Bibliography and
Research Manual
of the History of
Mathematics

Kenneth O. May

# Bibliography and Research Manual of the History of Mathematics

University
of Toronto
Press

Reprinted 2017
Printed in the United States of America

ISBN 978-1-4875-8718-5 (paper)
ISBN 978-0-8020-7077-7 (cloth)
Microfiche
LC 71-151379
AMS 1970 Subject Classification 00A15, 0100

SCHOLARLY REPRINT SERIES
LC 71-151379

CONTENTS

Appendices

# PREFACE

The purpose of this book is to assist mathematicians, users of mathematics, and historians in finding and communicating information required for research, applications, teaching, exposition, and policy decisions. It consists of a classified, indexed, and annotated bibliography of the secondary literature on the history of mathematics preceded by a brief manual on methodology. The bibliography contains about 31,000 entries under about 3,700 topics. There is appended a coded list of about 3,000 periodicals in which papers on mathematics and its history have appeared. The introductions to the parts and chapters give further descriptive details.

The scientific literature of mathematics (excluding popularizations and elementary textbooks) now consists of about half a million titles and is growing at the rate of more than 15,000 titles a year (see K. O. May, "Quantitative Growth of the Mathematical Literature," *Science*, 154 [1966], 1672-1673). This enormous collection is not indexed. No one knows its nature or contents. Preliminary studies suggest that there is a vast amount of duplication, and that the important information is contained in perhaps as little as 10 per cent of the titles (see K. O. May, "Growth and Quality of the Mathematical Literature," *Isis*, 59 [1968], 363-371). No one has suggested a way of measuring mathematical information, but it is certainly true that many publications contain several new mathematical results. It seems likely that the number of items of mathematical information is of the order of one million. It may well be most efficient in the long run to code and index the entire literature, but there is no possibility of doing this in the near future. The problem then remains of obtaining effective entry to this storehouse so as to find information, orientation, and enlightenment.

Often the best path to primary sources of mathematical information is through materials prepared with historical motives. These include bibliographies, bibliographical works, and historical studies of topics, periods, and other categories around which the mathematical enterprise can be organized conceptually. The historical literature in this broad sense is valuable in itself, is often the most useful point of entry in any information retrieval problem, and is necessary in planning further indexing of the mathematical literature. It is for this reason that I attack it first, leaving for the future the much bigger problem of indexing the substantive mathematical literature.

The manual for information retrieval and handling, which forms Part I of this book, is inspired by experience in research and teaching the history of mathematics over a number of years. I hope that it will save some readers the many hours that are often wasted by those unfamiliar with information sources, efficient procedures for information handling, and standards of scholarly communication. It is based in part on material in a syllabus prepared with support from the United States Office of Education and used at the University of Toronto (*Mathematics since 1800: Syllabus for a Course* [University of Toronto, 1967; 2nd ed., 1969]).

The bibliography in Part II developed from information-collecting activities that were at first quite personal. Turning to professional work in the history of mathematics rather late in my career, I was soon aware of the necessity for efficient, time-saving methods. Laborious searches of the historical literature would have been required to locate information on a number of related topics in which I was interested. It appeared most efficient to search for all of them at once and not much more difficult to index the entire historical literature simultaneously meeting my own needs and saving future scholars from similar tedious searches.

Thanks to a Science Faculty Fellowship from the National Science Foundation I was able to devote myself largely to historical research during the years 1962-1965. A grant from the American Philosophical Society supported the initial search of abstracting journals. Grants from the Society of Sigma Xi and the American Council of Learned Societies made possible special projects and studies. During the last six years, the Canada Council has generously supported the most tedious part of the work - completing bibliographic searches, coding, checking, and editing.

Many colleagues here and abroad supplied titles or helped identify obscure references. These include G. S. Andonie, G. Arrighi, K.-R. Biermann, P. Bockstaele, C. B. Boyer, R. P. Broughton, H. S. M. Coxeter, P. Delsedime, R. Dieschbourg, J. Folta, H. Freudenthal, Theo. Gerardy, A. Gloden, B. R. Goldstein, J. Itard, G. Maheu, R. M. MacLeod, I. Namioka, L. Novy, M. L. O'Malley, C. J. Scriba, D. J. Struik, R. Taton, K. Vogel, E. P. Wolfers, A. P. Yushkevich. Librarians at the University of Toronto and elsewhere cheerfully handled a heavy load of inquiries and inter-library loans, often finding answers in "impossible" cases.

The work of searching, recording, checking, and proofreading has been done largely by students and research associates. These include Charles Aronson, Holly Gardner, Charles V. Jones, Gregory H. Moore, Steve Regoczei, Dan Sunday, Henry S. Tropp, Dagney Wakfer, and James H. Willox. Harold P. Anderson identified and coded journal titles, a task whose difficulty can be appreciated only by someone who has attempted it. He and Henry S. Tropp assisted with the final editing of the bibliographic chapters.

Typing the manuscript for photographic reproduction was mainly the work of Jill Stiles, who performed this and many other secretarial tasks with her unusual diligence and efficiency. The final checking, editing, and typing of the coded serial list was carried through by Constance Gardner with a care and thoughtfulness that would have reduced error to zero if my impatience had not intervened. Others who assisted with clerical tasks were Gary Adamowicz, Edith Charlot, Sue L. Cline, Bob Holmes, Marilyn Luciano, Miriam May, Gail Pool, Nina Moore, Shirley Ann Sky, Fran Soule, and R. S. Winslow.

The College of Education at the University of Toronto supplied office space and photocopy services. I am responsible for planning, choice of sources, classification, selection of entries to be included, and, of course, for all faults and

mistakes.  I hope that users will
become collaborators by correcting
errors and omissions.

The present work is based on a
growing mathematical information
bank of about 200,000 slips containing data on the history, bibliography, terminology, and results
of mathematics.  I hope over the
next few years to make more of this
information generally available in
a mathematical dictionary and other
publications.


KENNETH O. MAY
Toronto, March 1972

PART I

RESEARCH MANUAL

# INTRODUCTION

The mathematician, user of mathematics, or student may approach the mathematical literature with a variety of motives. He may wish merely to discover the meaning of a word, to find a formula, a theorem, the bare facts about an individual, or the essentials of a topic or branch of mathematics. At a deeper level, he may seek insights provided by the history of an idea, an expository survey of a field, or an account of applications. To lay the basis for a substantive contribution to mathematics, he may wish to know what has already been done. He may be doing research on which to base a contribution of his own to the secondary literature. In every case, he will have to solve problems of retrieving, storing, handling, analysing, and presenting information. It is the purpose of the brief manual to suggest appropriate techniques for attacking such problems efficiently and effectively.

The goal of information retrieval is history in the broadest sense - an understandable account of past activity and results. In some cases, this may appear to involve no historical (time-dependent) component, such as when a person wants only a mathematical "fact." But even mathematical facts are dated, and an apparently timeless piece of mathematical information is simply one that has had a long and stable life. Accordingly I discuss all information retrieval from a historical point of view and consider any non-historical ("timeless") question as a special case.

Figure 1 suggests the general procedure to be followed in doing historical research. The goal is history. The raw material for historical analysis is chronology, that is a time-ordered list of events. Chronology is established by searching sources, and it is with this search that I deal in Chapter 1. The construction of a personal information file and the use of this file for analysis and writing are discussed in Chapters 2 and 3.

Figure 1 shows the investigator starting with topics, consulting reference sources for bibliographic and chronological information, then secondary and primary sources, and finally engaging in analysis and the production of history. To avoid a maze, I have not shown the constant feedback and recycling between sources on the left and records on the right. The investigator does not "start" anywhere. He comes to his problem with some information, he uses it to enter the flow where convenient, and he moves back and forth between different sources according to opportunity and knowledge. At every stage he records and uses mathematical, historical, and bibliographical information. From the beginning he critically analyses his stock of information (and misinformation), makes conjectures, draws conclusions, and reconsiders. There should be constant interaction between retrieval, storage and use.

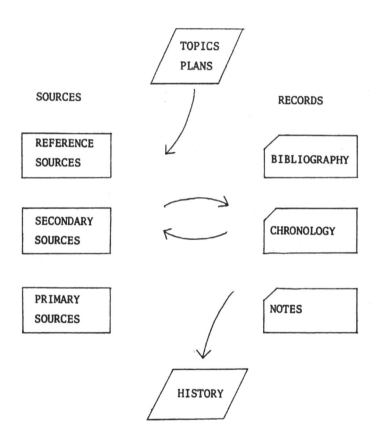

Figure 1

## 1

INFORMATION RETRIEVAL

The enormous amount of information now in storage in the world presents real problems for the person wishing to retrieve items of interest to him. In part the difficulties arise from the inadequacy of reference tools, but by far the most important factor is ignorance of the facilities already available. This ignorance often leads people to seek answers by picking accessible brains when more reliable information could be obtained easily from printed sources.

Information retrieval is not taught in our educational system, even in the education of scholars. It can be fully mastered only by experience, but one can benefit also from the experience of others and from learning general procedures. There are few profound difficulties for most people, and every increase in skill produces immediate rewards.

The following brief sketch is not intended as a complete guide but rather as a suggestive survey of activities. Combined with practice it should enable anyone to find the information he needs quickly and surely.

TYPES OF SOURCES

It is convenient to divide information sources according to their proximity to the events.

*Primary sources* are direct records or evidence. The most important examples in mathematics are original publications containing contributions to mathematical knowledge. Other primary sources include letters, manuscripts, artifacts, reports by witnesses (e.g. notes on lectures), films, pictures, and tapes.

*Secondary sources* are writings based on primary sources. These include historical publications, survey treatises, and some textbooks. It often happens that a single publication contains both primary and secondary materials. The same source may be primary for one purpose and secondary for another. For example, a historical article is a secondary source on its subject and a primary source on its author.

*Tertiary sources* are those based on secondary sources. Typical examples are dictionaries, encyclopaedias, bibliographies, and some textbooks. It is tempting to continue the classification, since for every n there appear to be n-ary textbooks, and one might even define ω-ary sources as those of an entirely speculative character.

Primary sources are the essential raw materials for the historian as well as for the mathematician. Nevertheless, secondary and tertiary sources are no less essential to the mathematical and scholarly enterprise as a whole. Frequently, the primary source on a mathematical topic is quite useless to the mathematician and of interest only to the historian. Moreover, when retrieving information, it is often most efficient to proceed from the tertiary to the primary sources in reverse historical order. Generally it is only from secondary and tertiary sources that one can find out what and where the primary sources are. In some cases, tertiary sources supply all the information required. The desirability and necessity of penetrating toward the primary sources is proportional to the depth and difficulty of the problem.

It is usually assumed that an n-ary source is always less reliable than an (n-1)-ary source. In fact, however, sources of all types are subject to error. Primary sources may intentionally or unintentionally misrepresent the facts, and tertiary sources may be entirely accurate and reliable. *Every source should be considered sceptically and critically.*

Because most materials are derived from others of the same type, mere agreement is no proof of accuracy. It is necessary to examine and compare sources with care, weighing the reliability of evidence as does the jury in a trial, and being particularly wary of generalizations and unsupported conjectures.

BRAINPICKING

What about persons as sources? Questioning an "expert" verbally or by correspondence is the best procedure when the information desired is known by an accessible individual and can be communicated easily. Experts can often give effective suggestions on reliable sources, good expositions, or individual facts. But consulting persons has severe limitations. Usually they enjoy answering a question that they can answer easily, but they seldom are willing to look up things for you, to be consulted very often, or to spend much time on *your* concerns. Moreover, even the most "expert" consultant is not infallible, especially in matters peripheral to his interests. Individuals are primary sources on their own opinions, observations, and direct knowledge. On other matters they may be secondary, tertiary, or n-ary. For individual facts, even about themselves, they are often very unreliable. Verbal communications are usually made with less care than written ones. They are also subject to faulty reception, recording, and remembering. Those who rely on brainpicking should ponder the possibility that their sources may have the same habit! The alleged expert opinion may be hardly more than verbal tradition, hearsay, or rumour, no longer, if ever, corresponding to reality.

There are also social reasons for

not relying mainly on personal communications. The store of information available and the part of it relevant to any particular problem are far too vast to be "known" by individuals or to be transmitted verbally. If everyone obtained his information orally, in a generation we should be reduced to the state of knowledge characterizing primitive societies, which had to depend on individual memory and verbal communication. Our complex culture, especially its technological and scientific base, requires a social bank of information from which individuals may retrieve information. The invention of writing first made this possible, and improvements, such as printing and electronic information processing, have simply increased capacity and accessibility. Verbal communication is as vigorous as ever, of course, but without interaction with recorded knowledge it soon becomes an exchange of misinformation.

## TOPICS

It seems obvious that one must choose a topic before beginning work, but it is equally true that one should consider a set of related topics and modify it as the study progresses. A little thought at the start and continued alertness will pay. Consider synonyms (including different names of persons, obsolete terms, near synonyms), antonyms, subtopics, related topics (including persons, mathematical and non-mathematical topics), and topics related to the mathematical, scientific, and social environment (including organizations, institutions, places, and general history of the times). A glance through the topics in Part II may be suggestive.

## REFERENCE MATERIALS

Few realize the enormous amount of information available in the reference departments of libraries. Often one can quickly acquire all the information needed for a specific purpose by consulting the reference shelves. For anyone making more than superficial investigations it is worth while to become quite familiar with the reference collection of the library he is using. A "walk through" the entire collection, careful examination of the catalogue or other guides, and discussion with the staff librarians will lay the basis for a never-ending increase in one's ability to find information.

Selected reference sources are listed in Part II, Chapters 1, 4, and 5, but one should take advantage of whatever is available in the library he is using. It would be impossible (and tiresome if it were possible) to describe the many skills of a craft that can be learned only by practice, but below are a few comments on particular types of reference materials.

We describe sources in a systematic order, but it is not suggested that the investigator make a thorough search in each category before going on to the next. As suggested in Figure 1, he should cycle back and forth among different sources from tertiary to primary, with emphasis on the latter in proportion to the seriousness and depth of his purpose.

To avoid difficulties later it is essential to keep good records from the beginning of an investigation, including precise information on the source of each note. This is discussed in detail in Chapter 2.

*General Dictionaries.* Dictionaries should be consulted as a first step. Recent editions of the unabridged *Webster's International* and the *Random House* are fairly up to date. The great *Oxford English Dictionary* contains much historical information. Do not overlook older dictionaries and those in foreign languages.

*Biographic Dictionaries.* For living mathematicians consult *American Men of Science, Who's Who in America,* and similar publications in other countries. In addition to general biographical dictionaries (which vary from directories with brief entries, like the *Century Dictionary of Names,* to multivolume encyclopaedic collections of long articles, like the *Dictionary of National Biography*), consult specialized scientific biographical dictionaries like *Poggendorff, Who's Who in World Science,* and the Russian *Biograficheskii Slovar Deyatelei Estestvoznaniya i Tekhniki,* and the *Dictionary of Scientific Biography.* Often older biographical dictionaries include people omitted later. (See Part II, Chapter 1.)

*Mathematical Dictionaries.* There is no satisfactory twentieth-century dictionary of mathematics, but consult whatever is available in order to find out something about word usage and related topics. Older dictionaries should not be neglected. (See Part II, Chapter 5.)

*General Encyclopaedias.* For general orientation and information about major topics and well-known individuals, the best encyclopaedias are very useful. Particularly good in mathematics is the *Encyclopedia Britannica* in its eleventh edition (1910) and in editions since 1965. The EB yearbooks contain authoritative brief accounts of main events in mathematics for each year. Do not ignore the great multivolume encyclopaedias in languages other than English, especially the *Bolshaya Sovetskaya Entsiklopediya,* to which the best Russian mathematicians and

historians have contributed. Several older encyclopaedias are excellent, such as the *Penny Cyclopaedia* (1833-1843) for which De Morgan wrote the mathematics and mathematical biography. *Always consult the index of the encyclopaedia you are using.* (See Part II, Chapter 5.)

*Mathematical Encyclopaedias.* There is at this writing no encyclopaedia of mathematics in English, though a very fine Japanese one-volume work is being translated. (See the list in Part II, Chapter 5.)

*Mathematical Handbooks.* These are often the quickest source for a formula or tabular value.

*The Library Card Catalogue.* After a first orientation in the reference materials, consult the card catalogue under author, title, and subjects. You may find a book that gives you just what you want. You may find a person's collected works, which will often include a biography and bibliography. If your library has a union catalogue, be thankful and use it.

*Printed Library Catalogues.* The great printed catalogues of the Library of Congress, British Museum, and Bibliotheque Nationale, and the many catalogues of other libraries are excellent sources of bibliographic information. Usually they can provide a fairly complete list of books published by an author. However, the Library of Congress and other national libraries do *not* keep or catalogue all books printed in their countries, so that failure to locate a book in their catalogues is not evidence of non-existence.

*Abstracting Journals.* The first abstracting journal in mathematics (*Jahrbuch ueber die Fortschritte der Mathematik*) was founded in 1868 and continued publication until World War II. It abstracted books and articles for an entire year in a single volume, but this completeness was paid for by a substantial delay in publication. In 1930, a second abstracting journal (*Zentralblatt der Mathematik*) was founded. It intended to publish abstracts as quickly as they were received without trying to cover the publications of a single year in one volume. In practice this speeded things up less than expected. Because of Nazi interference, these abstracting journals became less useful during the thirties, and the *Mathematical Reviews* was founded in 1940. This is now the most useful abstracting journal for English-speaking people. The Russian *Referativni Zhurnal* founded in 1935 is also good. The *Revue Semestrielle,* founded in 1893 and merged with the *Zentralblatt* in 1934, duplicated the *Jahrbuch* for the most part, but listed book reviews as well as books and articles. There is also a French abstracting service called the *Bulletin Signaletique*. All these abstracting journals arrange abstracts by subjects and are partially indexed by author and subject. Browsing through them can be very rewarding. They all have separate sections on history. Author indexes are included in each volume, and *Mathematical Reviews* has cumulative author indexes. These are excellent sources for a person's bibliography.

*Newspaper Indexes.* With the approximate date of death of an individual, you may be able to find an obituary through the indexes of the *New York Times* or the *London Times,* or by searching in local papers.

*Indexes of Individual Journals.* Consult the annual and general indexes of individual periodicals. Sometimes these are classified by subject (for example, the long-run indexes to the *American Mathematical Monthly* and the *Mathematics Teacher*).

*Bibliographies.* There are many specialized bibliographies and even bibliographies of bibliographies. (See Part II, Chapter 5.) Usually each country has one or more bibliographic services, such as *Books in Print* and the *Cumulative Book Index,* which record information on all books published.

*Serial Guides.* There are several very fine guides (e.g. *The Reader's Guide to Periodical Literature*), which index current journal articles by author, title, and subject (with cross-references). They do not cover technical mathematical publications but may often lead to useful biographical papers, popularizations, philosophical discussions, and so on, that are not usually covered by the mathematical abstracting services.

*Science Citation Index.* This is an important new source. Under each article are listed all those articles that cite the article. Thus from an article on a topic one can locate *later* articles on related topics.

*Serial Directories.* You will soon find that one of the vexing problems in historical research is identifying journals and locating libraries that hold them. The difficulty arises from the insufficient or inaccurate information given by authors, the many variations in names of the same journal, and the nasty habit of translating journal titles. Get acquainted with the *Union List of Serials* and other directories.

*Historical Treatises.* General histories and treatises on particular periods or topics often supply a surprising amount of information in historical context (use the index!). However, keep in mind that the information may be misinformation. Record sources and check with other and more direct sources.

*Reference Librarians.* Many library users do not realize that these professionals are trained to

help the searcher by answering inquiries, by assisting in information
retrieval, and even by carrying out
searches themselves on the user's
behalf.

Examples of appropriate questions
to ask a reference librarian are:
Where is the guide (catalogue, floor
plan) of the reference collection?
Where can I find a directory of
manuscript collections? I need to
know the year and place of publication of this book but I cannot find
it anywhere; can you help me? I cannot identify this journal; can you
help me? Can you suggest how I might
get started to find some information
on _____?

Of course, one should not impose
unduly on a reference librarian's
time, but assisting with information
retrieval is their work, not a dis-
traction. Usually they are glad to
take up a challenging problem.

SECONDARY SOURCES

From reference sources (including
Part II of this guide) you will have
constructed a bibliography of secondary sources. Read these critically for chronology, possible inter-
pretations, and indications of the
most important primary sources. Note
which are derivative from other
secondary sources and which are
based on direct knowledge of primary
sources. Distinguish facts from in-
ferences and speculations. If you
can find a carefully written and
comprehensive monograph or paper on
the topic of your interest, your
task may be limited to examining it
critically and bringing it up to
date. But this will seldom happen.
Surprisingly little satisfactory
work has been done.

In addition to titles listed in
bibliographies, you may find additional materials in older reference
works, in mathematical treatises
(the best of which have historical
and bibliographical sections, comments, or notes), and in publications
relating to a particular person (local newspapers, school publications,
institutional histories, commemora-
tive volumes, biographies of associates and relatives, etc.). Look
for obituaries. Often a man will
move in several circles, and valuable
information about him may be found
in publications ordinarily unrelated
to mathematics or the history of
science. Random search ("browsing")
in journals and books may be expected
to produce unexpectedly useful information, perhaps because the world
is really rather small and highly
interrelated.

PRIMARY SOURCES

Useful and essential as secondary
and reference sources are, it is the
primary sources that offer the
greatest joy and profit to the investigator. Here and here alone is
the original source of reliable information on what actually happened.

For a person, the most important
sources are his own writings. Collected works, translations, and
edited versions may be convenient,
but they are often incomplete or
unfaithful to the originals. Similarly, published letters may be expur-
gated, and volumes of correspondence
may be incomplete. *Primary sources
must be examined as critically as
all others*. If possible, one should
go direct to the first editions,
manuscripts, notebooks, diaries,
personal possessions (especially
books in which notes may have been
made), artifacts (inventions, instruments, equipment), mementos,
monuments, etc. In recent times,
photographs, films, sound tapes,
video tapes, etc. should be sought

out. Also important are contemporary records made by other people - biographies, related scientific publications, criticisms, reviews, commentaries, and documents (legal, financial, political, educational, medical, etc.). How far to go in studying all aspects of the man's environment depends on the depth of understanding desired.

For a subject, relevant material may be much harder to locate because information is more thoroughly indexed by persons than by topics. Most important are publications and other records of communications on the subject. An idea becomes part of history only when it is communicated. It may be necessary to search extensively. Skimming entire journal files and library collections will turn out to be more feasible and rewarding than the tyro imagines. Often one can centre the study on the greatest contributors, but this distorts the story by underestimating the role played by many "unimportant" and little-known participants.

Each project is different. When the easily available library resources have been exploited, the researcher must depend on his imagination, ingenuity, and perseverance. For example, he may track down elusive documents, interview principals or witnesses (producing the "oral history" on sound-tracks), or search for evidence in unexpected places. Such detective work is fascinating in itself and has many fringe benefits.

## OBTAINING MATERIALS

Even the greatest libraries do not have everything, and it is surprising how soon a thorough investigation will exhaust local resources. But this need not discourage the patient investigator. Through interlibrary loan or purchase of a photocopy, microfilm, or microfiche almost anything in any library or archive in the world can be obtained, *provided it can be identified and located. (The italicized condition is a prime reason for recording complete, accurate information and the sources of information.* Librarians do not care for wild goose chases and want to be sure the item you request really exists!) Fortunately for the travel-loving scholar there are always exceptions - items too rare to risk possible damage by copying, items in private hands, collections that are uncatalogued and too large to be copied in their entirety, etc. Usually, however, it is better to move copies of materials to one's place of work than it is to move oneself about. Here is where the reference librarian is invaluable. The essential thing is to supply full information and to be patient. It may take the library staff a very long time, involving much correspondence, to locate and secure an item. For this reason one should plan ahead and order needed items as soon as one knows of them, without waiting until they are to be used immediately. Sometimes materials can be secured directly through colleagues, who may be able to buy, lend, or copy an item. Often archivists, librarians, or individuals will be extraordinarily helpful if one shows serious interest in a matter that is valued by them. Correspondence may save a trip or make one more profitable.

## 2

PERSONAL INFORMATION STORAGE

Almost all the contents of the world's libraries and archives are available to one who knows how to ask, as was suggested in the last chapter. Thanks to modern transport and copying methods one ·can visit, borrow by interlibrary loan, or get photocopies or microfilms·. Automated storage and retrieval systems will undoubtedly increase the efficiency and universality of access still further. But this in no way decreases the importance of the individual's personal store of information with content and arrangement suited to his own needs.

## MEMORY

The centre of any personal information storage system is one's own memory, but it is foolish to rely on it too much. Human memories are very unreliable, and they have the dangerous habit of being often most certain of their least reliable contents. Moreover, the mind is used to manipulate information as well as to store it, and there is some evidence that too much remembering inhibits thinking. The whole point of a personal information storage system is to minimize the use of one's own mind for storage and routine handling of information so as to free it for functions not effectively performed by books, notes files, instruments, computers, and other devices. The individual who develops his own system of work along these lines will be following the historical pattern of communications technology, which tends to transform the intellectual worker from one who *knows* into one who *knows how*.

## BOOKS

A personal working library (I am not concerned here with collections for autotelic or commercial purposes) may consist of a single book (which should be a dictionary!) or a collection requiring the services of a librarian. If not adapted to the individual's needs it can be a burden. When carefully selected and organized as part of a system, it can be a powerful tool. The essential thing is to choose only books that are frequently used for reference or intensively studied over a long period of time. Do not try to duplicate institutional libraries. Instead, supplement your own library by a shifting collection of borrowings from them. Constantly weed your library of books that are no longer appropriate. Of every book, ask whether it would not be better to borrow it from a library when needed. If the book is rare and unavailable in libraries, you may find it better to place it in a library - to make it more accessible to others and also to save yourself the responsibility of caring for it.

How should a personal library be arranged? According to the whim of the owner, and in no other way! Like tools on a work-bench, books should be arranged according to convenience (usually frequency of usage) and not rigidly according to any logical schema. However, any collection that is very large had better be arranged in some systematic way if the owner does not wish to waste time searching or find himself buying or going to the library for a book "lost" on his own shelves! There is much to be said for straight alphabetical arrangement by author. Any subject matter schema fails to fit realities, gets out of date, and is ambiguous. In any case, a good library, like a good garden, requires constant cultivation, weeding, and replenishing.

EXCERPTS

Parts of books or journals in the form of reprints, photocopies, or tear sheets are essential. The publications from which they come would be too bulky to keep and too expensive to own. Excerpts can be collected and kept conveniently at hand while working on a topic, whereas the corresponding complete volumes would be physically unmanageable. But these materials can become just a pile of waste paper unless they are selected, stored systematically, and weeded. For short-run purposes, collections by subject matter do well. In the long run, it is better to file strictly by author in filing cabinets and to have a catalogue by authors (and subjects if required) on 3 by 5 inch slips according to standard library practices. This catalogue may be incorporated into the general file of slips to be described below, provided the slips for material owned have some special code.

Photocopying devices now make it feasible to incorporate in one's own files far more excerpt material than was possible when copying had to be done by hand. Many people have not yet fully adjusted to these possibilities and still copy by hand or paraphrase passages from books. If a book contains a passage that is wanted for later reference, make and file a photocopy. It will be much easier to retrieve than the book itself and more accurate than any note you could make by hand. *Be sure to write on the photocopy full descriptive information,* or else attach a photocopy of the title-page and other pages required for complete identication. In addition, write code words that identify the file location. For example, for filing by author (the most trouble-free system), a code such as "Smith 1928 abacus" will serve to indicate that it is to be filed under Smith and then by date 1928, the subject code serving for further identification.

## MANUSCRIPTS

Manuscripts (including preprints, ephemeral semi-publications, duplicated materials, notes, and letters) are very important and worth collecting. The remarks of the previous section on filing apply, but one should be much more cautious about discarding manuscripts than printed materials or copies. You should not keep in your working library materials unlikely to be of use in your work, but materials of possible historical value (including your own manuscripts) should not be thrown away thoughtlessly. They should be stored or placed in an institutional archive. Below is reprinted an appeal for the preservation of archival materials issued by the Advisory Committee on History, Conference Board of the Mathematical Sciences and published in a number of mathematical journals, including the *Notices of the American Mathematical Society* (October 1969, pp. 888-889).

*Our knowledge of mathematics and its history depends substantially on the preservation by our predecessors of manuscripts, notebooks, correspondence, apparatus, and other archival materials. This is obvious for ancient and medieval times, since what little we know depends on the few scraps that have survived. The invention of printing increased the diffusion and chance of survival for published materials, but private correspondence continued to be the most important mode of mathematical communication. Even after the founding of specialized journals, unpublished communications remained an essential part of the record.*

*Most highly creative mathematicians do not find the time to publish all their results, to say nothing of their ideas about mathematics and related matters. Gauss is an extreme case. Most of what he wrote has been published posthumously, and his famous journal did not come to light until 50 years after his death. If these ephemeral materials had not survived, many mathematical ideas would have been irretrievably lost and our picture of Gauss and of mathematics in the 19th century would be distorted. The situation is much less dramatic, of course, for most mathematicians, but it is not unimportant even for relatively "minor" figures. For example, only through the writings of others do we know who first made the four colour conjecture, because the discoverer did not write about it himself and his papers have not been preserved.*

*The twentieth century revolution in means of travel and communication has decreased the relative importance of printed material. The volume and slowness of publication has reduced the usefulness of journals as a means of communication. Private conversation and correspondence, notes, research reports, informal conference proceedings, preprints and various other ephemeral forms of communication play a greater role. One symptom of this is the number of "well-known results" that are not published anywhere. Another is the many "rediscoveries" of published results. If in the future what we are doing now is to be known and understood in its scientific and social context, we must do a better job of preserving mathematical records.*

*The initial and primary responsibility must rest with the individual mathematician. He should not destroy correspondence to and from his mathematical colleagues. He should preserve a file of his own unpublished and semi-published material, including letters, drafts and other manuscripts, notebooks, diaries, bibliographies, preprints, reports, syllabi, notes for his own lectures, notes on the lectures of others, preliminary edi-*

tions for local use, etc. He should keep photographs, sound recordings, apparatus, and mementos that throw light on the course of an investigation, the environment in which it took place, the organizations involved, etc.

Of course, not everything is worth preserving. But the mathematician should keep in mind that future generations will be interested in mathematicians as well as in their finished work, and that much of what is familiar to all of us will be unknown in the future unless some record of it is preserved. Not only historians of mathematics, but also a very wide scientific and lay public, are as interested in the origin and development of a mathematical theory as they are in specific mathematical results. Scholars generally are concerned with the whole range of mathematics as a component of civilization, and "personal" details of the lives of mathematicians frequently are of scientific and historical, as well as human interest.

We live in a period of rapid change in mathematics itself and in the many activities that make up the mathematical enterprise and its relations with other aspects of culture. Many of these changes will be of the greatest interest to future generations. Organizations should take care to preserve their own records and to arrange for their deposit and permanent accessibility. This applies especially to the many curricular reform groups, committees, and research groups whose records, memoranda, bulletins, and draft publications will be of the greatest interest to historians of the future and will enable them to understand events that today are familiar to all of us, but which may well appear quite mysterious some years from now.

It does little good for the mathematician to keep his papers unless he makes provision for their proper handling after his death. The best time to do this is in advance, by arranging for the deposit of papers in a library or other institution. In any case the mathematician should make provision in his will for materials that have not previously been provided for.

The families of mathematicians often play a key role in preserving documents. The best procedure for the layman is to consult mathematicians, historians of science, librarians, and archivists who can give advice based on knowledge of the materials and the means for preserving them.

An important responsibility rests also on departments of mathematics and mathematical organizations. They should establish their own archives which can serve as sources for the future historian. They should also see to it that the papers of deceased colleagues are not ignored until too late.

Where should materials be deposited? A natural place to consider first is an institution with which the mathematician has been associated. For example, government employees may approach the National Archives or the Library of Congress. Professors may find their own institutions appropriate, provided the library has an interest and adequate facilities. Often local museums, academies, or historical societies have archives. Many universities systematically collect materials on the history of science. Some private organizations, such as the American Philosophical Society in Philadelphia, the Niels Bohr Library of the American Institute of Physics and the New York Public Library, collect documents. A good source of information on depositories is A Guide to Archives and Manuscripts in the United States, by P. M. Hamer (Yale University Press, 1961).

It happens sometimes that prompt

*steps to preserve archival materials are not taken because the task of arranging and sorting appears formidable. Actually this task should be done by experts in any case. The best procedure for the possessor of a collection is to consult the institution where the materials may be preserved and to encourage its staff to do the work. Arbitrary removal of materials, editing or rearrangement can be very damaging. Materials that are appropriately deposited will be properly organized and catalogued. Listing in The National Union Catalog of Manuscript Collections, published by the Library of Congress, will make them available to scholars. The donor can, of course, place limitations on use in terms of time and circumstance, but restrictions are awkward to administer and should be kept to a minimum.*

*To sum up, we ask all those who are in possession of archival materials in mathematics to arrange for their proper preservation so that our mathematical work will be firmly linked both to the past and to the future. Members of the Advisory Committee would be glad to advise and assist in placing collections and would welcome information about materials not yet listed in the National Union Catalogue.*

## NOTES

Personal notes should be valuable and useful records of information peculiar to the individual's needs. Yet much scholarly effort is wasted through the failure to make good notes. Sometimes an important item (an idea, calculation, observation, citation) is not recorded and forgotten before it can be used. More commonly, a note is made so badly (incompletely, illegibly) or is so poorly stored and indexed that it is unusable or irretrievable. Even more often notes are made to no purpose at all. Most notes serve only to impress ideas on the memory and are never again consulted. Yet properly made notes can enormously enhance the effectiveness or research activity. By "properly made notes" I mean simply those that are made with a definite purpose and so as to meet this purpose effectively.

First of all one should not attempt to duplicate material by taking notes. Handwritten copies, detailed abstracts, or digests are less reliable than the original. When used as the basis for writing they may encourage the illusion of originality without helping the researcher to create his own synthesis. If one composes directly from such notes or from other people's writing without absorbing the ideas, he is simply "cutting and pasting" his own paraphrases of other people's expositions, often without being clear as to his sources. The result is hack work at best, plagiarism at worst. Such methods are reflected in the witticism: "To copy from one source is plagiarism; to copy from three is research."

The most common kind of note-taking arises from the habit, originating in school-days, of taking notes on lectures. Where the lecture represents a contribution not to be found in print, notes may be valuable or indispensable, but most such notes are poor records of material that appears more clearly and completely in print. Notes on lectures should be confined to general outlines plus a record of specific items of information or opinion that appear to be important or unique. But most students acquire the habit of "taking down" what the teacher says, in part exactly and in part paraphrased. The practice has been obsolete since the invention of printing. It becomes quite absurd when carried over from a lecture to printed matter. Nevertheless, when gathering material for a talk or paper, students often take notes that are merely diluted versions of the originals. From such notes on several sources they work toward what they imagine is their own product, but which is actually a hodge-podge whose sources are unclear to themselves.

If it is harmful to use notes as a substitute for the original or as an intermediate stage in the hidden copying of other people's ideas, it does not follow that note-taking is useless. Quite the contrary. Among the highly useful kinds of notes are: (1) records of one's own observations, calculations, conclusions, and plans; (2) abstracts, outlines, and commentaries by yourself of other people's work; and (3) records of "hard facts" (bibliographic, historical, mathematical) including the facts that certain people hold certain opinions. The best method of handling (3) is the classic 3 by 5 or 5 by 8 slip, whose use we consider below. The following general observations apply to all notes.

1. Do not take notes where the original or photocopy can be used.

2. Label all notes completely as to subject, meaning of notations, source, nature, and date. If there is more than one loose sheet, label each one.

3. Write so that someone else can understand. Later you will be that someone else to whom your own notes may have become meaningless!

4. If notes are not on individual slips filed for quick retrieval, index each note by making a 3 by 5 slip. For example, if you take notes in bound volumes, number the volumes systematically, number each item within the volume, and file index slips referring to volume and item numbers.

A note that cannot be found is waste paper. A note that is not important enough to record systematically and to index in some way is not worth keeping. It may still be worth making, of course, for temporary storage of information or to assist in thinking or remembering. Paradoxically, one often remembers just those items whose recording seems to have made remembering unnecessary! But the original record is still valuable for later checking, since memory is seldom complete or accurate.

SLIPS

The usefulness of 3 by 5 slips (and for some purposes 4 by 6 or 5 by 8) has been established by generations of librarians, bibliographers, historians, and office managers. For files (such as library catalogues) that are heavily used by many people, cards of substantial thickness are needed, but for personal files such cards are wastefully bulky. Thin paper slips are inexpensive, universally obtainable in convenient pads (usually called memo pads), fit a variety of filing equipment, and are quite strong enough if treated with a little care.

The basic feature of the slip, which makes it so useful for information storage and retrieval, is that each one contains a separate item of information that can be filed (and refiled!) for efficient retrieval. If index slips that give the location of materials in other forms are included, a file of slips can serve as the main personal information store and as the index to all other deposits in which the individual is interested. Each individual's system should be adapted to his own needs, but the following general guide-lines apply to all systems: (1) In terms of your needs, decide on definite procedures and modify them as your needs change. (2) Work through your files periodically, weeding and rearranging. (3) Record only one item on each slip. (One item may be defined as a message that cannot be split without being incomplete, e.g., the full bibliographic description of a publication, the definition of a word, or a person's name and address. Of course, the meaning of "complete" depends on the purpose.) (4) For each slip decide on a key word (or words) under which it will be filed. This (these) should be the word(s) under which *you* are most likely to think of looking for it. File one slip for each key word used, each with the complete message, or (more often) a master slip and other slips referring to it. Do not be concerned at inevitable variations in your choices, since when you come to recover slips you will also try various key words and achieve uniformity if you wish. (5) Be explicit, clear, complete, and legible. If you use abbreviations, make slips for them!

Where a personal information system is very large and subject to frequent manipulation, it may be advisable to *consider* automation, and especially the use of computer filing, manipulation, and storage. However, before information can be fed into a computer it must somehow be recorded and coded, so slips or equivalents must be made anyway. For most personal files, computerization is inconvenient and expensive. Only where the number of slips is very large (at least of the order of $10^5$) and subject to heavy manipulations, long calculations, and constant retrieval is computerization advantageous. Feeding information into a computer memory requires expensive operations (the preparation of cards, tapes, or other hardware), and retrieval requires that the information be in the computer memory for handling (subject to delay, unless one can afford arrangements for immediate access). In any case, computerization should be viewed as an addition to personal files rather than as a substitute for them. It is hard to imagine a personal information system more convenient than one based on a 3 by 5 memo pad that can be carried everywhere and on a file of slips that can be consulted instantly in order to find the original record.

The building of a personal file of slips is a very common practice among scholars. In 1908 one of the most creative men of the age gave the following advice:

*Moreover, procure, in lots of twenty thousand or more, slips of stiff paper of the size of postcards, made up into pads of fifty or so. Have a pad always about you, and note upon one of them anything worthy of note, the subject being stated at the top and reference being made below to available books or to your own note books. If your mind is active, a day will seldom pass when you do not find a dozen items worth such recording; and at the end of twenty years, the slips having been classified and arranged and rearranged, from time to time, you will find yourself in possession of an encyclopaedia adapted to your own special wants. It is especially the small points that are thus to be noted; for the large ideas you will carry in your head.* (From the *Collected Papers of Charles Saunders Peirce,* edited by C. Hartshorne and P. Weiss, Harvard University Press, 1933, vol. 4, paragraph 597.)

Each person should build his own system. What follows is the description of one that has proved very suitable for research in the history of mathematics. It is based on four types of slips (author-title slips, historical slips, chronology slips, and note slips) to be filed in a single alphabetical file under key words.

Such a file contains records of publications owned, consulted, or to be consulted, of dated events, and notes on facts of every description. The system is described in some detail, not under the illusion that anyone will wish to copy it exactly, but in order to give a specific example.

## Author-title slips

These are similar to the familiar cards that appear in library catalogues. The individual may wish to omit some of the information that usually appears on library cards and to add data of interest to himself. Typical slips are illustrated below. Uniformity of format is important for ease of filing and recovery.

Figure 2. Author slip for a book

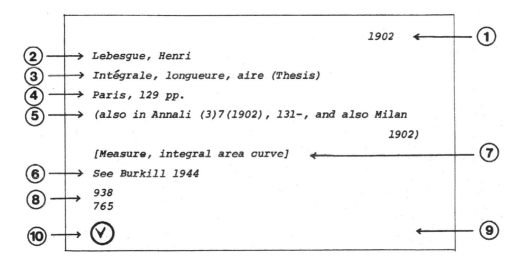

1. Year of publication. Useful at top, since these slips will be filed by author and then by year.
2. Author. Full first name often makes for easier location in a library catalogue. Where no author is known, use "Anon."
3. Title in full. (This may be abbreviated on other kinds of cards.)
4. City of publication. For recent books the publisher is most informative.
5. Information about reprinting and other publishings.
6. "Burkill 1944" is enough to identify a biographical card on the same man.
7. Subject classification, key words or codes.
8. Dates read (month and year).
9. Other information that might be recorded: library call numbers, brief characterization, location of one's more extensive notes, location of reviews, etc.
10. Code indicating that book is in one's own library.

Figure 3.  Author slip for an article

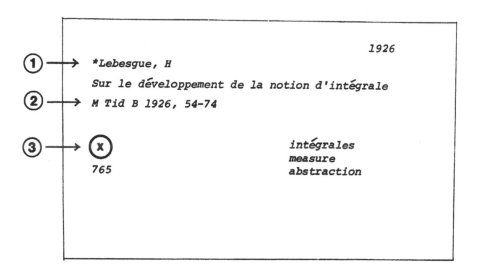

This is similar to a book slip except:
1.  Quality indication. The star means that this is a "best buy" item.
2.  Serial title abbreviated, volume, pages. Include any additional infor-
    mation such as series or number of issue.
3.  Indication that a xeroxed copy is owned.

## History slips (h-slips)

These slips describe publications *on* the history of a topic (or biography of a person). They are secondary sources. A publication that is an event *in* the history of mathematics is not recorded here but on a c-slip described in the following section.

For example, an article *about* a proof by Gauss in 1799 is recorded on an h-slip. The publication in which the proof occurred would be recorded on a c-slip. An h-slip names a publication on history, a c-card records an event that history describes.

Figure 4.  History slip for an article on a person

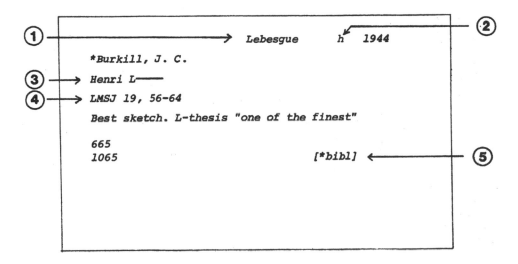

1. Person who is the subject of reference.
2. "h" indicates a history slip to be filed by subject and then by author and date.
3. Abbreviations may be used in title. If there is an author card giving details on this title in full, the information may be abbreviated more than it is here, as in Figure 5.
4. In abbreviating journal titles, follow a standard source or make slips showing your abbreviations.
5. Indicates that article contains good bibliography.

Figure 5.  History slip for an article on a mathematical topic

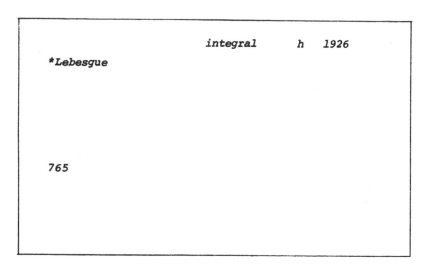

This example is highly abbreviated, but gives enough to locate the author card containing more information (see Figure 2).

Figure 6.  History slip on a period

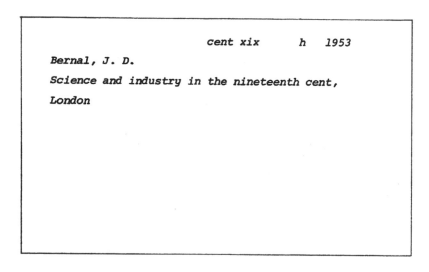

This slip is filed under "century," xix, and then by author and year.

Chronology slips (c-slips)

The chronology slip describes an event or a primary source, that is, a part of the development of the subject rather than a publication about its development. These slips are classified and filed by subject and within the subject by date, because their importance is primarily chronological. They are distinguished by a "c" after the subject.

Figure 7.  Chronology slip for a biographical event

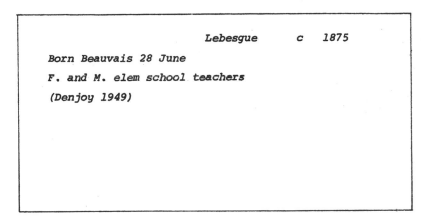

The parenthetical expression refers to a source described on an h-slip.

Figure 8.  Chronology slip for a publication on a topic

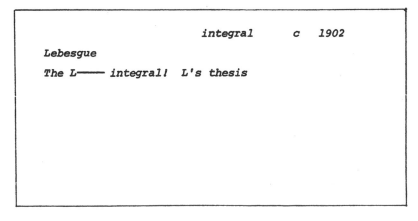

There is sufficient information to find the author card.

Figure 9.  Chronology slip on a mathematical event

non-Euclidean geom    c    1823

*J. Bolyai in letter to his father (3 Nov) describes*
*"new world" he has created "from nothing"*

*(Dict. Sci. Biog.)*

Figure 10.  Chronology slip on a country

England        c    1803

*Woodhouse Robert*
*Principles of analytical calculation,*
*Cambridge*

*First (?) effort to intro diff calc in England*

Note slips (n-slips)

The essence of historical information is the time dimension, and for this reason virtually all slips can be dated. However, it may be desirable to make notes of various kinds in order to assemble information by topics cutting across time. Such slips are indicated by the letter "n" following the subject. They may be used to file all sorts of information about a subject that you wish to collect by topic rather than by time. They may be filed at random or according to the subtopic. They may be used to index other slips by topics.

Figure 11. Note slip on an alias

*Fibonacci* *n*

*= Leonardo da Pisa*

Figure 12. Note giving a definition

*equipollences* *n*

*An alternative theory of complex and hypercomplex numbers due to Bellavitas*

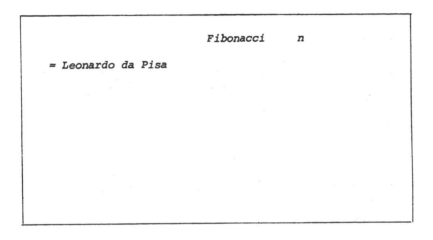

Figure 13.  Notes on subtopics

_Gauss_            n

<u>_Bolyai_</u>

_Closest college friend. They discussed foundations_
_of geometry_

_Gauss_            n

<u>_Family_</u>

_No known university-educated people among_
_ancestors_

Filing slips

A variety of containers are useful
for storing and filing slips: ordinary letter envelopes (for blanks,
for carrying slips, and for filing
a small collection), small boxes
holding several hundred slips (for
collecting slips on a topic or at a
convenient location), large card-
board boxes, filing drawers, and
cabinets. Be generous with dividers,
which can be made cheaply in a variety of colours by cutting 3 by 5
cards to 3 by 3 1/2. It pays to in-
sert dividers (alphabetical or
chronological) generously so as to
limit searches to a very small batch
of less than 30 cards that can be
picked out and riffled with finger
or eraser (the best procedure for
slips though not for cards).

Slips can be filed in many ways,
and it matters little as long as
some system is adopted. Usually the
best is a single alphabetical order
by author and topic, with an agreed
order for author slips, h-slips,
c-slips, and n-slips in the same location. But inevitably it will be
convenient to have separate special
purpose files.

HISTORICAL ANALYSIS AND WRITING

Having discussed the collection of information in Chapter 1 and its storage in Chapter 2, I now come to its utilization to produce history. Of course this division of the discussion and its order are not intended to correspond to separate, unrelated, and ordered activities. There should be constant interaction between collection, organization, and analysis of information.

Before writing or, better yet, before beginning work one should have in mind topics, audiences, places of publication, ends and means.

## APPROACHES TO HISTORY

History has far more facets than can be described here. One can focus on the development of ideas, biography, periods, civilizations, countries, cities, institutions, terms, symbols, topical fields, education, communication, bibliography, sociology of mathematics, and so on. No one approach can give a complete picture. Each one and every combination has value. But a choice will help to guide information collection as well as analysis and writing.

## CHRONOLOGY VERSUS HISTORY

Chronology, the time-ordered array of events, is the raw material of history, just as experimental data and observations are the raw material of natural science. History arises when chronology is selected, organized, related, and explained. The result may be, and usually is, a coherent descriptive account that attempts no very broad generalizations, or it may be a theory arising from the description and verifiable (or falsifiable) by further referenc to the record or by comparison of predictions with future experience.

To give rules for the production of history is as difficult as to tell someone how to discover a theorem or construct a new proof. It is easier to suggest some things to avoid.

## CREDULITY

The historical and mathematical literature contains many errors. Some result from repeating baseless traditions. Some are merely careless. Others arise from ignorance, misunderstandings, prejudice, or chauvinism. They appear in all sources, even the most respectable and seemingly primary.

For example, in 1866 Quetelet wrote that when he visited Gauss in 1829 the latter was studying Russian. Actually, the letters of Gauss himself and other documents show conclusively that Gauss did not begin learning Russian until ten years later, and Quetelet himself, in his description of the visit written immediately afterward in 1830, says nothing about Gauss learning Russian. Apparently, during the following 36 years Quetelet heard of the Russian studies and, forgetting his source, imagined that Gauss had told him in 1829. Whatever the explanation, this illustrates that witnesses are often unreliable, and the more so with the passage of time. (I am indebted to Kurt-R. Biermann for this example.)

It is therefore incumbent on the historian to examine everything critically. He should not *believe* anything automatically, but he should record what he finds and record also the source of information. There are advantages in recording information on the same matter from several sources. With properly organized notes, such information will be filed in juxtaposition. Then discrepancies can be observed and investigated. In general, a healthy scepticism backed by an analysis of sources of information is essential.

## PRIORITITIS

It is soon apparent to the historian that people often do not get the credit they deserve. But it is quixotic to try to provide everyone with his due. Questions of priority are worthy of historical consideration only when they involve historical issues, not merely efforts to dispense justice. Moreover, the question of who did something "first" usually turns out to be meaningless because of unremovable ambiguities in the definition of the "something" and what is meant by "doing" it. It is these ambiguities and the interrelations of mathematical topics that almost always make it possible to find a precursor for the work of anyone. And how could it be otherwise? No one descends from heaven, and the seeds of ideas must have come before the ideas themselves! Priority chasing for its own sake should be avoided.

## THE SOLOMON SYNDROME

Closely related to the illusion that historians are dispensers of priority is the idea that they should rank mathematicians and judge who was "right" and "wrong" by their own standards. This may be good fun, but it is not very illuminating. It tells little about what happened and why. The historian's tasks are description and explanation, not judging and moralizing. That is not to say that mathematical (or ethical) judgments need not be made, but rather that they should be made in historically meaningful and explicitly stated terms, not as personal awards.

## UNHISTORICAL ANALYSIS

One of the weaknesses of scientists who take up history is the tendency to judge historical events timelessly or, more often, as though they were occurring today. The amateur historian examines a mathematical paper of the past as if its author were taking an examination in mathematics today and amuses himself by finding "errors." Often he looks at past events in terms of the extent to which they appear to lead to the mathematics of today, as if that were the final form and criterion by which everything must be judged. Or he may force the ideas of one period into an inappropriate framework of another.

Of course, it is important to see how modern ideas originated and developed, but a full understanding of this development requires considering things *as they were at the time and in terms of conditions of that time*. In particular it must be remembered that words mean different things at different times. Failure to take this into account can lead to ludicrous errors. The matter was summed up well by Augustus De Morgan in a letter to Sir William Rowan Hamilton in 1853: "In reading an old mathematician you will not read his riddle unless you plow his heifer; you must see with his light, if you want to know how much he saw." (27 January. Quoted from R. P. Graves, *Life of Sir William Rowan Hamilton*, vol. III [Dublin, 1889], p. 438.)

## SPECULATIVE HISTORY

It is often necessary, and quite legitimate, to supply missing information by deduction from established facts, or even by plausible inference, if it is labelled as such. But there is a style of "history" that consists in simply fabricating an account of how something might have happened (for example, how counting might have first been invented) and then coming to regard this as a fact. Perhaps the most pernicious form of this aberration is the formulation of historical theories based on superficial information or prejudice and then the "verification" of them by making up a suitable historical narrative.

## THE MATTHEW SYNDROME

In the Gospel according to Matthew it is said that to those who already have much, more shall be given, and this seems to be true of the allocation of fame. The importance, ability, and achievements of well-known figures are exaggerated due to the fact that their papers and publications are more likely to be preserved, they are more often cited, and interest in them results in publications that create an audience and trigger further work. The result is a distorted picture of the history of mathematics as the work of a few "stars." The historian should be on guard against this "halo effect" and strive to give a balanced picture that includes "minor" as well as "major" personalities.

## PLAGIARISM

I close this incomplete catalogue of historical sins with one that is most distasteful to scholars. Plagiarism is "copying or imitating the language, ideas, and thoughts of another and passing off the same as one's original work" (*American College Dictionary* [Random House, 1959]), i.e., "to steal or purloin and pass off as one's own the ideas, words, artistic production, etc., of another; to use without due credit the ideas, expressions, or productions of another" (*Webster's International Dictionary,* 1947).

Clearly the offence involves both theft and fraud. It is to be distinguished from violations of copyright, which involves the improper use of copyrighted material whether or not any misrepresentation is involved. For example, extensive quotation from a copyright book with due credit but without permission, or a pirated printing of an entire book without permission, involves no plagiarism but is an infringement of copyright. There is no danger of copyright violation if the writer gets permission for any extensive quotations and gives credit to his sources, of if he uses his own words.

Most plagiarism does not involve copyright questions because it is only the form of presentation that can be copyrighted. Ideas are not subject to the copyright law. Nevertheless, ideas are of the greatest importance to the intellectual community. The intellectual worker wishes his ideas to spread and to be used by others. He does not want to restrict their use. However, he wishes full credit for what he has done, and this may be of great importance to him professionally and financially. Accordingly, the offence of plagiarism is above all an offence against the intellectual community and a violation of scholarly ethics. The offender forfeits the trust of this community and may be very severely penalized by it. Offending students may lose credit or be expelled. More mature offenders may find their careers cut short, and are certain to lose the respect of their fellow scholars.

A careful writer who is genuinely anxious to do his own work, to give full credit to others, to reveal his sources frankly, and to take credit only for his own contributions will not be guilty of plagiarism. One who is careless about such matters may commit the offence without intent. The following examples suggest the many possibilities.

The most blatant type of plagiarism is the *copying of an entire work,* substituting the offender's name for that of the true author. In previous centuries there are many examples, but this offence is rather rare now for a published book or paper, because the size and vigilance of the academic community as well as the copyright laws make it extremely dangerous. Today the offence is more often committed by students who wish to produce a paper in a hurry and who copy one from a printed source or from an old paper written by another student, or who hire a ghost writer. If the original author is agreeable, no theft is involved, but the fraud in this case is no less unethical. One of my students once copied most of an article from the *Encyclopedia Britannica* without even noticing that I was the author!

A rather common type of plagiarism is *quotation without quotation marks and acknowledgment.* To avoid this, use quotation marks for every quotation, no matter how short, and give the source accurately.

*Unacknowledged paraphrase* is also plagiarism. Common as the practice is, the rewriting of another's work in one's own words without acknowledgment clearly falls within the dictionary definitions given above. To avoid this, accompany a paraphrase by an acknowledgment. This may be done in a footnote or by a phrase such as "according to ____ (with exact reference)" or "the following exposition is based on...." Such acknowledgment does not relieve the writer of the obligation of using quotation marks when the original author is being quoted.

Extensive paraphrasing or quotation is seldom justified. An author should base his production on the assimilation of many sources, and he should not have to fall back on what is practically copying, even if he is honest about it. "Cutting and pasting" by quotation from several sources, even if they are properly acknowledged, is not to be confused with an original product based on reworking of assimilated material.

Equally a case of plagiarism is *unacknowledged use of translation.* A translated passage or a paraphrase of one must be treated no differently than the original. The actual source must be given, and the writer must indicate that he is the translator if that is not evident.

Another type of plagiarism is *false citation,* which usually involves citing a primary source for a quotation or idea when actually the source was a secondary source that cited the primary source! This kind of plagiarism involves both theft and fraud. The writer is making use of the work of the man who wrote the secondary sources without giving him credit. He is also pretending to have consulted the primary source, when in fact he has not done so. In addition to the violation of ethics, this practice is very undesirable because it perpetuates whatever errors appear in secondary sources.

When quoting an author, the exact citation must be given. This citation should be the source from which the writer actually got the quotation. If he got it from someone else's translation, from an anthology, or from another source that quoted the original source, this must be indicated by a footnote of the type "Quoted from..., page...."

A more subtle form of plagiarism is *failure to acknowledge the actual source of an idea.* This may appear as a judgment stated as if it were based on one's own research when actually it is taken from a secondary source. Such a misdemeanour may be avoided by including a phrase such as "Smith concludes that..." together with an exact reference. Another common device is to say at the beginning of a portion of the manuscript something like "The following conclusions are due to..." and then give the precise reference in a footnote.

Bibliographies serve both to indicate sources and to suggest possible additional titles of interest to the reader. The writer should make sure that his bibliography does not confuse these two functions. He may give a single list with indication of which sources have been consulted. He may list a reference with a notation that "this source was not consulted." The *essential thing is not to give a false impression of the materials that have actually been used.*

In order to avoid these and other kinds of plagiarism, the author may ask himself: "Is what I have just written my own idea, or did I get it from elsewhere?" In almost every case the idea comes from another source. If it is a matter of general knowledge obtainable in many sources, there is no need to give a specific source. The reader will understand that no claim for originality is being made. If the matter is one on which there may be some doubt or disagreement, or if the idea is apparently new, *the actual source used by the writer must be indicated.*

This discussion makes obvious the importance of taking notes and of recording the source of every note that is taken. Moreover, notes shoul indicate whether a listed source has actually been consulted directly. Quotations should be distinguished from paraphrase. *Unless such careful records are made, it is impossible for the author to present an honest and accurate picture of his work and results.*

## PLANNING A PAPER

Planning a paper involves choosing a topic, approach, style, and format. One should think about these matters over a period of time and while doing other things. One should revise one's plans while conducting research. Make outlines and possibly draft sample passages (e.g., an introduction), but do not begin the writing too soon. It is easier to change ideas in your head than on paper! If you are working systematically, much of the organizing work will be done automatically. You will have an ordered collection of slips giving sources, chronologies, and notes.

How should you present your material? Consider various possibilities. Dream about the topic and your audience. Ask such questions as: In this material what ideas, patterns, facts are of interest to my audience today? Or ought to be of interest? What light does this information throw on the development of mathematics and its state today? How is this matter related to other events of the time, to the development of science, to general social and political history? What do I find most interesting? What will my audience find most interesting? How would I tell a friend about it? Do I have a thesis or theses that I wish to assert and support? Do I have the evidence to back my hunches? What are the main ideas I wish to communicate? If you don't have some satisfying answers to these or similar questions, withhold your pen! Population control is needed also in this area.

## WRITING

Outline in your head or on paper. Think before you write, imagine what you are going to write, and write.

As you write, document or otherwise justify every claim. The extent to which you will include footnotes giving sources will depend on the nature of your writing and its place of publication, but you must possess a source or argument for every statement. In mathematics, rigor means precise proof from explicit axioms and definitions according to explicit methods (or the citation of such proofs). In history, rigor means documentation of every claim or its establishment by argument from documented facts. You should write nothing that cannot be so justified, even though you may refrain from complete documentation or argument in a popularization or other work where it is not expected. You should always give some indication of your sources and be prepared to give complete documentation if it is requested. One who fails to have proper documentation is comparable to an experimental scientist who claims certain results without giving sufficient information to enable another scientist to retrace his steps and without keeping the records on which his results are based.

When you have finished the first draft of your paper, reconsider everything, let it rest for a time, and show it to colleagues. Then be ready to rewrite, many times if required. Is what you have written going to be clear to your audience? Could it be cut without loss of content? Consider the writing really finished only when you feel that all that can be done now has been done. Then be ready to revise again when referees and editors report on it. Many writers revise a manuscript a half dozen times or more, and too

many publications would have bene-
fitted from several revisions.

The discussion above has stressed
revision. It is hard to imagine any
paper that could not be improved.
Still the striving for perfection
can become a fault. The scholarly
paper or book is supposed to be a
contribution to an ongoing social
process of collective research. It
can never be the final word. The
author should work only until he has
a worthy contribution that fulfills
limited objectives. Extraneous ideas
should be put aside for inclusion in
a later publication. When a compromise has been reached between unat-
tainable perfection and inexcusable
carelessness, the time has come to
send off the package and turn to
other projects.

The following bibliography is a
first rough approximation to a complete bibliography of published
writings on the history, biography,
and bibliography of mathematics. It
is directed toward historians and
mathematicians and focuses on util-
ity for information retrieval rather
than on matters of antiquarian interest.

COVERAGE AND SOURCES

The main sources are the *Jahrbuch
ueber die Fortschritte der Mathematik*
(1868-1940), *Revue Semestrielle des
Publications Mathématiques* (1893-
1932), *Zentralblatt der Mathematik
und ihre Grenzgebiete* (1930-1965),
*Mathematical Reviews* (1940-1966),
*Isis Bibliographies* (1913-1966), and
Cecil B. Read's bibliography cover-
ing seven expository mathematical
journals to 1965 (in *School Science
and Mathematics,* 66 [1966], 147-179).

Other sources are numerous special bibliographies, indexes of
mathematical journals, historical
treatises by M. Cantor, H. Wieleitner, F. Cajori, etc., encyclopaedia
articles, guides to the literature
by G. Sarton, G. Loria, N. G. Parke,
J. E. Pemberton, G. A. Miller, F.
Mueller, etc., and hundreds of other
books and papers.

The following have been *deliber-
ately excluded,* except in a few cases
of unusual merit or lack of other
sources:
1. Articles within reference
works. (But reference works are listed in Chapters 1 and 5.)
2. Collected and selected works,
and articles within them.

3. Chapters and parts of general
historical treatises. (But treatises
are listed in Chapter 4.)
4. Translations, revisions, and
reprints other than the major ones.
5. Ephemeral materials that are
effectively unavailable. These include many theses and "Program" publications of German gymnasia, separate reprints of journal articles,
privately published and circulated
materials, manuscripts, publications
not available in major libraries,
and unidentifiable items.

The first four categories are
valuable parts of the literature and
should be consulted in any investigation. They are excluded here because they can be located so easily.
The last would have led the user of
this guide to many a wild goose chase
of the kind I have "enjoyed" during
the last few years. In a few cases
titles have turned out to be antici-
pations never actually published. In
others, information given was insufficient to locate the item through
diligent search by reference librarians at several libraries and correspondence with scholars familiar
with the field, or else such a search
established that the publication was
practically inaccessible (e.g., not
available even in central libraries
of its country of origin!) and not
of sufficient importance to justify
the trouble of obtaining a copy.

I have not attempted to decide
whether an item is or is not relevant to mathematics. A topic is included if titles about it are cited
in mathematical sources. Similarly,
I have not excluded titles because
of judgments of quality. "Unimportant" writings are sometimes useful,

and works of low quality are part of the historiographic record.

It appears likely that the coverage of historical, biographical, and bibliographical publications is fairly complete, with the exception of biographical writings prior to 1868 and Asian historical publications not cited in the Occident. The first exception results from the fact that prior to the *Jahrbuch,* we have only those references that appear in treatises and miscellaneous sources. These appear to mention practically all titles on the history of mathematical topics, but biographical publications on only the most important mathematicians. The omission is not too important, since *Poggendorff* and the *Dictionary of Scientific Biography* fill the gap. The second exception is more serious. It is partially covered by Needham's great work on China, but Japan, India, and the Arab world are poorly represented. Nevertheless, I guess that 90 per cent of the historical, bibliographical, and biographical titles appear in this bibliography, and that 99 per cent of the rest can be found through the publications here listed.

Often important historical and bibliographical information appears incidentally in mathematical treatises. Coverage of such sources is not systematic, but they are included in so far as they appear in my files.

## ARRANGEMENT

For convenience in consultation, the bibliography is divided into five chapters: biography, mathematical topics, epimathematical topics (e.g., fields of application), historical classifications (e.g., time periods), and information retrieval (e.g., bibliography). *Each chapter is pre-ceded by an introduction containing information helpful to the user.* Two appendices give conventions and a coded list of journals.

Since practically every key word (including names of people!) has many synonyms and there are only a few suggestive cross-references, the user must expect some trial and error. However, I have chosen key words to achieve naturalness and efficiency rather than uniformity, "correctness," or adherence to any systematic scheme. Each title is classified under as many key words as seem appropriate.

## FORMAT OF ENTRIES

Under each key word, publications are ordered by author ("Anon" is used where the author is unknown to me) and date of publication (often intrinsically approximate). For books, the title (in italics), place of publication and/or publisher, and number of pages (where known) are given. For articles or passages within books, pages are sometimes omitted if the book contains an adequate index or table of contents. For serial publications, there appear the journal code (these are listed alphabetically in Appendix 2), series if any (in parentheses), volume, issue number (in parentheses), and beginning and ending pages. Readers are cautioned that many journals carry several different numbering systems. This is one of many reasons for making entries systematically redundant. Titles of journal articles are omitted. There may follow additional comments, such as reference to an abstracting journal entry (F = *Jahrbuch,* Z = *Zentralblatt,* MR = *Mathematical Reviews*) giving volume (sometimes year) and page number (review number in MR after 1958). Finally, additional key words (in square brackets) may be given to provide at least as much information as an omitted title. These key words also supplement the explicit cross-references. Where the publication includes a significant bibliography or a portrait, this is noted. An asterisk indicates an item that I consider to be especially worth consulting, but it does not imply a complete endorsement or that unmarked titles (with which I may be unfamiliar) are of lesser quality.

METHOD AND COST

Author slips (in the format described in Part I, Chapter 2) were prepared by the author and assistants (see the Preface) in the course of personal reading and in systematic searches of sources chosen by the author. The cost per slip for those done by asistants was about 10¢. The author then assigned topics, writing one (a person if possible, otherwise the most evident topic) at the top of the slip and the others in square brackets at the bottom. This required about 30 seconds per slip. Sorting, filing by topic and later refiling cost about 2¢.

After some years, there were imbedded in the author's files the slips from which the bibliography was prepared. Many duplicate slips appeared from different sources. The author then went through the file, selecting slips for inclusion, comparing information from different sources, discarding duplicates (after recording all information) or redirecting them by assigning new topics from the square brackets at the bottom, making additional topical cards where required, and dealing with discrepancies and doubtful assignments of topics. This operation took another 30 seconds on the average, though the time required to handle an entry varied from seconds

to the hours needed to visit a library in another city or correspond with a colleague.

Next the selected file was swept to identify journals and write their codes on the slips. Each journal that appeared was identified by library search, assigned a code, and entered on a card that eventually showed the information included in the list in Appendix 2. This file of serials was checked and augmented by searching the lists appearing in abstracting journals, the *Royal Society Catalogue, Poggendorff,* dealers' catalogues, etc. The result was a list of mathematical serials much larger than needed for the bibliography. However it seemed a shame not to include all serials with substantial mathematical content, even if not cited in this bibliography, since such a list appears to exist nowhere else and to be of considerable general use. The part of this work required for the bibliography added about 10¢ more per bibliographic entry (about $1 per serial cited).

The final editing for the typist by the author took about 12 seconds per entry. Typing and proof-reading (really assignable to publication cost) added about another 20¢ per entry. Materials and supplies came to about 3¢ an entry. This totals about 45¢ out-of-pocket costs per entry plus 1.2 minutes of professional time.

To those who have engaged in such work or paid for it, these cost figures may seem much too low. They are based on stop-watch studies of operations and on allocation of total salary payments, two independent procedures that produced approximately the same results. The secret, if one is needed, probably lies in the ability and motivation of the workers involved coupled with the fact that all work except the final typing was done on a part-time basis with constant effort to maximize efficiency. For example, slips were alphabetized in batches of several thousand and then filed by sweeping through the entire file. It is possible to do this kind of work very quickly for a few hours a day, but with full-time effort the hourly output drops sharply. The cost figures do not include the visible and hidden overhead for space, facilities, supplies, equipment, and so on. If these were added, the total cost would probably reach about 75¢ plus author's time per entry. For the whole bibliography this leads to a substantial total of about $23,000 plus about 620 hours of author's time, but this cost is quite low compared to the professional time that will be saved in the future.

The possibility of using computers was considered several times. But all the work we did up to the final typing of the manuscript would have been required in order to prepare copy to be keyboarded into the computer, and the cost of keyboarding would have been about the same as that of typing and proofreading the manuscript. In short, the cost of getting our information into the computer would have been about the same as the cost of the project.

Perhaps in a much larger project, in which full-time workers keyboarded information direct from sources into a computer programmed to do all indexing, identification, sorting, checking, and editing, much of the work we did might be automated, but at present such an approach is of greater interest to experimental computer scientists than to those wishing to do a job of this kind cheaply. Of course, this bibliography and serial list offer part of the information needed to construct an automated procedure.

## CORRECTIONS AND ADDITIONS

There are severe limitations to what can be done by one person, even with plentiful assistance, in one location. If this bibliography is to be improved substantially, it will be by the collaboration of others. I appeal to all readers to call attention to errors, omissions, further sources, and so on. Please give *complete* information (redundancy is helpful) and supply reprints, photocopies of title pages, etc., if possible. All help will be acknowledged gratefully.

## 1

BIOGRAPHY

This chapter contains systematic classifications in the first four sections, the remainder of the chapter being arranged alphabetically by name.

Where an individual has several names, we have chosen the most familiar one. Modern German names involving "von" are alphabetized under the name following the "von," but where the individual has anglicized his name and included "Von" as part of it, alphabetization follows this usage. Similar remarks apply to "de" and "De," but Dutch names are alphabetized under "Van." Medieval names such as Robert of England (and similar forms involving equivalents of "of" in other languages) are alphabetized under the preceding name, since the place name is not a family name.

The criterion for inclusion of a person is simply that biographical material about him has appeared in the literature of mathematics or history of mathematics. The number of titles on an individual is not my decision. In some cases, worthy mathematicians (e.g. Frobenius) have been given little or no biographical notice. In many others, quite minor figures in the development of mathematics have received attention for a variety of reasons. I do not attempt complete coverage of individuals whose main fields are non-mathematical, but exclude titles not of mathematical interest.

The reader is reminded that the best initial source for biographical information is usually a reference work. It should be noted that in addition to the biographical references and collections listed in this chapter, general reference works are listed in Chapter 5. With a few exceptions, material within reference works is not listed in this bibliography.

BIOGRAPHICAL COLLECTIONS

See also under countries, civilizations, and time periods in Chapter 4.

Ahrens W 1914 *Mathematiker-Anekdoten,*
Leipzig-Berlin, 86 p (F45, 56)
[Cauchy, Euler, Fermat, Gauss,
Grassman, Lagrange, Riemann, Riese,
Schellbach, portraits]
- 1920 *Mathematiker-Anekdoten,* 2nd ed
(Teubner), 42 p [portraits]
Amici N 1906 *Matematici, fisici, as-*
*tronomi delle Marche,* Macerata
Anon 1940 AMM 47, 109 [club topic]
Arago D F J 1854 *Notices biographiques,*
3 vols, Paris [Vols 1-3 of Oeuvres
complètes] [Ger tr 1856, Engl tr
1857]
Archibald R C 1932-1936 Scr 1, 173-181
265-274, 346-362; 2, 75-85, 181-187
282-292, 363-373; 3, 83-91, 179-190,
266-276; 4, 82-93, 176-188, 273-282,
317-328 [index 1932-1936]
- 1938 *A Semicentennial History of*
*the American Mathematical Society*
*1888-1838,* NY, 262 p
[*bib, presidents]
1945 Scr 11, 213-245 (MR8, 3)
[bib, port, table makers]
- 1946 Scr 12, 15-51 [con of 1945]
- 1948 *Mathematical table makers,* NY
(Scr Stu 3) [bib, port]
Aubrey John 1898 *Brief Lives,* Oxford
Baillaud B 1930 Rv Sc, 353-358
[founding of observatories]
Baldi Bernardino 1707 *Cronica de*
*Matematici overo epitome dell'is-*
*toria delle vite loro,* Urbino,
156 p
- 1872 Boncomp 5, 427-534 [arab, mid]
- 1879 Boncomp 12, 352-438 (comment
by Boncompagni)
- 1886 Vite inedite di Matematici
italiani, Boncomp 19, 383-406, 437-
489, 521-640 (also 1887 Rome)
(ed E Narducci)
Barbieri M 1906 *Notizie istoriche dei*
*matematici e filosofi del Regno di*
*Napoli,* Naples

Barinaga J 1937 *Miscelanea matematica,*
Madrid, 131 p (F63, 817) [port]
Barr E S 1964 Isis 55, 88- [biog in
Phi Mag to 1900]
*Becher O 1954 *Grundlagen der Mathemati.*
*in geschichtlicher Entwicklung,*
Freiburg
*Bell E T 1937 *Men of Mathematics,* NY,
613 p [Interesting, often unreliabl
See Miller G A 1937 M Mag 12, 389-
392]
Bense Max ed 1943 *Briefer grosser*
*Naturforscher und Mathematiker,*
Koeln, 246 p (Z28, 385)
Bertrand J 1902 *Eloges académiques,*
Paris, 462 p (F33, 35)
*Biermann K-R 1960 *Vorschlaege zur*
*Wahl von Mathematikern in die*
*Berliner Akademie,* Berlin, 75 p
(Ber Ab (1960) (3))
Bochner S 1966 *The Role of Mathematics*
*in the Rise of Science,* Princeton,
386 p [sketches in app]
Baffito Guiseppe 1933-1937 *Scrittori*
*barnabiti, o della Congregazione dei*
*Chierici Regolari di San Paolo 1533-*
*1933. Bigrafia, bibliografia,*
*iconografia,* Florence (Biblioteca
barnabitica), 4 vols
Cap P A 1857 *Etudes biographiques pour*
*servir a l'histoire des sciences,*
Paris
Capaccio J C 1608 *Illustrium mulierum*
*et illustrium virorum Elogia,* Naples
Catalan E 1895 Lieg Mm (2)18
(F26, 28) [letters]
Coolidge J L 1949 *The Mathematics of*
*Great Amateurs* (Clarendon Press),
211 p (MR11, 149) (Dover pap 1963)
Condorcet M J A 1773 *Eloges des Acadé-*
*miciens ...morts depuis 1666 jusqu-*
*en 1699,* Paris, 5 vols
Czistiakow W 1963 *Rasskazy e matie-*
*matikach,* Minsk, 343 p
Darboux G 1912 *Eloges académiques et*
*discours,* Paris, 524 p (F43, 52)
Dedron Pierre + 1959 Mathématiques et
*mathématiciens,* Paris, 443 p
Diogenes Laertius 1925 *Lives of Em-*
*inent Philosophers,* London
(tr R D Hincks)
Drohojowska 1887 *Les savants modernes*

*et leurs oeuvres: Cassini, Arago,*
*Le Verrier, Puiseux,* Lille, 192 p
(F21, 1254)

Eells W C 1925-1936 AMM 32, 258-259;
33, 274-276; 34, 141-142; 35, 437;
36, 99-100; 37, 150-151; 38, 100-
101; 39, 298-299; 40, 359-360; 41,
260-261; 42, 171-173; 43, 234-235
(F56, 3; 59, 4) [centennials]
- 1962 MT 55, 582-588
[100 eminent maths]

Eisenhart C 1965 Am Statcn 19(5), 21-
[anniversaires]
- 1967 Am Statcn (Ap), 31-33
[anniversaries]

Enestroem G 1901 Bib M (3)2, 326-350
(F32, 11) [necr 1881-1900]
- 1902 Bib M (3)3, 226-234 (F33, 48)
[anniversaries]

Ferroni P 1794 It SS 7, 319-

Fontenelle F L B de 1719 *Eloges des*
*Académiciens,* Paris, 3 vols

Fuss P H 1843 *Correspondence mathé-*
*matique et physique de quelques*
*celebres géometres de XVIII siecle,*
Pet

Gallego Armestro H 1925 *Matematicos*
*espanoles contemporaneos,* Santiago
(F51, 38)

Gloden A 1949 Thermik 2, 12-36
(MR11, 573) [Luxembourg]

Higgins T J 1944 AMM 51, 433; 56, 310-
312 (Z60, 13) [bib]

Hooper Alfred 1948 *Makers of Mathe-*
*matics,* NY [pop]

Isely L 1898 Neuc B 27, 167-172
[tombstone inscriptions]

Jones Bessie Z ed 1966 *The Golden Age*
*of Science* (Simon and Schuster)
[Laplace, Legendre, Poincaré]

Klein Felix 1926-1927 *Vorlesungen*
*ueber die Entwicklung der Mathe-*
*matik in 19 Jahrhundert,* Berlin
(repr 1950, 1952 Chelsea)

Kowalewski Gerhard 1938 *Grosse Mathe-*
*matiker,* Munich, 300 p
(F64, 1; Z17, 385)

Laurent C 1906 *Les grands ecrivains*
*scientifiques de Copernic à Berthe-*
*lot. Extrait, biographies et notes,*
2nd ed, Paris

Lodge O 1893 *Pioneers of Science,*
London, 420 p (F25, 1913)

Lorey W 1953 Leop NA (NS)16, 419-424
(Z53, 197)

Macfarlane Alexander 1916 *Lectures on*
*Ten British Mathematicians of the*
*Nineteenth Century,* NY, 148 p
[Peacock, De Morgan, Hamilton, Boole,
Cayley, Clifford, H J S Smith,
Sylvester, Kirkman, Todhunter]
- 1919 *Lectures on Ten British Phys-*
*icists of the Nineteenth Century,*
NY, 144 p [Tait, Babbage, Stokes,
Airy...]

Mahrenholz J 1936 *Anekdoten aus den*
*leben deutscher Mathematiker,*
Leipzig, 48 p (F62, 2)

Mansion P 1898 *Mélanges mathématiques*
*(1833-1898). I. Histoire, esquisses*
*biographiques,* Paris
- 1913 *Mathesis* (4)3(s)(July-Aug),
28 p

Martin Benjamin 1764 *Biographica*
*philosophica,* London, 565 p

Meschowski Herbert 1961 *Denkweisen*
*grosser Mathematiker,* Braunschweig,
94 p (Z114, 4)
- 1964 *Ways of Thought of Great Math-*
*ematicians,* (Tr of 1961 by J. Dyer-
Bennet) San Francisco, 118 p
(MR30, 373)

Midonik Henrietta O 1965 *A Treasury*
*of Mathematics,* NY [brief biogs]

Mueller Felix 1912 *Gedanktagbuch fuer*
*Mathematiker,* 3rd ed, Leipzig
[birthdays]
- 1913 BSM (2)37, 123-127 [birthdays]

Muir Jane 1963 *Of Men and Numbers, the*
*Story of Great Mathematicians,* NY,
249 p

Nielsen Niels 1910-1916 *Matematiken i*
*Danmark,* Copenhagen, 3 vols
- 1935 *Geómetres français du dix-*
*huitième siecle,* Copenhagen, 437 p

Ocagne Maurice d' 1927 Rv Q Sc 12.
257-283 [XIX, France]
- 1930 *Hommes et choses des sciences,*
Paris, 309 p (2nd ser 1932, 291 p,
F58, 993; 3rd ser 1936, 275 p,
F62, 1009)
- 1952 *Histoire abrégés des sciences*

*mathématiques,* R Dugas ed, Paris

Pekarskii P 1870 *Istoriya Imperator-skoi Akademii nauk v Peterburge,* Pet

Picard Emile 1922 *Discours et mélanges,* Paris

- 1924 *Mélanges de mathématiques et de physique,* Paris
1931 *Eloges et discours académiques,* Paris (F57, 4)

- 1936 *Discours et notices,* Paris, 374 p (F62, 1038)

Poincaré H 1910 *Savants et ecrivains,* Paris

*Prasad Ganesh 1933-1934 *Some Great Mathematicians of the Nineteenth Century: Their Lives and their Works,* Benares, 2 vols

Prielipp R W 1969 MT 62, 125-127 [Galois, Newton, Pascal]

Prudnikov V E 1956 *Russkie pedagogi-Matematiki XVIII-XIX vekov,* Moscow

Rebiere A 1893 *Mathématiques et mathématiciens,* 2nd ed, Paris

- 1899 *Les savants modernes, leurs vies et leurs travaux,* Paris

Rigaud S J 1841 *Correspondence of Scientific Men of the XVII Century,* Oxford, 2 vols (repr 1862)

Sedillot L A 1869-1870 *Boncomp* 2; 3 [France]

Smith David E 1936 *Portraits of Eminent Mathematicians with Brief Biographical Sketches,* (Scripta Mathematica) (a portfolio)

Stonaker Frances B 1966 *Famous Mathematicians,* Philadelphia, 118 p [pop]

Studnicka F J 1898 *Heroen des Geistes* Prague, 264 p (F29, 23)

Suter H 1900 Ab GM 10 [Arab]

Taylor E G R 1954 *The Mathematical Practitioners of Tudor and Stuart England,* Cambridge, 454 p

- 1966 *The Mathematical Practitioners of Hanoverian England 1714-1840,* Cambridge, 518 p (MR34, 2411)

Thébault V 1958 M Mag 32, 79-82 (Z83, 245) [XIX, France, geom]

Tricomi F 1940 Saggiat 1, 155-160 (F66, 19) [Scandinavia]

Turnbull H W 1929 *The Great Mathe-maticians,* London, 136 p [pop]

- 1961 *The Great Mathematicians* 2nd ed, NY, 156 p (MR23, 4)

Van Hée L 1926 Isis 8, 102-118 [China]

Zhatykov O A 1956 *Leading Russian Mathematicians* (in Kazak), Alma Ata, 248 p [port]

## PORTRAITS

Anon 1932 Nat 129, 910-911 [London
  Royal Society]
Archibald R C 1945-1946 Scr 11, 213-
  245; 12, 15-51 [table makers]
Bouvier E L 1932 Rv GSPA 43, 366-367
Ionescu I 1933-1934 Gaz M 39, 44-47,
  163-167, 201-204 (F59, 843); 40,
  193-196, 338-340, 392-395, 437-438
  (F61, 20)
Johnson R A + 1932 Scr 1, 183-184
  [coins, stamps]
Lacroix A 1932 Figures de savants,
  Paris, 2 vols (F58, 993)
Miller G A 1945 Sci 101, 223-224
  [false portraits]
Riesz M ed 1913 Acta mathematica 1882-
  1912, table générale des tomes 1-35,
  Upsalla

## REFERENCE SOURCES

Note: Not listed here are the very
useful national biographical diction-
aries such as Allgemeine deutsche
Biographie, Dictionary of American
Biography, Dictionary of National
Biography, American Men of Science.
See also under civilizations, coun-
tries, etc. in Chapter 4.

Academie des Sciences 1954 Index
  biographiques...1666-1954, Paris,
  534 p
American Statistical Association 1962
  Directory of Statisticians and
  Others in Allied Professions, Wash
  DC
Anon 1905 Mathesis (3)4, 217-218,
  250-251, 274-275; (3)5, 33, 57,
  89, 113, 145, 169, 201-202
  [calendar]
- 1912 Satzungen und Mitglieder-
  Verzeichnis der Deutschen-Mathemati-
  ker Vereiningung..., (Teubner)
  (Repr from DMV 21(1))
  1961 The Royal Society of London
  ...A Tercentenary Tribute, (Dawsons)
  282 p London [bib]
  1961 World Directory of Mathemati-
  cians, 2nd ed, Bombay (Tata
  Institute)
- 1966 McGraw-Hill Modern Men of
  Science [33 mathematicians]
Arnim Max 1930 Corpus Academicum
  Gottingense, Goettingen
*Becker O + 1951 Geschichte der Mathe-
  matik, Bonn (Athenaeum), 263-340
Bierman K B + 1960 Deutsche Akademie
  der Wissenschaften zu Berlin. Bio-
  graphischer Index der Mitglieder,
  Berlin
Brocard H 1915 Intermed 22, 129
  (Q2044) [pseudonyms]
*Debus A G ed 1968 World Who's Who in
  Science - A Biographical Dictionary
  of Notable Scientists from Antiquity
  to the Present (Marquis)
Gillispie C C 1970 Dictionary of

*Scientific Biography* (Scribners)
*Grimal P ed 1958 *Dictionnaire de Biographies,* Paris, 2 vols
Hottenberg G H 1788 *Nachrichten von dem Leben und den Erfindungen der beruehmstesten Mathematiker,* Munster, 308 p
Laisant C A + 1902 *Annuaire des mathématiciens pour 1901-1902,* Paris, 400 p
Mueller Felix 1904 *Gedenkentagebuch fuer Mathematiker* (Teubner) (2nd ed 1908, 3rd 3d 1912 (F35, 1)
- 1911 *Der Mathematische Sternenhimmel des Jahres 1811* (Teubner), 30 p
*Poggendorff J C 1863 *Biographisch-literarisches Handwoerterbuch zur geschichte der exacten Wissenschaften,* Leipzig (Many editions)
Riesz Marcel ed 1913 *Acta Mathematica, 1882-1912: Table générale des tomes 1-35,* Uppsala, 180 p
Teller J D 1943 MT 35, 369-371 [birthday calendar]
*Zvorykii A A ed 1958 *Biograficheskii slovar deyatelei estestvoznaniya i tekhnike,* Moscow, 2 vols

BIOGRAPHY, THEORY

Glenn O E 1953 M Mag 28, 229-302 [autobiography]
Favard A 1878 *Statistica degli scienzati vissuti nei due ultimi secoli,* Padova (F10, 23)
Kneser A 1900? ZM Ph 45, 113 [Minding's biog work]
Riccardi P 1882 Mod Mm 20, 299-310 (F14, 20-21) [ranking]

PERSONS

Abaco. See Paolo.

Abakanowicz, Bruno Abdank   1852-1900

Anon 1900 Wiad M 4, 266

Abbe, Ernst   1840-1905

Czapski S 1905 Deu Ph G 7, 89-121
  (F36, 33)
Knopf O 1905 DMV 14, 217-230
  (F36, 32) [port]
Voigt W 1905 Gott N, 34-44
Voit C 1905 Mun Si 35, 346-355
Wiener O 1906 Leip Ber 58, 631-646

Abbott, Edwin A.   1838-1926

Anon 1920 Nat (Feb 12)
Newman James 1956 *World of*
  *Mathematics,* NY

Abel, Niels Henrik   1802-1829

Anon 1902 Ber Si, 1001-1002 (F33, 21)
- 1902 DMV 11, 437-438
- 1903 Act M 27, 391   (F34, 12)
  [letter]
- 1907 Crelle 82, 308
- 1929 Nor M Tid, 11   (F55, 14)
Belga (pseud) 1920 Intermed 27, 54-56
  (F47,33) [Cauchy]
Bergh-Kragemo H 1929 Nor M Tid 11,
  49-52 (F55, 14) [letters]
Bertrand J L F 1885 BSM 9, 190-202
  [Bjerknes 1885]
Biermann K-R 1963 Nor M Tid 11, 59-63
  (Z117:1, 8) [Humboldt]
Biermann K-R + 1958 Nor M Tid 6,
  84-86, 96 (MR20, 738; Z82, 242)
Bjerknes C A 1885 Bord Mm (3)1, 1-365
- 1885 *Abel, Tableau de sa vie et de*
  *son action scientifique,*
  Paris, 368 p

- 1902 *Niels Henrik Abel: Mémorial*
  *publié à l'occasion du centenaire*
  *de sa naissance,* Kristiania, 450 p
  (also Paris, London, Leipzig)
  (F33, 19)
- 1929 *N. H. Abel,* Oslo  (F55, 612)
- 1930 *Niels Henrik Abel,* Berlin,
  141 p  (Ger tr of 1929)  (F56, 22)
Brun V 1949 Nor M Tid 31, 56-58
  (Z40, 1) [ms]
- 1953 Nor M Tid 1, 91-97, 143-144
  (MR15, 276; Z52, 3) [ms]
- 1953 Nor VSF 25, 25*-43*
  (MR14, 832; Z50, 243)
- 1954 Crelle 193, 239-249
  (MR16, 433; Z57, 3)
- 1955 Rv Hi Sc Ap 8, 103-106
  (MR17, 117) [ms]
Brun V + 1958 Nor M Tid 6, 21-24
  (Z80, 8) [letter, youth]
Brunel G 1885 BSM (2)9, 141-153
  [Bjerknes 1885]
Czuber E 1902 Z Realsch (27 Sept),
  17 p  (F33, 22)
Dickson L E 1930 *Modern Algebraic*
  *Theories,* NY, Ch 10, 178
  [Hamilton, quintic]
Fehr H 1902 Ens M 4, 445-447 (F33, 22)
Harkin D 1950 Nor M Tid 32, 68-78
  (Z40, 1) [Lie]
Heegard P 1935 Nor M Tid 17, 33-38
  (F61, 21) [Degen]
- 1937 Act M 68, 1-6 (Z16, 197)[Degen]
*Hille E 1953 M Mag 26, 127-146
  (Z52, 3)
Hliblowicky K 1903 Lem Sam 9, 1-88
Holmboe B M 1829 *Kort fremstilling av*
  *Niels Henrik Abels liv og videnska-*
  *belige virksomhet,* Oslo-Christiania
  (also Mag Nat 9)
Lange-Nielsen F 1929 Nor M Tid 11,
  13-17 (F55, 14) [Paris]
- 1929 Nor M Tid 11, 53-55  (F55, 14)
- 1953 Nor M Tid 1, 65-90, 143
  (MR15, 276; Z52, 3)
Lorey W 1929 Nor M Tid 11, 2-13
  (F 55, 14)
- 1930 Crelle 161, 65-72  (F55, 14)
Mansion P 1902 Rv Q Sc (3)2, 603-618
  (F33, 22)
- 1908 Brx SS 32A, 182-189  (F39, 42)

- 1909 Mathesis (3)9(2)  (F40, 20)
*Mittag-Leffler G 1907 *Niels Henrik
   Abel,* Paris, 48 p  (F40, 20)
- 1907 Rv Mois 4, 5-26, 207-229
Nielsen N 1928 Dan H 8(8), 31 p
   (F54, 32)
Ore O 1950 *N.H.Abel,* Basel, 23 p
   (Elem M S8)  (MR11, 707)
- 1954 *N.H.Abel, Et geni og hans
   samtid,* Oslo, 317 p  (MR16, 1)
- 1957 *Neils Henrik Abel; Mathemati-
   cian Extraordinary,* Minneapolis,
   257 p  (MR19, 826; Z82, 11)
Peslouan C L de 1906 *Abel, sa vie et
   son oeuvre,* Paris, 184 p  (F37, 9)
Picard E 1903 J Sav, 109-119
Prasad G 1933 *Some Great Mathemati-
   cians of the Nineteenth Century,*
   Benares 1, 111-165.
Smith D E 1922 AMM 29, 394
Sorokina L A 1959 Ist M Isl 12,
   457-480  (Z102, 5)  [equations]
Stormer C 1903 Chr Sk 5, 1-8
   (also Crelle 125, 237-240)
   [Kulp, letter]
- 1923 Nor M Tid 5, 33-34  (F49, 8)
- 1929 Nor M Tid 11, 85-142  (F55, 14)
Sylow L 1902 DMV 11, 377-382 (F33, 21)
- 1902 Wiad M 6, 311-316
Tambs L R 1929 Nor VSF 2, 28-31
   (F55, 15)
Taton R 1948 Rv Hi Sc Ap 1, 356-58
   [Acad Sci]
Warhaftman S 1930 Mathes Po 5, 62-68
   (F57, 1311)
Wilson E B 1902 AMSB (2)9, 154-157


Abraham.  See Ezra; Savasorda.


Abramescu, Niculea  1884-

Ionescu D V 1948 Math Tim 23, 139-140
   (MR10, 174; Z30, 101)


Abria, Jérémie Joseph Benoit
   1811-1892

Rayet G 1893 Bord Mm (4)3, 301-330
   (F25, 36)

Abu.  See word following abu in the nam

Ackeret, Jacob  1898-

Tank F + 1958 ZAMP 9b, 9-52
   (Z79, 244)


Ackermann, Wilhelm  1896-1962

Hermes H 1862 MP Semb 10, 11-13
   (Z113, 2)


Adam, Jehan  XV

Thorndike L 1926 AMM 33, 24-28
   (F52, 13)  [arith]


Adam, Pedro Puig  1900-1960

Fernández B 1960 Gac M 12, 5-8
   (Z97, 2)
Garcia R J 1960 Gac M 12, 3-4
   (Z97, 2)
Ibarra J R P 1960 Ens Sc 1(6), 2-5
   (Z99, 246)
- 1960 Gac M 12, 9-14  (Z97, 2)


Adams, Daniel  1773-1864

Sleight E R 1936 M Mag 10, 193-199
   (F62, 1036)
Sueltz B A 1943 MT 36, 183-85


Adelard of Bath  XII

   = Adelhard = Aethelard = Alard
   = Adeardus Bathoniensis

Bliemetzrieder F 1935 Adelhard von
   Bath, Munich, 402 p  (F61, 14)
Boncompagni B 1881 Boncomp 14, 1-134
   (F13, 7)
Clagett M 1953 Isis 44, 16-42
   (MR15, 275)  [Euclid, mid]

Guenther S 1873 Boncomp 6, 314-340
  [polygon]
Heiberg J L 1890 Hist Abt 35, 48-58,
  81-98 [Euclid]
Henry C 1880 ZM Ph 25(S), 129-139
  (F12, 7)
Sarton G 1931 Int His Sc 2(1), 168
  [bib]
Weissenborn H 1880 Hist Abt 25,
  141-166 [Campano, Euclid]

Adelbold of Utrecht 970?-1026

  = Albaldus = Adelbaldus
  = Athalbaldius = Adelberon

Sarton G 1927 Int His Sc 1, 714
  [bib, circle, sphere]
Tannery P 1904 Intermed 11, 254-257
  [Gerbert]

Adrain, Robert 1775-1843

Coolidge J L 1926 AMM 33, 61-76
  (F52, 22) [USA]
Gehman M 1955 MT 48, 409-410
Pettengil G E 1943 Berks HR 8,
  111-114
Seinin O B 1965 Ist M Isl 16, 325-336
  [least squares]

Adrianus Romanus 1561-1625

Gilbert P 1851 Notice sur le mathématicien louvaniste Adrianus
  Romanus, Louvain
Maupin M 1898 Rv M Sp 9, 3
  [equations]

Agnesi, Maria Gaetana 1718-1799

Amati A 1898 Lom Gen (2)31, 1380-1394,
  1493-1520 (F29, 6)
Anon 1915 Intermed 22, 215
Anzoletti L 1900 Maria Gaetana Agnesi,
  Milan, 495 p (F31, 11)
Boyer J 1898 Appleton 53, 403-409
  (F30, 9)

Coolidge J L 1951 Scr 17, 20-31
  (Z43, 5)
Fazzari G 1898 Pitagora 4, 88
Frisi A F 1799 Elogio storico di
  Domino Maria Gaetana Agnesi milanese
  Milan (Fr tr 1807 Paris)
Labrador J F 1951 Gac M 3, 175-178
  (MR13, 810; Z43, 5)
Masotti A 1940 Milan Sem 14, 89-127
  (MR8, 190; F66, 1187; Z27, 3)
- 1941 Saggiat 2, 279-281 (Z26, 195)
Sister M Thomas á Kempis 1938 Scr 6,
  211-217

Agnostini, Amedeo 1892-

Anon 1959 Pe M (4)37, 245-251
  (MR22, 1594; Z88, 244)

Ahmad ibn Yusef

  = Abu Jafar Ahmad ibn Yusuf ibn
  Ibraham ibn al Daya al Misri

Cantor M 1888 Bib M (2)2, 7-9
  [proportion]
Curtze M 1889 Bib M (2)3, 15-16
Sarton G 1927 Int His Sci 1, 598
Steinschneider M 1888 Bib M (2)2,
  49-52, 111-117 [names]

Ahmes -XVII

  See Rhind Papyrus under Manuscripts in Chapter 5

Ahrens, Wilhelm E. M. G. 1872-1927

Anon 1927 Ens M 26, 144
Lorey W 1927 Arc GMNT 10, 328-333
  (F53, 32)
Staude O 1928 DMV 37, 286-287
  (F54, 42)
Wangerin G 1927 ZMN Unt 58, 290
  (F53, 32)

Aida, Sanzaemon

Hayashi T 1914 Toh MJ 5, 69-82
 (F45, 16)
Watanabe T 1914 Toh MJ 5, 65-69
 (F45, 16)


Airey, John Robinson  1868-1937

Archibald R C 1948 *Mathematical Table
 Makers*, New York, 5-7  (Scr Stu 3)
 [biobib, port]


Airy, George Biddle  1801-1892

Airy G B 1896 *Autobiography* ...,
 Cambridge, 426 p  (F27, 18)
 [bib, port]
Anon 1892 Nat 55, 232-233  (F24, 27)
Budde E 1892 Pogg An 45, 601-604
 (F24, 27)
Faye H A E 1892 Par CR 114, 91-93
 (F24, 27)
Macfarlane A 1919 *Lectures on Ten
 British Physicists of the Nineteenth
 Century*, New York, 106-118
Loria G 1945 Scr 13, 120-122
 [numerical lunar thy]


Aiya, V. Ramaswami

Naraniengar M T + 1936 M Stu 4, 111-
 119, 120-129, 144-157 (F62, 1041)


Ajdukiewicz, Kasimier  1890-1963

Anon 1964 Log An (NS)7, 3
 (Z115, 246)
Jordan Z A 1967 Enc Phi


Akhiezer, Naum Ilich  1901-

Baltaga V + 1951 UMN (NS)6(2),
 191-194  (MR13, 1)
Krein M G + 1961 UMN 16(4), 223-
 239  (Z98, 9)


Al. See under word following "al" in
 the name.


Alasia, Chistoforo  1869-

Halsted G B 1902 AMM 9, 183-185


Albanese, Giacome  1890-1947

Dantoni G 1947 It UM (3)2,170-171
 (Z30, 338)


Albatenius.  See Battani


Albeggiani, Giusseppe  XIX

Gebbia M 1893 Paler R 7, 39-47
 (F25, 36)


Albert of Saxony  1316?-1390

= Albertus de Saxonia = Albertus
de Helmstede = Albertus de
Riemestorp (Riggensdorf)
= Albertutius = Albertus Novus

Boncompagni B 1871 Boncomp 4, 498-511
 (F3, 5)
Duhem P 1906-1913 *Etudes sur Leonardo
 da Vinci*, Paris, 3 vols
- 1906 *Origines de la statique*, Paris
 2, 1-185, 336
Ginsburg B 1936 Isis 25, 341-362
 (crit  Duhem)
Jacoli F 1871 Boncomp 4, 493-497
 (F3, 5)
Suter H 1884 Hist Abt 29, 81-102
- 1887 Hist Abt 32, 41-56


Alberti, Leon Battista  1404-1472

Favaro A 1884 Boncomp 16, 325-332
Gukovski M A 1935 Len IINT 7, 105-128
 (F61, 948)  [Leonardo da Vinci;
 mechanics]

Loria G 1895 Bib M (2)9, 9-12
(F26, 9)
Mancini G 1911 *Vita di Leon Battista
Alberti,* Firenze
Natucci A 1927 Pe M (4)7, 242 256
(F53, 11) [area, circle]
Schilling F 1922 ZAM Me 2, 250-262
(F48, 31) [art]
Wolff G 1936 Scientia 60, 353-359
S142-S147 (F62, 1027) [mechanics]
- 1937 Euc Gron 13, 234-245 (F63, 20)
- 1937 Z Kunst 5, 47-54 (F63, 20)
Zubov V P 1960 Bib Hum 22, 54-61


Albertus Magnus  1205-1280

  = Albert the Great = Ratisboneusis
  = Albertus Yeutonicus

De Raeymaeker L 1933 Rv Neosco (2)35,
1-36
Geyer B 1958 Angelic 35, 159
Greenwood T 1932 Arc Sto 14, 69-73
- 1932 Nat 129, 266-268
Hofmann J E 1960 Int Con 1958,
554-566 (MR22, 927)
Sister M Stephen 1962 MT 55, 291-295
[mid]
Steinschneider M 1871 ZM Ph 16,
357-396 (F3, 4)
Weisheipl J A 1958 Am C Phil 32,
124-139


Alcuin  735?-804

  = Albinus = Ealwhine = Alchvine

Anon 1930 Mathesis 44, 393 (F56, 14)
Lorentz F 1837 *The Life of Alcuin,*
London, 290 p (tr by J M Slee
from the German 1829 Halle)
Sarton G 1927 Int His Sc 1, 528-529
[bib]


Aleksandrov, Aleksandr Danilovich
1912-

Efimov N V + 1962 Rus MS 17(6),
127-141 (Z112-243)

1962 UMN 17(6), 171-184 (MR26, 1)
Vallander S V + 1963 Len MV (1), 7-9


Aleksandrov, Pavel Sergeevich 1896-

Kolmogorov A N + 1966 UMN 21(4), 4-7
(MR33, 2517)
Pontryagin L S + 1956 UMN 11(4),
183-192 (Z70, 5)


Alembert, Jean Babtiste le Rond d'
1717-1783

*Bertrand J 1889 *d'Alembert,* Paris
Broglie L de 1951 Rv Hi Sc Ap 4,
204-212
Delannoy 1895 Fr SMB 23, 262-265
(F26, 53) [probability]
Dobrovolskii V A 1968 *Dalamber,*
Moscow (Znanie), 31 p
Dugas R 1955 Par CR 241, 1437-1438
(MR17, 337)
Dupont P 1963 Tor FMN 98, 537-572
(MR27, 638) [center of rotation,
Euler, Poisson, Poncelet]
Favaro A 1884 Ven I At (6)1, 533-545
[Lagrange, letters]
Grimsley R 1963 *Jean d'Alembert,*
Oxford, 316 p [bib]
Henry C 1886 Boncomp 19, 136-148
[letters, Euler]
Janet P 1933 Par CR 196, 452-455
Jelitai J 1937 MP Lap 47, 173-199
[Clairault, La Condamine, Teleki]
Korner T 1904 Bib M (3)5, 15-62
Laissus Y 1954 Rv Hi Sc Ap 7, 1-5
[letter]
Liouville J 1837 JMPA 2, 245-
[Lagrange, letter]
Loria G 1936 Int Con HS (1934) 3,
146-160 [Encyclopédié]
Masotti A 1928 Pe M 8, 239-256
(F54, 8)
Maupin G 1895 Fr SMB 23, 185-190
(F26, 53)
Muller Maurice 1926 *Essai sur la
philosophie de Jean d'Alembert,*
Paris, 314 p
Sanford V 1934 MT27, 315

Stackel P 1910 Bib M (3)11,
   220-226  (F41, 11)  [Euler, letter]
Vahlen J 1899 Ber Si, 49
Valabrega E G 1962 Tor FMN 96, 644-654
   (Z108, 2, 249)  [probability]


Alexander, Andreas   XVII

Enestrom G 1903 Bib M (3)4, 290-291
   (F34, 7)


Alfred the Great   849-901

Clagett M 1954 Isis 45, 269-277
   [Euclid]
Jorg E 1933 Fors Fort 9, 275
- 1935 Des Boetius und des Alfredus
   Magnus Kommenter zu den "Elementen"
   des Euklid Buch 2,  Heidelberg,
   34 p   (diss)  (Z12, 97)


Alfsen, Magnus   1870-1943

Piene K 1943 Nor M Tid 25, 1-5
   (Z27, 290)


Alhazen   965?-1039?

   = Abu Ali al-Hasan ibn al-Hasan
   (Husain) ibn al-Haitham.

Amir-Moez A R 1956 M Mag 30, 93
   (MR18, 630)  [geom, problem]
Dilgan H 1955 Instanb UB 8, 36-41
   (Z67, 245)  [ms]
Goeje M J de 1901 Ned Arch (2)6, 668
Heiberg J + 1910 Bib M 10, 201-237,
   293-307  [parabola]
- 1911 Bib M 11, 193-208  [parabola]
Lindberg D G 1967 Isis 58, 321-341
   [optics]
Narducci E 1871 Boncomp 4, 1-48,
   137-139  (F3, 4)
Nesselmann H G F 1843 Essenz des
   Rechenkunst von Mohammed Beha-eddin
   ben Allhasain aus Amul arabisch und
   deutsch herausgegeben, Berlin

Padoa A 1932 Pe M, 113-118  [problem]
Ronchi V 1953 Int Con HS 7(1953), 516
   [optics]
Rosenfeld B A 1958 Ist M Isl 11, 733-78
   (MR1964, 3; Z100, 293)
Schoy C 1921 Deu Morg Z 75, 242-253
- 1926 Isis 8, 254-260
Sedillot L A 1834 J Asi 13, 435-458
   [tr of extracts]
Steinschneider M 1883 Boncomp 14,
   721-740
- 1884 Boncomp 16, 505-513  [astronomy
Suter H 1899 ZM Ph 44, 33-47
   [circle squaring]
- 1911 Bib M (3)12, 289-332  (F42, 63)
   [paraboloid]
Wiedemann E 1893 Erlang Si 24, 83
- 1910 Bib M (3)10, 293-307
- 1911 Jb Phot 25, 6-11
Winter H J J 1954 Centau 3, 190-210
   (Z59, 1, 5; 60, 5)  [optics]
Winter H J J + 1949 Beng Asi J 15,
   25-40  (Z41, 338)  [paraboloid]
- 1950 Beng Asi J 16, 1-16
   (MR13, 809)  [optics, paraboloid]
Woepcke F 1851 L'algèbre d'Omar
   Alkhayyami, Paris  [bib]


Alibrandi, Pietro  -1921

Ugolini G B 1921 Vat NLA a75, 69-70
- 1922 Vat NLA a76, 127-128


Alliaume, Maurice  1882-1931

La Vallée Poussin C J de 1931 Rv Q Sc
   (4)20, 387-390  (F57, 1311)
Van Hecke A + 1932 Rv Q Sc (4)21, 192-
   220  (F58, 992)


Allman, George Johnston  1824-1904

Anon 1904 Nat 70, 83  (F35, 34)
Lamb H 1904 Brt AAS 74, 421

Alsted, Johann Heinrich   1588-1638

Maria V 1939 Gaz M 44, 342-345,
  398-402   (F65, 9)


Alvarus Thomas   XV-XVI

Valentin G 1914 Bib M 3(B14), 249-252


Amaldi, Ugo   1875-1957

Anon 1937 Vat An 1, 96-101   (F63, 19)
- 1954 It XL An (1953), 209-213
  (MR15, 591)
Ghizzetti A 1958 Ren M Ap (5)16,
  511-514   (Z84, 3)
Terracini A 1958 Tor FMN 92, 687-695
  (Z82, 243)
Viola T 1957 It UM (3)12, 727-730
  (Z77, 242)
- 1958 Archim 10, 33-37   (MR19, 1150)


Amanzio, Domenico   1854-1908

Amodeo F 1912 Vat NLA (2)17


Amici, Giovanni Battista   1786-1863

Palermo F 1870 Boncomp 3, 187-248
  (F2, 21)


Ampère, Andre Marie   1775-1836

d'Adhémar R 1905 Quinzain (5), 15 p
  (F36, 22)   [Cauchy, Hermite]
Guillemin E A 1932 Isis 18, 118-126
  [electromagnetic thy letters]
Launay L de + 1936 Par CR 203, 1194-
  1197   (F62, 1036)
Lebesgue H 1937 Par CR 204, 925-928
  (F63, 16)
Valson C A 1876 La vie et les travaux
  de A. M. Ampere, Lyon

Amsler-Laffon, Jakob   1823-1912

Amsler A + 1912 Zur NGV 56, 1-17
  (sep Zurich, 17 p)   (F43, 37)
Anon 1912 Ens M 14, 139 140   (F43, 37)
- 1912 Rv GSPA 23, 129


Anaxagorus of Clazomenae   -499?--428

Gershenson D + 1964 Anaxagoras and
  the Birth of Scientific Method
  (Blaisdell, pap)


Anaximander   -610--547?

Tannery P 1882 Rv Phi 13, 500-529


Andalo di Negro   1260?-1340?

Boncompagni B 1874 Boncomp 7, 337-376
  (F6, 26)
Favaro A 1876 Ven I At 2, 545-558
  (F8, 13)
Sarton G 1947 Int His Sc 3(1), 645-648
  [bib]
Simoni C de 1874 Boncomp 7, 313-338
  (F6, 26)
Thorndike L 1928 Isis 10, 52-56


Anderson, Oskar Nikolai   1887-1960

Sagoroff S 1960 Metrika 3, 89-94
  (MR22, 619)
Tinter G 1961 ASAJ 56, 273-280
  (MR22, 1594)
Wold H 1961 An M St 32, 651-660
  (MR23, A2301)


Anding, Ernst Emil Ferdinand   1860-1945

Archibald R C 1948 Mathematical Table
  Makers, NY, 7-8   [biobib]


Andoyer, Henrie Marie   1862-1929

Anon 1929 Ens M 28, 136-137

Boucheny G 1920 Lar Men 5, 29-30
Picard E 1930 Fr Long(1930) (D), 5 p


**Andrade, Jules Charles Frederick**
**1857-1933**

Lecornu L 1933 Par CR 196, 821-822
(F59, 34) [chronom, eng, mech]


**Andreas, Alexander   XV-XVI**

Enestrom G 1902 Bib M (3)3, 355-360
[XVI, Germany]


**Andreev, Konstantin Alekseevich**

Chernyaev M P 1956 Ist M Isl 9,
723-756   (Z70, 243)
Egorov D F 1922 Mos Mo Sb 31, 337-340
(F50, 15)
Gordevskii D Z 1955 *K. A. Andreev*
*vydayushchiicya russkii geometr,*
Kharkov, 46 p   (MR17, 697)


**Andres de Li, Mossen Juan   XV-XVI**

Lecat M 1915 Intermed 22, 20-21
Torner J 1914 Esp SM 4, 33-36
(F45, 5)   [arith, finger reckoning,
Spain]


**Andronov, Aleksandr Aleksandrovich**
**1901-1952**

Aizerman M A + 1955 *Pamyati A. A.*
*Andronova,* Moscow   (MR17, 337)
[automatic control]
Anon 1961 Aut RC 22, 1018-1020
(Z101, 9)
Gorelik G S 1953 UFN 49, 449-468
(MR15, 89)


**Angelescu, Aurel A.   1866-1938**

Ionescu D V 1941 Math Tim 17, 111-
128   (Z25, 293)


**Angeli, Stefano   1623-1697**

Tenca L 1954 Ven MN 112, 1-15
(Z60, 7)   [Viviani]
- 1958 Bln Rn P (11)5(1), 194-257
(Z99, 244)


**Anianus, Magister   XV**

Cajori F 1928 Archeion 9, 31-42
Sarton G 1931 Int His Sc 2(2), 992
Smith D E 1928 *Le complet manuel de*
*Magister Anianus,*   Paris, 197 p
(F54, 22)


**Anissimoff, W.   1860-1907**

Mordoukhay-Boltowskoy D 1909 Wars PI
An, 114 p   (F40, 3)


**Anjema, Hendrik   -1765**

Beeger N G W H 1948 M.Tab OAC 3, 331
(Z37, 291)   [tables]


**Antiphon   -V**

Rudio F 1907 *Der Bericht des Simpliciu*
*ueber die Quadraturen des Antiphon*
*und Hippokrates,* Leipzig, 194 p
(Urk GMA1)   (F38, 64)


**Antoine, M.L.A.   1888-**

Errera A 1957 Bel SM 9, 30-58
(MR20, 1243; Z88, 6)


**Antomari, Xavier   1855-1902**

Laisant C A 1902 Nou An (4)2, 239


**Antonelli, Giovanni   1818-1872**

Stiattesi A 1872 Boncomp 5, 253-276,
277   (F4, 19)

Anzai, Hirotada   1919-1955

Kakutani S 1955 Osak JM 7, i-ii
   (MR17, 337)


Apian, Philipp   1531-1589

Guenther S 1882 *Peter und Philipp
   Apian, zwei deutsche Mathematiker
   und Kartographen,* Prag Ab,
   [XVI, cartog, Ger]
North J 1965 Physis Fi 7, 211-214
   [Pacioli's polyhedra]


Apollonius of Perga   -III

Agostini A 1931 Pe M (4)11, 293-300
   (F57, 15; Z3, 98)  [conic]
Archibald R C 1910 Edi S P 28, 152-
   157  [Heraclitus]
Arendt F 1914 Bib M (3)B14, 97-8
Barbarin P + 1912 Intermed, 234-237
   [hyperbola]
Bilimovitch A 1953 Beo Ak N (NS)5,
   49-56  (MR17, 445)
Bortolotti E 1924 Pe M(4)4, 118-130
- 1929 Archeion 11, 395-396
   (F55, 8)
- 1930 Archeion 12, 267-271  (F56, 18)
   [calculus, Torricelli]
Bosmans H 1924 Mathesis 38, 105-148,
   205-207
Brocard H + 1906 Intermed 13, 267
Chasles M 1842 Par CR 14, 547-
- 1853 Par CR 37, 553-  [irrationals]
- 1854 JMPA 19, 413-430
   [irrationals; Woepcke]
Drobisch M W 1856 Leip Ber 8, 103-
   [conics Bk V]
Czwalina A 1926 *Die Kegelschnitte des
   Apollonios,* Munich, 220 p (based on
   Heiberg 1890)  (F52, 8)
Flauti V 1852 It SSMF 25, 223-
Garcia Tranque T 1949 Gac M (1)1,
   3-10  (MR11, 707; Z36, 145)
Gardiner M 1860 Vict RS 5, 19-
   [porisms]
Giannattasio F 1825 Nap Barbo 2, 317-
Giovannozzi P 1916 Vat NL Mm (2)2,
   1-31

Gohlke P 1925 *Auswahl aus den Werken
   des Archimedes und Apollonius,*
   Frankfurt, 36 p
Heath T L ed 1896 *Apollonius of Perga.
   Treatise on conic sections,*
   Cambridge  (Based on Heiberg 1890)
Heiberg J L 1890-1893 *Apollonii Pergaei
   ...,* Leipzig, 2 vols
Housel 1858 JMPA 3, 153
Kliem F 1927 *Apollonius,* Berlin, 75 p
Loria G 1930 Archeion 12, 13-14
   (F56, 12)  [Arab]
Metzner K 1930 N Jb W Jug 6, 474-486
   (F56, 12)
Milankovitch M 1956 Beo IM 9, 79-92
   (MR19, 108)  [Aristarchos]
Milne J J 1894 M Gaz, 49  [conics]
Neugebauer O 1932 QSGM (B)2, 215-254
   (Z6, 2)  [alg]
- 1955 CPAM 8, 641-648  (MR17, 337)
   [planets]
- 1959 Scr 24, 5-21  (Z84, 242)
Sangro G 1825 Nap Barbo 2, 45-
Soldaini E 1934 Pe M (4)14, 90-100
   (Z8, 337)
Tannery P 1881 BSM (2)5, 124-136
   (F13, 4)
Terquem 1844 Nou An 3, 350-352,
   474-488  [bib]
Thaer C 1940 Deu M 5, 241-243
   (MR2, 114; Z23, 386)  [constructions]
Ver Eecke P ed 1924 *Les coniques
   d'Apollonius,* Bruges
   (Based on Heiberg 1890)
Woepcke F 1856 Par Mm 14, 658-720


Appell, Paul Emile   1855-1930

Anon 1927 *Cinquantenaire scientifique
   de Paul Appell,* Paris, 47 p
   (anniv, port]
- 1956 Par U An 26(1), 13-31  [anniv]
*Appell Paul 1923 *Souvenirs d'un alsacien,
   1858-1922,* Paris, 317 p  (F49, 23)
*- 1925 Act M 45, 161-285  (F51, 33)
*Buhl A 1927 Ens M 26, 5-11
*--1931 Ens M 30, 5-21 [portrait]
- 1934 Ens M 33, 229-231  (F61, 28)
*Levi-Civita T 1931 Linc Rn (6)13,
   241-242  (F57, 41)

Luzin N 1931 SSSR Iz, 319-323
   (F57, 1308)
*Poincaré R 1930 Par U An 5, 463-477
   [facs, autog, port]
Pompeiu D 1930 Buc EP 2, 48-49
   (F56, 28)
- 1930 Math Cluj 4, 186-7
*Lebon E 1910 *Paul Appell. Biographie*
   *et bibliographie analytique des*
   *écrits*, Paris, 79 p  (F41, 36)


Aquinas, Thomas  1225?-1274

Bodewig E 1931 Arc Phi 11, 1-34
   (Z3, 242)
Iseukrahe K 1920 *Die Lehre des hl.*
   *Thomas Aquinis neuzeitlichen mathe-*
   *matik*, Bonn, 230 p
Moreno A 1963 Not DJFL 4, 113-134
   (MR29, 3359)  [logik]
Winace E 1955 Rv Neosco 53, 482-510


Arbogast, Louis Francois Antoine
   1759-1803

Brocard H 1909 Intermed 16, 125 (Q949)
   (See also Intermed 3, 275; 4, 141,
   275; 5, 154; 8, 313; 10, 9; 11, 76-
   77, 949)  [ms]
De Morgan A 1846 Cam Dub, 238-255
- 1851 Cam Dub  (Feb)
Frechet M 1920 Rv Mois, 337-362
   (repr in Frechet 1940 and in
   Frechet 1955 *Les math. et la*
   *concret,*  Paris)
- 1940 Thales 4, 43-55
   (MR9, 74; Z61, 5)
Tanner H W L 1890 Mess M (2)20,
   83-101
Zimmerman K 1934 *Arbogast als Mathe-*
   *matiker und Historiker des Mathe-*
   *matik*, Heidelberg, 58 p  (diss)
   (Z12, 98)


Arbuthnot, John  1667-1735

Aitken, George A 1892 *The Life and*
   *Works...,* Oxford  [statistics]

Beattie Lester M 1935 *John Arbuthnot*
   *mathematician and satirist,* Cambridg
   Mass, 448 p    [Huygens]


Archibald, Raymond Clare   1875-1955

Adams C R + 1955 AMM 62, 743-745
   (MR17, 337)
- 1955 Scr 21, 293-295
Sarton G 1956 Osiris 12, 5-34
   (MR18, 710; Z70, 244)
Stahlman W D 1956 Isis 47, 243-246
   (Z70, 244)


Archilla y Espejo, D. S.

Rueda C J 1913 Esp SM 2, 213-219
   (F44, 17)


Archimedes  -287?--212

*Aaboe A 1964 *Episodes from the Early*
   *History of Mathematics,* (Blaisdell,
   Singer), Ch 3  [trisection]
Allendoerfer C B 1965 MT 58, 82-88
   [angle, arc]
Amodeo F 1920 Nap FM Ri (3)26, 170-178
   (F47, 26)  [Nicolo de Martino]
Anon 1900 Pitagora 7, 79  [polygons]
*Anon ed 1963 *Città di Siracusa,*
   *celebrazioni Archimedee del secolo*
   *XX, 11-16 aprile 1961,* vol 1,
   Gubbio, 256 p·  (Z106,1; 108:2, 244)
Arendt F 1915 Bib M 14, 289-311
Audisio F 1929 Tor FMN 65, 101-108
   [pi]
Babini José 1948 Archimedes, Buenos
   Aires, 155 p  (Z38, 146)
- 1948 Arc In HS 28, 66-75  (Z31, 241)
Bashmakova I G 1953 Ist M Isl 6, 609-
   658  (MR16, 660; Z53, 244)
   [calc]
   1956 Int Con HS 8, 120-122  [calc]
- 1956 Ist M Isl 9, 759-788
   (Z70, 242)
- 1963 Arc HES 2, 87-101  [calc]
Beumer M G 1946 N Tijd 33, 281-287
   (MR8, 2)  [trisection]

Blass F 1883 As Nach 104, 255
[father]
Bockstaele P 1956-57 Nov Vet 34,
299-312 [circle]
Bosmans H 1922 Rv Q Sc (Ap), 5-23
[Moerbeke]
Bromwich T J I A 1928 M Gaz 14,
253-257 (F54, 14) [roots]
Brun V 1935 Nor M Tid 17, 1-13
(Z11, 193) [sphere]
Buffon G L L C de 1947 Par Mm, 82-
101 [burning mirrors]
Carra de Vaux B 1900 Bib M (3)1, 28
[machines]
Carruccio E 1938 Pe M (4)18, 207-216
(F64, 908) [octagon]
Casara G 1942 It UM (2)4, 244-262
(MR5, 353) [cubic]
Child J M 1921 Studies in the History
of Science, Oxford 2, 490-520
[balance]
Christensen S A 1885 Zeuthen (5)4,
47-50 [exhaustion]
Clagett M 1952 Isis 43, 36-38
(MR15, 276; Z47, 1) [mid, sphere]
- 1952 Isis 43, 236-242 (Z47, 242)
[Johannes de Muris, Moerbeke]
- 1952 Osiris 10, 587-618 (Z46, 1)
[circle, mid]
- 1954 Osiris 11, 295-358
(MR16, 660; Z55, 2) [cylinder,
sphere]
- 1955 Isis 46, 281-282
- 1957 Isis 48, 182-183 [Oresme]
- 1959 Isis 50, 419-429 (Z94, 2)
- 1964 Archimedes in the Middle Ages.
Vol 1: The Arabo-Latin Tradition
(Univ Wisconsin Pr), 749 p
(MR29, 1068)
- 1964 Koyre 1, 40-60
Culum Z 1954 Beo DMF 6, 108-111
(MR17,1; Z57, 241) [roots]
Czwalina A 1925 Archimedes, Leipzig-
Berlin, 47 p (F51, 16)
- 1928 Arc GMNT (2)10, 464-466
(F54, 14)
Davis H T 1945 SSM 44, 136-145
Dijksterhuis E J 1938-1939 Archimeses,
Groningen-Batavia, 2 vols (also in
Euc Gron 15 (1938), 96-135, 248-
289; 16 (1939), 104-131; 17 (1940),

8-41; (Z21, 193; F64, 905; F65:2,
1079; F66, 10)
- 1952 Ab WGWL 1, 5-31
- 1954 Nor M Tid 2, 5-23 (MR15, 923;
Z56, 242) [calc, intgl]
- 1956 The Arenarius of Archimedes,
Leiden, 24 p (MR18, 268; Z73, 2)
- 1956 Archimedes, Copenhagen (also
Act HSNM 12, 1-422; also 1957 NY,
Humanities Pr) (MR18, 981)
Drachmann A G 1963 Centau 8, 91-146
Duhem P 1900 Bib M (3)1, 15
[hydrost]
Erhardt R v + 1942 Isis 33, 578-602
[Aristarchos, Copernicus, sand-
reckoner]
- 1943 Isis 34, 214-215 (reply to
Neugebauer 1942)
Evans G W 1928 MT 20, 243-252
Favard A 1908 Ven I At 67, 635-638
(F39, 10) [ms]
- 1912 Archimede, Genoa, 83 p (F43, 7)
- 1912 Ven I At 71, 953-975 (F43, 9)
[Leonardo da Vinci]
- 1940 Archimede, 2nd ed, Milano, 88 p
(F66, 10; Z26, 98)
Fleckenstein J 1956 Ens M (2)2,
324-326 (Z71, 1)
Gazis D C + 1960 Scr 25, 228-241
(MR22, 1341) [roots]
Gericke H 1962 MP Semb 8, 215-222
(Z107, 2) [lever]
Gibson G A 1897 Edi MSP 16, 2
[arith prog]
Gohlke P 1925 Auswahl aus den Werken
des Archimedes und Apollonius,
Frankfurt, 36 p (F51, 16)
Gould S H 1955 AMM 62, 473-476
(Z65, 243) [calc, heuristic]
Guzzo A 1952 Filo Tor 3, 149-168
(MR14, 609; Z47, 1)
Heath T L ed 1897 The Works of
Archimedes, Cambridge, 326 p
- 1920 Archimedes, London, 58 p
Heiberg J L 1879 Quaestiones
Archimedeae, Copenhagen (F11,2)
- 1879 Hist Abt 24, 177-182
(F11, 3)
1880 Hist Abt 25, 1-67 (F12, 1)
[conics]
1887 Op Phil Co, 1-8 (F21, 1255)

- 1890 ZM Ph 34(5), 1-85 (also in Ab GM 5, 1-85)
- 1907 Hermes 42, 235-303 (F38, 6) [method]
- 1907 Kagan (450-451) [method]
- 1909 *Eine neue Schrift von Archimedes,* Odessa, 46 p (F40, 6)
- 1909 Monist 19, 202-224 (F40, 6) [method]
- 1909 *Festskrift til H.G. Zeuthen,* Copenhagen, 63-66 (F40, 62) [method]
Heiberg J L + 1907 Bib M (3)7, 321-363 (F37, 1) [method]
- 1916 Scientia 20, 81-89
Heller S 1954 Mun Abh (NF)(63), 39 p (MR17, 1; Z55, 242) [mistake]
Hermelink H 1953 Arc In HS (NS)6, 430-433 (MR15, 591; 52, 1) [mistake]
Hjelmslev J 1934 M Tid B, 61-67 (F60, 823) [exhaustion]
- 1950 Dan M Med 25(15) (MR11, 571; Z36, 2) [number]
Hofmann J E 1930 Arc GMNT 12, 386-408 (F56, 805) [root]
- 1934 DMV 43, 187-210 (Z8, 337) [Heron, roots]
- 1935 Unt M 41, 37-40 (F61, 11) [pi]
- 1954 Archims 6(5) (Z56, 242)
- 1963 Arc M 14, 212-216 (MR27, 463; Z118:1, 2) [regular solids]
Hoppe E 1922 Arc GNT 9, 104-107 [pi]
Hultsch F 1877 Hist Abt 22, 106-108 (F9, 2)
- 1893 Gott N(1), 367-429 [roots]
- 1894 Hist Abt 39, 121-137, 161-172 (F25, 79) [pi]
Johnson M C 1933 Scientia 53, 213-217 [Leonardo da Vinci]
Juel C 1914 Dan Ov, 421-440 (F45, 83)
Kagan V F 1951 *Arkhimed,* Moscow-Leningrad, 56 p (MR14, 1050)
Kierboe T 1913 Bib M (3)14, 33-40 (F44, 49)
Kliem F + 1927 *Archimedes,* Berlin, 151 p
Lambossy P 1929 Freib NG 29, 20-39 [hydrost, method]

Lejeune A 1947 Brx SS 61, 27-47 (Z31, 1) [optics]
- 1947 Isis 38, 51-53 (Z29, 385) [optics]
Lenzen V F 1932 Isis 17, 288-289 [lever]
Lorent H 1955 Mathesis 64(1-2)(S) (Z64, 1)
Lorey W 1908 Z Latein 19, 1-8 (F39, 9)
Loria G 1928 *Archimede,* Milan, 72 p (F54, 45)
- 1928 Bo Fir 24, i-ii [sphere]
Lure S Ya 1945 Arhimed, Moscow-Lenin, 272 p (MR9, 3; Z60, 5) [pop]
Mansion P 1908 Brx SS 32, 188-189 (F39, 9)
Midolo P 1912 *Archimede e il suo tempo,* Syracuse, 548 p (F43, 7) [pop]
Milhaud G 1908 Rv Sc (5)10, 417-423 (F39, 9)
Miller G A 1928 Sci 67, 555-556 (F54, 16) [trig]
Muller C 1932 QSGM (B) 2-3, 281-285 [roots]
Neugebauer O 1942 Isis 34, 4-6 (Crit of Erhardt 1942) (MR3, 258) [Aristarchus]
Oldham R D 1926 Nat 117, 337-338
Onicescu O 1961 Buc U Log 4(4), 113-11 (MR24, 220) [real]
Painlevé P 1907 Rv GSPA 18, 911-913 (F38, 7) [method]
Paoli H J M 1925 "Spammites" di Archimedes..., Buenos Aires, 92 p
Pixley L W 1965 MT 58, 634-636
Procissi A 1953 It UM (3)8, 74-82 (MR14, 832)
Read C B 1961 SSM 61, 81-84
Reeve W D 1930 MT 23, 61-62
Reinach T 1907 Rv GSPA 18, 911-928, 954-961 (F38, 6)
Richardt T 1925 Nor M Tid 7, 73-88 [root]
Rigaud S P 1835-37 Ashmo T 1(9), 32 p
Rome A 1932 Brx SS 52A, 30-41 [Theon]
Rose V 1884 Deu LZ 5, 210-213 (F16, 8)
Rudio Ferdinand 1892 *Archimedes, Huygens, Lambert, Legendre, vier*

*Abhandlungen ueber die Kreismessung
...,* Leipzig, 173 p
[squaring circle]
Rufini R 1926 *Il "Metodo" di
Archimede e le origini dell'analisi
infinitesimale nell'antichita,*
Rome, 301 p  (F52, 8)
Schaeffer J J 1958 Montv Did 1,
57-93  (MR21, 484; Z90, 5)
- 1902 Bib M (3)3, 176
Schmidt W 1900 Bib M (3)1, 13
[ephodikon]
Schrek D J E 1942 N Tijd 30, 1-13
(F68, 6)  [lune]
Severi Francesco 1949 Archim 1, 3-6
(Z31, 97)
Shoen H H 1934 Scr 2, 261-264, 342-
347  (Z10, 98)
Sibirani F 1939 It UM 1, 160-172,
259-274  (Z21, 194)  [spiral]
Sierpinski W 1961 Scr 26, 143-145
Sister Anne Agnes 1954 MT 47, 366-67
Slichter C S 1908 AMSB (2)14, 382-
393  (F39, 10)  [method]
Smith D E 1900 Monist 19, 225-230
(F40, 6)
Stamatis E S 1946 *Archimedes' Me-
chanics...,* Athens, 31 p
(Isis 53, 583)
- 1946 *Archimedes' Quadrature of the
Parabola,* Athens, 46 p
(Isis 53, 583)
- 1950 *Archimedes: Kreismessung,*
Athens, 30 p  (Z45, 290)
- 1965 Gree SM (NS)6(2), 265-297
(MR34, 5609)  [reconstr of ms]
Stein W 1930 QSGM 1(B), 221-244
[cent grav]
Suter H 1894 ZM Ph 44(S), 491-500
(F30, 51)  [arab]
Tannery P 1881 Bord Mm (2)4, 313-339
(F13, 40)  [pi]
- 1881 BSM (2)5, 25-30  (F13, 32)
[numb thy]
- 1882 Bord Mm (2)5, 49-63
Thébault V 1927 Ed M 29, 105-107;
30, 104  (F54, 14)
- 1928 Mathesis 42, 199-205
Thiele J N 1884 Zeuthen (5)2, 151-
153  [roots]
Thurot C 1869 Rv Arch

Timchenko I Yu 1913 Kagan (598-600)
[Democrates]
Toeplitz O 1932 QSGM (B)2-3, 286-290
[approx roots]
Tropfke J 1928 Arc GMNT 10, 132-163
[trig]
- 1936 Osiris 1, 636-651  [polygons]
Vacca G 1914 Linc Rn 23, 850-853
(F45, 74)
- 1940 It UM (2)3, 71-73  (F66, 10;
Z25, 145)  [optics]
- 1940 It UM Con 2, 900  (MR3, 97;
F68, 6; Z26, 289)  [optics]
Vailati G 1896 Tor FMN 32, 500
- 1897 Tor FMN 32, 742-758
(F28, 49)  [cent grav]
Van der Waerden B L 1954 Elem M 9, 1-9
(Z56, 1)  [heuristic, sphere]
Veselovskiy I N 1957 *Archimedes,*
Moscow, 111 p  (Z80, 2)
Vetter Q 1920 Cas MF 49, 224-243
(F47, 23)
- 1921 Cas MF 50, 81-89, 250-254
(F48, 2)
Vietzke A 1912 MN Bl 9, 1-3  (F43, 8)
Vogel K 1932 DMV 41, 152-158
[approx roots]
Vollgraff J 1913 Ned NGC 14, 211-213
(F44, 39)
Vries H de 1926 N Tijd 13, 1-15
(F52, 8)
Weissenborn H 1883 Hist Abt 28, 81-99
[approx roots]
- 1894 Ber CPA 14  (F25, 80)
[Fibonacci]
Wieleitner H 1930 QSGM (B)1(2) 201-
220  [calc]
Yushkevich A P 1948 Mos IIET 2, 567-
572  [Euclid, Russia]
Zeuthen H G 1879 Zeuthen (4)3, 145-
155  (F11, 3)
- 1893 Bib M (2)7, 97-104  (F25, 61)
[cubic]
- 1909 Arc GNT 1, 320-327  (F41, 62)
Zubov V 1965 Ist M Isl 16, 235-272


Archytas of Tarentum  -V

Anon 1899 Pitagora 5(2), 93
Boeckh L 1841 *Ueber den Zusammenhang*

*der Schriften, welche der Pythag-*
*oraeer Archytas hinterlassen haben*
*soll,* Karlsruhe Lycemsprogramm
Fabricius 1716 Bib Graec 1, 833
Graesser R F 1956 MT 49, 393-395
 [Delian probl]
Skof F 1958 Pe M (4)36, 19-40, 76-92
 (Z88, 5) [Delian probl]
Tannery P 1878 Bord Mm (2)2, 277-283
 (F10, 25) [Delian probl, Eudoxos]
- 1886 BSM (2)10, 295-302
 [Democritos]
Thaer C 1909 Jena Sem, 13-15
 (F40, 66) [Delian probl]


Arentz, Friedrich Christian Holberg
 1736-1825

Arvesen O P 1941 Nor VSF 13, 54-57
 (F66, 18) [lin eqs]


Argand, Jean Robert  1768-1822

Enestrom G 1895 Bib M (2)9, 32, 64
Enestrom G + 1902 Intermed 9, 74
Fehr H 1902 Bib M (3)3, 145
 (F33, 19)


Argoli, Andrea  1570-1657

Fabris V 1895 Abruz BSS 8, 247-257
 (F26, 12)


Aristaeos the Elder  -IV

Conte L 1953 Pe M (4)31, 265-274
 (MR15, 383; Z53, 245)


Aristarchos of Samos  -III

Besso D 1889 Pe M 4, 14-17
Erhardt R v + 1942 Isis 33, 578-602
 [Archimedes, Copernicus]
*Heath T L 1913 *Aristarchus of Samos,*
 *the Ancient Copernicus,* London,
 Oxford

- 1920 *The Copernicus of Antiquity,*
 London, 64 p
Martin H 1870 Boncomp 3, 299-302
 (F2, 17) [false attribution]
Meyerhof M 1942-43 Eg IB 25, 269-74
Milankovitch M 1956 Beo IM 9, 79-92
 (MR19, 108) [Apollonius]
Miller M 1958-59 Dres Verk 6, 257-265
 (MR22, 259)
Neugebauer O 1942 Isis 34, 4-6
 (Z60, 5) [Archimedes]
Tannery P 1882 Bord Mm (2)5, 237-258
- 1888 Bord Mm (3)4, 79-96
Thirion J 1913 Rv Q Sc 24, 91-126


Aristophanes  -IV

Rudio F 1907 Bib M (3)8, 13-22
 (F38, 64) [circle sq]


Aristotle

Apostle H G 1952 *Aristotle's Philos-*
 *ophy of Mathematics,* Chicago,
 238 p (MR14, 831)
Bollinger Jenny 1925 *Die sogenannter*
 *Pythagoraeer des Aristoteles,*
 Zurich, 78 p (F51, 15)
Borkowski L 1957 Stu Log 5, 13-26
 (MR19, 518)
Boussinesq J 1915 Par CR 161, 21-27
 [mech]
- 1915 Par CR 161, 45-47
- 1915 Par CR 161, 65-70 [mech]
Boyer C B 1945 Sci Mo 60, 358-364
 [Archimedes]
Buerja A 1790-1791 Ber Mm (1790),
 257; (1791), 266
Dehn M 1936 Scientia 60, 12-21, 69-74
 (F62, 1016; Z14, 145)
Drabkin I E 1950 Osiris 9, 161-198
 (Z41, 337) [wheel paradox]
Dutordoir 1900 Brx SS 24(1), 52-
 [nat of math]
Einarson B 1936 Am J Phil 57, 33-54,
 151-172 (F62, 1016)
Goerland A 1899 *Aristoteles und die*
 *Mathematik,* Marburg, 249 p
 (F30, 3)

Greenwood T 1942 Ott UR 12, 65-94
- 1950 Can Rv Tri 36, 380-386
  (MR12, 311)
- 1954 Thomist 17, 84-94 [constr]
Heath T L 1899 Z M Ph 44(S), 153-180
  (F30, 50) [constr]
- 1949 Mathematics in Aristotle,
  Oxford, 305 p (MR10, 667)
Heiberg J L 1904 Ab GM 18, 1-49
  (F35, 6)
Heller S 1955 Centau 4, 34-50
  (Z66, 244)
Lacombe G 1939 Aristoteles latinus,
  Rome
*Lukasiewicz J 1951 Aristotle's
  Syllogistic from the Standpoint of
  Modern Formal Logic, Oxford
  (2nd ed 1957)
Lumpe A 1955 Arc Begrf 1, 104-116
  (Z65, 242)
Mansion P 1913 Bib M (3)14, 30-32
  (F44, 46) [geom]
Markovic Z 1961 Scientia 15, 37-41
  [Plato]
McCall Storrs 1963 Aristotle's Modal
  Syllogisms, Amsterdam, 108 p
  (MR27, 1078)
Mieli A 1932 Arc Sto 14(2), 169-182
Milhand G 1906 Etudes sur la penseé
  scientifique chez les Grecs et chez
  les modernes, Paris, 137-176
  [prob]
Mundolfo R 1934 Rv Fil 25, 210-219
  (F60, 823) [infinity]
Pastore A 1912 Tor FMN 47, 201-217
  [def, phi math]
- 1912 Tor FMN 47, 478-494 [def]
Poske F 1883 Hist Abt 28, 134-138
Reidemeister K 1943 Ham M Einz 37
  (MR8, 189)
Ruffini E 1920 Rass MF 1, 133-144
- 1923 Arc Sto 4(1), 78-92
  [Greek geom]
Stenzel Julius 1924 Zahl und Gestalt
  bei Platon und Aristoteles,
  Leipzig, 154 p [Plato]
- 1929 QSGM (B)1(1), 34-66
Struik D 1925 Nieu Arch 15, 121-37
Timpanaro-Cardini M 1961 Physis Fi 3,
  105-112 (Z101, 242)
Toeplitz O 1929 QSGM (B)1, 3-33
  [Plato]

Vailati G 1896 Tor FMN 32, 678
Vredenduin P G J 1960-61 Euc Gron 36
  (MR23A, 3) [number]
Zuercher Josef 1952 Aristoteles'
  Werk und Geist, Paderborn, 435 p
  (Z66, 244)

**Arkhangelskii,** Andrei Dmitrievich
1879-1940

Anon 1940 SSSR Vest (8-9), 78-85
  (MR2, 115)

**Armellini,** Giuseppe  1887-1958

Anon 1937 Vat An 1, 102-112
  (F63, 19)
Cecchini G 1958 It UM (3)13, 615-617
  (Z82, 243)
- 1959 Tor FMN 93, 587-596  (Z87, 6)
Cimino M 1959 Ren M Ap 5(17), 475-
  492  (MR23A, 136; Z89, 5)
Nobile V 1959 Linc Rn (8)27, 275-288
  (Z88, 294)

**Arnauld,** Antoine  1612-1694

= Arnau

Bopp K 1902 Ab GM 14, 187-336
  (F33, 16)
- 1929 Drei Untersuchungen zur
  Geschichte der Mathematik, Berlin,
  66 p (F55, 606)
Brocard H 1907 Intermed 14, 273-275
*Coolidge J L 1949 The Mathematics of
  Great Amateurs, Ch 7  (repr 1963
  Dover)
Schrecker P 1936 Thales 2, 82-90
  [neg numb]

**Arnoux,** Gabriel  1831-1913

Laisant C A 1913 Ens M 15, 337-339
  (F44, 28)

Arnovicevic, Ivan 1869-1951

= Arnov

Vrecko M 1952 Beo Ak N Z6 18(2), 1-8
(Z47, 6)

Aronhold, Siegfried Heinrich
1819-1884

Gundelfinger S 1901 Crelle 124, 59
Lampe Emil 1899 *Die reine Mathematik
in den Jahren 1884-1889,* Berlin,
33-48 [invar]

Arrest, H. L. d' 1882-1875

Kobell F V 1976 Mun Si, 124  (F8, 22)

Artin, Emil  1898-1962

*Brauer R 1967 AMSB 73, 27-43
(MR34, 2417) [bib]
Cartan H 1963 Ham Sem 28, 1-6
(MR30, 872)
Chevalley C 1964 Fr SMB 92, 1-10
(MR1965, 639; Z117:1, 9
Schoeneberg B 1963 MP Semb (NF)10,
1-10  (Z113, 2)
- 1966 Ham MG 9(3), 30-31
(MR33, 3876)
Zassenhaus H 1963 Not DJFL 5, 1-9
(MR30, 872) [list of students]

Artobolevskii, Ivan Ivanovich
1905-

Dikushin V I 1957 SSSR Mash 17(63),
29-36, 73-85  (Z88, 6)

Artus de Lionne  1583?-1663?

Hofmann J E 1939 M Mag 12  (F64, 16)
[area, circle sq]

Aryabhata  476-550?

Datta B 1927 Clct MS 18, 5-18
- 1932 Clct MS 24, 19-36  (F58, 19)
[diophantine anal]
Gânguli S K 1926 Bihar ORS, 78-91
- 1926 Clct MS 17, 195-202
(F52, 10) [Greek numer]
- 1927 AMM 34, 409-415 [arith]
- 1930 AMM 37, 16-29 [pi]
Kaye G R 1909 Bib M (3)10, 289-292
(F40, 7)
Mueller C 1940 Den M 5, 244-255
(MR2, 114; Z25, 146)  [area,
cone, vol]
Prasad B N + 1951 Alla UMAB 15, 24-32
(MR13, 420; Z43, 243)
Sengupta P C 1930 Clct MS 22, 115-120
(F56, 807)

Arzachel.  See Zarkali

Arzela, Cesare  1847-1912

Anon 1912 Pe MI (a)28, 10-11
Bortolotti E 1912 Mathes It 4, 66-73
(F43, 38)
Pincherle S 1911 Bln Rn 16, 159-176
Severini C 1930 It UM 9, 116-117
Sibirani F 1912 Pe MI (3)10, 45-48
(F43, 38)

Asachi, Gheorghe  1788-1869

Cazanacli V 1937 Gaz M 43, 64-69, 120-
123, 235-239, 344-347, 398-403,
453-457 (F63, 16, 812; 64, 19)
Cimpan F 1956 Iasi UM (NS)2, 333-340
(Z73, 241)

Ascoli, Giulio  1843-1896

Riboni G 1901 Pe MI (2)4, 144-151
(F32, 12)

Ascoli, Guido  1887-1957

Picone M 1958 Linc Rn (8)24, 614-625
  (Z80, 9)
Tricomi F G 1957 Tor Sem 16, 5-9,
  15-35  (MR19, 1247; Z77, 242)  [bib]
- 1957 It UM (3)12, 347-352
  (MR18, 519; Z77, 8)
- 1958 Tor FMN 92, 180-184  (Z82, 12)
Zin G 1957 Tor Sem 16, 11-14
  (Z77, 242)


Aubrey, John  1626-1697

Anon 1950 Lon Tim LS (Jan 13, 20)


August, Ernst Ferdinand  1797-1870

August F 1870 Arc MP Lit 51(204), 1-5
  (F2, 25)


Augustine, Saint  354-430

Mitchell S O 1959 Modern Sc 37, 49-52


Ausonius, Decimus Magnus  310?-390?

  = Ausone

Francon M 1951 Isis 42, 302-303
  (Z45, 145)  [perf numb]


Autolycos of Pitane   -IV--III

Falco V de 1930 QSGM (B)1, 278-300
  [Hypsikles]
Mogenet J 1948 Arc In HS 2, 139-164
  (MR10, 173; Z31, 1]
- 1950 Autolycos de Pitane, Louvain,
  336 p  (Z41, 337)
Schmidt O 1952 Sk Kong (1949),
  202-209  (MR14, 831; Z48, 241)
Tannery P 1886 Bord Mm (3)2, 173-199


Autonne, Léon  1859-1916

Autonne L 1913 Notice sur les recherches
  mathématiques de Léon Autonne,
  Paris, 36 p  (F44, 40)


Avicenna  980-1037

  = Abu Ali al-Hussein Ibn Abdallah
  ibn Sina = Aven Sina

Akhadova M A 1964 Bakh PIUZ 1(13),
  143-205  (MR34, 5611)  [geom]
*Carra de Vaux 1900 Avicenne, Paris,
  310 p
Lokotsch K 1912 Avicenna als Mathe-
  matiker..., (Diss, Bonn), 69 p
  (F43, 8)  [Euclid, planimetry]
Rescher N 1963 Not DJFL 4, 48-58
  (Z118:1, 247)  [logic]
Sarton G 1927 Int His Sc 1, 709-  [bib]
Tannery P Bu Sc M Ast 6, 142-144
  [casting out nines]
Wiedemann E 1925 Z Instr 45, 269-275
  [instruments]
Woepcke F 1863 J Asi 1, 502-504
  [casting out nines]


Azzavelli, Mattia  1811-1897

De Rossi M S 1898 Vat NLA 51, 49-55
  (F29, 18)


Babbage, Charles  1792-1871

Anon 1960 Elek Rech 2, 7-8  (Z87, 273)
Babbage Charles 1864 Passages from the
  Life of a Philosopher, London, 512 p
  [autobiog]
Babbage H B ed 1889 Babbage's Calcu-
  lating Engines..., London, 342 p
  [bib]
Bowden B V 1960 Think 26, 28-32
  [computer]
Buxton L H 1933-34 Newcomen 14, 43-65
Comrie L S 1946 Nat (Oct 26)
Fyvie John 1906 Some Literary Eccen-
  trics, London, 179-209  [port]

Lyons H G 1941 Lon RSNR 3, 146-148
[ophthalmoscope]
Macfarlane Alexander 1919 *Lectures
on Ten British Physicists of the
Nineteenth Century*, NY, 71-83
Menabrea L F 1842 Taylor Ms 3, 666-
731
*Morrison P + 1952 Sci Am (April),
66-73
*Morrison P + *Charles Babbage and His
Calculating Engines* (Dover), NY
Mullett C T 1948 Sci Mo 67, 361-71
Neville E H 1926 M Gaz 13, 163-164
(F52, 25)  [fun eq]
Quetelet A 1873 Bel Anr 149-169
(F6, 28)
Smith D E 1922 AMM 29, 114
[Mme. Laplace]
Sueltz B A 1965 MT 58, 446-7
[tables]


Bacaloglu, Emanoll  1830-1891

Cimpan F 1939 Gaz M 44, 393-398
462-468  (F65, 16)
- 1957 Iasi UM (NS)(1)3, 423-428
(Z82, 11)
Myller A 1935 Gaz M 41, 123-126
(F61, 953)


Bach, Carl von  1847-1931

Mises R v 1931 ZAM Me 11, 403-404
(F57, 39)


Bachet de Meziriac  1581-1638

Collet C G + 1947 Rv Hi Sc AP 1,
26-50  (MR10, 420)
Henry C 1879 Boncomp 12, 477-568,
619-740  (F11, 16)
[Fermat, Malebranche]
Matrot A 1891 Fr AAS 20, 185-191
(F23, 35)


Bachmann, Paul  1837-1920

Haussner Robert 1921 in Paul Bachmann,
*Grundlehren der neueren Zahlen-
theorie,* 2nd ed, Leipzig
Hensel K 1927 DMV 36, 31-73
(F53, 30)  [bib, port]
Lorey W 1936 Muns Semb 8, 70-76
(F62, 26)  [Rudolf Sturm]


Backlund, Albert Victor  1845-1922

Ossen C W 1929 DMV 38, 113-152
(F55, 17)


Bacon, Francis  1561-1626

Eiseley L F C 1931 Sci 133, 1197-
Milhaud G 1917 Scientia 21, 185-198
[Descartes]


Bacon, Roger 1214?-1294?

Anon 1915 Ens M 17, 53-56  (F45, 5)
- 1933 Nat 132, 809-810
Duhem P 1907 Par CR 146, 156-158
Forsyth A R 1905 Brt AAS 75, 307
Guareschi I 1916 Tor Mm (2)65(4, 9)
Keyser C J 1937 Scr 5, 177-180
Smith D E 1914 in A G Little, *Roger
Bacon Essays,* Oxford
Steele R 1933 Isis 20, 53-71
Thomson S H 1937 Isis 27, 219-224
[time, mech]


Bagdadinus  XI-XII

= Muhammad ibn abd al-Baqi
al-Baghdadi

Sarton G 1927 Int His Sc 1, 761-762
[bib]
Suter H 1903 Bib M 4, 19-27, 24-25
[ Euclid, Gerhard of Cremona]
- 1905 Bib M (3)6, 321-322  (F36, 64)
- 1907 Bib M 7, 234-251  [Euclid]

Bagnera, Giuseppe 1865-1927

Anon 1927 Ens M 26, 144 (F53, 34)
- 1929 Bo Fir (2)8, xlviii
  (F56, 815)
Severi F 1928 Linc Rn (6a)8(2),
  xii-xx


Bailey, Wilfred Norman 1893-1961

Slater L J 1962 LMSJ 37, 504-512
  (Z106, 3; MR1962, 967)


Baire, Louis René 1874-1932

Buhl A 1933 Ens M 31, 5-13 (F59, 41)
Lebesgue H 1932 Par CR 195, 86-88
  (F58, 992)
- 1957 Ens M (2)3, 28-30 (Z77, 8)


Baker, Henry Frederick 1866-1956

Edge W L 1957 Edi MSN (41), 10-28
  (MR19, 108; Z77, 8)
Hodge W V D 1957 LMSJ 32, 112-128
  (MR18, 710; Z77, 8)


Balakram, H.

Anon 1929 Indn MSJ 18, 17-72
  (F55, 32)


Balbin, Valentin 1851-1901

Babini J 1964 Isis 55, 82-85
  (Z117:2, 250) [Argentina, serial]


Baldi, Bernadino 1553-1617

Drake S + ed 1969 Mechanics in Sixteenth-century Italy (Univ of
  Wis Press), 428 p [bib]
Steinschneider M 1872 Boncomp 5, 427-
  534 (F4, 4) [Arab]
- 1874 Vite di matematici Arabi
  tratte da un'opera inedita di
  Bernardino Baldi con note, Rome
  (repr of 1872)
Zaccagnini G Bernardino Baldi nella
  vita e nelle opere, 2nd ed
  Pistoia, 372 p


Baliani, Giovanni Battista 1582-1666

Natucci A 1959 Geno AL 16, 13-27
  (MR24A, 3)
- 1960 Arc In HS 12, 267-283
  (Z98, 7)
- 1960 Int Con HS 9, 547-552
  (Z114, 7)
Tannery P 1885 Bib M, 200
  [curve, Fermat]
- 1896 Intermed 3, 213 (Q797) [helix]


Ball, Robert Stawell 1840-1913

Anon 1914 Lon RS (A)91, xvii-xxii
*Ball R S 1915 Reminiscences and
  Letters, London [ast, dyn, screws]
JLED 1913 Nat 92, 403-404


Ball, Walter William Rouse 1850-1925

Cajori F 1926 Isis 8, 321-324
Lefebvre B 1908 Rv Q Sc (3)12, 252-
  267, 558-578 (F39, 3)
- 1909 Mathesis (3)9(S), 1-72
Martin A 1892 NYMS 2, 10-11 (F24, 46)
  [mistake]
Miller G A 1902 AMM 9, 280-283
  [Ball's history]
Whittaker E T 1925 M Gaz 12, 449-454
  (F51, 29)


Ballo, Giuseppe XVI-XVII

Giacomelli R 1912 Nap FM Ri (3)18,
  166
- 1914 Nap FM At (2)15, 25 p

Ballore.  See Montessus.

Balmer, Johann Jakob  1825-1898

Balmer H 1961 Elem M 16, 49-60
  (MR, 135;  Z99, 5)

Baltzer, Heinrich Richard  1818-1887

Thaer A 1889 ZMN Unt 20, 312-314

Banach, Stefan  1892-1945

Anon 1946 UMN (NS)1(3, 4), 13-16
  (MR10, 174;  Z60, 13)
*Anon 1946 Colloq M 1(2), 67-192
  (special issue; papers by Orlicz,
  Steinhaus, Marcewski)
  (MR10, 174;  Z37, 291)
Mazur S 1961 Wiad M 4, 249-250
  (Z96, 4)
Steinhaus H 1960 Pol Ac Rv 5(3-4),
  82-90  (MR22, 1341; Z93, 9)
- 1961 Scr 26, 93-100
- 1961 Wiad M 4, 251-259  (Z96, 4)
- 1963 Stu M (S)(1), 7-15
  (Z107, 248)
Szökefalvi-Nagy B 1961 Wiad M 4, 269
  -270  (Z96, 4)

Banachiewicz, Tadeusz  1882-1954

Witkowski J 1959 Wiad M 2, 197-203
  (Z91, 6)

Bandini, Silvio

Vivanti G 1927 Pe M (4)7, 355-356
  (F53, 34)

al-Banna  1256?-1321?

  = Abul Abbas Ahmad ibn Muhammed
  ibn Uthman al-Azdi ibn al-Banna

Marre A 1864 Vat NLA 17, 289-
- 1865 Vat NLA 19, 1-
- 1879 Cron Cien 2, 329-332
  (F11, 36)  [mult]
- 1879 Nou An (2)18, 260-265
  (F11, 36)
Renaud H P J 1937 Isis 27, 216-218
  (F63, 12)
- 1938 *Ibn al-Banna de Merrakech...*
  (Hesperis), Paris, 30 p
Rodet L 1879 Fr SM B 7, 159-167
  [approx]
Steinschneider M 1877 Boncomp 10,
  313-314  (F9, 2)
Suter H 1901 Bib M 2, 12-40  [arith]

Banneker, Benjamin 1731-1806

Mulcrone T F 1961 MT 54, 32-37

Baqi.  See Bagdadinus.

Baraniecki, Marian Alexander

Dickstein S 1895 Wszech 14, 145-149
  (F26, 33)

Baranowski, Anton

Dickstein S 1903 Wiad M 7, 107-109
  (F34, 27)

Barbarin, Paul Jean Joseph  1855-1931

Buhl M A 1932 BSM (2)56, 72-78
  (F58, 46)
- 1932 Ens M 30, 287-8  (F58, 991)
CCD 1933 Arg SC 115, 46-49  (F59, 852)
Halsted G B 1908 AMM 15, 195-196

**Bari**, Nina Karlovna  1901-1962

Lavrentev M A + 1951 UMN (NS)6(6),
  184-185 (MR13, 612; Z43, 5)
Menshov D E  + 1962 Mos UM Me (1)
  (MR25, 389)
- 1962 UMN 17(1), 121-133
  (MR24A, 468)

**Barnes**, Ernest William  1874-1953

Bailey W N 1954 LMSJ 29, 498-503
Whittaker E T 1954 Lon RS Ob 9,
  15-25  (MR 16, 660)

**Barozzi**, Francesco  1537-1604

Boncompagni B 1885 Boncomp 17,
  795-848

**Barrau**, Johan Anthony 1873-1953

Popken J 1953 Nicw Arch (3)1, 89-91
  (MR14, 1050; Z50, 2)

**Barrell** F. R.  1873-1915

Anon 1915-1916 M Gaz 8, 214 (F46, 35)
- 1915 Nat 96, 402-403  (F45, 49)

**Barrow**, Isaac  1630-1677

Anon 1930 Nat 126, 658 (F56, 19)
- 1933 MT 25, 487-488
Azpeitia A G 1956 Gac M 9, 123-129
  (MR18, 982; Z72, 246)
Cajori F 1929 MT 22, 146-151
  [Hobbes, Wallis]
Child J M 1916 Monist 26, 251-267
  [geom]
- 1916 Open Ct 30, 65-69  [tangents]
- 1930 Sci Prog 25, 295-307
  [calc, Leibniz, Newton]
Lindemann F 1927 Mun Si 3, 273-284
  [Fabri, Leibniz]
Mathews G B 1916 Nat 100, 222
  (F46, 34)

Osmond Percy H 1944 *Isaac Barrow:
  His Life and Times*, London, 237 p
Reeve W D 1932 MT 24, 487-88
Schaaf W L 1955 MT 48, 565-66
Zeuthen H G 1897 Dan Ov, 565-606
  (F28, 40) [calc]
- 1897 Int Con 1, 274-280  [calc]

**Barsukov**, Aleksandr Nikolaevich

Andronov I K + Matv Shk (1), 72-74

**Bartolmeo** da Parma  XIII

Narducci E 1884 Boncomp 17, 1-120,
  165-218 [sphere]
Sarton G 1931 Int His Sc 2(2), 988
  [bib]

**Bassani**, Anselmo

Anon 1911 PE MI (2)9, 47-48

**Bassett**, Alfred Barnard  1854-1930

Lamb H 1931 LMSJ 6, 239-240
  (F57, 41)

**Basso**, Giuseppe  1842-1895

Ferraris 1895 Tor FMN 31, 3  (F26, 28)

**Bateman**, Harry  1882-1946

Bell, E T 1946 QAM 4, 105-111
  (Z60, 13)
Erdelyi A 1947 LMSJ 21
  (MR9, 74; Z61, 5)
- 1847 Lon RS Ob 5, 591-618
  (MR12, 311)
Murnaghan F D 1948 AMSB 54, 88-103
  (Z 31, 99)

Bath.  See Adelard.

Battaglini, Giuseppe  1826-1894

Amodeo F 1906 Nap Pont 36(3), 64 p
  (F37, 15)
- 1907 Batt 45, 229-274  (F38, 23)
Capelli A 1894 Batt 32, 205-208
  (F25, 44)
Fambri P 1893 Ven I At (7)5, 1419-
  1420  (F26, 30)
Galdeano Z G de 1894 Prog M 4, 195-
  196  (F25, 44)
Ovidio E d' 1894 Tor FMN 29, 678-679
  (F25, 43)
- 1895 Linc Mr (4)1, 558-610
  (F26, 29)
Pascal E 1894 Rv M 4, 91-96
  (F25, 43)
Pinto L 1894 Nap FM Ri (2)32, 49-54
  (F25, 43)
Torelli G 1894 Paler R 8, 169-179
  (F25, 43)


al-Battani  858?-929

  = Albategnius = Albatenius
  = Abu Abdallah Muhammad ibn Jabir
  ibn Sinan al-Battani al-Harrani
  al-Sabu

Nallino C A ed 1899, 1903, 1907 Al
  Battani sive Albatenii Opus
  astronomicum..., Milan, 3 vols
  [X, ast]
Sarton G 1927 Int His Sc 1, 602-603
  [bib]


Bauer, Gustav  1820-1906

Anon 1906 Leop 42, 82  (F37, 25)
- 1907 Loria 10, 64  (F38, 29)
Voit C v 1907 Mun Si 37, 249-257
  (F38, 29)
Voss A 1907 DMV 16, 54-75  (F38, 28)


Bauer, Mihaly  1874-1945

Rédei L 1953 M Lap 4, 241-262
  (MR15, 923)

Bauernfeind, Karl Maximilian von
  1818-1894

Anon 1894 ZMN Unt 25, 465-466
  (F25, 44)


Bauhuys, P.

Lecat M 1914 Intermed 21, 114, 132


Baur, Carl Wilhelm  1820-1894

Anon 1894 ZMN Unt 25, 388-392
  (F25, 44)
Cranz C 1898 Wurt Sch 1, 385


Bauschinger, Julius  1860-1934

Anon 1933 Ber Si, 23-32, 889-890
  [ast]
Archibald R C 1938 Mathematical Table
  Makers, NY, 10-11  (Scr Stu 3)
  [biobib]
Hopmann J 1934 Leip Ber 86, 299-306
Stracke G 1934 Leip As 69, 146-163
  [port]


Bayes, Thomas  1702-1761

Anderson J G 1941 M Gaz 25, 160-162
Barnard G A 1958 Biomtka 45, 293-315
  (Z85, 6)
Deming W E 1940 Facsimiles of Two
  Papers by Bayes, Washington, 71 p
  (F66, 16)


Beaugrand  XVII

De Waard C 1918 BSM (2)42, 157-177,
  327-328  [calc, Descartes, Fermat]
Lefner A 1909 Intermed 16, 263
  (Q3565)  [Torricelli]


Beccari, Nello

Levi G 1958 Linc Rn (8)24, 102-113
  (Z79, 5)

Beck, Rudolf Hans Heinrich 1876-1942

Salkowski E 1943 DMV 53(1), 91-103
(Z61, 5)

Becker, Ernst Emil Hugo 1843-1912

Archibald R C 1948 *Mathematical Table Makers,* NY, 11 (Scr Stu 3)
[biobib]
Jost E 1913 Leip St 48, 2-12 [port]
Valentiner W 1912 As Nach 192, 321-324

Beekman, Isaac 1588-1637

Brocard H 1908 Intermed 16, 59 (Q2778) (also Intermed 13, 106) [Descartes]
Dijksterhuis E J 1924 N Arc (2)14, 186-208 [mech]
- 1941 Arc Mu Tey 9, 268-342 [Stevin]

Behnke, Heinrich Adolph Louis 1898-

Lohmeyer H 1958 MN Unt 11, 227-229
(Z99, 246)

Behren, Arnold

Wunderlin W 1958 Schw Vers 58, 14-17
(Z79, 5)

Behrend, Felix A 1911-1962

Cherry T M + 1963 Au MSJ 4, 264-270 (MR30, 872; Z118:1, 111)
Neumann B H 1963 LMSJ 38, 308-310 (MR28, 2; Z112, 2)

Beldomandi, Prosdocimo de 1380?-1428

Favaro A 1879 Boncomp 12, 1-74, 115-251 (F11, 9)
- 1886 Boncomp 18, 405-423
- 1890 Bib M (2)4, 81-90 [astrolabe]

Riccardi P 1890 Bib M (2)4, 113-114 [astrolabe]
Sarton G 1947 Int His Sc 3(1), 740; 3(2), 1106 [bib]

Bell, Eric Temple 1883-1960

*Bell E T 1955 *Twentieth Century Authors, First Supplement,* NY [autobiog]
*Broadbent T A A 1964 Nat (11 Feb), 443
Smith E C 1938 Scr 5, 77-78

Bellacchi, G

Maroni A 1924 Bo Fir 3, 1-4
-1924 Pe M (4)4, 264-5

Bellavitis, Giusto 1803-1880

Abonné 1881 Nou An (2)20, 137-139 (F13, 24)
Alasia C 1906 Ens M 8, 97-117 (F37, 14) [letters]
Alasia C + 1902 Intermed 9, 30
Anon 1881 Linc At (3)5, 15-19 (F13, 23)
Brocard H 1906 Intermed 13, 145 (Q2137) (also Intermed 8, 189; 9, 30)
Favaro A 1881 ZM Ph 26-2, 153-169 (F13, 23)
Laisant C A 1880 BSM (2)4, 343-348 (F12, 15)
- 1887 *Theorie et applications des equipollences,* Paris
Laquiere E M 1881 Fr AAS, 76-84
Legnazzi E Nestore 1881 *Commemorazione del prof. Giusto Bellavitis,* Padova [bib]
Teixeira F G 1880 Teixeira 2 (F12, 15)
Turazza D 1882 Ven I At (5)8, 395-422

Beltrami, Eugenio 1835-1900

*Anon 1901 Bib M (3)2
Bonola R 1906 Loria 9, 33-38 [pseudosphere]

Frattini G 1900 Ens M 2, 173
Halsted G B 1902 AMM 9, 59-63
Levy M 1900 Par CR 130, 677
- 1900-1901 Kazan FMO 2(10), 32
Loria G 1901 Bib M (3)2, 392-440
  (F32, 17) [port]
Maggi G A 1935 Milan Sem 9, 75-85
  (F61, 953)
Mangolot H v 1904 Gott Anz, 341-348
  (F36, 16)
Mansion P 1901 Mathesis (3)1, 247-248
  (F32, 18)
Pascal E 1901 Lom Gen (2)34, 57-108
  (F32, 17)
- 1902 Wiad M 6, 1-55 (F33, 35)
- 1903 M Ann 57, 65-107


Benaflah, Cheber

Sanchez Pérez J A 1911 Esp SM 1,
  113-120   (F42, 3)


Benedetti, Giambattista   1530-1590

Anon 1933 Pe M 13, 191   (F59, 41)
Bordiga G 1926 Ven I At (2)85,
  585-757   (F52, 15)
Drake S + ed 1969 *Mechanics in
  Sixteenth-century Italy* (Univ of
  Wis Press)
Vailati G 1997 Tor FMN 33, 359


Benedictus Herbestus   1531-1593

Grabowski J 1913 Krak BI (A), 63-64


Benitez y Parodi, Manuel

Toledo L 1911 Esp SM (a)1, 508, 193-
  196   (F73, 34)


Bennett, Geoffrey Thomas   1868-1943

Baker H F 1944 LMSJ 19, 107-128
  (MR6, 254; Z60, 14)
- 1944 Lon RS Ob 4, 547-615
  (Z60, 14)

Benteli, Emmanuel Albrecht   1843-1917

Flukiger H 1919 Bern NG 221-7


Berardinis, Giovanni de

Marcolongo R 1938 Nap FM Ri (4)9,
  120-123   (F64, 919)


Berger, Alfred   1882-1942

Schönwiese R 1942 Bla Versi 5, 335-
  339   (F68, 18; Z27, 196)
Zwinggi E 1943 Schw Vers 43, 53-54
  (Z28, 99)


Berivald, Ludwig   1883-1942

Pinl M 1965 Scr 27, 193-203


Berkeley, George   1685-1753

Archibald R C 1935 Scr 3, 81-83
  (F61, 28)   [bib]
Evans W D 1914 M Gaz 7, 418-421
  [Newton]
Gibson G A 1898 Edi MSP 17, 9-32
  [limit, Moritz Cantor]
- 1899 Bib M 13, 65-70   [calc]
Johnston G A 1916 Mind 25, 177-192
  [phil]
- 1918 Monist 28, 25-46   [logic]
La Cinta Badillo M de 1952 Gac M (1)4,
  233-235   (MR14, 832)
Leroy A L 1956 Rv Syn 770, 135-169
  [mistakes]
Sanford V 1934 MT 27, 96-100
Stammier G 1922 Kantstu (55)
Synge J L 1957 *Kandelman's Krim: A
  Realistic Fantasy,* London, 175 p
  (MR21, 1)
Whitrow G J 1953 Hermath, 82   [calc]
Wisdom J O 1939 Hermath (54), 3-29
  (MR3, 258)
- 1953 Brt JPS 3

**Berliner,** Henoch 1883-1934

Anon 1936 Wiad M 40, 231-233
(F62, 27)

**Bermant,** Anisim Fedorovich 1904-1959

Kurosh H G + 1959 UMN 14(5), 117-121
(MR22, 619; Z87, 6) [port]

**Bernays,** Isaac Paul 1888-

Fraenkel A A 1958 Dialect 12, 274-279
(Z88, 244)

**Bernoulli** (family)

Behr-Pinnow C 1934 Natur W 22, 717-
721 (F60, 831)
Brunet P 1936 Archeion 18, 185-187
(F62, 23)
Conte L 1951 Pe M (4)29, 113-126
(Z42, 242)
Enestrom G 1880 Cron Cien 3, 329-335,
353-357, 377-382 [Euler]
Fleckenstein J O 1949 Elem M (S) 7)
- 1958 L'école mathématique baloise
des Bernoulli à l'aube du IVIIIe
siècle, Paris, 21 p (Z88, 243)
- 1964 Int Con HS 10
Fuss P H 1843 Correspondance mathé-
matique et physique de quelques
célèbres géomètres du XVIIIe siècle,
Pet, 2 vols (repr (1968) Johnson)
Hix 1912 Intermed, 237
Lemaire G 1906 Intermed 13, 237 (Q82)
[eq, approx]
Lick D W 1969 MT 62, 401-409
Merian P 1860 Die Mathematiker
Bernoulli, Basel
Ore O 1961 Isis 52, 586 [letters]
Sergescu P 1942 Tim Rev M 22, 24-26
(F68, 15)
Spiess O 1935 Schw NG, 278-279
(Z13, 193)
- 1937 Int Con (1936)2, 271
(F63, 20) [letters]
- 1948 Arc In HS, 356-62

- 1948 Die Mathematiker Bernoulli
Basel, 36 p
Truesdell C 1958 Isis 49, 54-62
(MR19, 826; Z79, 4)
Wolf R 1869 Boncomp 2 [letters]

**Bernoulli,** Daniel 1700-1782

Anon 1884 Basl V 7 [Euler]
Berthold G 1876 Pogg An 159, 659-661
(F8, 18)
Catalan E 1886 Boncomp 18, 464-468
[Goldbach]
Eneström G 1906 Bib M (3)7, 126-156
(F37, 7) [Euler, letters]
Guareschi I 1910 Tor FMN 45, 423-444
Huber Friedrich 1959 Daniell Bernoulli
(1700-1782) als Physiologe und
Statistiker, Basel-Stuttgart, 103 p
(MR22, 2052)
Jelitai J 1936 MP Lap 43, 142-160
(Z15, 290) [Johann Bernoulli,
Teleki]
- 1938 Magy MT 57, 501-508
(F64, 916) [Clairaut, Teleki]
Linder A 1936 Lon St (NS) 99, 138-141
(F62, 23; Z13, 193) [ins, Lambert]
Puppini U 1943 Bln Mm (9)10, 75-86
Rainov T I 1938 SSSR Vest (7-8),
84-93
Sheinin O B 1965 Vop IET 19, 115-117
Spiess O + 1941 Basel V 52, 189-266
(Z26, 195)
Staeckel P 1909 Arc GNT 1, 1-9
(F40, 59) [curves, Euler]

**Bernoulli,** Jacob 1654-1705

= Jacques, James, Jacob

Biermann K R 1956 Arc In HS 35, 233-
238 [induction]
Cinta Badillo M de la 1953 Gac M (1)5,
103-105 (MR15, 276)
Dietz P 1959 Basl V 70, 81-146
[calc vars]
Enestrom G 1905 Intermed 12, 151

- 1909 Bib M 3, 206-210
Fleckenstein J O 1949 Elem M (S6)
  (MR11, 707) [Johann Bernoulli]
- 1956 Int Con HS 8, 437-440
Hofmann J E 1956 Ens M 2, 61-171
  (MR18, 268; Z74, 6) [calc]
- 1957 *Ueber Jakob Bernoullis
  Beitraege zur infinitesimale Mathe-
  matik,* Geneva, 126 p (MR24A, 468)
  [calc]
Likhin V V 1959 Ist M Isl 12, 59-134
  [fun, Russia]
Sanford V 1933 MT 26, 382-4


Bernoulli, Johann I  1667-1748

  = Jean, John, Giovanni

Bachmann E 1948 Schw Verm 46, 125-
  128 (MR9, 485)
Barbensi G 1930 Rv Sto Cr 29, 168-180
Conte L 1948 Arc In HS, 611-622
  (MR10, 174) [B Taylor]
- 1949 Iasi IP 4, 36-53 (Z49, 289)
  [Fagnano, B Taylor]
Delorme S 1957 Rv H Sc Ap 10, 339-
  359 [Cury]
Enestroem G 1879 Boncomp 12, 313-314
- 1879 Sv Bi 5(21) (F11, 18)
  [Euler]
- 1894 Bib M (2)8, 65-72 (F25, 66)
- 1897 Bib M (2)11, 51-56
  [Euler, letters]
- 1903 Bib M (3)4, 344-388 (F34, 10)
  [Euler, letters]
- 1904 Bib M (3)5, 248-291 (F35, 10)
  [Euler]
- 1905 Bib M (3)6, 16 [Euler, letter]
Fleckenstein J O 1946 Elem M 1, 13-17
  (MR7, 354) [Taylor series]
- 1946 Elem M 1, 100-108 (Z61, 5)
  [Newton]
- 1949 *Johann und Jakob Bernoulli,*
  Basel, 24 p
- 1952 Basl V 63, 273-295 (Z49, 4)
  [N Bernoulli]
Grinwis C H C 1893 Ned NGC 4, 148-149
Hofmann J E 1967 Praxis 9(8), 209-212
  [calc]
Jelitai J 1936 MP Lap 43, 142-160
  (F62, 1035) [D Bernoulli]

Lambert J H 1784 *Deutscher gelehrter
  Breifewechsel von J Bernoulli,*
  Berlin, 5 vols
Lorey W 1938 Gies Hoch 12, 29-36
  (F64, 17) [Liebknecht, letters]
Nordenmark N V E 1940 Lychnos, 236-
  247 (F66, 1187)
Procissi A 1934 Pe M (4)14, 1-21
  (F60, 11)
Rebel O J 1934 *Der Briefwechsel
  zwischen Johann (I) Bernoulli und
  dem Marquis de L'Hospital in
  erlaeutender Darstellung,* Postberg,
  53 p
Sanford V 1933 MT 26, 486-89
Schafheitlin P 1920 Basl V 32, 230-
  235 [calc]
- 1922 Basl V 34 [calc]
- 1924 *Die Differentialrechnung von
  Johann Bernoulli aus dem Jahre
  1691-92...,* Leipzig, 56 p [calc]
Sieber L 1874 *Johannis Bernoulli ad
  Johanner Jacobum de Mairan ex
  autogropho Basileensi...,* Basle
  (F6, 27)
Stieda W 1926 Ber Ab A, 64 p
Van Geer P 1876 N Arc 2, 193-205
  (F8, 18)
Wollenschlager Karl 1933 Basl V 43,
  151-317 (Z6, 338) [De Moivre]


Bernoulli, Niklaus I  1662-1716

Fleckenstein J O 1952 Basl V 63, 273-
  295 [Johann]


Bernstein, Felix  1878-1956

Gini C 1957 Rv II St 25, 185-186
  (Z88, 6)


Bernstein, Sergei Natanovich  1880-

Akhiezer N I 1955 *Akademik S. N.
  Bernshteîn i ego raboty po konstruk-
  tivnoî teorii funktsii,* Kharkov,
  112 p (MR17, 697)
Akhiezer N I + 1961 UMN 16(2), 5-20
  (MR24A, 128) [part diff eqs]

Anon 1950 SSSRM 14, 193-198
(MR11, 707; Z36, 5)
Gelfond A O + 1960 SSSRM 24, 309-314
(MR??, 519; Z92, 246)
Goncharev V L 1950 UMN 5(3), 172-183
(Z36, 146)
Goncharoff V + 1940 SSSRM 4, 249-260
(MR2, 114; F66, 21; Z24, 244)
Kolmogorov A N + 1960 Teor Ver 5,
215-221 (MR24A, 222) [prob]
Kuzmin R O 1941 UMN 8, 3-7
(Z60, 14)
Linnik Yu V 1961 UMN 16(2), 25-26
(MR24A, 128; Z98, 9) [prob]
Videnskii V S 1961 UMN 16(2), 21-24
(MR24A, 128; Z98, 9) [constr thy
of functions]

Bernstein, Vladimir 1900-1936

Finzi B 1936 Milan Sem 10, xv-xvi
(F62, 1042)

Berra, Sagastume 1905-1960

Durañona Y Vedia A 1961 Arg UMR 19,
245-250 (Z98, 9)
Férnandez G 1961 Tucum Um (A)13
(1960), 223-230 (Z111, 3)

Bertelsen N P

Burrau C 1939 Sk Akt 22, 75-77
(F65:1, 17)

Bertini, Eugenio 1846-1933

Beltrami E ed 1935 *In memoria di
Eugenio Bertini a cura della
figlia Eugenia,* Florence, 35 p
(F61, 27)
Berzolari L 1933 It UM 12, 148-153
(F59, 32)
- 1933 Lom Gen (2)66, 67-68, 609-635
(F59, 32)
Castelnuovo G 1933 Linc Rn (6)17,
745-748 (F59, 32)
Conti A 1934 Bo Fir (2)13, 46-52
(F60, 15)

Fubini G 1933 Tor FMN 68, 447-453
(F59, 650)
Scorza G 1934 Esercit (2)7, 101-117
(F60, 836)

Bertrand, Joseph Louis François
1822-1900

Berthelot Marcelin 1901 *Science et
Education,* Paris, 113-119
Brillouin M 1901 Rv GSPA 11, 115-
Bryan G H 1899? Nat 61, 614
- 1900 Batt 38, 171-
Catalan E 1871 Boncomp 4, 127-134
(F3, 9)
*Darboux G 1902 in J. Bertrand,
*Eloges académiques,* Paris, vii-li,
387-399 [bib]
- 1902 Mathesis (3)2, 167
- 1904 Par Mm 47, 321-386
Levy M 1900 BSM (2)24, 69
Poincaré H 1910 *Savant et ecrivains,*
Paris, 157-161

Bertrand, Louis 1731-1812

Palazzo E 1960 Archim 12, 50-53
(MR23A, 272)
- 1960 Int Con HS 9, 558-563
(Z117:1, 5)
- 1961 Arc In HS 13, 223-236
(Z108:2, 249) [geometry]

Berwald, Ludwig 1883-1942

Pinl M 1965 Scr 27, 193-203

Berzolari, Luigi 1863-1949

Bausotti L 1936 in *Scritti matematici
offerti a Luigi Berzolari,* Pavia,
xxi-xxix, xxxi-xxxvii (F62, 1041)
[bib]
Bompiani E 1950 Linc Mr (8)9, 396-410
(Z39, 243)
Brusotti L 1950 IT UM (3)5, 1-19
(MR11, 708; Z36, 5)

Villa M 1954 Bln Rm 57 (1952-1953), 27-41 (MR15, 923)

Besant, William Henry 1828-1917

Anon 1918 LMSP (2)16, 50-53 (F46, 35)

Bessel, Friedrich Wilhelm 1784-1846

Anger C J 1846 *Errinerungen an Bessels Leben und Wirken*, Danzig
Anon 1900 Ber Si, 745 [Olbers]
Bierman K R 1966 Gauss Mt 3, 7-20
Boncompagni B 1884 Vat NLA 34, 241-244 [bib, Gauss]
*Busch August L 1848 Konig U As 24, xxi-lviii [bib]
*- 1848 *Verzeichniss saemmtlicher Werke, Abhandlungen, Aufsaetze und Bemerkungen von F. W. Bessel*, Koenigsberg
Encke J F 1846 Ber Ab, xxi-xlii [cel mech]
Foerster W 1894 *Ueber das Zusammenwirken von Bessel, Encke und A. von Humboldt unter der Regierung Friedrich Wilhelm III*, Berlin 21 p (F25, 1912)
Franz J 1884 Konig Ph 25
Herschel J F W 1847 *A Brief Notice of the Life, Researches and Discoveries of Friedrich Wilhelm Bessel*, London, 16 p
Przybyllok E 1931 As Nach 242, 365-368
Schoenberg E + 1955 Mun Abh (NF)(71) (MR17, 338) [Gauss]
Sommer J 1911 Z Ver Wien 40, 333-341 (F42, 12)

Besso, Davide 1845?-1906

Lazzeri G 1906 Pe MI (3)4, 48 (F37, 25)
Marcolongo R 1907 Pe MI (3)4, 147-156 (F38, 29)

Bessy. See Frenicle.

Beth, E. W.

Vredenduin P G J 1964 Euc Gron 39, 225 (Z118-1, 11)

Beth, Hermanus Johannes Elisa 1880-1950

Bottema O 1950 Euc Gron 25, 241-252 (Z37, 1)

Betti, Enrico 1823-1892

Basso G 1892 Tor FMN 28, 3-6 (F24, 27)
Beltrami E 1892 Paler R 6, 245-247 (F24, 27)
- 1893 Prag M 3, 30-32 (F25, 37)
Brioschi F 1892 Annali (2)20, 256 (F24, 27)
Cerruti V 1894 Paler R 8, 161-165 (F25, 37)
Padova E 1893 Ven I At (7)4, 609-622 (F25, 36)
Pascal E 1892 Rv M 2, 151-153 (F24, 27)
Pinto L 1892 Nap FM Ri (2)6, 143-144 (F24, 28)
Procissi A 1953 It UM (3)8, 315-328 (MR15, 276) [Galois]
Volterra V 1892 Nuo Cim (3)32, 5-7 (F24, 28)
- 1900 Int Con 2, 43-57 [anal]
- 1920 *Saggi scientifici*, Bologna, 33, 55-

Bettini, Bettino

Anon 1930 Pe M (4)10, 188 (F56, 29)

Bezout, Etienne 1730-1783

Bouligand G 1948 Rv GSPA 55, 121-123 (MR10, 174; Z35, 147) [alg]

White H S 1909 AMSB (2)15, 325-338
(F40, 68) [resultants, geom]

Bhaskara 1114-

= Bhaskaracarya

Banerji H C 1927 *Colebrooke's Trans-*
*lation of the Lilovati,* 2nd ed,
Calcutta, 324 p
Brockhaus H 1852 *Ueber die Algebra*
*des Bhaskara,* Leipzig
Datta B 1930 Indn HQ 6, 727-736
Gânguly S 1926 Clct MS 17, 89-98
(F52, 11) [numb thy]
- 1931 Indn MSJ 19, 6-9 (F57, 1293)
Krishnaswami Ayyangar A A 1929
Indn MSJ 18, 1-7 (F55, 8)
- 1930 Indn MSJ 18, 225-248 (F56, 13)
- 1951 M Stu 18(1950), 12 (MR13, 197)
[sine]
Nagarajan K S 1949 Aryan Pth, 310-14
Sanjana K J 1930 Indn MSJ 18, 176-188
(F56, 13)
Shukla K S 1963 *Bhaskara I and His*
*Works,* Lucknow, 3 vols
Sinha S R 1951 Alla UMAB 15, 9-16
(MR13, 420; Z43, 243)
Somayaji D A 1951 M Stu 18(1950),
1-8 (MR13, 197)

Bhatta, Kamalakara

Dwivedi P 1920 Benar MS 2, 68-81

Bianchi, Luigi 1856-1928

Anon 1929 Rv M Hisp A (2)4, 210-227
- 1937 Vat An 1, 115-124 (F63, 19)
Bedarida A M 1930 Bo Fir (2)8 (1929),
i-vi (F56, 815)
Blaschke W 1954 Pisa SNS (3)8, 43-52
(MR16, 1; Z56, 2) [diff geom]
Conti A + 1928 Bo Fir (2)7, 89-97
(F54, 40)
Fubini G 1929 Annali (4)6, 45-83
(F55, 19)
- 1929 Linc Rn (8)70 xxxiv-xliv
(F55, 19)

Hilton H 1929 LMSJ 4, 79-80
(F55, 19)
Krylov N 1929 SSSR Iz (7)2, 855-858
(F56, 815)
Pistolesi E 1928 Vat NLA 81, 378-380
(F54, 40)
· Scorza G 1930 Pisa SNS 16(2-5), 5-27
Vincensini P 1956 Tor Sem 16, 115-157
(MR20, 1325)

Bianchini, Giovanni XV

Birkenmajer L 1911 Krak BI (A), 268-
278 (F42, 69)
Bortolotti E 1942 Bln Mm (9)9, 81-90
(MR9, 486) [Regiomantanus, C
Roder, G Speir]
Thorndike L 1953 Scr 19, 5-17, 169-
180 (MR12, 311; 14, 832; Z50, 243)
- 1955 Scr 21, 136-137 (MR17, 337)
[instruments]

Bickmore, Charles Edward 1850?-1901

Elliott E B + 1902 LMSP 34, 129

Bienaymé, Irénée Jules 1796-1878

La Gournerie de 1878 Par CR 87, 617-
619 (F10, 18)

Bierens de Haan, David 1822-1895

Archibald R C 1948 *Mathematical Table*
*Makers,* NY, 12-13 (Scr Stu 3)
[biobib]
Enestroem G 1884 Sv Ofv 41, 191-197
[Girard, Spinoza, Stevin]
Kluyver J C + 1895 N Arc (2)2, i-xxviii
(F26, 34) [bib]
Korteweg D J 1896? N Arc (2)2, 16
[bib]
Schrek,D J E 1955 Scr 21, 31-41
(Z64, 3)
Van Haaften M 1929 N Arc (2)16(2), 1-3
[port]

Biernacki, Mieczystaw 1891-1959

Anon 1960, 1962 Lub (A) 14, 5-6
  Z(98, 9)
Bielecki A + 1962 Colloq M 9, 361-
  381 (MR25, 967; Z105, 3)
Krzyz J 1962 Wiad M 5, 1-14
  (Z100, 247)


Bigourdan, Guillaume 1851-1932

Anon 1932 ASP 44, 133
Brunet P 1932 Arc Sto 14, 256
Dyson F 1932 Nat 129, 643


Bikadze, Andrei Vasilevich

Kantorovic L V 1966 Sib MJ 7, 729-730
  (MR33, 3874)
Sobolev, S L + 1966 (MR33, 1215)


Bilharz, Herbert Emil Ludwig 1910-

Anon 1957 Arc M 8, i (MR19, 108)
Blenk H 1957 Jb WG, Flug (1956),
  232-233 (Z78, 3)
Koenig H 1958 DMV 61(1), 97-103
  (MR21, 750)


Billy, Jacques de 1602-1679

Brocard H 1911 Intermed 18, 56
  (Q1763) (See also 7, 77, 173,
  376; 8, 173) [bib]
Enestroem G 1895 Bib M (2)9, 32, 64,
  96
Schaewen P v 1908 Bib M (3)9, 289-300
  (F39, 58) [Fermat]
- 1910 Ens M 13, 77


Biot, Jean Baptiste 1774-1862

Lefort F 1867 Correspo (NS)36,
  955-995
Picard E 1928 Par Mm (2)59, i-xxxix
  1928 Rv I Ens 82, 129-145, 237-252
  (F55, 610)

Bird, John 1709-1776

Hellman C 1932 Isis 17, 125-53
  [instruments]


Birkeland, Richard 1879-1928

Ore O 1928 Nor M Tid 10, 81-89
  (F54, 42)
- 1929 Act M 53, i-iv (F55, 22)
- 1929 Oslo Sk (11), 7 p (F55, 22)


Birkhoff, George David 1884-1944

Anon 1945 Mex SM 2, 15-18 (Z60, 14)
- 1937 Vat An 1, 125-138 (F63, 20)
Etherington I M H 1947 Edi MSN (36),
  22-23 (MR9, 74)
Garcia G 1942 Lim Rev 44(440), 187-
  232 (MR4, 65; Z60, 14)
- 1944 Lim Rev 46, 675-677 (Z60, 14)
Hadamard J 1945 Par CR 220, 719-721
  (Z60, 14)
Kosambi D D 1945 M Stu 12, 116-120
  (Z60, 14)
Langer R E 1946 AMST 60, 1-2
  (Z60, 14)
Morse M 1946 AMSB 52, 357-391
  (MR8, 3; Z60, 14)
Moulton E J 1937 AMM 44, 185-186
Rey Pastor J 1945 Arg UMR 10, 65-68
  (MR6, 141; Z60, 14)
Vandiver H S 1963 JM An AP 7, 271-
  283 (Z109, 5; MR27, 900)
White H S 1937 Sci Mo 44, 191-193
Whittaker E T 1945 LMSJ 20, 121-128
  (Z60, 14)
Wilson E B 1945 Sci (NS)102, 121-
  128, 578-580 (Z60, 14)


Birnbaum, Walter 1897-1925

Anon 1925 ZAM Me 5, 278 (F51, 32)

al-Biruni   973-1048

= Abu Rachan Muhammad ibn
Ahmad al-Biruni   (Bairuni)

Anon 1951 *Al-Biruni Commemoration
  Volume,* Calcutta, 303 p  (Z45, 290)
Cassina U 1937 It UM Con, 457
  (F64, 24) [cubic eq]
- 1941 Pe M (4)21, 3-20   (MR3, 97;
  Z24, 242) [cubic eq]
- 1941 Pe M (4)21, 77-87 (MR3, 97;
  Z3, 97; Z24, 242) [trisection]
Courtois V 1952 *al-Biruni. A Life
  Sketch,* Calcutta, 42 p (Z48, 242)
Datta B 1925 Benar MS, 7-23
  [Arabic numer]
- 1926 Clct MS 17, 59-74  (F52, 10)
Davidian M L 1960 AM OSJ 86, 330-335
  (Z102, 245) [chronometry]
Hermelink H 1960 Sudhof Ar 44, 329-
  332 (Z95, 2)
Kennedy E S 1959 Scr 24, 251-255
  (Z89, 4) [ast]
- 1963 MT 56, 635-637  (MR30, 201)
  [ast]
- 1965 JNES 24, 274-  [calendar]
Kennedy E S + 1958 JNES 17, 112-121
  (Z82, 241) [ast]
Lesley M 1957 Centau 5, 121-141
  (Z82, 9) [ast]
Rozenfeld B A 1959 Ist M Isl 12,
  421-430 [interpolation]
Sachau E ed 1887 *alberuni's India...,*
  London
Sadykov H U 1950 As Zh 27, 73-80
  (MR11, 707)
- 1953 *Biruni i ego raboty po astro-
  nomii i mathematicheskoi geografia,*
  Moscow, 152 p  (MR17, 337)
Sarton G 1927 Int His Sc 1, 707-709
  [bib]
Schoy C 1925 *Die trigonometrischen
  Lehren des ostarabischen Astronomen
  Muhammed ibn Ahmad Abul Riban al-
  Biruni...,* Hannover, 160 p [trig]
  1926 AMM 33, 95-96  (F52, 37)
  [approx, trig]
- 1926 AMM 33, 323-325  [pi]
- 1927 *Die trigonometrischen Lehren
  des Persischen Astronom...,* Hannover

Suter H 1910 Bib M(3)(B)11, 11-78
  (F41, 63)
Tekeli S 1963 Belleten 27, 25-36
Tropfke J 1928 ZMN Unt 59, 193-206
  [Euc constr, nonogon]
Wiedemann E 1913 Islam 4, 5-13
  [instruments]
Wiedemann E + 1912 Mit GMNT 11, 313-
  321


Bjerknes, Carl Anton   1825-1903

Bjerknes V 1903 Phy R 17, 125-126
- 1903 *Till minde om C.A.Bjerknes,*
  Christiana, 24 p  (F34, 29)
  1904 *C.A.Bjerknes,* Leipzig, 31 p
  (F35, 24)
- 1925 *Carl Anton Bjerknes...,* Oslo
  246 p  (F51, 24)  [Ger transl
  1933 Berlin, F59, 848]
Bryan G H 1903 Nat 68, 133  (F34, 29)
Korn A 1904 DMV 13, 253-266  (F35, 23)


Bjerknes, Vilhelm Friman Koren
  1862-1951

Gold E 1951 Lon RS Ob 7, 303-317
  (MR13, 612)


Bjoergum, Oddvar   1916-1961

Godske C L 1962 Nor M Tid 10, 5-7
  (Z100, 247)


Bjoernbo, A. A.   1874-1911

Heiberg J L 1911 Bib M (3)12, 337-344
  (F42, 32)


Blaney, John Hugh   1917-1960

Clarke L E 1961 LMSJ 36, 499-500
  (Z98, 9)

Blaschke, Ernesto 1856-1926

F I 1926 Gi M Fin 8, 251-252
(F52, 32)

Blaschke, Wilhelm 1885-1962

Blaschke Wilhelm 1961 *Reden und
Reisen eines Geometers,* 2nd ed,
Berlin, 153 p (MR23A, 135)
Bompiani E 1962 It UB (3)17, 247-
248 (Z100, 247)
Burau W 1963 Ham MG 9(2), 24-40
(MR29, 639)
Kaehler E 1955 Fors Fort 29, 286-
287 (Z65, 244)
Mueller H R 1961 Ham Sem 25, 5-9
(MR23A, 272; Z95, 4)
Reichardt H 1966 DMV 69(1), 1-8
(MR34, 1150)
Sperner E 1963-64 Ham Sem 26, 111-128
(MR30, 569, Z114, 8)
Yaglom I M 1963 Rus MS 18(1), 135-
143 (Z112, 3)

Blaserna, Pietro 1836-1918

Cantone M 1918 Linc Rn (5)27, 262-
269 (F46, 20)

Blasius of Parma

= Biagio Pelecani

Thorndike L 1929 Archeion 9, 177-190

Blauner, Niklaus 1713-1791

Graf J H 1897 *Niklaus Blauner, der
erste Professor der Mathematik
an der bernischen Akademie,* Bern,
23 p (F28, 10)

Blichfeldt, Hans Frederik 1873-1945

Dickson L E 1947 AMSB 53, 882-883
(MR9, 74; Z31, 99)

Bliss, Gilbert Ames 1876-1951

Archibald R C 1938 *A semicentennial
history of the American Mathematical
Society,* NY, 201-206 [bib]
Bliss G A 1952 AMM 59, 595-606
(MR14, 343)
Graves L M 1952 AMSB 58, 251-264
(MR13, 810; Z46, 2)
McLane S 1951 APSY(1951-1952), 288-291
McShane E J 1931 Am NASBM, 32-45

Blissard, John 1803-1875

Bell E T 1938 AMM 45, 414-421
(F64, 20; Z19, 389) [numb thy]

Blondel, Nicolas François de
1618-1686

Mauclaire P + 1938 *Nicolas-François
de Blondel, ingénieur et architecte
du roi,* Laon, 299 p [bib]

Blumenthal, Ludwig Otto von 1876-1944

Behnke H 1958 M Ann 136, 387-392
(MR20, 945; Z82, 12)
Sommerfeld A + 1951 Aach THJ (1950)
21-26 (Z43, 245)

Boad, Henry XVIII

Bradley A D 1941 Scr 9, 101-104

Bobeck, Karl 1855-1899

Pick G 1900 Mo M Phy 11, 93-101
(F31, 19)
Weiss W 1901 DMV 9, 27-33

Bobillier, Etienne 1797-1834

Dupont P 1963 Tor FMN 98, 442-471
(MR29, 4232)

Fricker M + 1901-1902 Intermed 8
  330; 9, 163 (Q2138)
Mehmke R 1910 Intermed 17, 125
  (Q1104)


Bobynin, Viktor Viktorovich 1849-
  1919

Lukomskaya A M 1950 Ist M Isl 3,
  358-396 (Z41, 341) [bib]
Rybnikov K A 1950 Ist M Isl 3, 343-
  357 (MR13, 1; Z41, 341)
- 1950 UMN 5(1), 203-210 (Z35, 148)
Zubov V P 1956 Mos IIET 15, 277-322
  (MR19, 825; Z74, 246) [hist math]


Bôcher, Maxime 1867-1918

Anon 1918 Ens M 20, 226 (F46, 17)
- 1919 AMST 20, 1- [port]
Archibald R C 1938 A Semicentennial
  History of the American Mathematical Society, NY, 161-166
  [bib, port]
Birkhoff G D 1919 AMSB 25, 197-215
  (F47, 14) [bib]
Osgood W J 1919 AMM 26, 262-263
- 1919 AMSB (2)25, 337-350
- 1929 AMSB 35, 205-217 (F55, 22)


Bodio, Luigi

Foeldes B 1929 Allg St Ar 18, 426-
  436 (F57, 1307)


Boeklen, Georg Heinrich Otto
  1821-1900

Wölffing E 1901 Wurt MN (2)3, 1-16
  (F32, 18)


Boethius, Anicius Manlius Severinus
  480?-524

  = Boece = Boetius

Barret Helen M 1940 Boethius. Some
  Aspects of His Life and Work,
  Cambridge
Bubnov N 1907 Pet MNP (1), 1-33; (2),
  296-316; (3), 132-171 (F38, 62)
Chasles M 1836 Sur le passage du
  premier livre de la geométrie de
  Boéce, relativ à un nouveau système
  de numération, Brussells
- 1837 Par CR 4, [abacus]
Duerr Karl 1951 The Propositional
  Logic of Boethius, Amsterdam, 89 p
  (MR14, 524)
Friedlein G 1861 Gerbert, die Geométrie
  des Boethius..., Erlangen, 60 p
Gustafsson F 1880 Hels Act 11 (F12, 2)
Joerg Edgar 1935 Des Boetius und des
  Alfredus Magnus commentar zu den
  "Elementa" des Euklid... zweiter
  Buch, Postberg, 34 p (Z12, 97)
Koeppen F T 1892 Pet B 35, 31-42
  (F24, 41) [abacus]
Narducci E 1877 Linc At 1, 503-
  (F9, 16) [abacus, circle]
Paulson J 1885 Lund U 21, 30p
Tannery P 1900 Bib M (3)1, 39-50, 524
Weissenborn H 1879 Hist Abt 24, 187-
  240 (F11, 6)


Boev, Georgii Petrovich 1898-1959

Gnedenko B + 1960 Iz VUZM 1(14),
  245-248 (Z91, 6)


Bogdan, Constantin P 1910-1965

Cimpan F 1965 Iasi UM 11(a), xv-xvi
  (MR34, 4092) [bib]


Bogdan, Petru 1874-1944

Costeau G 1948 Iasi UM 30, i-iv
  (Z30, 338)

Bogolyubov, Nikolai Nikolaevich  1909-

Mitropolskii, Yu A + 1959 UFN 69,
  159-169  (MR22, 1131)
- 1959 Ukr IM 11, 295-311   (Z86, 5)
- 1959 UMN 14(5), 167-180
  (MR22, 619; Z87, 6)


Bohl, Piers G 1865-1921

Kneser A + 1924 DMV 33, 25-32
  (F50, 16)
Mishkis A D + 1965 Matematik Pirs
  Bol iz Riga, Riga, 99 p
  (MR34, 4095)
Rabinovich I M 1956 Sov Cong (1956),
  166-167


Bohlmann, Georg  1869-1928

Lorey W 1928 Bla Versi 1, 3-9
  (F54, 42)


Bohnicek, Stjepan  1872-1956

Kurepa D 1956 Zagr Gl (2)11, 275-277
  (Z71, 244)


Bohr, Harald  1887-1951

Bohr H 1947 M Tid A, 1-27   (Z31, 99)
Bochner S 1952 AMSB 58, 72-75
  (MR13, 420; Z46, 3)
Jessen B 1951 Act M 86, i-xxiii
  (MR13, 420; Z43, 6)
- 1951 M Tid, 1-18   (Z42, 243)
Neugebauer O 1952 APSY, 307-311
Perron O 1952 DMV 55(1), 77-88
  (MR13, 810; Z46, 2)
*Tambs-Lyche R 1951 Nor M Tid 33, 2-16
  (MR13, 1; Z42, 4)
 Titchmarsh E C 1953 LMSJ 28, 113-115
  (MR14, 324; Z50, 2)

Boltzmann, Ludwig Eduard  1844-1906

Broda E 1955 *Ludwig Boltzmann,* Wien,
  170 p   (Z66, 247)
- 1957 Vop IET (4), 47-54   (Z80, 3)
Classen  Johannes 1908 *Vorlesungen
  ueber moderne Naturphilosophen,*
  Hamburg, 108-128
Ehrenfest P 1906 MN Bl 3, 205-209
  (F37, 27)
Flamm L 1956 Ost Anz, 141-145
  (Z71, 3)
Kaller E 1906 Ens M 8, 484-485
  (F37, 27)
Meyer S ed 1904 *Festschrift Ludwig
  Boltzmann gewidmet zum sechzigsten
  Geburtstage,*  Leipzig, 942 p
  [math phys, mech, port]
Voit C 1907 Mun Si 37, 262-267


Bolyai family

Bell E T 1937 Scr 5, 37-44, 95-100
David L v 1951 *Die beiden Bolyai,*
  Basel, 24 p  (MR13,1; Z42, 4)
Jelitai J 1940 Magy MT 59, 812-845
  (F66, 19); Z24, 244)
Mansion P 1897 Mathesis (2)7, 194
Palffy Ilona + 1962 *Bibliographia
  Bolyaiana 1831-1960,* Budapest
Prosper V R 1894 Prag M 4, 37-40
  (F25, 25)
Schlesinger L 1903 Bib M (3)4, 260-270
  (F34, 16)
Schmidt F 1867 Bord Mm 5, 191-205
  (F1, 14)
Schmidt F 1868 Arc MP 48, 217-228
  (F68, 14)   (transl in Boncomp 1,
  277-299; Bord Mm 5)
Staeckel P + 1897 BSM (2)21, 206-228
  (F28, 42)  [Gauss]
Staeckel P 1913 *Wolfgang und Johann
  Bolyai Geometrische Untersuchungen,*
  Leipzig, 2 vols  (Urk GNG 2)
Study E 1914 Int Monat 8, 1231-1242

Bolyai, Janos    1802-1860

See also non-Euclidean geometry
in Chapter 2

Alexits G 1952 M Lap 3, 107-110
   (MR17, 2)
- 1953 Magy MF 3, 131-150
   (MR15, 383)
- 1954 Hun Act M 5(S), 1-20
   (MR16, 985; Z57, 243)
Anon 1896 M Gaz 1, 25-31
- 1902 Joannis Bolyai, (Hungarian
   Univ. of Kolozsvar), 169 p
   (F33, 24)
- 1902 BSM (2)33, 67-8
Darval I 1904 Int Con H 12, 45-49
   (F35, 14)
Emanuele M A 1952 Matiche 7, 18-20
   (MR14, 832)  [trig]
Gratsianska L M 1961 Kiev Vis 4,
   3-18  (MR33, 2506)
Halsted G B 1898 AMM 5, 35-38
- 1910 AMM 17, 31-33
Havlicek K 1960 Prag Pok 5, 345-357
   (Z100, 246)
Jelitai J 1939 M Ter Er 58, 35-39
   (F65, 15; Z21, 3)
Kreiling E 1932 Tim EP 4, 295-298
   (F58, 994)
Kurschak J + 1902 Johannes Bolyai de
   Bolya Appendix, Budapest 1902,
   Nieu Arch (2)6(1), 79-80
- 1900 MNB Ung 18, 250-279
Macauley F S 1897 M Gaz, 25-31, 50-60
   (F28, 53)
Mansion P 1902 Brx SS 26, 146
Mihaileanu N N 1960 Gaz MF(A)12,
   286-290  (MR23A, 4; Z102, 5)
Nagy G 1953 M Lap 4, 84-86 (MR15, 923)
Pavlicek J 1960 Cas M 85, 241-255
   (MR22, 770; Z100, 246)
Petkancin B 1960 Bulg FM 3(36),
   113-127  (MR23A, 433)
Petrosyan G B 1960 Arm FM 13(4), 69-
   78 (Z91, 5)
Popoviciu T 1960 Cluj UBBM (1)5(1),
   9-13 (Z113, 2)
Rényi A 1952 M Lap 3, 173-178
   (MR17, 2)
Sarloska E 1965 Magy MF 15, 341-387
   (MR33, 5438)

Schlesinger L 1902 MP Lap 11, 53-56
- 1903 DMV 12, 165-194  (F34, 16)
- 1903 MP Lap 12, 57-88  (F34, 17)
Schmidt F 1897 DMV 4, 107-109
   (F28, 42)
- 1898 Ab GM 8, 133-146  (F29, 9)
Staeckel P 1898 MNB Ung 16, 263
   [complex numb]
- 1899 MNB Ung 17, 1
- 1900 MNB Ung 18, 280-307
- 1902 M Ter Er 20, 180
- 1903 M Ter Er 21, 135-145
- (F34, 17)
- 1904 MNB Ung 19, 1-12
Szabo P 1910  M Ter Er 29, 135-164
Toth Imre 1956 Johann Bolyai. Leben
   und Werk des grossen Mathematikers,
   Bucarest, 73 p  (MR17, 117; Z64, 242)
Varicak V 1907 DMV 16, 320-321
Wiesner S 1920  DMV 29, 129-135
   (F47, 9)
Wigand K 1961 Praxis 3, 104-105


Bolyai, Wolfgang Farkas   1775-1856

Boncompagni B 1884 Vat NLA 34, 279
   [bib, Gauss]
Bujdosó E 1934 Deb M Szem 8, 1-34
   (F60, 833)  [educ]
David L 1959 Magy MF 9, 215-236
   (MR21, 1023; Z86, 4)
Halsted G B 1896 AMM 3, 1-5
Jelitai J 1937 MP Lap 44, 168-172
   (F63, 816)
- 1938 MP Lap 45, 200-203
   (F64, 919; Z19, 389)
Koncz Joseph 1887 in History of the
   Evangelical Reformed College of
   Maros Vasarhely, 271-388
   (Hungarian)
Mentovich F 1902 MP Lap. 11, 90-96
   [Gauss]
Schlesinger L 1902 MP Lap 11, 179-230
   (F33, 23)
Schmidt F + ed 1899 Briefwechsel
   zwischen Carl Friedrich Gauss und
   Wolfgang Bolyai, Leipzig, 220 p
Staeckel P 1897 Gott N, 1-12
   (F28, 42)  [Gauss]
Szabo P 1907 M Ter Er 25, 326-338
   (F38, 22)  [Gauss]

Szénássy B v 1937 Deb M Szem 13, 1-30
  (F63, 809; Z17, 290) [calculus]
Toth K K 1959 M Lap 10, 12-22
  (MR22, 618; Z86, 4) [educ]


Bolza, Oskar 1857-1942

  = F H Marneck

Anon 1943 Sci (NS)97, 108-109
  (MR4, 181; Z60, 14)
Bliss G A 1944 AMSB 50, 478-489
Bolza O 1936 Aus meinem Leben,
  Munich
Heffter L 1943 DMV 53, 2-13
  (MR8, 190; Z28, 100)


Bolzano, Bernard 1781-1848

Anon 1956 Cas M 81, 387 (Z74, 245)
Bar-Hillel Y 1950 Methodos 2, 32-55
  [logic]
- 1950 Theoria 16, 91-117 [logic]
- 1952 Arc MLG 1, 65-98 (MR14, 121)
Baumann J 1908 An Natphi 7, 444-449
  [Dedekind]
*Berg Jan 1962 Bolzano's Logic
  Stockholm, 214 p (Z107, 5) [bib]
Bergmann Hugo 1909 Das philosophische
  Werk Bernard Bolzano, Halle, 244 p
Bolzano B 1836 Lebensbeschreibung des
  Dr. Bolzano, Sulzbach, 338 p
  [autobiog]
Brzhetska V F 1949 UMN 4 (2), 15-21
  (MR11, 572; Z35, 147)
  [Bolzano fun]
Buhl G 1961 Ableitbarkeit and Abfolge
  in der Wissenschaftstheorie
  Bolzanos, Cologne
Capone-Braga C 1927 Bernardo Bolzano
  Padoa, 35 p (F52, 38)
Coolidge J L 1949 The Mathematics of
  Great Amateurs (Clarendon Press),
  ch 16 (Repr 1963 Dover)
Dubislav W 1931 Gorres Ja 44, 448-456
  (F57, 1304) [logic]
- 1931 Unt M 37, 340-44 (F57, 32)
Fels H 1926 Gorres Ja 39, 384-418
  (F52, 24)

- 1929 Bernard Bolzano, Leipzig 119 p
  (F55, 623)
Fesl J M 1836 Lebensbeschreibung des
  Dr. B. Bolzano und einige seiner
  ungedruckten Aufsatze, mit dem
  Bildniss des Verfassers, Subzbac
Folta J 1966 Act HRNT (2), 75-104
  [bib, geom]
Franzis Emerich 1933 Bernard Bolzano
  Der paedagogische Gehalt seiner
  Lehre, Muenster, 269 p
Hornich H 1961 Ost Anz 17-19
  (Z94, 4)
Jarnik V 1931 Cas MF 60, 240-262
  (F57, 32)
- 1961 Cz MJ 11, 485-489 (Z100, 7)
Jasek M 1921 Prag V (1), 1-32
- 1922 Cas MF 51, 69-76
  [Bolzano fun]
- 1924 Cas MF 53, 102-110
  [patho funs]
*Kolman E 1955 Bernard Boltsano,
  Moscow, 224 p (Mos IIET 19) (MR17,
  813) (also 1957 Prague) [bib]
- 1963 Bernard Bolzano, Berlin, 251 p
  (Z112, 2)
Kowalewski G 1923 Act M 44, 315-
  [patho funs]
Kreibig J K 1914 Arc Phi 27, 273-287
Laugwitz D 1965 Arc HES 2, 398-409
  (MR33, 11)
Lingua P 1964 Pe M (4)42, 209-214
  (MR30, 568) [real numbs]
Novy L 1961 Prag Dej 6 [anal]
Palagy 1902 Kant und Bolzano, Halle
Rychlik K 1932 Int Con 6, 503-505
  [fun]
- 1956 Cas M 81, 391-395 (MR19, 519;
  (Z74, 245) [real numbs]
- 1958 Cz MJ 8, 197-202 (MR20, 1041)
  [logic]
- 1958 Ist M Isl 11, 515-532
  (Z109, 238) [real numbs]
- 1961 Rv Hi Sc Ap 14, 313-327
  (tr of 1958)
1962 Rv Hi Sc Ap 15, 163- [Cauchy]
- 1962 Theorie der reelen Zahlen in
  Bolzanos handschriftlichem Nachlasse
  Prague, 103 p (MR28, 576; Z101, 247)
Scholz H 1937 Muns Semb 9, 1-53
  (F63, 17)

- 1937 Fries Sch (NS)6, 399-472
  (repr 1961 Mathes Un, 219-267)
Sebestik J 1964 Rv Hi Sc Ap 17, 129-
  164 (MR30, 201) [fund thm calc]
Seidlerová I 1956 Cas M 81, 388-390
  (Z74, 245) [political]
Smart H R 1944 Phi Rev 53, 513-533
  [logic]
Steele D A 1950 in *Paradoxes of the
  Infinite* by B.Bolzano, London
Stolz O 1880 Wien Anz, 91-92
  (F12, 34) [calc]
- 1881 M Ann 18, 255-279 (F13, 35)
  [calc]
Struik R + 1928 Isis 11, 364-366
  [Cauchy]
Talacko J V 1964 Cz Am Con 2, 11-13
Van Rootselaar B 1963 Arc HES 2,
  168-180 [real numbs]
Vetter Q 1923 AMM 30, 47-58
Vojtech J 1935 Slav Cong 2, 264-265
  (F61, 21)
Voltagio F 1965 in *I paradossi dell'
  infinite* by B.Bolzano, Milan
  (MR34, 11)
Winter E 1932 Gorres Ja 45, 325-346,
  483-499 (F58, 994) [phil]
- 1933 *Bernard Bolzano und sein Kreis,*
  Leipzig, 278 p (F59, 857; Z7, 147)
- 1933 Fors Fort 9, 93-94
  (F59, 28)
- 1933 Gorres Ja 46, 515-520
  (F59, 859) [mss]
- 1938 Gorres Ja 51, 29-60
  (F64, 918) [letters]
- 1949 Halle Mgr (14), 100 p
  (MR13, 1; Z40, 11)
- 1949 Int Con Ph 10, 1186-1190
- 1956 *Der boehmische Vormaerz in
  Briefen B. Bolzanos an K. Prihonsky
  (1824-1848),* Berlin, 314 p
  (Z71, 244)
Wisshaupt 1850 *Skizzen aus dem Leben
  Bolzano's,* Leipzig
Wrinch D M 1917 Monist 27, 83-104
Wussing H 1962 ZGNTM 1(3), 57-73
  (Z107, 4) [calc]
Yushkevich A P 1947 Mos IIET 1,
  388-389 [Cauchy, intgl]
Zeil W 1963 Fors Fort 37, 178-180
  (Z108-2, 250)

**Bombelli,** Rafael   XVI

Albarran F 1936 Rv M Hisp A (2)11,
  143-150 (F64, 911) [Heron, root]
Bortolotti E 1922-1923 Bln Rn, 14 p
  [cubic, trisection]
- 1927 Archeion 8, 49-64 (F53, 11)
  [variable]
- 1928 Pe M 8, 334-44 (F54, 24)
- 1929 Arc GMNT 11, 407-24 (F25, 11)
- 1932 Int Con 6, 393-400, 415-420
  (F58, 32)
- 1936 Osiris 1, 184-230 [alg, geom]
Enestroem G 1892 Bib M (2)6, 96
Favaro A 1893 Bib M (2)7, 15-18
  (F25, 16)
Jayawardene S A 1962-1963 Bln Rn P
  (11)10(2), 235-247 (MR29, 1069)
- 1963 Isis 54, 391-395 (Z117-2, 243)
- 1966 Isis 56, 298- [bib]
Loria 1932 Int Con 6, 415-420
  [Diophantus]
Vivanti G 1925 Pe M (4)5, 39-40
  (F51, 20) [alg]
Wieleitner H 1926 Arc Sto 7, 28-33
  (F52, 15) [variable]

**Bompiani,** Enrico  1889-

Anon 1954 It XL An (1953), 297-309
  (MR15, 591)

**Bonasoni,** Paolo  XVI-XVII

Bortolotti E 1924-1925 Bln Rn (2)29,
  90-105 (F51, 11) [anal geom]
- 1928 *Studi e ricerche sulla storia
  della matematica in Italia nei
  secoli XVI e XVII,*  Bologna

**Boncompagni,** Baldassarre  1821-1894

Cantor M 1892 ZM Ph 39, 201-203
  (F25, 44)
Catalan E 1884 Bel Bul (11 Oct),
  310-311)
Chasles M 1852 Par CR 34, 889-

Favaro A 1894 Ven I At (7)6, 509-
  521  (F26, 30)
Galli J 1894 Vat NLA 47, 161-186
  (F26, 30)
Genocchi A 1884 Tor FMN 20, 237-242
  [Gauss]
Mansion P 1894 Rv Q Sc (2)6, 262-264
  (F25, 45)
Narducci E 1892 *Catalogo Dei mano-
  scritti ora posseduti da Baldasarre
  Boncompagni* 2nd ed, Rome [library]
Rossi M S de 1894 Vat NLA 47, 131-
  134  (F26, 30)
Tannery P 1914 Intermed 21, 126
  [library]


Bonferroni  Carlo Emilio  1892-1960

Pagni P 1960 It UM (3)15, 570-574
  (Z93, 9)


Bonfils, Immanuel  XIV

Gandz S 1936 Isis 25, 16-45
  (F63, 20; Z15, 53) [decimal,
  exponential]
Sarton G 1936 Isis 25, 132-33
  [decimal]


Bongo, Pierre

Fontes 1893 Tou Mm (9)5, 371
  (F26, 12)


Bonnesen, Tommy  1873-1935

Mollerup J 1936 M Tid B, 16-24
  (F61, 26)


Bonnet, Pierre Ossian  1819-1892

Anon 1856 *Notice sur les traveaux
  mathématiques de M. Ossian Bonnet*,
  Paris, 16 p
- 1892 Par CR 114, 1509-1510
  (F24, 28)

- 1892 Par CR 115, 1113-1119
  (F24, 28)
Appell P 1893 Par CR 117, 1014-1024
  (F25, 37)
Tisserand 1892 Fr Long  (F24, 28)


Bonola, Roberto 1874-1911

Amaldi U 1911 Mathes It 3, 127,
  145-152  (F42, 33; 43, 35)
Kulisher A 1912 Kagan (568), 103-104
  (F43, 35)
Veneroni E 1911 Pe MI 26, 319-320


Boole, George 1815-1864

Anon 1867 Lon RSP 15, vi-xi
- 1960 Elek Rech 2, 57-  (Z87, 243)
*Boole Mary E 1878 Uni Mag (2)1, 105-
  114, 173-183, 326-336, 454-460,
  (repr in her Collected Works 1931)
- 1897 *The Mathematical Psychology of
  Gratry and Boole,* NY
Broadbent T A A 1964 M Gaz 48, 373-
  378
Enestroem G 1886 Bib M, 96 (Q10)
Feys R 1954 Dub Ac P 57(A), 97-106
Gridgeman N T 1964 New Sci 24(420),
  655-  [family]
Hackett F E 1954 Dub Ac P 57(A), 79-
  87 [education]
Halsted G B 1878 J Sp Phil 12, 81-91
- 1878 Mind 3, 134-137 [Jevons]
*Harley R 1866 Brt Q Rv 44, 141-181
  (repr in Boole 1952)
Hesse M B 1952 An Sc 8, 61-81
Jourdain P E B 1910 QJPAM 41
Kneale W 1948 Mind 57, 149-175
  (MR9, 485) [logic]
*- 1956 Lon RSNR 12, 53-63
Macfarlane Alexander 1916 *Lectures on
  Ten British Mathematicians of the
  Nineteenth Century,* NY, 50-63
Rhees R 1954 Dub Ac P 57(A), 74-78
  [Student, teacher]
Rosser J B 1955 Dub Ac P 57, 117-120
  (Z66, 7)  [fun]
Taylor G 1956 Lon RSNR 12, 1-
*Taylor G + 1955 Dub Ac P(A)57, 63-130
  (MR17, 337) [centenary tribute]

Thomas I 1954 Dub Ac P 57(A), 88-96
    [phil of sci]
Venn J 1876 Mind 1, 479-491 [logic]


Boone, James Jr 1743?-1795


Bradley A D 1938 Scr 6, 219-227


Booth, James 1816-1878

Glaisher J W L 1879 Lon As Mo N 39,
    219-225 (F11, 28)


Bopp, Karl 1877-1934

Lorey W 1905 MN B1 2, 194-195
    (F36, 34)
- 1935 DMV 45, 116-119 (F61, 26)


Borchardt, Karl Wilhelm 1817-1880

Graf J 1915 Bern NG, 50-69
    [Schlaefli]


Borda, Jean Charles 1733-1799

Archibald R C 1948 *Mathematical
    table makers,* NY, 13-14
    (Scr Stu 3)


Borel, Emile 1871-1956

*Anon 1940 *Selecta. Jubilé scienti-
    fique de...,* Paris, 418 p
    (Z60, 14) [*bib, port]
- 1940 *Jubilé scientifique...Sorbonne
    ...14 janvier 1940,* Paris, 38 p
    (also 1949 Par Not D 2, 324-359
- 1956 Par I St 5, 55-56 (Z71, 3)
*Borel E 1912 *Notice sur les travaux
    scientifiques de M. Emile Borel,*
    Paris, 79 p (F43, 61)
- 1921 *Supplement (1921) à la notice
    (1912) sur les travaux scientifiques
    de Emil Borel 1913-1921,* Paris
    (Repr in Anon 1940)

- 1936 Organon 1, 34-42 (F62, 1041)
    (Repr in Anon 1940; Frechet 1967)
    [heuristic, measure]
*Broglie Louis de 1957 Par Not D 4,
    1-24
*Collingwood E F 1959 LMSJ 34, 488-512
    (MR21, 1023) [bib]
- 1960 LMSJ 35, 384 (MR23A,4; Z88, 6)
Darmois G 1956 Rv II St 24, 154-156
    (Z72, 246)
Fréchet M 1953 Ecmet 21, 95-96
    [game thy]
- 1953 Ecmet 21, 118-124
- 1956 Rv Phi 146, 158-160
- 1961 Rv Phi 151, 397-416
    (Repr in Frechet 1965)
*- 1965 Ens M (2)11, 1-95
    (Repr as Ens M Mon 14) [bib]
- 1967 *Emile Borel, philosophe et
    homme d'action. Pages choisies et
    présentées par Maurice Fréchet*
    (Gauthier Villars), 406 p (Les
    Grands Problemes des Sciences 19)
Maurain C 1957 Par ENSB, 29-32
*Montel P 1956 Par CR 242, 848-850
    (MR17, 697; Z70, 5)
Murata T 1959 Paul UCM 7, 57-63,
    99-116 (MR21, 750) [foundations]
Pérès J 1956 Par U An 26, 213-217
    [port]
Pompeiu D 1935 Math Cluj 10, 209-211
    (F61, 28) [Rumania]
Stoilow S 1956 Gaz MF(A) 8, 169-175
    (MR18, 182; Z74, 8)


Borelli, Giovanni Alfonso 1608-1679

Barbensi G 1938 Rv Sto Cr 29, 168-180
    [J. Bernoulli]
Del Gaizo M 1890 Nat NLA 20, 1-48
- 1908 Rv FMSN 17, 385-402
    [Torricelli]
Derenzini T 1959 Physis Fi 1, 224-243
    (Z107, 246) [letter, Marchetti]
- 1960 Physis Fi 2, 235-241
    (Z107, 245) [letter, Cassini]
Giovannozzi G 1925 Vat NLA 79, 61-66
Itard J 1962 Arc In HS 14, 201-224
    (Z107, 246) [Apollonius]
Tenca L 1956 Lom Mr (3)21, 107-121
    (Z71, 244) [Viviani]

Bortkiewicz, Ladislaus von  1868-1931

Anderson O 1932 Z Natok 3, 242-250
 (F58, 992)
Anon 1932 Sk Akt 15, 95-101
 (F58, 995)
Lorey W 1932 Versich 3, 3-10
 (F58, 49)


Bortolotti, Enea  1896-1942

Blaschke W 1942 DMV 52(1), 173
 (Z27, 290)
Bompiani E 1942 Rm U Rn (5)3, 231-
 240  (Z27, 290;  MR8, 190)


Bortolotti, Ettore  1866-1947

Agostini A 1947 It UM (3)2, 87-88
 (Z30, 338)
Bompiani E 1947 Mod At (5)7, 185-202
 (MR9, 485)
Carruccio E 1948 Pe M (4)26, 1-13
 (MR10, 174; Z29, 241)
Segre B 1949 Bln Rn (NS)52(1947/1948),
 47-86  (MR12, 1)


Boruvka, Otakar  1899-

Anon 1959 Cz MJ 9, 309-313  (Z88, 6)
Koutsky K 1959 Prag Pok 4, 730-733
 (Z87, 7)
Novotny M + 1959 Cas M 84, 236-250
 (Z85, 6)


Boscovich, Ruggiero Giuseppe 1711-1787

Cermeli L 1929 Arc GMNT 11, 424-444
Costabel P 1961 Arc In HS 14, 3-12
 (Z109, 4)
- 1962 Rv Hi Sc Ap 15, 31-42
 (Z109, 4) [cycloid]
*Eisenhart C 1962 Actes du Symposium
 International R. J-B. 1961
 [abs value, regression]
Goditskii-Tsoirko A M 1959 Scientific
 Ideas of Roger Joseph Boscovitch,
 Moscow, 94 p   (in Russian)

Hercigonja M 1955 Beo DMF 7, 109-118
 (MR17, 813; Z67, 247) [pseudonym]
Hondl S 1950 Glasn MPA (2)5, 21-32
 (MR12, 311) [mech]
Langley E M 1894 M Gaz 1, 1
Markovic Z I 1956 Int Con HS 8,
 202-206
Nikolic D 1962 Arc In HS 14, 315-335
 (Z101, 9)
Sister M M Fitzpatrick 1968 MT 61, 167
 171 [atomic thy]
Stipanic E 1963 It UM Con 7
 [continuum]
- 1967 MV 4(19), 277-292  (MR37, 16)
Stoianovitch C 1918 Rv Sc, 456-460
 (F46, 35)
Varicak V 1910 Zagr Rad (181), 75-208
- 1914 Zagr Iz (1), 1-24  (F47, 20)
- 1927 Zagr Rad (234), 123-188
 (F53, 20)
Volta L 1937 It SPS 26
Whyte L L 1959 Lon RSNR 13(1958),
 38-48  (Z108-2, 248)
*Whyte L L ed 1961 Roger Joseph
 Boscovich...: Studies in His Life
 and Work on the 250th Anniversary
 of His Birth, London (repr 1964 NY)
 [bib]


Bosmans, Henri  1852-1928

Bernard-Maitre H 1950 Arc In HS 29,
 619-623  (Z38, 4)
- 1950 Arc In HS 29, 629-656 (Z36, 146)
 [bib]
Mansion P 1907 Brx SS 31(A), 242
 (F38, 58)
Mineur A 1928 Mathesis 42, 49
 (F54, 38)
Peeters P 1928 Rv Q Sc 13, 201-14
Pelliot P 1928 Toung Pao 26, 190-199
Rome A 1928 Brx SS(A)48, 57-59
 (F54, 39)
- 1929 Isis 12, 88-112 (repr in
 Brx SS(A)49, 88-112) [bib]
Sarton G 1949 Isis 40, 3-6
Sauvenier-Goffin E 1952 Lieg Bul 21,
 301-302  (Z48, 243) [Grégoire de
 Saint-Vincent]
Vetter Q 1928 Cas MF 57, 167 (F54, 38)

Bosse, Abraham   1611-1678

Valentin G 1912 Bib M(3)13, 23-28
  (F43, 12)  [Desargues]


Bossut, Charles   1730-1814

Doublet E 1914 BSM (2)38, 93-96,
  121-125, 128, 158-160, 187-190,
  220-224  (F45, 16)
Mulcrone T 1965 Jesuit SB 42, 16-19


Botea, Nicolae G

Ionescu-Bujor C 1937 Gaz M 43, 225-
  228  (F63, 816)


Bothvidi, J.  XVII

Eneström G 1889 Bib M (2)3, 64


Boulanger, August Henri Leon
  1866-1923

Anon 1923 Intermed (2)2, 73-75
  (F49, 13)


Bouquet, Jean Claude   1819-1885

Bertrand J + 1885 Par CR 101, 585-588
- 1885 BSM (2)9, 301-305
Halphen G H 1886 Par CR 102,
  1267-1273


Bour, Jacques Edmond Emile   1832-1866

Anon 1866 Philom Bu 3, 119-
- 1867 MOS Mo Sb 2-2, 44-
- 1867 Nou An 6, 145-
Résal H 1879 An Mines (7)16, 275-283
  (F11, 24)


Bourbaki, Nicolas  XX

Anon 1958 Le Monde (No.49, 6-12 March)

- 1959 Réalités (January), 61-
Bernard-Maitre H 1960 Rv Syn 81,
  348-349
Bouligand G 1953 Rv GSPA 60, 193-195
  [1in anal, fun anal]
Cartan Henri 1959 Nicolas Bourbaki
  und die heutige Math, Cologne, 27p
  (MR22, 260)
- 1959 Nor Wes AF 76, 5-8  (Z84, 3)
*Cavallari A 1964 Atlas 7(5), 311-
Choquet G 1962 MP Semb 9, 1-21
Halmos P R 1957 Parana M 4, 18-28
  (Z82, 12)
- 1957 Sci Am (May), 88-99  (MR18, 709)
Henney D R 1951 M Mag 36, 252-254
Lyapunov A A 1960 Prag Pok 5, 518-519
  (MR24A, 4)
Rychlik K 1959 Prag Pok 4, 673-678
  (MR23A, 698; Z87, 6)
Segre B 1955 In memorie di G. Peano
  ed by A Terracini, Cuneo  (Z66, 5)


Bourdin, Pierre   1595-1653

Jones P S 1945 Scr 13, 119-120


Bourlet, Carlo 1866-1913

Boulanger A 1920 Rv Sc, 42-45
Bricard R 1913 Nou An (4)13, 433-438
Castelnuovo G 1913 Mathes It 5, 85
  (F44, 30)
Fehr H 1913 Ens M 15, 417-418
  (F44, 30)
Laisant C A + 1913 Nou An (4)13, 337
  (F44, 30)


Bournous, Rombaut   1731-1788

Houzeau J C + 1877 Bel Bul (2)62,
  675-676  (F9, 7)
Mailly E 1877 Bel Mm 27  (F9, 7)


Boussinesq, Joseph 1842-1929

Mangin L 1929 Rv GSPA 40, 193-194
  (F55, 613)

Marcolongo R 1929 Nap FM Ri (3a)35,
114-115
Picard E 1934 Par Mm (2)61, i-xliii
(F60, 835)
- 1934 Rv Sc 72, 33-43, 76-84
(F60, 835)


Bouton, Charles Leonard 1869-1922

Osgood W F + 1922 AMSB 28, 123-124
(F48, 23)


Boutroux, Pierre Leon 1880-1922

Anon 1922 Ens M 22, 225-226
(F48, 19)
Brunschvicg L 1922 Rv Met Mor 29,
285-288 (F48, 18)


Bovillus, Caroli 1470?-1553?

= Charles Bovilles = Bouvelles

Brocard H 1902 Intermed 9, 207 (Q73)
Ericsson A P 1898 Intermed 5, 249
(Q73)
Fontes M 1894 Par IBL (9)6, 155
Setnoff 1894 Intermed 1, 26, 121-
123 (Q73)


Bowditch, Nathaniel 1773-1838

Archibald R C 1929 DAB 2, 496-98
Bowditch Nathanial (1773-1838) 1937
*A Catalogue of a Special Exhibition
of Manuscripts, Books, Portraits
and Personal Relics* (Peabody
Museum, Salem), 44 p
Bowditch H I 1839 "Memoire by his son"
in *La Place, Mécanique Celeste*,
Boston, vol 4, 1-169 (revis 1865;
repr 1966 Chelsea) [bib, port]
Pelseneer J 1930 Isis 14, 227-228
Pickering John 1838 *Eulogy of
Nathaniel Bowditch*, Boston
Stanford Alfred 1927 *Navigator. The
Story of Nathaniel Bowditch*, NY
320 p

Tozzer A M 1922 Am Ac Pr 57, 476-478
(F48, 23)
White D A 1838 *Eulogy on the Life and
Character of Nathaniel Bowditch*,
Salem
Young Alexander 1838 *Discourse on the
Life and Character of Nathaniel
Bowditch*, Boston


Bradwardine, Thomas 1290?-1349

Crosby H L ed 1955 *Thomas of
Bradwardine, His "Tractatus de
Proportionibus", Its Significance
for the Development of Mathematical
Physics*, Madison, 203 p
Eneström G + 1885 Bib M 94, 196 (Q3)
- 1921 Arc Sto 2, 133-136 (F48, 3)
Hofmann J E 1951 Centau 1, 293-308
(MR13, 420; Z43, 2)
Maier A 1949 *Die Vorlaeufer Galileis
im 14 Jahrhundert*, Rome
Murdoch J E 1957 Isis 48, 351-352
- 1960 Int Con HS 9, 538-542
(Z118:1, 2) [geom, continuum]
Stamm E 1936 Wars SPTN (3)28, 26-44
(F62, 1026; Z13, 193)
- 1937 Isis 26, 13-32 (F63, 12)
Zubov V P 1960 Ist M Isl 13, 385-440
(MR25, 1)


Brahe, Tycho 1546-1601

Burckhardt F 1887 *Aus Tycho Brahe's
Briefwechsel*, Basel, 24 p
(F21, 1254)
Dreyer J L E 1890 *Tycho Brahe: a
Picture of Scientific Life and Work
in the Sixteenth Century*, Edinburgh
(F22, 8)
- 1893 *Tycho de Brahe. Ein Bild
wissenschaftlichen Lebens im 16
Jahrhundert*, Karlsruhe, 446 p
(tr of 1890) (F25, 1912)
- 1916 Obs 39, 127-131 [trig]
Gade John A 1947 *The Life and Times
of Tycho Brahe*, Princeton, NJ
Hasner J V 1872 *Tycho Brahe und J.
Kepler in Prague*, Prague (F4, 10)

Schiller O 1932 Nor As Tid (2)13,
  16-18
Studnicka F J 1899 Prag Si (39)
  (F30, 62)
- 1901 *Prager Tychoniana*, Prague
Thirion J 1902 Rv Q Sc (3)1, 248-259
  (F33, 14)

**Brahmagupta**   598?-660?

Colebrooke H T 1817 *Algebra with
  Arithmetic and Mensuration, from
  the Sanscrit of Brahmagupta and
  Bhaskara,* London
Datta B 1930 Clct MS 22, 39-51
  [China]
Lorcy 1933 Euc Gron 9, 198-210
  [Euler, Gauss]
Sengupta P C 1931 Clct MS 23, 125-
  128  (F57, 18)  [interpolation]
- 1934 *The Khandakhadyaka. An astro-
  nomical treatise of Brahmagupta,*
  Calcutta, 234 p  (F60, 820)
Simon M 1913 Arc MP (3)20, 280-281
  [diophantine eqs]
Zeuthen H G 1876 Zeuthen (3)6, 168-
  174, 181-191  (F8, 1)

**Braikenridge,** William  XVIII

Brocard H 1908 Intermed 15, 56,
  (Q1761)  (See also Intermed 7, 76;
  14, 272)
Eneström G 1895 Bib M (2)9, 32, 64,
  96
Vacca G 1902 Bib M (3)3, 145
  (F33, 17)

**Brandley,** F. W.  1904-1953

Whittaker J M 1956 LMSJ 31, 251-252
  (Z70, 5)

**Brandt,** Heinrich Karl Theodor
  1886-1954

Eichler M 1955 M Nach 13, 321-326
  (MR17, 337; Z64, 242)

**Brasseur,** Jean Baptiste  1802-1868

Le Roi A 1869 Boncomp 2, 263-272
  (F2, 24)
Liagre J 1869 Bel Anr, 121-144
  (F2, 24)
Spring 1869 Arc MP Lit 198, 1-3
  (F2, 23)

**Bratu,** Gheorge  1881-1941

Sergescu P 1941 Math Tim 17, 137-142
  (Z26, 98)

**Braunmuehl,** Anton Edler von  1853-1908

Anon 1908 Leop 44, 45-46  (F39, 37)
Guenther S 1908 Mit GMNT 7, 362-367
  (F39, 37)
- 1909 Loria 11, 60-64  (F40, 31)
Wieleitner H 1910 Bib M (3)11, 316-330
  (F41, 20)

**Bravais,** Auguste  1811-1863

Anon 1866 Par Mm 35, xxiii-
- 1869 Smi R, 145-
Walker H M 1928 Isis 10, 466-484
  [correlation, Plana]

**Breguet,** Louis François Clement

Jonquierres E de 1886 Par CR 103,
  5-14

**Bremekamp,** Hendrik  1880-1963

Timman R 1963 Nieu Arch (3)11, 61-63
  (MR27, 463; Z112, 243)

**Bretschneider,** Carl Anton  1808-1878

Regel 1879 Hist Abt 24, 79-91
  (F11, 27)

Brewster, David   1781-1868

Martius V 1869 Arc MP Lit 194, 12-14
   (F2, 23)
Smith D E 1922 AMM 29, 157
   [stereoscope]


Brianchon, Charles Julien 1783-1864

Enestrom G + 1894 Intermed 1, 63, 121


Briggs, Henry   1561-1631

Abellan J 1952 Gac M (1)4, 39-41
   (MR14, 609)
Anon 1931 Nat 127, 133-134
   (F57, 23)
Archibald R C + 1943 M Tab OAC 1,
   129-130  [Vieta]
Thompson A J 1925 AMM 32, 129-131
   [log]
- 1925 Bo Fir (2)4, xxxiii-xxxv
   (F51, 21)  [log]
Whiteside D T 1961 M Gaz 45, 9-12
   [binom thm]


Brill, Alexander von   1842-1935

Anon 1912 Wurt MN (2)14, 33-35
   (F43, 52)
Berzolari L 1935 Lom Rn M (2)68,
   488-491  (F61, 954)
Finsterwalder S 1936 M Ann 112, 653-
   663  (F62, 26)
Loeffler E 1943 DMV 53, 82-89
   (MR8, 190; Z28, 100)
Schoenhardt E 1936 Deu M 1, 17-22
   (F62, 26)
Severi F 1922 DMV 31, 89-96
   (F48, 25)


Bring, Erlund Samuel   1736-1798

Cajori F 1907 Bib M 3, 417-420
   [Tschirnhaus transfs]
Montessus M R de 1898 Intermed 5, 40
   [Jerrard]

Brioschi, Francesco   1824-1897

Anon 1897? Nat 57, 279
- 1898 Rv GSPA 9, 49
- 1900 Mathesis (2)10, 112
Aschieri F 1899 Lom Gen (2a)31, 108
Beltrami E 1898 Pe MI 13, 33-36
   (F29, 13)
- 1900 *Francesco Brioschi...*, Milan,
   36 p  (F31, 17)
- 1900 Paler R 14, 262-274  (F31, 17)
Capelli A 1898 Batt 36, 51-54
   (F29, 13)
Colombo G + 1900 Annali (3)5, 141
Cremona L + 1898 Annali (2)26, 343-
   347  (F29, 13)  [ell fun, hyperb
   fun, mech, surf]
Cremona L 1898 LMSP 29, 721
Fuchs L 1898 Crelle 119, 259 (F29, 13)
Hermite C 1897 Par CR 125, 1139-1141
   (F28, 26)
- 1898 Kazn FMO (2)8, 4
- 1898 Loria 1, 62-73  (F29, 14)
Messedaglia A 1897 Linc Rn (5)6,
   353-355  (F28, 27)
Noether M 1898 M Ann 50, 477-491
   (F29, 12)
Pascal E 1899 Pavia U An  (F30, 19)
Prasad Ganesh 1934 *Some Great Mathe-
   maticians of the Nineteenth Century
   ...,* Benares 2, 94-115
Veronese G 1898 Ven I At (7)9, 144-
   145  (F29, 14)
Voit C v 1898 Mun Si 28, 449-452
   (F29, 14)
Volterra V 1900 Int Con 2, 43-57
   [analysis]


Broad, Henry      -1759

Bradley A 1943 Scr 9, 101-104


Brocard, Henri   1845-1922

Anon 1922 Intermed (2)1, 97-98
   (F48, 46)
Bricard R 1920 Nou An (4)20, 357-358
Emmerich A 1889 ZMN Unt 20, 259-260
Guggenbuhl L 1953 M Gaz 37, 241-243
   [triangle]

Lange J 1889 ZMN Unt 20, 181-183
  260 [B angle]
Mansion P 1890 Mathesis 10, 28-30
  [Crelle]
Schloemilch O 1889 ZMN Unt 20, 401-
  405 [Crelle]
Simon M 1905 Arc MP (3)9, 90, 181,
  206, 303 (F36, 66)

Broch, Ole Jacob  1818-1889

Mittag-Leffler G 1889 Act M 12

Brodetsky, Selig  1888-1954

Milne W P 1856 LMSJ 31, 121-125
  (MR17, 446; Z70, 5)

Bromwich, Thomas John I.  1875-1929

GNW 1929 Nat 124, 520  (F55, 617)
Hardy G H 1930 LMSJ 5, 209-220
  (F56, 31)

Brooks, Charles Edward

Anon 1937 Am I Actu R 26, 348-349
  (F63, 816)

Brooksmith, John  1824-1888

Tucker R 1888 LMSP 19, 591-592

Broscius, Joannes  1585-1652

Dickstein S 1882 Wars Pad E
  (F13, 25)
Franke J N 1884 Joannes Broscius,
  Krakow Scholar (in Polish), Kracow
  (Krakow Ac of Sci), 312 p
Stamm E 1936 Archeion 18, 174-183

Brouncker, William  1620-1684

Brun V 1951 Nor M Tid 33, 73-81
Coolidge J L 1949 The Mathematics of
  Great Amateurs (Clarondon Press),
  Ch 11  (Repr 1963 Dover)
Hofmann J E 1950 Mainz MN (3)
- 1960 Ber Mo 2, 310-314  (Z95, 2)
Scott J F + 1960 Lon RSNR 15, 147-157
  (Z107, 245)

Brouwer, Luitzen Egbertus Jan
  1881-1966

Dresden A 1924  AMSB 30, 31-40
Larguier E H 1940 Scr 7, 69-78

Brown, Ernest William  1866-1938

Archibald R C 1938 A Semicentennial
  History of the American Mathematical
  Society, NY, 173-183 [bib, port]

Browne, Edward Tankard  1894-1959

Lasley J W Jr + 1959 Elish Mit 75,
  81-85  (MR23A, 136)

Brozka, Jana

Opial Z 1958 Kwar HNT 3, 357-563

Brunacci, Vincenzio

Kneser A + 1901 Intermed 7, 181

Brunel, Georges  1856-1910

Barbarin P 1901 Ens M 3, 237-239
  (F32, 18)
Duhem P 1903 Bord Mm (6)2, i-lxxxix
  (F34, 24)

Brunhes, Edmund 1834-1916

Garnier F 1951 Isis 42, 234-237

Brunn, Herman Karl 1862-1939

Blaschke W 1940 DMV 50, 163-178
(MR2, 306; Z24, 49)

Bruno, Giordano -1600

Atanassievitch X 1923 *La doctrine
métaphysique et geométrique de
Bruno, ...,* Paris, 156 p
(F49, 23)

Bruno, Giuseppe 1828-1893

Segre C 1893 Tor U Ann, 16 p
(F25, 40)

Bruns, Heinrich 1848-1919

Hayn F 1920 As Nach 210, 15-16
(F47, 17)

Brusotti, Luigi 1877-1959

Chisini O 1960 Linc Rn (8)28, 731-
736 (Z92, 3)
Galafassi V E 1959 It UM (3)14,
287-294 (Z86, 5)
Masotti B G 1959 Milan Sem 29,
xiii-xvi (Z92, 3)

Bryan, George Hartley 1864-1928

Bairstow L 1933 Lon RS Ob 1, 139-42
Love A E H 1929 LMSJ 4, 238-240
(F55, 21)

Bubnov, Nicholas M XIX-XX

Lattin H P 1933 Isis, 19, 181-194
[numeration]

Buckley, William 1519-1571

Varieta 1900 Pitagora 6, 31-
[arith, Wallis]

Budan, F. D. XIX

Aubry V 1897 JM Sp 21, 61

Buddha, Gautama -563?--483?

= Siddhartha = Sakyamuni

Bell E T 1940 MT 33, 252-261
[advice to students and teachers]

Buée, Adrien Quentin 1748-1826

Kramar F D 1966 Ist M Isl 17,
309-316 (MR37, 1221) [alg, Wallis]

Buergi, Jobst 1552-1632

= Jost Buergl = Justus Byrgius
= Byrg Joost = Borgen = Burgi

Archibald R C 1948 *Mathematical
Table Makers,* NY, 15 (Scr Stu 3)
[biobib]
Gieswold 1856 *Justus Byrg als Mathe-
matiker und dessen Einleitung in
seine Logarithmen,* Danzig, 36 p
Mautz O 1920 Basl V 32, 104-106
[log]
Pajares E 1952 Gac M (1)4, 157-163
(MR14, 609)
Roessler G 1932 Z Instr 52, 31-38
(Z3, 242)
Voellmy E 1948 *Jost Burgi und die
Logarithmen,* Basel, 24 p Elem M (S)5
(MR12, 381; Z33, 2) [log]
Wolf R 1872 *Johannes Kepler und Jose
Buergi,* Zurich (F4, 11)
- 1893 Zur NGV 38, 1-9 (F25, 17)

**Buermann**

Cantor M 1872 ZM Ph 17, 428-430
(F4, 16)
Caspari 1873 ZM Ph 18, 120-122
(F5, 32)
Eneström G 1892 Bib M (2)6, 120
Eneström G + 1897 Intermed 4, 47
(Q909)

**Buffon,** George Louis Leclerc
Comte de  1707-1788

Brunet P 1931 Archeion 13, 24-39
(F57, 29) [infinity]
Coolidge J L 1949 *The Mathematics of
Great Amateurs* (Clarendon Press),
Ch 13 (Repr 1963 Dover)
Frechet M 1947 Eg IB 28, 185-202
(MR9, 485)
- 1954 in Boyer R ed, *Buffon* (Corpus
général des philosophes français,
Presses Univ) (repr in Frechet
1955 Les Mathematiques et la
concret, Paris) [phil, stat]
Gridgeman N T 1960 Fr SMB 25, 183-
185 [needle probl]
Thumm W 1965 MT 58, 601-7 [pi]
Weil F 1961 Rv Hi Sc Ap 14, 97-136
[letter, Cramer]

**Bugaev,** Nikolai Vasilevich  1837-1903

Anon 1905 Mos Mo Sb 25, 370
Braytzeff I 1904 Wars Na Tr, 14-15,
32-34
Bugaev N V 1959 Ist M Isl 12, 525-
558 (MR24A, 128; Z98, 7)
Egorov D T 1904 Kiev UFMO (10), 69-
73 (F35, 25)
Lakhtin L K 1905 Mos MO Sb 25, 251-
269 (F36, 23)
- 1905 Mos MO Sb 25, 322-330
(F36, 24)
Lopatine L M 1905 Mos MO Sb 25, 270
Minin A P 1905 Mos MO Sb 25, 293-321
(F36, 24) [numb thy]
Shevelev F Ya 1959 Ist M Isl 12,
551-558 (MR24A, 128)

**Buhl,** Adolphe  1878-1949

Fehr H 1951 Ens M 39, 6-8 (MR13, 1)
Viguier G 1950 Mak DMF 1, 83-87
(MR12, 311)

**Buka,** Felix  1852-1896

Hauck G 1899 DMV 6, 23-25 (F30, 17)

**Bukreev,** Boris Kakovlevich  1859-

Anon 1950 Ukr IM 2(1), 3-9
(Z41, 2)
Belousova V P 1959 Ukr IM 11, 312-314
(MR22, 619; Z86, 4)
Belousova V P + 1959 UMN 14(5), 181-
195 (MR22, 619; Z87, 6) [port]

**al-Buni**  XII-XIII

Ahrens W 1922 Islam 12, 157-177
- 1925 Islam 14, 104-110 [magic sq]
Sarton G 1931 Int His Sc 2(2), 595

**Bunickij,** Eugen  XX

Bily J 1953 Cas M 78, 287-290
(Z53, 340)

**Bunyakovskii,** Viktor Yakovlevich
1804-1889

Anon 1883 *Liste des travaux mathé-
matiques de Victor Bouniakowsky
membre effectif de l'Académie Im-
periale des Sciences a St.-
Petersburg,* St. Petersburg
Depman I J 1955 Ist M Isl 8, 630-635
(Z68, 6)
Dobrovolskii V A 1959 Ist M Isl 12,
505-511, 511-524 (MR24A, 4;
Z100, 246)
Lobanova T V 1959 Mos IIET 22, 289-
292 (Z95, 4)
Otradnykh F P 1955 Len M V 10(5),
49-54 (MR17, 2; Z64, 242)

Prudnikov V E 1953 Ist M Isl 6, 223-237
- 1954 *V. Ya. Bunyakovskii...,* *scientist and teacher,* Moscow

Burali-Forti, Cesare 1861-1931

Marcolongu R 1931 It UM 10, 182-185 (F57, 43)

Buratini, Tito Livio 1610?-1682

Favaro A 1896 *Intorno alla vita ed ai lavori di Tito Livio Buratini,* Venice

Burbury, Samuel Hawksley 1831-1911

Bryan G H 1913 Lon RSP (A)88, i-iv (F44, 20)
J L 1911 LMSP (2)10, iv-v (F43, 35)

Burckhardt, Johann Karl 1773-1825

Smith D E 1922 AMM 29, 297 [educ]
≈n.1922 AMM 29, 299 [publishing]

Burgatti, Pietro 1868-1938

Graffi D 1938 It UM 17, 145-151 (F64, 23)

Burgess, Alexander G XIX-XX

Comrie P 1933 Edi MSP (2)3, 300 (F59, 40)

Burgess, George Kimball 1874-1932

Anon 1932 Sci Mo 35, 183-184

Burkhardt, Heinrich 1861-1914

Liebmann H 1915 DMV 24, 185-195 (F45, 36)
Rudio F 1914 Zur NGV 59, 565-566 (F45, 36)

Burmester, Ludwig Ernst Hans 1840-1927

Dupont P 1963 Tor FMN 98, 489-513 (MR29, 807) [graphs, mech
Mueller R 1930 DMV 39, 1-21 (F56, 26)

Burnside, William 1952-1927

Forsyth A R 1927 Nat 120, 555-557 (F53, 34)
1928 Lon RS(A)117, xi-xxv (repr in Burnside 1928 *Theory of Probability)*
- 1928 LMSJ 3, 64-80 (F54, 38)

Burrau, Carl 1867-1944

Archibald R C 1948 *Mathematical Table Makers,* NY, 16 (Scr Stu 3) [biobib]
Nielsen Niels 1910-1916 *Matematiken i Danmark,* Copenhagen, 3 vols

Busche, Edmund 1861-1916

Hoppe E 1917 Ham MG 5, 217-229 (F46, 18)
Mangoldt H v 1917 Ham MG 5, 230-237
Riesebell P 1916 DMV 25, 283 (F46, 18)

Butchart, Raymond Keiller 1888-1931

J E A S 1931 Edi SP (A)51, 200-201 (F57, 47)

**Buteo,** Jean   1492-1565

= Buteon = Buteone

Procissi A 1946 Pe M (4)24, 141-151
[Cardano, lin eqs]
Wertheim G 1901 Bib M 3(2), 213


**Butrigarius,** Hercules   XVI

Ginsburg J 1932 Scr 1, 92


**Bydzovsky,** Bohamil   1880-

Bilek J 1960 Cas M 85, 226-227
(MR22, 770; Z100, 247)
Koutsky K 1950 Cas MF 75, D349-
D357 (Z39, 243)
Metelka J 1960 Prag Pok 4, 603-612
(Z98, 9)


**Byushgens,** Sergei Sergievich 1882-

Finikov S P 1953 UMN (NS) 8(4), 185-
192 (MR15, 89)


**Cabeo,** Niccolo   1585-1650

= Cabeus

Thirion J 1893 Rv Q Sc (2)4, 563-
572 (F25, 88)


**Caccioppoli,** Renato   1904-1959

Miranda C 1959 JMPA (4)47, v-vii
(MR22, 26; Z88, 6)
Scorza Dragoni G 1963 Linc Rn, 85-93
(MR29, 880)


**Cagwin,** Samuel Gardner   XIX

Burleson B F 1894 AMM 1, 374-375


**Caillier,** Charles Marc Elie   1865-1922

Wavre R 1922 Arc Sc Ph (5)4, 417-429


**Cairns,** William DeWeese   1871-1955

Carver W B 1956 AMM 63, 204-205
Hedrick E R 1943 AMM 50, 1


**Cajori,** Florian   1859-1930

Archibald R C 1932 Isis 17, 384-407
[bib]
- 1933 Am Ac Pr 68, 605-609 (F59, 853)
- 1959 Colo Stu (3), 5-9
Baidaff B T 1930 B Ai Bol 3, 125-126
(F56, 816)
Loria G 1897 Pe MI 11, 1-13 (F27, 1)
Miller G A 1919 AMSB 26, 79-85
(F47, 28)
- 1920 SSM 19, 830-835 [crit of
his *History*]
- 1921 SSM 20, 300-304
- 1922 Sch Soc 16, 449-454
- 1924 SSM 23, 138-149
- 1925 SSM 24, 939-947 [D E Smith]
Neville E H 1935 M Gaz 19, 134-135
(F61, 28) [crit of his Newton]
Schreiber E W 1932 SSM 32, 129-134
Simons L G 1930 AMM 37, 460-462
Smith D E 1920 AMM 27, 121
- 1898 Sch Rv 5, 184
- 1930 AMSB 36, 770-780
- 1930 Archeion 12, 369-371 (F56, 28)
- 1930 MT 23, 509-510 (F56, 816)


**Caldonazzo,** Bruto   1886-1960

Anon 1960 Parma Rv M (2)1, 294-296
(Z91, 6)
Sestini G 1960 It UM (3)15, 340-341
(Z91, 6)


**Caligny,** Anatole François Huë
Marquis de   1811-1892

Boussinesq J 1892 Par CR 114, 797-802
(F24, 29)

Callandreau, Pierre Jean Octave
1852-1904

Anon 1904 Ens M 6, 150  (F35, 35)
- 1904 Nat 69, 441
- 1904 Rv GSPA 15, 281-282


Callet, Jean Charles  1744-1799

Schols C M 1887 Delft EP 3, 130-139


Cambier, Augustin  1852-1909

Mansion P 1909 Mathesis (3)9, 57-58
  (F40, 34)


Campagne, Cornelius  1902-1963

Brans J A T M 1964 Actuar St (6),
  151-158  (Z115, 246)


Campano  XVII

  = Giovanni (John, Johannes)
  Campano (Campanus) da Novara

Benjamin F S 1954 Osiris 11, 221-246
  (Z55, 244)  [John of Gmunden]
Brocard H 1910 Intermed 17, 212
Sarton G 1931 Int His Sc 2(2), 985-
  987 [bib]
Weissenborn H 1882 Die Ueberzetzungen
  des Euklid durch Campano und
  Zaniberti,  Halle
Zapelloni M T 1928 Pe M (4)8, 175-184
  (F54, 5)


Campbell, John Edward  1862-1924

E B E 1925 Lon RSP (A)107 ix-xii
  (F51, 30)
H H 1925 LMSP (2)23, lxx-lxxi
  (F51, 30)

Campbell, John Robert  1827-1897

Anon 1897 LMSP 28, 587-588  (F28, 27)


Camus, Charles Étienne Louis 1699-1768

L'huilier T 1863 Essai biographique
  sur le mathématicien Camus, né a
  Grécy-en-Brie,  Meaux


Canacci, Rafaello  XIV

Procissi A 1953 Geno AL 9, 55-76
  (Z53, 196)  [alg]
- 1954 It UM (3)9, 300-326, 420-451
  (MR16, 433; Z37, 3)  [alg]
- 1955 I "Ragionamenti d'algebra" di
  R. Canacci..., Bologna, 81 p
Sarton G 1938 Isis 29, 99


Cantemir, Antioch  1709-1744

Ionescu I 1940 Gaz M 45, 337-339,
  393-396  (F66, 18)


Cantone, Michele  1857-1932

Anon 1932 It UM 11, 128
Carrelli A 1932 Nap FM Ri (4)2, 13-
  19  (F58, 46)
- 1932 Nuo Cim (2)10, 45-51
Perucca E 1932 Tor FMN 67, 558-565
Polvani G 1932 Lom Rn M (2)65, 307-308


Cantor, Georg Ferdinand Ludwig
  Philip 1845-1918


Anon 1918  Ens M 20, 68-69 (F46, 15)
Bellon W 1945 Colom Cul 3, 353-73
Crespo Pereira R 1952 Gac M (1)4,
  67-73  (MR14, 609; Z46, 3)
*Fraenkel A 1930 DMV 39, 189-266
- 1932 in Cantor, gesammelte Abhandlung, 452-483  (abr of Fraenkel 1930)

- 1935 Fund M 25, 45-50 [diag meth]
Guidice F 1892 Paler R 6, 161-164
Gutberlet C 1919 Gorres Ja 32, 364-
Hill L S 1934 Scr 2, 41-47 (F60, 15)
*Jourdain P 1910 Arc MP (3)16, 21-43
- 1914 Arc MP 22, 1-21
- 1915 in Cantor *Contributions to
the Founding of the Theory of Trans-
finite numbers,* Chicago-London,
1-81 (Repr 1952 Dover)
[Fourier series]
Lorey W 1915 ZMN Unt 46, 269-274
(F45, 55)
Meschkowski H 1965 Arc HES 6, 503-
519 (MR33, 3868)
*- 1965 *Evolution of Mathematical
Thought,* (Holden-Day), Ch. 5
* - 1967 *Probleme des Unendlichen.
Werk und Leben Georg Cantor,*
Braunschweig, 288 p
Mittag-Leffler M G 1929 Act M 50,
25-26
Prasad Ganesh 1934 *Some Great Mathe-
maticians of the Nineteenth Century
Century...,* Benares, 183-211
*Russell Bertrand 1951 *The Autobio-
graphy of Bertrand Russell 1872-
1914,* 356-361
Schoenflies A 1922 DMV 31, 97-106
(F48, 19)
- 1927 Act M 50, 1-24 (F53, 26)
Staeckel P 1900 Gott Anz 162, 251-
264
Ternus J 1929 Scholas 4, 561-571
(F55, 613) [Hontheim]
Wangerin A 1918 Leop 54, 10-13
(F46, 15)
Young W H 1926 LMSP 24, 412-426
[letter]

**Cantor,** Moritz Benedikt 1829-1920

Bobynin W W 1899 Fiz M Nauk (2)1,
76-83 (F30, 28)
Bopp K 1920 Heid Si, 16 p
- 1930 Heid Tat (F56, 24)
Bosmans H + 1908 Bib M (3)9, 71-80,
139-175, 237-265, 321-350
(F39, 3)
Cajori F 1920 AMSB 27, 21-28
(F47, 28)

Curtze M 1900 Bib M (3)1, 227
- 1900 ZM Ph 45, 41
Curtze M + 1899 *Festschrift zum 70
Geburtstage,* Leipzig (also ZM Ph
44(s)) (F32, 28)
Eneström G H 1903-1904 Bib M (3)4
Eneström G + 1909 Bib M (3)10,
53-83, 164-179, 260-277, 341-350
(F40, 1)
Gibson G A 1899 Edi MSP 18, 9-32
(F30, 46)
Gram J P 1893 Dan Ov, 18- [cubic]
Guenther S 1920 Mit GMNT 19, 222
Guenther S + 1909 *Zum 80. Geburtstage
Moritz Cantor,* Leipzig
Hoffmann J C V 1893 ZMN Unt 24, 170-
172 (F25, 60)
Junge G 1929 Unt M 35, 239-241
(F55, 16)
Loria G 1894-1896 Batt 32, 23-27,
353-357; 34, 11
- 1922 Scientia 31, 265-279 [Zeuthen]
- 1924 Arc Sto 5, 28-36 [bib]
Mikami Y 1909 Arc MP (3)15, 68-70
(F40, 2) [Chinese]
Neugebauer O 1956 Isis 47, 58
["A notice of ingratitude"]
Sarton G 1950 Int Con HS 6, 45-78
Schiaparelli E G V 1881 Lom Gen (2)14,
62-69 (F13, 25)
Smith D E 1933 Scr 1, 204-207
(F59, 30)
Zemaitis Z 1930 Lit UM 5(1), 177-219
(F57, 1306)
Zeuthen H G 1893 Dan Ov, 1-
[cubic eq]
- 1894 BSM (2)18, 163-169 (F25, 2)
- 1895 BSM (2)19, 183

**Capella,** Martianus Minneus Felix V

Narducci E 1883 Boncomp 15, 505-580

**Capelli,** Alfredo 1855-1910

Amodeo F 1910 Pc MI (3)7, 191-192
(F41, 26)
Gallucci G 1913 Nap Pont (2)18(3), 49
Miller G 1915 Batt 53, 313-315

Natucci A 1955 Pe M (4)33, 257-275
  (MR17, 813; Z70, 3)
- 1956 Batt (5)3, 297-300
  (MR17, 931; Z67, 247)
Ricci G 1910 Mathes It 2, 10-15
  (F41, 26)
Torelli G 1910 Batt (3)1, 5-15
  (F41, 26)
- 1910 Bo Fir (a)9, 1-10
- 1910 Nap FM Ri (3)16, 20-25


Caporali, Ettore  1855-1886

Loria G 1889 Batt 27, 1-32


Caraccioli, Giambattista B.

Loria G 1903 Loria 6, 33-38


Caramuel, Juan  1606-1682

  = J. Caramuel-Lobkowitz

Diéguez, D G 1919 Rv M Hisp A 1


Caratheodory, Constantin 1873-1950

Anon 1946 Gree SM 22, 198-207
  (MR8, 498; Z60, 14)
Errera A 1958 Brx U Rv 2  (MR20, 1137)
Hölder E 1950 Fors Fort 26, 290-293
  (Z38, 4)
Perron O 1952 DMV 55, 39-51
  (MR13, 810; Z46, 3)
Sakellariov N 1952 Gree SM 26, 1-13
  (MR14, 121; Z49, 291)
Schmidt E 1943 Fors Fort 19, 249-250
  (Z60, 14)
Tietze H 1950 Arc M 2, 241-245
  (MR12, 311; Z37, 2)
- 1950 Mun Si, 85-101  (Z41, 3)


Caravaggio, Pietro Paolo

Tenca L 1953 Lom Mr 86, 835-846
  (Z52, 3)  [Viviani]

Caravelli, Vito  1724-1800

Amodeo F 1902 *Dai fratelli di Martino
  e Vito Caravelli,* Naples 64 p
  (F33, 17)
Ionescu I 1902 Gaz M 6, 29
Ionescu I + 1900 Intermed 7, 323
Vereecke P 1935 Mathesis 49, 59-62
  (F61, 20)


Carcavi, Pierre de  ?-1684

Henry C 1884 Bomcomp 17, 317-391
Tannery P 1893 Bu Sc M Ast 28


Cardano, Geronimo 1501-1576

  = Girolamo C. = Hieronymus Cardanus
  = Jerome C.

Anon 1931 MT 23, 458
Battistini M 1952 Rv Sto Cr 42, 92-10
  (Z48, 242)
Bellini Angelo 1947 *Gerolamo Cardano
  e il suo tempo,* Milan 338 p
Bilancioni G 1930 Rv Sto Cr 21, 302-
  329 [Leonardo de Vinci]
Bortolotti E 1933 Bln Sto SM·12, 79 p
  (Z7, 387)
Boyer C B 1950 AMM 57, 387-390
  (MR12, 1)  [binom coeffs]
Cantor M 1904 Int Con H 12, 31-43
  (F35, 7)  (also in Heid N Jb 13)
Cardano G 1575 *De vita propria liber*
- 1821 *Vita di Girolamo Cardano...,*
  Milan  (tr of Cardano 1575 by M.
  Montovani)
- 1914 *Des Girolamo Cardano von
  Mailand...,* Jena (tr of Cardano
  1575 by H. Hefele)  (F45, 13)
- 1931 *The Book of My Life,* London-NY
  349 p (tr of 1821) (repr 1962 Dover)
Cassina U 1929 Pe M 9, 117-129
  [Tartaglia]
- 1932 Int Con 6, 443-448  (F58, 31)
  [Tartaglia]
Charadze A K 1943 Gruz Soob 4, 195-
  199  (MR6, 141)  [Hudde]

Crossley J 1836 *The Life and Times of Cardan*, London
Cunha F 1935 Am J Surg 30, 191-202
Drake S + ed 1969 *Mechanics in Sixteenth-century Italy* (Univ Wisc Pr)
Eckman James 1946 *Jerome Cardan*, Baltimore 133 p
Feldmann R W Jr 1961 MT 54, 160-63 [Tartaglia]
Firmiani 1904 *Girolamo Cardano, la vite e l'opere*, Naples
Gravelaar N L W A 1909 Nieu Arch (2) (2)8, 408-444 (F40, 54)
Harig G E 1935 Len IINT 7, 67-104 (F61, 949) [Tartaglia]
- 1936 Mos IINT 9, 23-67 (F62, 1029) [Tartaglia]
Larder D F 1968 Isis 59 74-77 [chem]
Lindsay R B 1948 AJP 16, 311-317 (MR9, 485)
Margolin J 1965 Bib Hum 27, 655-668
Mendelsohn C J 1939 Scr 6, 157-168 [crypt]
Morley Henry 1854 *Jerome Cardan: The Life of Gerolamo Cardano of Milan*, London
Mueller C H 1915 ZMN Unt 46, 500 (F45, 13) [Timerding]
Nordgaard M A 1938 M Mag 12, 327-346 [Tartaglia]
Ore Oystein 1953 *Cardano, the Gambling Scholar*, Princeton 263 p (MR14, 609; Z51, 243)
Procissi A 1946 Pe M (4)24, 141-151 (MR8, 497)
Reeve W D 1930 MT 23, 458
Sanford V 1952 MT 45, 368, 372 [medicine]
Vacca G 1937 Milan Sem 11, 22-40 (F63, 804)
Van de Velde A J J 1951 Bel Vla Jb 24
Vooys C J 1959-1960 Euc Gron 35, 162-166 (MR23A, 4) [complex numb]
Waters W C 1898 *Jerome Cardan*, London
Wieleitner H 1925 Arc Sto 6, 201-205 [quadratic]
- 1928 Erlang Si (58/59), 173-176 (F54, 24) [cubic eq]
Zeuthen H G 1893 Dan Ov 1-17, 303-341 (F25, 62)

**Carey,** Frank Stanton 1860-1928

Proudman J 1929 LMSJ 4, 139-140 (F55, 21)

**Carleman,** Torsten 1892-1949

Carlson F 1950 Act M 82, i-vi (MR11, 708)

**Carlini,** Francesco

Tosi A 1959 Pe M (4)37, 137-146 (MR23(B), 1545; Z88, 5)

**Carlson,** Fritz David 1888-1952

Frostman O 1953 Act M 90, ix-xii (MR15, 276)

**Carlyle,** Thomas 1795-1881

Wursthorn P A 1966 MT 59, 755-770

**Carnap,** Rudolph 1891-

McLane S 1939 AMSB 44
Schlilpp P A ed 1963 *The Philosophy of Rudolph Carnap*, La Salle, Ill [*bib]

**Carnot,** Adolphe 1839-1920

Breton J L 1922 Par Mm (2)58, xxix-xxxviii

**Carnot,** Lazare Nicolas Marguerite 1753-1823

Anon 1923 *Centenaire de L. Carnot, notes et documents inédits*, Paris [bib, ports]
Arago D F J 1850 Par Mm 22, 1-120 (Engl tr in Arago 1857 *Biographies of Distinguished Scientific Men*, London, 287-361)

Boyer, C B 1954 AMM 61, 459-463
  (MR16, 1; Z56, 243) [calc]
- 1956 MT 49, 7-14
Carnot Hippolyte 1861-1863 *Memoires
  sur Carnot par son fils*, Paris,
  2 vols (New ed 1893) [bib]
Carré Henri 1947 *Le grand Carnot*,
  Paris
Dupre Huntley 1940 *Lazare Carnot,
  Republicain patriot*, Oxford,
  Ohio 343 p
Fink K 1894 *Lazare-Nicolas-Marguerite
  Carnot. Sein Leben und seine Werke,
  nach den Quellen dargestellt*,
  Teubingen, 135 p (F25, 22)
Koerte Wilhelm 1820 *Das Leben Lazare
  N. M. Carnots: Mit einem Anhange
  enthaltend die ungedruckten Poesien
  Carnots*, Leipzig, 490 p
Kramar F D 1963 Ist M Isl 15, 225-290
  [XIX, vectors]
Reinhard Marcel 1950-1952 *Le grand
  Carnot*, Paris, 2 vols
Rioust Mathieu Noel 1817 *Carnot*,
  Paris, Brussels, Gand
Smith D E 1933 Sci Mo 37, 188-189
Watson S J 1954 *Carnot*, London


Carnot, Sadi Nicolas Leonhard
1796-1832

Anon 1932 Nat 130 266-267
Carnot H 1878 Par CR 87, 967-970
  (F10, 16) [letter]
Johnson E H 1933 Sci Mo 36, 131-137
  (F59, 847)
Mach E 1892 Wien Si 101, 1589-1612
  (F24, 54)
Plank R 1932 Deu Ing Z 76, 821-822


Carosio, Matteo

Favaro A 1919 Arc Sto 1, 28-38
  [Galileo]


Carroll, Lewis  1832-1898

  = Charles Lutwidge Dodgson

Anon 1898 Nat 57, 279-280  (F29, 14)
- 1932 *Catalogue of an Exhibition to
  Commemorate the One Hundredth
  Anniversary of the Birth of Lewis
  Carroll*, NY, 160 p
Archibald R C 1932 Scr 1, 172-173
  [bib]
- 1933 AMSB 39, 846-848  (F59, 31)
Becher, Florence 1945 *Victoria
  Through the Looking Glass*, NY
  (esp. Ch. 15)
Berneis B 1938 Sphinx 8, 110-112
  (F64, 23)
Braithwaite R B 1932 M Gaz 16,
  174-178  (F58, 995) [logic]
Collingwood S D 1898 *The Life and
  Letters of Lewis Carroll Dodgson*,
  London  (F30, 19)
- 1961 *Diversions and Digressions of
  Lewis Carroll* (Dover) [=*The Lewis
  Carroll Picture Book* 1899 +
  photographs]
De la Mare, Walter 1932 *Lewis Carroll*,
  London, 67 p
Eperson D B 1933 M Gaz 17, 92-100
  (F59, 31)
Gardner Martin 1960 *The Annotated
  Alice*, NY
- 1962 *The Annotated Snark*, NY
Lennon Florence B 1962 *The Life of
  Lewis Carroll*, (Collier Books)
  (Rev ed)
Madan F 1932 *The Lewis Carroll Cente-
  nary in London*, 1932, London, 169 p
Schiller F C S 1901 Mind, 87-101
  [Snark = the absolute!]
Stright R L 1964 Ari T 11, 571-573
Smith D E 1922 AMM 29, 15
Smith D E + 1932 MT 25, 38-43
Taylor A L 1952 *The White Knight*,
  Edinburgh
Weaver W 1938 AMM 45, 234-236
- 1954 APSP 98, 377-381  [mss]
- 1956 Sci Am 194(Ap), 116-128
Williams S H + 1931 *A Handbook of the
  Literature of the Rev. C. L.
  Dodgson*, Oxford, 360 p (suppl in 193

Willerding M F 1960 Scr 25, 209-219
(Z97, 244)

Carslaw, Horatio S 1870-1954

Houstoun R A 1956 Edi MSN (40, 26)
(MR18, 710; Z71, 3)
Jaeger J C 1956 LMSJ 31, 494-501
(MR18, 182; Z70, 244)

Cartan, Elie J. 1869-1951

Chern S + 1952 AMSB 58, 217-250
(MR13, 810; Z46, 3)
Hodge W V D 1953 LMSJ 28, 115-119
(MR14, 524; Z50, 2)
Finzi A 1954 Riv Lemat 8, 76-80
(MR16, 434)
Javillier M 1951 Par CR 232, 1785-
1791 (MR13, 1; Z42, 4)
Picard E 1939 BSM (2)63, 163-166
(F65, 17)
Saltykow N 1952 Feo DMF 4(3/4), 59-
64 (MR 14, 832; Z48, 243)
Simonart F 1950 Bel BS (5)36, 1010-
1025 (Z40, 289) [XIX, Gauss]
Whitehead J H C 1952 Lon RS Ob 8,
71-95 (MR14, 524)

Cartazar D. J.

Irveste J A 1912 Esp SM 1, 285-290
(F43, 23)

Carus, Mary Hegeler 1861-1936

Smith D E 1937 AMM 44, 280-283

Carus, Paul 1852-1919

Sheridan J F 1957 *Paul Carus: A Study
of the Thought and Work of the
Editor of the Open Court Publishing
Company*, Ann Arbor
Slaught H E 1923 AMM 30, 151
- 1925 MT 18, 183-184

Carver, Walter Buckingham 1879-1961

Walker R J 1962 AMM 69, 688

Casanova, Giovanni (Jacques) Giacomo
de Seingalt 1725-1803

Henry C 1883 Boncomp 15, 637-670

Casey, John 1820-1891

Anon 1891 LMSP 22 (F23, 29)
- 1891 Lon RSP 49, (F23, 29)
Hayashi T 1912 Toh MJ 1, 204-206
Zacharias M 1943 DMV 52-2, 79-89

Casorati, Felice 1835-1890

Bertini E 1892 Lom Gen 25, 1206-
(F24, 29)
Brioschi F 1890 Annali (2)18, 264
Galdeano Z G de 1891 Prog M 1, 22-24
(F23, 25)
Loria G 1891 Bib M (2)5, 1-12
(F23, 25)
- 1891 Paler R 5, 236-251 (F23, 24)
d'Ovidio E 1890 Tor FMN 26, 3-4
Vivant G 1935 Milan Sem 9, 127-138
(F61, 953)
Volterra V 1900 Int Con 2, 43-57
[anal]

Caspar, Max 1880-1956

Leibbrand W 1957 Arc In HS 10, 89-90
(Z88, 6)
Volk O 1960 DMV 62, 93-98; 63, 52
(Z88, 6)

Caspary, Ferdinand 1853-1901

Jahnke E 1903 DMV 12, 42-60
(F34, 25)

Cassani, Pietro 1832-1905

Ricci G 1907 Ven I At (8)9, 175-186
(F38, 27)

Cassini, Giovanni (Jean) Domenico
(1625-1712)

Antoniadi E M 1925 Astmie Fr 39,
417-434
Derenzini T 1960 Physis Fi 2, 235-
241 [Borelli]
Smith D E 1921 AMM 28, 123, 369
[Picard]
Tenca L 1955 Bln Rn P (11)2(2), 162-
177 (Z67, 246) [Viviani]

Castelli, Benedetto 1577-1644

Boncompagni B 1878 Boncomp 11, 587-
665 (F10, 11)
Favaro A 1884 Boncomp 16, 545-564

Castelnuovo, Guido 1864-1952

Anon 1954 Rm U Rn (5)13, 1-49
(MR15, 923)
Garnier R 1952 Par CR 234, 2241-2244
(MR13, 810)
Godeaux L 1953 Rv GSPA 60, 8-14
[alg geom]
Hodge W V D 1953 LMSJ 28, 120-125
(MR14, 524; Z50, 2)
Sebastiao e Silva J 1952 Gaz M Lisb
13(52), 1-3 (MR14, 243; Z42, 6)
Severi F 1952 Archims 4, 130-131
(Z47, 6)
Terracini A 1952 Tor FMN 86, 366-377
(MR15, 276; Z47, 245)

Castigliano, Alberto

Biadego G B 1885 Boncomp 18, 293-320

Castillon, Giovanni Francesco Mauro
Mel'chior 1708-1791

Anon 1792-1793 Ber Mm
Conte L 1942 Pe M (4)22, 70-90
[Fermat]
Court N A 1954 Scr 20, 118-120, 232-
235 (Z60, 10) [problem]

Caswell, John 1655-1712

Karpinski L C 1912 AMSB (2)19, 446
(F44, 39)
- 1912 Bib M (3)13, 248-249
(F43, 14)

Catalan, Eugene Charles 1814-1894

Boyer J 1894 Rv GSPA 5, 228
Galdeano Z G de 1894 Prog M 4, 58-60
(F25, 45)
Gloden A 1953 Int Con HS 7, 316-319
- 1952 Scr 19, 271
Longchamps G de 1894 JM Sp(4)3,
49-53 (F25, 45)
Mansion P 1885 Lieg Mm 12, 1-38
(Mathesis 5(S2), 1-38)
- 1896 Bel Ann (S), 62 p (F27, 19)
Mansion P + 1894 Mathesis (2)4, 33
(F25, 45)

Cataldi, Pietro Antonino 1522-1626

Bortolotti E 1930 Pe M (4)10, 227-228
Grunert J A 1858 Arc MP 30, 275-
[cont frac]
Wertheim G 1902 Bib M (3)3, 76-83
(F33, 54)

Cauchy, Augustin Louis 1789-1857

d'Adhemar R 1905 Quinzain (S), 15 p
(F36, 22) [Ampere, Hermite]
Belga 1920 Intermed 27, 54-56 [Abel]
Beman W W 1899 Ens M 1, 162
[complex numb]

Bertrand Joseph 1902 *Eloges acadé-
miques, nouvelle serie,* Paris,
101-120
- 1904 Par Mm 47, 183-205 (F35, 22)
Bertrand J + 1869 J Sav (Aug)
(F2, 19)
- 1870 BSM 1, 105-107 (F2, 19)
Biot J B 1857 *M. le Baron Cauchy.
Lettre à M. Falloux,* Paris, 14 p
Boncompagni B 1869 Boncomp 2, 1-96
(F2, 19) (Rev of Valson 1868)
Carruccio E 1957 It UM 12, 298-307
(MR20, 128) [calc]
David 1882 Tou Mm (8)4, 174-175
Ettlinger H J 1922 An M (2)23, 255-
270 [intgl]
Genocchi A 1870 Tor Atti 5 (F2, 20)
Jourdain P E B 1905 Bib M (3)6, 190-
207 (F36, 59) [Gauss, intgl]
- 1914 Isis 1, 661-703 [cont,
Fourier ser, intgl]
Loria G 1932 Int Con (1932) 2,
340-341 (F58, 52) [anal geom]
- 1933 Scr 1, 123-128 [anal geom]
Matzka W 1844 Arc MP 4, 357-
Medvedev F A 1961 Mos IIET 43, 264-
289 [fun]
Miller G A 1910 AMM 17, 162 [group]
- 1910 AMSB (2)16, 510-513
Narducci E 1869 Boncomp 2, 97-102
(F2, 19)
Ocagne Maurice d' 1930 *Hommes et
choses de science,* Paris, 111-125
*Pasch M 1927 *Mathematik am Ursprung,*
Leipzig, 47-73 [calc]
Peano G 1894 Tor FMN 30, 20-41
[limits]
Pelseneer J 1951 Arc In HS 30, 631-
633 (MR13, 197; Z42, 242)
Plakhowo N 1907 Intermed 14, 255
(Q3219)
Prasad Ganesh 1933 *Some Great Mathe-
maticians of the Nineteenth Century
...*Benares 1, 68-110
Procissi A 1947 It UM (3)2, 46-51
(Z31, 146) [letter, Libri,
numb thy]
Rychlik K 1957 Cas M 82, 227-228
(MR19, 826; Z98, 7)
- 1957 Cz MJ 7, 479-481 (Z90, 7)
- 1957 Rv Hi Sc Ap 10, 259-261

- 1958 Cz MJ 8, 619-631 (MR20, 129;
21, 485; Z80, 8)
- 1962 *Theorie der reellen Zahlen in
Bolzanos handschriftlichen Nachlass,*
Prague, 103 p (MR28, 576)
Smith D E 1922 AMM 29, 394
[Abel, Legendre]
Struik R + 1928 Isis 11, 364-366
[Bolzano]
Studnicka F J 1876 *Cauchy als formaler
Begruender der Determinantentheorie,*
Prag, 40 p
Teichmueller O 1939 Deu M 4, 115-116
(F65, 16) [complex anal, Riemann,
Weierstrass]
Terracini A 1957 It UM 12, 290-298
(Z77, 8)
- 1957 Tor Sem 16, 161-203 (Z78, 3)
- 1957/1958 Tor Sem 17, 81-82
(MR21, 485; Z82, 243)
Valson C A 1868 *La vie et les travaux
du baron Cauchy...avec une préface
de M. Hermite...,* Paris, 2 vols
(F1, 15)
Yushkevich A P 1947 Mos IIET 1, 373-
411 (MR11, 572) [integral]


Cavalieri, Bonaventura 1598-1647

Agnostini A 1925 It UM 4, 104-107
(F51, 8) [intgl, lim]
- 1940 It UM (2)2, 147-171
(MR 2, 114; Z23, 387) [center of
gravity]
Amodeo A 1909 Linc Rn (5)18(1), 661-
668 [conic]
Anon 1933 MT 25, 93-94
Artom E 1923 Pe M (4)3, 17-20
Bortolotti E 1930 Pe M (4)10, 227-228
Bosmans H 1922 Brx SS 42(1), 82-89
[indivisibles]
- 1922 Mathesis 36, 365-373, 447-456
Boyer C B 1941 Scr 8, 79-91
(MR3, 258)
Cellini G 1966 Pe M (4)44, 1-21
(MR34, 1139) [indivisibles]
- 1966 Pe M (4)44, 85-105
(MR34, 7319)
De Waard C 1919 Loria 2, 1-12
Evans G W 1917 AMM 24, 447-451
(F46, 52) [theorem]

Favaro A 1890 Romag Sto (3)6, 60 p
  (F21, 1255)
- 1904 Bib M (3)5, 415   (F35, 63)
  [spiral]
- 1905 Lom Gen (2)38, 358-372
  (F36, 60)  [spiral]
Jacoli F 1869 Boncomp 2, 299-312
  (F2, 10)
Koyré A 1953 Int Con HS 7, 405-410
Luria A 1935 Len IINT 5, 491-97
  [circle]
Morgan D 1958 MT 51, 473-74
Piola G 1844 *Elogio di Bonaventura
  Cavalieri. Con note e postille
  matematiche*, Milan
Procissi A 1960 Physis Fi 2, 321-324
  (Z106, 2)
Reeve W D 1932 MT 25, 93-94
Sittignani M G 1933 Pe M (4)13, 266-
  288  (F59, 842)  [indivisibles]


Cavalli, Ernesto

Ascione E 1913 Nap Pont (2)18 (2), 8
Resta O 1932 It UM 11, 126-127
  (F58, 49)


Caverni, Raffaello

Giovannozzi G 1919 Arc Sto 1, 267-271
Lungo C del 1919 Arc Sto 1, 272-282
Mieli A 1919 Arc Sto 1, 262-265
  (F47, 22)


Cayley, Arthur   1821-1895

Anon 1883 Nat 28, 481-485  [port]
- 1894 Nat 51, 323
- 1895? Crelle 115, 349
  [Schlaefli, Dienger]
- 1895 ZMN Unt 26, 394-395 (F26, 36)
Brioschi F 1895 BSM (2)19, 189-200
  (F26, 35)
- 1895 Linc Rn (5)4(1), 117 185
  (F26, 34)
Feldmann W Jr 1962 MT 55, 482-484
  [matrix]
Forsyth A R 1895 Lon RSP 58, 40 p
  (F26, 35)

Fuchs L 1895 Crelle 115, 349-350
  (F26, 36)
Graf H J 1905 Bern NG, 70-107
  (Also sep Bern, 42 p) [Schlaefli]
Halsted G B 1895 AMM 2, 96, 102
Hermite 1895 Par CR 120, 233-234
Lampe E 1897 Naturw R 12, 359-363
  (F28, 31)  [Sylvester]
MacFarlane A 1895 AMM 2, 99-102
- 1916 *Lectures on Ten British Mathe-
  maticians of the Nineteenth Century,*
  NY, 66-77
Mansion P 1895? Mathesis (2)5, 84
Matz F P 1895 AMM 2, 28-29
Miller G A 1930 Sci (2)72, 168-169
  (F56, 5)  [group]
Noether M 1895 M Ann 46, 462-480
  (F26, 34)
Novy L 1966 Act HRNT (2), 105-154
  [bib, group]
Prasad Ganesh 1934 *Some Great Mathe-
  maticians of the Nineteenth Century
  ...,* Benares 2, 1-33
Roberts S 1895 LMSP 26, 546-551
  (F26, 36)
Scott C A 1894 AMSB (2)1, 133-141
Vasiliev A 1895 Kazh FMO (9)5, 29
Watson E C 1939 Scr 6, 32-36  [port]


Caztelu, D. Luis de

Anon 1927 Rv M Hisp A (2)2, 118
  (F53, 34)


Cazzaniga, Camillo Tito   1872-1900

Viterbi A 1902 Loria 5, 87-90
  (F33, 35)


Cech, Eduard   1893-1960

Aleksandrov P S + 1961 UMN 16(1),
  119-126  (MR23A, 272; Z98, 9)
Anon 1953 Cas M 78, 195- 198
  (Z53, 340)
- 1960 Cas M 85, 217 (MR22, 770)
- 1960 Prag Pok 5, 341-342 (Z100, 247)
Katetov M + 1959/1960 Tor Sem 19, 57-
  88  (MR23A, 272; Z91, 6)

- 1960 Cz MJ 10, 614-630)
  (MR23A, 584; Z94, 5)
- 1961 Cas M 85, 477-491 (Z117:2, 251)
- 1968 in *Topological papers of Eduard
  Cech*, Prague, 7-19
Koutsky K 1965 Cas M 90, 104-118
  (MR34, 16) [topology]
Novák J 1953 Cas M 78, 185-194
  (Z53, 340)
- 1953 Cz MJ 3, 183-194 (MR15, 770;
  Z53, 247)
Villa M 1960 It UM (3)15, 342-343
  (Z91, 6)

## Celoria, Giovanni 1842-1920

Cerulli G 1920 Linc MR (5a)30(1),
  188-194
De Gaspapis 1880 Linc At (3)4, 105-
  106 (F12, 37)

## Cerruti, Valentino 1850-1909

Bortoletti E 1914 Mod Mm (3)11, 3-10
  (F45, 32)
Burgatti P 1910 Loria 12, 28-32
Garlando F 1909 Pe MI (3)7, 46-48
  (F40, 34)
Lauricella G 1912 Batt (3)3, 320-336
  (F43, 33)
Levi-Civita T 1909 Linc Rn 18, 565-
  575

## Cervenka, Ladislaw

Vetter Q 1934 Cas MF 63, D33-D41

## Cesaro, Ernesto 1859-1906

Alasia C 1907 Ens M 9, 5-23 (F38, 31)
- 1907 Rv FMSN 15, 23-46 (F38, 31)
Amodeo F 1906 Pe MI (3)4, 49-53
  (F37, 27)
Anon 1907 Ens M 8, 485 (F37, 27)
- 1907 Rv GSPA 18, 129-130
Bortolotti E 1909 Mod Mm (3)8, 77-82
Cerruti V 1906 Paler R 23, 221-226
- 1907 Linc Rn (5)16, 76-82 (F38, 31)

Del Pezzo P 1906 Nap FM Ri (3)12,
  358-375 (F37, 27)
Natucci A 1957 Batt (3)5, 126-127
  (Z77, 242)
Nunziato Cesàro C 1956 Archim 8,
  285-287 (MR18, 549; Z71, 3)
Pascal E 1906 Lom Gen (2)39, 916-920
  (F37, 27)
Perna A 1907 Batt 45, 299-332
  (F38, 30)
- 1956 It UM (3)11, 457-468
  (MR18, 549; Z71, 3)
Quint N 1907 Wisk Tijd 3, 152-153
  (F38, 31) [Planton]
Torelli G 1908 Nap Pont (2)13(2)

## Cesaro, Giuseppe 1849-1939

Buttgenbach H 1942 Bel Anr 108, 1-33

## Ceva, Giovanni 1647-1734

Fazzari G 1899 Pitagora 5(1), 7
  [Ceva thm]
Loria G 1915 Lom Gen, 3p (F45, 16)
Mase-Dari E 1935 *Un precursore dell'
  econometria. Il saggio di Giovanni
  Ceva "De re nummaria" edito in
  Mantova nel 1711*, Modena
Oettel H 1960/61 MN Unt 13, 257-260
  (MR22, 1129; Z109, 238)
Pascal A 1915 Annali (3)24, 287-310
  (F45, 11)
- 1915 Lom Gen (2)48, 173-181
  (F45, 80) [letter, Grandi]
Procissi A 1940 Pe M (4)20, 289-312
  (MR3, 97; Z23, 387)
  [letter, Magliabechi]
- 1942 It UM Con 2, 895-896
  (MR8, 499; Z26, 289)
Vivanti G 1915 Intermed 22, 175
  (Q267)
Vivanti G + 1899 Intermed 6, 177
  (Q267, Q585, Q1483)

## Ceva, Tomaso 1648-1736

Tenca L 1951 Lom MR (3)15, 519-537
  (MR14, 832; Z45, 147) [Grandi]

Chace, Arnold Buffum   1845-1932

Archibald R C 1933 AMM 40, 139-142
  (F59, 32)


Challis, James   1803-1882

Glaisher J W L 1883 Lon As Mo N 43,
  160-174


Chaplygin, Sergei Alekseevich
  1869-1942

Anon 1941 Pri M Me 5, 131-148
  (MR4, 65)
- 1942 SSSR Vest, 86-90   (MR4, 181)
Arzhanikov N S 1960 Vop IET (60),
  42-48  (Z117:1, 9)
Frodlin B N 1963 in Ocherki istorii
  matematiki i mekhaniki, Mos, 147-
Sretenskii L N 1953 SSSR Tek, 106-108
  (MR14, 609)
Topolyanskii D B 1966 Ist Met EN 5,
  75-79  (MR33, 5430)


Chappel, William   1718-1781

Candido G 1900 Pe MIS 3, 113   [Euler]


Charles, Jacques Alexandre Cesar
  1746-1823

Anon 1829 Par Mm 8, 73-


Charlier, Carl Ludwig 1861-1934

Wicksell S 1935 Lund F 5, 45-49
  (F62, 1041)


Charpit   XVIII

Brocard H 1902 Intermed 9, 123
Saltykow N 1930 BSM (2)54, 255-264
  (F56, 20)
- 1937 BSM (2)61, 55-64
  (F63, 16; Z16, 145)


Chasles, Michel   1793-1880

Anon 1917 Loria 19, 1-8  [Giogini]
Bertrand J 1873 Par C R 76, 909-911
- 1902 in his Eloges Académiques,
  nouvelle series, Paris 27-58
- 1904 Par Mm 47, xxxix-lxii
  (F35, 22)
Bertrand J + 1880 BSM (2)4, 433-436
  (F12, 15)
- 1880 Par C R 91, 1005-1015 and
  BSM (2)4, 433-436   (F12, 15)
Boncompagni B 1881 Boncomp 13, 815-828
  (F13, 23)
Brioschi F 1881 Annali (2)10, 158-160
  (F13, 23)
Brocard H 1906 Intermed 13, 117-118,
  202 (See also Intermed 12, 129,
  255)  [Vrain Lucas mss hoax]
Candido G 1935 Pe M (4)15, 58-62
  (Z10, 244)  [Fagnano, generalization,
  Pappos, Stewart]
Chernyaev M P 1957 Rostov Pe 4, 35-41
  (MR20, 6054)
Darboux G 1880 BSM (2)4, 436-442
  (F12, 16
- 1881 LMSP 12, 216-217 (F13, 23)
Kaiser F 1869 Arc MP 49, 81-
Loria G 1936 Osiris 1, 421-450
  (Z14, 243)  [conic, port]
Marianini 1808 Mod Mm (2)2   (F1, 2)
  [porism]
Pelseneer J 1936 Brx Con CR 2, 105-122
  (F62, 1536)
- 1955 Flambeau (3), 311-318
Rosenbaum R A 1959 MT 52, 365-6
Tannery P + 1897 Intermed 4, 156
  (Q961)


Chatelet, Madame de   1706-1749

  = Gabriele Emilie Le Tonn de
  Breteuil, Marquise du C.

Cajori F 1926 M Gaz 13, 252   (F52, 38)
  [calc]
Coolidge J L 1951 Scr 17, 20
  (MR13, 1)
Cox J F 1950 Ciel Ter 66, 1-11
Smith D E 1921 AMM 28, 368

Chaundy, Theodore William, 1889-

Ferrar, W L 1966 LMSJ 41, 755-756
  (MR33, 5432)


Chauveau, Jean Baptiste XVII

Tannery P 1895 BSM (2)19, 34-37
  (F26, 14)
- 1910 Intermed 17, 8


Chauvenet, William  1820-1870

Anon 1905 Wash UB 3, 123-128
Coffin J H C 1877 USNAS Bg 1, 227-244
Johnson W W 1917 USNI 43(10)
Littlehales G W 1905 USNI 31(3)
Matz F P 1895 AMM 2, 31-37
Roever W H 1925 Wash U St 12(7), 97-117
- 1926 Sci 64, 23-28  (Abr of 1925)
Van Vleck E B + 1925 AMM 32, 439
  [Chauvenet prize]


Chauvet, Jacques

Boyer J + 1896 Intermed 3, 146 (Q712)
  [Taillefer]
Brocard H + 1903 Intermed 10, 163
- 1903 Intermed 10, 313 (Q2239)


Chebotaryev, Nikolai Grigorevich
  1894-1947

Anon 1947 UMN 2(6), 68-71
Chebotarev N G 1948 UMN (NS)3(4),
  3-66 (MR10, 74)
- 1955 Ro Sov (3)9(3), 84-116
  (MR17, 1037)
Delone B 1948 SSSR M 12, 337-340
  (MR9, 174; Z30, 2)
Urazbaev B M 1952 Kazk Ak (7), 74-80


Chebuev, G. N.

Bolotoff E A 1901-1902 Mos MO Sb 22,
  vii-xv


Chebyshev, Pafnutii Lvovich  1821-1894

Anon 1895 Khar M So 4, 273-280
  (F26, 32)
- 1895 Nat 52, 345  (F26, 32)
- 1895 Pet B (5)2, 189-194  (F26, 32)
Artobolevskii I I + 1947 SSSR Mash 2,
  34-52  (MR12, 69)
Bernstein S N ed 1945 *Nauchnoe nas-
  ledie P. L. Chebysheva. Vypusk
  pervyi: Matematika,* M-L, 174 p
  (MR7, 355)
Brocard H 1902 Intermed 9, 158
  [transls]
- 1908 Intermed 15, 57  [transls]
Bytskov V P 1966 Ist Met EN (5), 86-
  92  (MR34, 2415)
Dakhiya S A 1953 Ist M Isl 6, 239-244
  (MR16, 660; Z52, 3)
Delaunay N 1899 ZM Ph 44, 101-171
- 1945 *P. L. Chebyshev and the Russian
  School of Mathematics,* M-L, 9 p
  (in Russ)  (MR16, 434)
Galchenov R I 1961 Ist M Isl 14, 355-
  392 [St. Pet Univ]
Gnedenko B V 1953 Ist M Isl 6, 215-
  222 (MR16, 660; Z53, 247)
Grave D 1895 Pet B (5)2, 131
Gurov S + 1961 *The Great Russian
  Scientist P. L. Chebyshev...,*
  Kaluga, 55 p (in Russ) (MR24A, 222)
Halsted G B 1895 AMM 2, 61-63
- 1898 AMM 5, 285-288
Kelbert S L 1956 Uzb FMN (3), 89-95
Kiselev A A + 1963 Ist M Isl 15, 291-
  318
Kryloff N M + 1921 Crim UMLP 3,
  xxii-liv  (F48, 15)
Liapunov A M 1895 Khar M So 4, 263-
  273  (F26, 32)
Maistrov L E 1961 Ist M Isl 14, 349-
  354  (Z118:1, 11)  [instrument]
Mansion P 1900 Mathesis 10, 67
Plakhowo N 1907 Intermed 14, 81
  [transls]
Prudnikov V E 1948 Ist M Isl 1, 184-
  214
- 1949 Mos IIET 3, 117-135
  (MR11, 572)
- 1949 UMN (NS)4(2), 173-175
  (MR11, 573)

- 1950 *P. L. Chebyshev as Student
  and Teacher,* Lehrer, Moscow, 144 p
  (Z41, 341)
- 1953 Ist M Isl 6, 223-237
- 1957 Ist M Isl 10, 639-648
  (Z119-1, 10) [astronomy]
Sochocki J L + 1899 Pet MO, 91-102
  (F30, 15)
Steklov V A 1946 UMN 1(2), 4-11
  (MR10, 174)
Stoermer C 1902 Intermed 9, 158
  [transls]
Urazbaev B M 1950 *Velikii russkii
  matematik P. L. Chebyshev,* Alma-
  Ata, 38 p  (in Kozak)
Vasiliev A V 1894 Kagan (203)
  (Repr from "Volzhskii Vestnik")
- 1898 Loria 1, 33-45, 81-92, 113-139
  (F29, 11)  (Also in German,
  Leipzig, 1900) (F29, 11)
Weyr E 1896 Cas MF 25, 38
Yuskevich A P 1965 Vop IET 18, 107-
  111


Chelini, Domenico 1802-1878

Beltrami E 1881 Chelini, I-XXXII
  (F13, 23)
Cremona L 1879 BSM (2)3, 228-238
  (F11, 27)
-.1879 Linc At (3)3, 54-58
  (F11, 27)
Cremona L + 1878 Batt 16, 345
  (F10, 20)
Cremona L + ed 1881 *In memoriam...
  Collectanea...,* Mediolani, 465 p
Foglini G 1879 Linc At 32, 152-165
  (F11, 27)


Cherbuliez, Emil 1837-1914

Rudio F 1914 Zur NGV 59, 566-571
  (F45, 37)


Chernae, Ladislaus  1740-1816

Jelitai J 1937 Deb Szeml (7/8)
  (Z17, 290)

Chernikov, Sergei Nikolaevich  1912-

Mal'cev A I + 1962 UMN 17(5), 177-181
  (MR26, 237; Z112, 3)


Chetaev, Nikolai Gurevich  1902-

Anon 1959 SSSR Mek (6), 3-6
  (Z88, 7)
- 1960 Pri MMe 24, 3-5, 171-200
  (MR22, 619)


Cheyne, Charles Hartwell Horne
  1838-1877

Anon 1877 Lon As Mo N 37, 147-148
  (F9, 16)


Chio, Felice 1813-1871

Boncompagni B 1871 Boncomp 4, 381-400
  (F3, 11)  [bib]
Genocchi A 1871 Boncomp 4, 363-380
  (F3, 11)


Chokuyen, Ajima Naonobu 1739-1798

Harzer P 1905 DMV 14(6)
Hayashi T 1916 Toh MJ 11, 17-37
  [Matsumaga]
Kikuchi D 1895 Tok M Ph 7, 114
  [length]
Mikami Y 1913 Arc MN 33, 8 p
- 1917 Tok M Ph (2)9, 186-193
  (F46, 29)


Christian of Prachatice  1368-1439

Sarton G 1948 Int His Sc 3(2), 1114
Studnika, F J 1893 Prag Si 24, 44
  (F24, 44)


Christoffel, Elvin  1829-1900

Anon 1901 Loria 4, 57-58  (F32, 19)

- 1901 M Ann 54, 344 [bib]
Geiser C F + 1901 M Ann 54, 329-341
   (F32, 19)
Winderband 1901 M Ann 54, 341


Chrystal, George  1851-1911

Black J S + 1913 Edi SP 32, 477-503
   (F43, 34; 44, 20)
McKinney T E 1909 AMM 16, 197  [alg]
Knott C G 1911 Nat 88, 47-49
   (F42, 34)


Chudakov, Nikolai Grigorevich  1905-

Anon 1965 UMN 20(2), 237-240 [bib]
- 1966 in *Studies in Number Theory*
   (1), 4-5  (MR34, 21)
Linnik J V + 1955 UMN 10(3), 213-215
   (Z64, 3)


Chunikhin, Sergei Antonovich

Maltsev A I + 1967 UMN 22(2), 189-
   197  (MR34, 5633)


Chuquet, Nicholas  1445-1500

Anon 1881 Boncomp 13, 543-659, 693-
   814  (F13, 8)
Gram J P 1882 Zeuthen (4)6, 126-138
Lambo C 1902 Rv Q Sc (3)2, 442-472
   (F33, 55)
*Marre A 1880 Boncomp 13, 555-659,
   693-814  (F13, 8)
Tannery P 1887 Bib M (2)1, 17-21
   [roots]


Chwalla, Ernest  1901-

Parkus H 1960 Ost Ing 14, 77-78
   (Z92, 3)


Cioranescu, Nicolae  1903-1957

Iacob D 1957 Gaz MF(A) 9, 214
   (MR21, 750)


Cipolla, Michele  1880-1947

Amato V 1948 Matiche 3, iii-xvi
   (MR10, 420; Z33, 4)
Mignosi G 1947 Annali (4)26, 217-220
   (Z30, 338)
- 1948 It UM (3)3, 94-95  (Z30, 338)
Sansone G 1956 Linc Rn (8)21, 507-523
   (MR20, 503; Z71, 3)


Cisotti, Umberto  1882-1946

Masotti A 1948 Milan Sem 18, i, 1-35
   (MR10, 667; Z32, 194)


Clairaut, Alexis Claude   1713-1765

Bigourdan G 1928 Par Sav CR, 26-40
   (F56, 812) [Euler, letters]
Boncompagni B 1892 Vat NLA 45, 157-291
   (F25, 21) [letter]
Boyer C B 1948 AMM 55, 556-557
   (MR10, 420) [anal geom, distance]
Brunet P 1951 Rv Hi Sc Ap 4, 13-40,
   109-153  (MR14, 832)
- 1952 *La vie et l'oeuvre de Clairaut*
   ..., Paris, 112 p  (Z68, 5)
- 1952 Rv Hi Sc Ap 5, 334-349
   (MR14, 832)
- 1953 Rv Hi Sc Ap 6, 1-17
   (MR14, 832)
Jelitai J 1937 MP Lap 44, 173-99
   [d'Alembert, La Condamine, Teleki]
- 1938 Magy MT 57, 501-508
   (F64, 916) [D. Bernoulli, Teleki]
Lefebvre B 1923 Rv Q Sc (4)3, 166-191,
   441-462 [Carnot]
Natucci A 1932 Pe M (4)12, 305-307
   (F58, 994) [pyramid]
Somigliana C 1941 Tor Sem 7, 19-24
   (MR3, 258) [Pizzetti, Stokes]
Speziali P 1955 Rv Hi Sc Ap 8, 193-
   237  (MR17, 446) [Cramer]

Clairaut le Cadet   1716-1732

Boyer C B 1964 Isis 55, 68-70
  (MR29, 219)   [Thabit ibn Qurra]


Clairin, Jean   1876-1914

Goursat E 1916 Fr SMB 44, 15-16
  (F46, 24)


Clasen, B. I.   1829-1902

Gloden A 1953 Rv Hi Sc Ap 6, 168-
  170   [lin eqs]


Clausius, Rudolf   1822-1888

Basso G 1888 Tor FMN 24, 3-4
Crew H 1940 Scr 8, 111-113
Fitzgerald G F 1888 Nat 38, 491
- 1890 Lon RSP 48, i-viii
Folie F 1890 Rv Q Sc 27, 419-487
Gibbs Josiah Willard 1906 The Scien-
  tific Papers..., NY 2, 261-267
Helmholtz H v 1889 Ber Ph 8, 1-6
Reinganum M 1910 ADB 55, 720-729
Riecke E 1888 Rudolf Clausius,
  Goettingen, 39 p [bib]
Tunzelmann G W de 1888 Nat 38, 438-
  439


Clavius, Christopher   1537-1612

  = Clau   = C. S. Clavio

Anon 1932 MT 23, 40
Phillips E 1939 Arc His SJ 8, 193-222
Reeve W D 1931 MT 24, 40


Clebsch, Alfred   1833-1872

Anon 1873 M Ann 8, 1-55 (F5, 29)
- 1874 Annali (2)6, 153-207
  (F6, 30)
Beltrami E 1872 Batt 10, 347-349
  (F4, 18)

Boernstein R 1872 Altpr Mo 9,
  653-656   (F4, 19)
Cremona L 1872 Lom Gen (3)5,
  1041-1042   (F4, 18)
Favaro O 1874 Batt 12, 28-74
  (F6, 30)
Kobell E 1873 Mun Si 2, 129
  (F5, 31)
Mansion P 1875 Boncomp 8, 121-184
  (F7, 10)   [bib]
Neumann C 1872 Gott N, 550-559
  (F4, 18)
- 1873 Batt 11, 44-48 (F5, 31)
- 1873 M Ann 6, 197-202   (F5, 29)


Clerke, Agnes Mary   1842-1907

Moreux  T 1907 Rv GSPA 18, 429-430


Clifford, William Kingdon   1845-1879

*Macfarlane Alexander 1916 Lectures on
  Ten British Mathematicians..., NY,
  78-91
*Newman J R 1946 Forward to  Common
  Sense of the Exact Sciences by
  W K Clifford, (Knopf).  [bib]
- 1953 Sci Am 188, 78-84 (MR15, 89)
Pollock F 1879 In W. K. Clifford,
  Lectures and Essays, London 1, 1-43
Smith D E 1922 AMM 29, 157
  [precocity]
Tait P G 1878 Nat 18, 89-91


Coates, W M

A D H L 1911 Edi SP 32, xl-xlviii


Coble, Arthur Byron   1878-

Archibald R C 1938 A Semicentennial
  History of the American Mathematica
  Society..., NY, 233-236
  [bib, port]

Cochin, C N

Brocard H + 1904 Intermed 11, 171-174
Lemaire G 1908 Intermed 15, 279
  (Q2690)

Cocker, Edward

Anon 1922 Lon Tim LS (Sept 7, Sept 14)
Sleight E R 1943 M Mag 17, 248-257

Cockle, James  1819-1895

Anon 1895 LMSP 26, 551-554
  (F26, 36)

Codazza, Giovanni  1816-1877

Cossa A 1878 Tor FMN 13, 25-33
  (F10, 20)

Cohn, Berthold  1870-1930

Archibald R C 1948 *Mathematical
  Table Makers*, NY, 16 (Scr Stu 3)
  [biobib]

Cohn-Vossen, Stefan  1902-1936

Aleksandrov A D 1947 UMN (NS)2(3),
  107-141  (MR9, 485)
Anon 1936 UMN 1, 5
- 1958 Biog Slov 1, 440

Coignet, Michel  1549?-1623

Bosmans H 1900 Brx SS 25(2), 91
  [sine]

Colbert, Jean Baptiste  1665-1746

Scheler L 1962 Rv Hi Sc Ap 15,
  351-365  (MR33, 1205)

Colburn, Warren

Richeson A W 1935 M Mag 10, 73-79
  (F61, 953)  [arith, USA]

Colburn, Zerah  1804-1840

Anon 1912 in *Encyclopedia of Vermont
  Biography*
Colburn Z 1833 *Memoir of Zerah Colburn
  Written by Himself*, Springfield
  Mass
Ellis S 1813 Nicholso 35: 9-  [methods]
Saint W 1813 Nicholso 34: 291-

Cole, Frank Nelson  1861-1926

*Archibald R C 1938 *A Semicentennial
  History of the American Mathematical Society...*, NY, 100-103
  [bib, port]
Fiske T S 1927 AMSB 33, 773-
  [biobib]

Colecchi, Attavio

Amodeo F 1917 Nap Pont (2)22(3) 17
Procissi A 1963 Physis Fi 5, 319-
  326  (Z117:2, 247)

Colin, le R. P. Elie  1852-1923

Lacroix A 1923 Par CR 177, 609-612

Colnet d'Huart, I. F. L. Alexandre de
  1821-

Grechen M 1904 Ven I At 44, 1-57

Combebiac, Gaston (Gustave) 1862-1912

Anon 1912 Ens M 14, 406-407 (F43, 38)
Macfarlane A 1913 Quatrns, 17-20
  (F44, 22)

Comessatti, Annibale Alessandro
1886-1945

Severi F 1947 Rv M Hisp A (4)7,
239-242

Commandino, Federigo 1509-1575

Baldi B 1714 Gi Let Ven 19
Drake S + ed 1969 *Mechanics in Six-
teenth-century Italy,* (Univ of
Wisc Press)

Como, G. A. de

Eneström G 1906 Bib M (3)7, 216
(F37, 3)

Comrie, Leslie John 1893-1950

Archibald R C 1948 *Mathematical Table
Makers,* NY, 16-18 (Scr Stu 3)
[bib, port]
Massey H S W 1952 Lon RS Ob 8, 97-107
(MR14, 524)
Porter J G 1951 Obs 71, 24-26

Comte, August 1798-1857

Bertrand J 1893 Nou An (3)12, 152-
163 (F25, 24)
Ducassé P 1935 Thales 1, 133-143
(F61, 953; Z13, 193)
- 1936 Thales 3, 39-43 [calc var]
Frances E K 1915 Pos Rv (Oct),
228-233
Laffite P 1894 Nou An (3)13, 65-80,
113-120, 405-428, 462-482
(F25, 24)
Ostwald W 1914 *Auguste Comte. Der
Mann und sein Werk,* Leipzig, 300 p
(F45, 19)
Vasiliev A 1900 Ens M 2, 157 [phil]
Zaharia N D 1935 Tim Rev M 14, 127-
129, 140-142 (F61, 958)
- 1936 Tim Rev M 15, 4-6, 16-18, 26-
28 (F62, 1043)

Comtino, Mordecai XV

Schub P 1932 Isis 17, 54-70 (F58, 30)
[XV Constantinople]

Concina, Umberto

Anon 1929 Pe M (4)9, 364 (F55, 623)

Condorcet, Marie Jean Antoine Nicolas
Caritat Marquis de 1743-1794

Arago D F J 1849 Par Mm 20, 1-
Burlingame Anne 1930 *Condorcet, the
Torchbearer of the French Revolution,*
Boston
Eneström G 1885 Bib M,191-192
[calc]
Granger Gilles-Gaston 1954 Rv Hi Sc
Ap 7, 197 [soc of sci]
- 1956 *La mathématique sociale du
Marquis de Condorcet,* Paris, 186 p
Henry C 1883 Boncomp 16, 271-324
(F16, 24) [bib]
Laboulle M J 1939 Rv HL Fr 46, 33-55
Pelseneer J 1950 Osiris 10, 322-327
[mss]
Sergescu P 1951 Rv Hi Sc Ap 4, 233-
237 (Repr 1952 in *L'Encyclopedie
et le progrès des sciences et des
techniques,* Paris, 30-34)
Taton R 1959 Rv Hi Sc Ap 12, 128-158,
243-262 [Lacroix]

Conforto, Fabio

Anon 1954 Archim 6, 107-122 (MR17, 2)
- 1954 Archim 6, 127-130 (MR15, 923)
- 1954 Rm U Rn (5) 13, 199-218
(MR15, 923)
Benedicty M + 1954 Archim 6, 95-96
(Z55, 4)
Segre B 1954 Archim 6, 91-94
(MR15, 923)
- 1954 Rm U Rn (5)14, 48-74
(MR16, 434; Z55, 4)

Congdon, Allen   1876

Reeves W D 1946 MT 39, 39


Conrad of Megenberg   1309?-1374

Sarton G 1947 Int His Sc 3(1), 817-821 [bih]
Noll-Husum H 1937 Isis 27, 324-325


Conta, Vasile   1846-1882

Sergescu P 1932 Gaz M (3), 3 p


Conti, Antonio   1677-1749

Brognoligo G 1893 Ven Atene (17)2, 162-179, 327-350 (F25, 21)
Popa I I 1938 Archeion 69-73
Tenca L 1954 Ven I At 112, 103-119 (MR16, 660; Z60, 7) [letter]


Conway, A. N.   1875-1950

Whittaker E T 1951 Lon RS Ob 7, 329-340 (MR13, 612)


Coolidge, Julian Lowell   1873-1954

Struik D J 1955 AMM 62, 669-682 (Z65, 244)


Copernicus, Nicolaus   1473-1543

= N. Mikolaj Kopernik
= Niklas Kippernigk

Africa T W 1961 Isis 52, 403-409 (Z108-2, 245) [Aristarchos, Pythagoras]
Anon 1873 Album Wydane staraniem Towarzystwa Przyjaciol Nauk W Poznian W czterechstn rocznie urodzin Nikolaja Kopernika, Poznan (F5, 13)
- 1873 Il quarto Centenario de Nicolo Copernico nell Universita di Padova, Padova (F5, 17)
- 1873 Die vierte saeculor feier des Geburt von Nicolaus Copernicus, Berlin (F6, 19)
Armitage Angus 1938 Copernicus, NY
Bender G 1920 Heimat und Voksthum des Familie Kippernigk, Breslau
Berti D 1876 Copernico e le vicende del sistema Copernicano in Italia nella seconda meta del Secolo XVI e nella prima del XVII con documenti inediti intorno a Giardano Bruno e Galileo Galilei, Rome (F8, 16)
Baranowski H 1958 Kopernikus-Bibliographie 1509-1955, Warsaw
Birkenmajer L 1897 Wiad M 1, 178
- 1902 Wiad M 6, 278-280
Birkenmajer L + 1909 Krak Bl (2), 20-36
Boncompagni B 1877 Vat NLA 30, 341-397 (F9, 3)
Braunmuehl A 1896 Biog Bl 2(4), 12 p (F27, 6)
Brohm R 1873 Nicolaus Copernicus. Skizze seines Lebens und Wirkens..., Thorn (F5, 16)
Cantor M 1876 Boncomp 9, 701-716 (F8, 14)
Chasles M 1872 Par CR 75, 893-894 (F4, 9)
Curtze M 1872 Altpr Mo 9, 187-189 (F4, 7)
- 1872 Altpr Mo 9, 1-2 (F6, 20)
- 1872 Arc MP Lit 54(216), 1-7 (F4, 8)
- 1873 Altpr Mo 10, 155-162 (F5, 20)
- 1874 Altpr Mo 11, 278-279 (F6, 23)
- 1874 Arc MP 56, 325-326 (F6, 20)
- 1874 ZM Ph 19, 76-82, 432-458 (F6, 21)
- 1875 Hist Abt 20, 60-62 (F7, 5)
- 1875 ZM Ph 20, 221-248 (F7, 5)
- 1878 Arc MP 62, 113-148, 337-374 (F10, 6)
- 1878 Boncomp 11, 167-171 (F10, 8)
- 1878 Boncomp 11, 172-176 (F10, 8)
Dick J 1953 Wiss An 2, 450-458 (MR15, 89)
Erhardt R + 1942 Isis 33, 578-602 [Archimedes, Aristarchos]

Favaro A 1877 Boncomp 10, 303-313 (F9, 3)
- 1878 Boncomp 11, 319-334 (F10, 9)
- 1879 Boncomp 12, 775-807 (F11, 12)
- 1890 Mex Alz Mm 3
Flammarion C 1872 *Vie de Copernic et histoire de la découverte du système du monde,* Paris (F5, 15)
- 1873 *Zycie Nikolja Kopernika przelozyl i przypiskam i dopelnil Filip Sulimierski,* Warsaw (F5, 15)
- 1891 *Copernic et la découverte du système du monde,* Paris (F23, 45)
Gabba L 1944 Lom Rn M (3)8, 321-327 (MR8, 306)
Gilbert P 1891 Rv Q Sc 29, 589-594 (F23, 45)
Guenther S 1879 *Malagola's und Curtze's neue Forschungen ueber Copernicus...,* Dresden (F11, 11)
- 1880 Copp Mit 5(2) (F12, 9)
Hipler F 1873 Altpr Mo 10, 193-218 (F5, 11)
- 1875 *Die Portaets des Nicolaus Kopernikus,* Leipzig (F7, 7)
- 1876 Boncomp 9, 320-325 (F8, 14) [Bologna]
Karlinski 1873 *Zywot Mikolaja Kopernika,* Krakow (F5, 16)
Karpinski L C 1943 Scr 9, 139-154 (MR5, 253)
- 1945 M Mag 19, 342-348 (MR7, 254)
Kience H 1943 Naturw 31, 1-12 (MR4, 181)
Knoetel A 1872 Rueb (2)11, 285-291, 334-339 (F4, 6)
Koyre A 1933 Rv Phi Fr E 116, 101-118
Lindemann 1892 Konig Ph, 4 p (F24, 47)
Lindhagen A 1882 Sv Han 6(12)
Losada y Puga de 1943 Peru U Rv 11, 149-178 (MR7, 355)
Lundmark K 1945 Lund F 14(3), 22-39 (MR7, 106)
Malagola C 1880 *Der Aufenthalt des Copernicus in Bologna,* Thorn, (transl by Curtze) (F12, 9)
Mansion P 1894 Brx SS 18(1), 12-15 (F29, 90)
Montanari A 1873 *Nicolo Copernico ed il Suo libro de Monetae Cudendae ratione Studio,* Padova (F5, 20)
Nesteruk F Ya 1953 SSSR Tek, 1341-1349 (MR15, 89)
Polkowski K H 1873 *Zywot Mikolaja Kopernika,* Gnesen (F5, 12)
Prowe Leopold 1870 Das Andenken des Copernicus bei der dankbaren Nachwelt, Thorn-Lambeck (F2, 8)
- 1873 *Festrede zur 4 ten Saecular feier des Geburtstages von Nicolaus Copernicus...,* Berlin (F5, 14)
- 1883 *Nicolaus Coppernicus,* Berlin
Przypkowski Tadeusz 1954 *Dzieje mysli Kopernikowskiej. Wydawnictwo Ministerstioa Obrony Nanodowej,* Warsaw, 115 p (MR16, 781)
Ramsauer R 1942 Fors Fort 18, 316-318 (Z27, 289)
Rhetikus G J 1943 *Des Georg Joachim Rhetikus erster Bericht ueber die 6 Buecher des Kopernikus* (tr by K. Zeller), Munich-Berlin, 208 p (Z28, 2)
Righini G 1932 *La Laurea di Copernico allo Studio di Ferrara,* Ferrara
Romer 1872 *Beitraege zur Beantwortung der Frage nach der Nationalitaet des Nicolaus Copernicus,* Breslau (F4, 6)
Rosen E 1962 Isis 53, 504-509 (Z118:2, 247)
Rosenblatt A 1943 Rv Cien 45, 409-442 (MR5, 253)
Schiaparelli G V 1873 *I precursori di Copernico nell'antichità...,* Milan-Naples (F5, 2)
- 1876 *Die Vorlaeufer des Copernicus im Alterthum...,* Leipzig (F8, 7)
Snell C 1873 *Nicolaus Copernicus. Rede,* Jena (F5, 14)
Sparagna A 1876 Boncomp 9, 315-319 (F8, 14)
Stamm E 1933 Int Con H 7(1933)2, 155-174 (F61, 949) [geom]
- 1934 Wiad M 37, 57-100 (Z9, 97) [geom]
Steinschneider M 1871 XM Ph 16, 252-253 (F3, 8)
Studnicka F J 1873 Cas MF 2, 1-56 (F5, 17)

- 1873 *Mikulas Kornik na oslavu 400
  leti pamatky jeho narozeni,* Prag
  (F5, 17)
Sturm R 1911 DMV 20, 152-167
Wolynski A 1873 *Cenni biografici di
  Niccolo Copernico,* Firenze
  (F5, 15)
- 1873 *Kopeanik w Italji czyli doku-
  menta italskie do Monografji
  Kopernika,* Posen  (F5, 15)
  (2nd ed 1874) (F6, 19)
Zeissberg H 1872 Altpr Mo 9, 377
  (F4, 7)
Zich O 1953 Cas M 78, 297-304
  (MR16, 434)
Zilsel E 1940 J H Ideas 1, 113-118
  (MR1, 129) [mech]
Zinner 1942 Fors Fort 18, 183
  (Z27, 3)


Corachan, Juan Batista  1661-1741

Bunge M 1943 Archeion 25, 289-290


Corbino, Orso Marlo  1876-1937

Amerio A 1937 Milan Sem 11, xvii-
  xviii  (F63, 816)


Coriolis, Gustave Gaspard  1792-1843

Dugas R 1941 Rv Sc 74, 267-270
Freiman L S 1961 Mos IIET 43, 478-489
  (Z117:1, 8)
Rees E L 1936 AMM 33, 510


Cornu, Alfred Marie  1841-1902

Cerruti V 1902  Linc Rn (5)11, 347
Laisant C A + 1902 Ens M 4, 212-215
  (F33, 40)
Poincaré H 1905 Par EP (2)10, 143-176
Raveau C 1903 Rv GSPA 14, 1023-1040

Cortes, Mariano Fernandez

Plans J M 1933 Rv M Hisp A (2)8, 266
  (F59, 855)


Cossali, P.

Bortolotti E 1930 Pe M 10, 85-91
Boncompagni Baldassare 1857 *Scritti
  inediti del P. D. Pietro Cossali..,*
  Rome, 417 p


Cotes, Roger  1682-1716

Braunmuhl A v 1905 Bib M (3)5, 355-365
  [intgl, Newton]
Edleston J 1850 *Letters Between Cotes
  and Newton,*  London
Huxley G 1963 Scr 26, 231-238
  (Z111, 4)
Terquem O 1850 Nou An 9, 195-


Cotterill, Thomas  1808-1881

Mayor J E B 1881 LMSP 12, 217-218
  (F13, 24)


Cotton, Emile Clement  1872-1950

Moret 1950 Gren IF 1, 1-4
  (MR11, 573; Z35, 148)


Cournot, Antoine Augustin  1801-1877

Anon 1905 Rv Met Mor 13 (May)
- 1939 *Cournot, nella economia e nella
  filosofia,* Padova, 245 p
Antonelli E 1935 Ecmet 3, 119-127
  [Walras]
Bompaire François 1931 *Du principe de
  liberté économique dans l'oeuvre de
  Cournot et dans celle de l'école de
  Lausanne,* Paris, 740 p
Bottinelli E P 1913 *A. Cournot, méta-
  physicien de la connaissance,*
  Paris

- 1913 *Souvenirs d'A. Cournot, 1760-77; precédés d'un introduction par E. P. Bottinelli,* Paris, 302 p [econ, phil, prob]
Edgeworth F Y 1925 in *Palgrave's Dictionary of Political Economy,* 1, 445-447
Evans G C 1929 AMSB 35, 269-
Floss S W 1941 *An Outline of the Philosophy of Antoine-Augustin Cournot,* Philadelphia, 168 p
Hecht Lilly 1930 *A.Cournot und L. Walras,* Heidelberg, 93 p
Lechalas G 1906 Rv Met Mor 14, 109 [prob]
Liefmann-Keil E 1938 Z Natok 9, 505-540 (F64, 915) [Leibniz]
Lurquin C 1939 Brx U Rv 44, 347-55
Mentré F 1905 Rv Met Mor 13, 485-508 (F36, 14)
  1908 *Cournot et la renaissance du probabilisme au XIXe siècle,* Paris, 658 p
- 1927 *Pour qu'on lise Cournot,* Paris
Milhaud Gaston 1906 *Etudes sur la pensée scientifique chez les Grecs et chez les modernes,* Paris, 137-176 [Aristotle, chance]
- 1911 Rv Mois 12, 70-72, 404-428
- 1911 Scientia 10 (20), 370-380 [pragmatism]
- 1927 *Etudes sur Cournot,* Paris
Moore H L 1905 Rv Met Mor 13, 521-543 (F36, 14)
Poincaré H 1905 Rv Met Mor 13, 293-306 (F36, 14) [calc]
Roy R 1933 Ecmet 1, 13-22

**Cousin,** Jacques Antoine-Joseph 1739-1800

Zero 1915 Intermed 22, 151 (Q3941)

**Couturat,** Louis 1868-1914

Anon 1915 Isis 3, 121 [biobib, logic]
Dassen C C 1934 Arg SC 118, 136-143 (F60, 840)

**Crabtree,** Harold -1915

Anon 1916 M Gaz 8, 280 (F46, 35)

**Craig,** Thomas 1855-1900

Matz F P 1901 AMM 8, 183-187

**Cramer,** Gabriel 1704-1752

Boyer C B 1966 Scr 27, 377-379 [Maclaurin]
Speziali P 1955 Rv Hi Sc Ap 8, 193-237 (MR17, 446) [Clairaut]
Weil F 1961 Rv Hi Sc Ap 14, 97-136 [Buffon]

**Cranz,** Carl Julius 1858-1945

Mises R v 1928 ZAM Me 8, 81 (F57, 43)
Schardin H 1932 Fors Fort 8, 32-36;9, 1-18, 31-32

**Crelle,** August Leopold 1780-1855

Biermann K-R 1959 Ber Mo 1, 67-72 (Z83, 245)
- 1960 Crelle 203, 216-220 (MR23A, 433) [journal, port]
Biermann K-R + 1958 Nor M Tid 6 [Abel, Otto Aubert ms]
Lorey W 1926 Crelle 157, 3-11 (F52, 26)
Mittag-Leffler G 1925 Crelle 157, 12-14
Schloemilch O 1889 ZMN Unt 20, 401-405 [Brocard]
Vagarie E 1890 JM El (3)4, 32-35

**Cremona,** Luigi 1830-1903

Anon 1903 Nat 68, 393-394
- 1903 Pe MI (3)1, 53-56 (F34, 30)
- 1905 Lon RSP 74, 277-279
- 1909 *Onorazeal Prof. Luigi Cremona,* Rome, 59 p (F40, 28)

- 1957 Archim 9, 137-139 (MR19, 1150)
*Bertini E 1904 Batt 42, 317-336
  (F35, 28)
- 1904 LMSP (2)1, 5-18  (F35, 28)
Berzolari, E L 1906 Lom Gen (2)39,
  95-155 (F37, 18)
Blaserna P 1901 Edi SP 24, 646-649
  [Chrystal]
Castelnuovo G 1930 Linc Rn (6)12,
  613-618  (F56, 24)
Celoria G 1903 Lom Gen (2)36, 753-754
  (F34, 30)
Enriques F 1903 Bln Rn (2)7, 37-51
  (F37, 19)
- 1915 Annali (3)24, 157-8
- 1917 Annali (3)26, 225-226
  (F46, 14)
Fergola E 1903 Nap FM Ri (3)9,
  174-175
Gabba A 1954 Lom Rn M 87, 290-294
  (MR16, 985; Z57, 243)
Jung G 1903 Annali (3)9, 91-92
  (F34, 30)
Lampe E 1903 Naturw R 18, 465-467
  (M34, 30)
Loria G 1904 Bib M (3)5, 125-195,
  311 (F35, 27) [synth geom]
- 1904 Geno ASL 15  (F35, 28)
Maggi G A 1931 Milan Sem 5, 59-62
  (F57, 1306)
Noether M 1904 M Ann 59, 1-19
  (F35, 26)
d'Ovidio E  1903 Tor FMN 38, 549-551,
  817-820
Prasad Ganesh 1934 *Some Great Mathe-
  maticians of the Nineteenth Century*,
  Benares, 2, 116-143
Rodriguez S 1930 Rv M Hisp A (2)5,
  233-238 (F56, 24)
Sturm R 1904 Arc MP (3)8, 11-29,
  195-213 (F35, 25)
Veronese G 1903 Linc Rn (5)12, 664-
  678 (F34, 29) [bib]
- 1904 Wiad M 8, 150-164  (F35, 28)
Voit C 1904 M Ann 59, 245-252
Voss A 1904 Mun Si 34, 249-252
  (F35, 28)
White H S 1917 AMSB 24, 238-243

Cremonese, Leonardo

Favaro A 1905 Bib M (3)5, 326-341
  (F35, 6)

Cristescu, Vasile

Ionescu I 1928 Gaz M 34, 441-449
  (F54, 42)
- 1929 Gaz M 35, 164-170
  (F55, 623)
- 1933 Gaz M 39, 241-8
- 1933 Gaz M 39, 289-294

Crocco, Gaetano Arturo  1877-

Anon 1937 Vat An 1, 246-257
  (F63, 20)

Crone, C

Juel C 1931 M Tid B, 1-2
  (F57, 1307)

Cuffeler, Abraham Jean  XVII

  = Cufaeler  = Kufaeler
  = Cufueler

Brocard H 1912 Intermed 19, 223
  (Q504)  (See also Intermed 2,
  93;5, 253)

Cullis, Cuthbert Edmund  1868-1954

Sen N R 1954 Clct MS 46, 139-140
  (MR17, 2)
Turnbull H W 1955 LMSJ 30, 252-255
  (MR16, 660)

Culmann, Karl  1821-1881

Favaro A 1882 Ven I At (5)8, 715-740
Tetmajer L 1882 *Ueber Culmann's blei-
  bende Leistungen*, Zurich

Cunningham, Allan Joseph Champneys
  1842-1928

Archibald R C 1948 *Mathematical Table
  Makers,* NY, 18-21 (Scr Stu 3)
  [biobib, port]
Western A E 1928 LMSJ 3, 317-318
  (F54, 37)


Cunradus.  See Dasypodius.


Curbastro.  See Ricci


Curtiss, David Raymond   1878-1953

Moulton E J 1953 AMM 60, 566-569


Curtze, E. L. W. Maximilian   1837-1903

Cantor M 1903 DMV 12, 357-368
  (F34, 31)
Eneström 1904 Bib M (3)5, 312
  (F35, 55)
Favaro A 1904 Ven I At 63, 377-395
  (F35, 62)
Gunther S 1903 Bib M (3)4, 65-76,
  76-81
Tannery P 1904 Bib M (3)5, 416
  (F35, 55)


Cusanus, Nicolas   1401-1464

  = Niklaus von Cues
  = Nicolaus de Cusa

Gensteenberghe E 1920 *Le Cardinal
  Nicolas de Cues 1401-1464,* Paris
Goldammer K 1965 Bei GTI 5, 25-41
Guenther S 1899 ZM Ph 44(S), 123-152
  (F30, 4)  [geography]
Hofmann J E 1941/42 Heid Si 1-36
  (Z28, 1)  [Lull]
Johnson R A 1920 AMM 27, 368-369
  [Ozanam]
Lampe E 1892 Mathesis (2)2, 230-231
  (F24, 46)

Loeb H 1907 *Die Bedeuting der Mathe-
  matik fuer die Erkenntnislehre des
  Nicolas Cusanos,* Berlin
Lorenz S 1926 *Das Unendliche be
  Nicolas Cusanus,* Fulda
Rotta P 1928 *Il cardinale Niccolò di
  Cusa,* Milan, 464 p  (Milan USC 12)
  (F52, 13)
Schanz 1872 *Der Cardinal Nicolaus von
  Cusa als Mathematiker,* Tuebingen
  (F4, 6; 5, 10)
- 1873 *Die astronomischen Anschauunge
  des Nicolaus von Cusa und seiner
  Zeit,* Rottweil  (F5, 10)
Simon M 1912 in *Festschrift Heinrich
  Weber,* Leipzig, 298-337  (F43, 8)
Wiedemann E 1909 Arc GNT 1, 157-158
  (F41, 65)  [pi]
Zimmermann R 1852 Wien Si PH 8, 306-
  328 [Leibniz]


Czapski, Siegfried   1861-1907

Rohr M v 1932 Naturw 20, 495-496


Czuber, Emanuel   1851-1920

Anon 1930 Bo Fir (2)8, XLVII
  (F56, 815)
Dolezal E 1927 ZAM Me 7, 85-87
  (F53, 30)
- 1928 DMV 37, 287-297  (F54, 38)
Guldberg A 1925 Sk Akt 8, 249
  (F51, 29)
Lietzmann W 1926 ZMN Unt 57, 233-234
  (F52, 30)
Lorey W 1926 Z Versich 26, 117-124
  (F52, 30)


Dagomari.  See Paolo dell'Abbaco.


Dainelli, Ugo   -1906

Frattini G 1907 Pe MI (3)4, 191-192
  (F38, 31)

Damascios  458?-533?

= Damaskios
= Damasius of Damascus

Sarton G 1927 Int His Sc 1, 421


Dandelin, Pierre  1794-1847

Boyer L E 1938 MT 31, 124-125
  [spheres]
Householder A S 1959 AMM 66,
  464-466  [Graeffe, Lobachevskii]


Daniell, Percy John  1889-1946

Stewart C A 1947 LMSJ 22, 75-80
  (Z29, 2)


Danti, Ignazio  1536-1586

= Egnazio Dante

Arrighi G 1963 Physis Fi 5, 464-473


Darboux, Gaston  1842-1917

Anon 1912 *Eloges académiques et dis-
  cours,* Paris  (Comité de Jubilé
  scientifique de M.Gaston Darboux)
  [port]
- 1912 Rv GSPA 23, 89
- 1917 Nou An (4)17, 96
- 1918 Lon RSP 94, xxi-xxiv
  (F46, 17)
- 1918 LMSP (2)16, 46-49  (F46, 16)
Arsonval M de 1917 Rv Sc, 187-188
Bouligand G 1950 Arc In HS 29, 103-
  113  (Z34, 147)  [geom]
Dehérain H 1917 J Sav 132-135
Eisenhart L P 1917 AMSB 24, 227-237
  (F46, 16)  [geom]
- 1919/20 Act M 42, 275-284
  (F47, 12)  [geom]
Fehr H 1917 Ens M 19, 87-88
  (F46, 16)
Galdeano Z G de 1916 Esp SM 6, 64-65
  (F46, 17)

Gilbert P 1880 BSM (2)4, 317-318
  (F12, 11)  [letter]
Guichard  C 1917 Rv GSPA 28, 198-200
Hilbert D 1917 Gott N, 71-75
  (F46, 16)
- 1919-1920 Act M 42, 269-273
  (F47, 12)
Labrador J F 1953 Gac M (1)5, 3-5
  (MR14, 1050)
Larmor J 1917 Nat 99, 28
Lebon Ernest 1910 Fr AAS, 10
  [Picard]
- 1910 *Gaston Darboux,* Paris, 80 p
  (F41, 36)  [bib, port, surfaces]
- 1913 *Gaston Darboux, biographie,
  bibliographie analytique des écrits*
  2nd ed, Paris, 97 p  (1st ed 1910)
Picard E 1915 Par Mm 55, 36 p
- 1917 An SENS (3)34, 81-93 (F46, 16)
- 1917 BSM (2)41, 97-107 (F46, 16)
- 1917 *Notice historique sur Gaston
  Darboux,* Paris, 36 p
- 1918 Fr Long, 20 p
- 1918 Rv Sc, 33-44
- 1922 in his *Discours et mélanges,*
  Paris
Prasad Ganesh 1934 *Some Great Mathematicians...,* Benares 2, 143-182
Sintsov D 1921 Khar Me M 16, 1-10
  (F48, 18)
Voss A 1919 DMV 27, 196-217 (F47, 13)


Darmois, Georges  1888-1960

Danjon A 1960 Par CR 250, 241-245
  (Z88, 7)
Dugué D 1961 An M St 32, 357-360
  (MR 23A, 272; Z94, 5)


Darwin, George Howard  1845-1912

A E H L 1913 LMSP (2)12, 55-56
Fehr H 1913 Ens M 15, 68 (F44, 40)
Darwin F + 1916 in *Scientific Papers
  of G. H. Darwin,* Cambridge, Vol 5
  [biog, port]
Jourdain P E B 1913 Open Ct 27, 193-
  201, 572-573  (F44, 22)

Dase, Johann Martin Zacharias
1824-1861

*Anon 1846 *Auszug aus dem in der Zeit
vom 5 Janvar 1845 bis zu 20 Juli
1846 gefuehrten Album von Zacharias
Dase,* Berlin, 32 p
Archibald R C 1948 *Mathematical Table
Makers,* NY, 21-22   (Scr Stu 3)
[biobib]
Barlow Fred 1951 *Mental prodigies...,*
London   (Repr 1952 NY)


Dasek, Vaclav

Servit R 1957 Prag M Ap 3, 71-74
(Z82, 12)


Dasypodius, Konrad   1532-1606

= Cunradus

Tannery P + 1902 Intermed 9, 112


Dauge, Felix 1829-1899

Mansion P + 1899 Mathesis (2)9, 177
(F30, 20)


Davidov, Avgust Yulevich   1823-1885

Gatlich A F 1912 M Obraz (1), 30-36
(F43, 40)
Zhukovskii N E + 1890 Mos MO Sb 15(1),
1-57
Simonov P A 1957 *Pedagogicheskoe
nasledie professora matematiki
Moskovskogo Universiteta A. Yu
Davidova,* Moscow
Simonov R A 1961 Mos UM Me (1)16(6),
67-73   (Z114, 243)


Davies, Charles   1798-1876

Kunkel R V 1934 MT 26, 471-476

Davis, Ellery Williams   1857-1918

Hedrick E R 1918 AMSB 25, 36-38
(F47, 20)


Davis, Harold Thayer   1892-

Archibald R C 1948 *Mathematical
Table Makers,* 22   (Scr Stu 3)
[bib, port]


Davis, Robert Frederick

Williams C E 1927 M Gaz 13, 405-406
(F53, 34)


de. For modern names see also the
following word, for medieval names
the preceding word.  For anglicized
and Dutch names see De.


De-Amicis, Enrico

Sforza G 1925 Bo Fir 4, 138-142
(F51, 30)


Dedekind, Richard   1831-1916

Anon 1901 *Festschrift zur Feier des
siebzigsten Geburtstag von Richard
Dedekind,* Braunschweig, 254 p
- 1902 Arc MP (3)3, 28, 34
- 1902 Ber MG 1, 28-29   (F33, 45)
- 1902 Ber Si, 329-331   (F33, 45)
Baumann J 1908 An Natphi 7, 444-449
[Bolzano, infinity]
Bonnesen T 1930 M Tid B, 41-62
(F56, 10)
Busulini B 1957 Ferra U An (NS)5,
79-83   (Z78, 242)  [reals]
Fehr H 1916 Ens M 18, 132-134
(F46, 15)
Grave D A 1925 Uk Zap FM 1, 6-7
[Arab, Euclid, false position,
irrationals]
Jourdain P E B 1916 Monist 26, 415-42

Landau E 1917 Gott N, 50-70
  (F46, 15)
Matthews G B 1916 Nat 96, 103-104
  (F46, 16)
Medvedev F A 1966 Ist Met En 5,
  192-199 (MR34, 5622) [set thy]
Weyl H 1919 DMV 28, 85-92
  [cut, foundations]


Dee, John  1527-1608

Heppel G 1894 M Gaz 40


Degen, Carl Ferdinand  1766-1825

Heegaard P 1935 Nor M Tid 17, 33-38
  (Z11, 385)
- 1937 Act M 68, 1-6  [Abel]


Dehn, Max  1878-1952

Magnus W + 1954 M Ann 127, 215-227
  (MR15, 591; Z56, 2)


Delambre, Jean Baptiste  1749-1822

Chartres R 1889 Nat 40, 644
Dupin 1822 Rv Enc (Dec)
Martin Jadraque V 1956 Gac M (1)8,
  191-193 (MR19, 108)
Smith D E 1921 AMM 28, 64
- 1921 AMM 28, 303  [Stanhope]
Vigarie E + 1900 Intermed 7, 311
  [chronometry]


Delaunay, Charles Eugene  1816-1872

Faye + 1872 An Mines (7)2, 193-205
  (F4, 23)
- 1872 BSM 3, 317-320  (F4, 23)
- 1872 Institut 11, 279-280
  (F4, 23)
- 1872 JMPA (2)17, 348-350
  (F4, 23)


Delgleize,  Augustin

Rozet O 1954 Leig Bul 23, 248-250
  (MR16, 207; Z56, 2)


Delisle, Joseph Nicolas  1688-1768

Doublet E 1933 Rv Sc 71, 714-720
  (F59, 845)


Della Faille, Jean Charles

Bosmans H 1927 Mathesis 41, 5-11


Delone, Boris Nikolaevich  1890-

Anon 1950 SSSRM 14, 297-302
  (Z36, 146)
- 1960 Kristal 5, 339-340  (Z89, 5)
Faddeev D K 1950 UMN (NS)5(6),
  159-163 (MR12, 577; Z38, 148)
Shafarevich I R 1961 Rus MS 16(3),
  151-156 (Z98, 9)
- 1961 UMN 16(3), 239-244


Delsaulx, Joseph  1828-1891

Mansion P 1891 Brx SS 15(A), 86-91
  (F23, 29)
- 1891 Rv Q Sc 29, 585-588  (F23, 29)


Demartes, Gustave  1848-1919

Anon 1919 Ens M 20, 450-452
  (F47, 19)


Demidovich, Boris Pavlovich  1906-1966

Levitan B M + 1966 UMN 21(6), 155-160
  (MR34, 18)


Democritos  -V

Enriques F 1917 Bln Rn 22, 106-110
  [mech]

Enriques Federigo + 1948 *Le Doctrine de Democritos d'Abdera, Testi e Commenti,* Bologna, 362 p (MR10, 419)

Loewenheim Louis 1914 *Die Wissenschaft Demokrits und irh Einfluss auf die moderne Naturwissenschaft,* Berlin, 255 p (F45, 83)

Luria S 1928 Doklady (A), 74-79 (F54, 45) [Protagoras]

Tannery P 1886 BSM (2)10, 295-302 [Archytas]


De Moivre, Abraham 1667-1754

Anon 1954 Nat 174, 950-951

Archibald R C + 1926 Nat 117, 551-552, 894 (F52, 19)

- 1930 M Gaz 14, 574-575 (See White 1930) [letter]

Aubry V 1896 JM Sp 20, 222

Bohren 1911 Intermed 18, 55 (Q1359) [ins]

Braunmuhl A v 1901 Bib M 3, 97

Deming W E 1933 Nat 132, 713 [normal dist]

Maty 1760 *Mémoire sur la vie et sur les écrits de M. Abraham de Moivre ...,* The Hague

Smith D E 1922 AMM 29, 340

Volterra V 1933 Nat 132, 848 (F59, 45)

Walker H M 1934 Scr 2, 316-333 (Z10, 98)

Walker H M + 1934 MT 26, 424-427

White F P 1930 M Gaz 15, 213-216 (F56, 20) [Halley]

Wollenschlaeger K 1933 Basl V 43, 151-317 (F59, 844) [letters, J I Bernoulli]


De Morgan, Augustus 1806-1871

Anon 1871 Nat (Mar 23)

Ball W W R 1915 M Gaz 8, 43-45 (F45, 60)

Cajori F 1920/21 AMSB 27, 77-81 (F47, 32) [series]

*Clifford W K 1873 Academy 4(78)

De Morgan Mary ed 1895 *Threescore Years and Ten...,* London [letters, mss]

De Morgan Sophia E F 1882 *Memoir of Augustus De Morgan,* London, 432 p [bib, port]

Graves R P 1882 *Life of Sir William Rowan Hamilton,* Dublin 3, 245-632 [letters]

Halsted G B 1884 J Sp Phil 18, 1-9 [logic]

- 1897 AMM 4, 1-5

- 1900 Monist 10, 188 [letters, Sylvester]

Hamilton W 1847 *Letter to Augustus De Morgan,* Edinburgh (also in appendices to De Morgan's *Formal Logic*)

*Jevons W S 1911 Enc Br [bib]

Ladd C 1879 AJM 2, 211

Lord R D 1958 Biomtka 45, 282

Macfarlane Alexander 1916 *Lectures on Ten British Mathematicians...,* NY, 19-33

Oppermann L 1879 Zeuthen (3)1, 94 (F3, 9)

Pade H 1906 Bord PV, 66-68 [calc]

Reynard A C 1872 Lon As Mo N 32(4), 112

Rosen E 1957 Am Bib SP 51, 111-118 [Maurolico]

Sanders S T 1934 M Mag 8, 93-95

Sister Mary Constantia 1963 MT 56, 31-37

Smith D E 1922 AMM 29, 115


Denza, Francesco 1834-1894

Anon 1894 Nat 51, 179 (F25, 47)

Giovannozzi G 1895 Vat NLA 48, 13-36 (F26, 31)


Deparcieux, Antoine 1703-1768

Smith D E 1921 AMM 28, 432

**Deruyts, François  1864-1902**

Godeaux L 1938 Bel Anr 114, 85-102
Le Piage C 1902 Bel Bul, 168, 210

**Deruyts, Jacques Joseph Gustave
1862-1945**

Godeaux L 1949 Bel Anr, 21-43

**Desargues, Gérard  1593-1661**

Amodeo F 1906 Nap FM Ri (3)12,
  232-262 [Chaveau, conic]
Brocard H 1913 *Analyse d'autographes
  et d'outres écrits de Gérard
  Desargues*, Bar-le-Duc, 30 p
Chasles M 1845 Par CR 20, 1550-
Chrzaszczewski S 1898 Arc MP (2)16,
  119 [proj geom]
Costabel P 1953 Int Con HS 7, 241-245
Court N 1954 Scr 20, 5-13
  (MR15, 923; 16, 434)
De Vries H 1927 N Tijd 14, 365-380
  (F53, 14)
- 1928 N Tijd 15, 1-14  (F54, 27)
Eneström G 1885 Bib M, 89-90
- 1895 Bib M (2)9, 64, 96
- 1914 Bib M 14, 253-258
Gerhardt C J 1892 Ber Si, 183-204
  (F24, 15) [conic, Pascal]
Ivins W M Jr 1941 Scr 9, 33-48
  (MR4, 181)
- 1942 NY Met MA 1, 33-35 (MR4, 181)
- 1947 Scr 13, 203-210  (MR9, 485)
Lenger F 1950 Brx Con CR 3, 27-30
  [involution]
Le Paige C 1888 Bib M (2)2, 10-12
Loria G 1895 Bib M (2)9, 51-53
  (F26, 62) [enumerative geom]
Sarton G 1950 Isis 41, 300
  [Japan]
Swinden B A 1950 M Gaz 34, 253-260
  (MR12, 382; Z39, 4) [geom]
Tannery P 1890 BSM (2)14, 248-250
Taton R 1949 Rv Hi Sc Ap 2, 197-224
  [Monge]
- 1950 Int Con HS 6, 151-161
- 1951 Arc In HS (NS)4, 620-630
  (MR13, 197)

- 1951 *L'oeuvre mathématique de G.
  Desargues*, Paris 232 p
- 1951 Rv Hi Sc Ap 4, 176-181
  [conic]
Valentin G 1912 Bib M 3, 23-8
  [Bosse]
Viola T 1946 Linc Rn (8)1, 570-575
  (MR8, 306; Z60, 11) [triangle]
Zacharias M 1920 ZMN Unt 51, 21-22
  (F47, 6) [quadrilateral]
- 1941 Deu M 5, 446-457 (MR2, 306)
  [proj geom]

**Descartes, René  1596-1650**

Adam C 1896 Rv Met Mor 4, 573-583
  (F27, 91) [mss]
* 1910 *Descartes. Sa vie et ses
  oeuvres*, Paris, 178 p
  (Repr 1937 Paris)
- 1933 Rv Phi Fr E 115, 373-401
  (F59, 24) [letters]
Adam H 1896 BSM (2)20, 221 [geom]
Allard Jean-Louis 1963 *Le mathématiques de Descartes*, Ottawa
Anon 1878/79 Harv Lib B, 246-250
  (F11, 42) [geom]
- 1896 Rv Met Mor (4) (spec 300
  anniv issue]
- 1933 MR 25, 173-175
- 1937 *Bibliothèque Nationale.
  Descartes. Exposition organisée
  pour le IIIe centenaire du "Discours
  de la methode"*, Paris, 190 p
- 1937 *Congrès Descartes (Paris, 1937)*
  Paris, 356 p (Act Sc Ind (530);
  (531); (532))
- 1950 Rv Hi Sc Ap 3, 179  [300 anniv]
Armitage A 1950 Lon RSNR 5, 1-19
  [Royal Society]
Auger L 1950 Thales 6, 59-67
  (MR13, 810) [Roberval]
Ayyangar A A K 1940 M Stu 8, 101-108
  (F66, 1187)
*Baillet Adrien 1691-1693 *La Vie de
  Monsieur Des-Cartes*, Paris, 2 vols
Beck L J 1952 *The Method of Descartes
  (A Study of the Regulae)* Oxford,
  326 p  (Z49, 4)
Belaval Yvon 1960 *Leibniz critique de
  Descartes* (Librarie Gallinard),

<u>Descartes,</u> René  (continued)

Paris, 559 p  (MR22, 2050)
Berthet J 1896 Rv Met Mor 4, 399-414
(F27, 9)
*Bessel-Hagen E 1939 Muns Semb 14,
39-70  (F65, 13)  [geom]
Beth E W 1937 Ned Ps 31, 41-49
(F63, 14)  [universal lang]
Bierens De Haan 1887 Hist Abt 32,
161-173  [Huygens, letters]
Bompiani E 1921 Pe M (4)1, 313-325
Bosmans H 1927 Mathesis 41(S), 1-29
- 1927 Rv Q Sc 4, 113-141 (F53, 14)
[Huygens]
Bosmans H + 1903 Intermed 10, 96
[letter]
Boutroux E 1894 Rv Met Mor 2, 247
- 1896 Rv Met Mor 4, 502 (F27, 9)
[phil]
- 1914 Rv Met Mor 22, 814-827
(F45, 76)  [geom]
Boyer C B 1944 MT 37, 99-105
[anal geom, Fermat]
- 1952 Isis 43, 95-98 [rainbow]
- 1953 Scr 18, 189-217 (Z50, 2)
[Fermat]
- 1959 AMM 66, 390-393 (Z92, 245)
[anal geom]
Brocard H 1909 Intermed 16, 12
(Q2778)  (See also Intermed 11,
115; 13, 106-107; 15, 59-60)
[Beekman, free fall]
- 1911 Intermed 18, 272  [works]
Brunschvigg L 1927 Rv Met Mor 34,
277-324  (F53, 14)  [phil]
- 1937 *Descartes,* Paris
Bunge E S 1937 Bo Fir 10, 153-165
(Z18, 50)  [anal geom]
Cajori F 1926 Ens M 25, 7-11
(F52, 21)  [Newton]
Carruccio E 1951 Parma Rv M 2, 133-
152  (MR13, 2; Z42, 241)
Cassirer Ernst 1938 Lychnos, 139-
179
- 1939 *Descartes. Lehre, Persoenlich-
keit, Wirkung,* Stockholm, 308 p
Christensen S A 1933 M Tid A, 64-71
(F59, 843)
Cochin D 1913 *Descartes,* Paris
De Giuli G 1931 Scientia 49, 207-220
(Z1, 322)  [Galileo]

Demissoff F 1952 Am C Phil 26,
179-184
De Vries H 1916 N Tijd 4, 145-167
[Fermat]
De Waard C 1905 Nieu Arch (2)7(1),
64 [refraction]
- 1905 Nieu Arch (2)7, 69-87
(F36, 7)  [letters]
- 1925 Rv Met Mor 32, 53-89
[Petit]
Dijksterhuis E J 1932 Euc Gron 9,
57-76  (F58, 34)
- 1950 Euc Gron 25, 265-270
(Z37, 1)  [Pascal]
Dimier L 1926 *La vie raisonnable de
Descartes,* Paris, 284 p
Dingle H 1950 Nat 165, 213-214
(Z34, 147)
Dubouis E 1911 Intermed 18, 55, 224
(Q1346)  (Also 5, 197; 17, 6)
[four square decomposition]
Dugas R 1953 Par CR 237, 1477-1478
(Z51, 3)  [Huygens, Newton]
Eneström G 1903 Bib M (3)4, 211
[Editions of *Geométrie*]
- 1911 Bib M (3)11, 241-243
Erim K 1952 Pak J Sc 4, 57-60
(Z48, 242)
Estiv E 1942 Universd 13, 31-53
[mechanics, physics]
Faggi A 1923 Tor FMN 58, 323-337
(F49, 6)  [Newton]
Finkel B F 1898 AMM 5, 191-195
Fleckenstein J O 1950 Gesnerus 7,
120-139 [physics]
- 1956 Fors Fort 30, 116-121
(Z70, 5)
Foucher de Cariel A L 1879 *Descartes,
la princesse Elisabeth et la reine
Christine...,* Paris-Amsterdam,
140 p
Garaudy R 1946 Let Fr Par 6(102)
Gibson B 1896 Rv Met Mor 4, 386-398
[geom]
Gilson E 1921 Rv Met Mor 28, 545-556
[Holland]
- 1923/24 Brx URv 29, 105-39
[scholastic phil]
Greenwood T 1948 Can Rv Tri 34,
166-179  [anal geom]
Gueroult M 1938 Rv Met Mor 45, 105-
126

Descartes, René (continued)

Haldane Elizabeth S 1905 *Descartes,
  His Life and Times,* London
Hall A R 1961 Arc IIES 1, 172-178
  (MR24A, 2)
Hannequin A 1896 Rv Met Mor 4, 433-
  458 (F27, 9) [Leibniz]
- 1906 Rv Met Mor 14(a), 755-774
Hervey H 1952 Osiris 10, 67-90
  (Z46, 1) [Hobbes]
Hoffmann A 1923 *René Decartes,* 2nd
  ed, Stuttgart, 198 p (F49, 23)
Itard Jean 1956 *La géométrie de
  Decartes,* Paris, 14 p
Jacobi C G J 1846 JMPA 12, 97-
Jonquieres E de 1890 Bib M (2)4,
  [mss]
- 1890 Par CR 110, 261-266
  [ms, polyhedra]
- 1890 Par CR 110, 315-317
  [ms, polyhedra]
- 1890 Par CR 110, 677-680
Jungmann K 1908 *René Descartes.
  Eine Einfuehrung in seine Werke,*
  Leipzig, 243 p (F39, 14)
Karpinski L C 1939 Sci (2)89, 150-
  152 (F65, 1084)
Keeling S V 1934 *Descartes,* London,
  293 p (F60, 840)
Korteweg D J 1896 Rv Met Mor 4, 489
  [Snellius]
Koyré Alexandre 1937 *Trois leçons
  sur Descartes,* Cairo, 57 p
- 1944 *Entretiens sur Descartes,*
  NY, 113 p
Krishnaswami A A A 1940 M Stu 8, 101-
  108 (MR2, 306)
Kuehn F R 1923 *Descartes' Verhaeltnis
  zu Mathematik und Physik,* Munich,
  173 p (F49, 23)
Kuznetsov B G + 1967 *Frantsuzkaya
  nauka i sovremennaya fizika,* Moscow
Langer R E 1937 AMM 44, 495-512
  (F63, 13; Z17, 290)
Lanson G 1896 Rv Met Mor 4, 517-550
  (F27, 9)
Launay L de 1923 *Les grands hommes
  de France: Descartes,* Paris,
  128 p (F49, 23)
Lefebre Henri 1947 *Descartes,* Paris,
  311 p

Leisegang G 1954 *Descartes Dioptrik,*
  Meisenheim am Glan 168 p
  (Z55, 3)
Le Lionnais F 1952 Rv Hi Sc Ap 5,
  139-54 [Einstein]
Lera Sor 1956 Gac M 8, 47-56
  (Z74, 5)
Lorey W 1938 Euc Gron 14, 285-291
  (F64, 15) [calc, Hudde, tan]
Loria G 1923 Bo Fir (2)2, i-ix
  [numb thy]
- 1937 Rv Met Mor 44, 199-220
  (F63, 13; Z16, 145)
Milhaud G 1906 Rv GSPA 17, 73
  [anal geom]
- 1908 Rv GSPA 18, 223-228 [sine law]
- 1910 *Descartes savant,* Paris,
  251 p (also 1921)
- 1916 Rv GSPA 27, 502-510
- 1916 Rv GSPA 28, 332-337 (F46, 51)
  [Fermat, tan]
- 1916 Rv GSPA 28, 464-469
  (F46, 51) [calc]
- 1916 Rv Met Mor 25, 163-175
  (F46, 51)
- 1917 Rv GSPA 28, 464-469 [calc]
- 1917 Scientia 21, 185-198
  [Bacon]
- 1918 Scientia 23, 1-8, 77-90
  (F46, 55) [1619-1620]
- 1937 Act Sc Ind 531, 21-26
  (F63, 818) [geom]
Millet J 1867-1870 *Histoire de
  Descartes,* Paris, 2 vols
Milne J J 1928 M Gaz 14, 413-414
  [anal geom]
Mitrovich R 1933 *La théorie des
  sciences chez Descartes d'après sa
  géométrie,* Paris, 64 p (F59, 857)
Monchamps 1886 Bel Mm Cou 39, 1-643
Moorman R H 1943 M Mag 17, 296-307
  (MR4, 18; Z61, 4) [phil]
Mouy P 1930 Scientia 48, 227-36
  [physics]
Mueller C 1956 Sudhof Ar 40, 240-258
  (MR18, 453) [anal]
Natorp P 1896 Rv Met Mor 4, 416-432
  (F27, 9)
Nunziante-Cesaro 1956 Pe M (4)34,
  169-170 (MR18, 368) [quartic eq]
Omont H 1923 BSM (2)47, 194-195
  [letter, Mersenne]

Descartes, René (continued)

Pelseneer J 1938 Isis 29, 24-28
(F64, 913) [Poincaré]
Picard E 1934 BSM (2)58, 13-21
(F60, 827)
Pierce J M 1875 Harv Lib B 1, 157-
158, 246-250, 289-290
Pogorzelski H A 1959/60 M Mag 33,
184 (MR 23A 4) [symbolism]
Read C B 1961 MT 54, 567-69
[anal geom]
Reeve W D 1932 MT 25, 173-175
Rey A 1938 Par Radio 9, 118-123
Roth Léon 1924 Par Mor Tr, 411-424
[letters, Huygens]
- 1926 Correspondence of Descartes
and Constantyn Huygens, 1635-1647,
Oxford, 426 p
- 1937 Descartes' Discourse on
Method, Oxford, 150 p
Saltykow N 1938 BSM (2)62, 83-96,
110-123 (F64, 913; Z18, 340)
[300 anniv]
- 1939 Beo M Ph (5), 71-74
(MR1, 290; F65:1, 13; Z22, 3)
[Pappus]
Schlesinger L 1895 Die Geometrie von
René Descartes, Berlin
Schneider H 1904 Die Stellung
Gassenis zu Descartes, Leipzig,
67 p (F36, 7)
Scholz Heinrich + 1951 Descartes
Drei Vortraege, Muenster 80 p
(Z42, 3)
Schrecker P 1937 Rv Phi Fr E 62, 336-
- 1936 Thales 3, 145-154 [bib]
Schwarz H 1896 Rv Met Mor 4, 459-477
(F27, 9)
Scott J F 1952 The Scientific Work
of René Descartes, London, 221 p
(Z48, 242)
Scriba C J 1961 Arc HES 1, 406-419
(MR25, 221)
Sebba Gregor 1964 Bibliographia
cartesiana. A critical guide to the
Descartes literature 1800-1960
The Hague, 525 p
Senn G 1945 Gesnerus 2, 16-22
[Theophrast of Eresos]
Sergescu P 1950 Rv Hi Sc Ap 3, 262-
265

Sirven J 1928 Les années d'appren-
tissage de Descartes (1596-1628)
Paris, 470 p
Slebodzinski W 1950 Kwar Fil 19, 67-
70 (MR12, 1)
Smith D E 1921 AMM 28, 166
[Huygens]
Smith Bunge E 1937 B Ai Bol 10, 153-
165 (F63, 817) [anal geom]
Snow A J 1923 Monist 33, 611-617
Stojanovich C + 1897 Intermed 4, 258
[Marinus Ghetaldud]
Studnicka F J 1879 Cas MF 26, 73-94
(F28, 7)
Takekuma R 1955 J HS (33), 35-36
Tannery P 1891 BSM (2)15, 69-75, 111
120, 202-212, 228-236, 260-274,
281-296, 301-308; (2)16, 32-40
(F23, 17) [mss]
- 1893 La correspondance de Descarte
dans les inedits du fonds Libri,
Paris
- 1896 Rv Met Mor 4, 478-488
(F27, 9) [physics]
- 1899 SM Ph 44(S), 501-513 (F30, 7)
- 1911 Intermed 18, 55 (Q1346)
[four squares]
- 1933 Archeion 15, 177-180
(F59, 24) [Huygens, letter]
Taton R 1951 Arc In HS 30, 620-630
(Z43, 4)
Titeica G 1937 Gaz M 43, 170-173
(F63, 817)
Tocco F 1896 Rv Met Mor 4, 568-572
[Vico]
Torrière E 1914 Isis 2, 106-124
[Leibniz, transcendance]
Tropfke J 1931 Archeion 13, 300-319
[variable]
Vuillemin Jules 1960 Mathématiques
et Métaphysiques chez Descartes,
Paris, 192 p (MR26, 930)
Wieleitner H 1913 Bay Bl G 2, 299-
313 [alg]
Zeuthen H G 1900 Nyt Tid (B) 11, 49
[tangents]

Deschales, Claude François Milliet
1621-1678

= Dechales = De Chales
= Des Chales

Rose J 1909 Intermed 16, 263 (Q3563)


Despian   XVIII-XIX

Cajori F 1924 AMSB 30, 387  (F50, 19)
[recreations]


Deweck, Maurice 1904-1953

Deaux R 1953 Mathesis 62, 81-84
(MR14, 1050)


De Witt, John  1625-1672

Barnwell Robert G 1856 Life and Times
of John De Witt, NY
Biermann K-R 1959 Fors Fort 33, 168-
173 [insurance, Leibniz]
Brocard H + 1911 Intermed 18, 272
[ins]
Easton J B 1963 MT 56, 632-635
[conics]
- 1956 AMM 72, 53-56 [ellipse,
Nasir al Din, Schooten]
Eneström G 1896 Sv Ofv 53, 41
[XVII]
- 1896 Sv Ofv 53, 157 [ins]
- 1897 Amst Verz 3, 62-68
(F28, 40) [ins]
Kist J 1908 Amst Verz 10, 337-344
(F39, 60)
Van Dorsten R H 1908 Amst Verz 10,
85-111  (F39, 59)
- 1909 Amst Verz 11, 29-36 (F40, 56)
- 1910 Amst Verz 11, 319-325
(F41, 60)
Van Geer P 1914 Nieu Arch (2)11,
98-126  (F45, 12)
Van Rooijen J P 1937 Verzek 18,
(41)-(85)  (Z17, 50)

Diadochos.  See Proclos


Diaz, Juan

Smith D E 1921 The Sumario compendioso of Brother Juan Diaz.  The
earliest mathematical work of the
New World, Boston, London


Dickson, James Douglas Hamilton
1849-1931

M M C F 1931 Edi SP 51, 205-206
(F57, 39)


Dickson, Leonard Eugene  1874-1954

Albert A A 1955 AMSB 61, 331-345
(MR17, 2; Z65, 244)
Archibald R C 1938 A Semicentennial
History of the American Mathemati-
cal Society 1888-1938,  NY, 183-
194  [bib, port]
Montel P 1954 Par CR 239, 1741-1742
(MR16, 434)


Dickstein, Samuel  1851-1939

Knaster B 1955 Prace M 1, 5-8, 9-12
(Z66, 7)
Mostowski A 1949 Prace MF 47, VII-XII
(MR11, 573; Z38, 148)


Diderot, Denis  1713-1784

Brown  B H 1942 AMM 49, 302-303
(MR3, 258)  [Euler, religion]
Gillings R J 1954 AMM 61, 77-80
(MR15, 591)  [Euler]
- 1955 Au MT 11, 2-4
Krakeur L G + 1941 Isis 33, 219-232
(MR3, 98; Z60, 11)
Lecat M 1919 Intermed 26, 105-106
(F47, 33)
Rabut C + 1907 Intermed 14, 9
Struik D J 1940 Isis 31, 431-432
[Euler]

**Didymos of Alexandria** -I

Heiberg J L 1927 Dan H 13(3), 107 p
Sarton G 1927 Int His Sc 1, 233-234
Ver Eecke P 1936 Brx SS (A)56, 7-17
  (F62, 14)
Ver Eecke P ed 1940 *Les opuscules
  mathématiques de Didyme, Diophane
  et Anthemius suivis du fragment
  mathématique di Bubolo,* Paris-
  Bruges, 95 p  (F66, 1186)

**Diekmann,** Franz Joseph Konrad
1848-1905

Thaer A 1905 ZMN Unt 36, 316-318
  (F36, 34)

**Dienes,** Paul 1882-1952

Cooke R G 1960 LMSJ 35, 251-256
  (MR22, 107; Z87, 7)

**Dienger,** Josef  1818-1894

Anon 1894 Arc MP (2), 13-    (F25, 45)
Fuchs L 1895 Crelle 115, 350
  (F26, 31)

**Digges,** Thomas 1543?-1595

Patterson L D 1951 Isis 42, 120-121
  [father]
- 1952 Isis 43, 124

**Di Legge,** Alfonso  1847-

Armellini G 1939 Linc Rn (6)29,
  344-352  (F65, 17)

**Dini,** Ulisse  1845-1918

Anon 1957 Archim 9, 263-266
  (MR19, 1150; Z78, 3)
Bianchi L 1918 Annali (3)28, 1
  (F46, 18)

- 1921 Pisa U 7(41), 155-169
Ford W 1919 AMSB (2)26, 173-177
- 1922 Rv M (2)1, xcvii-ci  (F48, 24)
Pascal E 1918 Batt 56, 220-222
  (F46, 19)
- 1918 Nap FM Ri (3a)24, 109-112
Sansone G 1939 It UM (2)1, 373-383
  (F65, 17; Z21, 196)

**Dinnik,** Aleksandr Nikolaevich
1876-1950

Anon 1951 Pri M Me 15, 124-136
  (Z42, 4)
Penkov O M 1955 Pri Me 1(4), 371-377
Savin G N 1951 Ukr IM 3, 123-127
  (MR14, 832; Z45, 148)

**Dino,** Nicola Salvatore

Torelli G 1919 Nap FM Ri (3a)25,
  21-23  (F47, 15)

**Diocles**  -II

Wellmann M 1912 Hermes 47, 160

**Dionysodoros of Amisos**  -II

Schmidt W 1963 Bib M (3)4, 321-325
  (F34, 5)

**Diophantos**  III

Agostini A 1929 Archeion 11, 41-54
  (F55, 11)
Auerbach M 1930 Mathes Po 5, 16-23
  (F57, 1291)
Bakmakova 1966 Rv Hi Sc Ap 19(4),
  289-306  [Fermat, num thy]
Bennett A A 1925 AMM 32, 78-85
Birkenmajer A 1935 Math Cluj 9, 310-
  320  (F61, 945)  [Euclid]
Bortolotti E 1927 Pe M 7, 42-50
Bosmans H 1926 Mathesis 40(S), 1-14
- 1926 Rv Q Sc (4)9, 443-456
  (F52, 9)

Cavazzoni L 1931 Pe M (4)11, 84-109
(Z1, 113)
- 1931 Rm Sem(2)7, 77 (F57, 16)
- 1932 Rm Sem (2)7, 109-173
(F58, 988; Z5, 2)
Curzon H 1924 AMM 31, 290-292
Eneström G 1884 Bib M, 47-48
Fermat Pierre de 1932 *Bemerkungen zu
Diophant*, Leipzig, 49 p (Ostw Kl
234) (F58, 25; Z5, 3)
Gollob E 1899 ZM Ph 44, 137-140
(F30, 3)
Gram J P 1909 in *Anon, Festskrift
til H. G. Zeuthen*, Copenhagen,
48-68 (F40, 55) [Fermat]
Haentzschel E 1916 DMV 24, 467-471
Heath T L 1910 *Diophantus of
Alexandria*, 2nd ed, Cambridge
Heiberg J L 1927 Dan H 13(3)
Jacobi C G J 1899 Ber Si, 265-
Loria G 1932 Int Con 6, 415-420
[Bombelli]
Lucas Edouard 1961 *Recherches sur
l'analyse indéterminée et
l'Arithmétique de Diophante*,
Paris, 97 p (MR24A, 4)
Milhaud G 1911 Rv GSPA 22, 749-752
(F42, 51)
Nesselmann G H F 1892 ZM Ph 37,
121-146, 161-193 (F24, 6)
Read C B 1964 SSM 64, 606-607
Rodero J 1950 Gac M (1)2, 69-72
(Z36, 145)
Stamatis E S 1961 Platon 13, 125-133
(Z106, 1)
Steinschneider M 1865 ZM Ph 10, 499
Swift J D 1956 AMM 63, 163-170
(Z70, 4)
Tannery P 1879 BSM (2)2, 261-269
(F11, 5)
- 1884 BSM (2)8, 192-206
- 1887 Bib M(2)1, 37-43, 81-88, 103-
108
- 1888 Bib M (2)2, 3-6
- 1892 ZM Ph, 41-46 (F24, 6)
Turrière E 1926 Ens M 25, 196-217
Ver Eeeke P 1926 *Diophante
d'Alexandrie*, Bruges, 340 p
- 1940 *Les opuscules mathématiques
de Didyme, Diophane, et Anthémius,
...*, Paris, 102 p

Wertheim G 1890 *Die Arithmetik und
die Schrift ueber Polygonalzahlen
des Diophantus von Alexandria*
(Teubner), 355 p
- 1897 ZM Ph 42, 121 [polyg numbs]
Zeuthen H G 1884 Zeuthen (5)1,
145-156

## Dirichlet, Peter Gustav Lejeune
## 1805-1859

Ahrens W 1905 MN Bl 2, 36-39, 51-55
(F36, 13)
Biermann K-R 1959 Ber Ab (2), 68 p
(Z86, 4) [mss]
- 1959 Ber Mo 1, 320-323 (Z84, 3)
- 1960 *Vorschlaege zur Wahl von
Mathematikeru in die Berliner
Akademie*, Berlin, 6 p
- 1960 Ber Mo 2, 386-389 (Z89, 5)
Bjerknes C A 1873 Gott N, 439-447
(F5, 52) [sphere, ellipsoid]
Brendel M + 1928 Crelle 159, 1-2
(F54, 34) [Gauss]
Cantor M 1877 ADB 5, 251-252
Davenport H 1959 M Gaz 43, 268-269
Dunnington G W 1938 M Mag 12, 171-
182 (F64, 24)
Kronecker L 1888 Ber Si, 439-442
Kummer E E 1861 Ber Ab, 1-36
[Attraction of ellipsoids,
Fourier series, intgl, numb thy]
Lampe E 1906 Naturw R 21, 482-485
(F37, 14) [teaching]
Minkowski H 1905 DMV 14, 149-163
(F36, 13)
Petrova S S 1965 Ist M Isl 16, 295-310
[Riemann]
- 1966 Ist Met EN 5, 200-218 (MR
(MR34, 1146) [D. principle]
Schering E 1885 Gott N, 361-382
[Kronecker]
Schroeder K + 1963 in *Bericht von
der Dirichlet-Tagung*, Berlin
(Akademie Verlag) (MR30, 871;
Z114, 243)
Tannery J 1908 BSM 32, 47-62, 88-95
[Liouville]
- 1909 BSM 33, 47-64 [Liouville]
Taton R 1954 Rv Hi Sc Ap 7, 172-174
[letter]

Titchmarsh E C 1929 Sci Prog 22,
  565-573   [divisor problem]


Disteli, Martin  1862-1923

Schur F 1927 DMV 36, 170-173
  (F53, 31)


Dixon, Alfred Cardew  1865-1936

Anon 1936 Lon RS Ob 2, 165-174
Whittaker E T 1937 LMSJ 43, 145-155
  (F63, 19)


Dixon, Arthur Lee 1867-1955

Ferrar W L 1956 LMSJ 31, 126-128
  (MR17, 446; Z70, 5)


Dobriner, Hermann 1857-1902

Mueller C H 1903 ZMN Unt 34, 179-180
  (F34, 27)


Doeblin, V.  1915-1940

Levy P W 1955 Rv Hi Sc Ap 8, 107-115
  (MR17, 337)


Doehlemann, Karl   1864-1926

Anon 1926 ZAM Me 6, 179 (F52, 32)
Faber G 1928 DMV 37, 209-212


Doerge, Karl  1899-

Anon 1938 Deu M 3, 326-335  (F64, 23)


Doergens, Richard  1839-1901

Lampe E 1902 DMV 11, 57-68 (F33, 36)


Dolbnia, Ivan P  1853-1912

Anon 1912 Pet Gor 4, xxix-xl  [bib]
Bernatzkij W A 1912 Ped Sb (3), 397-
  401 (F43, 37)
Krylov N M 1912 *I. N. Doblnya,* Pet
Mordukhai-Boltovskoi D A 1912 Mos
  MO Sb 28, 473-491  (F43, 39)


Dolezal, Eduard  1862-1955

Kracke 1932  Z Vermess 61, 146-147
Winter F 1932 Bildmess 7, 2-11
  (F58, 993)


Dolomieu, Deodat  1750-

Lacroix A 1918 Par Mm 56, i-lxxxviii


Domalip, Karl  1846-1909

Kucera B 1910 Cas MF 39, 387-395


Dominicus de Clavasio  XIV

  = Clavisio = Chivasso
  = Clavagio = Dominicis Parisienis

Bjoernbo A A 1912 Bib M 12, 203-218
Busard H 1965 Arc HES 2, 520-575
  (MR33, 1200)
Curtz M 1895 Bib M 9, 107-110
- 1896 Bib M 10, 69-72
Eneström G 1907 Bib M 7, 252-262


Dominis, Marcantonio de  1566-1624

Bottari A 1931 Bo Fir (2)10, 162-165
  (F57, 23)  [Descartes, Newton]


Domninos of Larissa  V

Sarton G 1927 Int His Sc 1, 408
Tannery P 1884 BSM (2)8, 288-298

Donati, Luigi  1846-1932

Anon 1934 It UM 4, 189  (F51, 33)
Foa E 1932 It UM 11, 127-128
  (F58, 44)
Rimini C 1932 Nuo Cim (2)9, 189-
  195


Dordea,  T. A.

Antoniv I S 1929 Gaz M 35, 361-362
  (F55, 618)


Dounot, Didier  XVI-XVII

  = Didier Dounot de Bar-le-Duc

Brocard H + 1899-1907 Intermed 6,
  127; 7, 33, 115, 150; 8, 115-116;
  14, 58 (Q1515)


Dovey  W. R.

Anon 1933 Lon Acta 64, 383


Dowling, Linneaus Wayland  1867-1928

Anon 1929 AMSB 35, 123


Dreetz, Werner  1887-1960

Schmeidler W 1960 MP Semb 7, 112-113
  (Z99, 246)


Dreyer, John Louis Emil  1852-1926

Sampson R A + 1934 Isis 21, 131-144


Drobisch, Moritz Wilhelm  1802-1896

Credaro L 1847 *M. G. Drobisch,* Rome,
  21 p (F28, 21)
Heinze M 1896 Leip Ber 48

Dronke, J.

Loetzberger P 1914 ZMN Unt 45, 642
  (F45, 38)


Droste, Johannes,  1886-1963

Zaanen A C 1963 Nieu Arch (3)11, 93-
  94  (Z113, 2)


Drude, Paul L 1863-1906

Falkenhagen H 1963 Fors Fort 37, 220-
  221  (Z107, 6)
Lave M 1906 MN Bl 3, 174-175
  (F37, 29)


Drummond, Josiah H.  1827-1902

Emch A 1902 AMM 9, 297-298


Drushel, Jacob Andrew  1872-1940

Deutsch J G 1941 MT 34, 1


Dubnov, Yakov Semenovich  1887-1957

Lopsic A M 1960 M Prosv 5, 3-16
  (Z117:1, 9)
Vagner V V + 1961 Mos Ve Ten 11, 3-17


Du Bois-Reymond, Paul  1831-1889

Boeklen O 1890 Wurt MN 3, 1-4
Cavailles Jean 1938 *Remarques sur la
  formation de la théorie abstraite
  des ensembles,* Paris
Hardy G H 1924 *Orders of infinity.
  The "Infinitor-Calcul" of Paul
  Du Bois-Reymond,* 2nd ed, Cambridge
Kronecker L 1889 Crelle 104, 352-354
Lampe E 1910 Ber MG 9, 53-56
  (F41, 17)
Stepanov V V 1918 M Sbor 30(4)
  [orders of inf]

Voit C 1890 Mun Si 20 415-418
Weber H 1890 M Ann 35, 457-469


Duclout, Jorge

Anon 1930 Rv M Hisp A (2)5, 23
  (F56, 34)


Duerer, Albrecht  1471-1528

Ahrens W 1915 Z Bild K 26, 291-301
  (F45, 69)  [magic sq]
Amodeo F 1907 Nap FM At (2)13, 16 p
  (F38, 66)  [Monge]
- 1907 Nap FM Ri (3)13, 322-323
  (F38, 67)  [Monge]
de Haas K H 1932 Rott NV (2)10(S1),
  32 p  (F58, 32)
Guenther S 1886 Bib M, 137-140
Hunrath K 1905 Bib M (3)6, 249
  [approx]
- 1906 Bib M (3)7, 120-125
  [trisection]
*Panofsky Erwin 1948 Albrecht Duerer,
  3rd ed, Princeton, 2 vols
Ransom W R 1964 SSM 64, 236
  [pentagon]
Schuritz Hans 1919 Die Perspective in
  der Kunst Duerers, Frankfurt, 50 p
Steck Max 1948 Duerers Gestaltlehre
  der Mathematik und der bildenden
  Kuenste, Halle, 188 p (Z31, 242)
- 1953 Leop NA 16, 425-434
  (Z53, 196)
- 1958 Fors Fort 32, 246-251
  (Z80, 7)
*- 1964 Int Con HS 10, 655-658
  [art, bib]
Steinbrenner G 1914 Braun Mo, 254-257
  (F45, 77)  [regular polygons]
Wolff G 1928 Unt M 34, 161-168
  (F54, 23)


Duhamel, Jean-Marie-Constant
  1797-1872

Jamin J C 1872 BSM 3, 314-317
  (F4, 22)
- 1872 JMPA (2)17, 324-327 (F4, 22)


Duhem, Pierre  1861-1916

Anon 1920 Bord Mm (7)1, 9-169
Bosmans R P H 1921 Pierre Duhem.
  Notices sur ses travaux...,
  Louvain (also 1921 Rv Q Sc (July,
  Sept))
Bryan G 1917 Nat 98, 130
Ginzburg B 1936 Isis 25, 341-362
  (F62, 19)  [Jordanus]
Hadamard J 1927 Bord Mm (7)1, 637-
  665  (F53, 29)
Marcolongo R 1916 Batt, 44, 365-368
Picard E 1922 Par Mm (2)57, c-cxlii
  (F48, 24)
- 1922 Rv Sc, 465-483  (F48, 24)


Dulac, Henri  Claudius, Rosario
  1870-1955

Anon 1932 Rv M Hisp A (2)7, 137-138
  (F58, 992)
Julia G 1955 Par CR 241, 913-916
  (MR17, 338; Z65, 245)


Dumas, Gustave  Alphonse  1872-1955

Rham Georges de 1955 Elem M 10, 121-
  122  (MR17, 338; Z65, 245)


Dunkel, Otto  1869-1951

Rider P R 1957 AMM 64, 1-2
  (MR19, 1150)


Duns Scotus, Joannes  1266?-1308

Balic K M 1936 Wiss Weis 3, 120-130
  (F62, 1042)
Sarton G 1931 Int His Sc 2(2), 967-
  970


Dupin, Charles  1784-1873

Fink K 1893 Wurt Sch, 1-27 (F25, 26)
Smith D E 1921 AMM 28, 121

Duporq, Ernest  1873-1903

Laisant C A 1903 Nou An (4)3, 97-98
  (F34, 32)


Durand, William Frederick  1859-1958

Karman T v 1958 J Aer Sci 25, 665-
  666 (Z80, 9)


Dwight, Herbert Bristol  1885-

Archibald R C 1948 *Mathematical
  Table Makers,* NY, 23-24
  (Scr Stu 3) [biobib, port]


Dyck, Walther von  1856-1934

Faber G 1935 DMV 45, 89-98
  (F61, 25)
- 1935 M Ann 111, 629-630
Lorey W 1935 Unt M 41, 28-29
  (F61, 26)


Dziobek, Otto  1856-1919

Haentschel E 1920 Ber MG 19, 31-33


Eberhard, Victor  1861-1927

Rosenthal A 1931 DMV, 41, 40-49
  (F57, 43)


Echegaray y Eizaguirre, Jose
  1832-1916

Carrasco P 1916 Esp SM 6, 2-6
  (F46, 36)
Galdeano Z G de 1905 Rv Trim 5, 33
Sanchez Perez J A 1932 Rv M Hisp A
  (2)7, 49-58  (F58, 993)


Eckhardt, Chistian Leonhard Phillipp
  1784-1866

Grunert J A 1869 Arc MP Lit 49(194),
  1-2  (F2, 22)


Edgeworth, Francis Ysidro  1845-1926

Bowley A L 1928 *Francis Ysidro
  Edgeworth's Contributions to Mathematical Statistics,* London, 146 p
  [econ, psych]
- 1934 Ecmet 2, 113-124 (F60, 835)
  [port]


Efimov, Nikolai Vladimirovich  1910-

Aleksandrov A D + 1960 UMN 15(6),
  175-180  (Z94, 5)


Egervary, Jeno  1891-1958

Anon 1959 M Lap 10, 1-4
  (MR22, 260; Z86, 5)
Rozsa P 1959 M Lap 10, 195-225
  (MR23A, 136; Z97, 244)
- 1960 Magy MF 10, 1-3
  (Z100, 247)


Ehrenfest, Paul  1880-1933

Klein M J 1959 Amst P(B)62, 41-62
  (MR21, 86, 459) [quantum stat]
Kramers H A 1933 Nat 132, 667
- 1933 Physica 13, 273-276
  (F59, 856)
Pauli W 1933 Naturw 21, 841-843


Einstein, Albert  1879-1955

Biermann K-R 1959 Ist M Isl 12,
  493-502  (Z98, 7) [bib]
Borel E 1922 Rv Hebdom 31(4), 195-202
  [Paris]
- 1967 in Frechet M, *Emile Borel,
  philosophe et homme d'action*
  (Gauthier-Villars) (Repr of 1922)

Born M 1955 Naturw 42, 425-431
  (MR17, 2)  [quantum thy]
- 1956/57 MN Unt 9, 97-105
  (MR18, 453)
Carnahan, W H 1951 SSM 50, 171-174
Eisenhart C 1964 Wash Ac 54, 325-
Kuznetsov B G 1963 *Einstein,* 2nd ed
  Moscow, 414 p  (Z106, 3)
Lanczos C 1955 Nuo Cim (10)2(S),
  1193-1220  (MR17, 931)
  [relativity]
Schaaf W L 1955 MT 48, 168-169
- 1955 MT 48, 416
*Schlipp Paul A 1949 *Albert Einstein:*
  *Philosopher Scientist,* Evanston,
  779 p [autobiog, 1-96, bib]
Talmey M 1933 Scr 1, 68-71
Temple G 1956 LMSJ 31, 501-507
  (MR18, 182)
Wheeler J A 1968 *Einsteins Vision,*
  Berlin
Whittaker E T 1943 Phi Mag (7)34,
  266-280  (Z60, 13) [Aristotle,
  Newton]


**Eisenhart,** Luther Pfahler  1876-

Archibald R C 1938 *A Semicentennial*
  *History of the American Mathemati-*
  *cal Society 1888-1938,*  NY, 228-
  233 [bib, port]


**Eisenlohr,** Wilhelm 1799-1872

Kobell E 1873 Mun Si, 131 (F5, 33)
  [astron, phys]


**Eisenstein,** Ferdinand Gotthold
  1823-1852

Biermann K-R 1958 Fors Fort 32, 78-81
  (Z79, 243)
- 1958 Fors Fort 32, 332-335
  (Z85, 6)
  1959 Ber Mo 1, 67-72  [Crelle]
*- 1959 Ist M Isl 12, 493-502
  (MR23A, 696)  [bib]
  1961 ZGNTM 1(2), 1-12 (Z100, 246)

- 1964 Crelle 214/215, 19-30
  (MR29, 1)
Hurwitz A 1895 Ab GM 7, 169-203
  (F26, 19)
Rudio F 1894 Ab GM 7, 143-168
  (F26, 19)


**Elie de Beaumont,** Jean Baptiste
  Armand Louis Léonce  1798-1874

Kobell F V 1875 Mun Si, 132-134
  (F7, 14)


**Elliott,** Edwin Bailey  1851-1937

Turnbull H W 1938 Lon RS Ob


**Elliott,** Ezekiel Brown

Harkness W 1891 Wash PS 11, 470-473
  (F23, 30)


**Ellis,** Alexander John  1814-1890

Tucker R 1890 LMSP 21, 453-461


**Elola,** Jose de

Plans J M 1933 Rv M Hisp A (2)8,
  265-266 (F59, 854)


**Emmanuel,** David  1854-1951

Coculescu N 1929 Math Cluj 2, iii-iv
Ionescu I 1929 Math Cluj 2, 164-175
  (F55, 24)
Pangrati E A 1929 Math Cluj 2, 137-139
Sergescu P 1929 Ro SSM 32, 206-207
Tzitzeica G 1929 Math Cluj 2, v-vi


**Emsmann,** August Hugo  1810-1889

Anon 1890 ZMN Unt 21, 155-156

**Encke,** Johann Franz   1791-1865

Bruhns Karl 1869 *Johann Franz Encke,
    sein Leben und Wirken,* Leipzig,
    360 p  (F2, 21)
Foerster W 1894 *Ueber das Zusammen-
    wirken von Bessel, Encke und A.
    von Humboldt unter der Regierung
    Friedrich Wilhelm III,* Berlin,
    21 p  (F25, 1912)
Jelitai J 1938 M Ter Er 57(1), 136-
    143  (Z18, 340)  [Gauss]


**Endo,** Toshisada   1843?-1915

Hayashi T 1915 Toh MJ 8, 127-129
    [bib]
- 1916 Toh MJ 9, 174  [bib]
Mikami Y 1915 Toh J M 8, 119-126
- 1919 Tok M Ph (3)1, 304-305
    [Kawakita]


**Eneström,** Gustav   1852-1923

Favaro A 1891 Ven I At (7)3, 637-
    644
- 1892 Ven I At (7)4, 403-409
Hagstroem K G 1953 Nor M Tid 1, 145-
    155, 182  (MR15, 59; Z53, 197)
Lorey W 1926 Isis 8, 313-320  [port]
Sarton G 1923 Isis 5, 421
Wieleitner H 1924 Unt M 30, 19
    (F50, 16)


**Engel,** Friedrich   1861-1941

Anon 1938 Deu M 3, 701-719
    (Z19, 389)
Heegaard P 1941 Nor M Tid, 23,
    129-131  (MR18, 190; Z25, 386)
Ullrich E 1945 Gies Sem (34), 15 p
    (MR11, 708; Z61, 5)
- 1951 Gies Hoch 20, 139-154
    (Z43, 245)
- 1951 Gies Sem (40)  (MR15, 276)


**Englehardt,** Valentin   1516-1562?

Richeson A W 1939 Scr 6, 26-31
    (F65, 9)


**English,** Harry   1865?-1933

Anon 1933 MT 26, 490


**Enriques,** Federigo   1871-1946

Agostini A 1952 Livo Rv (1)
Anon 1957 Ren M Ap (5)16, 1-22
    (MR19, 1248; Z89, 5)
Bottari A 1956 Pe M (4)34, 130-131
    (Z70, 244)
Campedelli L 1947 Pe M 4(25), 95-114
    (Z29, 195)
- 1956 Archim 8, 97-103
    (MR18, 453; Z70, 244)
- 1966 Pe M (4)44, 367-379
    (MR34, 5630)
Campedelli L + 1947 Pe M (4)25,
    124-151  (Z29, 195)
Castelnuovo G 1947 Linc Rn (8)2,
    3-21  (Z29, 195)
- 1947 Pe M (4)25, 81-94 (Z29, 195)
Chisini O 1947 Pe M (4)25, 117-123
    (Z29, 195)
- 1952 Pe M (4)30, 1-3 (Z46, 3)
Conforto F 1947 Ren M Ap (5)6,
    226-252  (MR9, 74; Z30, 338)
Cuzzer O 1956 Civ Mac 4(1), 73-76
    (MR17, 931)
Dassen C C 1931 Arg SC 111, 387-422
    (F57, 1310)
Deiman I Ya 1947 UMN 2(4), 207-208
Forti U 1950 Arc In HS 29, 915-918
    (Z38, 4)
Godeaux L 1951 Lieg Bul 20, 77-85
    (MR13, 197; Z43, 5)
- 1953 Rv GSPA 60, 8-14  [alg geom]
Levi B 1946 M Notae 6, 119-123
    (Z60, 14)


**Enskog,** David   1884-1947

Chapman S 1948 Nat 161, 193-194
    (Z29, 2)

Epicuros of Samos   -341 to -270

Keyser C J 1936 Scr 4, 221-240
  [inf]
Lasker E 1938 Scr 5, 122-123


Epiphanios   IV

Shaw A A 1936 AMSB 42, 821
  (F62, 1043)  [metrol]
- 1936 M Mag 11, 3-7 (F62, 1018)
  [metrol


Eratosthenes   -276? to 194?

  See also sieve in Ch. 2.

Bernhardy G 1822 *Eratosthenica,*
  Berlin
Jadraque V M 1957 Gac M 9, 159-161
  (Z88, 242)
Solmsen F 1942 Am Phlg TP 73, 192-
  213
Ver Eecke P 1956 Arc In HS 35, 217-
  226  [dup of cube, Eutocius]
Wilamowitz-Moellendorff U v 1894
  Gott Ph HN, 15
Wolfer E P 1954 *Eratosthenes von
  Kyrene als Mathematiker und
  Philosoph,* Groningen-Djakarta,
  172 p  (MR17, 117)


Erbiceanu, C. G.   1872-1904

Davidoglou A 1904 Gaz M 9, 173
  (F35, 35)


Erdös, Paul   1913-

Anon 1958 M Lap 9, 136-147
  (MR20, 848; Z80, 9)  [bib]
Juran P 1963 M Lap 14, 1-28
  (MR33, 13; Z113, 2)


Erini, Kerim

Anon 1953 Istanb (A)18, i-iv
  (MR14, 832)


Erlang, A. K.   1878-1929

Brockmeyer E + 1960 *The Life and
  Works of A. K. Erlang,* Act PSPM
  (287), 277 p  (MR22, 619; Z33, 51)
Nubolle H C 1929 M Tid B, 32-34


Erlenssön, Hauk   1264-1334

Eneström G 1885 Bib M (4), 199


Ermakov, V. P.   1845-1922

Dobrovolskii V A 1963 Ist M Zb 4,
  37-41  [series]
Gracianskaja L N 1956 Ist M Isl 9,
  667-690  (Z70, 5)
Latysheva K Ya 1955 Ukr IM 7, 231-
  238  (MR17, 117)  [bib, diff eqs]
Potapov V S 1953 Stal Ped (3), 3-8
  (MR17, 698)  [vector alg]


Errard, J.

Bosmans H 1906 Intermed 13, 183
  (Q3027, Q3028)


Errera, Alfred   1886-1960

Godeaux L 1960 It UM (3)15, 575-578
  (Z93, 9)
- 1961 Mathesis 69, 311-312
  (Z93, 9)


Erugin, Nikolai Pavlovich   1907-

Basov V P + 1958 UMN 13(2), 247-251
  (Z79, 5)

Escherich, Gustav von  1849-

Wirtinger W 1935 Mo M Phy 42, 1-6
(F61, 24)


Esson, William  1838-1916

Anon 1917 LMSP (2)15, 50-51
(F46, 35)


Etienne d'Espagnet  XVII

Brocard H 1902 Intermed 9, 268-269
(Q898) [instruments]
- 1903 Intermed 10, 261 (Q898)
Tannery P + 1896 Intermed 3, 199,
244 (Q898)


Euclid   -III

See also non-Euclidean Geometry

Agostini A 1928 Pe M 8, 185-188
- 1936 Gli elementi d'Euclide e la
critica antica e moderna Libri
XI - XIII, Bologna, 355 p
Alasia C 1909 Rv FMSN 20, 183-229
(F40, 62) [non-Euc geom]
Albaugh A H 1961 MT 54, 436-439
Alimov N G 1955 Ist M Isl 8, 573-
619 (MR17, 1; Z68, 3)
[quantity, proportion]
Archibald R C 1950 AMM 57, 443-452
(MR12, 311; Z41, 337)
[first Engl tr]
Bashmakova I G 1948 Ist M Isl 1,
296-328 [arith]
Baudoux C 1935 Brx Con CR 2, 73-75
(F62, 1017) [Syriac tr]
- 1937 Archeion 19, 70-71  (F63, 8)
[Arab, Ishaq]
Becker O 1936 QSGM 3, 533-553
(Z15, 147) [even and odd]
Belfroid J 1948 Et Class 16, 24-32
(Z37, 290)
Bergstraesser G 1933 Islam 21, 195-
222 [Pappus]
Besthorn R O 1892 Bib M, 65-67
[Simplicius]

Bilimović A 1955 Beo Ak N Zb 43(4),
67-71 (Z68, 3) [Bk VI]
Bini U 1950 Archim 2, 207-211
(Z41, 337)
Birkenmajer A 1935 Math Cluj 9, 310-
320 [Diophantos]
Bjoernbo A A + 1912 Al Kindi, Tideus
und Pseudo-Euklid. Drei optische
werke, Ab GM 26, 176 p
Bosmans H 1920 Rv Q Sc (Ap), 11p
[Zeuthen]
Brandt-Mueller 1930 M Tid A, 6-9
(F56, 11)
Breton de Champ P 1849 JMPA 2, 185-
[porisms]
- 1849 Par CR 29, 479 [porisms]
- 1858 JMPA 3, 89- [porisms]
Bretschneider C A 1870 Die Geometrie
und die Geometer von Euklides,
Leipzig (F2, 1)
Brusotti L 1938 Pe M (4)18, 133-150
(F64, 903)
Bunt L N H 1954 Van Ahmes tot Euclides,
Groningen-Djakarta, 178 p
(MR16, 551)
Burton H E 1945 Opt S Am 35, 357-372
(MR6, 253; Z60, 5)
Busard H L L 1968 Janus 54, 1-140
(MR37, 1216) [XII, Hermann the
Dalmation, translation]
Cabrera Emanuel S 1949 Los elementos
de Euclides como exponente del
"milagro griego," Buenos Aires,
150 p
Cajori F 1921 Sci 53, 414
[pseudo port]
Cantor M 1857 ZM Ph 2, 17- [porisms]
- 1867 ZM Ph 12(S)
- 1872 Boncomp 5, 1-74  (F4,2; 5, 3)
(tr of 1867)
Carruccio E 1963/64 Tor FMN 98, 226-
239 (MR29, 1068) [termin]
Chasles M 1838 Cor M Ph 10, 1-
[porisms]
- 1859 Par CR 48, 1033- [porisms]
- 1860 Par CR 50, 940, 997, 1007-
[porisms]
- 1860 Par CR 51, 1043-
(re Breton 1858)
Christensen S A 1888 Zeuthen (5)6,
161-192

Euclid   (continued)

- 1909 in *Festskrift til H. G.
  Zeuthen,* Copenhagen, 18-26
  (F40, 61)  [Denmark]
Clagett M 1953 Isis 44, 16-42
  (MR15, 275)  [Adelard, middle ages]
- 1954 Isis 45, 269-277  (MR16, 1)
  [Alfred]
Conte L 1941 Pe M (4)21, 113-127
  (MR3, 97; Z24, 242)  [Bk  XIV]
Curtze M 1868 ZM Ph (S), 45-104
  (F1, 5)  [mss]
- 1896 Bib M (2)10, 1-3  (F27, 3)
  [mid]
Dellac H + 1895 Intermed 2, 406  (Q432)
De Morgan A 1857 QJPAM 1, 47-
Depman I Ya 1950 Ist M Isl 3, 467-474
  (MR13, 1)  [Russ tr]
*Dijksterhuis E J 1929-1930 *Die
  Elementen van Euclides,* Groningen,
  2 vols
- 1950 Euc Gron 25, 43-54
  (Z34, 145)  [Proclos]
- 1955 *The First Book of Euclidis
  elementa,* Leiden, 59 p  (MR18, 182)
Dixon E T 1930 M Gaz 15, 1-4
  (F56, 11)  [relativity]
Eneström G 1884 Bib M, 79-80
- 1885 Boncomp 18  [Swedish tr]
*Enriques F 1925-1936 *Gli Elementi
  d'Euclide e la critica antica e
  moderna,*  4 vols, Rome (Z14, 49)
- 1947 Pe M (4)25, 66-72  (Z29, 193)
Evans G W 1927 MT 20, 127-141
  (Repr 1968 MT 61, 405-414)  [alg]
- 1928 MT 20, 310-320
Favaro A 1883 Ven I At (6)1, 393-397
Fine H B 1917 An M 19, 70-76
  [proportion]
Fladt K 1927 *Euklid,* Berlin, 80 p
Fletcher W C 1938 M Gaz 22, 58-65
Frajese A 1940 Pe M (4)20, 137-154
  (MR3, 97; Z24, 79)
- 1950 Int Con HS 6, 123-132
- 1950 Scientia 44, 299-305
  (MR12, 310)
- 1951 Arc In HS 30, 383-392
  (MR13, 1; Z42, 1)
- 1951 It UM (3)6, 50-54  (Z42, 2)
  [parallel]

- 1953 Archim 5, 45-48  (Z50, 241)
- 1954 Archim 6, 258-262
  (MR16, 433; Z57, 2)
- 1958 Archim 10, 130-135
  (MR20, 620)
Frankland W B 1902 *The Story of Euclid*
  London-NY, 176 p (F33, 10) [pop]
Gartz 1823 *De interpretibus et ex-
  planatoribus Euclidis Arabicis,*
  Halle
Ghose A K 1897 Nat 56, 224
  (F28, 3)  [lost books]
Gohlke P 1925 *Auswahl aus den Elemente
  des Euklid,* Frankfurt  (F51, 15)
Goldat G 1957 Isis 48, 351  [mid]
Gould S H 1962 M Gaz 46, 269-290
  (Z111, 2)  [axioms]
Guzzo A 1952 Filo Tor 3, 45-82
  (MR14, 609; Z47, 1)
- 1954 Tor Sem 13, 1-17 (MR16, 433;
  Z55, 243)
Halsted G B 1879 AJM 2, 46-48
  (F11, 2)  [first Engl tr]
*Heath T L 1925 *The Thirteen Books of
  Euclid's Elements Translated from
  the Text of Heiberg with Introduc-
  tion and Commentary* (Cambridge
  Univ Pr), 3 vols   (1st ed 1908;
  repr 1956 Dover paper)
Heiberg J L 1884 Hist Abt 29, 1-23
  [Arab]
- 1889 *Om scholierne til Euklids
  Elementer,* Dan Sk, 78 p (Fr summ)
- 1895 Dan Ov 1, 2, 117  [optics]
Heller S 1964 Janus 51, 277-290
  (MR 30, 870) [congruence, similarity
Hill M J M 1898 Cam PST 16, 227-261
  [Bk V]
- 1904 Cam PST 19, 157-172  [Bk V]
- 1922 Cam PSP 21, 474-476  [Bk V]
- 1923 Cam PST 22, 87-99  (F45, 95)
  [Bk V]
- 1923 Cam PST 22, 185-189  [Bk V]
- 1923 Cam PST 22, 449-462  [Bk V]
  1923 M Gaz 11, 213-220 [propor]
Hjelmslev J 1934 M Tid B, 33-39
  (F60, 825)  [exhaustion]
Hultsch F 1904 Bib M (3)5, 225-233
  (F35, 54)  [sexagesimal]

Euclid   (continued)

Ionescu I 1939 Gaz M 44, 228-230
(F65, 4)
Itard J 1961 *Les livres arithmétiques
d'Euclide,* Paris 230 p   (Z117:1, 5)
Jacoli F 1870 Boncomp 3, 297-298
(F2, 17)
Jones P S 1955 MT 48, 30-32
[ballad of Sir P. Spens]
Joerg E 1933 Fors Fort 9, 275
(F59, 45)   [Alfredus Magnus]
Junge G 1931 Fors Fort 7, 55
- 1934 QSGM (B)3, 1-17   (Z10, 244)
[Pappus]
- 1948 Osiris 8, 316-345
(Z32, 241)   [pentagram]
Junge G + 1932 *Codex Leidensis 339,
1. Euclidis Elementa ex inter-
pretatione Al-Hadschdschapschi cum
commentariis Al-Narizii,*
Copenhagen, 215 p   (Z5, 2)
Kapp A G 1934 Isis 22, 150-172
(Z11, 385)   [Arab tr]
- 1935 Isis 23, 54-99   [Arab]
Kugener A 1936 Brx Con CR 2, 70-72
(F62, 1018; Z13, 193)
Lacoarret M 1957 Rv Hi Sc Ap 10,
38-58   (Z93, 2)
Langer R E 1935 SSM 34, 412-423
Leclerq C 1936 Sphinx 6, 189-191
(F62, 13)
Lejeune A 1938 Rv Q Sc, 402-410
[optics]
- 1948 Arc In HS 28, 598-613
(MR11, 150; Z32, 193) [optics]
- 1948 *Euclide et Ptolémée. Deux
stades de l'optique géométrique
grecque,* Louvain, 196 p
(Z39, 241)
Levi Beppo 1947 *Leyendo a Euclides,*
Rosario, Argentina, 224 p
Levi ben Gerson 1958 Ist M Isl 11,
763-776   (Z100, 243)
(tr by I. G. Polskii)
Locke L L 1941 Scr 8, 34-42
[Malton's "Royal Road..." 1774]
Maggi A 1941 Pe M (4)21, 205-223
(MR8, 189; Z25, 385) [porisms]
Maistrov L E 1949 Ist M Isl 2, 505-
507 (MR12, 1)

Mansion P 1886 Brx SS 10(A), 46
- 1899 Brx SS 24, 47
Marian V 1939 Gaz M 44, 564-567,
621-624   (F65:1, 9)
Markushevich A I 1948 Ist M Isl 1,
329-342   (MR11, 150)
Martin T H 1874 Boncomp 7, 263-267
(F5, 13)
Molodshi V N 1949 Ist M Isl 2, 499-
504 (MR12, 1)   [Plato]
Mordoukhay-Boltovsky D 1933 Pe M
(4)13, 169-183   (F59, 18)
[Bk II]
Ofterdinger 1853 *Beitraege zur Schrift
des Euklid ueber die Theilung der
Figuren,* Ulm
Peters T 1936 Kantstu 40, 180-252;
41, 127-168   (F62, 13)
Petrosian G B 1964 Int Con HS 10
Plooij E B 1950 *Euclid's conception
of ratio and  his definition of
proportional magnitudes as
criticized by Arabian commentators,*
Rotterdam, 71 p   (diss) (Z45, 146)
Raik A E 1948 Ist M Isl 1, 343-384
(MR11, 150)   [Bk X]
Reeve W D 1930 MT 23, 332
Reinsoso N 1905 Rv Trim 5, 25-26
(F36, 63)   [span tr]
*Riccardi P 1887-93 *Saggio di una
bibliogafia Euclidea,* Bln Mm (4)8,
399-523; (4)9, 321-343; (5)1,
25-84; (5)3, 637-694
Robb A A 1929 M Gaz 14, 473-476
(F55, 26)   [relativity]
Ruddick C T 1927 AMM 34, 30-33
[circle, optics]
*Rutt N E 1937 M Mag 11, 374-381
(F63, 799)
Sabra A I ed 1961 *Explanation of the
Difficulties in Euclid's Postulates,*
Alexandria   (in Arabic)
Scholz H 1934 Muns Semb 5, 103-120
(F60, 8)
Shaw A A 1938 M Mag 13, 76-82
(F64, 904)   [pre-Euclid]
Shenton W F 1928 AMM 35, 505-512
[first Engl tr]
Simon Max 1901 *Euklid und die 6
planimetrischen Bucher,* Leipzig
(Also in Ab GM 11)

Euclid   (continued)

Smith D E 1935 Scr 3, 5-10
  (Z11, 193)   [Omar Khayyam,
  Saccheri]
Smith T 1902 *Euclid, his Life and
  System,*  London-NY, 231 p
  (F33, 10)
Spoglianti M 1960 Pe M (4)38, 175-
  186, 234-253, 265-292   (Z100, 6)
- 1961 Pe M (4)39, 17-36   (Z100, 6)
Stamatis Evangelos 1953 *Der Schluss
  von der vollstaendigen Induktion
  bei Euklid,* Athens, 6 p
  (Z51, 242)
- 1955 Athen P 30, 410-414
  (MR18, 368; Z67, 244) [circle]
- 1956 *Euklid: "Ueber inkommensurable
  Groessen",* Elemente Buch 10,
  Athens, 314 p   (Z74, 4)
- 1957 Athen P 32, 251-266
  (MR20, 620   [Bk X]
- 1962 Platon 14 (27/28), 310-313
  (Z107, 243)   [Bk V]
Steck M 1956 Fors Fort 30, 183-185
  (Z70, 4)   [Ger tr]
- 1957 Fors Fort 31, 113-117
  (Z77, 4)
Steinschneider M 1886 Hist Abt 31,
  81-110 [Arab]
Summent G A 1957 Isis 48, 66-68
  [lost work by Austin 1781]
Suter H 1907 Bib M (3)7, 234-251
  [Arab]
- 1922 Ab GN Med 1, 9-78
  [Bk X, Pappus, Arab]
Szabo A 1960 Arc HES 1, 37-106
- 1960 Magy MF 10, 441-468
  (MR24A, 127)   [termin]
Tannery P 1884 BSM (2)8, 162-175 [axioms]
- 1886 BSM (2)10, 183-194
- 1887 BSM (2)11, 17-28
  [technology]
- 1887 BSM (2)11, 86-96
- 1887 BSM (2)11, 97-108   [Heron]
Tartalea N 1926 Pe M (4)6, 121-122
  (F52, 7)
Thaer C 1936 QSGM (B)3, 116-121
  (F62, 14)   [al Tusi]
- 1940 ZMN Unt 71, 66-67
- 1942 Hermes 77, 197-205
  (MR9, 74; Z27, 194)   [Arab]

- 1962 *Data von Euklid,* Berlin-
  Goettingen-Heidelburg, 73 p
Thomas-Stanford C 1926 *Early Editions
  of Euclid's Elements,* London, 72 p
  (F52, 7)
Timchenko I Yu 1915 Kagan (643)
Tuman'ian T G 1953 Ist M Isl 6,
  659-671   (Z53, 244) [Armenia]
Vailati G 1897 Bo SBM,  21  [mech]
Valentin G 1893 Bib M (2)7, 33-38
  (F25, 76)
Van Deventer C M 1929 Euc Gron 6,
  170-184  [Tannery]
Vanhée L 1939 Isis 30, 84-88
  [China]
Vincent A J H 1844 Nou An 3, 5-
Vogt H 1912 Bib M (3)13, 193-202
  (F43, 7)
Vygodskii M Ya 1948 Ist M Isl (1),
  217-295  (MR11, 150)
Webb H E 1919 MT 12, 41-60
  [foundations]
Weissenborn H 1880 ZM Ph 25(S), 141-166
  (F12, 2)  [Adelhard]
Wilkinson J M 1868 Manc Pr 7, 68-72
  (F1, 1)  [porisms]
Wilson J M 1868 Batt 6, 361-368
  (F1, 1)
Yamamoto S 1953 Paul UCM 1, 59-66
  (MR15, 89; Z50, 241)
Yeldham F A 1927 Isis 9, 234-239
  (F53, 12)  [Engl tr]
Yushkevich A P 1948 Mos IIET 2, 567-
  572  (MR11, 150)  [Russ tr]
Zaharia R O 1930 B Ai Bol 3, 153-156
  (F56, 805)  [phil]
Zeuthen H G 1882 Zeuthen (4)6,
  97-101  [conics, geom progression]
  1910 Dan Ov (5), 395-435
  [irrat]
- 1918 Scientia 24, 257-269   [defs]


Eudemos   -IV

Spengel L ed 1870 *Eudemi Rhodii
  Peripatetici fragmenta quae
  supersunt,* Berlin, 111-143
Tannery P 1882 Bord Mm (2)5, 211-237
  [lunes]

Eudoxos   -IV

Becker O 1933 QSGM 2, 311-333,
   368-387 [proportion]
- 1936 QSGM (B)3, 236-244
   (Z13, 337) [continuity]
   1936 QSGM (B)3, 370-388
   (Z14, 145) [excluded middle]
- 1936 QSGM (B)3, 389-410
   (Z14, 145) [phil]
Bell E T 1945 Scr 11, 153
Bonnesen T 1930 M Tid B 41-62
   [Dedekind]
Franklin S P 1961 Scr 25, 353-355
   (Z100, 243) [con fractions,
   Omar Khayyam]
Heegaard P 1921 Nor M Tid 3, 49-58
   (F48, 2) [hippopede]
Hjelmslev J 1950 Centau 1, 2-11
   (MR12, 311; Z45, 290)
   [Archimedes ax]
Neugebauer O 1954 Scr 19, 225-229
   (MR15, 591; Z55, 243)
   [hippopede]
Rufini E 1921 Arc Sto 2, 222
   (F48, 2)
Santillana G de 1949 Isis 32, 248-262
   (Z38, 145) [Plato]
Schiaparelli G V 1875 Milan Ob 9,
   1-63 (F7, 1)
- 1877 ZM Ph 22(S), 101-198
   (F9, 28)
Tannery M 1876 Bord Mm (2)1, 441-451
   (F8, 4) [ast]
- 1878 Bord Mm (2)2, 277-283
   (F10, 25) [Archytas, Delian probl]
- 1882 Bord Mm (2)5, 129-149 [ast]


Euler, Johann Albrecht   1734-1800

Staeckel P 1910 Zur NGV 55, 63-90
   (F41, 12)
Stieda W 1932 Fors Fort 8, 244-287
   (F58, 994) [letters]
- 1932 Leip PH 84, 5-43
Thiersch H 1929 Gott N 1928/29,
   264-289+ [L Euler, ports]
- 1930 Gott Ph HN, 219-249
   [L Euler]

Euler, Leonhard   1707-1783

Ackeret J 1945 in *Festschift zum 60
   geburtstag... A Speiser,* 160-168
   (MR7, 354; Z60, 11)
Agostini A 1922 Pe M 2, 430-451
   [log, Mengoli]
Ahrens W 1909 Allg Z (s 94)
   (F38, 18) [Works]
- 1909 Int Woch 3, 1195-1204
   (F40, 18) [Works]
- 1909 MN Bl 6   (F38, 18)
Alasia C 1908 Rv FMSN 17, 74-83
   (F39, 18)
Anon 1907 Ab GM 25 [anniv]
- 1910 DMV 19, 104-116, 129-142
   (F41, 11) [Works]
- 1912 Z Realsch 37, 1-4
   (F43, 19) [Works]
- 1920-1922 Schw NG(1920) 68-70; (1921)
   48-51; (1922), 62-65   (F48, 6)
- 1933 B Ai Bol 6, 81-88
   (F59, 846) [anniv]
- 1935 *Leonard Euler. Recueil des
   articles et matériaux en commémora-
   tion de 150e anniversaire...,*
   Moscow, 239 p
- 1957 Pri M Me 21, 153-156
   (MR19, 826) [anriv]
- 1957 SSSR Tek (3), 3-9
   (MR19, 624; Z87, 5) [anniv]
Archibald R C 1917 AMM 25, 276-282
   [integrals, spiral]
Aubry A 1900 Prog M (2)2, 401
- 1909 Ens M 11, 329-356
   [numb thy]
Backlund O 1907 Pet B (6), 476
   (F38, 13) [anniv]
Ball W W R 1924 AMM 31, 83-4
   (F50, 12) [output]
Barbak, P M 1959 Ist M Zb 1,
   [technology]
Bashmakova I G 1957 Ist M Isl 10,
   257-304   [fund thm alg]
Bashmakova I G + 1954 Ist M Isl 7,
   453-512   (MR16, 781; Z59, 18)
Belyi Yu A 1961 Ist M Isl 14, 237-
   284   (Z118-1, 5) [elem geom]
- 1965 Ist M Isl (16), 181-186
   (MR33, 7228) [polyhedra]

Euler   (continued)

Beman W W 1888 Bib M (2)2, 120  [e]
- 1897 AAAS 46, 33-50  (F28, 46)
  [complex numb]
Bergman G 1966 M Ann 164, 159-175
  (MR33, 5553) [Fermat thm]
Bertrand J 1868 J Sav 133-
Biermann K-R 1957 Ens M 3(2), 251-
  262 [iteration]
- 1958 Ens M (2)4, 19-24  (Z89, 4)
  [iteration]
- 1963 Fors Fort 37, 236-239
  (Z112, 2)
Bigourdan G 1928 Par Sav CR, 26-40
  (F56, 812) [Clairaut, letters
Bjerknes C A 1888 Bib M (2)2, 1-2
  [Degen]
Bocher M 1893 NYMS 2, 107-109
  (F25, 76) [part diff eqs]
Boncompagni B 1879 Boncomp 12,
  808-811 (F11, 22)
Bopp K 1924 Ber Ab (2), 4-45
  [Lambert]
Bosmans H 1909 Brx SS (2)33, 265-289
  (F40, 17) [works]
Boyer C B 1951 AMM 58, 223-226
  (Z42, 3) [introduction]
Brauer E 1908 DMV 17, 39-46
  [engineering]
Brennecke R 1924 Z Ph 25, 42-45
  [potential]
Brill A 1907 DMV 16, 555-558
  (F38, 13)  [anniv]
Brown B H 1942 AMM 49, 302-303
  [Diderot]
Brown W 1965 AMM 72, 973-977
  [Pfaff-Fuss probl]
Brykczynski A 1937 AMM 44, 45-46
Burckhardt F 1908 Basl V 19, 122-136
  [family]
Cajori F 1929 AMM 36, 431-437
  [developable surf]
Candido G 1900? Pe MIS 3, 113
  [W Chappel]
Cantor M 1878 Boncomp 11, 197-21
  [Lagrange, letters]
- 1878 ZM Ph 23, 1-21 [Lagrange,
  letters]
Chovanskii A N 1957 Ist M Isl 10,
  305-326  (Z101, 245) [cont
  fractions]

Condorcet J-A 1786 Eloge de M. Euler,
  Paris  (Opera omnia (3)12)
Davis P J 1959 AMM 66, 849-869
  (Z91, 5)  [gamma fun]
De Vries H 1933 N Tijd 21, 188-226
  (F60, 832)
Dingeldey F 1935 DMV 45, 132-133
  (F61, 951)  [mistake]
Doublet E 1911/1912 Bord PV, 5-7
  (F43, 15)
- 1912 BSM (2)36, 310-319  (F43, 15)
  [works]
Duerr K 1949 Int Con Ph 10(1948)2,
  720-721  (Z31, 2)  [logic, Venn]
Dupont P 1962/63 Tor FMN 97, 927-962
  (MR28, 1)
- 1963 Tor FMN 98, 537-572
  (MR29, 638)
Eneström G 1879 Stoc Han (10)
  (F11, 38)  [E. summation]
- 1880 Cron Cien 3, 329-335, 353-357,
  377-382  (F12, 13)  [Bernoulli]
- 1897 Bib M (2)11, 51-56 (F28, 10)
  [J I Bernoulli, letters]
- 1904 Bib M (3)5, 209-210 (F35, 59)
  [E. summation]
- 1904 Bib (3)4, 344-388; (3)5, 248-
  291  [J I Bernoulli]
- 1905 Bib M (3)6, 16-87
  [J I Bernoulli]
- 1905 Bib M (3)6, 186-189  (F36, 54)
  [convergence]
- 1906 Bib M (3)7, 126-155
  [J.I Bernoulli]
- 1907 Bib M (3)8, 372-374  [port]
- 1908 Bib M (3)9, 175-176  (F39, 61)
  [calc]
- 1908 DMV 17, 405-407 (F39, 19)
  [Mueller 1907]
- 1909 Bib M (3)10, 308-316 (F40, 16)
*- 1910 Verzeichnis der Schriften
  Leonhard Eulers, (DMV 19(S4), 1-388,
  (F41, 10)  [*bib]
- 1911 Bord PV, 2-5  (F43, 15)
- 1912 Bib M (3)12, 238-241
  [diff eqs]
- 1913 DMV 22 (2), 191-205 (F44, 10)
  [mss]
Euler Karl 1955 Das Geschlecht Euler-
  Schoelpi, Geschichte einer alten
  Familie, Giessen,  320 p

Euler (continued)

Fabarius W 1914 *Leonhard Euler und
das Problem Fermat,* Cassel, 12 p
(F45, 95)
Faris J A 1955 JSL 20, 207
[Gergonne relations]
Fehr H 1921 Ens M 21, 343-344
[works]
Felber V 1908 Cas MF 37, 177-192
(F39, 18)
Finkel B F 1897 AMM 4, 297-302
Frankl F I 1950 UMN 5(4), 170-175
(MR12, 577) [hydro dyn]
- 1954 Ist M Isl 7, 596-624
(MR16, 781; Z60, 9)
[part diff eqs]
Freiman L S 1957 Vop IET (4), 164-
167 (Z80, 4) [mech]
Frobenius F G 1917 Zur NGV 62, 720-
722 (repr in Frobenius *Ges. Abh.*
3, 732-734)
Fueter R 1948 Elem M(S), 24 p
(MR11, 709; Z32, 51)
Fuss N 1783 *Eloges de M. Leonard
Euler,* St Petersbourg [*bib]
*- 1786 *Lobrede au Herrn Leonhard
Euler,* Basel
Fuss P H 1843 *Correspondance mathé-
matique et physique de quelques'
célèbres géomètres du XVIII
siecle, precede d'une notice sur
les travaux de Leonhard Euler,*
St Petersbourg, 2 vols
(MR37, 1220) (Repr 1968 Johnson)
[*bib, biog]
Gelfond A O 1957 UMN 12(4), 29-39
(MR19, 826) [anal]
Gillings R J 1954 AMM 61, 77-80
[Diderot]
Glaisher J W L 1872 Mess M 1, 25-
- 1873 Mess M 2, 64
Gnedenko B V 1959 Ist M Zb 1, 71-76
Graf J H 1907 *Der Basler Mathematiker
Leonhard Euler bei Anlass der
Feier seines 200 Gebrutstages,*
Bern, 24 p (F38, 18)
Grigoryan A T 1957 Mos IIET 17,
312-319 (Z112, 242)
- 1961 *O cherki po istorii mekhaniki
v Rossii,* Moscow

Grigoryan A T + 1957 Vop IET (4),
3-14 (Z99, 3)
Gussov V V 1953 Ist M Isl 6, 355-475
Haentzschel E 1913 DMV 22, 278-284
[ell fun , Weierstrass]
Hagen J G 1896 *Index operum Leionardi
Euleri,* Berlin, 88 p (F27, 11)
- 1897 DMV 5, 82-83 (F28, 10)
[works]
Hagenbach R R 1851 *Leonhard Euler
als Apologet des Christenthums,*
Basel
Hansted B 1879 BSM (2)3, 26-32
(F11, 18) [letters]
Henry C 1880 BSM (2)4, 207-256
(F12, 12)
- 1886 Boncomp 19, 136-148
[d'Alembert, letters]
Hocking W E 1909 Calif Ph (2)2, 31-44
[graphs, logic]
Hofmann J E 1948 Experien 4, 364-366
[notation]
- 1957 MZ 67, 139-146
(MR19, 108; Z87, 242) [summation]
- 1958 Phy Bl 14, 117-122 (Z78, 3)
- 1961 Arc HES 1, 122-159
(Z94, 3) [Fermat, numb thy]
Hoppe E 1907 DMV 16, 558-67
[optics]
- 1923 ZMN Unt 54, 181-184 (F49, 7)
[relativity]
Horner F 1822 "Memoir of the life
and character of Euler" in Euler
*Elements of Algebra* (transl John
Hewlett, 3rd ed), viii-xxiv
Isely L 1908 Neuc B 36, 57-65
Jonquieres E de 1890 Par CR 110,
169-173 [polyhedra]
Khovanskii A N 1956 Mos M Sez, 236-
237 [cont fractions]
Kiselev A + 1965 Ist M Isl 16, 145-180
- 1966 Ist Met EN 5, 31-34
[Goldbach, numb thy]
Kneser A 1907 Ab GM 25, 21-60
(F38, 15) [calc var]
Knoblauch J 1907 Ber MG 6, 69-72
(F38, 18) [works]
- 1912 Arc MP (3R)19, 2-3 [port]
- 1912 Ber MG 11, 2-3 [port]
Kolman E 1957 Vop IET (4), 15-25
(Z99, 4) [Russia]

Euler (continued)

Komarevskii V 1925 Tash NO Tr 2,
141-170 (F51, 10) [polyhedra]
Kopelevich Yu Kh 1957 Ist M Isl 10,
9-66 (Z99, 245)
- 1957 Ist M Isl 10, 95-116
(Z99, 244) [Bruce]
- 1959 Ist M Isl 5, 271-444
[Meyer]
Kopelevich Yu Kh + ed 1962-1965
*Ruk-opisnye materialy L. Eulera v
Arkhive Akademii Nauk SSSR,*
Mos-Len, 2 vols, 427 p, 574 p
(MR34, 5620-5621)
Korner T 1904 Bib M (3)5, 15-62
[mech]
Kostryukov K T 1954 Ist M Isl 7, 630-
640 (MR16, 781; Z59:1, 9; 60, 9)
Krasotkina T A 1957 Ist M Isl 10,
117-158 (Z100, 7) [Stirling]
Kravchuk M 1935 Uk Izv An, 46 p
(F61, 958; Z13, 193)
Krazer A 1925 DMV 33, 125 (F51, 39)
Kronecker L 1875 in Kronecker Werke
2, 3-10 [quad recip]
Kryloff A N 1936 Len IINT 8, 281-
299 (F62, 1011) [Euler-
Lambert thm]
Kushnik E A 1957 Ist M Isl 10,
363-370 (Z101, 9) [diff eqs]
Kuznetsov B G 1947 Mos IIET 1, 347-
371 [mech, space]
Lampe E 1907 Ab GM 25, 117-137
(F38, 16) [complex numbs, logs]
Langer R E 1935 Scr 3, 61-66,
131-138 (Z11, 193)
- 1957 AMM 64, 37-44 [fun eq]
*Lavrentiev M A + 1958 *Leonard Eiler:
Sbornik statei v chest 250-letiya
so dnya rozhdeniya,* Mos, 612 p
(MR23A, 3628)
Lebesgue H 1925 BSM 42, 273-276
[polyhedra]
Lefebvre B 1928 Rv Q Sc 14, 64-83
(F55, 13)
Likhin V V 1966 Ist Met EN 5, 35-44
[Bernoulli numbers, bib, fin
diffs, Lagrange]
Liouville, J 1874 JMPA 19, 189-
Lorey W 1907 Gorlitz 25(S)
(F38, 14)

- 1933 Crelle 170, 129-132
[mistake]
- 1933 Euc Gron 9, 198-210
[Euler identity, Gauss]
- 1933 Unt M 39, 265-267 (F59, 26)
[anniv]
Loria G 1923 Ens M 13, 142-147
[anal geom]
- 1935 in *Studi onore Salvatore
O. Carboni* Rome, 233-238
[figurate numbers, Pythagoras]
- 1942 Ens M 38, 250-275 (MR4, 65)
[optics]
Lusin N 1965 Ist M Isl 16, 129-143
(MR33, 2503) [Goldbach]
Lysenko V I 1966 Vop IET 20, 38-46
[Newton]
Mandryk A P 1957 Vop IET (4), 26-33
(Z98, 7) [ballistics]
- 1958 *Ballisticheskii
raboty Leonardo Eulera,* Mos-Len,
185 p (Z95, 3)
Maroni A 1921 Pe M (4)1, 337-346
[Descartes, polyhedra]
Matvievskaya G P 1959 Mos IIET 22,
240-250 (Z96, 3) [Diophantine
anal, mss, numb thy]
- 1960 Ist M Isl 13, 107-186
(MR23A, 696) [Diophantine anal,
mss, numb thy]
- 1961 Ist M Isl 14, 285-288
(Z118-1, 8) [Bertrand]
Meier J 1907 *Festekt der Universitet
Basel zur Fuer der Zucihundertstun
Geburtstager Leonhard Eulers,*
Basel, 21 p (F38, 13) [anniv]
Melnikov I G 1957 Ist M Isl 10, 211-
228 (Z99, 245) [arith]
- 1960 Ist M Isl 13, 187-216
- 1966 Ist Met EN 5, 15-30
(MR34, 2410) [Goldbach, numb thy]
Melnikov I G + 1957 Ist M Isl 10, 229-
256 (Z101, 246) [primitive roots]
Mikhailov G K 1955 SSSR Tek (1),
3-26 (MR16, 985)
- 1956 Mos M Sez, 232
- 1957 Ist M Isl 10, 67-94
(Z99, 245)
- 1957 SSSR Tek (3), 10-37
(MR19, 624; Z87, 5)
Minchenko L S 1959 Mos IIET 28, 188-20
(Z94, 4) [electricity]

Euler    (continued)

Mueller F 1907 Ab GM 25, 61-116
    (F38, 15)
- 1907 DMV 16, 185-195, 423-424
    (F38, 17)  [bib]
- 1907 Leop 43, 91-94  (F38, 17)
- 1907 Unt M 13, 97-104  (F38, 19)
- 1908 DMV 17, 36-39    (F39, 18)
- 1908 DMV 17, 313-318  (F39, 18)
- 1908 DMV 17, 333-339  (F39, 19)
    [works]
Oldfather W A + 1934 Isis 20, 72-160
    [elastic curves]
Pasquier L du 1909 Zur NGV 54, 116-
    148  [ins]
- 1910 Zur NGV 55, 14-22  [ins]
- 1927 *Léonard Euler et ses amis*,
    Paris, 134 p  (F53, 20)
Pelseneer J 1951 Bel BS (5)37,
    480-482  (MR13, 197; Z42, 242)
    [Rameau]
Petrem K 1933 Cas MF 63, 68-71
    (F59, 858)
Plakhowo N + 1906 Intermed 13, 232
    [alg, Lagrange]
Polak L S 1957 Mos IIET 17, 320-362
    (Z115, 242)  [least action]
Pringsheim A 1905 Bib M (3)6, 252
    [series]
Rabinovich I 1957 Vop IET (4), 163-
    164  (Z80, 4)
Raskin N M 1957 SSSR Tek (3), 38-48
    (Z87, 5)
Reidemeister K 1958 MP Semb 6, 4-9
    (Z86, 4)
Rocquigny G de 1895 Intermed 2, 175,
    364 [Fermat, four squares,
    heuristic]
Rozenfeld B A  1958 Ist M Isl 10,
    371-422 (Z102, 4) [geom transf]
Rudio F 1907 Zur NGV 52, 537-542
    (F38, 12)  [anniv]
- 1907 Zur NGV 52, 542-546 (F38, 12)
    [works]
- 1908 Zur NGV 53, 456-470
--1908 Zur NGV 53, 481-484  (F39, 19)
    [works]
- 1910 Int Woch 4, 86-94  (F41, 10)
    [works]
- 1912 Int Con (1912) (F43, 15)
    [works]

- 1912 Zur NGV 56, 552-557  (F43, 15)
    [works]
- 1913 Ens M 15, 59 [works]
- 1913 Zur NGV 58, 431-437
    (F44, 9)  [works]
- 1914 Zur NGV 59, 564-565 (F45, 14)
    [works]
- 1915 Zur NGV 60, 643  (F45, 14)
    [works]
- 1922 Zur NGV 67, 396-399 (F48, 6)
    [works]
- 1923 DMV 32, 13-32 [Staeckel,
    works]
- 1926 Zur NGV 71, 299-302 (F52, 21)
    [works]
Rudio F + 1913 Zur NGV 57, 596-604
    [Weber]
- 1921 Zur NGV 66, 347-360  (F48, 6)
    [works]
*Rybkin G F + ed 1957 Ist M Isl 10,
    1-424  (250th anniv series of papers)
Sanford V 1935 MT 27, 205-207
Schafheitlin P 1922 Ber MG 21, 40-44
    [ms]
- 1925 Ber MG 24, 10-13 (F51, 22)
Schering E 1880 Gott N, 489-491
    (F12, 13)  [Canterzoni, Lagrange,
    Laplace]
*Schroeder K ed 1959 *Leonard Euler.
    Sammelband der zu Ehren des 250
    Geburtstages*  Berlin, 346 p
    (MR23A, 1, 4, 433)
Schulz-Euler S 1907 *Leonhard Euler*,
    Frankfurt a M., 39 p  (F38, 19)
Sheinin O B 1965 Vop IET 19, 115-117
    (MR37, 15)  [D. Bernoulli]
Simonov N I 1954 Ist M Isl 7, 513-
    595  (Z59, 8; 60, 8) [diff eqs]
- 1955 UNM 10(4), 184-187  (Z65, 244)
    [diff eqs]
- 1956 Int Con HS 8, 146-148
    [diff eqs]
- 1956 Mos M Sez, 102-103 [diff eqs]
- 1956 Mos M Sez 3, 588-596
    (Z105, 3)  [diff eqs]
- 1957 Ist M Isl 10, 327-362
    (Z108-2, 249)  [part diff eqs]
    1957 *Methods of applied analysis in
    Euler*, Mos, 167 p  (MR19, 1247)
- 1959 Ist M Zb (1), 20-39
    [diff eqs]

Euler. (continued)

- 1959 Mos IIET 28, 138-187
  (Z99, 4) [diff eqs]
Smirnov V I 1956 Mos M Sez, 103
  [letters]
- 1957 SSSR Vest 27(3), 61-68
  (MR19, 518; Z85, 5)
Smirnov V I + ed 1963 *Leonard
  Euler, Pisma K uchenym*, Mos-Len
- 1967 *Leonard Euler. Perepiska.
  Annotirovannyi ukazatel*, Len
Sofonea T 1957 Verzek 34, 87-104
  (Z77, 242) [ins]
Sokolov Yu D 1959 Ist M Zb(1), 5-19
Speiser A 1926 Crelle 157, 105-114
  [phil, Riemann]
- 1929 Zur NGV 74, 309 (F55, 623)
  [works]
- 1934 Schw NG, 271-272 (F60, 840)
  [works]
- 1937 M Mag 12, 122-124 (F63, 817)
  [works]
- 1947 Helv CM 20, 288-318
  (MR9, 74; Z34, 147) [works]
Speziali P 1953 Stultif 10, 6-9
Spiess O 1929 *Leonhard Euler. Ein
  Beitrage zum Geistesgeschichte
  des XVIII Jahrhundert*, Leipzig,
  228 p (F55, 13)
Staeckel P 1907 Bib M (3)8, 37-68
  [sums]
- 1907 ZMN Unt 38, 300-307
  (F38,118) [elem math]
- 1909 Arc GNT 1, 1-9 [D Bernoulli]
- 1909 Arc GNT 1, 293-300 [curves]
- 1909 Zur NGV 54, 261-288
  (F40, 18) [works]
- 1910 DMV 19, 25-29 (F41, 11)
  [works]
- 1911 Bib M (3)11, 220-226
  [d'Alembert]
  1913 Annali 20, 193-200
  [arc length]
- 1913 Porto Ac 7, 207-213
  [arc length]
- 1907 Bib M (3)8, 233-306 (F38, 19)
  [Fuss, Jacobi]
-11908 *Der Briefwechsel zwischen
  C.G.J.Jacobi und P.H.von Fuss
  ueber die Herausgabe der Werke
  Leonard Eulers*, Leipzig, 196 p

Steinig J 1966 Elem M 21, 73-88
  (MR33, 5570) [idoneal numbs]
Stieda W 1931 Leip Ph 83(3)
  (F57, 29; Z2, 379) [Berlin,
  Petrograd]
Struik D J 1940 Isis 31, 431-32
  [Diderot]
Subbotin M F 1959 Vop IET 7, 58-66
  (MR22, 618) [ast]
Thiersch H 1929 Gott N 1928/29,
  264-289 + [J A Euler]
- 1930 Gott Ph HN, 193-218 [port]
- 1930 Gott Ph HN, 219-249
  [J A Euler]
- 1931 Fors Fort 7, 409-410
  (F57, 1302)
Thomas I 1957 JSL 22, 15-16
  [logic]
Timerding H E 1908 DMV 17, 84-93
  [hydrodyn]
Toeplitz O 1932 QSGM (B)2, 288
  [Euler identity]
Truesdell C 1953 Helv CM 27, 233-234
  [geom]
- 1957 Ens M (2)3, 251-262
  (MR20, 945; A79, 242) [mech]
Trusov Yu D 1963 Ivan GPI 34, 63-66
  (MR33, 5426) [diophantine anal]
- 1963 Ivan GPI 34, 67-70,
  (MR33, 5427) [diophantine anal]
Tyulina I A 1957 Vop IET (4), 34-46
  (Z98, 7) [hydrodyn]
Valentin G 1898 Bib M 12, 41-49
  (F29, 6) [bib]
- 1906 DMV 15, 270-271 (F37, 7)
  [Berlin home]
- 1907 Ab GM 25, 1-20 (F38, 121)
Van Den Broek J A 1947 AJP 15,
  309-318 (MR9, 74) [eng]
Van Veen S C 1943 Zutphen (B)12, 1-4
  (MR7, 354) [intgl]
Veslovskii N I 1957 Mos IIET 19, 271-
  281 [mech]
Vivanti G 1908 Bib M (3)9, 266
  (F39, 61) [cal]
- 1909 Bib M (3)10, 244-249
  (F40, 58) [complex var]
Wildschuetz-Jessen J L 1913 Nyt Tid
  (B)24, 69-71 (F44, 48)
Winter Eduard 1957 *Die Registres der
  Berliner Akad. der Wissenschaften
  1746-1766. Dokumente fuer das*

Euler (continued)

*Wirken Leonhard Eulers in Berlin,* Berlin, 406 p (Z85, 5)
- 1958 *Die deutsch-Gussische Begeg- nung und Leonhard Euler,* Berlin, 204 p (Z93, 6)
Yushkevich A P 1949 Mos IIET 3, 45- 116 [Prussia]
- 1954 Ist M Isl 7, 625-629 (Z59:1, 8) [Goldbach]
- 1957 Ist M Isl 10, 159-210 (Z102, 4) [circle sq]
- 1957 UMN 12(4), 3-28 (MR19, 826; Z99, 3)
Yushkevich A P + 1962-1965 *Manu- scripta Euleriana Archivi Scientiarum URSS,* 2 vols, Mos
- 1965 *Leonhard Euler und Christian Goldbach Briefwechsel 1729-1764,* Berlin, 429 p (Ber Ab (1)) (MR34, 1144)
Yushkevich A P + ed 1959-1961 *Die Berliner und die Petersburgh Akad. der Wiss. in Briefwechsel Leonhard Eulers,* Berlin, 2 vols (Quellen und Studien aus ges- chichte Osteuropas 3 (1)) (MR22, 12029; 23A1496; Z113, 1)
Zaharia N V 1933 Tim Rev M 13, 61-63 (F61, 958) [Legendre]
Zeuthen H G 1904 Bib M (3)5, 108- [Brahmagupta, Euler id]

Eutocius, V

Arendt F 1914 Bib M 14, 97-98 [Apollonius]
Sarton G 1927 Int His Sc 1, 427
Tannery P 1884 BSM (2)8, 315-329
Ver Eecke P 1954 Arc In HS 33, 131- 132
- 1956 Arc In HS 35, 217-226 [Delian probl, Eratosthenes]

Ezra, Abraham Ibn 1092-1167 .

Amir-Moez A R 1957 Scr 23, 173-178 [Karkhi]
Ginsburg J 1922 MT 15, 347-356 [combin]

Sarton G 1931 Int His Sc 2(1), 187- 189
Smith D E + 1918 AMM 25, 99-108 (F46, 53) [Arab numer]
Steinschneider M 1866 Deu Morg Z 20, 427-432 (Repr Ges Schr 1, 498-506)
- 1867 Z M Ph 12, 1-44 (Repr Ges Schr 1, 327-287)
*- 1880 Ab G M 25, 57-128 (Repr Ges Schr 1, 407-498)
Tannery P 1901 Bib M (3)2, 45-47 (F32, 4)
Terquem O 1841 JMPA

Faa di Bruno, Francesco 1825-1888

Berteu A 1898 *Vita dell'abate Francesco Faa di Bruno, fonditore del Conservatorio di N.S. del Suffragio in Torino,* Turin, 436 p (F29, 10)
Ovidio E d' 1888-9 Tor U Ann

Fabre, Jean Henri 1823-1915

Simons L G 1933 Scr 1, 208-221
- 1939 *Fabre and Mathematics and Other Essays,* NY, 101 p (F65, 15) (Scr Lib 4)

Fabri, Honoratus 1607-1688

= Honoré = Onorato Fabbri

Fellmann E A 1959 Physis Fi 1, 1-54, 73-102 (Z101, 8)
Fellmann E A + 1957 Schw NG 137, 57-61 (Z112, 242)
Lindemann F 1927 Mun Si, 273-284 (F53, 18) [Barrow, Leibniz]

Faddev, Dmitrii Konstantinovich 1907-

Venkov B A + 1958 UMN 13(1), 233-238 (Z78, 3)

Faerber, Karl

Mueller R 1912 Arc MP (3)19, 25-30
- 1912 Ber MG 11, 25-30   (F43, 40)


Fagnano, Giulio Carlo   1690-1760?

= Marquis De Toschi i S. Onorio
de Fagnani

Boncompagni B 1870 Boncomp 3, 27-46
(F2, 18)
Bopp K 1923 Isis 5, 400-402   [ell]
Candido G 1935 Pe M (4)15, 58-62
(Z10, 244)   [Chasles, generali-
zation, Pappus, Stewart]
Cantor M 1872 ZM Ph 17, 88
(F4, 15)
Conte L 1949 Iasi IP 4, 36-53
[J. Bernoulli]
- 1952 Archim 4, 214-217 (Z48, 243)
[Huygens]
- 1954 Archim 6, 37-41 (Z56, 2)
- 1956 Archim 8, 92-95, 236-238
(Z74, 7)
Court N A 1948 Scr 17, 147-150
Genocchi A 1869 *Di una formola del
Leibniz e di una lettere di
Lagrange al Conte Fagnano,*  Torino
(F2, 18)
Grunert J A 1869 Arc MP, 223-231
(F2, 18)   [Lagrange]
Harding P 1911 M Gaz 5, 68-78
[ell trammels]
Isely L 1908 Genv Arc 25, 288
Siacci F 1870 Boncomp 3, 1-26
(F2, 18)
Tenca L 1954 Bln Rn P(11)1(2),
77-87 (MR16, 985)  [Grandi]
Watson G N 1933 M Gaz 17, 5-17
(F59, 8)  [ell fun, Landen, length]


Fais, A

Fais A 1923 *Pagine autobiografiche,*
Sassari, 51 p  (F49, 13)
Usai G 1928 Bo Fir 7, xxv-xxix


Faisnier, Jean

Godeaux L 1914 Intermed 21, 161-164
(Q3976)


Fano, Gino   1871-1952

Anon 1950 Tor Sem 9, 33-45
(MR12, 382; Z37, 291)
Segre B 1952 Archim 4, 262-263
(Z47, 245)
Terracini A 1952 It UM (3)7, 485-490
(MR14, 524)
- 1953 Tor FMN 87, 350-360
(MR15, 923; Z53, 197)


Fantappiè, Luigi   1901-1956

Anon 1956 It UM (3)11, 641-645
(Z71, 3)
- 1957 Collect M 9, 3-5
(MR19, 1150; Z78, 242)
Arcidiacono G 1957 Rv M Hisp A (4)17,
14-17  (MR19, 108)
Fichera G 1957 Bari Sem 32, 24 p
(Z80, 9)
- 1957 Ren M Ap (5)16, 143-160
(MR19, 1248; Z89, 5)
Orts J M 1957 Rv M Hisp A (4)17,
3-9  (MR19, 108; Z77, 8) [anal]
Pellegrino F 1956 Archim 8, 282-285
(Z71, 3)
- 1956 Ren M Ap (5)15, 505-519
(MR19, 108; Z90, 8)
Severi F 1957 Ren M Ap (5)16, 140-142
(MR19, 1248; Z89, 5)


Farkas, Julius   1847-

Ortvay R 1927 MP Lap 34, 5-25
(F53, 35)


Faure, Henri Auguste   1825-

Rudi 1914 Intermed 21, 96 (Q4349)

**Favaro,** Antonio 1847-1922

Bortolotti E 1924 Romag Sto 14(1-3),
36p
Cajori F 1921 Nat 111, 368
Riccardi P 1893 Bib M (2)7, 64

**Favero,** Giovanni Battista 1832-

Guidi C 1907 Tor FMN 42, 318-320
(F38, 31)

**Fechner,** Gustav Theodor 1801-1887

Kuntze J E 1891 *Gustav Theodor
Fechner. Ein deutsches Gelehrten-
leben,* Leipzig, 383 p (F23, 23)
[port]
Lasswitz K 1896 *Gustav Theodor
Fechner,* Stuttgart, 215 p
(F27, 17)
Stevens S S 1961 Sci 133, 80
[psych]
Wundt W 1902 *Gustav Theordor
Fechner,* Leipzig

**Federhofer,** Karl 1885-

Girkmann K 1955 Ost Ing 9, 73-78
(Z64, 242)
Parkus H 1960 Ost Ing 14, 243-244
(Z90, 8)
Willers F A 1950 ZAM Me 30, 230
(Z36, 146)

**Federov,** Evgraf Stepanovich 1853-
1919

Ansheles O M 1956 Mos IIET 10, 13-18
(Z74, 246) [cryst]
Delone B N 1956 Mos IIET 10, 5-12
(Z74, 246) [geom]
Kavaǹko A S 1955 UMN 10(4), 193-196
(MR17, 338) [anniv]
Ratskin N M + 1956 SSSR Vest (1),
71-77
Shafranovskii I I 1956 Mos IIET 10,
28-65 (Z74, 246)

**Fehr,** Henri 1870-1954

Anon 1955 Ens M (2)1, 5-17
(MR17, 698)
Ruffet J 1955 Elem M 10, 1-4
(MR16, 434)

**Fejer,** Leopold 1880-1959

Aczél J 1961 Pub M 8, 1-24
(MR24A, 128)
Anon 1950 M Lap 1, 267-272
(MR12, 311) [bib]
- 1959 Bud UM 2, 3-4 (MR22, 770)
- 1959 Gaz MF (A)11, 687-688
(Z89, 5)
- 1959 Hun Act M 10, 249-250
(Z87, 243)
- 1959 Koz M Lap 19, 81-82
(MR24A, 571)
Perron O 1960 Mun Jb, 169-172
(MR22, 1131)
Polya G 1961 LMSJ 36, 501-506
(Z98, 9)
Szasz P 1960 Magy MF 10, 103-147
(MR24A, 129; Z100, 247)
Szego G 1960 AMSB 66, 346-352
(MR22, 929; Z89, 242)
- 1961 M Lap 11, 225-228
(Z96, 4)
Turan P 1950 M Lap 1, 160-169
(MR12, 1)
- 1960 M Lap 11, 8-18
(MR23A, 136; Z97, 2)
- 1960 UMN 15(4), 111-122 (Z93, 9)

**Fekete,** Michael 1886-

Anon 1958 M Lap 9, 1-5
(MR20, 848; Z80, 9) [bib]
Balazs J 1958 M Lap 9, 197-224
(MR21, 1023; Z84, 3)
Rogosinski W W 1958 LMSJ 33, 496-500
(MR20, 1138; Z82, 243)

**Felici,** Riccardo 1819-1902

Pitoni R 1902 Pe MI (2)5, 69-72
(F33, 41)

Felix, Vaclav

Nachtikal F 1933 Cas MF 62, 267-271
  (F59, 41)


Ferenc, Lukacs  1891-

Balint E 1962 M Lap 13, 1-8
  (Z115, 246)


Fergola, Emanuele  1830-1915

Jadanza N 1915 Tor FMN 50, 721-736
Millosevich E 1915 Linc Rn (5a)
  24(2), 411-417
Pinto L 1915 Nap FM Ri (3a)21, 121-
  126


Fergola, Nicolo  1752-1824

Amodeo F 1903 Nap Pont (2)8(11), 32 p
  (F34, 11)
Loria G 1892 Geno U, 144 p (F24, 18)
- 1893 BSM (2)17, 237-240
- 1893 Rv M 3, 179-186


Fermat, Pierre de  1601-1665

Baltzer R 1879 Crelle 172 (F11, 17)
Bell E T 1949 Scr 15, 162-163
  (Z34, 146)  [Wallis]
Bjerkeseth H 1943 Nor M Tid 25,
  40-45  (MR8, 189)  [quad indet eq]
Blanquiere Henri + 1957 *Un mathématicien de génie, Pierre de
  Fermat...*, Toulouse, 88 p
Bosmans H 1920-21 Brx SS 40(A),
  135-141  (F48, 4) [Giovannozzi]
- 1923 Rv Q Sc (4)3, 422-441  [mss]
Boyer C B 1944 MT 37, 99-105
  [anal geom, Descartes]
- 1945 M Mag 20, 29-32 (Z60, 10)
  [intgl]
- 1947 Scr 13, 133-53 [anal geom]
- 1952 Scr 18, 189-217
  (MR14, 609)  [Descartes]
Brassine E 1843 *Précis des Oeuvres
  mathématiques de Pierre Fermat et
  de l'arithmétique de Diophante*,
  Toulouse
- 1853 Tou Mm 3, 1-
Brocard H 1920 *Congres des mathé-
  maticiens de Strasbourg, comptes
  rendus*, 621
Bussey W H 1918 AMM 25, 333-337
  [induc]
Carcavi 1665 J Sav (9 Feb)
  [anal geom]
Child J M 1920 Isis 3, 255-262
  [Pellian eq]
Conte L 1938 Pe M (4)18, 50-54
  (F64, 916)
- 1942 Pe M (4)22, 70-90
  (F68, 15) [Castillon, Newton,
  parabola]
- 1952 Archim 4, 126-129 (Z47, 244)
Cowan R W + 1947 Scr 13, 123-127
  (Z29, 194)  [calc]
De Vries H 1916-17 N Tijd 4, 145-167
  (F46, 55)  [Descartes]
De Waard C 1918 BSM (2)42, 157-177,
  327-328  (F46, 38) [calc, Descartes]
- 1920 Tou Mm (11)5, 71-88
  [letters]
Dourmain C 1925 *Halley and Fermat*,
  Breslau, 118 p
Eves H 1960 MT 53, 195-196  [induc]
Galan G 1911 Esp SM 1, 1-3
  (F42, 5)
Genocchi A 1884 Mathesis 4, 106-108
  [ms]
Genty Abbé 1784 *L'influence de
  Fermat sur son siècle, relative-
  ment aux progrès de la haute
  géométrie et du calcul...*,
  Orleans
Giovannozzi G 1917 Vat NLA (a)71,
  51-56
- 1919 Arc Sto 1, 137-140  [calc]
- 1920 It SPS 10, 468-471 (F47, 5)
  [letter]
Gram J 1909 in *Festskrift til H. G.
  Zeuthen*, Copenhagen, 48-62
  [Diophantus]
Henry C 1879 Boncomp 12, 477-568,
  619-740 (F11, 16) [Bachet,
  Malebranche]
- 1881 Boncomp 13, 437-470  (F13, 16)
  [Bachet, Malebranche]
- 1883 Linc At (3)7, 39-40

Hoffmann J E 1944 Ber Ab(7), 19 p
  (MR8, 305; Z60, 10)
- 1943 Ber Ab(9), 52 p (MR8, 305)
- 1961 Arc HES 1, 122-159
  (Z94, 3) [Euler]
- 1963 M Ann 150, 45-49 [four cubes]
Itard J 1948 Arc In HS, 589-610
  [calc]
- 1950 *Pierre Fermat,* Basel, 24 p
  (MR13, 2; Z38, 147) (Elem M (S))
- 1950 Rv Hi Sc Ap 3, 21-26
  [numb thy]
Kummer E E 1850 Nou An 9, 386-
Loria G 1905 Bib M (3)6, 343-346
  (F36, 60) [contact transf]
- 1930 BSM (2)54, 245-254 (F56, 19)
  [competition, Frénicle]
Machaby Armand 1949 *La philosophie
  de Pierre de Fermat,* Liège, 127 p
Milhaud G 1917 Rv GSPA 28, 332-337
  [Descartes, tangents]
Mouchez 1882 Par CR 95, 399-402
Oblath R 1953 M Lap 4, 18-30
  (MR17, 1055)
Omont H 1918 J Sav 16, 321-323
  [d'Athenée, Théodoret]
Palamà G 1953 It UM (3)8, 414-422
  (Z51, 243)
Reeve W D 1931 MT 24, 512-513
Ritter F 1880 BSM (2)4, 171-182
  (F12, 11) [letter, Roomen]
Rocquigny G de 1894 Intermed 1, 220
- 1895 Intermed 2, 175, 364
  [Euler, four sq]
- 1906 Intermed 13, 264
Rocquigny G de + 1898 Intermed 5, 166
Sanford V 1932 MT 24, 512-513
Sarantopoulos S 1935 Gree SM 16,
  201-240 (F61, 940) [last thm]
Schaefer J 1926 DMV 35, 129-137
  (F52, 18)
Schaewen P von 1908 Bib M (3)9,
  289-300 (F39, 58)
Simon M 1913 Arc MP (3)21, 300
  (F44, 39)
Sister M Stephen 1960 MT 53, 193-195
Takekuma R 1953 JHS (26) (July)
  [Pascal, prob]
Tannery P 1883 BSM (2)7, 116-128
Vacca G 1894 Bib M (2)8, 46-48
  (F25, 64)

- 1927 Tor FMN 63, 241-252
  [irrat]
Walsh C M 1928 An M (2)29, 412-432
  [Pyth triples]
Wertheim G 1899 Hist Abt 44, 4-7
  (F30, 40)
- 1899 ZM Ph 44(S), 555-576
  (F30, 40) [Wallis]
Western A E 1957 M Gaz 41, 56-57
  (MR18, 784)
Wieleitner H 1929 DMV 38, 24-35
  [deriv, lim]


Fernel, Jean   1497-1558

Wilson W G 1947 Life Let (Oct), 18-26


Ferrari, Ludovico   1522-1565

Bortolotti E 1926 Bln Rn 30, 51-62
  (F52, 14) [compet, Tartaglia]
Giordani E 1876 *I sei cartelli di
  matematica disfida primamente
  intorno alla generale risoluzione
  delle equazioni cubiche di
  Ludovico Ferrari...,* Milan
  (F11, 20)
Notari V 1924 Pe M (4)4, 327-334
  [quartic eq]
Pasquale L di 1957 Pe M (4)35,
  253-278 (Z79, 241) [compet,
  Tartaglia]
- 1958 Pe M (4)36, 175-198 (Z79, 241)
  [compet, Tartaglia]
Vooys C J 1959 Euc Gron 34, 200-204
  (Z85, 4)


Ferrel, William 1817-1891

Abbe C 1895 Wash PS 12, 448-460
  (F26, 3) [meterology]
Anon 1891 Nat 64, 527-528   (F23, 29)


Ferro, Scipione dal   1465?-1525

Frati L 1910 Loria 12, 1-5 (F41, 4)
  [cubic]

Ferroni, Pietro XVIII

Anon 1830 Fir Atti 7, 33-


Fester, Diderich Christian 1732-1811

Lyche R T 1935 Nor VSF 7, 26-38
  (F61, 951) [Norway]


Feuerbach, Karl William 1800-1834

Cantor M 1910 Heid Si (25), 18 p
  (F41, 13)
Guggenbuhl L 1955 Sci Mo 81, 71-76
Kiefer A 1911 *Die Einfuhrung der
  homogener Koordinatan durch K. W.
  Feuerbach,* Strassburg, 55 p
  (F42, 70) [homog coord]
Quint N 1905-1908 Intermed 12, 265;
  13, 253; 15, 184 (Q2970)
  [F thm]


Feys, Robert

Anon 1949 Syn Dor 7, 447-452
  (MR11, 708)


Fibonacci XIII

  = Leonardo da Pisa

Agostini A 1949 Archim 1, 114-121
  (Z39, 4)
- 1949 It UM (3)4, 282-287
  (MR11, 572; Z35, 146) [letter]
- 1953 Archim 5, 205-206 (MR15, 276)
Babini J 1941 Archeion 23, 57-70
  (MR3, 97)
Bellacchi G 1892 Be MI 7, 81-88,
  113-118, 169-171 (F24, 48)
- 1893 Pe MI 8, 23-28, 113-116, 137-
  144 (F24, 48)
Boncompagni B 1851-52 Vat NLA 5, 5-,
  208-
Bortolotti E 1924 Pe M 4, 134-139
- 1930 Per M (4)10, 230-234 [bib]
- 1930 Bln Mm (8)7, 91-101
  (F57, 1298) [Arab]

- 1931 Pe M (4)11, 211-217 (Z2, 6)
  [cont frac, pi]
- 1939 It SPS 27(6), 521-533
  (F65, 1081)
Brooke M 1962 Recr M Mag (7), 42-46
  (Z102, 1)
Carruccio E 1939 Pe M (4)19, 189-197
  (F65, 8) [cube root]
Cassina U 1924 Tor FMN 59, 14-29
  [cubic]
Chasles M 1841 Par CR 12, 741-
  [alg, mistake]
Cossali P 1857 in Boncompagni B ed,
  *Scritti inediti del P. D. Pietro
  Cossali,* Rome, 1-288
Curtze M 1896 ZM Ph 41, 81 [alg]
Dunton M + 1966 Fib Q 4, 339-354
  (MR34, 7318) [Egypt]
Durazzini 1768 *Elogio di Leonardo
  Pisano,* Florence
Eisele C 1951 Scr 17, 236-259
  (MR13, 612; Z44, 246) [C.S. Pierce]
Eneström G 1905 Bib M (3)6, 214-215
  (F36, 65) [Jordanus Nemorarius]
Fazzari G 1895 Pitagora 1, 2, 10
  [quad eq]
Frajese A 1940 It UM (2)2, 363-365
  (F66, 12) [anal geom]
Gegenbauer L 1893 Mo M Phy 4, 402
  (F25, 58)
Genocchi A 1855 Tortolin 6, 161-,
  186-, 218-, 251-, 273-, 345-
Gerhardt C I 1842 Arc MP 2, 423-
Gram J P 1893 Dan Ov, 18-28
  (F25, 62)
Guglielmini G B 1813 *Elogio di
  Leonardo Pisano,* Bologna
King C 1963 Fib Q 1(4), 15-19
  (Z113, 1)
Lazzarini M 1903 Loria 6, 98-102
  (F34, 6, 35, 6)
- 1904 Loria 7, 1-7
Lebesgue V A 1876 Boncomp 9, 583-594
  (F8, 13)
Loria G 1924 Pe M (4)4, 131-134
McClenon R 1919 AMM 26, 1-8
Pfeifer G 1886 ZMN Unt 17, 250-254
Schutter C 1952 MT 45, 605-606
Terquem O 1856 Tortolin 7, 106-
Vacca G 1930 It UM 9, 59-63
  (F56, 15) [cubic eq, Euclid]

Vetter Q 1927 Tor FMN 63, 296-299
  [cubic eq]
- 1928 Cas MF 58, 149-151
  (F55, 600)  [cubic eq]
Vogel K 1940 Deu M 5, 219-240
Vooys C J 1958 Euc Gron 34, 108-09
  (Z85, 2)
Weinber J 1935 Scr 3, 279-281
  [John of Palermo]
Weissenborn H 1894 Ber CPA 14
  (F25, 80)  [Archimedes]
Woepcke F 1854 JMPA 19, 401-
  [cubic eq]
- 1856-1861 Vat NLA 10, 236; 12,
  230-, 399; 14, 211-, 241-, 301-,
  343-
Zeuthen H G 1893 Dan Ov, 1-17, 303-
  341  (F25, 62)  [cubic eq]


Fiedler, Otto Wilhelm  1832-1912

Brown B H 1925 AMM 32, 517
  ["cyclography"]
Fiedler E 1915 Deu Biog 17, 14-25
Grossman M 1913 Schw NG, 8 p
  (F44, 23)
Kollros S 1913 Ens M 15, 66-68
  (F44, 23)
Voss A 1913 DMV 22, 97-113
  (F44, 23)  [port]


Fields, John Charles  1863-1932

Fehr H 1933 Ens M 31, 127 (F59, 39)
- 1933 Ens M 31, 279 (F59, 45)
  [Fields medal]
Hopkins F G 1933 Lon RSP (A)139, 7
  (F59, 39)
McLennan J C 1932 Nat 130, 688-689
  (F58, 49)
Parks W A 1932 Univ.of Toronto
  Monthly 33(Nov), 56
P D 1932 It UM 11, 310-311
Synge J L 1933 LMSJ 8, 153-160
  (F59, 39)
*- 1933 Lon RS Ob(2), 131-138
  [bib, port]
Wallace W S ed 1949 Royal Canadian
  Institute Centennial Volume
  (1949-1949), Toronto, 163, 291

Fikhtengolts, Grigorii Mikhailovich
  1888-

Kantorovic L V + 1958 Len MV 13(7),
  5-13  (MR20, 1042; Z79, 5)
- 1959 UMN 14(5), 123-128
  (MR22, 260; Z87, 7)


Filipescu, G. E.

Ionescu I 1938 Gaz M 43, 281-289,
  337-342  (F64, 23)


Filippis, Vincenzo de  XVIII

Marcolongo R 1909 Int Con 3, 488-499
  [Lagrange, mech]


Filon, Louis Napoleon George
  1875-1937

Jeffery G B 1938 LMSJ 13, 310-318
  (F64, 23)
- 1938 M Gaz 22, 1-2  (F64, 919)


Fine, Henry Burchard  1858-1928

*Archibald R C 1938 A Semicentennial
  History of the American Mathemati-
  cal Society 1888-1938, NY, 167-
  [bib, port]
Veblen O 1929 AMSB 35, 726-730
  (F55, 20)


Finikov, Sergei Pavlovich  1883-

Laptev G 1954 UMN 9(3), 245-252
  (MR16, 207)


Fink, Karl  1851-1899

Fleischmann L 1899 DMV 7, 33-35
  (F30, 19)

Finkel, Benjamin Franklin
1865-1947

Cairns W D 1947 AMM 54, 311-312
(Z29, 2)


Finlaison, John

Anon 1863 Assur Mag 10, 147-


Finsterwalder, Sebastian   1862-1951

Clauss 1932 Z Vermess 61, 722-726
(F58, 993)
Graf H 1953 DMV 56, 27-31
(MR15, 89; Z50, 2)
Kneissl M 1942 Bildmess 17, 53-64
(F68, 18)


Fiorini, Matteo

Jadanza N 1901 Tor FMN 36, 416-418
(F32, 22)


Fisher, George  XVIII

Karpinski L C 1935 Scr 3, 337-339


Fisher, Irving   1867-1947

Lubin I + 1947 ASAJ 42, 1-4
Sasuly M 1947 Ecmet 15, 255-278
(Z29, 195)


Fisher, Ronald Alymer   1890-1962

Anon 1963 Clct St 12, 1-3 (Z112, 3)
- 1963 Lon St (A)126, 159-178
[*bib]
Bliss C I 1962 Biomtcs 18; 437-454
(Z106, 3)
*Kendall M G 1963 Biomtka 50, 1-15
(MR27, 900; Z109, 5)
Linder A 1962 Metrika 5, 141-144
(Z106, 3)

Mahalanobis P C 1938 Sankh 4(2),
265-272 (repr in R A Fisher 1950
*Contributions to Mathematical
Statistics* (Wiley)).
- 1964 Biomtcs 20, 238-251
(MR30, 373)
Neyman J 1938 Biomtka 30, 11-15
[K. Pearson]
Pearson E S 1968 Biomtka 55(3), 445-457
[Gosset, K. Pearson]
*Yates F + 1963 Lon RSBM 9  [*bib]
Youden W J 1962 ASAJ 57, 727-728
(MR25, 740)


Fiske, Thomas Scott   1865-1944

Archibald R C 1938 *A Semicentennial
History of the American Mathematical Society 1888-1938,* NY, 151-
153 [bib, port]


Fitting, Hans   1906-1938

Zassenhaus H 1939 DMV 49(1), 93-96
(MR1, 34; F65:1, 17; Z21, 196)


Fizeau, Hippolyte   1819-1896

Picard E 1922 Par Mm (2)58, i-lvi
[optics]


Fladt, Kuno   1889-

Raith F 1959 MN Unt 12, 84-85
(Z100, 247)


Flauti, Vincenzo

Amodeo F 1920 Nap Pont (2)25, 113-141
- 1921 Nap Pont (2)26, 55-57
[Trudi]
- 1921 Nap Pont 51, 81-118

**Flechsenhaar,** A.

Lietzmann W 1933 ZMN Unt 64, 87-88

**Fletcher,** William Charles  1865-1959

Siddons A W + 1959 M Gaz 43, 85-87

**Foeppl,** August Otto  1854-1924

Foeppl A 1925 *Lebenserinnerungen,*
  Munich, 160 p   (F51, 29)
Foeppl I 1924 ZAM Me 4, 530-531
Schlink W 1923 ZAM Me 3, 481-483

**Foglini,** P. G.

Carrara B 1907 *Il P. G. Foglini,*
  *S. J.,* Prato, 43 p  (F38, 36)

**Folie,** François  1833-1905

Godeaux 1942 Bel Anr 108, 1-33
Heen P de 1905 Bel Bul, 60-63
Le Paige 1905 Rv Q Sc (3)7, 698-700
  (F36, 35)

**Folkierski,** W

Dickstein S 1904 Wiad M 8, 164-169
  (F35, 36)

**Fontana,** Giovani 1395?-1455?

Birkenmajer A 1932 Isis 17, 34-53
Tenca L 1956 Lom Rn M 90, 547-558
  (Z71, 244)

**Fontené,** Georges  1848-1923

Bricard R 1923 Nou An (5)1, 361-363
  (F49, 13)

**Fontenelle,** Bernard Le Bovier
1657-1757

Petronievics B 1917 Linc Rn (5)26,
  309-316  (F46, 44)  [inf]

**Fontès,** M. 1842-1902

Rouquet V 1904 Tou Mm (10)4, 344-355
  (F35, 23)

**Forcadel,** Pierre  1560-1573

Fontes 1893 Fr AAS 22, 236-243
  (F25, 64)  [binom coeffs,
  Pascal, Stifel]
- 1894 Tou Mm (9)6, 282-296
  (F26, 11)
- 1895 Tou Mm (9)7, 316-346
  (F26, 12)
- 1896 Tou Mm (9)8, 361-382
  (F27, 7)
Mogenet J 1950 Arc In HS 29, 114-128
  (Z34, 145)

**Forchheimer,** Philipp  1852-1933

Terzaghi K 1932  Fors Fort 8, 288
  (F58, 995)

**Forestier,** Charles

Legoux A 1908 Tou Mm (10)8, 183-190

**Forni,** Luigi

Vivanti G 1915 Loria 17, 122-125
  (F45, 19)

**Forster,** Gustav

Anon 1932 Allg Verm 44, 113

Forsyth, Andrew Russell   1858-1942

A W S 1942 M Gaz 26, 117-118
Chapman S + 1942 Nat 150, 49-40
   (MR3, 258; Z60, 14)
Neville E H 1942 LMSJ 17, 237-250
   (F68, 18; Z60, 15)
Piaggio H T H 1930 M Gaz 15, 461-465
Whittaker E T 1942 Lon RS Ob 4, 209-
   227 (Z60, 15)


Forti, Angelo

Anon 1928 *Intorno alla vita e alle
   opere di Angelo Forti*, Pisa,   26 p
   (F54, 35)
- 1929 Bo Fir (2)8, xxxiv  (F56, 813)


Fotheringham, John Knight
1874-1936

Sampson R A 1937 Isis 27, 485-492
   (F63, 19)
Singer C 1937 Archeion 19, 172-174
   (F63, 816)


Foucault, Jean Bernard Léon
1819-1868

Anon 1879 BSM (2)3, 353-380
   (F11, 24)
Freudenthal H 1951 MP Semb 5, 230-238
   (MR19, 264)  [diff geom]
Moigno F 1878 Mondes (2)47, 181-182
   (F10, 18)


Fourier, Jean Baptiste Joseph
1768-1830

Arago D F J 1838 An Ch Ph 67, 337-
- 1838 Par Mm 14: lxix-;  (English
   tr in Arago 1857 *Biographies of
   Distinguished Scientific Men,*
   London, 242-286.  Repr 1871 Smi
   R, 137-176
Bose A 1817 Clct MS 7, 33-48

Budan de Bois-Laurent F F D 1832
   *Quelques observations sur l'oeuvre
   posthume de Fourier, et sur
   l'avertissement placé en tête par
   son éditeur,* Paris, 8 p
Cajori F 1911 Bib M (3)11, 132-137
   [approx, Mouraille, Newton-Raphson]
Cousin 1831 *Notes Biographiques,*
   Paris
De Morgan A 1869 Lon Actn 14, 89-
   [stat]
Doublet E 1930 Rv Sc 68, 405-407
   (F56, 22)
Gilbert P 1888 Rv Q Sc 24, 245-254
Jourdain P E B 1913 Int Con (5)2,
   526-527  (F44, 48)  [phil]
Labra M 1943 Cub SCPMR 2, 24-26
   (MR7, 355; Z60, 11)
Larmor J 1933 Phi Mag (7)17, 110-116
   [discontinuities]
- 1934 Phi Mag (7)17, 668-678
   (F60, 817)  [discontinuities]
*Ravetz J 1960 Int Con HS 9, 574-577
   (Z114, 7)  [physics]
- 1961 Arc In HS 13, 247-251
   (Z106, 3)
San Juan R 1930 Rv M Hisp A (2)5,
   134-137  (F56, 22)
*Van Vleck E B 1914 Mathes It 6,
   157-174  (F45, 75)  [series]
Wagner K W 1930 Elek Nach 7, 173
   (F56, 819)


Fowler, Alfred   1868-1940

Dingle H 1940 Obs 63, 262-267
   (MR2, 165)


Fowler, Ralph Howard   1889-1944

Milne E A 1944 LMSJ 19, 244-256
   (Z60, 15)


Fowler, Robert Nicholas   1828-1891

Anon 1891 LMSP 22, 476-481 (F23, 29)

**Français** (Family)

Eneström G 1903 Bib M (3)4, 241-292
(F34, 12)


**Francesca,** Piero della 1420?-1492

= Piero dei Franceschi

Coolidge J L 1949 *The Mathematics
of Great Amateurs* (Clarendon Pr),
Ch 3 (Repr 1963 Dover)
[perspective]
Pittarelli G 1904 Int Con H 12, 251-
266 (F35, 64)
- 1909 Int Con 3, 436-440
(F40, 7)
Vacca G 1920 Nap FM Ri (3a)26,
232-236 [alg, Archimedes]


**Francesco da Sole**

Riccardi P 1877 Boncomp 10, 407-432
(F9, 3)


**Francesco,** Galigai

Agostini A 1953 Pe M (4)31, 201-206
(MR15, 276; Z50, 242)


**Franchis,** Michele de 1875-1946

Anon 1946 Annali (4)25, xvii-xix
(Z61, 5)
Severi F 1947 Rv M Hisp A(4)7, 279-
281 (Z33, 4)


**Franck,** Paul Albert Richard 1874-1936

Franke W 1937 Ham MG 7, 357-362
(F63, 19)


**Franco of Liège** XI

Sarton G 1927 Int His Sc 1, 757-758
[bib]


**Francoeur,** Louis Benjamin 1773-1849

Brocard H 1910 Intermed 17, 107
(Q3177) (Also Intermed 14,
166-167; 15, 137)
Smith D E 1912 AMM 28, 254


**Franke,** Walter

Burau W 1966 Ham MG 9(3), 26-29
(MR33, 3872)


**Franklin.** See also Ladd Franklin.


**Franklin,** Benjamin 1706-1790

Anon 1891 Nat 43, 39-40 (F23, 20)
*Cohen I Bernard 1956 *Franklin and
Newton,* Philadelphia, 657 p
(APS Mm 46)
Feldmann R W Jr 1959 MT 52, 125-127
Govi G 1889 Linc Rn 5(1), 138-142
Marder Clarence C 1940 *The Magic
Squares of Benjamin Franklin,* NY,
36 p
Robbins C K 1950 M Mag 24, 55-57
[Collinson, letter]


**Franklin,** Fabean 1853-1939

Wilson A H 1939 Scr 6, 121-23


**Fraser,** Peter 1880-1958

Hodge W V D 1959 LMSJ 34, 111-112
(MR20, 945; Z82, 243)


**Frattini,** Giovanni 1852-1925

Marcolongo R 1926 Bo Fir (2)5, 41-
48, 122 (F52, 31)
Teofilato P 1926 Vat NLA 79, 114-116
(F52, 31)

**Fréchet,** Maurice René 1878-

Dubreil P 1956 Rv GSPA 63, 193-195
Fréchet M 1933 *Notices sur les travaux scientifique,* Paris, (supplement 1951)

**Frederick the Great** XVIII

Cajori F 1927 AMM 34, 122-130 (F53, 20)

**Fredholm,** Eric Ivar 1866-1927

Anon 1927 Ro SSM 30, 61 (F53, 23)
- 1928 Act M 51, i-ii
Birkeland R 1927 Nor M Tid 9, 127-135 (F53, 33)
Labrador J F 1958 Gac M 9, 217-220 (Z83, 245)
Norlund N E + 1928 Act M 51, I-II (F54, 42)
Pleijel Ake 1956 Nor M Tid 4, 65-75 (Z72, 246)
Scharing F 1927 Sk Akt 10, 160-163 (F53, 33)
Sretenskii L N 1966 Ist Met EN 5, 150-156 (MR34, 15)
Zeilow N 1930 Act M 54, i-xvi (F56, 27) [intgl eq]

**Freeman,** Alexander 1838-1897

Anon 1897 LMSP 28, 586-587 (F28, 28)

**Frege,** Gottlob 1848-1925

Anscombe G E M + 1961 *Three Philosophers,* Oxford
Beumer M G 1947 Sim Stev 25, 146-149 (MR9, 74; Z29, 195)
Biryukov B V 1964 *Two Soviet Studies on Frege* (Reidel), 101p
Black M 1950 Phi Rev 59, 77-93, 202-220, 332-345
Dummett M 1956 Phi Rev 65, 229 [fun]
*- 1967 Enc Phi [bib]

Egidi R 1962 Physis Fi 4, 5-32 (Z118-1, 10) [logic, phil]
- 1963 Physis Fi 5, 129-144 (Z118-1, 10) [logic, phil]
Jourdain P E B 1912 QJPAM 43, 219
Papst Wilma 1932? *Gottlob Frege als Philosoph,* Berlin, 51 p
Scholz H + 1936 Act Sc Ind (395), 24-30 (F62, 1038)
Steck M 1941 Heid Si (2) (MR11, 150) [geom]
Sternfeld R 1966 *Frege's Logical Theory* (S. Illinois Univ P), 200 p
Wells R S 1951 Rv Met 4, 537-573
Wienpahl P D 1950 Mind 59, 483-494

**Fremberg,** Nils Erik 1908-1952

Pleijei A 1953 Nor M Tid 1, 7-9 (MR14, 1050; Z51, 243)

**Frénicle de Bessy,** Bernard 1605-1675

Hofmann J E 1943 Ber Ab (9)
Loria G 1930 BSM 54, 10 p [Fermat]

**Fresenius,** Carl Remigius 1818-1876

Oppel 1876 ZMN Unt 7, 520-524 (F8, 24)

**Fresnel,** Augustin 1788-1827

Brillouin M 1928 An Physiq (10)9, 5-34 (F54, 33)
Fabry C 1927 Rv I Ens 81, 321-345 (F52, 25)
- 1928 Rv Sc, 33-42
Picard Emile 1927 *Ceremonie du centenaire de la mort de Fresnel,* Paris, 35 p

**Freud,** Geza 1922-

Anon 1959 M Lap 10, 142-144 (Z86, 5)

Fridman, Aleksandr Aleksandrovich
1888-1925

Anon 1959 Mos IIET 22, 324-388
   (Z95, 4)
Flicker H v 1925 ZAM Me 5, 526-527
   (F51, 32)
Gavrilov A F 1959 Mos IIET 22, 389-
   400  (Z95, 4)
Loskutov K N 1963 in *Ocherki istorii
   matematiki i mekhaniki,* Moscow,
   228-
Steklov V A 1927 Len Sb Gf 5(1),
   7-8  (F53, 35)


Friedl, Werner  1893-1936

Alder A 1938 Schw NG 119, 421-425
   (F65, 1091)
H W 1936 Schw Vers 32, III-XV
   (F62, 1042)
Lorey W 1937 Bla Versi 37, 70-74
   (F63, 19)


Friedlein, Johann Gottfried
1828-1875

Bobynin W W 1904 Fiz M Nauk (2)1(11),
   324-331  (F35, 39)
Cantor M 1873 Hist Abt 18, 85-86
   (F5, 1)
- 1875 Hist Abt 20, 109-113
   (F7, 13)
- 1876 Boncomp 9, 531-536  (F8, 22)
   [bib]


Frigyes, Riesz  1880-1956

Anon 1956 M Lap 7, 1-9  (MR20, 848)
Szokefalvi-Nagy B 1956 Magy MF 6,
   143-156  (Z74, 246)


Frisch, Ragnar

Arrow K J 1960 Ecmet 28(2), 175-192


Frisi, Paolo  1728-1784

Boffito G 1933-1935 *Scrittori
   Barnabiti...,* Florence 2, 73-96;
   4, 381-397
Enestrom G 1885 Bib M 3, 174  [ast]
Favaro A 1895 Tor FMN 31, 138
   [Lagrange, letter]
Massoti A 1943 Lom Rn M(3)7, 301-315
   (MR8, 306; Z60, 11)


Frobenius, Ferdinand Georg  1849-1917

Siegel C L 1968 in Frobenius 1968
   *Ges. Abh.,* New York (Springer 1,
   iv-vi


Frola, Eugenio

Geymonat L 1963 Tor FMN 97, 986-997
   (Z114, 8)


Frost, Percival  1817-1898

Anon 1898 Lon RSP 64, XVII
Halsted G B 1899 AMM 6,  189-191
Taylor H M 1898 LMSP 29, 726-727
   (F29, 19)


Frullani, Giuliamo  1795-1834

Agostini A 1926 Arc Sto 7, 209-215
   [series]
- 1931 Pe M (4)29, 241-248
   (MR13, 420)  [series]


Fubini, Guido  1879-1943

Dodge C W 1969 MT 62, 44-46
Segre B 1954 Linc Rn (8)17, 276-294
   (Z55, 244)
Terracini A 1944 Arg UMR 10, 27-30
   (MR6, 141; Z60, 15)
- 1950 Tor Sem 9, 97-123
   (MR12, 382; Z37, 291)  [diff geom]
- 1951 It UM Con 3, 41-44 (Z45, 147)
   [diff geom]

Fuchs, Immanuel Lazarus  1833-1902

Anon 1902 Nat 66, 156
Cerruti V 1902 Linc Rn (5)11, 397
Denizot A 1902 Wiad M 6, 245-251
  (F33, 42)
Hamburger M 1902 Arc MP (3)3, 177
- 1902 *Gedaechtnisrede auf I.L.Fuchs,*
  Leipzig-Berlin, 16 p  (F33, 41)
Heffter L 1950/51 Ber MG, 5-15
  (Z46, 3)
Jordan C 1902 Par CR 134, 1081-1083
  (F33, 42)
Loria G 1902 Loria 5, 126-127
  (F33, 43)
Schlesinger L 1905 Int Con 3, 543-544
  (F36, 23)
- 1920 DMV 29, 28-40  (F47, 10)
Voit C 1903 Mun Si 33, 512-515
Wilczynski E J 1902 AMSB (2)9, 46-49
  (F33, 42)


Fueter, Karl Rudolph  1880-1950

Burckhardt J J 1950 Zur NGV 95, 284-
  287  (MR12, 382; Z40, 290)
Fueter R 1950 Elem M 5, 99-104
  (MR12, 311; Z37, 291)
Speiser A 1950 Elem M 5, 98-99
  (MR12, 311)


Fuhrmann, Wilhelm  1833-1904

Kostka C 1905 ZMN Unt 36, 68-71
  (F36, 26)
Saalschuetz L 1905 DMV 14, 56-60
  (F36, 25)


Fujiwara, Matsusaburo  1881-1946

Anon 1942 Toh MJ 49, 133-138
  (Z60, 15)  [bib]
Kubota T 1949 Toh MJ (2)1, 1-2
  (MR11, 573; Z41, 3)
Sunouchi G 1952 Toh MJ (2)4, 316-317
  (MR14, 832)  [bib]

Funk, Paul  1886-

Basch A 1956 Ost Ing 10, 117-119
  (MR18, 182)


Furtwaengler, Phillip  1869-1940

Hofreiter N 1940 Mo M Phy 49, 219-225
  (MR2, 115; Z23, 196)
Huber A 1940 DMV 50, 167-178
  (MR2, 306; Z24, 49)


Fuss, Nikolaus Ivanovich  1755-1825

Lysenko V I 1960 Vop IET 9, 116-120
  [geom]


Fuss, Paul Heinrich  1797-1855

Staeckel P 1907 Gott N, 372-373
  (F38, 22)  [letter, Gauss]
Staeckel P + 1907 Bib M (3)8, 233-306
  [Euler, Jacobi]


Gabriel, Robert Mark  1902-1957

Verblunsky S 1958 LMSJ 33, 125-128
  (MR19, 1150; Z39, 5)


Gagaev, Boris Mikhailovich  1897-

Salechov G S + 1949 UMN 4(3), 177-179
  (Z32, 98)


Galdeano, Don Zoel Garcia de  XX

Gallego Armesto H 1925 *Matematicos
  espanoles contemporaneio,* Santiago
  (F51, 38)
Pineda P 1924 Rv M Hisp A 6, 97-103

Galerkin, Boris Grigorevich 1871-

Anon 1941 Pri M Me 5, 331-334
(Z60, 15)
Gregoriev 1941 SSSR Tek (4), 115-120
(Z25, 147)
Sokolovskii V V 1951 SSSR Tek, 1159-
1164 (MR13, 197)


Galileo Galilei 1564-1642

Amodeo F 1914 Nap Pont (2)44, 23 p
Anon 1879 Nat 21, 40, 58 [applics]
- 1893 ZM Ph 38, 197-198 [anniv]
- 1932 MT 24, 118-120
Artom E 1928 Bo Fir, 50-52 [ratio]
Bellacchi G 1891 Galileo ed i suoi
successori, Florence, 29 p
(F23, 12)
Berthold G 1896 ZM Ph 42, 5 [Eppur
si muove]
- 1897 Bib M, 57
Biancoli B 1922 Arduo 2, 18-22, 130-
135
Blaschke W 1943 Ham M Einz 39
(MR10, 668; Z28, 99) [Kepler]
- 1954 Batt (5)2, 309-334
(MR15, 923) [Kepler]
Boerma E 1964/65 MN Unt 17, 49-53
(MR29, 219)
*Boffito G 1943 Bibliografia
Galileiana: 1896-1940, Rome (Incl
additions to Favaro 1896 and index
1568-1940)
Boncompagni B 1873 Boncomp 6, 45-51,
52-60 (F5, 23) [Morin]
Bosmans H 1912 Rv Q Sc 22, 573-586
[Huygens]
Boutroux P 1922 Scientia 31, 279-290,
347-360 [Mersenne]
Boyer C B 1967 in McMullan 1967
Busulini Bruno 1963 Tor FMN 97, 809-
849 (Z117:2, 244)
Cajori F 1900 AMM 7, 111 [free fall]
Cantoni G 1892 Linc Rn (5)1(2),
405-410 (F24, 14)
*Carli A + 1896 Bibliografia
Galileiana 1568-1895 raccolta ed
illustrata, Rome, 512 p
(F27, 8) [bib]

Carruccio E 1939 It UM 2, 73-75
[isoperim]
- 1940 It UM Con 2, 901 (F68, 13)
[set]
- 1942 It UM (2)4, 175-187
(MR7, 354; Z27, 3) [set]
- 1957 It UM (3)12, 307-309
(Z77, 8)
Cassirer E 1937 Scientia 62, 121-130,
185-193 (Z17, 50)
Cattaneo C 1955 Archim 7, 68-73
(Z64, 1)
Clavelin M 1960 Thales 10,
1-26 (MR22, 260) [continuum, inf]
- 1968 La philosophie naturelle de
Galilée (Colin), 504 p
Costabel P 1955 Rv Hi Sc Ap 8, 116-128
(MR17, 338) [center of gravity]
De Giuli G 1931 Scientia 49, 207-220
(Z1, 322) [Descartes]
Drake S 1958 Osiris 30, 262-291
[mech]
- 1960 Physis Fi 2, 211-212 [mech]
- 1964 AJP 32(8), 601-608 [inertia]
- 1966 Isis 57(2), 269-271 [mech]
- 1968 in Singleton C ed, Art, Science
and History in the Renaissance,
(Johns Hopkins U P), 305-330
Drake S + ed 1969 Mechanics in Sixteenth-century Italy (U Wisc P)
[bib]
Einstein A 1964 Vop IET (16), 29-33
(MR30, 719)
Elia P M d' 1960 Galileo in China
(Harvard U P), 130 p
Favaro A 1886 Boncomp 19, 219-290
[lib]
- 1887 Bib M (2)1, 96 [letters]
- 1887 Boncomp 20, 372-376 [lib]
*- 1896 Bibliografia Galileiana 1568-
1895, Rome (also 1942 Venice)
- 1901 Bib M (3)2, 213 [Simon Marius]
- 1904 Tor FMN 39, 493-501 [Plana]
*- 1939 Galileo Galilei, 3rd ed,
Milan, 75 p (F65:2, 1082)
Galli M 1953 It UM (3)8, 328-336
(MR15, 276; Z51, 243)
[earth rotation]
- 1954 It UM (3)9, 289-300
(MR16, 434; Z55, 3) [mech]
- 1955 It UM (3)10, 77-96 (Z64, 1)
[mech]

- 1957 It UM (3)12, 80-82
(MR19, 518; Z77, 8) [mistake]
Gambera P 1873 *Di Galileo Galilei
considerato come fondatore de
metodo sperimentale e precursore
della moderna teoria dinamica,*
Novara (F5, 22)
*Garcia De Zuniga E 1945 Arg IM Ros 5,
171-174 (MR7, 106; Z60, 8)
*Geymonat Ludovico 1965 *Galileo
Galilei* (McGraw-Hill), 260 p
Gliozzi M 1942 It UM (2)4, 118-129
(F68, 12; Z26, 193)
Goldbeck E 1902 Bib M (3)3, 84
Govi G 1875 Linc At 2, 230-
- 1881 Boncomp 14, 351-379 (F13, 15)
[letter]
Guenther S 1878 Gaea, 474-480, 538-
543 (F10, 9)
Ito S 1964 Tok CGE 14, 279-292
(MR30, 720)
Jourdain P E B 1918 Monist 28, 629-
633 [Newton]
Kasner E 1905 AMS (2)11, 499-501
(F36, 49) [inf]
Kuznetsov B G 1964 *Galileo,* Mos,
326 p (MR33, 7)
Loria G 1925 Scientia 38, 361-370
- 1936 *Galileo Galilei nella vita e
nella opera,* Paris, 68 p
(F62, 20)
- 1938 *Galileo Galilei,* 2nd ed,
Milan, 155 p (Z18, 50)
- 1938 It UM Con 1, 447-449, 521-527
(F64, 16)
Losada y Puga C de 1942 Peru U Rv 10,
253-282 (MR4, 181)
Maistrov L E 1964 Vop IET (16), 94-98
(MR30, 719)
*McMullin E 1967 *Galileo. Man of
Science* (Basic Books), 557 p
[*bib]
Mieli A 1939 Archeion 21, 193-297
(F65, 1082; Z20, 196)
Miridonova O P 1964 Vop IET (16),
85-90 (MR30, 719)
[acoustics, Mersenne]
Nobile V 1954 Linc Rn (8)16, 426-433
(MR16, 434) [geodesy]
Ondemans J A C + 1903 Ned Arch (2)8,
115-167 (F34, 1026) [Marius]

Pogrebysskii I B 1964 Vop IET (16),
34-37 (MR30, 719)
Procissi A 1950 It UM (3)5, 170-194
[bib]
Reeve W D 1931 MT 24, 118-20
Rosen E 1952 Isis 43, 344-348
(Z48, 242) [moon dist]
Rubini G 1878 *Galileo Galilei e la
variabilità dei volumi reali dei
Corpi,* Bologna (F10, 11)
Tannery P 1901 Rv GSPA 12, 330
[mech]
Thompson D'A W 1915 Nat 95, 426
[sim]
Volkmann W 1932 Z Ph C Unt 45(1),
25-28
Volpe G della 1953 Int Con Ph 11, 14,
157-162 (Z51, 243) [logic]
Vonwiller O U 1943 NSWRSJ 76, 316-328
(Z60, 8) [Newton]
Wiener P P 1936 Osiris 1, 733-746
Wohlwill E 1899 ZM Ph 44(S), 577-624
(F30, 55) [free fall, parabola]


**Gallas, A    1783-1807**

Saar J du 1927 Verzek 8 (F53, 21)


**Gallatly, W.**

Anon 1914 Indn MSJ 6, 107 (F45, 38)


**Gallenkamp, Karl Wilhelm    1820-1890**

= G. L. Wilhelm

Schwalbe B 1890 Ber Ph 9, 71-73


**Galois, Evariste    1811-1832**

Adhemar R d' 1908 Rv Mois 6, 426-439
*Anon 1965 MT 58, 332
Arnoux Alexandre 1948, *Algorithme,*
Paris
Bertrand J 1899 BSM (2)23, 198-212
(F30, 9)
- 1899 J Sav (July), 389- (F30, 9)

- 1902 *Eloges Académiques,* Paris, 329-345
*Birkhoff G 1937 Osiris 3(1), 260-268 [group]
Bourgne R + 1962 *Ecrits et mémoires mathématiques d'Evariste Galois...,* Paris, 563 p (MR27, 4)
Boyer J + 1896 Intermed 3, 97 (Q609)
Brocard W 1896 Intermed 3, 189 (Q723)
Cipolla M 1934 Esercit (2)7, 3-9 (F60, 834)
Dalmas André 1956 *Evariste Galois revolutionnaire et géomètre,* Paris 175 p
- 1960 *Evarist Galya, revolyutsioner i matematik,* Moscow
Davidson G 1938 Scr 6, 95-100
Devisme J 1939 Sphinx 9, 113-114 (F65, 16)
*Dupuy P 1896 An SENS (3)13, 197-266 [mss, port]
*- 1903 Cah Quin (5)2 (Repr of 1896) [port]
*- 1921 Sci Mo 13, 363-375 (abr tr of 1896)
- 1945 *La vita di Evariste Galois...,* Rome, 167 p (Ital tr of 1896)
- 1947 Venzl Ac 10, 219-299 (MR10, 175) (Span tr of 1896)
Gaiduk Yu M 1961 Mat v Shk (4), 78-80
*Infeld Leopold 1948 *Whom the Gods Love,* NY (Italian tr 1957) [*bib, pop]
Korda D 1910 MP LAP 19, 183-191 (F41, 1052)
Kullros L 1949 Elem M (S7), 24 p (MR11, 708)
Malkin I 1963 Scr 26, 197-200 (Z111, 4) [anniv]
Mansion P 1910 Brx SS 34(A), 104-105 (F41, 12)
Mariani J 1932 Rv Syn 4, 7-14
Myrberg P J 1961 Arkhimed (2), 1-3 (MR24A, 333)
Picard E 1897 Rv GSPA 8, 339-340 (F28, 11)
Procissi A 1953 It UM (3)8, 315-328 (MR15, 276) [Betti]
Sarton G 1921 Sci Mo 13, 363-375

*Sarton George + 1937 Osiris 3, 241-268 (Z18, 197) (incl repr of 1921; repr in Sarton 1948 *The Life of Science)*
Sylow L 1920 Nor M Tid 2, 1-17 (F47, 9)
Tannery J 1908 Par C R 146, 674-676 (F39, 22) [mss]
- 1909 BSM (2)33, 158-164 (F40, 20)
- 1909 Rv Sc (5)12, 129-132
Taton R 1947 Rv Hi Sc Ap 1, 114-130 (MR10, 175; Z31, 2)
Tikhomandritzky M A 1897 Khar M So (2)6, 125-128 (F28, 11)
Verriest G 1934 *Evariste Galois et la théorie des équations algébriques,* Louvain, 58 p
- 1934 Rv Q Sc (4)25, 341-366 (F60, 12) [eqs]
- 1935 Rv Q Sc (4)26, 12-39 [eqs]

Galton, Francis   1822-1911

Anon 1911 Nat 85, 440-445 (F42, 34)
Galton Francis 1908 *Memories of My Life,* London
*Newman James R 1956 in *The World of Mathematics,* NY
Pearson Karl 1914-1930 *The Life, Letters, and Labours of Francis Galton* (Cambridge U P), 4 vols

Gambier, Bertrand   1879-1954

Anon 1956 It UM (3)11, 599-607 (MR18, 710; Z71, 4)
Vincensini P 1955 Ens M 40, 57-61 (MR16, 781; Z64, 3)

Gandz, Solomon   1883-1954

Anon 1954 N Y Times (1 April), 41
- 1954 Wilson Lib Bull (May)
Dienstag Jacob I 1954 Hadoar (Am Hebrew Weekly), (May)
Levey M 1955 Isis 46, 107-110

Gans, Richard Martin   1880-1954

Beck G 1955 Arg UMR 16, 150-158
   (Z66, 7)  [bib]

Ganter, Heinrich

Lorey W 1916 ZMN Unt 47, 583-585
   (F46, 23)
Tuchschild A 1914 Zur NGV 60, 644-646
   (F45, 49)

Garbasso, Antonio Giorgio   1871-1933

Amerio A 1933 Lom Rn M (2)66, 6-10,
   335-337  (F59, 40)
Anon 1933 Pe M (4)13, 2-3, 191-192
Brunetti R 1932 Nuo Cim (2)10, 1-4,
   129-152

Garbieri, Giovanni

Fazzari G 1904 Pitagora 11, 33-38
   (F35, 41)
Marcolongo R 1904 Pe MI (3)2, 118-122
   (F35, 41)

Garnier, Jean Guillaume   1766-1840

Quetelet A 1841 Notice sur Jean
   Guill. Garnier, Brussels

Garver, Raymond   1901-

Whyburn W M 1935 M Mag 10, 72
   (F61, 957)

Gasco, Luis-Gonzaga   1844-1899

Eneström G + 1900 Bib M (3)1, 255
Gutzmer A 1900 DMV 8, 26-27
   (F31, 19)

Gassendi, Pierre   1592-1655

Rochot B 1957 Rv Hi Sc Ap 10, 69-78
   (Z94, 2)
Smith D E 1921 AMM 28, 433   [letter]

Gatlich, A. T.

Tschistiakov I 1912 M Obraz (6), 323-
   327, 384   (F43, 41)

Gauss, Carl Friedrich   1777-1855

Ahrens W 1914 Braun Mo (2 Feb), 428-
   444
- 1925 Weltall 24(11), 205-213
   [children, planets]
- 1925 Weltall 24(12), 231 [children]
- 1926 Weltall 26, 8-11
   [Sophie Germain]
Anon 1934 Len IINT 3, 209-238
   (F61, 952)  [St. Pet Acad]
- 1954 Euc Gron 30, 276-281
   (MR17, 117)  [geom]
- 1955 ...Gedenkfeier der Ak. der
   wiss...und..zu Goettingen anlaesslich seines 100ten Todestages,
   Goetting (Musterschmidt), 31 p
Archibald R C 1920 AMM 27, 323-326
   [17-gon]
1935 Scr 3, 193-196, 386
   [Disq Arith, Fr Acad]
Aubry A 1909 Ens M 11, 430-450
   [Lagrange, Legendre, num thy]
Auwers G F J A v ed 1880 Briefwechsel
   zwischen Gauss und Bessel, Leipzig
   (Engelmann), 523 p
*Bachmann P 1911 Gott N, 455-508
   (repr Werke 10(2) (1))
   [numb thy]
Bagratuni G V 1955 K. F. Gauss.
   Kratkii ocherk geodezicheskikh
   issledovanii, Moscow, 43 p
   (MR17, 698)
Balmer H 1968 Gauss Mt (5), 3-12
   [letters]
Bell E T 1924 AMSB 30, 236-238
   [class numb]
- 1928 AMSB 34, 490-494 [class numb]
- 1944 M Mag 18, 188-204, 219-223
   (MR5, 253)  [alg numbs]

Gauss, Carl Friedrich   (continued)

Bidenkapp G 1877 *Gauss. Eine Umriss seines Lebens und Wirken,* Braunschweig

Bieberbach Ludwig 1938 *C. F. Gauss, ein deutsches Gelehrtenleben,* Berlin, 179 p (F64, 18)

*Biermann K-R 1958 Ber Hum MN 8, 121-130 [Humboldt]

- 1963 Ber Mo 5, 43-46 [Berlin Acad]

- 1963 Ber Mo 5, 241-244 [ciphers]

*- 1963 Ber Hum MN 12, 209-227 [geomag, Humboldt]

- 1964 Fors Fort 38, 44-46 [Russ language]

- 1965 Fors Fort 39, 357-361 [geodesy, Humboldt, ins]

- 1965 Fors Fort 39 (5), 142-144 [prob]

- 1966 Gauss Mt(3), 7-20 [Bessel]

- 1967 Fors Fort 41, 361-364 [Dase]

*- 1967 Gauss Mt (4), 5-18 [Humboldt]

*- 1969 Gauss Mt (6), 4-6 [Quetelet]

Biermann K-R + 1959 Fors Fort 33, 136-140 [Humboldt]

- 1962 Fors Fort 36, 41-44 [Humboldt]

Birck O 1926 Leip As 61, 46-74 [ast, mss]

Bobynin V V 1889 *Karl Friedrich Gauss,* Moscow

Bocher M 1895 AMSB 1, 205-209 [fund thm alg]

Bodenmueller H 1955 Z Vermess 80, 33-42

Bolza O 1922 *Gauss und die Variationsrechnung,* Leipzig, 95 p (Werke 10(2) (5))

Boncompagni B 1884 Vat NLA 34, 201-296 [*bib]

Brendel M 1903 DMV 12, 61-63 [Gauss Archive]

- 1929 *Ueber die Astronomischen Arbeiten von Gauss,* in Werke 11(2) (3), 258 p

Brocard H + Intermed 7, 383 [calendar]

Bruhns K ed 1877 *Briefe zwischen*

*A. v. Humboldt und Gauss,* Leipzig, 79 p

Cajori F 1889 Sci (NS)9(229)(19 May), 697-704 [children]

- 1912 Pop Sci Mo 81, 105-114 [children]

*Cantor M 1899 Heid N Jb 9, 234-255

Carslaw H 1910 Nat 84, 362

*Dedekind R 1901 Gott Bei, 45-59 (repr 1931 Dedekind, Ges M Werke 2, 293-306) [least sq, lectures]

Depman I Ja 1956 Vop IET (1), 241-245 (Z73, 242)

*Dieudonne J 1962 *L'oeuvre mathématique de C. F. Gauss,* Paris (Palais de la découverte), 18 p

Dubyago A D 1956 Ist M Isl 9, 101-106 [Simonov]

Dunnington G W 1927 Sci Mo 24, 402-414 (F52, 24) [pop]

- 1935 M Mag 9, 187-192 [error in Ball 1893 gen]

- 1938 Jean Paul Blaetter 13(1), 2-3

- 1955 *Carl Friedrich Gauss. Titan of Science,* New York, 479 p (based on Haenselmann 1878, Mack 1927, Sartorius v. Waltershausen 1856)

Dunnington G W ed 1937 *Inaugural Lecture on Astronomy and Papers on the Foundations of Mathematics,* Baton Rouge, 102 p (text and notes translated from Werke 12)

Emch A 1935 AMM 42, 382-383 (F61, 21) [error in Ball 1893 gen]

*Eymard P + ed Rv Hi Sc Ap 9, 21-51 [*Tagebuch,* Fr comm]

Feyerabend Ernst 1933 *Der Telegraph von Gauss und Weber im Werden der elektrischen Telegraphie,* Berlin, 228 p

*Fraenkel A 1920 Gott N (S), 1-58 [alg, numb]

- 1922 DMV 31, 234-238 [fund thm alg]

Freudenthal H 1954 Euc Gron 30, 276-281 [geom]

*Galle A 1918 *Gauss als Zahlenrechner,* Leipzig, 24 p (also Gott N (S4)) (F46, 12)

*- 1924 *Ueber die geodaetischen Arbeiten von Gauss,* Werke 11(2)(1), 165 p [least sq]

Gauss, Carl Friedrich    (continued)

*- 1925 Weltall 24, 194-200, 230
   (F51, 23)   (repr 1969 Gauss Mt (6),
   8-15)  [Kant, phil]
*Geppert H 1933 *Ueber Gauss' Arbeiten
   zur Mechanik und Potentialtheorie,*
   Berlin, 61 p (repr Werke 10(2)(7))
 - 1940 Deu M 5, 158-175 (MR2, 114)
   [ell fun]
 Geppert H + 1935 Scr 3, 285-286
   [*Disq Arith,* Fr Acad]
 Gerardy T 1959 Gott N, 64-66
   (MR21-2, 1023)  [Humboldt]
 - 1959 Gott N, 37-63  (Z88, 243)
   [Lecoq, geodesy]
 - 1966 Gauss Mt (3), 25-35
   [sons]
 Gerardy T ed 1964 *Christian Ludwig
   Gerling an Carl Friedrich Gauss...,*
   Goettingen, 123 p  (MR30, 201)
 Ginsburg J 1938 Scr 5, 63-66
   [eight queens]
 Goellnitz E 1937 Deu M 2, 417-420
   [lemniscate fun]
*Gresky W 1968 Gauss Mt (5), 12-46
   [Linden, politics]
*Gronweald W + ed 1955 *C. F. Gauss
   und die Landesvermessung in
   Niedersachsen,* Hannover, 191 p
 Guenther P 1894 Gott N, 92-105
   (tr JMPA (5) 3, 95-111)
 Guenther S 1876 ZM Ph 21, 61-64
   [magic sqs]
 - 1881 ZM Phy 16, 19-25  (F13, 19)
   [Sophie Germain]
*Haenselmann L 1878 *K. F. Gauss.
   Zwoelf Captiel aus seinem Leben,*
   Leipzig, 106 p
 Halsted G B 1904 AMM 11, 85-86
   [non-Euc geom]
 Hattendorff C 1869 *Die elliptischen
   Functionen in dem Nachlasse von
   Gauss,* Hannover
 Heil J 1968 Gauss Mt (5), 48-49
   [surveying]
*Herglotz G 1921 Leip Ber 73, 271-277
   (F48, 436)  [num thy]
 Hofmann J E 1942 Deu M 6, 576-685
   [surveying]
 - 1955 MN Unt 8, 49-60  (MR17, 117)

Hoppe E 1925 Naturw 13, 743-744
   (F51, 23)
Jaeger W 1933 Fors Fort 9, 143-144
   [absolute units]
Jelitai J 1938 Leip As 73, 44-52
   (F64, 917)  (Also Magy MT 57,
   136-143)
Jonquieres E de 1896 Par CR 122,
   829-830, 857-859
   [*Disq Arith,* Poinsot, Poullet-
   Delisle]
- 1911 Sphinx Oe 6, 23-24, 34-35
   [Poullet-Delisle]
Jourdain P E B 1905 Bib M (3)6, 190-
   207  [Cauchy intgl thm]
Kazarinoff N D 1968 AMM 75, 647
   [constr of polygs]
Klein F ed 1901 *(Gauss' wissenschaft-
   liches Tagebuch,* Berlin, 44 p
   (Also in M Ann 57(1), 1-34; Werke
   10(1), 485-574  (F34, 15)
   [contains false portrait]
- 1903 Gott N, 118-124
   [false portrait]
Klein F + ed 1911-1920 *Materialialien
   fuer eine wissenschaftlichen
   Biographie von Gauss,* 8 parts,
   Leipzig.  (Parts listed by author
   here)
Kneser H 1939 Deu M 4, 318-324
   [Laplace, fund thm alg]
Koerber H G 1958 Fors Fort 32, 1-8
   [geomag]
- 1959 Fors Fort 33, 298-303
   [Humboldt, geomag]
Kolman E 1955 Mos IIET 5, 385-394
   (Z68, 6)
Kuerschak J ed 1902 MP Lap 11, 90-96
   (F33, 23)
Krylov An 1934 Len IINT 3, 183-192,
   208
La Vallée Poussin C J de 1962 Rv Q
   Sc 133, 314-330  (Z108-2, 250)
   [potential]
Levasseur K 1955 Ost Verm 43, 1-17
   (MR16, 781)  [geodesy]
Levi F W 1956 Ber MG, 24-29
   (also in MP Semb 5, 191-199)
   (MR19, 624; Z87, 6)  [space]
Loewy A 1917 DMV 26, 101-109, 304-322
   [alg, calendar]

Gauss, Carl Friedrich (continued)

- 1917 DMV 27, 189-195 [cyclometry]
- 1920 DMV 30, 147-153 [space]
Lorey W 1933 MP Semb 3, 179-192
   (MR15, 276) [personal]
Luckey P 1930 ZMN Unt 61, 384
   [cipher]
MacDonald T L 1931 As Nach 241, 31
Mack Heinrich 1927 *Carl Friedrich
   Gauss und die Seinen*, Braunschweig,
   141 p (2nd ed same year, 162 p)
- 1930 Museumskunde (NS), 1
   [Gauss museum in Braunschweig]
Maennchen P 1918 *Die Wechselwirkung
   zwischen Zahlenrechnung und
   Zahlentheorie bei C. F. Gauss*,
   Leipzig (also Gott N (S 7), 1-47;
   Werke 10(2)(6)) (F46, 12)
- 1934 Unt M 40, 104-106 [anagram]
Mansion P 1899 DMV 7, 156
   [non-Euc geom]
- 1908 Mathesis (3)8(S Dec), 1-16
   (also in RV Neosco 15(4), 441-453)
Meder A 1928 Arc GMNT (2)11, 62-67
   (F54, 33) [Dorpat]
Michling H 1966 Gauss Mt (3), 24
- 1967 Gauss Mt (4), 27-30
   [heliotrope]
- 1969 Gauss Mt (6), 16-21
   [Gauss-Weber memorial]
Nevanlinna R 1956 Nor M Tid 4, 195-
   209, 229
Niessen L 1936 ZMN Unt 67, 220-223
   (F62, 24) [elem probl]
Norden A P 1956 Ist M Isl 9, 145-168
   (Z70, 243)
Oblath R 1955 M Lap 6, 221-240
Ostrowski A 1920 *Zum ersten und
   vierten Beweise der Fundamental-
   satzes der Algebra*, Leipzig, 8 p
   (also in Gott N (S), 50-58;
   rev ed in Werke 10(2)3, 16 p)
Peters C A ed 1860-1865 *Briefwechsel
   zwischen C. F. Gauss und H. C.
   Schumacher*, Altona, 6 vols
Pogorelov A V 1956 Vop IET 1, 61-63
   (Z74, 8) [surfs]
Pohl R W 1934 Gott Jb (1933/1934),
   48-56
Popken J 1954 Euc Gron 30, 282-292
   (MR17, 117)

Pringsheim A 1932 Mun Si (3), 193-
   200 [fun thy]
- 1933 Mun Si, 61-70 (F59, 847)
Reichardt H 1956 Ber MG, 11-23
   [alg, numb thy]
*Reichardt H ed 1957 *C. F. Gauss
   Gedenkband*, Leipzig, 251 p
   (repr 1960 under title *C F Gauss
   Leben und Werk*, Berlin) (11 papers)
Riesbesell P 1928 Ham MG 6, 398-431
   (F54, 33) [Repsold]
Roloff Ernst A 1942 *Carl Friedrich
   Gauss*, Osnabruck, 80 p (F68, 10)
Rybnikov K A 1956 Vop IET 1, 44-53
*Sartorius von Waltershausen 1856
   *Gauss zum Gedaechtniss*, Leipzig
   (tr 1966 Colorado Springs)
Schaaf W L 1964 *Carl Friedrich Gauss,
   Prince of Mathematicians*, New York,
   168 p [pop, high school level]
*Schaefer C 1929 *Ueber Gauss' physika-
   lische Arbeiten*, Berlin, 217 p
   (repr Werke 11(2)(2)) (F55, 611)
*- 1931 Nat 128, 339-341
   [electrodyn]
- 1933 Gott N, 57-75 [Fraunhofer,
   Pastorff]
Schaefer C ed 1927 *Briefwechsel
   zwischen...Gauss und...Gerling*,
   Berlin, 840 p (F53, 23)
Schering E 1877 Gott N (11), 229-
   237 (F19, 12) [anniv]
- 1877 *Carl Friedrich Gauss...Festrede*,
   Goettingen (Dieterich), 40 p
   (Gott Ab 22, 127-166; tr 1879
   Ann MPA (2) 9, 21-239) (F11, 22)
- 1879 Gott N, 381-384 (F11, 22)
   [Sophie Germain]
- 1880 Gott N, 367-369
- 1887 *Carl F.. Gauss und die Erforschung des Erdmagnetismus*, Goettingen
   (Dieterich), 79 p (Gott Ab 34(3),
   1-79)
- 1897 Par CR 124, 170-171 (F28, 13)
   [errors in Werke]
Schilling C + ed 1900-1909 *Briefwechsel
   zwischen Olbers und Gauss*, Berlin,
   2 vols
*Schlesinger L 1898 Ber Si 28, 346-360
   [ell fun, arith-geom mean]
- 1911 DMV 20, 396-403 [ell fun,
   arith-geom mean]

Gauss, Carl Friedrich (continued)

- 1912 Gott N, 512-543  (F43, 23)
- 1912 *Ueber Gauss' Arbeiten zur Funktionentheorie,* Leipzig, 140 p (Gott N, 1-140; Werke 10(2)(2))
Schmidt Franz + ed 1899 *Briefwechsel zwischen C. F. Gauss und Wolfgang Bolya,* Leipzig, 220 p  (F30, 10)
Schneider H 1926 Braun Mo (1,2), 7-15 [Goettingen seven]
Schoenberg E + 1955 Leip Ab 71, 7-58 (Z66, 246)  [Bessel]
- 1963 Leip Ab (N) 110
Seydler A 1877 Cas MF 6, 184-197 (F9, 13) [ast, phys]
Simon H 1877 ZM Ph 31, 99-101 [mistakes by Gauss]
Sofonea T 1955 Versich 32, 57-69 (Z65, 245) [ins]
Staeckel P 1897 Gott N, 1-12 (F28, 42)  [Bolyais]
- 1896 Gott N, 40-43 [Gerling]
- 1907 Gott N, 372-373 (F38, 22)
*- 1917 *C. F. Gauss als Geometer,* Leipzig, 128 p (Gott N 4, 25-140; Werke 10(2)(4)) (F49, 7)
*Staeckel P + 1897 M Ann 49, 149-206 (F28, 42)  (tr in 1897 BSM (2) 21, 206-228) [non-Euclidean geom]
Stein W 1965 Gauss Mt (2), 7-9
Stern M A 1877 *Denkrede...,* Goettingen (Kaestner), 16 p
Subbotin M F 1956 Top IET 1, 64-69 (Z74, 8) [ast, geodesy]
Svyatski D O 1934 Len IINT 3, 209-238 [Fuss, St. Pet]
Szabo P 1909 MNB Ung 25, 226-240 [W. Bolyai]
Timerding H E 1923 Kantstu 28, 16-40
Toth I 1955 Gaz MF (A7), 447-457 [non-Euc geom]
Unsoeld A 1953 Braun WG 5, 203-209 (Z53, 197)
Valentiner W 1877 *Briefe von C. G. Gauss and B. Nicolai,* Karlsruhe, 36 p  (F9, 13)
Van der Blij F 1954 Euc Gron 30, 293-298 [Gauss sums]
Van Veen S C 1918 Wisk Tijd 15, 140-146  (F46, 34)  [Holland]

- 1962 Euc Gron 37, 161-183 (Z107, 247) [personal]
Venkov B A 1956 Vop IET 1, 54-60 (Z74, 8)  [numb thy]
*Vinogradov I M ed 1956 *Karl Friedrich Gauss. Sbornik Statei. 100 Let so Dnya Smerti,* Moscow, 310 p (5 papers)
Voss A 1877 *Carl Friedrich Gauss,* Darmstadt
Winberg Margarete 1928 *Gauss-Erinnerung fuer die lieben Kinder,* Braunschweig
Wietzke A 1928 Bremen NV 27, 125-142 [Bremen]
- 1931 DMV 41, 1-2  (F57, 31) [port]
Winnecke F A T 1877 *Gauss, Ein Unriss seines Lebens und Wirkens,* Braunschweig, 34 p  (F9, 13)
Worbs Erich 1955 *Carl F....Gauss. Ein Lebensbild,* Leipzig, 236 p (Z64, 3)  (See Yushkevich 1956) [pop]
Yushkevich A P 1956 Vop IET 1, 299-301  (Review of Worbs)
Zimmermann P 1921 Braun Mo (11), 752-763 [college days]

Gauthier-Villars, Albert  1861-1918

Anon 1918 Ens M 20, 136-137 (F46, 28)
- 1918 Nou An (4)18, 241, 321-322 (F46, 28)

Gautschi, Werner  1927-1959

Blum J R 1960 An M St 31, 557 (MR22, 1101)

Gavarni, Guillaume Sulpice Chevallier 1804-1866

Brocard H 1911 Intermed 18, 225 (Q2366)

Gavra, Alexandru

Ionescu I 1929 Gaz M 35, 54-55
  (F55, 523)


Gazzaniga, Paolo

Cattaneo P 1930 Bo Fir (2)9, 164
  (F56, 817)
- 1931 Pe M (4)11, 68 (F57, 40)


Gegenbauer, Leopold 1849-

Fehr H 1903 Ens M 5, 296  (F34, 33)
Stole O 1904 Mo M Phy 15, 3-10,
  129-136  (F35, 29)


Geiser, Carl Friedrich  1843-1934

Emch A 1938 M Mag 12, 286-289
  (F64, 23)
Kollros L 1934 Schw NG, 522-528
  (F60, 835)


Gelfand, Israil Moiseevich  1913-

Kolmogorov A N 1951 UMN 6(4), 184-186
  (MR13, 197)  [fun anal]


Gelfond, Aleksandr Osipovich  1906-

Hille E 1942 AMM 49, 654-
  [Hilberts 7th probl]
Linnik J v  1956  UMN 11(5), 239-245
  (Z70, 244)


Geminos of Rhodes  -I

Sarton G 1927 Int His Sc 1, 212-213
  [bib]
Steinschneider M 1887 Bib M (2)1,
  97-99

- 1890 Bib M (2)4, 107-8
Tannery P 1885 BSM (2)9, 209-220
- 1885 BSM (2)9, 261-276
- 1885 BSM (2)9, 283-292
- 1901 Bib M (3)2, 9-11 (F32, 3)
  [Aganis]


Gemma-Frisius,  Cornelius  1535-1577

Bosmans H 1906 Brx SS 30-1, 165-168
  [Stifel]
Curtze M 1874 Arc MP 56, 313-325
  (F6, 23)


Genocchi, Angelo  1817-1889

Battaglini G 1889 Nap FM Ri (2)3, 79
Cassina U 1951 Int Con HS 6, 172-177
  (MR17, 3)
- 1951 Lom Rn M 83, 311-328
  (Z45, 148)  [Hermite, Schwarz,
  surf area]
Menabrea L F 1872 Boncomp 5, 301-305
  (F4, 23)
Peano G 1889/90 Tor U Ann, 195-202
  (tr in Kennedy 1966)
Siacci F 1889 Tor Mm (2)39, 463-495


Geöcze, Zoard   -1916

Szénássy B 1959 M Lap 10, 26-38
  (MR22, 260; Z86, 6)


George Pachyméres  1242-1310?

Laurent P 1934 Isis 20, 440
Narducci E 1891 Linc Rn 7, 191-196
- 1892 Linc Rn 8, 1
Sarton G 1931 Int His Sc 2(2), 972-973
Tannery Paul 1940 *Quadriviums de
  Georges Pachymére...*, Citta del
  Vaticano, 558 p  (F66, 11;
  Z25, 386)

## Gerard of Cremona   1114-1187

= Gherardo Cremonese

Bjornbo A A 1902 Bib M (3)3, 63
  [XIV]
- 1905 Bib M (3)6, 239-248 (F36, 52)
  [Euclid, al-Khwarizmi]
Boncompagni B 1854 *Della vita e delle
  opere di Gherardo Cremonese*,
  Rome  (also in 1851 Vat NLA)
Clagett M 1956 Osiris 12, 73-175
  (MR18, 630; Z72, 245)  [mech]
Eneström G 1921 Arc Sto 2, 133-136
Sarton G 1931 Int His Sc 2(1), 338-
  344  [bib]
Suter H 1903 Bib M (3)4, 19-27
  (F34, 6)


## Gerbert   950-1003

= Pope Silvester II  = Gerbert
von Aurillac  = Gerberti

Bubnov 1911 *Podlinnoe sochinenie
  Gerberta ob abake...*, Kiev, 508 p
  (Published serially in Kiev U Iz
  1905, 1909, 1910, 1911, 1912)
Carrara P B 1908 Vat NL Mm 26, 195-
  228  (F39, 10)
Chasles M 1843 Par CR 16, 156-, 218-,
  1393-; 17, 143-  [abacus]
Curtze M 1894 Bib M, 13-14
Friedlein G 1861 *Gerbert, die
  geometrie des Boethius und die
  Indischen Ziffern*, Erlangen
  [numer]
Greene H C 1931 *Gerbert, Man of
  Science. c. 945-1003*, Cambridge,
  Mass, 8 p
Knowles D 1967 Enc Phi
Lattin Harriet 1951 *The Peasant Boy
  Who Became Pope*, NY, 190 p
Miller G A 1922 SSM 21, 649-653
  [Adelbold]
Nagel 1888 *Gerbert und die Rechen
  kunst des 10. Jahrhunderts*,
  Vienna, 65 p  (F21, 1256)
Olleris A 1867 *La vie et les
  Oeuvres de Gerbert*, Clermont

Simon M 1911 Arc MP (3)18, 244-248
  (F42, 54)
Suter H 1894 Bib M, 83
Weissenborn H 1888 *Gerbert. Beitraeg
  zur Kenntniss der Mathematik des
  Mittelalters*, Berlin, 257 p
  1892 *Zur Geschichte der Einfuehrung
  der jetzigen Ziffern in Europa durch
  Gerbert*, Berlin, 123 p  (F24, 42)
  1893 Bib M, 21-24
Wuerschmidt J 1912 Arc MP (3)19, 315-
  320  (F43, 78)  [geodesy]


## Gercken, Albert Wilhelm   1854-

Thaer A 1911 Unt M 17, 32  (F42, 35)


## Geresi, Stefan

Toth A 1963 Gaz MF (A)15, 519-520
  (Z117:1, 7)
Tóth S 1962 Cluj UBBM (1)7(2), 45-61
  (Z118:1, 3)


## Gergonne, Joseph Diez   1771-1859

Faris J A 1955 JSL 20, 207  [Euler]
Guggenbuhl L 1959 MT 52, 621-629
  [serial]
Henry C 1880 Boncomp 13  [bib]
- 1881 Boncomp 14, 211-218
  (F13, 19)  [bib]


## Gerhardt, Carl Immanuel   1816-1899

Cantor M 1900 DMV 8, 28
Mueller F 1900 Bib M (3)1, 205-216
  (F31, 19)


## Gerlach, Hermann Carl Andreas   1826-

Ahrens W 1903 ZMN Unt 34, 583-590
  (F34, 33)

Gerland XI

≠ Gerland of Besançon XII

Boncompagni B 1877 Boncomp 10, 648-
656 (F9, 2)
Sarton G 1927 Int His Sc 1, 758
Treutlein P 1877 Boncomp 10, 589-
607

Germain, Sophie 1776-1831

Anon 1832 Ausland 5, 623
Biedenkipp G 1910 *Sophie Germain ein
weiblicher Denker,* Jena, 168 p
(F41, 12)
Boncompagni B 1879 Boncomp (Dec)
[Gauss, letters]
- 1880 Boncomp 15, 174- [Gauss,
letters]
- 1880 *Cinq lettres de Sophie
Germain a C. F. Gauss,* Berlin
[elastic surfaces, numb thy]
Genocchi A 1880 Tor FMN 15, 795-808
(F12, 14) [letters, Gauss]
Germain Sophie 1879 *Oeuvres philoso-
phiques, suivies de pensées et de
lettres inédites et precédées
d'une notice sur sa vie et ses
oeuvres par Hippolyte Stupuy,*
Paris, 375 p (2nd ed 1898,
Paris, 411 p)
Goering H 1888 *Sophie Germain und
Clotilde de Vaux, ihr Leben und
Denken,* Zurich, 270 p
Govi G 1880 Nap FM Ri 19, 113-114
(F13, 19) [Gauss, letters]
Guenther S 1882 Boncomp 15, 174-179
[Gauss, letters]
Harlor 1953 Rv Deux M, 134-145
Martinez J A F 1946 Sci Mo 63, 257-60
Noguera R 1956 Studia 1, 107-148
(MR18, 381) [cubic]
Schering E 1879 Gott N, 381-384
(F11, 22) [Gauss]
Simonart F 1957 Bel BS (5)43, 942-
957 (Z80, 8)
Sister M Thomas à Kempis 1939 M Mag
14, 81-90 (F65, 15)

Germay, Rodolphe Henri Joseph 1894-

Badillo M C 1954 Gac M (1)6, 147-151
(MR16, 660)
Bruwier L 1954 Mathesis 63, 268-270
(MR16, 207)
Dehalu M + 1954 Leig Bul 23, 251-266
(MR16, 207; Z56, 2)
- 1954 Lieg Bul 23, 340-359
(MR16, 434) [bib]

Gernardus, Magister XIII

Duhem P 1905 Bib M 6, 9-15
Eneström G 1904 Bib M 5, 9-14
- 1912 Bib M (3)13, 289-332
(F43, 67)
- 1914 Bib M 14, 99-149
Sarton G 1931 Int His Sc 2(2), 616
[bib]

Gernerth, 1825-1876

Schlenkrich A 1876 ZMN Unt 7, 423-426
(F8, 24)

Gerondi, En Bellsham Ephraim
1361-1440

Ginsburg J 1932 Scr 1, 60-62

Gerono, Camille Christophe 1799-1891

Galdeano Z G 1893 Prog M 3, 28
Rouche E 1892 Nou An (3)9, 10-12,
538-542

Gerschgorin, S. A.

Anon 1933 Pri M Me 1, 3 (F59, 857)

Gerson. See Levi.

Gerstner, Franz Joseph von  1756-1832

Bolzano B 1837 *Leben Franz Joseph Ritters von Gerstner,* Prague

Gertsen, Aleksandr Ivanovich 1812-1870

= Herzen

Gurianov V P 1953 Mos IIET 5, 379-386
Maistrov L E 1955 Ist M Isl 8, 481-488  (MR17, 3)

Getaldic.  See Ghetaldi

Ghaligai, Francesco  XVI

= Ghaligaj

Benedict S R 1928 AMM 36, 275-78
  (F55, 10)  [alg]

Gherardelli, Giuseppe

Conforto F 1947 Rm U Rn (5)6, 215-216  (MR9, 74; Z30, 338)

Gherardi, Silvestro  1802-1879

Basso G 1880 Tor FMN 15, 369-376
  (F12, 15)

Ghetaldi, Marino  1568-1626

= Martin Getaldic

Gelcich E 1883 Hist Abt 28, 130-143
Pavlovic D 1957 Beo Ak N Zb 55(6),
  77-87  (MR20, 264; Z80, 4)
Saltykow N 1938 Isis 29, 20-23
  (F64, 911)  [Viete]
Stipanić E 1961 *Martin Getaldic und
  seine Stellung enn der Mathematik
  und in der wissenschaflicher Welt,*
  Belgrad, 195 p   (Z107, 2)

Stajanovich C 1897 Intermed 4, 258
  [Descartes]
Wieleitner H 1912 Bib M (3)13,
  242-247  (F43, 73)  [anal geom]

Ghiron, Guido Fubini  1879-

Anon 1946 Annali (4)25, ix-xii
  (MR9, 485; Z61, 5)

Ghosh, Manindra Nath  1918-1965

Anon 1965 Clct St 14, 89-92
  (MR33, 5435)

Gibbs, Josiah Willard  1839-1905

Alasia C 1905 Rv FMSN 6(2), 21-30,
  111-125  (F36, 24)
Anon 1905 Lon RSP 74, 280-296
- 1928 Nat 121, 245-246
Ball R 1903 LMSP (2)1, v-xxix
Bryan G H 1903 Nat 68, 11
Bumstead H A 1903 Am JS (4)16, 187-202 (Repr in Gibbs 1928 Collected
  Works)  (F34, 33)
Carslaw H S 1925 AMSB 31, 420-424
  (F51, 9)  [Fourier ser, Gibbs
  phenomenon]
Duhem P 1907 BSM (2)31, 181-211
- 1908 *Josiah Willard Gibbs,* Paris,
  44 p
- 1908 Rv Q Sc (3)13, 5-43
  (F39, 28)
Garrison F H 1909 Pop Sci Mo 74,
  470-484, 551-561; 75, 41-48,
  191-203
Haas A E + ed 1936 *A commentary on
  the scientific writings of J. W.
  Gibbs (Yale Univ Press),* 2 vols
Hyde E W 1902 Ter Magn 7, 115-124
  (review of *Vector Analysis*)
Jaffe Bernard 1958 *Men of Science
  in America* (Simon & Schuster),
  Ch 13
Kraus C A 1939 Sci 89, 275-282
Langer R E 1939 AMM 46, 75-84
  (F65, 17; Z21, 3)

Larmor J 1904 LMSP (2)1, xix-xxi
(F35, 30)
Le Chatelier H 1903, Rv GSPA 14,
537-541
Moore C N 1925 AMSB 31, 417-419
(F51, 9)
Pereira Forjaz A 1939 Lisb Mr 2, 285-
291 (F65, 1091)
Richardson R G D 1923 AMSB 29, 385
(F49, 24)
Rukeyser M 1949 Phy Today, 6-13
Smith P F 1903 AMSB (2)10, 34-39
(F34, 34)
Studley D 1949 M Mag 23, 75-78
Verrill A E 1925 Sci 61, 41-42
Voit C 1904 M Ann 59, 245-252
[Cremona]
- 1904 Mun Si 34, 245-248
(F35, 30)
*Wheeler Lynde Phelps 1951 *Josiah
Willard Gibbs, the History of a
Great Mind* (Yale Univ Press),
264 p (MR12, 382) [bib]
Wilson E B 1931 AMSB 37, 401-416
(F57, 38)
- 1931 DAB 7, 248-251
- 1931 Sci Mo 32, 211-227


Gibson, George C 1853-1931

Bell R J T 1931 Edi MSP (2)2, 265-
267 (F57, 43)


Gierster, Joseph 1854-1893

Fricke R 1891-92 DMV 2, 44-46


Gigli, Duilio

Vivanti G 1933 Pe M (4)13, 255-256
(F59, 42)


Gilbert, Louis Philippe 1832-1892

Laisant C A 1892 Philom Bu (8)4,
138-146 (F24, 31)
Mansion P 1892 Brx SS 16(A), 102-110
(F24, 30)

1892 Rv Q Sc (2)1, 620-627, 627-
641 (F24, 30) [bib]
- 1893 *Notice sur les travaux
scientifiques de Louis-Philippe
Gilbert,* Paris, 86 p (F25, 37)
[port]
Mansion P + 1892 Mathesis (2)2, 57
(F24, 30)


Gill, Charles 1805-

Anon 1857 Assur Mag 6, 216- [ins]
McClintock E 1913 Actuar SA 14, 9-16,
211-238; 15, 11-40, 227-270
Newmark S 1934 Scr 2, 139-142


Ginsburg, Jekuthiel 1889-1957

Belkin S 1958 Scr 23, 7-9
(Z82, 243)
Boyer C B 1958 Isis 49, 335-336
(Z80, 9)


Giordano, Annibale

= Annibal Jourdain

Amodeo F + 1912 Nap Pont 42(2)17(13),
28 p (F43, 22)
Amodeo J + 1903 Intermed 10, 272
(Q2595)
Brocard H + 1904 Intermed 11, 219,
227, 273 (Q2595)
Fitz-Patrick J 1905 Intermed 12, 158
(Q2595)


Giorgi, Giovanni Leone Tito Carlo
1871-

Anon 1937 Vat Act 1, 356-372
(F63, 20)


Giorgini, Gaetano 1795-

Chicca T Del 1911 Pe MI 27, 24-32
Loria G 1893 Batt 31, 23-31
(F25, 29)

Giovanni. See Jean, Johann, John

Girard, Albert 1595-1632

Bierens de Haan D 1884 Nieu Arch 11,
83-152
Bosman R P H 1926 Brx SS 45, 35-42
["syncrese", Viete]
- 1926 Mathesis 40, 59-67, 100-109,
145-155 [eqs]
- 1926 Mathesis 40, 337-348, 385-392,
433-439 (F52, 16) [trig]
Brocard H 1895 Intermed 2, 241
Eneström G 1898 Bib M (2)12, 18
(F29, 5)
Korteweg D J 1896 Intermed 3, 88
(Q373)
Maupin G 1895 Fr SMB 23, 191-192
Tannery P 1883 BSM (2)7, 358-360


Giraud, Georges Julien Adolphe
1889-1943

Cartan E 1943 Par CR 216, 516-518
(Z28, 195)


Girkmann, Karl 1890-1959

Parkus H 1959 Ost Ing 13, 57-58
(Z88, 7)


Girshick, M. A. 1908-1955

Blackwell D + 1955 An M St 26, 365-
367 (MR17, 3; Z65, 245)


Glagolev, Nil Aleksandrovich
1888-1945

Anon 1946 UMN 1(2), 43-47
(MR10, 175; Z60, 15)
Bachvalov S V 1961 N. A. Glagolev,
Mos, 31 p (Z106, 3)

Glaisher, James 1809-1903

Anon 1903 LMSP 35, 466-471
Archibald R C 1948 Mathematical
Table Makers, NY, 24-25
(Scr Stu 3) [bib]


Glaisher, James Whitebread Lee
1848-1928

Anon 1929 Ens M 28, 136-137
Archibald R C 1948 Mathematical
Table Makers, NY, 25-29 (Scr Stu 3)
[biobib, port]
Forsyth A R 1931 Nat 123, 135-138
Hardy G H 1929 Mess M 58, 159-160
(F55, 17) [serials]
Zeitlinger H 1929 Nat 123, 206-207
(F55, 614)


Glaser, Konrad -1546

Richeson A W 1938 Isis 28, 341-348
(F64, 910)


Glivenko, Valerii Ivanovich
1897-1940

Kolmogorov A 1941 UMN 8, 379-383
(MR2, 306; Z60, 15)


Glizonios of Chios, Manuel XVII

Cimpan F 1960 Iasi UM (1)6, 203-209
(Z108:2, 250)


Gnedenko, Boris Vladimirovich 1911-

Gihman I I + 1962 UMN 17(4), 191-200
(Z105, 3)
Kolmogorov A N 1962 Teor Ver 7, 323-
329 (MR27, 463; Z113, 2 [probab]

**Gob,** Antoine 1868-1919

Anon 1919 Ens M 20, 383 (F47, 19)

**Godefroy,** Abraham Nikolaas

Korteweg D J 1900 Nieu Arch (2)5(1),
1 [bib]
Schoute P H 1899 Nieu Arch (2)4(4),
353

**Godfrey,** Charles 1873-1924

Siddons A W 1924 M Gaz 12, 137-138
(F51, 32)

**Goedel,** Kurt 1906-

Nagel E + 1956 Sci Am 194, 71-84;
195, 12
*- 1958 *Goedel's Proof,* NY, 127 p
- 1961 J Phil 58, 218-220
Neves Real L 1951 Gaz M Lisb 12(48),
1-8 (Z42, 243) [foundations]
Silvers S 1966 Phi M 3, 1-8

**Goepel,** Adolph 1812-1847

Cantor M 1879 ADB 9, 370
Jacobi C G J + 1847 Crelle 35, 313-
318 (repr Ostw Kl 67, 52-58)

**Goerl,** Jirik XVI

= Georg Gehrl

Vetter Q 1928 Sch Vit 3 (3-4), 82-83
- 1929 Prag Karl (94), 1-16

**Goethe,** Johann Wolfgang von 1749-
1832

Cazalas E 1932 Sphinx 2, 65-66
[magic sq]
Dyck M 1956 German R (Feb), 46-69
(Z72, 246)

- 1958 MLAAP 73, 505-515 (MR21, 230)
[phil]
Epstein P 1932 Fors Fort 8, 86-88
(F58, 994)
Helmholtz H v 1932 Naturw 20, 213-23
Lochner L 1937 Int Con (1936) 2, 272
(F63, 20)
Richter M 1932 Deu Opt Wo 18, 177-183
Schiff J 1932 Naturw 20, 223-240

**Goetting,** Eduard

Lietzmann W 1927 ZMN Unt 58, 177-180
(F53, 32)

**Gokai Ampon** 1796-1862

Shinomiya A + 1917 Toh MJ 12, 1-12
[Seki Kowa]

**Gold,** John Steiner

Eaves J C 1968 Pi Mu EJ 4, 319-321
[port]

**Goldbach,** Christian 1690-1764

Archibald R G 1935 Scr 3, 44-50
Catalan E 1886 Boncomp 18, 464-468
[D. Bernoulli]
Catalan E + 1894 Intermed 1, 202
Eneström G 1884 Bib M 1, 15-16
[series]
- 1886 Boncomp 18, 468
- 1887 Bib M (2)1, 23-24 [series]
Fuss P H 1843 *Correspondence mathé-
matique et physique de quelques
célèbres géometres de XVIIIe
siecle,* Pet
Kiselev A A 1966 Ist Met EN 5, 31-34
(MR34, 5617) [Euler, Goldbach,
numb thy]
Lusin N 1965 Ist M Isl 16, 129-143
(MR33, 2503) [Euler]
Yushkevich A P 1954 Ist M Isl 7, 625-
629 (MR16, 781) [Euler, letter]

Goldenberg, A J

Wolkovskii D 1912 M Obraz (6), 245-254  (F43, 30)

Goldovski, Yurii Onisimovich 1907-1931

Stepanoff W 1932 Mos MO Sb 39 (I - II), 127-128  (F58, 50)  [Levin]

Golovin, Mikhail Evsevevich 1756-1790

Bobynin V V 1912 M Obraz 1, 178-185, 217-223, 278-282, 313-323, 369-374 (F43, 20)

Golubev, Vladimir Vasilievich 1884-1954

Anon 1954 SSSR Tek (12), 3-18 (MR16, 781)

Goluzin, Gennalii Mikhailovich 1906-1952

Smirnov V I + 1952 UMN 7(3), 97-102 (MR14, 2; Z46, 3)

Gompertz, Benjamin  1779-1865

De Morgan A 1861 Assur Mag 9, 86- [ins]

Gonseth, Ferdinand  1890-

Bouligand G 1951 Rv Sc 89, 243-244 (MR13, 420)
Gagnebin S 1960 Dialect 14, 109-120, 267-274  (MR22, 424)

Goormaghtigh, René  1893-1960

Deaux R 1960 Mathesis 69, 257-273 (MR23A, 135; Z93, 9)

Gordon, Paul  1837-1912

Anon 1912 Ber Si (241/242)  (F43, 51)
- 1913 M Ann 73, 321-322 (F44, 23) [serials]
- 1913 Nat 90, 597
A Y 1913 LMSP (2)12, LI-LIV (F44, 23)
Noether M 1914 M Ann 75, 1-41

Gosiewski, Wladystaw  1844-1911

Dickstein S 1911 Wiad M 15, 275-282 (F42, 35)

Gosse, René  1883-1943

Janet M 1966 Ens M (2)12, 1-8 (MR33, 3873)

Gosselin, Guillaume  XVI

Bosmans H 1906 Bib M (3)7, 44-46

Gosset, William Sealy  1876-1937

= "Student"

Fisher R A 1939 An Eug 9, 1-9 (F65, 1091)
McMullen L 1939 Biomtka 30, 205-210 (F65, 18)
Neyman J 1938 ASAJ 33, 226-
Pearson E S 1939 Biomtka 30, 210-250 (F65, 18)

Gottigniez, Gilles-François  1630-1689

Bosmans H 1928 Rv Q Sc (4)13, 215-244 (F54, 29)

Goupillière, Haton de la 1832-1926

Anon 1926 Ens M 25 302

Gournerie. See Maillard

Goursat, Edouard Jean Baptiste
1858-1936

Anon 1936 *Jubilé scientifique de M.
Edouard Goursat. Allocations
prononcées à la cérémonie du 20
Nov, 1935,* Paris, 35 p
Dunnington G W 1937 M Mag 11, 190
Perrin J 1936 Par CR 203(2), 1105
Picard E 1936 Par CR 203(2), 1105-
1107

Govi, Gilberto 1826-1889

Basso G 1890 Tor FMN 25, 10-29
Legnazzi E N 1889-1890 Virgil AM, 101
101-153

Graaf, Isaak de

Van Rooijen J P 1933 Verzek 14,
(120)-(140) (F59, 843)

Grace, John Hilton 1873-1958

Todd J A 1959 LMSJ 34, 113-117
(MR20, 1138; Z82, 243)

Grace, Samuel Forster 1895-

Bradley F W 1938 LMSJ 13, 78-80
(F64, 23)

Graeffe, Carl Heinrich 1799-1873

Householder A S 1959 AMM 66, 464-466
[Dandelin, Lobachevskii]
Hutchinson C A 1935 AMM 42, 149 [eqs]

Wolf R 1874 *Carl Heinrich Graeffe.
Ein Lebensbild,* Zurich (F6, 33)

Graf von Wildberg, Johnn Heinrich
1852-1918

Anon 1918 Ens M 20, 138-139
(F46, 35)
- 1918 Ens M 20, 224-225 (F46, 21)

Graindorge, Louis Arnold Joseph
1843-1896

Anon 1896 Mathesis (2)6, 48

Grandi, Luigi Guido 1671-1742

Agostini A 1943 *Padre Guido Grandi,
matematico...,* Pisa, 27 p
- 1953 Arc In HS 32, 434-443
[letter]
Arrighi G 1931 Archeion 13, 320-24
[geom, letter]
Bottoni B 1954 Pe M (4)32, 150-171
(MR16, 207)
Paoli A 1912 Pisa U 28:29 [letters]
Pascal A 1915 Lom Gen (2)48 [Ceva]
Tenca L 1951 It UM (3)6, 143-149
(Z42, 241) [cubic]
- 1951 Lom Rn M 83, 493-510
(MR13, 612; Z45, 147)
- 1951 Lom Rn M 84, 519-537
(MR14, 832; Z45, 147) [Ceva]
- 1951 Pe M (4)29, 181-197
(Z43, 4)
- 1952 Bln Mm (10)9, 49-60
(MR16, 434; Z48, 243)
- 1953 It UM (3)8, 201-204
(MR14, 1050; Z50, 243)
- 1954 Bln Rn P (11)1(2), 77-87
(Z59-1, 7; 60, 7) [Fagnani]
- 1959 Bln Rn P (11)6, 139-148
(Z100, 244) [Marchetti]
- 1957 It UM (3)12, 458-460
(Z79, 242) [curves]
- 1960 Bln Rn P (11)7(2), 87-90
(Z100, 245) [Varignon]
- 1960 Physis Fi 2, 84-89
(Z103:2, 243) [religion]

Grandidier, Alfred  1836-1921

Lacroix A 1922 Par Mm (2)58, i-lviii

Grassmann, Hermann Guenther
1809-1877

*Anon 1878 M Ann 14, 1-46  (F10, 20)
- 1895 Rv M, 179
- 1909 Ber MG 8, 80-114 (F40, 22)
  [anniv]
Bottema O 1947 Euc Gron 23, 121-127
  [Jacobi, politics, Poncelet]
Cox H G 1882 QJPAM 25, 1-
- 1882 Cam PST 13, 68-
- 1891 QJPAM 19, 74-
Delbrueck B 1877 Allg Z 18, x
  (F10, 21)
Dickstein S 1875 Arc MP 57, 420-
- 1879 Herrmann Grassmann..., Niwa
  (F11, 26)
Engel F 1909 Arc MP (3)15, 79-88
- 1909 Ber MG 8, 79-88  (F40, 22)
- 1909 DMV 18, 344-356 (F40, 22)
- 1910 DMV 19, 1-13  (F41, 15)
Favaro A 1878 Boncomp 11, 699-756
  (F10, 21)
*Forder H G 1941 The Calculus of
  Extension   [bib]
Genese R W 1893 Nat 48, 517
- 1926 M Gaz 13, 373-391
*Heath A E 1917 Monist 27, 1-56
  [geom, Leibniz]
Jahnke E 1909 Arc MP (3)15, 89-99
Junghans F 1878 Hist Abt 33, 69-75
  (F10, 20)
- 1878 ZMN Unt 9, 167-169, 250-253
  (F10, 20)
Lotze A 1913 Enc MW 3(1)8
  [anal geom]
Mueller F 1909 Dres Isis  (F40, 22)
Muesebeck C 1909 MN Bl 6, 33-36
  (F40, 23)
Peano G 1895 Rv M 5, 179-182 [bib]
*Sarton G 1945 Isis 35, 326-330
  (MR7, 106)  [Hamilton]
Schlegel Victor 1878 Hermann
  Grassmann, sein Leben und seine
  Werke, Leipzig  (F10, 20)  [bib]
Schroder E + 1879 M Ann 14, 1-45

Staeckel P 1912 Int Monat 6(10)
  [psych of math]
Wolff G 1923 ZMN Unt 54, 51-53
  (F49, 14)

Graustein, William Caspar  1888-1941

Coolidge J L 1941 AMSB 47, 343-349
  (Z25, 2)

Grave, Dmitrii Aleksandrovich
1863-1939

Anon 1940 Mos MO Sb (2)7(2), i-ii
  (F66, 21)
- 1940 Sbornik, posvyashchennyi
  pamyati akademika D. A. Grave,
  M-L
- 1940 Ukr IM (4), 1-6 (F66, 21)
Breus K A 1963 Ukr IM 15, 235-239
  (Z115, 246)
Chebotarev N G 1937 UMN 3
- 1940 Sbornik posvjashchenii
  pamyati D. A. Grave, Moscow, 1-14,
  320-326  (MR2, 115; F66, 21; Z23,
  389)  [bib]
Delone B N 1940 SSSRM (4)4, 349-356
  (F66, 21)  [bib]
Dobrovolsky B A 1963 Ist M Isl 15,
  319-360
Lyusternik L 1941 UMN 8, 377-378
  (Z60, 15)

Gravelarr, Nicolas Lambertus Willem
Anthonie  1851-1913

Struik D J 1929 Euc Gron 6, 204-207
  (F55, 612)

Gray, James Gorden  -1934

Tweedie C 1925 Edi MSP 43, 70-80
  (F51, 22)

Green, Gabriel Marcus   1891-1919

ilczynski E 1919 AMSB (2)26, 1-13


Green, George 1793-1841

Anon 1850 Crelle 39, 74-75
Green H G 1947 in *Studies and Essays*
   *Offered to George Sarton*, NY,
   545-594


Greenhill, Alfred George   1847-1927

Anon 1927 M Gaz 14, 417-420
P A M 1927 Nat 119, 323-325
Baker H F 1928 LMSP (A)119, i-iv
   (F54, 38)
Fehr H 1927 Ens M 26, 141-142
   (F53, 34)
Love A E H 1928 LMSJ 3, 27-32
   (F54, 38)
Nicholson J W 1929 M Gaz 14, 417-420
   (F55, 17)


Greenstreet, William John   1861-1930

Macauley J F S + 1930 M Gaz 15, 181-
   185   (F56, 29)
Kirkman J 1913 M Gaz 7, 28-29
   [E M Langley]


Greenwood, Isaac 1702-1745

Simons L G 1933 Scr 1, 262-264
- 1934 Scr 2, 117-124


Gregory, David   1661-1710

Hiscock W G 1937 *David Gregory,*
   *Isaac Newton and Their Circle,*
   Oxford, 57 p


Gregory, Duncan Farquharson   1813-1844

Ellis R L 1865 in *The Mathematical*
   *Writings of Duncan Farquarson*
   *Gregory,* Cambridge, 291 p
   [bib, port]


Gregory, James   1638-1675

Anon 1896 Mathesis (2)6, 42
Dehn M + 1939 in Turnbull H W ed,
   468-478   (MR1, 33) [calc, conv]
- 1943 AMM 50, 149-163
   (MR4, 181; Z60, 9)
Gibson G A 1922 Edi MSP 41, 2-25
Heinrich G 1901 Bib M (3)2, 77-85
Hofman J E 1950 Centau 1, 24-37
   (Z40, 1) [approx, area, conic]
Inglis A 1933 M Gaz 17, 327-328
   [ell, length]
Larmor J 1936 Lon Tim LS (Jan 18),
   55 [Newton]
Loria G 1894 Intermed 1, 193
   [circle area]
- 1902 Bib M (3)3, 127
Scriba C J 1957 Gies Sem 55,
   (Z80, 3) [calc]
Siebel A 1956 MP Semb 5, 143-146
   (MR18, 368)   [pi]
Turnbull H W 1933 Edi MSP (2)3,
   151-172   (F59, 25) [interpol]
- 1938 Nat 142, 57-58   (F64, 24)
- 1938 Obs 61, 268-274 (F64, 919)
- 1939 in *University of St. Andrews*
   *James Gregory Tercentenary,*
   St. Andrews 5-11   (MR1, 33; Z60, 9)
- 1940 Lon RSNR 3, 22-38
   [Royal Society]
Turnbull H W ed 1939 *James Gregory*
   *Tercentenary Memorial Volume,*
   London, 536 p   (MR1, 129; Z26, 289;
   F65:2, 1084)


Gregory St. Vincent   1584-1667

   = Gregorius a Sancto Vincentio
   = Grégoire de Saint-Vincent

Anon 1873 Bel Bul (2)36, 89-96
   (F5, 23)

- 1901 Intermed 8, 12
- 1908 Brx SS 33, 169
Aubry A 1904 Mathesis (3)4, 129-130
  [area, hyperb, log]
Bopp K 1907 Ab GM 20, 85-314
  (F38, 65)
Bosmans H 1902 Brx SS 26(B), 22-40
  (F33, 16) [letter]
- 1903 Brx SS 27(B), 21-64
  (F34, 9)
- 1909 Brx SS 34, 174
- 1910 Brx SS 34(A), 174  (F41, 8)
- 1911 Bel Biog 21, 141-171
  (F42, 5)
- 1913 Brug Em An, 41-50
- 1924 Mathesis 38, 250-256  [calc]
- 1925 Brx SS 44, 17-22
  (F51, 21)  [mech]
Cajori F 1922 Sci Mo 14, 294-296
Hofmann J E 1941 Ber Ab (13), 1-80
  (Z26, 194)  [Leibniz]
Naux C 1962 Rv Hi Sc Ap 15, 93-104
  (MR34, 5618)
Neuberg J 1911 Bel BS, 922-932
  (F42, 5)
- 1912 *Vie et ouvrage de Gregoire de
  Saint-Vincent,* Mathesis (4)2(S)
  (F43, 12)
Sauvenier-Goffin E 1951 Lieg Bul 20,
  413-426, 427-436, 563-590, 711-732,
  733-737  (MR13, 612)
- 1952 Lieg Bul 21, 301-302
  (MR14, 343)


Gretschel, Heinrich  1830-1892

Papperitz E 1893 DMV 2, 42-43
  (F25, 39)


Grimsehl, Carl Ernst Heinrich  1861-

Hillers W 1915 ZMN Unt 46, 1-22
  (F45, 39)


Grinwis, Cornelius Hubertus Carolus
  1831-1899

Anon 1899 Amst Vs G 8, 326

Griss, G. F. C.

Van Rootselaar B 1954 Euc Gron 28,
  42-45  (Z36, 2)


Grofe, Gustav von 1848-1895

Kneser A 1896 Tartu Est 11, 186


Gromeka, Ippolit Stepanovich
  1851-1889

Anon 1951 Pri M Me 15, 393-395
  (MR13, 197; Z42, 243)
- 1951 Pri M Me 15, 396-408
  (MR13, 197)  [hydromech]
Vasilev O F 1956 Mos IIET 10, 245-268
  (Z74, 245)


Gronwall, Thomas H 1877-1932

Hille E 1932 AMSB 38, 775-786
  (F58, 992)
Ritt J F 1932 Sci (2)75, 657
  (F58, 995)


Grosseteste, Robert 1175-1253

  = Robert of Lincoln  = Robert
  Grosthead  = Robert Greathead
  = Robertus Grosse Capitis
  Lincolniensis

*Baur L 1917 *Die Philosophie des
  Grosseteste,*  Muenster, 314 p
*Callus D A ed 1955 *Robert Grosseteste,
  Scholar and Bishop,* Oxford  [bib]
Lindhagen A 1916 Ark MAF 11, 41 p
Pegge S 1793 *The Life of Robert
  Grosseteste...,* London, 390 p
Sarton G 1931 Int His Sc 2(2), 583-586
  [bib]
Stevenson F S 1899 *Robert Grosseteste,
  Bishop of London,* London
Thomson S H 1933 Isis 19, 19-25

Grossmann, Marcel  1878-1936

Saxer W 1936 Zur NGV 81, 322-326
  (F62, 1041)


Gruebler, Martin Fuerchtegott  1851-

Anon 1931 ZAM Me 11, 468  (F57, 40)
Dupont P 1963 Tor FMN 98, 397-417
  (MR29, 807)  [mech]


Gruenwald, Geza

Turan P 1955 M Lap 6, 6-26
  (MR17, 446)


Gruess, Gerhard Christian  1902-

Willers F A 1950 ZAM Me 30, 232
  (Z36, 146)


Grundy, Patrick Michael  1917-1959

Goddard L S 1960 LMSJ 35, 377-379
  (MR23A, 136; Z94, 5)


Grunert, Johann August  1797-1872

Curtze M 1872 BSM 3, 285-287 (F4, 22)
- 1873 Arc MP 55, 1-4  (F5, 34)
Kobell E 1873 Mun Si, 133 (F5, 34)


Gua de Malves, Jean Paul de 1714-1785

*Sauerbeck P 1902 Ab GM 15, 1-166


Guccia, Giovani Battista  1855-1914

Anon 1914 Pe MI (3)12, 48
Castelnuovo G 1914 Mathes F 6, 174-
  177  (F45, 41)
Franchis M de 1915 Paler R 39, i-xiv
Torelli G 1914 Nap FM Ri (3a)20,
  217-219


Gudermann, Christoph  1798-1852

Plakhowo N + 1906-1907  Intermed 13,
  259;14, 48, 139 (Q3121)
  [Pyth thm, spher geom]
Sturm R 1910 DMV 19, 160  (F41, 17)
  [Weierstrass]


Guensche, Richard 1861-1913

Jahnke E 1913 Arc MP (3)21, 96-100
- 1913 Ber MG 12, 96-110
  (F44, 31)


Guenther, Adam Wilhelm Sigmund
  1848-1923

Favaro A 1877 Ven I At (5)3, 913-958
  (F9, 30)
Reindl J 1908 Nurn NG, 17 p
  (F40, 42)
Schueler W 1925 ZMN Unt 56, 111-113
  (F51, 28)
Wagner H 1891 Gott N, 256-278
  (F23, 45)


Guenther, Paul  1867-1891

Gutzmer A 1892 ZM Ph 37, 46-50
  (F24, 31)
L L 1897 JMPA (5)3, 112 [ell fun,
  lin diff eqs, Weierstrass]


Gugle, Elizabeth Marie  1877-1960

Fawcett  H 1961 MT 54, 390


Guglielmini, Giovanni Battista
  1763-1817

Denizot A 1932 Int Con 6, 475-482


Guldberg, Alf  1866-1936

Meidell B 1936 Nor M Tid 18, 33-39
  (F62, 27)

- 1936 Sk Akt 19, 126-128
  (F62, 1041)

Guldin, Paul Habakuk 1577-1643

Miller G A + 1926 Sci (2)64, 204-206
  (F52, 17) [plagiarism]
Ver Eecke P 1932 Mathesis 46, 395-397
  (F58, 989)

Gundelfinger, Sigmund 1846-1910

Dingeldey F 1917 DMV 26, 75-99
  (F46, 18)

Gurev, Semen Emelyanovich 1766-1813

Prudnikov V E 1956 Mos IIET 10, 384-
  392 (Z74, 245)
Yushkevich A P 1947 Mos IIET 1, 219-
  268 (MR12, 311)

Gutberlet, Constantin 1837-1928

Hartmann E 1928 Gorres Ja 41, 261-266
  (F54, 37)

Guthrie, Francis, 1833-1886

Anon 1899 J Bot 37, 528
- 1899 Nat (Nov 23), 84
- 1912 in Boase F ed, *Modern English
  Biography*, (S)2, 530
*May K O 1965 Isis 56, 346-348
  (repr 1967 MT 60(5), 516-519)

Gutzmer, August 1860-1924

Krazer A 1924 DMV 33, 1-3
  (F50, 17; 51, 30)
Lietzmann W 1925 ZMN Unt 55, 224-227
  (F50, 17)
Salkowski E 1924 Unt M 30, 62-65
  (F50, 17)
Wangerin A 1926 Leop 1, 133-135

Gylden, Johan August Hugo 1841-1896

Anon 1896? Nat 55, 158
Bohlin K 1897 Act M 20, 397-404
  (F28, 23)
Callandreau O 1896 Par CR 123, 771
Ernst M 1897 Wiad M 1, 29
  [Tisserand, Gould]

Gyunter, Nikolai Maksimovich 1871-1941

Smirnoff V I + 1941 SSSR Iz 5, 193-
  202 (MR3, 98; Z60, 15) [bib]

Haantjes, J. 1909-1956

Schouten J D 1956 Nieu Arch (3)4,
  61-70 (MR18, 182; Z70, 5)
Van Veen S C 1956 Sim Stev 31, 3-4
  (MR18, 268; Z70, 244)

Haar, Alfred 1885-1933

Anon 1933 Szgd Acta 6, 65-66
  (F59, 43)

Habich, Eduard

Dickstein S 1911 Wiad M 15, 269-272
  (F42, 22)

Hachette, Jean Nicolas Pierre
  1769-1834

Anon 1834 Fr Soc Ag, 143

Hadamard, Jacques 1865-1963

Anon 1930 Arg SC 110, 66-80 (F56, 32)
- 1937 *Jubilé scientifique de Jacques
  Hadamard*, Paris, 108 p
Desforge J + 1936 Ens Sc 9, 97-117
  [sec ed]
Heilbronn H + 1963 Nat 200, 937-938
  (Z113, 2)

Maeder A M 1964 Parana MB 7, 5-7
(Z114, 8)
*Mandelbrojt S 1953 AMM 60, 599-604
(MR15, 276; Z51, 243)
*Mandelbrojt S + 1965 AMSB 71, 107-129
[bib, port]
Muir T 1926 So Af RS 13, 299-308
(F52, 36)
Tannery J 1899 Ens M 1, 333  [geom]
Tsortsis A 1938 Gree SM 18, 203-207
(F64, 23)


Hagen, Gotthilf Heinrich Ludwig
1797-1884

Mansion P 1914 Mathesis (4)4
(F45, 21)
Schiller L ed 1934 Drei Klassiker der
Stromungslehre: Hagen, Poiseuille,
Hagenbach, Leipzig, 97 p
[aerodyn, hydrodyn]


Hagen, Johann Georg    1847-1930

Armellini G 1931 Linc Rn (6)13,
301-312  (F57, 39)


Hahn, Hans  1879-1934

Frank P 1934 Erkennt 4, 315-316
J B 1935 Rv M Hisp A (2)10, 126-127
(F61, 956)
Mayerhofer K 1934 Mo M Phy 41, 221-
238  (F60, 839)
Menger K 1933 Mo M Phy 40, 233
- 1935 Erg MK (6), 40-44  (F61, 26)
- 1935 Fund M 24, 317-320  (F61, 26)


Haitham.  See Alhazen


Hajos, Gyorgy  1912?-

Réidei L 1962 M Lap 13, 217-227
(MR33, 2519)  [anniv]

Halley, Edmond 1656-1741

Archibald R C 1929 M Gaz 14, 574-475
(F55, 12)  [De Moivre]
- 1932 M Gaz 16, 213-216
Armitage A 1966 Edmond Halley
(Nelson), 220 p
Bateman H 1938 AMM 45, 11-17 [eqs]
Doormann C 1925 Halley and Fermat,
Breslau, 118 p  (F51, 38)  [stat]
Eneström G 1896 Sv Ofv 53, 41
[XVII, Finance, Witt]
- 1899 ZM Ph 44(S), 83  [stat,
Wargentin]
Forsyth A R 1905 Brt AAS 75, 307
[comet]
Huxley G L 1959 Scr 24, 265-273
(MR22, 929; Z92, 245)
Jones H S 1956 Lon RIGBP, 20 p
(MR20, 738)
MacPike E F 1943 Not Qu SD 184, 298-
302 [bib]
Sanford V 1933 MT 27, 243-246
Schaaf W L 1956 MT 49, 41-3  [ins]


Halphen Charles  1885-1915

Weill 1916 Fr SMB 44, 14
(F46, 24)


Halphen, Etienne  1911-1954

Morlat G 1954 Par I St 3, 203-205
(Z59:1, 12; 60, 12)


Halphen, Georges Henri  1844-1889

Anon 1889 Paler R3, 210-222  [bib]
- 1901 Par CR 133, 722
Brioschi 1890 BSM (2)14, 62-72
Hermite C 1889 Par CR 108, 1079-
1081
Jadraque V M 1958 Gac M 10, 3-5
(Z83, 245)
Jordan C 1889 JMPA (4)5, 343-359
Picard E 1890 Par CR 110, 489-497
- 1924 Mélanges de mathématiques et
de physique, i-ii

Poincaré H 1910 *Savants et écrivains,* 125-140
- 1890 Par EP 60, 137-161

Halsted, George Bruce 1853-1922

Cajori F 1922 AMM 29, 338-340
Dickson L E 1894 AMM 1, 337-340
Humphreys A M 1922 Sci 56, 160-

Hamburger, Arthur

Lampe E 1911 Ber MG 10, 70-74
(F42, 27)

Hamburger, Hans Ludwig 1889-1956

Grimshaw, M E 1958 LMSJ 33, 377-383
(MR20, 848; Z80, 4)

Hamburger, Meyer 1838-1903

Denizot A 1903 Wiad M 7, 208-210
(F34, 34)
Lampe E 1904 DMV 13, 40-53 (F35, 31)

Hamel, Georg 1877-1954

Faber G 1955 Mun Jb, 178-180
(MR18, 784)
Haack W 1952-54 Ber MG (Z98, 9)
Kucharski W 1952 ZAM Me 32, 293-297
(MR14, 343; Z47, 6) [mech]
Prandtl L 1948 ZAM Me 28, 129-131
(Z29, 241)
Schmeidler W 1952 Ber MG, 27-34
(Z48, 243)
- 1952-54 Ber MG, 7-16
(MR17, 931; Z98, 9)
- 1955 DMV 58, 1-5 (Z64, 3)
Vogelpohl G 1948 ZAM Me 28, 131-132
(Z29, 241)

Hamill, Christine Mary 1923-1956

Edge W L 1956 Edi MSN (40), 22-25
(MR18, 710; Z71, 4)
Todd J A 1957 LMSJ 32, 384 (Z77, 8)

Hamilton, William Rowan 1805-1865

Anderson R E 1891 Athm (3310) (April 4), 444
Anon 1866 Am JS 42, 293-
- 1867 Dub Ac P 9, 307-
*- 1878 in *A Compendium of Irish Biography,* Dublin
Bateman H 1944 Scr 10, 51-63
(MR6, 141) (Repr 1945 Scr Stu 2)
[bib, mech]
Chetaev N G 1960 Pri M Me 24, 33-34
(MR22, 618) [mech]
Child J M 1915 Monist 25, 615-624
[hodograph]
Colthurst J R 1945 Dub Ac P 50, 112-121 [Icosian calc]
Conway A W 1931 Dub S 20, 125-128
(F57, 35)
*- 1951 Can Cong 2, 32-41
(MR13, 197; Z42, 243)
Conway A W + 1929 Nat 123, 349
[optics]
De Morgan A 1866 Gentleman's Magazine
(January), 128-134
Dugas R 1941 Rv Sc 129, 15-23
(MR6, 254; Z60, 12) [mech, optics]
Eves H 1963 MT 56, 348-49
[The two William Hamiltons]
Graves R P 1882-1889 *Life of Sir William Rowan Hamilton...,* Dublin, 3 vols [*bib, literature, mss, poetry]
- 1891 *Addendum to the Life of Sir William Rowan Hamilton,* Dublin
(F24, 21)
- 1891 Athm (3308) (March 21), 380-381
Grigoryan A T 1955 Nauk Zh (12), 58
Hawkes H E 1900 AMSB 7, 306-307
[irrat]
Knott C G 1911 Nat 87, 77 (F42, 19)
[Tait]
Kargon R 1964 AJP 32, 792-795
[Boscovich, Faraday]

*Lanczos C 1967 Am Sci 55(2), 129-143
MacDuffee C C 1942 Scr 10, 25-35
  [alg]
Miller G A 1910 Bib M (3)11, 314-
  315 (F41, 58) [group]
McConnell A J 1945 Dub Ac 50A, 70-88
  (Z60, 13) [Dublin]
- 1958 Adv Sc 14, 323-332 (Z79, 5)
Milne E A 1941 M Gaz 25, 106, 298-
  299
O'Donoghue D J 1912 *The Poets of*
  *Ireland,* London, 182
Ore O 1960 AMM 67, 55
Piaggio H T H 1943 Nat 152, 553-555
  (Z60, 12) [quaternions]
Polak L S 1956 Mos IIET 15, 206-276
  (MR19, 826; Z74, 245)
- 1936 Mos IINT (2)8 (F62, 1036)
  [mech]
Sarton G 1932 Isis 17, 154-70
  [conical refraction]
Smith D E 1922 AMM 29, 209
  [quaternions]
- 1942 Scr 10, 9-11
Smith D E ed 1945 *A Collection of*
  *papers in Memory of Sir William*
  *Rowan Hamilton,* NY, 82 p
  (Scr Stu 2)
Steward G C 1932 M Gaz 16, 179-191
  (F58, 991) [optics]
Study E 1905 DMV 14, 421-424
  (F36, 13)
Synge J L 1942 Scr 10, 13-24
  (MR6, 141; Z60, 12) [bib]
- 1962 Sci 138, 13-
*Synge J L + 1945 Dub Ac P 50A (6),
  69-121 [port]
Tait P G 1866 North British Review
  45; 37-74
Van Heel A C S 1942 Ned Ti N 9, 342-
  350 (F68, 2)
Whittaker E T 1940 M Gaz 24, 153-158
- 1941 M Gaz 25, 106-108, 299-300
*- 1945 Dub Ac P 50A, 93-98
  (MR6, 254) [heuristic, quaternions]
*- 1954 Sci Am 190, 82-87
- 1957 in *Lives in Science,* NY, 61-74

Hammond, James  1850-1930

Elliot E B 1931 LMSJ 6, 78-80
  (F57, 39)


Hampl, Miloslav  1898?-

Spacek L + 1958 Prag M Ap 3, 75-78
  (Z82, 12)


Hancock, Harris  1867-

Moore C N 1944 AMSB 50, 812-815
  (MR6, 141; Z60, 13)


Hanegraeff, M. E.  XIX

Mansion P 1889 Bib M (2)3, 64  [phil]


Hankel, Hermann  1839-1873

Cantor M 1874 Hist Abt 20, 27-38
  (F5, 1)
Loria G 1889 Bib M (2)3, 120
Mansion P 1875 Boncomp 8, 185-220
  (F7, 15)
Zahn W v 1874 M Ann 7, 583-590
  (F6, 31)
- 1876 Boncomp 9, 290-308  (F8, 20)
  [bib]
- 1891 Wurt MN 4, 1-11  (F23, 22)


Hansen, C. C.

Nielsen J 1935 M Tid A, 1-2
  (F61, 956)


Hansteen, Cristoph

Kobell F v 1874 Mun Si, 71-72
  (F6, 31)

Happach, Max

Leman 1914 ZMN Unt 45, 244-245
  (F45, 33)


Hardy, Godfrey Harold  1877-1947

Anon 1948 Nat 161, 797-798
Bosanquet L S 1950 LMSJ 25, 102-106
  (Z36, 145)  [series]
Davenport H 1950 LMSJ 25, 119-125
  (Z36, 146)  [Waring probl]
Ingham A E 1950 LMSJ 25, 115-119
  (Z36, 146)  [numb thy]
Littlewood + 1949 AMSB 55, 1082
  (MR11, 573)
Milne E A 1948 Lon As Mo N 108,
  44-46  (MR10, 175; Z30, 338)
Mordell L J 1950 LMSJ 25, 109-114
  (Z36, 145)  [Diophantine approx]
Newman M H A + 1948 M Gaz 32, 49-51,
  98
Offord A C 1950 LMSJ 25, 136-138
  (Z36, 146)  [Fourier ser]
Perron O 1948 Mun Jb (1944/48),
  282-285  (MR11, 573)
Piaggio H T H 1931 M Gaz 15, 461-465
Rado R 1950 LMSJ 25, 129-135
  (Z36, 146)  [ineqs]
Smithies F 1950 LMSJ 25, 102-138
  (MR12, 69; Z36, 145)
*Snow C P 1967 Foreward to 1967 re-
  print of Hardy 1940 *A Mathe-
  matician's Apology*
Soddy F 1941 Nat 147(Jan 4)
Summerhayes V S 1948 Nat 161, 797-
  798 (Z29, 241)
Titchmarsh E C 1949 Lon RS Ob 6,
  447-461  (MR12, 311)
- 1950 LMSJ 25, 81-101 (MR12, 69;
  Z36, 145)
- 1950 LMSJ 25, 125-128  (Z36, 146)
  [Lattice point, zeta fun]
Vijayaraghavan T 1947 M Stu 15,
  121-122 (MR10, 668)
Wiener N 1949 AMSB 55, 72-77
  (MR10, 420; Z31, 99)

Haret, Spiru C  1851-1912

Joachimescu G 1935 Gaz M 40, 481-482
  (F61, 958)


Hargreave, Charles James  1820-1866

Anon 1868 Lon RSP 16, xvii-


Harley, Robert  1828-

Anon 1910 Nat 83, 210  (F41, 28)
- 1911 LMSP (2)9, xii-xv  (F42, 27)
- 1914 Lon RSP (A)91, 623-624


Harmuth, Friedrich Traugott Theodor
  1854-

Thaer A 1911 Unt M 17, 154-155
  (F42, 35)


Harnack, Carl Gustav Axel  1851-1888

Noether M 1888 Hist Abt 33, 121-124
Voss A 1888 M Ann 32, 161-174


Harriot, Thomas  1560-1621

Bradbrook M C 1936 *The School of
  Night,* Cambridge
Cajori F 1928 AMSB 34, 434  (F54, 48)
- 1928 Isis 11, 316-324
Dickstein S 1902 Wiad M 6, 259-260
Lohne J 1959 Centau 6, 113-121
  (MR22, 424)
- 1965 Centau 11, 19-45
- 1966 Arc HES 3, 185-205
  (MR34, 7321)  [alg]
Morley F V 1922 Sci Mo 14, 60-66
Scriba C 1965 Centau 10, 248-257
  [Wallis]
Shirley J W 1951 AJP 19, 452-454
Vacca G 1902 Loria 1-6  (F33, 15)
  [binom, spher tria]

**Harsdoerfer,** Georg Philipp
1607-1658

Rudel K 1894 *Georg Philipp Harsdorfer,*
Nurnberg  (rev in ZM Ph 40, 136)

**Hartree,** Douglas Rayner  1897-1958

Darwin C G 1959 LMSJ 34, 118-128
(MR20, 1138; Z82, 243)
Wilkes M V 1958 Comp J 1, 48
(Z87, 243)
- 1958 Nat 181, 808  (Z79, 5)

**Harvoy,** Jean

Boyer J 1898 Intermed 5, 102

Hasan.  See Alhazen.

**Haslam-Jones,** Ughtred Shuttleworth
1903-1962

Titchmarsh E C 1963 LMSJ 38, 311-312
(MR28, 2; Z112, 3)

**Hassar,** Muhammed al-  XII-XIII

= Abu Zakariya (Bakr) M. ibn
Abdallah al-Hassar (Hasir)

Sarton G 1931 Int His Sc 2(1), 400
[bib]
Suter H 1901 Bib M (3)2, 9, 12-40
[Arab]

**Hasseé,** Henry Ronald  1884-1955

Vint J 1956 LMSJ 31, 252-255
(Z70, 6)

**Hassler,** Ferdinand Rudolph  1770-1843

Cajori F 1928 *The Chequered Career of
F. R. Hassler,* Boston (F54, 47)

**Hatton,** John Leigh Smeathman
1865-1933

Le Beau G S 1934 LMSJ 9, 74-76
(F60, 16)

**Hauck,** Guido  1845-

Hessenberg G 1906 ZMN Unt 37, 71-76
(F37, 20)
Lampe E 1905 DMV 14, 289-311
(F36, 35)
- 1907 DMV 16, 155-164
- 1906 *Die Enthuellungsfeier des
Denkmals...,* Leipzig, 15 p
(F37, 20)
- 1907 DMV 17, 155-164  (F38, 28)
Lampe E + 1907 M Gaz 4, 136

**Hausdorff,** Felix  1868-1942

Bergmann G 1967 DMV 69(2-1), 62-75
(MR34, 7330)  [mss]
Blumberg H 1920 AMSB 27, 121-122
Lorentz G G 1967 DMV 69(2-1), 54-62
(MR34, 7329)

**Hawkins,** George Edmon  1901-1956

Warner G W + 1956 SSM 56, 605-606

**Hayashi,** Tsuruich  1873-1935

Anon 1955 Tokush J 6, 1-2
(Z66, 7)
Fujiwara M 1936 Toh MJ 41, 265-289
(F62, 27)

**Hayward,** Robert Baldwin  1829-1903

Anon 1950 M Gaz 34, 81
- 1903 LMSP 35, 466-471  (F34, 35)

**Hayyim.**  See Levi.

Heath, Thomas Little  1861-1940

Archibald R C 1940 M Gaz 24, 234-237
  (Z60, 15)
Sarton G 1950 Int Con HS 6, 45-78
  [M. Cantor]
Smith D E 1936 Osiris 2, v-xxvii
Thompson D W 1940 Nat 145, 578-579
  (Z61, 5)
- 1941 Lon RS Ob 3, 409-426


Heaviside, Oliver 1850-1925

Anon ed 1950 The Heaviside Centenary
  Volume, London, 98 p
Appleyard R 1930 in Pioneers of
  Electrical Communication, London,
  211-260
Berg  E J 1936 Heaviside's
  Operational Calculus (McGraw Hill)
*Carslaw H S + 1948 Operational
  Methods..., 2nd ed, London
  (Repr 1963 Dover)
*Cooper J L B 1952 M Gaz 36, 5-19
  (MR13, 612)  [oper calc]
Curry H B 1943 AMM 50, 365
Ettlenger H J 1941 Scr 8, 237-250
Hackett F E 1952 Dub S (NS)26, 3-7
  (MR14, 832) [Fitzgerald, letters]
Halstead P E 1950 Am Sc 78, 610-611
Jackson W 1950 Nat 165, 991-993
Josephs H J 1959 IEE (C)106, 70-76
  (MR23B, 199)
Kennelly A E 1936 Am Ac Pr 70, 544-
  545  (F62, 1040)
Levy J 1950 Lim Rev 52, 59-62
  (Z38, 148)
Macfarlane, A 1893 Phy R 2, 152-154
McLachan N W 1938 M Gaz 22, 255-260,
  485
Minchin G M 1894 Phi Mag (5)38,
  146-156
Murnaghan F D 1927 AMM 34, 234
  [Cauchy]
Poritsky H 1936 AMM 43, 331
  [oper calc]
Russell A 1925 Nat 115, 237-238
  (F51, 29)
Watson-Watt R 1950 Sci Mo 71, 353-358
  (MR12, 311)

Whittaker E T 1930 Clct MS 20, 199-
  220   (F56, 26)


Heawood, Percy John  1861-

Dirac G A 1963 LMSJ 38, 263-277
  (MR28, 2; Z112, 3)


Hecke, Erich  1887-1947

Maak W 1949 Ham Sem 16, 1-6
  (Z32, 98)
Perron O 1948 Mun Jb 1944/48, 274-
  276  (MR11, 573)
Petersson H 1949 Ham Sem 16, 7-31
  (MR11, 573; Z32, 98)


Hedrick, Earle Raymond  1876-

Archibald R C 1938 A Semicentennial
  History of the American Mathemati
  cal Society..., NY, 223-228
  [bib, port]
Ford W B 1943 AMM 50, 409-411
  (MR5, 58; Z60, 15)
Reeve W D 1943 MT 36, 129


Heffter, Lothar  1862-1962

Heffter Lothar  1937 Mein Lebensweg
  und meine mathematische Arbeit,
  Leipzig, 27 p  (F63, 816)
Tautz G L 1964 DMV 66, 39-52
  (Z112, 3)


Heiberg, Johan Ludwig  1854-1928

Junge G 1929 DMV 38, 17-23
  (F55, 19)
Mieli A 1928 Archeion 9, 157-158
  (F54, 39)
Spang-Hanssen E 1929 Bibliografi
  over J. L. Heibergs skrifter,
  Copenhagen, 30 p  (F55, 614)

Heilermann, Johann Bernhard 1820-
1899-

Diekmann J 1900 ZM Ph 45, 57


Heine, Heinrich Eduard 1821-1881

Wangerin A 1928 Mitdeu Lb 3, 429-436
(F54, 35)


Helmert, Friedrich Robert 1843-1917

Berroth A 1944 Z Geoph 18, 87-99
(Z28, 337)
Kruksal W 1946 AMM 53, 435-438
[Helmert distr]
Lancaster H O 1966 Au J St 8(3),
117-126 [chi square]


Helmholtz, Hermann Ludwig Ferdinand
von 1821-1894

Anon 1891 *Ansprachen und Reden
gehalten bei der am 2. November
1891 zu Ehren von Hermann von
Helmholtz veranstalteten Feier,*
Berlin, 63 p (F23, 32)
- 1892 Ber Si, 905-909 (F24, 39)
- 1895 Crelle 114, 353
Dingler H 1935 Z Ph 94, 674-676
[geom]
Du Bois-Reymond E 1897 Rv Sc (4)8,
321, 360
- 1897 *Gedaechtnisrede auf Hermann
v. Helmholtz,* Berlin, 50 p
(F28, 20)
Fuchs L 1895 Crelle 114, 353
(F26, 25)
Kenner M R 1957 MT 50, 98-104
[geom]
Koenigsberger L 1896 *Hermann von
Helmholtz's Untersuchungen ueber
die Grundlagen der Mathematik und
Mechanik,* Heidelberg & Leipzig,
57 p (F26, 26) (tr 1896 Smi R 93)
[port]
- 1902 *Hermann von Helmholtz,*
Braunschweig, 387 p, 399 p, 152 p
(F33, 28)

Ostwald W 1909 in *Grosse Maenner,*
Leipzig, 256-310
Poincaré L 1894 Rv GSPA 5, 771-772
Reiner J 1905 *Hermann von Helmholtz,*
Leipzig, 204 p (F36, 16)
Voit C v 1895 Mun Si 25, 185


Hendricks, Joel E. 1818-1893

Colaw J M 1894 AMM 1, 65-67


Henoch, Max 1841-1890

Lampe E 1891 F 20, 1-6


Henrici, Olaus Magnus Friedrich
Erdmann 1840-

Lindemann F 1927 DMV 36, 157-162
(F53, 28)
Wilson, E B 1904 AMSB (2)11, 17-21


Henrion, Denis XVII

Maupin G 1907 Intermed 14, 131
(Q2360) ['mecometre']


Hensel, Kurt 1861-1941

Hasse H 1932 Marb U Mit, 3-6
(F58, 50)
- 1949 Crelle 187, 1-13
(MR11, 573; Z33, 242)
- 1962 Crelle 209, 3-4 (MR25, 569)


Heraclitos of Ephesos -V

Archibald R 1910 Edi MSP 28, 152-177
[Apollonius, geom]
Sarton G 1927 Int His Sc 1, 85
[bib]

Herbrand, Jacques 1908- 1931

Anon 1932 Rv Met Mor 39 (S)16
  (F58, 50)
Chevalley C 1935 Ens M 34, 97-102
  (F61, 954)
Dreben B + 1963 AMSB 69, 699-706
Hadamard J 1934 Act Sc Ind (109),
  5-6 (F60, 16)


Hercules. See Butrigarius


Herglotz, Gustav 1881-1953

Tietze H 1953 Mun Jb, 188-194
  (MR15, 923; Z55, 5)
- 1953 Mun Si, 163-167
  (MR15, 923; Z56, 2)


Herigone, Pierre XVII

Lemaire G 1906 Intermed 13, 191
  (Q217)


Hermann the Dalmation XII

  = the Carinthian
  = the Slav

Busard H L L 1968 Janus 54, 1-140
  (MR37, 1216) [Euclid]
Sarton G 1931 Int His Sc 2(1), 173
  [bib]


Hermannus the Lame 1013-1054

  = H. Contractus

Yeldham F A 1928 Speculum 3, 240-245
  [frac, tables]


Hermes, Oswald 1826-1909

Brueckner M 1911 DMV 20, 299-306
  (F42, 35)

Hermite, Charles 1822-1901

Adhémar R d' 1905 Quinzain (S), 15 p
  (F36, 22) [Ampére, Cauchy]
Anon 1893 Jubilé de M. Hermite...,
  Paris, 45 p (F25, 51)
- 1901 Mex Alz Mm 16, 61
- 1905 Lon RSP 74, 142-145
Appell P 1901 Rv GSPA 12, 109-110
  (F32, 27)
Borel E 1902 Annuaire, xi-xxii
  (F33, 37) [port]
Brocard H 1907 Mathesis 7, 145
Bryan G H 1901 Nat 63, 350-351
  (F32, 28)
Buhl A 1933 Rv Sc 71, 545-50 [phys]
Capelli A 1901 Nap FM Ri (3)7, 53-54
  (F32, 27)
Cassina U 1951 Lom Rn M 83, 311-328
  (Z45, 148) [Genocchi, Schwarz,
  surf area]
Darboux G 1906 Par Mm (2)49, 1-54
- 1906 Rv Mois 1, 37-58
- 1912 Eloges académiques et discours
  Paris, 116-172
De Gasparis A 1881 Linc At (3)5, 217
  (F13, 38) [letter]
Dini U 1901 Linc Rn 10, 84-88
  (F32, 27)
Duran-Loriga J J 1901 Matiche 1, 2-4
  (tr in AMM 8, 130-133, port;
  MP Ap 1, 30-32)
Forsyth A R 1905 Lon RSP 45, 144
Fuchs L 1901 Crelle 123, 174
Jahnke E 1901 Arc MP (3)1, 184
Jordan C 1900 AMSB (2)7, 277
- 1901 JMPA (5)7, 91-95 (F32, 25)
- 1901 Loria, 16
- 1901 Par CR 132, 101
- 1901 Rv Sc (4)15, 129-131
  [alg, anal, ell fun, fun]
Jordan C + 1901 Mathesis (3)1(S), 47
- 1901 Rv Q Sc (2)19, 353-396
Joubert C 1900-1901 Vat NLA 54, 99
Krause M 1901 Dres Isis, 1-13
Lampe E 1901 Naturw R 16, 333-335,
  348-350 (F32, 26)
- 1914 Arc MP (3)24, 193-220, 289-310
  (F45, 27; 46, 13) [Du Bois Reymond]
Lapparent A de + 1893 Rv Q Sc (2)3,
  235-246 (F25, 51) [anniv]

Loria G 1901 Loria, 20 [bib]
Mathews G B 1901 LMSP 33, 405
Mittag Leffler G 1901 Act M 24, 395-396 (F32, 27)
Noether M 1901 M Ann 55, 337-385 (F32, 24)
Ogigova H 1967 Rv Hi Sc Ap 20, 2-32 [Chebyshev, Markov, Sylvester]
Ovidio E d' 1901 Tor FMN 36, 245
Painlevé P 1905 Nou An (4)5, 49-53 (F36, 22)
Pascal E 1901 Lom Gen (2)34, 171-175 (F32, 26)
Picard E 1901 Act M 25, 87
- 1901 An SENS (2)18, 9-34 (F32, 22)
- 1901 Faler R 15, 132
- 1924 in Mélanges de mathématiques et de physique, Paris, 53-84, 121-125 (Repr of 1901)
Poincare H 1910 Savants et écrivains, Paris, 97-101
Prasad Ganesh 1934 Some Great Mathematicians of the Nineteenth Century ..., Benares 2, 34-59
Pringsheim A 1902 Mun Si, 262-268 (F33, 36)
Schuermans H 1907 Mathesis 7, 145-146
Sonin N J 1901 Pet B (5)14(2), xvii-xviii (F32, 28)
Timchenko J 1901 Kagan 25, 97
Voit C 1902 Mun Si 32, 262

## Heron of Alexandria  I

= Hero

A B 1905 Pe MIS 8, 65-69 (F36, 63) [area]
Agostini A 1951 Pe M 29, 168-169
Albarran F 1936 Rv M Hisp A (2)11, 143-150 (F64, 911) [Bombelli, roots]
Baltzer R 1865 Leip B 17, 3-5
Bardis P D 1965 SSM 65, 535-542 [inventions]
Birkenmajer L 1917 Krak BI (A), 298-299 [triangles]
Boncompagni B 1871 Boncomp 4, 122-126 (F3, 1)
Bruins E M 1957 Bagd CS 2, 99-103 (MR 21, 2355) [roots]
- 1957 Janus 46, 173-182 [icosohedron, Pappus]
- 1964 Codex Constantinopolitanus palatii Veteris No. 1, Leiden 228 p, 151 p, 354 p [mss]
Conte L 1951 Pe M (4)29, 281-282 (Z43, 244) [formula, Huygens]
Curtze M 1897 ZM Ph 42, 113, 145 [Regiomontanus]
Drachman A G 1950 Centau 1, 117-131 [Ptolemy]
- 1963 Centau 8, 91-146 [Archimedes]
Favaro A 1894 Ven I At (7)5, 1117-1132 (F26, 6)
Friedlein G 1871 Boncomp 4, 93-121 (F3, 1)
- 1871 ZMN Unt 2, 173-191, 277-291 (F3, 2)
Gandz S 1949 Isis 32, 263-266 (Z38, 146)
Gessler R 1935 Unt M 41, 73-77 (F61, 12) [roots]
Hofmann J E 1934 DMV 43, 187-210 (F60, 8) [Archimedes, roots]
Hoppe E 1927 Hermes 62, 79-105
Hultsch F 1964 Heronis Alexandrini geometricorum et stereometricorum reliquial, Berlin
Jaglarz A 1891 Heron d'Alexandrie et son probleme sur l'aire d'un triangle, Cracovie
Letronne 1851 Recherches critiques, historiques et géographiques sur les fragments d'Heron d'Alexandrie, Paris
Loria G 1911 Bib M (3)12, 182 (F42, 62) [Jacobi]
Martin T H 1954 Par Mm Div (1)
Neugebauer O 1938 Dan H 26(2), 26 p (F64, 10) [geodesy]
- 1939 Dan H 26(7), 1-11 (F64, 905; 65, 1080) [dist, geodesy]
Padoa A 1934 Pe M 14, 114-18
Rome A 1923 Brx SS 42, 234-258 [geodesy]
Schmidt W 1898 Ab GM 8, 175-194 (F29, 40) [Dasypodius]
- 1898 Ab GM 8, 195-214 (F29, 39) [XVII]

- 1900 Bib M 3, 319    [trig]
- 1902 Bib (3)3, 180    [Leonardo
  da Vinci]
Smyly J G 1944 Hermath 63, 18-26
  (MR6, 253; Z60, 5) [roots]
Tannery P 1881 Bord Mm (2)4, 161-195
  (F13, 30)  [Greek]
- 1882 BSM (2)6, 99-108  [Proclus]
- 1883 Bord Mm (2)5, 305-327
  [stereometry]
- 1883 Bord Mm (2)5, 347-371
- 1884 BSM (2)8, 329-344, 359-376
- 1887 BSM (2)11, 97-108  [Euclid]
- 1887 BSM (2)11, 189-193
  [pseudo-Heron]
- 1894 BSM (2)18, 18-22  (F25, 57)
- 1894 Hist Abt 39, 13-15  (F25, 57)
Tittel K 1907 Bib M (3)8, 113-117
  (F38, 69)
Vailati G 1897 Tor FMN 32, 940-962
  (F28, 49)  [Aristotle]
Van der Waerden B L 1950 Amst Vs G
  59, 103-105  (Z39, 2)
Wertheim G 1899 ZM Ph 44(1A), 1-3
  [roots]
- 1899 ZMN Unt 30, 253-254
  (F30, 39)  [roots]


Herranz, D. A. V.

Anon 1927 Rv M Hisp A (2), 184-185
  (F53, 35)


Herschel, Caroline  1750-1848

Sister May Thomas a Kempis  1955
  Scr 21, 237-251


Hertz, Heinrich Rudolph  1857-1894

Bonfort H 1896 Smi R
Comberiac G 1902 Ens M 4, 247  [mech]
D E J 1894 Nat 49, 265-266  (F25, 46)
Garbasso A 1894 Nuo Cim (3)35, 5-11
  (F25, 46)
Helmholtz H v 1895 Z Ph C Unt 8, 22-
  29  (F26, 24)
- 1905 Rv GSPA 16, 1024

Hertz  Johanna 1927 Heinrich Hertz,
  Erinnerungen, Briefe, Tagebuecher,
  Leipzig, 270 p
Knott R 1905 ADB 50, 256-259
Koenig W 1892/93 Frank PVJ, 13 p
  (F25, 42)
Lodge Oliver 1894 Nat 50, 133-139,
  160-161  (F25, 46)
- 1894 The Work of Hertz and Some
  of his Successors, London,  58 p
  [mech, phys, port]
Manno Richard 1900 Heinrich Hertz fuel
  die Willensfreiheit? Eine kritische
  Studie ueber Mechanismus and
  Willensfreiheit,  Leipzig, 72 p
Mirow B 1957 MN Unt 9, 391-394
  (Z109, 5)
Planck M 1894 Ber Ph 13, 9-29
  (F25, 46)
- 1894 Pogg An 52  (F25, 46)
Poincare H 1897 Rv GSPA 8, 734
  [mech]
Trigoryan A T + 1968 Henry Hertz,
  Moscow (in Russ)
Vedenskij B A 1957 Vop IET (5), 3-8
  (Z80, 4)


Hertzer, Hugo Ottomar  1831-

Lampe E 1909 DMV 18, 417-433
  (F40, 32)


Herzog, Josef  1859-1915

Jahnke E 1915 Ber MG 14, 137-138
  (F45, 50)


Hess, Edmund  1843-1903

Anon 1904 Ens M 6, 146-147  (F35, 31)
Apel B 1904 ZMN Unt 35, 438-443
  (F35, 31)
Loeffler B 1904 MN Bl 1, 22-23
  (F35, 32)
Lorey W 1940 Lebensbi 2, 208-216
  (F66, 19)

Hesse, Emma Viola  1886?-1947

Reeve W D 1948 MT 41, 88


Hesse, Ludwig Otto  1811-1874

Bauer Gustav 1882 *Gedaechtnissrede
  auf Otto Hesse,* Munich, 36 p
Borchardt C W 1875 Crelle 79, 345-347
  (F7, 12)
Cantor M 1880 ADB 12, 306-307
Gundelfinger S 1901? Crelle 124, 80
  [Aronhold]
Klein F 1876 Boncomp 9, 309-314
  (F8, 20)
Kobell N F v 1875 Mun Si, 130-132
  (F7, 11)
Mansion P 1875 Nou Cor 1, 167-168,
  210-211   (F7, 12)
Noether M 1875 Hist Abt 20, 77-88
  (F7, 12)


Hessel, Johann Friedrich Christian
1796-1872

Hess E 1896 N Jb MGP, 107-122
  (F27, 11)


Hessenberg, Gerhard  1874-1925

Anon 1925 ZAM Me 5, 527  (F51, 32)
Leitzmann W 1926 ZMN Unt 57, 232-233
  (F52, 33)
Lorey W 1926 Unt M 32, 38-39
  (F52, 33)
Rothe R 1926 Ber MG 25, 26-44
  (F52, 32)
- 1927 DMV 36, 312-332  (F53, 31)
Salkowski E 1926 Ber MG 25, 45-52
  (F52, 33)  [descr geom]


Hettner, Hermann George  1854-

Hensel K 1914 Crelle 145
Lampe E 1914 Arc MP (3)23/24, 2-7
- 1915 Ber MG 14, 1-7  (F45, 41)
- 1915 DMV 24, 51-58


Heytesbury, William  XIV

  = Hethelbury = Hegterbury
  = Entisberus = Tisberius

Sarton G 1947 Int His Sc 3(1), 565-
  566 [bib]
Wilson Curtis 1956 *William Heitesbury:
  medieval logic and the rise of
  mathematical physics,* Madison,
  231 p


Hilb, Emil  1882-1929

Haupt O 1933 DMV 42, 183-198
  (F59, 42)


Hilbert, David  1862-1943

Aleksandrov N V 1960 Mos IIET 34,
  287-298  (MR33, 3152) [indep thm]
Anon 1922 *Festschrift David Hilbert
  zu seinem sechzigsten Geburtstag
  ...,* Berlin, 566 p  (F48, 26)
*- 1922 Naturw 10(1)  [bib]
*- 1922 Naturw 10(4), 27, 65-104
- 1935 Ber Si, 117-118  (F61, 27
Bachiller T R 1943 Rv M Hisp A (4)3,
  77-81  (Z28, 2)
Bieberbach L 1922 DMV 31, 3-10
  (F48, 26)
- 1922 ZAM Me 2, 79-80 (F48, 26)
Biermann K-R 1964 M Nach 27, 377-384
  (Z117:2, 250)
Brocard H + 1901 Intermed 7, 194
  (Q1630)
Caratheodory C 1943 Mun Si, 350-354
  (MR8, 3; Z60, 15)
Chakalov L 1962 Bulg FM 5, 126-135
  (MR26, 687)
Courant R 1932 Fors Fort 8, 39-40
  (F58, 52)
Freudenthal H 1957 Nieu Arch (3)5,
  105-142  (MR20, 738) [geom]
Hamel G 1935 DMV 45, 1-2  (F61, 27)
Heisenberg W 1943 Phy Z 44, 277-278
  (Z28, 337)
Hilbert D 1964 *Hilbertiana. Fuenf
  Aufsaetze,* 111p  (MR33, 1216)

Hoefling O 1932 Bildmess 7, 145-149
(58, 995)
- 1932 Unt M 38, 29-31
Lemoine E + 1900 Intermed 7, 105
Lietzmann W 1932 ZMN Unt 63, 37-39
(F58, 52)
Machado B 1943 Gaz M Lisb 4(14), 1-2
(Z60, 15)
Myller-Lebeden V 1962 Gaz MF (A)14,
47-50   (ZM 100, 247)
Sommerfeld A + 1943 Naturw 31, 213-
214   (MR5, 58; Z28, 2; 60, 15)
Steck M 1940 Heid Si (6), 8p
(MR2, 306)
Van Veen S C 1943 Zutphen 11, 159-169
(MR7, 355; Z60, 15)
Weyl H 1932 Naturw 20, 57-58
(F58, 50)
- 1944 AMSB 50, 612-654   (Z50, 15)
- 1944 Lon RS Ob 4, 547-553
(MR6, 142; Z60, 15)
- 1946-1947 Sao Pau SM 1, 76-104;
2, 37-60   (Tr of 1944)   (MR10, 175)
Weyl H + 1932 DMV 42, 67-68
(F58, 995)


Hill, George William   1838-1914

Anon 1914 Am JS (4)37, 486
*Archibald R C 1938 A Semicentennial
History of the American Mathemati-
cal Society,  NY, 117-
[bib, port]
Brown E W 1914 Nation 98, 540-541
*- 1915 Lon RSP (A)19, xlii-li
(also in AMSB (2)21, 499-511)
(F45, 41; 46, 25)
- 1916 Am NASBM 8, 273-309   [port]
Magnus W + 1966 Hill's Equation
(Interscience), 127 p [bib]
Woodward R S 1920 As J 28, 161-162


Hill, Micaiah John Muller   1856-
1929

Filon L N G 1929 LMSJ 4, 313-318
(F55, 20)


Himstedt, Franz   1852-1933

Konig W 1932 Fors Fort 8, 264


Hintikka, E. A.

Bonsdorff I 1936 Sk Akt 19, 129-130
(F62, 1042)


Hipparchos   -II

Aaboe A 1955 Centau 4, 122-125
[Babylonia]
Berger H 1869 Die geographischen
Fragmente des Hipparch, Leipzig
Filippoff L 1931 Rv Sc, 38-47
Hultsch F 1899 ZM Ph 44(S), 191-
209   (F30, 59)   [angle]
Marchand G + 1896 Intermed 3, 49
(Q625)   [Ptolomy]
Neugebauer O 1957 In Weinberg S  S
ed, the Aegean and the Near East.
Studies Presented to Hetty Goldman
Locust Valley, NY, 292-296
(Z92, 242)
Tannery P 1902 Intermed 9, 85


Hippasos   -V

Fritz K 1945 An M (2)46, 242-64
[irrat]
Heller S 1958 Ber Ab(6)


Hippias of Ellis   -V

Strycker E de 1937 in Mélanges Emile
Boisacq, Brussels, 317-326
- 1941 Antiq Cl 10, 25-36   [irrat]


Hippocrates of Chios   -V

Becker O 1936 QSGM (B)3, 411-419
(Z14, 145)   [area]
Bottari A 1954 Pe M (4)32, 223-230
(MR16, 433)
Rey Abel 1946 L'Apogée de la Science
Technique Grecque...Les Mathémati-
ques d'Hippocrate à Platon,

Paris, 331 p  (MR8, 497)
Rudio F 1902 Bib M (3)3, 7
  [Antiphon, area, Simplicius]
- 1905 Zur NGV 50, 177-200, 224
  (F36, 63)
- 1907 *Der Bericht des Simplicius
  ueber die Quadraturen des Antiphon
  und des Hippokrates,* Leipzig, 194 p
  (Urk GMA 1)  (F38, 64)
Tannery P 1878 Bord Mm (2)2, 179-184
  (F10, 2)  [lunes]
- 1886 BSM (2)10, 213-226


**Hirst,** Thomas Archer  1830-1892

Anon 1892 Nat 45, 399-400  (F24, 31)
- 1893 Lon RSP 52, xii-xviii
  (F25, 39)


**Hjelmslev,** Johannes T.  1873-1950

  = Johannes Petersen

Bohr H 1950 Act M 83, vii-ix
  (MR11, 708; Z35, 148)
Fog D 1950 M Tid A, 1-20
  (MR12, 311; Z37, 291)


**Hlavaty,** Vaclav  1894-

Truesdell C 1953 Int M Nach (29/30),
  2 p


**Hobbes,** Thomas  1588-1679

Cajori F 1929 AMSB 35, 13  [phil]
- 1929 MT 22, 146-151  [Wallis]
Hervey H 1952 Osiris 10, 67-90
  (Z46, 1)  [Cavendish, Descartes,
  Pell]
Hoffmann J C V 1901 ZMN Unt 32, 262-
  267 (F32, 7)
Kaminski S 1958 Stu Log 7, 43-69
  (MR20, 5119)  [defs]


**Hobson,** Ernest William  1856-1933

Anon 1934 Lon RS Ob 1, 237-249
Hardy G H 1934 Nat 133, 938-939
  (F60, 15)
- 1934 LMSJ 9, 225-237  (F60, 15)
Piaggio H T H 1931 M Gaz 15, 461-465
  (F57, 1310)
Prasad G 1933 Clct MS 25, 31-54
  (F59, 852)


**Hocevar,** Franc  1853-1919

Povshich J 1953 Beo MF 2, 226-232
  (Z53, 340)


**Hodgkinson,** Jonathan  1886-

Ferrar W L 1940 LMSJ 15, 236-240
  (MR2, 115; Z27, 196; F66, 1188)


**Hoeckner,** Georg  1860-1938

Schoenweise R 1935 in *Festschrift
  Georg Hoeckner* (Mittler), 1-23
  (F61, 957)


**Hoelder,** Ludwig Otto  1859-1937

Lichtenstein L 1930  Fors Fort 6,
  12-13  (F56, 817)
Van der Waerden B L 1938 Leip Ber 90,
  97-102  (F64, 23; Z20, 197)
- 1938 M Ann 116, 157-165  (F64, 919)


**Hoffmann,** J. C. Volkmar  1825-1905

Lietzmann W 1926 ZMN Unt 57, 34-35
  (F52, 26)
Schotten H 1902 ZMN Unt 33, 4-9
  (F33, 48)

**Hoffmann,** Johann Joseph Ignatz 1777-1866

Grunert J A 1869 Arc MP Lit 49(186), 1-5  (F2, 21)


**Hohenberg,** Herwart von   XVI-XVII

Glaisher J W L 1875 Cam PSP  2, 386-392


**Holmgren,** Erik Albert  1872-1943

Carleman T 1945 Act M 76, i-iii (Z60, 15)
Nagell T 1944 Nor M Tid 26, 1-2 (Z28, 337)


**Holroyd,** Ina Emma    -1951

Anon 1952 MT 45, 77


**Holst,** C. X

Anon 1873 Nou An (2)12, 433-435 (F5, 36)


**Holst,** Elling Bolt  1849-

Stoermer C 1919 Nor M Tid 1, 81-87 (F47, 34)
- 1949 Nor M Tid 31, 85-88 (Z40, 1)


**Holzmueller,** Gustav  1844-1914

Lorey W 1915 ZMN Unt 46, 117-124 (F45, 42)


**Hood,** Thomas  XVI

Johnson F R 1942 J H Ideas 3, 94-106


**Hooke,** Robert 1635-1703

Buchdahl G 1954 Scr 23, 77-82
Hall A R 1951 Isis 42, 219-230 (Z45, 147)


**Hoover,** William 1850-

Finkel B F 1894 AMM 1, 35-37


**Hopf,** Ludwig  1884-1939

Ewald PP 1940 Nat 145, 379-380 (MR2, 115)
Sommerfeld A + 1952-1953 Aach THJ, 24-26


**Hopkinson,** John  1849-1898

Glazebrook R T 1898 LMSP 29, 727-731 (F29, 19)


**Hoppe,** Ernst Reinhold Eduard 1816-1900

Archibald R C 1948 in *Mathematical Table Makers,* NY, 30  (Scr Stu 3) [bib]
Lampe E 1901 Arc MP (3)1, 4-19 (F32, 20)
- 1901 DMV 9, 33
Lorenz F 1901 DMV 9, 59  (F32, 16)
Meyer A 1928 Archeion 9, 513-514 (F54, 39)


**Horner,** William George  1786-1837

Coolidge J L 1949 *The Mathematics of Great Amateurs* (Clarendon Pr), Ch 15   (Repr 1963 Dover)


**Horsburgh,** Ellice Martin 1870-1935

Whittaker E T 1936 Edi MSP (2)4, 272-273  (F62, 27)

Horta, F. da Pnata

Teixeira F G 1900 Teixeira 14, 3


Hotelling, Harold 1895-

Madow W G + 1960 in Harold Hotelling
  *Contributions to Probability and
  Statistics,* Stanford, 3-   (MR22,
  11442-11443)
Pfouts R W ed 1960 *Essays in
  Economics and Econometrics: A
  Volume in Honor of Harold Hotelling,*
  U North Carolina P, 251 p
  (MR23B, 467)


Houel, Guillaume Jules 1823-1886

Barbarin P 1926 BSM 50, 50-64,
  74-88 [De Tilly]
Brunel G 1888 Bord Mm (3)4, 1-78
Halsted G B 1897 AMM 4, 99-101


Howard, J. L. XVIII

Lodge O J 1900 Phi Mag 49, 160


Howland, Raymond Clarence James

Filon L N G 1937 LMSJ 43, 158-160
  (F63, 19)


Hronec, Juraj 1881-1959

Anon 1960 Prag Pok 5, 114-115
  (Z87, 243)
Harant M + 1956 Brat Act M 1, 145-
  148 (Z73, 242)
Kolibiar M + 1960 Cas M 85, 218-225
  (MR22, 770; Z100, 247)


Hruska, Vaclav 1888-

Pleskot V 1956 Stroj ZI 3, 9-14
  (MR19, 1030)


Hube, Michat 1837-1907

Dickstein S 1937 Wiad M 43, 183-186
  (F63, 817)


Huber, Daniel 1768-1829

Spiess W 1939 Schw Verm 37, 11-17,
  21-23 (F65, 1090) [least sqs]


Huber, M. T. 1872-1950

Drobot S 1953 Zastos M 1, 55-65
  (MR14, 1050)
- 1954 Colloq M 3, 63-72
  (MR15, 770; Z56, 2)


Hudde, Jan 1628-1704

  = Johann Huddenius

Boyer C 1965 MT 58, 33-36
  [anal geom, space]
Brocard H 1911 Intermed 18, 272
  (Q1058)
Charadze E K 1943 Gruz Soob 4, 195-
  199 (Z60, 7) [Cardano, cubic]
Coolidge J L 1949 in *The Mathemati-
  matics of Great Amateurs* (Clarendon
  Press), Ch 10   (Repr 1963 Dover)
Haas K 1956 Centau 4, 235-284
Korteweg D J 1896 ZM Ph 41, 22-23
  (F27, 10)
Lorey W 1938 Euc Gron 14, 285-291
  [Descartes]


Hudson, Ronald William Henry Turnbull
1876-1904

Anon 1904? Nat 30, 533
Bryan G H 1904 Ens M 6, 484
  (F35, 36)
Cameron J F 1905 LMSP (2)2, xv-xvii
  (F36, 26)
Forsyth A R 1905 Brt AAS 75, 307
F S M 1904? M Gaz 3, 73-75

Hudson, William H H 1838-1916

Anon 1915 Nat 96, 118   (F45, 50)
- 1916 M Gaz 8, 245-246
- 1917 LMSP (2)15, App 52
  (F46, 35)


Hugh of Saint Victor   XII

   = Hugo de S. Victore
   = Hugonis

Baron R 1956 Osiris 12, 176-224
  (MR18, 630)  [geom]
Liouville R 1931 Par EP (2)28, 1-14
Sarton G 1931 Int His Sc 2(1), 193-
  194  [bib]
Tannery P 1901 Bib M 3, 41


Hugues d'Omerique   XVII

   = Hugo de Omerique

Berenguer P A 1895 Intermed 2, 204
  (Q604)  [Newton]
- 1895 Prog M 5, 116-121
Braid H + 1899 Intermed 6, 110
  (Q604)  [Newton]


Hukahara, Masuo   1905-

Anon 1966 Fun Ekv 9, 2-17
  (MR34, 7323)
Yosida K 1966 Fun Ekv 9, ii-v,
  ix-xvii   (MR34, 7324)
  [bib, port]


Hultsch, Friedrich Otto   1833-

Rudio F 1907 Bib M (3)8, 325-402
  (F38, 32)


Humbert, Georges   1859-1921

d'Adhemar R 1921 Rv GSPA 32, 97-98
Anon 1921 Par CR 172, 189-191
  (F48, 18)

Borel E 1926 Par Mm (2)58(1), i-xix
  (F52, 31)
Carvallo G 1921 Par EP (2)21, 1-3
Jordan C 1921 JMPA (8)4, 1-2 (F48, 18)
Lebesgue H 1922 BSM (2)46, 220-233
- 1922 Rv Sc, 249-262  [Jordan]
Lemoine G 1921 Rv Sc, 88-89
  (F48, 18)
Ocagne M d' 1930 Rv Q Sc 17, 77-82


Humbert, Pierre   1891-

Sergescu P 1954 Arc In HS (NS)7,
  181-183  (MR15, 923)


Humboldt, Alexander von   1769-1859

Anon 1959 Alexander von Humboldt:
  *Gedenkshrift zur 100 Wiederkehr*
  *seines Todestages,* Berlin
Bierman K-R 1958 Ber Hum MN 8, 121-
  130
- 1959 *Ueber die Foerderung deutscher*
  *Mathematiker durch Alexander von*
  *Humboldt,* Berlin, (Z85, 6)
  (repr from Anon 1959, 83-159)
- 1963 Ber Mo 5, 445-450
  (Z113, 2)  [Lagrange]
- 1963 Nor M Tid, 59-63   [Abel]
- 1966 Ber Mo 8, 33-37
  [Weierstrass]
Biermann K-R + 1959 Fors Fort 33,
  136-140  [Gauss]
- 1962 Fors Fort 36, 41-44  [Gauss]
Bruhns Ked 1877 *Briefe zwischen A. von*
  *H. und Gauss,* Leipzig, 79 p
Foerster W 1894 *Ueber das Zusammen-*
  *wirken von Bessel, Encke und A. von*
  *Humboldt unter der Regierung*
  *Friedrich Wilhelm III,* Berlin, 21p
  (F25, 1912)
Gerardy T 1959 Gott N(2), 64-66
  (Z88, 244)  [Gauss]
Hoffmann J E 1959 *Alexander von*
  *Humboldt in seiner Stellung zur*
  *reinen Mathematik und ihrer*
  *Geschichte,* Berlin (repr from Anon
  1959, 239-287)   (Z88, 5)
Maire A 1912 Fr AAS 40, 66-72
  (F43, 22)  [letters, Arago]

Mittag-Leffler G 1912 Act M 35,
29-65
Simons L G 1939 Scr Lib 4, 25-43
Théodoridès J 1961 Rv Hi Sc Ap 14,
329-330 [Lacroix, letter]


Hume, David 1711-1776

Atkinson R F 1960 Phi Q 10 (39),
127-137
Gossman L ed 1960 J H Ideas 21, 442-
449


Hume, James XVII

Boyer C B 1950 AMM 57, 7-8
[exponents]


Hunter, William 1899-1949

Bowman F 1950 LMSJ 25, 353-354
(Z37, 2)


Huntington, Edward Vermilye
1874-1952

Bridgman P W + 1953 AMSB 59, 399
(MR14, 832)


Hurewicz, Witold 1904-

Lefschetz S 1957 AMSB 63, 77-82
(MR19, 108; Z77, 8)


Hurwitz, Adolph 1859-1919

Anon 1919 Ens M 20, 452 (F47, 13)
Hilbert D 1920 Gott N, 75-83
(F47, 13)
- 1921 M Ann 83, 161-172
(F48, 15)
W H Y 1922 LMSP (2)20, 48-54
(F48, 15)

Hutton, Charles 1737-1823

Archibald R C 1948 *Mathematical
Table Makers,* NY, 30-32 (Scr Stu 3)
[bib]
Bruce J 1823 *Memoires of Charles
Hutton,* Newcastle
Gregory O 1823 Imperial Magazine (March)
Raine A 1946 M Gaz 30, 71-81
[letters, Harrison]


Huygens, Christian 1629-1695

= Huyghens

Anon 1886 *Liste alphabétique de la
corréspondance de Christiaan
Huygens qui sera publiée par
société hollandaise des sciences à
Harlem,* Harlem
- 1905 Par CR 140, 1377
Archibald R C 1921 AMM 28, 468-280
Beattie Lester M 1935 *John Arbuthnot,
Mathematician and Satirist,*
Cambridge, Mass, 448 p
[translator of Huygens]
Bell A E 1940 Nat 146, 511-514
(F66, 16; Z25, 2)
- 1941 *Christian Huygens and the
Development of Science in the
Seventeenth Century,* London
(also NY-London 1947)
- 1941 Nat 148, 245-248 (Z28, 99)
Bertrand J 1868 Nouv Ann (2)7, 229-
230 [log]
- 1868 Par CR 66, 565-567 [log]
Bierens de Haan D 1889 Fr AAS 18,
233-237
- 1892 Fr AAS 21, 159-166 (F24, 15)
Biernacki W 1895 Wszech 14, 417-420,
436-439 (F26, 14)
Bosmans H 1912 Rv Q Sc 22, 573-586
[Galileo]
- 1927 Rv Q Sc 11, 113-141
[Descartes]
Bosscha J 1895 *Christian Huygens.
Rede am 200. Gedaechtnistage...,*
Leipzig, 77 p (F26, 14)
- 1895 Ned NGC 5, 583
1896 BSM (2)20, 33-64 (F27, 10)
- 1896 Ned Arch 29, 352

- 1900 Bib M (3)1, 93-96 (F31, 10)
Boyer C B 1947 Isis 37, 148-149
  [graphs]
Brocard H 1897, 1902 Intermed 4,
  191; 9, 147
Bureau F 1947 Bel BS (5)32, 730-744
  (MR9, 74; Z61, 4)
Cavallaro V G 1939 Pe M (4)19, 45-49
  (Z20, 196)
Conte L 1951 Pe M (4)29, 281-282
  (Z43, 244) [Heron]
- 1952 Archim 4, 214-217 (Z48, 243)
  [Fagnano]
- 1953 Pe M (4)31, 145-157
  (MR15, 276) [mean propor]
Costabel P 1957 Rv Hi Sc Ap 10,
  120-131 (Z98, 6)
Crommelin C A 1931 Physica 11, 359-
  364 (Z3, 242) [pendulum]
- 1938 Wis Nat Ti 9, 1-19 (F64, 915)
- 1939 C Huyg 17, 247-270
  (F65, 1087)
- 1956 Sim Stev 31, 5-18
  (MR18, 268; Z71, 244; F48, 244)
  [calc, curve]
De Vleeschauwer H J 1941 Bel Vla Me
  3 (Z27, 290) [Tschirnhausen]
Dijksterhuis E J 1929 C Huyg 7, 161-
  180 [mech]
- 1937 Gids 101, 331-336 (F63, 807)
- 1951 Christian Huygens (Bij de
  Voltooiing van zijn Oeuvres
  Completes), Haarlem, 29 p
  (MR14, 343; Z45, 147)
- 1953 Centau 2, 265-282 (MR14, 1050;
  Z53, 197)
Dugas R 1953 Par CR 237, 1477-1478
  (MR15, 276; Z51, 3)
  [Descartes, Newton]
- 1954 Rv Hi Sc Ap 7, 22-33
  (MR15, 770)
Grigoryan A T 1963 in Ocherki Istorii
  Matematiki i Mekhaniki, Mos, 218-
  [oscillation]
Grimm G 1949 Elem M 4, 78-85
  (MR11, 572) [circle, pi]
Guenther S 1876 Vermischte Unter-
  suchungen zur Geschichte der
  mathematischen Wissenschaften,
  Leipzig, 352 p (F8, 30)
Harting P 1867 Christian Huygens in

zijn leven en werken..., Groningen
  77 p
Hofmann J E 1952 MN Unt 4, 321-323
  (MR13, 612; Z47, 5) [pi]
- 1966 Arc HES 3, 102-136
  (MR34, 5616) [area, circle, length]
Kiessling 1869 Chr. Huygens "De
  circuli magnitudine inventa" als
  ein Beitrag zur Lehre vom Kreise,
  Flensburg (F2, 12)
Koppe M 1901 Bib M (3)2, 224 [approx]
Korteweg D J 1887 Amst Vs M (3)3,
  253-283 [Descartes]
- 1887 Ned Arch 22, 422-466
  [Descartes]
- 1900 Bib M (3)1, 97 [catenary]
- 1908 Intermed 15, 84, 186 (Q3232)
Le Heaux J 1913 Wisk Tijd 10, 103
  [lemniscate]
Loria G 1898 Mathesis (2)8, 265
- 1942 Vat Com 6, 1079-1136 (MR10, 17
Mascart J 1907 La découverte de
  l'anneau de Saturne par Huygens,
  Paris
McCarthy J P 1936 M Gaz 20, 280-281
  (F62, 27) [Pythagoras]
Milne R M 1903 M Gaz 2, 309-311
  [length]
Monchamp G 1894 Bel Bul (3)27, 255-
  308 (F25, 20)
Nijland A A 1928 C Huyg 7, 192-208
  [ast]
Picard E 1929 Rv GSPA 40, 321-323
  (F55, 605)
Puiseux P 1905 J Sav, 596-605
  (F37, 5)
Roth L 1924 Par Mor Tr, 411-424
  [Descartes]
Rudio F 1892 Archimedes, Huygens,
  Lambert, Legendre, vier Abhandlunge
  ueber die Kreismessung...,
  Leipzig, 173 p (F24, 50)
Sanford V 1934 MT 26, 52-53
Schouten J A 1920 DMV 29, 136-144
  (F47, 5)
Schuh F 1921 C Huyg 1, 1-28
- 1928 De eerste uitingen van het
  genie van Christiaan Huygens
  C Huyg 7, 214-217
Smith D E 1921 AMM 28, 166 [Descartes
Struik D J 1958 Het land van Stevin
  en Huygens, Amsterdam, 147 p

Tannery Mme P 1933 Rv Sc 71, 481-482
  (F59, 843) [Descartes, letter]
Tannery P 1892 BSM (2)16, 247-255
  (F24, 15)
Thomas F 1868 Par CR 66, 661-664
  (also Nou An (2)7) (F1868, 13)
Van Geer P 1906 Nieu Arch (2)7,
  215-226, 438-454 (F38, 67)
- 1906 *Christiaan Huygens'
  Leesjahren,* Tijdspgl (S), 22 p
  (F37, 4)
- 1906 *Christiaan Huygens' Reis en
  Studiejhren,* Tijdspgl (S), 28 p
  (F37, 4)
- 1907 *Christiaan Huygens en Isaac
  Newton,* Tijdspgl (S), 23 p
  (F38, 11)
- 1907 *Christiaan Huygens' verblijf
  te Parijs (1666 bis 1681),*
  Tijdspgl (S)
- 1908 Nieu Arch (2)8, 1-24, 34-68,
  145-168, 289-314, 444-464 (F39, 16)
- 1908 *Christiaan Huygens en Gottfried
  Wilhelm Leibniz,* Tijdspgl (S) (F39, 16)
- 1908 *Christiaan Huygens laatste
  Levensjahren,* Tijdspgl (S) (F39, 16)
- 1909-1911 Nieu Arch (2)9, 6-38,
  202-230 338-358 (F40, 16; 41,
  10; 42, 65)
- 1912-1913 Nicu Arch (2)10, 39-60,
  178-198, 370-395 (F44, 54)
Van Tricht V 1877 Nou Cor 3, 209-210
  (F9, 6)
Vollgraff J A 1928 C Huyg 7, 181-191
- 1940 Janus 44, 198-201
  [Tschirnhaus]
- 1948 Arc In HS 28, 165-179
  (Z31, 97)
- 1951 Arc In HS (NS)4, 634-637
  (MR13, 197; Z42, 242) [letters]
Wieleitner H 1929 Unt M 35, 109-117
  (F55, 12)

## Hypatia  IV.

Coolidge J L 1949 Scr 17, 20-31
Kingsley Charles 1853 *Hypatia,*
  London [historical novel]
Mauthnes F 1892 *Hypatia,* Stuttgart,
  238 p

Richeson A W 1940 M Mag 15, 74-82
  (F66, 11)
Rome A 1926 Brx SS 46, 1-14
  (F52, 9)

## Hypsicles of Alexandria  -II

Falco V 1930 QSGM (B)1, 278-300
  [Autolycos]
Friedlein G 1873 Boncomp 6, 493-529
  (F5, 4)
Sarton G 1927 Int His Sc 1, 181-182
  [bib]

## Iacob, Caius

Gheorghitya S I 1962 Gaz MF (A)14,
  545-556 (Z112, 3)

## Iamblichos  III-IV

Hultsch F 1895 Gott Ph HN, 246-255
Sarton G 1927 Int His Sc 1, 351-352

## Ibanez de Ibero, Carlos  1825-1891

Bertrand J 1891 Par CR 112, 266-269
  (F23, 30)
Galdeano Z G de 1891 Prog M 1, 25-26
  (F23, 30)

## Ibbetson, William John  1861-1889

Anon 1889 LMSP 20, 424-427

## Imshenetskii, Vasilii Grigorevich
## 1832-1892

Andreev K A 1892 Khar M So (2)3,
  296-300 (F24, 31) [bib]
- 1896 *Zhizn i nauchnaya deyatelnost
  Vasiliya Grigorevicha Imshenetsgogo,*
  Mos
Andreev K A + 1896 Mos MO Sb 18, 347-
  367 (F27, 19)

Isnoskof I 1893 Kazn FMO (2)3(2)
Suvorov T 1892 Kazn FMO (2)2
(F24, 31)


Inaudi, Jacques 1867-1950

Flammarion C 1892 Astmie 11, 114
[prodigy]
Thévenot J 1948 Radio Actualité,
Lausanne, 16 July 1948


Ince, Edward Lindsay 1891-1941

Aitken A C 1941 Nat 148, 309-310
(Z27, 196)
Whittaker E T 1941 LMSJ 16, 139-144
(MR3, 98; F67, 970; Z28, 195)
Young A W 1941 Edi MSP (2)6, 263-
264 (Z60, 15)


Ingvarssoen. See Petrus


Ino, Tadataka

Otani R 1918 JP Ac P 1, 171-239
(F46, 29)


Ionescu, Ion

Campan F T 1950 Ro An MFC 3, 683-738
(Z53, 340)


Isaac ben Salomon

Eneström G + 1902 Intermed 9, 300
(Q1905)
Tannery P + 1903 Intermed 10, 159
(Q1905)
Sarton G 1948 Int His Sc 3(2), 1520


Ise, Ernesto

Brunelli P 1920 Vat NLA (2)25, 11-18

Ishaq ibn Hunain IX-X

= Abu Yaqub Ishaq ibn Hunain ibn
Ishaq al-Ibadi

Baudoux C 1937 Archeion 19, 70-71
[Euclid]
Sarton G 1927 Int His Sc 1, 600-601
[bib]


Ishlinski, Aleksandr Yulevich 1913-

Savin G N + 1963 Ukr IM 15, 299-302
(Z115, 246)


Isidore of Seville 560?-636

Sánchez Pérez J A 1929 Rv M Hisp A
(2)4, 35-53 (F55, 9)
Sarton G 1927 Int His Sc 1, 471-472
[bib]


Isserlis, Leon

Irwin J O 1966 Lon St (A)129, 612-
616 (MR34, 1148)


Iuga, George 1871-1958

Andonie G 1961 Gaz MF (A)13, 215-
217 (Z94, 5)


Ivanov, Ivan Ivanovich 1862-1939

Kuzmin R O 1940 SSSRM 4, 357-362
(F66, 21; Z24, 244; F66, 21)


Iwata, Kosan

Hayaski T 1895 Tok M Ph 6, 41
Mizuhara J 1891 Tok M Ph 4, 267
Terao H 1885 Tok M Ph 1, 151
Ueno K 1925 Tok S Ph S 34, 43-49
[geom]

Ja.  See Ya.

Jackson, Charles Samuel  1867-

Anon 1917 M Gaz 9, 1, 45-49
  (F46, 35)

Jackson, Dunham  1888-1946

Anon 1933 Am NAS Mm, 142-171
Hart W L 1948 AMSB 54, 847-860
  (Z31, 99)

Jackson, Frank Hilton  1870-

Chaundy T W 1962 LMSJ 37, 126-128
  (Z99, 246)

Jacob ben Machir  1236-1304

Ginsburg J 1932 Scr 1, 72-78, 153-155
  [Menelaus, spher trig]
Sarton G 1931 Int His Sc 2(2), 850-
  853  [bib]
Steinschneider M 1876 Boncomp 9,
  595-613  (F8, 13)
- 1888 Bib M, 13-16

Jacob of Florence  XIII-XIV

Karpinski L C 1929 Archeion 11,
  170-177  (F55, 10)

Jacobi, Carl Gustav Jacob  1804-1851

Ahrens W 1904 MN Bl 1, 165-172
  (F35, 13)
- 1906 Bib M (3)7, 157-192
  (F37, 11)
- 1906 MN Bl 3, 191-194, 209-212
  (F37, 11)  [Steiner]
- 1907 C G J Jacobi als Politiker,
  Leipzig, 45 p  (F38, 22)
Anon 1853 Arc MP 22, 158-
Appell P 1906 J Sav, 132-138
  (F37, 11)

Archibald R C 1948 *Mathematical
  Table Makers,* NY, 32-34  (Scr Stu 3)
  [bib]
Biermann K-R 1961 Crelle 207, 96-112
  (MR24A,4; Z97, 243)  [iter fun]
Bottema O 1947 Euc Gron 23, 121-127
  [Grassman, Poncelet]
Dirichlet G L 1851 Crelle 42, 91-
- 1852 Ber Ab
Dyck W v 1901 Mun Si 31, 203
  [F. Neumann]
Gaiduk Yu M 1956 Mos M Sez, 165-166
- 1959 Ist M Isl 12, 245-270
  (Z101, 249)  [Russ]
Gundelfinger S 1898 Ber Si, 342-345
  (F29, 31)
- 1908 Bib M (3)9, 211-226  (F39, 63)
  [ell fun]
Jahnke E 1902 Arc MP (3)4, 277-280
  [Steiner]
- 1903 Arc MP (3)4, 37-40  (F34, 13)
  [Leverrier]
*Koenigsberger L 1904 *C. G. J. Jacobi,*
  Leipzig, 572 p  (F35, 11)
- 1905 Int Con 3, 57-85  (F36, 11)
Krazer A 1910 Bib M (3)10, 250-9
  [ell intgl, Puiseux]
Lampe E 1901 Arc MP (3)2, 253
Loria G + 1906 Intermed 13, 39
  [Greek]
Meyer U H 1854 Arc MP 22, 474-
Natucci A 1952 Geno AL 9, 40-54
  [ell fun]  (MR14, 1049)
Prasad Ganesh  1933 *Some Great Mathe-
  maticians of the Nineteenth Century
  ...,* Benares 1, 166-219
Saltykow N 1939 BSM (2)63, 213-228
  (F65:2, 1090)  [part diff eq]
Schlesinger L 1905 Bib M (3)6, 88
  [complex anal]
- 1910 Bib M (3)11, 138-152, 275-276
  (F41, 61)
Schwarz H A 1904 DMV 13, 433-435
  (F35, 13)
Staeckel P + 1908 *Der Briefwechsel
  Zwischen C. G. J. Jacobi und
  P. H. v. Fuss ueber die Herausgabe
  der Werke Leonhard Eulers,*
  Leipzig, 196 p  (F39, 20)

Jacobsthal, Ernst Erich  1882-

Selbert S 1965 Nor VSF 38, 70-73
  (MR33, 17)

Jacquemet, Claude  1651-1729

Marre A 1879 Boncomp 12, 886-899
  (F11, 12)  [Bizance]
- 1880 BSM (2)4, 200-207
  (F12, 9)  [letters]
- 1883 Boncomp 15, 679-683  [letters]

Jadanza, Nicodemo  1847-1920

Berardinis, G de 1920 Nap Pont (2)25,
  39-53
Brocard H 1908 Intermed 15, 106
  (Q3074)  (Also Intermed 13, 142;
  14, 36)
Panetti M 1919 Tor FMN 55, 401-417

Jager, Robert  XVII

Patterson B C 1939 Isis 31, 25-31
  (F65:2, 1088)

Jahnke, Eugen  1863-

Baruch A 1922 Ber MG 21, 30-39
  (F48, 16)
St Jolles  1922 DMV 31, 177-184

James, Charles Gorden Fleming
1898-1926

Bath F 1931 LMSJ 6, 153-160  (F57, 47)

James, Glenn  1882-1961

Hyers D H 1959 M Mag 34, 310

Jamitzer, Wentzeln  1508-1588

Jamitzer W 1956 Elem M 11, 97-100
  (MR18, 268)

Janisch, Eduard  1868-1915

Anon 1915 Arc MP (3)34, 288
  (F45, 51)
Schmid T 1917 DMV 26, 158-160
  (F46, 18)

Janiszewski, Zygmunt  1888-1920

Anon 1920 Fund M 1,  vii  (F47, 16)
Dickstein S 1921 Wiad M 25, 91-98
Knaster B 1960 Wiad M 4, 1-9
  (MR22, 1131; Z91, 6)

Janko, Jaroslav  1893-

Truska L 1954 Cas M 79, 181-185
  (MR16, 434; Z59,:1, 12; Z60, 12)

Janssen, Pierre  1824-

Belopoliskij  1908 Pet B(3), 231-232
Nordmann C 1907 Rv GSPA 18, 223-225

Jarankiewicz, Kazimierz  1902-1959

Bergman S + 1964 Colloq M 12, 277-288
  (MR30, 872)

Jarnik, Vojtech  1897-

Anon 1958 Cz MJ 8, 155-161  (Z78, 3)
Knichal V + 1957 Cas M 82, 463-492
  (Z98, 9)
Korinek Vladimir + 1958 Prag Pok 3,
  1-8  (MR20, 264; Z83, 245)

Jasinski, Feliks  1856-1899

Jasinski S 1956 Arc Me Sto 8, 259-291
  (MR19, 81)
Wierzbicki W 1956 Arc Me Sto 8, 293-
  317  (MR19, 81)


Jayadeva, Acarya

Shukla K S 1954 Ganita 5, 1-20
  (MR17, 117; Z59:1, 3; 60, 3)


Jbikowsky, A. K.

Netschaief, N V 1900 Kazn FMO (2)10,
  39


Jean.  See also Johann, Giovanni
       John


Jean de Linières  XIV

Baldi B 1879 Boncomp 12, 352-438
Bigourdan G 1915 Par CR 161, 714-717,
  753-758  [bib, mss]
- 1916 Par CR 162, 18-23, 61-67
Boncompagni B 1879 Boncomp 12,
  420-428  (F11, 10)
Curtze M 1895  Bib M (2)9, 105-106
  (F26, 8)
Sarton G 1947 Int His Sc 3(1), 649-
  652  [bib]
Steinschneider M 1879 Boncomp 12,
  345-351  (F11, 9)  [Johannes
  Siculus]
Wittstein A 1895 Hist Abt 40, 121-125
  (F26, 4)


Jean de Meurs  XIV

  = John (Joannes) of Meurs (Murs,
  Muris, de Morys)

Clagett M 1952 Isis 43, 236-242
  (MR15, 277)  [Archimedes]
Eneström G 1908 Bib M 8, 216

Karpinski L C 1912 AMSB (2)18, 440
  (F43, 66)  [alg]
*- 1912 Bib M 13, 99-114  (F43, 66)
- 1913 AMSB (2)19, 294  (F44, 60)
- 1917 Sci 45, 663-665
  [deci frac]
Nagl A 1890 Ab GM 5, 135-146


Jeans, James Hopwood  1877-1946

Milne E A 1947 Lon As Mo N 107, 46-53
  (Z29, 242)
  1947 Lon RS Ob 5, 573-589
  (MR12, 311)
- 1952 Sir James Jeans--a Biography,
  London, 192 p  (Z48, 243)


Jefferson, Thomas  1743-1826

Martin Edward T 1952 Thomas Jefferson:
  Scientist, NY, 299 p
Smith D E 1932 Scr 1, 3-14, 87-90
  (Repr 1934 Scr Lib 1)


Jeffery, George Barker  1891-1957

Anon 1959 LMSJ 34, 251-256
  (Z83, 245)


Jeffery, Henry Martyn  1826-1891

Anon 1891 LMSP 22, 476-481
  (F23, 29)


Jellet, John Herwitt  1817-1888

Anon 1888 Nat 37, 396-397


Jensen, J. L. W. V.  1859-1925

Norlund N E 1926 M Tid B, 1-7
  (F52, 31)
Polya G 1927 Dan M Med 7 (17)
  [alg fun]

Jerabek, Vaclav

Roháchek J 1932 Cas MF 61, 105-108
  (F58, 51)


Jevons, William Stanley  1835-1882

Halsted G B 1878 Mind 3, 134-137
  [Boole]
Jevons H S + 1934 Ecmet 2, 225-237


Jimenez, D. E.

De Toledo L O 1912 Esp SM 2, 1-5
  (F43, 23)


Johann von Gemunden  1380?-1442

Benjamin Francis S 1954 Osiris 11,
  221-246  [Campanus of Novara]
Curtze M 1896 Bib M (2)10, 4
  (F27, 6)
Steinschneider M 1896 Bib M (2)10,
  96  (F27, 6)


John.  See also Jean


John of London  XIII

Boncompagni B 1879 Boncomp 12, 352-
  439
Fontès M 1897 Tou Mm (9)9, 382-386
  (F28, 3)  [Petrus Peregrinus]
- 1898 Tou Bu 1, 146


John of Saxony  XIV

Baldi B 1879 Boncomp 12, 352-438
Sarton G 1947 Int His Sc 3, 118


Joly, Charles Jasper  1864-1906

Anon 1906 Ens M 8, 159  (F37, 30)
G B M 1906 LMSP (2)4, xiii-xiv
  (F37, 30)

MacFarlane A 1908 Quatrns, 46-51
  (F39, 27)


Jones, Hugh  1692-1760

Phalen H R 1949 AMM 56, 461-465
  (Z33, 241)  [octaves]


Jonghe, Ignatius de 1632-1692

Bockstaele P 1967 Janus 54, 228-235


Jonquieres, Jean P. Ernest de
  Fauque de   1820-1901

Anon 1904 Mathesis (3)4, 224-225
Guyove 1903 Par CR 136, 1021-1031
  (F34, 26)
Loria G  1902 Bib M (3)3, 276-322
  (F33, 37)
- 1902 Loria 5, 71-82  (F33, 38)
- 1947 Scr 16, 5-15
  (MR9, 74; Z29, 194)


Jordan, Camille  1838-1922

Adhémar R d' 1922 Rv GSPA 23, 65-66
  [bib]
Bertin E + 1922 Par CR 174, 209-211
  (F48, 18)
Bianchi L 1922 Linc Rn (5)31, 398-
  404  (F48, 18)
Bosmans H 1926 Rv Q Sc 9, 165-166
  (F52, 27)
Buhl A 1922 Ens M 22, 214-218
  (F48, 18)
Lebesgue H 1922 Rv Sc (22 Ap), 249-
  262  (Repr 1957 Ens M (2)3, 188-)
- 1923 Notice sur la vie et les
  travaux de Camille Jordan, Paris,
  28 p  [port]
- 1926 Par Mm (2)58, xxxix-lxvi
  (F52, 27)  [port]
- 1957 Ens M (2)3, 81-106  (Z84, 245)
Picard E 1922 Par CR 175, 1257
Teofilato P 1921 Vat NLA (a)75, 122-
  123
Villat H 1922 JMPA (9)1, i-iv (F48, 18)

Jordan, Charles  1871-1959

Anon 1956 M Lap 7, 291-294
  (MR20, 848) [bib]
- 1960 Hun Act M 11, 1-2 (Z87, 243)
Kendall D G 1960 LMSJ 35, 380-383
  (MR23A, 136; Z94, 5)
Renyi A 1952 M Lap 3, 111-121
  (MR17, 3)
Takács L 1961 An M St 32, 1-11
  (MR22, 1131; Z94, 5)


Jordanus Nemorarius  1200?-1237

  = J. Teutonica
  = J. Saxo (Saxonia)
  = J. de Nemorel
  = Jordan of Namur

Curtze M 1880 Leop 16  (F12, 6)
- 1891 Hist Abt 36, 1-23, 41-62,
  81-95, 121-138 (F23, 6)
Duhem P 1904 Bib M (3)5, 321-325
  (F35, 64)
- 1905 Bib M (3)6, 9-15  (F36, 50)
  [mss]
Eneström G 1904 Bib M (3)5, 9-14
  (F35, 56)
- 1905 Bib M (3)6, 214-215
  (F36, 65) [Fibonacci]
- 1906 Bib M (3)7, 38-43
- 1907 Bib M (3)8, 135-153
  (F38, 57)
- 1912 Bib M (3)13
- 1913 Bib M (3)14, 41-54 (F44, 43)
  [frac]
Ginzburg B 1936 Isis 25, 241-362
  [calculus, Duhem]
Karpinski L C 1910 AMM 17, 108-113
  [Sacrobosco]
O'Connor J R 1931 Dominica 16, 128-
  137
Schreider S N 1959 Ist M Isl 12, 654-
  678 (MR24A, 127; Z100, 244)
- 1959 Ist M Isl 12, 679-688
  (MR24A, 127; Z10, 244) [alg]
Sterneck R D v 1895 Mo M Phy 7, 165-
  179 (F27, 5)
- 1896 Mo M Phy 7, 165
Treutlein P 1879 Hist Abt 24, 125-166
  (F11, 34)

Wertheim G 1900 Bib M 3, 417


Josephus Sapiens  X

Curtze M 1894 Bib M (2)8, 13-14
  (F25, 15) [Gerbert]
Suter H 1894 Bib M (2)8, 84
  (F25, 15)
Weissenborn H 1893 Bib M (2)7, 21-23
  (F25, 15) [Gerbert]


Joubert, P. C.

Carrara P B 1907 Vat NLA 60, 69-82
  (F38, 37)


Jourdain. See also Giordano


Jourdain, Philip E. B.  1879-1919

Anon 1920 Monist 30, 181-182
  [bib]
- 1921 Arc Sto 2, 167-184
  (F48, 24)
D M W 1921 LMSP (2)19, lix-lx
  (F48, 24)
Jourdain L + 1923 Isis 5, 126-136
  [bib, port]
Loria G 1921 Arc Sto 2, 167-184


Ju.  Se Yu


Juan y Santacilia, Don Jorge  1713-
1773

Leon y Ortez E 1912 Esp SM (a)2, 67-
  116


Judaeus.  See Philon

Judah ben Salomen na-Kohen    XIII

= Ibn Matqah

Sarton G 1931 Int His Sc 2(2), 603-
604    [bib]


Judd, Charles Hubbard    1873-1946

Reeve W D 1946 MT 39, 291-292
[psych]


Juel, Christian Sofus    1855-1935

Fog D 1935 M Tid B, 3-15    (F61, 24)
Jepsen H 1925 M Tid A, 1-3
(F51, 32)


Juergens, Enno    1849-1907

Krause M 1908 DMV 17, 163-170
(F39, 29)


Julia, Gaston Maurice    1893-

Anon 1954 It XL An (1953), 455-457
(MR15, 591)


Julius, Victor August    1851-1902

Bakhuyzen H G van d S 1902 Amst Vs G
11, 3


Jung, Franz    1872-

Basch A 1958 Ost Ing 11, 327-328
(Z79, 5)


Jung, Giuseppe    1845-1926

Maggi G A 1927 Milan Sem 1, 17
(F53, 34)
- 1927 Lom Gen (2)60, 291-307
(F53, 34)


Jung, Heinrich Wilhelm Ewald    1876-
1953

Keller O H + 1955 DMV 58, 5-10
(Z64, 3)
- 1955 Halle UM 4, 417-422
(Z64, 243)


Kaderavek, Frantisek    1885-

Havlicek K 1961 Prag Pok 6, 231-234
(Z96, 4)
Kepr B 1960 Prag M Ap 5, 479-485
(Z93, 9)
- 1962 Cz MJ 12(87), 157-158 (Z100, 247)


Kaestner, Abraham Gotthelf    1719-
1801

Bopp K 1928 Heid Si (18)    (F54, 47)
[Lambert]
Goe G 1964 Int Con HS 10, 659-661
[Gauss, geom, Hilbert, Pasch, rigor]
Peters W S 1962 Arc HES 1, 480-487
(Z101, 246)    [non-Eucl geom]


Kagan, Veniamin Fedorovich    1869-
1953

Dubnov Ya S + 1949 Mos Ve Ten 7, 16-
30    (MR12, 576; Z39, 5)
- 1956 Mos Ve Ten 10, 3-21
(MR18, 550; Z72, 246)
Efimov N V + 1949 UMN 4(2), 5-14
(MR10, 668; Z32, 98)
Rasevskii P K 1953 UMN 8(5), 131-138
(MR15, 591; Z51, 243)


Kalmar, Laszlo    1905-

Peter R 1955 M Lap 6, 138-150
(MR17, 446)

Kamil, abu    IX-X

= a. K. Shoja ben Islam ibn
Muhammad ibn Shiya al-hasib
al-Misri

Karpinski L C 1912 Bib M (3)12, 40-55
  [alg]
- 1914 AMM 21, 37-48  [alg]
Levey M 1958 Ens M (2)4, 77-92
  (Z82, 9)  [alg]
  1966 The "Algebra" of abu Kamil in
  a commentary by Mordecai Finzi,
  Madison, 237 p  (MR34, 5613)
Suter H 1910 Bib M (3)10, 15-42
  [polygon]
- 1910 Bib M (3)11, 100-120
  (F41, 53)  [arith]


Kant, Immanuel  1724-1804

Anon 1891 Kagan (120)  [numb]
Briedenbach W 1924 ZMN Unt 55, 112-
  114  [neg nums]
Couturat L 1904 Rv Met Mor 12, 321-
  383  [logic, phil]
Emden E 1934  Naturw 22, 533-535
  (F60, 11)  [surf]
Galle A 1925 Weltall 24, 194-200,
  230  (F51, 23)  [Gauss]
Mansion P 1907 Brx SS 31(A), 243-
  245  (F38, 73)  [mistakes]
- 1908 Mathesis (3)8(S)  [Gauss,
  non Eucl geom]
*- 1908 Rv Neosco 15(4), 441-453
  [Gauss]
Nelson L 1909 Gott Anz, 979-1001
  [Koenig]
Peters W S 1963 Arc HES 2, 153-167
Suppes P 1967 in Anton J P ed,
  Naturalism and Historical
  Understanding, NY, 108-120
Werner C 1913 Int Con Ph 2, 182-
  192  [space]


Kantorovich, Leonid Vitalevich  1912-

Akilov G P + 1962 UMN 17(4), 201-215
  (Z105, 9)

Anon 1962 Sib MJ 3, 5-6   (Z100, 247)


Kapinski, Stanislaus  1867-1908

Zorawski K 1908 Wiad M 11, 161-167


Karahisi.  See Omar.


Karapetoff, Vladimir  1876-1948

Ettlinger H J 1940 Scr 8, 237-250
Karapetoff V 1940 Scr 7, 63-67


al-Karkhi    X-XI

= Abu Bakr Muhammad ibn al-Hasan
(Husain) al-hasib al Karkhi

Amir-Moez A R 1957 Scr 23, 173-178
  (MR20, 945; Z84, 241)
  [Ezra, summation]
Sarton G 1927 Int His Sc 1, 718-719
  [bib]


Karman, Theodore von 1881-1963

Blenk H 1956 Z Flug 4, 161
  (MR18, 182)
Dryden H 1963 Astnt A En 1(6), 12-17
- 1963 Nat 199, 20-21   (Z109, 5)
Gabrielli G 1963/64 Tor FMN 98,
  471-485  (MR30, 872)
Ippen A T 1956 J Aer Sci 23, 438-443,
  499  (MR17, 931)  [hydrodyn]
Karman T v 1968 Die Wirbelstrasse.
  Mein Leben fuer die Luftfahrt,
  Hamburg
Karman I v + 1967 The Wind and
  Beyond, (Little Brown), 376 p
Malina F J 1963 Rv Fr Asq (NS)4,
  149-151
Millikan C B 1941 in Karman Anniver-
  Anniversary Volume (Calif Inst
  Tech)
- 1963 Lon Aer 67, 615-617
Pritchard J L + 1963 Lon Aer 67, 611-
  617

Rannie W D 1963 West Aer 43 (June),
10-11
Roy M 1963 Par CR 256, 4545-4551
(Z106, 3)
Schultz-Grunow F 1963 ZAM Me 43,
521-522   (Z113, 2)
Sears W R 1964 Phy Fl 7, v-viii
- 1965 SIAM J 13, 175-185 [port]
Taylor G I 1963 J Fl Me 16, 478-480
(Z109, 5)
Wattendorf F L 1956 Z Flug 4, 163-165
(Z70, 244)


**Karpinsky,** Louis Charles  1878-

Jones P S 1956 Sci 124, 19


**al-Kashi** XIII-XIV

= Jemshid ibn Medud ibn Mahmud
Giyat ed-din a.K.

Aaboe A 1954 Scr 20, 24-29
(MR14, 923; Z56, 1) [approx,
trig]
Dakhel Abdul-Kader 1960 *Al-Kashi on
root extraction,* Beirut, 49 p
(MR25, 739; Z111, 3) [roots]
Kennedy E S 1948 Isis 38, 56-59
(Z29, 241) [ast]
- 1950 Isis 41, 180-183
(Z37, 290) [ast]
- 1960 Oriental 29, 191-213
(Z102, 245) [letter]
Luckey Paul 1951 *Die Rechenkunst bei
Gansid b. Masud al-Kasi,* Wiesbaden,
153 p  (MR13, 611)
Struik D J 1959 Sim Stev 33, 65-71
(Z101, 8) [decimal]
Yushkevich A P + 1954 Ist M Isl 7,
380-449  (MR17, 1)


**Katetov,** Miroslav

Cech E 1953 Cas M 78, 277-281
(Z53, 340)

**Kavan,** Jiri

Nusl F 1933 Cas MF 62, 382-383


**Kawaguchi,** Akitsugu  1902-

Ide S 1963 Tensor (NS)13, i-xiv
(MR27, 900; Z109, 239)


**Kaye,** G. R. 1866-1929

Smith D E 1929 Archeion 11, 230-231
(F55, 22)


**Keck,** Wilhelm

Birk A 1902 Biog J 5, 185-186
(F33, 35)


**Keldysh,** Ludmilla Vsevolodovna  1904

Aleksandrov P S + 1955 UMN 10(2),
217-223  (MR17, 3; Z64, 4)


**Keldysh,** Mstislav Vsevolodovich
1911-

Anon 1961 Matv Shk (4) 70-71


**Kellogg,** Oliver Dimon 1878-1932

Birkhoff G D 1933 AMSB 39, 171-177
(F59, 42)
Coolidge J L 1933 Am Ac Pr 68, 642-
644  (F59, 856)


**Kelly,** Mary   1870-1941

Reeve W D 1942  MT 34, 326-327


**Kelvin,** William Thomson  1824-1907

Anon 1908 LMSP (2)6, xv-xix

Brillouin M 1908 Rv Mois 5, 257-271
  rebe L 1908 MN Bl 5, 38 (F39, 31)
  anssen P + 1908 Lon RSP (A)81,
  lxxvii-xciii
Negro C 1908 Rv FMSN 17, 61-70 (F39,31)
Picard E 1920 Rv Sc, 193-207
- 1922 Discours et melanges, Paris,
  41-73
- 1922 Par Mm (2)57, i-xxxix
  (F48, 24)
Poincaré H 1910 Savants et écrivains,
  Paris, 213-244
Sraer M G 1960 Mos II ET 34, 103-109
  (Z117-2, 9) [potential]


Kemmochi, Akiyuki

Hayashi T 1920 Toh MJ 18, 302-308
  (F47, 28)    [sphere]


Kempe, Alfred Bray   1849-1922

Anon 1919 Wisk Tijd 16, 242-4
Geikie A 1923 Lon RSP (B)94, i-x


Kepinski, Stanislaus Martin   1867-

Zorawski K 1908 Wiad M 12, 161-167
  (F39, 38)


Kepler, Johannes 1571-1630

Anon 1932 MT 24, 184-185
Anschuetz C 1886 Prag Si, 417-523
Archibald R C 1948 in Mathematical
  Table Makers, 34-42  (Scr Stu 3)
  [bib, port]
B A S 1954 M Gaz 38, 44-46
Baumgardt Carola 1951 Johannes
  Kepler, NY, 209 p  (MR12, 578;
  Z42, 2)
Blaschke Wilhelm 1943 Galilei und
  Kepler, Leipzig-Berlin, 14 p
  (MR10, 668; Z28, 99)
- 1954 Batt (2)5, 309-334 (MR15,
  923; Z56, 2) [Galileo]
Breitschwert J L E v 1831 Johann
  Kepplers Leben und Werken, Stuttgart

Bruhns C 1872 Leip Ber 24, 31-49
  (F4, 11)
Caspar M 1928 in Festschrift zur 30
  Hauptversammlung des mathematischen
  und naturwissenschaftlicher
  Unterichts zu Stuttgart, 44-53
- 1932 Unt M 38, 227-229  (F58, 34)
  [calc]
- 1938 Bla D Phi 12, 39-49  (F64, 913)
- 1948 Johannes Kepler, Stuttgart,
  179 p  (Z33, 2)
- 1962 Kepler (Collier), 466 p [pap]
  1930 Johannes Kepler in seinen
  Briefen, Munich, 424 p, 364 p
  (F56, 16)
- 1936 Bibliographia Kepler iana,
  Munich, 158 p  (Z14, 50) [bib]
Diessmann G A 1894 Johann Kepler und
  die Bibel, Ein Beitrag zur
  geschichte der Schriftautoritaet,
  Marburg, 34 p (F25, 1912)
Dittrich A 1913 Cas MF 42, 237-245
Eneström G 1913 Bib M (3)13, 229-241
  [intgl, trig]
Epstein P 1924 ZMN Unt 55, 142-151
  [log]
Frisch 1855 Arc MP 24, 286  [log]
Hasner J V 1872 Tycho Brahe und
  J. Kepler in Prague,  Prague
  (F4, 10)
Kubach F 1935 Johannes Kepler als
  Mathematiker, Karlsruhe, 83 p
  (Diss Heidelberg)  (F62, 1032;
  Z14, 147)
Maier J 1931 MN Bl 25, 1-10
  (F57, 1300)
Neugebauer O 1961 Comm PAM 14, 593-
  597  (MR24A, 221; Z106, 1)
Peinlich R 1874 Arc MP Lit 56, 15-17
  (F6, 24)
Rogner J 1872 Arc MP 54, 447-458
  (F4, 10)
Slovka H 1931 Cas MF 60, 49-56
  (F57, 24)
Steck Max 1941 Ueber das Wesen das
  Mathematischen und die Mathematische
  Erkenntnis bei Kepler, Leipzig,
  32 p  (Z25, 291)
Struik D J 1931 in Johann Kepler,
  A tercentenary Commemoration of His
  Life and Works, Baltimore

Taylor C 1900 Cam PST 18, 197
  [geom, Newton]
Tosi A 1950 Pe M (4)28, 159-168
  (MR12, 382; Z41, 338)
Tuck B H 1967 MT 60, 58-
Viola T 1946 Pe M (4)24, 68-83
  (Z60, 8) [conics]
Wagner H 1917 Gott Ph HN, 254-267
Weiss E A 1940 Deu M 5, 262-265
  (MR2, 114)
Wieleitner H 1922 DMV 31, 175
- 1930 Regensb 19, 279-313
  (F56, 811) [volume]
- 1930 Unt M 36(6), 176-185  [vol]
Wolf R 1872 *Johannes Kepler und
  Jost Buergi,* Zurich  (F4, 11)


Kerekes, Ferenc

Szenassy B 1957 Deb U Act 3(2), 3-12
  (MR19, 1030; Z77, 8)


Kerschensteiner, Georg

Anon 1932 Gree SM 13, 35-38
  (F58, 995)
Cramer H 1932 Unt M 38, 145-148
Lietzmann W 1932 ZMN Unt 63, 141-143


Kersseboom,  Willem  1691-1771

Brocard H 1911 Intermed 18, 272
  (Q1058)


Keyser, Cassius Jackson  1862-1947

Bell E T 1948 Scr 14, 27-33
Lasker E 1938 Scr 5, 121-123
  (F64, 24)  [Epicurus]


Khalilov, Zaid Ismaid  1911-

Ageev G N 1961 Az FRT (2), 3-12
  (Z95, 4)
- 1961 UMN 16(5), 231-237
  (MR24A, 221)

Kharkevich, Aleksandr Aleksandrovich

Bloh E L + 1966 Pro Per In 2, 3-13
  (MR33, 14)


Khayyam.  See Omar.


Khinchin, Aleksandr Yakovlevich
  1894-

Crámer H 1962 An M St 33, 1227-1237
  (Z105, 244)  [prob]
Doob J L 1961 Berk SMSP 2, 17-20
  (MR24A, 221)
Gnedenko B V 1955 UMN 10(3), 197-212
  (MR17, 3; Z64, 3)
- 1960 Teor Ver 5, 3-6 (MR24A, 222)
- 1961 Berk SMSP 2, 1-15
  (MR24A, 333; Z99, 246)
Gnedenko B V + 1960 UMN 15(4), 97-110
  (MR23A, 698; Z91, 6)
Khinchin A Ya 1968 *The Teaching of
  Mathematics* (Amer Elsevier)  [bib]


al-Khwarizmi   IX

  = al-Khowarizmi  = Alchwarazmi

Bjornbo A A 1905 Bib M (3)6, 239
  [translations]
- 1909 in Anon ed, *Festskrift til
  H. G. Zeuthen,* Copenhagen, 1-17
  (F40, 64)  [tables trig]
Burckhardt J J 1961 Zur NG/ 106, 213,
  231  (Z103-2, 242)  [ast]
Curtze M 1898 Ab GM 8, 1-27
Dumont M 1947 Rv GSPA (2)54, 7-13
  (MR9, 74; Z29, 1)
Dunlop D M 1943 Lon Asia, 248-250
Gandz S 1936 Osiris 1, 263-277
  (Z13, 193)  [alg]
- 1938 Osiris 5, 319-391
  (F64, 11; Z19, 387)  [finance]
Karpinski L C 1910 Bib M (3)11, 125-
  131  (F41, 53)  [Robert of Chester]
*- 1915 *Robert of Chester's Latin
  translation of the Algebra of Al-
  Khowarizmi,* NY, 172 p  [bib, biog,
  Robert of Chester]

Kennedy E S 1958 Scr 27, 55-59
[calendar, Hebrew]
Kennedy E S + 1965 Centau 11(2), 73-78
Marre A 1846 Nou An 5, 557-
[alg geom]
- 1865 Annali 7, 269-
Nagl A 1889 ZM Ph 34, 129-146, 161-170
Rodet L 1878 J Asi (F10, 6)
[alg, Greek, Hindu]
Ruska J 1917 Heid Si Ph, 1-125
Sarton G 1934 Isis 21, 209
[Roomen]
Simon M 1911 Arc MP (3)18, 202-203
(F42, 52)
Suter H 1902 Bib M 3, 350-354
- 1903 Bib M (3)4, 127-129
(F34, 5)
- 1914 *Die astronomischen Tafeln des Muhammed Ibn Musa Al Chwarizmi...,* Copenhagen, 290 p
Wieleitner H 1922 ZMN Unt 53, 56-67
(F48, 41)
Yushkevich A P 1954 Mos IIET 1, 85-127
(MR16, 660; Z59:1, 4; 60, 4)


Kiepert, Friedrich William August Ludwig   1846-1934

Lorey W 1934 Versich 5, 211-217
(F60, 835)


Kikuchi, Dairoku   1855-1917

Fujisawa R 1917 Tok M Ph (2)9, 171-177 (F46, 28)
- 1918 Jp Ac P 1, 246-259
(F46, 28)


Kilchevskii, Nikolai Aleksandrovich 1909-

Savin G N + 1959 Ukr IM 11, 431-433
(MR23A, 272; Z87, 7)


Killing, Wilhelm   1847-

Engel F 1930 DMV 39, 140-154
(F56, 27)


Killingworth, John   XV

Karpinski L C 1913 AMSB (2)20, 63
(F44, 60)   [algorism]
- 1914 Engl HR (Oct), 707-717
[algorism]


King, George   1846-1934

Braun H 1934 Bla Versi 3, 58-60
(F60, 836)
R C S 1933   Lon Actu 64, 241-263


Kinkelin, Hermann 1832-1913

Schaertlin G 1933 Schu Vers 28, 1-17   (F59, 848)


Kirchhoff, Gustav Robert   1824-1887

Kistner A 1925   Unt M 30, 51-54
(F50, 16)
Voigt W 1888 *Zum Gedaechtnis von G. Kirchhoff*, Goettingen, 10 p


Kiricov, Gheorghe

Ionescu I 1939 Gaz M 44, 225-227
(F65-1, 17)


Kirik Novgorodets   XII

Raik A E 1965 Ist M Isl (16), 187-189
Zubov V P 1953 Ist M Isl 6, 192-195
- 1953 Ist M Isl 6, 196-212


Kirilov, G   1845-1908

Joachimescu A 1908 Gaz M 14, 97-99

Kirkman, Thomas Penyngton 1806-1895

Macfarlane Alexander 1916 in *Lectures
on Ten British Mathematicians of
the Nineteenth Century,* NY, 122-
133


Kirkwood, Daniel 1814-1895

Aley R J 1894 AMM 1, 141-149


Kirsch, Ernst Gustav 1841-1901

Lorenz F 1902 DMV 11, 188


Kitao, Diro 1884-

Nakamur S 1907 Tok M Ph (2)4, 188-191


Klapka, Jiri 1900-

Havel V 1961 Prag Pok 6, 48-49
(Z98, 10)


Klein, (Christian) Felix 1849-1925

Anon 1919 DMV 27, 59-60 (F47, 21)
- 1925 It UM 4, 191-192 (F51, 28)
- 1925 M An 95, 1
- 1925 ZAM Me 5, 358-359 (F51, 28)
- 1926 ZAM Me 6, 84 (F52, 29)
- 1927 Rv M Hisp A (2)2, 153-154
(F53, 29)
Archenhold F S 1925 Weltall 25, 17-
18 (F51, 28)
Archibald R C 1914 AMM 21, 247-259
Bachiller T R 1925 Rv M Hisp A 7,
181-184 (F51, 28)
Baker H F 1926 LMSJ 1, 25-32
(F52, 30)
Behnke H 1960 MP Semb 7, 129-144
(MR23A, 135)
Berzolari L 1925 Lom Gen (2)58,
691-696 (F51, 22)
Bianchi L 1926 Linc Rn (6)4, xxvi-
xxxv (F52, 29)

Blumenthal O 1928 DMV 37, 1-3
(F54, 37)
Castelnuovo G 1926 Annali (4)3, 241-
245 (F52, 29)
- 1928 Rm Sem 5, 28-29 (F54, 37)
Courant R 1925 DMV 34, 197-213
(F52, 29)
- 1925 Naturw 13, 765-772 (F51, 28)
- 1925 Gott N, 39-46 (F52, 29)
Dyck W v 1925 Deu NA Mit 2, 21-22
(F51, 29)
Fano G 1934 Tor Sem, 151-171
(F60, 833)
Fehr H 1925 Ens M 24, 287-290
(F51, 29)
Fontanilla R 1927 Rv M Hisp A (2)2,
148-152 (F53, 29)
Halsted G B 1894 AMM 1, 417-420
Hamel G 1926 Ber MG 25, 69-80
(F52, 29)
Heegard P 1926 Nor M Tid 8, 45-51
(F52, 29)
Heger R 1903 ZMN Unt 34, 302-304
(F34, 28)
Klein F 1923 Gott UBM 5, 11-36
Kneser H 1949 Arc M 1, 413-417
(MR11, 573; Z32, 51)
Koenig J C 1927 Rv M Hisp A (2)2,
144-147 (F53, 29)
Lietzmann W 1925 ZMN Unt 56, 257-263
(F51, 28)
Lorey W 1926 Ber MG 25, 54-68
(F52, 29) [educ]
- 1926 Leop 1, 136-151 (F52, 29)
- 1926 Z Versich 26, 124-125
(F52, 29)
- 1950 Deu Vers M 1(1), 39-50
(MR15, 90) [ins, Laplace]
Manger E 1934 DMV 44, 4-11
Miller G A 1927 Am NASP 13, 611-613
(F53, 29) [historiog]
1933 M Stu 1, 49-51 (F59, 827)
[hist]
- 1933 Rv M Hisp A 8, 184-187
(F59, 827) [hist]
- 1933 Sch Soc 38, 120-121
(F59, 827) [educ]
- 1942 Sci 95, 353-354 [group, Lie]
Mises R v 1924 ZAM Me 4 86-92
(F50, 16)
Noerlund N E ed 1923 Act M 39, 94-132
[Poincaré]

Parfentiev N N 1926 Kazn FMO (2)25,
31-40 (F52, 30)
Pascal E 1925 Batt 63, 205-208
(F51, 29)
- 1926 Nap FM Ri 31, 144-148
(F52, 29)
Poske F 1925 Deu NA Mit 2, 22-23
(F51, 28) [educ]
Pranotl L 1926 Ber MG 25, 81-87
(F52, 29)
Prasad Ganesh 1934 in *Some Great
Mathematicians of the Nineteenth
Century...,* Benares 2, 245-
Timerding H E 1925 Unt M 31, 193-203
(F51, 28)
Walther A 1926 ZT Ph 7, 2-7
(F52, 29)
Weyl H 1930 Naturw 18, 4-11
(F56, 26)
Young W H 1928 Lon RSP (A)121, i-xix
(F54, 37)

Klingenstierna, Samuel 1698-1765

Heyman H J 1927 in *Symbola litteraria
Hyllninjsskrift till Uppsala
Universitet,* Uppsala, 187-209
Oseen C W 1925 *Samuel Klingenstiernas
Levnad och Verk,* Stockholm, 69 p

Klos, Thomas XV

Baraniecki M A 1889 *Algorithmus von
Thomas Klos aus dem Jahre 1538,*
Krakow, 80 p (in Polish)
Dziwinski P 1888 *"Algoritmus" von
Thomas Klos,* Lemberg, 24 p
(in Polish)

Kluyver, J. C. 1860-1932

Anon 1933 Nieu Arch (2) 18, 1-2
(F59, 36)

Knapman, Herbert 1880-1932

E H N 1932 Nat 130, 426-427

Kneser, J. C. Chr. Adolph 1862-

Koschmieder L 1930 Ber MG 29, 78-102
(F56, 29)

Knight, Jonathan 1787-1858

Dorwart H L 1950 Scr 16, 181-85

Knoblauch, Johannes 1855-1915

Hensel K 1915 Crelle, 146(3)
Reiher H 1932 ZT Ph 13, 297-298
Rothe R 1915 Arc MP (3)24, 15, 19-28
- 1915 DMV 24, 443-457
(F45, 51)
- 1915 MN Bl 12, 101-106 (F45, 51)

Knopp, Konrad Hermann Theodor
1882-1957

Anon 1957 MZ 67, i (MR19, 1030)
Kamke E + 1957 DMV 60, 44-49
(Z77, 242)
Loebell F 1958 Mun Jb, 187-189
(MR22, 425)

Knott, Cargill Gilston 1856-1922

Milne A 1922 Edi MSP 40, 50-51

Koch, Carl Ferdinand 1812-1891

Schubring G 1891 ZMN Unt 22, 633-635
(F23, 32)

Koch, Helge von 1870-

Anon 1925 Act M 45, 345-348
(F51, 33) [bib]

Kochanski, Adamus Adamandus XVII

Dickstein S 1901 Prace MF 11, 225-278
(F32, 7) [Leibniz]

- 1902 Prace MF 13, 237-283
  (F33, 17) [Leibniz]
- 1935 Slav Cong 2, 245-246
  (F61, 18)


**Kochin,** Nikolai Evgrafovich
1901-1944

Anon 1945 Pri M Me 9, 3-12
  (MR7, 106; Z60, 15)
- 1946 UMN (NS)1(1), 27-29
  (Z61, 5)


**Koehler,** Carl August  1855-1932

Heffter L 1932 Heid Tat, 7-15
  (F58, 991)
- 1934 DMV 44, 199-210  (F60, 837)
Levy L 1889 JM Sp (3)3, 68-70


**Koenig,** Johann Samuel  1712-1757

Graf J H 1889 *Der Mathematiker
Johann Samuel Koenig und das
Princip der kleinsten Action,*
Bern, 46 p


**Koenig,** Julius  1849-

Koenig D 1914 MP Lap 23, 291-302


**Koenigs,** Gabriel Xavier Paul  1858-
1931

Buhl A 1932 Ens M 39, 286-287
  (F58, 992)
D E P 1932 Rv M Hisp A (2)7, 85-87
  (F58, 995)


**Koenigsberger,** Leo  1837-

Anon 1910 Ber Si, 530-531  (F41, 46)
Bopp K 1923 DMV 33, 104-112
  (F50, 16)
Koenigsberger L 1919 *Mein Leben,*
  Heidelberg  (F47, 9)

**Koersma,** J.

Loria G + 1898 Intermed 5, 200
  [cardioide]


**Koessler,** Milos  1884-

Kopriva J 1961 Prag Pok 6, 226-230
  (Z96, 4)
Nozicka F 1962  Cz MJ 12, 153-156
  (Z100, 247)


**Kohn,** Gustav  1859-

Mueller E 1922 Mo M Phy 32, 281-293
  (F48, 19)


**Kolacek,** Frantisek  1851-1913

Anon 1932 Cas MF 62, V1-V8
Kucera B 1914 Cas MF 44, 129-141
  (F45, 34)
Novak V 1912 Cas MF 41, 432-442
  (F43, 51)
Zaviska F 1912 Cas MF 41, 273-303


**Kollros,** Louis  1878-

Saxer W 1959 Elem M 14, 97-100
  (Z88, 7)
- 1959 Helv SN 139, 423-426
  (MR21, 107; Z112, 243)


**Kolmogorov,** Andrei  Nikolaevich
1903-

Aleksandrov P S 1963 Mos UM Me
  (1)18(3), 3-6   (MR27, 4; Z117:2,
  25)
Alexsandrov P S + 1953 UMN 8(3),
  177-200  (MR15, 90; Z50, 2)
Anon 1953 SSSRM 17, 181-188
  (MR15, 90; Z50, 2)
- 1963 UMN 18(5), 121-123
  (MR29, 639)  [bib]
Gnedenko B V + 1963 Teor Ver 8, 167-
  174  (MR27, 165)  [prob]

Mihoc G 1954 Ro Sov (3)7(1), 105-118
(MR16, 207) [prob]


Kolosov, G. V. 1867-1936

Muskhelishvili N 1938 UMN 4, 279-
281


Komensky, J. A.

Cupr K 1935 Slav Cong 2, 244-245
(F61, 17) [geom, ms]


Koralek, Filip

Psota F 1958 Prag Pok 3, 361-365
(Z82, 243)
Rvchlik K 1960 Prag Pok 5, 472-478
(Z98, 7)


Korinek, Vladimir 1899-

Holubar J + 1959 Prag Pok 4, 724-
730 (Z87, 7)
Schwarz S 1959 Cas M 84, 222-235
(Z85, 6)


Korkin, Aleksandr Nicolaevich
1837-1908

Ozhigova E P 1968 *Alexandr
Nikolaevich Korkin,* Leningrad,
148 p [bib]
Posse K 1907 Khar M So (2)10, 217-230
- 1908 Pet MNP (11), 25-46
(F39, 38)
- 1909 Mos MO Sb 27(1), 1-27
(F39, 38)


Korn, Arthur 1870-1945

Anon 1928 Rv M Hisp A (2)3, 233-245
(F54, 44)


Kortum, Hermann 1836-1904

Anschuetz + 1906 DMV 15, 60-63
(F37, 19)


Koschmieder, Lothar Eduard 1890-

Herrera F E 1960 Tucum U (A)13,
41-46 (Z112, 3)


Koshlyakov, Nikolai Sergeevich 1891-

Smirnov V I + 1959 UMN 14(3), 115-122
(MR22, 619; Z88, 244)


Kossak, Ernst August Martin
1839-

Lampe E 1903 DMV 12, 500-504
(F34, 21)


Kotelnikov, Aleksandr Petrovich
1865-1944

Rosenfeld B A 1956 Ist M Isl 9, 317-
400 (Z70, 243)


Kotelnikov, P. J.

Kotelnikoff E 1887 Kazn OEFM 5, 225-
249
Suvoroff T 1887 Kazn OEFM 5, 250-254


Kotov, T. I.

Sintsov P 1924 Khar MUZ 1, I-II
(F51, 40)


Koutsky, Karel 1897-

Boruvka O 1957 Cas M 82, 493-497
(Z98, 10)

Kovalevskaya, Sofya Vasilevna
  1850-1891

= Sonya Kovalevski (Kowalewski)

Andreian-Cazacu 1962 Gaz MF (A)14,
  101-105   (Z100, 246)
Anon 1895 Sat Rv (13 July), 51-52
- 1951 Pamyati S. V. Kovalevskoi,
  Sbornik Statei, Moscow, 155 p
  (MR13, 810)
Archibald R C + 1902 Intermed 9,
  195
Banerji H 1917 Clct MS 8, 53-56
Bjerknes C A 1891 Nor VS Ov, 7-20
  (F23, 26)
Borisiak A 1928 Vladimir Onufreivich
  Kovalevskii, Leningrad,
  135 p   (Biog of Sofia's
  husband)
Depman I Ya 1954 Ist M Isl 7, 713-715
  (Z59:1, 11; Z60, 11)
Edgren A C 1884 Ny Ill Tid, 269-270
Galdeano Z G de 1891 Prog M 1, 88-90
  (F23, 27)
Golubev V V 1950 Pri M Me 14, 236-244
  (MR12, 1; Z37, 2)   [mech]
Holmberg A 1963 Lychnos, 250-255
Kerbedz E de 1891 BSM (2)15, 212-
  220   (F23, 27)
- 1891 Paler R 5, 121-128   (F23, 27)
Kovalevsky Sofia V 1895 Sonya
  Kovalevsky: Her recollections of
  Childhood, NY, 326 p [biog, port]
- 1895 Souvenirs d'enfance de
  Sophie Kovalevsky, Paris, 344 p
  (F26, 46)
- 1896 Jugenderinnerungen, Berlin,
  213 p (F27, 18)
- 1960 Vospominaniya detstva,
  Moscow, 239 p   (Z92, 246)
Kronecker L 1891 Crelle 108, 88
  (F23, 27)
Leffler A C 1891 Annali (2)19, 201-
  211   (F23, 26)
- 1892 Sonja Kowalewsky (in Swedish)
  (Ger tr 1895, F26, 46; Engl tr
  1895)   [sister of Sofya]
Manville D 1938 Bord U 2, 59-71
  (F65: 2, 1092)
Mittag-Leffler G 1893 Act M 16(4),
  385-392

- 1902 Int Con 2   [letters]
- 1923 Act M 39, 133-198
  [Weierstrass]
Novarese E 1891 Rv M 1, 21-22
  (F23, 27)
Papfentiev N N 1923 Kazn FMO (2)23,
  1-11   (F51, 40)
Polubarinova-Kochina P Ya 1950
  Pri M Me 14, 229-235   (MR12, 1;
  Z37, 2)
- 1950 UMN 5(4), 3-14   (MR12, 311;
  Z36, 146)
- 1950 Zhizn i deyatelnost S. V.
  Kovalevskoi, Mos-Len, 51 p
  (Czech tr 1951)   (MR13, 612)
- 1952 UMN 7(4), 103-125
  (MR14, 122; Z46, 3)
- 1954 Ist M Isl 7, 666-712
  (MR16, 781; Z59:1, 11; 60, 11)
- 1957 Vop IET (5), 156-162
  (Z80, 5)   [Sylvester]
*Rachmanowa Alja 1953 Sonja Kowalewski,
  Leben und Liebe einer gelehrten
  Frau, Zurich, (Rascher Verlag),
  350 p   (tr from Russ ms by A. v.
  Чoyer) [port, bib]
Russyan 1921 Khar Me M 15, 162-173
  (F48, 9)
Stoletov A G + 1891 Mos MO Sb 16,
  1-38   (F23, 27)
Voronets, L A 1957 Sofya Kovalevskaya
  1850-1891, Moscow, 334 p
  (Z82, 12)
Wentscher M 1909 DMV 18, 89-93
  [Weierstrass]
Zednik Jolla v 1898 Sophie Kowalewsky,
  ein weiblicher Professor, Prague,
  15 p (F29, 25)
Zhatykov O A 1950 Kazk Ak (1), 86-92


Kovanko, Aleksandre Sergeevich  1893-

Kostovskii A N 1954 UMN 9(2), 215-221
  (MR16, 434; Z56, 2)   [biog]


Kowa.  See Seki.

Kowalczyk, Jan  1833-1911

Dickstein S 1911 Wiad M 15, 285-287
   (F42, 36)


Kowaleski, (Hermann Waldemar)
   Gerhard  1876-1950

Kowalewski G 1950 *Bestand und Wandel,*
   Munich, 309 p  (Z37, 2)


Kowalski, Zdenek

Piska R + 1961 Prag Pok 6(3), 174


Kozakiewicz, Waclaw  1911-1959

Anon 1959 Can MB 2, 148-150
   (Z86, 6)


Krahe, Augusta

Lorente de Nó F  1931 Rv M Hisp A
   (2)6, 15-32  (F57, 45)'


Kraitchik, Maurice Borisovich  1882-

Archibald R C 1948 *Mathematical Table*
   *Makers,* NY, 43  (Scr Stu 3)
   [bib, port]
Errera A 1957 Mathesis 66, 303-309
   (Z78, 242)


Kramp, Christian  1760-1826

Aubry V 1897 JM Sp 21, 131  [eqs]
Mason Du Pré A  1938 Isis 29, 43-
   48  (F64, 917)  [normal prob
   intgl]


Kraus, Ludwig  1857-1885

Weyr E  1886 Cas MF 15, 49


Krause, Johann Martin  1851-1920

Herglotz G 1921 Leip Ber 72, 103-106
   (F48, 16)


Krause, Karl Christian Friedrich
   1781-1832

Hueniger H 1894 *Der Philosoph K.C.F.*
   *Krause als Mathematiker,*
   Eisenberg, 32 p  (F25, 1913)
Junge M E 1927 Leip Ber 79, 124-133


Krazer, Carl Adolph Joseph   1858-

Boehm K 1928 DMV 37, 1-33  (F54, 40)


Krein, Mark Grigorievich  1907-

Kolmogorov A N + 1958 UMN 13(3),
   213-224  (Z79, 5)


Kresa, P. J.

Jemelka A 1913 Cas MF 42, 501-509
   (F44, 39)


Kretkowski, Wladyslaw  1840-1910

Dickstein S 1911 Wiad M 15, 273-274
   (F42, 28)


Kretschmer, Walter  1897-

Anon 1925 ZAM Me 5, 447  (F51, 32)


Krogness, O. A.

Devik O 1934 Nor M Tid 16, 73-74
   (F60, 839)

Kronecker, Leopold   1823-1891

Anon 1892 Naturw Wo 8, 591-593
  (F24, 32)
- 1894 ZMN Unt 25, 225-233
  (F25, 34)
Bell E T 1924 AMSB 30, 236-238
  [class numbers]
- 1929 AMSB 35, 12  [alg numbs]
- 1936 Phi Sc 3, 197-209
Cantor M 1906 ADB 51, 393-395
Crespo P 1952 Gac M (1)4, 199-204
  (MR14, 609)
*Fine H B 1891 NYMS, 173-184
  [alg eqs]
- 1913 AMSB 20, 339-358
  [num eqs]
Francken E 1897 Intermed 4, 34 (Q294)
  [numb]
Frobenius F G 1893 Ber Ab (S), 22 p
  (F25, 33)
- 1893 Ber Si, 3-22 (also in
  Frobenius 1968 Ges Abh 3, 705-724)
- 1902 Ber Si, 329-331  (also
  Frobenius 1968 Ges Abh 3, 725-727)
Fujisawa R 1892 Tok M Ph 5
Hensel K ed 1910 Ab GM 29  [port]
Hermite C 1892 Par CR 114, 19-21
  (F24, 32)
Kneser A 1925 DMV 33, 210-228
  (F51, 25)
*Lampe E 1892 Leop 28, 94-
- 1892 Naturw R 7, 128-129
  (F24, 32)
- 1892 Pogg An 45, 595-601
  (F24, 32)
Mansion P + 1892 Mathesis (2)2, 18-
  19, 136-137  (F24, 32)
Miller G A 1910 Bib M (3)11, 182
  (F41, 58)  [Galois thy]
Netto E 1896 Chic Cong, 243-252
  (F28, 53)
Schering E 1885 Gott N, 361-382
  [Dirichlet]
Weber H 1893 DMV 2, 5-31
  (F25, 33)
- 1893  M Ann 43, 1-25

Kronland.  See Marci.


Krueger, Johann Heinrich Louis
  1857-1923

Haussman K 1925 DMV 34, 52-57
  (F51, 30)


Krumme, Wilhelm   1833-1894

Viereck L 1894   ZMN Unt 25, 461-464
  (F25, 47)


Kruse, Friedrich

Thaer A 1891 ZMN Unt 22, 468-470
  (F23, 32)


Krylov, Aleksei Nikolaevich   1863-
  1945

Anon 1956 Mos IIET 15, 4-168
  (MR19, 825)
- 1958 A. N. Krylov. In Memoriam,
  Mos-Len, 248 p  (Z82, 243)
Chpligin S A 1934 SSSR Stek 5, 5-18
  (F60, 837)
Geronimus J L 1953 A. N. Krylow,
  Berlin, 56 p  (MR15, 923)
Idelson N I 1956 Mos IIET 15, 24-31
  (Z74, 246)  [ast]
Ioffe A F 1956 Mos IIET 15, 6-12
  (Z74, 246)
Kolcov A V 1956 Mos IIET 15, 46-53
  (Z74, 246)
Kravec T P 1956 Mos IIET 15, 32-39
  (Z74, 246)
Krylov A N 1956 Vospominaniya i
  ocherki, Moscow, 884 p
  (MR18, 860; Z74, 8)
*Kryzhanovskaya N A  1952 A. N. Krylo
  Bibliograficheskii Ukazatel,
  Leningrad
Luchininov S T 1959 A. N. Krylov...,
  Moscow, 167 p  (Z87, 243)
Lyusternik L A 1946 UMN 1(1), 3-10
  (MR8, 498; Z60, 15)

Shatelen M A 1956 Mos IIET 15, 40-45
  (Z74, 246)
Shilov N I 1955 Mos IIET 5, 381-384
  (Z68, 6) [Vitkovskii]
Shtraih S Ya 1950 *A. N. Krylov...,*
  Mos-Len, 88 p (MR14, 1051)
Smirnov V I 1946 UMN 1(3/4), 3-12
  (MR10, 175; Z60, 15)
- 1956 Mos IIET 15, 13-23 (Z74, 246)
Vavilov S I 1956 Mos IIET 15, 4-5
  (Z74, 246)


Krylov, Nikolai Mitrofanovich 1879-
1955

Anon 1955 Ukr IM 7, 3-4 (Z64, 4)
- 1955 Ukr IM 7, 347-359
  (MR17, 814; Z65, 245)
Bogolyubov N N 1950 Ukr 2(3), 3-6
  (MR13, 810; Z45, 293)
- 1950 UMN 5(1), 230-233 (Z35, 148)
Bogoljubov N N + 1960 Ukr IM 12, 205-
  208 (Z93, 9)
Khanovich I G 1967 *Akadamik Aleksei*
  *Nikolaevich Krylov,* Len, 250 p
Krylov O V 1945 *Nikolai Mitrofanovich*
  *Krylov,* Mos


Kubota, Tadahiko 1885-1952

Anon 1949 Toh MJ (2)1, 3-12
  (MR11, 573) [bib]
Sasaki S 1952 Toh MJ (2)4, 318-319
  (MR14, 832; (Z48, 243)
- 1952 Toh MJ (2)4, 321-322
  (MR14, 832) [bib]


Kuehn, Heinrich 1690-1769

Beman W W 1897 AAAS 46


Kuepper, Karl Josef 1828-1900

Waelsch E 1905 DMV 14, 389-394
  (F36, 18)

Kuerschak, Joseph Andreas 1864-1933

Stachó T v 1936 MF Lap 43, 1-13
  (F62, 1041)


Kulakov, A. A. 1898-1946

Anon 1947 UMN 2(2), 185-187
  (MR10, 175)


Kulczycki, Stefan

Straszewicz S 1961 Wiad M (2)4, 151-
  154 (MR23A, 4; Z10, 247)


Kulik, Jacob Philipp 1793-1863

Depman I Ya 1953 Ist M Isl 6, 573-608
  (MR16, 660)


Kummer, Ernst Eduard 1810-1893

Anon 1881 Ber Mo (F13, 24)
  [anniv]
- 1910 *Festschrift zur feier des 100*
  *geburtstages Edouard Kummers...*
  (Teubner), 103 p (= Abh Gesch
  M W 29) [Hensel talk, letters]
- 1894 Mathesis (2)4, 40
- 1933 Crelle 170, 1-3 [Kronecker]
- 1940 Deu M 5, 337 (F66, 20)
  [H A Schwarz]
Cantor M 1906 ADB 51, 438-440
Galdeano Z G De 1893 Prog M 3, 234-
  236 (F25, 41)
Gudian F 1937 Z Gesm Nat 3, 70-80
  (F63, 810) [Jacobi]
Hancock H 1928 AMM 35, 282
  [chem, ideal numbs, Plato]
Hensel K ed 1910 *Festschrift zur*
  *Feier des 100. Geburtstages...,*
  Berlin, 107 p (Ab GM 29, 46-107)
  (F41, 15) [Kronecker, port]
Hensel K 1910 *E.E.Kummer und der*
  *grosse Fermatsche Satz,* Marburg
  (F41, 16)
Hermite C 1893 Par CR 116, 1163-1164
  (F25, 41)

Hudson R W H T 1905 *Kummer's
  quartic surface,* Cambridge
Lampe E 1892 DMV 3, 13-28
- 1893 Naturw R 8, 361-364
  (F25, 41)
O N H 1893 ZMN Unt 24, 310-313
  (F25, 42)


Kupradze, Viktor Dmitrievich 1903-

Anon 1965 Tbil M Me 110, 21-23
  (MR33, 5439) [bib]
Gokieli L P + 1965 Tbil M Me 110,
  7-19 (MR33, 5436)


Kuratowski, Kazimierz 1896-

Anon 1959 Prag Pok 4, 228-232
  (Z82, 12)
- 1960 Wiad M (2)3(3), 223-224
- 1960 Wiad M (2)3, 245-250
  (MR22, 1130; Z91, 6) [bib]
Bursuk K 1960 Wiad M (2)3, 231-237
  (MR22, 1130; Z91, 6) [topology]
Jarnik V 1960 Wiad M 3, 225-230
  (Z91, 6)
Marczewski E 1960 Wiad M (2)3, 239-
  244 (MR22, 1130; Z91, 6)
  [mea thy, set thy]


Kurinek, Vladimir

Anon 1959 Cz MJ 9, 305-308
  (Z88, 7)


Kuroda, Narikiyo

Mikami Y 1943 Toh MJ 49, 223-242
  (MR8, 497) [surveying]


Kurosh, Aleksandr Gennadievich 1908-

Aleksandrov P S + 1958 UMN 13(1),
  217-224 (MR20, 264; Z78, 4)


Kuzmin, Rodion Osieviz   1891-1949

Smirnov V I 1949 SSSRM 13, 385-388
  (MR11, 573; Z32, 4)
Venkov B A + 1949 UMN 4(4), 148-155
  (MR11, 573; Z32, 194)


Kwietniewski,  W.

Dickstein S 1903 Wiad M 7, 169-176


Labatie,  XIX

Brocard H 1911 Intermed 18, 271 (Q3026
  (Also Intermed 13, 60, 224;  14,
  86, 181; 15, 105; 16, 128)


Labosne, A.   XIX

Brocard H 1904 Intermed 11, 241
  (Q1822)  (See also Intermed 7, 125;
  8, 176)
- 1906 Intermed 13, 151 (Q2944)
  (Also Intermed 12, 172; 13, 28)


Lacaille, Nicolas Louis de  1713-1762

Boncompagni B 1872 Boncomp 5, 278-293
  (F4, 9)


La Chapelle,  Abbé de  1710-1792

Itard J 1952 Rv Hi Sc Ap 5, 171-175
  [educ]


La Condamine, Charles Marie de
  1701-1774

Jelitai J 1937 MP Lap 44, 173-199
  (F63, 817) [Alembert, Clairaut]


La Cour, Paul   1846-

Nordgaard M A 1938 Scr 6, 88-94
  [educ]

Lacroix, Sylvester François 1765-1843

Anon 1911 BSM (2)35, 273-275 (F42, 11) [letter]
Boyer C B 1947 Scr 13, 133-153 (MR9, 485) [anal geom, Fermat]
Riccardi P 1898 Mod Mm (3)1, 105-129 (F29, 7) [Paoli]
Taton R 1953 Int Con HS 7, 588-593
- 1953 Rv Hi Sc Ap 6, 350-360 [Laplace]
Théodorides Jean 1961 Rv Hi Sc Ap 14, 329-330 [Humboldt, letter]


Ladd Franklin, Christine 1847-1930

Reyes y Prosper V 1891 Prog M 1, 297-300 (F23, 33)


La Faille, Jean Charles de 1597-1652

Bosmans H 1914 Brx SS (B)38, 244-317 (F45, 85)
- 1927 Mathesis 41, 5-11 (F53, 14)
Bosmans H + 1906 Intermed 13, 15
Mansion P 1913 Brx SS 38, 151-154


Lagny, Thomas Fantet de 1660-1734

Nordgaard M A 1937 M Mag 11, 361-373 (F63, 808)


La outinsky, M. XIX

Pcheborsky A 1915 Khar M So (2)15, 77-80


Lagrange, Joseph Louis 1736-1813

non 1812 Par Mm, xxvii-
- 1873 Boncomp 6, 131-141 (F5, 28) [letters, Lorgna]
- 1877 Tid M 1, 129-
- 1879 J Sav, 572-574 (F11, 20) [letters]
- 1913 Annali (3)20

- 1937 Zhosef Lui Lagranzh (1736-1936) Sbornik statei k 200-letiyu so dnya rozhdeniya, Mos, 140 p
Aubry A 1909 Ens M 11, 430-450 [Gauss, Legendre, numb thy]
Beaujouan G 1950 Rv Hi Sc Ap 3, 110-132 (MR11, 708)
Beaumont E de 1867 Smi R
Bessel-Hagen E 1937 Proteus 2, 99-104
Biadego G 1873 Boncomp 6, 101-130 (F5, 24) [letter]
Bianco O 1913 Tor Mm (2)63, 59-110 [Comet, Gauss, Laplace, Schiaparelli]
Biermann K-R 1963 Ber Mo 5, 445-450 (MR29, 220; Z113, 2) [Humboldt]
Boncompagni B 1873 Boncomp 6, 142-150 (F5, 28) [letter]
- 1873 Boncomp 6, 539-543 (F5, 28) [letter]
Briano G 1942 Lagrange, Torino
Burzio Filippo 1941 Saggiat 2, 314-316 (Z26, 196) [infinitesimal]
- 1942 Lagrange, Turin, 275 p (F68, 16)
Cantor M 1869 Hist Abt 14 (56/57) (F2, 19)
- 1873 Hist Abt 18(86) (F5, 24)
- 1878 Boncomp 11, 197-216 (F10, 16) [Euler]
- 1878 Hist Abt 23, 1-21 (F10, 16) [Euler]
- 1879 Hist Abt 24, 182-184 (F11, 20)
Chebotarev N G O 1936 UMN 2, 17-31 [numb thy]
Chib F 1937 Bo Fir (2)16, xxix-xxxv (F63, 816)
Cossali P 1813 Elogio di Lagrange, Padova
De Gregori 1814 Necrologio di tre piemontesi illustri: Bodini, Denina, Lagrange, Vercilli
De Morgan A 1837 Differential calculus in Penny Encyclopedia, London, [bib, calc]
De Vries H 1934 N Tijd 22, 72-94 (F60, 832)
Dickstein S 1899 Prace MF 10, 178-192 [calc]
- 1899 ZM Ph 44(S), 65-74 (F30, 46) [calc]
Eneström G 1879 Zeuthen (4)3, 33-45 (F11, 21) [Euler, letters]

Lagrange, Joseph Louis (continued)

Fallex E 1868 *Lagrange,* Paris
(F1, 15)
Favaro A 1879 Pado Riv 29, 163-184,
193-201 (F11, 20) [letter]
- 1884 Ven I At (6)1, 533-545
[Alembert, letters]
- 1895 Tor FMN 182-194 (F26, 18)
[Frisi, letters]
Forti A 1868 *Intorno alla vita e
alle opere die Luige Lagrange,*
Pistoia (F1, 15)
Genocchi A 1877 Boncomp 10, 657-667;
12, 350-370 (F9, 8)
[Euler, letter]
- 1879 Tor FMN 14, 459-463 (F11, 20)
[letter]
- 1879 Tor FMN 14, 1138-1179
(F11, 20) [letter]
- 1882 Tor FMN 17, 531-533
[Alembert, letters]
Govi G 1880 Nou An (2)19, 421-428
(F12, 13) [letters]
Guareschi I 1914 Tor Mm (2)64(1)
(F45, 16)
Hamel G 1936 Naturw 24, 51-53
(F62, 1036; Z13, 193)
Henry C 1886 Boncomp 19, 129-136
Herrmann D 1963 Sterne 39, 58-63
Jourdain P E B 1905 Bib M (3)6, 350-
353 (F36, 55) [diff eqs, mech]
- 1913 Int Con 5(2), 540-541
(F44, 48) [anal fun]
Julia G 1951 Ens M 39, 9-21
(MR13, 2; Z42, 3)
Khovanskii A N 1961 Matv Shk (4),
81-82
Kilmister C W 1967 *Lagrangian
Dynamics...,* NY, 136 p
Korn A 1913 Arc MP (3)21, 90-94
- 1913 Ber MG 12, 90-94
(F44, 15)
Korner T 1904 Bib M (3)5, 15-62
Krylov A N 1936 UMN 2, 3-16
Likhin V V 1966 Ist Met EN 5, 35-
44 [Bernoulli numbs, bib,
Euler, fin diff]
Lorenzoni G 1880 Ven I At (5)5,
453-455 (F12, 13) [letter]
Lorey W 1936 Crelle 175, 224-239
(F62, 33; Z14, 245)

Loria G 1913 Annali (3)20, ix-lii
(F64, 14)
- 1913 Bib M 13, 333-338 (F43, 20)
- 1913 Intermed, 10 (Q4044)
- 1923 Linc Mr (5)14, 777-845
[anal geom]
- 1938 Isis 28, 366-375 [*bib]
- 1949 Isis 40, 112-117 (Z37, 291)
[bib]
Magistrini A 1916 *Discorso in lode
di G. de Lagrange,* Bologna
Marcolongo 1909 Int Con (4)3, 488-499
[Filippio, mech]
Menabrea L F 1873 Boncomp 6, 435-457
(F5, 28) [Genocchi]
Pidduck F B 1935 M Gaz, 206-
Pierpont J 1895 AMSB 1, 196-204
(F26, 51) [group]
Pittarelli G 1909 Int Con (4)3,
554-556 (F40, 19) [Caluso,
letter]
Plakhowo N + 1906 Intermed 13, 232
[alg, Euler]
Polak L S 1935 Len IINT 5, 155-181
[least action]
Revelli P 1918 *Un Maestro del
Lagrange, Filippo Antonio Revelli,*
Genoa, 34 p (F46, 11)
Riccardi P 1898 Mod Mm (3)1, 105-129
(F29, 7) [Paoli]
Sanford V 1935 MT 27, 349-351
Sarton G 1941 Isis 33, 55
- 1944 APSP 88, 457-496
(MR6, 141)
Sarton George + 1950 Rv Hi Sc Ap 3,
110-132
Schering E 1880 Gott N, 489-491
(F12, 13) [Canterzoni, Euler,
Laplace]
Schiaparelli G V 1877 Lom Gen (2)10,
185-188 (F9, 8) [Euler, letter]
Sclopis J 1872 Tor FMN 7, 428-434
(also Arc MP Lit 215, 2-6)
(F4, 16) [letter]
Serret J A 1877 Par CR 84, 1064-1084
(F9, 12)
Sintsof D M 1899 Kazn FMN (2)9, 44
[anal parallelogram, Newton]
Smith D E + 1921 MT 14, 362-66
Sokolov J D 1961 Ukr IM 13(2), 127-
135 (MR24A, 333; Z96, 4)
Tychsen C 1877 Zeuthen (4)1, 129-143
(F9, 8)

- 1879 Boncomp,12, 815-827
  (F11, 19)
Vacca G 1901 Loria 4, 1-4  (F32, 8)
Verriest G 1934 Wis Nat Ti 7, 1-5
  (F60, 817)  [eq]
Virey Julien J + 1813 *Précis
  historique sur la vie et la mort
  de J. L. Lagrange,* Paris
  [medical history]
Vuillemin J 1960 *La philosophie de
  l'algèbre de Lagrange,* Paris, 24 p
  (MR22, 1853; Z99, 4)
Zani V 1936 Bo Fir 32, xxi-xxiv
  (F62, 1040)


Laguerre, Edmond  1834-1886

Bertrand J + 1886 Par CR 103, 407,
  424-425
Poincaré H 1887 *Notice sur Laguerre,*
  Paris, 14 p  [funs, geom transfs,
  num eqs]
- 1887 Par CR 104, 1643-1650
Rouche E 1887 Nou An (3)6, 105-173
- 1887 Par EP 56, 213-277


La Hire, Philippe de 1640-1718

Curtze M 1888 Bib M (2)2, 65-66
- 1895 Bib M (2)9, 33-34  (F26, 63)
Le Paige C 1887 Bib M (2)1, 109
Taton R 1953 Rv Hi Sc Ap 6, 93-111
  (MR15, 90)  [geom]
Wieleitner H 1913 Arc GNT 5, 49-55
  (F45, 79)  [conic]


Laisant, Charles Ange  1841-1920

Anon 1920 Intermed 27, 81-83
  (F47, 15)
Boyer J 1920 Rv GSPA 31, 397-398
Bricard R 1920 Nou An (4)20, 449-454
  (F47, 15)
Buhl A 1920 Ens M 21, 57, 73-80
  (F47, 15)

La Lande, Joseph Jerome le François
  de  1732-1807

Anon 1932 Nat 130, 48
Archibald R C 1948 *Mathematical Table
  Makers,* NY, 43-45  (Scr Stu 3)
  [bib]
Doublet E 1932 Rv Sc 70, 451-458
  (F58, 994)
Wolf R 1883 Zur NGV 28, 65-68


Lalescu, Trajan  1882-1929

Abason E 1929 Buc EP 1, 72-73
  (F55, 22)
Anon 1929 Buc U Rev 1, 245-251
  (F55, 23)  [bib]
Ghermanescu M 1929 Gaz M 35, 121-124
  (F55, 618)
Lalescu T 1929 Buc U Rev 1, 245-251
  (F55, 23)  [bib]
Sergescu P 1929 Ro SSM 32, 101-103


Laloubere, Antoine de    1600-1664

  = Loubère
  = Lalouère
  = Lovera
  = Lalovera

Kropp G 1948 *Beitraege zur Philosophie,
  Paedogogik und Geschichte der
  Mathematik,* Berlin, 103 p
- 1951 Crelle 189, 1-76  (Z43, 243)
  [intgl]'
- 1959 MN Unt 12, 23-26  (Z107, 246)
Lefner A 1909 Intermed 16, 280
  (Q3567)


Lamarle, Anatole Henri Ernest
  1806-1875

De Tilly J M 1879 Bel An 45, 205-253
  (F11, 24)

Lamb, Horace   1849-1934

Glazebrook R T + 1935 Lon RS Ob 1,
   375-392
Love A E H 1937 LMSJ 12, 72-80
   (F63, 19)
Taylor G I 1935 Nat 135, 255-257
   (F61, 955)


Lambert, Johann Heinrich   1728-1777

Barthel E 1928 Arc GMNT 11, 37-62
   (F54, 31)
Bopp K 1914 Mun Si, 361-368
   (F45, 77)
- 1916 Mun Ab 27(6), 84 p  (F46, 11)
- 1924 Ber Ab (F52, 22)  [Euler]
- 1928 Heid Si (18)  (F54, 47)
   [Kaestner]
Busch W 1933 Gies Phil 30, 1-37
   (F59, 825)  [Ger, termin]
Duerr Karl 1945 in *Festschrift zum
   60 Geburtstag von Prof. Dr.
   A. Speiser,* Zurich
   (MR8, 497; Z60, 12)
Eberhard 1779 in J. H. Lambert,
   *Pyrometrie oder vom Maasse des
   Feuers und der Waerme,* Berlin
Eisenring M E 1941 *J. H. Lambert
   und die wissenschaftliche
   Philosophie der Gegenwart,*
   Zurich, 117 p  (Thesis ETH)
   (MR11, 150)
Halsted G B 1893 NYMS 3, 79
   [non-Euc geom]
- 1895 AMM 2, 209-211
Huber D 1829 *Lambert nach seinen
   Leben und Wirken,*  Basel
Krienelke K 1909 *J. H. Lamberts
   Philosophie der Mathematik,*
   Berlin
Labrador J F 1958 Gac M 10, 131-134
   (Z82, 11)
Lepsius J 1881 *J. H. Lambert. Eine
   Darstellung seiner Kosmologischen
   und philosophischen Leistungen,*
   Munich
Linder A 1936 Lon St 99, 138-141
   (F62, 23)  [D Bernoulli, ins]

Loewy A 1927 in *J. H. Lamberts
   Bedeutung fuer die Grundlagen des
   Versicherungswesens,*  Berlin,
   280-287  (F54, 31)
Lorey W 1928 Ber MG 28, 2-27
   (F55, 13)
Matthias Graf 1829 *J. H. Lamberts
   Leben,* Muehlhausen
Middleton W E K 1960 Isis 51, 145-149
   (Z93, 8)
Pasquier L G du 1929 Fr AAS 53, 80-88
   (F57, 1303)  [prob]
Pringsheim A 1932 Mun Si 3, 193-200
   (Z6, 145)  [Gauss, tangent]
Rudio F 1892 *Archimedes, Huygens,
   Lambert, Legendre, Vier Abhand-
   lungen ueber die Kreismessung...,*
   Leipzig, 173 p  (F24, 50)
   [circle sq]
Schur F 1904 *J. H. Lambert als
   Geometer,* Karlsruhe, 20 p
   (F35, 10)
- 1905 DMV 14, 186-198 (F36, 8)
Staeckel P 1899 Bib M (2)13, 107-110
   (F30, 50)  [non-Euc geom]
Steck Max 1943 *Bibliographia
   Lambertiana,* Berlin, 84 p
- 1951 Gesnerus 8, 245-49
   [Gesner]
- 1956 Fors Fort 30, 39-44   (Z70, 6)
- 1956 Fors Fort 30, 71-74   (Z71, 3)
   [letters]
- 1956 Int Con HS 8, 64-67
Wolf Rudolph 1860 *Biographien zur
   Culturgeschichte der Schweiz,*
   Zurich (3), 317-356
Wolff G 1928 Unt M 34, 234-237
   (F54, 47)  [anniv]


Lamé, Gabriel   1795-1870

Bertrand Joseph 1870 *Discours
   prononce aux funerailles de M. Lame,*
   Paris  (Repr in Bertrand 1890
   *Eloges académiques,*
   Paris, 131-158)
- 1878 An Mines (7)13, 236-259
   (F10, 18)

Bertrand J + 1872 An Mines (7)1,
274-282 (F4, 20)
de Fourcy E F 1872 An Mines (7)1,
271-273 (F4, 20)
De Vries H 1948 Scr 14, 5-15
[anal geom]
Gaiduk Yu + 1965 Ist M Isl 16, 337-
372 (MR34, 12)
Malkin I 1952 Scr 21, 44 [elast]
Sagnet L 1895 Grande En 21, 827-828


Lampaaridis, Grigorios

P V 1937 Gree SM 17, 182-184
(F63. 816)


Lampe, Karl Otto Emil 1840-1918

Hensel K 1919 Crelle 149
Jahnke E 1919 Deu Ph G 21, 33-42
(F47, 13)
- 1920 Arc MP (3)28, 1-16
Korn A 1919 Ber MG 18, 4-12


Landau, Edmund 1877-1938

Hardy G H + 1938 LMSJ 13, 302-310
(F64, 23; Z19, 389)
Knopp K 1951 DMV 54(1), 55-62
(MR12, 578; Z42, 4)


Landau, Ludwika 1901-1944

Anon 1961 Prz St 8(2), 211-212


Landen, John 1719-1790

Green H G + 1944 Isis 35, 6-10
(MR5, 253; Z60, 11)
Watson G N 1933 M Gaz 17, 5-17
[ell fun, Fagnano, length]
Winter H J J 1943 Peterbo (Dec)

Landré, C. L. 1838-1905

Paraira M C 1905 Niew Arch (2)7, 1-6
(F36, 36)


Lange, Oscar 1904-1965

Kowalik T 1964 in On Political Economy
and Econometrics. Essays in honour
of Oskar Lange, Warsaw [bib]


Langley, Edward Mann 1851-

Kirkman J P 1913 M Gaz 7, 28
Kirkman J P + 1933 M Gaz 17, 225-229
(F59, 857)
Lodge A + 1913 M Gaz 7, 25-27


Laon, Radulph von

Nagl A 1890 Ab GM 5, 85-134


Laplace, Pierre Simon 1749-1827

Andoyer Henri 1922 L'oeuvre scienti-
fique de Laplace, Paris, 162 p
(F48, 8)
Anon 1826 Hall Bij 2, 284-
- 1831 Par Mm 10, 81-
- 1949 Endeav (Ap), 49-50
Arago D F J 1854 Oeuvre 2, 593-671
- 1874 Smi R, 129-168
Bianco O 1913 Tor Mm 63, 59-110
Biot J B 1858 Mélanges scientifiques
et litteraires 1, 1-10
Boncompagni B 1883 Boncomp 15, 447-
463, 463-465
Danjon André 1949 Cérémonies du deux
centième anniversaire de la
naissance de Pierre-Simon Laplace...,
Paris, 20 p
David F N 1966 in LeCam, Lucien + ed,
Bernoulli-Bayes-Laplace Anniversary
Volume (Springer)
De Morgan A 1837 Dub Rv 1 (Ap),
2 (July)
Forsyth A R 1905 Brt AAS 75, 307
[cel mech]

Fourier J J 1831 Par Mm Div 10,
lxxi-cii (transl in Phi Mag
(2)6, 370-381)
- 1839 *Eloge historique de M. le Ms.
de Laplace,* Paris, 22 p
Hadamard J 1950 Arc In HS 29, 287-90
[anniv]
Henry C A 1886 Boncomp 19, 149-178
[comets, letters, Pingre]
Kneser H 1939 Deu M 4, 318-324
[fund thm alg, Gauss]
Korner T 1904 Bib M (3)5, 15-62
[mech]
Lemaitre G 1950 Astmie Fr 64, 89-97
[cel mech]
Lichtenstein L 1939 Wiad M 47, 1-86
(Z21, 3)   [XVIII]
Lilley S 1949 Nat 163, 468-469
(Z31, 99)   [ast]
Lorey W 1934 Allg St Ar 23, 398-410
[stat]
- 1950 Deu Vers M 1, 39-50
(Z36, 145)   [ins, Klein]
L S 1932 Isis 18, 333
Mansion P 1913 Brx SS 37(2), 107-117
[mistake]
Molina E C 1930 AMSB 36, 369-392
[prob]
Newman J R 1954 Sci Am 191(6), 76-81
Pearson K 1929 Biomtka 21, 202-216
(F55, 609)
Picard E 1927 Rv GSPA 38, 357-366
(F53, 18)   [Newton]
- 1931 in *Eloges et discours,* Paris,
167-206
Riccardi P 1898 Mod Mm (3)1, 105
[letter, Lacroix, Paoli, Ruffini]
Richeson A W 1942 M Mag 17, 73-78
(MR4, 65; Z60, 72)
Sanford V 1935 MT 28, 111-113
Schofield M 1949 Contem Rv (June),
355-358
Sergescu P 1949 Rv GSPA 56, 241-244
(Z34, 147)
Seydl O 1927 Cas MF 56, 296-298
(F53, 21)
Simon A G 1929 Biomtka 21, 217-230
Taton R 1949 Isis 40, 351
[Lacroix]
- 1949 Nat Par 77, 221-23
Van Dantzig D 1955 Arc In HS 8, 27-37
(Z68, 5)   [prob]

Whittaker E 1949 AMM 56, 369-372
(Z32, 51)
- 1949 M Gaz 33, 1-12   (Z33, 2)

Lappo-Danilevskii, Ivan Aleksandrovic.
1895-1931

Luzin N 1931 SSSR Iz, 729-732
(F57, 1304)

Larmor, Joseph   1857-1942

Birkhoff G D 1943 Sci (NS)97, 77-79
(MR4, 181; Z60, 13)   [phys]
Cunningham E 1943 LMSJ 18, 57-64
(Z28, 337)

La Roche, Estienne de    XVI

Anon 1884 Rv Sc 1, 561-563
[XVI, arith]

Laska, Vaclav

Vetter Q 1922 Cas MF 53

Laszlo, Csernak

Jelitai J 1937 Deb Szeml (7/8)
(F63, 16)

Latimer, Clairborn  G.  1893-

Parker E T 1965 AMM 72, 1127-1128

Lattès, A. B. Samuel   -1918

Anon 1918 Ens M 20, 138   (F46, 21)

Latysheva, Klavdiya Yakovlevna
1897-1956

Pavljuk I A 1956 Ukr IM 8, 342-344
(Z70, 244)

Laue, Max Theodor Felix von
1879-1960

Meissner W 1960 Mun Si, 101-121
(MR24A, 4; Z95, 4)
Papapetrov A 1959 Fors Fort 33, 316
317 (Z86, 6)


Lauffer, Rudolf 1882-

Pinl M 1963 DMV 65, 143-147
(Z106, 3)


Laura, Ernesto

Grioli G 1950 Pado Sem 19, 443-449
(MR12, 311; Z37, 291)


Lauremberg family XVI-XVII

William 1547-1612
Peter 1585-1639
Johann Wilhelm 1590-1658

Cantor M + 1906 Intermed 13, 82,
193-194 (Q860) (Also 3, 152)
[dioph eqs]


Laurent, Paul Mathieu Hermann
1841-1908

Brocard H + 1903 Intermed 10, 238,
312 (Q2102)
Maillet E 1910 Intermed 17, 36 (Q3396)


Lauricella, Giuseppe 1867-

Daniele E 1914 Cat Atti (5)7, 12 p
Silla L 1913 Mathes It 5, 34-40
(F44, 31)


Laussedat, Aimé 1819-

d'Ocagne M 1907 Rv GSPA 18, 341-342


Lavagna, Giovanni Marie 1812-1870

Agostini A 1951 Livo RV (3), 6 p


La Vallée Poussin, Charles Jean de
1866-

*Anon 1928 *Manifestation en l'honneur
de M. Ch. J. de la Vallée Poussin...,*
Louvain, 145 p [bib, port]
- 1928 Ens M 27, 145-146 (F54, 43)
- 1928 Rv Q Sc (4)14, 5-15
(F55, 25)
- 1937 Vat An 1, 281-293 (F63, 20)
- 1946 Rv Q Sc 7, 455-479
Burkill J C 1964 LMSJ 39, 165-175
(MR28, 756; Z118:1, 12)
Montel P 1962 Par CR 254, 2473-2476
(Z101, 9)
Simonart F 1926 Brx SS 45, 99-122
(F52, 34)
- 1962 Rv Q Sc (5)23, 161-165
(Z98, 11)
Valiron G 1929 Rv I Ens 83, 371-374
(F55, 619)
Verriest G 1928 Brx SS (A)48(2), 1-23
(F54, 43)


Lavrentiev, Mikhail Alekseievich
1900-

Anon 1960 JAM Me 24, 1475-1492
(Z96, 4)
- 1960 PMTF (3), 3-15 (Z95, 4)
- 1960 Sb TN (10), 3-4 (MR23A, 136)
- 1960 Sib MJ 1, 297-302 (Z91, 6)
Bicadze A V + 1961 Rus MS 16(4),
143-153 (Z98, 10)
- 1961 UMN 16(4), 211-221
(MR24A, 222) [port]
Ishlinskii A J 1960 PMTF (3), 16-19
(Z95, 4)
Keldysh M V 1951 SSSRM 15, 3-8
(MR12, 578; Z42, 4)
Lyusternik L A + 1951 UMN 6(1), 190-
192 (MR13, 2; Z42, 4)
Shtokalo I Z + 1960 Ukr IM 12, 490-
491 (Z109, 239)

**Lawson,** George    -1941

Comrie P 1941 Edi MSP (2)6, 261-262
  (Z60, 16)

**Lazar,** G.

Marian V 1936 Gaz M 41, 414-419,
  460-462  (F62, 25)

**Lazarus,** Wilhelm

Anon 1900 Ham MG 3, 426

**Lazzeri,** Giulio Giuseppe Giovanni
  Maria  1861-1935

Agostini  A 1931 Pe M (4)11, 170-171
  (F57, 40)
Anon 1935 Pe M (4)15, 261-262
  (F61, 955)
Sporza G 1895 Pe M 10, 154

**Lebesgue,** Henri Leon  1875-1941

*Burkill J C 1944 LMSJ 19, 56-64
  [bib]
*- 1944 Lon RS Ob 4, 483-490  [bib]
 Denjoy A 1946 Par CR 223, 61
*- 1949 Par Not D 2, 576-606
*Denjoy A + 1957 Ens M 3, 1-18
  (MR19, 108; Z77, 8)
 Fayet J 1941 Rv M Hisp A (4)1, 195-
  197  (MR7, 106)
 Fehr H 1942 Ens M 38, 330-332
  (MR4, 65)
 Felix L 1953 Rv GSPA 60, 265-276
  (MR15, 384)  [geom constr]
 Leconte T 1948 Rv Hi Sc Ap 1, 257-
  265   [historical work]
 - 1956 Ens M 2, 224-237  (MR18, 182)
 May K O 1966 in *Measure and the
  Integral* by Henri Lebesgue
  (Holden-Day), 1-7  [bib]
*Montel P 1941 Par CR 213, 197-200
 Perrin L 1948 in Le Lionnais, *Les
  Grands Courants de la Mathémati-
  ques Modernes,* Paris, 286-290

Rosenblatt A 1942 Rv Cien 44, 357-364
  (MR4, 65)
Sergescu P 1942 Monog M 7, 15-23
  (MR4, 66; F68, 18)
Stoilow S 1942 Math Tim 18, 13-25
  (MR4, 66; Z27, 196; F68, 18)
Vicente Gonçalves J 1942 Gaz M Lisb
  3(12), 2-3   (MR7, 106)

**Lebesgue,** Victor Amédée   1791-1875

Abria J J B + 1876 Boncomp 9, 554-
  582   (F8, 21)
- 1877 Nou An (2)16, 115-129
  (F9, 14)

**Lebon,** Désiré Ernest  1846-

Carnoy H 1910 *Sur les travaux mathématiques de M. Ernest Lebon,*
  Paris, 55 p   (F41, 46)

**Leclerc,** Sebastien  1637-1714

Graf J H 1899 ZM Ph 44(S), 115
  [Ozanam]
Lemaire G 1908 Intermed 15, 279
Sanford V 1953 MT 46, 348-354

**Lefschetz,** Solomon   1884-

Archibald R C 1938 *A Semicentennial
  History of the American Mathe-
  matical Society...,* NY, 236-240
  [bib]
Carli M 1957 Prin MS 12, 44-49
  (MR18, 784; Z77, 9)  [bib]
Hodge W V D 1957 in Fox R H + ed
  *Algebraic Geometry and Topology--a
  Symposium* (Princeton Univ P), 3-23
  (MR19, 173)
Lefschetz S 1968 AMSB 74, 854-880
Steenrod N E 1957 in Fox R H + ed,
  24-43  (MR19, 158)  [alg top]

Legendre, Adrien Marie   1752-1833

Anon 1833 Bib UN 52, 45-82   [bib]
Archibald R C 1948 *Mathematical
  Table Makers,* NY, 45-49
  (Scr Stu 3) [bib, port]
Aubry A 1909 Ens M 11, 430-450
  [Gauss, Lagrange, numb thy]
Beaumont L E de 1867 *Elogie
  historique de Adrian Marie
  Legendre,* Paris  (transl 1874 Smi
  R, 131-157)
Bourget H + 1896-1915 Intermed 3, 37;
  12, 100-101; 13, 100, 191-193;
  14, 55-56, 127; 20, 152; 22, 55-56
  (Q755)
Hellman C D 1936 Osiris 1, 314-40
  [metrol]
Neville E H 1933 M Gaz 17, 200-201
  (F59, 28)  [bib]
- 1934 M Gaz 18, 195-196
Nielsen Niels 1929 in *Geomètres
  français sous la Révolution,*
  Copenhagen, 166-174
Rudio F 1892 *Archimedes, Huygens
  Lambert, Legendre, Vier abhand-
  lungen ueber die Kreismessung...,*
  Leipzig, 173 p  (F24, 50)
Sanford V 1935 MT 28, 182-184
Smith D E 1922 AMM 29, 394
  [Abel, Cauchy]
Smith D E + 1932 MT 14, 362-366
Zaharia N U 1933 Tim Rev M 13, 61-63
  (F61, 958)  [Euler]


Legendre, François   XVII

Sanford V 1936 Osiris 1, 510-518


Lehmer, Derrick Norman   1867-1938

Archibald R C 1948 *Mathematical
  Table Makers,* NY, 50-51
  (Scr Stu 3) [bib, port]
Putnam T N 1939 AMSB 45, 209-212
  (F65, 18)


Leibenzon, Leonid Samuilovich
  1879-1951

Anon 1957 *L. S. Leibenzon...,* Mos,
  52 p  (Z77, 242)
Sedov L I 1952 UMN 7(4), 127-134
  (MR14, 122)


Leibniz, Gottfried Wilhelm   1646-1716

Agostini A 1952 Arc In HS 31, 3-5
  [letter, Marchetti]
- 1953 Arc In HS 32, 434-443
  (MR15, 591; Z53, 338)
  [Grandi, letters]
Aiton E J 1964 An Sc 20, 111-123
  (MR33, 8)  [cel mech]
Alexander H G 1956 *The Leibniz-Clarke
  Correspondence...,* Manchester,
  256 p  (Z72, 245)
Anon 1869 Boncomp 2, 375-376
  (F2, 19)
- 1903 J Sav, 172-179
- 1916 Monist 26, 481-629
- 1961 Ist M Isl 14, 607-610
  (Z118-1, 4)  [Perier, letter]
- 1969 *Akten des Internationalen
  Leibniz-Kongresses (1966),*
  Vol 2, Wiesbaden, 287 p
  (= Studia Leibnitiana Supplementa
  2(2))
Belaval Yvon 1960 *Leibniz Critique de
  Descartes,* Paris, 559 p
Biermann K-R 1954 Fors Fort 28, 357-
  363  (MR16, 434; Z59:1, 7)
  [combinatorics]
- 1955 Fors Fort 29, 110-113
  (Z65, 243)  [prob]
- 1955 Fors Fort 29, 205-208
  (Z65, 244)  [ins]
- 1956 Fors Fort 30, 169-172
  (Z71, 244)  [combin]
- 1959 Fors Fort 33(6), 168-173
  [de Witt, ins]
Biermann K-R + 1957 Fors Fort 31,
  45-50  (Z77, 7)
Bodemann E 1889 *Der Briefwechsel des
  Gottfried Wilhelm Leibniz in der
  Koenigl. Oeffentlichen Bibliothek
  zur Hannover,* Hannover, 419 p

Leibniz, Gottfried (continued)

Boehm A 1938 Le "Vinculum sub-
stantiale" chez Leibniz. Ses
origines historiques, Paris, 130 p
Boncompagni B 1869 Boncomp 2, 273-
274 (F2, 19) [deriv]
Bopp K 1929 Strasb W G 10, 5-18
[XVII, Arnauld, De Nonancourt]
Borchardt C G 1869 Boncomp 2, 277-
278 (F2, 19) [deriv]
Budylina M V 1954 Mos IIET 1, 309-
316 (MR16, 660; Z59:1, 7; 60, 7)
[letter]
Cajori F 1916 Monist 26, 557-565
- 1919 AMM 26, 15-20 [calc]
- 1921 AMSB 27, 453-458 (F48, 6)
[calc, notation]
- 1923 AMM 30, 223-234 (F49, 20)
[lim]
*- 1925 Isis 7, 412-429 [notation]
Campori 1901 Carteggio far Leibniz
e Muratori, Modena
Cantelli G ed 1958 La disputa
Leibniz-Newton sull'analysi
scelta da documenti degli anni
1672-1716, Turin, 239 p
(Z84, 245)
Cantor M 1869 Hist Abt 14, 30-31
(F2, 19) [der]
Carr E W 1929 Leibniz, London, 228 p
(F55, 623)
Carruccio E 1927 Pe M (4)7, 285-301
(F53, 18)
Cassiner E 1903 Leibniz' System in
seinen Wissenschaftlichen
Grundlagen, Marburg Rv Met Mor
11, 83-99
Child J M 1916 Monist 26, 577-629
[calc]
- 1917 Monist 27, 238-294, 411-454
[calc]
- 1930 Sci Prog 25, 295-307
[Barrow, calc, Newton]
Costabel P 1949 Rv Hi Sc Ap 2, 311-
332 (MR11, 572) [Reyneau]
- 1956 Int Con HS 8, 25-28
[mech]
- 1960 Leibniz et la dynamique: les
textes de 1692, Paris, 128 p
(MR24A, 3; Z93, 5) [mech]

*Couturat Louis 1901 La logique de
Leibniz, d'après de documents
inédits, Paris (Repr 1961)
- 1902 Rv Met Mor 10, 1
- 1961 Opuscules et fragments
inédits de Leibniz, Hildesheim
699 p (MR29, 220)
Dalbiez R 1937 Int Con Ph (9)6, 3-7
[logic]
Degel C 1909 Intermed 16, 201
(Q1274)
Dickstein S 1896 Krak BI, 208-
[Kochanski]
- 1897 Krak Roz 33, 1-9 (F28, 9)
[Kochanski]
- 1897 Prz Fil 1, 70-85 (F28, 9)
[Kochanski]
- 1901 Prace MF 11, 225-278
[Kochanski]
- 1902 Prace MF 13, 237-283
[Kochanski]
Duerr Karl 1930 Neue Beleuchtung
einer Theorie von Leibniz.
Grundzuege des Logikkalkuels,
Darmstadt, 192 p
Durdik J 1869 Leibniz und Newton,
Halle (F2, 13)
Eneström G 1896 Bib M (2)10, 32
[Newton]
- 1909 Bib M 3(9), 309-320 [calc]
- 1909 Bib M (3)10, 43-47
(F40, 67)
- 1910 Bib M (3)11, 354-355
(F41, 60) [Newton]
Engelhardt W V 1947 Naturw 34, 97-
114
Erdmann B 1916 Naturw 4, 673-675
(F46, 34)
Ferrari G M 1921 Arduo 1, 134-140
[Vico]
*Fischer Kuno 1920 Gottfried Wilhelm
Leibniz. Leben, Werke und Lehre,
5 ed, Heidelberg
Fischer R 1967 Fors Fort 41(3), 83-85
[name]
Fleckenstein J O 1956 Die·Prioritaet-
streit zwischen Leibniz und Newton,
Basle-Stuttgart, 27 p
- 1964 MP Semb 11, 129-143
(MR30, 720)
Franke O 1928 Deu Morg Z 7, 155-78
[China]

Leibniz, Gottfried  (continued)

Freudenthal H 1954 in *Homenaja a Millas-Vallicrosa,* Barcelona. (Consejo sup. de invest. cien.) 1, 611-621 [topology]
Gagnebin S 1947 Dialect 1, 77-97 (MR9, 75)
Galego-Diaz J 1946 Gaz M Lisb 7(30), 3-4  (Z60, 9)
Galli M 1956 It UM (3)11, 445-446 (MR18, 368)
Garcia de Zuniga E 1947 Montv H Ci 1, 207-210
Gent W 1926 Kantstu 31, 61-88 (F52, 20) [space]
Gerhardt K J 1848 *Die Entdeckung der Differentialrechnung durch Leibniz mit benutzung des Leibnizischen Manuscripte auf der Koeniglichen Bibliothek zu Honnover,* Halle, 64 p
- 1856 Arc MP 27, 125-132 (re Weissenborn 1856)
- 1885 Ber Si, 19-23, 133-143
- 1891 Ber Si, 157-176  (F23, 19)
- 1891 Ber Si, 407-423  (F23, 19) [det]
- 1891 Ber Si, 1053-1068  (F23, 20) [Pascal]
- 1898 Ber Si, 417
- 1917 Monist 27, 524-560
- 1918 Monist 28, 530-566  [Pascal]
Gerhardt K J ed 1963 *Briefwechsel zwischen Leibniz und Christian Wolff,* Hildesheim, 192 p (MR29, 220)
- 1863 *Ausgabe der Mathematischen Schriften von Leibniz VII,* Halle
- 1875-1890 *Die philosophischen Schriften von G. W. Leibniz,* Berlin, 7 vols
Gerland E 1900 Bib M 3, 421
- 1906 Ab GM 21, 253
Gibson G A 1896 Edi MSP 14, 148-174 [Newton]
Guhrauer G E 1842-1846 *G. W. Leibniz,* Breslau, 2 vols + s
Guitton J 1949 Int Con Ph 10, 1145-1147 [Pascal]
Halbwachs L 1906 *Leibniz,* Paris, 124 p  (F37, 6)

Hannequin Arthur 1906 Rv Met Mor 14(a), 775-795 [mech]
- 1908 in *Etudes d'histoire des sciences et d'histoire de la philosophie,* Paris 2, 240-
Hofmann J E 1941 Ber Ab (13) (MR8, 190) [Gregory of St. Vincent]
- 1943 Ber Ab (2), 130 p (MR8, 190; Z61, 5) [Newton]
- 1947 in *Berichte Math-tagung Teubingen 1946,* 13-35 (MR9, 75; Z29, 2)
- 1948 *Leibniz Mathematischen Studien in Paris,* Berlin, 70 p (MR12, 382)
- 1949 *Die Entwicklungsgeschichte der Leibnizschen Mathematik waehrend des Aufenthaltes in Paris (1672-1676),* Munich, 261 p (MR12, 382; Z32, 193)
- 1957 Leop (3)3, 67-72  (Z105, 2)
- 1966 Praxis 8(10), 253-259 [calc]
- 1966 Ost Si M (2)175, 208-254 (MR37, 13) [calc]
- 1966 Sudhof Ar 50(4), 375-391
Hofmann J E + 1931 Arc GNMT 13, 277-292  (F57, 27) [calc, Tschirnhaus]
- 1931 Ber Si 26, 562-600 (Z3, 242) [fin dif]
Hoppe E 1928 DMV 37, 148-187 (F54, 6) [calc, Newton]
Houeel J 1869 *Sur une formule de Leibniz,* Bordeaux  (F2, 18)
Huber Kurt 1951 *Leibniz,* Munich, 451 p  (MR14, 524; Z45, 292)
Isely L 1903 Neuc B 32, 173-214 [Bourguet]
Jacobi C G J 1850 Ber Ber, 426- [letters]
Jourdain P E B 1916 Monist 26, 504-523 [logic]
Koser R 1902 Ber Si, 546
Koyré A + 1961 Isis 52, 555-566 (Z107, 246) [Clarke, Newton]
Langen L 1939 Muns Semb 14, 103-114 (F65, 12) [continuity]
Laurent H 1900 Intermed 7, 62 (Also 6, 138-139) [compound interest]
Lechalas G 1912 Rv Met Mor 20, 718-721 [line, Lobachevskii plane]

Leibniz, Gottfried (continued)

Liefmann-Keil E 1938 Z Natok 9,
505-540 (F64, 915) [Cournot]
Lerner A 1910 Intermed 17, 79
(Q1274) [bib]
Lindemann F 1927 Mun Si 3, 273-284
[Barrow, Fabri]
Lindemann H A 1946 Arg SC 142, 164-
176 (Z60, 9) [logic]
Locke L L 1933 Scr 1, 315-321
(Z7, 147) [calc mach]
Lorey W 1936 DMV 46 (F62, 27)
[Giessen]
Mahnke D 1912 Bib M (3)13, 29-61,
260 (F43, 68) [prime]
- 1912 Bib M (3)13, 250-260
(F43, 69) [exponent, notation]
* 1925 Leibnizens Synthese im
Universal mathematik und Indivi-
dual metaphysik, Jb Phi Pha 7,
308 p [phil]
- 1927 Isis 9, 278-293
- 1932 Marb Si 67, 31-69
(F59, 843; Z6, 337) [calc]
Mangeron D I 1946 Rv St Adam 32
83-99 (Z60, 9)
Martin Gottfried 1960 Leibniz:
Logik und Metaphysik, Cologne,
McClenon R B 1923 AMM 30, 369-374
[complex numb]
Mesnard J + 1963 Rv Hi Sc Ap 16,
11-22 (MR28, 1) [letter, Perier]
Miller M 1957 Dres Verk 5(2), 285-291
(MR22, 107) [calc]
Mitropolskii Yu A + 1967 Ukr IM
19(2), 90-94 (MR34, 4094)
Mollat G 1893 Mitteilungen aus
Leibnizens ungedruckten Schriften.
Neue Bearbeitung, Leipzig, 147 p
(F25, 21)
Montagu M F A 1941 Isis 33, 65
[Newton]
Moorman R H 1945 M Mag 19, 131-140
[phil]
Paul C Y 1954 Scr 20, 37-50
(Z56, 243) [calc]
Petronievics B 1934 Isis 22, 69-76
[deriv]
Planck M 1946 Z Natfor 1, 298-300
(Z60, 9)

Politano M L 1957 Archim 9, 178-180
(Z78, 241) [top]
Rafael Verhulst R P E de 1947 Mod
Rv 41, 5-29 (Z29, 241) [phil]
Ravier E 1937 Bibliographie des
oeuvres de Leibniz, Paris,
710 p (F63, 808) (See Schrecker
1938)
Rescher N 1954 JSL 19, 1-13
(MR15, 591) [logic]
- 1955 Phi Rev 64, 108-114
[inf, numb, quantity]
Ritter P 1904 Ber Ab, 44 p
- 1909 Ber Si, 897-901 (F40, 16)
- 1932 Fors Fort 8, 31
Rivaud A 1906 J Sav 4, 370-389,
431-441 [bib]
- 1913 Rv Met Mor 21(a)6, 94-120
Runge C 1905 Int Con 3, 737-738
(F36, 54) [calc mach]
Russell Bertrand 1900 A Critical
Exposition of the Philosophy of
Liebniz, Cambridge U Pr, 328 p
Sanford V 1934 MT 26, 183-185
Schmalenbach Hermann 1921 Leibniz,
Munich, 626 p
Schmidt F 1960 Kantstu 52, 43-58
Schneider H 1953 Isis 44, 266-272
(Z51, 3) [letter]
Scholz H 1942 DMV 52, 217-244
[foundations]
Schrecker P 1936 Thales 2, 130-135
- 1938 Rv Phi Fr E 126, 324-346
(F64, 915) (suppl to Ravier 1937)
[bib]
- 1947 JH Ideas 8, 107-116
[algorisms]
Sloman H 1857 Leibnitzens Anspruch
auf die Erfindung der Differential-
rechnung, Leipzig
- 1860 The claim of Leibniz to the
invention of the differential
calculus, London (rev transl
of 1857)
Sofonea T 1958 Verzek 35(s), 62-79
(Z83, 244) [ins]
Stamm E 1913 Wiad M 17, 43-90
Stammler G 1930 Leibniz, Munich,
183 p (F56, 19)
Studnicka F J 1889 Cas MF 18, 97
Tardy P 1868 Boncomp 1, 177-186
(F1, 13)

- 1868 Mondes (2)18, 687
  (F1, 13)
Tauc C Y 1951 Scr 20, 37-50
Torriere E 1914 Isis 2, 106-124
  [Descartes]
Vacca G 1899 Loria 2, 113-116
  (F30, 9)
Vahlen J 1905 Ber Si, 653-671
  (F36, 8)
Vleeschauwer H J de 1953 Phi Nat
  2, 358-375 [Newton, C Wolff]
Weissenborn H 1856 ZM Ph 1, 240-244
  (re Gerhardt 1855)
Wieleitner H 1931 ZMN Unt 62, 18-20
  (Schooten]
Wiener P P 1939 Phi Rev 48, 567-586
  [logic]
*Wiener Philip P ed 1951 *Leibniz
  Selections,* NY, 606 p
Wilde H 1909 Manc Mr 53(13), 1-9
Yushkevich A P 1948 UMN 3(1),
  150-164 (Z31, 145) [calc]


Le Lasseur

Gerardin A 1922 Fr AAS 46, 133
  [letters]


Lemaitre, Georges

Anon 1937 Vat An 1, 489-493
  (F63, 20)


Leman, Alfred Karl Hugo  1855-

Goedseels E 1922 Brx SS 42, 63-65
  [calc, Mansion]


Lemoine, Emile (Michel-Hyacinthe)
  1840-1912

Goormaghtigh R 1949 Scr 18, 182-184
Hagge K 1912 ZMN Unt 43, 422-423
  (F43, 43)
Jahnke E 1912 Ber MG 11, 57-62
  (F43, 43)

Laisant C A 1912 Ens M 14, 177-183
  (F43, 43)
Loria G 1912 Mathes It 4, 60-65
  (F43, 43)
Smith D E 1896 AMM 3, 27-33


Lenard, Philipp Eduard Anton
  1862-1947

Gehrcke E 1932 Fors Fort 8, 215-216


Lenhart, William  1785-1840

Kirkwood D 1875 Analyst H 2, 181-182
  (F7, 17)


Lennes, Nels Johann  1874-

Montgomery D + 1954 AMSB 60, 264-265
  (MR15, 770)


Leonardo. See also Cremonese;
  Fibonacci.


Leonardo da Vinci  1452-1519

Agostini A 1952 It UM (3)7, 321-327
  (MR14, 344; Z47, 4)
  [barycenter]
Albenga G 1932 Int Con 6, 431-432
Anon 1902? Nat 67, 440-441
  [hydraulics]
- 1954 Gac M (1)6, 105-107
  (MR16, 434)
Balta-Elias J 1955 Stu Gen 8, 626-
  636  (Z65, 243)
Beltrami Luca 1929 *Documenti e
  memorie riguardanti la vita e le
  opere di Leonardo da Vinci,*
  Milan
Beltrami L + 1891 *Il Codice di
  Leonardo da Vinci nella biblioteca
  del Principe Trivulzio,* Milan
  (F23, 7)
Berthelot D 1902 J Sav, 116
  [military]

Leonardo da Vinci (continued)

Bilancioni G 1930 Rv Sto Cr 21, 302-
329  [Cardano]
Cantor M 1890 Ham MG 2, 8-15
[constr]
*Coolidge J L 1949 The Mathematics
of Great Amateurs (Clarendon P)
Ch 4  (Repr 1963 Dover)
Duhem Pierre 1904 Bib M (3)4, 338-
343  [vector]
- 1906 BSM (2)31, 52-57
- 1906-1913 Etudes sur Leonard de
Vinci..., Paris, 3 vols
- 1906 Par CR 143, 946
Elsasser W 1899 ZM Ph 45, 1
Favaro A 1911 Ven I At 71, 953-975
[Archimedes]
- 1919 Arc Sto 1, 313-323
(F47, 2)
Feldhaus F M 1922 Leonard, der
Techniker und Erfinder, Jena
Giocomelli R 1952 Aerot 22, 178-191
(MR14, 344)  [dyn]
Gukovski M A 1935 Len IINT 7, 105-
128  [Alberti, mech]
Ivanov I + 1963 Gaz MF (A)15, 686-
693  (MR29, 219)
Johnson M C 1933 Scientia 53, 213-
217  (F59, 23)  [Archimedes]
Marcolongo R 1924 Nap FM Ri (36)30,
65-72  (F50, 13)
- 1925 It UM 4, 13-21  (F51, 18)
- 1926 Act M 49, 69-94  [ms]
- 1926 Scientia 40, 277-286,
"93-101"  [mech]
- 1927 Scientia 41, 245-254,
"99-106"  [inventions]
- 1929 Int Con 8, 275-293  (F55, 601)
- 1929 Linc Rn (6)9, 259-261
(F55, 10)  [geom, mech]
- 1929 Nap FM Ri, 35
[Alhazen, instruments]
- 1930 Scientia 47, 1-8, "1-10"
[mech]
- 1931 Nap FM Ri (4)1, 7-15
[Boffito, propor compass]
- 1932 Int Con 6, 431-432  [mss]
- 1932 Le invenzioni di Leonardo
da Vinci, Florence, 20 p
- 1932 La meccanica di Leonardo da
Vinci, Naples, 150 p

- 1934 Nap FM At (2)20(9)
(F60, 826)
- 1935 Rv FMSN 9, 281-298
(Z11, 193)
- 1937 Memorie sulla geometria e la
meccanica di Leonardo da Vinci,
Naples, 376 p
- 1939 Leonardo da Vinci artista-
scienziato, Milan  (2nd ed 1943)
McCurdy E 1920 Nat 105, 307-309,
340-342
Natucci A 1952 Archim 4, 209-213
(Z47, 244)  [geom]
- 1952 Batt 81, 89-103 (MR14, 832;
Z48, 242)  [geom]
Patroni A 1951 It UM (3)6, 159-162
(MR13, 2; Z43, 2)
Pignedoli A 1952 Mod At (5)10, 120-
133  (MR14, 1051)
Pirenne M H 1952 Brt JPS 3, 169-185
[perspective]
Sarton G 1919 Scribners Mag 65, 531-
540 (repr in Sarton 1948 The Life
of Science)
Satterly John 1952 MT 45, 576-577
[cen grav tetrahedron]
Schmidt W 1902 Bib M (3)3, 180
[Heron]
Severi F 1953 Scientia 88, 41-44
(MR14, 832)
Signorini A 1952 Archim 4, 221-227
(MR14, 524; Z47, 244)  [mech]
- 1952 Ric Sc 22, 2267-2274
(MR14, 524)  [mech]
- 1953 Scientia (6)88, 1-10
(MR15, 591)  [mech]
Speziali P 1953 Bib Hum, 295-305
[Pacioli]
Stites R S 1968 Am Sc 56, 222-243
Thayer W R 1893 Monist 4, 507
Venturi J B 1797 Essai sur les
ouvrages physico - mathématiques
de Leonardo de Vinci, Paris
(Repr 1912 Milan, 1937 Paris)
(F43, 10)
Verga E 1931 Bibliografia Vinciana,
Bologna, 2 vols
Wohlwill E 1888 Bib M (2)2, 19-26
[mech]
Zubov V P 1954 Mos IIET 1, 219-248
[Z59-1, 6)  [perspective, Witello]

Zubov V P 1968 *Leonardo da Vinci* (Harvard Univ Press), 325 p (Tr by D H Kraus)

Leonelli, Giuseppe Zecchini 1776-1847

Berkeley L M 1931 USNI 57, 207-209
Porro F 1886 Boncomp 18, 652-671

Leon y Ortiz

de Toledo L 1914 Esp SM (a)4, 1-5

Lepaige, Constantin-Jerome 1852-1929

Godeaux L 1939 Bel Anr 105, 239-269

Lepaute, Nicole-Reine Etable de la Brière 1723-1788

Connor E 1944 ASP (189)

Le Poivre, Jacques F.  -1710

Godeaux L 1935 Brx Con CR 2, 94-95
  [proj geom]

Lerch, Matyas (Mathias) 1860-1922

Boruvka O + 1957 Brno AS 29, 417-540
  (MR21, 1023; Z89, 242) [anal]
- 1959 in Schoeder K ed *Sammelband...
  Eulers,* 78-86 (MR23A, 584; Z111 5)
  [gamma fun]
  1961 Brno U,  352-360, 362-372
  (Z98, 10)
Cupr K 1923 Cas MF 52, 301-313
Cupr K + 1925 Cas MF 54, 140-151
  (F51, 30) [bib]
Frank L 1953 Cas M 78, 119-137
  (Z53, 340)
- 1953 Cz MJ 3(78), 111-122
  (MR15, 770; Z53, 247)

- 1960 Prag Pok 5, 764-771
  (Z102, 6)
Skrazek J 1953 Cz MJ 3(78), 111-122
  (MR15, 770; Z53, 340) [bib]
- 1960 Cas M 85, 228-240
  (MR22, 770; Z100, 247)
- 1960 Cz MJ 10(85), 631-633
  (MR23A, 584; Z194, 4)

Lesage, George-Louis 1724-1803

Rossier P 1941 Genv CR 58, 204-207
  (Z27, 3) [circle sq]

Leslie, John 1766-1832

Fraser-Harris D F 1932 Nat 130, 651-
  652 (F58, 42)
Glaisher J W L 1890 Nat 41, 9
Horsburgh E M 1933 Edi MSN 28, i-v

Lesniewski, Stanislaw 1886-1939

Luschei E C 1962 *The logical Systems
  of Lesniewski,* Amsterdam

Lesser, Oscar

Wolff G 1921 ZMN Unt 52, 131-136
  (F48, 22)

Letnikov, Aleksei Vasilievich 1837-1888

Shostak R J 1952 Ist M Isl 5, 167-
  238 (MR16, 434; Z49, 5)
Sludskii T 1889 Mos MO Sb 14,
  i-xxxiii

Lettenmeyer, Fritz 1891-1953

Schmidt H 1958 DMV 61(1), 2-6
  (MR20, 848; Z80, 4)

Leudesdorf, Charles   1853-1924

E B E 1925 LMSP (2)23, lxviii-lxix
  (F51, 29)


Le Verrier, Urbain Jean Joseph
  1811-1877

Anon 1911 *Centénaire de U. J. J.*
  *Le Verrier,* Paris, 128 p
  (F43, 23)
Aoust Barthelemy 1877 *Leverrier, sa*
  *vie et ses travaux,* Marseille
Bertrand J 1877 BSM (2)1, 116-124
  (F9, 15)
- 1879 *Eloge historique de*
  *Le Verrier,* Paris  (Repr 1890 in
  Eloges..., 159-192)  (F11, 28)
Brault L 1880 Rv Sc 25, 944-948
  [meteorology]
Guillot 1878 BSM (2)2, 29-41
  (F10, 17)
Jahnke E 1903 Arc MP (3)5, 37-40
  [Jacobi]
Pechule C 1878 Zeuthen (4)2, 93-96
  (F10, 17)
Smith D E 1921 AMM 28, 255
Souza Pinto R R da 1878 Lisb JS 1,
  86-89  (F10, 17)
W B 1877 Par CR 85, 579-596
  (F9, 15)


Levett, Rawson   1844-1923

Godfrey C 1923 M Gaz 11, 328-329
Mayo C H P 1923 M Gaz 11, 325-328


Levi, Beppo   1875-

Anon 1956 Arg UMR 17, vii-xvi
  (MR18, 550; Z71, 4)
Peano G 1955 *In memoria di. Studi*
  *di Beppo Levi...,* Cuneo, 115 p
Pla C + 1963 M Notae 18, iii-xxviii
  (Z109, 239)  [bib]
Terracini A 1963 Linc Rn (8)34, 590-
  606  (Z111, 3)


Levi, Eugenio Elia   1883-1917

Fubini G + 1918 Loria 20, 38-45
  (F46, 19)


Levi ben Abraham ben Hayyim
  XIII-XIV

Davidson I 1936 Scr 4, 57-65
  (Z13, 337)
Sarton G 1931 Int His Sc 2(2), 885
  [bib]


Levi ben Gerson   1288-1344

  = Gersonide

Carlebach J 1910 *Levi ben Gerson als*
  *Mathematiker...,* Berlin
Curtze M 1898 Bib M (2)12, 97-112
  (F29, 32)  [trig]
Lange C 1900 *Sefer Maassei Choschen*
  *...,* Frankfurt
Rozenfeld B A 1958 Ist M Isl 11,
  733-742, 777-782
Sarton G 1947 Int His Sc 3(1),
  594-607  [bib]
Steinschneider M 1870 Heb Bib 9,
  162-164  (F2, 5)
- 1890 Bib M (2)4, 107-108


Levi-Civita, Tullio   1873-1941

Amaldi U 1946 Linc Rn (8)1, 1130-
  1155  (Z60, 16)
Anon 1937 Vat An 1, 496-511 (F63, 20)
- 1941 Rv Cien 43, 683-685
  (MR3, 258)
- 1946 Annali (4)25, iii-viii
  (Z60, 16)
Buhl A 1942 Ens M 38, 350-351
  (Z60, 16)
Cartan E 1942 Par CR 215, 233-235
  (F68, 18; Z60, 16)
Cisotti U + 1942 Lom Gen (3)6, 110-
  112 (F68, 18)
Gomes R L 1943 Porto Fac 28, 5-7
  (MR9, 170; Z61, 5)

F L U 1929 Rv M Hisp A (2)4, 184-188
 (F55, 25)
Hodge W V D 1942 Lon RS Ob 4, 151-165
 (Z60, 16)
- 1943 LMSJ 18, 107-114 (Z60, 16)
Krall G 1953 Civ Mac 1(4), 33-57
 (MR15, 276) [mech]
- 1953 Civ Mac 1(6), 42-48
 (MR15, 276) [relativity]
Levi B 1942 M Notae 2, 155-159
 (MR4, 66; Z60, 16)
Masotti A 1946 Milan Sem 17, 16-61
 (Z61, 6) [bib, Volterra]
Roth L 1942 Nat 149, 266 (MR3, 258;
 Z60, 16)
Ruse H S 1943 Edi MSN (33), 19-24
 (Z60, 16)
Somigliana C 1947 Milan Sem 17, 16-
 61 (MR9, 485; Z61, 6)


Levin, Boris Yakovlevich 1906-

Akhiezer N I + 1957 UMN (NS)12(2),
 237-242 (MR19, 826; Z78, 4)


Levin, S. S.

Stepanoff W 1932 Mos MO Sb 39,
 127-128


Levitzki, Jacob

Amitsur S A 1957 Riv Lemat 11, 1-6
 (MR20, 739)


Levy, Lucien 1853-

Anon 1912 Rv Sc (5)18, 186
Bricard R 1913 Nou An (4)13, 355-363
 (F44, 24)
Buhl A 1912 Ens M 14, 404-405
 (F43, 43)


Levy, Maurice 1838-1910

Fantoli G 1912 Linc Rn (5)21, 517-525

Hadamard J 1911 Rv GSPA 32, 141-143
Lecornu L 1915 Par Mm 53
Picard E 1910 Rv Sc 14, 529-530


Lewis of Caerleon XV

Kibre P 1952 Isis 43, 100-108
 (MR15, 276; Z47, 4)


Lexell, Anders Jean 1740-1784

Lysenko V I 1960 Vop IET 9, 116-120
 (Z119:1, 10) [Fuss, geom]


Lexis, Wilhelm 1837-1914

Klein F 1914 DMV 23, 314-317
 (F45, 44)
Lietzmann W 1915 ZMN Unt 41, 161
 (F45, 44)
Lorey W 1925 Nor St Tid 4, 31-41
 (F51, 26) [ins]


L'Hospital, Guillaume François de
 1661-1704

Bernardi J 1807 *Essai sur la vie,
 les écrits et les lois de Michel
 de L'Hopital,* Paris [bib]
Conte L 1955 Archim 7, 132-135,
 228-230 (Z64, 242)
Coolidge J L 1949 in *The Mathematics
 of Great Amateurs* (Clarendon Pr),
 Ch 12 (Repr 1963 Dover)
Costabel P 1965 Rv Hi Sc Ap 18, 29-
 43 [cubic eqs]
Marre A 1879 Par CR 88, 76-77
 (F11, 17)
Rebel Otto Julius 1934 *Der Brief-
 wechsel Zwischen Johann (I)
 Bernoulli und den Marquis de
 L'Hospital...,* Heidelberg, 46 p
 (Diss) (Z9, 97)
Sanford V 1934 MT 26, 306-307
Spiess O ed 1955 *Der Briefwechsel
 von Joh. Bernoulli,* Basle
Turrière E 1937 Ens M 36, 179-194
 (Z17, 50) [curve]

L'Hoste, Jean  1570-1631

Gérardin A 1930 Par Sav CR (1928),
  9-10  (F56, 820)
Xardel P 1924 Fr AAS, 74 (F51, 39)


L'Huilier, Simon Antoine Jean
  1750-1840

  = Lhullier

Dickstein S 1930 Slav Cong, 111-118
  (F56, 22)
Guenther S 1887 Wurt MN 2, 1-9


Liagre, Jean Baptiste Joseph
  1815-1891

Brialmont A 1892 Bel Anr 58, 323-
  376  (F24, 33)


Liard, Louis

Dauriac L 1926 Rv Met Mor 33, 379-
  423  (F52, 27)


Libri, Guillaume B. I. T.
  1803-1869

  = L.-Carrucci della Sommaia

Brocard H + 1912 Intermed 19, 183
  (Q3811)  (Also 18, 25-26,
  161-163)
Candido G 1937 It UM Con 1, 478-502
  (F64, 20)
Corsini A 1956 Int Con HS 8, 163-173
Loria G 1918 Geno ASL 28, 35 p
Natucci A 1953 It UM Con 2, 663-673
  (Z50, 243)
Procissi A 1947 It UM (3)2, 46-51
  (MR9, 74; Z31, 146)
  [Cauchy, numb thy]
Stiattesi A 1879 *Commentario storico-
  scientifico sulla vita e le opere
  del conte Guglielmo Libri,* 2nd ed,
  Florence

Vetter 0 1928 Arc Sto 9, 175-176
  [XIV, ms]


Lichnerowicz, André Léon  1915-

Ehresmann C 1956 It UM Con 5, 21-26
  (Z72, 246)


Lichtenberg, Georg Christoph
  1742-1799

Elestein E 1907 ZM Ph 54, 224


Lichtenstein, Leon  1878-1933

Anon 1933 Rv Phi Fr E 116, 478
- 1935 Wiad M 38, 131-136  (F61, 956)
Holder O 1934 Leip Ber, 307-314
Schauder J 1933 Mathes Po 8, 149-156
  (F59, 856)  [part diff eqs]
Steinhaus H 1933 Mathes Po 8, 131-
  137, 137-142  (F59, 856)


Lie, Marius Sophus  1842-1899

Anon 1899 Kazn FMO (2)9(2)
- 1900 Mathesis (2)10, 228
- 1905 Lon RSP 74, 60-68
Beltrami 1899 Linc Rn (5)8, 281
  (F30, 23)
Bianchi L 1899 Linc Rn (5)8, 360-366
  (F30, 23)
Braunmuehl A v 1901 Biog J ℓ, 324-
  325  (F32, 16)
Burnside W 1899 LMSP 30, 332-336
  (F30, 23)
Cremona 1899 Linc Rn (5)8, 281
  (F30, 23)
Darboux G 1898 AMSB (2)5, 367-378
- 1899 AMM 6, 99-101
- 1899 Par CR 128, 525-529
  (F30, 23)
Dodsanmeldelse 1899 Nyt Tid (B)10,
  48
Engel F 1899 Leip Ber 51, xi-lxi
  (F30, 22)
- 1900 Bib M (3)1, 166-204

- 1900 DMV 8, 30-36
- 1902 Batt 40, 325-363
- 1914 DMV 5, 14-79 (F45, 81)
  [contact transf]
- 1922 Nor M Tid 4, 80-84 (F48, 11)
- 1922 Nor M Tid 4, 97-114 (F48, 12)
- 1929 Nor M Tid, 97-102 (F55, 17)
Engel F + 1934 Nor M Tid 16, 105-113
  (F60, 834)
Forsyth A R 1899 Nat 59, 445-446
  (F30, 24)
Freudenthal H 1968 Rv Syn (3),
  49-52, 223-243 [bib, top alg]
Guldberg A 1913 Chr Sk 1(5), 40 p
- 1913 *Verzeichnis ueber den
  wissenschaftlichen Nachlass von
  Sophus Lie,* Kristiana, 43 p
  (F44, 40)
Halsted G B 1899 AMM 6, 97-98
Harkin D 1950 Nor M Tid 32, 68-78
  (MR12, 311) [Abel]
Isely L 1907 Neuc B 33, 138-151
  (F38, 24)
Johansson I 1942 Nor M Tid 24,
  97-106 (MR8, 140; F168, 17;
  Z27, 196)
Klein S 1893 in his *Lectures on
  Mathematics...1893,* NY (Macmillan),
  9-24
Klein F 1896 Nou An (3)15, 1
  [geom]
- 1898 M Ann 40, 583
Lampe E 1899 Naturw R 14, 216-218
  (F30, 24)
Miller G A 1896 AMM 3, 295-296
- 1899 AMM 6, 191-193
- 1942 Sci 95, 353-354 [group, Klein]
Mira Fernandez A de 1942 Gaz M Lisb
  3(12), 1-2 (MR7, 106; Z60, 13)
Noether M 1900 Leip Ber 52, 1
- 1900 M Ann 53, 1-41
- 1903 Batt 41, 145-179 (F34, 24)
Scheffers G 1903 DMV 12, 525-539
  [intgl]
Segre C 1899 Loria 2, 68-75
  (F30, 23)
- 1899 Tor FMN 34, 363-366 (F30, 23)
Stoermer C 1904 Chr Sk (1)(7), 1-31
  (F35, 21)
Study E 1908 DMV 17, 125-142
  [group]

Sylow L 1899 Arc MN 21, 1-22
  (F30, 24)
- 1899 Chr Sk (1)(9), 1-15 (F30)
Voit C 1900 Mun Si 30, 339
Weiss E A 1936 Deu M 1, 23-37
Zorawski K 1899 Prace MF 10, 85
- 1899 Wiad M 3, 85-119 (F30, 24)


Lies-Bodart    XIX

Laurent H 1900 Ens M 2, 346


Lietzmann, Walter    1880-1959

Proksch R 1959 MN Unt 12, 227-228
  (Z100, 247)
Stender R 1960 MP Semb 7, 1
  (Z99, 246)
Wansink J H 1959 Euc Gron 35, 81-82
  (Z88, 7)


Ligowski, Wilhelm    1821-1893

Anon 1897 DMV 4, 46 (F28, 20)


Lilienthal, Reinhold von    -1935

Behnke H 1935 Muns Semb 7, 120-123
  (F61, 955)


Lilio, Luigi    XVI

= Aloysius  Giglio (Lilius)

Mayr J 1933 As Nach 247, 429-444


Lill, E.    XIX

Jones, P S 1953 MT 46, 35-37
  [polynomials]


Lindeberg, Jarl Waldemar    1876-

Nevanlinna F 1933 Sk Akt 11, 197-198
  (F59, 856)

**Lindeloef,** Ernest Leonard  1870-

Myrberg P J 1947 Act M 79, i-iv
  (MR9, 75; Z29, 195)
Orts J M 1948 Rv M Hisp A (4)8,
  19-22  (Z33, 4)


**Lindeloef,** Lorenz Leonard  1827-

Mittag-Leffler G 1908 Act M 31,
  407-408 (F39, 39)
Sonin N J 1908 Pet B (6), 476-480
  (F39, 39)


**Lindemann,** Carl Loyis Ferdinand von
  1852

Carathéodory C 1940 Mun Si, 61-63
  (F66, 22; Z24, 244)
Volk O 1932 Fors Fort 8 , 145
  (F58, 995)


**Linder,** Koenraad  1874-1961

Brans J A T M 1962 Actuar St (4),
  105-108  (Z99, 246)


**Lindner,** Paul

Zuehlke P 1917 Ber MG 16, 55-57
- 1917 ZMN Unt 48, 424-427
  (F46, 24)


**Lindsay,** William A  -1926

R W 1926 Edi MSP 44, 55-56
  (F52, 34)


**Lindstedt,** Anders

Akesson O A 1940 Sk Akt 23, 58-61
  (F66, 22)


**Linebarger,** Charles Elijah  1867-1937

Anon 1938 SSM 37, 770


**Linnik,** Yuri Vladimirovich
  1915-

Prokhorov Yu V 1965 Teor Ver 10, 117-
  129  (MR30, 560) [prob, stat]


**Lionnet,** François Joseph Eugene
  1805-1884

Marre A 1886 Boncomp 18, 424-440


**Liouville,** Joseph  1809-1882

Ahrens W + 1902 Intermed 9, 215
Bell E T 1947 Scr 13, 177-185
  [quad forms]
Brocard H 1906 Intermed 13, 13-15
  (Q2285)  (See also Intermed 9,
  36, 215-217; 14, 59)
Jahnke E 1903 Arc MP (3)5, 41
  [Jacobi]
Laboulaye F 1882 Par CR, 467-471
Loria G 1936 Archeion 18, 117-139
  (F62, 25)
- 1936 Scr 4, 147-154, 257-262,
  301-306  (F62, 1036)
Tannery J 1908 BSM (2)32, 47-62,
  88-95; 33, 47-64  (F39, 23)
  [Dirichlet]


**Liouville,** R.  1856-1930

Levy P 1931 Par EP (2)29, 1-5
  (F57, 42)


**Lipka,** Joseph  -1924

Graustein W C 1924 AMSB 30, 352-356
Wiener N 1924 MITM Cont (2)(73)

Lippmann, Gabriel 1845-

Lebon E 1911 BSM (2)36, 153-154
Leduc A 1921 Rv GSPA 32, 565-570
Volterra V 1921 Linc Rn (5)30, 388-
389 (F48, 47)


Lipschitz, Rudolf Otto Sigismund
1832-1903

Anon 1903 Gott Jb, 116-117
(F34, 36)
Kortum H 1906 DMV 15, 56-59
(F37, 19)


al-Lith XI

= Abul Jud Muhammad ibn
al-Lith

Sarton G 1927 Int His Sc 1, 718
[bib]
Schoy C 1925 Isis 7, 5-8 [area]


Livens, George Henry 1886-1950

Morris R M 1951 LMSJ 26, 156-160
(Z42, 4)


Livet, Jean Joachim 1783-1812

Dickstein S 1903 Wiad M 7, 225-243
(F34, 11)


Lobachevskii, Nikolai Ivanovich
1792-1856

Aleksandrov P S 1943 SSSR Vest
(11/12), 52-62 (Z60, 13)
- 1946 UMN 1(1), 11-14
Aleksandrov P S + 1943 Nikolai
Ivanovich Lobachevskii, Mos-Len,
100 p (MR7, 355; Z60, 13)
Andronov A A 1956 Ist M Isl 9, 9-48
(Z70, 6) [birth]
Anon 1893 Nou An (3)12, 188-192

- 1894 Kazn FMO (2)4, 31-43
- 1896 Compte rendu du bureau local
du Comité Lobatschefskij. 1893-1895
Kasan, 25 p (F27, 14)
- 1896 Kazn FMO (2)6(2), 33-51
[port]
- 1927 Ad Annum MCMXXVI Centesimum a
Geometra Kazaniensi N. I. Lobacewski
Nonedklideae Geometriae Systematis
inventi concelebrandum, Kazan,
2 vols (F53, 24)
- 1956 Matv Shk (3), 1-2 (MR17, 932)
- 1961 Ist M Isl 14, 623-627
(Z118:1, 8) [letter]
Archibald R C 1954 Isis 45, 199
[bib]
Bespamyatnyh N D 1950 Ist M Isl 3,
154-170 (MR13, 2) [neg numb]
Bondarenko I 1893 Kagan (173)
*Bonola Roberto 1906 La geometria non-
euclidea..., Bologna, 220 p
(Ger transl 1908, Engl transl 1911
+ appendices, repr 1955 Dover)
Bronshtein I N 1950 Ist M Isl (3),
171-194 (MR13, 2) [educ]
Chistyakov V D 1956 Ist M Isl 9, 247-
270 (Z70, 5) [educ]
Deaux R 1948 Mathesis 56, 106-109;
57, 40-45, 128-133 [*bib]
Depman I Ya 1948 Mos IIET 2, 561-
563 (MR11, 573)
- 1950 Ist M Isl 3, 475-485
(MR13, 197) [M F Bartel]
- 1956 Ist M Isl 9, 111-122
(Z70, 6) [Lettrov]
Dittrich E 1913 An Natphi 12, 62-87
Efimov N V 1956 UMN 11(1), 3-15
(MR17, 814; Z70, 6)
*Engel F 1898 Nikolai Ivanovich
Lobachewsky, Leipzig
*- 1899 Zwei geometrisches Abhand-
lungen aus den Russischen ueberzetzt
mit Anmerkung und mit einer
Biographie der Verfassers, Leipzig
Engel F + 1898-1913 Urkundeu zur
Geschichte der nichteuklidischen
Geometrie, Leipzig, 2 vols.
Fedorenko B V 1956 Ist M Isl 9, 65-
75 (Z70, 6)
- 1957 Mos IIET 17, 163-228
(Z115, 244) [geom]

Lobachevskii, Nikolai I. (continued)

Gaiduk Yu M 1954 Ukr IM 6, 476-478
(Z55, 4) [geom]
- 1956 Ist M Isl 9, 215-246
(Z70, 5)
Gerasimov V M 1952 Ukazatel litera-
tury po Geometrii Lobachevekogo i
razvitiyu ee idei, Mos, 192 p
Gnedenko B V 1949 Ist M Isl (2), 129-
136 (MR12, 1) [prob]
Gutman D S 1956 Ist M Isl 9, 77-100
(Z70, 6) [org]
Halsted G B 1895 AMM 2, 137-139
- 1905 Sci 22, 161-167
Hatzikakis N 1926 Gree SM 7, 81-88
(F52, 26)
*Houel J 1870 BSM 1, 66-71
(F2, 22)
Householder A S 1959 AMM 66, 464-466
(Z89, 242) [Dandelin, Graeffe]
Idelson N I 1949 Ist M Isl (2),
137-167 (MR12, 1) [ast]
Ilin A S 1957 Matv Shk (3), 1-4
(MR19, 624) [geom, phil]
Kagan V F 1894-1898 Kagan (174-276)
(Repr 1900 Odessa)
- 1943 SSSR Vest (7/8), 44-83
(MR5, 253)
- 1944 Lobachevskii, Mos-Len, 347 p
(MR6, 254: Z60, 13) (2nd rev ed
1948; Z33, 3)
- 1955 Lobachevskii i ego geometriya,
Mos, 303 p (Z66, 5)
*- 1957 N. Lobachevsky and His
Contribution to Science, Moscow,
92 p [bib, chron, pop]
Kagan V F + 1956 in Lobachevskii N I,
Tri sochineniya po geometrii,
Mos, 5-29, 221-413
Karteszi F 1953 Magy MF 3, 189-197
(MR15, 384)
- 1954 Hun Act M 5(S), 127-136
(MR16, 986; Z57, 243)
- 1956 Magy MF 6, 157-161
(MR20, 620; Z74, 245)
Khilkevich E K 1949 Ist M Isl 2,
168-230 (MR12, 1)
- 1956 The geometry of Lobachevskii
and experience, Tyumen, 16 p
(MR19, 1030) [geom, phil]

Kolesnikov M 1965 Lobachevskii,
Mos, 318 p [facs, ports]
Kolman E 1956 Int Con HS 8, 134-137
[geom]
- 1956 Velikij russkii mystitel
N. I. Lobachevskii, 2nd rev ed,
Mos, 102 p
Kotelnikov A P 1950 Nekotorye
primeneniya idei Lobachevskogo v
mekhanike i fizike, Mos-Len
Kurshak J + 1902 MNB Ung 18, 250-279
Lakhtin L 1894 Mos MO Sb 17, 474-493
(F25, 26)
Laptev B L 1951 Ist M Isl 4, 201-229
(MR14, 524; Z44, 246) [parallels]
- 1951 UMN 6(3), 10-17
(MR13, 197; Z54, 3)
Lechelas G 1912 Rv Met Mor 20, 718-
721 [Leibniz, line, plane]
Licis N 1960 Lat Ves (3), 33-44
(MR22, 2051; Z92, 3)
[geom, phil]
- 1960 Lat Ves (8), 5-18 (MR22, 2051)
[space, time]
Litvinov E 1894 N. I Lobachevskii,
Pet (F25, 26)
Lunc G L 1949 Ist M Isl (2), 9-71
(MR12, 1) [anal]
- 1950 UMN (NS)5(1), 187-195
(MR11, 573; Z36, 5) [anal]
- 1951 UMN 6(1), 163-164 (Z42, 4)
Mansion P 1897 Brx SS 22, 44
Modzalevskii L B 1902 Kazn FMO
(2)12(2), 86-101 (F33, 24)
Modzalevskii L B ed 1948 Materialy
dlya biografii N. I. Lobachevskii,
Mos-Len, 827 p (MR11, 573)
Morozov V V 1951 Ist M Isl 4, 230-
234 (MR14, 524; Z44, 246) [alg]
Mozhalevski B 1902 Kazn FMO (2)12(1),
86-101 [letters, Welikpolsky]
Nagaeva V M 1949 Mos IIET 3, 368-377
[educ]
- 1950 Ist M Isl 3, 76-153 [educ]
Nasimov P S 1896 Kazn Uch Z (1),
111 [plane]
*Norden A P 1952 Sto dvadtsat pyat
let naevklidovoi geometrii
Lobachevskogo 1826-1951, Mos
(Z47, 245; 49, 290)
Norden A P + ed 1956 N.I. Lobachevski.
Tri sochineniya po geometrii, Mos,
415 p

Lobachevskii, Nikolai I. (continued)

- 1956 Mos M Sez, 101-102
- 1958 Ist M Isl 11, 97-132
  (MR23A, 584; Z102, 5) [geom]
Parfentiev N N 1932 Int Con 6, 483-
  488 (F58, 41) [phil]
Pavlicek J B 1953 Zaklady Neeuklei-
  dovske geometrie Lobacevskeho,
  Prague, 224 p
- 1956 Cas M 81, 376-385
  (Z74, 245)
Piccard Sophie 1957 Lobatchevsky,
  grand mathematicien russe. Sa
  vie, son oeuvre, Paris, 39 p
  (MR19, 108; Z77, 9)
Privalova N I 1956 Ist M Isl 9,
  49-64 (Z70, 6) [birth]
Reyes y Prosper V 1893 Prog M 3,
  321-324 (F25, 25)
Rogachenko V F 1953 Ist M Isl 6,
  477-494 (MR16, 660; Z53, 246)
  [approx, eqs]
Rozenfeld B A 1956 Mos M Sez, 234
  [geom]
Rybkin G F 1950 Ist M Isl 3, 9-29
- 1951 UMN (NS)6(3), 18-30
  (MR13, 197; Z54, 3)
- 1956 SSSR Vest (4), 151-153
- 1956 Vop IET (2), 50-60
  (MR18, 784)
Rybkin G F + 1956 Ist M Isl 9, 107-
  110 (Z70, 6) [org]
Shirokov P 1964 A Sketch of the
  Fundamentals of Lobachevskian
  Geometry, Groningen, 88 p
Sluginov S P 1927 Perm Sem 1, 11-19
  (F53, 25) [geom]
Sokolow N P 1894 Kiev U Iz (6),
  1-39 (F25, 26) (Also in Kiev
  UFMO 1893)
Sommerville D M Y 1911 Bibliography
  of Non-Euclidean Geometry, London,
  310-, 361- [bib]
Urazbaev B M 1948 Velikii russkii
  matematik N. I. Lobachevskii,
  Alma Ata, 31 p (in Kozak)
  (2nd ed 1950 32 p)
Vasilev Aleksandr Vasilevich 1894
  Nikolai Ivanovich Lobachevskii.
  Address Pronounced at the Commem-
  orative Meeting of the University

of Kasan Oct. 22, 1893, (transl
  G. B. Halsted), Austin, Tex, 48 p
  (French tr 1896, German tr 1895,
  Spanish tr 1896) (F25, 26)
- 1894 NYMS 3, 231-235 (F25, 27)
  [alg, anal]
- 1895 Prog M 5, 12-16, 33-34
  (F26, 20)
- 1895 ZM Ph 40(S), 205-244
  (F26, 19)
- 1896 Kazn Uch Z (10), 199-213
  (F27, 14)
- 1897 DMV 4, 88-90 (F28, 42)
  [geom]
  1914 N. I. Lobachevskii, Pet
- 1924 Ens M 23, 218-220
- 1925 Pe M 5, 121-122 (F51, 23)
  [alg]
Verkhunov V M 1958 Mos UM Me(6),
  77-89 (MR22, 618; Z86, 4) [mech]
*Vucinich A 1962 Isis 53, 465-481
  (MR26, 930; Z112, 243) [geom]
Weyr Ed 1896 Cas MF 25, 1-38
  (F27, 15)
- 1897 Cas MF 26, 249-254 (F28, 13)
Yakunin P F 1956 Ist M Isl 9, 129-144
  (Z70, 6)
Yanichevsky E 1869 Boncomp 2, 223-
  262 (F2, 22)
Yanovskaya S A 1950 Peredovye
  idei N. I. Lobacevskogo, Mos-Len,
  183 p (MR14, 609)
- 1951 Ist M Isl 3, 30-75;4, 173-200
  (MR14, 524; Z44, 246)
Yushkevich A P 1960 Int Con HS 9,
  622-623 (Z114, 7) [alg]
Yushkevich A P + 1949 Ist M Isl (2),
  12-128 (MR12, 1) [alg]
Zabotin I P 1953 Lobachevskii,
  Kazan
Zemaitis Z 1931 Kosmos 12, 212-216
  [geom]
Zhatykov O A 1954 Kazk Ak (2), 67-76
Zubov V P 1956 Ist M Isl (9),
  123-128 (Z70, 6)


Lobatie, Antide Gabriel Marguerite
1786-1866

Brocard H 1909 Intermed 17, 277-278
  (See also Intermed 13, 60, 224;
  14, 86-87, 181; 15, 105; 16, 128-129)

Lobatto, Rehuel   1797-1866

Matthes C J 1868 Arc MP 49,
  332-334  (F1, 15)


Lodge, Alfred   1854-1937

Archibald R C 1948 *Mathematical
  Table Makers,* NY, 51-52
  (Scr Stu 3)  [bib]
Tuckey C O 1938 M Gaz 22, 3-4
  (F64, 919)


Loeffler, Eugen 1883-

Toepfer H 1963 Praxis 5, 74


Loewy, Alfred   1873-1935

Fraenkel A 1938 Scr 5, 17-22
  (F64, 23)
Lorey W 1935 Bla Versi 3, 298-304
  (F61, 956)


Loewy, Maurice   1833-1907

Backlund O A 1907 Pet B (6)(16), 698
Hamy M 1907 Rv GSPA 18, 949-950


Logan, James   1674-1741

Brasch F E 1942 APSP 86, 3-12


Lohmeyer, Hans   -1960

Steiner H G 1961 MP Semb 8, 107
  (Z97, 244)


Lohse, Wilhelm Oswald   1845-1915

Archibald R C 1948 *Mathematical
  Table Makers,* NY, 52  (Scr Stu 3)
  [bib]


Lommel, Eugen Cornelius Joseph
  1837-1899

Archibald R C 1948 *Mathematical
  Table Makers,* NY, 52-53
  (Scr Stu 3)  [bib]
Boltzmann L 1900 DMV 8, 47


Lomonosov, Mikhail Vasilievich
  1711-1765

Kotov V F 1955 Mos IIET 5, 52-68
  (Z68, 5)  [mech]
Kuznetsov B G 1961 *Tvorcheskii put
  Lomonosov,* Moscow, 375 p
  (MR23A, 136)
Likholetov I I 1961 Mos UM Me
  (1)16(5), 16-24  (Z112, 2)
Maistrov L E 1961 Vop Fil (5),
  93-102
- 1962 Sov Rv 3(3), 3-18
  (Tr of 1961)
Zubov V P 1954 Mos IIET 1, 5-52
  (Z60, 9)


London, Franz   1863-

Study E 1917 DMV 26, 153-157
  (F46, 18)


Longchamps, Gaston Albert Gohierrede
  1842-

Lazzeri G 1906 Pe M 22, 53-59
  (F37, 31)


Loomis, Elias   1811-1889

Newton H A 1890 Smi R, 741-770
  (F23, 24)


Loomis, Elisha Scott   1852-1940

Anon 1945 SSM 41, 255
Finkel B F 1894 AMM 1, 219-222

**Lopatinskii,** Yaroslav Borisovich

Daniljuk I I + 1967 Ukr IM 19(2),
95-99 (MR34, 4093)


**Lorent,** Henri 1871-1956

Godeaux L 1956 Mathesis 65, 406
(Z71, 4)


**Lorentz,** Henrik Antoon 1853-1928

Anon 1900 Ned Arch (2)5
[phys, relativity]
Bridgeman P W 1936 Am Ac Pr 70,
550-552 (F62, 1041)
De Haas-Lorentz G L 1957 *H. A.
Lorentz,* Amsterdam, 716 p
(Z77, 242)
Langevin P 1928 Rv Sc, 157, 158
Page L 1928 Am JS (5)15, 374
Schidlof A 1929 Arc Sc (5)11, 7-9
(F57, 40)
Verschaffelt J E 1928 Wis Nat Ti
4(1), 19-20
Volterra V 1928 Nuo Cim (2)5, 41-43
(F54, 39)


**Lorenz,** Ludwig Valentin 1829-1891

Valentiner H 1896 Dan Ov, 440-445
(F27, 18)


**Lorenzini,** Lorenzo

Tenca L 1959 Lom Rn M (A)92, 292-
306 (Z93, 6)


**Lorey,** Wilhelm 1873-1955

Behnke H 1956 MP Semb 5, 1-3
(MR18, 453; Z74, 246)


**Lorgna,** Antonia Maria 1735-1796

Jacoli J 1877 Boncomp 10, 1-74
(F9, 8)


**Loria,** Gino 1862-1939

Archibald R C 1939 Osiris 7, 5-30
(MR1, 34; Z22, 297)
Natucci A 1954 Rv Hi Sc Ap 7, 372-
374
Pascal E 1892 Rv M 2, 179-186
(F24, 18)
Terracini A 1954 Tor FMN 88, 387-
392 (MR16, 434; Z55, 244)
- 1955 Ukr IM 7, 3-4 (Z64, 4)
Vetter Q 1932 Cas MF 61, 369-370
(F58, 50)
Vollgraff J A 1954 Synthese 9, 485-491
(MR17, 117)


**Loriga,** Juan Jacobo Duran

Dieguez D F 1912 Esp SM 1, 237-242
(F43, 35)
Gallego Armestro H 1925 in *Matemati-
cos espanoles cortemporaneos,*
Santiago (F51, 38)


**Losada y Puga,** Cristobal de
1894-

Anon 1940 *Notice sur ses travaux
scientifiques,* Paris, 58 p


**Love,** Augustus Edward Hough 1863-
1940

Anon 1929 Nat 123, 850 (F55, 615)
Milne E A 1941 LMSJ 16, 69-80
(MR3, 98; Z60, 16; 28, 337)
- 1941 Lon RS Ob 3, 467-482
(Z60, 16)

**Lowan,** Arnold Noah  1898-

Archibald R C 1948 *Mathematical Table Makers,* NY, 53-55 (Scr Stu 3)  [bib]

**Loyd,** Sam 1841-1911

Gardner M 1959 in *The First Scientific American Book of Math Puzzles...,* NY, Ch 9
Bain G G 1907 Strand 34, 771-777

**Loys de Cheseaux,** Jean Philippe de 1718-1751

Dumas S 1926 Schw Vers 22, 107-109 (F52, 22)

**Lozinskii,** Sergei Mikhailovich 1914-

Mysovskikh I P + 1964 UMN 19(6), 207-212  (MR30, 872)

**Lubbock,** John William 1803-1865

Anon 1866 Lon As Mo N 26, 118-
- 1867 Lon RSP 15, xxxii-

**Lubelski,** Salomon 1902-

Anon 1958 Act Ari 4, 1-2 (MR20, 129; Z79, 5)

**Lucas,** François Eduard Anatole 1842-1891

Harkin D 1957 Ens M (2)3, 276-288 (MR20, 620; Z78, 4)

**Lucchesse,** Smeraldo Borghetti

Boncompagni B 1880 Boncomp 13, 1-87, 121-201, 245-308  (F12, 11)

**Lucretius,** Titus Caras  -I

Keyser C J 1918 AMSB 24, 321-327 (F46, 45)  [inf]  (repr in The Classical Weekly Jan 27, 1919)

**Ludolph van Ceulen**  1540-1610

Bosmans H 1910 Brx SS 34(2), 88-139 [Viete]
- 1925 Mathesis 39, 352-360

**Lueroth,** Jakob  1844-

Brill A + 1911 DMV 20, 279-299 (F42, 28)
Neumann L 1911 Loria 13, 49-57 (F42, 29)
Voss A 1911 Mun Si 41, 21-33 (F42, 28)

**Lugli,**  A.

Frattine G + 1896 Pe M 11, 77

**Lukasiewicz,** Jan  1878-1956

Anon 1957 Stu Log 5, 7-11 (MR19, 519)
Borowski L + 1958 Stu Log 8, 7-62 (MR21, 1024)
Mostowski A 1957 Rund M 44, 1-11 (Z77, 242)
Scholz H 1957 Arc MLG 3, 3-18 (Z77, 242)
Sobocinski B 1956 Phil St Ire 6, 3-49 [bib]
Tarski A 1953 J Comp Sys 1, 3 [modal logic]
- 1956 *Logic, Semantics, Metamathematics,* Oxford U P (Pap)

**Lull,** Ramon  1232-1316
  = Raimondus Lullus
  = Lilly

*Anon 1885 *Histoire littéraire de la*

*France* 29, 1-386　　[*bib]
*Gardner Martin 1958 *Logic Machines
　and Diagrams* (McGraw Hill), 1-25
　[pop]
Hillgarth J N 1967 Euc Phi　[bib]
Hofmann J E 1941/42 Heid Si Ph (4)
　(F68, 9)　[circle sq]
*Peer E A 1929 *Ramon Lull,* London

Luqa.　See Qusta.

Lurie, Anatolii Isakovich　1901-

Anon 1961 JAM Me 25, 887-897
　(Z101, 9)
- 1961 Pri M Me 25, 593-599

Luzin, Nikolai Nikolaevich
1883-1950

Anon 1950 UMN (NS)5(4), 15-18
　(MR12, 311)
- 1955 Ist M Isl 8, 55-76
　(Z68, 7)
Bari N K + 1951 UMN (NS)6(6), 28-46
　(MR14, 122)　[metric thy of fun]
Egorov D 1953 UMN (NS)8(2), 105-110
　(MR14, 1051)　[intgl, series]
Fedorov V S 1952 UMN (NS)7(2), 7-16
　(MR13, 810)　[complex anal]
Goltsman V K + 1952 UMN (NS)7(2), 17
　17-30　(MR13, 810)　[diff eqs,
　num meth]
Keldysh L V + 1953 UMN (NS)8(2), 93-
　104　(MR14, 1051)　[set]

Lyapunov, Aleksandr Mikhailovich
1857-1918

Anon 1948 Pri M Me 12, 467-468
　(MR10, 420)
- 1948 Pri M Me 12, 553-560
　(Z30, 338)　[bib]
Bilimovic A D 1956 Beo IM 9, 1-7
　(MR18, 982; Z71, 4)
Gnedenko B V 1959 Ist M Isl 12,
　135-160　(MR23A, 697; Z98, 8)
　[prob]

Kazarinoff D 1923 AMSB 29, 440
　(F49, 12)
Krylov A N 1919 SSSR Iz (6)13, 389-
　394
Krylov N M + 1921 Crim UMLP 3, xxii-
　liv　[Chebychev]
Kudela F 1933 Cas MF 62, V73-V74
　(F57, 858)　[central lim thm, prob]
Lukomskaya A M 1953 *A. M. Lyapunov.
　Bibliografiya,* Mos-Len, 268 p
　(MR16, 986)
Lyapunov B 1930 SSSR Iz (7)3, 1-24
　(F56, 814)
Medvedev F A 1961 Ist M Isl 14, 211-
　234　[Stieltjes intgl]
Polak L S 1957 Vop IET (5), 31-38
　(Z80, 4)
Pressland A J 1931 Nat 128, 138-140
　(F57, 42)
Rumyancev V V 1957 SSSR Vest (6),
　44-49　(MR19, 518; Z85, 6)
Smirnov V I 1948 Pri M Me 12, 469-560
　(MR10, 420; Z38, 148; 30, 338)
　[bib]
Sobolev S L 1957 Pri M Me 21, 306-308
　(MR19, 826)　[pot thy]
Stepanov V V + 1949 SSSR Tek, 161-167
　(MR10, 420)
Yataev M Ya 1957 Kazk Ak (9), 118-121
Yushkevich A P 1965 Ist M Isl (16),
　375-388 (MR34, 7322)
Zhatykov O A 1950 *Velikii russkii
　matematik Aleksandr Mikhailovich
　Lyapunov,* Alma-Ata, 25 p

Lyle, John Newton　1836-

Matz F P 1896 AMM 3, 95-100

Lyte, Henry　XVII

Ockenden R E 1936 Isis 25, 135
　[decimal]

Lyusternik Lazar Aronovich　1899-

Aleksandrov P S + 1960 Rus MS 15(2),
　153-168　(MR22, 1594; Z89, 243)
- 1960 UMN 15(2), 215-230　(MR22, 424)

Kolmogorov A N 1950 UMN 5(1), 234-
235   (Z35, 148)

Macaulay, F. S.   1862-1937

Baker H F 1938 LMSJ 13, 157-160
(F64, 23)
Baker II F + 1938 Lon RS Ob 2, 357-361

MacCullagh, James   1809-1847

Platts C 1893 DNB 35, 15

MacDonald, Hector Munro   1865-1935

J A C   1936 Lon As Mo N 96,
295-296   (F62, 104)
Whittaker E T 1935 LMSJ 10, 310-318
(F61, 955)
- 1935 Nat 135, 945   (F61, 955)

Macfarlane, Alexander   1851-1913

Colaw J M 1895 AMM 2, 1-4
Knott C G 1913 Nat 92, 103-104
(F44, 32)
Shaw J B 1913 Quatrns (June), 11-16
(F44, 32)

Mach, Ernst   1838-1916

Bavink D 1916 Unt M 22, 41-45
Bouvier R 1923 *La pensee d'Ernest
Mach. Essai de biographie
intellectuelle et critique,*
Paris
Helm G 1916 Dres Isis, 45-54
*Pittenger H W 1965 Sci 150, 1120-
1122

Machir.   See Jacob ben Machir.

Machovec, Franz   1855-1892

Czuber E 1892 Mo M Phy 3, 403-406
(F24, 33)   [Winckler]

Machytka, Bohumil

Kaderavek K 1930 Cas MF 59, 3-8

MacIntyre, Sheila Scott   1910-1960

Cartwright M L 1961 LMSJ 36, 254-256
(MR23A, 136; Z95, 5)
Cossar J 1960 Edi MSN (43), 19
(Z94, 5)

Mackay, J. S.

Gibson G A 1914 Edi MSP 32
(F45, 44)
- 1914 M Gaz 7, 309-310

Mackie, John

Allen T M M 1956 Edi MSN (40), 27-28
(Z71, 4)

MacLaren, S. B.   1876-1916

Anon 1918 LMSP (2)16, 33-37

Maclaurin, Colin 1698-1746

Dingeledey F 1919 DMV 28, 158-162
(F47, 6)
Maclaurin Colin 1748 *Newton's
discovery,* London   [incl biog]
Masson R 1916 Ens M 18, 56-59
(F46, 9)
Sanford V 1935 MT 27, 155-156
Schlapp R 1949 Edi MSN 37, 1-6
(MR10, 420; Z35, 148)
Turnbull H W 1947 AMM 54, 318-322
(MR8, 498; Z29, 2)

- 1951 *Bi-centenary of the Death of
  Colin Maclaurin (1698-1746)...,*
  Aberdeen, 20 p   (Z43, 4)
Tweedie C 1916 Edi SP 36, 87-150
- 1915 M Gaz 8, 132-151   (F46, 9)
- 1917 M Gaz 9, 303-306
- 1918 M Gaz 10, 209   [Lagrange]


**MacMahon,** Percy Alexander   1854-1929

A R F 1930 Nat 125, 243-245
  (F56, 28)
Baker H E 1930 LMSJ 5, 307-318
  (F56, 28)


**MacRobert,** Thomas Murray   1884-

Erdelyi A 1963 Glasg M 6, 57-64
  (MR27, 5674; Z111, 3)
Rankin R A 1964 LMSJ 39, 176-182
  (MR28, 756; Z118:1, 12)


**Maggi,** Gian Antonio   1856-

Cisotti V 1937 Milan Sem 11, xix-xx
  (F63, 816)
- 1938 Milan Sem 12, 167-189
  (F64, 919)
Finzi B 1958 Milan Sem 27, xi-xiv
  (Z79, 5)
Somigliana C 1938 Tor FMN (1)73,
  519-528   (F64, 919)


**Maggi,** Pietro 1809-1854

Biadego G 1879 Boncomp 12, 839-846,
  847-862   (F11, 26)
  [Fusinieri, mech]
- 1879 *Pietro Maggi, Matenatico e
  Poeta Veronese,* Verona (F11, 25)


**Magini,** Giovanni Antonio   1555-1617

Ginsburg J 1932 Scr 1, 168-189


**Magliabechi,** A.

Procissi A 1940 Pe M (4)20, 289-312
  (Z23, 387)   [Ceva, letter]


**Magnitskii,** Leontii Filippovich
  1669-1739

Deaux R 1940 Mathesis 54, 139-42
Denisov A P 1967 L. F. M., 1669-1739,
  Mos
Prudnikov V E 1953 Mat v Shk (2),
  12-15
Shvetsov K I 1962 Ist M Zb 3,
  [arith]


**Mahavira**   IX

  = Mahaviracarya

Aiya S B 1954 MT 47, 528-533
Datta B 1928 Clct MS 20, 267-294
  [quad, triangle]
- 1932 Isis 17, 25-33   (Z4, 4)
  [Sridhara]
Kaye G R 1919 Isis 2, 325-356
Kangacarya M ed 1912 *The Ganita-Sara-
  Sangraha of Mahaviracarya...,*
  Madras, 352 p
Smith D E 1908 Bib M (3)9, 106-110
  (F39, 55)
- 1909 Int Con 4(3), 428-431
  (F40, 7)


**Mahnkopf,** Heinrich Wilhelm August
  1892-1932

Brennecke E 1933 Z Vermess 62, 97-101
  (F59, 857)


**Maievskii,** Nikolai Vladimirovich
  1823-1892

Klussman 1893 Arc Artil 99, 339-346
  (F25, 39)
Mandryka A P 1954 Mos IIET 1, 146-192
  (MR16, 660; Z59:1, 11)
  [ballistics]

Maillard de la Gournerie,, Jules
  Antoine René de 1814-1883

Bertrand J 1883 Par CR 97, 6-9


Mailly, Nicolas Edouard 1810-

Terby F 1893 Bel Anr 59, 377-404
  (F25, 33)


Maimonides 1135-1204

Frank E 1946 Jew Q Rv 37, 149-164
Neugebauer O 1949 Heb UC An 22, 321-
  363 [ast]
Sarton G 1931 Int His Sc 2, 369-380
  [*bib]


Mainardi, Leonardo

Favaro A 1903 Bib M (3)4, 334-337
  (F34, 7)


Maisano, Giovanni 1851-1929

Cipolla M 1932 Batt 70, 100-105
  (F58, 46)


Malebranche, Nicolas de 1638-1715

Brunet P 1933 Isis 20, 367-396
Henry C 1879 Boncomp 12, 477-568,
  619-740 (F11, 16)
  [Bachet, Fermat]
Mouy P 1938 Rv Met Mor 45, 411-435
  (F64, 914) [Newton]
Robinet A 1960 Rv Hi Sc Ap 13, 287-
  308 (MR22, 1341) [calc]
- 1961 Rv Hi Sc Ap 14, 205-254
  [phil]
Schrecker P 1935 Thales 2, 82-90
  [neg numb]
- 1937 Int Con Ph 9, 33-40


Malfatti, Gianfrancesco (Giovanni
  Francesco Giuseppe) 1731-1807

Anon 1876 Boncomp 9, 393-480
  (F8, 19) [letters]
Biadego G B 1876 Boncomp 9, 361-387
  (F8, 19) [bib]
Bortolotti E 1931 It SPS 19(2), 45-55
  (F57, 30)


Malkin, I. G.

Anon 1959 Iz VUZM (5)(12), 223-226
  (MR24A, 128; Z87, 243)


Mallet, Bernard 1859-1932

Anon 1932 Nat 130, 728-729
  (F58, 46)


Malmsten, Carl Johan 1814-1886

Eneström G 1886 Ny Ill Tid, 75-76


Malton, Thomas 1725-1801

Locke L L 1940 Scr 8; 34-42 [Euclid]


Maltsev, Anatolii Ivanovich 1909-1968

Aleksandrov P S + 1968 UMN 23(3), 159-
  170 (MR37, 19) [port]
Anon 1967 Alg Log 6(4), i-xi
  (MR37, 22) [port]
Kurosh A G 1959 UMN 14(6), 203-211
  (MR22, 260; Z87, 7)

Malves. See Gua de Malves.


Mancini, P. Nazarene

Marchetti F 1870 Boncomp 3, 429-438
  (F2, 24)

**Manfredi,** Eustachio  1674-1739

Arrighi G 1931 Archeion 13, 320-324
  (F57, 28)  [Grandi]
- 1962 Physis Fi 4, 125-132
  (Z109, 238)


**Manfredi,** Gabriele  1681-1761

Bortolotti E 1933 Bln Mm (8)10,
  47-58, 103-114  (F59, 845)
  [geom]


**Mangoldt,** Hans von Carl Friedrich
  1854-1925

Anon 1930 Bo Fir (2)8, xxiii-xxiv
  (F56, 815)
Knopp K 1927 DMV 36, 332-348
  (F53, 31)


**Mannheim,** Amédée  1831-1906

Anon 1902 Par EP (2)7, 221
- 1902 Rv GSPA 13, 1
- 1906 Nou An (4)6, 529  (F37, 31)
- 1907 Loria 10, 59-60  (F38, 34)
Bricard R 1907 Nou An (4)7, 97-111
  (F38, 33)
Buhl A 1907 Ens M 9, 66-68
  (F38, 33)
Darboux G 1907 LMSP (2)5, 13
  (F38, 33)
Duporcq E 1902 Nou An (4)2, 25-27
  (F33, 45)
Laisant C A 1907 Ens M 9, 169-179
  (F38, 33)
*Loria G 1908 Paler R 26, 1-63
  (F39, 28)  [geom]
- 1934 Scr 2, 337-342
Quint N 1907 Wisk Tijd 3, 198-199
  (F38, 34)
Reveille J 1907 Rv GSPA 18, 49-50
Rouché 1902 Nou An (4)2, 145-150
  (F33, 45)

**Mannoury,** Gerrit

Beth E W 1957 Euc Gron 32, 298-300
  (Z108-2, 250)
Van Dantzig D 1957 Nieu Arch (3)5,
  1-18 (MR18, 784; Z77, 9)
  [found]


**Mansion,** Paul  1844-1919

Demoulin A 1929 Rv Q Sc 15, 217-250
Goedseels E 1922 Brx SS 42, 63-65
  [calc, Leman]


**Mansson,** Peder  XVI

Eneström G 1885 Bib M
- 1888 Bib M (2)2, 17-18


**Mansur.**  See Nasr.


**March,** Lucien  1859-1933

Huber M 1933 Par St 74, 269-280
  (F59, 854)


**Marchetti,** Alessandro  1633-1714

Agostini A 1952 Arc In HS 31, 3-5
  (MR14, 344)  [Leibniz, letter]
Derenzini T 1959 Physis Fi 1, 224-243
  (Z107, 246)  [Borelli, letter]
Tenca L 1959 Bln Rn P (11)6, 139-148
  (Z100, 244)  [Grandi]


**Marci de Kronland,** Johann Marcus
  1595-1667

Hoppe E 1927 Arc GMNT (NF)10, 282-
  290  (F53, 14)
Laska W 1890 Hist Abt 35, 1-3
Studnicka F J 1875 Prag Si, 1-8
  (F8, 17)
- 1891 *Ioanne Marcus Marci a Cronland
  sein Leben und gelehrtes Wirken.
  Festvortrag,* Prague, 32 p  (F23, 16)

Marcinkiewicz, Joseph

Zygmund A 1960 Wiad M (2)4, 11-41
  (MR22, 1131; Z91, 6)
Zygmund A ed 1964 *Collected Papers*...,
  Warsaw   [bib, biog, port]


Marcolongo, Roberto  1862-

Pascal M 1948 Nap FM Ri (4)15
  (MR14, 344)


Mardini, Abdallah ibn Khalil al-
  XIV-XV

Worrell W H + 1942 Scr 10, 170-180
  [quadrant]
Sarton G 1948 Int His Sc 3(2), 1530-
  1533   [bib]


Mariantoni, F.

Anon 1899 Pe MI 2(1), 88


Marie, Charles François Maximilien
  1819-1891

Boncompagni B 1886 Bib M 43-45,
  87-90


Marie, Frère Gabriel  1834-1916

Archibald R C 1917 AMM 24, 280-281
  (F46, 25)
Sarton G 1950 Isis 41, 199


Marinos of Sichem   V

Hopfner F 1946 Wien Anz 83, 77-87
  (MR11, 150)   [cartog, Ptolemy]
Michaux M 1947 Louv HPRT (3)25
  (MR9, 74)   [Euclid]
Sarton G 1927 Int His Sc 1, 405
  [bib]

Marius, Simon  1570-1624

Bosscha J 1907 Ned Arch 12, 258-307
- 1907 Ned Arch 12, 490-527
Favaro A 1901 Bib M (3)2, 213
  [Galileo]
Ondemans J A C 1903 Ned Arch (2)8,
  168-172   (F34, 1026)
Ondemans J A C + 1903 Ned Arch (2)8,
  115-167 [Galileo]


Markov, Andrei Andreevich  1856-1922

Archibald R C 1948 *Mathematical
  Table Makers*, NY, 55-56
  (Scr Stu 3)   [bib]
Bezikovic A 1923 SSSR Iz (6)17, 45-52
  [prob]
Gunther N M 1923 SSSR Iz (6)17, 35-44
  (F50, 13)   [educ]
Koltsov A V 1956 Vop IET, 204-207
  (Z73, 242)
Mihoc G 1955 Ro Sov (3)9(2), 95-106
  (MR17, 338)
Minkowskii V L 1952 M v Shk (5)
  [educ]
Novlianskaya, M G 1952 UMN 7(6),
  213-215   (Z47, 6)
Ogigova H 1967 Rv Hi Sc Ap 20, 2-32
  [Hermite, letters, series]
Otradnykh F P 1953 Ist M Isl 6, 495-
  508   (MR16, 660; Z52, 3)
Steklov V 1922 SSSR Iz (6)16, 169-184
Uspenskii Ya V 1923 SSSR Iz (6)17,
  19-34   (F50, 15)


Markov, Andrei Andreevich  1903-

Delone B N 1948 UMN 3(5), 3-6
Linnik Yu V + 1954 UMN (NS)9(1), 145-
  154   (MR15, 770)


Markov, Vladimir Andreevich  1871-
  1897

Otradnykh F P 1954 UMN 9(4), 256-258

Markushevich, Aleksei Ivanovich
1908-

Gelfond A O + 1958 UMN 13(6), 213-
220  (MR20, 1243; Z85, 6)


Marletta, Giuseppe  1878-

Calapso R 1950 Cat  Atti (6)6, 24 p
(MR11, 708; Z37, 291)


Marneck.  See Bolza.


Martin, Artemus  1835-1918

Archibald R C 1948 Mathematical
Table Makers, NY, 56-57
(Scr Stu 3)  [bib]
Finkel B F 1894 AMM 1, 109-111


Martinovics, Ignaz Joseph
1755-1795

Szénássy B 1956 M Lap 7, 277-290
(MR20, 848)


Martins da Silva, Joaquim Antonio
1859-1885

Teixeira G 1885 Teixeira 6, 194-196


Martinus Rex    1422-1460

  = Martini de Zorawica
  = Martinus Polonus

Birkenmaier L 1893 Krak Roz 25, 1-64
(F25, 16)
- 1895 "Martinus Rex de Premislia"
vocitati, Geometriae practicae seu
Artis mensurationum Tractatus,
Warsaw, 91 p  (F26, 9)
(Repr 1897 Mo M Phy 8, 30)
Dickstein S 1895 Wszech 14, 742-747
(F26, 10)  [geom]


Marx, Karl    1818-1883

Bernal J D 1952 Marx and Science,
New York, 48 p
Gokieli A P 1947 Matematicheskie
rukopis Marksa, Tiblis
Kolman E 1932 Int Con (1932)2, 349-
351  (F58, 52)  [calc]
- 1934 Archeion 15, 379-384
(F60, 840)  [calc]
Rybnikov K A 1955 UMN (NS)10(1),
197-199  (MR16, 782)
Struik D 1948 Sci Soc 12, 181-196
- 1948 in Bernstein S ed, A Centenary
of Marxism, NY, 181-196
(MR9, 485)


Mascheroni, Lorenzo  1750-1800

Anon 1881 Bibliografia Mascheroniana
..., Bergamo  (F13, 18)
- 1904 Contributi alla biografica
di Lorenzo Mascheroni..., Bergamo
- 1912 Intermed 19, 92  [bib]
Cajori I 1929 AMM 36, 364-365
Cheney W F Jr 1953 MT 46, 152-156
De Vries H 1925 N Tijd 12, 379-394
(F51, 22)
Van Veen S C 1951 Passermeetkunde,
Gorinchem, 184 p


Maschke, Heinrich  1853-1907

Bolza O 1908 AMSB (2)15, 85-95
(F39, 39)
- 1908 DMV 17, 345-355
Van Vleck E B + 1908 AMSB (2)14, 425
(F39, 39)


Maseday Méndez.  Miguel Angel

Gran M F 1957 Cub SCPMR 4, 45-48
(Z80, 9)


Massau, Junius  1852-1909

Anon 1909 Mathesis (3)9, 89-91  [bib]

Bouny F 1954 Bel Oc 28, 52-73
(Z59-1, 11) [vectors]
Demoulin A 1909 Bel BS, 305-308
(F40, 36)
Demoulin A + 1909 Mathesis (3)9, 57,
89-91 (F40, 36)
Rose J 1910 Ens M 12, 187-200
(F41, 23)


**Mathews,** George Ballard 1861-1922

Gray A 1923 M Gaz 11, 133-135
- 1923 Nat 109, 712


**Mathews,** Robert Maurice 1884-1929

Reynolds C N 1930 AMSB 36, 161


**Mathieu,** Emile Léonard 1835-1890

Duhem P 1892 NYMS 1, 156-168


**Mathy,** Ernest Joseph 1855-1923

Godeaux L 1950 Brx Con CR 3, 34-35


**Matqah.** See Judah


**Matsunaga,** Sapanojo 1751-1795

Hayashi T 1917 Toh MJ 11, 17-37
[Ajima]


**Matteson,** James 1819-1876

Matz F P 1894 AMM 1, 373-374


**Matthiessen,** Heinrich Friedrich
Ludwig 1830-1906

Anon 1906 Leop 42, 158 (F37, 31)

**Mauduit,** Antoine René 1731-1815

Loria G + 1903 Intermed 10, 144


**Maunsell,** Frederick George 1898-1956

Davies E T 1958 LMSJ 33, 255-256
(Z80, 4)


**Maupertuis,** Pierre Louis Moreau de
1698-1759

Bois-Reymond E du 1892 Ber Si, 393-44
(F24, 16)
Brunet P 1929 *Maupertuis*, Paris,
493 p (F55, 606)
Dugas R 1942 Rv Sc 80, 51-59
(MR7, 106; Z60, 9) [least action]
Keller-Zschokke J V 1935 *Pierre
Louis Moreau de Maupertuis von
St. Malo...*, Basel, 105 p
(Z13, 338)
Schumaker J A 1957 Scr 23, 97-108
[comets]
Smith D E 1921 AMM 28, 430
[Frederick the Great]


**Maurer,** Hans Theodor Julius Christian
Karl 1868-1945

Anon 1938 Deu M 3, 109-119
(F64, 23)


**Maurolico,** Francesco 1494-1575

Amodeo F 1908 Bib M (3)9, 123-138
[conic]
Anon 1876 Boncomp 9, 23-1121
(F8, 15)
- 1896 *Commemorazione del 4°
centenario di Francesco Maurolico*,
Messina, 258 p (F27, 6)
Bussey W H 1917 AMM 24, 199-207
[induc]
De Marchi L 1885 Bib M, 141-144,
193-195 [mss]
- 1886 Bib M, 90-92
Flauti V 1856 Nap FM Ri 5, 112-
[Apollonius, Archimedes]

Fontana M 1808 It INFM 2, 275-
[arith]
Macri G 1901 *F. Maurolico nella
vita e negli Scritti,* 2nd ed
Messina, 367 p (F32, 4)
Martines D 1863 *Cenno analitico
intorno gli Studi e le opere di
Francesco Maurolico,* Messina
Napoli F 1876 Boncomp 9, 1-22
(F8, 15)
Rosen E 1956 Arc In HS 9, 349-350
(MR18, 981)
- 1957 Am Bib SP 51, 111-118
[De Morgan]
- 1957 Scr 22, 285-286 [death]
- 1959 Scr 24, 59-76 (Z85, 3)
Vacca G 1909 AMSB 16, 70-73
[induc]


Maxwell, James Clerk 1831-1909

Anon 1931 *A Commemoration Volume...,*
Cambridge, 152 p [port]
Campbell Lewis + 1882 *The Life of
James Clerk Maxwell,* London, 678 p
(abr ed 1884 London, 436 p;
repr 1969 Johnson) [port]
Kuchar K 1959 Prag Pok 4, 501-514
(Z87, 6) [Fields]
Tait P G 1898 *Scientific Papers,*
Cambridge, 1, 396-401
Turner J 1955 Brt JPS 6, 226-238
[quaternions]


Mayer, Adolf 1839-1908

Holder O 1908 Leip Ber 80, 353-373
(F39, 40)
Liebmann H 1908 DMV 17, 355-362
(F39, 41)
Muehl K v d 1908 M Ann 65, 433-434
(F39, 40)


Mayer, Johann Tobias 1723-1762

Loria G 1934 Isis 20, 441 [geom]
Ofterdinger C F 1887 Wurt MN 2, 116-
132

Mazurkiewicz, Stefan 1888-1945

Kuratowski C 1947 Fund M 34, 316-331
(MR10, 175)


McClintock, John Emory 1840-1916

*Archibald R C 1938 *A Semicentennial
History of the American Mathematical
Society...,* NY, 112-117
[bib, port]
Fiske T S 1917 AMSB 23, 353-357
(F46, 24)


McGiffert, James

Merrill L L 1944 M Mag 18, 142-144


Medigo, Joseph Salomo 1591-1655

Heilbronn J 1912 *Die Mathematischen
und naturwissenschaftlichen
Anschavungen des Josef Salomo
Medigo...,* Erlangen 92 p
(Thesis) (F43, 39)
Lewittes M H 1932 Scr 1, 56-59
[prosthaphaeresis]


Medler, Nicolaus 1502-1551

Neumueller G 1868 *Elemente der
praktischen Arithmetik von D.
Nicolaus Medler,* Naumberg
(F1, 10)


Mehler, Gustav Ferdinand 1835-1895

Krause M 1897 M Ann 48, 603-606
(F28, 21)


Meissel, Daniel Friedrich Ernst
1826-1895

Archibald R C 1949 M Tab OAC 3,
449-451

Meissner, Ernst    1883-1939

Kollros L 1939 Schw NG, 290-296
  (F65:2, 1092)


Melanchthon, Phillip  1497-1560

Bernhardt P 1865 *Phillip Melanchthon
  als Mathematiker und Physiker*,
  Wittenberg, 80 p  (F27, 7)
Marian V 1941 Gaz M 46, 396-400,
  460-466  (Z25, 1)


Melanderhjelm, Daniel   1726-1810

Eneström G 1885 Bib M 3, 174 (Q4)
  [ast, Frisi]


Mellin, R. H. 1854-

Lindelöf E 1933 Act M 61, i-vii
  (F59, 34)


Menabrea, Federigo Luigi  1809-1896

Genocchi A 1872 Boncomp 5, 535-542
  (F4, 24)
- 1873 Boncomp 6, 530-532  (F5, 28)


Menaechmos,   -IV

Sarton G 1927 Int His Sc 1, 139
Schmidt M C P 1884 Philolog 42,
  72-81
Schub P 1960 MT 53, 278-279
Vetter Q 1925 Cas MF 54, 348-350
  [hyperb]


Menelaos of Alexandria    I

Ginsburg J 1932 Scr 1, 72-78,
  153-155  (Z5, 242, 377)
  [J. ben Machir, trig]
Krause M 1936 Gott Ab H (3)17, 374 p
  (F62, 1019)  [Abu Nasr Mansur,
  sphere]

Schmidt O 1955 Nor M Tid 3, 81-95
  (Z65, 243)  [Ptolemy]
— Schwartz J 1934 Scr 2, 243-346
  (Z9, 388)  [sphere]


Mengoli, Pietro  1625-1686

Agostini A 1922 Pe M (4)2, 430-451
  [Euler, log]
- 1925 Pe M 5, 18-30  [calc, lim]
- 1925 Pe M 5, 137-146  [calc, intgl]
- 1940 Pe M (4)20, 313-327
  (MR3, 97; F66, 16; Z23, 387)
  [geom]
- 1941 It UM (2)3, 231-251
  (MR3, 97; Z25, 146)  [series]
- 1950 Archim 2, 165-170  (Z38, 3)
- 1950 Arc In HS 29, 816-834
  (MR12, 311; Z39, 4)
Karamata J 1959 Ens M (2)5, 86-88
  (MR23A, 2663; Z88, 243)
Vacca G 1915 Linc Rn (5)24, 617-620


Mennher, Valentin  1520-1573

Hairs E de 1930 Wis Nat Ti 5, 58-65
  (F56, 80)


Menshov, Dmitrii Efgenevich  1892-

Aleksandrov P S + 1962 UMN 17(5),
  161-175 (MR26, 449; Z111, 3)
Bari N K + 1951 UMN (NS)6(4), 187-189
  (MR13, 198)  [series]
- 1952 UMN (NS)7(3), 145-150
  (MR14, 2; Z46, 3)


Meray, Hugh Charles Robert  1835-1911

Bourlet C 1904 Fr AAS, 62  [geom]
Chevallier 1904 Fr AAS, 66
  [educ, geom]
Laisant C A 1901 Ens M 3, 98  [geom]
- 1911 Ens M 13, 151, 181-186
  (F42, 37)
Loria G 1911 Mathes It 3, 127-131
  (F42, 36)

Pionchon J 1912 Rv Bourg 22, 1-I58
(F44, 40) [irrat]

Mercator, Gerard 1512-1594

= Gerhard Kremer

Averdunk H + 1914 Peterman (182),
100 p
Breusing 1869 *Gerhard Kremer genannt
Mercator, der deutsche Geograph...,*
Duisburg (F2, 9)
Carlslaw H S 1924 M Gaz 12, 1-7
(F51, 14) [cartog]
Doederlein 1879 Bay Bl GR 15,
193-199 (F11, 15)
Fiorini M 1890 It Geog Bo (3)3,
94-110, 182-196, 243-256,
340-380, 550-556 [globe]
Grunert J A 1869 Arc MP Lit 200,
1-12 (F2, 9)
Rodero J 1953 Gac M (1)5, 143-146
(MR15, 592)
Van Durme M 1959 *Correspondance
Mercatorienne,* Antwerp, 285 p
(MR22, 1853)

Mercator, Nicolaus 1620-1687

Hofmann J E 1938 Deu M 3, 445-466
(F64, 16; Z19, 1) [log]
- 1938-1940 Deu M 3, 598-605; 4,
556-562; 5, 358-370 (MR1, 33)
[log, series]
- 1950 Mainz MN (3), 45-103
(MR11, 708)

Mercer, James 1883-1932

Hobson E W 1933 LMSJ 8, 79-80
- 1933 Lon RS Ob 1, 164-165
Hopkins F G 1933 Lon RSP (A)139, 3
(F59, 43)

Mercogliano, Domenico

Florio S 1936 Bo Fir 32, 124-128
(F62, 1041)

Merinoy Melchor, Miguel

Aguilar M 1912 Esp SM 1, 388-401
(F43, 32)

Merlani, Adolfo

Pincherle S 1925 Bln Rn (2)29, 57-58
(F51, 31)

Mersenne, Marin 1588-1648

Anon 1900 Pitagora 6, 32 [Fermat]
- 1932 MT 24, 369
Boncompagni B 1875 Boncomp 8, 353-
381 (F7, 21) [Torricelli]
Boutroux P 1922 Scientia 31, 273-291,
347-361 [Galileo]
Boyer J 1901 Intermed 7, 227
[letters]
Brocard H + 1902 Intermed 9, 101
(Q419) [letters]
Coste, Hilaron de 1649 *La vie du
R. P. Marin Mersenne,* Paris
De Waard D 1927 BSM (2)51, 65
(F53, 15)
- 1931 Archeion 13, 175-186
(F57, 24; Z2, 243) [Descartes,
letter, pendulum]
- 1948 Rv Hi Sc Ap 2, 13-28
(MR10, 420)
Duhem P 1906 Rv GSPA 17, 769-782,
809-817
Humbert P 1933 Rv Q Sc (4)24, 79-84
(F59, 842)
Lenoble Robert 1943 *Mersenne, ou la
naissance du mécanisme,* Paris
- 1948 Arc In HS 28, 583-597
(MR10, 608; Z31, 242)
Loria G 1922 Arc Sto 3, 272-276
[Torricelli]
Miridonova O P 1964 Vop IET (16),
85-90 (MR30, 719)
[acoustics, Galileo]
Omont H 1923 BSM (2)47, 194-195
(F49, 5) [Descartes, letter]
Plá C 1949 Lim Ac 12, 44-57
(MR11, 573; Z39, 4)
Reeve W D 1931 MT 24, 369

Sergescu P 1948 Rv GSPA (NS)55,
193-195   (MR10, 420)
- 1948 Rv Hi Sc Ap 2, 5-12
(MR10, 420)
Tannery Mme P 1926 Arc Sto 7, 256
(F52, 37)
- 1927 Brx SS (A)47(1), 33-39
(F53, 15)

Meshcherskii, Ivan Vsevolodovich
1859-1935

Grigoryan A T 1959 Vop IET (7),
127-130  (Z117:1, 8)
Tyvlina I A 1960 Mos IIET 34, 264-
272  (Z117-1, 9)

Metzburg, Georg Ignatz von
1735-1798

Ionescu I 1937 Gaz M 42, 234-235
(F63, 15)

Meusnier de la Place, Jean Baptiste
Marie Charles

Darboux G 1910 Par Mm (2)49, 1-54
- 1912 *Eloges académiques et
discours,* Paris, 218-262

Meyer, Arnold  1844-1896

Lang A 1897 DMV 5, 18-20  (F28, 24)

Meyer, Eugen  1868-1930

Fromm H 1931 ZAM Me 11, 163-164
(F57, 45)
Lorey W 1911 MN Bl 8, 3-5  (F42, 23)
Salkowski E 1910 Arc MP (3)16, 9-13
- 1910 Ber MG 9, 9-13  (F41, 23)

Meyer, Friedrich  1842-1898

Riehm G 1900 DMV 8, 59  (F31, 18)
[bib, port]

Meyer, Friedrich Wilhelm Franz
1856-1934

Anon 1928 MN Bl 22, 42-43  (F54, 48)
Arnot B 1935 DMV 45, 99-113
(F61, 25)
- 1935 ZMN Unt 66, 303-304
(F61, 25)

Meziriac.  See Bachet.

Michal, Aristotle, Demetrius
1899-1953

Hyers D H 1954 M Mag 27, 237-244
(MR15, 923; Z156, 2)

Michell, John Henry  1863-1940

Michell, A G M 1941 Lon RS Ob 3,
363-382

Michelsen, Johann Andreas Christian
1749-1797

Haentzschel E 1912 DMV 21, 102-103
(F43, 20)

Michelson, Albert Abraham  1852-1931

Anon 1932 Mathes Po 7, 59-60

Middelburg, Paul von  1445-1534

Struik D J 1925 Linc Rn (6)1, 305-308
(F51, 19)
- 1925 Pe M 5, 337-347  (F51, 19)

Mignosi, Gaspare  1875-1951

Mineo C 1952 Matiche 7, iii-xii
(MR14, 833; Z48, 243)

**Mikami,** Yoshio    1875-

Hayashi T 1906 DMV 15, 586
  [Harzer]
- 1909 Tok M Ph (2)4, 446-453
  (F40, 65)


**Mikan,** Milan    1892-

Havlicek K 1957 Cas M 82, 497-499
  (Z98, 10)


**Mikhlin,** Solomon Grigorevich   1908-

Bakelman I J + 1958 UMN 13(5),
  215-221   (Z85, 7)


**Milankovitch,** Milutin   1879-

Angelitch T P 1960 Arc In HS 12, 176-
  178   (Z89, 243)
Michkovitch V V 1958 Beo IM 12,
  vii-xii   (Z82, 12)


**Milhaud,** Gaston   1858-1918

Anon 1918 Ens M 20, 226-227
  (F46, 21)
Nadal A 1959 Rv Hi Sc Ap 12, 97-110


**Mill,** John Stuart    1806-1873

Wedgewood H 1875 New Eng 34, 488
  [foundations]
Whitmore C E 1945 JH Ideas 6, 109-
  112   (Z60, 16)


**Miller,** George Abram   1863-1951

Anon 1958 Ill JM 2, (4B)
Brahana H R 1951 AMM 58, 447-449
  (MR13, 198; Z42, 243)
- 1951 AMSB 57, 377-382   (Z43, 5)
Cajori F 1924 Sch Soc 19, 225
Dunnington G W 1938 M Mag 12, 384-387
  (F64, 23)

- 1952 DMV 55, 52-53   (Z46, 3)


**Miller,** Jeffrey Charles Percy 1906-

Archibald R C 1948 *Mathematical
  Table Makers,* NY, 57-58
  (Scr Stu 3)   [bib, port]


**Miller,** John

Arthur W 1957 Edi MSN (41), 29
  (Z77, 9)


**Miller,** Oscar von   1855-1934

Miller W v ed  1932 *Oskar von Miller.
  Nach eigenen Aufzeichnungen,*
  Munich, 192 p


**Miller,** W. J. C.   1832-

Finkel B F 1896 AMM 3, 159-163


**Milne,** Edward Arthur   1896-1950

McCrea W H 1951 Lon RS Ob 7, 421-443
  (MR13, 612)


**Milne,** William Proctor    1881-1967

Edge W L + 1969 LMSJ 44, 565-570


**Milnor,** John Willard    1931-

Whitney H 1963 Int Con (1962),
  xlviii-l    (Z112, 244)


**Mimori,** Mamoru  1859-1932

Mikami Y 1933 Scr 1, 254-255
- 1949 Arc In HS 28, 1140-1143
  (Z33, 242)

**Minchin,** George Minchin 1845-1914

A E H L 1914 LMSP (2)13, xliii-xlv
    (F45, 45)
R A G 1914 Nat 93, 115-116
    (F45, 45)


**Minding,** Ferdinand 1806-1885

Bierman K-R 1961 Ber Mo 3(2),
    120-133
Kneser A 1900 ZM Ph 45, 113


**Mineo,** Corradino Augusto 1875-

Boaga G 1961 Linc Rn (8)30, 576-591
    (Z94, 5)
Chiara L + 1960 It UM (3)15,
    460-466   (Z93, 9)


**Mineur,** Adolphe 1867-1950

Godeau R 1950 Bel SM 2, 10-14
    (MR12, 311; Z40, 290)
- 1950 Mathesis 59, 5-9   (Z36, 146)


**Minich,** Serafino Rafaeles
    1808-1883

Favaro A 1884 Ven I At (6)1, 1095-
    1173
Minich S R 1884 Ven Atene (7)1, 395


**Minkowski,** Hermann 1864-1909

Anon 1909 M Ann 66, 417-418
    (F40, 37)
Delone B N 1936 UMN 2, 32-38
Dumas G 1909 Ens M 11, 140-141
    (F40, 37)
Hancock H 1939 *The Development of
    the Minkowsky Geometry of Numbers,*
    2 vols   (repr 1964 Dover)
Hilbert D 1909 Gott Jb, 72-101
    (F40, 36)   [relativity]
- 1910 M Ann 68, 445-471 (F41, 23)


**Kagan** W 1909 Kagan (481), 16-17
    (F40, 38)
Malkin I 1959 Scr 24, 79-81
    (Z82, 234)
Rudio F 1909 Zur NGV 54, 505-506
    (F40, 37)
Wisk A 1909 MN Bl 6, 36-37 (F40, 38)


**Minto,** Walter 1753-1796

Eisenhart L P 1950 APS Lib B 94,
    282-294
Smith D E + 1934 *A History of Mathe-
    matics in America Before 1900,*
    Chicago, 30-31   (F60, 3)


**Mirimanoff,** Dmitri 1861-1945

Vandiver H S 1953 Ens M 39, 169-179
    (Z50, 2)


**Mises,** Richard von 1883-1953

Basch A 1953 Ost Ing 7, 73-76
    (MR14, 1051; Z50, 2)
Cramer H 1953 An M St 24, 657-662
    (MR15, 276; Z51, 244)
    [prob, stat]
Frank P 1953 Mun Jb, 194-197
    (MR15, 923)
Grunbaum A 1955 Scr 20, 109-110
Khinchin A Ya 1929 UFN 9(2)
    [prob, stat]
Saver R 1953 Mun Jb, 194-197
    (MR15, 923; 18, 784)


**Misrachi,** Elia XV-XVI

Wertheim G ed 1896 *Die Arithmetik
    des Elia Misrachi,* 2nd ed,
    Braunschweig


Misri.  See Ahmad.

Mister, Jean-Nicolas 1832-1898

Mansion P + 1898 Mathesis (2)8, 242

Mitchell, Oscar Howard 1851-1889

Anon 1889 LMSP 20, 424-427

Mitchell, Ulysses Grant 1872-1942

Dunnington G W 1942 M Mag 16, 240-242
Reeve W D 1942 MT 35, 134

Mitropolskii, Yurii Alekseevich

Erugin N P + 1967 Dif Ur 3, 158-166
 (MR34, 5632)
Glushkov V M + 1967 Ukr IM 19(1),
 3-8 (MR34, 2418)

Mittag-Leffler, Mangus Gosta 1846-
 1927

Anon 1915 Wisk Tijd 12(4), 236-7
- 1916 Act M 40, III-X (F46, 29)
- 1916 BSM (2)40, 316-320
 (F46, 30)
- 1916 Ens M 18, 130-132 (F46, 30)
- 1916 Ens M 18, 274-277 (F46, 30)
- 1916 Nou An (4)16, 337-338
 (F46, 30)
- 1923 Act M 39, 257-258
 [three body]
- 1926 Sk Kong 6, 27-44 (F52, 34)
- 1927 Manc Mr 72, vi
- 1927 Nat 120, 626
- 1927 Par CR 185, 93-95
- 1927 Ro SSM 30, 59-60 (F53, 33)
- 1928 Bo Fir (2)7, xlviii
 (F54, 38)
*- 1969 AMSN 16, 496-497
 Birkeland R 1927 Nor M Tid 9, 127-
 135 [Fredholm]
 Carleman T 1934 Sk Kong 8, xiii-xvi
 (F61, 958)
- 1935 Act M 64, i-iv (F61, 28)
 Courant R 1927/28 Gott Jb, 67-68
 (F54, 38)

Cramér H 1927 Sk Akt 10, 157-159
 (F53, 33)
Dassen C C 1930 Arg Sc 109, 417-422
 (F56, 27)
Fehr H 1927 Ens M 26, 140-141
 (F53, 33)
Frostman O 1966 in *Festschrift zur*
 *Gedaechtnisfeier fuer Karl*
 *Weierstrass,* Koeln-Opladen, 53-56
 (MR33, 2505)
Hardy G H 1928 LMSJ 3, 156-160
 (F54, 38)
 1928 Lon RSP (A)119, v-viii
 (F54, 38)
Jourdain P E B 1918 Sci Prog 12, 647
 (F46, 30)
Mongeron D I 1946 Tim Rev M 26, 8 p
 (MR8, 498; Z60, 16)
*Noerlund N E 1927 Act M 50, i-xxiii
 (F53, 33) [bib]
Sintsov D 1915 Khar M So (2)15, 296-
 300

Miyai Antai XVIII

Mikami Y 1914 Toh MJ 5, 176-179
 [approx]

Mlodzeevskii, Boleslav Korneliebich
 1858-1923

Egorov D F 1925 Mos MO Sb 32, 449-452
 (F51, 40)
Lapko L A + 1967 UMN 22(6)(133)
Rossinskii S D 1950 *B. K. Mlodzeevskii,*
 *1858-1923,* Mos (MR13, 2)

Moebius, August Ferdinand 1790-1868

Allardice R E 1891 Edi MSP 10, 2-21
 (F24, 52) [barycenter]
Anon 1889 Leip Ber, 14-21
Baltzer R 1885 Leip Ber, 1-6
 [Weiske]
Cantor M 1885 ADB 22, 38-43
De Vries, H K 1929 N Tijd 16, 201-227
 (F55, 14) [geom, mech]
- 1938 N Tijd 25, 226-231 (F64, 918)

- 1940 *Historische Studien III,*
Groningen-Batavia,  261 p
(Z22, 296)  [Pluecker]
Gretschel H 1869 Arc MP Lit 49(195),
1-9  (F2, 22)
Liebmann H 1910 Leip Ber 62, 189-196
(F41, 14)


**Moerbeke.**  See William.


**Mohr,** Georg  1640-1697

Cajori F 1929 AMM 36, 364-365
Eneström G 1909 Bib M (3)10, 71-72
- 1912 Bib M (3)12, 77
Geppert H 1929 Pe M 9, 149-160
(F55, 602)  [Mascherini constr]
Hellerberg A E 1960 MT 53, 127-132
Hjelmslev J 1928 M Tid B, 1-7
(F54, 28)
- 1931 Dan M Med 11(4), 1-22
(F57, 26; Z2, 243)
Marcolongo R 1929 Nap FM Ri (4)35,
25-31  (F55, 13)  [Euclid]
Schogt J H 1938 M Tid A, 34-36
(F64, 920; Z19, 388)
Zuehlke P 1956 MP Semb 5, 118-119
(MR18, 549)


**Moiseev,** Nikolai Dmitrievich
1902-1955

Rakcheev E N 1961 Mos UM Me (1),
71-77  (MR24A, 333; Z96, 4)


**Molk,** Jules  1857-1914

Eneström G 1915 Bib M (3)14,
336-340  [historiog, port]
Vogt H 1914 Ens M 16, 380-383, 387
(F45, 45)


**Moll,** Gerrit Gerhard  1785-1838

Quetelet A 1839 *Notice sur Gerard
Moll,* Brussels

**Mollame,** Vincenzio  1848-

Cipolla M 1912 Cat Atti (5)5
(F43, 44)


**Mollerup,** Johannes   1872-1937

Petersen R 1938 M Tid B, 1-6
(F64, 23)


**Monge,** Gaspard    1746-1818

*Arago F 1865 Oeuvres 2, 426-592
Archibald R C 1915 AMM 22, 6-12
[center of simil, Greek]
Aubry Paul V 1954 *Monge, le savant
ami de Napoléon Bonaparte,*
Paris, 375 p  (Z59:1, 11;60, 11)
Beumer M G 1945 Scr 13, 122-123
[chem]
Brisson B 1818 *Notice historiques
sur Gaspard Monge,* Paris
Brocard H 1908 Intermed 16, 60-63
(Q2948)  (Also Intermed 12, 172;
13, 47, 118, 202; 14, 13, 60-64)
Cayley A 1883 LMSP 14, 139-142
[deblais et remblais]
De Vries H 1914 N Tijd 3, 255-269
(F45, 78)  [descr geom]
- 1937 Euc Gron 14, 137-179
(F63, 809)
- 1939 C Huyg 17, 182-237  (F65, 14)
[part diff eqs]
*Dupin Charles 1819 *Essai historique
sur les services et les travaux
scientifiques de Gaspard Monge,*
Paris  [*bib, port]
Fink K 1892 Wurt Sch 39, 263-289,
339-359  (F24, 18)
Gafney L 1965 MT 58, 338-344
[descr geom]
Grevy A + 1906 Intermed 13, 48
Launay L de 1932 Rv Deux M 10, 640-70
813-39; 11, 127-55
- 1933 *Un grand Français, Monge,
fondateur de l'Ecole Polytechnique*
Paris, 280 p  (F59, 857)
Lefebvre,B 1923 Rv Q Sc (4)4, 333-368
[Desargues, Frezier, de l'Orme,
Pascal]

Loria G 1925 *Da Descartes e Fermat a Monge e Lagrange...,* Rome, 74 p (F51, 38) [anal geom]
Morand L 1904 *Généalogie de la famille Monge,* Dijon
Resal H 1874 Par CR 79, 821-822 (F6, 44) [sound]
Roever W H 1933 *The Mongean Method,* NY, [descr geom]
Sanford V 1935 MT 28, 238-240
Segre C 1907 Bib M (3)8, 321-324 (F38, 68) [Malus]
Sergescu P 1947 Sciences (54), 288-310
Smith D E 1921 AMM 28, 166 [US]
- 1921 AMM 28, 208
- 1932 Scr 1, 111-22 (repr Scr Lib 1) [politics]
Taton R 1947 Rv Sc 85, 963-989 (MR9, 486; Z30, 2)
- 1948 Par CR 226, 36-37 (MR9, 485; Z29, 290)
- 1948 Thales, 43-49 [soc of math]
- 1950 Elem M(S)(9),24 p; (MR13, 2)
- 1950 Osiris 9, 44-61 (MR16, 660; Z41, 341) [fin diff eqs]
- 1950 Rv Hi Sc Ap 3, 174-179 [phys]
*- 1951 *L'oeuvre scientifique de Monge,* Paris, 441 p
- 1951 Par CR 232, 198-200 (MR12, 383; Z42, 3) [geom, oriented areas]
- 1952 Elem M 7, 1-5 (Z46, 2) [axial coords, Pluecker]
Vygodski M J 1935 Len IINT 6, 63-96 (F61, 951) [diff geom]

**Montel,** Benedetto Luigi 1872-1932

Codegone C 1932 Nuo Cim (2)9(8), 237-239

**Montel,** Paul 1897

Pompeiu D 1937 Math Cluj 13, 272-274 (F63, 816)

**Montesano,** Domenico 1863-1930

Marcolongo R 1931 It UM 10, 54-55 (F57, 44)
Scorza G 1929 Nap FM Ri (3a)35, 145-154
Tummarello A 1931 Bo Fir (2)10, 1-2 (F57, 44)

**Montessus de Ballore,** Robert de 1870-1937

Anon 1937 JMPA (9)16, 425-426 (F63, 816)
Dassen C C 1937 Arg SC 124, 355-365 (F63, 816)

**Montmort,** Pierre Rémond de 1678-1719

Eneström G 1886 Bib M, 143 (Q11) [series]
Lefner A 1909 Intermed 16, 280 (Q3566) [geom]

**Montucla,** Jean Etienne 1725-1799

Sarton G 1936 Osiris 1, 519-567 (F62, 23)
Smith D E 1921 AMM 28, 207
- 1922 AMM 29, 253

**Mookerjee,** Asutosh 1864-1924

Prasad G 1925 Clct MS 15, 51-56 (F51, 26)
Sen R N 1964 Clct MS 56, 49-62 (MR33, 3877)

**Moore,** C. L. E. 1876-1931

Franklin P 1932 AAAS 67, 606-608 (F58, 992)
Struik D J 1932 AMSB 38, 155-156 (F58, 995)
- 1932 JM Ph 11, 1-11 [bib, port]

Moore, Eliakim Hastings   1862-1932

Anon 1933 Bo Fir (2)12, lxxii
   (F59, 38)
*Archibald R C 1938 *A Semicentennial
   History of the American Mathe-
   matical Society,* NY, 144-150
   [bib]
Bliss G A 1933 AMSB 39, 831-838
   (F59, 38)
Bliss G A + 1935 Am NASBM 17(5)
   83-99
Dickson L E 1933 Sci (2)77, 79-80
   (F59, 854)
Slaught H E 1922 AMM 29, 207-209
- 1933 AMM 40, 191-195  (F59, 38)
Smith D E 1934 MT 26, 109-110


Moore, Robert Lee 1882-

Archibald R C 1938 *A Semicentennial
   History of the American Mathematical Society,*  NY, 240-244
   [bib, port]


Morales, Carlos M.

Dassen C C 1929 B Ai Bol 2, 97-98
   (F55, 20)


Mordukhai-Boltovskoi,  D. D.
   1876-1952

Chernyaev M P + 1953 UMN (NS)8(4),
   131-139  (MR15, 90; Z51, 244)


Morera, Giacinto    1856-

Maggi G 1909 Khar M So (2)11, 243-
   248
- 1910 Batt 48, 317-324


Moret, Theodor   1602-1667

Bosmans H 1928 Guld Pass
Hoffmann H 1935 Schles Jb 107, 118-
   155   (F61, 17)


Morgan, Williams  1750-1833

W P E 1933 Lon Actu 64, (F59, 27)


Morin, Jean Baptiste  1583-1656

Boncompagni B 1873 Boncomp 6, 45-51,
   52-60  (F5, 23)   [Galileo]
Iwanicki Joseph 1936 *Morin et les
   demonstrations mathématiques de
   L'existence de Dieu,* Paris, 144 p


Morley, Frank  1860-1937

Archibald R C 1938 *A Semicentennial
   History of the American Mathe-
   matical Society,* NY, 194-201
   [bib, port]
Philips H R 1939 Am Ac Pr 73, 138-139
   (MR2, 115; Z60, 16)
Richmond H W 1939 LMSJ 14, 73-78
   (F65, 18; Z20, 197)
Slaught H E 1937 AMM 44, 677


Moschopulos, Manuel   XIII

McCoy J C 1941 Scr 8, 15-26
   (MR3, 97; Z61, 2)  [magic sq]
Tannery P 1884 BSM (2)8, 263-277
   [Rhabdas]


Moser, Christian   1861-1935

Lorey W 1935 Bla Versi 3, 304-308
   (F61, 955)


Moses of Leon   XIII

   = Moses ben Shem-Tob

Heiberg J L 1887 Bib M (2)1, 33-36
   [Byzantine math]
Sarton G 1931 Int His Sc 2(2),
   878-881   [bib]

Mossotti, Ottaviano Fabrizio
1791-1863

Anon 1864 Lon As Mo N 24, 87-


Motte, Andrew

Cajori F 1929 Nat 124, 513


Moulton, Forest Ray    1872-1952

*Anon 1946 Curr Biog (Jan), 421-423
   [bib, port]
Carlson A J 1953 Sci 117, 545-546
   (MR14, 833)


Mountaine, William   1700?-1779

Chaplin W R 1960 Am Ncp 20, 185-190


Mourraille, J. Raym    XVIII

Cajori F 1911 Bib M (3)11, 132-137


Moutard, Theodore Florentin
1827-1901

Darboux G 1901 Par CR 132, 614


Moxon, Joseph    1627-1700

Hallerberg A E 1962 MT 55, 490-492


Mozzi, Giulio

Marcolongo R 1905 Loria 8, 1-8
   (F36, 9)


Mueller, Carl Heinrich    1855-

Flechsenhaar A 1927 Unt M 33, 157-159
   (F53, 32)
Lorey W 1927 ZMN Unt 58, 353-360
   (F53, 32)


Mueller, Conrad Heinrich   1878-1953

Quade W 1954 DMV 57, 1-3  (MR15, 592)
Toeplitz O 1932 QSGM (B)2, 286-290
   (Z6, 2)


Mueller, Emil 1861-

Jarosch J 1928 ZMN Unt 59, 132-135
   (F54, 41)
Krames J 1948 Ost Ing 2, 317-318
   (Z32, 242)
Kruppa E 1928 Mo M Phy 35, 197-219
   (F54, 41)
- 1931 DMV 41, 50-58   (F57, 43)
Scmid T 1928 ZAM Me 8, 81-83


Mueller, Felix    1843-

Mansion P 1905 *Mémorial Mathématique
   d'après le Professeur Dr. Félix
   Mueller*, Gand, 16 p   (F36, 2)


Mueller, Friedrich Carl Georg   1848-

Curio O 1932 Z Ph C Unt 45, 171-177


Mueller, Heinrich

Kullrich E 1915 ZMN Unt 46, 435-439
   (F45, 52)


Mueller, Johann Heinrich Jacob
1809-1875

Anon 1876 ZMN Unt 7, 85-90
   (F8, 23)
Fiedler W 1875 Zur NGV 20, 151-157
   (F7, 12)
Hermann 1875 Zur NGV 20, 187-191
   (F7, 12)


Mueller, Johannes. See Regiomontanus.

**Mueller-Breslau,** Heinrich
1851-1925

Anon 1912 *Festschrift*, Leipzig, 218 p
(F43, 55)
- 1925 ZAM Me 5, 277-278  (F51, 29)


**Muettrich,** Johann August  1799-1858

Mueller C H 1909 ZMN Unt 40, 506-510
(F40, 28)


**Muhammad ibn Ishaq.**  See Yaqub Nadim.


**Muir,** Thomas    1844-1934

Aitken A C 1936 Edi MSP (2)4, 263-267
(F62, 26)
E H N 1934 M Gaz 18, 257  (F60, 15)
Loram C T 1916 AMM 23, 74-75
*Turnbull H W 1934 Lon RS Ob 1, 179-
184
- 1935 LMSJ 10, 76-80  (F61, 24)


**Muirhead,** R. F.   1860-

Dougall J 1941 Edi MSP (2)6, 259-260
(Z60, 16)


**Mukhopadkyaya,** Syamadas   1867-

G B 1938 Clct MS 29, 115-120
(F64, 23)


**Mullendorff,**  Auguste

Grechen M + 1911 Lux Arch 6, 1-22


**Muller,** John   1699-1784

  = Mueller

Escott E B + 1904 Intermed 11, 88

**Musa,**  Banu  IX

  = Benu Musa  = Sons of Musa

Rosen 1831 *The Algebra of Mahommed
ben Musa,* London
Sarton G 1927 Int His Sc 1, 560-561
[bib]
Steinschneider M 1887 Bib M 1, 44-48,
71-75
Suter H 1902 Bib M 3, 259-272


**Muskhelishvili,** Nikolai Ivanovich
1891-

Anon 1951 Pri M Me 15, 265-278
(MR13, 2)
Gokieli L P 1961 Matv Shk (4), 72-74
Keldysh M V + 1951 UMN 6(2), 185-190
(MR13, 2)
Savin G M 1961 Pri Me 7, 223-227
(MR26, 1)
- 1961 Ukr IM 13, 119-123
(Z109, 239)
Vekua I N 1961 Rus MS 16(2), 91-109
(Z98, 10)
- 1961 UMN 16(2), 169-188 (in Russian
(MR23A, 1505)  [port]


**Muth,** Peter 1860-

Pasch M 1909 DMV 28, 454-456
(F40, 31)


**Mydorge,** Claude  1585-1647

Busard H L 1965 Janus 52, 1-39
[construction, polygon]
Eneström G 1912 Bib M 12, 353
[conic]
Henry C 1881 Boncomp 14, 271-278
(F13, 13)
Valentin G 1912 Bib M (3)13, 82-83
(F43, 72)  [conic]


**Myers,** Sumner B   1910-1955

Bott R + 1958 Mich MJ 5, 1-4 (Z82, 243

Myller, Alexander  1879-1965

Anon 1965 Iasi UM (1)11(a), i-xiv
   (MR34, 4097)
- 1966 Gaz M (A)71, 115-118
   (MR33, 15)
Popa I 1955 Iasi M 6, 1-12
   (MR17, 932)
- 1955 Iasi UM (NS)1, xi-xxiv
   (MR18, 710) [Vera Myller]


Nagumo, Mitio  1905-

Yosida K + 1966 Fun Ekv 9, 6-8,
   18-22  (MR34, 7326-7328)
   [bib, port]


Nagy. See also Szokefalvi


Nagy, Albino

Anon 1901 Rv M 7, 111


Naimark, Mark Aronovich  1909-

Krein M G 1960 Rus MS (15)2, 169-174
   (MR22, 1594; Z89, 243)


al-Nairizi   X

   = Anaritius = Abul Abbas
   al-Fadl ibn Hatim al-Nairizi

Mansion P 1900 Brx SS 24, 47-49
Sarton G 1927 Int His Sc 1, 598-599
   [bib]
Schoy C 1922 Mun Si, 55-68
Suter H 1906 Bib M (3)7, 396
   (F37, 2)


Nakayama, Tadasi  1912-1964

Anon 1966 Nagoya MJ 27, 1-7
   (MR33, 16)  [bib]


Nanda, D. N.

Chand V 1952 Indn Ag St 4, 109-112
   (MR14, 609)


Napier, John   1550-1617

Anon 1912/13 Merchist, 14 p
   (F44, 7)
- 1913 Nat 91, 20-21
- 1914 BSM (2)38, 125-126
- 1950 M Gaz 34, 1   [logs]
- 1932 MT 24, 310
Archibald R C 1916 AMSB 22, 182-187
   (F46, 8)
*- 1948 Mathematical Table Makers,
   NY, 58-63  (Scr Stu 3) [bib, port]
Biot 1835 Analyse des ouvrages
   originaux de Napier relatifs à
   l'invention des logarithmes,
   Paris
Bose A 1916 Clct MS 6, 13-31
Bosmans H 1919 Brx SS 39, 104-111
   (F47, 32)
Cairns W D 1928 AMM 35, 64-67
   (F54, 46) [logs]
Cajori F 1914 AMM 21, 321-323
- 1916 AMM 23, 71-72 [logs]
- 1927 AMSB 33, 270  (F53, 37)
- 1927 Sci 65, 547   [logs]
Carslaw H S 1914 NSWRSJ 48, 42-72
   (F45, 73) [logs]
- 1915 M Gaz 8, 76-84, 115-119
   (F45, 73) [logs]
- 1916 AMM 23, 310 [logs]
- 1916 NSWRSJ 50, 130-142   [logs]
- 1916 Phi Mag (6)32, 476-486
   [logs]
Coolidge J L 1949 The Mathematics
   of Great Amateurs (Clarendon Pr),
   Ch 6  (Repr 1963 Dover)
   [*logs, rods, negatives, trig]
De Vries H K 1914 N Tijd 3, 1-19
   (F45, 72)  [logs]
Fazzari G 1895 Pitagora 1, 4
Gibson G A 1914 Glasg PS  (F45, 71)
   [logs]
Gravelaar N L W A 1898 Amst Vh (1)6,
   159 p  (F30, 8)

Hobson E W 1914 *John Napier and the invention of logarithms 1614,* Cambridge, 48 p    (F45, 71) [logs]

Inglis A 1936 M Gaz 20, 132-134 (F62, 20)

Johnson W W 1919 Mess M 48, 145-153 [circular parts]

Jones P S 1954 MT 47, 482-487 [rods]

Jourdain P E B 1914 Open Ct 28, 513-520 [logs]

Knott C G 1913 M Gaz 7(105)
- 1914 Nat 93, 572-573  (F45, 72)
- 1915 Sci Prog 10, 198-203

Knott C G ed 1915   *Napier Tercentenary Memorial Volume,* London, 441 p [logs]

*Macdonald W R 1889 Catalogue of the Works of J. Napier of Merchiston, in *The construction of the Wonderful Canon of Logarithms,* London    (Repr 1966 Dawson)

Miller G A 1926 Sci Prog 21, 307-310   [logs]

Moritz R 1915 AMM 22, 220-222 [Napier rules]

Mueller C 1914 Naturw 2, 669-676 [logs]

Napier Mark 1834 *Memoirs of John Napier of Merchiston...with a history of the invention of logarithms,*   London

Read C B 1960 MT 53, 381-384 [logs]

Reeve W D 1931 MT 24, 310

Sarton G 1914 Isis 2, 166-167

Sleight E R 1944 M Mag 18, 145-152 (MR5, 253; Z60, 7)   [logs]

Smith D E 1914 AMSB (2)21, 123-127 (F45, 72)

Stewart David + 1787 *An Account of the Life, Writings, and Inventions, of John Napier of Merchiston,* Perth

Thomas W R 1935 M Gaz 19, 192-205 (F61, 16; Z11, 385)

Vanhée L 1936 AMM 33, 326 [China, rods]

**Napoleon Bonaparte**   1769-1821

Aubrey P V 1954 *Monge le savant ami de Napoleon Bonaparte,* Paris, 365 p

Brocard H + 1912 Intermed 19, 177-178 223  (Q1193)  (also 5, 2; 15, 242; 16, 32, 125)

Lecat M 1919 Intermed 26, 107-108 (F47, 33)

Miels C N 1953 MT 46, 344-345

**Naqqash.**   See Zarkali.

**Narayana**   XIV

Datta B 1931 Clct MS 23, 187-194 (F57, 18)  [approx, root]
- 1933 Isis 19, 472-485 (F59, 20; Z2, 387)  [alg]

**Nardi,** Antonio

Tenca L 1955 Lom Rn M (3)19, 491-506 (Z66, 245)

**Narducci,** Enrico   1832-1893

Boncompagni B 1893 *Catalogo dei lavori di Enrico Narducci,* Rome, 20 p   (F25, 42)

Cremona L 1877  Linc At (3)1, 129 (F9, 16)

Siacci F 1893 Tor FMN 28, 811 (F25, 42)

**Naronski,** Joseph

Stamm E 1935 Slav Cong 2, 246-248 (F61, 17)

**al-Nasawi,** Abul-Hasan Ali ibn Ahmad XI

Sarton G 1927 Int His Sc 1, 719 [bib]
Suter H 1906 Bib M (3)7, 113-119

Nasimov, P. S.

Anon 1902 Kazn FMO (2)12(1)
Vassiliev A 1902 Kazn FMO (2)12,
1-6    (F33, 38)


Nasir al-Din    1201-1274

= Nasir Eddin Mohammed ibn-
Hassan al-Tusi

Braunmuehl A v 1897 Leop NA 71, 61-
67  (F28, 44)  [Regiomontanus]
Castillon G 1788 Ber Mm 18, 175-183
[non-Euc geom]
Dilgan Hamit 1956 Buyuk Turk a limi
Nasireddin Tusi,  Istanbul
- 1956 Int Con HS 8, 183-191
(MR19, 825)
Easton J 1965 AMM 72, 53-56
[de Witt, ellipse, Schooten]
Kasumkhanov F A 1954 Mos IIET 1,
128-145  (Z59:1, 4; 60, 4)
[numbers, reals]
Kubesov A 1963 Az FMT (4), 147-152
(Z117:2, 242)  [calc]
Mamedbeili G D 1959 Muhammed
Nasirèddin Tusi on the theory of
parallel lines and the theory of
ratios, Baku 100 p  (MR23A, 4)
Mamedov K M 1963 Az IMM 2(10), 147-
158  (MR27, 899)  [Euclid]
Rozenfeld B A 1951 Ist M Isl 4,
489-512 (MR14, 524; Z44, 242)
Rozenfeld B A + 1960 Ist M Isl 13,
475-482  (MR27, 3)  [non-Euc
geom]
Sabra A I 1959 Alex UAB 13, 133-170
[non-Euc geom]
Sarton G 1931 Int His Sc 2, 1001-1013
Suter H 1892 Bib M (2)6, 3-6
(F24, 48)
- 1893 Bib M (2)7, 6
Thaer C 1936 QSGM (B)3, 116-121
(Z13, 338)  [Euclid]
Wiedemann E 1928 Erlang Si 58, 228-
336 (F54, 17)  [Euclid]
- 1928 Erlang Si 58, 363-379
(F54, 18)
- 1928 Erlang Si 60, 289-316
Winter H J J + 1951 Isis 42, 138-142
(Z42, 241)  [optics]

Nasr Mansur, Abu    X-XI

= A. N. M. ibn Ali ibn Iraq

Krause M 1936 Gott Ab H (3)17
(F63, 9) [Menelaos, sphere]
Maya Julia Samso 1969 Estudios
sobra abu Nasr Mansur b. ali b.
Iraq, Barcelona, 161 p
Sarton G 1927 Int His Sc 1, 668
[bib]


Natanson, Isidor Pavlovich  1906-

Kantorovich L V + 1956 UMN 11(4), 193-
196   (Z70, 6)
Vulikh B Z + 1965 UMN 20(1), 171-175
(MR30, 569)


Negro.  See Andalo.


Nekrasov, Aleksandr Ivanovich
1883-1957

Sekerzh-Zenkovich Ya I 1960 UMN
15(1), 153-162  (MR22, 424; Z87,
243)


Nelson, Leonard    1882-1927

Bernays P 1928 Naturw 16, 142-144
[phil]
Meyerhof O 1928 Naturw 16, 137-142
[phil]


Nemenyi, Paul Felix  1895-1952

Truesdell C A 1952 Sci (NS)116, 215-
216  (MR14, 122)
- 1953 Wash Ac 43, 62-63


Nemorarius.  See Jordanus.

Nemytskii, Viktor Vladimirovich
1900-

Vainberg M M + 1961 UMN 16(1),
201-212 (MR23A, 272; Z98, 10)


Nernst, Walther   1864-1941

Bodenstein M 1934 Naturw 22, 437-439
Dehn E 1964 MN Unt 17, 114-115
(Z117-1, 9)


Nestrovich, Nikolai Mikhailovich
1891-1955

Chernyaev M P + 1956 UMN 11(4),
117-118 (Z70, 6)


Neuberg, Joseph  1840-1926

Cristescu V 1929 Gaz M 35, 201-202
(F55, 623)
Gloden A 1949 Lux Arch 18, 19-23
(Z41, 341)
Godeaux L 1926 Mathesis 40, 241-244
(F52, 28)
Labrador J F 1950 Gac M (1)2, 217-219
(Z39, 5)
Mineur A 1926 Mathesis 40, 97
(F52, 27)


Neugebauer, Otto   1899-

Dunnington G W 1937 M Mag 11, 1-2
(F62, 1042)   [port]


Neumann, Carl   1832-1925

Anon 1925 M Ann 94, 177-178
(F51, 27)
Archenhold F S 1925 Weltall '24,
167-168   (F51, 27)
Duerll 1926 Unt M 32, 222-223
(F52, 27)
Hoelder O 1922 M Ann 80, 161-162
(F48, 25)
- 1925 Leip Ber 77, 154-172 (F51, 27)

- 1926 M Ann 96, 1-25  (F52, 27)
Liebmann H 1927 DMV 36, 174-178
(F53, 28)
Lorey W 1926 ZMN Unt 57, 417-423
(F52, 27)


Neumann, Franz  Ernst   1798-1895

Amsler-Laffon 1904 Zur NGV 49, 142-
158  (F35, 19)
Anon 1895 Nat 52, 176  (F26, 37)
- 1896 Lon RSP 60, 4
Bertrand J 1895 Par CR 120, 1189-119
(F26, 37)
Biermann K-R 1960 Fors Fort 34, 97-
101  (Z87, 243)  [phys]
Dyck W v 1901 Mun Si 31, 203-208
(F33, 22)  [Jacobi]
- 1902 M Ann 56, 252-256  [Jacobi]
Knott R 1906 ADB 52, 680-684
Lorey W 1904 MN Bl 1, 150-152
(F35, 19)
Neumann C 1917 Leip Ab 33, 195-458
(F46, 5)  [cryst]
- 1950 Mun Abh (NF)(59), 27 p
(MR12, 383)
Neumann Luise 1904 *Franz Neumann:*
*Erinnerungsblaetter von seine*
*Tochter,* Tuebingen, 459 p
(F35, 18)
- 1929 *Franz Neumann,* Leipzig, 48 p
(F55, 611)
Voigt W 1895 Gott N, 248-265
(F26, 37)
- 1909 Int Woch 3, 371-378, 397-410
(F40, 23)
Voit C 1896 Mun Si 26, 338
Volkmann Paul 1896 *Franz Neumann,*
Leipzig (Teubner), 75 p
(F26, 37)  [port]
- 1899 Konig Ph 40, 41
Wangerin A 1897 DMV 4, 54-68
(F26, 37; 28, 21)
- 1907 *Franz Neumann und sein Wirken*
*als Forscher und Lehrer,*
Braunschweig, 195 p  (F38, 23)
- 1907 Loria 11, 3-4
- 1910 Phy Z 11, 1066-1072  (F41, 13

Nevanlinna, Rolf Herman    1895-

Kuenzi H P 1955 Elem M 10, 97-100
   (MR17, 3; Z64, 243)
*Kuenzi H P + 1966 in *Festband 70
   Geburtstag R. Nevanlinna,*
   Berlin (Springer), 1-6, 135-149
   (MR34, 2419-2420)


Neville, Eric Harold    1889-

Broadbent T A A 1962 LMSJ 37, 479-
   482   (MR25, 967; Z106, 3)
Broadbent T A A + 1964 M Gaz 48, 131-
   163    (Z117-1, 9)


Newcomb, Simon    1835-1909

Anon 1909 Am JS 28, 290-292
- 1909 Clct MS 1, 219-221
- 1909 Ens M 11, 403-404
   (F40, 39)
- 1910 Wash PS 15, 133-167
- 1911 Lon RSP (A)84, xxxii-xxxviii
- 1936 Scr 4, 51-56
Archibald R C 1906 Can PT 11, 79-110
   (F37, 33)   [bib]
- 1916 Sci 44, 871-878
- 1924 Am NAS Mm 17, (1)2   [*bib]
Baillaud B 1909 Rv GSPA 20, 725-727
Ball R S 1909 Nat 81, 103-105
   (F40, 39)
Belopolsky A 1909 Pet B(6)15, 1013-
   1014
Berberich A 1909 Naturw 24, 453-455
   (F40, 39)
Brown E W 1910 AMSB (2)16, 341-355
   (F41, 24)
Campbell W W 1924 Am NAS Mm 17, 1-18
Colaw J M 1894 AMM 1, 253-256
Loewy 1899 Nat 60, 1-3   (F30, 29)
Millosevich E 1909 Linc Rn (5)18(2),
   409-414
Newcomb Simon 1903 *The Reminiscences
   of an Astronomer,* Boston, 434 p
   [port]
Seeliger H v 1910 Mun Si, 24-26

Newell, Marquis J.

Anon 1945 MT 38, 345-349


Newman, Francis William    1805-1897

Archibald R C 1945 M Tab OAC 1,
   454-459


Newton, Hubert Anson    1830-1896

Anon 1898 Lon RSP 53, i
*Gibbs J W 1897 Am JS 3(4), 359-
   [bib, port]
- 1902 Am NASBM 4, 101-124
   (repr of 1897)
Phillips A W 1896 AMSB (2)3, 169-
- 1897 AMM 4, 67-71


Newton, Isaac    1642-1727

Allen F 1944 Sci 99, 299
Andrade E N da C 1935 Nat 135, 360
- 1943 Lon RSP (A)181, 227-243
   (MR5, 57; Z60, 9)
- 1950 *Isaac Newton* (Anchor pap),
   111 p   (MR14, 344)
- 1954 *Isaac Newton,* London, 140 p
   (MR15, 923; Z59:1, 17)
- 1954 Rv Hi Sc Ap 6, 289-307
   (MR15, 592)
Anning N 1926 AMM 33, 211   [cubic]
Anon 1888 *Portsmouth Papers. A
   Catalogue of the Portsmouth
   Collection... Newton, the Scientific
   portion of which has been presented
   by the Earl of Portsmouth to the
   Univ of Cambridge,* Cambridge
   (F21, 1254)
- 1927 Nat (Mar 26)(S), 21-48
- 1927 *Newton 1727-1927,* Leningrad,
   73 p   (F53, 17)
- 1942 Nat 150, 654-655   [anniv]
*- 1943 *Isaak Nyuton, Sbornik statei
   ...,* Mos-Len
- 1947 *Newton Tercentenary Celebra-
   tions 15-19 July, 1947,* Cambridge,
   107 p   (MR9, 75)

Newton, Isaac (continued)

- 1950 M Gaz 33, 233
  1967 The Annis Mirabilis..., in
  *The Texas Quarterly,* 10(3) (Sp),
  287 p
Antoniadi E M 1931 Nat 127, 484
  [Greek phil]
Ball W W R 1892 LMSP 23, 226-231
  (F24, 52) [mech]
- 1893 *An essay on Newton's Principia.*
  London, 185 p (F25, 84)
  [Principia]
- 1914 M Gaz 7, 349-360 (F45, 4)
Barnes E 1926 Nat (5), 21-24
Barr E S 1966 Sci 154, 338
Bashmakova I G 1959 Ist M Isl 12,
  431-456 (MR23A, 697)
  [eqs, Waring]
Bell E T 1942 AMM 49, 553-575
  (MR4, 65; Z60, 9)
Berenguer P A 1895 Intermed 2, 204
  (Q604) (See also Intermed 6, 110)
  [Hugues]
Bertoldi I 1957 Pe M (4)35, 14-43
  (Z79, 242)
Beth H J E 1932 *Newton's Principia,*
  Groningen, 179 p, 154 p
  (F58, 990)
Biot J B 1829 *Life of Sir Isaac
  Newton,* London
- 1832 J Sav, 193-, 263-
  [Brewster 1855]
- 1852 J Sav, 133-, 217-, 269-
  [Cotes]
- 1855 J Sav, 589-, 662-
Bjerknes C A 1877 Chr Avh, 1-27
  (F9, 6)
Blaquier J 1947 B Ai Ci EFN 12, 9-32
  (MR11, 573)
Blueh O 1935 Nat 135, 658
  [Spinoza]
Bolza O 1912 Bib M (3)13, 146-149
  (F43, 76) [calc of var,
  least resistance]
Bosscha J 1909 Ned Arch (2)14, 278-
  288
Boyer C B 1949 AMM 56, 73-78
  (Z32, 51) [polar coords]
- 1950 Scr 16, 141-157, 221-258
  [Euler]

Brasch F E 1927 Isis 9, 427
- 1928 *Sir Isaac Newton,* Baltimore,
  340 p (F55, 623)
- 1928 Pop As 36, 14 p
- 1941 Scr 8, 199-227 [port]
- 1947 Sci 106, 102-103 [letters]
- 1952 Scr 18, 53-67 (Z47, 5)
- 1954 Scr 20, 224-225 [port]
Brasch F E ed 1928 *Sir Isaac Newton:
  A bicentenary evaluation of his
  Work,* Baltimore, 360 p
Braunmuehl A v 1904 Bib M (3)5, 355-
  365 (F35, 60) [intgl, Cotes]
Brewster D 1855 *Memoirs of the Life,
  Writings and Discoveries of Sir
  Isaac Newton,* Edinburgh, 2 vols
  (2nd unchanged ed 1860)
- 1867 Par CR 65, 261-, 537-, 653-,
  717-, 769-, 770-, 825-, 925-
  [Pascal]
Broad C D 1927 *Sir Isaac Newton,*
  London, 32 p
Brodetsky S 1927 *Sir Isaac Newton:
  a Brief Account of his Life and
  Work,* London, 173 p (F55, 605)
- 1942 Nat 150, 698-699
Brown B H 1926 AMM 33, 155-157
  (F52, 21) [pasturage problem]
Bunge Mario 1943 *El tricentenario de
  Newton,* Buenos Aires, 8 p
Cajori F 1894 AMSB (2)1, 52-54
  [binom thm]
- 1922 Arc Sto 3, 201-204 [grav]
- 1924 Sci 59, 390-392 [legends]
- 1926 Ens M 25, 7-11 [Descartes]
- 1926 Pop As 34 [religion]
- 1928 Sci Mo 27, 47-53 [grav]
- 1929 M Gaz 14, 415-416
- 1929 Nat 124, 513 [Motte]
Cantelli G ed 1958 *La disputa Leibniz-
  Newton sull'analisi,* Torino, 239 p
Carra De Vaux 1907 *Newton,* Paris,
  61 p (F38, 54)
Carruccio E 1938 Pe M (4)18, 1-32
  (F64, 915) [area]
Cassirer E 1943 Phi Rev 52, 366-391
  [Leibniz]
Chant C A 1943 Can As 37, 1-16
  (MR4, 181; Z60, 8)
Charlier C V L 1931 *Naturvetenskabens
  matematiska principer av Isaac
  Newton II-III,* Lund (F57, 1310)

Newton, Isaac   (continued)

Chasles M 1867-71 Par CR, 65-72
   [Pascal]
Cherry T M 1937 *Newton's Principia*
   *in 1687 and 1937...,* London, 28 p
   (F63, 817)
Cherubino S 1930 Pe M (4)10, 21-30
   (F56, 20)
Chevreul M E 1869 Par CR 69, 305-
   [Pascal]
Cohen I B 1943 Sky Tel 2(3), 3-5
- 1956 *Franklin and Newton,*
   Philadelphia, 657 p   (APS Mm 46)
- 1960 Isis 51, 489-514   (Z97, 2)
Conte L 1942 Pe M (4)22, 70-90
   (F68, 15)   [Castillon, Fermat,
   parab]
- 1947 Pe M (4)25, 1-15   (MR9, 169)
- 1947 Pe M (4)25, 165-180
   (MR9, 486)
Craig V J 1901 AMM 8, 157-161
Crew H 1941 Scr 8, 197-199
Crewe W H 1966 Sci 153, 1336   [death]
Crompton S 1867 Manc Pr 6, 1-; 7, 3-
   [port]
Dale H H 1946 Mus Bk, 400-463
   [letter]
David F N 1957 An Sc 13, 137-147
   [Dyse, Pepys]
De Losada Y Puga Cristóbal 1942
   Peru U Rv 10, 479-480
   (MR7, 355)
De Morgan A 1846 PT, 107-109
   [Newton-Leibniz]
- 1848 Phi Mag (3)32, 446-456
- 1852 Phi Mag (4)3, 440-444
- 1852 Phi Mag (4)4, 321-330
- 1855 Nort Br Rv 23, 307-338
   [Brewster 1855]
- 1885 *Newton: His Friend: and his
   Niece,* London, 161 p   (rep 1968
   with intro E A Osborne, London
   (Dawsons))
- 1914 *Essays on the Life and Work of
   Newton, Chicago (Open Court),*
   198 p   (Ed: P E P Jourdain
   w. notes and appendices)   (F45, 14)
   [*bib, mss]
Dessauer Friedrich 1945 *Weltfahrt
   der Erkenntnis, Leben und Werk*

*Isaac Newtons,* Zurich, 430 p
   (MR9, 169)
Dugas Rene 1953 *De Descartes à Newton
   par l'école anglaise,* Paris, 19 p
   (MR16, 433)
Dyson F 1926 Nat (S), 30-33   [ast]
Edleston 1850 *Correspondence of Sir
   I. Newton and Prof. Cotes...,*
   Cambridge
Einstein A 1927 Naturw 15, 273-276
   [phys]
- 1927 Smi R, 201-207
Emanuelli P 1927 It As (2)4, 57-85
   (F57, 1301)
Eneström G 1910 Bib M (3)11, 276,
   354-355   [calc]
- 1911 Bib M (3)12, 268   (F42, 61)
   [calc]
Evans M G 1955 J H Ideas 16, 548-557
   [real numb]
Evans W 1914 M Gaz 7, 418-421
   [Berkeley]
Faggi A 1923 Tor FMN 58, 323-337
   (F49, 6)   [Descartes]
Favaro A 1881 Boncomp 13, 481-514
   (F13, 12)
Ferguson A 1942 Phi Mag (7)33, 871-
   888   (MR4, 65; Z60, 9)
Fesenkov V G 1927 Rus As Zh 4, 91-101
   (F53, 18)
Findlay Shirras G 1951 Arc In HS(NS)4,
   897-914   (MR13, 612)
Fleckenstein J D 1946 Elem M 1, 100-
   108   (MR8, 190)   [J I Bernoulli]
- 1956 *Die Prioritaetstreit zwischen
   Leibniz und Newton...,* Basel---
   Stuttgart, 27 p   [Leibniz]
Fraser Duncan C 1927 *Newton's Inter-
   polation Formulas,* London
Fueter E 1937 Zur NGV 82(28)
   (F63, 807)
Fujiwara M 1941 Toh MJ 47, 322-338
   (Z61, 4)
Gager W A 1953 SSM 52, 258-262
Garcia de Zuniga E 1940 *Newton,*
   Montevideo, 12 p
Gibson G A 1921 Edi MSP 40, 9-20
   [Jurin, lim, Robins]
Glazebrook R 1926 Nat(S), 43-48
   [optics]
Graham R H 1890 Nat 52, 139-142
   [geom]

Newton, Isaac (continued)

Gray G J 1888 *Bibliography of the Works of Sir Isaac Newton Together with a List of Books Illustrating his Life and Works*, Cambridge, 40 p (2nd ed 1907, repr 1966 Dawsons)

Greenhill G 1923 Nat 111, 224-226 [mech]

Greenstreet W J 1927 *Isaac Newton..., A Memorial Volume...*, London, 189 p

Gridgeman N T 1966 New Sci (Aug 18)

Groat B F 1930 AMSB 36, 194 (F56, 35) [simil]

Hackett F E 1924 Nat 111, 395-396 [mech]

Haentzschel E 1915 ZMN Unt 96, 190-194

Hall A R 1948 Cam HJ 9, 239-250

- 1958 Osiris 13, 291-326 (Z90, 7)

Hall A R + 1963 Arc In HS 16, 23-28 (Z117:1, 7) [ell]

Henrotenu F 1916 Intermed 23, 81-84 (Q4385) [bib]

Herivel J W 1962 Isis 53, 212-218 (Z105, 2) [rot]

- 1962 Rv Hi Sc Ap 15, 105-140 (Z108-2, 246) [dyn]

- 1963 Arc In HS 16, 13-22 (Z117-1, 7)

Hessen Boris M 1931 in *Science at the Cross Roads*, London [econ, soc]

*- 1946 *The Social Economic Roots of Newton's Principia*, Sydney, Australia

*Hill, Christopher 1967 Tex Q 10(3), 30-51 [bib]

Hofmann J E 1943 Ber Ab (2), 130 p (Z61, 5) [calc, Leibniz]

- 1951 MP Semb 2, 45-70 (MR12, 578; Z42, 241)

Hoskia M 1961 Listener 66, 597-599

Hussain Z 1951 AJP 19, 197-202

Jacoli F 1887 An Ast Met 5, 153-157 (F21, 1255)

Jeans J H 1926 Nat (S), 28-30

- 1943 Lon RSP (A)181, 251-262 (MR5, 57)

Jones P S 1958 MT 51, 124-127

Jourdain P E B 1914 Monist 24, 515-564 [mech]

- 1915 Monist 25, 79-106, 234-254, 418-440 [ether, grav]

- 1918 Monist 28, 629-633 [Galileo]

- 1920 Monist 30, 19-36, 183-198, 199-202

Kirwan C de 1893 Rv Q Sc (2)3, 168-189 (F25, 84)

Klose A 1927 Lat U Rak 16, 623-633 [mech]

Korner T 1904 Bib M (3)5, 15-62 [mech]

Koyré A 1950 Arc In HS 29, 291-311 (Z36, 4)

- 1952 Isis 43, 312-337 (MR15, 276) [Hooke, letter]

- 1955 Rv Hi Sc Ap 8, 19-37 (MR16, 986)

- 1965 *Newtonian Studies* (Harvard U Pr)

Koyré A + 1962 Arc In HS 15, 63-126 (Z108:2, 247) [Clarke, Conti, Leibniz, des Maizeaux]

Kramar F D 1950 Ist M Isl 3, 486-508 [Wallis]

Kryloff A N 1923 Lon As Mo N 84, 392-395

- 1925 Lon As Mo N 85, 571-575 (F52, 21) [mech]

- 1925 Lon As Mo N 85, 640-656 (F52, 21) [ast]

Kudrjavcev P S 1955 Mos IIET 5, 33-51 (Z68, 5) [Lomonosov]

Laemmel Rudolf 1957 *Isaac Newton*, Zurich, 308 p (Z77, 242)

Lamb H 1926 Nat (S), 33-36 [mech]

Langer R E 1936 Scr 4, 241-255

Larmor J 1925 Isis 7, 110-112

- 1936 Lon Tim LS (Jan 18), 55 [Gregory]

Laue M V 1927 Naturw 15, 276-280 [optics]

Laugel L 1897 Intermed 4, 286

Lefebvre B 1899 Intermed 5, 63, 71 (Q1185) [exponents]

- 1924 Rv Q Sc (4)6, 115-140 [grav]

- 1925 Rv Q Sc (4)7, 126-180 [grav]

Lenzen V F 1937 Isis 27, 258-60 [mech]

Newton, Isaac   (continued)

Lerner A 1910 Intermed 17, 206
(Q3655)
Lippmann E O v 1937 Naturw 25, 238
[alchemy]
Littlewood J E 1948 M Gaz 32, 179-181
(MR10, 420) [mech]
Lohne J A 1960 Centau 7, 6-52
(MR22, 424) [Hooke]
*- 1965 Lon RSNR 20(2), 125-139
- 1961 Arc HES 1, 389-405
[color, sine]
- 1967 Hist Sci 6, 69-89
Loria G 1920 It SPS 10, 471-473
(F47, 6)
- 1920 Newton, Rome, 69 p  (F47, 6)
- 1926 Scientia 39, 323-334,
"104-114"  [Galileo]
- 1926 Scientia 40, 205-216,
"61-70"  [Galileo]
- 1939 Newton, Milan, 80 p
(F65, 1086)
Losada Y Puga C de 1942 Peru U Rv 10,
479-480 (Z60, 9)
Lysenko V I 1966 Vop IET 20, 38-46
(MR36, 4946) [Euler]
Macauley W H 1897 AMSB (2)3, 363-371
(F28, 49) [mech]
Macomber Henry P 1950 A Descriptive
Catalogue of the Grace K. Babson
Collection of the Works of Sir
Isaac Newton..., NY, 242 p
[bib]
- 1951 Isis 42, 230-232 (Z43, 142;
44, 244)
MacPike E F 1940 Not Qu Lon, 133
Marvin F 1926 Nat (S), 24-28
McColley G 1938 Isis 28, 94
Miller J E 1962 AMM 69, 624-631
(Z112, 2) [grav]
Miller J 1939 Brt As J 50, 2
[mech]
Miller M 1953 Dres Verk 1(1), 5-32
(MR22, 260) [cubic curves]
- 1954 Dres Verk 2(2), 1-16
(MR22, 260; Z55, 4) [series]
- 1954 Dres Verk 2(3) (MR22, 260;
Z60, 7) [calc]
Milne J J 1935 M Gaz 19, 139-140
(F61, 12)

Miner P 1961 Not Qu Lon 8, 15-16
Moore Patrick 1957 Isaac Newton,
London, 96 p  (Z77, 242)
Mordell L J 1921 Nat 119(S), 40-43
(F53, 17)
More L T 1934 Isaac Newton. A
Biography, NY, 687 p  (Repr 1962
Dover)
Mouy Paul 1938 Rv Met Mor, 411-435
[Malebranche]
Natanson Wladyslaw 1927 Newton,
Czerwiec, 53 p
Nordström J 1936 Lychnos 1, 225-229
[Mencke]
Ohlsson J 1928 Nor As Tid 9, 56-65,
88-99  (F57, 1301)
- 1929 Nor As Tid 10, 1-14, 56-63,
93-106  (F57, 1301)
- 1930 Nor As Tid 11, 21-32; 12, 6-18
(F57, 1301)
Ortvay R 1943 MP Lap 50, 262-289
(MR8, 190)
Oseen C W 1936 Lychnos 1, 217-23
(F62, 1035)
Parfentiev N N 1927 Kazn Ped 2, 27-43
(F53, 18)
Peddie W 1924 Nat 111, 395 [mech]
Pelseneer J 1929 Isis 12, 237-254
[letter]
- 1930 Isis 14, 155-165 [anal geom,
Hugh d'Omerique]
- 1932 Isis 17, 331
- 1936 Osiris I, 497-499 [letter,
Pepys, prob]
- 1939 Osiris 7, 523-555 (MR1, 33;
F65:2, 1086) [letters]
- 1952 Bel BS (5)38, 219-220
(Z47, 5)
Picard Emile 1927 Un double centenaire,
Newton et Laplace, Paris, 26 p
(Repr Rv GSPA 38, 357-366)
- 1927 Rv Fr, 753-761
Pighetti C 1963 Arc In HS 15, 291-
302
Pla C 1945 Isaac Newton, Buenos
Aires, 64 p
Popp K R 1935 Jacob Boehme und Isaac
Newton, Leipzig, 109 p
Pullin V E 1927 Sir Isaac Newton: a
Biographical Sketch, London, 80 p
Raphson 1715 History of Fluxions,
London

Newton, Isaac   (continued)

Rigaud Stephen Peter 1838 *Historical Essay on the First Publication of Sir Isaac Newton's Principia,* Oxford

Robinson H W 1939 An Sc 4, 324 (F65, 1087)   [letters]

Rollett A P 1965 M Gaz 49, 86-87

Rosenfeld L 1961 Isis 52, 117-120 [letters]

Rufus W C 1941 Scr 8, 228-231 (MR4, 181)   [Rittenhouse]

Sampson R A 1923 Lon As Mo N 84, 378-383

Sanford V 1934 MT 26, 106-109

- 1952 MT 45, 598-599

Santalo L A 1942 M Notae 2, 61-72 (MR4, 65; Z60, 9) [binom thm]

Schofield B 1937 Brt Mus Q 11, 66 [alchemy]

Scriba C J 1963 Arc HES 2, 113-137 [calc, Leibniz]

Shirras G F 1951 Int Con HS 6, 212-229  [bib, letters]

Sister M. Thomas a Kempis 1935 SSM 34, 569-573

Slichter C S 1937 AMM 44, 433-444 (F63, 808)  (repr in Slichter 1938 *Science in a Tavern)*

Smith B 1908 Nat 77, 510, 534 (F39, 17)

- 1909 Nat 79, 130

Smith D E 1909 Bib M (3)9, 301-308 [port]

- 1912 *The Portrait Models of Isaac Newton,* NY, 14 p

Snow A J 1927 Scientia 42, 1-10 [phys]

Strong E W 1951 J H Ideas 12, 90-110 (MR12, 383; Z45, 147)

Stukeley William 1936 *Memoirs of Sir Isaac Newton's Life,* London, 103p (F62, 1034)

Suchting W A 1967 Isis 58, 186- [Berkeley, mech, relativity, space]

Sullivan J W N 1938 *Isaac Newton...,* London, 295 p  (F64, 915)

Tait P G 1885 Edi SP(A)13, 72-78

Tannery P 1896 BSM (2)20, 24-28 [calc]

Taylor C 1900 Cam PST 18, 197-219 [geom, Kepler]

Teich N 1944 Nat 153, 42-45

Tibiletti C 1947 Pe M (4)25, 16-29 (MR9, 169) [Apollonius problem, Pluecker, Viete]

Tischer E 1896 *Ueber die Begrundung der Infinitesimalrechnung durch Newton und Leibniz,* Leipzig

Trevelyan G M 1927 Lon Tim LS (Mar 24

*Truesdell C 1967 Tex Q 10(3), 238-258

*Turnbull H W 1945 *The Mathematical Discoveries of Newton,* London, 68 p  (2nd ed, 1947)

Van Geer P 1907 Tijdspgl(s), 23 p [Huygens]

Van Heel A C S 1953 Nat 171, 305-306 (MR14, 609)  [optics]

Vavilov S I ed 1943 *Isaac Newton...,* Mos, 439 p

- 1947 Mos IIET 1, 315-326 (MR12, 311)  [optics]

- 1948 *Isaac Newton,* Vienna, 176 p

- 1951 *Isaac Newton,* Berlin, 224 p (Z43, 244)

- 1961 *Isaac Newton,* Moscow, 294 p (MR24A, 333)

Villamil, Richard de 1931 *Newton: The Man,* London, 117 p

Villat H 1939 *Newton,* Paris

Volkmann P 1898 Konig Ph 39, 1

Von Kármán T 1942 J Aer Sci 9, 521-522 548  (MR4, 65)  [aerodynamics]

Vonwiller O U 1943 NSWRSJ 76, 316-328 (MR5, 57)  [Galileo]

Wargny C + 1903 Intermed 10, 17

Westfall R S 1962 Isis 53, 339-358 (Z100, 7)  [color]

Whiteside D T 1961 M Gaz 45, 175-180 (Z103:2, 242)  [binom thm]

*- 1964 Lon RSNR 19, 53-62

*- 1967 Tex Q 10(3), 69-85  [*bib]

Wieleitner H 1914 Bib M 14, 55-62 [Stirling]

- 1927 Unt M 33, 103-107 (F53, 17)

Wilson E B 1946 Sci 104, 276

Winter H J J 1938 Lincolns 3, 294-298

Witting A 1911 Bib M (3)12, 56-60 (F42, 61)  [calc]

Zaviska 1927 Cas MF 56, 295-296 (F53, 18)

Zeuthen H G 1895 Dan-Ov 1, 2, 37,
193, 257-278 (F26, 65)

Neyman, Jerzy 1894-

*David F N ed 1966 *Festschrift for
J. Neyman*, NY, 468 p

Nicephoros Gregoras 1295-1359

Biedl A 1948 Wurz Jb (1), 100-106
(Z31, 97) [phil]
Sarton G 1947 Int His Sc 3(1),
949-953

Nichols, Irby Coghill 1882-1952

Sanders S T 1954 M Mag 27, 118

Nicholson, James W. 1844-

Anon 1894 AMM 1, 183-187
Bell E T 1893 AMM 1, 183

Nicolescu, Miron 1903-

Dinculeanu N + 1963 Gaz MF (A)15,
538-555 (Z118:1, 12)

Nicoletti, Onorato 1874-1929

Anon 1930 It UM 9, 121-124
(F56, 30)
Puccianti L 1930 Bo Fir (2)9, 81-84
(F56, 817)

Nicollic

Brocard H + 1902 Intermed 9, 127
(Q987)
Lemoine E + 1897 Intermed 4, 167

Nicolo de Martino XVII

Amodeo F 1920 Nap FM Ri (3a)26,
170-178 [Archimedes]

Nicomachos of Gerasa I-II

Becker O 1938 QSGM (B)4, 181-192
(F64, 10; Z18, 49)
Kutsch Wilhelm 1959 *Thabit B. Qurra's
arabische uebersetzung der
Arithmetike Eisagoge des Nicomachos
von Gerasa...*, Beyrouth, 369 p
Pistelli H 1894 *Jamblichi in
Nicomachi arithmeticam introduc-
tionem liber*, Leipzig
Robbins F E + 1926 Studies in Greek
Arithmetic in *Introduction to
Arithmetic by Nichomachus*, tr
M. L. D'Ooge, NY
Simon M 1909 Arc GNT 1, 163-171
(F40, 52)
Spezi G 1868 Boncomp 1, 57-62
(F1, 2)
Tabuenca O 1949 Gac M (1)1, 257-262
(MR11, 707; Z36, 3)
Wertheim G 1898 ZM Ph 43, 41

Nicomedes -II

Seidenberg A 1966 Arc HES 3, 97-101
(MR33, 7224) [Delian probl]

Nielsen, Jakob 1890-

Fenchel W 1960 Act M 103, vii-xix
(MR22, 425; Z92, 3)
- 1960 Nor M Tid 8, 5-10, 63
(MR23A, 4; Z112, 244)
Schieldrop E B 1960 Nor VSF 33, 1-6
(MR23A, 135; Z84, 5)

Nielsen, Niels 1865-1931

Anon 1932 It UM 11, 64
- 1939 Isis 30, 514
Archibald R C 1948 *Mathematical Table
Makers*, NY, 63-64 (Scr Stu 3) [bib]

Bohr H 1931 M Tid B, 41-45
 (F57, 1308)


Nikliborc, Wladyslaw    1899-1948

Slebodzinski W 1948 Colloq M 1, 322-
 330  (MR10, 668; Z37, 291)


Nikoladze, Giorgi

Mushelishvili N 1947 Tbil MI 15,
 1-17  (MR14, 525)


Nikolai, E. L.  1881?-

Anon 1941 Pri M Me 5, 3-10
 (Z60, 16)
- 1950 Pri M Me 14, 117-120
 (Z37, 2)


Nikolskii, Sergei Mikhailovich
 1905-

Kolmogorov A N + 1956 UMN 11(2),
 239-244  (Z70, 6)


Niven, William Dawson  1843-1917

Anon 1918 LMSP (2)16    (F46, 24)


Nobel, Emanuel  1859-

Riesenfeld E H 1929 Naturw 17,
 531-533  (F55, 24)


Nobile, Arminio    1838-1897

Montesano D 1899 Nap Pont 29,
Pinto L 1897 Nap FM Ri (3)3, 138-141
 (F28, 29)


Noéther, Emmy    1882-1935

Aleksandrov P S 1936 UMN 2, 254-265
Barinaga J 1935 Rv M Hisp A(2)10,
 162-163  (F61, 956)
Einstein A 1935 NYT (May 4)
 [ideals, non-comm]
Korinek V 1935 Cas MF 65, D1-D6
 (F61, 956)
Sagastume Berra A 1935 La Pla Pub 1C
 95-96  (F62, 1041)
Van der Waerden P L 1935 M Ann 111,
 469-476  (F61, 26)
Weyl  H 1935 Scr 3, 201-220
 [port]


Noether, Max    1844-1921

Anon 1922 Linc Rn (5)31, 38-39
 (F48, 17)  [H Schwarz]
- 1922 M Ann 85  (F48, 17)
Berzulari L 1921 Lom Gen (2)54,
 600-603  (F48, 17)
Brill A 1923 DMV 32, 211-233
Castelnuovo G 1922 Linc Rn (5a)31,
 404-414  (F48, 17)
Castelnuovo G + 1925 M Ann 93, 161-
 181  (F51, 27)
Segre C 1921 Tor FMN 57, 89-91
 [H Schwarz]


Norden, Aleksandr Petrovich  1904-

Kopp V G + 1964 UMN 19(5), 171-179
 (MR29, 879)


Novak, Josef    1905-

Fischer O + 1965  Cas M 90, 236-246
 (MR33, 1214)


Novalis    1772-1801

 = Friedrich von Hardenberg

Dyck Martin 1960 *Novalis and Mathematics,* Chapel Hill, 109 p
 (MR22, 1)

Novara da Ferrara, Domenico Maria
1454-1504

Boncompagni B 1871 Boncomp 4, 340-
341    (F3, 8)
Curtze M 1869 Altpr Mo 6, 735-743
(F2, 7) [Copernicus]
- 1870 Altpr Mo 7, 253-256  (F2, 7)
- 1870 Altpr Mo 7, 515-521  (F2, 7)
- 1870 Altpr Mo 7, 726-727  (F2, 8)
- 1870 Rv Europ 2(3), 1-3  (F2, 7)
[Copernicus]
- 1871 Boncomp 4, 140-149  (F3, 7)
Jacoli F 1877 Boncomp 10, 75-89
(F9, 2)


Novarese, Enrico  1858-1892

Anon 1892 Rv M 2, 35    (F24, 34)


Novgorodets.  See Kirik


Novikov, Peter Seegeevich  1901-

Lyapunov A A 1952 UMN (NS)7(2), 193-
196   (MR13, 810; Z47, 6)
Anon 1961 SSSRM 25, 629-634
(MR24A, 333; Z98, 10)


Nunes, Pedro   1519-1578

  = Nunez   = Nonius

Bosmans H 1907 Bib M (3)8, 154-169
(F38, 58) [alg]
- 1908 Porto Ac 3, 222-271
(F39, 57) [alg]
Escobar T M 1932-1933 Rv M Hisp A
(2)7,  269-281; (2)8, 26-40
(MR58, 989; 59, 841)
Fontoura da Costa A 1938 Petrus No 1,
337-356  (F64, 910) [sphere]
Guimaraes R + 1896 Intermed 3, 102,
263-266  (Q826)
Guimaraes R 1914 Porto Ac 9, 54-64
96-117   (F45, 6)
- 1915 Porto Ac 10, 20-36

- 1915 Sur la vie et l'oeuvre de
Pedro Nunes, Coimbra, 87 p
(F45, 6)
- 1918-1919 Porto Ac 13, 61-71
(F46, 9)
- 1930 P. Nonius, Coimbra
Guimaraes R + 1915-1916 Intermed 22,
56; 23, 21, 103  (Q915)  (F46, 9)
[mss]
Rueda C J 1911 Esp SM 2(4), 43-41
(F42, 4)


Nunn, Percy  1870-1944

C O T + 1945 M Gaz 29, 1-3
Reeve W D 1945 MT 38, 134


Nystroem, Evert Johannes  1895-

Stenij S E 1960 Nor M Tid 8, 105-109
(MR23A, 4; Z112, 244)


Oblath, Richard  XX

Anon 1959 M Lap 10, 192-194
(MR23A, 136; Z97, 244)  [bib]
Oláh G 1960 M Lap 11, 19-25
(MR23A, 136; Z97, 2)


Obradovici, G.

Marian V 1937 Gaz M 42, 337-341
(F63, 15)


Ocagne, Maurice Philbert d'
1862-1938

Buhl A 1933 Ens M 32, 87-89
(F59, 36)
Couffignal L + 1939 Rv Sc 77, 70-76
(F65, 1092)
Glagolev N A 1940 UMN 7, 322-326
(F66, 22; Z23, 196)
Humbert P 1959 Rv Q Sc (5)5, 17-27
(F65, 1092)
Monoide S P 1962  Parana MB 5, 10-11
(Z98, 10)

Rozenfeld B A 1963 Vop IET (14), 130-132

Rudio F 1899 ZM Ph 44(s), 385
[anal geom, tangential coords]

Oettinger, Ludwig 1797-1869

H M 1870 Arc MP Lit 51(201), 1-3
(F2, 24)

Ofterdinger, Ludwig Felix 1810-1896

Kunssberg H 1896 Bib M, 50

Ohara, Rimei

Hayashi T 1930 Toh MJ 36, 182-188,
395-397 (F58, 994)

Ohm, Georg Simon 1787-1854

Bauernfeind C M v 1882 *Gedaechtnis-srede auf Georg Simon Ohm den Physiker,* Munich
Hartmann L ed 1927 *Aus Georg Simon Ohms handschriftlichen Nachlasse,* Munich, 263 p (F53, 21)
Lemmel E 1889 *Georg Simon Ohms wissenschaftliche Leistungen. Festrede...,* Munich, 23 p
- 1893 Smi R, 10 p (F25, 1913)
Poske F 1889 Z Ph C Unt 2, 196-198
Vondermuehll K 1892 Pogg An 47, 163-168 (F24, 53)

Ohm, Martin 1792-1872

Kobell E 1873 Mun Si, 132
(F5, 34)

Okada, Yoshitomo 1892-1957

Fukamija M 1958 Toh MJ (2)10, 1-2
(Z88, 244)

Okamoto, Norifumi -1931

Smith D E 1931 Sci (2)73, 468-469
(F57, 45)

Okamura, Hiroshi

Matsumoto T 1950 Kyo CS 26, 1-3
(MR12, 311; Z45, 148)

Olbers, Heinrich Wilhelm Matthias 1758-1840

Boncompagni B 1884 Vat NLA 34, 206
[bib, Gauss]
- 1884 Vat NLA 34, 238 [bib, Gauss]
Favaro A 1884 Ven I At (6)3, 1-10
[Gauss, letter]
Focke Wilhelm Olbers 1912 in *Bremische Biographie des neunzehnten Jahrhunderts,* Breman, 359-376
[cel mech]
Schilling C ed 1894-09 *Wilhelm Olbers, sein Leben und seine Werke,* Berlin, 2 vols in 3 (F30, 9) [port]

Olds, George Daniel 1853-1931

Esty T C 1931 AMSB 37, 644

Olivier, Louis 1854-1910

Anon 1910 Ens M 12, 424 (F41, 32)

Ollero y Carmona, D. Diego

Grinon T 1912 Esp SM 2, 149-154

Olsson, R. H. G.

Selberg A 1957 Nor VSF 30, 71-79
(MR19, 1030; Z78, 3)

Oltramare, Gabriel  1816-1906

Fehr H 1906 Ens M 8, 378-382
   (F37, 32)
- 1906 Rv GSPA 17, 725


Omar al Karabisi   X

Bessel-Hagen E 1931 QSGM (B)1, 502-
   540  (Z2, 325)
Gandz S 1932 QSGM (B)2, 98-105
   (Z5, 4)


Omar Khayyam   XI

   = al-Hajjan

Amir-Moez A R 1959 Scr 24, 275-303
   (MR22, 929; Z90, 5)  [Euclid]
- 1963 Scr 26, 323-337  (MR29, 218)
Arberry Arthur J 1952 Omar Khayyam.
   A New Version Based upon Recent
   Discoveries, London, 159 p
Archibald R C 1953 Pi Mu EJ 1, 350-
   358  (MR15, 276)
Dilgan Hamit 1959 Omar Hayyam,
   der grosse Mathematiker, Istanbul,
   139 p  (Z92, 244)
Erani T 1936 Commentary on Euclid
   by Omar Khayyam,  Teheran
Eves H 1958 MT 51, 285-286
   [cubic eq]
Franklin S P 1961 Scr 25, 353-355
   (Z100, 243)  [cont frac, Eudoxos]
Jacob G + 1912 Islam 3, 42-62
Lamb Harved 1934 Omar Khayyam,
   Garden City, 319 p  (Ger tr 1939)
Ogannisyan V A 1966 Arm Ped FM 3,
   89-98  (MR37, 1217)
Rozenfeld B A 1953 UMN 8(3), 170-171
   (Z50, 1)
Rozenfeld B A + 1953 Ist M Isl 6,
   11-172  (MR16, 986; Z53, 195)
- 1965 Omar Khayyam, Mos, 191 p
   (MR33, 7225)
Ruska J 1929 Arc GMNT 11, 256-264
   (F55, 8)  [zero]
Salat P 1927 Life of Omar Khayyam,
   Paris

Shirozi J K M 1905 Life of Omar
   al-Khayyam, Edinburgh
Struik D J 1958 MT 51, 280-285
Winter H J J + 1950 Beng Asi J 16,
   27-78  (MR13, 809)  [alg]
Wittstein A 1895 Hist Abt 40, 1-6
   (F26, 4)  [Eudoxos]
Woepcke F 1850 Crelle 40, 160-
   [cubic eq]
- 1857 Algebra of Omar Khayyam,
   Paris
Yushkevich A P 1948 Mos IIET 2, 499-
   534  (MR11, 572)  [alg]


Omerique.  See Hugues d'Omerique


Omori, Fusakich  -1923

Palazzo 1924 Linc Rn (5)33, 542-546
   (F50, 16)  [seismology]
Volterra V 1924 Linc Rn (5)33, 43
   (F50, 15)


Onnes, Heike Kamerlingh  1853-1926

F G D 1927 Lon RSP (A)113,  i-vi
   (F53, 33)  [phys]


Oppermann, Ludwig Henrik Ferdinand
   1817-1883

Gram J P 1884 Zeuthen (5)1, 137-144


Ore, Oystein  1899-1968

Aubert K E 1949 Nor M Tid 31, 81-84
   (Z40, 1)


Oresme, Nicole  1323?-1382

Borchert E 1934 Die Lehre von der
   Bewegung bei Nicolaus Oresme,
   Muenster, 128 p  (F62, 1026)
Boyer C 1943 M Mag 18, 81-  [exponents]
Clagett M 1957 Isis 48, 182-183

Clagget M ed 1968 *Nicole Oresme and the Medieval Geometry of Qualities and Motions* (Univ of Wisconsin), 713 p [bib]

Curtze M 1868 *Der Algorithmus proportionum des Nicole Oresme...,* Berlin (F1, 8)

- 1870 *Die Mathematischen Schriften des Nicole Oresme,* Berlin (F2, 6)

Dingler H 1929 Archeion 11, xv-xxiii (F55, 10)

- 1932 Gorres Ja 45, 58-64 (F58, 988)

Droppers G 1957 Isis 48, 351

Duhem P 1909 Rv GSPA 20, 866-873 [Copernicus]

Durand D B 1941 Speculum 16, 167-185 [mid]

Grant E 1957 Isis 48, 351 [propor]

- 1960 Isis 51, 213-314 (MR22, 928; Z97, 241) [propor]

- 1961 Arc HES 1, 420-458 (Z112, 1) [irrat]

- 1966 Isis 56, 327- [propor]

Grant Edward ed 1966 *Nicole Oresme. De proportionibus proportionum and Ad pauca respicientes* (U Wisc P), 466 p

Itard J 1961 Ens Sc 2(13/14), 29-31 (Z118:1, 3)

*Krazer A 1915 *Zur Geschichte der graphischen Darstellung von Funktionen,* Karlsruhe, 31 p

Murdoch John 1964 Scr 27, 67-91 [Euclid]

Pederson Olaf 1956 *Nicole Oresme und sein naturwissenschaftlichen System,* Copenhagen, 290 p (Z73, 3)

Suter H 1882 Hist Abt 27, 121-125

Wieleitner H 1912 Bib M (3), 115-145 (F43, 73)

- 1917 Nat Kult 14, 529-536 (F46, 47) [anal geom]

- 1925 Isis 7, 486-489

Zubov V P 1958 Ist M Isl 11, 601-719 (MR25, 221; Z96, 1)

- 1958 Ist M Isl 11, 720-731 (MR25, 221; Z96, 1)

- 1959 Arc In HS 11, 377-378 (Z96, 2)

- 1959 Isis 50, 130-134

- 1960 Ist As (6), 301-400 (Z118-1, 248) [ast, irrat]


Orlando, Luciano 1877-1915

Marcolongo R 1916 Loria 18, 1-10
Teixeira F 1916 Porto Ac 11, 125-126


Orr, William McFadden 1866-1934

Conway A W 1935 Lon RS Ob 1, 559-562


Ortega, Juan de

Albarrán F 1936 Rv M Hisp A (2)11, 139-142 (F64, 911) [approx]
Barinaga J 1932 Rv M Hisp A (2)7, 194-207 (F58, 995)
- 1932 Rv M Hisp A (2)7, 244-245 (F58, 995)
Vera F 1932 Archeion 14, 554 [irrat]


Ortu-Carboni, S.

Anon 1940 It Attuar 11, 138-139 (F66, 22)


Osgood, William Fogg 1864-1943

*Archibald R C 1938 *A Semicentennial History of the American Mathematical Society...,* NY, 153-158 [bib, port]
Birkhoff G D 1943 Sci Mo 57, 466-469 (MR5, 58; Z60, 17)
Coolidge J L 1943 Sci 98, 399-400 (Z60, 17)
Koopman B O 1944 AMSB 50, 139-142 (Z60, 17)


Osipovskii, Timofei Fedorovich 1765-1832

Bachmutskaya E Ya 1952 Ist M Isl 5, 28-74 (Z49, 4)

Prudnikov V E 1952 Ist M Isl 5, 75-83 (Z49, 4)


Ostrogradsky, Mikhail Vasilevich
1801-1861

Anon 1902 *Feier veranstaltet von der Physiko-math. Gesellschaft...,* Poltawa, 138 p (F33, 24)
- 1951 Ukr IM 3, 235-239 (MR14, 833; Z45, 147)
- 1952 UMN 7(1), 203-205 (Z46, 3)
Antropova V I 1955 Mos IIET 5, 304-320 (Z68, 6)
- 1956 Ist M Isl 16, 97-126 (MR33, 2507 d) [thermodyn]
- 1957 Mos IIET 17, 229-289 (Z117:2, 247) [divergence thms]
- 1957 Vop IET 3
Chilkevich E K 1957 Vop IET (5), 162-164 (Z80, 4) [geom]
Depman I Ya 1951 Ist M Isl 4, 160-170 (Z44, 246)
Geronimus J L 1954 *M. W. Ostrogradski. Zum Prinzip der kleinsten Wirkung,* Berlin, 112 p (Z59-1, 11)
Gnedenko B V 1951 Ist M Isl 4, 99-123 (MR14, 525) [prob]
- 1951 UMN 6(5), 3-25 (MR13, 420; Z43, 245)
- 1952 *M. V. Ostrogradsky,* Moscow, 332 p (MR15, 277; Z49, 290)
Gnedenko B V + 1963 *Mikhail Vasilevich Ostrogradskii,* Moscow
Grigoryan A T 1959 Mos IIET 28, 250-258 (Z95, 4) [mech]
- 1960 *Sonjet Beitraege zur Geschichte der Naturwissenschaften,* Berlin, 192-200 (Z99, 246) [mech]
Kagan B T 1901 Kagan (305), 97-101 (F32, 8)
Kropotov A I + 1961 *M. V. Ostrogradskii i ego pedagogicheskoe nasledie,* Mos
Lakhtin L C 1901 Mos MO Sb 22, 540-554 [anal]
Lewicky W 1927 Lemb Si 5, 4-5 (F57, 1366)
Makarova V I 1954 Mos IIET 1, 317-319 (Z59-1, 11) [port]

Maron I A 1950 Ist M Isl 3, 197-340 (Z41, 241)
- 1951 Ist M Isl 4, 124-159 (Z44, 246)
Pogrebysskii I B + ed 1961 *Pedagogicheskoe nasledie. Dobumenty o zhizni i deyatelnosti,* Mos, 399 p
Polak L S 1960 Int Con HS 9, 564-565 (Z119, 8) [anal mech]
Prudnikov V E 1953 Ist M Isl 6, 223-237
- 1954 Ist M Isl 7, 716-719 (Z60, 11)
Putyata T V + 1951 *Mikhailo Vasilovich Ostrogradskii,* Kiev
Rabinovich Vu L 1951 UMN 6(5), 26-32 (MR13, 420; Z44, 246) [intgl]
Remez E Ya 1951 Ist M Isl 4, 9-98 (MR14, 525; Z45, 292)
- 1951 UMN 6(5), 33-42 [cont frac, irrat]
Rybkin G F 1952 UMN 7(2), 123-144 (Z48, 243) [Osipovsky, phil]
Sabinin E T + 1902 Mos MO Sb 22, 499-531, 532-539, 540-554, 555-573 (F33, 25)
Shtokalo I Z 1952 Ukr IM 4, 3-24 (Z48, 243) [phys]
Steklov V A 1952 UMN 7(1), 203-205
- 1953 UMN 8(1), 102-103 (Z51, 4)
Sveshnikov A G 1953 UMN (NS)8(1), 101-102 (MR14, 833; Z51, 4)
Tripolskii P I 1902 *Mukhail Vasilevich Ostrogradskii,* Poltava
Tyulina I A + 1963 in *Ocherki istorii matematiki i mekhaniki,* Mos, 125-146
Urazbaev B 1951 Kazk Ak (9), 52-58
Vasiliev A Y 1901 Kazn OEFM (2)11(4), 2-10 (F32, 8)
Verebrusov A 1901 Kagan 25, 97
Yushkevich A P + 1963 *Mikhail Vasilevich Ostrogradskii,* Mos
- 1965 Ist M Isl 16, 11-48
- 1965 Vop IET 18, 103-107 [divergence]
Zhukovsky N E 1901 Mos MO Sb 22, 532-539
- 1901 Mos MO Sb 22, 555-573 [mech]

Ostwald, Wilhelm    1853-1932

Baur E 1932 Naturw 20, 321   [chem]
Findlay A 1932 Nat 129, 750-751
Kistyakovskii V 1934 SSSR Iz (7),
   431-442   (F60, 15)
Koerber H G ed 1961 *Aus dem Wissen-
   schaftlichen Briefwechsel
   Wilhelm Ostwalds,*  Berlin, 167 p
   (MR24A, 222)
Le Blanc M 1932 Fors Fort 8, 174-
   175
Luther R 1933 Leip Ber 85, 57-71
Michaelis L 1932 Sci Mo 34, 567-570
Rodnyi N I + 1969 *Vilgelm Ostvald,*
   Mos, 275 p
W V 1932 Z Ph C Unt 45, 169-171


Otter, Christian   1598-1660

Peters T 1933 QSGM (B)2,   352-363
   (F59, 25)   [military]
Reidemeister K 1933 Konig Ph 10(5),
   155-182   (F59, 24)   [mech]


Oughtred, William   1575-1660

Anon 1949 M Gaz 33, 161
- 1960 Discov 21, 280
Bosmans H 1911 Brx SS 35, 24-78
   (F42, 56)   [Descartes]
Cajori F 1912 Calif Pu 1, 171-186
   [notat]
- 1915 Monist 25, 441-465   (F45, 12)
- 1915 Monist 25, 495-530   [educ]
- 1916 *William Oughtred...,*
   Chicago-London, 106 p   (F46, 34)
- 1920 Calif Pu 1(8), 171-186
   [notation]
Glaisher J 1915 QJPAM 46, 125-198
   [notation]
Reeve W D 1931 MT 24, 457-458
Sanford V 1932 MT 24, 457-458


Ovidio, Enrico d'   1843-1933

Anon 1918 *Scritti matematici offerti
   ad Enrico d'Ovidio,* Torino, 401 p
   (F46, 30)

- 1957 Archim 9, 218-220   (Z77, 242)
   [biog]
Battaglini G 1893 Tor FMN 29, 458-460
Brusotti L 1933 Lom Gen (2)66, 6-10,
   338-340
Fano G 1933 It UM 12(3), 153-156
Loria G 1933 Linc Rn (6)17, 996-
   1009
Scorza G 1933 Nap FM Ri (4)3, 1-8,
   93-100   (F59, 32)
Somigliana C 1934 Tor FMN 69, 119-138
   (F60, 835)


Ozanam, Jacques   1640-1717

Graf J H 1899 ZM Ph 44(S), 115
   [geom, Le Clerc, plagiarism]
Le Paige C 1890 Mathesis 10, 34-36
   [Snell]
Schaaf W L 1957 MT 50, 385-388
Tannery P 1900 An In Hist (5), 297-
   310   [mss]


Pachymeres.   See George.


Pacinotti, Antonio   1841-1912

Boccara V 1913 Pe MI (3)10, 188-190
   (F44, 24)
Grassi G 1913 Tor Mm (2)63, 205-211
   (F44, 24)


Pacioli, Luca   XV-XVI

Agostini Amodeo 1924 *Il "de viribus
   quantitatis" di Luca Pacioli,*
   Bologna   28 p
- 1924 Pe M (4)4, 165-192
- 1925 Arc Sto 6, 115-120   (F51, 18)
   [plagiarism]
Baldi B 1879 Boncomp 12, 352-439
   (F11, 10)   [John of Saxony,
   Jean de Linières]
Cajori F 1924 AMSB 30, 387
   (F50, 19)
- 1924 Arc Sto 5, 125-130
Kheil C P 1896 *Ueber einige aeltere*

*Bearbeitungen des Buchhaltungstract-
ates von Luca Pacioli,* Prague,
136 p   (F27, 6)
Mancini G 1913 Linc Mor (5)14
Masotti Biggiogero G 1960 Lom Gen
(A)94, 3-30   (MR23A, 270; Z99, 3)
North J 1965 Physis Fi 7, 211-214
[Apian, polyhedra]
Pittarelli G 1909 Int Con 4(3), 436-
440   [P. della Francesca,
plagiarism]
Ricci D I 1940 *Fra Luca Pacioli
L'uomo e lo scienziato,* Sansepalcro,
54 p   (F66, 12; Z25, 1)
Speziali P 1953 Bib Hum, 295-305
[Leonardo da Vinci]
- 1953 Genv Mus (May), 1
- 1953 Stultif 10, 84-90
Staigmueller H 1889 Hist Abt 34,
81-102, 121-129
Taylor R E 1942 *No Royal Road.
Luca Pacioli and his Times,*
Chapel Hill
Vianello V 1896 *Luca Pacioli nella
storia della ragioneria: con
documenti inediti,* Messina, 174 p
(F27, 6)

Padeletti, Dino   1852-1895

Pezzo P del 1895 Nap Pont 25(4)
(F26, 24)
Pinto L 1892 Nap FM Ri (2)6, 49-50
(F24, 35)
Torelli G 1892 Paler R 6, 68-72
(F24, 35)   [bib]

Padova, Ernesto   1845-1896

Ricci G 1897 *Commemorazione
del Prof. Ernesto Padova...,*
Padua, 41 p   (F28, 24)

Padula, Fortunato   1815-1881

Amodeo F 1921 Nap Pont 51, 81-118
[Flauti]
Rubini R 1881 FM Ri 20, 181-198
(F13, 24)

Painlevé, Paul   1863-1933

Anon 1933 Ens M 32, 92
- 1933 Rv Sc 71, 669-670
Basch V + 1933 Cah Dr Hom 33, 651-655
Borel E + 1933 Par CR 197, 953-958
((F59, 40)   [diff eqs, mech]
Chazy J 1934 Math Cluj 8, 201-204
(F60, 15)
Greenwood T 1933 Nat 132, 738-740
(F59, 40)
*Hadamard J 1934 Rv Met Mor 41, 289-
325   (F60, 837)
Hesse Germaine A 1933 *Painlevé, grand
savant, grand citoyen,* Paris,
246 p
Orts J M 1934 Rv M Hisp A (2)9, 34-40
(F60, 15)
Picard E 1933 Par CR 197, 955-958
(F59, 40)
P L 1933 Cah Ratnl (26), 230-260
Titeica G 1933 Gaz M 39, 161-162
Vivanti G 1933 Lom Gen (2)66, 16-20,
1051-1052   (F59, 855)
Whittaker E T 1935 LMSJ 10, 70-75
(F61, 26)

Painvin, Louis Felix   1826-1875

Anon 1875 BSM 9, 188-191   (F7, 11)
[bib]

Palama, Giuseppe   1898-1959

Petraria V 1960 It UM (3)15, 468-469
(Z93, 9)

Palatini, Francesco   1865-

Artom E 1929 Bo Fir (2)8, 38-40
(F56, 818)

Paley, Raymond Edward Alan Christopher
1907-1933

Hardy G H 1934 LMSJ 9, 76-80
(F60, 16)

Wiener N 1933 AMSB 39, 476  (F59, 43)

**Palmquist,** Reinhold

Hagstroem K-G 1940 Sk Akt 23, 62-65
   (F66, 21; Z23, 196)

**Palmstrøm,** Arnfinn  1867-1922

Meidell B 1923 Nor M Tid 5, 1-7
   (F49, 13)
- 1924 Chr Avh    (F51, 31)

**Panek,** Augustin   1843-

Petr K 1912 Cas MF 41, 1-8
   (F43, 33)

**Paola,** Gabrio

Vacca G 1903 Loria 6, 1-4

**Paoli,** Pietro  1759-1839

Agostini A 1938 Lib Civ 11, 3-7
   (F64, 919)
Catalan E + 1894 Intermed 1, 247
Riccardi P 1898 Mod Mm (3)1, 105-129
   (F29, 7)  [Lacroix, Lagrange,
   Laplace, letters, Ruffini]

**Paolis,** Riccardo de  1854-1892

Segre C 1892 Paler R 6, 208-224
   (F24, 34)

**Paolo dell'Abbaco**   XIV?

  = P. d. Abaco    P. Dagomari

Frizzo G 1883 *Le regoluzze di*
   *mastro Paolo dell'Abbaco, mate-*
   *matica del secolo XIV,* Venice
Narducci E 1882 Linc At (3)4, 324-326

- 1882 Boncomp 15, 111-135

**Papaioannou,** G I   XX

Papaioannou K 1932 Gree SM 13, 48
   (F57, 1309)

**Pappos of Alexandria**   III

  = Pappus   = Pappo

Bergstraesser G 1933 Islam 21, 195-
   222  [Euclid]
Candido G 1935 Pe M (4)15, 58-62
   (Z10, 244)  [Chasles, Fagnano,
   generalization, Stewart]
- 1935 Pe M (4)15, 286-289
   (F61, 941)
Cantor M 1876 Hist Abt 21, 37-42
   (F8, 7)
Chasles M ed 1860 *Les trois livres*
   *de porismes d'Euclide retablis pour*
   *la première fois d'opres la notice*
   *et les lemmes de Pappus...,*  Paris
Conte L 1954 Pe M (4)32, 125-141
   (Z57, 2)  [reg polyh]
Czwalina A 1927 Mit GMNT 26, 233-234
   [mistakes]
Eves H 1958 MT 51, 544-546
   [Pyth thm]
Gerhardt C J 1871 *Der Sammlung des*
   *Pappus von Alexandrien Siebentes*
   *und achtes Buch,* Halle  (F4, 3)
- 1874 *Die Sammlung des Pappus von*
   *Alexandrien,* Eisleben  (F8, 7)
Guenther S 1888 Bib M (2)2, 81-87
   [Kepler]
Heiberg J L 1878 Hist Abt 23, 117-128
   (F10, 3)
Hultsch F 1879 Boncomp 12, 333-344
   (F1879, 5)
Junge G 1934 QSGM (B)3, 1-17
   (Z10, 244)  [Euclid]
MacKay J S 1888 Edi MSP 6, 48-58
   [seqs]
Maroger A 1925 *Le probleme de Pappus*
   *et ses cent premières solutions,*
   Paris, 392 p
Milne J J 1911 *An Elementary Treatise*

on *Cross-ration Geometry, with Historical Notes,* Cambridge, 146-149 [locus]
- 1928 M Gaz 14, 413-414 [Descartes]
Mogenet J 1951 Bel BL 37, 16-23 [div]
Occella F 1897 *Intorno ad un probleme di Pappo,* Casale-Monferrato, 30 p (F28, 43)
Pelseneer J 1933 Mathesis 47, 328-333 (F59, 838)
Rigaud S P 1822 Edi PJ 7, 56-, 219- [mss]
Rome A 1927 Brx SS(A), 46-48
- 1934 Isis 20, 440
Saltykow M N 1939 Beo M Ph (5), 71-74 (Z22, 3) [Descartes]
Scardapane N M 1935 Pe M (4)15, 116-122 (F61, 7)
Tannery P 1880 Bord Mm (2)3, 351-378 (F12, 28) [numb thy]
Taylor C 1881 Mess M 10, 112- [thm]
Tibiletti C 1946 Pe M (4)24, 100-111 (Z60, 6) [Apollonius]
Ver Eecke P 1933 Scientia 54, 114-121 (F59, 18; Z7, 147) [mech]
- 1933 *Pappus d'Alexandre. La Collection Mathématique,* Paris-Bruges, 2 vols [hist intro 118 p, notes]
Weaver J H 1915 AMSB (2)21, 492-493 (F45, 62)
- 1916 AMSB 23, 127-135 (F46, 8) [plagiarism]
Zacharias M 1927 Mit GMNT 27, 1-2 (F54, 46) [mistakes]


Pardies, Ignace Gaston 1636-1673

Wieleitner H 1910 Arc GNT 1, 436-442 (F41, 64) [geom]


Parent, Antoine 1666-1726

Sergescu P 1938 Sphinx Br 8, 196-197 (F64, 17) [mech]


Pareto, Vilfredo 1848-1923

Keyser C J 1936 Scr 4, 5-23


Parmenides of Elea -VI

Taylor A 1916 Aristot 16, 234-289 [Socrates, Zeno]


Parseval-Deschenes, Marc Antoine -1836

Brocard H 1908 Intermed 15, 55 (Q971) (See also Intermed 4, 5; 13, 162, 263)
Eneström G 1892 Bib M (2)6, 64


Partenie, Iacob Petru -1790

Marian V 1958 Cluj MF 8, 291-301 (Z85, 5)


Pascal, Alberto 1894-1918

Marcolongo R 1918 Batt 56, 42-46 (F46, 20)
Pascal C 1918 Loria (2)20, 97-99 (F46, 20)


Pascal, Blaise 1623-1662

Anon 1962 *Pascal présent, 1662-1962,* Clermont-Ferrand, 292 p
Arov R 1926 Rv Met Mor 33, 85-91
Bacon H M 1938 MT30, 180-185
Bernard 1837 Par CR 65, 203-
Bertrand J 1912 *Blaise Pascal,* Paris
Bianco 1886 Tor FMN 21, 686-697 [hexagram]
*Bishop Morris 1936 *Pascal: The Life of genius,* NY
- 1938 *Pascal,* Berlin, 536 p (F64, 916)
Boberil R du 1902 *Pascal et Riemann,* Paris, 14 p (F33, 16)

Pascal, Blaise (continued)

Borel E 1909 Rv Mois 7, 98-100
[barometer]
Bosmans H 1922 Brx SS 42, 337-345
(F49, 6) [four dim space]
- 1923 Arc Sto 4, 369-379 (F49, 6)
[infinitesimal]
- 1923 Mathesis 37, 455-464
[Pascal tria]
- 1924 Mathesis 38, 250-256 [cal]
- 1924 Arc Ph Par, 21 p
- 1924 Rv Q Sc 5, 130-161, 424-451
Boyer C B 1943 Scr 9, 237-244
(Z60, 9) [summation]
- 1963 Scr 26, 283-307 (MR29, 638)
Brocard H 1906 Intermed 13, 31
[bib]
Brunschvicg Leon 1923 Rv Met Mor 30,
165-180
- 1924 Rv Phi Fr E, 1-29
- 1926 Le Génie de Pascal, Paris,
205 p
- 1932 Pascal, Paris, 86 p
Cantor M 1873 Preus Jb 32, 212-237
(F5, 23)
Carevyge 1914 Intermed 21, 85
(Q2959) [bib]
Chapman S 1942 Nat 150, 427, 508-509
(MR4, 63) [calc mach]
Coumet E 1965 Arc In HS 18, 245-272
[prob]
Delegue 1869 Dunkerq 14, 137-
[diff geom]
Delorme A 1960 Rv Syn 81, 353-355
- 1963 Rv Syn 84, 511-520
Desboves A 1878 Etude sur Pascal et
les géomètres contemporains suivie
de diverses notes scientifiques
et litéraires, Paris (F10, 11)
Dijksterhuis E J 1950 Euc Gron 25,
265-270 (Z37, 1) [Descartes]
- 1954 Amst L Med 14, 339-374
Dreydorff J G 1870 Pascal, sein
Leben und seine Kampfe, Leipzig
(F2, 12)
Duclaux Mary 1927 Portrait of Pascal,
London, 232 p
Duhem P 1905 Rv GSPA 16, 599
Eastwood D M 1936 The Revival of
Pascal, Oxford

Eneström G 1904 Bib M (3)5, 72-73
(F35, 58) [binom thm]
Faugere 1867-1868 Par CR 65, 202-,
340-, 455-, 643-, 702; 67, 497-
[mss]
Fonsny J 1952 Et Class 20, 181-191
[calc mach]
Fontes 1893 Tou Mm (9)5, 459-475
(F26, 52) [div numb thy]
Freudenthal H 1953 Arc In HS (22),
17-37 [induc]
Gennaro S di 1932 Pe M (4)12, 104-112
(F58, 35) [conic]
Gerhardt K J 1892 Ber Si, 183-204
[conic, Desargues]
- 1918 Monist 28, 530-566 [Leibniz]
Giraud Victor 1923 La vie héroique
de Blaise Pascal, Paris, 257 p
Gradstein S 1962-63 Philip TR 24,
101-105 [calc mach]
Guitton J 1949 Int Con Ph 10,
1145-1147 [Leibniz]
*- 1951 Pascal et Leibniz, Paris
Hankel H 1869 ZM Ph 14, 165-173
(F2, 13)
Hartog Philip 1927 Blaise Pascal,
Lahore
Hatzfeld A 1901 Pascal, Paris
Heilbronner P 1931 Par CR 192, 998-
1000 (F57, 26) [cycloid]
Hoecken K 1914 Ber MG 13, 8-29
(F45, 73) [calc mach]
Hofmann J E 1962 Praxis 4, 309-414
(Z107, 3)
Humbert Pierre 1944 Ciel Ter 60,
122-25 [ast]
- 1947 Cet effrayant génie...
l'oeuvre scientifique de Blaise
Pascal, Paris, 264 p
Ignacius G I 1961 Ist M Isl 14, 611-
622 (Z118-1, 4) [conics,
Leibniz, Périer]
Itard J 1940 Ens M 38, 27-38
(MR1, 290; Z23, 388) [geom]
Jasinski R 1933 Rv Hi Phi 1, 134-154
[infinity]
Josey E 1932 Les antécédents de
l'infiniment petit dans Pascal,
Paris, 229 p
Keyser C J 1938 Scr 5, 83-94
(F64, 16)

Pascal, Blaise  (continued)

Klinckowstroem Carl v 1922 Ges Bl
  TI 8, 16-22   [calc mach]
Lindeloef L L 1868 Hels Ofv 10, 17-
  [letters]
Lloyd A H 1925 Sci Mo 20, 139-152
Maire Albert 1909 Fr AAS, 71-79
  [bib]
- 1912-1921 L'oeuvre scientifique
  de Blaise Pascal. Bibliographie
  critique et analyse de tous les
  travaux qui s'y rapportant,
  Paris, 5 vols  (F43, 12)
- 1912 Fr AAS 40, 37  (F43, 12)
  [bib]
- 1925-1927 Bibliographie générale
  des oeuvres de Blaise Pascal,
  5 vols, Paris
Maupin G + 1902 Intermed 9, 323
  [geom]
M E P 1942 Nat 150, 527
Mesnard Jean 1951 Pascal, l'homme et
  l'oeuvre,  Paris, 192 p
- 1953 Sarav PL 2, 3-30
- 1963 Rv Hi Sc Ap 16, 1-10
  (Z117:2, 246)
Milne J J 1924 M Gaz 12, 37-49, 53-
  56   (F51, 20)
Molinéry 1932 Par Med 84, 159-160
  [illness]
*Montel P 1951 Pascal Mathématicien,
  Paris
Ocagne M d' 1930 Par CR 190, 1163-
  1164  [calc mach]
Ore O 1959 Colo Stu (3), 11-24
  [prob]
- 1960 AMM 67, 409-423  (MR22, 1129;
  Z92, 245)  [prob]
Paven J 1963 Rv Hi Sc Ap 16, 161-178
Perier 1876 Vie de Pascal écrite
  par Mme. Perier sa soeur,  Paris
Perrier L 1901 Rv GSPA 12, 482-490
  [calc, prob]
Picard E 1923 BSM (2)47, 257-267
  (F49, 6)
- 1923 Rv Fr 4, 272-283
- 1924 Pascal mathématicien, Paris,
  23 p   (F50, 18)
Roland P 1937 Rv Sc 75, 148   [calc]
Sanford V 1932  MT 25, 229-231

Scholtz H 1945 in Festschrift zum
  60 Geburtstag  von Prof. Dr. A.
  Speiser, 19-33  (MR7, 354; Z60, 9)
Sergescu Pierre 1950 Pascal et la
  science de son temps, Paris, 18 p
Servien P 1930 Rv Cou Con 32, 283-288
Shaw J B 1941 Scr 8, 69-77
  [hexagram]
Soltau R H 1927 Pascal: the Man and
  his Message,  London, 230 p
Stewart H F 1941 The Secret of Pascal
  (Cambridge U P), 118 p
- 1942 Brt Ac P 28, 197-215
Stuyvaert M 1907 Bib M (3)8, 170-172
  (F38, 61)
Suter R 1946 Sci Mo 62, 423-428
Takekuma R 1953 JHS (26)
  [Fermat, prob]
Tannery P 1889 Bord Mm (3)5, 55-84
  [Lalouvere]
- 1894 Bord Mm (4)4, 251-259
  (F25, 20)   [Lalouvere]
Taton René 1955 Rv Hi Sc Ap 8, 1-18
  (MR16, 986)  [conics]
*Taton R + 1962 Rv Hi Sc Ap 15 (3-4),
  190-392  (Special issue)
  (MR33, 1202; Z106, 2)
  [Chronology, port]
- 1963 Rv Hi Sc Ap 16(1), 1-83
- 1964 L'oeuvre scientifique de
  Pascal, Paris, 319 p
Teske A 1963 Kwar HNT 8, 3-21
Van Dantzig D 1950 Euc Gron 25,
  203-232 (MR11, 707; Z37, 1)
  [soc sci]
Vekerdi C 1963 Magy MF 13, 269-285
  (MR29, 638)
Yushkevich A P 1959 Vop I ET 7, 75-85
  (MR22, 618)


Pascal, Ernesto   1865-

Berzolari L 1940 Lom Gen (3)4, 162-
  170  (F66, 1188)
Colucci A 1939 Batt (3)76, vii-xxvii
  (F66, 21; Z25, 3)

Pasch, Moritz   1843-1930

Dehn M 1928 Naturw 16, 813-815
  (F54, 43)
Engel F + 1931 *Moritz Pasch,*
  Giessen, 27 p   (F57, 38)
- 1934 DMV 44, 120-142   (F60, 835)
Fritzsche R A 1931 Gies Hoch 8(2),
  22-26   (F57, 1307)
Vaerting M 1928 ZMN Unt 59, 461-463
  (F54, 47)


Pastor.   See Rey Pastor.


Paul of Middleburg   1445-1533

Struik D J 1925 Linc Rn (6&)2(1),
  305-307   [decimals]


Peacock, George   1791-1858

Clark J W 1895 DNB 44, 138-140
*Macfarlane Alexander 1916 *Lectures
  on Ten British Mathematicians of
  the Nineteenth Century,* NY, 7-18


Peano, Giuseppe   1858-1932

Anon 1928 *Collectio descripto in
  honore di prof. G. Peano...,*
  Milan, 96 p   (F54, 43)
- 1932 It UM 11, 187
- 1928 Sch Vit (S)
- 1932 Sch Vit 7(3)
*Ascoli Guido 1958 Archim 10, 263-266
  (Z82, 243)
 Boggio T 1933 It SPS 21(2), 100-102
  (F59, 853)   [geom]
- 1933 Tor FMN 68, 436-446
  (F59, 852)
 Burali-Forti C 1894 Tor FMN 30, 129-
  145
*Cassina U 1932 Sch Vit 7, 117-148
  (F58, 995)   [bib]
- 1933 It UM 12, 57-65   (F59, 853)
  [logic]
- 1933 Milan Sem 7, 323-389
  (F59, 582)

- 1933 Rv Met Mor 40, 481-491
  (F59, 34)
- 1950 Parma Rv M 1, 275-292
  [curve]
- 1952 Lom Rn M (3)16, 337-362
  (MR15, 276; Z49, 4)
*- 1955 It UM (3)10, 244-265
  (MR17, 3; Z66, 245)
- 1955 It UM (3)10, 544-574
  (MR17, 698; Z66, 245)
- 1957 It UM (3)12, 310-312
  (Z77, 9)
*- 1961 *Dalla geometria egiziana alla
  matematica moderna,* Rome, Ch 16-17
 Catalan E 1894 Rv M 4, 104-105
  [Catalan]
 Couturat L 1899 Rv Met Mor 7(4),
  616-646   [logic]
 Dickstein S 1934 Wiad M 36, 65-70
  (F60, 836)
- 1959 It UM (3)14, 109-118
  (Z86, 4)   [logic]
 Gliozzi M 1932 Archeion 14, 254-255
  (F58, 46)
 Hao Wang 1962 *A Survey of Mathemati-
  cal Logic,* Peking, 72-79
  Dedekind priority   [Peano axioms]
 Jourdain P E B 1912 QJPAM 43, 270-
  314
 Kennedy H C 1960 Pi Mu EJ 3, 107-113
*- 1963 Phi Sc 30(3), 262-266
  [phil]
- 1968 MT 61, 703-706
 Kozlowski W M 1934 Wiad M 86, 57-64
  (F60, 836)
 Levi B 1932 It UM 11, 253-262
  (F58, 46)
- 1933 It UM 12, 65-68   (F59, 853)
 Medvedev F 1965 Ist M Isl 16, 311-323
  (MR33, 5428)   [set fun]
 Natucci A 1932 Bo Fir (2)11, 52-56
  (F58, 46)
 Padoa A 1898 Rv M 6, 90
- 1899 Rv M 6, 105   [logic]
- 1933 Pe M (4)13, 15-22
  (F59, 853)   [logic]
- 1936 Act Sc Ind (395), 31-37
  (F62, 1038)   [logic]
- 1936 Int Con US 8, 31-38   [logic]
- 1939 It SPS 21(2), 99   (F59, 853)
  [logic]

Stamm E 1934 Wiad M 36, 1-56
(F60, 836)
Terracini A ed 1955 *In memoria di
Giuseppe Peano. Studi raccolti,*
Cuneo, 114 p (MR17, 338; Z64,
243; 66, 6)
Vacca G 1933 It SPS 21(2), 97-99
(F59, 853)
Vivanti G 1932 Lom Gen (2)65, 11-15,
497-498 (F58, 995)
Whitehead A N 1901 AJM 23, 139-140,
297 [bib]


Pearson, Karl    1857-1936

Archibald R C 1948 *Mathematical
Table Makers,* NY, 64-67
(Scr Stu 3) [bib, port]
Camp B H 1933 ASAJ 28, 395-461
(F59, 855) [stat]
G H T 1936 Edi SP (A)56, 274-275
(F62, 1041)
Haldane J B S 1957 Biomtka 44, 303-
313 (MR19, 1150; Z78, 4)
Morant G M 1939 *A Bibliography of
the Statistical and other Writings
of Karl Pearson,* Cambridge, 127 p
(F65, 1092)
Pearl R 1936 ASAJ 31, 653-664
(F62, 1041)
Pearson E S 1936 Biomtka 28, 193-
257 (F64, 20) [port]
- 1937 Biomtka 29, 161-248
(F64, 20) [port]
- 1938 *Karl Pearson. An Appreciation
and Some Aspects of his Life and
Works,* Cambridge, 178 p
(repr of 1936, 1938) (F79, 5)
Pearson K 1936 M Gaz 20, 27-36
Stouffer S A 1958 ASAJ 53, 23-27
(MR22, 1)
Thompson G H 1937 Edi SP (A)56, 274-
275
Walker H M 1958 ASAJ 53, 11-22
(MR22, 1; Z79, 5)
Yule G U 1938 Biomtka 30, 198-203
Yule G U + 1936 Lon RS Ob 2, 73-110

Peet, T. E.

Meyerhof M 1935 Archeion 17, 76-77
(F61, 957)


Peirce, Benjamin    1809-1880

Anon 1880 Nation (Oct 14)
- 1881 Am Ac Pr 8, 443-454
- 1913 Am JS 36, 208
Archibald R C 1925 AMM 32, 1-30
[bib]
*- 1925 *Benjamin Peirce,* Chicago-
Oberlin, 35 p (F51, 24)
(Revis of 1925 AMM)
- 1927 AMM 34, 525-527 (F53, 26)
[alg, C S Peirce]
Ginsburg J 1934 Scr 2, 278-282
[alg, letter]
*Hawkes H E 1902 AMJ 24, 88-95
[lin'alg]
*Hill T 1880 Harv Reg (May)
*King Moses ed 1881 *Benjamin Peirce
...a Memorial Collection,*
Cambridge, Mass
Matz F P 1895 AMM 2, 173-179
Newton H A 1881 Am JS (3)22, 167-178
Peterson S R 1955 J H Ideas 16,
89-112 (MR16, 434)


Peirce, Benjamin Osgood    1854-1914

*Archibald R C 1948 *Mathematical
Table Makers,* NY, 67-68
(Scr Stu 3)
Hall E H 1918 Am NASBM 8


Peirce, Charles Santiago Sanders
1839-1914

Anon 1914 Am JS (4)37, 566
- 1916 J Phi Ps SM (Dec 21)
- 1965- Peirce Tr
Archibald R C 1927 AMM 34, 525-527
[B Peirce]
Cohen M R 1923 *Love and Logic,* NY
[anthology]
Eisele C 1951 Scr 17, 236-259
[Fibonacci]

- 1954 APSY, 353-358
- 1956 Int Con HS 8, 1196-1200
*- 1956 Scr 24, 305-324
- 1957 APSP 101, 409-433
  [infinitesimals, Newcomb]
- 1958 Int Ac HS 11, 55-64
  [Galileo, logic, soc of sci]
- 1959 Int Con HS 9 [phil, Poincaré]
- 1959 Scr 24, 305-324  (MR22, 929;
  Z87, 6)
- 1963 APSP 107, 299-  [cartog]
- 1963 Physis Fi 5(2), 120-128
  [induction, infinity, infinitesimal,
  logic, sets]
Goudge Thomas A 1950 *The Thought of*
  *C. S. Peirce,* Toronto
Keyser C J 1935 Scr 3, 11-37
- 1941 Scr Lib 5, 87-112
  (MR3, 98; Z60, 17)
*Lewis C I 1918 *A Survey of Symbolic*
  *Logic,* Berkeley, Ch. 1(7)
Moore E C + ed 1964 *Studies in the*
  *Philosophy of Charles Sanders*
  *Peirce,* Amherst
Nagel E 1933-1936 J Phil, 30; 31; 33
- 1940 Phi Sc 7, 69-80
Royce J + 1916 J Phi Ps SM 13(26)
*Wiener P P 1947 APSP 91(2)
  [Hume, Langley]
Wiener P P + 1950 *Studies in the*
  *Philosophy of Charles Sanders*
  *Peirce,* Cambridge, Mass
Young F H + 1952 *Studies in the*
  *Philosophy of Charles S. Peirce*
  (Harvard U Pr)


Peirce, James Mills      1834-1906

Byerly W E 1925 Am Ac Pr 59, 650-651
  (F51, 25)
Byerly W E + 1906 Harv G Mag (June)
  [port]
Whittemore J K 1906  Sci (July 13)


Peletier, Jacques  1517-1582

Bosmans H 1907 Rv Q Sc (3)11, 117-173
  (F38, 59)  [alg]
Thébault V 1948 M Mag 21, 147-150
  (MR9, 486; Z29, 241)

Thureau M 1935 *Jacques Peletier*
  *mathématician manceau du XVI*
  *siècle,* Lavel, 27 p


Pelicani, Biagio

Amodeo F 1909 Int Con 3, 549-553
  (F40, 62)


Pelisek, Miroslav      1855-1940

Piska R 1956 Mat v Shk 6(6), 363-366
- 1958 Prag Dej 4, 96-101
  (Z112, 243)


Pell, John      1610-1685

Dijksterhuis E J 1932 Euc Gron 8,
  286-296   (F58, 36)  [circle, pi]
Hervey H 1952 Osiris 10, 67-90
  (Z46, 1)  [Cavendish, Descartes,
  Hobbes]


Pena, Fernando

Arregui J 1960 Gac M 12, 238-239
  (Z97, 2)


Pendry,  F. M. A.  1892-1958

V W G 1959 M Gaz 43, 37


Pennachietti, Giovanni    1850-

Daniele E 1917 Cat Atti (5)10(S)


Pepys, Samuel      1633-1703

Anon 1960 M Gaz 44, 299-300 [Wallis]


Pereira, J. R.   XVIII

Quintas Castans V 1954 Calc Au Ci
  3(8), 29-32   (MR16, 435)

Peres, Joseph Jean Camille
1890-1962

Costabel P 1962 Arc In HS 15, 137-
140 (Z100, 247)
Tricomi F G 1963 Tor FMN 97, 750-752
(Z114, 8)


Perez-Cacho, Teofilo

Garcia Rua J 1957 Gac M (1)9, 3-5
(MR19, 1030; Z79, 6)


Perigal, Henry 1801-1898

Anon 1898 LMSP 29, 721, 732ᵢ735
(F29, 19)


Perks, John XVII

Pedersen O 1963 Centau 8, 1-18
[lune]


Perry, John 1850-1920

Anon 1922 Enc Br
- 1926 Lon RSP (A), 111


Persiani, Odoardo 1848-1924

Sanctis P de 1923 Vat NLA 77, 101-102


Persidskii, Konstantin Petrovich
1903-

Zhatykov O A 1953 Kazk Ak (11), 46-
50 (MR17, 698; Z56, 2)
- 1954 UMN 9(1), 151-154


Pervushin, Ivan Mikheevich
1827-1900

Raik A E 1953 Ist M Isl 6, 535-572
(Z53, 339)

Pestalozzi, Johann Heinrich
1746-1827

Fehr H 1927 Ens M 25, 298 (F53, 22)
Sleight E R 1937 M Mag 11, 310-317
(F63, 810) [elem educ]


Peter. See also Petrus


Peter, Rozsa XX

Domolki B + 1965 M Lap 16, 171-184
(MR34, 5531)


Peters, Johann Theodor 1869-1941

Archibald R C 1948 *Mathematical
Table Makers,* NY, 68-71
(Scr Stu 3)


Petersen, Julius 1889-

Heegard P 1911 Ens M 13, 49-59
(F42, 29)


Peterson, Karl Mikhailovich
1828-1881

Depman I Ya 1952 Ist M Isl 5, 134-
164 (MR16, 435; Z49, 5)
Egorov D T 1903 Mos MO Sb 24, 22-29
(F34, 18) [part diff eqs]
Egorov D T + 1903 Tou FS (2)5, 459-
679 (F34, 18)
Mlodzhelovski B K 1903 Mos MO Sb
24, 1-21 (F34, 18) [geom]
Rossinskii S D 1949 UMN 4(5), 3-13
(MR11, 573; Z33, 4)
- 1952 Ist M Isl 5, 113-133
(MR16, 435) [surf]
Staeckel P 1901 Bib M (3)2, 122-132
(F32, 10)


Petiscus. See Pitiscus.

Petr, Karel    1868

Fusl F + 1928 Cas MF 57, 169-182
    (F54, 44)
Korinek V 1938 Cas MF 67, D245-
    D253    (F64, 919)
- 1948 Cas MF 73, D9-D18
    (Z33, 242)
Koutsky K 1950 Cas MF 75, 341-345
    (MR13, 2; Z40, 2)
Schwarz S 1960 Prag Pok 5, 598-603
    (Z98, 10)


Petrie, Flinders    1853-1942

Kendall D G 1963 II St B 40,
    657-681    (MR30, 5419)    [stat]


Petrovic, Mihail    1868-1943

Anon 1938 Beo U (6/7), xii-xxx
    (F64, 23)    [bib]
Mitrinovic D 1955 Nauk Pri 8, 271-
    284    (MR17, 932; Z68, 7)
Saltykow N 1939 L'oeuvre scienti-
    fique  du Professeur Dr. M.
    Petrovitch, Belgrade, 8 p


Petrovich, Michel   1894-1921

Milankovitch M 1922 Notice sur les
    travaux scientifiques de Michel
    Petrovich, Paris, 161 p
    (F48, 25)


Petrovskii, Ivan Georgevich   1901-

Aleksandrov P S + 1961 Rus MS 16(3),
    131-149    (Z98, 10)
Galpern S A + 1961 Mos UM Me (1),
    3-8  (MR23A, 272; Z96, 4)
Kolmogorov A N 1951 UMN 6(3), 160-164
    (MR13, 198; Z42, 243)
Landis E M + 1957 I. G. Petrovskii,
    Mos, 44 p   (Z80, 4)
Sobolev S L 1951 SSSRM 15, 201-204
    (MR13, 2; Z42, 4)

Teodorescu N 1954 Ro Sov (3)7(2),
    9-31   (MR16, 207)


Petrus Peregrinus    XIII

    = P. P. de Mahorn  = Maricurtia

*Bertelli T 1868 Boncomp 1
Fontes 1897 Tou Mm (9)9, 382-386
Sarton G 1931 Int His Sc 2(2),
    1030-1032   [bib]
*Schlund E 1911 Arc Fr Hi 4, 436-455,
    633-643
*Thompson S 1907 Petrus Peregrinus de
    Maricourt and his Epostola de
    Magnete, London, 32 p


Petrus Philomeni    XIII

    = Peter of Dacia
    = Peter Ingvarssoen

Curtze M 1897 Petri Philomeni de
    Dacia in Algorismum vulgarem
    Johannis de Sacrobosco
    commentarius, una cum Algorismo
    ipso,  Copenhagen
Eneström G 1885 Sv Ofv 42(3),
    15-27; (8), 65-70
Ionesco I 1909 Intermed 16, 46-48
    (Q3434)
Sarton G 1931 Int His Sc 2(2), 996-
    997   [bib]


Petzval, Joseph    1807-1891

Anon 1891 Mo M Ph 2, 479-480
    (F23, 31)
Gegenbauer L 1903 DMV 12, 324-344
    (F34, 19)


Peuerbach, Georg von 1423-1461

    = Purbach  = Peurbach

Alasia C + 1903 Intermed 10, 321, 322

Czerny A 1888 *Aus dem Briefwechsel des grossen Astronomen G. v. Peuerbach,* Vienna, 24 p (F21, 1254)
Gassendi P 1655 *Life of George Purbach,* Hague

**Peverone,** Giovan Francesco

Bianco O Z 1882 Tor FMN 17, 320-324

**Pezzo,** Pasquale Duca del    XIX

Gallucci G 1938 Nap FM Ri (4)9, 162-167   (F64, 919)

**Pfaff,** Johann Friedrich  1765-1825

Anon 1835 *Neuer nekrolog der Deutschen* 13, 575-578  [*bib]
Dunnington G W 1937 M Mag 11, 263-267  (F63, 809)  [port]
Pfaff D C ed 1853 *Sammlung von Breifen gewechselt zwischen J. F. Pfaff und...,* Leipzig

**Phillips,** Andrew Wheeler    1844-1915

Burgess H T 1916 AMM 23, 165-166 (F46, 25)

**Philomeni.**  See Petrus.

**Philon Judaeus**   -I to I

  = Philon of Alexandria

Eves H 1959 Scr 24, 141-148
Robbins F E 1931 Clas Phil 26, 345-361   (F57, 1292)
Sarton G 1927 Int His Sc 1, 236-237
Staehle K 1931 *Die Zahlenmystik bei Philon von Alexandrien,* Leipzig, 98 p   (F57, 15)

**Philon of Byzantium**    -II

Carra de Vaux B 1900 Bib M (3)1, 28 [Arab, mach]
Sarton G 1927 Int His Sc 1, 195-196

**Philoponus,** Johannes    VI

  = John the Grammarian

Haas A E 1906 Bib M 6, 337-342
Sarton G 1927 Int His Sc 1, 421
Steinschneider M 1869 Pet Mm 13, 152-176
Tannery P 1888 Rv Phi LHA 12, 60-73 [astrolabe]

**Phragmen,** L. Edvard  1863-1937

Carleman T 1938 Act M 69, xxxi-xxxiii (F64, 23)
H C 1937 Schw Vers 34, iii (F63, 816)
Lange-Nielsen F 1937 Nor M Tid 19, 49-51   (F63, 19)

**Piaget,** Jean  1896-

*Bunt L N H 1951 *The Development of the Ideas of Number and Quantity According to Piaget,* Groningen
Cohin A 1967 Minn CR 5(2), 12-13
Dienes Z P 1960 *Building Up Math,* London   [math education]
Lovell K 1961 *The Growth of Basic Math Concepts in Children,* London
Mays W 1967 Enc Phi
Parsons C 1960 Brt J Psy 51
Piaget J 1952 in E. G. Boring + ed *Psychology in Aubobiography* (Clark U P), 237-256  [autobiog]

**Picard,** Alfred   1844-1913

Appel P 1913 Rv GSPA 24, 378-380

Picard, Charles Emile 1856-1941

Anheluta T 1942 Cluj Sem 7, 3-14
  (F68, 18)
- 1942 Math Tim 18, 1-12
  (Z27, 196)
Anon 1928 *Cinquantenaire scienti-*
  *fique d'Emile Picard*, Paris, 123 p
  (F54, 48)
- 1928 Rv I Ens 82, 321-361
  (F55, 618)
- 1937 Vat An 1, 632   (F63, 20)
- 1939 Par CR 209, 849-865
Bouligand G 1942 Rv GSPA 52, 1-3
  (F68, 18; F68, 18; Z60, 17)
Broglie L de 1943 Par Mm (2)66
  (Z60, 17)
Buhl A 1928 Ens M 27, 5-13
  (F54, 43)
- 1942 Ens M 38, 348-350
  (MR4, 66; Z60, 17)
Dunnington G W 1942 M Mag 16, 186-
  187
Escobar T M 1928 Rv M Hisp A (2)3,
  123-126   (F54, 43)
Hadamard J 1942 Lon RS Ob 4, 129-150
  (Z60, 17)
- 1943 LMSJ 18, 114-128
  (MR5, 58; Z60, 17)
Lalescu T 1929 Buc U Rev 1, 49-57
  (F55, 24)
*Lebon E 1910 *Emile Picard*, Paris,
  88 p   (F41, 37)  [bib]
- 1914 *Savants du jour Emile Picard*,
  Paris, 88 p   (F45, 62)
Lefort G 1945 Rv M Hisp A (4)5, 147-
  151 (Z60, 17)
Mandelbrojt S 1942 AMM 49, 277-278
  (MR3, 258; Z60, 17)
Montel P 1942 BSM (2)66, 3-17
  (MR4, 253; Z26, 196; F68, 18)
Ocagne M d' 1931 Rv GSPA 42, 449-
  451 (F57, 42)
Rosenblatt A 1942 Rv Cien 44,
  311-356  (Z60, 17)
Smith D E 1921 AMM 28, 123
  [Cassini]

Picard, Jean 1620-1682

Doublet E 1920 Rv GSPA 31, 561-564

Picart, Lauro Clariana

Aracil J M O 1916 Esp SM 6, 58-64
  (F46, 36)

Picht, Karl Wilhelm Johannes
  1897-

Klebe J 1957 Pots Ped 3, 1-2
  (Z77, 242)

Picone, Mauro 1885-

Anon 1953 It XL An, 357-370
  (MR15, 592)
- 1956 *Onoranze a Mauro Picone*,
  Rome, 79 p   (Z72, 246)

Picquet, Louis Didier Henry 1845-

Liouville 1925 Par EP (2)25, 1-41
  (F51, 27)

Pieri, Mario 1860-

Castelnuovo G 1913 Mathes It 5, 40-
  41   (F44, 32)
Levi B 1913 Loria 15, 65-74
  (F44, 32)   [*bib]
Skof F 1960 It UM (3)15, 63-68
  (MR23A, 4; Z90, 8)

Pierpont, James 1866-1932

Ore O 1939 AMSB 45, 481-486
  (F65, 18)

Pietzker, Friedrich 1844-1916

Anon 1916 Unt M 22, 82-85

Schotten H 1911 ZMN Unt 42, 337-342
(F42, 41)
- 1916 ZMN Unt 47, 500-501 (F46, 23)


Pigott, Edward and Nathaniel XVIII

Bourget H + 1899 Intermed 6, 207


Pillai, Subbayya Sivasankaranarayana
1902-

Chandrasekharan K 1951 Indn MSJ
(NS)15, 1-10 (MR13, 198; Z42, 243)


Pincherle, Salvatore 1853-1936

Amaldi U 1937 Linc Rn (6)26, 418-429
(F63, 816)
- 1938 Annali (4)17, 1-21 (F64, 23)
Anon 1925 Act M 46, 341-362
(F51, 33)
Belardinelli G 1937 Milan Sem 11,
xv-xvi (F63, 816)
Berzolari L 1936 It UM 15, 149-152
(F62, 1041)
Bortolotti E 1937 It UM 16, 37-60
(F63, 19)
Conti A 1936 Bo Fir 32, 87
(F62, 1041)
Natucci A 1954 Batt (5)2, 335-342
(MR15, 924; Z56, 2)
Segre B 1953 Parma Rv M 4, 3-10
(MR15, 90; Z50, 243)
Tonelli L 1937 Pisa SNS (2)6, 1-10
(F63, 19)
Vivanti G 1936 Lom Gen (2)69, 956-
961 (F62, 1041)


Pinkerton, Peter 1870-1960

Dougall J 1931 Edi MSP (2)2, 268-271
(F57, 45)


Pinsker, Aron Grigorevich 1905-

Vladimirov D A + 1966 UMN 21(6),
169-170 (MR34, 20)


Pinto, Luigi

Campanella G 1922 Nap Pont (2)27, 215-
220
Cantone M 1920 Nap Pont (2)25, 27-37
Marcolongo R 1920 Nap FM Ri (3)26,
89-93


Pinto. See Sousa.


Piola, Gabrio 1794-1850

Martin Jadraque V 1955 Gac M 7, 57-
59 (MR17, 338)
Masotti A 1950 Lom Rn M (3)14, 695-
723 (MR13, 810; Z44, 247)


Pirie, George 1843-1904

Davidson W L 1905 LMSP (2)2, xviii-
xix (F36, 27)


Pirondini, Geminiano 1857-

Lazzeri G 1915 Pe MI 30(3)(12),
93-96 (F45, 46)
Teixeira F 1915 Porto Ac 10, 125-126


Pitcher, Arthur Dunn 1880-1923

Anon 1924 AMSB 30, 155


Pitiscus. 1561-1613

= Petiscus

Archibald R C 1949 M Tab OAC 3, 390-
392 (MR10, 420; Z41, 338)
Cajori F 1923 Arc Sto 4, 313-318
[dec point]

- 1924 AMSB 30, 13   (F50, 19)
Gravelaar N L W A 1898 Nieu Arch
   (2)3, 253-278   (F29, 32)  [trig]


Pitot, Henri   1695-1771

Humbert P 1953 Rv Hi Sc Ap 6, 322-
   328


Piuma, Carlo Maria   1837

Loria G 1913 *Prof. C. M. Piuma*,
   Genoa, 8 p   (F44, 25)


Pizzetti, Paolo   1860-

Jadanza N 1917 Tor FMN 53, 671-676
Reina V 1918 Linc Rn (5)27(1), 336-
   345
Somigliana C 1941 Tor Sem 7, 19-24
   (MR3, 258)  [Clairaut, Stokes]


Plana, Giovanni Antonio Amedeo
   1781-1864

Anon 1864 Lon As Mo N 24, 89-
- 1866 Edi SP 5, 293-
- 1873 BSM 5, 65-79   (F5, 36)
   [bib]
- 1873 Par Mm 38, cvii-
- 1887 It SS 6, lxxxvi-
Beamont E de 1873 Institut (2)1,
   45-49, 61-63, 69-73, 77-80, 85-88
   93-96   (F5, 36)
Bobynin V V 1899 Fiz M Nauk (2),
   14-25, 67-70   (F30, 13)
Favaro A 1904 Tor FMN 39, 493-501
   [Galileo]
Pearson K 1928 Biomtka 20(A), 295-
   299   (F54, 32)  [stat]
Realis S 1886 Boncomp 19, 121-128
Walker H M 1928 Isis 10, 466-484
   [Bravais, correlation]

Plarr, Gustav   1819-1892

Anon 1892 Nat 45, 419   (F24, 35)


Plateau, Joseph Antoine Ferdinand
   1801-1883

Van der Mensbrugghe G 1885 Bel Anr
   51, 389-486


Plato   -428? to - 348?

Andrissi G L 1940 It UM Con 2,
   912-920   (F68, 5)
Anon 1900 Pitagora 6, 58, 111
- 1900 Pitagora 7, 106
   [mean prop]
Becker O 1931 QSGM (B)1, 464-501
   (Z2, 324)
Benecke 1867 *Ueber die geometrische
   Hypothesis in Platons Menon*,
   Elbing
Bernays P 1935 Ens M 34, 52-69
Bettica-Giovanni R 1942 Pe M (4)22,
   129-144   (MR8, 2; F68, 5; Z28, 99)
   [geom]
Blass C 1861 *De Platone mathematico*
   Bonn, 31 p
Boncompagni B 1851 *Delle versioni
   fatte da Platone Tribustino,
   traduttore del secolo duodecimo*,
   Rome
Brocard H 1895 Intermed 2, 338 (Q138)
   [Rabelais]
Bruins E M 1957 Janus 46, 253-263
   [Egypt]
Brumbaugh R S 1954 *Plato's Mathe-
   matical Imagination* (Indiana U P),
   315 p   (MR16, 207) (repr 1968 Kraus
Butcher S H 1888 J Phlol, 12
   [geom]
Cherniss H 1951 Rv Met 4, 395-425
   (Z45, 145)
Cornford F M 1932 Mind 41, 37-52,
   173-190   (F58, 13; Z4, 193)
   [Marxism]
- 1937 *Plato's Cosmology*, London
Delastelle F 1896 Intermed 3, 87,
   155 (Q138)  [Rabelais]

Plato    (continued)

Demel S 1929 *Platons Verhaeltnis zur Mathematik*, Leipzig, 151 p (F55, 597)

Depuis J 1887 Boncomp 19, 641-645 [geom]

- 1887 Boncomp 19, 645-650 [geom]

Enriques F + 1932 Scientia 51, 5-20 (Z3, 241)

Erhardt R v + 1942 Isis 34, 108-110 [ast, helix]

Favaro A 1875 *Sulla ipotesi geometrica nel Menone di Platone*, Padua (F7, 2)

Frajese Attilio 1940 It UM (2)3, 62-70 (MR3. 97; Z23, 386; F 66, 10)

- 1940 It UM Con 2, 902 (F68, 5)

- 1943 It UM (2)5, 182-189 (MR7, 353; Z60, 5) [geom]

- 1950 Archim 2, 89-95 (MR12, 310; Z38, 1) [polyhedra]

- 1954 It UM (3)9, 74-80 (MR15, 591; Z56, 1) [irrat]

- 1961 Archim 13, 185-194 (Z105, 243) [conics]

- 1963 *Platone e la matematica nel mondo antico*, Rome, 218 p (Z112, 241)

Frank E 1923 *Plato und die sogenannten Pythagoreer*, Halle

Fritz K V 1932 Philolog 87, 40-62, 136-178 (F58, 13) [Theaetet]

Gaiser K 1964 Arc Phi 46, 241-292 (MR33, 1197)

Guenther S 1883 Bay Bl GR 19, 115-124

Heller S 1956 Centau 5, 1-58 (MR18, 453) [Theodoros]

Hofmann J E 1942 Deu M 7, 117-120 (MR8, 189; F68, 5) [Theodoros]

Krafft F 1965 Bei GWT 5, 5-24 [phys]

Lasserre F 1964 *The Birth of Mathematics in the Age of Plato*, London, 191 p (MR34, 7)

Lutfil Maqtul Molla 1940 *La duplication de l'autel (Platon et le problème de Delos)*, Paris, 66 p

Magnus L 1932 Nat 129, 473

Markovic Z 1939 Zagr BI 32, 28-48 (F64, 905; 65:1, 5) [Aristotle]

- 1939 Zagr Rad (261), 83-131 (F64, 905; 65:1, 5) [Aristotle]

- 1940 Zagr BI 33, 1-25 (F66, 1186; Z27, 289) [measure]

- 1954 Zagr BS 2(2), 57

- 1955 Rv Hi Sc Ap 8, 289-297

Milhaud G 1900 *Les philosophes géométres de la Grece. Platon et ses prédecesseurs*, Paris

Molodshi V N 1949 Ist M Isl (2), 499-504 (MR12, 1) [Euclid]

Mugler Charles 1948 *Platon et la recherche mathématique de son époque*, Strasbourg, 455 p

Pihl M 1951 M Tid A, 19-38 (Z43, 1) [Theodoros]

Places E des 1935 Rv Et Grec 48, 540-550 [irrat]

Reeve W D 1930 MT 23, 268-269

Reidemeister K 1942 Ham M Einz 35 (MR5, 253; Z27, 193) [logic]

Rodier 1902 Arc GPS 15

Santillana G de 1949 Isis 32, 248-262 (Z38, 145) [Eudoxos]

Sarton G 1927 Int His Sc 1, 113-116 [*bib]

Sedillot L A 1873 Boncomp 6, 239-248 (F5, 1) [spiral]

Solmsen F 1929 QSGM 1, 93-107

Stamatis E 1958 Athen P 33, 298-301 (Z84, 241) [set]

- 1962 Platon 14(27/28), 315-320 (Z109, 3) [geom]

Steele D A 1951 Scr 17, 173-189 (Z44, 243)

Stenzel  Julius 1924 *Zahl und gestalt bei Platon und Aristoteles*, Leipzig, 154 p (2nd ed 1933, 196 p Leipzig-Berlin, F59, 17: Z7, 145)

Strycker E de 1950 Rv Et Grec 63, 43-57

Tannery P + 1895 Intermed 2, 102 [Rabelais]

Taylor A E 1927 Mind 36, 12-33

Toeplitz O 1931 QSGM (B)1, 3-33 [arithmetisation]

- 1933 QSGM (B)2, 334-346

Van Deventer C M 1927 Euc Gron 4, 242-244 [irrat]

Vogt H 1910 Bib M(3)10, 97-155 [irrationals]

Wedberg Anders 1955 *Plato's Philosophy of Mathematics,* Stockholm, 154 p
Wex F C 1867 Arc MP 47, 131- [Plutarch, Pythagoras]


Platzman, Martin    1760-1786

Myrberg P J 1963 Arkhimed (1), 22-23 (Z111, 4)


Playfair, John    1748-1819

Funkhouser H G + 1935 Ec Hi 3, 103-109


Plemelj, Josip    1873-

Anon 1963 in *Neki nereseni Problemi u Matematici,* Belgrad, 3-10 (Z117, 9)


Plesner, Abram Iezekiilovich    1900-

Lyusternik L A + 1961 UMN 16(1), 213-218    (MR23A 272; Z98, 10)


Pluecker, Julius    1801-1868

Bertrand J 1867 J Sav, 269-
Clebsch A 1871 Gott Ab 16    (F3, 10)
- 1872 Boncomp 5, 183-212    (F4, 17)
- 1872 BSM 3, 59-64    (F4, 18) [bib]
- 1873 Batt 11, 153-180    (F5, 29)
De Vries H 1930 N Tijd 18, 56-84, 217-241; 19, 61-92, 283-302 (F57, 34)
Drohnke A 1871 *Julius Pluecker,* Bonn, 31 p    (F3, 10)    [dual, geom]
*Ernst Wilhelm 1933 *Julius Pluecker,* Bonn, 91 p
Loria G 1926 Ens M 25, 67-71 [surf]
Mansion P 1873 BSM 5, 313-319 (F5, 29)

Schoenflies A 1904 M Ann 58, 385-403 (F35, 14)
Schoenflies A + 1903 Gott N , 279-28 (F34, 42)
Tibiletti C 1947 Pe M (4)25, 16-29 (MR9, 169) [Apollonius problem, Newton, Viete]


Poe, Edgar Allen    1809-1849

Wylie C R 1946 Sci Mo 63, 227-235


Poggendorff, Johann Christian 1796-1877

W B 1877 Pogg An 160, i-xxiv (F9, 15)


Pogorzelski, Witold 1895-1963

Anon 1962 Wiad M 5, 29-37 (Z100, 247)
Wolska-Bochenek J + 1964 Pol An M 16, 1-16    (MR30, 872)


Poignard, François Guillaume

Fichefet J 1957 Lieg Bul 26, 396-405 (Z79, 243)


Poincaré, Henri    1854-1912

Adhemar Robert d' 1912 *Henri Poincar* Paris, 41 p (2nd ed 1914, 64 p)
Anon 1911 Rv GSPA 22, 665
- 1912 Ens M 14, 391-392 (F43, 45)
- 1912 Kagan (506), 33-64 (F43, 47)
- 1912 Mathesis (4)2, 233-238
- 1912 Nat 90, 353-356    (F43, 45) [port]
- 1912 Rv Sc (5)18, 90
- 1913 Paler R 8(S), 13-32 (F44, 26)
- 1915 Lon RSP (A)91, vi-xvi (F46, 5)

Poincaré, Henri (continued)

- 1919 Isis 2, 398-399
- 1921 Act M 38, 1-385 [bib]
- 1955 Le livre du centenaire de Henri Poincaré, 1854-1954, Paris, 305 p
Appell P 1913 Rv Sc 20, ii, 144-146
- 1925 Henri Poincaré, Paris, 120 p (F51, 25) [port]
Appell P + 1912 Rv Mois 14, 129-134
Archibald R C 1915 AMSB 22, 125-136
Bellivier A 1956 Henri Poincaré, ou la vocation souveraine, Paris, 247 p (MR18, 268)
Borel E 1907 Rv Mois 4, 115-118 [relativity]
*- 1909 Rv Mois 7, 360-362 [heuristic]
- 1924 Rv Sc, 321-324
- 1925 Fr SMB, 49-54
*Boutroux P 1913 Rv Mois 1, 155-183 [phil]
Brunschvicg L 1913 Rv Met Mor 21, 585-616
Buhl A 1913 Ens M 15, 9-32 (F44, 28)
- 1917 Ens M 19, 4-19 (F46, 5)
Cailler C 1914 Arc Sc Ph (4)38, 164-188 [ast]
- 1914 Henri Poincaré, Paris, 64 p (F45, 33)
Cajori F 1929 AMM 36, 162 [Institut H. P.]
Cath P G 1954 Euc Gron 30, 265-275 (MR17, 117)
Chatelet A + 1951 in Congrès International de Philosophie de Sciences, (1949), Paris, Vol. 1 (MR13, 421)
Chazy J 1951 Bu As (2)16, 145-160 (MR13, 198) [cel mech]
Chistyakov I 1912 M Obraz 5, 197-199 (F43, 47)
Dantzig Tobias 1954 Henri Poincaré Critic of Crisis, NY-London, 160 p (MR16, 2)
Darboux G 1914 Par Mm 52, lxxxi-cxlviii [port]
- 1914 Rv Sc 2, 97-110
Dickstein S 1912 Wiad M 15, 249-260 (F43, 47)
Dugas R 1951 Rv Sc 89, 75-82 (Z43, 5) [mech]

Echegaray J 1912 Esp SM 2, 33-39 (F43, 45)
*Furman L 1941 in Les Grand Savants Française, NY, 101-107
Gomes R L 1955 Gaz M Lisb 15, 1-3 (MR17, 338)
Guisthan + 1913 Fr Long (D) (F43, 45)
Hadamard J 1913 Rv Met Mor 21, 585-616 (F44, 25)
- 1913 Rv Mois 15, 91-96, 385-418 [three body probl]
- 1921 Act M 38, 203-287
- 1922 Rice Pam 9(3)
- 1933 Rice Pam 20, 1-130 (F59, 850)
- 1954 Rv Hi Sc Ap 7, 101-108 (MR15, 924)
Ingraham M H 1928 MT 20, 253-264 [William James]
Jourdain P E B 1912 Monist 22, 611-615 (F43, 46)
Juvet G 1924 Rv GSPA 35, 69-71 [relativity]
K 1912 Pe M (3)10, 42-45 (F43, 47)
Korn A 1913 Ber MG 12, 2-13 (F44, 27)
Kuznetsov B G + ed 1967 Frantsuzkaya nauka i sovremennaya fizika, Moscow
Langevin P 1913 Rv Met Mor 21, 675-718 [phys]
- 1913 Rv Mois 15, 91-96, 419-463
la Rive L de 1914 Arc Sc Ph (4)38, 149-163
- 1914 Arc Sc Ph (4)38, 189-201
Lebeef A 1913 Rv Met Mor 21, 659-674
Lebon Ernest 1909 Fr AAS, 29-30
*- 1909 Henri Poincaré, Paris, 88 p (F40, 41) (2nd ed 1912, 115 p)
Lippman G 1912 Par CR 155, 1277-1283 (F43, 36)
Love A E H 1913 LMSP (2)11, xli-xlviii (F44, 27)
Mansion P 1912 Mathesis (4)2, 233-238 (F43, 46)
Margaillan L 1913 Int Monat 7, 545-556 (F44, 27)
Masson F 1912 Rv Sc 18(2), 628-629
Meyer C 1912 Arg SC 74, 125-147
Mieli A 1913 Rv Fil 5, 44-48

Milhaud G 1912 Grande Rv (Dec 10)
Miller G A 1912 Sci 36, 425-429
Mooij J J A 1966 *La philosophie des mathématiques de Henri Poincaré,* Paris, 180 p (MR34, 1149)
Natucci A 1955 Batt (5)3, 115-130 (MR17, 117; Z64, 4)
Naville E H 1921 Nat 106, 661-662
Nordmann C 1912 Rv Deux M (Sept 15)
- 1912 Smi R, 741-763 (F44, 26)
Noerlund N E 1923 Act M 39, 94-132 (F49, 10) [Klein, letters]
Otradnykh F P 1955 UMN 10(2), 224
Papp Desiderio ed 1944 *El legado de Henri Poincaré al siglo XX,* Buenos Aires, 187 p
Pascal E 1912 Batt 50, 305-309
- 1912 Nap FM Ri (3)18, 309-313 (F43, 45)
Pelseneer J 1938 Isis 29, 24-28 [Descartes]
Picard E 1913 An SENS (3)30, 463-482 (F44, 25)
- 1913 Rv Sc 2, 705-713
- 1922 *Discours et mélanges,* Paris, 201-220
- 1931 *Eloges et discours,* Paris, 289-296
Rougier Louis 1920 *La philosophie géométrique de Henri Poincaré,* Paris, 208 p
Rudnicki J 1913 Wektor 2, 271-281, 312-321 (F44, 40)
Sageret J 1911 *Henri Poincaré,* Paris, 80 p (F42, 48)
Saltykov N 1955 Beo Ak N Sb 43(4), 1-13 (Z67, 247)
*Sarton George 1913 Ciel Ter [port]
- 1913 Isis 1, 95-96
Schlesinger L 1923 Act M 39, 240-245 (F49, 10) [Fuchsian, Weierstrass]
Shaw J B 1913 Pop Sci Mo 82, 209-224
Smith W B 1912 Monist 22, 611-617
Somigliana C 1914 Tor FMN 49, 45-54 (F45, 33)
Subbotin M F 1956 Vop IET (2), 114-123 (MR18, 784) [cel mech]
Torres Torya M 1916 Mex Alz Mm 34, 294-402 (F46, 5)
Toulouse Edouard 1910 *Henri Poincaré,* Paris, 204 p [psych of math]

Tzitzeica G 1912 Gaz M 17, 441-445
Veblen O 1912 APSP 51, iii-
Volterra V 1912 Rv Mois 15, 129-154 (F43, 47)
Volterra V + 1914 *Henri Poincaré,* Paris, 270 p
Weyl H 1912 MN Bl 9, 161-163 (F43, 46)
Zahara N O 1932 Tim Rev M 12, 61-67 (F58, 995)


Poinsot, Louis     1777-1859

Bertrand Joseph 1902 *Eloges académiques,* (NS), Paris, 1-27 [mech]
Boyer J + 1894 Intermed 1, 106-107
Jonquières de 1896 Par CR 122, 829, 857- [Gauss, letter]
Poinsot L 1873 Boncomp 6, 536-538 (F5, 47)
Reveille M 1920 Par EP (2)20, 107-114 [mech]


Poisson, Raimond

Brocard H 1912 Intermed, 224 (Q2838)


Poisson, Siméon Denis 1781-1840

Anon 1841 Phil Mag (2)18, 74-77
- 1857 *Catalogue des ouvrages et mémoires scientifique de S. D. Poisson,* Paris
Dupont P 1963 Tor FMN 98, 537-572 (MR29, 638) [d'Alembert, Euler, mech, Poncelet]
Gliozzi M 1939 Pe M (4)19, 90-91 (F65, 15) [elec]
Hermite C 1890 BSM (2)14, 9-14
Hostinsky B 1924 Par CR 179, 1199-1201 [magn]
Pajares E 1955 Gac M (1)7, 105-108 (MR17, 932; Z68, 6)
Polak L S 1957 Mos IIET 17, 450-472 (Z118-1, 6; Z118:1, 6)

**Pokorny**, Martin   1836-

Panek A 1901 Cas MF 30, 81-100
  (F32, 29)

**Pokrovsky**, Petr S.   1857-1901

Przeborski A P 1901 Mos MO Sb 22(2),
  1-33   (F32, 29)
- 1902 Kiev U Iz (9c)
- 1903 DMV 12, 117-119   (F34, 26)

**Poleni**, Giovanni   1683-1761

Anon 1963 *Giovanni Poleni (1683-1761)
  nel bicentario della morte,
  Padova, 17 dicembre 1961,* Padua,
  147 p   (MR29, 219)
Cossali 1813 *Elogio di Poleni,*
  Padua

**Pollard**, Samuel   1894-1905

Burkill J C 1945 LMSJ 20, 189-192
  (MR8, 3; Z60, 17)

**Polya**, Georg   1887-

Hartokopf 1963 Praxis 5, 100-103
  [heuristic]

**Pompeiu**, Dimitrie   1873-1954

Anon 1954 Ro IM 5, 7-10
  (MR16, 434; Z57, 4)
- 1954 Ro IM 5, 11-17   (MR16, 434)
  [bib]
- 1954 Ro IM 5, 421-422   (MR16, 660)
Onicescu O 1943 Math Tim 19, 12-15
  (Z60, 17)
Sergescu P 1930 Buc EP 1, 200-201
  (F56, 34)
- 1955 Ens M 40, 70-71   (Z64, 4)
Stoilov S 1954 Ro IM 5, 19-24
  [anal fun]
Teodorescu N 1959 Gaz MF (A)11, 685-
  686   (Z89, 5)

**Poncelet**, Jean Victor   1788-1867

Bapst G 1897 Par CR 124, 1135-1137
  (F28, 17) [Russ]
Bertrand Joseph 1879 Par Mm 41(2),
  i-xxv   (repr 1890)
- 1890 *Eloges académiques,* 105-129
  (Repr of 1879)
Bottema O 1947 Euc Gron 23, 121-127
  [Grassman, Jacobi]
Camberousse C de 1874 Nou An (2)13,
  174-185   (F6, 29)
Cantor M 1869 Hist Abt 14, 53-56
  (F2, 24)
Cassina U 1923 Esercit 3, 32-39
  [proj geom]
Didion Isidore 1869 *Notice sur la
  vie et les ouvrages du général
  Jean Victor Poncelet,* Paris, 59 p
  [proj geom]
Dupont P 1963 Tor FMN 98, 537-572
  (MR29, 638) [d'Alembert, center
  of rotation, Euler, mech, Poisson]
Holst E 1879 *Om Poncelet's Betydning
  for Geometrien,* Christiania, 162 p
  (F11, 23)
Hunyady 1877 Magy MT   (F11,23)
Kravec T P ed 1955 SSSR Tek (4),
  120-130   (MR17, 3)
Loria G 1889 Bib M (2)3, 67-74
  [polyg]
Morin G B 1874 Par CR 78, 229-236
  (F6, 29) [mech]
Toride 1913 Intermed 30, 130-131
  (F44, 60) [polyg]
Tribout H 1936 Rv Sc 74, 389-395
  (F62, 25)
- 1936 *Un grand savant le général
  Jean Victor Poncelet,* 1788-1867
  Paris, 225 p   (F62, 1036)
Vuckic M 1951 Glasn MPA (2)6, 115-
  121   (MR13, 421) [approx]
Weiss E A 1939 Deu M 4, 126-127
  (F65, 16)

**Pontryagin**, Lev Semenovich   1908-

Aleksandrov P S + 1959 UMN 14(3),
  195-202   (MR22, 619; Z88, 244)
- 1960 Ro Sov (3)14(1), 203-207
  (MR22, 929)

Poole, Edgar Girard Croker
1891-

Chaundy T 1941 LMSJ 16, 125-130
  (F67, 970; Z28, 195)


Popov, V. V.

Anon 1956 BSSRFMT (1), 3-10
  (MR20, 264)
Bychkov V P 1957 Tiras Ped 5


Popoviciu, Tiberiu    1906-

Nicolescu M G 1958 Cluj MF 8(1),
  7-19   (Z82, 12)


Poretzky, P. S.    1846-1897

Dubjago D 1908 Kazn FMO (2)16, 3-7
Slezhinski J 1909 Kagan (487), 145-
  148  (F40, 31)


Porta, Giambattista della
1538-1615

  = Giovanni Battista

Hofmann J E 1953 Arc In HS 23, 193-
  208  (Z52, 2)  [area]
- 1953 Arc In HS 26, 16-   [area]
Loria G 1916 AMSB 22, 340-343
  (F46, 8)  [geom]
Riccardi P 1888 Boncomp 20, 605-
  [circle sq]


Posidonius    II-III

  = P. of Rhodes  = P. of Apamea

Sarton G 1927 Int His Sc 1, 204
  [bib]
Sepp B 1882 Bay Bl Gr 18, 397-399


Pospisil, Bedrich   1912-1944

Cech E 1947 Cas MF 72, D1-D9
  (MR9, 75; Z33, 242)


Postma, O.

Bottema O 1960 Euc Gron 35, 230-233
  (Z105, 3)


Potter, Mary A.   XX

Anon 1949 MT 42, 374


Poudra, Noel Germinal    1794-

Anon 1868 Boncomp 1, 302-308
  (F1, 14)  [bib]


Poulet, Paul

Kraitchik M 1953 Int Con HS 7, 415-
  417 [bib]


Poullet-Delisle,    A. C. M.

Boncompagni B 1883 Boncomp 15, 670-
  678


Prachatice.  See Christian


Prandl, Ludwig   1875-1956

Ackeret J 1954 ZAMP 5, 175-177
  (Z56, 2)
Blenk H 1953 Z Flug 1, 49-51
  (Z51, 4)
Hoff W 1935 Luftf 12, 1-3
  (F62, 1041)
Leibbrand W 1957 Arc In HS 10, 91-92
  (Z88, 7)
Sommerfeld A 1935 ZAM Me 15, 1-2
  (F61, 28)
Tollmien W 1953 Jb WG Flag, 22-27
  (Z59-1, 12; 60, 12)

Trefftz E 1925 ZAM Me 5, 87-88
(F51, 33)


Prange, Heinrich Friedrich Wilhelm
  Georg   1885-1941

Koppenfald W v 1941 DMV 51(1), 1-14
(Z25, 3)


Prasad, Badri Nath   1899-

Sinha S R 1965 Indn JM 7(2), 1-11
(MR34, 4096)


Prasad, Ganesh   1876-1935

Anon 1935 Nat 135, 644   (F61, 956)
Bagchi S C 1935 Clct MS 27, 93-98
(F61, 956)
Narayan L 1939 Benar MS 1, 107-114
(MR1, 290; Z60, 17)


Prediger, Johann Carl   1822-1895

Meyer F 1897 DMV 4, 51-52
(F28, 21)


Prestet, Jean 1648-1691

Bouligand G 1960 Rv Hi Sc Ap 13,
95-113
Schrecker P 1935 Thales 2, 82-90
[neg numb]


Prete, Guelfo del   1873-1901

Conti A 1901 Bo Bolog 2, 297-300
(F32, 29)


Price, Bartholomew   1818-1898

Elliot E B 1899 LMSP 30, 332-334
(F30, 20)


Priestley, Henry James   1883-1932

McCarthy J P 1932 M Gaz 16, 305


Prihonsky, F.  XIX

Winter Eduard  1956 *Der boehmische
Vormarz in Briefen B. Bolzanos
an F. Prihonsky,* Berlin, 316 p


Prince, Thomas   XVIII

Simons L G 1935 Scr 3, 94-96


Pringsheim, Alfred   1850-1941

Perron O 1930 Fors Fort 6, 315
(F56, 817)
- 1952 DMV 56(1), 1-6
(MR14, 344; Z47, 6)


Privalov, Ivan Ivanovich   1891-1941

Stepanov W 1941 SSSR Iz 5, 389-394
(MR3, 258; Z60, 17)


Proca, Alexandru   1897-1955

Andonie G S 1959 Gaz MF (A)11, 516-
528   (Z87, 244)


Prochazka, Friedrich   1855-

Vetter Q 1925 Sphinx Oe 20, 56-57


Proclus   410-485

  = Proculus = Proclos Diadochos

Boncompagni B 1874 Boncomp 7, 152-166
(F5, 13)  [Euclid]
Dijksterhuis E J 1951 Arc In HS (NS)4,
602-619  (MR13, 2; Z43, 1)

Frankland W B 1905 *The First Book of Euclid's Elements with a Commentary Based Principally upon That of Proclus Diadochus* (Cambridge Univ Press)
Hartmann N 1910 Arc MP (3)16, 344-345
Knoche 1865 *Untersuchungen ueber die neu aufgefunden scholien des Proklus Diadochus zur Euklids Elementen Commentariis...*, Herford
Knoche + 1856 *Ex Procli successoris in Euclidis elementa commentariis ...*, Herford
Martin T H 1874 Boncomp 7, 145-152 (F6, 12)
Ritzenfeld A 1912 *Procli Diadochi Lycii Institutio physica*, 11, Leipzig, 94 p
Steck M 1943 Leop NA (NF)13, 131-149 (MR14, 1051; Z61, 2)
Tannery P 1885 Bib M, 199 (Q8) [Nicomachus]
- 1886 BSM (2)10, 49-64


Protagoras  -V

Luria S 1928 Doklady (A), 74-79 (F54, 45) [Democritos]


Prudhomme.  See Sully.


Pruefer, Heinz  1896-1934

Behnke H + 1934 Muns Semb 5, 1-14 (F60, 839)
- 1935 DMV 45, 32-40  (F61, 27)


Prym, Friedrich  1841-1915

Krazer A 1916 DMV 25, 1-15 (F46, 17)
- 1916 Wurz Ver 44, 167-171 (F46, 18)
Rudio F + 1916 Zur NGV 61, 727-751 (F46, 17)


Psellos, Michael Constantine  XI

Redl G 1929 Byz Par 4, 197-236 [chronom]
- 1929 Byz Z 29, 168-187
- 1930 Byz Par 5, 229-246


Ptaszycki, Jan  1854-1912

Dickstein S 1912 Wiad M 16, 241-247 (F43, 48)
Posse K A 1912 Pet MNP 6, 95-101 (F43, 48)
- 1913 Khar M So 13


Ptolemy, Claudius  II

Aaboe A 1960 Isis 51, 565 [Fourier ser]
*- 1964 *Episodes from the Early History of Mathematics* (Blaisdell, Singer), Ch 4 [tables, trig]
Archibald R C 1918 AMM 25, 94 [trig]
Boll F 1894 *Studien ueber Claudius Ptolemaus*, Leipzig
Boncompagni B 1871 Boncomp 4, 470-492 (F3, 2) [optics]
- 1873 Boncomp 6, 159-170 (F5, 5) [optics]
Brendan T 1965 MT 58, 141-149 [trig tables]
*Cajori F 1926 Archeion 1, 25-28 [fourth dim]
Czwalina A 1927 Arc GMNT 10, 241-249 [trig]
- 1958 Centau 5, 283-306
Diller A 1941 Isis 33, 4-7 (MR3, 97; Z61, 2) [cartog]
Doederlein 1879 Bay Bl GR 15, 396-400, 433-441 (F11, 11) [Muenster]
Drachmann A G 1950 Centau 1, 117-131 (MR12, 311) [Heron]
Goldstein B R 1967 APST 57(4) [ast]
Govi G 1885 *L'ottica di Claudio Tolomeo da Eugenio*, Torino-Paravia, 220 p
Guzzo A 1952 Filo Tor 3, 351-370 (MR14, 610; Z47, 2)
Heiberg J L 1895 ZM Ph 40(S), 1-30 (F26, 7)

Ptolemy, Claudius (continued)

Heiberg J L 1899 *Claudii Ptolemaei opera quae exstant omnia. Syntaxis mathematics*, Leipzig
Hopfner F 1946 Wien Anz 83, 77-87 (MR11, 150) [cartog, Marinos]
Jacobi C G J 1849 Ber Ber (Aug), 222-226 [Egypt]
Lejeune A F 1947 Antiq Cl 15, 241-56 [optics]
- 1950 Scriptor 4, 18-27 [optics]
Lynn W T 1896 Nat 53, 488-490 (F27, 5)
Manitius Karl 1912-1913 *Des Claudius Ptolemaus Handbuch der Astronomie*, Leipzig, 2 vols
Marchand G + 1896 Intermed 3, 49
Martin H 1871 Boncomp 4, 464-469 (F3, 2) [optics]
Milankovitch M 1953 Beo Ak N Zb 35(3), 11-14 (MR15, 591) [pi]
Mogenet J 1956 Bel Let Mm 51(2)
Narducci H 1888 Bib M(2)2, 98-102
Neugebauer O 1959 Isis 50, 22-29 (MR20, 1041) [geog]
Ottema J G 1844 Amst NWNV i, 235- [geog]
Rome A 1936 *Commentaires de Pappus et de Theon d'Alexandrie sur l'Almageste*, Rome, 2 Vols
- 1939 Brx SS(1)59, 211-224 (MR1, 33; F65:1, 6; Z22, 2)
Schmidt O 1955 Nor M Tid 3, 81-95, 127 (MR17, 117) [Menelaus]
Shively L S 1946 MT 39, 117-120 [reg polyg]
Stahlman W D 1960 Int Con HS 9, 593-605 (Z114, 5) [tables]
Steinschneider M 1892 Bib M, 53-62 [Arab]
Thaer C 1935 Unt M 41, 117-119 (F61, 12)
- 1950 MN Unt 3, 78-80 [trig]
Van der Waerden B L 1953 Mun Si, 261-272 (MR15, 923; Z56, 1) [tables]


Puiseux, Pierre Henri 1855-

Anon 1928 Ens M 27


Puiseux, Victor Alexandre 1820-1883

Bertrand J 1884 BSM (2)8, 227-234
Krazer A 1910 Bib M (3)10, 250-259 [ell intgl, Jacobi]
Tisserand F 1884 BSM (2)8, 234-245


Pupin, Michael 1872-1935

Ettlinger H J 1940 Scr 8, 237-250


Purser, William XVII

Conte L 1953 Pe M (4)31, 1-6 (MR14, 833; Z50, 243) [Ireland]


Puzyna, Joseph 1856-1919

Lomnicki A + 1921 Wiad M 25, 113-119


Pythagoras -VI

Badrau D 1962 Buc U Log 5(5), 87-110 (Z117:2, 241) [numb]
Ball W W R 1915 M Gaz 8, 5-12 (F45, 60)
- 1930 MT 23, 185-186
Baltzer E 1867 *Pythagoras, der Weise von Samos*, Nordhausen, 188 p
Brunschwig L 1937 Act Sc Ind (447) (F63, 817) [mysticism]
Burkert Walter 1962 *Weisheit und Wissenschaft, Studien zu Pythagoras, Philolaus, und Platon*, Nuremberg
Cardini Maria T ed 1958 *Testimonianze e frammenti*, Florence, 197 p
Delatti A 1922 *Essai sur la politique Pythagoricienne*, Liege
Fettweis E 1952 Z Phi Fors 5, 179-196
Frank Erich 1923 *Plato und die sogenannten Pythagoreer*, Halle, 318 p
Fritz K v 1940 *Pythagorean Politics in Southern Italy*, NY
G A 1954 Gac M (1)6, 3-7
Haebler A 1957 Hist Abt 45, 161-

Pythagoras   (continued)

Heidel W A 1940 Am J Phil 61, 1-33
   (MR1, 129)
Heller Siegfried 1958 *Die Entdeckung
   der stetigen Teilung durch die
   Pythagoreer*, Berlin 28 p
Hopper G M 1936 AMM 43, 409-413
   (F62, 10)  [numb thy, seven]
Iamblichos 1818 *Life of Pythagoras*,
   London  (transl Thomas Taylor)
- 1963 *Pythagoras. Legende, Lehre,
   Lebensgestaltung*,  Zurich-
   Stuttgart,  280 p
Junge G 1940 Deu M 5, 341-357
   (F66, 9; Z24, 97)
- 1948 Clas Med 9, 183-194
Keyser C J 1939 Scr 6, 17-22
Levi B + 1942 M Notae 3, 74-100
   (Z60, 5)
Lévy Isidore 1926 *Recherches sur
   les sources de la légende de
   Pythagore*, Paris, 152 p
- 1927 *La Légende de Pythagore de
   Grèce en Palestine*, Paris, 352 p
Lloyd William 1966 *A Chronological
   Account of the Life of Pythagoras
   and of Other Famous Men His
   Contemporaries*, London
Loria Gino 1935 "Eulero e i neo-
   pitagorici? Una questione di
   priorita," in *Studi in onore del
   S. D. Carboni*, Rome, 8 p
Maddalena Antonio 1954 *I Pitagorici*,
   Bari, 366 p  (Z57, 1)
Martin T H 1872 Boncomp 5, 99-126
   (F4, 2)  [ast]
Michel P H 1958 *Les nombres figurés
   dans l'arithmétique Pythagoricienne*,
   Paris, 23 p  (MR20, 830)
   [figurate numb]
Miller G A 1935 Sci (NS), 82-129
Naber H 1908 *Das Theorem des
   Pythatoras, wiederhergestellt in
   seiner ursprunglichen Form und
   betrachtet als Grundlage der ganzen
   pythagoreischen Philosophie*,
   Haarlem, 2 vols
Narducci E 1887 Boncomp 20, 197-308
Obenrauch F J 1903 Mo M Phy 14, 187-
   205  [curve]

Papadakis K N 1957 *Ein Beitrag zum
   Studium der Seiten und Durchmesser-
   zahlen der Pythagoreer*, Athens,
   30 p  (Z84, 241)
Peirce C S 1892 Open Ct (Sept 8)
Raven J E 1948 *Pythagoreans and
   Eleatics*, Cambridge
Reeve W D 1930 MT 23, 185-86
Riecke A 1882 *Pythagoras. Zeit- und
   Lebensbild aus dem alten Griechen-
   land*, Leipzig-Berlin
Schroeder Leopold v 1884 *Pythagoras
   und die Inder*, Leipzig
Shulte A P 1964 MT 57, 228-232
Stamatis E 1955 Athen E 30, 262-282
   (MR17, 337; Z65, 243)  [alg]
Stapleton H E 1958 Osiris 13, 12-53
   (Z87, 3)
Struyk A 1954 MT 47, 411-413
Szabo A 1956 Elem M 11, 101-105
   (MR18, 182; Z70, 3)
- 1963 M Ann 150, 203-217  (Z115, 241
   [Euclid]
Tannery P 1885 BSM (2)9, 69-89
- 1886 BSM (2)10, 115-128
Taylor T 1934 *The Theoretical
   Arithmetic of the Pythagoreans*,
   Los Angeles
Ver Eecke P 1938 Antiq Cl 7, 271-273
   [geom]
Vogt 1908 Bib M (3)9, 15-54
   (F39, 68)  [geom]
Van Der Waerden B L 1941 Himmelsw
   (MR3, 257; Z61, 4)  [ast]
- 1941 M Ann 118, 286-288
   (MR3, 257; Z25, 385)  [roots]
- 1943 Hermes 78, 163-199
   (MR8, 189; Z60, 5)
- 1948 M Ann 120, 127-153, 676-700
   (Z30, 98; 32, 49)  [irrat]
- 1951 Amst Vh (1)20(1)
   (MR13, 611)  [ast]
Wellmann M 1919 Hermes 54, 225-248
*Zeller E 1881 *A History of Greek
   Philosophy From the Earliest
   Period to the Time of Socrates* 1,
   306-533

Quetelet, Ernest   1825-1878

Mailli E 1878 Mondes (2)47, 225-227
  (F10, 17)
- 1880 Bel An, 169-216   (F12, 14)


Quetelet, Lambert Adolphe Jacques
  1796-1874

Cox J F 1948 Bel BS 34, 799
Freudenthal H 1951 Arc In HS (NS)4,
  25-34   (MR12, 577)
  [Arbuthnot, soc sci, stat]
- 1966 Bel Vla Ve 28
Kobell F v 1874 Mun Si, 88-91
  (F6, 34)
Lottin J 1912 Louv ISPA 1, 97-173
  [soc]
- 1912 Quetelet, statisticien et
  sociologue, Paris-Louvain, 594 p
Lurquin C 1924 Rv Sc, 390-393
Mailli E 1874 Brx Ob Anr 52, 45-268
  (F6, 35)
- 1874 Bel Bul (2)38, 816-844
  (F6, 35)
- 1875 BSM (2)2, 240-246
  (F10, 18)
Pelseneer J 1935 Brx Con CR 2, 105-
  112 (F62, 1036) [Chasles]
Reichesberg N 1896 Z Schw St 32
  (F27, 16)


Quint, N.

Anon 1907 Wisk Tijd 4, 129-130


Qurra.  See Thabit


Qusta ibn Luqa   IX-X

Gabrieli G 1912 Linc Rn Mo 21, 341-
  382  [bib]
Sarton G 1927 Int His Sc 1, 602
  [bib]
Suter H 1908 Bib M (3)9, 111-122
  [false position]

Radl, Frantisek   1876-

Rychlik K + 1957 Cas M 82, 378-382
  (Z98, 10)


Rado, Tibor   1895-1965

Anon 1950 Parma Rv M 1, 239-273
  (MR12, 383; Z39, 243)


Radon, Johann Karl August   1887-1956

Funk P 1958 Mo M Phy 62, 189-199
  (Z80, 5)
Hornich H 1960 DMV 63(1), 51-52
  (MR22, 425; Z92, 3)


Raets, Willem

Smeur A J E M 1960 Sci Hist 2, 22-23
  [arith]


Raffaele, Rubini

Torelli G 1890 Nap FM Ri (2)4,
  134-135


Raffy, Louis   1855-

Appel P + 1910 Fr SMB 38(S), 241-248
  (F41, 33)


Rahn, Johann Heinrich   1622-1676

Cajori F 1924 AMM 31, 65-71
  [notat]
- 1925 AMSB 30, 225   (F50, 19)
  [notat]
Wertheim G 1902 Bib M (3)3, 113-126
  [alg]


Raitenau, P. A. E. von   1605-1675

Schwab P F 1898 P.A.E. von Raitenau,
  1605 bis 1675..., Salzburg, 105 p
  (F29, 6)

Rakhmaninov, I. I.

Sousloff G K 1898 Kiev U Iz 5, 32


Rakhmatullin, C. A.

Anon 1959 Uzb FMN (3), 3-4
(Z84, 3)


Ramanujan, Srinivasa    1889-1920

Aiyar P V S 1920 Indn MSJ 12, 81-86
Anon 1928 Bo Fir (2)7, lxxi-lxxii
(F54, 43)
Hardy G H 1917 Indn MSJ 9, 30-48
1920 Nat 105, 494-495
1921 LMSP (2)19, xl-xlviii
(F48, 19)
- 1937 AMM 44, 137-155
(F63, 19; Z16, 145)
*- 1940 *Ramanujan. Twelve lectures
on subjects suggested by his life
and work,* Cambridge, 236 p
(MR21, 893; Z27, 196)
Levin V I 1960 Ist M Isl 13, 335-
378   (MR23A, 584)
Littlewood J E 1929 M Gaz 14, 425-428
(repr 1929 Nat 123, 631-633)
Miller J S 1952 SSM 51, 637-645
Myrberg P J 1965 Archim (2), 1-6
(MR34, 2421)
Naylor V 1930 M Gaz 15, 260-261
(F56, 32)   [Fuerstenau's process]
Neville E H 1942 Nat 149, 292-294
(MR3, 258; Z28, 385)
Rao R R 1920 Indn MSJ 12, 87-90
Srikantia B M 1928 AMM 35, 241-246
(F54, 43)
Stoermer C 1934 Nor M Tid 16, 1-13
(F60, 839)
Watson G N 1931 LMSJ 6, 137-153
(F57, 47)
Wilson B M 1930 M Gaz 15, 89-94
(F56, 31)


Ramsey, Frank Plumpton    1903-1930

Braithwaite R B 1931 LMSJ 6, 75-78
(F57, 47)


Ramus, Petrus  1515-1572

  = Pierre de la Ramée

Delcourt M 1934 Bude Bu 44, 3-15
[Rhedicus]
Desmaze C 1864 *P. Ramus, Professeur
au college de France, sa vie,
ses ecrits, sa mort,*  Paris, 135 p
Eneström G 1909 Bib M (3)10, 180
(F40, 51)
Lebesgue H 1957 Ens M (2)3, 188-215
(Z78, 3)  [Humbert, Jordan,
Roberval]
Verdonk J    1966 *Petrus Ramus en de
Wiskunde,* Assen, 465 p  [bib, 27 p]
Waddington C 1855 *Ramus, Pierre de
la Ramée. Sa vie, ses ecrits,
ses opinions,*  Paris


Rankine, William J. M . 1820-1872

Anon 1932 Nat 130, 175
Henderson James 1932 *M. Rankine,*
Glasgow, 28 p
Macfarlane Alexander 1919 *Lectures
on Ten British Physicists of the
Nineteenth Century,*  NY, 22-37
Mansion P 1875 Nou Cor 1, 167-168,
210-211  (F7, 12)
Quercia M 1874 Boncomp 7, 1-61
(F6, 36)


Ranyard, Arthur Cowper   1845-1894

Anon 1894 Nat 51, 179  (F25, 47)
- 1895 LMSP 26, 554-557  (F26, 31)


Rashevsky, P. Konstantinovich 1907-

Norden A P + 1958 UMN (NS)13(1),
225-231  (MR20, 265; Z78, 4)


Rasp, Carl von

Dorn H 1927 Z Versich 27, 248-250
(F53, 33)

Rayleigh, Lord    1842-1919

= John William Strutt

Anon 1922 LMSP (2)20  [A. Hurwitz]
Strutt Robert J 1924 *The Life of
John William Strutt, Third Baron
Rayleigh,* London (enlarged ed
1968 U Wisc P) (MR37, 20)
Thomason J J 1919 Nat 103, 365-369


Razmadze, A. M.    1929-

Gokieli L P 1937 Tbil MI 1, 1-10
(F63, 19)
Tonelli L 1937 Tbil MI 1, 11-16
(F63, 19)


Re, Alfonso del    1859-1921

Montesano D 1922 Nap FM Ri (3a)28,
187-189
Signorini A 1925 Nap Pont 55, 133-135
(F51, 31)


Realis, Savin    1818-1886

Catalan E 1886 Nou An (3)5, 200-203
Genocchi A 1886 Batt 24, 56
- 1886 Boncomp 19, 55-58
- 1886 Tor FMN 21, 549-551


Réaumur, René Antoine Ferchault
1683-1757

Taton R 1958 Rv Hi Sc Ap 11, 130-133


Recorde, Robert    1510-1558

Anon 1947 M Gaz 31, 129 [biog]
Cajori F 1922 MT 15, 294-302
Clarke F M 1926 Isis 8, 50-70
Easton J B 1962 Scr 27, 339-355
- 1966 Isis 57, 1, 121  [birth]
- 1967 Isis 58, 515-
Ebert E R 1937 MT 30, 110-121

Johnson F R + 1935 Hunt LB (7), 59-87
(F62, 1029) [logic]
Patterson L D 1951 Isis 52, 208-218
Sanford V 1952 MT 45, 546
[pasturing probl]
- 1957 MT 50, 258-266


Reeve, William David    1883-1961

Rosskopf M 1961 MT 54, 389-390


Regiomontanus    1436-1476

= Johannes Mueller

Blaschke W 1953 Matiche 8(1), 50-58
(MR16, 207; Z50, 242)
Blaschke W + 1956 Mainz MN, 445-529
(MR19, 108)
Bortolotti E 1942 Bln Mm (9)9, 81-90
(MR9, 486) [Bianchini, Roder,
Speir]
Braunmuhl A v 1898 *Nassir Eddin Tusi
und Regiomontan,* Leipzig
Cantor M 1874 Hist Abt 19, 41-53
(F6, 16)
Curtze M 1897 ZM Ph 42, 145
[Heron]
- 1897 ZM Ph 42(s), 143  [root]
Fuller A W 1957 M Gaz 41, 9-24
(MR18, 982) [chronom]
Guenther S 1885 Bib M, 137-140
Hughes Barnabas 1967 *Regiomontanus on
Triangles* (U Wisc Pr), 298 p
[biog, commentary]
Magrini S 1916 *Johannes de Blanchinis
Ferrariensis e il suo carteggio
col Regiomontano,* 1463-64, Ferrara
Sister Mary C Zeller 1946 *The
Development of Trigonometry from
Regiomontanus to Pitiscus,*
Ann Arbor
Ziegler A 1874 *Regiomontanus ein
geistiger Vorlaeufer des Columbus,*
Dresden  (F6, 16)
Zinner E 1936 Fors Fort 12, 210-212
- 1936 Philb Wi 9, 89-97
(F62, 1027)
- 1937 M Ter Er 55, 280-288
(F63, 13)

- 1938 *Leben und Wirken des Johannes Mueller von Koenigsberg, genannt Regiomontanus,* Munich, 307 p (F64, 910)

**Rehorovsky,** Vaclav Karel  1849-1911

Sobotka I 1913 Cas MF 42, 129-145 (F44, 21)

**Reichenbach,** Georg von  1772-1926

Dyck W v 1908 Int Woch 2, 1474-1486 (F39, 21)
- 1912 Int Con 5(2), 528

**Reichenbach,** Hans  1891-1953

Grunbaum A 1963 *Philosophical Problems of Space and Time,* NY, Ch 3

**Reina,** Vincenzo  1862-

Jadanza N 1919 Tor FMN 55, 138-142

**Reinhardt,** Karl  1895-1941

Maier W 1942 DMV 52(1), 75-83 (F68, 18; Z27, 196)

**Reinhold,** Erasmus  1511-1553

Wilkening W 1961 Z Vermess 86, 46-53 (Z101, 8)  [XVI, mens]

**Reissner,** H. J.

Harrington R P + 1949 *Reissner Anniversary Volume: Contributions to Applied Mechanics,* Ann Arbor, 1-12  (Z39, 243)

**Reitz,** Henry Lewis  1875-1943

Smith C D 1944 M Mag 18, 182-184

**Rellich,** Franz  1906-1955

Courant R 1957 M Ann 133, 185-190 (MR19, 108; Z77, 9)

**Remez,** Ergenii Yakovlevich  1896-

Gnedenko B V + 1956 Ukr IM 8, 218-222 (Z71, 4)

**Remundos,** Georgios I  1878-1928

Anon 1928 Gree SM 9, 3-4  (F54, 47)
Sakellariu N 1929 Gree SM 10, 73-89 (F55, 618)

**Renaldini,** Carlo  1615-1698

Gunther S 1897 ZMN Unt 28, 239-240 (F28, 43)

**Resal,** Henri  1828-1896

Cailler C 1896 Rv GSPA 7, 893
Jordan C 1896 JMPA (5)2, 453
Levy M 1896 JMPA (5)2, 455
- 1896 Par CR 123, 435

**Reuleaux,** Franz  1829-

Mises R v 1929 ZAM Me 9, 519 (F55, 15)
Weihe Carl 1925 *Franz Reuleaux und seine Kinematik,* Berlin, 108 p

**Reusch,** Friedrich Eduard  1812-1891

Boeklen O 1892 Wurt MN 5, 1-18 (F24, 35)

Reuschle, Carl   1847-1909

Anon 1909 Ens M 11, 404   (F40, 40)
Woelffing E 1910 Wurt MN (2)12, 34-
  56   (F41, 25)
Zech P 1876 Hist Abt 21, 1-4
  (F8, 21)


Revelli, F. A.   1716-1801

Revelli P 1918 *Un maestro del
  Lagrange: F. A. Revelli.* Genova,
  34 p   (F46, 11)


Reye, Carl Theodor   1838-1919

Castelnuovo 1922 Linc Rn (5)31,
  268-272   (F48, 21)
Geiser C F 1921 Zur NGV 66, 158-180
  (F48, 21)
Schur F 1921 M Ann 82, 165-167
  (F48, 21)
Segre C 1922 Linc Rn (5)31(1), 268-
  272
Timerding H E 1922 DMV 31, 185-203


Reymond, Arnold   1874-1958

Delorme S 1958 Arc In HS 11, 32-34
  (Z93, 9)


Reyneau, Charles René   1656-1728

Costabel P 1949 Rv Hi Sc Ap 2, 311-
  332 [Leibniz]


Rey Pastor, Julio   1888-1962

Anon 1946 Arg IM Ros 6, 355-377
  (MR8, 3; Z60, 17)
- 1962 Arg UMR 22, 201
Babini J A 1962 Arc In HS 15, 361-
  364   (Z106, 3)
- 1963 Isis 54, 259
Babini J A + 1962 Rv UM Arg 21(1),
  3-22, 35-55   (MR26, 687; Z106, 3)

Orts J M 1962 Rv M Hisp A 22, 96-101
  (Z101, 10)
Penalver P 1962 Rv M Hisp A 22, 102-
  105   (Z101, 10)
P P A 1959 Gac M 11, 123-129
  (Z95, 5)
Rios S 1962 Rv M Hisp A 22, 106-113
  (Z101, 10)
San Juan Llosa R 1962 Rv M Hisp A 22,
  60-93   (Z101, 10)
Vidal Abascal E 1962  Rv M Hisp A
  22, 116-120   (Z101, 10)


Rhabdas, Nicolaos Artabasdos   XIV

Sarton G 1947 Int His Sc 3(1), 681-
  682
Tannery P 1884 BSM (2)8, 263-277
  [Moschopulos]
- 1920 in his *Mémoires Scientifique*,
  4, 1-19, 61-198


Rheticus, Georg Joachim   1514-1576

Archibald R C 1949 M Tab OAC 3, 552-
  561   (MR11, 573; Z41, 338)
- 1953 M Tab OAC 7, 131
  (MR14, 833; Z50, 2)
Bernoulli J 1786 Ber Mm 1, 10
  [Pitiscus]
Burmeister Karl H 1967 *Georg Joachim
  Rhetikus,* Wiesbaden, 2 vols
  [bib]
Delcourt M 1934 Bude Bu 44, 3-15
  [Ramus]
Hipler F 1876 Hist Abt 21, 125-150
  (F8, 14)
Hunrath K 1899 ZM Ph 44(S), 213
  [Viète]


Riabovchinsky, M. D.

Demtchenko B 1954 Fr Air NT, i-xxxv
  (MR15, 924)

Riboni, Gaetano

Piazza S 1932 Pe M (4)12, 187-188
   (F58, 991)


Riccardi, Geminiano

Lodi L 1875 Boncomp 8, 1-50
   (F7, 9)   [bib]


Riccardi, Pietro   1828-1898

Bortolotti E 1932 Int Con 6, 407-408
   (F58, 4)
Cavani F 1899 *Della vita e delle
   opere del....,* Bologna, 66 p
   (Bln SA Ing) (F30, 20)
Favaro A 1881 Ven I At (5)7, 47-64
   (F13, 26)
Pantanelli D 1899 Loria 2, 23-29
   (F30, 20)


Riccati, Jacopo Francesco   1676-1754

Kristhoff A + 1905 Intermed 12, 162
   [bib]
Michieli  A A 1944 Ven ML 102, 535-
   587  (MR9, 74; Z61, 5)
- 1945 Ven ML 103, 69-109
   (MR9, 74)
- 1946 Ven ML 104, 771-859
   (MR9, 74)


Riccati, Vincenzo   1707-1775

Conte L 1952 Pe M (4)30, 125-128
   (MR14, 344)   [cubic eq]


Ricci, Matteo   1552-1610

Bosmans H 1921 Rv Q Sc (3)29, 135-
   151
Hofmann J E 1963 Centau 9, 139-193
   (MR28, 962)
Masotti A 1952 Lom Rn M 85, 415-445
   (MR15, 277; Z49, 5)


Tenca L 1954 Lom Rn M (3)18, 212-228
   (Z57, 243)   [Viviani]
Tonolo A 1961 Annali (4)53, 189-207
   (MR23A, 5)
Vella P F S 1910 Vat NL Mn 28, 51-71
   (F41, 5)


Ricci Curbastro, Gregorio   1853-

Anon 1954 *Celebrazione in lugo del
   centenario della nascita di...,*
   Lugo, 93 p  (MR16, 207)
- 1958 Archim 10, 230-234
   (Z82, 243)
Bortolotti E 1937 Mos Ve Ten 4, 17-2
   (F63, 816)
Labrador J F 1956 Gac M 8, 3-5
   (Z71, 4)
Levi-Civita T 1926 Linc Mm (6)1, 555
   564  (F52, 31)
Natucci A 1954 Batt (5)2, 437-442
   (MR16, 660; Z57, 3)
Palatini A 1926 Rv M Hisp A (2)1,
   49-51  (F52, 31)
Teofilato P 1925 Vat NLA 79, 195-196
   (F52, 31)
Tonolo A 1954 Pado Sem 23, 1-24
   (MR15, 592; Z56, 2)


Richardson, Archibald Read   1881-1954

Turnbull H W 1956 Edi MSN (40), 31-32
   (Z71, 4)
- 1956 LMSJ 31, 376-384   (Z70, 244)


Richardson, George   1841?-1904

Anon 1904 M Gaz 3, 25-26


Richardson, Lewis Fry   1881-1953

Gold E 1954 Lon RS Ob 9, 217-235
   (MR16, 660)
Sutton O G 1954 Listener (Mar 25),
   522-524   [meteorol]
Todd J 1954 M Tab OAC 8, 242-245
   (MR16, 207)

Richardson, Roland George Dwight
1878-1949

Archibald R C 1938 *A Semicentennial
History of the American Mathematical Society*, NY, 103-105
[bib, port]
- 1950 AMSB 56, 256-265
(MR11, 708; Z36, 146)

Richelot, Friedrich Julius
1808-1875

Aronhold S 1901 Arc MP (3)1, 38
Kobell F v 1876 Mun Si, 125
(F8, 22)

Richmond, Herbert William 1863-1948

Milne E A 1948 Nat 161, 877-878
(Z29, 386)
Milne E A + 1948 Lon RS Ob 6,
219-230
- 1949 LMSJ 24, 68-80
(MR10, 420; Z31, 99)

Riebesell, Paul    1883-1950

Lorey W 1951 Deu Vers M 1(2), 43-50
(Z42, 5)

Riemann, Georg Friedrich Bernhard
1826-1886

Anon 1867 Gott N (1), 305
- 1868 Mos MO Sb 3(2), 153-
- 1868 Lon RSP 16, lxix-
- 1895 Intermed 2, 415 (Q548)
B H T 1932 Muns Semb 1, 93-95
(F58, 994) [Siegel]
Boberil R du 1902 *Pascal et Riemann*,
Paris, 14 p  (F33, 16)
Burkhardt H 1892 *Bernhard Riemann*,
Goettingen (Math. Verein), 12 p
(F24, 21)
Conforto F 1951 Experien 7, 476-477
*Courant R 1926 Naturw, 814-818, 1265-
1277

Drenckhahn F 1927 Weltell 26, 140-144
(F53, 37)
Goncharov V L 1948 in *B. Riemann
Sochineniya*, Mos-Len
- 1959 Wiad M 2, 155-196
(MR24A, 5; Z91, 5)
Hancock H 1887 Brt AAS [Abelian fun]
Kagan V F 1933 *The Geometric Ideas
of Riemann...*, Mos-Len, 74 p
(F61, 958)
Klein Felix 1894 Deu NA Tag 66, 57-
72, 212-221  (F25, 77)
- 1894 *Riemann und seine Bedeutung
fuer die Entwicklung der modernen
Mathematik*, Leipzig, 18 p
- 1895 ZMN Unt 26, 55-66  (F26, 57)
*- 1895 AMSB 1, 165-180  (F26, 57)
- 1895 Annali (2)23, 209-224
(F26, 57)
- 1897 Gott N(1), 189-190  (F28, 17)
- 1897 DMV 4, 71-87  (F28, 53)
Kulczycki S 1955 Wiad M (2)1, 180-193
(MR21, 892)
Lampariello G 1957 Ber FIM 1, 222-234
(MR18, 860)  [phys]
Mira Fernandes A de 1956 Lisb (A)(2)5,
329-342  (MR19, 108; Z71, 3)
Neumann A 1865 *Vorlesungen ueber
Riemann's Theorie der Abelschen
Integrale*, Leipzig (2nd ed 1884)
Noether M 1900 DMV 8, 177-178
(F31, 16)  [Abelian fun]
- 1909 Gott N (1)23-25  (F40, 21)
Petrova S S 1965 Ist M Isl (16),
295-310  (MR33, 12)
[Dirichlet's princ]
Schering E 1870 Boncomp 3, 409-428
(F2, 22)
Schroeder K 1957 Ber FIM 1, 14-26
(MR18, 860)
Siegel C L 1932 QSGM (B)2  [numb thy]
Speiser A 1926 Crelle 157, 105-114
(F52, 26)  [Euler, phil]
Teichmueller O 1939 Deu M 4, 115-116
(F65, 16)  [Cauchy, complex anal,
Weierstrass]
Tricomi F G 1965 Tor Sem 25, 57-72
(MR34, 2422)  [Italy]

Riese, Adam   1489?-1559

Albrecht G 1894 *Adam Ries und die*
*Entwickelung unserer Rechenkunst,*
Prague, 18 p  (F25, 1912)
Anon 1892 ZMN Unt 23, 237  (F24, 8)
Berlet B 1855 *Ueber Adam Riese,*
Annaberg
- 1860 *Die Coss von Adam Riese,*
Annaberg
- 1892 *Adam Riese, sein Leben, Seine*
*Rechenbuecher und seine Art zu*
*rechnen. Die Coss von Adam Riese.*
*Mit dem Brustbild und der*
*Handschrift von Adam Riese,*
Frankfurt, 70 p  (F24, 8)
Carpenter D 1965 MT 58, 538-543
Darmstaedter L 1925 Fors Fort 1,
25-26  (F51, 38)
Deubner Fritz 1962 ZNTM 1(3), 11-
44  (Z102, 1)
Falckenberg H 1938 Deu M 3, 3-8
(F64, 12; Z18, 195)
Gruber A 1929 Bay Bild 3, 240-248
Mueller K 1959 MN Unt 11, 450-452
(Z100, 244)
Roch Willy 1959 *Adam Ries. Ein*
*Lebensbild des grossen, Rechen-*
*meister,*  Frankfurt, 79 p
Vogel Kurt 1959 *Adam Riese der*
*Rechenmeister,* Munich, 47 p
(MR23A, 692; Z97, 1)
- 1959 Praxis 1, 85-88  (Z86, 3)


Riesz, Frederic   1880-1956

Anon 1950 M Lap 1, 273-277
(MR12, 312)
- 1956 Hun Act M 7, 1-3   (Z70, 7)
- 1957 UMN 12(4), 155-166
(MR22, 770; Z78, 4)
Julia G 1956 Par CR 242, 2193-2195
(MR17, 932; Z70, 6)
Kalmar L + 1956 Szgd Acta 17, 1-3
(MR18, 550)
Rogosinski W W 1956 LMSJ 31, 508-512
(MR18, 182; Z70, 244)
Stsillarda K S + 1957 UMN 12(4),
155-160
Szokefalvi-Nagy B 1950 M Lap 1, 170-
182  (MR12, 1)

Rietti, Teofilo

E P 1926 Pe M (4)6, 72  (F52, 33)


Rietz, Henry Lewis   1875-1943

Crathorne A R 1944 An M St 15, 102-
108  (Z60, 17)
Smith C D 1944 M Mag 18, 182-184
(Z60, 17)


Rigaud, Stephen Peter   1774-1839

Rigaud G 1875 Lon As Mo N 36, 54-57
(F7, 10)


Righi, Augusto   1850-

Amaduzzi L 1920 Scientia 28, 467-472


Rignano, Eugenie

Anon 1930 Bo Fir (NS)9, 85-90


Rimini, Cesare   1882-

L C 1960 It UM (3)15, 349-351
(Z91, 7)


Rinaldini, Carlo

Tenca L 1956 Bln Rn P (11)3(1), 197-
208  (Z73, 3)


Rinonapoli, Michele

Marcolongo R 1912 Nap Pont 42(4)
(F43, 32)


Riquier, Charles Edmond Alfred   1853

Anon 1929 Ens M 28, 136-137

Ritchie, William Irvine    -1903

Anon 1903 LMSP 35, 466-471


Ritt, Joseph Fels    1893-1951

Lorch E R 1951 AMSB 57, 307-318
  (MR13, 2; Z42, 343)
Smith P A 1956 Am NASBM 29, 253-258


Rittenhouse, David    1732-1796

Babb M J 1932 Sci Mo 35, 523-
  [port]
Brooke Hindle 1964 *David Rittenhouse,*
  Princeton, 394 p
Cope T D 1933 Franklin 215, 287-297
Rufus W C 1941 Scr 8, 228-231
  (MR4, 181; Z60, 9) [Newton]
Struik D 1966 Isis 57, 282-284


Ritter, Ernst    1867-1895

Klein F 1897 DMV 4, 52-54
  (F28, 21)


Ritter, Robert Bernhard Alwin   1905-

Pinl M 1961 DMV 63(1), 137-140
  (MR23A, 584; Z94, 5)


Ritter, Wilhelm    1847-1906

Guidi C 1907 Tor FMN 42, 163-166
  (F38, 34)
Timerding H E 1907 DMV 16, 244-248
  (F38, 34)


Rivard, Dominique François
  1697-1778

Archibald R C 1948 *Mathematical
  Table Makers,* NY, 71
  (Scr Stu 3)   [bib]


Robert of Chester    XII

  = Robertus Castrensis, etc

Karpinski L C 1911 Bib M (3)11, 125-
  131 [al Khwarizmi]
- 1915 *Robert of Chester's Latin
  Translation of the Algebra of
  Al-Khwarizmi,* NY
Sarton G 1931 Int His Sc 2(1), 175-
  177   [bib]


Robert of England    XIII

  = Robertus Anglicus
  = Johannes Anglicus

Alliaume M 1926 Brx SS 45(1), 139-148
  (F52, 13)
Curtze M 1899 ZM Ph 44(S), 41-63
  (F30, 60)
Mansion P 1898 Mathesis (2)8, 83-87
  (F29, 42)
Sarton G 1931 Int His Sc 2(2), 993-
  994
Steinschneider M 1896 Bib M (2)10,
  102-104   (F27, 5)
Tannery P 1897 Bib M (2)11, 3-6
  (F28, 3)


Roberts, Samuel    1827-1913

Roberts S 1914 Lon RSP (A)89, xx-xxi


Roberts, Samuel Oliver    XIX

Anon 1899 M Gaz 1, 18-20, 278


Robertson, H. P.    1903-1961

Taub A H 1962 SIAM J 10, 737-746
  [port, bib]
- 1963 SIAM J 11, 741-750
  (Z106, 3)

Roberval   Gilles Pérsonne de
1602-1675

Auger Léon 1949 Thales 6, 59-67
  [Descartes]
- 1962 *Un savant méconnu: Gilles
  Personne de Roberval*, Paris, 215 p
  (MR27, 4; Z101, 8)
Costabel P 1950 Rv Hi Sc Ap 3, 80-
  86   [Diophantine anal]
de Waard C 1921 BSM 45, 206-216,
  222-230 [cycloid, letter Mersenne]
Hofmann J E + 1952 Rv Hi Sc Ap 5,
  312-333 (MR14, 833) [Viete]
Jacoli F 1875 Boncomp 8, 265-304
  (F7, 21) [calc, Torricelli]
Lebesgue H 1922 Rv Sc, 249-262
  [Ramus]
Rakhmaninov I I 1884 Kagan (1)
Walker E 1932 T Col Cont (446)
  [calc]
- 1938 *A Study of the Traité des
  indivisibles of Gilles Persone
  de Roberval*, NY, 272 p
  (F58, 988)


Robin, Victor Gustave  1855-1897

Raffy L 1899 *Gustave Robin...*,
  Paris
- 1902 BSM (2)26, 87 [thermodyn]


Robins, Benjamin  1707-1751

Hartenberg R S 1956 Z Flug 4, 213-
  217 (Z70, 7)


Robins, Sylvester   1834-1900

Smith D E 1900 AMM 7, 179


Robson, Alan  1888-

Anon 1958 M Gaz 42, 203-204
  (Z80, 9)

Rocca, Giannantonio   1667-1656

Favaro A 1902 Bib M (3)3, 328-412
  (F33, 16)


Rode, H. H.

Schieldrop E 1930 Nor M Tid 12,
  85-86   (F56, 816)


Rodenberg, Karl Friedrich   1851-

Korteweg D J 1893 Nieu Arch 20(1),
  63-97   [cubic surf]


Rodriguez, C. A. de Campos

Anon 1905 Teixeira 15, 181


Roe, Edward Drake Jr 1859-1929

Campbell A D 1930 AMSB 36, 161


Roemer, Ole   1644-1710

Meyer-Bjerrum K 1910 BSM (2)34,
  73-96
Van Biesbroeck G 1913 Ciel Ter 34,
  152-167 [ast]
Van Biesbroeck G + 1913 Dan Ov (4),
  213-324   [ast]


Rogers, Leonard James  1862-1933

Dixon A L 1934 LMSJ 9, 237-240
  (F60, 15)
- 1934 Lon RS Ob 1, 299-301
N R C 1933 Nat 132, 701-702
  (F59, 37)


Rohn, Karl F. W.   1855-1920

Hoelder O 1921 Leip Ber 72, 109-127
  (F48, 16)

Lorey W 1920 Leip HSAN 2, 70-71
   (F47, 13)
Schur F 1923 DMV 32, 201-211


Rohrbach, Karl Ernst Martin   1861-
   1932

Lietzmann W 1933 ZMN Unt 64, 138


Rollandus   XV

Karpinski L C 1914 AMSB (2)20, 305-
   306  (F45, 96)  [alg]


Rolle, Michel  1652-1719

Anon 1895 Intermed 2, 210 (Q82)
   [cascades]
Cajori F 1910 Bib M (3)11, 300-313
   (F41, 57)
- 1911 AMSB (2)18, 55-56  (F42, 70)
   [thm]
- 1918 AMM 25, 291-292
   [Rolle curve]
Cantor M  1895 Intermed 2, 96
   [cascades]
Sergescu Petric 1942 *Eine Episode
   in dem Kampf fuer den Triumph
   der Differentialrechnung...*,
   Bucharest, 17 p  (F68, 15)
   [Sourin]
Shain J 1937 AMM 44, 24-29
   [cascades]
Yanovskaya S A 1947 Mos IIET 1,
   327-346  (MR12, 312)  [calc]


Rollin

Hendle P 1916 Intermed 23, 134-135
   (Q4557)  [alg]


Romagnosi, Giovanni Domenico
   1761-1835

Stiattesi A 1878 *Notizia storica di
   G. D. Romagnosi Considerato*

*precipuamcute come matematico,*
Firenze   (F10, 17)


Romain, Adrien  1561-1615

   = Adrianus Romanus = van Roomen
   = Adraen oder Romain

Bosmans H 1904 Brx SS 28(B), 411-429
   (F35, 56)
- 1905 Brx SS 29A, 68-79  (F36, 65)
- 1906 Brx SS 30(2), 267-287
   [al-Khwarizmi]


Romanov, Nikolai Pavlovich   1907-

Sarymsakov T A + 1957 UMN 12(3),
   251-253  (Z78, 4)


Romanovskii, Vsevolod Ivanovich
   1879-1954

Nikolaev A N 1939 Tash UA 5(S), 9-32
   (F65;1, 18; Z23, 388)  [anal]
Sarymsakov T A 1950 UMN 5(3), 184-
   186  (Z36, 146)
- 1955 UMN 10(1), 79-88  (Z64, 4)


Romanus.  See Adrianus.


Roomen, Adriaan van  1561-1615

   = Adrianus Romanus

Ritter F 1880 BSM (2)4, 171-182
   (F12, 11)  [Fermat, Viète]
Sarton G 1934 Isis 21, 209
   [al-Khwarizmi]
Vetter Q 1930 BSM 54, 7 p  [eq]


Roos, C. F.

Davis H T 1958 Ecmet 26, 580-589
   (Z82, 12)

Rosa, Paolo    1825-1874

Marchetti F 1875 Boncomp 8, 305-320
  (F7, 12)    [bib]


Rosati, Carlo   1876-1929

Conti A 1929 Bo Fir (2)8, 120
  (F56, 816)
Puccianti L 1929 Bo Fir 8, 121-124
Scorza G 1929 Bo Fir 8, 125-128
- 1929 Pe M (4)9, 289-291
  (F55, 617)


Rosenbach, Joseph Bernhardt   1897-

Anon 1952 MT 45, 77


Rosenberger, Johann Karl Ferdinand
  1845-1899

Anon 1900 Wiad M 4, 134
Gunther S 1900 Bib M (3)1, 217


Rosenblatt, Alfredo   1880-1947

Anon 1936 Lim Rev 38(418), 21-28
  (F62, 1040)
Valentinuzzi M 1948 Arg UMR 13, 97-
98   (Z30, 338)


Rosenhain, Johann Georg   1816-1887

Cantor M 1889 ADB 29, 209   double
  theta functions   [theta fun]
Hermite C 1887 Par CR 104, 891


Rosenthal, Arthur . 1887-

Haupt O 1960 DMV 63(1), 89-96
  (MR23A, 136; Z91, 7)

Rosius, Jacobus   XVII

Burchhardt F 1903 Basl V 16, 376-387


Roth, K. F.   XX

Davenport H 1960 Int Con (1958),
  lvii-lx   (MR22, 929)
  [Fields medal]


Rothe, Hermann   1882-1923

Radon J 1924 DMV 35, 172-175
  (F52, 33)


Rothe, Rudolph   1873-1942

Gruess G 1942 ZAM Me 22, 302-303
  (F68, 18; Z27, 291)


Rouché, Eugène   1832-1910

Anon 1910 Ens M 12, 425   (F41, 33)
- 1910 Mathesis (3)10, 241-242
- 1910 Rv GSPA 21, 761
Cavallaro V G 1938 Bo Fir 34, xxxvi-
  xxxviii   [tria]


Routh, Edward John   1831-1917

Anon 1912 Lon RSP (A)84, xii-xvi
  (F43, 32)
Forsyth A R 1907 LMSP (2)5, xiv-xx
  (F38, 41)
Larmor J 1907 Nat 76, 200-202
  (F38, 41)


Rowe, Charles Henry   1893-1943

Semple J G 1944 LMSJ 19, 241-244
  (Z60, 17)

Rowe, John XIX

Hendricks F 1858 Assur Mag 7, 136-
[ins]

Rubini, Raffaele 1817-1890

Capelli A 1891 Nap Pont 21, 275-281
(F23, 25)

Rudd, Thomas

Archibald R C 1915 Nieu Arch (2)11,
191-195 (F45, 78) [Cardinael]

Rudio, Ferdinand 1856-

Anon 1929 Ens M 28, 136-137
Schroeter C + 1926 Zur NGV 71, 115-
135 (F52, 34)

Rudnicki, Juliusz 1881-1948

Anon 1948 Colloq M 1, 268-271
(Z37, 291)

Rudolff, Christof 1499?-1545?

Eneström G 1886 Bib M, 243 (Q12)
Pringsheim A 1886 Bib M, 239-244

Ruffini, Paolo 1765-1822

Agostini A 1926 Arc Sto 7, 209-215
[Frullani, series]
Anon 1929 Bln Rn 2, 33
Barbensi Gustavo 1956 *Paolo Ruffini*,
Modena, 138 p (MR17, 814)
Bompiani E 1955 Archim 7, 145-149
(Z66, 5)
Bortolotti Ettore 1902-1903
*Influenza dell'opera matematica
di Paolo Ruffini sullo svolgiamento
delle teorie algebriche*,
Modena, 57 p (F33, 55)

- 1906 It SS (3)14, 291-325
(F38, 20) [eq]
- 1907 Paler R 24, 403-411 (F38, 20)
- 1916 Mod Mm (3)12, 179-193
[group]
- 1928 Bln Rn (NS)33, 83-87 [group]
- 1932 Int Con 6, 401-405
(F58, 40)
- 1943 It UM (2)5, 114-120
(MR7, 354; Z60, 8)
- 1947 Bln Mm (10)3, 215-224
(MR10, 175)
Burkhardt H 1892 Ab GM 6, 119-159
[group]
- 1894 Annali (2)22, 175-212
(F25, 63) [group]
Cajori F 1911 AMSB (2)17, 409-414
[Horner meth]
Gonçalves J V 1951 Lisb (A)(2)1, 408-
409 (MR14, 1051)
Lombardi 1824 *Notizie sulla vita...di
Paoli Ruffini*, Modena
Mieli A 1924 Arc Sto 5, 332-334
Miller G A 1909 Bib M 10, 318
Pascal E 1903 Lom Gen (2)36, 159-161
(F34, 11)
Riccardi P 1898 Mod Mm (3)1, 105-129
(F29, 7) [Paoli]
Vacca G 1947 Mod At (5)7, 203-204
(MR9, 486)

Rumi, Qadi Zade al- XIV-XV

= Kazi zade al-R.

Rosenfeld B A + 1960 Ist M Isl 13,
533-538 [sine]
- 1960 Ist M Isl 13, 552-556

Runge, Carl David Tolmé 1856-1927

Anon 1926 Zt Ph 7, 360a-361a
(F52, 34)
- 1927 Nat 119, 533
- 1927 Naturw (10), 225-248
Courant R 1927 Naturw 15, 229-231
(F53, 31)
Lietzmann W 1927 ZMN Unt 58, 482-483
(F53, 31)

Prandtl L 1927 Gott N(1), 58-62
  (F53, 31)
- 1927 Naturw 15, 227-229
  (F57, 31)
Ramser H 1952 Experien 8, 237
Runge I 1949 Gott Ab (3)(23)
  (MR11, 708; Z41, 2)
Trefftz 1926 ZAM Me 6, 423-424
  (F52, 34)


Runkle, John Daniel    1822-1902

Newcomb S 1903 AMM 10, 130-
Tyler H W 1903 AMM 10, 183-185
- 1902 Tec Rv 4, 277-


Russell, Bertrand    1872-

Ayuso A 1958 Gac M 10, 71-75
  (Z80, 9)
Darbon André 1948 La philosophie
  des mathématiques. Etude sur
  la logistique de Russell,
  Paris,  215 p
Hardy G H 1942 Bertrand Russell and
  Trinity..., Cambridge, 62 p
Jourdain P E B 1912 Monist 22, 149-
  158
Leggett H W 1950 Bertrand Russell
  O. M., NY, 79 p  (MR12, 1;
  Z38, 148)
Russell B 1967 The Autobiography of
  B. R. 1872-1914, Boston, 356
- 1968 The Autobiography...1914-
  1944 (Little, Brown), 404 p
Schilpp Paul A ed 1944 The Phil-
  osophy of BR, Evanston, Ill,
  830 p  [*bib]


Russo, Giovanni

Anon 1929 Pe M (4)9, 140
  (F55, 614)


Rychlik, Karel    1885-

Korinek V 1960 Cas M 85, 492-498
  (Z117:2, 251)


Rymarenko, Boris Aleksandrovich
  1906-1966

Videnskii V S ɤ 1966 UMN 21(6), 153-
  154   (MR34, 19)


Saadia Gaon   892-942

Gandz Solomon 1943 in American
  Academy for Jewish Research,
  Saadia Anniversary volume, NY,
  141-195
Hirsch M 1951 MT 44, 591
Polachek H 1933 Scr 1, 245-246
  (F59, 19)  [X, arith]
Sarton G 1927 Int His Sc 1, 627
  [bib]


Saalschuetz, Louis   1835-

Anon 1913 Leop 49, 64
  (F44, 33)


Saavedra y Moragas, Eduardo

Echegaray D J 1912 Esp SM 1, 333-353
  (F43, 48)


Sabinin, E. T.

Wolkov A 1911 Mos MO Sb 27, 407-412
  (F42, 24)


Saccheri, Girolamo   1667-1733

Agostini A 1931 It Ac Mm 2(2), 1-20
  (Z3, 242; F57, 28)  [letter]
Allegri L 1964 Int Con HS 10
Beltrami E 1889 Linc Rn Mo 5, 441-448
*Bonola R 1955 Non-Euclidean Geometry
  (Dover)
Bosmans H 1925 Rv Q Sc (4)7, 401-430
  (F51, 21)
Brusotti L 1952 Geno AL, 155-164
  (MR14, 1050; Z53, 197) [Filippa]
Emch A F 1935 Scr 3, 51-60, 143-152,
  221-233  (Z10, 386; 11, 193; 12, 98)

Favaro A 1903 Rv FMSN 4, 424-434
(F34, 9) [letter, Viviani]
Katsoff L O 1962 MT 55, 630-636
[quadril]
Langenkamp O 1907 *Ueber Saccheris
Untersuchungen des Parallelaxioms,*
Munster, 30 p (F38, 74)
Musatti C L 1926 Pe M (4)6, 313-329
(F52, 20) [logic]
Pascal A 1914 Batt 5, 229-251
(F45, 13)
Russo F 1956 Int Con HS 8, 29-32
Segre C 1903 Tor FMN 38, 351-163
Sister M M Fitzpatrick 1964 MT 57,
323-332
Smith D E 1935 Scr 3, 5-10
[Euclid, Omar Khayyam]
Vailati G 1903 Rv Fil 5, 13 p
(F34, 9) [logic]
Vassiliev A 1893 Kazn FMO (2)3(2)

Sacrobosco    XIII

= John of Halifax (Holywood,
Halyfax, Holywalde) = Sacro Busto

Bosmans H 1925 Brx SS, 458-462
Curtze M 1895 Bib M (2)9, 36-37
(F26, 48)
Eneström G 1894 Bib M 8, 63
- 1897 Bib M (2)11, 97-102
(F28, 38) [limits]
- 1899 Bib M 13, 32
Eneström E + 1901-1915 Intermed 7,
199, 268; 8, 263-265; 9, 275; 10,
16, 82, 261, 16, 271; 22, 10-11
(Q1906) [death]
Karpinski L C 1910 AMM 17, 108-113
[Jordanus Nemorarius]
Riccardi P 1894 Bib M (2)8, 73-78
(F25, 58)
Sarton G 1931 Int His Sc 2(2), 617-
619 [bib]
Thorndike Lynn 1949 *The Sphere of
Sacrobosco and its Commentators*
(U Chicago P), 506 p (MR10, 419)

Sadd, Grace Dorothy

H H 1925 LMSP (2)23, lxix   (F51, 32)

Saint Germain, Alberti Leon de
1839-1914

Bioche C 1914 Ens M 16, 481-483
(F45, 46)

Saint-Venant, Adhémar Jean Claude
Barre de    1797-1886

Higgins T J 1942 AJP 10, 248-259
(F68, 2) [torsion]
Pade H 1904 Rv GSPA 15, 761-767
[mech]
Phillips E 1886 Par CR 102, 141-147
Todhunter Isaac 1886-1893 *A History
of the Theory of Elasticity and of
the Strength of Materials,*
Cambridge, 2 vols [physics,
vectors, biog]

Saladini, Girolamo    1731-1813

Agostini A 1938 It UM Con 1, 453-456
(F64, 17) [propor]
- 1940 It UM Con 2, 886-899
(MR8, 497; Z26, 291; F68, 16)
[calc]

Salee, Achille    1883-1932

Kaisin F 1932 Rv Q Sc (4)22, 1-29

Saligny, Anghel

Anon 1925 Gaz M 30, 401-406
(F51, 29)

Salmon, George    1819-1904

Anon 1904 M Gaz 32, 273
- 1904 Nat 69, 324-326   (F35, 37)
- 1905 Lon RSP 74, 347-355

Ball R S 1904 LMSP (2)1, v-xxviii
(F35, 38)
Fehr H 1904 Ens M 6, 232 (F35, 38)
Lamb H 1904 Brt AAS 74, 421
Noether M 1905 M Ann 61, 1-19
(F36, 27)
Quint N 1908 Wisk Tijd 4, 129-130
(F39, 31)

Sanchez Perez, A.

Anon 1959 Gac M 11, 3-5 (Z85, 7)

Sanderson, Mildred Leonora

Dickson L E 1915 AMM 22, 264

Sang, Edward 1805-1890

Archibald R C 1948 *Mathematical Table
Makers,* NY, 71-72 (Scr Stu 3)
[bib]
Peebles D B 1896 Edi SP 21, XVIII
[bib]

Sannia, Achille 1823-1892

Masoni U 1894 Nap Pont 24 (F25, 33)
Torelli G 1892 Paler R 6, 48-52
(F24, 36)

Sannia, Gustavo 1875-1930

Pizzuti M 1933 Nap Pont 62, 533-540
(F59, 856)
Scorza G 1931 Batt 69, 227-231
(F57, 46)
- 1931 It UM 10, 181-182
(F57, 46)

Santini, Giovanni 1787-1877

Lorenzoni G 1877 *Giovanni Santini.
La sua vita e le sua opere,*
Padua (F9, 14)

Millosevich E 1878 Boncomp 11, 1-110
(F10, 19)
Turazza D 1878 Ven I At (5)4, 5-21
(F10, 19)

Sarpi, Paolo

Cassani P 1882 *Paolo Sarpi e le
scienze matematiche e naturali,*
Venice
Favaro A 1884 Ven I At (6)1, 893-913

Sarrau, Jacques Rose Ferdinand
Emile 1837-1904

Anon 1904 Ens M 6, 311 (F35, 38)
Vieille P 1905 Rv GSPA 16, 7

Sarton, George 1884-1956

*Anon 1957 Isis 48(3)
Ashley-Montagu M F ed 1947 *Studies
and Essays...in Homage to George
Sarton...,* NY, 608 p
Rome A 1958 Osiris 13, 5-10
(Z84, 3)

Sasayama, Hiroshi

Anon 1958 Sasayama 1, 188-190
(Z80, 9)

Sauri, Abbé 1741-1785

Rossier P 1948 Arc In HS 27, 297-311
(MR9, 486)

Saurin, Joseph 1659-1737

Sergescu P 1942 *Eine Episode aus dem
Kampf um den Triumph der Differ-
entialrechnung. Der Streit Rolle-
Saurin 1702-1705* Bucarest, 17 p
(F68, 15)

Sauvage, Louis Charles    1853-

Fabry E 1924 Mars Ann 24, 98-103
Payot M J + 1924 Mars Ann 24, 104-110


Savasorda, Abraham    XII

    = Abraham bar Hiyya (Chyja)
    ha-Nasi

Cohen B 1918 Mo Ges W Ju 62, 186-194
    [encyclopedia]
Levey M 1952 Isis 43, 257-264
    [encyclopedia]
- 1954 Osiris 11, 50-64
    (MR16, 660; Z55, 3) [algorism]
Sarton G 1931 Int His Sc 2, 206-208
*Steinschneider M 1864 Heb Bib 7, 84-
    95 [encyclopedia]
- 1866 ZM Ph 11, 235-
- 1867 ZM Ph 12, 1-44
    [Abraham ibn Ezra]


Saville, Henry    1549-1622

Anon 1950 Lon Tim LS (Jan 27)
Wright E M + 1960 Scr 25, 63-65
    (Z92, 3)


Savin, Gurii Nikolaevich    1907-

Ishlinskii A Yu + 1957 Ukr IM 9, 225-
    229 (Z77, 242)


Saxony.  See Albert.


Sbrana, Francesco    1891-

Graffi D 1958 It UM (3)13, 618-620
    (Z82, 243)
Togliatti E 1959 Geno AL 15, 466-471
    (MR22, 619)


Schaar, Mathias    1817-1867

Anon 1868 Bel Anr, 99    (F1, 14)


Schachenmeier, Wilhelm Friedrich
    1882-1927

Foeppl L 1928 ZAM Me 8, 83    (F54, 43)


Schaeffer, Albert Charles    -1957

Duffin R J 1957 Sci 126, 156
    (MR19, 518)


Schafheitlin, Paul    1861-

St. Jolles  1925 Ber MG 24, 14-20
    (F51, 31)


Schapira, Hermann Hirsch    1840-1898

Cantor M 1898 Loria 1, 106-109
    (F29, 20)
Koehler C 1900 DMV 8, 61 (F31, 18)


Schauder, J. P.

Leray J 1959 Wiad M (2)3, 13-19
    (MR22, 1131; Z91, 7)


Scheefer, Ludwig    1859-1885

Cantor G 1885 Bib M, 197-199
Dyck W 1886 Hist Abt 31, 50-55


Scheffelt, Michael    1652-1720

Keefer H 1911 MN Bl 8, 81-84
    (F42, 59)


Scheibner, Wilhelm    1826-

Neymann C 1908 Leip Ber 60, 375-390
    (F39, 42)

Scheiner, Christoph 1575-1650

Braunmuel A v 1891 *Christoph
   Scheiner als Mathematiker,
   physiker und Astronom,* Bamberg
Huber G 1925 Weltall 24, 149-151
   (F51, 20)


Schell, Wilhelm Joseph Friedrich
   Nicolaus 1826-1904

Anon 1904 Ens M 6, 232-233 (F35, 38)
Hatzidakis N 1905 DMV 14, 389-394
   (F36, 29)
Lueroth J 1905 DMV 14, 113-121
   (F36, 29)


Schellbach, Karl Heinrich 1805-1892

Anon 1892 ZMN Unt 23, 315-317, 637-
   638 (F24, 36)
Mueller Felix 1892 *Karl Heinrich
   Schellbach. Gedächtnisrede...,*
   Berlin, 35 p (F24, 36)
- 1904 Arc MP (3)8
- 1905 Ab GM 20, 1-86
   (F35, 17; 36, 16; 41, 15)
- 1905 Ber MG 4, 8-10 (F36, 16)
- 1906 BSM (2)30, 317-322
Poske F 1892 Z Ph C Unt 5, 301-303
   (F24, 36)


Schering, Ernst Christian Julius
   1833-1897

Anon 1897 Crelle 118, 86
Baerwald H 1925 Phy Z 26, 633-635
   (F51, 30)
Fuchs L 1898 Crelle 119, 86
   (F29, 16)
Klein F 1899 DMV 6, 25-27
   (F30, 19)
Loria G 1898 Loria 1, 26-29
   (F29, 16)
Mansion P 1892 Brx SS 16(A), 51-53
   (F24, 47) [geom]
Schur W 1898 Leip AS 33, 2-5
   (F29, 16)

Young G C + 1898 Nat 57, 416
   (F29, 16)


Scheubel, Johannes 1494-1570

Day Mary S 1926 *Scheubel as an
   Algebraist,* NY, 175 p (F52, 15)
Staigmueller H 1899 ZM Ph 44(s),
   429-469 (F30, 4) [alg]


Scheuchzer, Johann Jacob 1672-1733

Steiger R 1933 Zur NGV 78(S21)
   (F59, 845) [bib]


Schiaparelli, Giovanni Virginio 1835

Anon 1910 Rv Sc 13, 58
- 1910 Lom Gen (2)43, 525-532
Bianchi E 1935 Milan Sem 9, 87-111
   (F61, 953)
Celoria G 1910 Linc Rn (5)19(2),
   528-555
Fergola E 1910 Nap FM Ri (2)16,
   203-205
Gabba A 1954 Lom Rn M (3)18, 290-294
   (MR16, 985) [Cremona]
Gabba L 1924 Arc Sto 5, 57-58
- 1928 Milan Sem 2, 121-138
- 1941 Milan Sem 15, 29-51 (Z27, 4)
Jadanza N 1912 Tor Mm (2)62, 361-385
Loria G 1910 Bib M (3)10, 330-340
Maggi P + 1900 Rv FMSN 2, 289
Moreux T 1910 Rv GSPA 21, 629-630
Seeliger H v 1911 Mun Si, 17-21


Schiavoni, Federico 1810-1894

Jadanza N 1894 Nap Pont 24
   (F25, 47)


Schickard, Wilhelm 1592-1635

Anon 1935 Nat 136, 636 (F61, 17)

Schiff, P. A.

M P 1910 Mos MO Sb 27, 259-262
   (F41, 25)


Schilling, Georg Friedrich
   1868-1950

Graf 1951 DMV 55(1), 1-4  (Z42, 243)


Schimmack, Rudolph    1881-1912

Weinrich 1913 ZMN Unt 44, 113-116
   (F44, 28)


Schissler, Christoph  XVI

Bobinger Maximillian 1954 *Christoph
   Schissler der Aeltere und der
   Juengere,* Augsburg-Basel, 148 p
   (Z56, 243)
Wunderlich H 1955 Dres THWZ 4, 199-
   227  (MR16, 986)


Schjellerup, Hans Carl Friedrich
   Christian    1827-1887

Dreyer J L E 1888 Nat 37, 154-155
Hjort V 1887 Tid M (5)5, 148-153
   (F21, 1255)


Schlaefli, Ludwig  1814-1895

Brioschi F 1895 Linc Rn (5)4(1), 310-
   312  (F26, 39)
Burckhardt J J 1942 Bern NG, 1-22
   (MR9, 170; F68, 17; Z61, 5)
- 1948 Elem M (S4), 23 P  (MR11, 708)
Fuchs L 1895 Crelle 115, 350
   (F26, 39)
Graf J H 1895 Bern NG 120
- 1896 *Ludwig Schlafli,* Bern
- 1905 Bern NG, 70-107 [Cayley,
   letters]
- 1915 Bern NG, 50-69
   [Borchardt]

- 1915 Loria 17, 36-40, 81-86, 113-
   123  (F45, 22)
- 1916 Loria 18, 21-35, 49-64, 81-83,
   113-121   (F46, 14, 35)
- 1917 Loria 19, 9-14, 43-49, 64-73
   [Italy]
Linder A 1931 Helv CM 3, 148-150
   (F57, 35)  [relativity]
Schlaginhaufen O 1932 Bern NG, 35-66


Schlauch, William S.    1873-1953

Anon 1948 MT 41, 299-301
Reeve W D 1953 Scr 19, 91-92
Schlauch W S 1949 MT 41, 299-301


Schlegel, Stanislaus Ferdinand Victor
   1843-

Fehr II 1906 Ens M 8, 55  (F37, 23)


Schleiermacher, Ludwig    1855-

Borkowski H 1898 Arc MP (2)16, 337-
   346  (F29, 29)


Schlesinger, Ludwig  1864-

Dunnington G W 1935 Scr 3, 67-68
Varicak V 1907 DMV 16, 320-321
   (F38, 67)


Schlick, Friedrich Albert Moritz
   1882-1936

Juhos B 1957 Stu Gen 10, 81-87
   (Z77, 9)


Schloemilch, Oskar Xavier   1823-1901

Anon 1901 Loria 4, 124-125
   (F32, 30)
Cantor M 1901 Bib M (3)2, 260-263
   (F32, 29)
Helm G 1901 ZM Ph 46, 1-7 (F32, 30)
Krause M 1901 Leip Ber 53, 507-520
   (F32, 30)

Schmalz, Johann Jakob    1820-1892

Wolf R 1892 Zur NGV 37, 228-232
  (F24, 36)


Schmehl, Christoph    1853

Pfersdorff F W 1930 ZMN Unt 61,
  182-184  (F56, 30)


Schmeidler, Werner Johannes    1890-

Hellwig G + 1960 Z Flug 8, 177
  (Z95, 5)


Schmid, Hermann Ludwig    1908-1956

Hasse H 1958 M Nach 18, 1-18
  (MR20, 265; Z79, 244)
Schmidt H 1958 DMV 61(1), 7-11
  (MR20, 848; Z80, 5)


Schmidt, Adolf  Friedrich Karl
1860-

Anon 1932 Ber Si 1-12, 75-76


Schmidt, Erhard    1876-1959

Julia G 1959 Par CR 249, 1676-1677
  (Z88, 7)
Nevanlinna R 1956 Fors Fort 30, 60-62
  (Z70, 7)
- 1956 M Nach 16, 1-6
  (MR17, 814; Z70, 244)
Rohrback H 1967 DMV 69(4, 1), 209-
  244 (MR37, 2570)
Schroeder K 1963 M Nach 25, 1-3
  (MR27, 4; Z106, 3)
Tietze H 1960 Mun Jb 176-177
  (MR22, 1131)


Schmidt, Franz    1827-1901

Halsted G B 1902 Amm 8, 107-110


Staeckel P 1902 DMV 11, 141-146
  (F33, 38)


Schmidt, Wilhelm    1862-1905

Rudio F 1905 Bib M (3)6, 354-386
  (F36, 37)


Schober, Karl    1859-

Wirtinger W 1900 DMV 8, 66


Schoener, Johann    1477-1547

Pelseneer J 1932 Isis 17, 259
  [notat]
Sarton G 1927 Int His Sc 1, 569, 759
Smith D 1910 Bib M (3)11, 79-80


Schoenflies, Arthur Moritz    1853-

Bieberbach L 1923 DMV 32, 1-6
  (F49, 14)
Reinhardt K 1928 MN Bl 22, 87-88
  (F54, 47)


Scholz, Arnold    1904-1942

Trussky-Todd O 1952 M Nach 7, 379-
  386  (MR14, 122; Z46, 3)


Scholz, Heinrich    1884-1956

Anon 1957 MP Semb 5, 181
  (MR19, 1150)
Hermes H 1955 MP Semb 4, 165-170
  (MR17, 338)


Schopenhauer, Arthur    1788-1860

Rostand F 1953 Rv Hi Sc Ap 6,
  203-230 (MR15, 384) [proof]

Schorling, Raleigh    1887-1950

Breslich E R 1951 MT 44, 67-68
Edmonson J B 1951 MT 44, 76
Jones P S 1951 MT 44, 79-80, 99, 107,
    134, 148
Mayor J + 1951 SSM 50, 523-524
Reeve W D + 1952 MT 44, 81-82
Rickard R B 1952 MT 44, 69-75
Wickham J J 1952 MT 44, 77-78


Schottky, Friedrich Hermann  1851-

Bieberbach L 1936 Ber Si, cv-cvi
    (F62, 27)
Schottky F 1903 Ber Si, 714-716
    (F34, 39)


Schou, Eric    1873-1928

Anon 1929 Ens M 28, 136-137
Bonnesen T 1928 M Tid B, 55-56
    (F54, 42)


Schoute, Pieter Hendrik  1846-

Anon 1913  Amst Vs G 21, 1396-1400
    (F44, 34)
Fehr H 1913 Ens M 15, 256-257
    (F44, 34)
Teixeira F G 1913 Porto Ac 8, 125


Schoy, Carl  1877-1925

Smith D E 1926 AMM 33, 28-31
    (F52, 33)  [Arab]
Spies O 1926 Deu Morg Z 5, 319-327
Wieleitner H 1927 DMV 36, 163-167
    (F53, 32)


Schreiber, Edwin W.    1890-1951

Ayre H G 1953 MT 45, 243-244
Gingery W G 1953 SSM 52, 173-174


Schreier, Otto    1901-1929

Menger K 1930 Mo M Phy 37, 1-6
    (F56, 32)


Schrentzel, Wilhelm    1861-1896

Schlesinger L 1897 Hist Abt 42, 1-5
    (F28, 25)


Schroeder, Ernst    1841-1902

Crespo R 1951 Gac M (1)3, 211-214
    (MR13, 810; Z43, 245)
Haussner R 1902 Rv M 8, 54-56
    (F33, 43)
Lueroth J 1903 DMV 12, 249-265
    (F34, 28)
Mueller E 1905 Int Con 3, 216-218
    (F36, 25)


Schroeter, Heinrich Eduard    1829-1892

Sturm R 1892 Bresl U Ch  (F24, 37)
- 1892 Crelle 109, 358-360
    (F24, 37)
- 1893 DMV 2, 32-41  (F25, 40)
Vogt H 1892 ZMV Unt 23, 230-232
    (F24, 37)


Schroeter, Jens Fredrik Wilhelm
1857

Lous K 1927 Nor M Tid 9, 53-54
    (F53, 33)


Schubert, Hermann Caesar Hannibal
1848

Burau W 1966 Ham MG 9(3), 10-19
    (MR33, 3871)


Schuessler, Rudolf    1865-

Horninger H 1944 Deu M 7, 598-601
    (Z61, 6)

Schuette, Fritz    1864-

Loria G 1914 DMV 23, 429-430
  (F45, 46)


Schulhof, Leopold   1847-1921

Bigourdan G 1922 Astmie Fr 36, 84-87


Schultz, Heinrich

Behnke H 1935 Muns Semb 6, 1-3
  (F61, 27)


Schultz, Henry

Douglas P H 1939 Ecmet 7, 104-106
  (F65, 18)
Hotelling H 1939 Ecmet 7, 97-103
  (F65:1, 18)


Schur, Friedrich Heinrich  1856-1932

Engel F 1935 DMV 45, 1-31
  (F61, 24)
Fladt K 1957 MP Semb 5, 182-185
  (MR19, 1150)


Schur, Wilhelm   1846-1901

Becker E 1902 DMV 11, 292-301
  (F33, 39)


Schuster, Max

Lietzmann W 1911 ZMN Unt 42, 88-92
  (F42, 30)
Thaer A 1910 Unt M 16, 111-112
  (F41, 34)


Schwalbe, Georg Bernhard  1841-1901

Pietzker F 1901 Unt M 7, 42-44
  (F32, 31)


Schwarz, Hermann Amandus   1843-1921

Anon 1915 *H. A. Schwarz-Festschrift,*
  Berlin, 459 p  (F45, 55)
- 1922 Linc Rn 31, 38-39
- 1923 Crelle 152
- 1940 Deu M 337  (Z23, 388)
Bieberbach L 1922 Ber MG 21, 47-52
  (F48, 17)
Caratheodory C 1927 Deu Biog 3, 236-
  238
Cassina U 1951 Lom Rn 83, 311-328
  (Z45, 148)  [Genocchi, Hermite,
  surf area]
Hamel G 1923 DMV 32, 6-13  (F49, 11)
Mises R v 1921 ZAM Me 1, 494-496
  (F48, 17)
Segre C 1922 Tor FMN 57, 161-163
  (F48, 17)  [Noether]
Vivanti G 1922 Lom Gen (2)55, 118-120
  (F48, 17)  [Jordan]


Schwarz, Maria Josepha de

Caligo D 1957 It UM (3)12, 732-734
  (Z77, 242)


Schwarzschild, Karl    1873-1916

Blumenthal O 1917 DMV 26, 56-75
  (F46, 22)


Schwatt, Isaac-Joachim    1867

J R P 1935 Rv M Hisp A (2)10, 125-
  126   (F61, 957)


Schweizer, Kaspar Gottfried    1816-
1873

Struve O v 1873 Leip As 8, 163-165
  (F5, 37)


Schwerd, Friedrich Magnus    1792-1871

Kobell V  1872 Mun Si (F4, 20)

Schwering, Karl Maria Johann Gerhard
1846-

Heinrichs J 1926 Unt M 32, 84-86
(F52, 28)
Lietzmann W 1926 ZMN Unt 57, 271-272
(F52, 28)

Scorza, Gaetano    1876-

Anon 1940 Esercit (2)12, iii-xvii
(F66, 22)
Berzolari L 1939 It UM (2)1, 401-408
(MR1, 130; F65:1,18)
- 1940 Lom Gen (3)4, 125-143
(F66, 1188)
Bompiani E 1939 Rm Sem 3, 139-152
(MR1, 290; Z22, 297; F65:1, 18)
Severi F 1941 Annali (4)20, 1-20
(Z25, 2)

Scott, Charlotte Angas   1858-1931

Macaulay F S 1932 LMSJ 7, 230-240
(F58, 46)

Scotus.  See Duns Scotus.

Secchi, Angelo   1818-1878

Bricarelli C 1888 Vat NL Mm 4, 41-106
Ferrari S 1891 Vat NLA 64, 240-245
(F23, 22)
Van Tricht 1878 Rv Q Sc 4, 353-402
(F10, 19)

Sédillot, Louis Pierre Eugine Amelie
1808-1875

Sédillot C E 1876 Boncomp 9, 649-700
(F8, 20)  [bib]

Segarceanu, Niculae C.

Ionescu I 1928 Gaz M 34, 171-176
(F54, 43)

- 1929 Gaz M 35, 179-185   (F55, 623)

Segner, Jean Andre de   1704-1777

Guenther S 1876 Boncomp 9, 217-229
(F8, 18)  [meteorol]

Segre, Carrado   1863-1924

Anon 1924 Annali (4)1, 319-320
(F50, 14)
- 1959 Archim 11, 304-308 (Z87, 6)
Baker H F 1926 LMSJ 1. 263-271
(F52. 32)
- 1927 It UM 6, 276-284 (F53, 34)
Berzolari L 1924 Lom Gen (2)57,
528-532  (F50, 14)
Boggio T 1927 Tor FMN 63, 299-348
Castelnuovo G 1924 Linc Rn (5)33,
353-359 (F50, 14)
Coolidge J L 1927 AMSB 33, 352-357
(F53, 34)
Loria G 1924 Annali (4), 1-21
(F50, 14)  [geom]
Pascal E 1924 Batt 63, 203-204
(F51, 30)
- 1924 Nap FM Ri (3a)30, 114-116
(F50, 14)
- 1925 Batt (3)63, 203-204
Terracini A 1926 DMV 35, 209-250
(F52, 32)
- 1953 It UM Con 4(1), 252-262
(Z50, 3)
Tricomi F 1941 Tor Sem 7, 101-117
(Z25, 386)
Viglezio E 1924 Rass MF 5, 1-2
(F50, 13)
Volterra V 1924 Linc Rn (5)33, 459-
461  (F50, 15)

Seguim, Marc   1786-1875

Picard E 1923 BSM (2)17, 291-298
(F49, 24)

Seidel, Philip  Ludwig von
1821-1896

Lindemann F 1899 DMV 7, 23-33
  (F30, 17)
- 1899 *Gedaechtnisrede auf Philipp
  Ludwig von Seidel...*, Munich,
  84 p  (F30, 17)


Seidel, Wolfgang  1492-1562

Poehlein Hubert 1951 *Wolfgang Seidel,*
  Munich, 255 p     (Z43, 2)


Seifert, Ladislav  1883-

Klapka J 1956 Cas M 81, 370-376
  (Z74, 246)


Seitz, Enoch Beery    1846-1883

Finkel B F 1894 AMM 1, 3-6


Seki Kowa    XVII

  = Seki Takakazu

Endo 1907 Tok M Ph (2)4,  94-96
  (F38, 12)
Fujiwara M 1941 Toh MJ 48, 201-214
  (MR10, 419)
Hayashi T 1906 Tok M Ph 3, 127-141
  (F37, 5)
- 1907 Tok M Ph 3, 183-201
  (F38, 11)
- 1908 Tok M Ph (2)4, 264-265
  [port]
- 1908 Tok M Ph (2)4, 446-453
- 1911 Tok M Ph (2)6, 144-152
Hosoi So 1959 Sugaku 10, 134-138
  (MR25, 389)
Kato H 1941 Toh MJ 48, 1-24
  (MR7, 353; Z60, 1)
- 1959 Sagaku 10, 138-141
  (MR25, 389)
Kawakita C 1895 Tok M Ph 7, 127
Kawakita-Tyorin 1907 Tok M Ph (2)4,
  88-94    (F38, 11)

Kikuchi D 1897 Tok M Ph 8, 179
  [circle]
*Mikami Y 1908 DMV 17, 187-196
  [Shibukawa Sukezayemen]
- 1908 Tok M Ph (2)4, 442-446
  (F39, 69)  [circle]
- 1910 Nieu Arch (2)9, 158-171
  (F41, 8)
Minoda T 1940 Toh MJ 47, 99-109
  (MR1, 289; F66, 14)
- 1941 Toh MJ 48, 167-173
  (MR7, 353; Z23, 196, 293)
- 1943 Toh MJ 49, 220-222
  (MR8, 497; Z60, 1)
Oya S 1959 Sugaku 10, 141-145
  (MR25, 389)
Shinomiya A + 1917 Toh MJ 12, 1-12
  (F46, 48)  [Gokai]
Yubuuchi K 1959 Sugaku 10, 133-134
  (MR25, 389)


Selivanov, Dmitri Fedorovitch
  1855-1932

H K 1932 Cas MF 62, 79-82
  (F58, 46)
Rothe R 1934 DMV 44, 210-214


Semeijns, Meindert

Bierens de Haan D 1872 Boncomp 5,
  213-220   (F4, 15)
Boncompagni B 1872 Boncomp 5, 221-
  228   (F4, 15)


Sen, Nikhilranjan R.   1894-1963

Burman U R 1963 Clct MS 55, 103-111
  (MR30, 872; Z119:1, 10)


Sereni, Carlo  1786-1868

Perassi R 1937 Pe M (4)17, 14-24
  (F63, 809)  [curvature]

Serenos Antinoeia          IV

Conte L 1938 Pe M (4)18, 70-86
  (F64, 905) [conic]
- 1939 Pe M (4)19, 16-35
  (MR1, 289) [conic]
- 1940 Pe M (4)20, 1-23, 218-239
  (MR3, 97) [conic]
Heiberg J L 1894 Bib M (2)8, 97-98
  (F25, 13)
Sarton G 1927 Int His Sc 1, 353-354
Tannery P 1883 BSM (2)7, 237-244


Sergescu, Petre  1893-1954

Andonie G S 1968 Janus 55(1), 3-11
*Andonie  G S + 1968 *Pierre Sergescu,*
  Leiden (E J Brill), 74 p [bib]
Reymond A 1955 Ens M (2)1, 21-29
  (Z66, 7)
Taton R 1955 Par BS Rou 3, 3-12
Taton R + 1968 Janus 55(1)
  (Special issue)


Serret, Paul Joseph  1827-1898

Darboux G 1898 Par CR 127, 37-38
  (F29, 20)
Jordan C + 1885 BSM (2)9, 123-132
- 1885 Par CR 100,647-681
Loria G 1898 Loria 1, 57 (F29, 20)


Servais, Clement  1862-1935

Godeaux L 1950 Bel Anr, 1-22


Servois, François  Joseph
  1767-1847

Balitrand F 1917 Intermed 24, 22-23
  (Q4668) (F46, 36)
Boyer J 1896 *Le mathématicien franc-
  comtois François Joseph Servois,
  ancien conservateur du musee
  d'artillerie, d'apres des documents
  inedits (1767-1847),* Paris, 26 p
  (F27, 12)
- 1896 JM E1 20, 39

Dickstein S 1894 Intermed 1, 116; 2,
  58-59, 220-221; 3, 182; 9, 119,
  331  (Q220)


Sevastiyanov, Ya. A.  1796-1849

Gusev V A 1952 Mos IIET 4, 183-194
  [descr geom]


Severi, Francesco  1879-1961

Anon 1931 It Annu 2, 305-313
  (F57, 46)
- 1953 It XL An, 13-30  (MR15, 592)
- 1962 It UM (3)17, 243-246
  (Z100, 247)
Garnier R 1962 Par CR 254, 777-782
  (Z115, 246)
Marchionna E 1963 Tor FMN 97, 556-
  566  (Z114, 8)
Papi G U + 1962 Ren M Ap (5)20, 482-
  486  (Z101, 10)
Rey Pastor J 1928 Rv M Hisp A 3, 41-
  49  (F54, 44)
Roth L 1963 LMSJ 38, 282-307
  (MR26, 221; Z111, 3)
Segre B 1962 Ren M Ap (5)21, 524-584
  (MR26, 1142; Z106, 3)
- 1963 Annali (4)61, i-xxxvi
  (MR29, 880; Z107, 248)


Severini, Carlo  1872-

Straneo P 1952 It UM (3)7, 98-101
  (MR14, 344)


Severus Sebokht          VII

Nau F 1894 J Asi 10(16), 219-228
  [arabic numer]
Ruska J 1896 *Das Quadrivium aus
  Severus Bar Sakku's Buch der
  Dialoge,* Leipzig
Sarton G 1927 Int His Sc 1, 493
  [bib]

Seydler, August  J. F.   1849-1891

Strouhal V 1892 Cas MF 21, 193
 (F24, 37)


Sforza, Giuseppe, 1858-

Conti A 1928 Bo Fir (2)7, 43
 (F54, 41)


Shanks, William   1812-1882

Hoffmann J C V 1895  ZMN Unt 26,
 261-264 [pi]   (F26, 55)


Sharp, Abraham  1651-1742

Archibald R C 1948 *Mathematical
 Table Makers,* NY, 72-73
 (Scr Stu 3) [bib]
Connor E 1942 ASP 54, 12-18
Cudworth W 1889 *Life and Correspond-
 ence of Abraham Sharp, the York-
 shire Mathematician and Astronomer.
 With memorial of his family,*
 London


Sharp, William Joseph Curran
 1856?-1891?

Anon 1891 LMSP 22, 476-481
 (F23, 29)


Shattuck, Samuel Walker  1841-1915

Miller G A 1916 AMM 23, 45-46
 (F46, 25)


Shatunovskii, Samnil Osipovich
 1859-1929

Bakhmutskaya E Ya 1965  Ist M Isl
 16, 207-216   [foundations]
- 1966 Ist Met EN 5 (MR33, 3870)
 [foundations]


Chebotarev N G 1940 UMN 7, 316-321
 (F66, 22)


Shebuev, Georgi Nikolayevitch
 1850-1900

Bolotov E A 1901 Mos MO Sb 22(1),
 XII-XV  (F32, 21)


Sheppard, William Fleetwood
 1863-1936

A C A + 1936 Edi SP 56, 279-282
 (F62, 1041)
Archibald R C 1948 *Mathematical
 Table Makers,* NY, 73-74
 (Scr Stu 3)   [bib]
Fisher R A 1937 An Eug 8, 11-12
 (F63, 816)
Sheppard N F 1937 An Eug 8, 1-9
 (F63, 816)


Sherman, Frank Asbury  1841-1915

Hahen J V 1916 AMM 23, 114-115
 (F46, 25)


Shigetomi, Hazama Gorobei  1756-1816

 = Hazama Jufu

Kikuchi D 1912 Jp Ac P 1, 88-95
 [ell]


Shiller, Nikolai Nikolaevich  1848-
 1910

Kagan W 1910 Kagan (528), 314
 (F41, 25)


Shiragooni VII

 = Anania of Shirak

Shaw A A 1932 AMSB 38, 638 (F57, 1310)

Vogel K 1935 Scr 3, 283-284
(F61, 12)


Shmidt, Otto Yulevich 1891-1956

Anon 1956 SSSR Gf, I-VII
(MR18, 784)
- 1959 *Sbornik Otto Yulevich Schmidt.
Zhizn i deyatelnost,* Mos, 470 p
(Z87, 7)
Breus K A 1966 Ukr IM 18(5), 3-6
(MR33, 5432)
Kurosh A G 1951 UMN (NS)6(5), 197-
199 (MR13, 421; Z43, 5)
- 1956 UMN (NS)11(6), 227-233
(MR18, 709; Z72, 246)


Shnirelman, Lev Genrikhovich
1905-1938

Anon 1939 UMN 6, 3-8 (MR1, 33)


Shtokalo, Iosif Zakharovich 1897-

Sokolov J D + 1958 Ukr Im 10, 105-
106 (Z79, 6)


Siacci, Ugo Aldo de Francesco
1839-1907

Anon 1940 *Ugo Aldo de Francesco
Siacci. Uno scienziato...,*
Naples, 147 p (F66, 20; Z25, 2)
Loria G 1929 Linc Rn (6)9(1), 596-
597
Marcolongo R 1906 Loria 9, 107-116
- 1908 Batt 46, 375-380 [bib]
Milone F 1909 Nap Pont (2)14, 39-49
Morera G 1908 Tor FMN 43, 568-578
(F39, 33)


Siber, Thaddaeus 1774-1854

Ruttmanner M ed 1927 Oberbay 65, 83-
225


337 / Biography

Sibirani, Filippo 1880-

Varoli G 1957 It UM (3)12, 125-130
(MR19, 518; Z77, 9)


Siddons, A. W. 1876-

Snell K S 1960 M Gaz 44, 35-36


Sidler, Georg Joseph 1831-

Anon 1907 Ens M 9, 493 (F38, 42)
Rudio F 1908 Zur NGV 53, 1-32
(F39, 34)


Sierpinski, Wastawa 1882-

Anon 1949 *Polski Zjazd. Matematyczny
Jubileusz 50-Lecia Dziatalnosci na
katedre Universyteckiej Professora
Wactawa Sierpinskiego,* Warsaw, 95 p
(MR11, 573; Z35, 147)
Fryde M M 1963 Pol Rv 8(1), 1-8
(MR27, 4)
- 1964 Scr 27, 105-111
(MR29, 639; Z117:1, 9)
Melnikov 1 1968 in *Sierpinski W,
250 zadach po elementarnoi teorii
chisel,* Mos, 3-15


Signorini, Antonio 1888-1963

Agostinelli C 1963 It UM (3)18, 327-
330 (Z111, 3)
Anon 1952 Parma Rv M 3, 301-306
(MR14, 833; Z48, 243)
- 1963 Annali (4)62, i-iv (Z113, 2)
Cattaneo C 1963 Ren M Ap (5)22,
345-350 (Z113, 2)


Siliceo, Juan Martinez 1477-1557

= Silicius, Dubois, Guisco, Guijeno

Reyes Y Prosper V 1911 Esp SM 1, 153-
156 (F42, 4)

Silva, Daniel Augusto da    1814-1878

Almeida e Vasconcelos Fernando de
  1934 Archeion 16, 73-92
Quesada C A de 1909 Porto Ac 4, 166-
  192  [numb thy]
- 1914 Porto Ac 9(2), 65-95


Silvan y Gonzalez, D. Graciano

Pineda P 1934 Rv M  Hisp A (2)9,
  95  (F60, 838)


Simandl, Wenzel 1887-1918

Pelisek M 1919 Cas MF 48, 203-206


Simerca, P. Wenzel  1819-1877

Cupr K 1914 Cas MF 43, 482-489
Panek A 1888 Cas MF 17, 253
Petrzilka V 1926 Cas MF 55, 352-360


Simon, Max  1844-

Lorey W 1918 Leop 54, 31-32
  (F46, 15)


Simonov, Nikolai Vasilievich

Anon 1966 Teor Ver 11, 495-496
  (MR33, 5440)


Simons, Lao Genevra   1870-

Eisele C 1950 Scr 16, 22-30
  (Z37, 292)


Simony, Oskar 1852-1915

Mueller Ernst 1926  *Bibliographisches*
  *Verzeichnis der Schriften Oskar*
  *Simonys,* Vienna, 9 p  (F52, 39)

Simplicios   VI

Besthorn R O 1892 Bib M (2)6, 65-66
  (F24, 6)
Rudio F 1903 Bib M (3)4, 13-18
- 1905 Zur NGV 50, 213-223
  (F36, 64)
- 1907 *Der Bericht der Simplicius*
  *ueber die Quadraturen des Antiphon*
  *und des Hippokrates,* Leipzig, 194 p
  (Urk GMA 1)  (F38, 64)
Schmidt W 1903 Bib M (3)4, 118-126
Steinschneider M 1892 Bib M (2)6, 7-9
  (F24, 6)
Tannery P 1902 Bib M (3)3, 342-349
  [circle sq]


Simpson, Thomas   1710-1761

Buicliu C G 1933 Tim Rev M 13, 6
  (F61, 958)
Clarke F M 1929 *Thomas Simpson and*
  *His Times,* NY
Hutton 1792 *Memoirs of the Life and*
  *Writings of Thomas Simpson,*
  London
Smith E 1910 Nat 84, 254-255


Simson, Robert   1687-1768

Archibald R C 1933 Scr 2, 73-75
Gibson G A 1926 Edi MSP 44, 39-46
  (F52, 7)  [Heath]
Mackay J S 1903 Edi MSP 21, 2-39
  [Stewart, Stirling]
Trail William  1812 *Account of the*
  *Life and Writings of Robert Simson*
  *MD late professor of mathematics*
  *in the University of Glasgow,*
  Bath-London


Sina.  See Avicenna.


Singh, Aradhesh Narayam   1901-

Sinvhal S D 1955 Ganita 5, i-vii
  (MR18, 710; Z60, 12)

Sinram, Heinrich Theodor   1840-1895

Gutzmer A 1897 DMV 5, 17-18
(F28, 21)


Sintsov, Dmitri Matveevich   1867-1946

Anon 1933 Khar Me M 6, 80 (F59, 40)
- 1948  Khar M Za (4)19, 5-9
(MR12, 1)
Bernshtein S N + 1947 UMN 2(4),
191-206   (MR10, 420)
Naimov J A 1955 D.M. Sintsov,
Kharkov, 72 p   (Z66, 7)


Sipos, Paul   1759-1816

Jelitai J 1934 Archeion 16, 298-306
(F61, 20; Z11, 194)
- 1934 MP Lap 41, 45-54
(F60, 11, Z9, 8)
Woyciechowsky J v 1932 Deb M Szem 6
(F58, 38; Z5, 242)


Sitter, Willem de 1872-1934

Anon 1934 Amst P 37, 734-737
A S E 1934 Nat 134, 833-835
Esclangon E 1935 Par CR 200, 21-22
(F61, 957)


Skiba, E. W.   1843-1911

Dickstein S 1912 Wiad M 16, 237-240
(F43, 37)


Skinner, Ernest Brown   1863-1935

Langer R E 1935 AMM 42, 535-537
(F61, 955)


Skolem, Thoralf Albert 1887-

Crespo R 1952 Gac M (1)4, 109-112
(MR14, 610)


Fenstad J E 1963 Nor M Tid 11, 145-
153  (Z118:1, 12)
Ljunggren W 1963 Math Scan 13, 5-8
(MR29, 1; Z115, 246)
Nagell T 1963 Act M 110, 303
(MR27, 463; Z111, 3)
Ore O 1947 Nor M Tid 29, 33-34
(Z30, 101)
Selberg S 1963 Nor VSF 36, 165-168
(MR28, 756; Z115, 246)


Slaught, Herbert Ellsworth   1862-1937

Anon 1938 SSM 37, 770-771
Bliss G A 1937 AMSB 43, 597
- 1938 AMM 45, 5-10   (F64, 23)
Cairns W D 1938 AMM 45, 1-4
(F64, 23)
Dark H J 1948 The Life and Works of
Herbert Ellsworth Slaught,
Nashville, 152 p  [bib]
Dickson L E + 1937 Sci 86, 73
Greene N L 1937 Ed Screen 16(6), 186
Reeve W D 1937 MT 30, 293


Sloginov, N. P.

Kasankin N P 1897 Kazn FMO (2)7, 79


Slonimski, C. Z. 1860-1904

Anon 1904 Wiad M 8, 337-338
(F35, 38)


Sludskii, Fedor Alekseevich   1841-
1897

Joukovsky N E 1898 Mos MO Sb 20, 337


Sluse, René François Walter de
1622-1685

Le Paige C 1884 Boncomp 17, 427-554,
603-726 [letters]
- 1887 Ciel Ter (2)2,  (F21, 1256)
Marre A 1884 Lisb JS 38  [letter]

Slutskii, Evgenii Evgenevich
    1880-1948

Allen R G D 1950 Ecmet 18, 209-216
    (MR12, 1; Z36, 146)
Kolmogorov A N 1948 UMN (NS)3(4),
    143-151  (MR10, 175)
Smirnov N 1948 SSSRM 12, 417-420
    (MR10, 175; Z30, 101)


Smadecki, Jan

Chamcowna Miroskawa 1963 *Jan Smadecki*,
    Kracow, 136 p


Smirnov, Nikolai Vasilevich  1900-

Gnedenko B V 1960 Teor Ver 5, 436-
    440  (MR24A, 221)  [stat]
Kolmogorov A N + 1951 UMN (NS)6(4),
    190-192  (MR13, 198)


Smirnov, Vladimir Ivanovich  1887-

Aleksandrov P S + 1957 UMN (NS)12(6),
    197-205  (MR19, 1030; Z78, 4)
Fok V A 1962  Rus MS 17(6), 146-148
    (Z115, 246)
Ladyzhenska Ya O A + 1957 Len MV
    12(7), 5-14  (Z89, 243)
- 1968 UMN 23(4)(142)
Radovskii M I 1962 Rus MS 17(6),
    143-146  (Z115, 246)
Sobolev S L + 1949 *Vladimir
    Ivanovich Smirnov*, Mos-Len
Sobolev S L 1957 SSSRM 21, 449-456
    (Z77, 242)


Smith, David Eugene  1860-1944

Anon 1926 MT 19, 257-281, 297-305
- 1936 Sci 83, 424-426
- 1940 MT 33, 43
- 1943 Scr 11, 364-369
- 1945 Scr 11, 207-380
Brasch F E 1933 Sci 78, 384
- 1944 Sci 100, 257-259  (Z60, 17)

Breslich E R 1945 SSM 44, 838-839
Fite W B 1945 AMM 52, 237-238
    (Z60, 17)
Frick B M 1936 Isis 25, 140
- 1936 Osiris 1, 9-84  [*bib]
Ginsburg J 1926 MT 19, 306-311
Miller G A 1924 SSM 24, 939-947
    [Cajori]
- 1930 Indn MSJ 18, 265-270
    (F56, 2)
Reeve W D 1944 MT 37, 278-279
Simons L G 1945 AMSB 51, 40-50
    (MR6, 142; Z60, 17)
Taton R 1948 Arc In HS 28, 741-742
    (Z31, 242)


Smith, Henry John Stephen  1826-1883

Cayley A 1895 AMSB 1, 94-96
    (F26, 57)
Clerke A M 1898 DNB 53, 50-53
Cremona L 1883 Linc At (3)7, 162-163
Glaisher J W L 1883 Cam PSP 4,
    320-322
- 1884 Lon As Mo N 44, 138-149
*Macfarlane Alexander 1916 *Lectures on
    Ten British Mathematicians of the
    Nineteenth Century*, NY, 92-106
Mansion P 1898 Mathesis (2)8(S)
- 1898 Rv Q Sc 43, 219-227  (F29, 9)


Smith, William Benjamin  1850-1934

Keyser C J 1934 Scr 2, 305-311
    (F60, 836)
- 1935 Scr Lib 2, 108-119
- 1937 Scr Lib 3, 1-32


Smogorzhevskii, Aleksandr Stepanovich
    1896-

Dobrovolskii V A + 1966 Ukr IM 18(5),
    94-96  (MR33, 5433)

Smoluchowski, Marien   1872-

= Marjan Ritter von Smolan

Godlewski T 1919 Wiad M 23, 1-36
Malarski T 1932 Mathes Po 7, 118-130
  (F58, 992)


Snell, Willebord   1591-1626

= Snellius   = Snell Van Roijen

Bosmans H 1899 Brx SS 24(2), 111
  [geod]
Hofmann J E 1933 Unt M 39, 45
  (F59, 24)
Korteweg D J 1896 Nieu Arch (2)3, 57
  [Descartes]
- 1896 Rv Met Mor 4, 489-501
  (F27, 9)  [Descartes]
Lampe E 1892 Mathesis (2)2, 230-231
  (F24, 46)  [Cusanus]
Le Paige C 1890 Mathesis 10, 34-36
  [Ozanam formula]
Mansion P 1884 Mathesis 4, 64
Reeve W D 1931 MT 24, 244
Sanford V 1932 MT 24, 244
Tannery P 1886 Bib M 47-48, 143, 243
  (Q9)
Van Geer P 1883 Ned Arch 18, 453-468
Verdam G J 1842 Arc MP 2, 81-


Sniadecki, Jan   1756-1830

= Johann Baptist Sniedecki

Birkenmajer A + 1938 Organon 2, 95-
  132  (F64, 18)
Dickstein S 1903 Wiad M 7, 22-31
  (F34, 12)
- 1931 Wiad M 33, 1-14 (F57, 1303)
Drobot S 1955 Wiad M 1(1), 95-111
  (Z66, 8)


Snyder, Virgil   1869-1950

Archibald R C 1938 A Semicentennial
  History of the American Mathe-

matical Society, NY, 218-223
  [bib, port]
Coble A B 1950 AMSB, 468-471
  (Z38, 4)


Sobolev, Sergei Lvovich   1908-

Vishik M I + 1959 UMN 14(3), 203-214
  (MR22, 619; Z88, 244)
- 1960 Ro Sov (3)14(1), 208-215
  (MR22, 929)


Sobotka, Jan   1862-1931

Seiffert L 1935 Zagr Ljet 47, 223-228
  (F61, 958)
- 1936 Zagr BI, 29-30, 133-134
  (F62, 1041)


Sofronov, Mikhael   XVIII

Smirnov Vladimir L + 1954 Mikhail
  Sofronov--russkii matematik
  serediny XVIII veka, Mos, 54 p
  (Z59:1, 9)


Soha, Hatono   XVII   1641-1697

Mikami Y 1910 Nieu Arch 9, 158-171
  [Seki Kowa]


Sohncke, Leonard   1842-1897

Voit C v 1898 Mun Si 28, 440-449
  (F29, 17)


Sokhotskii, Yulian Vasilevich
  1842-1929

Markushevich A I 1950 Ist M Isl (3),
  399-406  (MR13, 2; Z64, 242)
  [anal fun]

Sokolov, Yurii Dmitrievich

Mitropolskii Y A 1966 Ukr IM 18(4),
94-101 (MR33, 3875)
Putyata,T V + 1956 Ukr IM 8, 223-230
(Z71, 4)
Tsyganova N I 1959 Mos IIET 22, 202-
213 (Z95, 4) [anal mech]


Somerville, Mary 1780-1872

Richeson A W 1941 Scr 8, 5-13
Somerville Mary 1874 *Personal
Recollections of Mary Somerville,*
Boston, 383 p


Somigliana, Carlo 1860-1955

Agostinelli C 1954 Tor Sem 14
(MR17, 698; Z65, 245)
- 1955 It UM (3)10, 650-656
(Z66, 8)
- 1956 Tor FMN 90, 217-222
(Z71, 4)
- 1961 Tor Sem 20, 15-38
(Z98, 11)
Anon 1953 It XL An, 9-11
(MR15, 592)
Finzi B 1958 Milan Sem 27, xv-xvi
(Z79, 6)
Signorini A 1956 Linc Rn (8)21,
343-351 (MR20, 503)


Sommerfeld, Arnold Johannes Wilhelm
1868-1951

Born M 1952 Lon RS Ob 8, 275-296
(MR14, 525)
Pauling L 1951 Sci (NS)114, 383-384
(MR13, 198)
Whittaker E T 1953 LMSJ 28, 125-128
(Z50, 243)


Sommerville, Duncan McLaren Young
1879-1934

Halsted G B 1912 AMM 19, 1-4

Turnbull H W 1934 Edi MSP (2)4, 57-
60 (F60, 16)
- 1934 LMSJ 9, 316-318 (F60, 16)


Somov, Osip Iosif Ivanovich
1815-1876

Boncompagni B 1878 Boncomp 11, 482-
486 (F10, 22) [letter]
Geronimus J L 1954 *O. I. Somow.
Einige Probleme der hoehere
Dynamik,* Berlin, 55 p
(MR15, 924; Z60, 11)
Kramar F D + 1958 Kazk Ak (3), 44-49
(Z79, 5)
Kramar F D + 1965 *Iosif Ivanov
Somov...,* Alma-Ata
Prudnikov V E 1953 Ist M Isl 6,
223-237
Somov A 1878 Boncomp 11, 453-481
(F10, 22) [bib]


Sonin, Nikolai Yakovlevich 1849-1915

Anon 1915 Ens M 17, 128 (F45, 53)
Kropotov A L 1967 *Nikolai Yakovlevich
Sonin,* Len, 135 p [bib, port]
Posse K 1913 Khar M So (2)14, 275-
293


Soper, H. E.

Greenwood M 1931 Lon St 64, 135-141
(F57, 44)


Soto, Dominic 1494-1560

Dugas R 1948 Rv Sc 86, 131-134
(Z31, 2)


Sourek, Anton Wenzel 1857-

Sobotka J 1927 Cas MF 56, 1-6
(F53, 35)

Sousa, Joaquim Gomez de 1829-1863

Basseches,B 1955 Parana M 2, 18-25
   (Z67, 247) [bib]


Souza Pinto, Rodrigo Ribeiro de
   1811-1893

Anon 1894 Teixeira 12, 3-10
Teixeira A J 1894 Teixeira 12, 3-10
   (F25, 42)


Sovero, bartolomeo

Busulini B 1957 Pada Pat 70(2),
   35-88   (MR22, 928; Z97, 241)
Favaro A 1882 Boncomp 15, 1-48
- 1886 Boncomp 19, 99-114


Spacek, Antonia   1911-1961

Hans O 1962/63 Z Wahr, 315-318
   (MR26, 1142; Z107, 6)
Winkelbauer K 1962 Aplik M 7, 161-170
   (Z101, 10)
- 1962 Cz MJ 12, 314-321 (Z100, 248)


Spataruc, N. M.

Cristescu V 1929 Gaz M 35, 16-19
   (F55, 623)


Spencer, Herbert   1820-1903

Mackay J S 1907 Edi MSP 25, 95-106
   (F38, 23)


Spengler, Oswald   1880-1936

Riebesell P 1920 Naturw 8, 507-509
   (F47, 21)
Schaaf W L 1955 MT 48, 262-266


Spieker, Theodor

Otte P 1914 ZMN Unt 45, 194-198
   (F45, 34)


Spinoza, Benedict   1632-1677

Bierens De Haan D 1884 Nieu Arch 11,
   49-82
Blueh O 1935 Nat 135, 658 [Newton]
Brunschvicg L 1905 Rv Met Mor 13, 673
- 1906 Rv Met Mor 14, 35
Dutka J 1953 Scr 19, 24-32
   (Z50, 243) [prob]
Keyser C J 1938 Scr 5, 33-35
Moorman R H 1944 M Mag 18, 108-115
Serouya H 1933 Spinoza, sa vie et
   sa philosophie, Paris, 83 p
Vanos C H 1946 Tijd man ein getal,
   naar aan leiding van Spinoza's
   brief over het oneindige,
   Leiden, 20 p
Vleeschauwer H J 1942 Tijd Phi 4,
   345-396  (F68, 14) [Tschirnhaus]
Wolf A 1933 Phi Lon 8, 3-13


Spottiswoode, William   1825-1883

Anon 1883 Nat 27, 597-  [bib, port]
- 1884 Lon RSP 38, xxxiv-xxxix


Sretenskii, Leonid Nikolaevich   1902-

Anon 1962 J Ap M Me 26, 585-600
   (Z111, 4)
- 1962 Mos UM Me 17(2), 77-80
   (Z105, 244)


Sridhara   X-XI

Datta B 1932 Isis 17, 25-33
   [Mahavira]
Ramanujacharia N + 1912 Bib M (3)13,
   203-217  (F43, 68)
Shankar Shukla K 1951 Ganita 1, 1-12,
   53-64  (MR13, 420) [numb thy]

Shankar Shukla K ed 1959 *The Patiganita of Sridharacarya With an ancient Sanskrit Commentary*, Lucknow

Staeckle, Paul Gustav     1862-

Anon 1920 Ens M 21, 63 (F47, 12)
- 1922 Crelle 151
Lorey W 1921 ZMN Unt 52, 85-88
  (F48, 22)
Perron O 1920 Heid Si (7)
  (F47, 12)
Rudio F 1923 DMV 32, 13-32
  (F49, 13)  [Euler]

Stahl, Herman Bernhard Ludwig  1843

Brill A v 1911 Wurt MN (2)13, 1-8
  (F42, 24)

Stahl, Wilhelm  1846-1894

Reye T 1894 Crelle 104, 45-46
  (F25, 47)
- 1897 DMV 4, 36-45    (F28, 20)

Staude, Ernst  Otto  1857-

Wangerin G 1928 ZMN Unt 59, 278-279
  (F54, 40)

Staudigl, Rudolph 1838-1891

Anon 1891 Mo M Phy 2, 479-480
  (F23, 31)

Staudt, Carl Georg Christian von
  1798-1867

Archibald R C 1918 AMSB 25, 132-134
  (F47, 31)
Cantor M 1893 ADB 35, 520
Hofmann J E 1960 Fraenkis 7(6), 536-
  548  (Z91, 6)

Martius V 1869 Arc MP Lit 49(193), 1-
  (F2, 23)
Noether Max 1901 in *Festschift dem Prinzregenten Luitpold dargebracht* Erlangen, Vol 4(2), 63-86
- 1901 *Zur Errinerung an K. G. C. von Staudt,* Erlangen-Leipzig, 24 p
  (F32, 9)
- 1923 DMV 32, 97-119  (F49, 12)
Segre C 1888 *C. G. C. von Staudt ed i suoi lavori,* Turin, 17 p

Steen, Adolph  1816-1886

Zeuthen H G 1886 Zeuthen (5)4, 65-70

Stefan, Josef  1835-1893

Subic J 1902 Matica S1 4, 62

Steffensen, J. F. 1873-1961

Norlund N E 1962 Nor M Tid 10, 105-
  107  (Z100, 248)

Stegemann, Wilhelm

Sondheimer 1915 ZMN Unt 46, 487-490
  (F45, 53)

Steggall, John Edward Aloysius
  1855-1935

H  W  T  1936 Edi SP 56, 284-285
  (F62, 1041)
Peddie W 1936 Edi SP (2)4, 270-271
  (F62, 27)

Stegmann, Johann Gottlieb  1725-1795

Merczyng H 1907 Krak BI, 1075-1079
  (F38, 59)

Steiner, Jacob    1796-1863

*Archibald R C + 1948 Scr 14, 187-264
  Scr 14, 187-264  (Scr Stu 4)
  (Z33, 2) [bib, constr, port]
Biermann K-R 1963 Arc In HS 16, 167-
  171  (Z113, 2)
- 1963 Fors Fort 37, 125-126
  (Z107, 6)
  1963 Leop NA (NF)27(167), 31-45
Bussey W H 1914 AMM 21, 3-12
Buetzberger F 1896 ZMN Unt 27, 161-
  171  (F27, 15)
- 1913 Ueber bizentrische Polygone,
  Steinersche Kreis-und Kugelreihen
  und die Erfindung des Inversion,
  Leipzig, 60 p
Cantor M 1874 Hist Abt 19, 65-67
  (F6, 30)
De Vries H 1925 N Tijd 12, 260-289
  (F51, 23)
Emch A 1929 AMM 36, 273-275
  (F55, 14) [mss]
Geiser C F 1874 Zur Erinnerung an
  Jakob Steiner, Zurich, 37 p
  (F6, 29)
- 1875 Annali (2)7, 65-68  (F7, 12)
Graf J H 1896 Bern NG 61 [Schaefli]
- 1897 Der Mathematiker Jakob
  Steiner von Utzensdorf,  Bern,
  54 p  (F28, 16) [port]
- 1898 Bern NG, 8  (F28, 16)
  [Schaefli]
- 1905 Bern NG, 59-69  (F36, 13)
Hesse O 1863 Crelle 62, 199-200
Jahnke E 1903 Arc MP (3)4, 268-277
  (F34, 13) [Jacobi]
Kollros L 1936 Schw NG 117, 245
  (F62, 1043; Z15, 290)
- 1947 Elem M (2) (MR11, 708; Z33, 2)
Lampe E 1900 Bib M (3)1, 129-141
  (F31, 15)
Lange Julius 1899 Jakob Steiner's
  Lebensjahre in Berlin 1821-1863
  Berlin, 70 p [port]  (F30, 12)
Longchamps G de 1898 Intermed 5, 44
  [inversion]
Lorey W 1933 MN Bl 27, 30-32
  (F59, 847)
Loria G 1897 Bo SBM, 1-2, 5-6, 9-11
  (F28, 17)

Oettingen A J v 1913 Die geometrischen
  Constructionen...von Jacob Steiner
  (1833), Ostw Kl 60
  [Comment, 81-84]
Stuloff N 1963 Praxis 5, 313-315
Sturm R 1900 ZM Ph 45, 235-239
  (F31, 15)
- 1903 Bib M (3)4, 160-184
  (F34, 17)


Steinmetz, Charles Proteus    1865-1923

Ettlinger E J 1940 Scr 8, 237-250
Leichliter V H 1967 MT 60, 448
Sutton R M 1957 MT 50, 434-435


Steklov, Vladimir Andreevich
1864-1926

Anon 1928 Sbornik pamyati V. A.
  Steklova, Leningrad
- 1930 Bo Fir (2)9, xlviii
  (F56, 816)
Bachmutskaja E J 1953 Ist M Isl 6,
  529-534  (Z52, 3)
Depman I J 1953 Ist M Isl 6, 509-528
  (MR16, 660; Z53, 340)
Gyunter N M 1932 Khar Me M 4(5),
  3-5
- 1946 UMN (NS)1(3/4), 23-43
  (MR10, 175; Z60, 17) [phys]
Ignatsiyus G I 1967 Vladimir
  Andreevich Steklov 1864-1926,
  Mos [bib]
Kneser A 1929 DMV 38, 206-231
  (F55, 21)
Koltsov A V 1959 Vop IET (7), 107-
  112  (Z117-2, 9)
Kryloff A 1926 Nat 118, 91-92
Nikiforov P M 1926 Priroda (9/10),
  3-20
Parfentiev N N 1926 Kazn FMO (3)1,
  2-13  (F57, 1311)
Smirnov V I 1946 UMN (NS)1(3/4),
  17-22  (MR10, 175; Z66, 17)
- 1964 Stekl MI 73, 5-13 (MR30, 373)
Uspenskii Ya V 1926 SSSR Iz (6)20,
  837-856  (F52, 39)

Stengel, J. P.    XVII

Eneström G 1885 Bib M (Q2)


Steno, Nicolaus    1631-1686

= Stensens = Stenone = Stenonius

Scherz G 1958 *Nicolaus Steno and 'his
   Indice,* Copenhagen, 314 p
   (Z79, 243)


Stepanov, Vyacheslav Vasilevich
   1889-1950

Alexsandroff P S 1950 UMN 5(5),
   3-10  (MR12, 312; Z37, 292)
Aleksandroff P S + 1956 *v.v.
   Stepanov,* Moscow, 60 p
   (MR18, 550; Z70, 7)
Anon 1950 Pri M Me 14, 565-572
   (MR12, 312; Z37, 292)


Stephanos, Kyparissos   1857-1917

Anon 1918 Ens M 20, 138-139
Varopoulos T 1951 Gree SM 25, 7-22
   (Z42, 5)


Stern, Maritz Abraham   1807-1894

Rudio F 1894 *Erinnerung an Maritz
   Abraham Stern,*  Zurich
- 1894 Zur NGV 39, 133-143
   (F25, 48)
   1895 ZMN Unt 26, 392-394
   (F26, 31)
   1897 DMV 4, 34-36   (F28, 20)


Stevin, Simon   1548-1620

Anderhuss J H 1943 Deu M 7, 299-304
   (Z28, 99)
- 1901 Intermed 8, 318   [loxodrome]
Archibald R C 1948 *Mathematical
   Table Makers,* NY, 74-76
   (Scr Stu 3)  [bib]

Aubry A 1928 Ens M 27, 106-123
   (F54, 8)  [mech, vector]
Bobynin W W 1950 Fiz M Nauk (2)1(6),
   167-180  (F31, 8)
Bosmans H 1911 Brx SS 35, 292-313
   (F42, 55)
- 1913 Brx SS 37, 171-199  [calc]
- 1920 Rv Q Sc, 35 p
- 1922 Mathesis 36, 167-174, 226-232,
   275-281
- 1923 Mathesis 37, 12-18, 55-62,
   105-109 (F49, 19)  [calc]
- 1923 Mathesis 37, 246-254, 304-311,
   341-347  [cubic eq]
- 1924 Bel Biog 23, 888-938
- 1925 Mathesis 49-55, 99-104, 146-
   153 [quartic eq]
- 1926 Pe M (4)6, 231-261  (F52, 16)
Depau R 1942 *Simon Stevin,* Bruxelles,
   126 p  (F68, 14)
Dijksterhuis E J 1932 Unt M 38, 148-
   150  (F58, 51)
- 1941 Arc Mu Tey 9, 268-342
   [Beeckman]
- 1943 *Simon Stevin,* The Hague, 291 p
- 1947 Sim Stev 25, 1-21
   (Z29, 289)
- 1949 Euc Gron 142-155 (Z34, 146)
- 1951 Sim Stev 28, 129-139
   (Z45, 148)
Gloden A 1948 Lux SNBM (NS)42, 70-73
   (MR11, 573)
Gothals J V 1841 *Notice historique...
   de Stevinus,* Brussels
Gravelaar N L W A 1901 Nieu Arch
   (2)5(2), 106
Loria G 1948 Bo Fir (5)2, xii-xiii
   (Z41, 338)
- 1950 Archim 2, 253-255   (Z38, 147)
Mansion P 1912 Brx SS 37, 66-67
   [calc]
Maupin G 1902 *Opinions et curiosités
   touchant la mathématique,*  Paris
Minnaert M 1961 Amst Vs G 70, 11-12
   (Z91, 5)
Pajares E 1955 Gac M (1)7, 3-6
   (MR17, 118)
Pelseneer J 1937 Lychnos 2, 373-377
   [letter]
Saedeleer A de 1936 *Een krachtige
   Figur, Simon Stevin...,* Bruge, 46 p
   (F62, 1031)

Sanford V 1921 MT 14, 321-333
[deci]
Sarton G 1934 Isis 21, 241-303
- 1935 Isis 23, 153-244 [deci]
- 1941 Isis 33, 55
Schor D 1902 Bib M (3)3, 198
[hydrostat]
Schouteet A 1937 Brug Em An 80,
137-146
Steichen 1846 *Mémoire sur la vie et
les travaux de Simon Stevin*,
Brussels
Struik D J 1958 *The Principal Works
of Simon Stevin*, Amsterdam
(MR34, 1141)
- 1959 MT 52, 474-478 [deci]
Tesch J W 1897 Nieu Arch (2)3, 94
(F28, 4)
Van de Velde A J J + 1948 *Simon
Stevin*, Brussels, 112 p (Z41, 338)
Van Hercke J J 1942 Ciel Ter 58,
31-34
Vollgraff J A 1936 Isis 25, 136
[loxodrome]
Wieleitner H 1928 Erlang Si 58,
177-180 (F54, 26) [quad eq]


Stewart, Matthew   1717-1785

Anon 1892 Mathesis (2)2, 63-64
- 1901 Bu SMP El 7, 177
Chanzy L 1924 Fr AAS, 7 (F51, 21)
Mackay J S 1891/92 Edi MSP 10, 90-94
[tria]
- 1903 Edi MSP 21, 2-39 [Stirling]
Mannheim A + 1900-1904 Intermed 7,
128; 7, 176; 11, 241 (Q1822)
Morris R 1928 MT 21, 465-478


Stickelberger, Ludwig   1850-

Heffter L 1937 DMV 47, 79-86, 197
(F63, 19)


Stieltjes, Thomas Jan 1856-1894

Archibald R C 1948 *Mathematical
Table Makers*, NY, 76-77
(Scr Stu 3) [bib]

Carmichael R D 1921 AMSB 27, 170-178
(F48, 44)
Cosserat E 1895 Tou FS 9, (1)-(64)
(F26, 40)
Hamburger 1920 M Ann 81, 235-319;
82, 120-164, 168-187
Van Vleck E V 1903 AMST 4, 297-332


Stifel, Michael   1487-1567

Fontes 1893 Fr AAS 22, 236-243
(F25, 64) [binom, coeff, Forcadel,
Pascal]
Giesing J 1879 *Stifel's arithmetica
integra. Ein Beitrag zur Geschichte
der Arithmetik des 16 Jahrhunderts*,
Doebeln (F11, 31)
Glaisher J W L 1927 Mess M 56, 33-172
(F53, 11) [alg, notat]
Hofmann J E 1968 Essling J 14, 30-60
Hoppe E 1900 Ham MG 3, 411
Mueller T 1897 *Der Esslinger Mathe-
matiker Michael Stifel*, Esslinger,
39 p (F28, 4)


Stirling, James   1692-1770

Krasotkina T A 1957 Ist M Isl 10,
117-158 (Z100, 7) [Euler]
Mackay J S 1903 Edi MSP 21, 2-39
(F34, 11) [Simson, Stewart]
Tweedie C 1920 M Gaz 10, 119-128
- 1922 *James Stirling. A Sketch of
His Life and Works Along With His
Scientific Correspondance*,
Oxford, 225 p (F48, 7)
*Wieleitner H 1913 Bib M (3)14, 55-62
(F44, 52) [Newton]


Stodola, Aurel   1859-1942

Koener K 1929 ZAM Me 9, 256-259
(F55, 24)


Stoica, Alexandru

Grigorescu V 1925 Gaz M 31, 19-20
(F51, 39)

**Stoilov,** Simion   1887-1961

Anon 1961 Ro IM 12, 7-19
  (MR23A, 584; Z96, 4)
- 1961 Ro Rv M 6, 413-427
  (MR28, 221; Z102, 6)
Ghika A 1961 Gaz MF (A)13, 282-292
  (Z98, 10)

**Stokes,** George Gabriel   1819-1903

Anon 1875 Nat 12, 201-   [port]
- 1899 Nat 60, 125-129
  (F30, 29)
- 1900 *Memoirs Presented to the
  Cambridge Philosophical Society
  on the Occasion of the Jubilee
  of Sir George Gabriel Stokes,*
  Cambridge
- 1904 Nat 70, 247
- 1905 Lon RSP 74, 199-216
Brillouin M 1904 Rv GSPA 15, 22-29
  (F35, 33)
E W B 1904 Phy R 18, 58-62
Kelvin W T 1902 Nat 67, 337-338
Lamb H 1904 Brt AAS 74, 421
Larmor J 1907 *Memoirs and Scientific
  Correspondence of the Late Sir
  G. G. Stokes,* Cambridge [port]
Macfarlane Alexander 1919 *Lectures
  on Ten British Physicists of the
  Nineteenth Century,* NY
Mascart J 1903 Par CR 136, 841-846
Somigliana C 1941 Tor Sem 7, 19-24
  (MR3, 258)   [Clairaut, Pizzetti]
Voigt W 1903 Gott Jb, 70-80
  (F34, 35)
Voit C v 1903 Mun Si 33, 512-515
  (F34, 36)
- 1903 Mun Si 33, 550-556

**Stoll,** Franz Xaver   1834-1902

Kiefer 1902 ZMN Unt 33, 143-144
  (F33, 44)

**Stolz,** Otto   1842-

Gmeiner J A 1906 DMV 15, 309-322
- 1906 Mo M Phy 17, 161-178
  (F37, 24)
Kaller E 1906 Ens M 8, 55   (F37, 24)
Voit C 1906 Mun Si 36, 477

**Stone,** Charles Arthur

Kinney J M + 1945 SSM 44, 699-700

**Stone,** Ormond   1847-1933

Luck J J 1933 AMSB 39, 318-319
  (F59, 33)
Matz F P 1895 AMM 2, 299-301
Mitchell S A 1933 Sci (2)77, 107-108
  (F59, 850)

**Størmer,** Fredrik Carl Muerlertz
1874-1957

Brun V 1957 Nor M Tid 5, 169-175,
  213   (MR19, 1150; Z78, 4)
- 1958 Act M 100, i-vii
  (MR20, 739; Z80, 9)
Chapman S 1957 Nat 180, 633-634
  (Z77, 242)

**Story,** William Edward   1850-1930

Anon 1930 Sci (2)71, 409   (F56, 27)
Franklin F 1936 Am Ac Pr 70, 578-580
  (F62, 1040)

**Strack,** Otto

Treutlein P 1899 ZMN Unt 30, 316-318
  (F30, 25)

**Strnad,** Alois   1747-1799

Sobotka J 1912 Cas MF 41, 553-557
  (F43, 49)   [port]

Strode, Thomas   XVII

Schrek D J E 1950 Euc Gron 25, 169-172   (Z37, 1)   [constr]


Stromer, Heinrich

Gunther S 1881 Prag Ab (6)10
    (F13, 10)


Stroud, William   1860-1938

Copley G N 1960 Nat 188, 254
    [calc]


Strouhal, C.

Novak V 1910 Cas MF 39, 363-383
    (F41, 38)


Struve, W. B. 1854-1912

Alexandrov I 1912 M Obraz (3), 134-136   (F43, 49)


Struyck, Nicolas   1687-1769

Brocard H 1911 Intermed 18, 272
    (Q1058)
Van Haaften M 1924 Verzek 5, 203-233
    (F51, 21)
- 1925 Nicolaas Struyck, The Hague,
    74 p   (F51, 22)
- 1925 Verzek 6, 49-86   (F51, 22)
    [bib]
Vollgraf J A ed 1912 Les oeuvres de
    Nicolas Struyck qui se rapport...,
    Amsterdam   [prob]


Stuart, George Henry   -1904

Greenhill A G 1904 LMSP (2)1, 29
    (F35, 33)


"Student."   See Gosset.


Studnicka, Frantisek Josef   1836-1903

Anon 1899 Titel der Buecher und
    Abhandlungen..., Prague, 24 p
    (F30, 29)
- 1903 Cas MF 32, 297
Panek A 1904 Cas MF 33, 369-480
    (F35, 39)
Petr K 1928 Cas MF 57, 163-164
    (F54, 36)


Study, Eduard   1862-1930

Barinaga J 1931 Rv M Hisp A (2)6, 177-178   (F57, 44)
Bodewig E 1930 Batt 68, 233-235
    (F56, 30)
Engel F 1908 DMV 17, 143-144
- 1931 DMV 40, 133-156
    (F57, 44)
Weiss E A 1930 Ber MG 29, 52-77
    (F56, 30)
- 1931 Ens M 29, 225-230
    (F57, 44)
- 1933 DMV 43, 108-124, 211-225
    (F59, 38; 60, 834)
- 1936 Deu M 1, 711-715   (F62, 1041)


Sturm, Charles Jacob Karl Franz
1803-1855

Bocher M 1911 AMSB (2)18, 1-18
    [alg and diff eqs]
- 1914 Rv Mois 17, 88-104
    (F45, 57)
Lorey W 1919 ZMN Unt 50, 289-293
    (F47, 13)
- 1936 Muns Semb 8, 70-76
Loria G 1938 Ens M 37, 249-274
    (F64, 918; Z21, 3)
Ludwig W 1925 DMV 34, 41-51
    (F51, 27)
Speziali Pierre 1964 C.-F. Sturm
    (1803-1855) Documents inédits,
    Paris, 32 p   (Z117:1, 8)

Stuyvaert, Modeste 1866-1932

Godeaux L 1937 Bel Anr 103


Subic, Simon 1830-1903

Anon 1904 Matica Sl 6, 203


Sucharda, Anton 1854-

Sobotka J 1908 Cas MF 37, 353-359
(F39, 28)


Suess, Wilhelm 1895-1958

Behnke H + 1958 MP Semb 6, 1-3
(Z86, 6)
Bompiani E 1958 It UM (3)13, 464
(Z80, 9)
Gericke H 1967 DMV 69, (4, 1), 161-
183
Raith F 1958 MN Unt 11, 229
(Z99, 246)


Suiseth, Richard XIV

= calculator = Swinshed
= Swineshead = Suisset, etc.

Sarton G 1947 Int His Sc 3(1),
736-738 [bib]
Thorndike L 1932 Speculum 7, 221-230


Sulima, P. A. 1779-1812

Shatunova E S 1959 Ist M Isl 12, 179-
184 (MR23A, 696; Z10, 247) [anal]


Sully Prudhomme, René François Armand
1839-1907

Poincaré H 1909 Rv GSPA 20, 659-662
(F40, 20)


Sundman, Karl Frithiof 1873-

Järnefelt G 1950 Act M 83, i-vi
(MR11, 708; Z35, 148)
- 1953 Hels Arsb 30(2), 12 p
(Z53, 197)


Suprunenko, Dimitrii Alekseevich
1915-

Berman S D + 1966 UMN 21(6), 170-171
(MR34, 17)


Sushkevich, Anton Kazimirovich
1889-

Gluskin L M + 1959 UMN 14(1),
255-260 (MR20, 1138; Z88, 244)


Suslov, Gavril Konstantinovich
1857-1935

Anon 1911 *Susloff-Festschrift*,
Kiev, 412 p (in Russ) (F42, 41)
Sluginov S P 1927 Perm Zh 4, 89-93
(F53, 35)


Suter, Heinrich 1848-1922

Hankel H 1872 Boncomp 5, 297-300
(F4, 24)
Schoy C 1922 Zur NGV 67, 407-413
(F48, 21)


Suto, Onosaburo 1871-1915

Hayashi T 1916 Toh MJ 9, 71-76


Suvorov, F. M.

Olonichev P M 1956 Ist M Isl 9, 271-
316 (Z70, 243)
Parfentev N 1911 Kazn FMO (2)17(3),
31-45 (F42, 38)
Sintsov D 1911 Kagan (541), 19-21
(F42, 38)

Suyehiro, Kyoji   1877-1932

Anon 1932 Jp Ac P 8, xv


Swain, Lorna Mary   1891-1936

Kennedy M D 1937 LMSJ 12, 155-157
  (F63, 19)


Swedenborg, Emanuel   1688-1772

Eneström G 1889 Sv Ofv 46, 529-531
- 1890 Sv Bi 15
Nordenmark N V E 1933 Ark MAF 23(13)
  [ast]


Swenson, John A.   XIX-XX

Reeve W D 1945 MT 37, 133-134


Swift, Jonathan      1667-1745

Merrill C F 1961 MT 54, 620-625
  [Gulliver]
Nicholson M + 1937 An Sc 2
Patter G R 1941 Philol Q 20


Sylow, Ludvig   1832-1918

Anon 1918 Ens M 20, 227   (F46, 17)
Kragemo Helge B 1933 Ludvig Sylow,
  Oslo, 27 p [group]
- 1933 Nor MFS 2(3), 25-30
  (F59, 31) [bib]
- 1933 Nor M Tid 15, 73-99
  (F59, 30)
Matthews G B 1918 Nat 103, 49
  (F46, 17)
Skolem T 1919 Nor M Tid 1, 1-13
  (F47, 11)
- 1921 Chr Sk 2(18)
  1932 Nor M Tid 14, 117-123
  (F58, 991)
- 1933 Nor MFS 2(2), 14-24
  (F59, 31)
Stormer C 1933 Nor MFS 2(1), 7-13
  (F59, 30)


Sylvester, James Joseph   1814-1897

Anon 1889 Nat 39, 217-
- 1897 AMJ 19(S), 1-8   (F28, 31)
- 1897 Bu SBM 16
- 1898 Lon RSP 53, ix
- 1948 M Gaz 32, 225
Archibald R C 1936 Osiris 1, 85-154
  (F52, 25)
- 1937 Int Con (1936)2, 272-273
  (F63, 20)
- 1947 in Ashley-Montagu M F ed,
  Studies and Essays...George Sarton
  ..., NY, 209-217 (MR8, 498)
Blondheim D S 1921 Johns HAM (Jan),
  22 p
Capelli A 1897 Nap FM Ri (3)3, 165-
  168   (F28, 30)
Dickstein S 1897 Wiad M 1, 75-77
  (F28, 31)
Edwards B 1924 M Gaz 12, 168-169
  (F51, 39)
Franklin F 1897 AMSB (2)3, 299-309
  (F28, 30)
- 1897 Johns. HC (June)
Gerardin A 1921 Fr AAS 45, 225
  [letters]
Halsted G B 1894 AMM 1, 295-298
- 1897 AMM 4, 159-168
- 1897 Sci 5, 597-604
- 1916 Johns NAM (Mar)
Kelvin W T 1897 Edi SP 22, 9
Lampe E 1897 Naturw R 12, 359-363
  (F28, 31) [Cayley]
Lemoine E + 1900 Intermed 7, 375
MacDuffee C C 1943 AMM 50, 360-365
Macfarlane Alexander 1916 Lectures
  on Ten British Mathematicians of
  the Nineteenth Century, NY, 107-121
MacMahon P A 1897 Nat 55, 492-494
  (F28, 30)
- 1898 Lon RSP 63, ix-xxv   [port]
Mansion P 1897 Mathesis (2)y, 245-246
  (F28, 31)
- 1913 Rv Q Sc 73, 568-579
- 1914 Mathesis (4)4(S), 12 p
  (F45, 27)
Matheson P E + 1898 Johns HC 18, 29
  (F29, 17)
Matz F P 1893 AMM 1, 383 .[det]
Miller G A 1926 Sci (2)64, 576-577
  (F52, 38)

Noether M 1897 M Ann 50, 113-156
  (F28, 30)
Picard E 1897 Rv GSPA 8, 689-690
  (F28, 31)
- 1924 *Mélanges de mathématiques
  et de physique,* Paris, 29-34
Polubarinova-Kochina P Ya 1957
  Vop IET (5), 156-162  (Z80, 5)
  [Kovalevskaya]
Rosenbaum R A 1960 MT 53, 35-38
Smith D E 1922 AMM 29, 14
  [poetry]
Vassilief A 1897 Kazn FMO (2)7, 89
Walker J J 1897 LMSP 28, 581-586
  (F28, 31)
Werebrusof A + 1901 Intermed 7, 172
  (Q1760)
Yates R C 1937 AMM 44, 194-201
  (F63, 17)  [U Virginia]


Szasz, Otto    1884-

Szego G 1954 AMSB 60, 261-263
  (MR15, 770; Z56, 2)


Szasz, Pal    1901-

Karteszi F 1962 M Lap 13, 9-21
  (Z115, 246)


Szele, Tibor    1918-1955

Anon 1954 Pub M 3, 193-194
  (MR17, 118)
Fuchs L 1955 M Lap 6, 97-129
  (MR17, 446)
Kertész A 1956 Pub M 4, 115-125
  (Z70, 7)
Varga O 1955 Deb U Act 2, 5-6
  (Z67, 247)


Szokefalvi-Nagy, Gyula    1887-

Anon 1953 M Lap 4, 81-83 (MR15, 924)
- 1954 Szgd Acta 15, 97-98
  (MR15, 592)


Szuecs, Adolf André    1884-

Kosa A 1961 M Lap 12, 1-9
  (Z98, 10)


Tabit.    See Thabit.


Tacquet, André    1612-1660

Bosmans Henri 1925 Guld Pass 3, 63-87
- 1925 *Le jésuite mathématicien
  anversois André Tacquet,* Anvers,
  25 p
- 1927 Brx SS 47 (A)1, 39-42
  (F53, 15)
- 1927 Isis 9, 66-82


Taillefer,  P.

Boyer J + 1898 Intermed 5, 254 (Q712)


Taisnier, Jean    1508?-1562

Godeaux L 1915 Loria 17, 33-36
  [bib]


Tait, Peter Guthrie  1831-1901

Anon 1903 Loria 6, 28-29  (F34, 27)
Chrystal G 1901 Nat 64, 305-307
Hayashi T 1905 Bib M (3)6, 323
  (F36, 53)  [Japan]
Knott C G 1911 *Life and Scientific
  Work of P. G. Tait,* Cambridge,
  390 p  (F42, 18)  [port]
Macfarlane Alexander 1919 *Lectures
  on Ten British Physicists of the
  Nineteenth Century,* NY, 38-54
- 1902 Phy R 15, 51-64
  (F33, 39)
- 1903 Bib M (3)4, 185-200
  (F34, 26)
Mackay J S 1905 Ens M 7, 5-10
  (F36, 23)
Stark J 1901 Naturw R 16, 462
  (F32, 32)

Thomson William 1901 Edi SP (A)23,
498-504  (F32, 32)
- 1911 *Mathematical and Physical
Papers*, Cambridge 6, 363-369


Takagi, Teiji  1875-1960

Anon 1960 Jp MS 12, 225  (Z91, 7)
- 1960 Sugaku 12, 135-136
(Z114, 244)  [bib]


Takano, Kinsake   1915-1958

Matusita K 1958 Tok I Sta M 10, i-ii
(Z82, 243)


Tallquist, Axel Henrik  Hjalmar 1870-

Anon 1958 Arkhimed (2), 52-54
(MR22, 425)
Archibald R C 1948 *Mathematical
Table Makers, NY,* 79-80
(Scr Stu 3)  [bib, port]


Tamarkin, Jacob David   1888-

Hille E 1947 AMSB 53,440-457
(MR8, 498; Z31, 99)


Tamaru, Takuro

Shimizu T 1932 Tok M Ph (3)14,
450-451


Tanaka, Yoszane

Fujiwara M 1942 Toh MJ 49, 90-105
(MR7, 353)


Tanner, Henry W. Lloyd  1851-1915

Anon 1915 M Gaz 8, 215
R H P 1915 Nat 95, 707  (F45, 53)


Tannery, Jules   1848-1910

Anon 1910 BSM 34, 193-197
- 1910 Ens M 13, 56-58
Borel E 1911 Rv Mois, 5-17
Cahen E 1909 BSM (2)33, 157
[letter]
Chatelet A 1911 Ens M 13, 56-58
(F42, 31)
Darboux G + 1910 BSM (2)34, 193-197
(F41, 34)
Hovelaque E 1911 Rv Par (Jan)
Lelieuvre M 1911 Rv GSPA 32, 49-50
Mascart J 1911 Rv GSPA 22, 49-50
Picard Émile 1910 Rv Sc 14, 689-690
- 1925 *La vie et l'oeuvre de Jules
Tannery,* Paris, 32 p  (F51, 25)
- 1926 Par Mm (2)58, i-xxxii
(F52, 29)
- 1926 Rv Deux M, 858-884
Sarton G 1948 Isis 38, 33-51
(Z29, 195)  [Marie Tannery]


Tannery, Paul   1843-

Anon 1905 BSM (2)29, 102-109
(F36, 29)
- 1905 M Gaz 3, 168
- 1905 Rv GSPA 16, 97-99
- 1929 Archeion 11, lxxv-cviii
- 1930 Bo Fir (3)11, 80-92
- 1931 Archeion 13, 49-54
[letters, Zeuthen]
- 1936 Isis 25, 57-59 [Hultsch,
letter]
- 1938 Osiris 4, 633-709  [bib]
- 1942 Rv Sc 80, 99-103  (Z60, 5)
[letters, Zeuthen]
Bosmans H 1905 Mathesis (3)5(S)
- 1905 Rv Q Sc (3)7, 544-574
(F36, 31)
Boutroux P 1938 Osiris 4, 690-702
(F64, 22)
Ducasse P 1938 Osiris 4, 708-709
(F64, 22)  [Mme Tannery]
Duhem P 1908 Bord Mm (6)4, 295-298
(F39, 27)
Enriques F 1924 Rv Met Mor 31, 425-
434  (F50, 6)
Fehr H 1905 Ens M 7, 51-52 (F36, 31)

Loria G 1905 Loria 8, 27-30
  (F36, 31)
- 1929 Archeion 11
- 1947 Scr 13, 155-162
  (MR9, 486; Z33, 4)
Rivaud A 1912 Rv Met Mor 21, 177-210
Sarton G 1938 Osiris 4, 703-705
  (F64, 22)  [*bib]
Tannery J 1908 Bord Mm (6)4, 269-293
  (F39, 26)
Tannery V 1908 Bord Mm (6)4, 299-382
  [bib]
- 1931 Archeion 13, 49-54
  (F57, 38)  [letters, Zeuthen]
Teichmuller G 1880 Rv Phi Fr E
Van Deventer C M 1929 Euc Gron 6,
  170-184  (F55, 599)  [Euclid]
Zeuthen H G 1905 Bib M (3)6, 257-304
  (F36, 30)


Tardy, Placido   1816-

Loria G 1915 Linc Rn 24, 505-531
  (F45, 47)


Tartaglia, Nicolo   1506?-1559

Boncompagni B 1881 in Cremona L + ed
  *In Memoriam Domenico Chelini,*
  363-412  (F13, 10)  [mss]
Bortolotti E 1926 Bln Rn, 12 p
  [Ferrari]
Cassina U 1932 Int Con 6, 443-448
  [Cardano]
Cossali P 1857 in Boncompagni B ed,
  *Scritti inediti del P. D. Pietro
  Cossali...,* Rome, 289-316
Drake S 1961 Isis 52, 430-431
Drake S + ed 1969 *Mechanics in
  Sixteenth-century Italy* (U Wis
  P),  428 p
Favaro Antonio 1881 *Intorno al
  testamento inedito di Nicolò
  Tartaglia pubblicato da
  D. B. Boncompagni,* Padua
  (F13, 11)
- 1913 Isis 1, 329-340
- 1913 *Per la biografia di Niccolò
  Tartaglia,* Rome, 40 p

Feldmann R W 1961 MT 54, 160-163
  [Cardano]
Loria G 1932 Bo Fir 32, lxi-lxiii
  (F62, 1029)  [geom]
Masstti A 1960 Lom Rn M (A)94, 31-41
  (MR23A, 270)  [Ferrari]
- 1960 Lom Rn M (A)94, 42-46
  (MR23A, 270)
- 1959 *Niccolò Tartaglia quesiti et
  inventioni diverse,* Brescia, 597 p
  [port]
Montagnana M 1958 Archim 10, 135-139
  (MR20, 620; Z80, 7)
Natucci A 1956 Batt (5)4, 261-271
  (MR18, 710; Z71, 243)
- 1956 Int Con HS 8, 75-83
  [Archimedes]
- 1956 It UM (3)11, 594-598
  (MR18, 710; Z71, 243)  [root]
- 1956 Pe M (4)34, 294-297
  (MR19, 108)
Nordgaard M A 1938 M Mag 12, 327-346
  (F64, 12)  [Cardano]
Oliva A 1909 *Sulla soluzione dell'
  equazione cubica di Tartaglia,*
  Milan, 36 p  (F40, 54)
Pasquale L di 1957 Pe M (4)35, 79-93
  (MR19, 825; Z80, 3)  [cubic]
- 1957 Pe M (4)35, 253-278
  (MR20, 945)  [competition, Ferrari]
- 1958 Pe M (4)36, 175-198
  (MR20, 945)
Reeve W D 1930 MT 23, 385
Santalo L A 1941 M Notae 1, 26-33
  (MR3, 97)  [cubic]
Tonni-Bazza V 1901 Linc Rn (5)10(2),
  39  [letter]
- 1904 Int Con H 12, 293-307
  (F35, 7)
- 1904 Lin Rn 13, 27-30
  (F35, 7)
Villa M 1960 Archim 12, 161-167
  (Z98, 5)
- 1961 Archim 13, 51-61  (Z98, 5)
- 1963 in *Città di Siracusa Celebr.
  Archimedee del secola XX, 1961,*
  Vol 1(1), 61-62  (Z107, 2)
Zeuthen H G 1893 Dan Ov 3, 303, 330
  [Cardano]

Tartakovskii, Vladimir Abramovich 1901-

Linnik Yu V + 1961 UMN (16(5), 225-230 (MR24A, 221; Z100, 8)

Taurinus, Franz Adolph 1794-1874

Staeckel P 1899 ZM Ph 44(S), 397-427 (F30, 13)

Taylor, Brook 1685-1731

Auchter H 1937 *Brook Taylor, der Mathematiker and Philosoph,* Wuerzberg, 112 p (F63, 14, (Z16, 197)
Bateman H 1906 Bib M (3)7, 367-371 (F37, 7) [letters]
Conte L 1948 Arc In HS, 611-622 (MR10, 174) [Bernoulli]
- 1949 Iasi IP 4, 36-53 (MR20, 264) [Bernoulli, Fagnano]
Eneström G 1894 Sv Ofv 51, 177-187 (F25, 65)
Jones P S 1951 AMM 58, 587-606 (Z44, 244) [perspective]
Kirby John J 1754 *Dr. Brook Taylor's Method of Perspective...,* Ipswich
Langley E M 1907 M Gaz 4, 97-98 (F38, 12) [letter]
Sanford V 1935 MT 27, 60-61
*Young W 1793 in Taylor, Brook, *Contemplatio Philosophicae,* London

Taylor, Charles 1840-

Love,A E H 1909 LMSP (2)6, xx-xxi (F40, 33)

Taylor, Henry Martin 1842-

Lamb H 1928 LMSJ 3, 110-112 (F54, 38)

Tedone, Orazio 1870-

Sbrana F 1954 Bari Sem (4), 17pp (Z57, 243)
Somigliani C 1923 Linc Rn (5)32(1), 173-180

Teixeira, Francisco Gomez 1851-1933

Almeida e Vasconcellos F de 1933 Archeion 15, 164-167 (F59, 33)
Appell P 1917 Porto Ac 12, 126-128 (F46, 36)
Freire L 1951 Gaz M 12(50), 109-111 (MR14, 2)
J R P 1911 Esp SM 1, 77-80 (F42, 41)
Leite D 1934 Porto Fac 18, 193-207 (F60, 836)
Sarmento de Beires R 1951 Porto Fac 35, 173-192 (MR13, 810)
Toledo L O de 1933 Rv M Hisp A (2)8, 50-55 (F59, 850)
Vilhenna Henrique de 1936 *O Professor Doctor F. Gomas Teixeira. Elogio, notas de biografia, bibliografia, documentos,* Lisbon, 336 p

Teleki, Josef and Samuel 1739-1822

Gulyas K 1912 MP Lap 21, 194-223 (F43, 20) [letters]
Jelitai J 1936 MP Lap 43, 142-160 [J II Bernoulli]
- 1937 MP Lap 44, 173-198 (F63, 817) [d'Alembert, Clairaut La Condamine]
- 1938 Magy MT 57, 501-507 [D Bernoulli, Teleki]
Marian V 1939 Gaz M 45, 113-117, 229-231, 289-292 (F65:1, 13)

Temperley, Ernest 1849-1889

Anon 1889 LMSP 20, 424-427

Tenca, Luigi 1877-1960

Procissi A 1960 It UM (3)15, 466-468
  (Z93, 9)
- 1960 Physis Fi 2, 358-359
  (Z98, 10)


Tendlebury, Charles 1854-1941

Punnett M + 1942 M Gaz 26, 1-4


Teofilato, Pietro 1879-1952

Sestini G 1952 Parma Rv M 3, 291-296
  (MR14, 833; Z48, 243)


Termier, Pierre 1859-1930

Kaisin F 1932 Rv Q Sc (4)21, 1,
  5-32


Tesch, Johann Wendel 1840-1901

Schoute P H 1901 Nieu Arch (2)5(3),
  310


Tetmajer, Ludwig von 1850-1905

Anon 1905 Ens M 7, 143-148
  (F36, 39)


Thabit ibn Qurra IX

  = Abu-1-Hasan Thabit ibn Qurra
  ibn Marwan al-Harrani

Bessel-Hagen E + 1932 QSGM (B)2,
  186-198 (Z5, 4)
Björnbo A 1924 Ab GN Med 7, 91 p
Boyer C B 1964 Isis 55, 68-70
  (MR29, 219) [Clairaut le Cadet]
Burger H + 1925 *Thabits Werk ueber
  den Transversalen satz*, Erlangen,
  99 p (F51, 38)
Carmody F J 1941 Thabit b. Qurra:
  *Four Astronomical Tracts in Latin*,
  Berkeley, 28 p (MR5, 57)

- 1955 Isis 46, 235-242 (MR17, 118)
  [ast]
Garbers K ed 1937 QSGM (A)4, 84 p
  (F63, 8)
Karpova L M + 1966 Ist Met EN 5, 126-
  130 (MR34, 5612) [ratio]
Kennedy E S 1960 UAR M Ph (24), 71-74
  (Z107, 244)
Krasnova S A 1965 Ist M Isl (16), 437-
  446 (MR34, 2406) [conics]
Kutsch Wilhelm 1959 *Tabit ibn
  Qurra's arabische Uebersetzung...
  des Nikomachos...*, Beyrouth, 372 p
  (Z102, 245)
Luckey P 1937 QSGM (B)4, 95-148
  (F63, 8: Z17, 289)
- 1941 Leip Ber 93, 93-114
  (MR11, 150) [quad eq]
Nix L M Ludwig 1889 *Das V Buch der
  Conica des Apollonius von Perga in
  der arabischen Uebersetzung des
  Tabit ibn Corrah*, Leipzig
Rozenfeld B A + 1961 Ist M Isl 14,
  587-602 [Euclid]
Sayili A 1960 Isis 51, 35-37
  (Z89, 241; Z89, 241) [Pyth thm]
Steinschneider M 1873 ZM Ph 18,
  331-338 (F5, 7) [bib]
Suter H 1916 Erlang Si, 48-49,
  65-88, 186-227 [paraboloid]
- 1918 Zur NGV 63, 214-228 (F46, 37)
  [parab]
Wiedemann E 1912 Bib M (3)12, 21-39
  (F42, 65) [mech]
- 1922 Erlang Si 52, 189-219
Wiedemann E + 1922 Dan M Med (4)9,
  24 p
Yushkevich A P 1966 Ist Met EN 5,
  118-125 (MR34, 1136) [area,
  parab]


Thales of Miletus -624? to-548?

Anon 1930 MT 23, 84-86
Bortolotti E 1930 Per M (4)10, 228-
  230 [Thales thm]
Decker 1865 *De Thalete Milesio*,
  Halle
Dicks D R 1959 Clas Q (NS)9, 294-309
Frajese A 1941 It UM (2)4, 49-60
  (Z26, 97) [geom]

Harding P J 1913 Int Con 5(2), 533-
  538 (F44, 50) [geom]
Loria G 1928 Bo Fir 24, iii-v
Plummer W E 1890 Nat 42, 390-391
Stamatis E 1959 Epist Tec 116
  (MR21, 230)


Theaetetos of Athens    -IV

Anderhub J H 1941 *Genetrix
  irrationalium. Platonis
  Theaetetus...*, Frankfurt, 222 p
  (Z26, 97)
Fritz K v 1932 Philolog 87, 40-62,
  136-178
Sachs Eva 1917 *Die fuenf platonis-
  chen Koerper*, Berlin
Sarton G 1927 Int His Sc 1, 116
  [bib]
Stamatis E 1956 Athen P 31, 10-16
  (MR18, 268; Z70, 242)
- 1957 Athen P 32, 87-90
  (Z77, 241)


Thebault, Victor    1882-1960

Byrne W E 1947 AMM 54, 443-444
Court N A 1947 AMM 54, 445-446
  [geom]
Deaux R 1690 Mathesis 69, 377-395
  (MR23A, 698; Z97, 2)
Guillotin M R 1961 Scr 25, 331-333
  (Z97, 244)
Starke E P 1947 AMM 54, 444-445
  [numb thy]


Theodoros of Cyrene    -V

Heller S 1956 Centau 5, 1-58
  (Z71, 1)    [Plato]
Hofmann J E 1942 Deu M 7, 117-120
  (F68; 5; Z27, 193)    [Plato]
Omont H 1918 J Sav 16, 321-323
  [Fermat]

Theodosios of Bithynia    -II to -I

  = Theodosios of Tripoli

Bosmans H 1927 Arc Sto 8, 465-476
  [sphere]
- 1928 Rv Q Sc (4)13, 82-85
  (F54, 14)
Hultsch F 1887 Leip Ph Ab 10, 381-
  446 (F21, 1255) [sphere]


Theon of Alexandria    IV

Halma 1821 *Théon d'Alexandrie,*
  Paris
Halma ed 1821 *Commentaire de Théon
  sur la composition Mathématique de
  Ptolémée,* Paris
Kuensberg H 1884 Bay Bl GR 20, 368-
  372 [geog]
Rome A 1926 Brx SS 46, 1-14
  (F52, 9) [Hypatia]
- 1932 Brx SS 52(A), 30-41
  (Z5, 2) [Archimedes, optics]
- 1933 Brx SS(A)53, 39-50
  (F59, 838; Z7, 147) [Menelaos]
- 1934 Isis 20, 440
- 1938 Brx SS (1)58, 6-26
  (F64, 905) [ast]
- 1950 Brx Con CR 3, 21-23
- 1952 Int Con (1950)1, 209-219
  (MR13, 419) [ast]
- 1953 Bel BL 39, 500-521


Theon of Smyrna    II

Martin T H 1849 *Theonis Smyrnaei
  liber de astronomie,* Paris
Vedova G C 1951 AMM 58, 675-683
  (MR13, 611; Z44, 242)


Thibaut, Bernhard Friedrich    1775-
1832

Dunnington G W 1937 M Mag 11, 318-323
  (F63, 809) [educ]

Thiele, Thorrald Nicolai
1838-1910

Burrau C 1929 Nor St Tid 8, 340-348
  (F56, 814)
Gram J 1910 Nyt Tid (B)21, 73-78


Thieme, Karl Gustav Hermann  1852-

Lietzmann W 1926 ZMN Unt 57, 322-324
  (F52, 31)


Thirring, Hans  1888-

Flamm C 1963 Fors Fort 37, 93-94
  (Z106, 3)


Thom, René  1923-

Hopf H 1960 Int Con (1958), lx-lxiv
  (MR22, 929)


Thomae, Johannes  1840-1921

Herglotz G 1922 Leip Ber 74, 159-160
  (F48, 16)
Liebmann H 1921 DMV 30, 133-144
  (F48, 17)


Thomas of Namur, Antoine  1644-1709

Bosmans H 1924 Brx SS 44, 169-208
  (F52, 19)
- 1926 Brx SS 46, 154-181  (F52, 19)


Thomé, Ludwig Wilhelm  1841-

Engel F 1911 DMV 20, 261-278
  (F42, 31)


Thompson, Alexander John  1885-

Archibald R C 1948 *Mathematical
  Table Makers,* NY, 78-79
  (Scr Stu 3)  [bib, port]


Thompson, D'Arcy Wentworth  1861-1940

Turnbull H W 1952 Edi MSN 38, 17-18
  (Z49, 291)


Thomson, James  1822-1892

Anon 1891 Lon RSP 53  (F23, 30)
- 1892 Nat 46, 129-130  (F24, 38)


Thomson, Joseph John  1856-1940

Novak V 1927 Cas MF 56, 215-218
  (F53, 36)
Price D J 1956 Nuo Cim (10)5(S),
  1609-1629  (MR19, 236)
Thomson G 1956 Nat 178, 1317-1319
  (Z71, 4)


Thorndike, Edward Lee 1874-1949

Bailey M A 1923 MT 16, 129-140


Thorne, Roger Chapman  1929-1959

Ursell F + 1959 Au MSJ 1, 255-256
  (MR22, 107; Z89, 5)


Thue, Axel  1863-1922

Bjerknes V + 1922 Nor M Tid 4, 40-49
  (F48, 20)  [bib]
Bohr H 1922 M Tid B (2), 33
  (F48, 20)
Brun V + 1923 Chr Sk (12)
Stoermer C 1922 Nor M Tid 4, 33-39
  (F48, 20)


Thureau-Dangin,  F.

Rey A 1940 Thales 4, 227-234
  [Babylonia]

Thysbaert, J. F. XVIII

Lorent H 1956 Bel BS (5)42, 83-91
(Z71, 245)


Tikhomandritzky, M. A. 1844-1921

Anon 1921 Tauric ML 2, xx-xxxii
(F48, 24)


Tikhonov, Andrei Nikolaevich 1906-

Aleksandrov P S + 1956 UMN (NS)11(6),
235-245 (MR18, 709; Z72, 246)
- 1967 UMN 22(2), 133-138
(MR34, 5623, 5624) [fun anal, top]
Anon 1967 UMN 22(2), 185-188
(MR34, 5629) [bib]
Dmitriev V I + 1967 UMN 22(2), 138-
149 (MR34, 5625) [phys]
Ilin V A 1967 UMN 22(2), 168-175
(MR34, 5627)
- 1966 Dif UR 2, 1408-1412
(MR34, 1147)
Samarskii A A 1967 UMN 22(2), 176-
184 (MR34, 5628)
Vasileva A B + 1967 UMN 22(2), 149-
168 (MR34, 5626) [diff eqs]


Tilly, Joseph Marie de 1837-1906

Anon 1914 Mathesis (4)4(Feb)(S), 54 p
Barbarin P 1909 Bord PV 145-155
(F40, 29)
- 1926 BSM (2)50, 50-64, 74-88
(F52, 27) [Houël]
Mansion P 1895 Rv Q Sc 37, 584-595
and in Mathesis (2)5(S)
(F26, 6) [geom]
- 1906 Bel Anr, 622-629 (F37, 32)
- 1906 Mathesis (3)6(S)
- 1906 Rv Q Sc (3)10, 353-361
(F37, 32)
- 1914 Bel Anr, 203-285
(F45, 31)
Quint N 1907 Wisk Tijd 3, 106-107
(F38, 33)


Tilser, Frantisek 1825-1913

Prochazka B 1914 Cas MF 43, 1-25
(F45, 34)
Zedek M 1957 Olom Pri V 3, 89-108
(Z91, 5)


Timerding, Heinrich Carl Franz Emil
1873-

Rehbock F 1943 Deu M 7, 252-254
(Z28, 100)


Timmermans, J. A. 1801-1869

Anon 1868 Bel Anr, 99 (F1, 14)
Quetelet A 1869 Arc MP Lit 49(194),
2-12 (F2, 21)


Timoshenko, Stepan Prokotevich
1878-

Anon 1958 SSSR Tek (12), 3-4
(Z87, 7)
Lessels J M 1938 *Timoshenko 60th
Anniversary Volume, 1-8,* (in Russ),
(F64, 919) [bib]


Tisserand, François Felix 1845-1896

Bertrand J 1904 Par Mm 47, 259-282
Conu A 1896 Par CR 123, 623-625
Ernst M 1897 Wiad M 1, 29
Poincaré H 1896 Rv GSPA 7, 1230


Titchmarsh, Edward Charles 1899-

Cooper J L B 1963 Nat 198, 1039
(Z107, 248)
Levitan B M 1964 UMN 19(6), 123-131
(MR30, 872)


Titeica, Gheorghe 1875-1939

Anon 1939 Tim Rev M 18, 137-138
(F65, 18)

Cioranescu N 1939 Ro SP 53, 170-199
  (F65, 1092)
Gheorghiu G T 1939 Gaz M 45, 169-173,
  281-288, 339-344, 396-405, 452-457
  (F65, 18)
- 1956 in *Lucrarile constatuirii de
  geometrie differentiala din 9-2
  Iunie 1955,* 55-62  (Z74, 246)
Onicescu O + 1941 Ro SSM 43, 147-161
  (MR6, 254; Z60, 18)
Zervos P 1939 Intbalkn 2, 1
  (F65, 18)


Todhunter, Isaac    1820-1884

Macfarlane Alexander 1916 *Lectures
  on Ten British Mathematicians of
  the Nineteenth Century,* NY,
  134-146
*Pearson K 1926 Nature 117(17 April),
  552


Toepler, August Joseph Ignaz
  1836-

Anon 1910 Ber Si, 532-533
  (F41, 46)


Toeplitz, Otto   1881-1940

Behnke H 1949 MP Semb 1, 89-96
  (MR11, 573; Z31, 99)
Behnke H + 1963 DMV 66, 1-16
  (MR29, 1069; Z108:2, 250)
Born M 1940 Nat 145, 617
  (MR2, 115; Z60, 18)


Toledano, Eduardo Leon de

De Toledo L O 1914 Esp SM 4, 1-5
  (F45, 47)


Toledo y Zulueta, Don Luis
  Octavio de

Sanchez Perez J A 1934 Rv M Hisp
  A(2)9, 49-53

Tolstoi, Lev Nikolaevich    1828-1910

Bunyakovskii V Ya 1959 Ist M Isl 12,
  505-511 (MR24A, 4)


Tonelli, Alberto   1849-

Gradara E 1920 Rass MF 1, 77-79


Tonelli, Leonida   1885-

Anon 1946 Annali (4)25, xiii-xvi
  (Z61, 6)
- 1952 *Leonida Tonelli in Memoriam,*
  Pisa, 193 p  (MR15, 90)
Cinquini S 1950 Pisa SNS (2)15,
  1-37   (MR11, 708; Z36, 5)
Faedo S 1947 Rm U Rn (5)6, 217-225
  (MR9, 75; Z30, 338)
Mambriani A 1950 Parma Rv M 1, 157-
  188   (MR12, 312; Z39, 243)
Sansome G 1948 Linc Rn (8)4, 594-
  624   (MR10, 175; Z36, 146)


Torelli, Gabriele   1849-1931

Cipolla M 1932 Batt 70, 62-78
  (F58, 45)
Marcolongo R 1931 Nap FM Ri (4)1,
  109-118   (F57, 39)
- 1932 Batt 70, 55-61
- 1932 It UM 11, 62-63  (F58, 45)
- 1932 Nap FM Ri (4)1, 9-12
  109-118
Ricci C L 1935 Nap Pont 63, 561-538
  (F61, 955)


Torelli, Ruggiere   1884-1915

Severi F 1916 Loria 18, 11-21
  (F46, 35)


Torricelli, Evangelista   1608-1647

Agostini Amodeo 1930 Archeion 12, 33-
  37 [calc, intgl]

Torricelli, Evangelista (continued)

- 1930 Bo Fir (2)9, xxv-xxviii
  (F56, 812) [length, spiral]
- 1930 Pe M 10, 143-151 [spiral]
- 1931 Archeion 13, 55-59
  (F57, 25) [intgl]
- 1931 Archeion 13, 65 (Z2, 6)
  [calc]
- 1949 *La memoria di Evangelista
  Torricelli sopra la spirale
  logarithmica riordanata e
  completata,* Livorno, 45 p
  (Z41, 339) [spiral]
- 1950 Pe M (4)28, 141-158
  (Z39, 243) [calc]
- 1951 It UM (3)6, 149-159
  (Z42, 241) [barycentric calc]
- 1951 It UM (3)6, 319-321
  (MR13, 613; Z43, 243)
- 1951 It UM Con 3, 250-251
  (Z45, 147)
- 1951 Livo Na PS (10), 15 p
  (Z45, 292) [cycloid]
- 1951 Livo Na PS (11), 14 p
  (Z45, 292)
  1951 Parma Rv M 2, 265-275
  (MR13, 612; Z44, 244) [max]
- 1951 *Sul moto dei proiettili
  di Evangelista Torricelli,*
  Leghorn, 14 p
- 1953 It UM Con 4(2), 629-632
  (Z50, 242) [max]
Anon 1925 Pe M (4)5, 43-46
  [mss]
- 1932 Archeion 14, 12-14 (F58, 35)
  [letter, spirals]
Boncompagni B 1875 Boncomp 8, 353-
  456 (F7, 21) [letter, Mersenne
  Verdus]
Bortolotti E 1923 Pe M 3, 317-319
  [stereometry]
- 1923 Pe M 3, 429-430 [stereometry]
- 1924 Int Con 2, 943-958
  (F54, 48)
- 1924 M El Rm 2, 11 p
- 1925 Arc Sto 49-58, 139-152
  (F51, 19) [calc]
- 1928 Bln Rn (2)32 (F54, 27)
  [length]
- 1928 Pe M 8, 19-59 [calc]

- 1928 Pe M (4)8, 205-206
  (F54, 27)
- 1930 Archeion 12, 267-271
  [Apollonius]
- 1930 Pe M 10, 85-91
  [definitions]
- 1931 Archeion 13, 60-64 (Z2, 6)
  [intgl]
- 1938 It UM Con 1, 19 p
  [infinitesimal]
- 1939 Mo M Phy 48, 457-486
  (MR1, 130; F65:1, 10; Z12, 2)
  [geom]
Bosmans H 1920 Brx SS (A)40, 141-148
  (F48, 4)
Del Gaizo M 1908 Rv FMSN 17, 385-402
  (F39, 12) [Borelli]
de Waard C 1919 Loria (2)2, 33-35
Duhem P 1906 Par CR 143, 809
  [mech]
Eremeeva S I 1959 Mos IIET 22, 281-
  288 (Z96, 2)
Favaro A 1914 Loria 16, 1-6
  (F45, 11)
- 1921 Arc Sto 2, 46-50
  (F48, 3) [Ciampoli]
Ghinassi G 1864 *Lettere fin qui
  inedite di Evangelista Torricelli
  precedute dalla vita di lui,*
  Florence
Govi 1886 Nap FM Ri 25, 163-169
  [letter]
Jacoli F 1875 Boncomp 8, 265-304
  (F7, 21) [calc, Roberval]
Lefner A 1909 Intermed 16, 263
  (Q3565) [Beaugrand]
Lombardo-Radice L 1959 Gaz MF (A)10,
  259-261 (Z88, 243)
Loria Gino 1897 Linc Rn (5)6, 318
  [length]
- 1900 Bib M (3)1, 75 [log curve]
- 1909 *In memoria di Evangelista
  Torricelli,* Genoa, 12 p
  (F40, 12)
- 1919 Linc Rn (5)28(2), 409-415
  (F47, 3) [geom]
- 1922 Arc Sto 3, 272-276 (F48, 3)
  [letter, Mersenne]
- 1922 Bo Fir (2)1, i-vii (F48, 4)
  [geom]
- 1938 Archeion 21, 62-68
  (F64, 15; Z19, 100) [plagiarism]

Mezzetti P 1908 Rv FMSN 17, 516-527
  (F39, 13)
Michl F 1909 Cas MF 38, 257-262
  (F40, 13)
Opial Z 1960 Wiad M (2)3, 251-265
  (MR22, 1853; Z118:1, 4)
  [spiral]
Podetti F 1914 Loria 16, 65-76
  (F45, 68) [propor]
Procissi A ed 1951 *E. T. nel terzo
  centenaio della morte. Conferenze
  ...,* Florence [port]
- 1953 Pe M (4)31, 34-43
  (MR14, 833) [envelope]
Segre B 1958 Archim 10, 170-188
  (Z89, 4)
Surico L A 1929 Archeion 11, 64-83
  (F55, 11) [intgl]
Tannery Mme P 1933 Rv Sc 71, 33-38
  (F59, 843) [letter]
Tenca L 1956 It UM (3)11, 258-259
  (Z70, 242)
- 1956 Pe M (4)34, 109-112
  (Z70, 7)
- 1958 Pe M (4)36, 251-263
  (MR20, 1137; Z83, 244)
- 1960 Pe M (4)38, 87-94
  (MR22, 2052; Z91, 5) [Viviani]
Torricelli E 1925 Pe M (4)5, 43-46
  (F51, 19) [will]
Turrière E 1919 Ens M 20, 245-268
  (F47, 4)
Vassura G 1909 Loria 11, 104-105
- 1909 *La pubblicazione delle opere
  di Evangelista Torricelli, con
  alcani documenti inediti,*
  Faenza, 59 p (F40, 13)
Weis F 1927 Arc GNMT 10, 250-281
  (F53, 19)


**Tortolini,** Barnaba   1808-1874

Anon 1875 Annali (2)7, 63-64
  (F7, 10)
Diario V 1875 BSM 8, 273- and Vat
  NLA 28, 93-106 (F7, 10)
Mansion P 1875 Nou Cor 1, 167-168,
  210-211   (F7, 12)
d'Ovidio E 1887 It SS (40)3, vi

**Tosca,** T. V.

Crespo Pereira R 1953 Gac M (1)5, 53-
  60 (MR15, 277)


**Toth,** Laszlo Fejes   XX

Anon 1957 M Lap 8, 281-283
  (MR20, 848)


**Toyos,** Daniel Marin

Guzman S A 1948 Rv M Hisp A (4)8,
  243-245 (Z33, 4)


**Tralles,** Johann Georg   1763-1822

Graf J H 1886 *Der Mathematiker Johann
  Georg Tralles,* Bern, 21 p


**Transon,** Abel   1823-1876

An Mines (7)14, 433-499 (F10, 18)


**Trefftz,** Erich Immanuel   1888-1937

Grammel R 1938 ZAM Me 18, 1-11
  (F64, 23)
Rehboc, F 1937 Deu M 2, 581-586
  (F63, 19)


**Treitschke,** Heinrich von

Weyrauch J J 1894 Schwa Mer (285)(S)
  (F25, 87)
- 1895 ZMN Unt 26, 226-233 (F26, 66)


**Trenchant,** Jean   XVI

Anon 1901 Intermed 7, 111
Bosmans H 1909 Brx SS 33(A), 184-192
  (F40, 53)
Ockenden R E 1935 Isis 24, 113
Sarton G 1944 Isis 35, 331

Treutlein, Josef Peter 1845-

Behm H W 1912 ZMN Unt 43, 521-530
  (F43, 50)
Cramer H 1912 Unt M 18, 121-123
  (F43, 50)
Staeckel P 1912 DMV 21, 384-386
  (F43, 50)


Trier, Viggo

Heegard P 1916 Nyt Tid 27(A), 1-5
  (F46, 24)


Tropfke, Johannes 1866-1939

Hofmann J E 1941 Deu M 6, 114-118
  (MR3, 98; Z25, 147)
Vogel K 1953 Arc In HS 6, 86-88
  (Z53, 340)


Trudi, Nicola 1811-1884

Amodeo F 1921 Nap Pont (2)26, 55-75
  [Flauti]
Pittarelli G 1923 Batt (3)61, 92-108
  [Sannia]
Torelli G 1884 Batt 22, 304-307


Truhlar, J

Studnicka F J 1898 Prag V (39)
  [tables]


Truksa, Ladislav 1891-

Bily J 1961 Cas M 86, 492-496
  (Z96, 4)


Tschirnhausen, Ehrenfried Walter von
1651-1708

  = Tschirnhaus

De Vleeschauwer H J 1941 Bel Vla Me
  3, 69 p  (Z27, 290) [Huygens]

- 1942 Tijd Phi 4, 345-396  (F68, 14)
  [Spinoza]
Hofmann J E + 1931 Arc GNT 13, 277-292
  [alg, Leibniz]
Oettel H 1958 MN Unt 11, 193-198
  (MR20, 810; Z99, 245)
Reinhardt C 1911 *Briefe an Ehrenfried
  Walther von Tschirnhaus von Pieter
  van Gent,* Freiberg, 32 p
Rychlik K 1959 Prag Pok 4, 232-234
  (Z82, 12)
Vollgraff J A 1940 Janus 44, 198-201
  [Huygens]
Weissenborn H 1866 *Lebensbeschreibung
  des ehrenfr. W. von Tschirnhausen
  aus Kissingswalde, und Wuerdigung
  seiner Verdienste,* Eisenach
Winter E 1959 Ber Si, 16 p


Tschuprow, Alexander Alexandrovich
1874-1926

Chetverikov N S 1926 Metron 6, 314-
  320  (F52, 33)
F I 1926 Gi M Fin 8, 194-195
  (F52, 33)
Kohn S 1926 Nor St Tid 5, 171-184
  (F52, 33)
- 1926 Nor St Tid 5, 481-502
  (F52, 33)
L I 1926 Lon St(A)(2)89, 619-622
  (F53, 33)
Slutsky E 1926 ZAM Me 6, 337-338
  (F52, 33)


Tsinger, Vasili Yakovlevich 1836-
1907

Andreev K N 1907 in *Otchet
  imperatorskogo Moskovskogo universiteta za 1907,* Moscow (F39, 35)
- 1911 Mos MO Sb 28, 3-39  (F42, 20)
Lopatin L M 1908 Vop F Ps (95), 219-
  227  (F39, 35)
Mlodzhizovski B 1911 Mos MO Sb 40-49
  (F42, 20)
Zhukovski N 1911 Mos MO Sb 28, 50-53
  [mech]

Tsuji, Masatsugu

Anon 1959 Jp JM 29, 185-189
(MR22, 1131; Z96, 5)

Tswetkoff, Y. Y.

Anon 1907 Nat 75, 165  (F38, 42)

Tucker, Robert  1832-

Halsted G B 1900 AMM 7, 237-239
Hill J M 1905 LMSP (2)3, xii-xx
(F36, 39)

Tunstall,  Cuthbert

Anon 1951 M Gaz 35, 1  [first English
printed arithmetic]
Sturge Charles 1938 *Cuthbert Tunstal,
Churchman, Scholar, Statesman,
Administrator,*  London, 413 p

Turan, Pal  1910-

Renyi A 1961 M Lap 11, 229-263
(Z96, 5)

Turanski, W. T.  1925-1960

Anon 1960 ACMC 3(A), 14  (MR22, 1131)

Turazza, Domenico  1813-1892

Favaro A 1892 *Della vita e delle
opere del Senatore Domenico
Turazza,* Padua, 82 p  (F24, 38)
Paladini E 1892 Politec 40, 170-180
(F24, 38)

Turing, Alan Mathison  1912-1954

Anon 1954 Lon Tim (Jan 14)
- 1954 Nat 174, 535-536

Newman M H A 1955 Lon RSBM 1, 253-
263
Taussky O 1956 M Tab OAC 10, 180-181
(Z70, 7)
Turing Sarah 1959 *Alan M. Turing,*
Cambridge, 173 p  (MR21, 1023)
[bib]

Turk, Abdal al Hamid ibn

Sayih Aydin 1962 *Logical Necessities
in Mixed Equations by Abd al Hamid
ibn Turk and the Algebra of His
Time,* Ankara  181 p  (MR26, 1)

Turnbull, Herbert Western  1885-1961

Ledermann W 1963 LMSJ 38, 123-128
(Z107, 6)
Rutherford D E 1963 Edi MSP 13, 273-
276  (Z111, 4)

Turner, Herbert Hall  1861-1930

Archibald R C 1948 *Mathematical
Table Makers,*  NY, 79-80
(Scr Stu 3)  [bib]

Tuschel, Ludwig  1886-1913

Mueller E 1914 Mo M Phy 25, 177-178
(F45, 35)

Tusi.  See Nasir.

Tweedie, Charles  -1925

Horsburgh E M 1925 M Gaz 12, 523
(F51, 31)
- 1925 Edi MSP 45, 131-135
(F51, 31)
- 1925 Edi MSP 45, 381-383
(F51, 31)

Twisden, John   XIX

Greenhill G 1914 Nat 94, 427
(F45, 47)

Tycho.  See Brahe.

Tzitzeica, Georges   1873-

N C 1930 Buc EP 1, 199-200
(F56, 34)
Tzitzeica Georges 1941 *Oeuvres.*
I. *Notes et Mémoires,*
(Academie Roumaine), 116 p
(F67, 967)

Uhler, Horace Scudder   1872-

Archibald R C 1948 *Mathematical
Table Makers,* NY, 80-81
(Scr Stu 3) [bib, port]

Ullrich, Egon Leopold Maria
1902-1957

Nevanlinna R + 1958 DMV 61(1), 57-65
(MR21, 750; Z82, 12)

Ulrich, Georg Karl Justus   1798-1879

Anon 1879 Gott N (1), 339-341
(F11, 26)

Unverzagt, Karl Wilhelm   1830-1885

Schmidt A 1886 Hist Abt 31, 41-50

Urbanski, Wojcierch   1820-1903

Zakrzewski I 1904  Wiad M 8, 145-150

Uryson, Pavel Samuilovich   1898-1924

Alexandrov P S 1925 Fund M 7, 138-140
(F51, 26)
- 1950 UMN (NS)5(1), 196-202
(MR11, 573; Z35, 148)

Ushinskii, K. D.

Fridman L M 1956 Matv Shk (1), 12-14

Utrecht.  See Adelbold.

Vacca, Giovanni Enrico Eugenio
1872-

Bigourdan G 1916 Par CR 162, 679-681
(F46, 9)
Carruccio E 1953 It UM (3)8, 448-456
(MR15, 592; Z51, 244)
Cassina U 1953 Lom Gen (3)17, 185-200
(MR16, 2)

Vagner, Viktor Vladimirovich   1908-

Liber A E + 1958 UMN 13(6), 221-227
(MR20, 1243; Z85, 7)

Vaidyanathaswamy, R.   1894-1960

Narasinga Rao A + 1961 M Stu 29, 1-14
(MR25, 967; Z98, 11)

Vailati, Giovanni   1863-1909

Loria G 1909 Ens M 11, 296-299
(F40, 41)
Vailati G 1911 *Scritti di G. Vailati*
Leipzig-Florence, 1035 p
(F42, 25)
Volterra V 1909  Mathes It, 60-63
- 1909 Pe MI (3)6, 289-292
(F40, 41)

Vaisala, Yrjo

Oterma L 1961 Hels M (A)(3)61, 15-30
(Z99, 247)
Virtanen A I 1961 Hels M (A)(3)61,
7-14 (Z99, 247)


Valentin, Georg Hermann 1848-1926

Archibald R C 1932 Int Con 6, 465-
472 (F58, 44)


Valentiner, Hermann 1850-

Anon 1913 Ens M 15, 513 (F44, 34)
Juel C 1913 Nyt Tid (B)24, 65-69
(F44, 34)


Valerio, Luca 1552-1618

Bosmans H 1913 Brx SS 37(2)
(F44, 60) [infinitesimal]
Tosi A 1957 Pe M (4)35, 189-201
(MR19, 1150; Z79, 242)


Valiron, Georges 1884-1954

Milloux H 1956 Ens M (2)2, 217-223
(MR18, 182; Z70, 7)


Valyi, Gyula 1855-1913

Obláth R 1956 M Lap 7, 61-70
(MR20, 848)


Van Amringe, John Howard 1835-1915

*Archibald R C 1938 A Semicentennial
History of the American Mathemati-
cal Society, NY, 110-112
[bib, port]
Keyser C J 1916 AMM 23, 15-16
(F46, 24)


Van Ceulen, Ludolph 1540-1610

Bierens de Haan D 1873 Mess M (2)3,
24-26 (F5, 45) [pi]
Bosmans H 1925 Mathesis 39, 352-360
(F51, 19)
Catalan E 1874 Boncomp 7, 141-144
(F6, 25)
Jong C de + 1938 M Gaz 22, 281-282
(Z19, 101)
Oijen V v 1868 Boncomp 1, 141-157
(F1, 9)


Van Dantzig, David 1900-1959

Anon 1959 Nieu Arch (3)7, 90
(MR23A, 136)
- 1959 St Ned 13, 415-432
(MR24A, 222)
- 1959 Synthese 11, 319-328
(MR22, 1)
Freudenthal H 1960 Ned Ja (Z107, 6)
- 1960 Nieu Arch (3)8, 57-73
(MR23A, 4)
Hemelrijk J 1959 Synthese 11, 335-
351 (MR22, 1) [stat]
- 1960 An M St 31, 269-275
(MR22, 929) [stat]
Wansink J H 1959 Euc Gron 35, 1-2
(Z88, 7)


Van den Berg, Franciscus Johannes
1833-1892

D B d H 1894 N Arc (2)1, 1-10
(F25, 44)
Schoute P H 1897 Amst Ja


Vandermonde, Alexandre Alexis
Theophile 1735-1796

Birembaut A 1953 Fr AAS 72 (Congress
July 1953), 530-533
Bottema O 1956 Nieu Arch (3)4, 2-12
(Z70, 244)
Lebesgue H 1940 Thales 4, 28-42
(MR9, 75; Z61, 5)
- 1956 Ens M (2)1, 203-223 (Z70, 244)

Loewy A 1919 DMV 27, 189-195
(F47, 7) [cyclotomy, Gauss]
Rocquigny G de 1898 Intermed 5, 184
Simon H 1896 Hist Abt 41, 83-85
(F27, 11)

Van der Pol, Balthasar    1889-1959

Anon 1959 Nieu Arch (3)7, 89
(MR23A, 135)
Bremmer H 1960 Philip TR 22, 36-52
(MR22, 1131)
Cartwright M L 1960 LMSJ 35, 367-376
(MR23A, 135; Z94, 5) [bib]
De Claris N 1960 IRECTT 7, 360-361
(MR22, 1341)
Parodi H 1959 Par CR 249, 1420-
1422 (Z86, 6)
Smith-Rose R L 1959 Nat 184, 1020-
1021 (Z86, 6)

Van der Schuere

Anon 1925 Euc Gron 2, 48-53
(F52, 15)
Hallema A 1924 Euc Gron 1, 161-193
[arith]

Van der Waals, J. D.    1837-1923

Maggi G A 1924 Linc Rn (5)33(1),
152-159
Onnes H K 1924 Nat 111, 609-610

Van der Waerden, Bartel Leendert
1903-

Maggi G A 1924 Linc Rn (5)33(1),
106,152-159  (F50, 10)

Van der Woude, Willem  1876-

Bottema O 1956 Nieu Arch (3)4,
2-12   (MR17, 932)

Vanecek, Matthias N.    1859-

Hruska V 1923 Cas MF 52, 313-319

Van Lansbergen, Philip
1561-1632

Bosmans H 1928 Mathesis 42, 5-10
(F54, 26)

Van Rees, Richard    1797-1875

Mailly E 1877 Bel Anr 43, 227-240
(F9, 14)

Van Schooten, Frans Jr. 1615-1661

Easton J 1965 AMM 72, 53-56
[ell, Nasr al Din, J de Witt]
Hofmann J E 1962 Frans van Schooten
der Juenger, Wiesbaden, 54 p
(Z111, 4)

Van Swinden, Jan Hendrick   1746-1823

Dijksterhuis E J 1931 Euc Gron 8,
265-285  (F58, 38)
Korteweg D J + 1899 Amst Vs G 8, 389-
523.  [bib]

van Vleck, Edward Burr  1863-1943

*Archibald R C 1938 A Semicentennial
History of the American Mathematical Society, NY, 170-173
[bib, port]
*Langer R E + 1957 Am NAS Mm 30, 399-
409 [bib, port]
Rosenbaum R A 1956 Wesl UA (Nov), 2-3

Varignon, Pierre  1654-1722

Boyer J + 1901 Intermed 8, 311
Brocard H + 1902-1903 Intermed 9, 69;
297; 10, 157, 259, 305 (Q264) [mss]

Costabel P 1965 *Pierre Varignon (1654-1722) et la diffusion en France du calcul différentiel et integral,* Paris, 28 p. (MR34, 5619)

Fedel E J 1933 *Der Briefwechsel Johann Bernoulli-Pierre Varignon aus Jahren 1692 bis 1702 in erlaeuternder Darstellung,* Heidelberg, 42 p (Z12, 98) [Thesis]

Fleckenstein J O 1948 Arc In HS 28, 76-138 (MR10, 420; Z31, 98) [XVII]

Tenca L 1960 Bln Rn P (11)7(2), 87-90 (Z100, 245) [Grandi]


Vashchenko-Zakharchenko, M. E.

Gracianskaya L N 1961 Ist M Isl 14, 441-464 (Z118:1, 9)


Vasilescu, Florin 1897-1958

Andonie G S 1959 Gaz MF (A)11, 160-164 (Z86, 6)


Vasiliev, Aleksandr Vasilievich 1853-1929

Fehr H 1930 Ens M 28, 317-318 (F56, 28)

Halsted G B 1897 AMM 4, 265-267

Korzybski A 1929 Sci 70(20 Dec), 599-600

Loria G 1930 Archeion 12, 46-47

Miller H A 1900 AMM 7, 215-216

Parfentev N N 1925 Kazn FMO (2)24, 27 (F51, 40)

- 1929 Kazn FMO (3)4

- 1930 Kazn Uch Z 90, 943-956 (F57, 1308)

Rainoff T 1930 Isis 14, 342-348

Sincov D 1926 Khar MUZ 2, 161-163 (F52, 35)


Veblen, Oswald 1880-1960

Archibald R C 1938 *A Semicentennial History of the American Mathematical Society,* NY, 206-211 [bib, port]

Castrucci B 1961 Parana MB 4, 29-31 (Z98, 11)

Hodge W V D 1961 LMSJ 36, 507-510 (Z98, 11)

Montgomery D 1963 AMSB 69, 26-36 (MR26, 1142; Z107, 6) [bib]


Vega, Georg Freiherr-von 1756-1802

Anon 1903 Matica S1 5, 217

Depman I Ya 1953 Ist M Isl 6, 573-608 (MR16, 660; Z53, 339) [Kulik]

Dohlemann K 1894 ZM Ph 39, 204-211

Kaucic F 1902 ZMN Unt 33, 525-528 (F33, 19)


Vekua, Ilya Nestorovich 1907-

Anon 1967 Tbil MI, 33, 9-14 (MR37, 1222) [bib]

Lavrentev M A + 1957 UMN (NS)12(4), 227-234 (MR19, 826; Z78, 4)


Velmin, Volodimir Petrovich 1886?-

Dobrovolskii V O 1961 Kiev Vis, 125-132 (MR33, 2516)


Venema, Pieter 1711-1748

Bradley A D 1949 Scr 15, 13-16 (Z34, 147)


Venkov, Boris Alexeyevich 1900-

Malyshev A V + 1961 UMN 16(4), 235-240 (MR24A, 222; Z98, 11)

Venn, John    1834-1923

Heath P L 1967 Enc Phi
Keynes J N 1906 *Studies and
  exercises in Formal Logic,*
  4th ed, London


Verbiest, Ferdinand  1623-1688

Bosmans H 1912 Brug Em An, 15-61
- 1912 Rv Q Sc 21, 195-273,
  375-464
- 1927 Brx SS 47, 14-19  (F53, 16)
  [mss]
- 1913 Rv Q Sc 24, 272-298


Vercelli, Francesco Maggiorino
1883-

Anon 1937 Vat An 1, 750-758
  (F63, 20)


Verdam, Gideon Jan  1802-1866

Anon 1866 Amst Ja, 56-
- 1866 Arc MF Lit 46(183)


Ver Eecke, Paul Louis  1867-1959

Mogenet J 1960 Arc In HS 12, 296-297
  (Z93, 9)
Pelseneer J 1948 Osiris 8, 5-11
  (Z31, 242)
Roma A 1960 Isis 51, 202-203
  (Z93, 9)


Verhulst, R. P. Enrique de Rafael

Sanchez Perez J A 1955 Mad Rv 49,
  213-222  (Z65, 245)


Vernic, Radovan

Markovic Z 1958 Glasn MPA (2)13, 287-
  290  (MR21, 750; Z83, 245)


Vernier, Pierre    1580-1637

Pereira A 1916 Porto Ac 11, 147-154


Veronese, Giuseppe  1854-1917

Bordiga G 1931 Pado Sem 2, 63-79
  (F57, 41)
Castelnuovo G 1958 Archim 10, 165-169
  (Z80, 5)
Lazzeri G 1917 Pe MI 32, 214-216
  (F46, 19)
Schoenflies A 1897 Linc Rn (5)6, 362
  [inf numb]
Segre C 1917 Linc Rn (5)26(2), 249-
  258  (F46, 19)


Versluys, Jan    1845-

Anon 1919 Wisk Tijd 16, 244-247
Schlesinger L 1920 DMV 29, 236-237
  (F47, 15)


Vessiot, Ernest Paulin Joseph  1865-

Cartan E 1947 Fr SMB 75, 1-8
  (MR9, 486; Z31, 242)


Vetter, Quido Karl Ludwig    1881-

Novy L 1961 Arc In HS 13, 270-273
  (Z106, 3)


Vicaire, Joseph Marie Hector Eugene
1839-1901

Andre D 1903 Philom Bu (9)4, 123-126
  [bib]


Victor.  See Hugh of St. Victor.


Victorius of Aquitania  V

Friedlein G 1871 Boncomp 4, 443-469
  (F3, 6)

- 1871 ZM Ph 16, 42-79, 253-254
  (F3, 6)
Sarton G 1927 Int His Sc 1, 409-410
  [bib]


Vicuna, Gumersindo    1840-1890

Eneström G 1891 Bib M (2)5, 33-34
  (F23, 25)


Viète, François    1540-1603

  = Franciscus Vieta

Allegret M 1867 Eloge de Viète,
  Paris, 28 p
Anon 1910 Intermed 17, 32 (Q2407)
Archibald R C 1948 Mathematical
  Table Makers, NY, 81-82
  (Scr Stu 3)  [bib]
Archibald R C + 1943 M Tab OAC 1,
  129-130  [Briggs]
Bigourdan G 1916 Par CR 162, 237-240
  (F46, 9)  [ms]
Bosmans H 1910 Un émule de Viète:
  Ludolph Van Ceulen, Louvain,
  56 p  (F41, 4)
- 1926 Brx SS 45(1), 35-42
  (F52, 17)  [Girard]
Bosmans H + 1901 Intermed 8, 265
Boyer C B 1962 MT 55, 123-127
  [deci frac]
Chasles M 1841 Par CR 12
Eneström G 1886 Bib M 226-239,
  244 (Q13)  [root]
- 1912 Bib M (3)13, 177-178
  (F43, 68)
- 1915 Bib M 14, 354  [anal geom]
Fitz-Patrick J + 1905-1906
  Intermed 12, 286; 13, 46 (Q2943)
Gambier G 1911 Le Mathématicien
  François Viète, La Rochelle, 31 p
  (F42, 48)
Hofmann J E 1953 Archims 5, 113-116
  (Z53, 245)
- 1953 Pyramide 11/12, 201-204
  (Z53, 338)
- 1954 Arc M 5, 138-147
  (MR15, 770; Z56, 1)  [spiral]

- 1956 Centau 4, 177-184  [constr]
- 1962 MP Semb 8, 191-214
  (Z102, 1)  [geom]
Hunrath H 1899 ZM Ph 44(S), 211-240
  (F30, 52)  [Rheticus]
Karpinski L C 1939 Scr 7, 133-140
Pierce J M 1875 Harv Lib B 1, 157-
  158, 246-250, 289-290  [bib]
Reeve W D 1930 MT 23, 508
Ritter E 1868 Boncomp 1, 223-244
  (F1, 9)
Ritter F 1892 Fr AAS 21, 17-25
  (F24, 8)  [alg]
- 1892 Fr AAS 21, 177-182
  (F24, 8)  [alg]
- 1892 Fr AAS 21, 208-211  (F24, 8)
  [trig]
- 1895 Viète. Notice sur sa vie et
  ses oeuvres,  Paris, 102 p
  (F27, 7)
Saltykow N 1938 Isis 29, 20-23
  [Ghetaldi]
Sanford V 1931 MT 23, 508
Smith D E 1940 Bo Fir 13, 221-223
  (MR2, 114; Z60, 7; 61, 4)
Tannery P 1897 Intermed 4, 204 (Q978)
- 1902 Intermed 9, 329
Vacca G 1916 Par CR 162, 676-679
  (F46, 9)


Vigenère, Blaise de    1523-1596

Condit M C 1963 MT 56, 160-163
Mendelsohn C J 1940 APSP 82, 103-129


Vignola, Iacopo Barozzi da
1507-1573

Casotti M W 1953 Pe M (4)31, 73-103
  (MR15, 89)  [perspective]


Vijayaraghavan, Tirukkannapuram
1902-1955

Chandrasekharan K 1957 M Stu 24, 251-
  267  (MR19, 108)
Davenport H 1958 LMSJ 33, 252-255
  (Z80, 5)

Villacampi, Bernardus de

Eneström G 1907 Bib M (3)8, 215-216
(F38, 8)


Villarreal, Federico 1850-1923

Losada y Puga C de 1933 *La
personalidad y labobra de Federico
Villarreal, matematico peruano,*
Lima, 19 p


Vincent. See Gregory St. Vincent.


Vinogradov, Ivan Matveevich 1891-

Delone B N 1951 SSSRM 15, 385-394
(MR13, 198; Z42, 243)
Linnik Yu V + 1962 UMN 17(2), 201-214
(MR25, 222; Z100, 248)
Mardzhanishvili K K 1951 UMN (NS)6(5),
190-196 (MR13, 421; Z43, 5)
Postnikov A G 1961 SSSRM 25, 621-628
(MR25, 221; Z98, 11)


Virgil -70 to -19

= Publius Vergilius Maro

Tannery P + 1904 Intermed 11, 255


Vitale da Bitonto, Giordano

Bonola R 1905 Loria 8, 33-36
(F36, 66) [paral]


Vitali, Giuseppe -1932

Bortolotti E 1933 Batt 71, 201-236
(F59, 856)
Pincherle S 1932 Bln Rn (2), 36, 58-
60 (F58, 995)
- 1932 It UM 11, 11-12, 125-126
(F58, 50)
Severini C 1932 Bo Fir (2)11, xxv-
xxxii, lxi-lxiv (F58, 49)


Tonolo A 1932 Pado Sem 3, 67-81
(F58, 49)
- 1959 Archim 11, 105-110
(MR23A 4; Z88, 244)


Vitelo 1225?-1280?

= Witelo = Vitellius = Vitellio
= Vitello etc

Boncompagni B 1871 Boncomp 4, 78-81
(F3, 5) [Pacioli]
Curtze M 1871 Boncomp 4, 49-77
(F3, 4)
Czerminski Adrian 1964 *Swiatlo
Witelona,* Katowice, 123 p (Engl,
Ger, Russ abstrs) [facs, bib]
Sarton G 1931 Int His Sc 2(2), 1027-
1028 [bib]
Zebrawski T 1879 Boncomp 12, 315-317
(F11, 9)


Viterbi, Adolfo 1873-1917

Vivanti G 1918 Loria 20, 33-37
(F46, 19)


Vitruvius -I

= Marcus Vitruvius Pollio

Luce J H 1953 Rv Hi Sc Ap 6, 308-321
[perspective]
- 1953 Rv Hi Sc Ap 6, 308-321
(MR15, 591) [perspective]
Pottage J 1968 Isis 59, 190-197
[pi]
Schmidt W 1900 Bib M (3)1, 297
[Heron]


Vivanti, Giulio 1859-

Pastori M 1950 Milan Sem 20, xv-xix
(MR12, 578; Z41, 341)

Viviani, Vincenzo  1622-1703

Conte L 1952 It UM (3)7, 334-340
  (Z47, 5)  [Torricelli]
- 1952 Pe M (4)30, 185-193
  (MR14, 525; Z48, 243)
  [mean propor]
Favaro A 1903 Rv FMSN 4, 424-434
  (F34, 9)  [letter, Saccheri]
- 1912 Ven I At 72, 1-155
  [Galileo]
Ferroni P 1821 It SS 19, 187-
Loria G 1929 Bo Fir (2)8, xlix-
  (F56, 812)  [curve]
Pascal A 1915 Annali (3)24, 287-310
  [G. Ceva]
Procissi A 1953 It UM (3)8, 74-82
  (Z50, 242)  [Archimedes]
Tenca L 1952 It UM (3)7, 328-334
  (Z47, 4)
- 1953 It UM (3)8, 456-459
  (MR15, 592; Z51, 243)
- 1953 Lom Rn M (3)17, 113-126
  (MR16, 207; Z53, 245)
- 1953 Lom Rn M (3)17, 835-846
  (MR16, 207; Z52, 3)  [Caravaggi]
- 1954 Lon Rn M (3)18, 212-228
  (Z57, 243)  [Ricci]
- 1954 Ven I At 112, 1-15
  (MR16, 660)  [Angeli]
- 1955 Bln Rn P (11)2(2), 162-177
  (Z67, 246)  [Cassini]
- 1960 Pe M (4)38, 87-94
  (MR22, 2052)  [Torricelli]


Vlasov, Aleksei Konstantinovich
1868-1922

Glagolev N A 1925 Mos MO Sb 32, 273-
  275  (F51, 40)


Vogel, Hermann Karl  1841-1907

Belopoliskij A 1907 Pet B 6, 487-488


Vogler, Christian Wilhelm Jacob
  August  1841-

Blumenberg H 1925 Allg Verm 37, 217-
219  (F51, 27)


Voigt, Woldemar  1850-1919

Runge D 1920 Gott N (1)1, 46-52


Voit, C. von  XIX

Anon 1894 Mun Si 24, 113-148
  (F25, 40)
- 1895 Mun Si 25, 161-196
  (F26, 27)


Vojtech, Jan  1879-

Vychichlo F 1953 Cas M 78, 283-286
  (Z53, 340)


Volkmann, Karl Bernhard Wilhelm
  1872-

Matthee H 1932 Z Ph C Unt 45, 241-247


Vols, E.  1650-1720

Marian V 1938 Gaz M 43, 449-452
  (F64, 17)


Volta, Alessandro  1745-1827

Berzolari L 1927 Lom Gen (2)60, 449-
  583  (F53, 21)
Furlani G 1927 Pe M (4)7, 154-156
  (F53, 22)
Libicky V 1927 Cas MF 56, 219-221
  (F53, 21)
Patetta F 1926 Tor FMN 62, 252-281
  [letters]
- 1927 Tor FMN 62, 712-741
  (F53, 21)  [letters]

Somigliana C 1926 Tor FMN 62, 94-113
- 1927 Tor FMN 62, 210-229
  (F53, 21)
Volta L 1928 Bo Fir (2)7, 1-19
  (F54, 47)


Voltaire    1694-1778

Brocard H 1898 Intermed 5, 263
  [trisection]
Kasner E + 1950 Scr 16, 13-21
  (MR12, 312) [horn angles]
Smith D E 1921 AMM 28, 303


Volterra, Vito    1860-1940

Agostinelli C 1961 Tor Sem 20, 15-38
  (Z98, 11) [Somigliana]
Allen E S 1941 AMM 48, 516-519
  (Z60, 18)
Anon 1937 Vat An 1, 759-775
  (F63, 20)
- 1941 Archeion 23, 325-359
- 1958 Archim 10, 29-32
  (MR19, 1150; Z79, 6)
Armellini G 1951 Ric Sc 21, 3-12
  (MR12, 578)
Cardoso J M 1960 Parana MB 3, 28-29
  (Z89, 242)
Castelnuovo G 1943 It SS (3)25,
  87-95  (MR9, 75; Z61, 6)
Colonnetti G 1956 Tor Sem 16, 95-100
  (MR19, 1247; Z77, 242)
Krall G 1955 Cir Mac 3(1), 64-77
  (MR16, 600)
Labrador J F 1957 Gac M 9, 39-42
  (Z79, 6)
Levi B 1941 Arg IM Ros 3, 25-26,
  37-48  (MR2, 306; Z60, 18)
  [bib]
Lorente de Nó F 1925 Rv M Hisp A 7,
  117-130 (F51, 33)
Masotti A 1946 Milan Sem 17, 16-61
  (Z61, 6) [bib]
Pérard A 1941 Cah Phy (3)51, 58
  (Z60, 18)
Picard E 1940 Par CR 211, 309-312
  (MR3, 98)
Picone M 1956 Ric Sc 26, 3277-3289
  (MR18, 453)

Rosenblatt A 1942 Rv Cien 44, 423-442
  (Z60, 18)
Somigliana C 1942 Vat Act 6, 57-85
  (MR10, 175; Z60, 18)
- 1942 Vat Act 6(9), 85 p
  (F68, 18)
Thompson D W 1941 Nat 147, 349-350
  (MR2, 306; Z60, 18)
Wavre R 1942 Ens M 38, 347-348
  (MR4, 66; Z60, 18)
*Whittaker E T 1941 Lon RS Ob 3, 691-
  729 (repr. in Volterra 1959
  *Theory of Functionals* (Dover)
  (Z60, 18) [bib]
- 1941 LMSJ 16, 131-139
  (F67, 970; Z28, 195)


von. See also following word
(modern names) or preceding word
(medieval names). See also Von
for anglicized names.


Von der Muehll, Karl    1841-

Fueter R 1912 M Ann 73    (F43, 51)
Knapp M 1912 Basl V 23, 1-5


Von Neumann, John    1903-1957

Anon 1957 M Lap 8, 1-7   (Z94, 5)
- 1957 M Lap 8, 210
  (MR20, 848) [bib]
- 1959 Elek Rech 1, 57   (Z87, 243)
Behnke H + 1957 MP Semb 5, 186-190
  (Z88, 7)
Birkhoff G 1958 AMSB 64(3)(2), 50-56
  (Z80, 4) [lattice]
Bochner S 1958 Am NASBM 32
*- 1958 Am NASP 32, 438-457
*Burks A W 1966 in J Von Neumann,
  *Theory of Self-reproducing Auto-
  mata* (U Ill P), 1-28   [bib]
Freudenthal H 1958 Ned Ja, 1-7
  (Z92, 246)
*Goldstine H H + 1957 Sci 125, 683-684
  (MR18, 784)
Hajnal A 1959 M Lap 10, 5-11
  (MR22, 3679) [set thy]

*Halmos P R 1958 AMSB 64, 86-94
  (Z80, 4) [ergodic thy, measure thy]
Halperin I 1960 Foreword to Von
  Neumann, *Continuous Geometry*,
  Princeton U P
Jungk R 1958 *Brighter Than a Thousand
  Suns. A personal history of the
  Atomic Scientists* (Engl tr of
  Ger orig of 1956) NY, 360 p
Kadison R V 1958 AMSB 64, 61-85
Kuhn H W + 1958 AMSB 64, 106-122
  (MR20, 503; Z80, 4)
  [econ, games]
Murray F L + 1958 AMSB 64, 57-60
  (MR20, 620) [operators]
Oxtoby J C + ed 1958 AMSB 64 (3:2)
  (May)(S), 129 p
Penney W G 1957 Nat 179, 510
  (Z108:2, 250)
Redei L 1959 M Lap 10, 226-230
  (MR23A, 136; Z97, 244)
  [alg, numb thy]
Shannon C E 1958 AMSB 64, 123-129
  (MR19, 1084; Z80, 4) [automata]
Smithies F 1959 LMSJ 34, 373-384
  (MR21, 893; Z85, 7)
Szökefalvi-Nagy B 1957 M Lap 8,
  185-210 ((MR20, 848; Z94, 5)
  [operators]
Tarjan R 1958 M Lap 9, 6-18
  (MR20, 848) [computer]
Tompkins C B 1957 M Tab OAC 11, 127-
  128 (MR19, 128; Z78, 242)
*Ulam S 1958 AMSB 64, 1-49
  (MR19, 1030; Z80, 4) [*bib]
*Van Hove L 1958 AMSB 64, 95-99
  (MR19, 1131; Z80, 4) [quantum thy]
Von Neumann K 1958 Preface to Von
  Neumann, *The Computer and the
  Brain*


Voronets, P. V.

Fradlin B N 1960 Vop IET (10),
  73-76 (Z117-2, 9) [mech]
- 1961 Ist M Zb 2, 104-127
  (MR33, 3869)


Voronoi, Georgi Feodosievich
1868-1908

Braytzeff I 1908 Wars PI An, 15 p
- 1909 Khar M So (2)11, 197-210
Markov A 1908 Pet B(6)(17), 1247-
  1248    (F39, 42)


Voss, Aurel Edmund   1845-

Dingler H 1915 Allg Z 118, 691-692
  (F45, 56)


Vranceanu, Gheorghe   1900-

Teleman O 1961 Gaz MF((A)13, 641-659
  (Z106, 3)


Vries, Jan de    1858-

Anon 1940 Zutphen (A)9, 1
  (MR2, 115)


Vycichlo, Frantisek   1905-1958

Babuska I + 1958 Cas M 83, 374-387
  (MR20, 1137; Z80, 9)
Havlicek K 1959 Prag Pok 4, 497-501
  (Z87, 7)


Vygodskii, M. Ya.   1898-1965

Rosenfeld B A 1967 in *Arifmetika i
  algebra v drevnem miri,* 350-362


Vyshnegradskii, Ivan Alekseevich
1831-1895

Andronov A A 1949 SSSR Tek, 805-818
  (MR11, 574)   [automatic control]


Waard, Cornelius de

Hoovkaas R 1960 Arc In HS 12, 173-175
  (Z89, 243)

Wachter, Friedrich Ludwig   1792-1817

Lorey W 1934 Archeion 16, 307-315
  (Z11, 194)
Staeckel P 1900 M Ann 53, 49
  [non Euc geom]


Wada Yenzo Nei     1787-1840

Hayashi T 1916 Toh MJ 10, 232-233
Mikami Y 1916 Toh MJ 10, 229-231


Waelsch, Emil   1863-

Fanta E 1928 ZAM Me 8, 245-248
  (F54, 41)


Wafa, Abul   X

Bertrand J 1873 *La théorie de la
  Lune d'Aboul-Wefa,* Paris
  (F5, 8)
Bertrand J + 1871 J Sav, 457-474
  (F3, 3)   [moon]
- 1871 Par CR 73, 581-589, 637-647,
  765-766, 805-808, 889-890, 932-934
  (F3, 3)   [moon]
Chasles M 1673 Par CR 76, 859-864,
  901-909   (F5, 8)
Marre A 1874 Boncomp 7, 267-277
  (F5, 14)
Medovoi M I 1958 Ist M Isl 11, 593-
  598   (Z96, 1)   [negatives]
- 1959 Vop IET (8), 101-106
  (Z91, 5)   [frac]
- 1960 Ist M Isl 13, 253-324
  (MR25, 739)
Nadir N 1960 MT 53, 460-463
  (Z119:1, 6)
Sédillot L A 1868 Par CR 66, 286
  (F1, 5)   [Brahe]
- 1873 Par CR 76, 1291-1293
  (F5, 8)   [ast]
Suter H 1922 Ab GN Med 4, 94-109
  [constr]
Wiedemann E 1879 Hist Abt 24, 121-
  122   (F11, 8)

Wakeford, Edward Kingsley   1894-1916

Anon 1918 LMSP (2)16, 54-57
  (F46, 35)


Wald, Abraham   1902-1950

Anon 1952 An M St 23, 29-33
  (Z46, 3)
Finetti B de 1951 Stat Bln 11, 185-
  192
Menger K 1952 An M St 23, 14-20
  (MR13, 613; Z46, 3)   [geom]
Morgenstern O 1951 Ecmet 19, 361-367
  (MR13, 421; Z43, 245)
Roy S N 1951 Clct St 3, 133-138
  (MR13, 421)
Schmetterer L 1951 Sta Vier 4, 69-74
  (MR13, 2)
Tintner G 1952 An M St 23, 21-33
  (MR13, 613)   [bib, econometrics]
Wolfowitz J 1952 An M St 23, 1-13
  (MR13, 613; Z46, 3)


Waldo, Clarence Abiathar   1852-1926

Roever W H 1927 AMSB 33, 613-614
  (F53, 34)


Walfisz, Arnold   1892-1962

Lomadze G A + 1963 UMN 18(4), 119-128
  (MR29, 880)
Lomadze G A   1964 Act Ari 10, 225,
  227-237, 239-244   (MR30, 201-202)
  •


Walker, John James   1825-1900

Tucker R 1900 LMSP 32, 439


Walkingame, Francis   1723?-1783

Wallis P J 1963 M Gaz 47, 199-208

**Wallenberg,** Georg Jacob    1864

Haentzschel E 1925 Ber MG 24, 38-41
(F51, 32)


**Walling,** Henry Francis    1825-

Boutelle C O 1891 Wash PS 11, 492-
496  (F23, 31)


**Wallingford,** Richard    1292-1335

Bond J O ed 1923 Isis 5, 99-115,
339-363
Gunther R T 1926 Nat 118, 773-774
(F52, 13)


**Wallis,** John    1616-1703

Archibald R C 1936 AMM 43, 35-37
(F62, 22)  [religion]
Aubry A 1900 Prog M (2a)2, 16
Bell E T 1947 Scr 15, 162-163
[Fermat]
Brun V 1951 Nor M Tid 33, 73-81
(Z44, 245)  [Brouncker, pi]
Cajori F 1929 AMSB 35, 13  (F55, 26)
[Barrow, Hobbes]
- 1929 MT 22, 146-151  (F55, 588)
[Barrow, Hobbes]
Cassina U 1956 Int Con HS 8, 33-38
[non-Euc geom]
- 1956 Pe M (4)34, 197-219
(Z72, 245)  [non-Euc geom]
Cayley A 1889 QJPAM 23, 165-169
[pi]
Cherkalova L I 1964 Jaros Ped 2, 153-
160  (MR33, 2502)  [ratio]
Dickinson G A 1937 M Gaz 21, 135-
139  (F63, 807)  [pi]
Eneström G 1907 Bib M (3)7, 263-269
[complex numb]
Kearns D A 1953 MT 51, 373-374
[complex numb]
Koppe M 1903 Arc MP (3)3, 56-60
Kramer F D 1950 Ist M Isl (3), 486-
508  (MR13, 1)  [anal, Newton]
*- 1961 Ist M Isl 14, 1-100

- 1966 Ist M Isl 17, 309-316
(MR37, 1221)
- 1967 Vop IET 21, 103-108
(MR37, 14)  [mech, vector alg]
Kutta W 1901 Bib M (3)2, 230
[ell intgl]
Marco G de 1956 *Il calcolo dell'
infinitamen grande, Vol. II. Sulle
orme di Wallis ed oltre,* Naples,
126 p  (Z74, 244)
Nunn T P 1910 M Gaz 5, 345-356, 377-
386  [intgl]
*Prag A 1930 QSGM (B)1(3), 381-412
Sanford V 1933 MT 25, 429-431
Scott J F 1936 An Sc 1, 335-357
(F62, 1033; Z14, 245)
- 1938 *The Mathematical Work of John
Wallis,* London, 251 p  (F64, 916)
- 1960 Lon RSNR 15, 57-67
(Z107, 244)
Scriba C 1965 Centau 10, 248-257
[Harriot]
- 1966 *Studien zur Mathematik des
John Wallis...,* Wiesbaden, 144 p
(Boethius 6)
Smith D 1918 AMSB (2)34, 82-96
[cryptography]
Tenca L 1955 It UM (3)10, 412-418
(MR17, 118; Z64, 243)
Wertheim G 1899 ZM Ph 44(S), 557
[Fermat]
Wieleitner H 1930 Weltall 29, 56-80
[anal geom]
Yule G U 1939 Lon RSNR, 73-82


**Wallner,** Karl Raimund    1881-1934

Thiersch F K F 1935 DMV 45, 113-116
(F61, 21)


**Walras,** Leon    1834-1910

Antonelli E 1935 Ecmet 3, 119-127
[Cournot, Jevons]
Hecht Lilly 1930 *A. Cournot und L.
Walras,* Heidelberg, 93 p
Pareto V 1910 Ec J (Mar)

**Walsh,** Joseph Leonard   1895-

Sewell W E 1966 SIAM (B)3, 344-348
(MR33, 5907)
Widder D V 1966 Siam(B)3, 171-172
(repr in *Studies in Approximation
and Analysis* [port])

**Wangerin,** Friedrich Heinrich Albert
1844-

Lorey W 1914 ZMN Unt 46, 53-57
(F45, 55)
- 1933 Unt M 39, 861   (F59, 32)

**Wantzel,** Pierre Laurent   1814-1848

Cajori F 1918 AMSB 24, 339-347
(F46, 13)
Kazarinoff N D 1968 AMM 75, 647
[reg polyg]

**Wappler,** Hermann Emil   1852-1899

Eneström G + 1900 Bib M (3)1, 255
[port]

**Ward,** J.

Cajori F 1890 Colo Stu 1, 27-33
Hooker J H 1894 M Gaz 1, 29

**Wargentin,** Pehr Vilhelm   1717-1783

Eneström G 1899 ZM Ph 44(S), 81-95
Nordenmark N V E 1939 *P.W. Wargentin,
Sekretae der koeniglichen Akademie
der Wissenschaften und Astronomie
1749-1783,* Uppsala, 464 p
(F65:2, 1089)

**Waring,** Edward   1736-1798

Crawford L 1942 So Af RS 29, 69-74
(MR5, 57; Z60, 11)

Mayer F X 1924 *E. Waring's
"Meditationes algebricae,"* Zurich,
60 p (Thesis)   (F50, 18)
Smith E C 1934 Nat 133, 13-15

**Washington,** George   1732-1799

Ingalls E E 1954 MT 47, 409-410
[educ]

**Watson,** George Neville   1886-

Rankin R A 1966 LMSJ 41, 551-565
(MR33, 2518)

**Watson,** Henry Charles   1842-1925

H H T 1925 M Gaz 12 433-434
(F51, 25)

**Wavre,** Rolan   1896-1949

Ammann A 1950 Philom Bu  57, 7-8
(Z35, 148)
Fehr H 1950 Helv SN 130, 420-428
(MR13, 2)
Tiercy G 1950 Lyon (A)(3)13, 5-6

**Weber,** Constantin Heinrich   1885-

Willers F A 1950 ZAM Me 30, 230-232
(Z36, 146)

**Weber,** Heinrich   1842-1913

Anon 1912 *Festschrift Heinrich Weber
zu seinem siebzigsten Geburtstag...,*
Leipzig, 508 p  (F43, 54)
- 1913 M Ann 74, 1-2  (F44, 35)
Fehr H 1913 Ens M 15, 339-340
(F44, 35)
Frobenius F G 1913 Ber Si, 248-249
Rudio F + 1913 Zur NGV 57, 596-604
[Euler]
Rudio F 1913 Zur NGV 58, 437-453
(F44, 35)

Voss A 1914 DMV 23, 431-444
(F45, 35)
- 1914 Mun Jb 20, 90-108   (F45, 35)


Wedderburn, Joseph Henry Maclagan
1882-1948

Aitken A C 1952 Edi MSN 38, 19-22
(MR14, 344; Z49, 291)
Taylor H S 1949 Lon RS Ob 6, 619-623
(MR12, 312)


Weierstrass, Karl Theodor Wilhelm
1815-1897

Ahrens W 1907 MN Bl 4, 41-47
(F38, 24)
Anon 1894 DMV 4, 20
- 1897 Bo SBM 7, 16   (F28, 35)
[Sylvester]
- 1897? Crelle 117, 357
- 1897 Nap FM Ri (3)3, 63-64
(F28, 32)
- 1899 Ber Si, 79
- 1902 Ber Si, 100
- 1923 Act M 39, 257-258
(F49, 10) [Mittag-Leffler,
three body]
- 1925 Ber Si, LX   (F51, 39)
Auwers A 1904 Ber Si, 235
Behnke H + 1966 Festschrift zur
Gedaechtnisfeier fuer Karl
Weierstrass, Koeln-Opladen
(MR34, 2412)
Biermann K-R 1966 Ber 8, 33-37
[Humboldt]
- 1966 Crelle 223, 191-220
(MR33, 1210)   [bib, mss]
- 1967 Int Con HS (1965), 235-237
Chevalley C 1935 Rv Met Mor 42, 375-
384   (F62, 7)
Dantscher V 1908 Vorlesungen ueber
die Weierstrassche Theorie der
irrationalen Zahlen, Leipzig
Dickstein S 1897 Wiad M 1, 53, 58
(F28, 35)
Flaskamp F 1961 Fors Fort 35, 236-239
(Z95, 4)
Haentzschel E 1913 DMV 22, 278-284
[ell fun, Euler]

Hancock H 1895 An M 9, 179-
Hermite C 1897 Kazn FMO (2)7, 85
- 1897 Par CR 124, 430-433
(F28, 32)
Hilbert D 1897 Gott N, 60-69
(F28, 32)
Kiepert L 1926 DMV 35, 56-65
(F52, 26)
Killing Wilhelm 1897 Karl Weierstrass,
Munster, 21 p
- 1897 Nat Offen 43   (F28, 33)
Koenigsberger L 1916 DMV 25, 393-424
(F46, 14)   [ell fun]
Kossak Ernst 1872 Die Elemente der
Arithmetik, Berlin, 29 p
Lampe Emil 1897 Karl Weierstrass,
Leipzig, 24 p
- 1897 Ber Ph 16, 50-71   (F28, 32)
- 1899 DMV 6, 27-44   (F30, 19)
- 1910 Arc MP (3)16, 53-56
[Bois-Reymond]
- 1915 DMV 24, 416-438   (F45, 24)
- 1916 Ber MG 15, 35-59
(F45, 24)
Lorey W 1915 ZMN Unt 46, 597-607
(F45, 25)
- 1916 ZMN Unt 47, 185-188
(F46, 15)
Mittag-Leffler G 1897 Act M 21, 79-82
(F28, 32)
- 1902 Int Con 2, 131-153   (F32, 15)
- 1911 Act M 35, 29-65   (F42, 17)
- 1923 Act M 39, 1-57   (F49, 8)
- 1923 Act M 39, 133-198   (F49, 9)
[Kovalevsky]
Ocagne M d' 1897 Mathesis (2)7(S)
- 1897 Rv Q Sc (2)12, 484-507
(F28, 34)
Picard E 1897 Rv GSPA 8, 173-174
(F28, 35)
Pincherle S 1880 Batt 18, 178-254,
317-357   [complex anal]
- 1896 Bln Rn (NS)1, 101
Poincaré H 1898 Act M 22, 1-18
(F29, 17)
Pokrovsky P M 1897 Kagan (255), 62-66
(F28, 34)
- 1898 Kiev U Iz 5, 63
Polubarinova-Kocina P J 1966 UMN 21,
213-224  (MR33, 2510)
Prasad G 1923 Benar MS 5, 35-42

- 1925 Clct MS 15, 111-118
  (F51, 25)
Rothe R 1916 DMV 24, 438-442
  (F45, 23)
- 1928 DMV 37, 199-208   (F54, 35)
- 1937 Deu M 2, 17   (F63, 811)
Runge C 1926 DMV 35, 175-179
  (F52, 26)
Schubert H 1897 ZMN Unt 28, 228-231
  (F28, 33)
Siacci F 1897 Nap FM Ri (3)3, 63-64
  (F28, 32)
Simon H 1915 Voss Z (557)   (F45, 24)
Sleschinsky J 1897 Kagan (255) 59-62
  (F28, 34)
Sturm R 1910 DMV  19, 160
  [Gudermann]
Teichmueller O 1939 Deu M 4, 115-116
  (F65, 16) [Cauchy, complex anal,
  Riemann]
Tikhomandritskii M A 1897 Khar M So
  (2)6, 35-56
Van Emelen 1914 Intermed 21, 151-152
  (Q2279) [bib, sigma and rho funs]
Vasiliev A W 1885 Fiz M Nauk 1, 225-
  231, 257-264
Voit C v 1897 Mun Si 27, 402-409
  (F28, 34)
Wentscher M 1909 DMV 18, 89-93
  (F40, 25)  [Kovalevsky]


Weigel, Erhard   1625-1699

     1868 ZM Ph 13(S), 1-44
Spiess E 1881 Erhard Weigel, der
  Lehrer von Leibniz und Pufendorf,
  Leipzig   (F13, 17)


Weingarton, Leonhard Johannes,
  Gottfried Julius   1836-

Lueroth J 1910 Loria 12, 65-70
  (F41, 35)
St. Jolles 1910 Arc MP (3)17, 8-14
- 1910 Ber MG 10, 8-14   (F42, 32)
- 1911 BSM (2)35, 142-148


Weinmeister, Johann Phillip   1848-

Weinmeister P 1910 DMV 19, 321-327
  (F41, 36)


Weinmeister, Paul Franz Wilhelm
  1856-1927

Lorey W 1927 Unt M 33, 330-331
  (F53, 32)


Weiss, Ernst August Karl Hermann
  1900-1942

Blaschke W 1942 DMV 52(1), 174-176
  (Z27, 291)
Strubecker K 1943 Deu M 7, 254-298
  (Z28, 100)


Weiss, Mary   1930-1966

Zygmund A 1968 in Haimo Deborah H ed,
  Orthogonal Expansions and Their
  Continuous Analogues (Southern Ill
  U P),  xi-xviii  [bib]


Weiss, Wilhelm   1859-

Waelsch E 1905 Mo M Phy 16, 3-6
  (F36, 32)
Waelsch E + 1905 DMV 14, 171-175
  (F36, 32)


Weitbrecht, Wilhelm

Schmelz 1932 Z Vermess 61, 1-3


Wendelin, Hermann   1895-

Anon 1938 Deu M 3, 336-338
  (F64, 23)


Wentworth, George Albert   1835-1906

Finkel B F 1908 SSM 7, 485-488

Gwinner H 1935 M Mag 9, 165

Werner, Johann  1468-1528

  = Jean Verneris

Anon 1909 M Gaz 5, 107-108
Bjoernbo A A 1907 *Joannis Verneri de triangulis sphaericis libri quator...,* Leipzig, 200 p
  (Ab GM 24(1), 1-184)  (F38, 66)
Dickstein S+ 1902-1908 Intermed 9, 158; 15, 181 (Q2072) [trig]
Wuersshmidt J 1913 *Ioannis Verneri de triangulis sphaericis libri quator...,* Leipzig-Berlin, 266 p
  (F44, 50)

Wernicke, Friedrich Alexander  1857-

Lietzmann W 1915 ZMN Unt 46, 386-389
  (F45, 54)

Wertheim, Gustav  1843-1902

Eneström G 1902 Bib M (3)3, 395-402
  (F33, 44)

Wessel, Caspar  1745-1818

Beman W W 1897 AAAS 46, 33-50
  [complex numb]
- 1899 Ens M 162-184  (F30, 54)
  [Cauchy]
Brun V 1959 Rv Hi Sc Ap 12, 19-24
  [complex numb]
Budon J 1933 BSM (2)57, 175-200, 220-232  (F59, 7)
Cajori F 1912 AMM 19, 167
Floyd W F 1935 Nat 136, 224
Juel C 1895 Nyt Tid 6, 25-35
  (F26, 50)
Laurent H 1897 JM Sp (5)21, 128-130
  (F28, 46)

Wessell, E

Backlund R J 1933 Sk Akt 11(3), 199-200

Westergaard, H. L.  1853-1937

Pietra G 1938 Metron 13, 173-174
  (F64, 23)

Western, Alfred Edward  1873-

Miller J C P 1963 LMSJ 38, 278-281
  (Z111, 4)

Weyer, Georg Daniel Eduard  1818-1896

Anon 1897 Leop 33, 49  (F28, 26)
- 1897 Nat 55, 299  (F28, 26)
Pochhammer L 1899 DMV 6(1), 44-45
  (F30, 18)

Weyl, Claus Hugo Hermann 1885-1955

Anon 1957 Lon RSBM 3, 305-328  [bib]
Beisswanger P 1966 Ratio 8, 25-45
  (MR33, 2512)
Chevalley C + 1957 Ens M (2)3, 157-187  (MR20, 620; Z78, 4)
Denjoy A 1955 Par CR 241, 1665-1667
  (MR17, 446; Z65, 245)
Dyson F J 1956 Nat 177, 457-458
  (MR17, 814)
Freudenthal H 1955/56 Ned Ja, 1-8
  (MR20, 503; Z72, 246)
Hoelder D 1955 Fors Fort 29, 350-351
  (Z65, 245)
Koenig R 1956 Mun Jb, 236-248
  (MR18, 784)
Newman M H A 1958 LMSJ 33, 500-511
  (MR20, 1138; Z82, 243)

Weyr, Eduard  1852-1903

Anon 1904 Cas MF 33, 1
- 1904 Mo M Phy 15, 137

Petr K + 1905 Cas MF 34, 457-516
  (F36, 25)


Weyr, Emil 1848-1894

Kohn G 1895 Mo M Phy 6, 1-4
  (F26, 33)
- 1895 Paler R 9, 260-262
  (F26, 33)
- 1897 DMV 4, 24-33 (F28, 20)
Panek A 1895 Cas MF 24, 161-224
  (F26, 32)


Wheatstone, Charles 1802-1875

Kobell F v 1876 Mun Si, 117-118
  (F8, 23)


Wheeler, A. H. -1950

Anon 1951 MT 44, 364


Whewell, William 1794-1866

Anon 1867 Lon As Mo N 27, 110-
- 1868 Lon RSP 16, 1i-
Cajori F 1926 Arc Sto 7, 25-28
  (F52, 2) [ptolemy, space]
Todhunter I 1876 W. Whewell, An
  Account of His Writings,
  Cambridge, 2 vols


Whipple, Francis John Welsh 1876-

Bailey W N 1943 LMSJ 18, 249-256
  (Z60, 18)


Whiston, William 1667-1752

Whiston W 1749 Memoirs of His Own
  Life and Writings, Cambridge,
  2 vols


White, Henry Seely 1861-

*Archibald R C 1938 A Semicentennial
  History of the American Mathematical Society, NY, 158-161
  [bib, port]


Whitehead, Alfred North 1861-1947

Broad C D 1948 Mind 57, 139-145
Crespo Pereira R 1949 Rv M Hisp A
  (4)9, 49-52 (MR11, 574, Z33, 242)
Emmett D 1948 Brt Ac P
Hammerschmidt W W 1948 Scr 14, 17-23
  (Z33, 4; MR9, 175) [port]
Palter R M 1960 Whitehead's Philosophy
  of Science (U Chicago P), 263 p
  (MR22, 929)
Price L 1954 Dialogues of Alfred
  North Whitehead, London, 386 p
Russell B 1948 Mind 57, 137-138
- 1952 Portraits from Memory and
  other Essays, London
*- 1967-1968 The Autobiography of
  Bertrand Russell, London, vols 1
  and 2
Schaaf W L 1955 MT 48, 347- [educ]
*Schlipp P A ed 1941 The Philosophy
  of Alfred North Whitehead,
  Evanston, Ill (2nd ed 1951 New York)
  [bib]
Whitehead A N 1936 Atlantic (June),
  672-679
- 1948 Science and Philosophy, NY
  (Philosophical Lib), Pt 1, 9-84
*Whittaker E T 1948 Lon RS Ob (17),
  281-296
Whittaker E T + 1948 Nat 161, 267-
  268


Whitehead, John Henry Constantine
  1904-1960

Hilton P J 1962 Ens M (2)7, 107-124
  (MR25, 570; Z101, 10)
James I M 1960 Nat 186, 932
  (Z87, 244)
Wylie S 1962 LMSJ 37, 257-273
  (Z100, 8)

Whittaker, Edmund   1873-1956

Aitken A C 1956 Nat 177, 730-731
  (MR17, 932)
- 1958 Edi MSP 11, 31-38
  (MR20, 503; Z83, 245)
  [alg, numer anal]
Anon 1937 Vat An 1, 776-781
  (F63, 20)
- 1956 Edi Actu 23, 454-456
  (Z70, 7)
Erdélyi A 1957 M Tab OAC
  (MR19, 108; Z109, 5)
Julia G 1956 Par CR 242, 2493-2495
  (Z70, 7)
Martin D 1958 Edi MSP 11, 1-9
  (MR20, 738; Z83, 245)
McConnell J 1958 Edi MSP 11, 57-70
  (Z83, 245)  [phys]
McCrea W H 1957 LMSJ 32, 234-256
  (MR19, 236)
Rankin R A 1958 Edi MSP 11, 25-30
  (Z83, 245; MR20, 738)
  [automorphic fun]
Synge J L 1958 Edi MSP 11, 39-55
  (Z83, 245)  [relativity]
Temple G 1958 Edi MSP 11, 11-24
  (Z83, 248)  [harmonic fun]


Whitworth, William Allen   1840-1905

Irwin J O 1968 Lon St (A)130


Wicksell, Johan Gustav Knut
1851-1926

Akerman J 1933 Ecmet 1, 113-118
  (F59, 850)  [econometrics]


Wicksell, Sveg Dag   1890-1939

Larsson T 1939 Sk Akt 22, 78-80
  (F65, 18)


Widman, Johannes   1460?-1500?

Boncompagni B 1876 Boncomp 9, 188-210
  (F8, 15)


Drobisch M W 1940 *De Joannis Widmanni
  ...*, Leipzig


Wiedemann, Eilhard Ernst Gustav
1852

Ruska J 1928 Archeion 9, 158
  (F54, 39)


Wiedemann, Gustav Heinrich   1826-1899

Ostwald W 1899 Leip Ber 51, lxxvii


Wieland, L. W.   1901-1945

van der Corput J G + 1948 Nieu Arch
  (2)22, 217-219   (Z29, 2)


Wieleitner, Heinrich Karl 1874-1931

Bortolotti E 1932 It UM 11, 63-64
  (F58, 49)
Hofmann J E 1932 Unt M 38, 58-62
  (F58, 992)
- 1933 DMV 42, 199-223   (F59, 41)
Lietzmann W 1932 ZMN Unt 63, 91-93
  (F58, 995)
Lorey W 1932 MN Bl 26, 19-23
  (F58, 49)
Ruska J 1932 Isis 18, 150-165
Tropfke J 1932 Mit GMNT 31, 97-101
  (F58, 49)
Vogel K 1932 Archeion 14, 112-115
  (F58, 49)
- 1932 Bay Bl GR 68, 90-94


Wien, Wilhelm Carl Werner Otto Fritz
Franz   1864-

Anon 1929 Ens M 28, 136-137
Baier O 1928 MN Bl 22, 112-114
  (F54, 41)

Wiener, Ludwig Christian 1826-1896

Brill A + 1899 DMV 6(1), 46-69
(F30, 18)
Wiener O 1926 Naturw, 81-84

Wiener, Norbert 1894-1964

Dalbear K 1910 J Gen Psy 19, 452-465
*Levinson N + 1966 AMSB 72(1,II), 1-
145 [special issue, 9 papers,
bib, port]
Pitt H R 1964 Nat 202, 540-541
(Z118:2, 247)
Sartorius H 1964 Regel tec 12, 241
(Z117:1, 9)
*Struik D J 1966 Am Dialog 3(1),
34-37
Wiener Norbert 1953 *Ex-prodigy.*
*My Childhood and Youth,* NY,
321 p (MR15, 277)
- 1956 *I Am a Mathematician. The*
*Later Life of a Prodigy.* Garden
City NY, 380 p (MR17, 1037;
Z71, 4)

Wijdenes, P.

Streefkerk N 1950 Euc Gron 26, 3-8
(Z37, 292)

Wilczynski, Ernest Julius 1876-1932

Anon 1936 Am NASBM 16, 295-319
Bliss G A 1932 Sci (2)76, 316-317
(F58, 995)
Lane E P 1932 AMM 39, 567-569
- 1933 AMSB 39, 7-14 (F59, 41)
- 1933 Annali (4)11, 363-364
(F59, 41)
Sperry P 1933 AMSB 39, 203
(F59, 45)

Wildberg. See Graf.

Wilks, Samuel Stanley 1906-1964

Anderson T W 1965 An M St 36, 1-23,
24-27 (MR30, 872)
Gulliksen H 1964 Psychmet 29, 103-
104 (MR29, 639)

Willers, Friedrich Adolf 1883-1959

Heinrich H 1958 Fors Fort 33, 190-
191 (Z83, 245)
- 1959 MTW 6, 43 (Z85, 7)
Sauer R + 1960 ZAM Me 40, 1-8
(MR22, 260; Z87, 244)

William of Lunis XIII

= Wilhelmus de Lunis Apud Neapolim

Eneström G 1891 Bib M (2) (F23, 4)
[alg]
Sarton G 1931 Int His Sc 2(2), 563
[bib]

William of Moerbeke 1215?-1286?

Bosmans H 1922 Rv Q Sc, 5-23
[Archimedes]
Clagett M 1952 Isis 43, 236-242
(Z47, 242) [Archimedes, Johannes
de Muris]
Heiberg J L 1892 Hist Abt 37, 81
(F24, 7)
Sarton G 1931 Int His Sc 2(2), 829-
831 [bib]

William of Ockham XIII-XIV

Birch T B 1936 Phi Sc 3, 494-505
[continuum]
Burns C 1916 Mind 25, 506-512
[continuity]
Henry D P 1964 Not DJFL 5, 290-292
(MR32, 1114) [logic]

Williamson, Benjamin  1827-1916

Anon 1917 LMSP (2)15, 48-49
  (F46, 35)


Williamson, John  1901-

Turnbull H W 1952 Edi MSN 38, 23-24
  (MR14, 344; Z49, 291)


Wilson, Bertram Martin  1896-1935

Turnbull H W 1936 Edi SP (A)55,
  176-177  (F62, 1042)


Wilson, Cook

Furlong E J 1941 Mind (Ap), 122-139
  [non-Euc geom]


Wilson, Jean

Glaisher J W L 1876 Nou Cor 2, 110-
  114  (F8, 19)


Wiltheiss, Ernst Eduard
  1855-1900

Wirtinger W 1901 DMV 9(1), 59-63
  (F32, 21)


Wilton, John Raymond 1884-

Carslaw H S + 1945 LMSJ 20, 58-64
  (Z60, 18)


Wiman, Anders    1865-

Nagell T 1959 Sv Arsb, 17-19
  (Z92, 3)
- 1960 Act M 103, i-vi
  (MR22, 1131; Z87, 7)


Winckler, Anton 1821-1892

Czuber E 1892 Mo M Phy 3, 403-406
  (F24, 33)


Wind, Cornelis Harm  1867-1911

Lorentz H 1911 Amst P 14(1), 172-174


Wingate, Edmund    1593-1656

Hammer E 1911 Z Vermess 40, 27-28
  (F42, 57)  [slide rule]
Hooker J H 1896 M Gaz 1, 35-38


Winkler, Johann Jakob    1831-1893

Emch A 1909 Int Con 4(3), 538-540
  [calculating prodigy]


Winlock,  William Crawford  1859-1896

Eastman J R 1900 Wash PS 13, 431-434
  (F32, 13)


Wintner, Aurel   1903-1958

Anon 1958 AJM 80, 1  (MR19, 1248)
Hartman P 1962 LMSJ 37, 483-503
  (MR25, 967; Z106, 4)


Wirtinger, Wilhelm    1865-

Carathéodory C 1948 Mun Jb (1944/48),
  256-258  (MR11, 574)
Hornich H 1948 Mo M Phy 52, 1-12
  (MR9, 486; Z30, 101)


Wishart, John   1898-1956

Pearson E S 1957 Biomtka 44,  1-8
  (MR18, 982)


Witelo.  See Vitelo.

Witkowski, August Victor   1853-1913

Dickstein S 1913 Wiad M 17, 189-193
  (F44, 35)


Witte

Maupin G 1908 *"Le comte Witte mathe-maticien"* Rv Sc (5)9, 586-594


Wittgenstein, Ludwig Joseph Johann
1889-1951

Malcolm N 1967 Enc Phi  [bib]
Pears D F 1969 New York Review of
  Books (Jan 16), 21-30
Russell B 1969 *The Autobiography of
  Bertrand Russell (1914-1944)*
  (Bantam Books), 64-65, 132-135,
  156-165, 282-288  [letters]


Woepcke, Fran   1826-1864

Biermann K-R 1960 Ber Mo 2, 240-249
  (Z89, 5)
Narducci E 1869 Boncomp 2, 119-152
  (F2, 21)


Wojewodka, Bernard XVI

Dziwinski M 1889 *Ueber den Algorith-mus von Bernard Wojewodka aud dem
  Jahre 1553,* Lemberg (in Polish)


Wolf, Max   1863-1932

A C D C 1933 Nat 131, 353
Kopff A 1932 Naturw 21, 181-183
Plassman J 1932 Himmelsw 42, 215-216
Vogt H 1933 As Nach 247, 313-316


Wolf, Rudolf   1816-1893

Doublet E 1930 Rv Sc 68, 658-660
  (F56, 23)

Graf J H 1894 Bern NG, 193-231
  (F25, 43)
- 1895 *Professor Dr. Rudolf Wolf
  1816-1893,* Bern
Weilenmann A 1894 Zur NGV 39, 1-64
  (F25, 43)
W J L 1894 Nat 49, 266-267
  (F25, 43)


Wolff, Christian   1679-1754

Caesar J 1879 *Christian Wolff in
  Marburg,* Marburg  (F11, 18)
Marian V 1936 Gaz M 42, 120-123,
  179-183
Tonelli G 1959 Arc Phi 9
- 1967 Enc Phi  [bib]
Vleeschauwer J J de 1932 Bel Rv PH
- 1953 Phi Nat 2, 358-375


Wolfram, Isaac   XVIII

Archibald R C 1950 M Tab OAC 4,
  185-200  (MR12, 312)


Wolibner, Witold   1902-1961

Charzynski Z + 1963 Colloq M 10, 353-
  360  (MR28, 2; Z112, 244)


Womersley, John Ronald   1907-1958

Smithies F 1959 LMSJ 34, 370-372
  (Z85, 7)


Wood, De Volson   1832-1897

Matz F P 1895 AMM 2, 253-256
- 1897 AMM 4, 197-199  [port]


Wood, Hudson A.   1841-

Matz F P 1895 AMM 2, 343-345

Wood, Philip Worsley 1880-1956

White F P 1959 LMSJ 34, 486-487
(Z88, 7, 244)


Woodward, Robert Simpson 1849-1924

*Archibald R C 1938 *A Semicentennial
History of the American Mathematical Society,* NY, 139-144
[bib, port]
Sarton G 1925 Isis 7, 112-114
Wright F E 1926 Am Geol SB 37, 115-
134


Woolhouse, Westley Stoker Barker
1809-1893

Anon 1893 LMSP 25, 1-5 (F25, 43)


Worden, Orpha -1938

Reeve W D 1938 MT 31, 136


Worpitzky, Julius Daniel Theodor
1835-1895

Lampe E 1895 Ber Ph 14, 33-39
(F26, 40)
- 1897 DMV 4, 47-51
(F28, 21)


Wozniakowski, J.

Dobrowolski W O 1963 Kwar HNT 8,
519-524


Wren, Christopher 1632-1723

Anon 1933 MT 25, 368
Huxley G L 1960 Scr 25, 201-208
(MR22, 929; Z97, 243) [geom]
Lena Milman 1908 *Sir Christoph Wren,*
London
Reeve W D 1932 MT 25, 368

Simons L G 1934 Scr 2, 362
Whiteside D T 1960 Lon RSNR 15, 107-
111 (Z107, 245)


Wren, Walter 1831-1898

Edwards J 1898 LMSP 29, 731-732
(F29, 27)


Wright, Edward 1558-1615

Heppel G 1894 M Gaz 1, 11


Wronski, Josef Maria Hoëné de
1778-1853

Archibald R C 1948 *Mathematical
Table Makers,* NY, 29-30
(Scr Stu 3) [bib]
Benes J 1891 Cas MF 20, 24 (F23, 46)
Bertrand J 1897 J Sav
- 1897 Rv Deux M (Feb)
Dickstein S 1887 *Hone-Wronski*
(in Polish)
- 1889 Krak Roz 19, 167-192
- 1890 Krak Roz 20, 287-291
- 1890 *Die logarithmische Tafel von
Hoene-Wronski,* Warsaw, 15 p
- 1890 *Die logarithmischen Canones
von Hoene-Wronski,* Warsaw, 30 p
- 1890 Poz Rocz 17, 1-10
- 1890 Prace MF 2, 145-168
- 1892 Bib M (2)6, 48-52, 85-90
(F24, 19)
- 1892 Krak BI, 64- [numb thy]
- 1893 Krak Roz 4, 73-, 396
[numb thy]
1894 Bib M (2)7, 9-14; (2)8, 49-54
85-87 (F25, 25)
- 1896 Bib M (2)10, 5-12 (F27, 13)
- 1896 *Hoene Wronski. Sein Leben und
Seine Werke.* Cracaw, 372 p
(F27, 13) (in Polish)
- 1896 Krak BI, 165
- 1897 Wiad M 1, 22-26 (F28, 12)
[Bertrand]
- 1905 Int Con 3, 515-525
(F36, 11)

Echols W H 1893 NYMS 2, 178-184
(F25, 69)
Kozlowski W 1908 Rv Met Mor (S Sept),
29
*Levi B + 1943 M Notae 3, 74-100
(MR5, 57; Z60, 5) [Pythagoras]
Mongardon L + 1915-1918 Intermed 22,
68; 23, 113, 164-167, 181-183,
199; 25, 55-57 (F46, 11-12)
[bib]
Montessus M R de + 1898 Intermed 5,
264
Montessus M R de 1935 Rv Sc 73, 329-
333 (F61, 952)
Warrain F 1933-1938 *L'oeuvre
philosophique de Hoëné Wronski,*
Paris, 2 vols (F59, 857)
West E 1886 *Exposé des méthodes
générals en mathématiques...
d'après Hoëné Wronski,* Paris, 314 p


Wullner, Friedrich Hugo Anton Adolf
1835-

Anon 1905 Arc MP (3)11, 97-98


Wylie, Alexander 1815-1887

Wang Ping 1962 in *Proceedings, Second
Biennial Conference, International
Association of Historians of Asia,*
Taipei


Ximenes, Leonardo 1716-1786

Mori A 1904 Int Con H 12


y. See preceding word in Spanish
names.


Yamabe, Hidehiko 1923-1960

Goto M 1961 Osak JM 13, i-ii
(MR23A, 698; Z95, 5)

Yang, Hsia-Hou VI

Vanhee L 1924 AMM 31, 235-237


Yanovskaya, Sofiya Aleksandrovna
1896-

Anon 1956 UMN (NS)11(3), 219-222
(MR18, 182; Z70, 6)


Yaqub Nadim X

= Abul-Faraj Muhammad ibn Ishaq
ibn abi Yaqub al-Nadim al-Warraq
al-Baghdadi

Suter H 1892 ZM Ph 37(S), 1-89
- 1893 ZM Ph 38, 126-128
Sarton G 1927 Int His Sc 1, 662


Yoshikawa, Jitsuo 1878-1915

Anon 1955 Tokush J 6, 3-4
(Z66, 8)
Hayashi T 1915 Toh MJ 7, 116-122


Young, A. A.

W C M 1929 ASAJ 24(166), 200-201
(F55, 617)


Young, Alfred 1873-1940

Turnbull H W 1941 LMSJ 16, 194-208
(F67, 970; Z28, 195)
- 1941 Lon RS Ob 3, 761-778
(Z60, 18)
Wilson G H A 1941 Nat 147, 229
(Z60, 18)


Young, Grace Chisholm 1868-

Cartwright M L 1944 LMSJ 19, 185-192
(Z60, 18)

**Young,** John Wesley  1879-1932

Beetle K D + 1932 AMSB 38, 603-610
  [bib]
Decker F F 1932 AMSB 603-610
  (F58, 992)
Hopkins E M + 1932 AMM 39, 309-315
  (F58, 50)
Hopkins F G 1933 Lon RSP A 139, 7
Sanford V 1932 MT 25, 232-234
Smith D E 1932 ZMN Unt 63, 195-196
  (F58, 995)


**Young,** Thomas  1773-1829

Larmor J 1934 Nat 133, 276-279
  (F60, 12)
Nichols E L 1933 Am Opt J 23, 1-6
  (F59, 846)  [Fresnel]
Wood Alexander 1954 *Thomas Young,*
  *Natural Philosopher,* Cambridge,
  374 p  (Z60, 11)


**Young,** William Henry  1863-1942

Hardy G H 1942 LMSJ 17, 218-237
  (F68, 18)
- 1943 Lon RS Ob 4, 307-323
  (Z60, 18)
Sutton G 1963 M Gaz 49, 16-21
Ursell H D 1963 Nat 200, 1274-1275


**Yuan Yuan**  1764-1849

Mikami Y 1928 Isis 11, 123-126
Van Hee P L 1926 Isis 8, 106


**Yule,** George Udny 1871-1951

Yates F 1952 Lon RS Ob 8, 309-323
  (MR14, 525)


**Yushkevich,** Adolf-Andrei Pavlovich
  1906-

Bashmakova I G + 1967 UMN 22(1), 187-
  194  (MR34, 2416)

Markushevich A I + 1956 UMN 11(4),
  197-200  (Z70, 6)


Yusuf.   See Ahmad


**Yvon-Villarceau,** Antoine Joseph
  François      1813-

Aubry A + 1896 Intermed 3, 18
  (Q456)
Blanchard E + 1883 Par CR 97, 1454-
  1463
Clery A 1895 Intermed 2, 274, 300


**Zach,** François Xavier de  1754-1832

Houzeau J C + 1877 Bel Bul (2)42,
  475-479  (F9, 12)
Wolf R 1874 BSM 6, 258-272
  (F6, 28)


**Zahradnik,** Karel  1848-

Vojtech J 1917 Cas MF 46, 235-303


**Zajaczkowski,** Wladyslaw  1837-1898

Dickstein S 1898 Wiad M 2, 258-261
  (F29, 21)


**Zallinger,** Franz  1743-1828

Gunther S 1904 Bib M (3)3, 208-225
  (F33, 19)


**Zamorski,** Jan  1927-1964

Charzynski Z + 1963 Colloq M 10, 361-
  364  (MR28, 2; Z112, 294)

Zantedeschi, Francesco    1797-1873

Kobell F v 1874 Mun Si, 70-71
   (F6, 31)


Zaremba, Stanislaw    1863-1942

Szarski J 1962 Wiad M 5, 15-28
   (Z100, 247)


Zarkali, al    XI

   = Arzachel = Al Zarqali = Abu
Ishaq   Ibrahim ibn Yahya al-Naqqash

Eneström G 1896 Bib M (2)10, 53-54
   (F27, 5) [Ziegler]
- 1901 Intermed 8, 12
Steinschneider M 1881 Boncomp 14,
   171-182  (F13, 6)
- 1883 Boncomp 16, 493-504
- 1884 Boncomp 17, 765-794
- 1885 Boncomp 18, 343-360
- 1887 Boncomp 20, 1-36
- 1888 Boncomp 20, 575-604
- 1890 Bib M (2)4, 11-12
Wittstein A 1894 ZM Ph 39, 41


Zbikowski, A.

Anon 1900 Wiad M 4, 266


Zebrawski, P.

Curtze M 1880 Arc MP 64, 432-434
   (F12, 6)


Zeising, Adolf    1810-1876

Guenther S 1876 Hist Abt 21, 157-165
   (F8, 25)


Zeitz, Hermann

Braver A 1934 Ber MG 23, 2-6
   (F60, 16)


Zelbr, Karl    1854-1900

Waelsch E 1901 DMV 9, 63


Zeller, Christian Julius Johannes
   1822-1899

Anon 1899 Wurt MN (2)1, 52


Zeno.  See Zeno paradoxes in
   Chapter 3.


Zenodoros    -II

Goulard A + 1896 Intermed 3, 140
   (Q696) [geom]
Mueller W 1953 Sudhof Ar 37, 39-71
   (MR15, 383) [isoperimetric
   problem, Theon]


Zermelo, Ernst    1871-1953

Anon 1953 Int M Nach 9(27/28)
Pinl M 1969 DMV 71, 221-222
Serpinski V K 1922 M Sbor 31(1)


Zero    XVIII

Anon 1915 Intermed 22, 82-83
   (Q3940)


Zerr, George B. McClennan
   1862-1910

Finkel B F 1911 AMM 18, 1-2


Zenner, Gustav Anton 1828-

Krause M 1908 Leip Ber 60, 341-351


Zeuthen, Hieronymus Georg    1839-

Anon 1909 *Festskrift til H. G. Zeuthen*
   Copenhagen, 156 p  (F40, 42)

- 1919 Ens M 20, 452   (F47, 12)
- 1920 Rv Met Mor 27 (4)(S), 9-10
Bierman K R 1966 Sudhof Ar 50, 335
Bohr H 1949 M Tid A 60-66
   (Z36, 4)
- 1952 Sk Kong 11, 195-200
   (MR14, 523; Z48, 241)
Cantor M 1895 BSM (2)19, 64-69
   (F26, 2)
Heegaard P 1920 Nor M Tid 2, 33-37
   (F47, 11)
Hjelmslev J 1939 M Tid A, 1-10
   (F65, 1092)
H W R 1921 LMSP (2)19, xxxvi-xxxix
   (F48, 20)
Loria G 1919 Arc Sto 1, 447-451
   (F47, 11)
- 1922 Scientia 31, 265-279
   [M Cantor]
Noether M 1921 M Ann 83, 1-23
   (F48, 19)
Picard E 1923 BSM (2)47, 366-369
   (F49, 12)
- 1923 Par CR 177, 565-568
   (F48, 20)
Segre C 1919 Tor FMN 55, 177-178
- 1920 Tor FMN 55, 327-328
   (F47, 12)


Zhukovskii, Nikolai Egorovich
   1847-1921

Anon 1947 JAM Me 11, 9-40   (MR9, 74)
- 1947 Pri M Me 11, 3-40   (Z29, 195)
Artobolevskiy I I 1952 SSSR Mash
   12(46), 5-14 (MR16, 2) [appl]
Golubev V V 1947 UMN 2(3), 3-17
   (MR9, 486)
- 1951 SSSR Tek, 1152-1158
   (MR13, 198) [aerodyn]
Kalinin N N + 1952 *N. E. Zhukovskii
   otets russkoi aviatsii,* Alma Ata,
   28 p  (in Kozak)
Khristianoviche S A 1951 SSSR Tek,
   1137-1151   (MR13, 198)
Kostitzin V A 1922 Mos MO Sb 31, 5-6
Sintsov D M 1924 UKUZ 1
Yurev B I 1947 Pri M Me 11, 3-8
   (MR9, 74)

390 / Biography

Ziegler, Jakob   1480-1549

Eneström G 1896 Bib M 10, 53-54
   [Zarkali]
Guenther S 1896 *Jakob Ziegler,
   ein Bayerischer Geograph and Mathematiker,* Ansbach, 64 p  (F27, 5)


Zillmer, August   1831-1893

Anon 1897 DMV 4, 23-24   (F28, 20)


Zindler, Konrad   1866-

Wirtinger W 1935 Mo M Phy 42, 215-220
   (F61, 956)


Ziwet, Alexander   1853-

Anon 1929 Ens M 28, 136-137
Karpinski L C + 1929 AMSB 35, 259-
   260


Zmurko, Wawrzyniec (Lorenz)   1824-
   1889

Dziwinski P 1890 Prace MF 2, 433-448


Zolotarev, Egor Ivanovich   1847-1878

Bashmakova IG 1949 Ist M Isl (2),
   231-351   (MR12, 1) [cong, div]
Kuzmin R O 1947 UMN 2(6), 21-51
   (MR10, 420)
Nalbandian M B 1965 Ist M Isl 16,
   295-310 [ell fun]
- 1966 Ist Met EN 5, 96-104
Ozhigova E P 1966 *Egor Ivanovich
   Zolotarev,* M-L, 143 p  (MR33, 2509)


Zorawski, Paulin Kazanierz Stephan
   1866-1953

Slebodzinski W 1956 Colloq M 4, 74-88
   (MR17, 932; Z70, 244)

- 1956  Prace M 2, 79-93
  (MR18, 710; Z74, 246)

Zottu, J. G.   1870-1902

Anon 1902 Gaz M 7, 269

Zsigmondy, Karl   1865-

Schmid T 1927 DMV 36, 167-170
  (F53, 31)

Zuehlke, Faul   1877-1957

Leumann H 1957 MN Unt 10, 234-235
  (Z100, 8)

Zurria, Giuseppe   1810-1896

Pennachietti G 1897 Cat Bol 46
  (Mar), 32

Zuse, Konrad   1910-

Anon 1959 Elek Rech 1, 5-6
  (Z87, 243)

## 2

MATHEMATICAL TOPICS

This chapter lists historical studies of topics within mathematics. Key words are names of branches of mathematics or of subtopics. Generally a title is classified under the key word of smallest scope. For example, an article on the history of cubic equations is classified under that topic and does not appear under "algebraic equations" or "algebra." Because of the many synonyms in mathematical terminology and the inevitable vagueness of many key words, no unique classification scheme is possible. However, if the user begins with one or several key words related to his interests and follows cross-references and key words that appear in the comments on each entry, he will be able to locate appropriate references without excessive difficulty. Often entries will lead to other chapters, notably to the biography of people who worked on a topic. Titles on a theorem or concept named after a man may appear under his name in Chapter 1 if there are too few to justify a separate entry here.

## Addition logarithms

Berkeley L M 1931 USNI 57, 207-209
[addition logs, Leonelli]
Bunyakovskii V A 1863 Pet B 5, 471-
[addition logs]
Eneström G 1896 Bib M (2)10, 64
[addition logs, Cavalieri,
Leonelli]
Schroeter J F 1923 Nor M Tid 5, 35-
43   (F49, 19)


## Agnesi curve

= witch of Agnesi

Arcais J d' + 1901 Intermed 7, 111,
227
Beran H 1940 Scr 8, 135
Larsen H D 1947 SSM 46, 57-62
Loria G 1897 Bib M (2)11, 7-12,
33-34   (F28, 47)
Mulcrone T F 1957 AMM 64, 359-361
(MR18, 982)
Spencer R C 1940 Am Opt J 30, 415-
419
Vacca G 1901 Loria 4, 33   [Grandi]


## Algebra

See also Associativity, Commutative,
Distributive, Division algebra,
Equations, Field, Forms, Four and
eight square, Fundamental theorem,
Galois theory, Geometric algebra,
Invariants, Jordan algebra, Lattice,
Matrix, Partial fractions,
Quaternions, Rings, Summations,
Symmetry, Universal algebra.

Anon 1939-1940 AMM 46, 234; 47, 107
[club topic]
Aubry A 1907 Mathesis (3)7, 257-261
[complex anal, series]
Bashmakova I G 1966 Ist M Isl 17,
317-323   [XVIII]
Baumgart J K 1961 MT 54, 155-160
[axioms]
Bell E T 1938 in *AMS Semicentennial
Publications,* NY 2, 1-34 [XX, USA]

Berzolari L 1934 Esp Prog 1, 25-28
(Z12, 243)
Boguslawski A 1894 *Die Axiome der
Arithmetik und Algebra nach
Helmholtz und Lobatschewsky,* Mos,
8 p (in Russ)   (F25, 61)
Bortolotti Ettore 1903 *Influenza
dell'opera matematica di Paolo
Ruffini sulla svolgimento delle
teorie algebriche,* Modena
- 1918/19 Mod U An, 100 p (F46, 44)
[Italy]
- 1923 Esercit 3, 69-87, 161-164
(F49, 18) [Bologna lib]
- 1925 Pe M (4)5, 147-184
[XVI, Bologna]
- 1929 Pe M 9, 161-166 [geom, Italy]
- 1934 Bln Mm 2, 27 p  [anc]
- 1936 Osiris 1, 184-230  (Z14, 243)
[anc]
- 1937 Archeion 19, 192-195
[Babylonian]
- 1941 Bln Mm (9)8, 3-11
(MR9, 483; Z61, 2; F67, 968)
[arith]
Bouligand G 1948 Rv GSPA 35, 121-123
[Bezout]
Bruins K M 1959 Euc Gron 34, 131-
159  (Z88, 4)  [anc, mid]
Carnahan W H 1947 SSM 46, 7-12,
125-130
Chasles M 1841 Par CR 13, 497-601-
Chebotarev N G 1961 Ist M Isl 14,
539-550   [ideals]
Chiari A 1900 Pitagora 7, 39, 107
[elem]
Cossali Pietro 1797-99 *Origine,
trasporto in Italia, primi progressi
in essa dell' algebra,* Parma,
2 vols [obsolete but important]
Curtze M 1895 ZM Ph 40(S), 31-74
(F26, 49)  [XV Ger]
Delone B N 1952 UMN 7(3), 155-178
(Z49, 290)
Eneström G 1911 Bib M (3)12, 181-182
(F42, 55)   [XVI]
Fazzari G 1904 Pitagora 10, 87-92,
131-134; 11, 14, 55   [anc]
Fleckenstein J O 1946 Experien 2,
321-323  [XVI-XIX]
Frajese A 1940 It UM (2)2, 363-365
(MR2, 114)  [Fibonacci, geom]

Algebra    (continued)

Franchini P 1827 *La storia dell'
  algebra e dei suoi principali
  scrittori sino al secolo XIX,
  illustrata ed estesa col mezzo
  degli originali documenti...,*
  Lucca   (Also in Lucca 3, 211-)
  [obsolete but important]
Gandz S 1926 AMM 33, 437-440
  [word]
Gerhardt C J 1843 Arc MP 3, 284-
  [Italy]
- 1870 Ber Mo, 141
*Harkin D 1951 Nor M Tid 33, 17-26
  (MR13, 1)
Hayashi T 1935 Toh MJ 40, 317-369
  (Z10, 386)   [Japan]
Hyde E W 1891 AAAS, 51-65
Karpinski L C 1916 Scientia 26, 89-
  101  (F46, 54)
- 1923 SSM 23, 54-64
- 1934 Isis 22, 104-105  (Z10, 243)
  [Greek]
- 1944 Scr 10, 149-169
  (MR6, 141; Z60, 7) [bib to 1700]
Klein J 1934 QSGM 3(1), 18-105;
  (2), 122-235  (Z10, 248; 14, 49)
  (Engl tr 1968 MIT P)
Kramar F D 1966 Ist M Isl (17), 309-
  316  (MR37, 1221)  [Buée]
Kurosh A G 1948 *Algebraischeskii
  referativnyi sbornik za 1941-1946,*
  Moscow, 615 p
Landsberg G + 1910-1911 Enc SM
  1(2)(2), 233-328; 1(2)(3), 329-385
  [alg geom, field]
Libri G 1841 Par CR 13, 559-
Loria G 1930 Pe M 10, 152-154
  [XIII-XV]
- 1938 Ens M 37, 77-82  (F64, 4)
  [word]
Lure S Ya 1946 UMN 1, 248-257
  [origin]
Mansion P 1899 *Mélanges Mathématiques,*  Paris, vol 2
Miller G A 1928 Sch Soc 28, 363-364
- 1929 SSM 29, 404-410  (F54, 45)
- 1932 Sci (2)76, 14-15
  (F58, 986)  [word]
- 1940 Sci 91, 571-572

Natucci Alpinolo 1932 Pe M (4)12,
  173-179  (F58, 32)  [XVI]
- 1954 Batt (5)2, 429-436
  (MR16, 660; Z57, 2)  [origin]
- 1955 *Sviluppo storico dell'arit-
  metica generale e dell'algebra,*
  Napoli, 367 p  (Z67, 243)
*Nesselmann G H F 1842 *Vessuch einer
  Kritischen Geschichte der Algebra
  I. Die Algebra der Griechen,*
  Berlin
- 1855 Nou An 14, 445-  [word]
Netto E + 1907 Enc SM 1(2)1, 1-232
  [polyn, rat fun]
Neugebauer O 1932 QSGM (B)2, 1-27
  (Z4, 2)  [anc]
- 1936 QSGM (B)3, 245-259
  (Z13, 337)   [geom]
Nielsen N 1928 Dan Sk 8(8), 32 p
  [to Abel]
Parmentier G 1881 Nou An (2)20, 139-
  140  (F13, 26)   [word]
Queysanne M + 1955 *L'Algèbre moderne*
  Paris, 136 p  (MR16, 1081)
  [elem survey]
Reves G E 1952 SSM 52, 61-69
Sanford V 1927 *The History and
  Significance of Certain Standard
  Problems in Algebra,* T Col Cont
  (251), 110 p
Segre B 1963 Cah H Mond 7, 383-406
  [geom]
Smith D E 1936-1937 Scr 4, 111-125;
  5, 15-16  (Z15, 147)  [anc]
Stamatis E 1955 Athen P 30, 263-282
  (Z65, 243)  [geom, Pythagoreans]
Strachey E 1816 Asi R 12, 158-
Sushkevich A K 1951 Ist M Isl 4, 237-
  451  (MR14, 523)  [XIX, Russ]
*Taton Rene 1965 in *Science in the
  Nineteenth Century,* NY, 7-43
Thureau-Dangin F 1937 Archeion 19,
  1-12  [Babyl]
- 1940 Par IBL, 292-318  (MR48, 73)
  [origin]
Tropfke Johannes 1933 *Geschichte der
  Elementar-Mathematik...,*  vol. 2,
  part 3, Berlin-Leipzig, 270 p
  (F59, 3)
Van der Waerden B L 1948 in
  Friedrichs K O + *Studies and Essays,*

## Algebra    (continued)

*Courant Anniversary Volume,* NY,
   437-449
*- 1966 DMV 68(1), 155-165
   [XIX, XX, field, group, ideal]
Vashchenko-Zakharchenko M E 1884
   Kagan 1 [notation]
Vaux  C de 1897 Bib M (2)11, 1-2
   (F28, 38) [word]
Verriest G 1957 Civ Mac 5(4), 57-65
   (MR19, 1029) [Galois]
*Vivanti G 1924 Pe M (4)4, 277-306
   (F50, 7) [bib, to 1800]
Vogel K 1931 ZMN Unt 62, 266-271
   [Greek]
- 1936 Muns Semb 9, 107-123
   (F63, 5)
Vuillemin Jules 1960 *La philosophie
   de l'algébre de Lagrange.
   Réflexions sur le Mémoir de
   1770-1771,* Paris, 24 p
   (MR22, 1853)
Vygoskii M Ya 1967 *Arifmetika i
   algebra v drevnem mire,* 2nd ed,
   Mos, 367 p (MR37, 1215) [anc]
*Wallis John 1657 *Mathesis Universalis*
   Oxford
- 1685 *Treatise of Algebra both
   Historical and Practical,* London
- 1693 *Tractatus de Algebra
   historicus et practicus* in *Works,*
   Oxford
Wappler E 1899 ZM Ph 44(S), 537-554
   [Ger]
Woepcke F 1854 Par CR 39, 162 [Arab]
Zeuthen H G 1919 Dan Sk 2(4)
   (F47, 23) [origin]

## Algebraic functions

See also Elliptic functions.

Baker H F 1897 *Abel's Theorem and
   and the Allied Theory Including
   the Theory of Theta Functions*
   (Cambridge U P)
Bourbaki Nicolas 1960 *Eléments
   d'histoire des mathématiques,*
   Paris, 107-

*Brill A + 1894 *Die Entwickelung der
   Theorie der algebraischen Functionen
   in aelterer und neuerer Zeit,*
   DMV 3, 107-566 (F25, 70)
Clebsch R F A 1864 Crelle 63, 189-243
   [geom]
Eneström G 1908 Bib M (3)9, 206-210
   (F39, 63) [theta fun, Jacob
   Bernoulli]
*Hancock H 1897 Brt AAS, 246-286
   (F28, 41) [Abelian funs to
   Riemann]
Hensel K + 1902 *Theorie der
   algebraischen Funktionen einer
   variabeln...,* Berlin, 707 p
*Klein Felix 1882 *Ueber Riemanns
   Theorie der algebraischen Functionen
   und ihrer Integrale,* Leipzig,
   82 p
*- 1890 M Ann 36, 1-83 [Abelian fun]
- 1894 in *The Evanston Colloquium,*
   NY-London (MacMillan), Lecture 10
*- 1963 *On Riemann's Theory of
   Algebraic Functions...,* NY
   (Dover)  (tr of 1882)
Lange-Nielsen F 1927 Nor M Tid 9,
   35-43, 205  (F53, 25)
   [Abel thms]
Noether E 1919 DMV 28, 182-203
   [survey]
Pexider J V 1903 Bib M (3)4, 52-64
   [Abel thms]
Pokrowsky P M 1886 Fiz M Nauk 2, 47-
   65  [ell fun]
Segre C 1894 Tor FMN 30, 91-93

## Algebraic geometry

See also Algebraic functions,
   Analytic geometry, Cramer paradox,
   Cremona transformation, Curves,
   Surfaces.

*Benedicty M 1953 *La geometria
   algebrica astratta e il concetto
   di varietà algebrica,* Archim 4-5
Bertrand J 1867 J Sav, 644-
   [surf]
Calleri P 1938 Pe M (4)18, 33-42
   [ell point, hyperb point]

## Algebraic geometry (continued)

Castelnuovo G 1929 Int Con (1929),
1, 191-201 (F55, 589) [Italy]
Coolidge J L 1931 *A Treatise on
Alg Plane Curves* (Oxford U P),
513 p (repr 1959 Dover) [bib]
Enriques F + 1915-1934 *Lezioni sulla
teorica geometrica delle equazioni,*
Bologna, 3 vols
Godeaux L 1947 Bel BS (5)33, 901-918
(MR9, 485)
- 1953 Rv GSPA 60, 8-14
[Castelnuovo, Enriques]
Hodge W V D 1957 in Fox R H + ed
*Algebraic Geometry and Topology
--a Symposium* (Princeton U P)
3-23 (MR19, 173) [Lefschetz]
Klein F 1893 in his *Lectures on
Mathematics...1893,* NY (Macmillan),
25-32
Marre A 1846 Nou An 5, 557-
[al-Khwarizmi]
Picard E 1895 Paler R 9, 159-
(F26, 65) [surf]
Rosenblatt A 1912 Prace MF 23, 51-
192 [surf]
*Samuel Pierre 1967 *Méthodes d'algèbre
abstraite en géometrie algébrique*
(Springer), 2nd ed, 124-129
[bib]
Scorza G 1930 Bo Fir (2)8, 41-53
(F56, 798) [Italy]
Segre B 1963 Cah H Mond 7, 383-406
Severi F 1951 Tor Sem 10, 67-95
(MR14, 2)
Snyder V + 1928 Am NRCB (63), 359 p
[*bib]
- 1934 Am NRCB (96) (S to 1928)
[bib]
Teixeira F 1905 Porto Ac 1, 7
[Anastacio da Cunha, Monteiro
da Rocha]
Van der Waerden B L 1948 in *Studies
and Essays Presented to R. Courant
on his 60th Birthday,* NY, 437-449

## Algebraic numbers

Bell E T 1929 AMSB 35, 12 (F55, 26)
[Kronecker]

*- 1944 M Mag 18, 188-204, 219-223
[Gauss]
Dickson L E 1917 Am M 18(2), 161-187
[Fermat last thm]
*Dickson L E + 1923 Am NRCB (62)
(repr 1967 Chelsea)
Furtwaengler P 1953 Enz MW 1(2)(8)(2)
[*bib]
Hancock H 1925 Sci (2)61, 5-10,
30-35 (F51, 9)
- 1931-1932 *Foundations of the Theory
of Algebraic Numbers* (Macmillan),
2 vols (repr 1964 Dover)
*Hecke Erich 1923 *Vorlesungen ueber
die Theorie der algebraischen
Zahlen,* Leipzig, 266 p [*bib]
*Hilbert D 1894 DMV 4, 175-546
(Tr 1909, 1910, 1911 Tou FS (3)1,
257-328; 2, 226-456; 3, 1-62)
[bib]
Mordell L J 1923 AMSB (2)29, 445-463
*Vandiver H S + 1928 Am NRCB (28)
(repr 1967 Chelsea) [bib]

## Algorithm

See also Arithmetic.

Bortolotti E 1913 Loria 15, 97-98
(F94, 45) [word]
- 1914 Loria 16, 33-38 (F45, 67)
[word]
Grosse H 1901 *Historische rechen-
buecher des 16. und 17. Jh.,*
Leipzig [bib]
Karpinsky L C 1914 AMSB (2)21, 69
(F45, 96)
- 1932 Int Con 6, 455-458 [France]
Kennedy E S + 1956 AMM 63, 80-83
[iteration]
Kovach L D 1964 M Mag 37, 159-165
(Z117:1, 4) [anc, comp]
Hunrath K 1887 Bib M (2)1, 70 [word]
Rybnikov K A 1956 Int Con HS 8, 142-
145 [influence]
Schrecker P 1947 J H Ideas 8, 107-116
[heuristic, Leibniz]
Spasskii I G 1952 Ist M Isl 5,
269-420

## Alhazen problem

*Baker M 1881 AJM 4, 327-332
   (F14, 24)  [bib]
*Bode P 1893 *Die Alhazensche Spiegel-Aufgabe*, Frankfurt, 50 p
Marcolongo R 1929 Nap FM Ri (4)35,
   22-24  (F55, 10) [Leonardo da Vinci]
Thebault V 1955 Scr 21, 148-150

## Alligation

Eveland K C 1955 MT 48, 339-341
Katra A E 1940 MT 33, 225-228

## Almost periodic functions

Besicovitch A S 1932 *Almost Periodic Functions* (Cambridge U P)
   (repr 1954 Dover) [bib]

## Amicable numbers

Curtze M 1895 Bib M 10, 33-42
*Escott E B 1946 Scr 12, 61-72
   [survey]
Hunrath K 1909 Bib M (3)10, 80-81
Rolf H L 1967 MT 60, 157-
Rudio F 1915 Bib M (3)14, 353-354

## Analysis

See also Calculus, Complex analysis,
   Integral equations, Limits,
   Operators, Pathological curves,
   Potential, Real analysis, and in
   Ch. 3 Analysis and synthesis.

Boutroux Pierre 1914-1919 *Les principes de l'analyse mathématique exposé historique et critique*,
   Paris, 2 vols
Boyer C B 1954 MT 47, 450-462
   [word]
Buarque de Gusmao A 1955 Rv Min Eng
   18(65), 19-23  (MR17, 2)
*Delachet André 1949 *L'Analyse*

*Mathématique*, Paris, 119 p
   (MR11, 150) [continuity, function]
Fichera G 1949 Iasi IP 4, 63-107
   (MR20, 264)
Gelfond A O 1957 UMN 12(4), 29-39
   (Z99, 4) [Euler]
*Gerhardt C F 1855 *Die Geschichte der hoheren Analysis*, Halle
Hagen J G 1900 AMSB (2)6, 381-390
*Hobson E W 1926-1927 *Theory of Functions of a Real Variable...*,
   (Cambridge U P), 2 vols
   (repr 1957 Dover)
Kowalewski G 1910 *Die Klassischen Probleme der Analysis des unendlichen*, Leipzig
   [Bernoullis, Euler, series]
Lepage T 1932 Brx U Rv 37, 376-384
   (F57, 1287)  [XIX]
Mahnke D 1925 Ber Ab (1), 1-64
   (F52, 20)
Manheim Jerome H 1964 *The Genesis of Point Set Topology* (Pergamon),
   166 p
Mansion P 1899 *Mélanges Mathématiques II*, Paris
*Novy L 1964 Int Con HS 10    [XIX]
Picard E 1905 *Sur le developpement de l'analyse et ses rapport avec diverses sciences*, Paris
   (also in BSM (2)28, 267-278, 282-295; Sci (NS)20, 857-872; AMSB
   (2)11, 404-426  (F36, 56); M Gaz
   3, 193-201, 217-228  (F35, 58);
   Howard J Rogers, *Congress of Arts and Sciences*, NY, 497-517)
*Rybnikov K A 1954 Ist M Isl 7,
   643-665  (Z59:1, 12, 60, 12)
Schlesinger L 1912 Gott N, 1-140
   (F43, 23)  [Gauss]
Tropfke Johannes 1924 *Geschichte der Elementar-Mathematik in Systematisches Darstellung...*, vol. 6,
   Berlin, 173 p
Volterra V 1900 Int Con 2, 43-57
   [Betti, Brioschi, Casorate]
*Whittaker E T + 1927 *A Course of Modern Analysis* (Cambridge U P)
   (repr 1965)
Yushkevich A P 1959 in Schroeder K
   ed, *Sammelband der zu Ehren des*

Analysis    (continued)

*250 Geburtstages des Deutschen Akademie der Wissenschaften zu Berlin...,* Berlin, 224-244 (MR23A, 271)   [Euler, Lagrange]

Analytic geometry

See also Algebraic geometry, Coordinates, Solid geometry.

Baltzer R 1865 Leip B 17, 5-6
*- 1882 *Analytische Geometrie,* Leipzig
Berenguer P A 1895 Prog M 5, 116-121   [XVII, Hugo de Omerique]
Bortolotti E 1923 *Lezioni di geometria analytica,* Vol I [XVII, Italy]
- 1924 Bln Rn 29, 90-125 [Bonasoni]
*- 1930 Pe M (4)10, 89-
Bosmans H 1906 Mathesis (3)6, 260-264
Bouligand G 1951 Arc In HS (NS)4, 884-896  (MR13, 611)
Boyer Carl B 1944 MT 37, 99-105 [Descartes, Fermat]
- 1947 Scr 13, 133-153 (MR9, 485; Z33, 50)   [Fermat, Lacroix]
- 1949 Sci Am 180(Jan), 40-45
- 1951 Scr 17, 32-54, 209-230
- 1953 Scr 19, 230-238  (MR15, 591)
- 1954 Scr 19, 97-108  (Z52, 2)
- 1954 Scr 20, 30-36, 143-154 (Z55, 243)   [Greek]
- 1955 Scr 21, 101-135 (MR17, 337; Z66, 4)   [Descartes]
*- 1956 *History of Analytic Geometry,* NY, 300 p  (Scr Stu 6-7) (MR18, 368; Z73, 2)  [*bib]
- 1959 AMM 66, 390-393 (MR21, 750)   [Descartes]
- 1965 MT 58, 33-36  [Hudde, Lahire]
Bunt Lucas + 1963 *Van Ahmes tot Euclides,* 4th ed, Groningen [Greek]
Chasles M 1838 Cor M Ph (NS)10, 1-20  [Euclid porisms]
Conte L 1947 Pe M (4)25, 1-15, 165-180  (Z29, 289)

- 1948 Pe M (4)26, 133-152 (Z32, 97)
Coolidge J L 1936 Osiris 1, 231-250 (Z14, 147)   [origin]
Cournot A A 1847 *De l'origine et des limites de la correspondance entre l'algèbre et la geométrie,* Paris
Das S R 1928 AMM 35, 535-540 [coord, Hindu]
Dassen C C 1930 Arg SC 110, 270-272 (F56, 6)
De Vries H 1948 Scr 14, 5-15 (MR10, 174; Z33, 337)   [XIX]
Eneström G 1911 Bib M (3)11, 241-243 [vs Oresme]
- 1915 Bib M 14, 354   [Viete]
Frajese A 1940 It UM (2)2, 363-365 (F66, 12)   [Fibonacci]
Funkhouser H G 1926 Osiris I, 260-262 [mid]
- 1937 Osiris 3, 269-404  (F63, 796) [bib, statistics]
Gagnebin S 1950 Gesnerus 7, 105-120
Gelcich E 1882 Ab GM 4, 191-232 [Rogusaer]
Gerhardt C I 1856 *Etudes historiques sur l'arithmétique de position,* Berlin
Gloden A 1948 Lux SNBM (NS)42, 3-5 (MR11, 150)   [origin]
Greenwood T 1948 Can Rv Tri 34, 166-179  (MR10, 174)   [origin]
Guenther S 1877 Boncomp 10, 363-406 (F9, 22)   [Greek]
- 1877 Nurnb NG 6, 3-50  (F9, 22)
Ionescu I 1930 Gaz M 36, 289-291 (F56, 23)   [Rumania]
Karpinski L C 1933 SSM 33, 34-39
*- 1937 Isis 27, 46-52   (F63, 8) [Greek, philosophy]
*Klein F 1926 *Vorlesungen ueber hoehere geometrie,* 3rd ed, Berlin [XIX]
Krazer A 1915 DMV 24, 340-363 (F45, 76)   [fun]
Lattin H 1948 Isis 38, 205-225 [XI, coord, logic]
Longcaire de 1898 Intermed 5, 273 [axes]
Loria Gino 1902 Monist 13, 80-102, 218-234   [Greek, axes]

## Angle     (continued)

Cajori F 1923 AMM 30, 65-66
  (F49, 20) [notation, units]
Crathorn A R 1912 AMM 19, 166
  ["radian"]
Gandz S 1929 Isis 12, 452-481
  [Babylonian, Hebrew]
Guentert H 1928 Wort Sach 11, 124-
  142  [anc]
Hultsch F 1899 ZM Ph 44(S), 193
  [Hipparchus]
Itard J 1964 Koyre 1, 346-360
*Jones P S 1953 MT 46, 419-426
Lemaire G + 1911-1912 Intermed
  18, 50; 19, 256-257 ["radian"]
Meringer R 1928 Wort Sach 11, 114-123
Siderski D 1929 Rv Assyr 26, 31-32
  (F59, 834)
Stohler H 1951 Elem M 6, 84-86
  (Z42, 241) [anc]
Thureau-Dangin F 1931 Rv Assyr 28,
  23-25 [Arab]
- 1931 Rv Assyr 28, 111-114
  [ast, Babylonia]
Vacca G 1902 Bib M (3)3, 191
  [spher trig]


## Antiparallel

James W J 1890 Nat 41, 10  [word]
Langley E M 1889 Nat 40, 460-461
  [word]
- 1890 Nat 41, 104-105  [word]


## Apollonius problem

*Ahrens J T 1832 *Apollonisches Problem,*
  Augsburg
Altshiller N 1915 AMM 22, 261, 304
Court N A 1961 MT 54, 444-452
Coxeter H S M 1968 AMM 75, 5-15
Milne J J 1931 M Gaz 15, 142-144
Simon Max 1906 *Ueber die Entwicklung*
  *der Elementar-Geometrie im*
  *XIX Jahrhundert, 97-105*  [bib]
Study E 1897 M Ann 69, 497-542
  [survey]
Tibiletti C 1946 Pe M (4)24, 100-111

  (MR8, 305) [Pappus]
- 1946 Pe M (4)24, 152-161
  (MR8, 497) [Gaultier]
- 1947 Pe M (4)25, 16-29
  (MR9, 169) [Newton, Pluecker,
  Viete]


## Approximation

See also Equations, Interpolation,
  Numerical analysis, Pi, Roots,
  Weierstrass approximation theorem.

Aaboe A 1954 Scr 20, 24-29
  [al-Kashi, trig]
Aubry A 1916 Porto Ac 11, 5-35
Cajori F 1910 Colo CPS (Science
  Series) 12(7), 171-286  (F41, 1052)
  [eq]
Fujiwara M 1939 Jp Ac P 15, 114-115
  (Z22, 3) [Japanese]
Guenther S 1879 Prag Ab (6)9
  (F11, 37) [anc]
Hofmann J E 1942 Deu M 6, 453-461
  (MR5, 57) [roots]
- 1950 Centau 1, 24-37 [Gregory]
Hunrath K 1905 Bib M (3)6, 249
  [Duerer]
Kennedy E S + 1956 AMM 63, 80-83
  (Z72, 244) [mid]
Koppe M 1901 Bib M (3)2, 224
  [Huygens]
Kowalewski A 1917 *Newton, Cotes,*
  *Gauss, Jacobi.  Vier grundlegende*
  *Abhandlungen ueber Interpolation*
  *und genaeherte Quadratur (1711,*
  *1722, 1814, 1826),* Leipzig, 110 p
  (F46, 38)
Lemaire G 1906 Intermed 13, 237
  [eq]
Luria S J 1934 Len IINT 4, 21-46
  [Greek]
Mikami Y 1913 Porto Ac 8(4), 210-216
  [Aida Ammei]
- 1914 Toh MJ 5, 176-179
  [Japan, Miyai Antai]
Miller G A 1939 M Stu 6, 137-142
  (F65, 2)
Mueller C 1932 QSGM (B)2, 281-285
  (Z6, 2) [Archimedes, roots]

Approximation    (continued)

Neugebauer O 1931 Arc Or 7, 90-99
  [Babylonian, roots]
*Nordgaard M A 1922 *A Historical
  Survey of Algebraic Methods of
  Approximating the Roots of
  Numerical Higher Equations up to
  the Year 1819,* NY, 70 p
Pressland A J 1891 Edi MSP, 23-24
  [geom]
Rodet L 1879 Fr SMB 7, 159-167
  (F11, 37) [anc]
- 1879 Fr SMB 7, 98-102  (F11, 36)
  [anc, roots]
Rogachenko V F 1953 Ist M Isl 60,
  477-494  [eq, Lobachevsky]
Schoy C 1926 AMM 33, 95-96
  [al-Biruni, trig]
Stamatis E S 1961 Platon 13
  (25/26), 318-325  (Z107, 243)
  [Diophantus]
Vahlen T 1911 *Konstruktionen und
  Approximationen* (Teubner), 285 p
Vaux C de 1898 Bib M 12, 1-2
  [Musa]
Vogel K 1932 DMV 14, 152-158
  [Archimedes, roots]
Weissenborn H 1883 Hist Abt 28, 81-
  99 [Archimedes, roots]


Approximation of functions

Gusak A A 1961 Ist M Isl 14, 289-348
  (Z117:2, 248)
Lorentz G G 1966 *Approximation of
  Functions,* NY, 188 p    [bib]


Arc length.  See Curve, Length.


Archimedean and non-Archimedean

See also Horn angles, Infinitesimals,
  Non-standard Analysis.

Graves L M 1956 MT 52, 72-77
  [ax on non-A]
Hjelmslev J 1950 Centau 1, 2-11
  [Eudoxos]

Kagan V F 1907 Kagan (439-441)
Kasner E + 1950 Scr 16,  13-21
  (MR12, 312)  [Voltaire]
Levi B 1942 M Notae 2, 109-141
  (MR4, 65; Z60, 5)
Robinson Abraham 1966 *Non-Standard
  Analysis*  (North-Holland)
Stolz O 1882 Innsbr B 12
- 1883 M Ann 22, 504-520  [geom]
Veronese G 1905 Linc Rn (5)14(1),
  347  [non-Arch geom]


Archimedes cattle problem

Archibald R C 1918 AMM 25, 411-414
Bell A H 1882 M Mag 1, 163
- 1895 AMM 2, 140-141
Dickson Leonard E 1919 *History of
  the Theory of Numbers,* Washington
*  2, Ch 12
*Doerrie Heinrich 1965 *100 great
  problems of elementary mathematics*
  (Dover), 3-7  [bib]
Krummbiegel B + 1880 Hist Abt 25,
  121-136, 153-171  (F12, 30)
Merriman M 1905 Pop Sci Mo 67, 660
Mita H 1951 JHS (18), 16-28
Vincent 1856 Nou An 15


Area

See also Integration, Lunes,
  Mensuration, Oriented areas,
  Squaring the circle, Surface area.

Archibald R C 1922 AMM 29, 29-36
  [quadril]
Artom E 1925 Pe M (4)5, 88-106,
  280 [conic]
Becker O 1954 Philolog 98, 313-316
  [lune]
Boyer C B 1954 MT 47, 36-37
  [parab]
Bruins E M 1957 Janus 46, 4-11
  [Babyl]
Carruccio E 1938 Pe M (4)18, 1-32
  (F64, 915)  [Newton]
Chakravarti G 1934 Clct Let 24(7a),
  19-22  (F62, 1024)

Area    (continued)

Coolidge J L 1939 AMM 46, 345-347
  (Z22, 2) [quadril]
Eneström G 1887 Bib M (2)1, 32, 64
  [Neil, Van Heuraet]
Fuije A de la 1930 Rv Assyr 27, 65-71
  (F59, 834)
Gandz S 1927 AMM 34, 80-86
  (F53, 8) [words]
Hadamard J 1896 Bord PV, 25
Hofmann J E 1893 M Mag 12
  (F64, 16) [Artus de Lione,
  circle sq]
- 1953 Arc In HS 32, 193-208
  [Porta]
- 1953 MP Semb 3, 59-79
  (Z50, 1) [hyepb, parab]
- 1954 Arc In HS 26, 16-34
  (Z56, 243) [Porta]
Hofmann J 1965 Fors Fort 39, 228-235
  [XVII, calc]
Hofmann J + 1941 Deu M 5, 571-584
  (Z25, 2) [cissoid]
Itard J 1955 Ci Tecnol 5(17), 53-64
  [spiral]
Loerchner H 1912 Ost Bau 18, 770-775,
  793-799, 812-816  (F43, 76)
Loria G 1895 Intermed 2, 188
  [quadril, tria]
Loria G + 1901 Intermed 8, 30
  [cycloid]
Miller G A + 1921-1922 AMM 28, 256-
  258; 29, 303-307 [triag]
Natucci A 1927 Pe M (4)7, 242-256
  [Alberti]
Schoy C 1925 Isis 7, 5-8 [al-Lith]
Stamatis E S 1946 Archimedes'
  Quadrature of the Parabola,
  Athens, 46 p
Suter H 1911 Bib M (3)12, 289-332
  [Alhazen]
Thureau-Dangin F 1932 Rv Assyr 29,
  26-28 (Z5, 1) [circle segment]
Zeuthen H G 1895 Dan Ov 1(2),
  37, 193, 257 [cal, Fermat]

Arithmetic

See also Numeration (Ch. 3),
  Numerical analysis, Multiplication

(and other operations), Alligation,
  Duplication, Finance (Ch. 3),
  Finger reckoning (Ch. 3), Fractions,
  Percent, Tables (Ch. 3).

Arithmetic: Bibliographies

Anon 1889 Elenco cronologico delle
  opere di computisteria e ragioneria
  venute alla luce in Italia dal
  1202, 4th ed Rome, 291 p  [*bib]
De Morgan Augustus 1847 Arithmetical
  Books, From the Invention of
  Printing to the Present Time...,
  London, 124 p (repr 1966) [*bib]
Smith D E 1908 Rara Arithmetika
  Boston, 507 p    [*bib]
*- 1939 Addenda to Rara Arithmetice,
  Boston

Arithmetic:  General

Adam W 1891 Geschichte des Rechnens
  and des Rechenunterrichts. Zum
  Gebrauch an gehobenen und hoeheren
  Lehranstalten, sowie auch bei der
  Vorbereitung auf die Mittelschul-
  lehrer- und Rectoratspruefung
  bearbeitet, Quedlinburg, 190 p
  (F23, 35)
Andreer N 1908 Ped Sb (1), 31-48
  (F39, 54)
Bellyustin V 1907  Kak postepenno
  doshli lyudi do nastoyashchei
  arifmetiki, Mos,  206 p
  (F39, 54)  (also 1941, 119 p)
Bergold L 1881 Arithmetik und
  Algebra, nebst einer Geschichte
  dieser Disciplinen, Karlsruhe
  (F13, 26)
Bobynin V V 1893 Fiz M Nauk 11, 1-30
  (F25, 61)
Boncompagni B 1862-63 Vat NLA 16, 1-,
  101-, 300-, 389-, 503-
Bortolotti E 1941 Bln Mm (9)8, 3-11
  (Z61, 2) [alg]
Brocard H 1908 Intermed 15, 56
  (Q1658) (See also Intermed 6, 243;
  11, 145-148; 12, 16)
Brun Viggo 1962 Regnekunsten i det

Arithmetic: General   (continued)

*gamle Norge, fra arilds tid til
Abel,* Oslo, 125 p   (Engl summ)
[Norway to XIX]
Bubnov N M 1908 *Arifmeticheskaya
samostoyatelnost evropeiskoi
kultury,* Kiev, 418 p   (F39, 53)
- 1914 *Arithmetische selbstaendigkeit
des europaeischen kultur,* Berlin,
293 p   (F45, 64)
Buchanan H E 1928 M Mag 2, 6-11
Buckingham B R 1956 Ari T 2, 1-5
Burkhardt H 1905 ZMN Unt 36, 9-20
(F36, 50)
Burchuladze G 1948 *Vozniknovenie i
razvitie sovremennoi arifmetiki,*
Tiblis, 99 p (in Gruzin)
Carnahan W 1947 SSM 46, 209-213
Cehakaja D G 1949 Tbil MI 17, 315-
339   (Z39, 242) [alg, Near East]
Cossali Pietro 1857 in Boncompagni
B ed, *Scritti inedite del P.D.
Pietro Cossali...,* Rome, 317-398
Cunningham S 1904 *The Story of
Arithmetic...,* London, 256 p
(F35, 71)
Dadic Z 1958 Zagr Gl (2)13, 281-286
(Z84, 2) [Silobod, Zoricic]
De Morgan A 1851 Brt Alm C
Depman I Ya 1959 *Istoriya arifmetiki,*
Moscow   [bib, chron]
Ferrari G 1875 Lom Gen 8, 5-, 227-,
289-, 624-, 1006-
- 1876 Lom Gen 9, 199-
Gerhardt 1856 *Etudes historiques
sur l'arithmétique de position,*
Berlin
Karpinski L C 1925 *The History of
Arithmetic,* Chicago, 200 p
(F51, 5)
- 1933 Scr 2, 34-40   (Z8, 99)
[ed]
Klimpert R 1885 *Kurzgefasste
Geschichte der Arithmetik und
Algebra. Eine Ergaenzung zu jedem
Lehrbuche der Arithmetik und
Algebra,* Hannover, (F17, 24)
Krysicki W 1958 *Wie man ehemals
rechnete und wie man jetzt
rechnet,* Warsaw, 111 p (Z82, 7)
Lemoine E + 1899-1900 Intermed 6,

207; 7, 137   (Q1477)
[problem collections]
Leslie John 1820 *Philosophy of
Arithmetic,* Edinburgh [unreliable]
Loria G 1920 Scientia 28, 77-93
[devices]
Lupton S 1910 M Gaz 5, 273-279
[long calculations]
Mueller F J 1964 Ari T 11, 386-390
Myers G W 1927 MT 20, 93-100
[re phil]
Natucci A 1924-1926 Rass MF 4, 161-
174; 5, 183-198; 6, 25-33
(F50, 10; 51, 8) [alg]
- 1955 *Suiluppo storico dell'
arithmetica generale e dell'algebra,*
Naples, 368 p [alg]
*Peacock George 1826 Enc Met
*- 1845 Enc Met
Popov L 1873 *Ocherk razvitiya
arifmetiki,* Kazan, 72 p
Reves G E 1952 SSM 51, 611-617
Sanford V 1951-1953 MT 43, 292-294,
368-370; 44, 29-30, 135-137;
45, 198, 204
Seidenberg A 1960 Calif Pu 3
[counting]
- 1961 Arc HES 2, 1-40   [counting]
Sergescu P 1932 Cuget Clar (11-12),
7 p
Smeur A J E M 1960 *De Zestien de-
eeuwse Nederlandse Rekenboeken,*
The Hague, 175 p [bib]
Sterner M 1891 *Geschichte der
Rechenkunst,* Munich-Leipzig, 556 p
Stiattesi A 1870 Boncomp 3, 389-408
(F2, 29)
Tannery J + 1904 Enc SM 1(1)(1), 1-62
Taton R 1946 *Histoire du calcul,*
Paris, 128 p
Thurion J 1885 *Histoire de l'arith-
métique,* Bruxelles, 172 p
*Tropfke Johannes 1921-1922 *Geschichte
der Elementar-Mathematik,*
Leipzig, vols 1 and 2   (3rd ed
1930, 1933 Berlin)   (F59, 3)
See Isis 5, 182-186
Unger F 1888 *Die Methodik der
praktischen Arithmetik in histor-
ischer Entwicklung vom Ausgange des
Mittelalters bis auf die Gegenwart*

Arithmetic: General   (continued)

nach den Originalquellen bear-
beitet, Leipzig, 252 p
Vetter Q 1928 Z Minulosti proctar-
skych zakladu, Brno, 16 p
(F54, 44) [foundations]
Villicus F 1897 Die Geschichte der
Rechenkunst von Altertum bis zum
XVIII Jahrhundert, 3rd ed, Vienna,
122 p    (F28, 53)
Wieleitner Heinrich 1912 Die Sieben
Rechnungsarten mit allgemeinen
zahlen, Leipzig, 74 p
- 1927 Rechnen und Algebra,
Berlin, 83 p [source book]
Wilt M L 1953 MT 46, 475-477
Yusupov N 1932 Ocherki po istorii
razvitiya arifmetiki na
Blizhnem Vostoke,   Kazan, 177 p
Zepf K 1907 Grundzuege des Geschichte
des Rechnens nebst kurzen
Erlaeuterung verschiedener
Rechnungsarten der Kulturvoelker
alter und neuer Zeit, Karlsruhe
102 p


Arithmetic: Ancient

Bobynin V V 1894 Bib M (2)8, 55-60
(F25, 63)
- 1895 Fiz M Nauk 12, 97-110
(F26, 48) [operations]
Bruins E M 1949 Amst P 52, 161-163
(Z32, 49)
- 1949 Euc Gron 24, 169-185
(Z34, 145)
Buzzi O 1900 Bo Bologl, 19, 51, 114,
168, 209, 330, 362
Conant L L 1906 SSM 5, 385-394
Dagobert E B 1955 MT 48, 557-559
De Morgan A 1850 Brt Alm C
Drieberg F v 1819 Die Arithmetik der
Griechen, Leipzig
Fazzari G 1903 Pitagora 10, 49-54
[alg]
Fettweis Ewald 1923 Wie men einstens
rechnete, Leipzig, 56 p (F49, 18)
(Russ tr 1925)
- 1927 Das Rechnen der Naturvoelker,
Leipzig-Berlin, 100 p

- 1953 Scientia 88, 235-249
Filon L N G 1925 M Gaz 12, 401-414
(F51, 6)
Freudenthal H 1948 Euc Gron 24, 12-34
(Z31, 337)
Gilain O 1827 La science egytienne:
l'arithmétique au Moyen Empire,
Paris
Gillet 1903 Intermed 10, 272
[Greek, Roman]
*Hawkes H E 1898 AMSB (2)4, 530-535
(F29, 27)   [Greek]
Loria G 1939 Porto Fac 24, 3-7
(MR1, 33)  [Greek]
Neugebauer O 1930 Arc GMNT 13, 92-99
(F56, 83)  [Egypt]
- 1930 QSGM (B)1(3), 301-380
[Egypt]
Raum O F + 1938 Arithmetic in Africa,
London (Evans Bros), 94 p
Reidemeister Kurt 1940 Die Arithmetik
der Griechen, Leipzig-Berlin, 32 p
(Z25, 145)
Richardson L J 1916 AMM 23, 7-14
(F46, 51)
Robbins F E 1929 Clas Phil 24, 321-
328 (F56, 804) [Egypt, probls]
- 1934 Isis 22, 95-103  (F60, 824)
[Egypt, Greek, probls]
Sanchez Perez J A 1943 La aritmetica
en Babilonia y Egipta, Madrid,
72 p (Z60, 3)
- 1946 Arithmetia in Greece, Madrid,
260 p (Z60, 5) (in Spanish)
- 1949 Arithmetic in Rome, India,
and Arabia, Madrid-Granada, 263 p
(in Spanish) (Z41, 338)
Smith D E 1910 Arc GNT 1, 301-309
(F41, 54)
Stamm E 1935 Archeion 17, 149-170
(F61, 940)
Tannery P 1905 Bib M (3)6, 225-229
(F36, 50)  [Greek]
Vogel Kurt 1929 Die Grundlagen der
aegyptischen Arithmetik in ihrem
zusammenhang mit der zin-tabelle
des Papyrus Rhind, Munich, 217 p
- 1936 Mun Si, 357-457  [Greek]
Vygodskii M Ya 1967 Arifmetika i
algebra v drevnem mire, 2nd ed,
Mos, 367 p (MR37, 1215)

Arithmetic: Medieval

Beaujouan G 1947 Par EC, 17-22
(Z31, 2)
- 1948 Rv Hi Sc Ap 1, 301-313
(MR10, 174) [X-XII]
- 1954 in Homenaje a Millas-Valli-
crosa, Barcelona 1, 93-124
Benedict Susan R 1914 A Comparative
Study of the Early Treatises
Introducing into Europe the
Hindu Art of Reckoning, Concord,
132 p
Carra de Vaux B 1899 Bib M, 33
[Arab]
Datta B 1928 Beng Asi J 23, 261-267
(F54, 46) [Hindu, checks]
Eneström G 1890 Bib M 4, 32 (Q29)
[tables]
Friedlein Gottfried 1864-1865
ZM Ph 9, 297; 10, 241-
- 1869 Die Zahlzeichen und das
elementare Rechnen der Griechen und
Roemer und des christlichen
Abendlandes vom 7 bis 13 Jahr-
hundert, Erlangen
Ganguli S 1927 AMM 34, 409-415
[Aryabhata]
Goodstein R L 1956 M Gaz 40, 114-129
(Z71, 1)
Houghtaling A E + 1925 MT 19, 179-
183 [France]
Karpinski L C 1930 Bo Fir 170-177
[XIV, Italy]
Larney B M 1932 SSM 31, 919-930
Luckey Paul 1951 Die Rechenkunst bei
Gamsid b Masud al-Kasi mit
Rueckblicken auf die aeltere
Geschichte des Rechnens, Wiesbaden,
150 p (Z44, 242)
Massignon L 1932 Archeion 14, 370-
371 (F58, 15) [Arab]
Mortet V 1908 Bib M (3)9, 55-64
[France]
Rath E 1912 Bib M 13, 17-22
[XV, Germany]
- 1914 Bib M (3)14, 244-248
[XV, Germany]
Sarton G 1933 Isis 20, 260-262
[Arab, finance]
*Schrader D V 1967 MT 60, 264 [univ]

Sleight E R 1940 MT 32, 243-248
[Engl]
- 1942 M Mag 16, 198-215, 243-251
(MR3, 258) [Engl]
*Spasskii I G 1952 Ist M Isl 5, 269-
420 [Russia]
Suter H 1911 Bib M (3)11, 100-120
[X, Ahmad]
Szily C v + 1895? Die Arithmetik des
Magisters Georgius de Hungaria aus
dem Jahre 1499, Berlin
Thorndike L 1926 AMM 33, 24-28
(F52, 13) [XV, Adam]
Unger F 1888 Die Methodik der prak-
tishchen Arithmetik in Historischer
Entwicklung vom Ausgange des
Mittelalters bis auf die Gegenwart,
Leipzig, 252 p
Vogel K 1954 Die Practica des
Algorismus Ratisbonensis, Munich,
294 p (Z57, 2)
Waters E G R 1929 Isis 12, 193-236
[France]
Yeldham Florence A 1926 The Story of
Reckoning in the Middle Ages,
London, 96 p
Zapilloni M T 1928 Pe M 8, 175-184
[Campanus]

Arithmetic: Modern

Anon 1884 Rv Sc 1, 561-563
[XVI, Roche]
Bockstaele P 1960 Isis 51, 315-321
(MR22, 769; Z96, 2) [ed, Engl,
Holland]
Bradley A D 1934 Scr 2, 235-241
[ms, 1727]
Cajori F 1927 Isis 9, 391-401
[America]
Davis N Z 1960 JH Ideas 21, 18-48
[finance]
DeGrodte H L V 1960 Sci Hist 2, 161-
172 [Holland]
Flett T M 1961 M Gaz 45, 1-8
[XIX, ed]
Grosse H 1901 Historische Rechen-
buecher des 16 und 17 Jahrhunderts,
Leipzig (repr 1965) [abacus]
Hallema A 1924 Euc Gron 1, 161-193
[XVII]

## Arithmetic: Modern  (continued)

Karpinsky L C 1912 Bib M (3)13, 223-228  (F43, 67)  [XVI]
- 1929 Archeion 11, 331-335 (F55, 10)  [Italy, XVI]
- 1936 Osiris 1, 411-420  (F62, 20) [Climent, Spain]
Keefer 1915 MN Bl 12, 137-140 (F45, 68)  [Germany]
Lady C H 1947 MT 40, 38
Perott J 1882 Boncomp 15, 163-170 [XVI, Spain]
Popa I 1955 Iasi M 6(1/2), 115-121 (Z67, 246)  [Moldavia]
Read C B 1959 MT 52, 366-367 [probl]
- 1961 MT 54, 361-363  [probls]
Richeson A W 1947 Isis 37, 47-56 (Z29, 194)  [Engl]
Shaw A A 1937 M Mag 11, 117-125 [alg, Armenia]
Shvetsov K I 1962 Ist M Zb 3 [XVII, Magnitskii]
Simons L G 1935 Scr 3, 94-96
Smeur A J E M 1960 Sci Hist 2, 22-23 [Raets]
Smith D E 1924 Isis 6, 311-331 [XV]
Steele R 1922 *The Earliest Arithmetics in English,* London, 102 p
Treutlein P 1877 ZM Ph 22(S), 1-100 (F9, 18)  [XVI]
Williamson R S 1928 M Gaz 14, 128-133 (F54, 32)  [XIX, ed]

## Arithmetic-geometric mean

Schlesinger L C 1911 DMV 20, 396-403  [Gauss]
- 1912 Gott N (1), 33p  (F43, 23) [Gauss]

## Associativity

Hammond J 1890 Mess M (2)20, 3-4 [Euclid]
Schafer R D 1955 AMSB 61, 469-484 [bib, non-asso]
- 1961 *An Introduction to Nonassocia-* *tive Algebra,* Stillwater, Oklahoma [bib]

## Automorphic functions

*Ford Lester R 1929 *Automorphic Functions,*  NY  (2nd ed 1951) [bib]
Fricke R 1904 Enc MW 2(2), 3
Klein Felix 1927 *Vorlesungen ueber Entwicklung der Mathematik im 19 Jahrhundert,* Berlin, Ch. 8
Kuznetsov V M 1966 Ist Met EN 5, 185-191  [modular functions]

## Axiom of choice

See also Continuum hypothesis.

Blumenthal L 1940 AMM 47, 346- [paradoxes]
Goedel K 1947 AMM 54, 515-525 [continuum hyp]
*Rubin H + 1963 *Equivalents of the Axiom of Choice,* Amsterdam  [bib]
Sierpinski W 1918 Krak BI, 97-152
- 1924 M Sbor 31, 94-128
Zlot W L 1960 Scr 25, 105-123 (MR22, 7934)

## Ballot theorems

*Takacs L 1967 *Combinatorial methods in the  theory of stochastic processes*

## Banach algebra

Goffman C 1962 MAA Stu 1, 171-180
*Wermer J 1961 Adv M 1(1), 51-

## Barycentric calculus

Agostini A 1951 It UM (3)6, 149-159 (MR13, 2)  [Torricelli]
- 1952 It UM (3)7, 321-327 [Leonardo da Vinci]

Barycentric calculus   (continued)

Allardice R E 1892 Edi MSP 10, 2-21
   (F24, 52)
De Vries H 1928 N Tijd 16, 201-227
Gibbs J W 1886 AAAS, 32 p  (F18, 52)
   (also in Sci Pap 2, 911-117)


Bernoulli numbers

*Ely G S 1883 AJM 5, 228-236  [bib]
Likhin V V 1959 Ist M Isl 12, 59-134
   (MR24A, 4)  [Russ]
- 1966 Ist Met EN 5, 35-44
   [bib, Euler, Lagrange]


Bernoulli series

Anon 1897 Pet B (5)6, 337


Bertrand curves

Dorwart H L 1928 AMM 35, 478-479
   (F54, 34)


Bertrand paradox

Czuber E 1908 Wahrscheinlichkeits-
   rechnung, Leipzig, 106-109


Bertrand postulate

Archibald R C 1945 Scr 11, 109-120
Dickson L E 1919 Dickson 1, Ch. 8
Matvievskaya G P 1961 Ist M Isl 14,
   285-288  (Z118:1, 8)  [Euler]


Bessel functions

Bocher M 1892 NYMS 2, 107-109
   [Euler]
Prasad G 1930-1932 A Treatise on
   Spherical Harmonics and the
   Functions of Bessel and Lamé,
   Benares, 2 vols

Rutgers J G 1907 Eenige beschouwingen
   over de Besselsche fucties,
   Alkmaar, 29 p  (F38, 62)


Big numbers

Jones P S 1952 MT 45, 528-530
   [big numbs]
Miller G A 1929 Sci 70, 282
   [big numbs]
Struyk A 1953 MT 46, 266-269


Binary systems

Bruck R H 1958 A Survey of Binary
   Systems (Springer)   [*bib]


Binomial coefficients

See also Figurate numbers.

Bosmans H 1906 Brx SS 31(1), 65-72
   [Pascal tria]
- 1923 Mathesis 37, 455-464
   (F49, 18)  [Pascal]
Boyer C B 1950 AMM 57, 387-390
   (MR12, 1)  [Cardan]
Cassina U 1923 Bo Fir 19, xxxiii-
   xxxix
   1961 in Dalla geometria egiziana
   alla matematica moderna, Rome,
   Ch. 2
Dickson L E 1919 Dickson 2, Ch. 1
Fontes 1893 Fr AAS 236-243
   (F25, 64)  [Forcadel, Pascal,
   Stifel]
*Gould H W 1959 Combinatorial
   Identities (US Defense Dept),
   100 p  [formula collection]
*- 1962 West Va AS 34, 158  [bib]


Binomial theorem

Aubry A 1899 JM El 24, 24, 39, 72,
   87 [before Newton]
- 1907 Mathesis (3)7, 62-66
   [Maskelyne thm]

Binomial theorem   (continued)

Bourgoin L 1874 Rv Sc 7, 453-
Burrow R 1790 Asi R 2, 487
  [Hindu]
Cajori F 1894 AMSB (2)1, 52-54
  (F25, 21)  [Newton]
Coolidge J L 1949 AMM 56, 147-157
  (MR10, 419; Z33, 242)
Delegue 1869 Dunkerq 14, 163-
  [Pascal]
Eneström G 1904 Bib M (3)5, 72-73
  (F35, 58)  [Pascal]
Godefroid 1898? Mathesis (2)9, 39
Hayashi T 1932 Toh MJ 35, 345-397
  [Japan]
Luckey P 1948 M Ann 120, 217-274
  [Arab, roots]
Marre A 1846 Nou An 5, 488-
Santalo L A 1942 M Notae 2, 61-72
  (MR4, 65)  [Newton]
Tannery P + 1896 Intermed 3, 98, 99,
  233 (Q615)
Tytler J 1820 Asi R 13, 457-
  [Arab]
Vacca G 1902 Loria 5, 1  [Harriot]
Whiteside D T 1961 M Gaz 45, 9-12
  [Briggs]
- 1961 M Gaz 45, 175-180
  [Newton]


Bisectors

Court N A 1953 Scr 19, 218-219
  [three bis probl]
Sister Mary Constantia 1964 MT 57,
  539-541  [Hopkins]


Bolzano function

Brzhetska V F 1949 UMN (NS)4(2), 15-
  21  (MR11, 572)
Kowlawski G 1923 Act M 44, 315-319


Boolean algebra

See also Logic, Lattice, Set theory.

Biermann K-R + 1958 JSL 23, 129-132
  [anc]
Carvallo M 1966 *Monographie des
  treillis et Algebre de Boole,*
  Paris, 2nd ed, 142 p  [bib]
*Flegg Graham 1964 *Boolean Algebra*
  (Blackie), 261 p  [*bib 1847-1964]


Brachistochrone

See also Cycloid, Tautochrone.

Albrecht B 1912 *Vom Problem der
  Brachistochrone. Eine geschicht-
  liche Skizze,* Frankfurt
  (F43, 82)
Carnahan W H 1948 SSM 47, 507-511
  [isochrone]
Compère C 1899 Lieg Mm (3)1, 128 p
  (F30, 56)
Kimball W S 1955 Pi Mu EJ 2(2), 55-77
Phillips J P 1967 MT 60, 506-508
  [cycloid, tautochrone]


Braids

Artin E 1950 Am Sc 38(1)
- 1959 MT 52, 328-333


Calculus

See also Analysis, Derivative,
  Integral, Limit, L'Hospital's rule,
  Maxima and minima, Mean value
  theorem, Partial derivative.

Agostini A 1930 Archeion 12, 33-37
  (F56, 18)  [Torricelli]
- 1951 Archim 3, 168-173  (MR13, 1)
  [Italy]
Anon 1939 AMM 46, 233  [club topic]
Aubry A 1899 Prog M (2)1, 129-137,
  164-169  (F30, 53)  [deriv]
- 1911 Porto Ac 6(2), 82-89  [XVII]
- 1912 Porto Ac 7, 160-185
  [before Descartes]
Babini J 1953 Imago Mun (1), 23-41

Calculus    (continued)

Bashmakova I G 1953 Ist M Isl 6,
    609-658 [Archimedes]
- 1964 Arc HES 2, 87-107
    (MR33, 5423) [Archimedes]
Bertrand J 1863 J Sav, 465-483
- 1864-1870 Traité de calcul
    différentiel et de calcul intégral,
    Paris, intro   (Russ tr 1912;
    F43  70)
Bieberbach L 1935 Z Gesm Nat 1, 171-
    177 (F61, 6; Z11, 385)
Bockstaele P P 1966 Janus 53, 1-16
    (MR34, 4090)  [XIX, Belgium,
    foundations]
Bohlmann G 1897 Deu NA Ver 69, 6
    [Euler, text]
- 1899 DMV 6, 91-110
    [bib; texts]
Bortolotti E 1921 Pe M (4)1, 263-
    276
- 1925 Arc Sto 6, 49-58, 139-152
- 1928 Pe M (4)8, 19-59  (F54, 27)
    [Torricelli]
- 1930 Archeion 12, 267-271
    (F56, 18) [Apollonius, Torricelli]
- 1938 It UM Con 1
    [infinitesimal, Torricelli]
- 1939 Bln Mm (9)6, 113-141
    (Z26, 289) [infinity, Italy,
    limit]
Bosmans H 1913 Brx SS 37(2)
    (F44, 60) [infinitesimal, Valerio]
- 1923 Mathesis 37, 12-18, 55-62,
    105-109  [Stevin]
- 1924 Fr AAS, 75-76  (F51, 39)
    [XVII]
- 1924 Mathesis 38, 250-256
    [Gregory St Vincent]
- 1927 Mathesis 41, 5-11  [Faille]
- 1929 Rv Q Sc (5)4, 136, 424
    [Pascal]
Boutroux Pierre 1914-1919 Les
    principes de l'analyse mathémati-
    ques, exposé historique et
    critique, Paris
Boyer C B 1946 MT 39, 159-167  [ed]
- 1947 MT 40, 267-275  [sine]
- 1948 Scr 16, 141-157, 221-258
- 1949 The Concepts of the Calculus,

A Critical and Historical
    Discussion..., NY
*- 1959 The History of the Calculus
    and Its Conceptual Development
    (Dover), 346 p (repr of 1949 with
    foreward by R Courant)
    (MR23A, 271; Z95, 3) [bib]
Brunschvicg L 1909 Rv Met Mor 17,
    309-356
Busulini B 1956 Ferra U An (NS)5, 59-
    67 (Z78, 241)
- 1957 Ferra U An (NS)7, 9-52
    (MR21, 750; Z92, 2)
Cajori Florian 1890 USO Ed CI 3,
    395-400 [bib, textbooks, US]
- 1917 AMM 24, 145-154
    [Berkeley, Woodhouse]
- 1917 MT 24, 146-151  [Barrow,
    Hobbes, Wallis]
- 1919 A History of the Conceptions
    of Limits and Fluxions in Great
    Britain from Newton to Woodhouse,
    Chic-Lon, 308 p
- 1919 AMM 26, 15-20  (F47, 32)
    [Leibniz]
- 1921 AMSB 27, 453-458 [notation]
- 1922 AMSB 28, 45 (F48, 45)
- 1922 AMSB 28, 95 (F48, 45)
    [notation]
- 1923 An M (2)25, 1-46 [notation]
- 1923 AMM 30, 223-234  (F49, 20)
    [Leibniz, limit]
- 1926 M Gaz 13, 252  [Chatelet]
Cantor M 1901 Int Con Ph 3, 3-25
Carruccio E 1957 It UM (3)12, 298-
    307  (Z77, 8)  [Cauchy]
Caspar M 1932 Unt M 38, 227-229
    [Kepler]
*Castelnuovo G 1938 Le origini del
    calcolo infinitesimale nell'era
    moderna,  Bologna, 164 p
    (F64, 5; Z21, 2)
*- 1962 Le origini del calcolo
    infinitesimale nell'era moderna,
    con scritti di Newton, Leibniz,
    Torricelli,  Milan, 236 p
    (MR34, 2409)
Child J M 1916 Monist 26, 577-629;
    27, 238-294, 411-454 [Leibniz, mss]
- 1930 Sci Prog 25, 295-307 (F57,
    1302)  [Barrow, Leibniz, Newton]

Calculus (continued)

Christensen S A 1933 M Tid A, 89-105
(F60, 817)
Clarke F M 1934 Scr 2, 155-160
Cohen H 1883 *Das Princip der
Infinitesimalmethode und seine
Geschichte*, Berlin
Cohn J 1896 *Geschichte des Unend-
lichkeitsproblems in abenlaen-
dischen Denken bis Kant*, Leipzig
Cowan R W + 1945 Scr 13, 123-127
[Fermat]
Crommelin C A 1956 Sim Stev 31, 5-18
(MR18, 268) [curves, Huygens]
De Morgan A 1952 Brt Alm C
[Leibniz, Newton]
De Waard C 1918 BSM (2)42, 157-177,
327-328 [Beaugrand, Descartes,
Fermat]
Dickstein S 1899 Ab GM 9, 65-79
[Lagrange]
- 1899 Prace MF 10, 178-192
(F30, 47) [Lagrange]
- 1898 Bib M (2)12, 50-52 (F29, 30)
- 1908 Bib M (3)9, 200-205 [trig]
- 1908 Bib M (3)9, 309-320
(F39, 60) [Leibniz]
- 1910-1911 Bib M (3)11, 276; 12,
268 (F41, 60) [Newton]
Fellman E A 1957 Schw NG 137, 57-61
*Fleckenstein J O 1950 Arc In HS 29,
542-555 (MR12, 69; Z38, 2)
Fort Osmar ·1846 *Andeutungen zur
Geschichte der Differential-
Rechnung*, Dresden
Friedlein G 1871 ZM Ph 16, 42-, 253-
[Victorius of Aquitania]
Gabba A 1825 Brescia, 89-
Gerhardt C I 1840 *Historische
Entwickelung des Princips der
Differentialrechnung bis auf
Leibniz*, Halle
- 1846 *Historia et origo calcuii

differentialis a Leibnizio con-
scripta*, Hannover
- 1848 *Die Entdeckung der Differen-
tialrechnung durch Leibnitz*,
Halle
- 1855 *Die Geschichte der hoeheren
Analysis, Erste Abt*, Halle,
163 p (all published)
- 1856 Arc MP 27, 125-132
Geymonat Ludovico 1947 *Storia e
filosofia dell'analisi infinitesi-
male*, Torino 352 p
Giesel F 1868 *Entstehung des Newton-
Leibnizschen Prioritaetsstreites
...*, Delitzsch
Ginsburg B 1936 Isis 25, 340-362
[Jordanus]
Giovannozzi P G 1919 Arc Sto 1, 137-
140 (F47, 5) [Fermat]
Gloden A 1966 Benelux H S 3, 190-194
[Holland]
Goldbeck E 1902 Bib M (3)3, 84-112
[Galileo, indivisibles]
Graves G H 1911 MT 3, 82-89
Grunsky H 1960 MN Unt 13, 1-12
(Z112, 2)
Guallart E 1895 Prog M 5, 73-85
Guili G de 1931 Scientia 49, 207-220
[Galileo, Descartes]
Gunther S 1888 Bib M (NS)2, 81-87
[Kepler, Pappus]
Hadamard J 1954 Bras Ac 26, 19-23
(Z55, 4)
Hathaway A S 1904 Indn NAS (A), 237-
240 [Newton]
- 1919 Sci 50, 41-43 [XVII]
- 1920 Sci 51, 166-167
Hofmann J E 1950 MP Semb 1, 220-225
(MR11, 707) [XVII]
- 1956 Ens M (2)2, 61-71
(MR18, 268) [Jacob Bernoulli]
- 1964 Arc HES 2, 271-343
(MR34, 7320) [XVII]
*- 1966 Ost Si M 2, 175, 209-254
[catenoid, cycloid, Leibniz]
Hoppe E 1928 DMV 37, 148-187
[Leibniz, Newton]
Itard J 1948 Arc In HS 27, 589-610
(MR10, 175) [Fermat]
- 1950 Rv Hi Sc Ap 3, 210-213
[Archimedes, Euclid]

Duhamel J M C 1864 Par Mm 32, 269-
330 [Descartes, Fermat, max, tan]
Eneström Gustaf 1878 *Differens-
kalylens Historia I*, Upsala
(F10, 27)
- 1894 Sv Ofv 51, 297-305
(F25, 66)

Calculus    (continued)

Jacoli F 1875 Boncomp 8, 265-304
    [Roberval, Torricelli]
Jourdain P E B 1899 Nat 60, 245
    (F30, 48) [founds]
Kaestner A G 1796-1800 *Geschichte
    der Mathematik seit der Wiederher-
    stellung der Wissenschaft bis
    das Ende des achtzehnten Jahr-
    hunderts,* 4 vols
Kolman E 1932 Int Con (1932) 2,
    349-351 (F58, 52) [Marx]
Kondo Y 1951 JHS (19);(20)
Kowalewski G 1919 *Einfuehrung in de
    Infinitesimalrechnung,* Leipzig-
    Berlin, 100 p
Kramar F D 1950 Ist M Isl 3, 486-508
    [Newton, Wallis]
Kubesov A 1963 Az FTM (7), 147-152
    [Nasir al-din]
Lacroix S 1810-1819 *Traité du
    calcul differentiel et integral,*
    Paris, 3 vols [bib]
Lasswitz K 1890 *Geschichte der
    Atomistik vom Mittelalter bis
    Newton,* Hamburg, 2 vols
Leibniz G W 1846 *Historia et origo
    calculi differentialis (1714),*
    Hannover (C. I. Gerhardt ed)
Lorenzen P 1960 *Die Entstehung der
    exacten Wissenschaften* (Springer)
Lueroth J 1889 Freib NG [XVII]
Mahnke D 1932 Marb Si 67, 31-69
    [Leibniz]
Mansion P 1884 Mathesis 4, 163, 177
- 1887 *Esquisse d d'histoire du
    calcul infinitesimal,* Gand, 38 p
Mau Juergen 1957 *Zum Problem des
    Infinitesimalen bei der antiken
    Atomisten,* 2nd ed, Berlin, 48 p
    [Z98, 5)
Milhaud G 1917 Rv GSPA 28, 332-337
    [Descartes, Fermat]
Mueller C 1930 Unt M 36, 381-382
    (F56, 35) [Leibniz]
Muller F 1904 DMV 13, 247-253
    [anal geom, bib]
- 1904 DMV 13, 247-253
    [anal geom, Euler]
*Pasch M 1927 *Mathematik am Ursprung,*
    Leipzig, 47-73   [Cauchy]

Poincaré H 1905 Rv Met Mor 13, 293
    [Cournot]
Raphson Joseph 1715 *The History of
    Fluxions, Showing in a Compendious
    Manner the First Rise of, and
    Various Improvements Made in that
    Incomparable Method,* London
    [pro Newton]
Robinet A 1960 Rv Hi Sc Ap 13, 287-
    308 [Malebranche]
Rosenthal A 1951 AMM 58, 75-86
    (MR12, 577; Z42, 2)
Ross R 1919 Sci Prog, 13-52, 634-635
Rufini Enrico 1926 *Il Metodo di
    Archimede e le origini dell'analisi
    infinitesimale nell'antichità,*
    Rome, 302 p   (2nd ed 1961;
    MR34, 1137)
Rybnikov K A 1958 Ist M Isl 11, 583-
    592 (Z95, 4)
Savèrien Alexandre 1753 *Histoire
    critique du calcul des infiniments
    petits,* Paris
Schafheitlin P 1920 Basl V 32, 230-
    235 (F47, 6; 48, 5)
Schrader D V 1962 MT 55, 385-396
    [Leibniz, Newton]
Scriba C J 1957 Gies Sem (55)
    (MR20, 738) [James Gregory]
Sergescu Petre 1938 Sphinx 8, 125-129
    (F64, 17)
- 1942 in *Studies dedicated to the
    memory of the great Nicolas Iorga,*
    Bucharest (MR4, 65)
    [Rolle, Sourin, XVIII]
- 1949 *Les recherches sur l'infini
    matématique jusqu'à l'établisse-
    ment de l'analyse infinitésimale,*
    Paris, 32 p
*Simon M 1896 Deu NA Ver 2, 257-263
- 1898 Ab GM 8, 115-132 (F29, 31)
Simons L G 1936 Scr 4, 207-219
    [ed, USA]
Stolz O 1880 Wien Anz, 91-92
    (F12, 34) [Bolzano]
Studnicka F J 1879 Cas MF 8, 1-10,
    97-109, 272-295 (F11, 39)
Szenassy B V 1937 Deb M Szem 13, 1-
    30 (F63, 809) [W Bolyai]
Tenca L 1953 It UM (3)8, 201-204
    (MR14, 1050)   [Grandi]

Calculus    (continued)

- 1956 Int Con HS 8, 3-9  [Italy]
Timerding H E 1912 ZMN Unt 43,
  343-347, 347-350
Tobiesen L H 1793 *Principa atque
  historia inventionis calculi
  differentialis et inegralis nec
  non methodi fluxioneum,* Goettingen,
  28 p
Toeplitz Otto 1949 *Die Entwicklung
  der Infinitesimalrechnung,* Berlin,
  190 p  (Z35, 146)
- 1963 *The Calculus - a Genetic
  Approach* (Univ of Chicago Pr)
  (tr of 1949)
Turnbull H W 1951 Nat 167, 1048-1050
  (Z42, 242)  [XVII]
Vacca G 1925 Rm Sem (2)2, 13-14
  (F51, 8)  [Cavalieri, Galileo,
  Kepler, Napier]
Van Geer P 1913 Wisk Tijd 10, 23-29,
  81-86, 129-133, 206-210
  (F44, 60; 45; 74)
Vivanti G 1930 Milan Sem 3, 1-12
  (F56, 5)
*Volterra V 1912 Rv Mois 13, 257-275
Voss A 1899 Enc MW 2A(2)(2), 54-134
Vredenduin P G J 1960 Euc Gron 25,
  305-328  (Z108-2, 246)
Walker E 1932 T Col Cont (446)
  [Roberval]
Wallner C R 1904 Bib M (3)5, 113-124
  (F35, 59)
Weissenborn H 1856 *Die Principien der
  hoeheren Analysis in ihrer
  Entwickelung von Leibniz bis auf
  Lagrange...,* Halle
Wieleitner Heinrich 1925 *Die Geburt
  der modernen Mathematik. II Die
  Infinitesimalrechnung,* Karlsruhe
  72 p
Wieleitner H 1930 QSGM (B)1(2), 201-
  220  [Archimedes]
Wisdom J D 1941 Hermath (57), 49-81
  (MR3, 258)
Witting A 1912 Bib M 12, 56-60
  [Newton]
Wittstein T 1851 *Drei Vorlesungen
  zur Einleitung in die Differential
  und Integralrechnung,* Hannover

Wren F L + 1933 AMM 40, 269-281
  (F59, 8; Z6, 337)
Yanovskaya S A 1947 Mos IIET 1, 327-
  346  [Rolle]
Zavagna I 1923 Pe MI (4)3, 408-427
  [bib to Cauchy]
Zeuthen H G 1895 Dan Ov 1, 2, 37, 193
  256  (F26, 54)


Calculus of variations

See also Isoperimetric problem.

Alexandrov N V 1959 Mos IIET 28, 219-
  236
Bar A 1914 Intermed 21, 108 (Q3917)
  [bib]
Bliss G A 1925 *Calculus of Variations*
  Chicago, 202 p  [bib, survey]
- 1936 AMM 43, 598-609  [bib]
- 1946 *Lectures on the Calculus of
  Variations* (Univ Chicago Pr)
  [bib, Bolza problem]
*Bolza Oscar 1904 *Lectures on the
  Calculus of Variations* (Univ
  Chicago Pr) [1870-1904, bib]
*Caratheodory C 1937 Osiris 3, 224-240
  (F63, 795; Z18, 196)
- 1945 in *Festschrift zum 60
  Geburtstag von Dr. A. Speiser,*
  Zurich, 1-18  (Z60, 12)
Dorofeva A V 1961 Ist M Isl 14, 101-
  180  (Z118-1, 5)
- 1963 Ist M Isl 15, 99-128  [XIX]
Dresden A 1926 AMSB 32, 475-521
  [recent work, bib]
Ducassé P 1936 Thales 3, 39-43
  [Comte]
Duren W 1930 in *Contribution to the
  Calculus of Variations* (Univ
  Chicago Pr)  [suff cond]
Forsyth A R 1926 *Calculus of Vari-
  ations* (Cambridge U P)  (repr
  1960 Dover)
Graefe C 1825 *Historia Calculi
  variationum,* Goettingen
Guirandet 1863 *Aperçu historique zur
  l'origine et les progrès du calcul
  des variations jusqu'aux travaux
  de Lagrange,* Lille, 43 p

## Calculus of variations (continued)

Hadamard J 1910 *Leçons sur le calcul des variations,* Paris

Kneser J C A 1900 *Lehrbuch der Variationsrechnung,* Braunschweig
- 1904 Enc MW 2(1)8
- 1907 Ab GM 25, 21-60 (F38, 15) [Euler]

*Lecat M 1913 *Bibliographie du calcul des variations,* Gand-Paris, 115 p
- 1916 *Bibliographie du calcul des variations depuis le origines jusqu'a 1850 comprenant la liste des travaux qui ont preparé ce calcul,* Gand-Paris
- 1921 *Bibliographie des Series trigonometriques avec une appendice sur le calcul des variations,* Brussels
- 1924 *Bibliographie de la relativité,* Brussels, app

Pascal E 1899 *Calcolo delle variazioni,* Milan [to Weierstrass]

Reiff R 1887 Wurt MN 2, 90-98

Rybnikov K A 1949 Ist M Isl 2, 355-498 (MR12, 1)

Todhunter I 1861 *A History of the Progress of the Calculus of Variations During the Nineteenth Century,* London, 532 p (repr 1961 Chelsea)

Woodhouse Robert 1810 *A Treatise on Isoperimetical Problems and the Calculus of Variations,* 163 p Cambridge, (rep 1965 Chelsea under title *A History of the Calculus of Variations in the Eighteenth Century)*

Zermelo E + 1904 Enc MW 2(1)8

## Cardioid

Loria G 1898 Intermed 5, 200

Yates R C 1959 MT 52, 10-14

## Categorical algebra

*Freyd Peter 1964 *Abelian Categories*
(Harper and Row), 164 p [bib]

*MacLane S 1965 AMSB 71, 40-106

Mitchell Barry 1965 *Theory of Categories,* NY [bib]

Pumpluen D 1968 Ueberbl 1, 177-192

## Catenary and catenoid

Brocard H + 1895 Intermed 2, 114, 115, 358, 405 (Q302)

Hofmann J E 1966 Ost Si M 2, 175, 209-254 [calc, cycloid, Leibniz]

Kato H 1940 Toh MJ 47, 279-293 (F66, 14) [Japanese]

Korteweg D J 1900 Bib M (3)1, 97 [Huygens]

Sestini G 1936 Pe M (4)16, 168-183 (Z14, 243)

## Cauchy integral theorem

Gould H W + 1965 Math Mngl (10) [*bib]

## Cayley algebra

Kleinfeld E 1963 in Albert A A ed; *Studies in Modern Algebra* (MAA)

## Cayley-Hamilton theorem

Mitchell A K 1933 AMM 40, 153

## Center of gravity and centroid

See also Mean.

Agostini A 1940 It UM (2)2, 147-171 (F66, 15) [Cavalieri]

Costabel P 1955 Rv Hi Sc Ap 8, 116-128 (MR17, 338) [Galileo]

Court N A 1960 MT 53, 33-35

Hofmann J E 1956 MN Unt 9, 252-255 (Z71, 242)

Juel C 1914 Dan Ov (5)6, 421-441 [Archimedes]

## Center of gravity and centroid
### (continued)

Piani D 1868 Boncomp 1, 41-42
  (F1, 15)
- 1870 Bln Mm (2)10  (F2, 29)
Vailati G 1897 Tor FMN 32, 500,
  742-758  (F28, 49)  [Archimedes]

## Center of similitude

Archibald R C 1915 AMM 22, 6-12
  [Monge]
- 1916 AMM 23, 155-161  (F46, 52)
Brown B H 1916 AMM 23, 155-159
  (F46, 52)

## Central limit theorem

Feller W 1945 AMSB 51, 800-832
Saxer W 1950 Elem M 5, 50-55
  (MR11, 707)

## Cevian

Court N A 1956 Scr 22, 193-202
Wayne A 1951 MT 44, 496-497
  [nedians, redians]

## Chain rule

Simons L G 1935 Scr 3, 93-94

## Chebyshev systems

Karlin Samuel + 1966 *Tchebycheff
  Systems...,* NY, 604 p  [*bib]

## Chi square

Cochran W G 1952 An M St 23, 315-345
Lancaster H O 1966 Au J St 8(3),
  117-126

## Chinese remainder theorem

Dickson L E 1919 Dickson 2, 57-64
Mahler K 1958 M Nach 18, 120-122
  (MR20, 503)
Matthiessen L 1881 Crelle 91, 251-262
  (F13, 35)
- 1881 Par CR 92, 291-294  (F13, 35)

## Chromatic graphs

See also Four color conjecture,
  Graph theory.

Coxeter H S M 1957 Scr 23, 11-
*Ringel G 1959 *Aarbungsprobleme auf
  Flachen und Graphen,* Berlin

## Circle

See also Cyclotomy, Squaring the
  circle, Center of similitude.

Agostini A 1930 Pe M (4)10, 36-38
  (F56, 18)  [cycloid, length]
Betazzi R + 1899 Intermed 6, 158
  (Q1382)  [terms]
Brown B H 1925 AMM 32, 517
  [Fiedler]
Chatley H 1911 Monist 21, 137-141
  (F42, 70)  ["magic circle"]
Dijksterhuis E J 1931 Euc Gron 8,
  286-296  [arc length, Pell]
Eneström G 1911 Bib M (3)12, 84
  [XVI, inscribed quadrilateral,
  Simon Jacob]
Goodrich L C 1948 Isis 39, 64-65
  (Z30, 1)  [Chinese]
Hofman J E 1966 Arc HES 3, 102-136
Kutta M 1896 Bib M, 16  [anc]
Lietzmann W 1951 *Altes und neues vom
  Kreis,* Leipzig 54 p
Luria A 1935 Len IINT 5, 491-497
  (F61, 949)  [Cavalieri]
Mackay J S 1893 Edi MSP 11, 104-106
  (F25, 82)  [Adam, hexagons]
Merrifield C W 1869 LMSP 2, 175-177
  [Arab, radical axis]
Mikami Y 1908 Tok M Ph (2)4, 442-446
  [Seki]

## Circle    (continued)

Natucci A 1931 Pe M (4)11, 69-83
    (Z1, 322) [VIII, sphere]
Neugebauer O ⊢ 1929 QSGM (B)1(1),
    81-92 [Babylonian]
Rossier P 1941 Genv CR 58, 170-171
    (Z27, 3) [length]
Ruddick C T 1927 AMM 34, 30
    [Euclid, optics]
Rupp O 1931 Arc GMNT 13, 293-326
    [hyperb]
Stamatis E 1955 Athen P 30, 410-414
    (MR18, 368) [Euclid]
Steele A D 1936 QSGM (B)3, 287-363
    [Greek]
Steiner J 1931 *Allgemeine Theorie
    ueber das Beruehren und Schneiden
    der Kreise und der Kugeln,*
    Zurich, 363 p (F57, 1305)
Steinschneider M 1891 Bib M (2)5,
    113-116 (F23, 3) [mid]
Terquem O 1846 Nou An 5, 636-
    [quadril]
Thureau-Dangin F 1932 Rv Assyr 29,
    26-28 (Z5, 1) [area]
Wiedemann E 1912 Mit GMNT 11, 252-
    255 [Hindu]
Wilton J R 1928 Mess M 58, 67-80
    (F54, 6)

## Cissoid

Hofmann J + 1941 Deu M 5, 571-584
    (MR2, 306) [area]
Roeser H 1915 SSM 14, 790-795

## Class field theory

Bell E T 1924 AMSB 30, 236-238
    [Gauss]
- 1928 AMSB 34, 490-494
    [ell theta fun, Gauss]
Hasse H 1966 DMV 68(1), 166-181
    (MR34, 165)

## Clumping

Roach S A 1968 *The Theory of Random
    Clumping,* London [bib]

## Cluster sets

Collingwood E F + 1966 *The Theory of
    Cluster Sets,* NY [bib]

## Combinatorial geometry

*Hadwiger H + 1964 *Combinatorial
    Geometry in the Plane,* NY, 120 p
    [bib]

## Combinatorial logic

Biermann K R + 1958 JSL 23, 129-132
    (MR21, 750; Z85, 2)
*Curry H B + 1958 *Combinary Logic*
    Vol 1, Amsterdam [bib]
Feys R 1953 Int Con Ph 5, 70-72
    (MR15, 90) [Burali-Forti, Peano]

## Combinatorics

See also Ballot theorems, Binomial
    coefficients, Figurate numbers,
    Graph theory.

Biermann K-R 1954 Fors Fort 28, 357-
    363 (Z60, 7) [Leibniz]
- 1956 Fors Fort 30(6), 169-172
    [Leibniz]
Boutroux P 1910 Rv Mois 9, 50-62
Chakrabartti G 1932 Clct MS 24, 79-88
    (F58, 18) [India]
Das L R 1932 Pe M (4)12, 133-140
    (Z5, 148) [Hindu]
Gandz S 1943 in *American Academy of
    Jewish Research, Saadia Anniversary
    Volume,* 141-195
Ginsburg J 1923 MT 15: 347-356 [Ezra]
Hayashi T 1931 Toh MJ 33, 328-365
    (Z1, 323) [Japanese]
J N 1900 Mathesis (2)10, 158 [Tarry]

## Combinatorics (continued)

Kaucky J 1963 MF Cas 13, 32-40
(MR27, 3563) [Chinese]
Kutlumuratov D 1964 *O razvitii
kombinatornykh metodov matematiki,*
Nukus
*Netto E 1927 *Lehrbuch der Combin-
atorik,* Leipzig, 2nd ed
(repr Chelsea)
Riordan J 1968 *Combinatorial
Identities* (Wiley) [bib]
Rome A 1930 Brx SS 50(A), 97-104
(F56, 15) [anc]
Ryser Herbert J 1963 *Combinatorial
Mathematics* (Carus Monograph 14)
[bib, diffc sets, latin sq]
Sorge F 1932 ZT Ph 13, 223-225
Turetsky M 1923 MT 16, 29-34
[XVI, Arab, Hebrew]
Vogt H + 1904 Enc SM 1(1)(1),
64-132
Vajda S 1967 *Mathematics of
Experimental Design* (Hafner)
[bib]
- 1967 *Patterns and Configurations
in Finite Spaces* (Hafner) [bib]

## Commutative laws

Mode E B 1945 MT 38, 108-111
Zapelloni M T 1929 Pe M (4)9,
19-24 (F55, 588)

## Compactness

*Hewitt E 1960 AMM 67, 499-
Pier J P 1961 Rv Hi Sc Ap 14,
169-179

## Compass (geometry)

See also Construction, Proportional
compass (Ch. 3).

Brocard H + 1901 Intermed 7, 229
(Q914) [Baradelle, Macquart,
Meynier]

*Cheney W F 1958 MT 46, 152-156
(repr SMSG RS 4)
*Court N A 1958 MT 51, 370-372
Drake S 1960 Physis Fi 2, 211-212
[Galileo, mechanics]
Hallerberg A E 1959 MT 52, 230-244
[fixed compass]
Nippoldt A 1932 Naturw 21, 306-307
(F59, 45)
Shafranovskaya T K 1960 Vost IINT (1),
56-63 (Z118:1, 245) [Chinese]
Schueck A 1915 *Der Kompass,*
Hamburg, 58 p (F45, 96)
Wiedemann E 1923 Z Ph 13, 113-116
(F49, 24)

## Complex analysis

See also Conformal mapping, Riemann
surface, Zeta function.

Aubry A 1907 Mathesis (3)7, 257-261
Belozerov S E 1956 Mos IIET 15, 169-
- 205 (MR19, 825)
- 1956 Mos M Sez, 229-230 [Russia]
- 1962 *Osnovnye etapy razvitiya
obshchei teorii analeticheskikh
funktsii,* Rostov
Beltrami E 1869 Batt 7, 29-41
Borel E 1917 Rice Pam 4(1)
Casorati Felice 1868 *Teorica delle
funzioni di variabili complesse,*
Pavia, 1-143
Dienes P 1931 *The Taylor series...,*
(Oxford U P), 552 p (repr 1957
Dover) [*bib]
Jourdain P E P 1905 Bib M (3)6, 190-
207 [Cauchy, Gauss]
Julia G 1933 *Essai sur le développement de la théorie des fonctions
de variables complexes,* Paris,
53 p
Look K H 1959 Sci Sin 8(11), 1229-
1237 (repr 1960 AMSN 7(2),
155-163) [China, several variables,
1948-1958]
Markushevich A I 1950 Ist M Isl 3,
399-406 (Z41, 341)
*- 1951 *Ocherki po istorii teorii
analitcheskikh funktsii,* Mos-Len,

Complex analysis    (continued)

127 p  (MR14, 2)  (Ger tr 1955
Berlin; MR 17, 445)
Nevanlinna R 1966 in Behnke H +
*Festschrift zur Gedaechtnisfeier
fuer Karl Weierstrass,* Koeln-
Opladen (MR33, 4284)
*Osgood W F 1914 *Topics in the Theory
of Several Complex Variables*
(AMS Col 4)
Petrovich M 1929 Beo Ak 134 (63),
87-89  (F55, 18)  [Poincaré,
limits at inf]
Picard E 1926 An SENS 43, 363-368
Pincherle S 1880 Batt 18, 178-254,
317-357  [Weierstrass]
Riabouchinsky D 1946 Par CR 222,
426-428  (MR7, 354)      [mech]
Schlesinger L 1905 Bib M (3)6, 88-96
(F36, 60)  [Jacobi]
Staechel P 1900 Bib M (3)1, 109-
128   [intgl]
- 1901 Bib M (3)2, 111-121  [XVIII]
Teichmueller O 1939 Deu M 4, 115-116
(F65, 16)  [Cauchy, Riemann,
Weierstrass]
Timchenko I Yu 1899 *Istoricheskie
svedeniya o razvitii pomyatii i
metodov, lezhashchikh v osnovanii
teorii analiticheskikh funktsii,*
Odessa, 670 p  (Pt. 1 of
*Osnovaniya teorii analiticheskikh
funktsii),* (Also 1892-1899 in
Novor M 12; 16; 19)  (F30, 48)
Valiron G 1949 BSM (2)73, 152-162
(MR11, 572)  [Ahlfors, Bloch]
Vivanti G 1909 Arc MP (3)25,
318-343   [bib]


Complex numbers

See also Imaginary elements, Numbers.

Agostini A 1924 It UM 3, 2 p
(F50, 10)
Allen E S 1922 AMM 29, 301
Archibald R C 1921 AMM 28, 116
[e, i pi]
Aubry V 1897 JM Sp 21, 83, 114

Baltzer R 1883 Crelle 94, 87-92
Bell E T 1934 Phi Sc 1, 30-49
(See also Phi Sc 2, 104-111)
Beman W W 1897 AAAS 46, 33-50
(F28, 46)  [geom, Kuehn, Wessel]
- 1898 AMSB (2)4, 274, 551  [i]
Bierens de Haan D 1858 Amst Vs M 8,
248-
Bortolotti E 1923 Scientia 33, 385-395
Brun V 1959 Rv Hi Sc Ap 12, 19-24
[Wessel]
Budon J 1933 BSM (2)57, 175-200,
220-232  (Z7, 388)  [Wessel]
Cajori F 1912 AMM 19, 167-171,
222 (F44, 60)   [before Wessel]
*Cartan E 1908 Enc SM (1)1(5), 329-
488
*Cayley A 1887 QJPAM 22, 270-308
[bib, linear alg]
Coolidge J L 1924 *Geometry of the
Complex Domain*  (Oxford U P),
242 p
Diamond L E 1955 M Mag 30, 233-249
Eneström G 1899 Bib M (2)13, 46
(F30, 49)  [e, i, pi]
- 1906 Bib M (3)7, 263-269  [Wallis]
Hankel H 1867 *Theorie der complexen
Zahlensysteme,* Leipzig, 71-73
Houel G J 1867-1870 Bord Mm 5, i-iv,
1-64; 6, 1-144; 8, 97-175
Jones P S 1954  MT 47, 106-114, 257-
263, 340-345
Kearns D A 1958 MT 51, 373-374
[Wallis]
Kiessling 1882 *Ueber die Entwicke-
lung des Imaginären in der Analysis,*
Hamburg
Kossak E 1872 *Elemente der Arithmetik*
Berlin  [Weierstrass]
Lampe E 1907 Ab GM 25, 117-137
(F38, 16)   [Euler, log]
Lange-Nielsen F 1923 Nor M Tid 5, 7-
23  (F49, 19)
Loria G 1917 Scientia 21, 99-121
(F46, 40)
McClenon R B 1923 AMM 30, 369-374
(F49, 6)  [Leibniz]
*Nagel E 1935 Stu HI 3, 429-476
[bib]
Peacock G 1833 Brt AAS, 185-352
[XIX]

Complex numbers (continued)

Studnicka F J 1884 Cas MF 13, 49,
   254
*Study E 1896 Chic Cong, 367-381
   (F28, 41)
Uhler H S 1921 AMM 28, 114
   [i to i power]
Van Raay W H L J 1894 Arc Mu Tey
   (2)4, 53-118  (F26, 50)
Vivanti G 1910 Bib M (3)10, 244-249
   [Euler]
Vooys C J 1960 Euc Gron 35, 162-166
   (Z105, 2)  [Cardan]
Wieleitner H 1927 DMV 36, 74-88
   (F53, 12)  [early]
Wigner E P 1960 CPAM 13, 1-14,
   [appl]
Wijdenes P 1933 Euc Gron 10, 1-22
   [i]
Windred G 1929 M Gaz 14, 533-541
   (F55, 26)


Conchoid

Loria G 1897 Mathesis (2)7, 5
Roeser H 1915 SSM 14, 790-795
   [Nicomedes]
Rudio F 1907 Mathesis (3)7, 261-262
   (F38, 68)
Wolffing E 1900 Loria, 97  [bib]


Cone

Mueller C 1942 Deu M 5, 244-255
   (MR2, 114)  [area, Aryabhatta]
Stoley R W 1927 Anc Eg, 16-17
   [parabolic]


Conformal mapping

See also Complex analysis.

Ahlfors L V 1953 An M Stu (30), 3-13
   (MR14, 1050)  [Riemann surf]
Carathéodory C 1932 Cam Tract (28)
Kober H 1952 *Dictionary of Conformal
   Representations* (Dover), 208 p

Kuenzi H P 1956 Elem M 11, 1-15
   (MR17, 473)
Wangerin A 1897 DMV 4, 126  (F28, 49)
   [Gauss, Lagrange, Lambert]


Congruence (geometry)

Kieffer L 1957 Lux Arch 24, 161-171
   [equivalence]
Murdoch J 1964 Koyre 1, 416-441
   [mid]


Congruence (number theory)

Alasia C 1903 Rv FMSN 4(2), 149-208
   [to 1852]
Heller S 1964 Janus 51, 277-290
   (MR30, 870)  [Euclid]
Miller G A 1936 Toh MJ 42, 362-365
   (F62, 1011)  [factor congruences]
Zuhlke P 1905 Arc MP (3)9, 59


Conics

See also Circle, Focus, Ellipse,
   Hyperbola, Parabola.

Alekseev V G 1893 *Teoriya chislovykh
   kharakteristik sistem krivykh
   linii...,* Mos, 214 p
Amodeo F 1908 Bib M 9, 123-138
   (F39, 70)  [Maurolico]
- 1909 Linc Rn (5)18, 661-668
   (F40, 67)  [Cavalieri]
Artom E 1925 Pe M (4)5, 88-106, 280
   (F51, 8)    [area]
- 1930 Pe M (4)10, 204-226, 296-316
   (F56, 11)    [anc]
Bompiani E 1939 It UM (2)1, 372-373
   (Z21, 194)  [constr]
Bopp K 1917 Ab GM 20, 87-310
   (F41, 66)  [Gregory St Vincent]
Braunmuehl A v 1890 Hist Abt 35,
   161-165
Chapin M L 1926 MT 19, 36-45
Conte L 1939 Pe M (4)19, 16-35
   (F65:1, 6)
Coolidge J L 1945 *A History of Conic*

## Conics (continued)

*Sections and Quadric Surfaces,*
(Oxford U P), 225 p (MR8, 1)
Easton J B 1963 MT 56, 632-635
[de Witt]
Eneström G 1911 Bib M (3)12, 353
(F42, 65) [Mydorge]
Eves H 1960 MT 53, 280-281
[names]
Fladt Kuno 1965 *Geschichte und*
*Theorie der Kegelschnitte und der*
*Flaechen zweites Grades,*
Stuttgart, 374 p
Frajese A 1961 Archim 13, 185-194
(Z105, 243) [Plato]
Gerhardt K J 1892 Ber Si, 183-204
[Desargues, Pascal]
Hayashi T 1906 AMM 13, 171-181
[Japanese]
Heath T L 1896 *Introduction to*
*Apollonius of Perga, A Treatise*
*on Conic Sections,* Cambridge
(F27, 4)
Heiberg J L 1880 Hist Abt 25, 41-67
(F12, 1) [Archimedes]
Hofmann J E 1950 Centau 1, 24-37
[approx, Gregory St Vincent]
Krasnova S A 1966 Ist Met EN (5),
140-149 (MR34, 2407) [constr, mid]
Lez H 1895 Intermed 2, 123
[trilin coord]
Loria G 1936 Osiris 1, 412-450
[Chasles]
Milne J J 1930 M Gaz 15, 142-144
(F56, 7) [Apollonius]
Neugebauer O 1932 QSGM (B)2, 215
*- 1948 APSP 92, 136-138 (MR10, 174;
Z38, 145) [ast]
Roberts S 1881 LMSP 12, 120-122
(F13, 43) [confocal, Grave]
Schoute P H 1881 *De Kegelsneden in*
*de projectivische meetkunde,*
Groningen (F13, 43) [proj geom]
Soldaini E 1934 Pe M (4)14, 90-100,
158-177 (F60, 6) [Apollonius]
*Taylor C 1881 *An Introduction to*
*the Ancient and Modern Geometry*
*of Conics,* Cambridge
- 1900 Cam PST 18, 197-219
[Kepler, Newton]

Valentin G 1912 Bib M 13, 82-83
Ver Eecke P 1925 *Introduction to*
*Apollonius les Coniques,* Bruges
Viola T 1946 Pe M (4)24, 68-83
(MR8, 306) [Kepler]
de Vries H 1923 N Tijd 11, 1-24,
188-210, 295-330 [proj geom]
- 1946 N Tijd 33, 100-164 (MR8, 2)
[contact, inters]
Weaver 1916 AMSB 23, 353-365
[Apollonius, foci]
Zeuthen H G 1885 *Kegelsnitslären i*
*Oldtiden,* Copenhagen
- 1886 *Die Lehre von den Kegelschnitten*
*ten im Altertum,* Copenhagen, 527 p
(2nd ed 1565 Hildesheim; MR34,
1138)

## Conjugate coordinates

Carver W B 1956 AMM 63(9)2
[Morley]

## Constructions

See also Compass, Cyclotomy,
Euclidean constructions, Existence,
Mascheroni constructions, Ruler,
Trisection.

Bieberbach L 1952 *Theorie der*
*geometrischer Konstruktionen,*
Basel
Bompiani E 1939 It UM (2)1, 372-373
(F65, 8) [conics]
Bortolotti E 1929 Pe M 9, 161-166
[algebra, Italy]
Courant R + 1941 *What is Mathematics?*
NY
Geppert H 1929 Pe M (4)9, 292-319
(F55, 602) [fixed compass]
Gibson B 1896 Rv Met Mor 4, 386-398
[Descartes]
Greenwood T 1954 Thomist 17, 84-94
[Aristotle]
*Hallerberg A E 1959 MT 52, 230-244
[fixed compass, bib]
Hofmann J 1962 Mathunt, 7-47
Hunrath K 1905 Bib M (3)6, 249-251
(F36, 65) [Duerer, reg polyg]

Constructions  (continued)

Krasnova S A 1964 Kolom Ped 8, 184-
204  (MR33, 2500)
Kubota T 1928 DMV 37, 71-74
(F54, 7)
Mackay J S 1887 Edi MSP 5, 2-22
[fixed compass]
Mehmke R 1910 Intermed 17, 125
(Q1104)  [Bobillier]
*Petersen J 1879 *Methods and Theories
of Problems of Geometrical
Constructions,* Copenhagen  (tr in
Danish, French, German, Italian,
Hungarian, Russian.)
Rogachenko V F 1965 Lvov U Vis (1),
34-45  [non-Euc geom]
Schramm M 1965 Hist Sci 4, 70-103
[mid]
Steele D A 1934 QSGM (B)3, 287-369
[Plato]
Tannery P 1895 Intermed 2, 29
[Greek]
Vahlen K T 1902 Arc M P (3)3, 112
[cubic]
- 1911 *Konstruktionen und Approxi-
mationen* (Teubner), 285 p
Wieleitner H 1917 ZMN Unt 48, 328
(F46, 47)  [approx, Duerer]
Zeuthen H G 1892 Nyt Tid 3(A), 105-
113  (F24, 48)  [Greek]


Contact transformations

Liebmann H + 1914 DMV (S)5, 93 p
(F45, 81)  [Lie]
Loria G 1905 Bib M (3)6, 343-346
(F36, 60)  [Fermat]


Continued fractions

Bortolotti E 1919 Mathes It 11, 157-
188  Cataldi
- 1920 Mathes It 12, 152-162
1931 Pe M (4)11, 133-148
(F57, 7)  [irrat]
1931 Pe M (4)11, 211-217
(F57, 7)  [Leonardo da Pisa, pi]
1932 Bln Mm 9, 85-93  [Egypt]

- 1932 Scientia 52(S), 133-
Cantor M 1872 Hist Abt 17, 102
(F3, 26)  [review of Guenther 1872
Chovanskii A N 1957 Ist M Isl 10,
305-326  (Z101, 245)  [Euler]
De Vries H 1939 Euc Gron 16, 44-84
(F65, 16)
Favaro A 1874 Boncomp 7, 451-502,
533-589  (F6, 39)  [XIII-XVII]
Franklin S P 1961 Scr 25, 353-355
(Z100, 243)  [Eudoxos, Omar]
Gonzolez J 1914 Intermed 21, 115
(Q4327)  [series]
Guenther S 1872 *Beitraege zur
Erfindungsgeschichte der Ketten-
brueche,* Weissenberger
- 1874 Boncomp 7, 213-254  (F6, 38)
[to Euler]
- 1876 *Vermischte Untersuchungen zur
Geschichte der Mathematischen
Wissenschaften,* Leipzig, ch.2
- 1882 Bord Mm (2)5, 91-107
[roots]
Hayashi T 1914-1915 Toh MJ 5, 188-
231; 7, 1-17  [Japan]
Isely L 1903 Neuc B 32, 72-79
Itard J 1954 Rv GSPA 61, 5-18
(MR15, 591)
*Jolliffe A E 1910 Enc Br  [bib]
Jones P S 1953 MT 46, 190-192
Khovanskii A N 1957 Ist M Isl 10,
305-326
Maccaferri E 1931 Piac IT An 6
(F57, 6)
Mehta D M 1932 *Theory of Simple
Continued Fractions with Special
Reference to the History of Indian
Mathematics,* Bhavaghar, 168 p
Molk J + 1907 Enc SM 1(1)(2), 282-328
Mongardon L 1916 Intermed 23, 180-182
(Q4423)  [roots]
Muir T 1876 Nat (Ap 6), 448-449
[XVII]
Perron O 1920 Heid Si (5), 9 p
(F47, 27)
*- 1929 *Die Lehre von den Ketten-
bruechen* 2nd ed, Leipzig,  536 p
[bib]  (repr Chelsea 1950)
(3rd ed 1954)
Plakhowo N 1907 Intermed 14, 260-263
(Q3243)

Continued fractions (continued)

Stieltjes T J 1894 Tou FS 8, J1-J122
*Wall H S 1948 *Analytic Theory of
  Continued Fractions* (Van Nostrand),
  433 p (rep 1967 Chelsea) [bib]
Wertheim G 1898 Ab GM 8, 147-160
Woelffing E 1908 Wurt MN (2)10, 18-
  32, 35-38 (F39, 58)

Continuity

See also Analysis, Functions,
  Limits.

Burns C 1916 Mind 25, 506-512
  [William of Ockham]
Dupont P 1965 Linc Rn (8)39,
  255-262 (MR34, 2404) [Galileo]
Frajese A 1939 It UM (6)2, 76-79
  (F65:2, 1079) [geom, Greek]
- 1951 Archim 3, 98-104
  (Z43, 241) [geom, Greek]
Heath T L 1925 *The Thirteen Books
  of Euclid's Elements...*,
  (Cambrdige U P) 1, 234-240
  [geom]
Jourdain P E B 1914 Isis 1, 661-703
  [Cauchy, intgl]
*Korner S 1967 Enc Phi
Lagage R 1929 Wis Nat Ti 4, 153-160
  (F56, 805)
Langen L 1939 Muns Semb 14, 103-114
  (F65, 12) [Leibniz]
Murdoch J 1964 Koyre 1, 416-441
  [mid]
Rosenfeld B 1965 Arc In HS 18, 3-22
  [geom]
- 1965 Ist M Isl 16, 273-294
  [geom]
Taylor C 1881 Cam PSP 4, 14-17
  (F13, 39) [geom]
Zariski O 1926 Rv M Hisp A (2)1,
  161-166, 193-200, 233-240,
  257-260 (F52, 3)

Continuous geometry

*Halperin I 1959 M Lap 10
  [von Neumann]

*von Neumann J 1960 *Continuous
  Geometry* (Princeton U P), 299 p

Continuum

See also Real numbers.

Noi S di 1951 Archim 3, 112-116
  (Z44, 246)
Simon M 1909 Int Con 3, 385-390
Stipanič E 1967 MV 4(19), 277-292
  [Boscovich]
Vredenduin P G J 1960 Euc Gron 36,
  1-6 (Z108:2, 244) [Aristotle]

Continuum hypothesis

See also Set theory.

*Cohen Paul J + 1967 Sci Am (Dec),
  104-116
*Gödel K 1947 AMM 54, 515-525
*Rubin Jean E 1967 *Set Theory for
  Mathematicians* (Holden-Day),
  Ch. 13)
*Smullyan R M 1967 Enc Phi [bib]

Convexity

*Hardy G H + 1934 *Inequalities*
  (Cambridge U P), Ch. 3)
Valentine F A 1964 *Convex Sets,* NY
  [bib]

Coordinates

See also Analytic geometry, Conjugate
  coordinates, Homogeneous coordinates,
  Polar coordinates.

Eneström G 1903 Bib M (3)4, 344-388
  [space]
- 1910 Bib M (3)10, 43-47
  [Leibniz]
Fiser R 1900 *Die Methoden der
  Analytischen Geometrie in ihrer
  Entwicklung im 19 Jahrhundert,*
  Braunau, 51 p

## Coordinates    (continued)

Guenther S 1877 Nurnb NG 6
   (tr 1877 Boncomp 10, 363-406)
Lattin H P 1948 Isis 28, 205-225
   [logic ms, mid]
Lez H + 1895 Intermed 2, 123; 3, 64-
   226 [conic, trilinear]
Loria G 1942-1948 Math Tim 18,
   125-145; 20, 1-22; 21, 66-83
   (MR6, 141)
- 1948 Osiris 8, 218-288
   (1942-1948)
Taton R 1952 Elem M 7, 1-5
   (MR13, 420)  [Pluecker coords,
   Monge]
Williamson R S 1942 Nat 150, 460-461
   [Egypt]
Woelffing E 1900 Bib M (3)1, 142-159
*Woolard E W 1942  ASP 54, 77-90
   [anc ast]

## Correlation

See also Statistics.

Walker H M 1928 Isis 10, 466-484
   [Bravais, Plana]

## Correspondence principle

See also curves.

Segre C 1892 Bib M (2)6, 33-48
   (F24, 51)   [bib]

## Cramer paradox

*Scott C A 1897 AMSB 4, 260-273

## Cremona transformation

*Snyder V + 1928 Am NRCB (63)   [bib]
White H S 1918 AMSB 24, 242
   [Cremona, Magnus, Noether]

## Cross ratio

Milne J J 1911 *An Elementary Treatise
   on Cross Ratio  Geometry With
   Historical Notes,* Cambridge

## Cubic curves and surfaces

Ball W W R 1891 Bib M (2), 35-40
   (F23, 39)   [Newton]
- 1891 LMSP, 104      [Newton]
De Vries H 1939 C Huyg 18, 73-109
   [F65, 1087)
Henderson A 1911 *The 27 lines upon
   cubic surfaces,*  Cambridge
   [bib, surf]
Miller M 1953 Dres Verk 1(1), 5-32
   (Z53, 196)
White H S 1925 *Plane Curves of the
   Third Order* (Harvard U P), 168 p

## Cubic equations

Anning N 1926 AMM 33, 211 [Newton]
Bortolotti Ettore 1926 *I contributi
   del Tartaglia, del Cardano, del
   Ferrari, e della scuola matematica
   bolognese alla teoria algebrica
   delle equazioni cubiche,* Imola,
   54 p
- 1934 Bln Mm 9(1), 79-94   (Z10, 243)
   [Babylonia]
Bosmans H 1923 Mathesis 37, 246-254,
   304-311, 341-347 (F49, 20)
   [Stevin]
Campan F 1938 Gaz M 43, 403-409
   (F64, 5)
Cantor M 1878 Boncomp 9, 177-196
   (F10, 26)   [Ferrari]
Casara G 1942 It UM (2)4, 244-262
   (F68, 6)   [Archimedes]
Cassina U 1924 Tor FMN 59, 14-29
   (F50, 10)  [Fibonacci]
- 1941 Pe M (4)21, 3-20   (MR3, 97)
   [al-Biruni]
Charadze E K 1943 Gruz Soob 4, 195-
   199 (Z60, 7)   [Cardan, Hudde]
Conte L 1952 Pe M (4)30, 125-128
   (Z47, 6)   [Riccati]
Costabel P 1965 Rv Hi Sc Ap 18, 29-
   43   [L'Hospital]

## Cubic equations  (continued)

Dassen C C 1942 Arg SC 134-170
(MR5, 57)
Eves H 1958 MT 51, 285-286
[Omar Khayyam]
Forestier C 1874 Tou Mm 6, 254-
[Cardan]
*Frati L 1910 Loria 12, 1-5
[Scipione dal Ferro]
Guilbeau L 1930 M Mag 5, 8-12
Harig G 1935 Len IINT,7, 67-104
[Cardan, Tartaglia]
Henderson A + 1930 AMM 37, 515-521
[Viete]
Korteweg D J 1893 Nieu Arch 20, 63-96
[F25, 83]  [Rodenberg]
Miller G A 1944 Sci 100, 333-334
Neugebauer O 1933 Gott N(2)
[Babylonian]
Noguera R 1956 Stvdia 1, 107-148
(MR18, 381)  [S. Germain]
Pasquale L di 1957 Pe M (4)35, 79-93
(Z80, 3)  [Tartaglia]
Procissi A 1935 Pe M (4)15, 197-219
(Z11, 385)
- 1951 Pe M (4)29, 263-280
(MR13, 420)  [Cardan]
Sedillot L P 1837 Recherches
nouvelles...orientaux...ou notice
de plusieur opuscules...,  Paris
[Arab]
Studnicka F J 1875 Cas MF 4  (F8, 12)
[Arab]
Tenca L 1951 It UM (6)3, 143-149
[Grandi]
Vacca G 1930 It UM 9, 59-63
[Euclid, Fibonacci]
Verdam G J 1846 Amst Inst, 163-
Vetter Q 1928 Tor FMN 63, 296-299
[Fibonacci]
- 1935 Math Cluj 9, 304-309
(F61, 943)
Vogel K 1934 Mun Si 1, 87-94
[Babylonian]
Wieleitner H 1926-1927 Erlang Si
58/59, 173-176  [Cardan]
Zeuthen H G 1893 Bib M (2)7, 97-104
(F25, 61)  [Archimedes]
- 1893 Dan Ov, 1-  [M. Cantor]

## Curvature

See also Differential geometry.

Christensen S A 1884 Zeuthen (5)1,
97-127
Coolidge J L 1952 AMM 59, 375-379
(MR13, 809; Z47, 4)
Dupont P 1963 Tor FMN 98, 397-417,
489-513  (MR29, 4231, 4233)
Haas August 1881 Versuch einer
Darstellung der Geschichte des
Krummungsmasses, Tubingen
Le Corbeiller P 1954 Sci Am 191,
80-86  [Riemann]
Perassi R 1937 Pe M (4)17, 14-24
(F63, 809)  [Serini]

## Curves: General

See also Algebraic geometry, Analytic
geometry, Correspondence principle,
Curves: special, Curvature,
Envelope, Singular points, Linkages
(Ch. 3), Pursuit problems.

Amodeo Federico 1945 Sintesi storico-
critica della geometria delle
curve algebriche, Napoli, 420 p
[alg, geom]
Aubry A 1909 Porto Ac 4, 65-112
(F40, 67)
Baker H B 1912 LMSP 12
Berzolari L 1906 Enc MW 3(2)(1)
[bib]
Bliss G A 1923 AMSB 29, 161-183
[birat transf]
Boyer C B 1945 M Mag 19, 294-310
(MR6, 253)
Braunmuehl A v 1892 in Katalog der
Mathematische Ausstellung,
Nuernberg,  (F24, 51)
Christianson S A 1887 Bib M (NS)1,
76-80  [length]
Coolidge J L 1953 AMM 60, 89-93
(Z50, 1)  [length]
Crommelin C A 1956 Sim Stev 31, 5-18
[Huygens]
Guenther S 1886 Bib M 3, 137-140
[Duerer]

## Curves: General    (continued)

Hayashi T 1934 Toh MJ 39, 125-179
  (Z9, 98) [Japanese]
Hill J E 1896 AMSB (2)3, 133
  [bib, space]
Hofmann J E 1956 Ost Ing 10, 190-195
  (MR18, 182; Z74, 7) [XVIII]
La Chapelle J B 1750 *Traite des
  sections coniques...*, Paris,
  320 p [survey]
*Loria G 1898 Int Con (1897)1,
  289-298 (F29, 35)
- 1898 Wiad M 2, 202-213
  (F29, 35)
- 1899 Bib M (2)13, 10-12
  (F30, 53)
- 1936 Milan Sem 10, 1-14  (F62, 4)
Lowd F H 1893 An M 8, 29-37
  [constr]
Lysenko V I 1961 Ist M Isl 14, 517-
  526 (Z118:1, 6) [turning point]
Obenrauch F J 1903 Mo M Phy 14, 187-
  205 [Pythagoras, space]
*Sauerbeck P 1902 Ab GM 15, 1-166
  [XVII-XIX]
*Scott C A 1897 AMSB (2)4, 260
  [inters]
Staeckel P 1909 Arc GNT 1, 293-300
  (F41, 61) [D Bernoulli, Euler]
Tannery P 1883 BSM (2)7, 278-291
  [anc]
Turrière E 1913 Intermed 20, 106-107
  [anallagmatic]
Vargas y Aguirre J D de 1908 Mad Mm
  26 [catalog]
Wieleitner H 1905 Arc MP 12, 83-
  (F36, 68) [bib, 1890-1904]
Wilczynski E J 1916 AMSB 22, 317-329
  [diff geom]
*Wilder R L 1962 MT 55, 462-
  [topology]

## Curves: Special

See also Brachistochrone, Cardioid,
  Catenary, Cevian, Circle,
  Cissoid, Conchoid, Conics, Cubic,
  Cycloid, Devil's curve, Ellipse,
  Dido curve, Folium, Helix, Hessian,
  Hyperbola, Isochrone, Lemniscate,
  Limaçon, Logistic, Lunes, Ovals,
  Pathological, Quadratrix, Sinusoid,
  Spiral, Strophoid, Tautochrone,
  Tractrix, Trochoid.

Anon 1891 Bib M (2)5, 64, (Q34)
  (F23, 4) [names]
- 1901 Intermed 8, 318 [loxodrome]
- 1910 Intermed 17, 252 (Q3641)
  [catalog]
Brocard H 1897 *Notes de bibliographie
  des courbes géométriques*,
  Bar-le-Duc, 344 p (F28, 46)
- 1899 *Notes de bibliographie des
  courbes géométriques, partie
  complementaire*, Bar-le-Duc, 243 p
  (F30, 53)
- 1916 Intermed 23, 103-104 (Q1755)
  [quartic]
Brocard H + 1919 *Courbes géométriques
  remarquables*, Paris, 3 vols
  (repr 1967-1970 Blanchard) [bib]
Cady W G 1965 AMM 72, 1065-
  [tractrix, bib]
Crommelin C A P 1954 Sim Stev, 17-24
Dijksterhuis E J 1933 Euc Gron 9,
  233-265 (F59, 10)
Juel C 1895 Intermed 2, 279
  [constr ratio of powers re circles]
*Lockwood E H 1963 *A Book of Curves*,
  Cambridge
*Loria G 1902 *Spezielle algebraische
  und transcendente ebene Kurven*,
  Leipzig (2nd ed, 2 vols, Leipzig,
  1910-1911, Italian 1930 Milan)
- 1907 Bib M (3)7, 270-281
- 1914 Porto Ac 9, 193-196
  [Sluze, spiric]
- 1925 *Curve sghembe speciali,
  algebriche e transcendenti*,
  Bologna, 257 p
- 1929 Bo Fir 8, xlix-li [Viviani]
Maupin G 1895 Intermed 2, 290
  [Clelies, Rhodonees]
*Teixeira F G 1908-1915 *Traité des
  courbes spéciales remarquables
  planes et Gauches*, Coimbra,
  3 vols
Tenca L 1957 It UM (3)12, 458-460
  (MR22, 260) [Grandi]
Yates R C 1947 *A Handbook on Curves
  and their Properties*, Ann Arbor

Delian problem   (continued)

*problème de Delos.)* Paris, 84 p
  (F66, 1186)
Reimer N T 1758 *Historia problematis
  de cubi duplicatione sive de
  inveniendis duabus mediis continue
  proportionalibus inter duas datas,*
  Goettingen.
Rodriguez A 1953 Rv M El 2, 8-12
  (Z51, 2)
Seidenberg A 1966 Arc HES 3(2), 97-
  101 [Nicomedes]
Skof F 1958 Pe M (4)36, 19-40,
  76-92 (MR20, 944) [Archytas]
Sturm A 1895 *Das Delische Problem,*
  Linz
Tannery P 1878 Bord Mm (2)2, 277-283
  (F10, 25) [Archytas, Eudoxos]
Thaer C 1940 Deu M 5, 241-243
  (F66, 11) [Apollonios]
Weaver J H 1915 SSM 15, 216-217
  [Pappus]
- 1916 AMM 23, 106-113 (F46, 52)
  [to Descartes]


Derivative

See also Calculus, Chain rule,
  Taylor series and theorem.

Agostini A 1950 Pe M (4)28, 141-158
  (MR12, 383) [Torricelli]
Aubry A 1889 Prog M 1(2), 129-164
Boncompagni B 1869 Boncomp 2, 273-274
  (F2, 19) [Leibniz]
*Bruckner A M + 1966 AMM 73, 24-
  [bib, XX]
*De Morgan A 1837 Differential
  Calculus in *Penny Encyclopedia.*
Eneström G 1909 Bib M (3)9, 200-205
  [trig]
Hoppe E 1919 Arc MP (3)28, 100-101
  (F47, 25)
Jourdain P E B 1912 Int Con 5(2), 540
  [Foncenex, Lagrange]
May K O 1965 M Mag 38, 307-308
  [det]
Miller M 1954 Dres Verk 2(3)
  (Z59:1, 7) [Newton]

Peano G 1892 Mathesis (2) [def]
Stephens E 1924 Wash U St 12, 149-152
  [bib, fractional]
Wieleitner H 1929 DMV 38, 24-35
  [Fermat]


Descriptive geometry

See also Perspective (Ch. 3),
  Projective geometry.

Adamo M 1953 Cagliari 22(S), 1-12
  (MR15, 591; Z52, 244)
Amodeo F 1939 *Origine e sviluppo
  della geometria proietliva,*
  Naples
Appell P 1909 Rv Mois 8, 728-729
  [1612]
Bourlet C 1907 Ens M 9, 89-93 [Paris]
Court T H + 1935 Phot J, 54-66
  (F61, 8) [instru]
De Vries H 1914 N Tijd 3, 255-269
  (F45, 78; 46, 34) [Monge]
Gafney L 1965 MT 58, 338-344
  [Monge]
Kaderavek F 1949 Prag V (15), 7 p
  [MR12, 311) [Czech]
Lalanne L 1884 Par CR 98, 1466-1470
Lavicka W 1878 Cas MF 7 (F10, 28)
Laussedat A 1892 Fr AAS (2)21, 215-
  238 [photography]
Lietzmann W 1933 ZMN Unt 64, 34-35
  (F59, 16) [anc]
Loria G 1899 Pe MI (2)14(1), 1
- 1909 Arc GNT 1, 335-346 (F41, 13)
  [Olivier]
*- 1921 *Storia della geometria
  descrittiva dalle origini sino
  ai giorni nostri,* Milano, 584 p
  (F48, 30)
Mannheim A 1875 LMSP 6, 35, 36
  (F7, 349) [Sylvester]
Mueller E 1919 Ost IAZ (F47, 29)
  [ed college]
Neder L 1937 ZMN Unt 68, 8-16
  (F63-3)
*Obenrauch F J 1894 *Geschichte der
  darstellenden und projectii-
  vischen Geometrie...,* Bruenn,
  448 p

Descriptive geometry    (continued)

Papperitz E 1901 *Ueber die Wissen-*
  *schaftliche Bedeutung der*
  *darstellenden Geometrie...,*
  Freiberg, 24 p
Salkowski E 1926 Ber MG 25, 45-52
  [Hessenberg]
Schilling F 1904 *Ueber die Anwendun-*
  *gen der darstellenden Geometrie*
  *inbesondere ueber die Photogram-*
  *metrie,* Leipzig, 204 p
Taton René 1954 *L'historie de la*
  *géométrie descriptive,* Paris,
  25 p
Timerding H E 1910 *Uber Ursprung und*
  *Bedeutung der darstellende*
  *Geometrie,* Braunschweig, 16 p
  (F41, 66)
Torelli G 1875 Batt 13, 352-356
  (F7, 21)
Vaiman A A 1960 Ist M Isl 13, 379-
  382  [Babylonia]
Wiener Christian 1884 *Geschichte der*
  *Darstellenden Geometrie,* Leipzig


Determinants

See also Hessian, Jacobian, Matrix,
Permanents.


Brocard H + 1903 Intermed 10, 164
  (Q2307)  [Gasparis 1861]
Cayley A 1862 Brt AAS 32, 184-
  [lin transf]
Echols W E 1893 An M 7, 109-142
Fontené G 1900 Nou An (3)19, 188
  [Rouché]
Frobenius G 1905 Crelle 129, 179-180
  [def]
Gerhardt K I 1891 Ber Si, 407-423
  [Leibniz]
Guenther S 1874 Deu NA Tag, 78-
Hayashi T 1910 Tok M Ph (2)5, 254-
  271  [Japan]
- 1912 Batt 50, 193-211  (F43, 69)
  [Japan]
Kato H 1939 Toh MJ 45, 338-353
  (F65:1, 8)  [Japan, Wasan]
Lecat Maurice 1911 *Histoire de la*

*théorie des déterminants à*
*plusieurs dimensions,* Ghent
- 1913 Intermed, 9  (Q3719)
  [bib, n-dim]
- 1913 Intermed, 163 (Q4166)
  [bib, geom]
- 1919 Intermed 26, 79-  [bib, n-dim]
- 1920 Intermed 27, 21-23
  [bib, Hadamard]
- 1924 *Bibliographie de la relativité,*
  Brussels, 439 p  (extract re n-dim
  1924 Louvain, 16 p)  (F50, 19)
*- 1929 Brx SS(A)49, 23-, 87-
  [n-dim]
- 1929 Rv GSPA 40, 231-  [n-dim]
Locke L L 1941 MT 34, 183-184
  [notation]
*MacMahon P A 1927 LMSJ 2, 273-
  [n-dim]
Matz F P 1893 AMM 1, 383
  [Sylvester]
May K O 1965 M Mag 38, 307-308
  [deriv]
Mikami Y 1910 Tok M Ph (2)5, 292-394
  (F41, 58)  [re Hayashi 1910]
- 1914 Isis 2, 9-36  [Japan]
Miller G A 1930 AMM 37, 216-219
  (F56, 785)
Molk J + 1904 Enc SM 1(1)(1), 133-
  160
Muir Thomas 1884 Phi Mag (5)18,
  416-
- 1886-1889 Edi Sp 13, 547; 14, 452-;
  15, 481-; 16, 207-, 389-, 748-
- 1887 Nat 37, 246-, 343, 344, 438-
  439, 445  [error in Baltzer]
- 1890 *The Theory of Determinants in*
  *the Historical Order of Development*
  1(1), London, 278 p (2nd ed 1906)
- 1899-1905 Edi SP 22, 441-; 23, 93-
  132, 181-, 423-; 24, 151-, 244-;
  555-571; 25, 61-91, 129-159, 648-,
  908-947; 26, 357-389
- 1903 Nat 67, 512
*- 1905 *The Theory of Determinants in*
  *the Historical Order of Development,*
  London , 504 p  (Pt 1= 1890)
  [to 1841]
- 1907-1908 Edi SP 27, 135-166; 28,
  197-209, 676-702
- 1907 QJPAM 38, 237-264  [bib]

## Determinants (continued)

- 1909 Edi MSP 27, 5-9 [Waring]
- 1910 Edi SP (A)31, 296-303 [Wronskians]
- 1910 Edi SP 31, 304-, 311-
- 1911 *The Theory of Determinants in the Historical Order of Development* 2, London, 475 p [1841-1960]
- 1913 Edi SP 33, 49-63; 34-59
- 1919 So Af RS 8, 101- [Koch]
- 1920 *The Theory of Determinants in the Historical Order of Development* 3, London, 530 p [1861-1880]
- 1923-1927 Edi SP 43, 127-148; 44, 218-241; 45, 51-55, 187-212; 46, 46-70; 47, 11-33; 252-282
- 1923 *The Theory of Determinants in the Historical Order of Development* 4, London 539 p [1880-1900] (repr vols 1-4, Dover)
- 1926 So Af RS 13, 299-308 [Hadamard]
- 1927 QJPAM 50, 333-349 [bib]
- 1927 So Af RS 14, 367-371 [Hessioms]
- 1928-1929 Edi SP 48, 37-57; 49, 1-15, 264-288 (F54, 6)
- 1929 So Af RS 18, 219-227 [bib]
- 1930 *Contributions to the History of Determinants 1900-1920,* London, 408 p [index to vols 1-4]

Nekrasov P 1886 Fiz M Nauk (A)2, 169-178

Relewicz S + 1968 *Equations in Linear Spaces,* Warsaw [bib]

Schlegel V v 1896 Hist Abt 41, 1, 41

Schuchmann H 1914 MN Bl 11, 65-68, 86-89 [Japan]

Studnicka F J 1876 Cas MF 5, 1-, 8-, 193-, 279- [Cauchy]
- 1877 Prag Ab 8(3) [Cauchy]

Vogt H + 1903 Intermed 10, 53 [Gasparis]

## Devil's curve

Borel E 1927 AMM 34, 365

## Dido curve

Lemoyne T 1914 Intermed 21, 85-86 (Q3549)

## Differential

See also Calculus, Infinitesimals.

Nashed M Z 1966 AMM 73(3), 63-
*Peano G 1913 Tor FMN 48, 47-69
Poincaré H 1899 Ens M 1, 106

## Differential algebra

*Kaplansky Irving 1957 *An Introduction to Differential Algebra,* Paris (Act Sc Ind 1251) [survey, bib]
Ritt J F 1950 *Differential Algebra,* AMS Col 33 (repr 1966 Dover)

## Differential equations

See also Partial differential equations, Lie groups, Linear differential equations, Singular solutions, Mechanics (Ch.3), Pursuit problems

Abraham Ralph + 1967 *Transversal Mappings and Flows* (Benjamin), 161 p [1880-1962]
Baily P B + 1968 *Nonlinear Two Point Boundary Value Problems* (Academic P) [bib]
*Bennett A A + 1931 *Numerical Integration of Differential Equations,* (Wash DC (NRC Div Phys Sci Bull 92) (repr 1956 Dover) [bib]
Bocher M 1911 AMSB (2)18, 1-18 [Sturm]
- 1912 Int Con (5)1, 163- [exist thms]
Boehmer P E 1936 Bl Versi 3, 399-403 (F62, 24)
Braunmuehl A v 1905 Int Con 3, 551-555 (F36, 54)

## Differential equations (continued)

Burkhardt H 1908 DMV 10(2), 1393-
1804 [period funs, phys]
Cartwright M L 1952 M Gaz 36, 81-88
[non-lin vib]
Fiorentini P 1906 Batt 44, 25-88,
291-313 [ordin first order]
Forsyth A R 1923 M Gaz 11, 73-81
[mech, phys]
Frankl F I 1951 UMN 6(2)(42)
[char eq]
Graig T 1893 NYMS 2, 119-134
(F25, 68) [ordin, 1878-1893]
Haimovici A 1954 Iasi UM 1, 413-426
(MR18, 710; Z59:1, 9; 60, 9)
Ince E L 1927 *Ordinary Differential
Equations*, NY (repr 1956 Dover)
[bib]
Kaplansky Irving 1957 *An Introduc-
tion to Differential Algebra*,
Paris (Act Sc Ind 1251)
[Galois thy]
Knapp H + 1968 Ueberbl 1, 87-114
[numer intgl]
Kushnir E A 1957 Ist M Isl 10, 363-
370 (Z101, 9) [Euler]
Miwa K 1888 Tok M Ph 3, 245-248
Murray F J + 1954 *Existence Theorems
for Ordinary Differential
Equations* (NY U P)
Rothenberg S 1908 *Geschichliche
Darstellung der Entwicklung der
Theorie der singulaeren Loesungen
totaler Differentialgrleichungen
von der ersten Ordnung miz zwei
variablen Groessen*, Leipzig-Berlin,
90 p (Ab GM 20(3))
Rubini R 1879 Batt 17, 149-159
(F11, 40) [2nd order]
Rybnikov K A 1958 Ist M Isl 11,
583-592 [alg]
Simonov N I 1954 Ist M Isl 7, 513-
595 (MR16, 781) [Euler]
- 1958 UMN 13(5), 223-228
(Z85, 5) [Euler]
- 1959 Mos IIET 28, 138-187
[Euler]
- 1963 Ist M Zb (4) [Pet]
- 1966 Ist Met EN (5), 157-174
(MR33, 7233) [XVIII-XIX]

## Differential geometry

See also Curvature, Geodesic, Tensors.

Blashke W 1954 Pisa SNS (3)8, 43-52
(Z56, 2) [Bianchi]
Boyer C B 1954 AMM 61, 459-463
(Z56, 243) [Carnot, deviation]
*Chern S S ed 1967 *Studies in Global
Geometry and Analysis*, (MAA Stu 4)
197 p [bib]
Darboux G 1908 BSM (2)32, 106-128
(F39, 71)
- 1909 Int Con 1, 105-122 (F40, 68)
Delegue 1869 Dunkerq 14, 137-
[Pascal]
Gheorghiu G T 1956 in *Tucrarile
consfatuirii de geometrie differen-
tiala din 9-12 Iunie 1955*, 55-62
(Z74, 246) [Titeica]
Liouville J 1850 in G. Monge,
*Applications de l'analyse à la
géométrie*, 5th ed, Paris, Note I
Pinl M 1956 MP Semb 5, 34-48
(MR21, 892) [1856-1956]
Ruse H S + 1961 *Harmonic Spaces*,
Rome (Cremonese), 240 p
[bib, classical diff geom 1940-1960]
Sperry P 1931 Calif Pu 2(6), 119-
127 [*bib, proj]
Staeckel P 1914 Heid Si 5(A), 27 p
*Struik D J 1922 *Grundzuge der
mehrdimensionalen Differential-
geometrie in director Darstellung*
(Springer) [*bib]
- 1925 *Ueber die Entwicklung der
Differential geometrie*, Utrecht
(DMV 14, 14-25)
- 1926 Rm Sem (2)3
- 1933 Isis 19, 92-120; 20, 161-191
(Z7, 388) [(Russ tr 1941)
- 1926 Su Buchen + 1959 Sci Sin 8(11),
1238-1242 (repr 1960 AMSN 7(2),
163-168 [China, 1948-1958]
Svec A 1960 Cas M 85, 389-409
(MR23A, 1299) [Czech]
Terracini A 1935 Pe M (4)15, 1-21
(F61, 8; Z10, 386) [origin]
- 1950 Tor Sem 9, 97-123 (MR12, 382)
[Fubini, proj dif geom]
Varga O 1951 M Lap 2, 190-218
(MR13, 611) [Russ]

## Differential geometry   (continued)

Veblen O 1928 Int Con 8(1), 181-
  [dif invar]
Villa M 1956 It UM (3)11, 591-593
  (MR18, 710)   [Rumania]
Vranceannu G +1966 Ro Rv M 11, 1147-
  1156   [Rumania]
Vygodskii M J 1935 Len IINT 6, 63-96
  [Monge]
Wilczynski E 1916 AMSB (2)22, 317-
  329   [curve]

## Dimension theory

Brouwer L E J 1928 Amst P 31, 953-
  957   (F54, 7)
Favaro J 1950 *Espace et dimension,*
  Paris, 302 p
Menger K 1930 Mo M Phy 37, 175-182
  (F56, 6)   [Brouwer]

## Diophantine equations

See also Number theory, Four cube
  problem.

Archibald R C 1933 Scr 2, 27-33
  [quad]
Carmichael R D 1924 MT 16, 257-265
  [empirical thms]
Cattaneo P 1938 Bo Fir (2)17, 60-62
  (F64, 906)   [Hindu]
Chatelet F 1960 Ens M (2)6, 3-17
  (Z96, 3)
Datta B 1931 Archeion 13, 401-407
  (F57, 18)   [Hindu]
- 1932 Clct MS 24, 19-36
  (Z5, 148)   [Aryabhata]
- 1938 Archeion 21, 28-34
  (F64, 11)   [ast]
Dickson L E 1919 Dickson 2
Gandz S 1948 Osiris 8, 12-40
  [Babyl]
Ganguli S 1931 Indn MSJ 19, 110-120,
  129-142   (Z5, 149)   [Hindu]
- 1932 Indn MSJ 19, 153-168
  (Z5, 149)   [Hindu]
Kamalamma K N 1948 Clct MS 40, 140-
  144 (Z41, 338)   [Hindu]

Katscher F 1965 Physis Fi 7, 107-114
  [cubic]
Schutter C 1952 MT 45, 605-606
Sen-Gupta P C 1919 Clct MS 10, 73-80
  (F47, 26)   [Hindu]
Shankar Shukla K 1951 Ganita 1, 1-12,
  53-64 (MR13, 420)   [Sridhara]
Simon M 1913 Arc M 20, 280-281
  [Brahmagupta]
Smith D E 1917 AMM 24, 64-71
Van der Waerden B L 1955 Zur NGV 100,
  153-170   [ast, Hindu]
Van Hee L 1914 Toung Pao 15, 203-210
  [China]
Woepke F 1853 *Extrait du Fakhri,*
  *précédé d'une memoire sur l'algebre*
  *indétermineé chez les Arabes,*
  Paris

## Dirichlet problem

See also Potential.

Bouligand G 1960 BSM (2)84, 111-115
  (MR22, 11226)   [origin]
Courant R 1950 *Dirichlet's Principle*
  (Interscience)
Kellogg O D 1926 AMSB 32, 601-625
  [recent]
- 1928 AMSB 34, 154
La Vallée Poussin C J de 1962 Rv Q
  Sc 23, 133, 327-   [Dirichlet]
Manheim Jerome H 1964 *The Genesis of*
  *Point Set Topology* (Pergamon P),
  179 p   (MR37, 2561)

## Discrete distributions   (statistics)

Patil G P ed 1965 *Classical and*
  *Contagious Discrete Distributions,*
  Calcutta   [*bib]

## Dissection

Archibald R C 1915 *Euclid's Book on*
  *Divisions of Figures,* Cambridge
  96 p
Brown W G 1965 AMM 72, 973-977

## Distance

Boyer C B 1948 AMM 55, 556-557
[Clairaut]
Natucci A 1933 Pe M (4)13, 47-52,
161-165, 301-310  (F59, 827)
[between skew lines]
Wieleitner H 1931 ZMN Unt 62, 18-20
(F57, 25)  [point to line, Leibniz,
Schooten]

## Distributive law

Pincherle S 1899 Bib M (2)13, 13-18
(F30, 45)  [bib]

## Divergent series

Burkhardt H 1910 M Ann 70, 169-206
[1750-1860]
Cajori F 1923 AMSB 29, 55
(F49, 20)  [name]
Chirikov M V 1960 Ist M Isl 13, 441-
472  (MR24A, 3)
Gloden A 1950 Int Con HS 6, 178-186
*Hardy G H 1949 *Divergent Series,*
Oxford
Karamata J 1959 Ens M (2)5, 86-88
(MR23A, 2662)  [Mengoli]
Pastori M 1927 Pe M (4)7, 302-320
(F53, 2)
*Smail Lloyd L 1925 *History and
Synopsis of the Theory of Summable
Infinite Processes* (U oregon P),
175 p  [bib]

## Divisibility

Bashmakova I G 1949 Ist M Isl 2
Dickson L E 1919 Dickson 1
Fontes 1893 Tou Mm (9)5, 459-475
[7, 9, Lagrange, Pascal]
Gillings R J 1956 Scr 22, 294-296
[tests]

## Division

See also Fractions, Ratio and
Proportion.

Beman W W 1887 Bib M (2)1, 96
(Q17)  [sign]
- 1900 Intermed 7, 86  [sign]
Bobynin V V 1899 Bib M (2)13, 81-85
(F30, 38)  [frac]
- 1899 ZM Ph 44(S), 1-13  (F30, 36)
[Egypt]
Cajori F 1914 Nat 94, 477  (F45, 95)
[sign]
- 1921 AMSB 27, 150-151  (F48, 44)
[sign]
Gandz S 1936 Isis 25, 426-432
[Babyl recip tables]
Gillings R J 1955 Au JS 18 (Oct),
43-49  [Greek]
Gillings R J +1964 Au JS 27, 139-141
(MR30, 1)  [Babyl recip tables]
Glaisher J W L 1927 Mess M 57, 38-
71  [in alg]
Lam L Y 1966 Brt JHS 3-1 (9), 66-69
(MR34, 4087)  [China, galley
method]
*Leslie John 1817 *The Philosophy of
Arithmetic,* Edinburgh, 169 p
[galley meth]
Mathieu 1896 JM El 20, 97
[Legendre, XVIII]
Palatini F 1895 Pe MI 10, 169
[def]
*Peacock 1829 Enc Met 1, 433
[galley meth]
Romig H G 1924 AMM 31, 387-389
(F50, 9)  [by zero]
Simons L G 1933 Scr 1, 362-363
Vetter Q 1921 Prag V 14, 1-25
(F48, 37)  [Egypt]

## Division algebra

Cartan E 1908 Enc SM 1(1), 360-362
*Curtis C W 1963 in A A Albert ed,
*Studies in Modern Algebra*
(MAA Stu 2), 100-125)
*Hankel H 1867 *Theorie der Complexen
Zahlen,* Leipzig, 106-108

Division algebra  (continued)

Kurosh A G 1963 *Lectures on General Algebra,* NY, Ch. 5
May K O 1966 AMM 13, 289-291 [vectors]
*Natucci Alpinolo 1923 *Il ancetto di numero...,* Turin, 482 p
- 1925 Bo Fir, 15 p


Duality

See also Projective geometry.

Bourbaki Nicolas 1960 *Eléments d'histoire des mathématiques,* Paris, 78-, 89-
Mueller J H T 1860 Arc MP 34, 1-
Nuvoli L 1960 Physis Fi 2, 101-120 (Z102, 6)


Duplation

Hirsch M 1952 MT 44, 591 [Saadia Gaon]


e

See also Irrational, Transcendental.

Archibald R C 1921 AMM 28, 116-120 [complex expo]
Beman W W 1894 Bib M (2)8, 32 (Q23) (F25, 65) [Euler]
Coolidge J L 1950 AMM 57, 591-602 (MR12, 381; Z41, 340)
*Duarte F J 1949 *Monografien uber die Zahen π und e,* Caracas, 246 p (Z41, 340) [bib]
Klein F 1893 in his *Lectures on Mathematics,* NY (Macmillan), 51-57 [transc]
Mitchell U G + 1936 Osiris 1, 476-496 (F62, 3; Z14, 244)
Pall G 1949 AMM 56, 682

Eigenvalues

Gould S H 1966 *Variational Methods for Eigenvalue Problems* (U Toronto P), 2nd ed [bib]
*Wilkinson J H 1965 *The Algebraic Eigenvalue Problem* (Clarendon P), 662 p [bib]


Eight square problem

See Four and eight square.


Ellipse

Ameline 1934 Rv GSPA 45, 243-247, 323-324 (F60, 840) [Egypt]
Boyer C B 1947 Isis 38, 54-56 (MR9, 169) [Copernicus, epicycle, Lahire]
Easton J 1965 AMM 72, 53-56 [Nasir al-Din, Schooten, J de Witt]
Endo T 1906 Tok M Ph 3, 72-74 [Japan, length]
Hermann 1928 Unt M 34, 359-360 (F54, 26)
Hofmann J E + 1930 Nieu Arch (3)16, 5-22 (F56, 6)
Inglis A 1933 M Gaz 17, 327-328 (F59, 843) [Gregory, length]
Johnson R A 1930 AMM 37, 188-189 [length]
Mazkewitsch D + 1961 MT 54, 609-612 [trammel constr]
Mikami Y 1912 Bib M 12, 225-237 [Japan, length]
Wieleitner H 1928 Unt M 34, 276-277 (F54, 26) [polar coord]
Wilson C 1968 Isis 59, 5-25 [Kepler]
Yanagihara K 1915 Toh MJ 7, 74-77 [Japan]


Ellipsoid

Grube F 1883-1888 *Zur Geschichte des Problems der Anziehung der Ellipsoide,* Schleswig, 2 vols

Ellipsoid   (continued)

- 1869 ZM Ph 14, 261-266
  (F2, 29) [MacLaurin]


Elliptic functions

See also Algebraic functions.

Adams O S 1925 *Elliptic Functions
  Applied to Conformal World Maps,*
  Wash D C (US Dept Commerce 112)
  [C S Peirce]
Anon 1913 Isis 1, 245
Bellacchi G 1894 *Introduzione
  storica alle teoria delle
  funzioni ellittiche,* Florence,
  320 p  (F25, 75)
Brown E W 1893 AJM 15, 321
  [lunar thy]
Dantzig T 1935 AMSB 41, 795
  (F61, 958)
*Enneper Alfred 1890 *Elliptische
  Funktionen. Theorie und Geschichte,*
  2nd ed, Halle, 617 p  (1st ed
  1875) [bib]
Gunther P 1894 Gott N, 92-105
  [Gauss]
- 1897 JMPA (5)3, 95-112  [Gauss]
Haentzschel E 1913 DMV 22, 278-284
  [Euler, Weierstrass]
Klein F 1893 in his *Lectures on Mathe-
  matics...,* 75-84 [hyperelliptic,
  Abelian]
Koenigsberger L 1879 *Zur Geschichte
  der Theorie der elliptischen
  Transcendenten in den Jahren
  1826-1829,* Leipzig, 104 p
  (F11, 40)
Nalbandyan M B 1965 Ist M Isl (16),
  191-206  (MR33, 1212)  [Zolotarev]
Natucci A 1953 Geno AL 9, 40-54
  (MR14, 1049; Z53, 197)  [Jacobi]
Pokrowskii P M 1886 Fiz M Nauk 2,
  47-65 [Abel fun] [ultra-elliptic
  and Abelian]
Russel W H L 1869-1873 Brt AAS 39,
  334-; 40, 102-; 42, 334-; 43,
  307-   (F5, 48)  [recent]
Siegel C L 1959 in Schroeder K ed,
*Sammelband der zu ehren des 250
  geburtstages des deutschen akademie
  der wissenschaften...,*  Berlin,
  315-318 [Euler]
Watson G N 1933 M Gaz 17, 5-17
  [arc length, Fagnano, Landen]


Elliptic integrals

Hancock H 1917 *Elliptic Integrals*
  (Wiley) (repr 1958 Dover)
Krazer A 1909 Bib M (3)10, 250-259
  (F40, 60) [Jacobi, Puiseux]
Mikami Y 1911 Bib M (3)12, 225-237
  (F42, 63) [Japan]


Elliptic point

Calleri P 1938 Pe M (4)18, 33-42
  (F64, 917)


Enumerative geometry

Loria G 1888 Bib M (2)2, 38-48,
  67-80
- 1895 Bib M, 51  [Desargues]
Severi F 1948 *Grundlagen der
  abzaehlenden geometrie,*
  Wolfenbuettel


Envelope

Bortolotti E 1921 Pe M (4)1, 263-276
  [calc]
Mackay J S 1905 Edi MSP 23, 80-88
  [bib, Wallace line]
Procissi A 1953 Pe M (4)31, 34-43
  (Z51, 3) [Torricelli]


Equations   (algebraic)

See also Cubic, False position,
  Fundamental theorem, Galois theory,
  Horner method, Linear, Newton-Raphson
  method, Quadratic, Quartic,
  Quintic, Symmetric functions,
  Tschirnhausen transformation.

Equations (algebraic)   (continued)

Aubry A 1894 JM Sp 18, 225-228,
   245-253, 276-279   (F25, 75)
- 1895 JM Sp 19, 14-17, 36-40,
   67-70, 81-85, 111-113, 127-131,
   153-156, 181-185, 197-200, 228-231,
   245-248, 269-76 (F26, 49)
- 1896 JM Sp 20, 76-, 254-
- 1897 JM Sp 21, 17-30, 61-62,
   83-88, 114-115, 131-132, 155-159
   (F28, 39)
Bashmakova I G 1960 Int Con HS 9,
   427-428   (Z114, 7)
Bateman H 1938 AMM 45, 11   [Halley]
Bocher M 1911 AMSB 18, 1-18
   (F42, 58)   [Sturm]
Bortolotti E 1906 It SS (3)14, 291-
   325 (F37, 8)   [Ruffini]
Bosmans H 1926 Mathesis 40, 59-67,
   100-109, 145-155   (F52, 17)
   [Girard]
Boyer C B 1943 Scr 11, 5-19
   [nomog]
Burnside William S + 1899-1901
   Theory of Equations, London
   (7th ed 1912; repr 1960 Dover)
Cajori F 1910 Colo CPS 12(7), 171-
   286 [approx]
Dehn E 1930 Algebraic Equations.
   An Introduction to the Theories
   of Lagrange and Galois,  NY,
   (repr 1960  Dover)
Dressler H 1905 Unt M 11, 82-83
   (F36, 51)   [appl]
Drobisch M W 1834 Grundzuege der
   Lehre von den hoeheren numerischen
   Gleichungen, Leipzig
Dickson L E 1903 Introduction to the
   Theory of Algebraic Equations,  NY,
   104 p  (repr 1967 Chelsea)
Favaro 1878 Mod Mm 18, 127-330
   [*bib]
*Fine H B 1892 NYMS 1, 173-184
   (F23, 32)   [Kronecker]
Foulkes H O 1932 Sci Prog 26, 601-608
   (F58, 986)   (1771-1932)
Gloden A 1958 Janus 47, 73-78
   (MR20, 945)
Gongalves J V 1958 Lisb (A)(2)7,
   57-58   (MR20, 847)

Hayashi T 1931 Toh MJ 34, 145-185
   (Z2, 243)   [Japan]
Karpinski L C + 1929 Sci (Sept 27),
   311-314 [Greek]
Kiro S N 1957 Vop IET (4), 169-171
   (Z80, 3)   [calcn mach]
Klein F 1893 in his Lectures on
   Math..., 67-74
Lemaire G 1906 Intermed 13, 237 (Q82)
   (See also Intermed 1, 34; 2, 96,
   210) [Bernoulli, methods of
   cascades]
Loria G 1938 Mathesis 52, 129-131
   [irrat]
Mattheissen L 1878 Grundzuege der
   antiken und modernen Algebra der
   Literalen Gleichungen, Leipzig
   [bib]
Miller G A 1911 in Young J W A ed,
   Monographs on Topics of Modern
   Mathematics..., Ch. 5
   (repr 1955 Dover)
- 1924 SSM 24, 509-510   (F50, 6)
- 1926 Toh MJ 26, 386-390   (F52, 3)
- 1939 Sch Soc 49, 178-179   (F65, 6)
   [anc]
Nordgaard M A 1922 A Historical
   Survey of Algebraic Methods of
   Approximating the Roots of
   Numerical Higher Equations up to
   the Year 1819, NY, 70 p
Rogachenko V F 1953 Ist M Isl 6,
   477-494   (Z53, 246) [approx,
   Lobachevsky]
Salmon George 1876 Lessons Intro-
   ductory to the Modern Higher
   Algebra, 3rd ed, Dublin, 298-392
   [Tchirnhausen]
Shain J 1937 AMM 44, 24-29
   [Rolle's meth of cascades]
Sorokina L A 1959 Ist M Isl 12,
   457-480   [Abel]
*Tropfke Johannes 1922 Geschichte der
   Elementar-Mathematik 3, 20-151
Verriest G 1934 Rv Q Sc (4)25, 341-
   366   [Galois]
Vetter Q 1928 Archeion 9, 175-176
   (F54, 46)   [XIV, Libri]
- 1930 BSM (2)54, 277-283
   (F56, 15)   [van Roomen]
Wieleitner H 1925  Arc Sto 5, 46-48
   [Arab]

## Ergodic

Birkhoff G D + 1932 Am NASP, 281
*Birkhoff G D 1942 AMM 49, 222-
Halmos Paul R 1949 AMSB 55, 1015-
1034 [*bib]
- 1958 AMSB 64, 86-94
(MR20, 620) [meas, von Neumann]
- 1956 *Lectures on Ergodic Theory*
Chelsea [bib, mech, top]
- 1958 AMSB 64(3)(2), 86-94
(Z80, 4) [von Neumann]
Kakutani S 1952 Int Cong (1950) 2,
128-142 [bib]
Oxtoby J C 1952 AMSB 58, 116-136

## Erlanger Program

Hoefling D 1932 ZMN Unt 63, 153-159
(F58, 995)

## Euclidean algorithm

Junge G 1932 Fors Fort 8, 331-332
Szabo A 1963 M Ann 150, 203-217
[Pythagoras]

## Euclidean constructions

See also Constructions, Geometro-
graphy, Malfatti problem, Regular
polygons.

## Euclidean constructions: General

Hess A L 1954 M Mag 29, 217-221
Hofmann J E 1956 Centau 4, 177-184
[Viete]
Horadam A F 1960 M Gaz 44, 270-276
Kazarinoff N D 1968 AMM 75, 647-
[Gauss, reg polyg, Wantzel]
*Niebel Eckhard 1959 *Untersuchungen
ueber die bedeutung der
geometrischen konstruktion in der
antike,* Cologne, 147 p
[existence]
*Pagliano C 1902 Bo Fir 1, 94
[compet, Ferrari, Tartaglia]

Schrek D J E 1950 Euc Gron 25, 169-
172 [Strode]
Steele A D QSGM (B)3, 287
[Greek]
Tannery Paul 1881 Bord Mm (2)4,
395-417 (F13, 26) [before Euclid]
Tropfke J 1928 ZMN Unt 59, 193-206
[Archimedes, al-Biruni, heptagon,
nonogon]
Yates R C 1954 MT 47, 231-233
*Zeuthen H G 1895 M Ann 46, 222
[exist, Greek]

## Euclidean constructions: Three classical problems

See also Delian problem, Squaring
the circle, Trisection.

Carrara B 1901 Rv FMSN 2, 36,
115, 208, 304, 492
- 1902 Rv FMSN 3, 296, 481, 696,
761, 926-939, 1056-1071
- 1903 Rv FMSN 4, 39-60, 142-156,
337-351, 442-453
- 1904 *I tre problemi classici degli
antichi,* Pavia
De Morgan Agustus 1872 *A Budget of
Paradoxes* (Athenaeum)
Dickson L E 1911 in J W A Young ed
*Monographs on Topics of Modern
Mathematics...,* (repr 1955 Dover)
Gambioli D 1903 *Breve sommario della
storia delle matematiche colle due
appendici sui matematici italiani
e sui tre celebri problemi
geometrici dell'antichita,*
Bologna, 241 p
Gibbens G 1930 AMM 37, 343-348
(F56, 813) [ruler]
Hudson H P 1953 in *Squaring the
Circle and Other Monographs*
(Chelsea), 9-31
Klein F 1930 *Famous Problems of
Elementary Geometry,* 2nd rev ed.
92 p (repr 1956 Dover)
(1st ed 1895) [Notes by R C
Archibald]
Saalschuetz L 1899 Konig Ph 39, 8-14
(F30, 51) [Delian problem,
trisection]

## Euclidean constructions: Three classical problems (continued)

Teixeira Francisco G 1915 *Sur les problèmes célèbres de la géométrie élémentaire non résolables avec la règle et la compas,* Coimbra, 132 p

## Euclidean geometry

See also Angle, Antiparallel, Apollonius problem, Bisectors, Center of gravity, Circle, Congruence, Curves, Dissection, Euclidean constructions, Line, Menelaos theorem, Parallelogram, Parallels, Porisms, Pythagorean theorem, Triangle.

Allman G J 1877 *Greek geometry, from Thales to Euclid,* Dublin (F9, 21) [Also 1880, F12, 35)
Bobynin V V 1907-1908 Pet MNP (11), 53-113; (1), 1-50 (F38, 64) [XVIII]
Bouligand G 1951 *L'accès aux principes de la géométrie Euclidienne,* Paris, 86 p
Cajori F 1924 AMSB 30, 387 (F50, 19) [symbols]
*Court Nathan A 1952 *College geometry,* 2nd ed (Barnes and Noble)
- 1964 MT 57, 163-166 [origin]
Dodgson C L 1878 *Euclid and His Modern Rivals,* Oxford
Enriques F 1927 *L'évolution des idées géometriques dans la pensée grecue: point, ligne, surface,* Paris, 56 p
Frankland W B 1902 *The Story of Euclid,* London [pop, non-Euc geom]
Hofmann Joseph E 1959 *Streifzug durch die Entwicklungsgeschichte des elementar geometrischen Methoden,* Frankfurt, 28 p (Z83, 243)
Hauser G 1955 *Geometrie der Griechen von Thales bis Euklid,* Luzern, 176 p

Huber T 1917 Bern NG, 53-57
Jones P S 1945 MT 37, 3-11 [U S]
Karpinski L C + 1925 SSM 24, 162-167 [termin]
Milhaud G 1899 Rv GSPA 10, 847-854 (F30, 50) [Greek]
*Mugler Charles 1958-1959 *Dictionnaire historique de la terminologie géométrique des grecs,* Paris, 2 vols (Z88, 241)
Natucci A 1933 Pe M (4)13, 47-50 (Z6, 145) [min dist]
- 1933 Pe M (4)13, 161-165 (Z6, 337) [irrat]
- 1942 It UM Con 2, 966-976 (MR8, 497) [-III to XX]
Ruska J + 1939 Bo Fir 35, i-ii
Segner Andrea 1739 *Elementa Arithmetica et Geometrial* Goettingen
Simon Max 1906 *Ueber die Entwicklung der Elementar-Geometrie im XIX Jahrhundert,* Leipzig, 286 p (DMV 15(S) [bib]
- 1906 *Ueber die Entwicklung der Elementar Geometrie im XIX. Jahrhundert,* Leipzig [XIX]
Siriati L 1948 Pe M (4)26, 65-73; 27, 133-139 (Z36, 4) [XIX]
Stephanos C 1894-1895 Intermed 1, 210; 2, 231, 364, 420 (Q347) [existence]
Tannery P 1887 *La géométrie grecque. Comment son histoire nous est parvenue et ce que nous en savons,* (Gauthier-Villars)
Taylor C 1900 Cam PST 18, 197 [Kepler, Newton]
Tropfke Johann 1923 *Geschichte der Elementar-Mathematik in systematischer Darstellung..,* Berlin, 242 p
Vereecke P 1937 Mathesis 51, 11-14 (F63, 799) [Greek]
Zeuthen H G 1913 Nyt Tid (A)24, 105-124 (F44, 3) [Plato]

## Exponent

Bortolotti E 1927 Arc Sto 8, 49-63 [Bombelli]

## Exponent    (continued)

Boyer C B 1944 M Mag 18, 81-86
  [frac, Oresme]
- 1950 AMM 57, 7-8 (MR11, 572;
  Z35, 146) [Hume]
Cajori F 1923 AMSB 29, 195
  (F49, 23) [nota]
- 1924 SSM 23, 573-581 [nota]
Enestrom G 1905 Bib M (3)6, 324-
  325, 410 (F36, 52) [fifth power]
Lefebre B 1899 Intermed 5, 63, 71
  [frac, neg, Newton]
Mahnke D 1913 Bib M (3)13, 250-260
  [Leibniz]
Smith D E 1915 in C G Knott ed,
  *Napier Tercentenary Memorial
  Volume,* London, 81-89 [XVI]
Tannery P + 1897 Intermed 4, 60
  (Q579) [nota]
Vincent A J H 1847 Nou An 6, 35
Wieleitner H 1922 Unt M [Stifel]
- 1924-1925 Isis 6, 509-520; 7,
  490-491 [frac]

## Exponential function

See also Logarithm.

Archibald R C 1914 AMM 21, 253
Burkhardt H + 1913 Bib M 13,
  150-153 [interpol]
Cajori F 1913 AMM 20, 5-14, 35-47,
  75-84, 107-117, 148-151, 173-182,
  205-210 [and log]
Sanford V 1952 MT 45, 451, 454
  [political wager 1828]

## Factor analysis

See also Statistics.

Harman Harry H 1967 *Modern Factor
  Analysis* 2nd ed (U Chicago P)
  [*bib]

## Factorial

See also Gamma function.

Cajori F 1921 AMSB 27, 298   (F48, 44)
  [nota]
- 1921 Isis 3, 414-418  [nota]
Dickson L E 1919 Dickson 1, Ch 9

## Factoring

Aubry A 1913 Ens M 15, 202-231
Dickson L E 1919 Dickson 1, Chs 11,
  13, 14, 16
Glaisher J W L 1878 Cam PSP 3, 99-138,
  228-229
- 1880 *Factor Table for the Fourth
  Million,* London
Kraitchik M 1952 Scr 18, 39-52
  (MR14, 121)
Lehmer D N 1918 Sci Mo 7, 227-234
Seelhoff P 1884 Arc MP 70, 413-427
  [tables]
Shankar Shukla K 1966 Ganita 17,
  109-117 (MR37, 2563)  [Hindu]
Wisner R J + 1959 MT 52, 600-603
  [unique prime fact]

## False position

See also Linear equations.

Eves H 1958 MT 51, 606-608
Matthiessen L 1870 ZM Ph 15, 41-47
  (F2, 25) [Arab, Hindu, mid]
Natucci A 1931 Bo Fir (2)10, 166-172
  (F57, 23)
Porcu L 1960 Pe M (4)38, 95-112,
  149-154, 213-227 (MR22, 1593)
Sanford V 1951 MT 44, 307-310
Suter H 1908 Bib M (3)9, 111-122
  (F39, 56) [Qusta ben Luqas]
Thureau-Dangin F 1938 Rv Assyr 35,
  71-77 (F64, 8) [alg]
Vogel K 1960 MP Semb 7, 89-95
  (Z105, 1) [Babyl]
Vygodskii M Ya 1960 Ist M Isl 13,
  131-252 (MR29, 221)

## Farey series

Dickson L E 1952 Dickson 1, 155-158
Glaisher J W L 1878 Cam PSP 3, 194
- 1879 Phi Mag (5)7, 321-336

## Fermat last theorem

Bachmann Paul 1919 *Das Fermat-
problem in seiner bisherigen
Entwicklung,* Berlin, 168 p
- 1919 *Der Fermatsche Satz,*
Berlin
*Bell E T 1961 *The Last Problem,*
NY
*Dickson L 1917 An M (2)18, 161-187
[alg numb, bib]
- 1919 Dickson 1, Ch. 4
Fabarius W 1914 *Leonhard Euler und
das problem Fermat,* Cassel, 12 p
(F45, 95)
Gambioli D 1900 Pe MT (2)3, 145
[bib]
- 1900 Pe MI (2)15, 4, 48  [bib]
Hensel Kurt W S 1910 *Ernst Eduard
Kummer und der grosse Fermatsche
Satz,* Marburg, 22 p
Lietzmann W 1930 *Der Pythagorische
Lehrsatz: mit einem Ausblick auf
das Fermatsche Problem,* 4th ed,
Leipzig, 75 p
Lind B 1910 *Ueber das letzte
Fermatsche Theorem* (Teubner)
[bib]
Mordell L J 1920 *Three Lectures on
Fermat's Last Theorem,* Cambridge
(repr 1955 in Klein, *Famous
Problems* (Chelsea))
Nagel T 1921 Nor M Tid 3, 7-21
Noguès R 1932 *Théorème de Fermat.
Son histoire,* Paris, 177 p
*Vandiver H S 1946 AMM 53, 555-578
[bib]

## Fibonacci sequences

See also Golden section.

Archibald R C 1918 AMM 25, 235-238
[bib]

Brother V Alfred 1965 *Introduction to
Fibonacci Discovery* (St. Mary's
C, Calif)
Dickson Leonard E 1919 Dickson 1,
Ch. 17
Guirao P 1957 Gac M 9, 43-53
(Z87, 241)

## Field

Baumgart J K 1961 MT 54, 155-160
[1966 SMSGRS 1]
Landsberg G + 1910-1911 Enc SM
1(2)(2-3), 233-385 [alg, alg geom]

## Figurate numbers

See also Binomial coefficients.

Biermann K-R 1961 Fors Fort 35, 195-
198  (Z99, 3)
Diophantus 1926 *Les six livres
arithmétique et les livres des
nombres polygons,* Bruges
Hayashi T 1932 Toh MJ 35, 171-226
[fin diffc, Japan, series]
Loria G 1935 in *Studi in onore
Salvatore Ortu Carboni,* 233-238
(Z13, 193)
Michel P H 1958 *Les nombres figurés
dans l'arithmétique pythagoricienne,*
Paris, 23 p  (Z90, 5)
Wertheim G 1897 Hist Abt 42, 121-126
(F28, 39)  [Diophantus]

## Finite differences

Adams C R 1931 AMSB 37, 362-400
[lin diffc eq]
Eneström G 1894 Sv Ofv, 177
[Nicole, Taylor]
Hayashi T 1932 Toh MJ 35, 171-226
(F58, 26)  [Japan]
Hofmann J E + 1931 Ber Si, 562-590
(F57, 27)  [Leibniz]
Lacroix S F 1819 *Traité du Calcule
Différentiel et du Calcule
Intégrale,* Paris, 2nd ed, Vol 3, 75-
[bib]

## Finite differences (continued)

Likhin V V 1966 Ist Met EN 5, 35-44
(MR34, 9) [Euler, Lagrange]
Mikami Y 1911 Toh MJ 1, 98-105
[Japan, Shoko]
*Noerlund N E 1924 *Vorlesungen ueber Differenzrechnung,* Berlin
[bib]
Taton R 1950 Osiris 9, 44-61
[Monge]

## Finite geometries

Dembrowski P 1968 *Finite Geometries* (Springer), 368 p [bib, survey]
Mitchell U G 1921 AMM 28, 85-87
Vajda S 1967 *Patterns and Configurations* (Hafner) [bib]

## Fixed point theorems

See also Topology.

Myshkis A D + 1955 UMN (NS)10(3), 188-192 (MR17, 2) [first proof, Bohl]
Walt T van der 1963 *Fixed and Almost Fixed Points,* Amsterdam, (Mathematisch Centrum), 128 p (MR34, 917-918) [bib]

## Focus

See also Conics.

Weaver J 1916 AMSB 23, 357-365

## Folium

Barbarin P + 1898 Intermed 5, 102, 104, 128 (Q1023) [Carre]
Hofmann J E 1954 Centau 3, 279-295 (MR16, 551; Z57, 242)
Loria G + 1897 Intermed 4, 238 [Carre, Huygens]
Tannery P 1897 Intermed 4, 19 (Q816)

## Forms

See also Determinants, Invariants.

Dickson Leonard E 1919 Dickson 1, 361-366, 417-418 [sum of squares]
Meyer F W + 1911 Enc SM 1(2)(3-4), 386-520 [invar]

## Foundations   See Chapter 3.

## Four and eight square problem

*Curtis C W 1963 in A A Albert ed, *Studies in Modern Algebra* (MAA Stu 2), 100-125 [bib]
Dickson L E 1918 An M (2)20, 155-171
Tannery P 1911 Intermed 18, 55 (Q1346) [Descartes]

## Four color conjecture

Anon 1936 AMM 43, 18
Ball W W Rouse 1892 *Mathematical Recreations and Essays,* NY, Ch. 8
Baltzer R 1840 Leip Ber 37, 1-6
Brahana H R 1923 AMM 30, 234-243
*Coxeter H S M 1959 MT 52, 283
Dynkin E B + 1968 *Mehrfarbenprobleme,* Berlin
*Errera Alfred 1921 *Du coloriage des cartes...,* Brussels
- 1927 Pe M 7, 20-41 (F53, 3)
*Franklin P 1939 Scr 6, 149-156, 197-210 (also in Scr M Lib 5) [bib]
Hasse Helmut 1967 in *Probe mathematische Forschung* (Salle), 55-101
Maddison I 1897 AMSB (2)3, 257 (F28, 43)
May, K O 1965 Isis 56, 346-348 (repr 1967 MT 60(5), 516-619)
*Ore Oystein 1967 *The Four-Color Problem* (Academic Press), 274 p [bib]
Ringel G 1959 *Farbungsproblem auf Flaechen und Graphen,* Berlin
Saaty T L 1967 M Mag 40, 31-36

## Fourier and trigonometric series
(continued)

Ravetz J R 1961 in *Essays Presented to M. Polanyi on His Seventieth Birthday, 11 March 1961*, London, 71-88
*Riemann B 1867 Gott Ab 13, 47p
Sachse A 1880 BSM (2)3, 43-64, 83-112 (F12, 35)
- 1880 ZM Ph 25(S), 229-276 (F12, 35)
Van Vleck E B 1914 Sci 39, 113-124
*Zygmund O 1935 *Trigonometric Series*, Warsaw (repr 1955 Dover) [bib]
*- 1959 *Trigonometric Series* (Cambridge U Pr), 2 vols

## Functional analysis

See also Schauder bases, Spectral theorem.

Andreoli G 1958 Batt (5)6, 86, 133-186 (MR23A, 697)
*Bernkopf M 1966 Arc HES 3(1), 1-96 [bib, intgl, eq]
Dunford N + 1958 *Linear Operators*, NY, 2 vols
Fichera G 1963 Cah H Mond 7, 407-417 [Italy]
*Fréchet Maurice 1925 Rv Met Mor 32 (repr in Frechet 1955 *Les mathématiques et la concrète*, Paris, Ch. 3)
- 1928 *Les espaces abstraits ...*, Paris, 305 p [bib]
Goffman C 1962 MAA Stu 1, 176-178
*Hadamard J 1912 Ens M 14, 5-
Wilansky A 1964 *Functional Analysis*, NY

## Functional equations

Aczel J 1969 *On Applications and Theory of Functional Equations*, (Birkhaeuser), 64 p
Babbage C 1820 *Examples of Solutions of Functional Equations*, London [bib]
Jurkat W B 1965 AMSP 16, 683-686 [Cauchy]
Kuczma Marck 1968 *Functional Equations in a Single Variable*, Warsaw, 383 p [*bib, 63 p]
Langer R E 1957 AMM 64 (8-2), 37-44 (MR20, 620) [Euler]
Neville E H 1926 M Gaz 13, 163-164 (F52, 25) [Babbage]

## Functions

See also Algebraic, Analysis, Bessel, Elliptic, Exponential, Factorial, Generalized, Hypergeometric, Lambda, Lamé, Logarithm, Pathological, Periodic, Recursive, Special, Spherical harmonics, Symmetric, Transformations, Variable (Ch. 3), Zeta.

*Bernstein S N 1912 Kagan (559)
*Bortolotti E 1926 It UM 5, 43-44 (F53, 3) [word]
Boyer C B 1946 Scr 12, 5-13 [eq, propor]
*Brill A + 1894 DMV 3, 107-566
Bychkov B P 1955 Matv Shk (6), 1-8 [XX, Russ]
*Dickson H 1967 *Variable, Function, Derivative*, Gothenburg
Dummett M 1956 Phi Rev 65, 229 [Frege]
Eneström G 1891 Bib M (2)5, 89-90 (F23, 37) [nota]
Goulard A + 1895-1900 Intermed 2, 279; 3, 22-23, 279; 7, 52 (Q531) [word]
*Hevesi J 1953 Int Con HS 7, 366-371 [variable before Galileo]
Jourdain P E B 1913 Int Con 2, 526-527, 540-554, [Fourier, Lagrange]
- 1914 Isis 1, 661-703
Klein F 1883 M Ann 22,
Krazer A 1915 *Zur Geschichte der graphischen Darstellung von Funktionen*, Karlsruhe, 31 p (F45, 76) (and DMV 24, 340-363)

Functions    (continued)

Likhin V V 1959 Ist M Isl 12,
   59-134 [Jacob Bernoulli, Russ]
Marcus S 1958 Gaz MF (A)10,
   416-423
Medvedev F A 1959 Ist M Isl 12, 481-
   492 (MR23A, 697) [measurable]
- 1961 Mos IIET 43, 264-289
   [Cauchy]
Miller G A 1925 Ens M 24, 59-69
   (F54, 48)
- 1925 Int Con 7(2), 959-967
   (F54, 48)
- 1928 SSM 28, 506-516  (F54, 6)
- 1929 SSM 28, 829-834  [graph]
Picard E 1936 in *Discours et notices*
   Paris, 189-225
Podetti F 1913 Loria 15, 1-8, 34-41
   [XVII, proportion]
Pringsheim A 1933 Mun Si 1, 61-70
   (Z7, 50)  [Gauss]
*Rosser J B 1955 Dub Ac P (A)57,
   117-120  (MR17, 337; Z66, 7)[Boole]
Rychlik K 1932 Int Con 6, 503-505
   (F58, 41)  [Bolzano]
Schramm 1965 Hist Sci 4, 70-103
   [Arab, mid]
Settle T 1961 Sci (6 Jan)
   [Galileo]
Spengler Oswald *Decline of the
   west,* Vol 1, Ch. 2
Staeckel P 1901 Bib M (3)2, 111
   [XVIII]
Stipanic E 1956 Int Con HS 8, 115-
   119
Valiron G 1948 in F. Le Lionnais ed,
   *Les Grands Courants de la pensée
   mathématique* (Cahiers du sud),
   157-172
Videnskiy V S 1961 Rus MS 16(2), 17-
   20 (Z98, 9) [Constructive thy
   fun, S. N. Bernstein]
Wampler J F 1960 MT 53, 581-583
*Wieleitner H 1912 Bib M (3)13, 115-
   145 [Oresme]
- 1914 Bib M 14, 193-243 [Oresme]
- 1916 DMV 25, 66 (F46, 47)
   [anal geom]
Wolff G 1933 Unt M 39, 296-306,
   332-340  (F59, 9)

*Yushkevich A P 1966 Ist M Isl 17,
   123-150

Fundamental theorem of algebra

Agostini A 1924 Pe M 4, 307-327
   (F50, 8)
Aubry V 1897 JM Sp 21, 17
   [d'Alembert, Gauss]
Bashmakova I G 1957 Ist M Isl 10,
   257-304  (Z101, 246)  [Euler]
- 1960 Arc In HS 13, 211-222
   [fields]
Bocher M 1895 AMSB (2)1, 205-209
   [Gauss]
Fine H B 1914 AMSB 20, 339-358
   [Kronecker]
Fraenkel A 1922 DMV 31, 234-238
   [Gauss]
Gasapina U 1957 Pe M (4)35, 149-163
   (Z79, 242)
Kneser H 1939 Deu M 4, 318-324
   [Gauss, Laplace]
Loria G 1891 Bib M (2)5, 99-112
   (F23, 35)
- 1891-1893 Rv M 1, 185-248; 2,
   37-38; 3, 105-108 (F23, 34; 24, 44;
   25, 61)
- 1893 Bib M (2)7, 47-50  (F25, 61)
   [fundamental theorem of alg]
- 1909 Loria 11, 33-37  (F40, 54)
Moritz R E 1903 AMM 10, 159
Netto E + 1907 Enc SM 1(2)(1), 189-
   205
Ostrowski A 1920 Gott N 50-58
   [Gauss]

Fundamental theorem of **arithmetic.**
   See Factoring

Fundamental theorem of **calculus**

See also Integral.

Bortolotti E 1924 Arc Sto 5, 205-227
Sebestik J 1964 Rv Hi Sc Ap 17,
   129-164

## Galois theory

*Artin Emil   1959 *Galois Theory*
   2nd ed Notre Dame, Ind
Dehn E 1930 *Algebraic Equations,*
   NY, Ch. 11   (repr 1960 Dover)
Miller G A 1910 Bib M (3)11, 182
   (F41, 58)   [Kronecker]
Pierpont J 1896 An M 1(2)113-;
   2(2)22-
- 1898 AMSB (2)4, 332-340
   (F29, 29)
- 1900 *Galois Theory of Algebraic
   Equations,* Salem
Procissi A 1953 It UM (3)8, 315-328
   (Z53, 246)   [Betti, Libri]


## Game theory

Epstein R A 1967 *The Theory of
   Gambling and Statistical Logic*
   (Academic Press)
*Fréchet M 1953 Ecmet 21, 95-96
   (Z50, 2)   [Borel]
*Kuhn H W + 1958 AMSB 64, 100-122
   (MR20, 503)   [von Neumann]
*Luce R Duncan + 1957 *Games and
   Decisions--Introduction and
   Critical Survey* (Wiley), 528 p
   [bib]
*McKinsey J C L 1952 AMSB 58, 591-611
Tucker A W + 1959 An M Stu 40
   [bib]
*- 1965 Games, theory of, Enc Br
Uspensky J V 1927 AMM 34, 516-525
*Von Neumann J 1953 Ecmet 21, 124-125


## Gamma function

See also Factoring

Boruvka O 1959 in *Sammelband zu
   Ehren des 250 Geburstages Leonhard
   Eulers,* Berlin, 78-88   (Z111, 5)
   [Lerch]
*Davis P J 1959 AMM 66, 849-869
   (MR21, 1024)   [Euler]
Godefroy M 1901 *La fonction gamma,*
   Paris

Gussov V V 1952 Ist M Isl 5, 421-472
   (MR16, 433)   [Russia]
Schenkel H 1894 *Kritisch-historische
   Untersuchung uber die Theorie der
   Gammafunction und Euler'schen
   Integrale,*   Zurich


## Generalized functions

See also Operational calculus.

Bochner S 1959 *Lectures on Fourier
   Integrals* (Ann M Stu 42)   Ch. 6
Bohr N 1950 Int Con 11, 127-134
   [Schwartz]
Gelfand I M + 1958-1966 *Generalized
   Functions,* NY, 5 vols   [since 1900]
Jammer Max 1966 *The Conceptual
   Development of Quantum Mechanics,*
   NY, 391 p
Schwarz L 1950 *Théorie des distri-
   butions,* Paris, 148 p   (Act Sc Ind
   1091)   [bib]
Van der Pol B + 1955 *Operational
   Calculus...,*   NY, 62-66
   [delta-fun]


## Geodesic

See also Differential geometry.

Eneström G 1899 Bib M (2)13, 19-24
   (F30, 54)


## Geometric algebra

Favaro 1878 Mod Mm 18, 127-330
Loria G 1939 Rv Met Mor 46, 57-64
Matthiessen L 1878 *Grundzuege der
   aantiken und modernen Algebra...,*
   Leipzig
Neugebauer O 1936 QSGM (B)3, 245-259
Zeuthen H G 1904 Bib M (3)5, 97-112
   (F35, 54)   [Greek, Hindu]

## Geometric number theory

Davenport 1952 Int Con 11, 166-174
Delone 1956 in Vinogradov I M ed,
  *Karl Friedrich Gauss, Sbornik
  statei. 100 Let so dnya smerti,*
  Mos  [Minkowsky, Venkov, Voronoi]
*Koksma Jurjen 1937 *Diophantische
  Approximationen,* Berlin, 163 p
Urazbaev B M 1951 Kazk Ak (3),
  132-138

## Geometric progressions

See also Progressions.

Aubry A 1894 JM Elem (4)3(18), 49-53
  (F25, 75)  [sum]
Cajori F 1922 SSM 22, 734-737
  [arith prog, propor]
Darzens G 1893 JM Elem (4)2, 246-248
  (F25, 75)  [sum]

## Geometrography

See also Constructions.

Archibald R C 1918 AMM 25, 37-38
- 1920 AMM 27, 323-326
*Hess A H 1956 M Mag 29, 217-221
  [bib]
Loria G 1908 Pe MI 24(6), 114-122
*Mackay J S 1893 Edi MSP 12, 2-16
  [Lemoine]

## Geometry

See also Algebraic, Combinatorial,
  Congruence, Continuous, Cross
  ratio, Enumerative, Erlanger,
  Euclidean, Finite, Geometric
  algebra, Global, Helly theorem,
  Integral geometry, Inversive,
  Isogonal centers, Lehmus-Steiner,
  Line geometry, Paper folding (Ch. 3),
  Point, Polygon, Polyhedra, Solid
  geometry, Spaces, Space (Ch. 3),
  Symmetry, Transformations.

## Geometry: General

Agostini A 1940 Pe M 20, 213-227
  [Mengoli]
Aley R J 1893 AMM 1, 42  [bib]
Baron R 1955 Rv Hi Sc Ap 8, 298-302
  *"Geometria theorica et practica"*
Bogomolov S A 1928 *Evolyutsiyu
  geometricheskoi mysli,* Len, 221 p
Bonola R 1902 Loria 5, 33-41, 65-71
Bopp K 1923 Isis 5, 406-408
Borel E 1922 *D'Euclide et de Descartes
  à Gauss et à Darboux,* Le Temps
  15 Aug
Bouligand G 1951 Arc In HS 30, 884-
  896
Brill A 1891 Wurt MN 4, 12-29
  (F23, 37)
Brocard H + 1896 Intermed 3, 50
  (Q648)
Cajori F 1911 *Notes on the History
  of Geometry and Algebra,* Boston
  25 p  (F42, 70)
Caporali Ettore 1888 *Memorie di
  Geometria,* Naples, 385 p
Cassina U 1950 Pe M (4)28, 1-12,
  73-84  (MR12, 64; Z37, 290)
Chernev C 1939 Sof FMD 2, 84-89
  (F66, 1185)
Chasles Michel 1837 *Aperçu historique
  sur l'origine et le développement
  des méthodes on géométrie,* Paris
  (2nd ed 1875, 3rd 1889, Russ tr
  1883, Ger transl 1939)
Coleman Robert Jr 1942 *The Develop-
  ment of Informal Geometry,* NY,
  190 p
Colerus Egmont 1935 *Vom Punkt zur
  vierten Dimension,* Vienna
  [pop]
- 1937 *Il Romanzo della Geometria,*
  Milan, 395 p  (transl of 1935)
  [pop]
Coolidge J L 1940 *A History of
  Geometrical Methods,* NY, 469 p
  (MR2, 113; Z113, 1)  (repr 1963
  Dover)  [bib]
Coxeter H S M 1961 *Introduction to
  Geometry* (Wiley)
Coxeter H S M + 1967 *Geometry
  Revisited,* New M Lib 19

Geometry: General   (continued)

Craft M G W 1753 *Institutiones Geometriae Sublimioris,* Tubingen
Depman I Ya 1955 Ist M Isl 8, 620-629 [appl]
Dupin Charles 1813 *Développements de Géométrie,* Paris
Enriques F 1907-1911 *Fragen der Elementargeometrie* (Teubner), 2 vols
Eves Howard 1965 *A Survey of Geometry* (Allyn and Bacon), 2 vols
Galdeano Z G de 1894 Prog M 4, 132-139, 216-218, 260-264
- 1895 Prog M 5, 5-8, 38-40, 85-88
- 1895 Prog M 5, 57-67
Godeaux L 1952 *Les géométries,* Paris, 215 p [pop]
- 1952 *La naissance et la développement de la géométrie,* Paris, 24 p
Gould S H 1957 NCTMY 23, Ch.9
Grzepski Stanislaw 1957 *Geometria to jest mirnicka nauka,* Wroctaw, 144 p [XVI]
Kaderavek F 1935 Slav Cong 2, 249-250 (F61, 8) [geom repres]
Klein F 1939 *Elementary Mathematics from an Advanced View Point,* 3rd ed (repr Dover) (1st ed 1908 Goettingen, 3rd ed 1925)
Kline M 1956 Sci Am 194, 104-114
Lame G 1927 *Examen des différentes méthodes employées pour résoudre les problemes de géométrie,* Paris, 136 p (F53, 21)
Lazarski M 1889 Kosmos 14, 271-281
Lenz Hanfried 1968 Ueberbl 1, 63-68
Limpert 1888 *Geschichte der Geometrie* Stuttgart
Loria Gino 1887 *Il passato e il presente delle principlai teorie geometriche,* 1st ed, Turin (2nd ed 1897, 3rd ed 1907, 4th ed 1931, F57, 8, 1287; transl Ger, Polish)

- 1888 *Die Hauptsachlichsten Theorien der Geometrie...,* Leipzig, 135 p
- 1902 Monist 13, 80-102, 218-234 [to 1850]
Marchal P E 1942 *Histoire de la géométrie,* Paris, 126 p
Meserve B E 1956 MT 49, 372-382
Mikan M 1954 *Jak se vyvinula matematika a geometrie,* Prague, 136 p
Natucci A 1933 Pe M (4)13, 301-310 (Z7, 387)
Pressland A J 1891 Edi MSP, 23-34 [approx]
Quint N + 1904 Intermed 11, 101 (Q2558) [space and plane]
Reves G E 1953 SSM 52, 299-309
Rey Pastor J 1918 Scientia 23, 413-422 [group]
Reye T 1902 DMV 11, 343-353
Rothe H 1914-1931 Enc MW 3(1), 1277-1423
Schell W 1895 Karlsruh 11, 136 [mech]
Steiner J 1936 *Systematische Entwicklungen...,* Leipzig, 126 p (F62, 1043)
Taylor C 1880 Cam PSP 4, 14-17 (F12, 35) [continuity]
Thom A 1961 M Gaz 45, 83-92
Vashchenko-Zakharchenko M E 1883 *Istoriya matematiki. Istoricheskii Ocherk rozvitiya geometrii,* 1, Kiev
Vera Francesco 1948 *Breve historia de la geometria,* Buenos-Aires, 200 p
Wolf Wilhelm 1865 *Drei Vortraege ueber die geschichte der praktischen Geometrie, Gehalten in den Versammlungen des Vereins praktischer Geometer,* Dresden, 98 p
Worrell W H 1943 Scr 9, 195-196 [ast]
Zeuthen H G 1893 Dan Ov, 1-17, 303-341 (F25, 62) [word]
- 1909 Int Con 4(3), 422-427 (F40, 61)

## Geometry: Foundations

See also Space (physical) in Ch. 3.

Blumenthal L M 1961 *A Modern View of Geometry* (Freeman)
Baker A 1906 Can T(3)12(3), 111-126
Bonola R 1900 Loria 3, 2-3, 33-60, 70-73 [bib]
*Borsuk Karol + 1960 *Foundations of Geometry*, Amsterdam
Bottema 0 1962 MP Semb 9, 164-168 (MR27, 3)
Cassina U 1931 Milan Sem 4, 18-37 (F57, 1287) [line, surf, solid]
Dehn Max 1926 App to Pasch M, *Vorlesungen ueber neuere geometrie*, Berlin, 184-275
Evans G W 1927 MT 19, 195-201 [heresy]
Haantjes J 1948 Euc Gron 23, 258-270 (Z32, 51)
Kagan V F 1901 Kagan (38), 308-312
- 1904 Kagan (380-384) (F35, 61)
- 1905 Kagan (387), 49-57; (391), 153-156; (392), 169-176, (395), 248-253; (396), 272-278 (F36, 66)
- 1907 Novor M 108, 1-320
- 1907 Novor M 109, 321-567
- 1907 *Osnovaniy Geometrii*, Odessa, 561 p
- 1909 Kagan (490)
- 1949 *Osnovaniya Geometrii. Uchenie ob obosnovanii geometrii v khode ego istoricheskogo razvitiya*, Mos-Len, 2 vols
Kerekjarto Bela 1955 *Les Fondements de la Géométrie*, Budapest, 2 vols
Osborne R 1950 M Mag 24, 77-82 [phil]
Pogorelov A V 1966 *Lectures on the Foundations of Geometry*, Groningen, 137 p
Raffalli 1896 JM El 20, 152
Redei L 1968 *Foundation of Euclidean and Non-Euclidean Geometries According to F. Klein* (Pergamon), 405 p
Ricci C G 1902 DMV 11, 382-403
Rossier P 1948 Arc In HS 27, 363-364 [elem ed]

Russel Bertrand 1897 *An Essay on the Foundations of Geometry* (Cambridge U P), 217 p (repr 1956 Dover)
Sommerville Duncan M Y 1911 *Bibliography of Non-Euclidean Geometry, Including the Theory of Parallels, the Foundation of Geometry, and Space of n Dimensions*, London
Sos E 1905 Bib M (3)6, 408-409 (F36, 70) ["natural geom"]
Viola T 1957 Sci Tec Pi (NS)1 (MR20, 128; Z73, 2)
Weyl F J 1962 SIAMR 4, 197-201
Whittaker Edmund 1949 *From Euclid to Eddington* (Cambridge U P), 222 p
Wiener J 1892 DMV 3, 70-80

## Geometry: Ancient and medieval

Adamo M 1952 Cagliari 22(S), 13-88 (MR15, 591) [anc]
August E F 1843 *Zur Kenntniss der geometrischen Methode der Alten...*, Berlin, 35 p
Blaschke W 1953 *Griechische und anschauliche Geometrie*, Munich, 60 p (MR15, 275)
Bretschneider C A 1870 *Die Geometrie und die Geometer von Euklides, ein historischer Versuch*, Leipzig (F3, 1)
Bruins E M 1958 Euc Gron 33, 264-284 (Z82, 9) [Greek]
- 1959 Sim Stev 33, 38-60 (MR21, 5156; Z89, 3) [Babyl geom without angle or parallel]
Chatterjee B 1949 Beng Asi J 15, 41-89 (MR12, 381) [ast, Hindu, Ptolemy]
Deacon A B 1934 Lon Anthr 64, 129-175 [New Hebrides]
Diesterweg W A 1828 *Geometrische Aufgaben nach der Methode der Griechen*, Berlin, 2 vols
Fettweis E 1929 Arc GMNT 12, 113-121 (F55, 6) [anc]
- 1929 Archeion 11, 366-374 (F55, 7) [anc]
- 1930 Bo Fir 11, 336-374 [anc]
Godeaux Lucien 1952 *La naissance et le développement de la géométrie,*

## Geometry: Ancient and medieval (continued)

Paris, 24 p [anc]
Gunn B + 1929 J Eg Arch 15, 167-185 [Egypt]
Hertz A 1929 Rv Syn Sc 47, 29-54 [anc]
- 1933 Pologne 1, 137-140 (F59, 832) [Babyl]
- 1934 Wiad M 36, 81-92 (F60, 817) [anc]
Hirsch 1896 Wurt Sch 3, 158 [anc]
Klinkenberg L 1916-1917 Wisk Tijd 13, 161-180; 14, 7-19, 145-160 [anc]
Levi B 1908 Bib M (3)9(2), 97-105 [Hindu]
Lietzmann W 1933 Isis 20, 436-439 [anc, art]
- 1934 ZMN Unt 65, 313-319 (F61, 5) [anc]
- 1940 *Fruehgeschichte der Geometrie auf germanischem Boden,* Breslau, 94 p (F66, 5)
Luckey P 1933 Isis 20, 15-52 [Egypt]
Luke F 1845 *Geometrische Aufgaben nach der Methode der Alten,* Thorn
Mansion P 1899 ZM Ph 44(S), 275-292 (F30, 59) [ast]
Seidenberg A 1959 Scr 24, 107-122 (Z88, 4) [Greek]
- 1962 Arc HES 1, 488, 527
Smith D E 1913 Isis 1, 197-204 [Hindu]
Timchenko I Yu 1914 Kagan (618) [anc phil]
Vincent A J H 1857 *Sur un point de l'histoire de la géométrie chez les grecs,* Paris
Woeckel L 1853 *Die Geometrie der Alten in einer Sammlung von 848 Aufgaben...,* 3rd ed, 1845

## Geometry: Modern

Amaldi U 1911 It SPS 5, 415-461 [Italy, XIX]
*Bachmann F 1959 *Die Entwicklung der geometrie aus dem Spiegelungsbegriff* Berlin
Bally Emile 1922 *Principes et premiers développments de géométrie générale synthétique moderne,* Paris, 226 p
Charles M 1870 *Rapport sur les progrès de la géométrie,* Paris (F3, 12)
*Coolidge J L 1929 AMSB 35, 19-37 (F55, 5) [XIX, "golden age"]
Couturat L 1901 *La logique de Leibniz d'après des documents inédits,* Paris, Ch. 9 [Leibniz]
Darboux J G 1896 (also in 1904 BSM (2)28, 234-263 F35, 61; 1905 AMSB (2)11, 517-543 F36, 63; 1905 M Gaz 3, 100-106, 121-128, 157-161, 169-173 F36, 63; Pe MI (3)7, 145-163, 193-197; F41, 62) [XIX]
- 1911 *Etyud o razvitii geometricheskikh metodov,* Kazan, 37 p
Dingler H 1935 Z Ph 94, 674-676 [Helmholtz]
Eisenhart L P 1918 AMSB 24, 227-237 [Darboux]
*Fano G + Enc SM 3(3), 185-259 [XIX]
Freudenthal H 1957 Nieu Arch 5, 105-142 [found]
*- 1960 MP Semb 7, 2-25 (MR23A, 135) [XIX, found]
*- 1962 in Nagel Ernest + ed, *Logic Methodology and Philosophy of Science* (Stanford U P), [XIX, found]
Galdeano G de 1900 Prog M (2)2, 310
Gruenbaum A 1959 Conventionalism in Geometry in *Intl Symp U Calif, Berkeley, Dec. 26, 1957-Jan 4, 1958* 204-222 (MR22, 107)
Hahn H 1929 Naturw 17, 916-919 (F55, 5) [set thy]
Hankel H 1876 Boncomp 9, 267-289 (F8, 30)
- 1885 *Esquisse historique sur la marche du développement de la nouvelle géométrie* (Gauthier-Villars)
Kenner M R 1957 MT 50, 98-104 [Helmholtz]
Klein Felix 1872 Gott Anz, 1-12 (F4, 28) [Chasles]

## Geometry: Modern   (continued)

- 1892 NYMS 2, 215-249    [XIX]
*- 1895 Prace MF 6, 27    [XIX]
- 1896 Kazn FMO (2)6(2), 131
   [XIX]
*Koetter E 1898 DMV 5(2), 1-128
   (F29, 36) [XIX to 1847]
 Kokomoor F W + 1928 Isis 10, 21-32,
   367-415    [XVII]
 Loria G 1924 Linc Mr M (5)14, 777-
   845    (F50, 5) [Descartes, Fermat,
   Lagrange, Monge]
- 1927 Bo Fir 6, xlix-liii
*Mansion P 1895 Rv Q Sc 37, 584-595
   [Tilly]
 Nagel E 1939 Osiris 7, 142-224
   [logic]
 Nielsen Niels 1935 Gécmètres
   francais du dix-huitieme siècle,
   Paris, 444 p  (Z12, 243)
 Segre B 1932 An M 11, 1-16
   [Italy, XIX-XX]
- 1958 Archim 10, 53-60  (MR20, 620)
   [XX]
*Taton R 1949 Rv Hi Sc Ap 2, 197-224
   (MR11, 150) [Desargues, Monge]
- 1951 Par CR 232, 198-200
   (MR12, 383) [Monge]
- 1957 in Hommages à Gaston
   Bachelard, Paris, 93-101
   [graphic meths]
 Taylor C 1900 Cam PST 18, 197
   [Kepler, Newton]
 Thebault V 1958 M Mag 32, 79-82
   (MR20, 847) [XIX, France]
 White H 1908 AMSB (2)15, 325-388
   [Bezout, resultants]
 Wilczynski E J 1914 Loria 16, 97-109
   (F45, 81)

## Global geometry

 Chern S S ed 1967 Studies in Global
   Geometry and Analysis (MAA Stu 4)
   [bib]

## Graph theory

See also Chromatic graphs, Combinatorics, Four color conjecture.

 Harary Frank + 1953 Graph Theory as
   a Mathematical Model in Social
   Science, Ann Arbor
- 1967 A Seminar on Graph Theory
   (Holt) [bib]
 Koenig D 1935 Theorie der endlichen
   und unendlichen Graphen, Leipzig,
   258 p  (repr 1950 Chelsea)
   [*bib]
 Moon J W 1968 Topics on Tournaments,
   NY  [bib]
 Ore Oystein 1959 MT 52, 367-370
- 1962 Theory of Graphs (AMS Col 38)
   [bib]
- 1963 Graphs and Their Uses (Random
   House) [pop]
 Polya G 1937 Act M 68, 145-254
   [trees]
 Rapaport E S 1959 Scr 24, 51-
   [Cayley color groups, Hamilton
   lines]
*Turner J 1969 Key-word indexed
   bibliography in Harary Frank ed,
   Proof Techniques in Graph Theory,
   (Academic), 189-330
*Zykov A A 1964 in Theory of Graphs
   and its Applications.  Proceedings
   of the Symposium held in Smolenice
   in June 1963, Prague, 171-234
   [*bib]

## Gregory series

 Marar K M + 1945 M Stu 13, 92-98
   (MR8, 2; Z60, 1) [Hindu]
 Rajagopal C T + 1951 Scr 17, 65-74
   (Z45, 146) [Hindu]

## Group theory

See also Erlanger program, Lie
   groups.

 Alasia C 1908 Rv FMSN (a)9(2), 630-
   638    [bib]

Group theory   (continued)

*Birkhoff G 1937 Osiris 3, 260-268
    (F63, 810)   [Galois]
Bortolotti E 1916 Mod Mm (3)12,
    179-193  [Ruffini]
- 1929 Bln Rn (2)33, 82-87
    (F56, 812)  [subst, Ruffini]
Burkhardt H 1892 ZM Ph 37(S), 119-159
    159   (F24, 44)  [Ruffini]
- 1894 Annali 22, 174-213   [Ruffini]
Burns J E 1913 AMM 20, 141-148
    (F44, 50)  [founding]
Cartan E 1914 Rv Mois 17, 438-468
Coxeter H S M 1961 Introduction to
    Geometry, NY  [bib]
Davis Constance 1969 A Biblio-
    graphical Survey of Simple Groups
    of Finite Order, 1900-1965 (Courant
    Int NYU) 209 p  [indexed]
Dickson L E 1899 AMSB 6, 13-27
Dubisch R 1947 AMM 54, 253
    [Wedderburn thms]
*Easton B S 1902 The Constructive
    Development of Group Theory
    Philadelphia, 89 p  (See AMSB(2)9,
    557-558)
Kaplansky I 1954 Infinite Abelian
    Groups  (Univ of Mich P),  [bib]
Klein F 1896 Chic Cong, 136-
    (F28, 53)
Litvak B 1964 MT 57, 30-32
    [inf Abel grps]
Maillet E + 1906 Intermed 13, 25
Malcev A 1941 Ivan GPI, 3-9
    (MR17, 823)  [local thms]
Miller G A 1895 AMSB (2)2, 138
    [subst grps table, bib]
- 1899-1902 AMSB (2)5, 227-249,
    6, 106-   (F30, 41)  [recent]
- 1901 AMM 8, 213-216
- 1903 AMM 10, 87
- 1909 Bib M (3)10, 317-329
    (F40, 55)  [fin]
- 1910 AMM 17, 162-165  [founding]
- 1911 Bib M (3)11, 314-315
    [Hamilton]
- 1912 Bib M 13, 62-64   [word]
- 1913 AMM 20, 14-20  [mistakes]
- 1919 AMM 26, 290-291  [indicator,
    simple group]
- 1920 AMSB 27, 459-462 (F48, 41)
    [reviews in F]
- 1921 Sci Mo 12, 75-82
- 1922 AMM 29, 319-328  [mistakes]
- 1925 Ens M 24, 59-69
    (F54, 48)
- 1925 Int Con 7(2), 959-967
    (F54, 48)
- 1927 Indn MSJ 17, 87-102
    (F53, 2)  [fin]
- 1927 Sci Prog 22, 225-230
- 1928 An M (2)29, 223-228  [def]
- 1930 Sci 72, 168-169  [defs, Cayley,
    Kronecker, Weber]
- 1931 Indn MSJ 19, 169-172
    (F57, 1287)  [log]
- 1932 Indn MSJ 19, 205-210
    (F58, 986)  [fin]
- 1933 Sci Mo 36, 146-147
- 1934 Sci (30 Mar), 291-292)  [def]
- 1934 Toh MJ 39, 60-65  (F60, 817)
    [fin subst grps]
- 1935 Sci Mo 41, 228-233  (F61, 958)
- 1938 Sci Mo 47, 124-127
- 1939 Sci 90, 234  [US]
- 1942 Sci 95, 353-354  [dilemma,
    Klein, Lie]
- 1943 Sci 97, 90-91
- 1947 Am NASP 33, 235-236  (MR9, 74)
    [Hamilton, quaternions]
- 1964 MT 57, 26-30
*Miller G A + 1916 Theory and Appli-
    cations of Finite Groups  (Wiley),
    407 p  (repr 1961 Dover)  [bib]
Novy L 1966 Act HRNT 2, 105-151
    [Cayley]
Pierpont J 1895 AMSB (2)1, 196
    [Lagrange]
Rey Pastor J 1918 Scientia 23, 413-
    422  [geom]
Schoenflies A 1896 Chic Cong, 341-349
    (F28, 43)  [cryst]
Severi F 1951 Archim 3, 45-55  [XX]
Vogt H + 1909 Enc SM 1(1)(4), 532-
    616
'Wussing H 1969 Die Genesis des
    abstrakten Gruppenbegriffes,
    Berlin, 258 p  [*bib]

## Harmonic analysis

See also Fourier series, Spectral theorems.

*Edwards R E 1967-1968 *Fourier series, a Modern Introduction* (Holt), 2 vols  [bib]
*Weiss G 1965 MAA Stu 3, 124-178
Wiener N 1930 Act M 55, 117-258 (repr 1964 Dover)  [*bib]
- 1938 in *AMS Semicentennial Publications* 2, 56-68  (F64, 6)
- 1964 *Generalized Harmonic Analysis and Tauberian Theorems,* (Dover), 246 p  bib  (repr of 1930 and 1932) (MR34, 23)

## Harmonic functions

Temple G 1958 Edi MSP 11, 11-24 (Z83, 248)  [Whittaker]

## Harmonic progression

Rocquigny G de + 1898 Intermed 5, 72, 164 (Q1223)  [word]

## Harmonic series

Karamata J 1959 Ens M (2)5, 86-88 (MR23A, 507)  [Mengoli]

## Hebel theorem

Pihl M 1950 M Tid B, 123-127 (Z38, 147)

## Helix

Erhardt Rudolf v 1942 Isis 34, 108-110  (MR4, 65)  [ast, Plato]
Tannery P 1896 Intermed 3, 213 (Q797) [Baliani]

## Helly theorem

Danzer L + 1963 Symp PM 7, 101-

## Heptagon

See also Seventeen-gon.

Aaboe A 1964 in *Episodes from the Early History of Mathematics* (Blaisdell, Singer), 88- [Archimedes]
Samplonius Y 1963 Janus 50, 227-249
Tropfke J 1928 ZMN Unt 59, 195 [Archimedes]
*- 1936 Osiris 1, 636-  [Archimedes]

## Hessian

Muir T 1927 So Af RS 14, 367-371 (F52, 39)  [det]
Wieleitner H 1922 Porto Ac 14, 223-228

## Hexagon

Carruccio E 1938 Pe M (4)18, 207-216 (Z19, 387)  [Archimedes]
Mackay J S 1893 Edi MSP 11, 104-106 (F25, 82)  [Adam, circle]

## Hilbert problems

*Aleksandrov P S 1969 *Problemy Gilberta,* Mos, 241  [bib]
Bieberbach L 1930 Naturw 18, 1101-1111  (F56, 3)
- 1930 Unt M 36, 381  (F56, 35)
Demidov S S 1966 Ist Mat Isl 17, 91-121  [*bib]
Freudenthal H 1968 Rv Syn (3), 49-52, 223-243  (esp 233-234)
Hille E 1942 AMM 49, 654-661 [7, Gelfond]
Montgomery D + 1955 *Topological Transformation Groups,* NY  [5th]
Rybnikov K A + 1970 Ist Met EN 9, 150-154  [bib]

## Homogeneous coordinates

De Vries H 1928 N Tijd 15, 224-259
(F54, 31)
Kiefer A 1910 *Die Einfuhrung der
homogein Koordinaten durch K. W.
Feuerbach*, Strassburg, 55 p
[Feuerbach]

## Homology

Maclane Saunders 1963 *Homology*
(Springer), 422 p [bib]

## Horn angle

Kasner E 1945 Scr 11, 263-267
Kasner E + 1948 Scr 16, 13-21
(MR12, 312) [Voltaire]
Vivanti Guilio 1894 Bib M 8, 1-10

## Horner method

Cajori F 1911 AMSB (2)17, 409-414
(F42, 58) [Ruffini]
- 1911 Loria 13, 81-86 [Ruffini]
Conkwright N B 1944 MT 37, 31-32
Ling Wang + 1955 Toung Pao 43(5),
345-401 [China]
Ruchonnet C + 1897 Intermed 4, 259;
5, 108 (Q1088) [priority]

## Hyperbola

Clagett M 1954 Osiris 11, 359-385
(Z55, 3) [Arab]
Hofmann J E 1953 MP Semb 3, 59-79
(Z50, 1) [area]
Inglis A 1935 M Gaz 19, 142-144
[arc length]
Mikami Y 1913 Toh MJ 3, 29-37
(F44, 52) [Japan]
Rupp O 1931 Arc GMNT 13, 132-166,
226-258 (F57, 8) [circle]
Vetter Q 1925 Cas MF 54, 348-350
(F51, 17) [Menaechmos]

## Hypergeometric function

Schulze K 1889 Ham MG 1, 110-

## Icosahedron

Bruins E M 1957 Janus 46, 173-182
[Heron, Pappos]
Coxeter H S M 1936 Scr 4, 156-157
[sphere]
Vereecke P 1935 Mathesis 49, 59-62
(Z11, 385) [Caravelli]

## Ideals

Chebotarev N G 1930 AMM 37, 117
[Zolotarev]
- 1961 Ist M Isl 14, 539-550
(MR32, 120)
Klein F 1893 in his *Lectures on
Mathematics* .., 58-66
Hancock H 1928 AMM 35, 282-290
[chem, Plato]
Kloosterman H D 1930 *Geschiedenis
der Idealtheorie*, Leiden, 31 p
(F56, 5)
- 1934 Euc Gron 10, 110-124
(F60, 817)
Krull Wolfgang 1968 *Idealtheorie*,
2nd ed (Springer), 160 p
(1st ed 1935) [bib]
Macduffee C C 1931 AMSB 37, 841-853
[lin alg]
Ore Oystein 1934 *Les corps algébrique
et la théorie des idéaux*,
Paris

## Imaginary elements

See also Complex numbers, Ideals.

Court N A 1951 Scr 17, 55-64
De Vries H 1949 N Tijd 36, 82-96
Gillet J 1898 Intermed 5, 254 [bib]
Loria G 1917 Scientia 22, 1-15
(F46, 41)
Romorino A 1897 Batt 35, 242-258;
36, 317-345 (F29, 36)

## Imaginary elements (continued)

Windred G 1935 M Gaz 19, 280-290
(F61, 942) [time]


## Indivisibles

See also Calculus, Infinitesimals

Bosmans H 1922 Brx SS 42(A), 82-89
(F48, 4) [Cavalieri]
- 1923 Arc Sto 4, 369-379 [Pascal]
Cajori F 1925 Scientia 37, 301-306
Hoppe E 1928 DMV 37, 148-187
[Leibniz, Newton]
Wallner C R 1903 Bib M (3)4, 28-47
[Cavalieri, Wallis]
Zubov V 1957 Rv Hi Sc Ap 10, 97-109
(Z94, 3) [Russia]


## Inequalities

Eves H 1964 MT 57, 481 [nota]
*Hardy G H + 1934 Inequalities
(Cambridge U P) (2nd ed, 1952)
Smith C L 1964 MT 57, 479-481
[nota]
Stamatis Evangelos 1953 Der Schluss
von der volstaendigen Induction bei
Euklid, Athens, 6 p (Z51, 242)


## Infinite products

Molk J + 1907 Enc SM 1(1)(2), 270-281


## Infinitesimals

See also Calculus, Exhaustion, Horn
angle, Indivisibles, Infinity,
Non-Archimedean.

Agostini A 1942 It UM Con, 886-894
(F68, 16; Z26, 291) [Saladini]
Bobori A 1964 Int Con HS 10
[Japan]
Bockstaele P B 1966 Janus 53(1),
1-16 [XIX, Belgium]

Bortolotti E 1937 Bln Mm (9)5, 147-
159 [anc]
- 1937 It UM Con 1, 459-477
(F64, 15) [calc, Torricelli]
- 1939 It UM (2)2, 57-63 (F65, 4)
[infinity]
Bosmans H 1913 Brx SS 37, 211-218
[Valerio]
- 1923 Arc Sto 4, 369-379
[Pascal]
Burzio F 1941 Saggiat 2, 314-316
(Z26, 196) [Lagrange]
Cajori F 1925 Scientia 37, 301-306
(F51, 9)
Cantor G 1884 Deu LZ 5, 266-268
(re Cohen 1883)
Cohen Hermann 1883 Das Princip der
Infinitesimal Methode und seine
Geschichte, Berlin, 162 p
De Morgan A 1852 Phi Mag (4)4, 321-
330 [England]
Guimaraes R 1918 Porto Ac 13, 61-71
[Nunes]
Heath T L 1923 M Gaz 11, 248-259
[Greek]
- 1923 Nat 111, 152-153
[geom, Greek]
Hobson E W 1903 LMSP 35, 117-140
Josey E 1932 Les antécédents de
l'infiniment petit dans Pascal,
Paris, 229 p
Levi-civita T 1892 Ven I At (7)4,
1765-1815
Loria G 1915 Scientia 18, 357-368
[anc]
- 1916 Scientia 19, 1-18 [XVIII]
Lure S Ya 1932 QSGM (B)2(2), 106-185
[anc]
- 1935 Teoriya beskonechno malykh u
drevnikh atomistov, Mos-Len, 197 p
Mau J 1954 Ber Hel Rm (4), 48 p
(MR16, 659) [Greek]
Mondolfo R 1934 Rv Fil 23, 210-219
[Aristotle]
*Robinson Abraham 1966 Non-Standard
Analysis (North-Holland), Ch 10
Vivanti G 1891 Bib M (2)5, 97-98
(F23, 36) [Newton]
- 1894 Bib M (2)8, 1-12 (F25, 68)
*- 1894 Il concetto d'infinitesimo e
la sua applicazione alla matematica,

Infinitesimals   (continued)

*saggio storico,* Mantova, 134 p
(F25, 67)  (2nd ed 1901 Naples,
163 p)
*- 1900 Batt 38, 265
- 1901 Batt 39, 317
Yushkevich A P 1959 in *Sammelband zu*
*ehren des 250 geburstages*
*Leonhard Eulers,* Berlin, 224-244
[Euler, Lagrange]
Zubov V P 1950 Ist M Isl (3), 407-
430  (MR13, 1) [Russia]


Infinity

See also Infinitesimals.

Baltzer R 1887 Bib M (2)1, 32, 64
[nota, Wallis]
Baumann J 1908 An Natphi 7, 444-449
[Bolzano, Dedekind]
Bell E T 1934 Phi Sc 1, 30-49
Bergmann Hugo 1913 *Das Unendliche*
*und die Zeit,* Halle, 88 p
Bochenski I M 1934 Bu Thom 11, 240-
248
Bogomolov S A 1934 *Aktualnaya*
*bezkonechnost,* Mos
Bortolotti E 1938 Bln Mm (9)5, 147-
159  (F64, 904) [anc]
- 1939 It UM 1, 47-60   [anc, limit]
- 1939 It UM 1, 275-286  [limit]
- 1939 It UM 1, 351-371  (Z21, 194)
[XVII, limit]
- 1939 It UM 2, 57-63
[infinitesimal]
Brunet P 1931 Archeion 13, 24-39
(Z2, 6)  [Buffon]
Burger D 1951 Int Con HS 6, 145-150
(MR17, 1)
Cassina U 1921 Pe M (4)1, 326-337
[persp, proj geom]
Cassirer E 1912 in *Philosophische*
*Abhandlungen. Hermann Cohen zum*
*70sten Geburtstag...,* Berlin
[Renouvier]
Cohn Jonas 1896 *Geschichte des*
*Unendlichkeitsproblems im*
*abendländischen Denkens bis Kant,*

Leipzig, 261 p  (repr 1960 Olms)
Datta B 1927 Clct MS 18, 165-176
[Hindu, zero]
Dempf A 1926 *Das Unendliche in der*
*Mittelalterlichen Metaphysik und*
*in der Kantischen Dialektik,*
Muenster
De Vries H 1949 N Tijd 36, 82-96
(MR10, 420)
Edel Abraham 1934 *Aristotle's Theory*
*of the Infinite,* NY, 102 p
[J Phil]
Eisele C 1963 Physis Fi 5(2), 120-
128 [infinitesimal, logic, Peirce,
recur, set]
Enriques F 1913 Int Con Ph 2, 357-
378
- 1933 Scientia 54, 381-401
(Z8, 98)
- 1935 Scientia 57, 310-314
(F61, 344)  [Greek]
Hahn Hans 1934 in *Alte Problem-*
*neue Loesungen in den exacten*
*Wissenschaften,* Leipzig, 93-116
Hardy G H 1924 *Orders of infinity.*
*The "Infinitar-Calcul" of Paul Du*
*Bois-Raymond,* 2nd ed, Cambridge
Hilbert D 1938 M Ann 95, 161-190
Hobson E W 1903 LMSP 35, 117
[anal, infinitesimal]
Janssen van Raaij W H L 1897 Ned
NGC 6, 211   [Bolzano, Cantor]
Jasinski R 1933 Rv Hi Phi 1, 134-159
[Pascal]
Kaestner A G 1799 *Aufangsgrunde der*
*Analysis des Unendlichen,* 3rd ed,
Goettingen
Kasner E 1905 AMSB (2)11, 499-501
[Galileo]
Kaufmann F 1930 *Das Unendliche in*
*der Mathematik und seine*
*Ausschaltung,* Leipzig, 213 p
Keyser C 1918 AMSB (2)14, 268, 321-
327 [Lucretius]
- 1937 Scr 4, 221-240
[cosmology, Epicurus]
Kovalevski G 1910 *Die Klassische*
*Probleme der Analysis des*
*Unendlichen,* Leipzig
Loria G 1915 Scientia 18, 358-368,
"227-239" [anc, infinitesimal]

Infinity   (continued)

- 1915 Scientia 18, 358-368, "227-
239" [anc, infinitesimal]
Mondolfo Rodolfo 1934 *L'infinito
nel pensiero dei greci,* Florence,
440 p  (Z10, 243)
- 1956 *L'infinito nel pensiero
dell'antichita,* Florence
Mordukhai-Boltovskoy D 1932 Scr 1,
132-134, 252-253
Nunn T P 1910 M Gaz 5, 345-356
Reymond A 1908 *Logique et mathématiques. Essai historique et
critique sur le nombre infini,*
Saint-Blaise, 219 p
- 1956 Int Con HS 8, 45-51
[anc]
Russell Bertrand 1914 *Our Knowledge
of the External World,* (Chic
(Open Court) (2nd ed 1926)
Schroeder E 1898 Leop NA 71, 303
[Cantor, Dedekind, Peirce, def]
Sergescu Pierre 1947 *Le développment
de l'idée de l'infini mathématique
du XIVe siècle,* Paris, 15 p
- 1949 Act Sc Ind (1083) (MR11, 571;
Z36, 4)
Ternus J 1926 Gorres Ja 39, 217-231
Thomas I 1958 JSL 23, 133-134
(MR21, 484; Z85, 2) [mid, paradox]
Weyl Hermann 1931 *Die Stuten des
Unendlichen...,* Jena
Winter M 1913 Int Con Ph 4(5), 455-
460
Wolfson H A 1929 *Crescas' Critique of
Aristotle,* Cambridge, Mass.
Zippin Leo 1967 *The Uses of Infinity*
(Random House)


Integral

See also Calculus, Elliptic integrals,
Exhaustion, Fundamental theorem
of calculus, Measure, Simpson
rule.

Agostini A 1925 It UM 4, 104-107
[Cavalieri, lim, Mengoli]
- 1925 Pe M 5, 137-146  (F51, 20)
[Mengoli]

- 1930 Bo Fir 12, 33-37  [Torricelli]
- 1931 Archeion 13, 55-59  (Z2, 6)
[Torricelli]
Antopova V I 1957 Mos IIET 17, 229-
269 [diverg thms, Ostrogradskii]
Ascoli G 1895 Annali (2)23, 67-71 [def]
Aubry V 1896 JM El 20, 18, 38, 62,
87, 114, 138, 162, 177, 194, 227,
248, 271  [area]
Bierens de Haan D 1875 Nieu Arch 1
(F7, 20)  [Van der Eycke]
Bortolotti 1924 Arc Sto 5, 205-227
(F50, 10)  [fund thm]
- 1930 Bo Fir 12, 267-271
[Apollonius, Torricelli]
- 1931 Archeion 13, 60-64  (F57, 25)
[Torricelli]
Boyer C B 1945 M Mag 20, 29-32
(MR7, 106)  [Fermat, power fun]
Braunmuehl A v 1904 Int Con H 12,
271-284  (F35, 60)
- 1905 Bib M (3)5, 355-365
[Cotes, Newton]
Brocard H 1917 Intermed 24, 140-142
(F46, 55)
Bryan N R 1922 AMM 29, 392-394
[table]
Cajori F 1915 Bib M 14, 312-319
Catalan E 1878 Nou Cor 4, 53-58
(F9, 27)
Dijksterhuis E J 1954 Nor M Tid 2,
5-23 (MR15, 923)  [Archimedes]
*Douglas J 1941 *Survey of the Theory
of Integration,* Scr Lib 5
[Denjoy, Lebesgue, measure,
Riemann, Stieltjes]
Duran-Loriga J J + 1899-1904 Intermed
6, 219; 8, 314; 11, 18  (Q1632)
[restricteurs, bib]
Eneström G 1889 Bib M (NS)3, 65-66
[Kepler, trig]
- 1912 Bib M (3)13, 229-241 (F43, 70)
Ettlinger H J 1922 An M (2)23, 255-
270 [Cauchy]
Fikhtengolts G M 1952 Ist M Isl 5,
241-268  (MR16, 433)
Fujisawa R 1902 Tok M Ph 7, 88
[Japan]
Hayashi T 1933 Toh MJ 36, 346-394
(Z6, 145)
Hofmann J E 1965 Arch HES 2(4), 270-
343 [*bib, Leibniz, Tschirnhaus]

Integral     (continued)

Jourdain P E B 1905 Bib M (3)6, 190-
    207 [Cauchy, Gauss, functions]
- 1913 Isis 1, 661-703
    [cont, Fourier series]
Kramar F D 1961 Ist M Isl 14, 11-100
    (Z119, 7) [Wallis]
Krazer A 1908 Zur Geschichte des
    Umkehrproblems der Integrale,
    Karlsruhe, 35 p (F39, 64)
- 1909 DMV 18, 44-75 (F40, 59)
Kropp G 1951 Crelle 189, 1-76
    (MR13, 612) [Lalouvère]
- 1959 MN Unt 12, 23-26
    (Z107, 246) [XVII]
Lebesgue H 1926 M Tid B, 54-74
    (repr 1927 Rv Met Mor 34, 149-167;
    transl 1967 in Lebesgue, Measure
    and the Integral, Holden-Day
    [L. integral]
*Loève M 1965 Enc Br
Nalbandyan M B 1966 Ist Met EN 5, 96-
    104 (MR34, 14) [XIX, Russia]
Nunn T P 1910 M Gaz 5, 345-356, 377-
    386 [Wallis, XVII]
Olds C D 1949 AMM 56, 29-30
    [by parts]
Perrier L 1901 Rv GSPA 12, 482-490
    [prob]
*Pesin I N 1966 Razvitie ponyatiya
    integrala, Mos, 207 p (MR34, 7764)
    [Cauchy, Lebesgue, Young, Denjoy,
    Perron, Daniel]
Rabinovich Ya L 1951 UMN 6(5), 26-
    32 (Z44, 246) [Ostrogradsky]
Rey Pastor J 1934 Esp Prog 1, 13-24
    (Z12, 243) [XX]
*Ritt J F 1948 Integration in Finite
    Terms, Liouville's Theory of
    Elementary Methods, NY [bib]
Saks ·Stanislaw 1937 Theory of the
    Integral, 2nd ed (Stechert)
    (repr 1964 Dover) [bib]
Scheffers G 1903 DMV 12, 525-539
    [Lie]
Sturiales S 1923 Il concetto
    d'integrale dei tempi autichi ai
    nostri giorni, Messina, 55 p
    (F49, 20)
Surico L A 1929 Archeion 11, 64-83
    [Torricelli, power fun]

Takeda U 1912 Toh MJ 2, 74-99,
    182-207 [Japan, table, Wada]
Tannery P 1905 Int Con 3, 502-514
    (F36, 54)
Turrell F M 1960 AMM 67, 656-658
    [nota]
Van Veen S C 1943 Zutphen (B)12,
    1-4 (Z28, 194) [prob intgl]
Yushkevich A P 1947 Mos IIET 1, 373-
    411 [Cauchy]
Zeuthen H G 1895 Dan Ov, 37-80
    (F26, 54) [Fermat]
- 1898 Int Con 1, 274-280 (F29, 30)
    [Barrow]
Zhirnov F P 1964 Mos OPI 40
    [approx, Euler]

Integral equations

*Bateman H 1910 Brt AAS
- 1917 Sci Prog 11, 508-512
Bocher Maxime 1909 An Introduction
    to the Study of Integral Equations,
    Cambridge (2nd ed 1914, repr 1926)
Davis H T 1929 Inda U Stu 17, 76 p
    [Volterra]
Hahn H 1911 DMV 20, 69-117
    [lin]
Heywood B + 1912 L'équation de
    Fredholm et ses applications à la
    physique mathématique,   Paris
Lalesco 1912 Introduction à la
    theorie des équations intégrales,
    Paris
Lauricella G 1912 It SPS 5, 217-236
Loria G 1929 Della tavola pitagorica
    alla equazioni integrali. Schizzo
    Storico, Padua, 51 p (F55, 587)
Volterra V 1959 Theory of Functionals
    (Dover) 226 p [bib]

Integral geometry

Stoka Marius I 1968 Géométrie
    Intégrale (Gauthier-Villars),
    1-2, 62-64 (Mem Sci M 165) [bib]

## Interpolation

Braunmuhl A v 1901 Bib M (3)2, 86
[early]
Burkhardt H + 1912 Bib M (3)13, 150-
153 (F43, 70) [expo fun]
Hamadanizadeh J 1963 Centau 9, 257-
[mid]
Kowalewski A ed 1917 *Newton; Cotes;
Gauss; Jacobi. Vier grundlegend
Abhandlungen ueber Interpolation
und genaeherte Quadratur (1711,
1722, 1814, 1826),* Leipzig, 112 p
Mansion S + 1886 Bib M, 141-144
[Newton]
Merrifield C W 1880 Brt AAS, 321
Peano G 1917 Tor FMN 53, 693-716
[numer tables]
Quiquet A 1894 *Aperçu...interpolation
...tables survie...mortalité,*
Paris
Rozenfeld B A 1959 Ist M Isl 12, 421-
430 (MR23A, 583; Z101, 243)
[al-Biruni]
Sengupta P C 1931 Clct MS 23, 125-
128 (Z2, 241) [Brahmagupta]
Turnbull H W 1933 Edi MSP (2)3,
151-172 (Z6, 145) [James Gregory]
Yen L 1956 Int Con HS 8, 70-72
[China]

## Invariants

Fisher C S 1966 Arc HES 3, 137-159
(MR34, 2413) [soc of sci]
Forsyth A R 1905 Lon RSP 45, 144
[Hermite]
Gundelfinger S 1902? Crelle 124, 83-
Hilbert D 1896 Chic Cong, 116-124
Meyer W F 1891 Halle NG 64, 5-7
(F23, 35) [proj, XIX]
- 1892 DMV 1, 79-292 [*bib]
- 1894-1899 Batt 32, 319-320, 321-
347; 33, 260-319; 35, 284-332; 36,
306-316; 37, 186-211 (F25, 63;
26, 51; 28, 53; 29, 29; 30, 41)
- 1894-1895 BSM (2)18, 179-196,
213-220, 284-308; 19, 87-110, 213-
224, 246-264 (F25, 63; F26, 51)
- 1897-1899 Prace MF 8, 139-177; 9,

222-241; 10, 193-268 (F28, 53;
29, 29; 30, 40)
- 1897 *Sur les progrès de la théorie
des invariants projectifs,* Paris,
136 p (F28, 53)
- 1900 *Rapporto sui progressi della
teoria proiettiva degli invarianti
nell' ultimo quarto di secolo,*
Naples
- 1904 Enc MW 1, 320-
- 1911 Enc SM 1(2)(3-4), 386-520
[forms]
*Pedoe Daniel 1965 Enc Br
Thomas T Y 1944 *The Concept of In-
variance in Mathematics* (U Calif
P), 14 p
Turnbull H W 1926 M Gaz 13, 217-221
- 1928 *The Theory of Determinants...,*
Lon, 354 p [bib]
White H S 1899 AMSB (2)5, 161-175
(F30, 41) [proj]

## Inversive geometry

Court N A 1962 MT 55, 655-657
Emch A 1914 AMSB (2)20, 412-415
(F45, 79) [discov]
- 1914 AMSB (2)21, 206 [discov]
Longchamps G de 1898 Intermed 5, 44
[Steiner]
Mikami Y 1929 Toh MJ 32, 173-192
[Japan]
Patterson B C 1933 Isis 19, 154-180
(F59, 10) [origin]

## Irrationals

See also e, Pi, Roots.

Archibald R C 1914 AMM 21, 253
[expon fun]
Bonnesen T 1920 M Tid B, 2-15, 55
(F47, 25) [anc]
- 1921 Pe M (4)1, 16-30 [anc]
Bortolotti E 1931 Pe M (4)11, 133-148
(Z1, 321) [cont frac]
- 1935 Pr M (4)15, 220-229 (Z12, 97)
[Babyl]
Bosch F 1931 DMV 41, 59-72 (Z2, 379)
[Greek]

## Isogonal centers

Neuberg J + 1892 Mathesis (2)2,
162-163, 274-275  [bib]

## Isoperimetric problem

See also Calculus of variations.

Anton L 1888 *Geschichte des
isoperimetrischen Problems, eine
geschichtliche Darstellung der
Variationsrechnung von Bernoulli
bis Lagrange,* Leipzig, 77 p (diss)
Carruccio E 1939 It UM (2)2, 73-75
(F65, 11)  [Galileo]
Eneström G 1876  Upp U, 77 p
(F9, 6)
- 1888 Bib M (2)2, 38
Kelvin W T 1894 Lon RIGBP 14, 111-
119
- 1894 Nat 49, 515-518  (F25, 76)
Mueller W 1953 Sudhof Ar 37,  39-71
[Theon, Zenodoros]
*Porter T I 1941 A history of the
classical isoperimetric problem
in Bliss G. A., *Contributions to
the Calculus of Variations,*
(U Chicago P)
Schmidt W 1901 Bib M (3)2, 5  [anc]

## Iteration

Biermann K-R 1957 Ens M (2)4, 19-24
[Euler]
- 1961 Crelle 207, 96-112
(MR24A, 4)  [Jacobi]
Kennedy E S + 1956 AMM 63, 80-83
[mid]

## Jacobian

Muir T 1909 Edi RSP 29, 499-516

## Jordan algebra

Paige L J 1963 in Albert A. A. ed,

*Studies in Modern Algebra* (MAA
Stu 2)  144-186   [bib]
Schafer R D 1955 AMSB 61, 469-484
[bib]

## Kirkman problem

Doerrie H 1965 *One Hundred Great
Problems,* NY, 14-19
*Eckenstein O 1911 Mess M 41, 33-36
[bib]

## Knot theory

*Crowell R H + 1963 *Introduction to
Knot Theory,* Boston, 182 p
[*bib]

## Lambda function

Archibald R C 1934 Scr 2, 300

## Lamé functions

Prasad G 1930-1932 *A Treatise on
Spherical Harmonics... Lamé,*
Benares, 2 Vols

## Latin squares

Kendall M G 1948 Am Stacn 2, 13
[XVI]
MacMahon Percy A 1915 *Combinatorial
Analysis,* (Cambridge U P), Pt. 5,
Ch. 3.
Turgeon J 1967 Jeune Sci (Jan), 89

## Lattice

Birkhoff Garrett 1940 *Lattice theory,*
(AMS Col 25)  [bib]  (2nd ed 1948;
3rd 1967)
*- 1965 Enc Br
*- 1966 MH En St
MacLane S 1939 AMM 46, 3

## Law of large numbers

Anon 1919 Isis 2, 395
  [Jacques Bernoulli]
Invrea R 1936 It Attuar 7, 229-230
  (F62, 12)
Markov A A 1914 Kagan (603)
Vassiliev A 1914 Ens M 16, 92-100
  (F45, 69)

## Law of tangents

Bradley H C + 1921 AMM 28, 440-443
  [bib]

## Least squares

See also Correlation.

Chebotarev A S 1961 Vop IET 11, 20-
  28
Eisenhart C 1964 Wash Ac 54, 24-33
Gerling C L 1861 Gott N, 273-
Gore J H 1887 USCS, 311-512
  [bib, geodesy]
Hall A 1872 Nat (6 June)
Merriman M 1877 Analyst H 4, 140-143,
  176  (F9, 21)
*- 1882 Conn T 4(1), 151-232  [*bib]
Plackett R L 1949 Biomtka 36, 458-
  460
Sheinin O B 1965 Ist M Isl 16, 325-
  336  (MR33, 325-336)  [Adrain]
Spiess W 1939 Schw Verm 37, 11-17,
  21-23  (Z22, 296)  [D. Huber]

## Lehmus-Steiner theorem

Henderson A 1937 Elish Mit 53(2),
  246
- 1955 Scr 21, 223-232, 309-312
- 1956 Scr 22, 81
MacKay D L 1940 SSM 39, 561-572
Thebault V 1955 MT 48, 97-98

## Lemniscate

Fouret G 1892 Fr SMB 20, 38-40
  (F24, 53)  [mech]
Wittstein A 1895 ZM Ph 40, 1-6
  [Omar]

## Length

Agostini A 1930 Bo Fir (NS)9, 25-28
  [spiral, Torricelli]
- 1930 Pe M 10, 36-38  [cycloid]
Bettazzi R 1892 Annali (2), 19-40
Bortolotti E 1927 Bln Rn (NS)32, 127-
  139  [Spiral, Torricelli]
Boyer C 1964 Koyre 1, 30-39
*Christensen S A 1887 Bib M (2)1, 76-
  80
- 1887 Zeuthen (5)5, 121-126
Coolidge J L 1953 AMM 60, 89-93
  (MR14, 523)
Endo T 1895 Tok M Ph 7, 103
  [cycloid, Japan]
- 1906 Tok M Ph 3, 72-74
  [ellipse, Japan]
Hofmann J E 1941 Deu M 6, 283-303
  (MR8, 189; Z26, 194)  [log]
Inglis A 1933 M Gaz 17, 327-328
  [ellipse, Gregory]
Kikuchi D 1895 Tok M Ph 7, 114
  [Ajima, circle, Japan]
Loerchner H 1912 Ost Bau 18, 770-775
  593-799, 812-816  (F43, 76)
  [area]
Loria G 1897 Linc Rn (5)6(2), 318-
  323  (F28, 48)  [Torricelli]
- 1938 It UM Con 1, 521-527
  (F64, 16)
Mikami V 1912 Bib M 12, 225-237
  [ellipse, Japan]
Milne R M 1903 M Gaz 2, 309-311
  [Huygens]
Neugebauer O 1937 QSGM (A)3
  (F63, 6)
Rossier P 1941 Genv CR 58, 170-171
  (MR3, 258)  [circle]
Staeckel P 1913 Porto Ac 7, 207-213
  [Euler]
Watson G N 1933 M Gaz 17, 5-17
  [ell fun, Fagnano, Landen]

## L'Hospital rule

Eneström G 1894 Sv Ofv, 297
[Bernoulli]
Struik D J 1963 MT 56, 257-260
[origin]


## Lie groups

*Dickson L E 1924 An M (2)25, 287-
378 [dif eqs]
Eisenhart Luther P 1933 *Continuous
Groups of Transformations,*
Princeton [bib]
Freudenthal H 1968 Rv Syn (3), 49-52,
223-243 [bib]


## Limaçon

Archibald R C + 1900 Intermed 7, 106
[Pascal]


## Limits

See also Calculus, Tauberian
theorems.

Agostini A 1925 It UM 4, 104-107
[Cavalieri]
- 1925 Pe M (4)5, 18-30 (F51, 20)
[Mengoli]
Bortolotti E 1939 Bln Mm (9), 113-
141 (F65, 1083) [Italy]
- 1939 It UM (2)1, 47-60
(F65:1, 4; Z20, 196) [inf]
- 1939 It UM (2)1, 275-286
(F65:1, 8) [XVI, Italy]
- 1939 It UM (2)1, 351-371
(F65:1, 10) [XVII, Italy]
Bosmans H 1913 Brx SS 37, (F44, 60)
[Stevin]
Cajori Florian 1915 AMM 22, 1-6, 39-
47, 77-82, 109-115, 143-149, 179-
186, 215-220, 253-258, 292-297
[Zeno]
*- 1919 *A History of the Conceptions
of Limits and Fluxions in Great
Britain, From Newton to Woodhouse,*
Chic-Lon, 307 p

- 1922 Nat 109, 477 [Newton]
- 1923 AMM 30, 223-234
[calc, Leibniz]
Cassina U 1936 Pe M (4)16, 1-19,
82-103, 144-167 (F62, 4; Z14, 244)
*- 1961 *Dalla geometria egiziana alla
matematica moderna,* Rome, Ch.7
Gibson G A 1899 Bib M (2)13, 65-70
(F30, 46) [Berkeley]
*McShane E J 1952 AMM 59, 1-11
[Moore-Smith] partial order
Medvedev F A 1960 Mos IIET 34
Molk J + 1904-1907 Enc SM 1(1)(1-2),
133-208
Mordukhai-Boltovskoi D 1933 Archeion
15, 45-72 (Z7, 49)
[mid to present]
Vogt H 1885 *Der Grenzbegriff in der
Elementar-Mathematik,* Breslau
Wallner C R 1903 Bib M (3)4, 246-259
Wieleitner H 1929 DMV 38, 24-35
(F55, 11) [Fermat]


## Line

Kline M 1956 Sci Am 194, 104-
Steele A D 1936 QSGM (B)3, 287-370
(F62, 11) [circle, Greek]


## Line geometry

Balitrand F 1918 Intermed 25, 25-26
(Q4815) (F46, 53) [Chasles,
Pluecker]
Rudio F 1899 ZM Ph 44(S), 383-397
(F30, 54) [line coordinates]
Taton R 1952 Elem M 7, 1-5
[Cayley, Monge, Pluecker]


## Linear algebra

See also Quaternions, Vectors,
Tensors.

Bork A M 1964 Isis 55(3)
[fourth dim, phys, quaternions,
vectors]
Boutroux P 1908 Int Con 4(3), 381-
384 [role in anal]

## Linear algebra   (continued)

Cayley 1887 QJPAM 22, 270-308
Crowe Michael J 1967 *A History of*
  *Vector Analysis* (U Notre Dame P),
  270 p  [bib]
Dickson Leonard E 1906 AMM 13, 201-
  205
- 1914 *Linear Algebras*, NY, 81 p
  [bib]
*- 1923 *Algebras and Their Arith-*
  *metics*, Chic, 249 p  (repr 1960
  Dover)  [bib]
*Gibbs J W 1886 AAAS 35, 37-66
Hankel H 1869 *Theorie der Complexen*
  *Zahlenapteme*, Leipzig, 167 p
*Pickert G 1953 Enc MW 1(1)(3)(1)
*Shaw J B 1907 Carn Pub (78)  [bib]
Study E 1896 Chic Cong, 367-
Wills A P 1931 *Vector Analysis...*,
  NY, 285 p  (repr 1958 Dover)
  [bib]


## Linear differential equations

Birkhoff G D 1941 Am  NASP 27, 65-67
Eneström G 1897 Bib M (2)11, 43-50
  (F28, 40)  [Bernoulli, Euler,
  const coef]
- 1911 Bib M (3)12, 238-241 (F42, 61)
  [Euler, var coeff]
Fields J C + 1885 AJM 7, 353  [bib]
Fry T C 1929 AMM 36, 499  [elec]
Heffter L 1896 Chic Cong, 96-104
  (F28, 40)  [Fuchs]
Moulton F + 1911 AJM 33, 63-96
  [perio coeff]
Moulton F R 1930 *Differential*
  *Equations*, NY [homog, perio coefs]
Nixon H B + 1885 AJM 7, 353-363
  [bib]
Schlesinger L 1895 *Handbuch der*
  *Theorie der linearen Differential-*
  *gleichungen*, Leipzig 1
- 1909 DMV 18, 133-266  (F40, 58)
  [*bib, since 1865]
Simonov N T 1966 Ist M Isl 17, 333-
  338  [Euler to Peano]


## Linear equations

See also False position.

Gloden A 1953 Rv Hi Sc Ap 6, 168-170
  [Clasen]
Kattsoff L O + 1965 MT 58, 295-297
  [recur sol syst]
Kieffer L 1959 Lux Arch 26, 105-124
  [syst]
Kloyda ⅃ 1937 Osiris 3, 165-192
  (Z18, 197; F63, 805)  [1550-1660]
Procissi A 1946 Pe M (4)24, 141-151
  [Z10, 7)  [Buteone, Cardan, Syst]
Rolewicz S + 1968 *Equations in Linear*
  *Spaces*, Warsaw [inf dim spaces]
Vogel K 1940 Deu M 5, 217-240
  (MR2, 114; F66, 11; Z23, 193)
  [Fibonacci]


## Linear programming and extensions

See also Operations research (Ch.3)

Arnoff E L + 1961 Prog OR 1, Ch. 4
  [bib]
Dantzig George B 1963 *Linear*
  *Programming and Extensions*
  (Princeton U P),  Ch. 2  [bib]
Dorfman R + 1958 *Linear Programming*
  *and Economic Analysis* (McGraw-Hill),
  1-7  [bib]
*Gass S I 1964 *Linear Programming*,
  2nd ed. (McGraw-Hill)  [appl, bib]
*Riley V + 1958 *Bibliography on Linear*
  *Programming and Related Techniques*
  (Johns Hopkins P),  623 p
Zangwill Willard I 1969 *Nonlinear*
  *Programming. A Unified Approach*
  (Prentice-Hall), 356 p  [bib]
Zwicky F + ed 1967 *New Methods of*
  *Thought and Procedure* (Springer),
  99-131  [dyn program]


## Logarithm

See also Addition logarithms,
  Exponential, Mantissa.

Abelson V B 1948 *Rozhdenie logarifmov*,
  Mos-Len, 231 p

Logarithm   (continued)

*Agostini A 1922 Pe M (4)2, 135-150
  [discov]
- 1922 Pe M (4)2, 430-451
  [Euler, Mengoli]
- 1923 Pe M (4)3, 177-190    [XVIII]
Andrews F E 1929 SSM 28, 103-130
Anon 1900 Intermed 7, 95
Archibald R C 1955 M Tab OAC 9, 62-
  63  (MR16, 985) [first table]
Aubry A 1906 Ens M 8, 417
  [before Napier]
Barbour J M 1940 Scr 7, 21-31
  [music]
Bertrand J 1868 Nou An (2)7, 229
  (F1, 13)  [Huygens]
- 1868 Par CR 66, 565-567 (F1, 13)
  [Huygens]
Bierens de Haan D 1873 Boncomp 6,
  203-238  (F5, 43)
  [Holland, tables]
- 1875 Amst Vh 15, 1-35
  (F7, 24)  [tables]
Boehm 1918 Ost IAZ 70, 516
  (F46, 46)  [Briggs]
Bohren A 1914 Bern MG, 318-324
  (F45, 72)  [anniv]
Bonardi J 1932 Rv Sc 70, 174-178
  (F58, 51)
Boys C V 1931 M Gaz 15, 367-368
  [Briggs, Hutton]
Cairns W D 1928 AMM 35, 64-67
  [Napier]
Cajori F 1899 ZM Ph 44(S), 32-39
  (F30, 49)
- 1913 AMM 20, 5-14, 35-47, 75-84,
  107-117, 148-151, 173-182, 205-
  210  (F44, 60)  [expo fun]
- 1916 AMM 23, 71-72  (F46, 51)
  [Napier]
- 1927 Sci (2)65, 547  (F52, 16)
  [Napier]
- 1930 Archeion 12, 229-233
  (F56, 15)  [XVI, Peucers]
Carslaw H S 1916 AMM 23, 310-313
  [Napier]
- 1916 M Gaz 8, 76-84, 115-119
  [Napier]
- 1916 Phi Mag (6)32, 476-486
  (F46, 46)  [Napier]

D G 1871 Dillner 4, 28    (F3, 9)
  [Napier]
Dieguez D 1917 Esp SM 6, 94-103
  (F46, 55)
Dubois E 1872 Mondes (2)27, 651-652
  (F4, 13)
Eneström G 1884 Bib M, 121-124
  [Sweden, tables]
Epstein P 1924 ZMN Unt 55, 142-151
  (F50, 11)  [Kepler]
Euler L 1751 Ber HARS 5(1749), 139-
  179  [Leibniz, Bernoulli, imag
  logs]
Eves H 1953 MT 53, 384-385
Ferguson A 1912 Sci Prog 7, 147-170
  [disc]
Fletcher A 1941 Nat 148, 728
  [early]
Gehler I S F 1776 Historiae
  logarithmorum naturalium primordia,
  Leipzig
Gibson G A 1914 Glasg PS  [Napier]
Gieswald 1856 Arc MP 26, 316-
Girshvald L Ya 1952 Istoriya
  otkrytiya logarifmov, Kharkov,
  32 p
Glaisher J W L 1872 Lon As Mo N 32,
  255-262, 288-290  (F4, 13)
  [mistakes, table, Vlacq]
- 1913 QJPAM, 249-301
  [calculation by geom means]
- 1915 QJPAM 46, 125-198
  [Oughtred]
- 1916 QJPAM 47, 249-301 Numerical
  Analysis   [calculation without
  series]
- 1918 QJPAM 48, 151-192
  (F46, 45)  [tables]
Guenther S 1876 Vermischte Unter-
  suchungen zur Geschichte der
  Mathematischen Wissenschaften,
  Leipzig, Ch. 5   [XVII, XVIII]
Gutzmer A 1914 DMV 23, 235-248
  [anniv]
- 1915 ZMN Unt 46, 69-81  [anniv]
Hayashi T 1922 Toh MJ 21, 148-190
  (F48, 40)  [Japan]
Hejzlar F 1874 Cas MF 3, 49-61
  (F6, 41)  [tables]
Henderson J 1930 M Gaz 15, 250-256
  (F56, 16)  [tables]

Logarithm    (continued)

Hobson E W 1914 *John Napier and the Invention of Logarithms,* Cambridge, 48 p
Hofmann J E 1938 Deu M 3, 446-466 (Z19, 100) [Mercator]
- 1939 M Mag 14, 37-45    (MR3, 97) [series]
- 1939 Deu M 4, 556-562 (F65, 13) [Mercator series]
- 1940 Deu M 5, 358-375 (MR7, 354) [Cotes, Halley, de Moivre series]
- 1941 Deu M 6, 283-304 (Z26, 194)    [intgl]
- 1940 Deu M 3, 598-605 [Mercator series]
- 1963 Praxis 5, 225-232  [natural]
Hohn F E 1943 AMM 50, 115 [exist]
Hoppe E 1901 Ham MG 4, 52 [tables]
Kewitsch 1896 ZMN Unt 27, 321 [Buergi, Napier]
*Knott C G 1915 *Napier Tercentenary Volume,* London
- 1915 Sci Prog (Oct)   [Napier]
Koppe M 1904 Ber MG 3, 48-52 (F35, 58)   [Napier]
Lampe E 1907 Ab GM 25, 117-137 [complex, Euler]
Loria G 1900 Bib M (3)1, 75-89 [curve, Torricelli]
Lupton S 1914 M Gaz, 7, 147-150, 170-173  [radix method]
Matzka W 1860 Arc MP 34, 341-
Miller G A 1926 Indn MSJ 16, 209-213  (F52, 16)
- 1928 Sci 70, 97-98    (F54, 6) [Napier]
- 1928 Toh MJ 29, 308-311   (F54, 6)
- 1932 Indn MSJ 19, 164-172 (Z4, 194)    [group]
- 1947 M Stu 15, 1-3   (Z31, 241) [Babyl]
Monkevich 1891 Kagan (111) [Bronski, tables]
Mueller C 1914 Naturw 2, 660-676 [Napier]
Naux Charles 1966 *Histoire des logarithmes de Neper à Euler,*

(Blanchard), Tome I.  La découverte ... premieres tables,  158 p
d'Ocagne M 1914 Nat Par (11 July), 114-118 [anniv]
Read C B 1953 MT 53, 381-385 [Napier]
Rychlik K 1960 Cas M 85, 37-43 (MR22, 928)  [calculation]
Schaaf W S 1952 MT 45, 361-363
- 1957 MT 50, 295-297
Schlepps F 1882 *Die Logarithmen,* Leipzig
Simon M 1913 Ost PZ 10, 157-159 (F44, 48)  [tables]
Sleight E R 1944 M Mag 18, 145-152 (Z60, 7)  [Napier]
Teixeira F G 1918 Intermed 25, 47-48  (F46, 55)  [neg numbs]
Thibaut Bernard F 1797 *Historiae controversiae circa numerorum negativorum et impossibilium logarithmos,*  Goettingen
Thoman F 1868 Nou An (2)7 (F1, 13)  [Huygens]
- 1868 Par CR 66, 661-664 (F1, 13)  [Huygens]
Thompson A J 1925 AMM 32, 129-131 [Briggs]
Timchenko I 1935 Odess UTM 1, 7-33 (Z14, 244)
Toledo L O de 1915 Esp SM 4, 175-177 (F45, 96)  [Spain, tables]
Uspenskii Ya V 1923 *Ocherk istorii logarifmov,*  Pet, 78 p
Valia G 1915 Tor FMN 50, 183-186 (F45, 72)
Van Haaften M 1925 Nieu Arch (2)15(1), 49-54 [Decker, table, Vlack]
- 1941 Nieu Arch (2)21, 59-64 (MR7, 354)  [Netherlands, tables]
Van Hée 1914 Toung Pao 15, 454-457   [China]
Vetter Q 1933 Cas MF 63, D41-D49 (F59, 827)
Viglezio E 1923 Tor FMN 58, 67-75 [calculation]
Voellmy E 1948 *Jost Buergi und die Logarithmens,* Basel, 24 p [Buergi]
Wackerbarth A D 1871 Lon As Mo N 31, 263-264  (F4, 13)  [natural]

Logarithm    (continued)

- 1871 Mondes (2)26, 626-627
  (F3, 8)
Wargentin Peter 1752 Sv Han 13, 1-11
  (also Sv Han Deu 14, 3-15)
Yoneyama K 1926 Tok S Ph S 36, 1-23
  [disc]


Logic

See also Combinatorial logic, Many
  valued logic, and in Ch. 3:
  Analysis and synthesis, Automata,
  Logic diagrams and machines,
  Proof.

Anon 1959 Mathunt (April)
- 1961 Mathunt (Jan)
*- 1966 JSL 26, 151-337
  [index of reviews]
Becker O 1957 Zwei Untersuchungen
  zur antiken Logik, Weisbaden,
  55 p
Beth Evert W 1940 Ned Ps 34, 53-68
  (MR7, 355)
- 1944 Geschiedenis der Logica,
  The Hague, 96 p  (MR7, 354)
- 1947 Dialect 1, 311-346
  [XIX, XX]
- 1948 J S L 13, 62-
- 1948 Symbolische Logic und
  Grundlegung der exakten Wissenschaften. Bibliographische
  Einfuehrung...,   Bern, 28 p
- 1950 Les fondements logiques des
  mathématiques, Paris
  1956 Int Con HS 8, 1104-1106
Blakey R 1848 Essay on Logic,
  London   [bib]
- 1851 Historical Sketch of Logic,
  London
Blanché R 1953 Rv Phi Fr E 143,
  570-598    [XX]
*Boehner P 1952 Medieval Logic, an
  Outline of its Development from
  1250  to ca. 1400, Chicago
Boll Marcel + 1920 Histoire de la
  logique (Presses Universitaires),
  7th ed, 127 p  (1st ed 1946)

Bochenski I M 1936 Angelic 13, 109-
  123
- 1939 Int Con Ph 10, 1062-1064
- 1951 Ancient Formal Logic
  (North Holland), 131 p
  (MR13, 419)
*- 1956 Formale Logik, Frieburg
  [bib]
*- 1959 A Precis of Mathematical
  Logic NY, 100 p
*- 1961 A History of Formal Logic
  (Notre Dame Press),  589 p
  [bib]
Braithwaite R B 1932 M Gaz 16, 174-
  178  [Lewis Carroll]
Calker J F A van 1822 Denklehre
  oder Logik und Dialektik, nebst
  einem Abriss der Geschichte und
  Literatur derselber,  Bonn,
  554 p
Calogero G 1931 Logos Tub 2, 414-429
  (F57, 5)  [anc]
Carruccio E 1948 It UM 3, 1-16
  [Leibniz]
*Church A 1936 JSL 1, 121-218    [*bib]
- 1938 JSL 3, 178-212  [*bib]
- 1952 Am Ac Pr 80  [bib]
*- 1956 Introduction to Mathematical
  Logic,  Vol I, Princeton    [bib]
- 1958 Int Con Ph 12(4), 77-81
- 1964 Int Con L  [exist import
  categ prop]
Clark Joseph T 1952 Conventional
  Logic and Modern Logic, Wookstock,
  Md,  117 p
*Copi J M + 1967 Readings in Logical
  Theory (Macmillan)
Couturat L  1901  La logique de
  Leibniz d'après des documents in
  inédits, Paris
- 1905 L'algèbra de la logique,
  Paris, 100 p
Curry H B 1957 Theory of Formal
  Deducibility,  Notre Dame, 129 p
  [bib]
Duerr K 1938 Erkennt 7, 160-168
  (F64, 909)  [mid]
Eberstein W L G 1794,1799 Versuch
  einer Geschichte der Logik und
  Metaphysik den Deutschen von
  Leibniz bis auf gegenwarlige Zert.
  Halle,  2 vols

Logic    (continued)

Enriques Federigo 1922 *Per la storia della logica: I principii e l'ordine della scienza nel concetto dei pensatori matematici* Bologna   (Fr transl 1926, Ger, transl 1927)   (F48, 29)
- 1926 *L'evolution de la logique,* Paris   (Engl transl 1929; F55, 587)
Eves H 1959 MT 52, 33
Fabricius J A 1699 *Specimen elencticum historiae logicae,* Hamburg
Feys R 1950 *Logistique, Chronique 1939-1945,* Paris
Fraenkel A A 1951 Scr 17, 5-16 [excluded middle]
Glanville J J 1954 New Schol 28, 187-198
Guggenheimer H 1966 in *Confrontations with Judaism,* Philip Longworth ed, London, 171-196 [Hebrew]
Gunther J 1952 *Die Anaprueche der Logistiker auf die Logik und ihre Geschichts-schreibung,* Stuttgart
Haldane J B S 1957 Sankh 18, 195-200 (Z99, 242) [Hindu]
Hedrick E R 1933 Sci (NS)77, 335-
Heiberg J L 1929 *Anonymi logica et quadrivium, cum scholiis antiquis edidit,* Copenhagen, 164 p
Henkin L 1953 AMST 74, 410- [alg]
*- 1962 Sci 138, 788-794 [logicism, Russell]
Herbrand J 1968 *Ecrits logiques,* Paris, 247 p
Hermes H + 1952 Eng MW (1)1(A), 12A
Johnson F R + 1935 Hunt LB 7, 59-87 (F62, 1029) [Recorde]
Joja A 1960 Buc U Log 3(1), 7-47 (MR24A, 220) [Greek]
Jordan A A 1945 *The Development of Mathematical Logic and of Logical Positivism in Poland Between the Two Wars,* Oxford
Joergensen J 1931 *A Treatise of Formal Logic,* Copenhagen-London, Vol 1

Jourdain P E B 1910 QJPAM 41, 324-352; 43, 219-314; 44, 113-128 (F41, 50)
- 1916 Monist 26, 522   [Lambert]
King G 1962 Not DJFL 3, 5-40 [XX, bib, Russia]
*Kleene S C 1967 *Mathematical Logic* (Wiley)   [bib]
Kneale W 1948 Mind 57, 149- [Boole]
Kneale W + 1962 *The Development of Logic,* Oxford, 762 p
Korcik A 1954 Stu Log 1, 247-253 (MR16, 781)
Korfhage Robert R 1966 *Logic and Algorithms,* NY, 194 p
Kotarbinski T 1959 *La logique en Pologne,* Rome
- 1964 *Lecons sur l'histoire de la logique,* Paris   (tr of *Wyklady z Dziejow Logiki,* Lodz 1957)
Kreisel G + 1967 *Elements of Mathematical Logic* (North-Holland) [model thy]
Kuypers K 1943 Ned Ps 37, 31-40 (MR7, 355) [Aristotle]
Lattin H 1948 Isis 38, 205-225 [XI, anal geom]
Leblanc H 1961 J Phil 58, 553-558
*Lewis C I 1918 *A Survey of Symbolic Logic* (U Calif P), 327 p   (repr less Chs.5-6, 1960 Dover)   [bib]
Lewis Clarence I + 1932 *Symbolic Logic,* NY-Lon, 515 p
Liard Louis 1878 *Les logiciens anglais contemporains,* Paris (5th ed 1907)
Loria G 1894 BSM 18, 107-112 (F25, 54) [before Leibniz]
Lukasiewicz J 1934 Prz Fil 37, 417-437 [predicate logic]
- 1935 Erkennt 5, 111-131 (=1934)
- 1957 *Aristotles' Syllogistic,* 2nd ed (Oxford U P)
MacLane S 1939 AMM 46, 289-296
Michalski Konstanty + ed 1937 *Mysl katolicka wobec logiki wspolcquesnej,* Poznan, 196 p
Moody E A 1953 *Truth and Consequences in Medieval Logic,* Amsterdam, 113 p

Logic    (continued)

Nagel E 1939 Osiris 7, 142-224
  (F65:2, 109)  [geom]
Nagi A 1892 Rv M 2, 177-179
Nelson E J 1934 AMSB 40, 478-
  [Whitehead, Russell]
*Nidditch P N 1962  Development of
  Mathematical Logic,  London, 88 p
  (Z115, 243)
Peano G 1894 Fr AAS 2, 222-226
Perez B J 1954 Teoria 2, 171-176
*Prandtl Carl 1855-1870 Geschichte
  der Logik im Abendlande,  Leipzig,
  4 vols (2nd ed vol 2, 1885, repr
  1927, 1955)  [anc, mid]
Primakovsky A P 1955 Bibliografiya
  po logike, Moscow, 96 p
  [XVIII, XX, Russia]
Prior A N 1960 Formal Logic,  2nd ed,
  Oxford  (1st ed 1953)
Prosper V·R 1893 Prog M 3, 41-43
  (F25, 55)  [Italy]
Rabus L 1895 Logik und System der
  Wissenschaft,  Erlangen-Leipzig
  [bib]
Ramsey F P 1926 M Gaz 13, 185-194
Reiffenberg F A F T 1833 Principes
  de logique, suivis de l'histoire et
  de la bibliographie de cette
  science,  Brussels, 83 p
Reimann J F 1699 Critisierender
  Geschichts-Calendar von der
  logica,  Frankfurt
Rescher N 1963 Not DJFL 4, 48-58
  (MR26, 1141)  [Avicenna]
Reymond A 1936 in Int Con PS
  (1935) 8  (=Act Sc Ind 395)
  [Greek, Russell]
Rychlik K 1958 Cas M 83, 230-235
  [Bolzano]
Scholz Heinrich 1931 Geschichte der
  Logik,  Berlin, 78 p  (F57, 1286)
  - 1961 Concise History of Logic
  NY,
Schrecker P 1937 Rv Phi Fr E 62,
  336-367  [Descartes]
Schroeder E 1890 Vorlesungen ueber
  die Algebra der Logik,  Leipzig,
  3 vols
  - 1895 Algebra und Logik der Relative

der Vorlesungen ueber die Algebra
  der Logic,  Leipzig, 657 p
  1909 Abriss der Algebra der Logik,
  Leipzig and Berlin, 50 p
Shearman A T 1906 The Development of
  Symbolic Logic, London
Staal J F 1960 Synthese 12, 279-286
  (MR23A, 221)  [Hindu]
Stegmueller W 1963 Kantstu 54, 317-
  334  [recent]
Styazhkin N I + 1962 Kratkii ocherk
  istorii obshchei i matematicheskoi
  logiki v Rossii, Mos, 87 p
  (Z105, 244)
Styazhkin N I 1964 Stanovlenie idei
  matematischeskoi logiki, Mos,
  304 p
  - 1969 History of Mathematical Logic
  from Leibniz to Peano  (MIT Pr),
  341 p  (tr of 1964)
Surma S J 1967 Buc U Log 10, 127-138
  (MR38, 571)  [Lindenbaum alg]
Thomas I 1957 JSL 22, 15-16
  [Euler]
Ueberweg F 1857 System der logik
  und Geschichte der logischen
  Lehren,  Bonn  (5ed 1882)
Vacca G 1899 Rv M 6, 121-125, 183
  [Pell 1668]
*Van Heyenoort Jean 1967 From Frege
  to Goedel, a Source Book in
  Mathematical Logic 1879-1931
  (Harvard U P)
Venn J 1881 Symbolic Logic, London
Volpe G della 1953 Int Con Ph 11, 14,
  157-162  (Z51, 243)  [Galileo]
Wilson Curtis 1956 William Heytesbury:
  Medieval Logic and the Rise of
  Mathematical Physics (U Wisc P),
  231 p  (MR18, 267)
Ziehen T 1920 Lehrbuch der Logik auf
  positivestischer Grundlage mit
  Berueck sichtigung der Geschichte
  der Logik,  Bonn    [bib]


Logistic curve

Miner J R 1933 Hum Bio 5, 673-689
  [demography, Verhulst]

## Lorentz transformation

Hill E L 1966 MH En ST

### Lunes

Becker O 1954 Philolog 98, 313-316
  [Eudemos, Hippocrates]
Eneström G + 1898 Intermed 5, 180
Landau E 1903 Ber MG  2
Schrek D J E 1942 N Tijd 30, 1-13
  (F68, 6)  [Archimedes]
Wieleitner H + 1906 Intermed 13,
  133-135, 223  (Q3009)
- 1934 Zur Geschichte der quadrier-
  baren Kreismonde, Munich, 78 p
  (ed by J E Hofmann)  (F10, 98;
  60, 5)

### Magic squares etc.

Ahrens W 1914 Himm Erde 27, 281-297,
  325-341  (F45, 70)
- 1915 Z Bild K 26, 291-301
  (F45, 69)  [Duerer]
- 1917 Islam 7, 186-250  [Arab]
- 1922 Islam 12, 157-177
  [al-Buni]
- 1925 Islam 14, 104-110  [al-Buni]
Andrews W S 1917 Magic Squares and
  Cubes, Chic
Aubry A 1927 Porto Fac 15, 193-200
Ball W W R 1960 Mathematical
  Recreations, Ch. 7
Bergstraesser G 1923 Islam 13, 227-
  235 [Arab]
Cammann S 1960 Am OSJ 80, 116-124
  (Z102, 244)   [China]
- 1961 His Relig 1, 37-80
  (Z102, 244)   [China]
- 1962 Sinolog 7, 14-53
  (Z102, 244)   [Chinese]
Carra de Vaux B 1948 Rv Hi Sc Ap 1,
  206-212  (Z37, 290)  [Arab]
Cazalas E 1922 Sphinx 2, 65-66
  [Goethe]
- 1934 Carrés magiques au degré n
  séries numérales de G. Tarry...,
  Paris, 192 p  [bib]

- 1934 Rv Hi Rel 110, 66-82
  [Agrippa]
Fazzari G 1895 Pitagora 1, 7, 13
Fontes J 1892 Fr AAS 21(1), 158
  (F24, 46)
Gerardin A 1928 Fr AAS 52, 40-41
  (F57, 1303)  [XVIII, Medrano]
Guenther S 1876 Vermischte Unter-
  suchungen..., Leipzig, Ch.4
Haas K H de 1935 Frenicle's 880
  Basic Magic Squares of 4 x 4 cells,
  Normalized, Indexed and Inventoried,
  Rotterdam
Hermelink H 1958 Sudhof Ar 42(3),
  199-217  (Z84, 244)
- 1959 Sudhof Ar 43, 351-354
  (Z92, 2)  [Arab]
Kowalewski Gerhard 1937 Magische
  Quadrate und magische Parkette,
  Leipzig, 78 p
Kraitchik M 1930 Traité des Carrés
  magiques, Paris, 108 p
McCoy J C 1941 Scr 8, 15-26
  [Moschopoulos]
Nassip Tou Mm (9)4, 423-454
Parmentier T 1894 Fr AAS 2
  [Knight's tour]
Sanford V 1923 MT 16, 348-349
  [circles]
Schubert H 1891 Monist 2, 487
- 1898 in Mathematical Essays...,
  (Open Court), 39-63
Schwartz J J 1933 Scr 1, 44-52
  [Arab]
Singh A N 1937 Int Con 10(2),
  275-276  (F63, 20)  [Hindu]

## Malfatti problem

Anon 1876 Boncomp 9, 388-397
  (F8, 19)  [bib]
Archibald R C 1933 Scr 1, 170-171
Baker M 1880 Wash PS 2, 113-123
  (F12, 35)
Bellacchi G 1895-1896 Pe M 10, 93,
  156; 11, 25
Derousseau J 1895 Lieg Mm (2)18
  (F26, 62)
Gérard Louis 1929 Sur le problème
  de Malfatti, le pendule de
  Foucault..., Paris, 64 p

## Malfatti problem    (continued)

Goldberg M 1967 M Mag 42, 241-247
Hayashi G 1937 Toh MJ 43, 127-132
    (Z17, 290)    [Japan]
Hirayama A 1936 Toh MJ 42, 67-74
    (Z14, 147)    [Japan]
Procissi A 1932 Pe M (4)12, 189-205
    [bib]
Scardapane N M 1931 Pe M (4)11,
    281-292    (Z3, 99)
Wittstein A 1871 *Geschichte des*
    *Malfatti'schen Problems,*
    Munich    (F3, 12)
- 1878 *Zur Geschichte des*
    *Malfatti'schen Problems,*
    Nördlingen    (F10, 27)

## Mantissa

Berdelle C + 1899-1900   Intermed 6,
    88, 181-182; 7, 61   (Q1336)
    [word]
Grove C C 1935 AMSB 41, 187
    (F61, 28)    [word]
Miller G A 1926 Sci (2)64, 279
    (F52, 16)    [mistakes]

## Many valued and modal logic

Carvallo Michel 1968 *Logique à trois*
    *valeurs. Logique à seuil,*
    Paris    [bib]
Fish M 1966 Peirce Tr 2, 71-85
    [C S Peirce]
Lewis C I 1918 *A Survey of symbolic*
    *Logic,* Berkeley [repr 1960 Dover]
McCall Storrs 1963 *Aristotle's Modal*
    *Syllogisms,* Amsterdam, 108 p
    (MR27, 1078)
Rescher N 1969 *Many-Valued Logic*
    (McGraw-Hill)   [*bib]
Rosser J B 1960 Log An 3, 137
    [inf val]
Turquette A A 1962 Act Phi Fe 16,
    261   [modal, many val]
Zawiriski Z 1932 Rv Met Mor 39,
    503

## Mascheroni constructions

Cajori F 1929 AMM 36, 364-365
    [Mohr]
Court N A 1958 MT 51, 370-372
Geppert H 1929 Pe M 9, 149-160
    [Mohr]
Hlavaty J H 1957 MT 50, 482-487

## Matrix theory

See also Eigenvalues.

Bellman R 1960 *Introduction to*
    *Matrix Analysis* (McGraw-Hill)
    [bib]
Cullis C E 1913-1925 *Matrices and*
    *Determinoids,* (Cambridge U P)
    3 vols
Feldmann R W 1962-1963 MT 55, 482-
    484, 589-590, 657-659; 56, 101-
    102, 163-164
MacDuffee C C 1943 AMM 50, 360-365
Miller G A 1910 AMM 17, 137-139;
    201-202
Muir T 1893 AJM 20, 225-228
    [bib]
- 1930 So Af Rs 18, 219-227
    [bib]
Rinehart R F 1955 AMM 62, 395
Ryser H J 1958 Can JM 10, 57-65
    [rank]
Taber H 1890 AJM 12, 337-396
- 1899 Clark University Decennial
    Publications, 73-83
*Trunbull H W + 1952 *An Introduction*
    *to the Theory of Canonical*
    *Matrices,* London   (repr 1961
    Dover)
*Wedderburn J H M 1934 *Lectures of*
    *Matrices,* NY, 200 p   (AMS Col 17)
    (repr 1964 Dover)   [bib]

## Maxima and minima

See also in Ch. 3: Least action,
Operations research.

Agostini A 1951 Parma Rv M 2, 265-
    275   (MR13, 612)   [Torricelli]

## Maxima and minima    (continued)

Aubry A 1900 Prog M (2)2, 41
Borel E 1914 Intermed 21, 105-106
  (Q1241)  [words]
Eneström G 1909 Bib M (3)10, 84
  (F40, 57)
Golovensky D I 1932 Scr 1, 53-55
  [Hebrew]
Hagge + 1898 Intermed 5, 141
  [min dist]
*Hancock Harris 1917 *Theory of Maxima
  and Minima*  (Ginn)  (repr 1960
  Dover)
Hayashi T 1906 Tok M Ph 4, 64
  [Nakagawa]
- 1931 Toh MJ 34, 349-396  [Japan]
Hoffmann E 1905 Bib M (3)5, 366-397
*Hutton Charles 1815 *A Philosophical
  and Mathematical Dictionary...*,
  London 2, 27-30
Mansion P 1882 Mathesis 2, 193-202
  [Fermat]
*Pitcher E 1965 Critical Points, Enc Br
Whittemore J 1917 An M (2)19, 1-20
  [min surf]
Wilde D J + 1967 *Foundations of
  Optimization* (Prentice-Hall)

## Maximum likelihood

Kendall M G 1961 Biomtka 48, 1-2
  (Z99, 244)  [Bernoulli]
LeCam L 1953 Calif Sta 1, 277
Sheinin O B 1965 Vop IET 19, 115-117
  (MR37, 15)  [Bernoulli, Euler]

## Mean

See also Barycentric calculus,
  Center of gravity.

Aubry A 1908 Mathesis (3)8, 206-210
  (F39, 58)
Britzelmayr W 1942 Allg St Ar 31,
  177-178  (F68, 3)        [Egypt]
*Dodd E L 1945 *Lectures on Probability
  and Statistics*,  Austin
  (U Texas P), 44 p    [bib]

*Dunnett C W 1960 Lon St (B)22, 1-40
  [bib, largest mean]
Gini C 1952 Metron 16(3-4), 3-26
  (Z47, 244)
Jones P S 1950 MT 43, 65-67, 164
Mansion P 1888 Bib M (2)2, 36
  [mediation]
Muirhead R F 1903 M Gaz 2, 283-287
  [arith, geom]
Plackett R L 1958 Biomtka 45, 130-135
Szabo A 1963 M Lap 14, 277-306
  (MR30, 372  [geom]

## Mean proportional

Anon 1900 Pitagora 7, 106  [Plato]
Carra de Vaux 1898 Bib M (2)12,
  3-4  (F29, 32)
Cavallaro V G 1929 Pe M (4)9, 269-
  273    (F55, 599)
Conte L 1952 Pe M (4)30, 185-193
  (Z48, 243)  [Viviani]
- 1953 Pe M (4)31, 145-157
  (MR15, 276)  [Huygens]

## Mean value theorem

Eneström G 1914 Bib M 14, 180
  [intgl, Newton]
Prasad G 1931 *Six Lectures on the
  Mean Value Theorem of the
  Differential Calculus,* Calcutta,
  117 p

## Measure

See also Ergodic, Integration,
  Mensuration, Metrology  (Ch.3).

Arton E 1925 Pe M (4)5, 255-264
  (F51, 11)  [anc]
Aubry A 1896 JM El 20, 173, 201
*Borel E 1936 Organon 1, 34-42
  (F62, 1041)    [heuristic]
Denjoy A 1940 in *Selecta, jubilé
  scientifique de M. Emile Borel,*
  Paris
Dinculeanu N 1966 Ro Rv M 11, 1075-
  11J2   [bib, Rumania]

## Measure    (continued)

Gunderson C 1901 *On the Content or Measure of Assemblages of Points,* NY
Hadwiger H 1954 Bern NG 11, 13-41 (MR16, 1)
Marczewski E 1948 Colloq M 1, 93-102 (Z37, 291)  [Banach]
Markovic Z 1940 Zagr BI 33, 1-25 (MR8, 497)  [Plato]
Medvedev F A 1959 Ist M Isl 12, 481-492   (MR23A, 697)

## Menelaos theorem

Burger H + 1924 Ab GN Med 7, 47-49, 80
Bresse G 1936 Ens Sc 9, 243
de Vries H 1924 N Tijd 12, 331-337 (F51, 17)
- 1936 N Tijd 24, 1-21   (F62, 1010)
Rome A 1933 Brx SS 53, 39-50 [Theon]

## Mensuration

See also Areas, Length, Measure, and in Ch. 3: Metrology, Surveying.

Aubry 1897 JM El (5)21, 18-22, 38-40, 62-64, 87-91, 114-118, 138-140, 162-166, 177-179, 194-198 (F28, 48)
Natucci A 1948 Pe M (4)26, 153-156 (MR10, 419)  [China]
Wiedemann E 1918 Erlang Si 50-51, 264-271

## Mersenne numbers

Archibald R C 1935 Scr 3, 112-119
*Ball W W R 1920 *Mathematical Recreations,* Ch. 17
Lehmer D H 1947 AMSB 53, 164-167 [survey]
Uhler H S 1952 Scr 18, 122-131 (MR14, 343]

## Modulus

Rocquigny G de 1898 Intermed 5, 277 [word]

## Monte Carlo Methods

Anon 1955 *Bibliography of the Monte Carlo Method...,* (U Fl P), 73 p
Hammersley J M + 1964 *Monte Carlo Methods* (Wiley), 178 p   [bib]

## Multigrades

Gloden A 1949 *Aperçu historique des multigrades,* Luxembourg,  41 p (Z38, 147)
Gloden A + 1948 *Bibliographie de Multigrades avec quelques notice bibliographiques,* Luxembourg

## Multiplication

Bobynin W W 1897 Fiz M Nauk 13, 77-80 (F28, 53)  [Egypt]
Bowden J 1912  MT 5, 4-8
Cajori F 1914 Nat 94, 363  (F45, 95) [nota]
- 1922 M Gaz 11, 136-143
- 1923 AMSB 29, 12  (F49, 22) [nota]
Colin G S 1933 Hesperis 16, 151-155 [Arab]
Curtze M 1895 ZM Ph 40, 7-13 (F26, 49) [abbreviated]
Eneström G 1891 Bib M (2)5, 96 (Q35)  (F23, 4)  [nota]
Frisone R 1917 Tor FMN 53, 420-427 [def]
Gandz S 1929 Heb UC An 6, 247-250 [Arab, Hebrew]
Glaisher J W L 1878 Phi Mag 5, 331
*Gravelaar N 1910 Wisk Tijd 6, 1-25  [nota]
Kleiber I 1887 Kagan (34)  [div]
Laisant C A + 1896 Mathesis (2)6, 85  [def]
Listray A 1897 Mathesis (2)7, 17 [def]

## Multiplication (continued)

Polachek H 1933 Scr 1, 245-246
[XII]
Sanford V 1952 MT 44, 256-258
[rule of signs]
Stern C + 1966 Ari T 11, 254-257
[Egypt]
Taton R 1948 Nat Par 76, 268-271
Vetter Q 1922 Cas MF 51, 271-278
[div, Babyl]
- 1923 Arc Sto 4, 233-240
[div, Babyl]

## Natural numbers

See also Big numbers, Figurate
numbers, Number theory, Numbers,
and in Ch.3: Billion, Fifty,
Million, Numeration, Seven,
Thousand.

Anon 1891 Kagan (120)   [Kant]
Fraenkel A A 1939 Scr 6, 69-79
[cardinal]
- 1940 Scr 7, 9-20  [ordinal]
- 1955 Integers and Theory of
Numbers, NY, 102 p  (Scr Stu 5)
- 1938 Die Zahlbegriffe der
Australien, Papua und Bahtunegen,
nebst einer Einleitung uber die
Zahl. Ein Beitrag zur Geistesgeschichte der Menschen, Berlin
(F64, 902)
Kluge Theodor 1937 Die Zahlbegriffe
der Sudansprache. Ein Beitrag
zur Geistesgesichte der Menschen,
Berlin, 260 p  (F64, 901)
- 1939 Die Zahlbegriffe der
Volker Amerikas, Nordeurasiens
der Munde und der Palaioafrikener
..., Berlin, 736 p  (Z23, 385)
- 1941 Die Zahlbegriffe der Dravida,
der Hamiten, der Semiten und der
Kaukasier; ein vierter Beitrag zur
Geistesgeschichte der Menschen,
Berlin,  65 p  (Z24, 242)
- 1941 Die Zahlbegriffe der Sprachen
Central und Südaniens, Indonesiens,
Micronesiens, Melanesians und

Polynesiens mit Nachtraegen zur den
Baenden 2-4. Ein fuenfter Beitrag
zur Geistesgeschichte der Menschen
nebst einer principiellen Unter-
suchungen ueber die Tonsprachen,
Berlin, 300 p, 200 p  (Z27, 1)
Koehler O 1951 Bu An Beh (9)(Mar)
[birds]
Lorey W 1924 Unt M 30 , 130
(F50, 10)
Menninger Karl 1957 Zahlwort und
Ziffer. Eine Kulturgeschichte der
Zahl. Bd. 1. Zahlreihe und
Zahlsprache, 2nd ed   Goettingen,
221 p  (MR19, 517)
Miller G A 1925 Ens M 24, 59-69
(F54, 48)
- 1925 Int Con 7(2), 959-967
(F54, 48)
Molodshii V N 1950 Ist M Isl (3),
431-466  (MR13, 1; Z41, 340)
[XVIII]
Nehring A 1929 Wort Sach 12, 253-288
Seidenberg A 1960 Calif Pu 3, 215-
299  (MR22, 1129)  [anc]
- 1962 Arc HES 2, 1-40  (Z109, 1)
[anc]
Solomon B S 1954 Harv J Asi 17, 253-
260  [one]
Vasiliev A V 1891 Kazn FMO (2)1,
1-21  (F23, 34)  [phil]
- 1919 Tseloe chislo. Istoricheskii
ocherk, Pet, 272 p  (F50, 19)

## Negative numbers

Bespamiatnykh N D 1950 Ist M Isl 3,
154-170 [Lobachevskii]
Briedenbach W 1924 ZMN Unt 55, 112-
114   [Kant]
Cajori F 1913 Sci 28, 51-52
[minus, plus signs]
De Morgan A 1842 Phi Mag
[Leonardo da Vinci]
Halsted G B 1913 Sci 27, 836-837
[minus, plus]
Mansion P 1888 Bib M (2)2, 63 (Q21)
[nota, XVII]
Medovoi M I 1958 Ist M Isl 11, 593-
598  (MR23A, 583)  [Abu Wafa]

Non-Euclidean geometry (continued)

*literatury po geometrii Lobachevskogo i razvityu ee idei,* Mos, 192 p (MR16, 1)
Grigorian A T 1960 Scientia (6)54, 347-350 (MR22, 1130) [mech, Russia]
Halsted G B 1878 AJM 1, 261-276, 384-385 (F10, 343) [bib, n-space]
- 1879 AJM 2, 65-70 (F11, 357) [bib, n-space]
- 1893 Monist 4, 483
- 1893 NYMS 3, 79-80 (F25, 79) [Lambert]
- 1893-1897 AMM 1, 70-72, 112-115, 149-152, 188-191, 222-223, 259-260, 301-303, 345-346, 378-379, 421-423; AMM 2, 10, 42-43, 67-69, 108-109, 144-146, 181, 214, 256-257, 309-313, 346-348; AMM 3, 13-14, 35-36, 67-69, 109, 132-133; AMM 4, 10, 77-79, 101-102, 170-171, 200, 247-249, 269-270, 307-308; AMM 5, 1-2, 67-68, 127-128, 290-291
- 1896 Chic Con, 92-95 (F28, 42)
- 1899 AAAS 53
- 1899 AMM 6, 166-172 [re Engel]
- 1900 AMM 7, 123-133 [Gauss]
- 1902 AMM 8, 216-230
- 1904 AMM 11, 85-86 [Gauss]
Holling N 1931 Gorres Ja 44, 41-78 (F57, 1304) [Bolyai]
Hoppe E 1925 Naturw 13, 743-744 (F51, 23) [Gauss]
Kagan V F 1933 *Geometricheskie idei Rimana i ikh sovremennoe razvitie,* Mos-Len, 76 p
- 1948 Mos IIET 2, 323-389 (MR11, 150) (repr in Kagan V F 1955 *Lobachevskii i ego geometriya,* Moscow, 193-294)
Kagan V F + 1950 *Stroenie neevklidovoi geometrii,* Moscow-Leningrad
Karagiannides A 1893 *Die nicht-euklidische Geometrie von Altertum bis zur Gegenwart,* Berlin, 44 p (F25, 78)
*Klein F 1871 M Ann 4, 573
- 1893 *Lectures on Mathematics...,* 85-98

Kuznetsov B G 1947 Mos IIET 1, 347-371 (MR11, 572) [Euler, mech]
Mansion P 1896 Mathesis (2)6(S), 1
- 1898 Mathesis (2)8, 44-45 (F29, 34) [Bolyai, Gauss, Lobachevsky]
- 1899 DMV 7, 156-158 (F30, 50) [Gauss]
- 1908 Mathesis (3)8(S)(3) (F39, 64) [Gauss vs Kant]
McClintock E 1893 NYMS 2, 144-147 (F25, 78)
Nevanlinna R 1956 Nor M Tid 4, 195-209, 229 (MR18, 710) [Gauss]
*Norden A P 1951 UMN (NS)6(3), 3-9 (MR13, 197; Z49, 290)
- 1956 Ist M Isl 9, 145-166 [Gauss, Lobachevsky]
Norden A P + 1952 *Sto dvadsat pyat let neevklidovoi geometrii Lobachevskogo, 1826-1951,* Mos-Len, 207 p (MR14, 832)
d'Ovidio E 1889 Tor FMN 24, 512-513 [re Beltrami 1889]
Pati T 1951 Alla UMAB 15, 1-8 (MR13, 420)
Petronievics B 1929 Rv Phi Fr E 54, 190-214 [Bolyai, Lobachevski]
Pierpont J 1930 AMSB 36, 66-76
Prosper V R 1894 Prog M 4, 13-16 (F25, 78)
Raaij W H L J van 1900 Kazn FMO (2)10(1), 1 [Holland]
Rényi A 1953 Magy MF 3, 253-273 (MR15, 383) [Bolyai, Lobachevskii]
- 1954 Hun Act M 5(S), 21-42 (Z57, 243)
Rose J 1911 Wisk Tijd 7, 183-192; 8, 21-29, 103-112 (F42, 62)
Rosenfeld B A 1956 Int Con HS 8, 138-141 [Lobachevskii]
Russo F 1956 Int Con HS 8, 29-32 [Saccheri]
- 1963 Rv Q Sc (5)24, 457-473 (Z117:1, 8)
Sommerville Duncan M Y 1910 Nat 84, 172
*- 1911 *Bibliography of Non-Euclidean Geometry, inclucing the Theory of Parallels, the Foundation of Geometry, and Space of n Dimensions,* London, 415 p

## Non-Euclidean geometry (continued)

- 1914 *The Elements of Non-Euclidean
  Geometry* (Bell), 290 p
  (repr 1958 Dover)
Staeckel P 1900 M Ann 53, 49
  [Wachter]
Staeckel P + 1897 BSM (2)21, 206-228
  [Bolyai, Gauss]
- 1897 M Ann 49, 149-206   (F28, 42)
  [Bolyai, Gauss]
- 1902 M Ter Er 19, 40
  [Bolyai, Lobachevskii]
Toth I 1955 Gaz MF (A)7, 447-456
  (MR17, 338) [Gauss]
Varga O 1953 Magy MF 3, 151-171
  (MR15, 383)
- 1954 Hun Act M 5(S), 71-94
  (Z58, 2)
Varicak V 1907 Zagr Rad 169, 110-194
Winger R M 1925 AMSB 31, 356-358
  (F51, 23) [Gauss]
*Wolfe Harold E 1945 *Introduction to
  Non-Euclidean Geometry* (Dryden),
  260 p
Zacharias Max 1937 *Das parallelelen-
  problem und seine Loesung Eine
  Einfuehrung in die hyperbolische
  nichteuklidische Geometrie*,
  Leipzig, 44 p

## Non-standard analysis

See also Infinitesimals,
  Archimedean.

Luxemburg W A J 1969 *Applications of
  Model Theory to Algebra, Analysis,
  and Probability* (Holt), 307 p
  [bib]
*Robinson Abraham 1966 *Non-Standard
  Analysis*, Amsterdam

## Nuisance parameters

See also Statistics.

Linnik Yu V 1968 AMS Trls 20,
  [*bib]

## Number theory

See also Algebraic numbers, Chinese
  remainder theorem, Congruence,
  Diophantine equations, Divisibility,
  Euclidean algorithm, Factoring,
  Fermat last, Fibonacci sequences,
  Geometric number theory, Magic
  squares, Mersenne numbers,
  Multigrades, Parity, Partitions,
  Pell equation, Perfect numbers,
  Prime numbers, Pythagorean triples,
  Quadratic reciprocity, Six square
  problem, Waring problem, Wilson
  theorem, Wolstenholm theorem,
  Nim (Ch.3).

Aubry A 1909 Ens M 11, 329-356
  (F40, 17) [Euler]
*- 1909 Ens M 11, 430-450   [Gauss,
  Lagrange, Legendre]
Bachmann P 1902 *Niedere Zahlentheorie*,
  Leipzig, 412 p
- 1911 Gott N, 455-508   [Gauss]
*Bell E T 1927 AMM 34, 55-75
  [bib, generalization]
- 1938 AMM 45, 414-421   [Blissard]
- 1943-1946 Scr 9, 209-231; 10, 81-
  147; 11, 21-50, 139-171; 12, 53-60
  (MR7, 353; 8, 189; Z60, 5)
Bespamyatnikh N D 1957 BSSR Ped 2,
  3-42 (MR20, 264) [XIX, Russia]
Bobynin V V 1894 Bib M, 55-60
  [elem meths]
Boncompagni B 1875 Boncomp 8, 51-62
  (F7, 19)
Bricard R 1936 Ens Sc 9, 240-242
Carmichael R D 1914 SSM 13, 392-399
Curtze M 1895 Bib M (2)9, 37-42,
  77-88, 110-114 (F26, 52) [XV]
Delone B N 1947 *Petersburgskaya
  shkola teorii chisel*, Mos-Len,
  421 p (MR19, 1029)
*Dickson Leonard E 1919-1923 *History
  of the Theory of Numbers*, Wash DC
  3 vols (repr 1934 NY) (abbreviated
  Dickson in this bibliography)
- 1930 *Studies in the Theory of
  Numbers*, Chicago, 240 p
Fraenkel A A 1955 *Integers and Theory
  of Numbers*, NY

## Number Theory    (continued)

Ganguly S 1927 Clct MS 18, 65-76
  [Bhaskara]
Gerardin A 1912 Int Con 5(2)(4a),
  539  (F44, 48)  [Plan for history]
Goldstein B R 1964 Centau 10, 129
  [mid]
Hardy G H 1922 Nat 110, 381-385
Hardy G H + 1938 *An Introduction
  to the Theory of Numbers*,
  Oxford  (3rd ed 1954)
Henry C 1880 BSM (2)4, 268-272
  (F12, 32)  [problems]
Hofmann J E 1961 Arc HES 1, 122-159
  (MR23A, 433)  [Euler, Fermat]
Itard J 1950 Rv Hi Sc Ap, 3 21-26
  (MR11, 572)  [Fermat]
Junge G 1940 Deu M 5, 341-357
  (MR7, 353)  [Pythagoras]
Kiselev A A 1966 Ist Met EN 5, 31
  [Euler, Goldbach, Waring, probl]
Landau E 1937 *Ueber einige neuere
  Fortschritte der additiven
  Zahlentheorie*, Cambridge, 94 p
Lehmer D N 1933 Scr 1, 229-235
  [meth]
Lehmer D H 1941 *Guide to Tables in
  Theory of Numbers*, Wash DC , 191 p
Loria G 1913 Ens M 15, 193-201
- 1923 Bo Fir    [Descartes]
Marre A 1880 BSM (2)3, 27-31
  (F12, 31)    [mss]
Matvievskaya G P 1960 Ist M Isl 13,
  107-186  (MR23A, 696)
  [Diophantine anal, Euler]
Miller G A 1907 AMM 14, 6    [probls]
- 1923 MT 16, 247-248
  [sums of digits]
Min S 1955 Shux Jinz 1(2), 397-402
  [China]
Minin A P 1905 Mos MO Sb 25, 293
  [Bugaev]
*Ore Oystein 1948 *Number Theory and
  its History*,  NY
 Poulet P 1929-1934 *La Chasse aux
  nombres*, Brussels, 2 vols
 Procissi A 1947 It UM (3)2, 46-51
  (MR9, 74)  [Cauchy, Libri]
 Quesada C A de 1909 Porto Ac 4, 166-
  192  [da Silva]

Rocquigny G de 1895 Intermed 2, 269
  [sum of four squares]
Shanks Daniel 1962 *Solved and
  Unsolved Problems in Number Theory*,
  Wash DC
Siegel C L 1932 QSGM (B)2(1), 45-80
  [anal numb thy, Riemann]
Studnicka F J 1876 Cas MF 4,
  (F8, 29)  [origin]
Turan P 1950 M Lap 1, 243-266
  (MR12, 311)  [USSR]
Yushkevich A P 1957 Mos IIET 17,
  300-311  [XVII, Russia]


## Numbers

See also Algebraic, Complex,
  Irrationals, Natural, Negative,
  Numeration (Ch.3), Real, Tran-
  scendental, Transfinite, Zero.


Badarav D 1962 Buc U Log 5(5), 87-
  110  (Z117, 241)  [Pythagoras]
Baltzer R 1865 Leip Ber 17, 1-2
Barry Frederick 1927 *The Scientific
  Habit of Thought* (Columbia U P),
  207-
Benacerraf P 1965 Phi Rev 74, 47-73
  (MR30, 1929)
Bouligand G 1965 Rv Syn 86, 11-41
Brix H 1889 *Der Mathematische
  Zahlbegriff und seine Entwickelung
  Formen*, Leipzig
Carmichael R D 1935 Sci Mo 41, 490-
  (F61, 940)
Cassina U 1929 Pe M (4)9, 238-250,
  320-335  (F55, 588)  [sq circle]
*Conant L L 1896 *The Number Concept*,
  NY  (repr 1923)
*Dantzig Tobias 1954 *Number, the
  Language of Science*, 4th ed
  (Doubleday), 357 p    [pop]
 Dijksterhuis E J 1930 Euc Gron 7,
  97-112  (F56, 804)  [Greek]
 Eneström G 1902 Ens M 4, 126-127
  (F33, 53)
 Fabinger F 1904 Cas MF 33, 74-93,
  198-209, 297-307  (F35, 71)
 Fehr H 1902 Ens M 4, 16-27  (F33, 53)
 Fine H B 1891 *The Number-System of*

Numbers    (continued)

*Algebra, Treated Theoretically and
Historically,* Boston and NY,
140 p  (F23, 34)  (2nd ed 1907;
repr 1937 Chelsea)
Fraenkel A A 1920 Gott N(S)
(F47, 8)  [Gauss]
Freitag H T + 1960  *The Number Story*
Washington, 80 p  [pop]
Hayashi T 1931 Toh MJ 33, 292-327
(F57, 19)  [Japan]
Helmholtz Hermann 1930 *Counting and
Measuring,* NY, 73 p
Itard J 1953 Rv Sc 91, 3-14
(Z52, 243)  [axiomatics]
Karpinski L C 1911 AMM 18, 97
Klein J 1934 QSGM (B)3(1), 18-105,
122-235  (Z10, 248; 14, 49)
Martell P 1926 Allg Verm 38, 281-285
(F52, 36)
Medvedev F R 1959 Mos IIET 28, 237-
249  [first textbooks]
Michel W 1941 *Die Entstehung der
Zahlen,* Bern, 31 p  (Z25, 1)
Molodshii V N 1963 *Osnovy ucheniya
O chisle v XVIII i nachale XIX
veka,*  Mos
Murnaghan F D 1949 Sci Mo 68, 262-
269  (MR10, 419)
Natucci Alpinolo 1923 *Il concetto di
Numero e le sue estensioni,*
Turin, 472 p  [bib]
Pasch M 1930 *Der Ursprung des
Zahlbegriffs,*  Berlin, 56 p
Pasquier L G du 1921 Neuc U Mm 3
(F48, 46)
Peano G 1928 Archeion 9, 364-366
(F54, 5)
Popken J 1940 Euc Gron 16, 225-239
(F66, 3)
Scriba Christoph J 1968 *The Concept
of Number,* Mannheim-Zurich, 216 p
Sergescu Pierre 1949 *Histoire de
nombre* (U Paris), 44 p
(Z51, 2)
Slaught H E 1928 MT 21, 305-315
(F54, 5)
Slocum S E 1918 Sci Mo 7, 68-79
Smeltzer Donald 1958 *Man and Number,*
NY, 122 p

Stammler Gerhard 1925 *Der Zahlbegriff
scit Gauss,* Halle, 208 p
(repr 1965; MR 34, 4091)
Vasiliev A V 1922 *Tseloe chislo,*
Len, 246 p
Volkov M 1889 *Evolyutsiya ponyatiya
o chisle,* Pet, 120 p
*Wieleitner Heinrich K 1911 *Der
Begriff der Zahl in seiner
logischen und historischen
Entwicklung,* Berlin, 66 p
(3rd ed 1927 Leipzig; F53, 36)
Young M H 1964 Ari T 11, 336-341
[bib]


## Numerical analysis

See Approximation, Computer (Ch. 3),
Equations, Interpolation, Iteration,
Monte Carlo methods (Ch. 3).

Aitken A C 1958 Edi MSP 11, 31-38
(Z83, 245)  [Whittaker]
Biermann K-R 1958 Ens M (2)4, 19-24
(MR20, 945)  [Euler]
Householder A S 1956 ACMJ 3, 85-100
[bib]
Popoviciu T 1966 Ro Rv M 11, 1139-
1146  [Cluj, Roumania]
Sergescu P 1932 Cuget Clar
(F59, 857)
Todd John 1962 *Survey of Numerical
Analysis,* NY  [bib]


## Operators

Berezanskii Yu M 1968 AMS Trls 17
[*bib, expansions]
Berg  Lothar 1967 *Introduction to
the Operational Calculus* (Wiley)
[bib]
*Cooper J L B 1952 M Gaz 36, 5-19
[Heaviside]
Dixmier J 1957 *Les algèbres
d'operateurs dans l'espace
Hilbertien,* Paris  [*bib]
*Dunford N + 1958 *Linear Operators,* NY,
2 vols
Garding L 1963 Int Con (1962), XLIV-
XLVII (Z112, 243)  [Hoermander]

## Operators    (continued)

Higgins T J 1949 El Engin 68, 42-45
  (MR10, 420)  [electronics]
Orlicz W 1948 Colloq M 1, 81-92
  (Z37, 291)   [Banach]
Pincherle S 1899 Bib M 13,
  [bib, distributive]
Putnam C R 1967 *Commutation Proper-
  ties of Hilbert Space Operators,*
  NY
Schwartz J T 1968 *W\*- Algebras,* NY,
  [bib]
Szokefalvi-Nagy B 1957 M Lap 8, 185-
  210  (Z94, 5)  [Neumann]
*Young G S 1958 AMM 65, 37-38


## Oriented area and volume

Taton R 1951 Par CR 232, 198-200
  [Monge]


## Orthogonal functions and expansions

Birkhoff G + 1960 AMM 67, 835-841
  [completeness of Sturm-Liouville]
Gagaev B M 1957 UMN (NS)12(4),
  251-262  (MR19, 825)  [Kazan]
Haimo Deborah H ed 1968 *Orthogonal
  Expansions and Their Continuous
  Analogues* (Southern Ill U P)
Geronimus Ya L 1950 *Teoriya
  ortogonalnykh mnogochlenov,*
  164 p  [Russ]
*Shohat J A + 1940 *A Bibliography
  on Orthogonal Polynomials,* Wash DC
  (Am NRCB 103, 203 p) [bib]


## Ovals

Ball N H 1930 AMM 37, 348-353  [bib]
Genocchi A 1884 Mathesis 4, 49-52
  [Descartes]
Liguine V 1882 BSM (2)6, 40-49
  [bib, Descartes]
Tannery P 1900 Intermed 7, 169
  [Descartes]


## p-adic numbers

MacDuffee C C 1938 AMM 45, 500-508
  [Hensel]


## Parabola

Boyer C 1954 MT 47, 36-37  [area]
Conte L 1942 Pe M (4)22, 70-90
  (F68, 15)  [Castillon, Fermat,
  Newton]
Heibert J + 1911 Bib M (3)11, 193-
  208  [Arab, mirrors]
Hofmann J E 1953 MP Semb 3, 59-79
  (MR14, 609)  [area, hyperbola]
Mikami Y 1913 Toh MJ 3(1), 29-37
  [Japan]
Suter H 1918 Zur NGV 63, 214-228
  [area, Thabit]
Wohlwill E 1899 ZM Ph 44(S), 577-
  624  (F30, 55)  [Caverni,
  Galileo, trajectory]


## Paraboloid

Suter H 1912 Bib M 12, 289-332
  [Alhazen]
Winter H J J + 1949 Beng Asi J 15,
  25-40 (Z41, 338)  [Alhazen]
- 1950 Beng Asi J 16, 1-16
  (MR13, 809)  [Alhazen]


## Parallelogram

Sintzow D 1899 Kazn FMO (2)9, 11,
  44-46  (F30, 49)    [analytic
  parallelogram, Lagrange, Newton]
Yates R C 1940 MT 33, 301-310


## Parallels

See also Non-Euclidean.

Al-Dhahir M W 1958 Bagd CS 3, 60-65
  (MR20, 944)
Cassina U 1956 Int Con HS 8, 33-38
  [Wallis]

## Partial differential equations
(continued)

Mansion P 1891 Brx SS 15, 32-37, 60
(F23, 37) [bib, sing sol]
- 1892 *Theorie der partiellen
Differentialgleichungen 1 er
Ordnung,* Leipzig
Prasad G 1924 *The Place of Partial
Differential Equations in Mathe-
matical Physics...,* U Patna, 49 p
Saltykow M N 1939 BSM 63, 213-228
(MR1, 33) [Jacobi]
- 1958 in *Calcutta Math Soc Golden
Jubilee Commemoration,* (2), 287-
316 (MR17, 678)
Simonov N I 1957 Ist M Isl 10,
327-362 (Z108, 249) [Euler]
Zhatykov O A 1955 Kazk AK (7), 4-19
(MR17, 931)

## Partial fractions

Macmahon Percy A 1916 *Combinatory
Analysis* (Cambridge U P) 2(7),
Ch. 5 (repr 1960 Chelsea)
Metzler G F 1915 MT 7, 159-166

## Partitions of numbers

Bachmann P 1910 *Niedere Zahlen-
theorie* 2, 102-283 [bib]
Biermann K-R 1961 Fors Fort 35, 71-
74 (Z95, 2)
Csorba G 1902 MP Lap 10, 257-281
[bib]
Dickson Leonard E 1920 Dickson 2,
Ch.3 (repr 1934 NY)
Enestrom G 1912 Bib M (3)13, 352
(F43, 68) [anc]
Kempner A J 1923 AMM 30, 354-369,
416-425 [Hardy, Littlewood,
Ramanujan]

## Pasturage problem

Brown B H 1926 AMM 33, 155
Doerrie Heinrich 1965 *100 Great
Problems...,* (Dover), 9-10

## Pathological curves, functions, etc.

Hildebrandt T H 1933 AMM 40, 547
Kowalewski G 1923 Act M 44, 315-319
[Bolzano, discon]
*Mansion P 1887 Mathesis 7, 222-225
[Weierstrass]
*Pascal E 1921 *Esercizi critici di
calcolo differenziale,* Milan
[*bib, non-diff cont fun]
Voss A + 1912 Enc SM (2)1(2)3,
259-262
Young G C 1916 QJPAM 47, 127
[Weierstrass]

## Pell equation

Brocard H 1898 Mathesis (2)8, 112
[bib]
Child J M 1920 Isis 3, 255-262
[Fermat]
Conte L 1938 Pe M (4)18, 50-54
(F64, 916) [Fermat]
Datta B 1928 Clct MS 19, 87-94
[Hindu]
Dickson Leonard E 1919 Dickson 1,
Ch. 14; 2, Ch. 12
Eneström G 1902 Bib M (3)3, 204
[name]
Ganguli S K 1928 Clct MS 19, 151-
176 (F54, 21) [Hindu]
Konen H 1901 *Geschichter der
Gleichung...,* Leipzig, 132 p
Selenius C O 1963 Turku Ac 23(10),
3-44 (MR28, 962; (Oct-Dec), 962
Z117:1, 6) [cont fra]
*Wentford Edward E 1912 N.Y., 197 p
(Thesis) [*bib]

## Perfect numbers

Archibald R C 1921 AMM 28, 140-153
Brown A 1954 Scr 20, 103-106
[multiperfect]
Carmichael R D + 1911 Inda Ac, 257-
270 [list to date]
Curtze M 1895 Bib M (2)9, 37-42,
77-88, 110-114 (F26, 52)
Dickson Leonard E 1919 Dickson 1,
Ch. 1

Perfect numbers    (continued)

- 1921 Sci Mo 12, 349-354
Fontes 1894 Tou Mm (9)6, 155-167
  (F26, 10)  [XVI, Bouelles]
Fraenkel A A 1943 Scr 9, 245-255
Francon M 1951 Isis 42, 302-303
  (Z45, 145)  [Ausone]
Gerardin A 1916 LMSP (2)15,
  xxii
Gridgeman N T 1963 New Sci (11 Ap)
Hanawalt K 1965 MT 58, 621-622
*Reid C 1953 Sci Am 188, 84-86
  [SMSG RS 2]  [lists 17]
Rosenthal E 1962 MT 55, 249-250
  [binary numer]
Uhler H S 1953 Scr 19, 128-131
  [16th, 17th]


Periodic functions

Bell A E 1941 Nat 147, 78-80
  [early]
*Burkhardt H 1901 DMV 10, 1-1392
  (F36, 56)
Capparelli Vincenzo 1928 *L'ordine
  dei tempi e delle forme in natura.
  Introduzione allo studio generale
  delle funzioni periodicke,*
  Bologna, 256 p


Permanents

Marcus M + 1962 An M 75, 47-62
  [bib]
- 1965 AMM 72, 577-591    [bib]


Phi

Sister Marie Stephen   1956 MT 49,
  200-204


Pi

See also Squaring the circle.

Anon 1919 AMM 26, 209   [club topic]

Archibald R C 1921 AMM 28, 116-121
  [e, i]
Audisio F 1930 Linc RN (6)11, 1077-
  1080  (F56, 19)  [Leibniz]
- 1930 Tor FMN 65, 101-108
  (F56, 11)  [Archimedes]
- 1931 PeM (4)11, 11-42, 149-150
  (F57, 7; Z1, 322)
Ball W W R 1894 Bib M (2)8, 106
  (F25, 65)  [nota]
Ballantine J P 1939 AMM 46, 499-501
  [calcn]
Baravalle H v 1952 MT 45, 340-348
  (repr 1967 MT 60, 479-487)
Barbour J M 1933 AMM 40, 69-73
  (F59, 20; Z6, 145)  [China, XVI]
Bardis P D 1960 SSM 60, 73-78
  (SMSG RS 6)
Becker O 1961 Praxis 3, 58-62
  [Babyl]
Beretta L 1938 Pe M (4)18, 193-206
  (Z19, 388)  [irrat]
Bierens de Haan D B 1858 Amst Vh 4,
  22  [mid]
- 1873 Mess M 24-26  (F5, 45)
  [Van Ceulen]
- 1875 Nieu Arch 1, 70-, 206-
  [Eycke]
Bortolotti E 1931 Pe M (4)11, 110-113
  (Z1, 322)
Bruins E M 1945 Amst P 48, 206-210
  (Z60, 5)  [Egypt]
Brun V 1951 Nor M Tid 33, 73-81
  (Z44, 245)  [Brouncker, Wallis]
- 1955 Nor M Tid 3, 159-166
  [Leibniz]
Ceretti U 1903 Rv FMSN  4(2),
  520-527    [China]
Datta B 1927 Beng Asi J 22, 26-42
  [Hindu]
- 1929 Clct MS 21, 115  [Hindu]
Dickinson G A 1937 M Gaz 21, 135-139
  [Wallis]
Dijksterhuis E J 1932 Euc Gron 8,
  286-296  [circle, Pell]
*Duarte F J 1949 *Monografien ueber die
  Zahlen* π *und e,* Caracas 246 p
  (Z41, 340)   [bib]
Dudley U 1960 M Mag 35, 153-154
Eneström G 1889 Bib M (2)3, 28, 4,
  22   [nota]

Pi (continued)

- 1890 Kagan (94)
Eneström G + 1902 Intermed 9, 166
    (Q2189)  [nota]
Eves H 1962 MT 55, 129-130
    [SMSG RS 7]
Ferguson D F 1946 Nat 157, 342
    [Shanks]
Ganguli S 1930 AMM 37, 16-22
    [Aryabhata]
Ginsburg J 1944 Scr 10, 148
    [rat approx]
Glaisher J W L 1873-1874
    Mess M (2)2, 119-128, (2)3, 27-46
    [mid]
Godefroy M 1902 Intermed 9, 52
    [nota]
Gould S C 1888 Bibliography on the
    Polemic Problem: What is the
    Value of π ? Manchester N.H.,
    32 p  (F21, 1255)
Gridgeman N T 1960 Scr 25, 183-195
    [Buffon needle]
Grimm G 1949 Elem M 4, 78-85
    (MR11, 572 [Huygens]
Gutenaecker J 1928 Kreis-Messung
    des Archimedes von Syrakus nebst
    dem dazu gehoerigen Kommentare
    des Eutokius von Askalon,
    Wurtemberg, 166 p
Halstead G B 1908 AMM 15, 84  [Asia]
Hawkins G S + 1965 Stonehenge
    Decoded, NY, 150 p
Hayashi T 1902 Bib M (3)3, 273-275
    [XVII, XVIII, Japan]
- 1909 Tok M Ph (2)5, 43-57
    [China, series]
- 1910 Tok M Ph (2)5, 407-414
    [China, series]
Heading J 1959 Nat 184, 78 (Z87, 3)
    [Hebrew]
Heyman H J 1925 Arc Sto 6, 113-114
    (F51, 9)  [calcn]
- 1925 Ark MAF 19B(5)  (F51, 9)
    [Klingenstiernas]
Hoffmann J C V 1895 ZMN Unt 26, 261-
    264  (F26, 55)  [Shanks]
Hofmann J E 1935 Unt M 41, 37-40
    (F61, 11)  [Archimedes]
- 1953 MP Semb 3, 193-206  (Z51, 242)
    [Hindu]

Hoppe E 1922 Arc GNT 9, 104-107
    [Archimedes]
Hultsch F 1892 ZM Ph 39, 121-137,
    161-172  [Archimedes]
Jablonski E 1915 BSM (2)39, 166-168
    [Egypt]
*Jones P S 1950 MT 43, 120-122, 208-
    209  [SMSG R5]  [Recent]
- 1957 MT 50, 162-165  [Babyl]
Kikuchi D 1895 Tok M Ph 7, 47
    [Japan, series]
Klein F 1893 in his Lectures on
    Mathematics, 51-57  [transc]
*Lawson D A 1945 Pentagon 4, 15-24
    [SMSG RS6]  [pop]
Mahler E 1882 Hist Abt 27, 207-210
Mansion P 1908 Mathesis (3)8, 236-
    242  (F39, 69)  [calcn]
- 1909 Mathesis (3)9, 14-17
    (F40, 65)  [Metius, Montucla]
Milankovitch M 1953 Beo Ak N Zb 35(3),
    11-13  (Z53, 245)  [Ptolemy]
Mitchell U G 1919 AMM 26, 209-212
Niven I 1939 AMM 46, 469  [transc]
Pottage J 1968 Isis 59, 190-197
    [Vitruvius]
Quint N 1907 Wisk Tijd 4, 7-9
    (F38, 65)
- 1908 Mathesis (3)8, 126-128
    [calcn]
- 1908 Wisk Tijd 4, 68-71
    (F39, 69)
Rajagopal C T 1947 Scr 15, 201-209
    [Hindu]
Rajagopal C T + 1949 Beng Asi J
    15(1), 1-13
- 1951 Scr 17, 65
- 1952 Scr 18, 25-30  (Z47, 244)
    [approx, Hindu]
Read C B 1960 SSM 60, 348-350
    ["oddities"]
- 1964 SSM 64, 765-766  [Hebrew]
- 1967 MT 60, 761-762
    [Shanks mistake]
Rice D 1928 M Mag 2, 6-8
*Schepler H C 1949 M Mag 23, 165-170,
    216-228, 279-283  (Z36, 1)
    [chronology]
Schorik S S 1932 Euklid die Quadratur
    des Kreises und das problem π,
    Plowdiw, 138 p  (F58, 994)

## Pi (continued)

Schoy C 1926 AMM 33, 323-325
  (F52, 12)  [al-Biruni]
Selander K E I 1868 *Historik
  oefver Ludolphska Talet,* Uppsala
  [calcn]
Sherk W H 1910 MT 2, 87-93
  [approx]
Siebel A 1956 MP Semb 5, 143-146
  (MR18, 368)  [J Gregory]
Smith D E 1895 AMM 2, 348-351
  [calcn]
- 1911 in Young J W A ed, *Monographs
  on Topics of Modern Mathematics...*
  (Longmans), Ch 9  (repr 1955 Dover)
Soddy F 1943 SSR (Feb), 171-178
  [Leibniz, Newton]
Stamatis Evangelas 1950 *Archimedes'
  Measurement of the Circle,*
  Athens, 32 p
Stengel C 1904 ZMN Unt 35, 508-509
  (F35, 62)  [approx]
Terquem O 1854 Nou An 13, 253-
  [Ptolemy]
Todd J 1949 AMM 56, 517-528
  [formulas]
Vasconcellos F de 1923 Esp Prog
  3, 5-13  [Greek]
Whish C M 1835 As So GB 3, 509
  [area]
Wiedemann E 1909 Arc GNT 1, 157-158
  (F41, 65)  [Cusanos]
*Wrench J W 1960 MT 53, 644-650
  [bib, chronology]
Zackova J 1966 Prag Pok 11, 240-250
  (MR34, 8)  [Lambert, irrat]


## Plateau problem

*Douglas J 1931 AMST 33, 263-321
  [bib]
- 1938 Scr 5,  159-164


## Point

Juel C 1896 Nyt Tid (8)7, 7
Korner T 1904 Bib M (3)5, 15-62
  [XVIII, Alembert, Euler, Lagrange,
  Laplace, Newton]

Simon M 1909 Int Con 4(3), 385-390
  (F40, 49)  [continuum]


## Poisson distribution

*Haight Frank A 1967 *Handbook of the
  Poisson Distribution* (Wiley), 168 p
  [*bib of 700]


## Polar`coordinates

Boyer C B 1949 AMM 56, 73-78
  (MR10, 420)  [Newton]
Coolidge J L 1952 AMM 59, 78-85
  (MR13, 611; Z47, 4)
Fontenelle 1704 Par Mm, 47-57
  [Bernoullis, Varignon]
Wieleitner H 1928 Unt M 34, 276-277
  [ellipse]


## Polygon

See also Regular polygons, Heptagon,
  Hexagon, Quadrilateral, Star
  polygons.

Anon 1898 Intermed 5, 276   [Nieuport]
- 1900 Pitagora 7, 79
  [Archimedes]
Brueckner M 1900 *Vielecke und
  Vielflaeche, Theorie und Geschichte,*
  Leipzig, 234 p
Lysenko V I 1959 Ist M Isl 12, 161-
  178  (MR24A, 221)  [XVIII, Russia]
Merriell D 1965 AMM 72, 960-965
  [concentric]
Suter H 1909 Bib M (3)10, 15-42
  (F40, 62)  [Kamil]
Tropfke J 1936 Osiris 1, 636-651
  [Archimedes]
Wolffing E 1899 Wurt MN (2)1, 87
  [min dist]


## Polyhedra

See also Icosahedron, Pyramid,
  Regular polyhedra, Stellated
  polyhedra.

## Polyhedra   (continued)

Brocard H + 1906 Intermed 13, 99,
  (Q653, 685) (Also Intermed 2,
  316, 401; 5, 151) [termin]
Brueckner M 1900 *Vielecke und*
  *Vielflaeche. Theorie und Geschichte,*
  Leipzig, 234 p
Court N A 1959 MT 52, 31-32
  [altitudes of tetrahedron]
Jonquieres E de 1890 Par CR 110, 315-
  317 [Descartes]
Lebesgue A 1924 Fr SMB 52, 315-336
  [Euler formula]
Lindemann F 1934 Mun Si, 265-275
  (F60, 817; Z10, 244)
Maroni A 1921 Pe M (4)1, 337-346
  [Descartes-Euler]
Phillips J P 1965 MT 58, 248-250
  [dodecahedron]
Schirmer O 1958 Bay Jb OK, 61-80
  (Z80, 6)
Ver Eecke Paul 1935 Mathesis 49, 59-
  82 (Z89, 241) [Vito Caravelli]
Wenninger M 1965 MT 58, 244-248

## Porisms

Breton P 1873 *Question des porismes.*
  *Notices sur les débats de priorité*
  *auxquels a donné lieu l'ouvrage*
  *de Chasles sur les porismes*
  *d'Euclide,* Paris (F5, 47)
Cantor M 1857 ZM Ph 2, 17-
Chasles M 1838 Cor M Ph 10, 1-20
  [anal geom]
*- 1860 *Les trois livres de*
  *porismes d'Euclide,* Paris
Davies T S 1850 Mathcian 3, 75-, 140-,
  225-, 311-; 3(S), 42-
Mordukhai-Boltovskoi D D 1948
  *Istorii estestvoznaniya*
  (Soveshchaniya, 1946), 161-172
  (MR11, 571)
Playfair J 1794 Edi ST 3, 154

## Positivity

*Karlin Samuel 1968 *Total Positivity,*
  (Stanford U P)   [*bib]

## Potential

See also Dirichlet problem.

*Bacharach, Max 1883 *Abriss der*
  *Geschichte der Potentialtheorie,*
  Goettingen, 78 p   [bib]
Baltzer R 1878 Crelle 86, 213-216
  (F10, 28)
Becker G F 1893 Am JS (3)45, 97-100
  (F25, 86) [Bernoulli, word]
Bocher M 1891 *Ueber die Reihent-*
  *wicklungen der Potentialtheorie,*
  Goettingen (Dietrich)
Brennecke R 1924 Z Ph 25, 42-45
  [Euler]
Carleson L 1967 *Selected Problems*
  *in Exceptional Sets* (Van Nostrand)
  [*bib]
Hathaway A S 1891 NYMS 1, 66-74
  (F23, 41) [early]
Kellog O D 1929 *Foundations of*
  *Potential Theory* (Ungar)
LaVallée Poussin C J de 1962 Rv Q Sc
  (23)133, 314-330 [Gauss]
Sneddon Jan N 1966 *Mixed Boundary*
  *Value Problems in Potential*
  *Theory,* NY   [bib]
Sologub V S 1962 Kiev Vis (5)2, 121-
  135   (MR34, 4517) [origin]
Zanotti-Bianco O 1893 Rv M 3, 56-60,
  114 (F25, 87) [origin]
Zanotti-Bianco O + 1893 Nat 47, 510
  (F25, 87) [origin]

## Pothenot problem

Brocard H 1908 Intermed 15, 81
  (Q2974, Q2976) (Also Intermed 12,
  266-267; 13, 122, 219-222; 14; 15)
Doerrie Heinrich 1965 *100 Great*
  *Problems...* (Dover), 193-195

## Prime numbers

See also Factoring, Mersenne numbers,
Sieve of Eratosthenes.

Banachiewicz T 1909 Wars SPTN 2, 7-
  10 (F40, 55) [China, Fermat]

Prime numbers  (continued)

Burali-Forti C + 1900 Intermed 7,
  38, 244 (Q1633)
Dickson Leonard E 1919 Dickson 1,
  Ch. 13, Ch.18
Dudley V 1969 AMM 76, 23-28
  [formula]
Eves H 1958 MT 51, 201-203
Glaisher J W L 1883 *Factor Table
  for the Sixth Million,* London
Hasse Helmut 1967 in *Proben Mathe-
  matischer Forshung* (Salle),
  1-13 [pop]
*Landau E 1909 *Handbuch der Lehre von
  der Verteilung der Primzahlen,*
  Leipzig
- 1927 *Vorlesungen ueber Zahlen-
  theorie,* Leipzig
Lehmer D N 1933 Scr 1, 229-235
Lorey W 1933 Crelle 170, 129-132
  (Z7, 388)  [Euler, mistake]
Mahnke D 1912 Bib M 13, 29-61, 260
  [Leibniz]
Poletti L 1939 Isis 30, 281
  [bib, ms]
- 1951 Parma Rv M 2, 417-434
  (MR14, 121   [Italy, tables]
Schwarz W 1968 Ueberbl 1, 35-61
  [prime numb thm]
Studnicka F J 1879 Cas MF 8, 36-37
  (F11, 35)
*Torelli G 1902 Nap FM At (2)11,
  1-222


Probability

See also Central limit theorem,
  Clumping, Law of large numbers,
  Poisson distribution, Stochastic
  processes.

Bayes T 1763 PT 53 (repr 1958
  Biomtka 45)
Bernstein S N 1940 Len MZ (10), 3-11
  (F66, 1)  [Pet, Russia]
Bianco O Z 1878 Batt 16, 26-30
  (F10, 27)  [re Todhunter 1865]
Bierens de Haan D 1884 Nieu Arch 11,
  75-  [Spinoza]

Biermann K R 1955 Fors Fort 29,
  110-113  (Z65, 243) [Leibniz]
- 1956 Wiss An 5, 542-548  (MR18, 453)
- 1957 Centau 5, 142-150
  [Bernoulli]
- 1957 Prag Pok 2, 31-35  (MR20, 204)
- 1965 Ber Mo 7, 70-75   [dice]
*- 1965 Fors Fort 39(5), 142-144
  [chronology, Gauss]
Biermann K R + 1957 Fors Fort 31,
  45-50  [Leibniz]
Birziska V 1931 Kosmos 12, 81-104
Blom Siri 1955 Theoria 21, 65-98
  [term, foundations]
Bosmans H 1923 Brx SS 43, 318-326
Boutroux P 1908 Rv Mois 5, 641-654
  [origins]
Brown S 1857 Assur Mag 6, 134
  [ins]
Buch K R 1955 Nor M Tid 3, 19
Byrne Edmund F 1968 *Probability and
  Opinion. A Study of the Medieval
  Presuppositions of Post-Medieval
  Theories of Probability* (Nijhoff)
  359 p
*Cantor G 1873 Halle NG, 34-42
- 1874 *Historischen Notizen,*
  Halle, 8 p
Carlini L 1900 Pitagora 7, 65
  [origin]
Carnap R 1950 *Logical Foundations
  of Probability,* Chicago  [*bib]
Castro G de 1954 Portu IA (10), 1-15
  (Z58, 2)
Coolidge J L 1925 *An Introduction
  to Mathematical Probability,*
  Oxford, 214 p  (repr 1962 Dover)
Cramer H 1962 An M St 33, 1227-1237
  (MR25, 570)  [Khinchin]
*Czuber Emanuel 1899 DMV 7(2), 1-279
  (F30, 43)  [*bib] (see Woelffing
  1899)
- 1968 *Warscheinlichkeitsrechnung und
  ihre Anwendung auf Fehleraus-
  gleichung, Statistik und Lebensver-
  sicherung.*  2 vols  (Johnson repr)
Darmois G 1965 in R. Taton *Histoire
  Génerale des Sciences* 3
David F N 1955 Biomtka 42, 1-15
  (MR16, 781)  [games]
- 1962 *Games, Gods and Gambling*
  (Hafner)  [gen hist]

Probability    (continued)

Dutka J 1953 Scr 19, 24-33
  (MR14, 833)  [Spinoza]
Eisenhart C 1964 Wash Ac 54, 24-33
Fisher R A 1934 M Gaz 18, 294-297
  [cards]
Fréchet M 1938 Exposé...recherches
  récentes..., Colloque sur la
  théorie des probabilities,
  (Hermann)
- 1946 Rv Phi, 129-169  [defs]
- 1964 Par CR 258, 4877-4878
  (MR28, 1055)  [zero prob]
Freudenthal H 1951 M Nach 4, 184-192
  [Petersburg probl]
Gabba A 1840 Brescia 203
- 1842 Considerazioni storico
  critiche sulla teorica e sul
  calcolo della probabiletà, Brescia
Gnedenko B V 1948 Mos IIET 2, 290-425
  (MR11, 150)  [Russia]
- 1949 Ist M Isl (2), 129-136
  (MR12, 1)  [Lobachevskii]
  1956 Int Con Hs 8, 128-131
  [stages]
- 1959 Ist M Isl 12, 135-160
  [Lyapunov]
- 1967 The Theory of Probability
  (Chelsea)
Gnedenko B V + 1961 An Elementary
  Introduction to the Theory of
  Probability (Freeman)
- 1963 Th Pr Ap 8, 157-174
  (Z117, 251)  [Kolmogorov]
Gouraud Charles 1848 Histoire du
  calcul des probabilités depuis
  ses origines jusqu'à nos jours,
  avec une thèse sur la legitimité
  des principes et des applications
  de cette analyse, Paris, 148 p
Hagstroem K-G 1932 Les Préludes
  antiques de la Théorie des
  Probabilités, Stockholm, 54 p
  (F58, 986; Z5, 241)
Hasofer A M 1967 Biomtka 54, 316-321
  [Hebrew]
Jecklin H 1949 Dialect 3, 5-15
  (MR11, 150)  [def]
Kendall M G 1956 Biomtka 43, 1-14
  [origin]

- 1957 Biomtka 44, 260-262
  (Z77, 242)  [cards]
- 1961 Biomtka 48, 1-2  (Z99, 244)
  [D. Bernoulli, max likelihood]
- 1963 Biomtka 50, 204-205
  [Todhunter 1865]
Keynes John M 1921 A Treatise on
  Probability (Macmillan), 477 p
  [*bib 29 p]
Khotimskii V 1936 PZM (1); (6)
  [See Maistrov 1967, 19]
King Amy C + 1962 SSM 62, 165-176
  [biog]
- 1963 Pathways to Probability...,
  NY, 139 p  [based on Todhunter]
Kolmogorov A N 1947 Mos UZ 91, 59-
  [Russia]
- 1962 Th Pr Ap 7, 313-319
  (Z113, 2)  [Gnedenko]
Kosten L 1957 Teletek E 1, 32-40
  [telephone traffic]
Kyburg H E 1963 Theoria 29, 27-55
  [randomness]
Le Cam M + 1963 Bernoulli 1713,
  Bayes 1763, Laplace 1813
  Anniversary Volume, NY, 280 p
Lechalas G 1906 Rv Met Mor 14, 109
  [Cournot]
Lefebvre B 1922 Rv Q Sc (4)2, 342-
  364
*Levy P 1940 in Borel Selecta, 310-
  315
Linnik J V 1961 Rus MS 16(2), 21-22
  (Z98, 9)  [Bernshtein]
Maistrov L E 1962 Deb U Act 7(2)
  (1961)  [origin]
- 1964 Vop IET 16  [Galileo]
- 1967 Teoriya veroyatnostei.
  Istoricheskii ocherk, Mos, 320 p
Mahalanobis P C 1957 Sankh 18(1-2)
Martic L 1953 Beo MF 2, 198-203
  (Z53, 339)
Milhaud G 1906 Etudes sur la penseé
  scientifiques chez les Grecs et
  chez les modernes, Paris, 137-176
  [Aristotle, Cournot]
Molina E C 1930 AMSB 36, 369-392
  [Laplace]
Moritz R E 1923 AMM 30, 14-18, 58-65
  [mistakes]
Mrochek V R 1934 Len IINT 2, 45-60
  (F61, 941)

## Probability (continued)

Muller M 1960 Schw Vers 60, 115-129
(Z87, 5) [betting]
Nagel Ernst 1939 *Principles of the
Theory of Probability*, Chic (Intl
Enc Unified Sci 1(6))
Nicod J 1930 *Foundations of Geometry
and Induction*, London [Keynes]
Ore O 1960 AMM 67, 409-419
(Z92, 245) [Pascal]
Pasquier L G du 1926 *Le calcul des
probilités, son évolution mathé-
matique et philosophique*, Paris,
325 p (F52, 3)
- 1929 Fr AAS 53, 80-88 (F57, 1303)
[Lambert]
Patil G P + 1968 *A Dictionary and
Bibliography of Discrete
Distributions* (Stechert-Hafner)
280 p
Read C B 1963 MT 56, 637-638
[origin]
Renyi Alfred 1954 Magy MF 4, 447-466
(MR16, 659; Z59:1, 8; 60, 8)
- 1969 *Brief ueber die Wahrschein-
lichkeit* (Birkaeuser), 94
[Fictitious letters Pascal-Fermat]
Sambursky S 1956 Osiris 12, 35-48
[Greece]
Struik D J 1934 Phi Sc 1, 50-70
[found]
Takekuma R 1953 JHS 26 (July)
[Fermat, Pascal]
Thatcher A R 1957 Biomtka 44, 515-
518 (MR19, 1247) [duration of
play]
Todhunter Isaac 1865 *History of the
Mathematical Theory of Probability
from the Time of Pascal to that of
Laplace*, Cambridge, 640 p
Usai G 1929 Cat ISECA (F55, 589)
Vacca G 1936 It Attuar 7, 231-234
(Z14, 243) [Greek]
Van Dantzig D 1955 Arc In HS 34,
27-37 [Laplace]
- 1955 Sta Neer 9, 233-242
[1945-1955]
Woelffing E 1899 Wurt MN (2)1, 76-
84 (F30, 45) [bib suppl Czuber]
- 1901 Wurt MN (2)3, 57, 93
[suppl 1899]

Wright G H v 1941 *The Logical Problem
of Induction*, Helsinki (2nd ed
1957) [*bib]

## Programming

See Linear programming, and in Ch. 3,
Computer, Operations research.

## Progressions

See also Arithmetic, Geometric,
Harmonic, Sequences.

Bortolotti E 1941 It UM (2)3, 395-401
(Z25, 145) [anc]
Cajori F 1922 SSM 22, 734-737
[names]
- 1923 AMSB 29, 12 (F49, 22)
[names]
- 1928 Isis 10, 362-366 [Cirvelo]
Curtze M 1895 Bib M (2)9, 37-42,
77-88, 105-114 (F26, 52) [mid]
Guenther S 1876 ZM Ph 21(S), 57-
[Arab]
MacKay J S 1888 Edi MSP 6, 48-58
[Pappus]
Mohan B 1958 Benar JSR 9(1), 19-28
(MR22, 769) [Hindu]
Vetter G 1923 Bo Fir 19, xcvii-xcix
[Egypt]

## Projective geometry

See also Descriptive geometry,
Duality, Perspective (Ch. 3)

Amodeo F 1908 Nap FM At 13 (16)
[Duerer, Monge]
- 1939 Arg IM ROS 1(3) (F65, 1090)
[origin]
- 1939 Batt (3)75, 1-68 (F65, 15)
- 1939 *Origine e sviluppo della
geometria proiettiva*, Naples,
174 p (F65, 14; Z22, 3)
- 1940 Batt 76, 1-88 (F66, 17)
Cassina U 1921 Pe M (4)1, 326-337
[inf, perspective]

## Projective geometry    (continued)

- 1923 Esercit 3, 32-39
  [cyclic pts, abs circle, Poncelet]
- 1957 Par PDC (D)(50)  (MR20, 264;
  Z79, 242)
Conforto F 1949 Archim 1, 7-17
  (Z31, 337)
*Coolidge J L 1934 AMM 41, 217-228
  (F60, 6; Z9, 98)  ["rise and fall"]
Frajese A 1940 It UM (2)2, 481-492
  (F66, 17; Z23, 388)  [origin]
Galdeano Z G de 1891 Prog M 1, 3-8,
  26-30, 54-57, 106-108, 256-258
  (F23, 39)
Godeaux L 1935 Brx Con CR 2, 94-95
  (F62, 1034; Z13, 194)  [le Poivre]
Levy Harry 1964 Projective and
  Related Geometries, NY
Libois P 1935 Brx Con CR 2, 96-104
  (Z13, 194)
Lodge A + 1909 M Gaz 5, 81-88
Massera J L 1956 Montv Bol 5, 405-
  458  (MR18, 227)  [found]
Obenrauch F J 1897 Geschichte der
  darstehenden und projectiven
  Geometrie..., Bruenn, 448 p
  (F28, 54)
Sakellariou N 1938 Gree SM 18, 123-
  141  (F64, 903)
Schollmeyer G 1928 DMV 37, 123-148
  [arithmetization]
Sperry P 1931 Calif Pu 2(6), 119-127
  [*bib, proj dif geom]
St. Chrzaszczewski 1898 Arc MP (2)16,
  119-149  (F29, 35)  [Descartes]
Taton René 1951 La géométrie
  projective en France de Desargues
  à Poncelet, Paris, 21 p
  (MR14, 1050; Z49, 290)
Uhden R 1937 Imag M Ber 2, 8
  [XV]
de Vries H 1923 N Tijd 11, 1-24,
  188-210, 295-330; 14, 124-144
- 1926 Historische Studien I,
  Groningen, 196 p [Pascal,
  Brionchon]
Zacharias M 1941 Deu M 5, 446-457
  (Z24, 98)  [Desargues]

## Pyramid    (volume problem)

Archibald R C 1932 Scr 1, 91
Bortolotti E 1935 Pe M (4)15, 87-92
  (F61, 10)  [Egypt]
Bruins E M 1945 Amst P 48, 209-210
Cassina M 1942 Pe M (4)22, 1-29
  [Egypt]
Drenchkhahn F 1934 Unt M 40, 98-104
  (F60, 7)  [Egypt]
Frajese A 1934 Pe M (4)14, 3-4, 211-
  234
Gillings R J 1964 MT 57, 552-555
  [Egypt]
Natucci A 1932 Pe M (4)12, 305-307
  (F58, 994; Z6, 4)  [Clairaut]
Neugebauer O 1933 QSGM (B), 347-351
  (Z7, 145)  [anc]
Vetter Q 1935 Math Cluj 9, 304-309
  (F61, 943)
Vogel K 1930 J Eg Arch 16, 242-249
  [Egypt]
- 1933 Arc Or 8, 22  [Babyl]
Wieleitner H 1925 Unt M 31, 91-92
  (F51, 8)
Zeuthen H G 1885 Zeuthen (S)4, 175-
  179

## Pythagorean theorem

Bortolotti E 1930 Pe M 10, 85-91
- 1933 Pe M (4)13, 151-152
  (F59, 15)
Boettcher J E 1927 ZMN Unt 52, 153-
  160  (F48, 40)  [Arab]
Cajori F 1899 AMM 6, 72-73
Campan F 1937 Gaz M 42, 410-413
  (F63, 8)
Eneström G 1898 Bib M (2)12, 113-114
  (F29, 33)  [analogs]
Eves H 1958 MT 51, 544-546
  [Pappus extension]
Haentzschel E 1916 ZMN Unt 47, 183-
  185  (F46, 46)  [proofs]
Heuchamps E 1936 Et Class 5, 61-72
Jones P S 1951 MT 43, 162-163, 208
Lidonnici A 1933 Pe M (4)13, 74-86,
  137-143, 193-211  (F59, 15;
  Z6, 337)  [anc]
- 1935 Pe M (4)15, 22-57 (Z10, 386)
  [proofs]

## Pythagorean theorem (continued)

Lietzmann W 1930 *Der Pythagorische Lehrsatz: mit einen Ausblick auf das Fermatsche Problem,* 4th ed, Leipzig, 75 p [1st ed 1912]
Loeschhorn K 1902 ZMN Unt 33, 369
Loomis Elisha S 1927 *The Pythagorean Proposition,* Cleveland, 214 p (2nd ed 1940, repr 1968) [256 proofs]
Marre A 1887 Boncomp 20, 404-406
Martin A 1892 M Mag Mart 2, 97 (F25, 79) [proof]
McCarthy J P 1936 M Gaz 20, 280-281 (F62, 27) [Huygens]
Mikami Y 1912 Arc MP (3)22 [China]
Miller G A 1935 Sci 8, 152 [Egypt]
Mueller E 1913 An Natphi 12, 170-186
- 1913 An Natphi 12, 234-235
Naber H A 1908 *Das Theorem des Pythagoras...,* Haarlem, 251 p (F39, 65)
Natucci A 1954 Archim 6, 156-161, 229-234 (MR16, 207) [proofs]
Neugebauer O 1928 Gott N (1), 45-48 (F54, 10)
Reinhardt H 1932 Bay Bl G 68(5), 321-322 [proof]
Sarton G 1943 Isis 34, 513 [Arab]
Sayili A 1960 Isis 51, 35-37 (MR22, 260; Z89, 241) [Thabit]
Schaaf W 1951 MT 44, 585-588
Vipper Yu 1876 *Sorok pyat dokazatelstv pifagorovoi teoremy s prilozheniem kratkikh biograficheskikh svedenii o Pifagore,* Mos, 46 p
Vogt H 1906 Schles Jb 84, 3-4 [Hindu]
- 1906 Bib M (3)7, 6-23 [Hindu, irrat]
Vos K 1920 N Tijd 8, 265-268
Weidner E F 1916 Orient Lz 19 [anc]
Winter M 1923 Rv Met Mor 30, 23-29
Young G C 1926 Ens M 25, 248-255 [heuristic]
Zeuthen H G 1904 Int Con Ph 2

## Pythagorean triples

Bruins E M 1949 Amst P 52, 629-632 (Z33, 49) [Babyl]
- 1957 M Gaz 41, 25-28 (Z77, 3) [Babyl]
Dantzig T 1955 *The Bequest of the Greeks,* Ch.9
Gillings R J 1958 M Gaz 42, 212-213 (Z82, 7) [Babyl]
Hart P J 1954 MT 47, 16-21
Price D J de S 1965 Centau 10, 1-13 (MR30, 568) [Babyl]
Yanagihara K 1914 Toh MJ 5, 120-123 [Japan]

## Quadratic equations

Anon 1951 MT 44, 193-194 [sol]
Artom E 1924 Pe M (4)2, 326-342 [Greek]
Berriman A E 1956 M Gaz 40, 185-192 (Z72, 243) [Babyl]
Bortolotti E 1936 Pe M 16, 34 [Babyl]
Carnahan W H 1948 SSM 47, 687-692 [geom sol]
Eells W 1911 AMM 18, 3-14 [Greek]
Fazzari G 1895 Pitagora 1, 22
Gandz S 1936 It UM Con 1, 528-531 [Arab, Babyl, Greek]
- 1937 Scientia 62, 249-257 (Z17, 289) [Arab, Babyl, Greek]
- 1938 Osiris 3, 405 [Babyl]
- 1947 Isis 32, 101-115 (MR8, 189) [Babyl]
Gundlach K B + 1964 Ham Sem 26, 248-263 (Z119, 6) [Babyl]
Jones P S 1950 MT 43, 279-280 [geom sol]
- 1957 MT 50, 570-571
Kloyda M T K 1937 Osiris 3, 165-192 (F63, 805)
- 1938 *Linear and Quadratic Equations 1550-1660,* Ann Arbor, 141 p (F64, 911)
Luckey P 1941 Leip Ber 93, 93-114 [Thabit ben Qurra]
Miller G A 1923 Indn MSJ 15(2), 153-155 [Greek]

## Quadratic equations (continued)

- 1934 Sch Soc 39, 211-212
  (F60, 816)
Milne J J 1927 M Gaz 13, 318-320
  (F53, 15) [nomog]
Raik A E 1953 Perm M 8(1), 31-63
  (Z68, 1) [Babyl]
Schaaf W 1956 MT 49, 618-621
Schuster H S 1930 QSGM (B)1, 194-200
  [Babyl]
Shaw J M 1928 MT 21, 121-134
  [aesthetics]
Thureau-Dangin F 1936 Rv Assyr 33,
  27-48 [Babyl]
Tropfke J 1933-1934 DMV 43, 98-107;
  44, 26-47, 95-119 (F59, 7; 60,
  5; Z9, 388)
Vera F 1934 Archeion 15, 321-325
  (F60, 816) [mid]
Vogel K 1933 Unt M 39, 76-81
  [Babyl]
- 1936 Osiris 1, 703-717
  (Z14, 243) [Babyl]
Wieleitner H 1925 Arc Sto 5, 201-205
  [Cardan]
- 1928 Erlang Si 58, 177-180
  [Stevin]

## Quadratic reciprocity and extensions

*Bachmann P 1902 *Niedere Zahlentheorie*
  1, Ch. 6 [cites 52 proofs]
Baumgart O 1885 Hist Abt 30, 169-236,
  241-277
Doerrie Heinrich 1965 *100 Great
  Problems...*, (Dover), 104-108
Eneström G 1908 Bib M (3)9, 265-266
  [cubic recip, mid]
Goulard A + 1895 Intermed 2, 33, 56,
  216, 403 (Q168) [geom]
Hancock Harris 1931 *Foundations of
  the Theory of Algebraic Numbers*
  1, Ch. 10 (repr 1964 Dover)
  [bib]
Kronecker L 1875 Ber Mo, 267-275
  (F7, 18)
Melnikov I G + 1959 Mos IIET 28, 201-
  218 [proofs]

## Quadratrix

Freeman K 1949 *The Pre-Socratic
  Philosophers*, 381-391 [=trisectrix]

## Quadric surfaces

See also Sphere.

Coolidge J L 1945 *A History of the
  Conic Sections and Quadric
  Surfaces* Oxford U P, 225 p
  (MR8, 1) (repr Dover 1968)
Fladt Kuno 1965 *Geschichte und
  theorie der Kegelschnitte und der
  Flaechen zweiten Grades*, Stuttgart,
  374 p
Tenca L 1952 It UM (3)7, 445-447
  (Z47, 4) [hyperb paraboloid]

## Quadrilateral

Archibald R C 1922 AMM 29, 29-36
  [area]
Coolidge J L 1939 AMM 46, 345
  [area]
Datta B 1930 Clct MS 20, 267-294
  (F56, 13) [Mahavira, rat]
Eneström G 1911 Bib M (3)12, 84
  (F42, 63) [inscribed]
Jones P S 1957 MT 50, 570-571
  [Babylonian, trapezoid]
Loria G 1895 Intermed 2, 188
  [area]
Weissenborn H 1879 Hist Abt 24, 167-
  184 (F11, 40) [Brahmagupta,
  Euclid, Heron, trapezoid]
Zacharias M 1906 Arc MP (3)9, 24
  [perp diags]
- 1920 ZMN Unt 51, 21-22 (F47, 6)
  [Desargues]

## Quartic equation

Bortolotti E 1926 Pe M (4)6, 217-230
  (F52, 15) [XVI]
Bosmans H 1925 Mathesis 39, 49-55,
  99-104, 146-153 (F51, 21)
  [Stevin]

## Quartic equation   (continued)

Candido G 1941 Pe M (4)21, 21-44,
88-106, 151-176   (MR3, 97;
Z24, 243; 25, 2)
Christensen S A 1889 Hist Abt 34,
201-217 [Euclid]
Conte L 1943 Pe M (4)23, 1-11
(MR8, 2; Z60, 7) [Descartes]
Desboves A 1878 Par CR 87, 925
(F10, 26) [Lagrange, Lebesgue]
Eneström G 1910 Bib M (3)11, 182-183
(F41, 56) [XVI]
Henderson A + 1930 AMM 37, 515-521
[Viete]
Lorey W 1937 M Mag 11, 216-220
[Dieffenbach]
Notari V 1924 Pe M (4)4, 327-334
(F50, 8)
Nunziante-Cesaro C 1956 Pe M (4)34
169-170 (MR18, 368) [Descartes]
Raik A E 1955 Perm M 8(3), 11-14
(Z68, 1) [Babyl]
Witting A 1934 Unt M 40, 53-54
(F60, 8) [Greek]
Woepcke F 1863 JMPA 8, 57-
[Arab]

## Quaternions

See also Hamilton, Vector, Linear
algebra.

Best R I 1944 Dub Ac P 50, 69-70
[anniv]
Birkhoff G D 1944 Dub Ac P 50, 72-
75 [anniv]
Bork A M 1965 Isis 55, 326-338
[XIX, physics]
- 1966 AJP 34, 202-211 [vector]
Byerly W E 1925 AMM 32, 5
[Benjamin Peirce]
Crowe Michael J 1967 A History of
Vector Analysis (U Notre Dame P),
270 p [bib]
*Dickson L E 1919 An M 20, 155
[eight square probl]
Dillner G 1876 M Ann 11, 168-
Fischer Otto F 1951 Universal
Mechanics and Hamilton's Quater-
nions, A Cavalcade, Stockholm

- 1957 Five Mathematical Structural
Models in Natural Philosophy with
Technical Physical Quaternions,
Stockholm
Fleuri G 1894 NSWRSJ 38, 65-93
Hawkes H E 1906 AMSB 13, 30-32
[bib]
Johnson W W 1880 Analyst H 7, 52
[name]
Kramar F D 1966 Ist Met EN (5), 175-
184 (MR34, 13) [Hamilton]
MacFarlane Alexander 1904 Biblio-
graphy of Quaternions and Allied
Systems of Mathematics, Dublin,
86 p (F35, 57) (Supplements in
Quatrns 1905, 1908-1910, 1912-1913)
- 1905 Vector Analysis and
Quaternions (Wiley), 50 p
*Piaggio H T H 1943 Nat 152, 553-555
(MR5, 57) [Hamilton]
Rastall P 1964 Rv Mod Ph 36, 820-
[relativity]
Synge J L 1944 Dub Ac P 50, 71-72
[anniv]
Tait P G 1900 Edi SP (A)23, 17-23
[Gauss]
Turner J 1955 Brt JPS 6, 226-238
[Maxwell]

## Quintic equation

Burkhardt H 1892 ZM Ph 37(S), 121-159
[group, Ruffini]
Dickson L E 1930 Modern Algebraic
Theories, NY, Ch.10
*Foulkes H O 1932 Sci Prog 26, 601-
608 [1771-1932]
*Klein Felix 1884 Vorlesungen ueber
das Ikosaeder und die Aufloesung
der Gleichungen vom fuenften
grades, Leipzig, 260 p (Engl
transl 1913; repr 1956 Dover)
[esp. Pt 2, Ch. 5]
Pierpont J 1895 Mo M Phy 6, 15-68
(F26, 51)
Prasad G 1924 Benar MS 6, 40-45

## Ratio and proportion

See also Division, Real numbers.

Agostini A 1938 It UM Con 1, 453-
453-456 [Saladini]
Artom E 1928 Bo Fir 7, 50-52
[Galileo]
Becker O 1933 QSGM (B) 2, 369-387
(Z7, 146) [Greek]
Boyer C B 1946 Scr 12, 5-13
Cajori F 1914 Nat 94, 477
[nota]
- 1922 SSM 22, 734-737 [name]
-1923 AMSB 29, 12 (F49, 22) [name]
-1928 Isis 10, 362-366 [name]
Cantor M 1888 Bib M [2] 2, 7-9
[Ahmed b-Yusuf]
Conte L 1953 Pe M (4) 31, 145-157
(Z51, 3) [Huygens]
Eneström G 1911 Bib M (3) 12, 180-181
(F42, 55) [ratio subduplicata]
Evans G W 1927 AMM 34, 354-357
[Greek]
Fine H 1917 An M (2) 19, 70-76
[Euclid]
Gandz S 1934 Isis 22, 220-222

[Arab, Hebrew, rule of three]
- 1936 Isis 26, 82-94 [rule of
three]
Grant E 1957 Isis 48, 351 [Oresme]
- 1960 Isis 51, 293-314
(MR22, 928) [Oresme]
Hill M J M 1912-1913 M Gaz 6; 7,
324-332, 360-368, 401
- 1914 The Theory of Proportion,
London, 128 p
- 1923 M Gaz 11, 213-220
[Euclid BL5]
- 1928 M Gaz 14, 36-56 (F54, 13)
[Euclid]
Lodge A 1928 M Gaz 14, 57-59
(F54, 13) [Euclid]
Moessel E 1926 Die Proportion in
Antike und Mittelalter, Munich
Natucci A 1953 It UM Con 4(2),
646-662 (MR14, 1049; Z51, 3)
[Italy renai]
Neugebauer O 1935 An Orient (12) 235-
258 (F61, 9) [Babyl]

Plooij E B 1950 Euclid's Conception
of Ratio and his Definition of
Proportional Magnitudes as Critic
Criticized by Arabian Commentators,
Rotterdam, 77 p (diss) (Z45, 146)
Podetti F 1913 Loria 15, 1-8, 33-41
(F44, 46) [XVII fun]
- 1914 Loria 16, 65-76 [Torricelli]
Schuhl P M 1939 Rv Et Grec 52, 19-22
[Plato Gorgias]
Stamatis E 1963 Platon 15, 290-293
(Z117, 5) [mirror]
Thomae W 1933 Das Proportionenwesen
in der Geschichte der gotischen
Baukunst und die Frage der
Triangulation, Heidelberg, 52 p
[mid archit, surveying]
*Tropfke Johannes 1922 Geschichte der
Elementar-Mathematik, 3, 3-19

## Real analysis

See also Analysis.

*Borel E + 1912 Enc S M 2(1)(2),
113-241
Lauricella G 1912 It SPS 5, 212-236
[Italy]
Luzin N N 1927 Sovremennoe sostoyanie
teorii funktsii deistvitelnogo
peremennogo (Doklad na 1-m
Vseross. cesde mat. v Moskve 29
apr 1927), Mos-Len, 59 p
Marcus S 1966 Ro Rv M 11, 1123-1138
[bib, Rumania]
*Natanson I P 1955-1960 Theory of
Functions of a Real Variable
(Ungar), 2 vols
Pringsheim A 1899 Enc MW 2(1)(1),
1-53
*Pringsheim A + 1909 Enc SM 2(1), 1-
112
*Rosenthal A + 1923 Enc MW 2(3), 851-
1187
Schlesinger L 1912 Ueber Gauss'
Arbeiten zur Funktionentheorie,
Gott N(1), 143 p [Gauss]
Tonelli L 1929 Int Con 8(1), 247-254
(F55, 584) [Italy]

## Real numbers

See also Number, Ratio and
proportion.

*Alimov N G 1955 Ist M Isl 8, 573-
619 [Euclid]
*Bashmakov I G 1948 Ist M Isl 1, 296-
328 [Euclid]
Becker O 1933-1936 QSGM (B)2, 369-
387; 3, 236-244 [Eudoxos]
*Busulini B 1955 Ferra U An (7)5, 79-
83 (MR19, 825) [Dedekind]
Chatelet A 1954 Rv GSPA 61, 272-277
Evans M G 1955 JH Ideas 16, 548-557
[Aristotle, Newton]
Goussinsky B 1959 *Continuity and
Number,* Israel, 31 p
*Guggenheimer H 1965 Dialect 19, 136-
143 [XVII, Euclid, transc]
Hjelmslev J 1934 M Tid B, 61-67
(F60, 823) [Archimedes, calc]
- 1950 Dan M Med 25(15) (MR11, 571)
[Archimedes]
- 1950 M Tid, 21-52 (Z39, 2)
[anc, modern]
Itard J 1953 Rv Sc 91, 3-14
(MR15, 770) [axiomatics]
*Jourdain P E B 1906-1914 Arc MP (3)10,
254-281; 14, 287-311; 16, 21-43;
22, 1-21 [defs]
Kasumhanov F A 1954 Mos IIET 1, 128-
145 (MR16, 660; Z60, 4) [al Tusi]
*Loria G 1939 Rv Met Mor 46, 57-63
(F65, 3) [geom]
Murnaghan F D 1949 Sci Mo 68, 262-269
(MR10, 419)
Natucci A 1914 Mathes It 6, 23-29
(F45, 68, 96)
Noi S di 1951 Archim 3, 56-61,
112-116 (Z42, 4) [continuum]
Onicescu O 1961 Buc U Log 4(4),
113-116 (MR24, 220) [Archimedes]
*Peano G 1922 Tor FMN 57, [units]
Rychlik Karel 1956 Cas M 81, 391-395
(MR19, 519) [Bolzano]
- 1957 Cz MJ 7, 553-567
(Z89, 242) [Bolzano]
- 1958 Ist M Isl 11, 515-532
(MR23, 583; Z109, 238)
[Bolzano]

- 1962 *Theorie der rellen Zahlen im
Bolzanos handscriften Nachlasse,*
Prague, 103 p (Z101, 247)
*Van Rootselaar, B 1963 Arc HES 2,
169-180 [Bolzano]

## Recursive functions

See also Recursion (Ch.3).

*Davis M 1967 *Recursive function
theory,* Enc Phi [bib]
Kleene S C 1935 AJM 57, 153-173,
219-244
Mikami Y 1912 Toh MJ 1, 98-105
(F43, 71) [Japan, Sohko]
*Peter Rozsa 1967 *Recursive Functions,*
Budapest and NY, 300 p [bib]

## Regular polygons

See also Construction, Cyclotomy,
Heptagon, Hexagon, Polygons,
Seventeen-gon.

Archibald R C 1914 AMM 21, 247-262
[Klein 1895 falsely attrib
constr proof to Gauss]
[construction]
- 1920 AMM 27, 323-326 [Gauss]
Bruins E M 1959 Janus 48, 5-23
[Babylonian, Greek]
Dickson L E 1911 in Young J W A ed,
*Monographs on Topics of Modern
Mathematics Relative to the
Elementary Field* (Longmans),
Ch.8 (repr 1955 Dover)
[construction]
- 1914 AMM 21, 259-262
Eves H 1969 MT 62, 42-44
Hayashi T 1932 Toh MJ 36, 135-181
(Z5, 150) [Japan]
Kazarinoff N D 1968 AMM 75, 647
[Gauss, Wantzel]
Klein F 1895 *Famous Problems of
Elementary Mathematics,*
[see Archibald 1914]
Lebesgue H 1924 Ens M 24, 264-275
[Gauss, see Archibald 1914]

Regular polygons (continued)

Richter M 1897 ZMN Unt 28, 252-255
  (F28, 43) [Renaldini]
Shively L S 1946 MT 39, 117-120
  [Ptolemy]
Steinbrenner G 1914 Broun Mo, 254-
  257 (F45, 77) [Duerer]


Regular polyhedra

See also Polyhedra.

Brueckner M 1907 Unt M 14, 104-110,
  120-127 (F38, 68)
Conte L 1954 Pe M (4)32, 261-278
  (Z57, 241 [Pappus]
Franceschi P di B dei 1915 Linc
  Mor (5)14
Hofmann J E 1963 Arc M 14, 212-216
  (MR27, 463) [Archimedes]
Lebesgue H 1938 Par CR 207, 437-439
  [equiv]
Lindemann F 1897 Mun Si 26, 625-758
  [cryst]
Loria G 1912 Bib M (3)13, 14-16
  (F43, 71) [semi-regular]
Sachs Eva 1917 Die fuenf platonischen
  Koerper, Berlin, 252 p
  (Phi Untgn 24)
Stolz O 1896 Mo M Phy 7, 296
Struik D J 1926 Nieu Arch (2)15(2),
  121-137 [tesselation]
Vacca G 1939 It UM (2)2, 66-70
  (F65, 5) [pentagon, dodecahedron]
Weaver J H 1915 MT 7, 86-88


Relations

See also Functions.

De Vries H 1949 N Tijd 37, 21-28,
  99-109 (MR11, 150)


Riemann surface

See also Complex analysis.

493 / Mathematical Topics

Klein F 1893 in his Lectures on
  Mathematics...1893, NY (Macmillan)
  33-40
Tonelli A 1895 Linc Rn (5)4, 300
*Weyl H 1955 Die Idee der Riemannschen
  Flaeche, 3rd ed, Stuttgart
*   1964 The Concept of a Riemann
  Surface, 3rd ed (Addison-Wesley),
  191 p


Rings

Reitberger H 1968 Ueberbl 1, 155-175
  [local rings]


Roots (of numbers)

See also Approximation, Equations,
  Square roots.

Albarran F 1936 Rv M Hisp A (2)11,
  143-150 (F64, 911)
  [Bombelli, Heron]
Barbour J M 1957 AMM 64, 1-9
  (MR19, 124) [approx, geom]
Bruins E M 1957 Bagd CS 2, 99-103
  (Z78, 3) [4th, 5th]
Carruccio E 1939 Pe M (4)19, 189-197
  (MR1, 130) [cube, Fibonacci]
Chakravarti G 1934 Clct Let 24(8),
  29-58 (F62, 1021) [Hindu]
Culum Z 1954 Beo DMF 6, 108-110
  (Z57, 241) [Archimedes]
Curtze M 1897 Hist Abt 42, 145-152
  (F28, 38) [Heron, Regiomontanus]
- 1897 Hist Abt 42, 113-120
  (f28, 38) [cube, sq, Heron]
Czwalina A 1927 Arc GMNT 10, 334-335
  [Greek]
Dakhel A 1960 Al-Kashi on root
  extraction, Beirut, 53 p
  (MR25, 739)
Datta B 1927 AMM 34, 420-423
  [Hindu word]
- 1931 AMM 38, 371-376 (Z2, 326)
  [Hindu word]
- 1931 Clct MS 23, 187-194 (Z4, 194)
  [approx, Narayana]
Demme C 1886 Hist Abt 31, 1-27
  [Archimedes, Hero]

## Roots (of numbers) (continued)

Eneström G 1913 Bib M (3)14, 83-84
  (F44, 44)  [cube, mid]
Gandz S 1926 AMM 33, 261-265  [word]
- 1928 AMM 35, 67-75  [word]
Gauchet L 1914 Toung Pao 15, 531-550
  [China]
Glaisher J W L 1927 Mess M 57, 38-71
  (F53, 11)  [div by radical]
- 1936 Muns Semb 8, 65-69
  (F62, 6)  [Babyl]
Kitao D 1892 Tok M Ph 5, 175
  [cube, Japan]
Lemeray E M 1898 Intermed 5, 69
  [word]
Luckey P 1948 M Ann 120, 217-274
  (MR9, 484)  [Arab binom thm]
Ma C C 1928 AMM 35, 29-30
  [Chinese, word]
Miller G A 1917 MT 9, 154-157
  [nota]
Natucci A 1956 It UM (3)11, 594-598
  (Z71, 243)  [Tartaglia]
Noordgaard M 1924 MT 17, 223-238
  [algorithms]
Rodet L 1879 Fr SMB 7, 98-102
  (F11, 36)  [approx]
Singh A N 1927 Clct MS 18, 123-140
  [Hindu]
- 1936 Math Cluj 12, 102-115
  (F62, 1022)  [Hindu]
Tytler J 1832 Asi R 17, 51-
  [Arab]

## Ruler

See also Constructions.

Archibald R C 1918 AMM 25, 348-360
  [double edged]
- 1946 Scr 14, 189-197  [Steiner]
Archibald R C ed 1950 *Jacob Steiner's
  Geometry, Constructions with a
  Ruler,* NY  (Scr Stu 4)
Gibbens G 1930 AMM 37, 343-348
  [marked ruler]
Juel C + 1895 Intermed 2, 124, 363
  [bib]
Mordukai-Boltovskoi D 1933 Archeion
  15, 45-72  (F59, 8)

Smogorzhevskii A S 1961 *The Ruler in
  Geometrical Constructions*
  (Blaisdell)

## Schauder bases

See also Functional analysis.

Sanders B L 1967 *A Bibliography of
  Papers Relating to (Schauder)
  Bases* (Texas Christian U)

## Schlicht functions

See also Complex analysis.

Bernardi S D 1966 *Bibliography of
  Schlicht Functions,* NY, 166 p
  (MR34, 508-509)  [*bib]

## Sequences

See Fibonacci sequences, Finite
  differences, Progressions, Series.

Dickson Leonard E 1919 Dickson 1,
  Ch.17
Frechet M + 1908 Enc SM 1(1)(3),
  469-488  [complex]
Hultsch F 1900 Bib M (3)1, 8
  [Greek, Pythagoras]
Molk J + 1908 Enc SM 1(1)(2), 162-
  328

## Series

See also Sequences, Divergent series,
  Farey series, Fourier series,
  Gregory series, Harmonic series,
  Orthogonal functions, Tauberian
  theorems, Taylor series.

Agostini A 1926 Anc Sto 7, 209-215
  (F52, 25)  [Frullani, Ruffini]
- 1941 It UM (2)3, 231-251
  [Mengoli]
- 1951 Pe M (4)29, 241-248
  (Z44, 245)  [Frullani]

## Series (continued)

Aubry A 1907 Mathesis (3)7, 257-261

Baire R + 1909 Enc SM 1(1)(4), 489-531

Bari N K + 1951 UMN (NS)6(4), 187-189 (MR13, 198) [Menshov]

Broadbent T A A 1930 M Gaz 15, 5-11 (F56, 21)

Bromwich T J I 1908 *An Introduction to the Theory of Infinite Series,* (Macmillan) (2nd ed 1926)

Busard H L L 1962 Ens M (2)8, 281-290 (Z115, 241) [mid]

- 1964 Arc HES 2, 387-397 (MR33, 2499) [mid]

Cajori F 1891 NYMS 1, 184-189 [mult]

- 1892 NYMS 2, 1- [conv]

- 1920 AMSB 27, 77-81 (F47, 32) [De Morgan]

Chakravarti G 1934 Clct Let 24(6), 1-18 (F62, 1021) [Hindu]

Dobrovolskii V A 1963 Ist M Zb 4, 37-41 [V P Ermakov]

Du Bois-Raymond 1913 Ostw Kl (185) 115 p

Eneström G 1879 Sv Han (9) (F11, 38) [conv]

- 1894 NYMS 3, 186-187 [XVIII, conv]

- 1902 Bib M (3)3, 243 (F33, 54) [mid]

- 1909 Bib M (3)10, 83-84 (F40, 57) [XVII, recip sq]

- 1909 Bib M (3)10, 350 (F40, 57) [pow ser]

- 1912 Bib M (3)12, 135-148 [XVII]

Gloden A 1947 Lux Arch (NS)17, 113-120 (MR9, 485)

- 1950 Lux Arch (NS)19, 205-220 (MR12, 577) [XIX-XX]

Hayashi T 1932 Toh MJ 35, 345-397 (Z5, 4) [binom, Japan]

Hofmann J E 1939 Deu M 4, 556-562 (MR1, 33) [log, Mercator]

- 1940 M Mag 14, 37-45 [Cotes, England, log]

Krasotkina T A 1957 Ist M Isl 10, 117-158

Lecat M 1924 *Bibliographie de la rélativité,* Bruxelles

Likhin V V 1966 Ist Met EN 5, 219-227 [Euler]

Miller M 1954 Dres Verk 2(2), 1-16 (Z55, 4) [Newton]

Molk J + 1907 Enc SM 1(1)(2), 162-270

Petersen G M 1966 *Regular Matrix Transformations* (McGraw-Hill), [bib] [transf of ser]

Pringsheim A 1905 Bib M (3)6, 252-256 (F36, 54) [conv, Euler]

Rajagopal C T + 1949 Beng Asi J 15(1) [Hindu]

- 1951 Scr 17, 65-74 (Z45, 146) [Gregory, Hindu]

*Reiff R 1889 *Geschichte der unendlichen Reihen,* Tuebingen, 217 p

Singh A N 1936 Osiris 1, 606-628 [Hindu]

Smail L L 1925 *History and Synopsis of the Theory of Summable Infinite Processes,* Oregon, 182 p (F51, 9)

Staeckel P 1907 Bib M (3)8, 37-60 [Euler, recip sq]

Vanhèe L 1930 Archeion 12, 117-125 (F56, 20) [Asia]

Wanner G 1968 Ueberbl 1, 133-153 [Lie ser]

Waschow H + 1932 QSGM (B)2, 298-304 (Z6, 1) [Babyl]

Wieleitner H 1914 Bib M 14, 150-168 [mid]

Williamson R S 1942 J Eg Arch 28, 67 [geom ser]

## Set theory

See also Foundations, Infinity, Logic diagrams, Number, Topology, Transfinite.

Baire R 1909 Enc SM 1(1)4, 490-531

*Borel E 1909 Rv GSPA 20, 315-324

Carruccio E 1942 It UM (2(4, 175-187 (MR7, 354) [Galileo]

*Cavailles Jean 1938 *Remarques sur la formation de la théorie abstraite

## Seventeen-gon

See also Constructions.

## Sieve of Erastosthenes and analogs

## Similarity and similitude

See also Geometry.

## Similarity and similitude

Thierry G de 1927 Am NASP 13, 684-688 [hydraulics]

## Simpson rule

See also Integration.

Braasch Johann 1908 *Historisches ueber die Simpsonsche Regel...,* Hamburg, 13 p (F39, 61)
Cattaneo P 1914 Pe MIS 17, 56-57 (F45, 78) [Vercellin]
Eneström G 1910 Bib M (3)11, 185 (F41, 62)
Heinrich G 1900 Bib M (3)1, 90
Peano 1915 Tor FMN 50

## Sine

Bioche C + 1898-1899 Intermed 5, 264; 6, 252 (Q1301) [word]
Boyer C B 1947 MT 40, 267-275 [calc]
Gordon C B 1952 M Gaz 36, 288 [word]
Gravelaar N L W A 1905 Wisk Tijd 2, 12 [word]
Inamder M G 1950 M Stu 18, 9-11 (MR13, 197) [Bhaskara, calcn]
Kanfinan H 1967 Scr 28, 29-36 [bib genzn]
Krishnaswami Ayyangar A A 1950 M Stu 18, 12 (MR13, 197) [Bhaskara, calcn]
Lohne J 1961 Arc HES 1, 389-405 [Newton]
Milhaud G 1908 Rv GSPA 18, 223-228 [Descartes, sine law]
Rozenfeld B A + 1960 Ist M Isl 13, 533-538 (MR24, 467) [al-Rumi]
Ruska J 1895 Hist Abt 40, 126-128 (F26, 64)

## Singular points

Brocard H + 1896 Intermed 3, 250 (Q387) [bib]

## Curetti Bovolo A 1959 Archim 11, 317-321 (Z92, 242)

## Singular solutions

See also Differential equations.

Lindeberg K M 1875 *Historisk oefversigt of teorierna foer singulaere solutioner till ordinaera differential equationer,* Stockholm (F7, 19)
Mansion P 1891 Brx SS 15, 32-37, 60 (F23, 37) [bib, part dif eq]
Rothemberg S 1908 Ab GM 20, 317-404 [total dif eq]

## Sinusoid

Goupillière Haton de la 1898 Nou An (3)17, 153-155 (F29, 37) [bib]

## Six square problem

See also Number theory.

Hofmann J E 1958 M Nach 18, 152-167 (MR20, 503; Z89, 241)

## Solid geometry

See also Spaces (Ch. 3).

Boyer C B 1965 MT 58, 33-36 [coords, Hudde]
Coolidge J L 1948 AMM 55, 76-86 (MR9, 485; Z30, 99) [anal geom]
Frajese A 1934 Pe M (4)14, 211-234 (F60, 6) [trihedral]
Obenrauch F J 1903 Mo M Phy 14, 187-205 [curve, Pythagoras]
Tannery Paul 1912 *Mémoires Scientifiques,* Toulouse 2, 1-47
Wieleitner H 1928 ZMN Unt 59, 357-358 (F54, 7)

## Spaces (mathematics)

See also Four dimensional space,
Linear algebra, Solid geometry,
Space    (Ch.3).

Freudenthal H 1965 Adv M 1, 145-190
Halsted G B 1878-1879 AJM 1; 2
[bib, n-dim]
- 1896 Chic Cong, 92-95
Libois P 1935 Brx Con CR 2, 96-104
[metric, proj]
Piaggio H T H 1930 Nat 125, 897-898
(F56, 7)
Schlegel V 1900 Ens M 2, 77-114
[n-dim]
*Sommerville Duncan M Y 1911 *Biblio-
graphy of Non-Euclidean Geometry
Including the Theory of Parallels
the Foundations of Geometry, and
Space of n Dimensions,* London
- 1929 *Introduction to Geometry of
n-Dimensions* (Methuen) (repr 1958
Dover)  [bib]
Van der Waerden B L 1950 *Over de
rμimte* (Noordhoff), 18 p
Veronese G 1894 *Grundzuge der
Geometrie von mehreren Dimensionen
und mehreren Arten gradliniger
Einheiten in elementarer Form
entwickelt,* Leipzig
Wieleitner H 1925 Isis 7, 486-489
[Oresme]

## Special functions

*Lebedev N N 1965 *Special Functions
and Their Applications,* Englewood
Cliffs, 308 p

## Spectral theorem

Lorch E R 1962 MAA Stu 1, 88-137
Riesz F + 1953 *Leçons d'analyse
fonctionelle,* Budapest [bib]

## Sphere

Bortolotti E 1935 Pe M (4)15, 87-92
(F61, 10)  [Egypt]
Braunmuhl A v 1898 Bib M 65
[tria]
Brun V 1935 Nor M Tid 17, 1-13
(F61, 12)  [Archimedes, vol]
Clagett M 1952 Isis 43, 36-38
[Archimedes, mid]
- 1954 Osiris 11, 295-358
(Z55, 2)  [Archimedes, Tinemue]
Gillings R J 1967 Au JS 30, 113-116
[area]
Krause M 1936 Gott Ab H 3(17)(7)
(Z15, 52)  [Menelaos]
Leech J 1956 M Gaz 40, 22-23
(MR17, 888)  [thirteen sph probl]
Loria G 1928 Bo Fir 24, I-II
[Archimedes]
Luckey P 1941 Deu M 5, 405-446
[tria]
Mueller C 1940 Deu M5, 244-255
(F66, 11; Z25, 146)
[area, Aryabhata]
Natucci A 1931 Pe M (4)11, 69-83
(F57, 31)  [mid]
Schwartz J J 1934 Scr 2, 243-246
[Menelaos ms]
Steiner J 1931 *Allgemeine Theorie
ueber das Beruehren und Schneiden
der Kreise und der Kugeln,*
Zurich, 363 p  (F57, 1305)
Suter H 1909 Bib M (3)9, 196-199
[area, Hindu]
Whyte L L 1952 AMM 59, 606-611

## Spherical harmonics

Heine Eduard 1878-1881 *Theorie der
Kugelfunctionen,* 2nd ed, Berlin,
2 vols
Hobson E W 1910 Enc Br
Nielsen N 1929 Dan M Med 10(5)
Prasad G 1930-1932 *A Treatise on
Spherical Harmonics...,* Benares,
2 vols
Rabut C + 1897 Intermed 4, 189
(Q1043)

## Spherical trigonometry

Braunmuehl A v 1898 Bib M (2)12,
65-72  (F29, 33)  [polar tria]
Buerger H + 1924 Ab GN Med 7, 40-91
[Arab]
Czwalina Λ 1958 Ens M (2)4, 292-299
(MR21, 893)  [Ptolemy]
Eneström G 1888 Bib M (2)2(Q22)
[area, Cavallieri, Girard]
Ginsberg J 1932 Scr 1, 72-78
[Jacob ben Machir Menelaus]
Gloden A 1949 Lux SNBM (NS)43, 1-17
(MR11, 571)
Haller S 1899 Bib M (2)13, 71-80
(F30, 52)  [tria sol]
Johnson W 1922 Mess M 50, 76-80
[circ parts]
Luckey P 1941 Deu M 5, 405-446
(MR2, 306; Z25, 290)
Mogenet J 1947 Brx SS (1)61, 235-241
(Z31, 1)  [anc def]
Moritz R 1925 AMM 22, 220-222
[Napier]
Schidlowski W 1905 Kagan (389),
106-113  (F36, 67)
Schubert F T 1801 Pet Nov Ac 12, 165-
[Ptolemy]
Sengupta P C 1928 Clct Let 21(4)
(F57, 1295)  [ast, Greek, Hindu]
Sperry Pauline 1928 *Short Course in
Spherical Trigonometry* (Johnson),
48-51
Todhunter J 1873 Phi Mag  (F5, 46)
Tropfke Johannes 1923 *Geschichte
der Elementar-Mathematik*, Berlin-
Leipzig 5(2)
Vacca G 1902 Loria 5, 1  [Harriot
mss]

## Spiral

Agostini Amodeo 1930 Bo Fir 9, xxv-
xxviii  [length, Torricelli]
1930 Pe M (4)10, 143-151  (F56, 19)
[Torricelli]
- 1949 *La Memoria di Evangelista
Torricelli sopra la spirale
logaritmica riordinata e completata*,
Livorno, 45 p

Archibald R C 1918 AMM 25, 189-193
[log]
- 1918 AMM 25, 276-282
[Cornu, Euler, Fresnel]
- 1920 in Jay Hambridge *Dynamical
Symmetry* [log]
Favaro A 1891 Bib M (2)5, 23-25
(F23, 39)  [log]
- 1905 Lom Gen (2)38, 358
[Cavalieri]
Hofmann J E 1954 Arc M 5, 138-147
(Z56, 1)  [Viete]
Itard J 1955 Ci Tecnol 5(17), 53-64
[area]
Opial Z 1960 Wiad M 3, 251-265
(Z118, 4)  [length, Torricelli]
Sedillot L A 1873 Boncomp 6, 239-248
(F5, 1)  [Plato]
Sibirani F 1939 It UM (2)1, 160-172,
259-274     (Z21, 194)
[Archimedes]
Weyer G D E 1894 *Ueber die para-
bolische Spirale*, Leipzig

## Square roots

See also Roots.

Anon 1899 Pitagora 5(2), 53
Bierens De Haan D 1877 Nieu Arch 3,
208-210  (F9, 19)  [Van Schooten]
Bobynin V V 1896 ZM Ph 41, 193
[Greek]
Bortolotti E 1935 Pe M 15, 220-229
[Babyl]
Bromwich J T 1928 M Gaz 14, 253-257
[Archimedes]
Bruins E M 1948 Amst P 51, 332-341
(Z30, 97)  [Babyl, Greek]
Ceretti U 1899 Pitagora 5(2), 30
[Arab]
Eneström G 1886 Bib M, 236-239, 244
[Alkalsadi]
Garver R 1932 Clct MS 24, 99-102
(F58, 24)
Gazis D C + 1960 Scr 25, 229-241
(Z101, 243)  [Archimedes]
Guenther S 1882 Ab GM 4, 1-134 [anc]
Gupta J 1940 Benar MS (NS)2, 33-37
(MR2, 306; F66, 1186)

Square roots (continued)

Gurjar L V 1942 Bombay (NS)10(5),
  6-10 (Z60, 1) [Hindu]
Heilermann H1881 Hist Abt 26, 121-126
  (F13, 28) [Archimedes]
- 1884 ZMN Unt 14, 81-89
Hofmann J E 1930 Arc GMNT 12, 387-
  408 [Archimedes]
- 1934 DMV 43, 187-210 (F60, 8)
  [Archimedes, Heron]
Hultsch F 1893 Gott N, 367-428
  (F25, 57) [Archimedes]
Hunrath K 1888 Hist Abt 33, 1-12
Joco-Seria 1941 Aus dem Papieren
  eines reisenden Kaufmannes,
  Wiesbaden (Kalle-Werke)
  [J H Anderhub, Theodorus]
Jones P S 1949 MT 42, 307-310
  [Babyl US]
Le Paige C 1892 Mathesis (2)2, 273-
  274 [def]
Loria G 1913 Int Con 2, 518-525
  (F44, 44) [Greek]
Mueller C 1932 QSGM (B)2, 281-285
  [Archimedes]
Neugebauer O 1931 Arc Or 7, 90-99
  (F58, 8; Z3, 97) [approx, Babyl]
Neugebauer O + 1932 QSGM (B)2,
  291-297 [Babyl]
Paev M 1965 Ist M Isl 16, 219-233
  [approx, Greek]
Raik A E 1954 Perm M 8(3), 49-58
  (Z68, 1) [anc]
Richardt T 1925 Nor M Tid 7, 73-88
  (F51, 16) [Archimedes]
Schoenborn W 1883 Hist Abt 28, 169-
  179 [Archimedes, Heron]
- 1885 Hist Abt 30, 81-90
  [Diophantus]
Smyly J G 1944 Hermath 63, 18-26
  (Z60, 5) [Heron]
Stamatis E 1953 Platon 5(2)
  (Z52, 2) [anc]
- 1955 Athen P 30, 255-262 (Z65, 243)
  [Archimedes]
Tannery P 1887 Bib M (2)1, 17-21
  [Chuquet]
Uhler H S 1953 Scr 19, 78-79
  [names]
Van der Waerden B L 1941 M Ann 118,

286-288 (MR3, 257)
  [Pythagoras]
Vogel K 1932 DMV 41, 152-158
  (F58, 13) [approx, Archimedes]
Weissenborn H 1884 Die irrationalen
  Quadratwurzeln bei Archimedes und
  Heron, Berlin
Wertheim G 1898 Ab GM 8, 147-160
  (F29, 28) [cont fra]
- 1899 Hist Abt 44, 1-3 (F30, 39)
  [cube, Heron]
- 1899 ZMN Unt 30, 253 [cube, Heron]
Woepcke F 1874 Boncomp 7, (F6, 39)
  [approx]
Zeuthen H G 1911 Nyt Tid (A)22, 77-81
  (F42, 52) [Bhaskara]


Squaring the circle

See also Construction, Pi.

Anon 1903 Rv Sc (4)21, 91
  [Egypt]
Aubry A 1900 Prog M (2)2, 281
Ball W W R 1960 Mathematical
  Recreations, Ch. 12.
Beutel E 1913 Die Quadratur des
  Kreises, Leipzig, 79 p
Bierens de Haan D 1874 Boncomp 7,
  99-140 [Holland, 1584-1754]
Bockstaele P 1956 Nov Vet 34, 299-
  312 [Archimedes]
Brun V 1941 Nor M Tid 23, 41-53
  (MR3, 97; Z25, 1)
Cassina V 1929 Pe M (4)9, 238-250,
  320-335 (F55, 588)
  [alg transc numbs]
Christensen A A 1894-1895 Nyt Tid 6,
  52-56, 63-67, 84-89 (F26, 61)
  [Greek]
Clagett M 1952 Osiris 10, 587-618
  (MR14, 524) [Archimedes, mid]
Curtze M 1901 Bib M (3)2, 48
  [XV]
Demme C 1886 Hist Abt 31, 132-135
  [Egypt]
Dianni J 1956 Int Con HS 8, 132-133
  [Poland]
Drenckhahn F 1936 DMV 46(1), 1-13
  (F62, 17) [Hindu]

Squaring the circle    (continued)

Eneström G 1889 Bib M (2)3, 96 (Q27)
  [Charles V, Rabelais]
- 1912 Bib M 12, 268 (Q157)
Glaisher J W L 1873 Mess M (2)3, 27-
  46 (F5, 46) [1580-1630]
Goering W 1899 *Die Auffindung der*
  *rein geometrischen Quadratur des*
  *Kreises...,* Dresden, 13 p
Goodrich L C 1948 Isis 39, 64
  [China]
Gurjar L V 1942 Bombay 10(5),
  11-16 (Z60, 1) [Hindu]
*Hobson E W 1913 *Squaring the Circle*
  *--a History of the Problem,*
  Cambridge, 57 p (F44, 50)
  (repr 1953 Chelsea)
Hofmann J E v 1938 M Mag 12, 223-230
  [XVI-XVII, Artus de Lionne]
- 1941 Heid Si Ph 4, 37 p
  [Lull]
- 1952 MN Unt 4, 321-323   (MR13, 612)
  [Huygens]
- 1954 Arc In HS (NS)7, 16-34
  (MR15, 924)  [Della Porta]
- 1961 Mathunt 7(3), 47-103
  [in schools]
Jansen F 1909 *De cirkelquadratuus*
  *bij de Grieken Hippokrates van*
  *Chios, Archimedes van Syrakuse,*
  Haarlem, 54 p (F40, 64)
Kikuchi D 1895 Tok M Ph 7, 24
  [Japan]
Kropp G 1951 Crelle 189, 75 p
  [area, Lalouvère]
Marar K M + 1944 Bom RASJ 20, 65-82
  (Z60, 1) [Hindu]
Menger Karl 1934 in *Alte Problem*
  *neue Loesungen in der Exakten*
  *Wissenschaften,* Leipzig, 1-29
Mikami Y 1909 Bib M (3)10, 193-200
  (F40, 65) [China]
Mitzscherling A 1913 *Das Problem*
  *der Kreisteilung,* Leipzig-Berlin,
  220 p
Montel P 1939 Arg SC 128, 321-330
  (F65:1, 2)
*Montucla 1754 *Histoire des recherches*
  *sur la quadrature du cercle...,*
  Paris

- 1831 *Histoire des recherches sur*
  *la quadrature du cercle,* 2nd ed,
  Paris (ed by S. F. Lacroix)
Neovius E 1884 Hels Tids 16, 325-339
Rossier P 1941 Genv CR 58, 204-207
  (MR3, 258) [Lesage]
Rudio F 1890 Zur NGV 35, 1-51
*- 1892 *Archimedes, Huygens, Lambert,*
  *Legendre, Vier Abhandlungen ueber*
  *die Kreismessung...mit einer*
  *Uebersicht ueber die Geschichte des*
  *Problemes von der Quadratur des*
  *Cirkels, von den aeltesten Zeiten*
  *bis auf unsere Tage,* Leipzig, 173 p
  (F24, 50)  (Russ tr 1934 Mos-Len)
- 1907 *Der Bericht des Simplicius*
  *ueber die Quadraturen...* Leipzig
- 1907 Bib M (3)8, 13-22
  [Aristophanes]
Schubert Hermann 1888 *Die Quadratur*
  *des Zirkels...,* Hamburg, 40 p
  (Engl transl 1890 Smi R, 97-120;
  1891 Monist 1, 197-; 1896 Chic
  Cong, 112-143)
Suter H 1899 ZM Ph 44, 33-47
  [Alhazen]
Tannery P 1902 Bib M (3)3, 342-349
  [Simplicios]
Teixeira F G 1917 Scientia 22, 169-
  178
Travnicek J 1889 *Das Problem der*
  *Kreismessung. 1. Die Zeit vor*
  *Archimedes,* Bruenn, 28 p
Vacca G 1909 Loria 11, 65-67
  (F40, 64) [Egypt]
Wasserstein A 1959 Phronsis 4, 92-100
  [Greek]
Williamson R S 1945 J Eg Arch 31,
  112 [Egypt]
*Wolff T 1933 Verm R 10, 116-122
  (F59, 828)
Yushkevich A P 1957 Ist M Isl 10,
  159-210  (Z102, 4) [Euler]


Star polygons

Boncompagni B 1873 Boncomp 6, 341-
  358, 544 (F5, 7) [Boethius]
Guenther S 1873 Boncomp 6, 313-340
  (F5, 5) [anc, mid]

Star polygons    (continued)

Hankin R H 1934 M Gaz 18, 165-168
(F60, 840)  [art]


Statistics

See also Chi square, Correlation,
Discrete distributions, Factor
analysis, Insurance (Ch.3),
Latin squares, Least squares,
Maximum likelihood, Mean,
Nuisance parameters, Probability.

Andersson T 1925 Nor St Tid 4, 1-24
(F51, 36)  [Sweden]
Anon 1930 *List of References on
Statisticians...,* Wash DC,
(Library of Congress), 5 p
Baten W D 1938 Scr 5, 165-170
[XX]
Boyer C B 1947 Isis 37, 148-149
(Z29, 194)  [Huygens]
*Buckland William R + 1963 *Biblio-
graphy of Basic Texts and Mono-
graphs on Statistical Methods
1945-1960* (Hafner), 304 p
Buros O K 1938 *Research and Statisti-
cal Method; Books and Reviews of
1933-1938* (Rutgers Univ Pr)
- 1941 *Second Yearbook of Research
and Statistical Method; Books and
Reviews,* Highland Park, NJ
- 1951 *Statistical Methodology
Reviews 1941-1950* (Wiley)
Buttenbaugh Grant I 1946 *A Bibli-
ography on Statistical Quality
Control* (U Washington P), 122 p
(Suppl 1951)
Darmois G + 1952 *Bibliographie sur
la méthode statistique...,*
(Institute internationale de
statistique), 49 p
*Deming L S 1960 USNBS (B)64(1),
55-82  [bib, regression, time
series]
- 1963 USNBS (B)67, 91-134  [bib]
Demster A P 1968 Sci 160, 661-663
Deutsch Ralph 1965 *Estimation theory,*
Englewood Cliffs, 269 p  [bib]

Doorman Carl 1925 *Halley und Fermat
Beitraege zur Geschichte der
Statistik mit einen Anhang ueber
die Fermatsche Problem,* Breslau,
122 p
Duhem Pierre 1905 *Les Origines de la
statistique,* Paris
Eisenhart C 1967 Am Statcn (Ap),
32-34  [annivs]
Eneström G 1899 ZM Ph 44(S), 81-95
(F30, 45)  [Halley, Wargentin]
Fabian F 1959 Ber Hum MN 5, 699-703
[Czech]
Finetti B de 1951 Stat Bln 11, 185-
192  (MR13, 421)  [Wald]
Fisher R A 1953 in A E Heath ed,
*Scientific Thought in the
Twentieth Century,* NY
- 1954 Am Sc 42, 275-282
Fitzpatrick P J 1956 Am Statcn 10,
14-19  [educ, USA]
- 1958 ASAJ 53, 689-701  [biog, USA]
*Freudenthal H 1951 Arc In HS 14, 25-
34  [Arbuthnot]
Funkhouser H G 1937 Osiris 3, 269-
404  (F63, 746; Z18, 195)
[graphs]
Gini C 1926 Lon St (2)89, 703-724
(F52, 39)  [Italy]
*Gordon R D 1940 Phi Sc 7, 389-
[inverse prob]
Good Irving J 1965 *The Estimation
of Probabilities. An Essay on
Modern Bayesian Methods* (MIT P),
[bib]
Gore J H 1887 USCS, 311-512  (2nd ed
1902 USCS, 427-787)  [bib]
Greenwood J A + 1962 *Guide to Tables
in Mathematical Statistics*
(Princeton U P), 1072 p
Haight F A 1961 USNBS (B)65(1),
23-60  [bib, distributions]
*Healy M J R 1963 Lon St(A)126, 270
[subj index to Kendall 1965]
[bib]
Irwin J O 1935 M Gaz 19, 18-30
(F61, 7)
John V 1884 *Geschichte der Statistik*
Stuttgart, 391 p  [descr]
*Kendall Maurice G 1951 Biomtka 38,
11-25  [regression]

Statistics (continued)

- 1958-1966 *The Advanced Theory of Statistics*, London, 3 vols [encyclopedic]
*- 1960 Biomtka 47, 447-449 [origins]
- 1961 Biomtka 48, 1-2 (MR23A, 433) [D. Bernoulli, max likelihood]
- 1961 Biomtka 48, 220-222 (MR23A, 433) ["the book of fate"]
*Kendall Maurice G + 1960 *A Dictionary of Statistical Terms*, 2nd ed, Edinburgh (1st ed 1957)
*- 1962-1968 *Bibliography of Statistical Literature*, Edinburgh, 3 vols (See Healy 1965 for index) [I 1950-1958; II 1940-1949; III to 1949]
Koren John 1918 *The History of Statistics, Memoirs to Commemorate the Seventy-fifth Anniversary of the American Statistical Association*, NY, 785 p
Kotz S 1965 Am Statcn (June), 16 [termin]
- 1965 Survey (Oct) [USSR]
Kozlov T I 1951 Ves St (4), 59-67 [Russia, XIX]
Kruskal W H 1957 ASAJ 52, 356-360 [Wilcoxon unpaired two-sample test]
*Lancaster H O 1969 *The Chi-Squared Distribution* (Wiley), 365 p [*bib]
Lehmann E L 1958 *Some Early Instances of Confidence Statements*, (Univ Calif Off. Naval Res. Tech. Report)
Lieblein J 1954 Biomtka 41, 559-560 [engineering]
Lord R D 1958 Biomtka 45, 282 [De Morgan, literature]
Lorey W 1934 Allg St Ar 23, 398-410 (F60, 832) [Laplace]
Mason Du Pré A 1938 Isis 29, 43-48 [C Kramp]
Meitzen A 1886 *Geschichte, Theorie u Technik der Statistik*, Berlin, 223 p [descr]
- 1891 *History, Theory and Technique of Statistics*, Philadelphia, 2 vols

Miller R G 1966 *Simultaneous Statistical Inference* (McGraw-Hill) [bib]
Mood Alexander M 1950 *Introduction to the Theory of Statistics* (McGraw-Hill) [hist notes]
Morice E + 1968 *Dictionnaire de Statistique*, Paris, 208 p
National Physical Lab 1951 Lon St (B)114, 497-558 [bib recent work]
Neyman J 1955 Sci 122, 401-406 [and science]
- 1962 Rv IT St 30(1), 11-27 [decisions]
- 1966 in *On Political Economy and Econometrics, Essays in honour of Oskar Lange*, Warsaw, 445-462 [phil]
Owen D B 1962 *Handbook of Statistical Tables*, Reading, Mass [bib]
Patino Emilio A + 1938 *Bibliografia del metodo estadistico y sus aplicaciones*, Mexico (Secretaria de la economia naconal), 169 p [1920-1938]
Pearson E S 1965 Biomtka 52, 3-18 [biometry]
*- 1966 in *Festschrift for J Neyman* ["Neyman-Pearson story"]
- 1967 Biomtka 54, 341-355 [1885-1920]
- 1968 Biomtka 55, 445-457 [Fisher, Gosset, K Pearson]
- 1924 Biomtka 16, 402-404 [normal curve]
- 1928 Biomtka 20(A), 295-299 (F54, 32) [Plana]
Plackett R L 1958 Biomtka 45, 130-135 [mean]
Ptukha M V 1945 *Ocherki po istorii Statistiki XVII-XVIII vekov*, Mos, 351 p
- 1958 *Ocherki po istorii statistiki v SSSR*, Mos
Quadri Antonii 1824-1826 *Storia della Statistica...*, Venice, 2 vols
Rietz H L 1922 AMM 29, 333-337 [termin]
Royston E 1956 Biomtka 43, 241-247 [graphs]
Savage I R 1953 ASAJ 48, 844-849 [non-par stat]

Statistics   (continued)

- 1962 *Bibliography of Nonparametric Statistics* (Harvard U P)
Seal H L 1967 Biomtka 54, 1-24 [bib, Gauss]
Seng Y P 1951 Lon St (A)114, 214-231 [sampling]
Shoen H H 1938 Osiris 5, 276-318 (F64, 918) [XIX, Prince Albert]
Sterling R E 1952 Hum Bio 24, 145-166
Stringfellow T L + 1961 SSM 61, 1-4
Teichroew D 1965 ASAJ 60(Mar), 27-49 [distribution sampling]
Thatcher A R 1957 Biomtka 44, 515-518 (MR19, 1247)
Van Dantzig D 1950 Sta Neer 4, 233-247 (MR13, 1) [descr]
Varburg D E 1963 MT 56, 252-257, 344-348
Vetter Q 1948 Arc In HS, 684-696 [Czech]
Walker Helen M 1929 *Studies in the History of Statistical Methods,* Baltimore, 237 p (F55, 622)
Westergaard H 1932 *Contributions to the History of Statistics,* London, 287 p (F58, 4) (repr 1968)
Wilcox W F 1935 Rv II St, 12 p [defs]
*Wilks S S 1948 AMSB 54, 6-50 [bib, order stat]
Williams C B 1956 Biomtka 43, 248-256 [literature]
*Wold H O A 1965 *Bibliography on Time Series and Stochastic Processes* (MIT P)


Stellated polyhedra

Guenther S 1876 *Vermischte Untersuchungen zur Geschichte der mathematischen Wissenschaften,* Leipzig, Ch. 1 (F8, 30)
Simon M 1904 Arc MP (3)7, 109 (F35, 62)


Stieltjes integral

Lebesgue Henri 1928 *Leçons sur l'intégration,* Paris
Medvedev F A 1961 Ist M Isl 14, 211-234 [Lyapunov]
- 1963 Ist M Isl 15, 171-224


Stochastic processes

*Bharucha-Reid A T 1960 *Elements of the Theory of Markov Processes and Their Applications* (McGraw-Hill), 468 p [bib]
*Cox D R + 1966 *The Statistical Analysis of Series of Events* (Wiley) [*bib]
Doig A 1957 Biomtka 44, 490-514 [bib, queues]
Hostinsky B 1950 Cas MF 74, 48-62 [Markov chains 1935-1948 bib]
Rozanov Yu A 1968 *Stationary Random Processes* (Holden-Day) [bib, notes]
Saaty T L 1961 *Elements of Queueing Theory with Applications* (McGraw-Hill) [*bib]
*Takacs Lajos 1967 *Combinatorial Methods in the Theory of Stochastic Processes* (Wiley), 273 p [*bib]
Wold H O A ed 1965 *Bibliography on Time Series and Stochastic Processes* (MIT Pr)


Strophoid

Aubry V 1897 JM Sp (5)21, 133 (F28, 48)
Loria G 1898 Loria 1, 1-7 (F29, 35)
- 1898 Mathesis (2)8, 265 [Delian problem Huygens]


Summations

See also Series.

Amir-Moez A R 1958 Scr 23, 173-178 (Z84, 241) [ibn Ezra, Karkhi]

## Summations    (continued)

Boyer C B 1941 Scr 9, 237-244
  [Pascal, powers]
Emmerich A 1896 ZMN Unt 27, 16
  [squares]
Eneström G 1890 Bib M (2)4, 22-24
  [recip sq]
Likhin V V 1966 Ist Met EN 5, 219-
  227 (MR34, 10) [funs]
Spiess O 1945 in Festschrift Zum 60
  Geburtstages von Prof. Dr. A.
  Speiser, Zurich, 66-86
  (MR7, 354; Z60, 12) [recip sq]
Staeckel P 1907 Bib M (3)8, 37-68
  (F38, 61) [Euler, recip sq]
Tannery P 1902 Bib M (3)3, 257-258
  [anc, cubes]


## Surface area

Bortolotti E 1935 Pe M (4)15, 87-92
  (F61, 10) [Egypt, sphere]
*Cassina U 1950 Lom Rn M (3)14,
  311-328 (MR13, 612) [Genocchi,
  Hermite, Schwartz]
Frechet M 1925 Pol SM An 3, 1-3
Gabba A 1957 Lom Rm M 91, 857-883
  (Z82, 11) [Lacorati, Peano,
  Schwartz]
Gillings R J 1967 Au JS 30, 113-116
  [Egypt, sphere]
Lebesgue Henri 1902 Integrale,
  longuerre, aire, Paris, Ch. 4
  (also 1902 Annali (3)7, 231-
  and 1902 Milan, 129 p)
- 1926 Fund M 8, 160-165
Mueller C 1940 Deu M 5, 244-255
  (F66, 11; Z25, 146) [Aryabhata,
  sphere]
Rado T 1943 AMM 50, 139
Saks S 1937 Theory of the Integral,
  Ch. 5
Suter H 1909 Bib M (3)9, 196-199
  [Hindu, sphere]


## Surfaces

See also Cone, Cubic, Curvature,
Ellipsoid, Paraboloid, Quadric
surfaces, Riemann surface, Sphere.

Aleksandrov A D 1955 Die innere
  Geometrie der konvexen Flaechen,
  Berlin (Engl transl 1967 AMS
  Trls 15) [bib]
Cajori F 1929 AMM 36, 431-437
  [developable surf, Euler]
- 1929 AMSB 35, 596 (F55, 26)
  [developable surf]
Calleri P 1938 Pe M (4)18, 33-42
  (Z18, 340) [ell, hyperb pt]
Clagett M 1954 Osiris 11, 295-358
  [Archimedes]
Emden R 1934 Naturw 22, 533-535
  [Kant]
Godeaux L 1933 Questions non résolues
  de géométrie algébrique. Les
  involutions de l'espace et les
  variétés algébriques à trois
  dimensions, Paris, 24 p
Hayashi T 1934 Toh MJ 39(1), 125-179
  [Japan]
Hill J E 1897 AMSB (2)3, 133-146
  (F28, 48) [bib]
Koetter 1896 DMV 5 (2), 65 p
  [XIX, Monge]
Morehead J C + 1902 in Gauss F K,
  General Investigations of Curved
  Surfaces.., Princeton, 115-126
  [*bib, 343 titles] (not in the
  reprint 1965)
*Purser F + 1910 Enc Br 26, 117-125
Staeckel P 1899 M Ann 52, 598-600
  (F30, 50) [one sided]
Tannery P 1884 BSM (2)7, 278-291
  [anc]
- 1896 Intermed 3, 143 (Q706)
Whittemore J 1917 An M (2)19, 1-20
  [min surf]


## Symmetric functions

See also Equations.

Agostini A 1925 Pe M 5 [eqs]
Bennett A A 1923 AMM 30, 180 [genzn]
Eneström G 1905 Bib M (3)6, 409-410
  (F36, 53) [eqs]

## Symmetric functions  (continued)

Funkhouser H G 1930 AMM 37, 357-365
  (F56, 5)  [eqs]
Saalshuetz L 1908 Bib M (3)9, 65-70
  (F39, 57)  [eqs]


## Symmetry

Anon 1949 Stu Gen 2 (July), 203-278
Ciamberlini C 1895 Pe M 10, 163
  [in proofs]
Fiala F 1949 Act Sc Ind (1066),
  81-102
Garrido J 1952 Par CR 235, 1184-1186
  (Z48, 241)  [art, Mexico]
Lemaire G 1915 Intermed 22, 184
  (Q4281)  [equiv sym figures]
Mansion P 1909 Bib M (3)10, 278
  (F40, 63)  [equiv]
- 1914 Intermed 21, 89 (Q4322)
  [equiv sym figures]
Nicolle J 1955 La Symetrie dans la
  nature et dans les travaux des
  hommes, Paris
Shaw J B 1943 Scr 9, 129-138
Shepard A O 1948 Carn Pub (574)
  [art]
*Shubinkov A V + 1964 Colored
  Symmetry, NY [art, *bib]
Speiser A 1956 Theorie der Gruppen
  von enlichen Ordnung,  4th ed,
  Basel
*Weyl Hermann 1952 Symmetry,
  Princeton, 171 p [bib]


## Tangents

See also Calculus.

Aubry A 1899 Prog M (2)1, 130, 164
  [XVII]
Bortolotti E 1921 Pe M (4)1, 263-276
  [envelopes]
Cajori F + 1923-1924 SSM 22, 463-464,
  715-717; 23, 64-66, 320-322
  [Greek]
Coolidge J L 1951 AMM 58, 449-462
  (MR13, 197; Z43, 4)

## Scriba

Scriba C J 1961 Arc HES 1, 406-419
  (Z102, 4)  [Descartes]
Zeuthen H G 1900 Nyt Tid (B)11, 49
  [Descartes]


## Tauberian theorems

See also Limits, Series.

Pitt H R 1918 Tauberian Theorems
  (Oxford U P)
Wiener N 1932 Ann M 33, 1-100
  [repr 1964 Dover]  [*bib]


## Tautochrone

Dijksterhuis E J 1928 Euc Gron 5,
  193-227  (F54, 29)
Ohrtmann C 1872 Das Problem der
  Tautochronen,  Berlin  (F4, 30)
- 1875 Le Problème des Tautochrones,
  Rome (Transl of 1872)  (F7, 23)
      Phillips J P 1967 MT 60, 506-508


## Taylor series and theorem

Dienes P 1931 The Taylor Series...,
  (Clarendon), 562 p  (repr 1957
  Dover)  [*bib]
Eneström G 1912 Bib M 12, 333-336
  (F42, 59)
Fleckenstein T O 1946 Elem M 1, 13-
  17  (Z60, 12)  [J. I. Bernoulli]
Gibson G A 1920 Edi MSP 39, 25-33
  [Bernoulli theorem]
Hadamard Jacques 1901 La série de
  Taylor.., Paris, 102 p (Scientia 12)
Moritz R E 1937 AMM 44, 31-33
*Pringsheim A 1900 Bib M (3)1, 433
Turnbull H W 1938 Nat (9 July)
  [Gregory]


## Tchirnhausen transformation

Burnside W S + 1928 Theory of
  Equations, Ch. 19
Dickson L E 1930 Modern Algebraic
  Theories, NY, 186-197

## Tensors

*Appell Paul + 1926 *Traité de mécanique rationelle*, Paris, vol 5

Rothe Herman 1923 *Einfuehrung in die Tensorrechnung*, Vienna, 183 p

*Wills A P 1931 *Vector Analysis With an Introduction to Tensor Analysis* (Prentice Hall) (repr 1958 Dover)

## Topological algebra

*Freudenthal H 1968 Rv Syn (3), 49-52, 223-243 [bib, Lie group]

## Topology

See also Cluster sets, Fixed point theorems, Graph theory, Knot theory.

Aleksandrov P S 1965 Rus MS 19, 1-39; 20, 177-178 [gen top]

Cantor G 1878 Crelle 84, 242-258

Dehn Max + 1907 Analysis Situs, Enc MW 3(1)(AB3), 153-220 [*homology]

*Feigl G 1928 DMV 37, 273-286 (F54, 7)

Franklin P 1935 Phi Sc 2, 39-47

Freudenthal H 1954 in *Homenaje a Millas-Vallicrosa*, Barcelona 1, 611-621 (MR16, 782) [Leibniz]

Gleason A M 1964 Sci 145 (3631), 451-

*Godeaux L 1913 Intermed, 10 (Q3977) [bib]

Hocking J G + 1961 *Topology* (Addison-Wesley) [homology]

Hopf H 1966 DMV 68 (1), 182-192 (MR33, 7996)

*- 1966 UMN 21, 8-16 (MR33, 2513) [autobiog]

Kuratowski K 1955 Shux Jinz 1(3), 601-607 [top spaces]

*Lebesgue Henri 1923 *Notice sur la vie et les travaux de Camile Jordan*, Paris, 28 p (repr 1957)

- 1924 Fr SMB 52, 315-336 [Euler, polyhedra]

*Lefshetz Solomon 1949 *Introduction to Topology* (Princeton U P), 226 p

Manheim Jerome H 1964 *The Genesis of Point Set Topology* (Pergamon; Macmillan), 179 p (MR37, 2561)

- 1966 MT 59, 36-41 [from 1964]

Massey William S 1965 Enc Br

McAllister B L 1966 AMM 73, 337-350 [bib, cyclic elems]

Nekrasov V L 1907 *Stroenie i mera lineinykh tochechnykh oblastei*, Tomsk, Ch. 1 and 3

Poutano M L 1957 Archim 9, 178-180 [Leibniz]

*Whyburn C T 1964 *Topological Analysis* Princeton, 2nd ed [bib]

*Wilder R L 1962 MT 55, 462-

## Transcendental numbers

See also e, Irrationals, Pi.

Dickstein S 1895 Kosmos 20, 359-365 (F26, 56)

Gelfond A O 1960 *Transcendental Algebraic Numbers* (Dover) (tr L F Boron) [bib]

Guggenheimer H 1965 Dialect 19, 136-148

Lang Serge 1966 *Introduction to Transcendental Numbers* (Addison-Wesley), 105 p [bib]

Lohnstein T 1892 Act M 16, 141-143

Popken J 1954 Bel SM, 71-82 (MR17, 117; Z66, 245)

Schneider T 1957 *Ein fuehrung in die Transzendenten Zahlen*, Berlin, 150 p (Fr tr 1959 Paris) [bib]

Torriere E 1914 Isis 2, 106-124 [Descartes, Leibniz]

Veblen O 1904 AMM 11, 219-223 [e, pi]

## Transfinite numbers

Jourdain P E B 1906-1910 Arc MP (3)10, 254-281; 14, 289-311; 16, 21-43; 22, 1-21 (F40, 60; 41, 51; 44, 49)

## Transfinite numbers (continued)

Keisler H J + 1964 Fund M 53, 225-308 [bib]
Petronievics B 1917 Rn 26, 309-316 [Fontenelle]
Schrecker P 1947 in *Studies and Essays in the History of Science Dedicated to George Sarton*, NY, 359-373 (MR8, 497)
*Sierpinski W 1965 *Cardinal and Ordinal numbers*, Warsaw [bib]
Tannery P 1894 Rv Met Mor 2, 465
Vivanti G 1892 Bib M, 9-25

## Transformations

De Vries H 1949 N Tijd 37, 21-28, 99-109 (MR11, 150)
Liebmann H + 1914 *Die Beruehrungs-transformationen. Geschichte und Invariantentheorie*, Leipzig
Rozenfeld B A 1960 Int Con 9, 583-584 (Z114, 7) [Euler]

## Triangle

See also Nine point circle, Pythagorean theorem, Wallace line.

Alasia C 1901 Pitagora 8, 43, 73 [termin]
- 1902 *Saggio terminologico-bibliografico sulla recente geometria del triangolo*, Bergamo, 49 p
Anon 1909 Mathesis (3)9, 153 [bib]
Birkenmajer L 1917 Krak BI (A), 298-299 [Heron, Hindu, rational]
Brocard H 1907 Fr AAS 35, 53-56 (F38, 67) [bib, 1895-1905]
- 1907 Mathesis 7 [bib 1895-1905]
Bruins E M 1953 Indagat 56, 412-422 (Z51, 242) [Babylonian]
Cajori F + 1922 AMM 29, 303-307 [area]
Cavallaro V G 1938 Bo Fir 34, xli-xliv [Italy]

Couturier C 1894 Intermed 1, 140-255
Datta B 1928 Clct MS 20, 267-294 [Mahavira, rat]
Delahaye G + 1894-1895 Intermed 1, 254; 2, 212 (Q120) [bib, sqs on sides]
Feuerbach Karl W 1908 *Eigenschaften einiger merkwurdigen Punkte des geradlinigen Dreiecks,* Haarlem
Galdeano Z G de 1891 Prog M 1, 223-228, 269-274, 317-318 (F23, 38)
Guggenbuhl L 1953 M Gaz 37, 241-243 [Brocard]
Hermelink H 1964 Sudhof Ar 48, 240-247 (MR30, 568)
Hofmann J E 1960 MZ 74, 105-118 (Z94, 3) [quadrisection]
Loria G 1895 Intermed 2, 188 [area]
- 1903 Bib M (3)4, 48-51 [quadrisection]
Mackay J S 1883 Edi MSP 1, 4 [circles]
- 1887 Edi MSP 5, 62-78 [circum and inscribed circles]
- 1890 Edi MSP 8, 93-94
- 1893 Edi MSP 11, 92-103 (F25, 82) [symmedian pt]
- 1893 Edi MSP 12, 86-105 [excircle, incircle]
*- 1894 Edi MSP 13, 37 [bisectors]
- 1894 Edi MSP 13, 166 [isogonals]
- 1895 Edi MSP 14, 37 [symmedian pt]
- 1902 Edi MSP 20, 18 [isosceles]
Miller G A 1921 AMM 28, 256-258 [area]
Neuberg J 1907 Mathesis 7, 184-185 [pseudo-isosceles]
- 1925 *Bibliographie des triangles spéciaux*, Bruxelles, 54 p (F51, 37)
Palamà G 1948 It UM (3)3, 49-66 (Z31, 97)
Stipanić E 1952 Beo DMF 4(3/4), 65-66 (Z48, 242)
Struyk A 1951 MT 44, 498-500 [nedian]
- 1954 MT 47, 116-118 [quasi-right]
Vigarie E 1887 Fr AAS, 87-112 [survey]

## Trigonometry (continued)

*vershiedenen Grundlegungen in der Trigonometrie,* Leipzig, 8 p (F28, 45)

Hayashi T 1926 Toh MJ 26, 408-419 (F56, 809) [Japan, tables]

Ibadov R I 1968 Uzb FMN 12(2), 23-26 (MR37, 2564) [tables]

Ideler C L 1812 Zach M C 26, 3-

Karpinski L C 1945 Scr 11, 268-272 (MR8, 2; Z60, 7) [appl]

- 1946 Scr 12, 267-283 (MR8, 497; Z60, 7) [*bib to 1700]

Khmelevskii B 1963 Lit M Sb 2(2), 319-341 (Z117;2, 249) [Lithuania]

Krishnaswami Ayyangar A A 1923 Indn MSJ 15, 121-126 [Hindu table]

Leon y Ortiz E 1913 Esp SM 2, 309-329; 3, 1-16, 65-73 [analogies]

Loria G 1912 Mathes It 4, 13-28 (F43, 71) [Greek]

Marian V 1937 Gaz M 42, 617-620 (F63, 15) [XVIII, Gooden]

- 1940 Gaz M 45, 561-564 (Z23, 194) [XVIII, Binder]

Miller G A 1928 Sci 67, 555 [Archimedes]

Naraharayya S N 1923 Indn MSJ 15(2), 105-113 [Hindu table]

Paplauskas A B 1961 Ist M Isl 14, 181-210 [uniqueness]

Pesci G 1912 Pe MIS 15, 43-45 [sine, termin]

Playfair J 1798 Edi ST 4 [Hindu]

Reves G E 1953 SSM 53, 139-145

Richeson A W 1952 Scr 18, 94 [re Karpinski 1946]

Rome A 1932 Brx SS 52A, 271-274 (F58, 987) [Greek]

- 1933 Antiq Cl 2, 177-192 (Z6, 337) [Greek]

Schoy C 1923 Isis 5, 364-399 [Arab]

- 1926 AMM 33, 95-96 [approx, al-Biruni]

Simon M 1911 Arc MP (3)18, 202 (F42, 63) [Hindu]

Singh A N 1939 Benar MS 1, 77-92 (MR1, 289) [Hindu]

Somayajulu D A 1934 M Stu 2, 12-21 (Z9, 98) [Hindu]

Suter H 1893 Bib M (2)7, 1-8 (F25, 82)

- 1909 Bib M (3)10, 156-160 (F40, 64) [Arab]

- 1914 Dan H Sk (7)3, 1-255 [Arab, Al-Khwarizmi, table]

Tannery P 1902 Intermed 9, 76, 278 (Q2072) [Werner]

Thureau-Dangin 1928 Rv Assyr 23, 187-188 [anc]

Tropfke J 1923 *Geschichte der Elementar-Mathematik in systematis-cher Darstellung...,* vol 5

- 1928 Arc GMNT (2)10, 432-461 (F54, 15) [Archimedes]

Van Veen S C 1936 Zutphen (A)4, 184-190 (F62, 1030) [XVII]

Vetter Q 1925 Cas MF 54, 281-283 [Egypt]

Wieleitner Heinrich 1927 *Geometrie und Trigonometrie,* Berlin, 76 p [source book]

Woepcke F 1854 JMPA 19, 153-, 302- [Arab]

- 1854 Nou An 13, 386 [Hindu]

Wolfe Harold E 1945 *Introduction to Non-Euclidean Geometry* (Dryden), Ch. 5

Zeller Mary C 1944 *The Development of Trigonometry from Regiomontanus to Pitiscus* (U Michigan thesis), 125 p (MR4, 2; Z60, 7)

Zeuthen H G 1900 Bib M (3)1, 20 [anc]

## Trisection

See also Constructions, Quadratrix.

Aubry A 1896 JM Sp 20, 76-106

Beumer M G 1946 N Tijd 33, 281-287 (Z60, 5) [Archimedes]

- 1948 Euc Gron 23, 230-236 (Z30, 337)

Bortolotti E 1922 Bln Rn 27, 125-139 [Bombelli]

Breidenbach W 1933 *Die Dreiteilung des Winkels* Leipzig, 42 p

Brocard H 1898-1899 Intermed 5, 263; 6, 68 (Q1298) [Voltaire]

## Trisection    (continued)

Carrara B 1904 *I tre problemi classici degli antichi, problema terzo: trisezione dell'angolo,* Paris

Cavallaro V G 1952 Archim 4, 259-261  (Z47, 244)

Conte L 1953 Archim 5, 77-80  (Z50, 242)  [Bernoulli, Comiers, Huygens, Kinner]

Curtze M 1874 ZMN Unt 5, 226-227  (F6, 41)  [conchoid]

Daniells M E 1941 MT 33, 80-81  [Amadori]

Dickson L E 1895 AMM 2, 71  [mistakes]

- 1914 AMM 21, 259-262

Escott E B 1901 Intermed 8, 304  (Q2168)  [bib]

Gardner M 1966 Sci Am (June), 116-

Goering W 1899 *Die Auffindung der rein geometrischen...die Teilung jedes beliebigen Winkels...,* Dresden, 13 p

Gruber M A 1895 AMM 2, 112

Hake P 1953 MT 46, 524

Hippauf K 1872 ZMN Unt 215-240  (F4, 28)  [conchoid]

Hunrath K 1906 Bib M (3)7, 120-125  [Duercr]

Kohl K 1925 Erlang S1, 54-55, 180-189 (F51, 10)

Loria G 1925 Mathesis 39, 348-350

Meserve B E 1951 MT 44, 547-550

*Montucla 1754 *Histoire des recherches sur la quadrature du cercle, avec une addition concernant les problèmes de la duplication du cube et de la trisection de l'angle,* Paris  (2nd ed 1831)

Nagy G 1953 M Lap 4, 84-86

Richardson E T 1953 MT 46, 344

Robusto C C 1959 MT 52, 358-360

Shields J 1953 MT 46, 344

Silvertzen O 1929 Nor M Tid 11, 18-19  [approx, Egypt]

Stamatis E 1949 *To Delion problema kai e trichotomesis gonias,* Athens, 32 p  (See Isis 53, 583)

Tandberg J 1935 Tid El M, 85-94  (F61, 941)

- 1936 Tid El M, 145-159 (F61, 941)

Todd W S 1950 MT 43, 278-279

Valentin G 1893 Bib M (2)7, 113-114  (F25, 80)

Weaver J H 1915 SSM 15, 590-595

Woelffing E 1900 Wurt MN (2)2, 21-27

Yates R C 1940 M Mag 15, 129-142, 191-202, 278-293  [bib]

- 1941 M Mag 16, 20-28, 171-182  [bib]

## Trochoid

Hoza F 1872 Cas MF 1, 54-60  (F4, 28)

## Universal algebra

*Birkhoff G 1946 Can Cong 1, 310-326  [bib]

Cohn P M 1965 *Universal Algebra,* NY, 333 p  [bib]

Graetzer George 1968 *Universal Algebra* (Van Nostrand), 268 p  [bib]

*Kurosh A G 1965 *Lectures in General Algebra* (Pergamon), 364 p  [bib]

*Neumann B H 1962 *Special Topics in Algebra: Universal Algebra,* NY, 78 p  [survey]

## Vectors

See also Linear algebra, Mechanics (Ch. 3), Quaternions.

Aubry A 1927 Mathesis 41, 251-261  (F53, 2)  [mech]

- 1928 Ens M 27, 106-123  (F54, 8)  [mech, Stevin]

*Beman W W 1897 AAAS 46, 33-50

Bork A M 1967 Isis 58, 210-  [elec, Maxwell, pot]

Bouny F 1954 Bel Oc 28(6), 52-73  (MR16, 434; Z60, 11)  [Massau]

Cassina V 1929 Linc Rn (6)9, 962-969  (F55, 5)

*Crowe Michael J 1967 *A History of*

Vectors   (continued)

  *Vector Analysis* (U Notre Dame P),
  270 p  [bib]
Duhem P 1903 Bib M (3)4, 338-43
  [composition, Leonardo da Vinci,
  mech]
- 1905-1906 *Les origines de la
  statique,* Paris, 2 vols (esp 2,
  245-265, 347-348)
Freudenthal H 1954 in *Homenaje a
  Millas-Villicrosa,* Barcelona 1,
  612-621  [Leibniz]
*Kramar F D 1963 Ist M Isl 15, 225-
  290  [XIX]
- 1967 Vop IET 21, 103-108
  (MR37, 14)  [mech, Wallis]
Lotze A + 1914-1931 Enc MW 3(1)
  1425-1595
Mikami Y 1914 Nieu Arch (2)11(1),
  76-78  [mech]
Moon P + 1965 *Vectors* (Van Nostrand)
  [bib]
Pawlikowski G J 1967 MT 60, 393
Rothe H 1916 Enc MW 3(1), 1277-1423
Siacci F 1899 Nap FM Ri (3)5, 69
  [mech]
- 1899 Nap FM Ri (3)5, 147 [mech]
Wilson E B 1931 Sci Mo 32, 211-227
  [Gibbs]


Volume

See also Mensuration, Oriented
  volumes, Pyramid.

Bellachi G 1893 Pe MI 8, 25, 57, 113,
  137 [polyh]
Bortolotti E 1923 Pe M 3, 317-319,
  429-430 [Torricelli]
Endo T 1895 Tok M Ph 7, 123
  [Japan, sphere]
Gerstinger H + 1932 Wien Papy (NS)1,
  11-76 (F58, 12) [Greek]
Hadamard J 1896 Bord PV, 25
Junge G 1926 Unt M 32, 240-244
  (F52, 7) [Egypt]
Mueller C 1940 Deu M 5, 244-255
  (MR2, 114) [Aryabhatta, cone]
Tenca L 1953 It UM (3)8, 337-342
  (Z51, 3)

Thorndike L 1949 Isis 40, 106-107
  (Z38, 2)
- 1957 Isis 48, 458  [termin]
Thureau-Dangin F 1935 Rv Assyr 32,
  1-28  [Babylonian]
Tropfke Johann 1924 *Geschichte ler
  Elementar-Mathematik* 7, Berlin, 133
Weaver J H 1916 AMSB 23, 127-135
  [Pappus thm]
Wieleitner H 1928 Regensb 19, 279-313
  [Kepler]
- 1930 Unt M 36, 176-185  (F56, 17)
  [Kepler]


Wallace line and point

See also triangle.

Archibald R 1910 Edi MSP 28, 64, 179
Candido 1899 Pe MIS 2, 85-86
  (F30, 51)
MacKay J S 1891 Edi MSP 9, 83-91
  (F23, 38)
- 1905 Edi MSP 23, 80-88
  (F36, 66)  [hypocycloid]


Waring problem

Archibald R C 1940 Scr 7, 33-48
  (Z61, 5)
Batchelder P M 1936 AMM 43, 21-27
  [1770]
Hardy G H 1920 *Some Famous Problems
  of the Theory of Numbers and in
  Particular Warings Problem,*
  Oxford, 35 p
Hardy G H + 1954 *An Introduction to
  the Theory of Numbers,* 3rd ed
  Oxford, 335-  [bib, survey 1914-
  1954]
Saalschutz L 1907 Arc MP 12, 199-207
  [Girard]


Weierstrass approximation theorem

Franklin P 1925 MITM Cont (2)(94)
Stone M H 1948 M Mag 21, 167-184
  [genzn]

## Wilson theorem

Cantor M 1902 Bib M (3)3, 412
(F33, 19)
Dickson Leonard E 1919 Dickson 1,
Ch. 3

## Wolstenholme theorem

Rao N R 1938 Clct MS 29, 167-170

## Wren's problem

Hall A 1965 Lon RSP 20, 140-144

## Zero

Boncompagni B 1884 Boncomp 16, 673-
686
Boyer C B 1943 AMM 50, 487-491
(MR5, 57) [div]
- 1944 M Mag 18, 323-330
(MR5, 253)
Bryan G H 1923 M Gaz 11, 334-336
(F49, 24) [symbol]
Cajori F 1903 AMM 10, 35
[China, symbol]
- 1929 AMSB 35, 597 (F55, 26)
[div]
Cazanacli P V 1934 Tim Rev M 14,
9-10 (F61, 957)
Cipolla M 1937 Esercit (2)10, 1-10
(Z16, 197)
Coomaraswamy A K 1934 Lon SOAS 7, 87-
97 [space]
Datta B 1926 AMM 33, 449-454
(F52, 10) [Hindu]
- 1927 Clct MS 18, 165-176
[Hindu]
*- 1931 AMM 38, 566-572 (Z4, 4)
[Hindu]
Datta B + 1935-1938 *History of Hindu
Mathematics,* Lahore
Halsted G B 1903 AMM 10, 89-90
[symbol]
Hunrath K 1887 Bib M (2)1, 120 (Q18)
[Ramus, symbol]
Jones P S 1957 MT 50, 162-165
[Babyl]

Neugebauer O 1941 Am OSJ 61, 213-215
(MR3, 97; Z61, 1) [Babyl]
Romig H G 1924 AMM 31, 387-389
[div]
- 1924 AMSB, 15-16 (F50, 19)
[div]
Ruska J 1929 Arc GMNT 11, 256-264
[Arab]
Stammler G 1928 An Phi 75, 146-164
Vygodskii M Ya 1959 Ist M Isl 12,
393-420 (MR24, 2) [Babyl, symbol]
Wieleitner H 1917 ZMN Unt 48, 213-215
(F46, 46) [Maya]

## Zeta function

Bohr N + 1913 M Ann 74, 3-30
Cahen Eugene 1894 *Sur la fonction
(zeta) de Riemann et sur des
fonctions analogues,* Paris, 93 p
Haselgrove C 1960 *Tables of the
Riemann Zeta Function,*(Cambridge
U P), 80 p [bib]
Landau E G H 1909 *Handbuch der
Lehre von der Verteilung der
Primzahlen,* Leipzig, 2 vols.
[*bib]
*Titchmarsh E C 1951 *The Zeta Function
of Riemann* (Clarendon), 346 p
[*bib]

## Zorn lemma

Cuesta N 1955 Gac M (1)7, 174-176
(MR17, 931)

# EPIMATHEMATICAL TOPICS

In this chapter are grouped titles referring to various external aspects of mathematics and related fields. These include fields of interaction and application, mathematical technology (including hardware and also software such as numeration systems), and various metamathematical topics such as heuristic and the philosophy, psychology, and sociology of mathematics. In every case it is to be understood that the key word refers to the subject in relation to mathematics. For example, under "education," there appear only titles on the history of mathematical education, and under "physics" only titles referring to the history of physics in relation to mathematics. Because of the vague criteria for deciding when a publication on the history of another science is relevant to mathematics, we have avoided the problem by including only those cited in the mathematical literature. Undoubtedly our coverage of the interplay of mathematics and other fields is weak. Systematic coverage of epimathematical topics must await comprehensive indexing of the entire scientific literature.

## Abacus

See also Arithmetic   (Ch.2).

Anon 1920 AMM 27, 180   [China]
Barnard F P 1916 *Casting-Counter and Counting Board,* Oxford, 357 p (F46, 56)
Bourget H + 1898 Intermed 5, 129 [origin]
Bubnov N M 1911 *Die echte Schrift Gerberts ueber dem Abakus oder das System der elementaren Arithmetik des klassischen Altertums,* Kiev, 517 p   (F42, 53)
Chasles M 1843 Par CR 16, 156-, 218-, 281-, 1393-; 17, 143- [Gerbert]
- 1867 Par CR 64, 1059-
Chin-Te Cheng D 1925 AMM 32, 492-499 (F51, 6)   [China, rods]
Curtze M 1898 ZM Ph 43, 122-130 (F29, 28)   [XII, XIII]
Datta B 1928 AMM 35, 520-529 [Hindu]
Faddegon J M 1932 J Asi 220, 139-148 [Arab]
Flewelling R W 1961 Ari T 7, 104-106
Gandz S 1921 AMM 34, 308-316
- 1927 AMM 34, 308-316   [Arab]
- 1930 Isis 14, 189-214 [Hebrew, knot, numer]
- 1931 Isis 16, 393-424   [Arab]
Grosse H 1901 *Historische Rechen-buecher des 16 und 17 Jahrhunderts,* Leipzig   (repr 1965)
Haskins C H 1912 Engl HR, 101-106 [England]
Hitomi C 1926 Tok M Ph 35, 55-58 [Japan]
*Howe H 1964 Colliers   [current use]
Iyer R V 1954 Scr 20, 58-63 (MR15, 923)   [Hindu]
*Knott C G 1886 Jp Asi So 14, 18- [Japan]
Kojima T 1956 *The Japanese Abacus. Its Use and Theory,* Tokyo, 102 p
Li S T 1959 ACMJ 6, 102-110 (MR20, 1137)   [China]
Locke L L 1932 Scr 1, 37-43 [Peru]

Marducci E 1882 Linc At 6, 324- [XII]
Mikami Y 1911 DMV 20, 381-393 [China, Japan]
Nagl A 1899 ZM Ph 44(S), 335-357 (F30, 37)   [Greek]
Narducci E 1885 Linc Rn 1, 563-566 [XII, Italy]
*Pullman J M 1968 *The History of the Abacus* NY (Praeger), 127 p   [bib]
Scesney F C 1944 *The Chinese Abacus,* Buffalo, 68 p
Schaaf W L 1955 MT 48, 417, 440, 567-568
Simons L G 1936 Scr 4, 94
Smith D E 1921 *Computing Jetons,* NY, 70 p   (F48, 45)
*- 1958 *History of Mathematics* (Dover) 2, Ch. 3, Pt 1   [bib]
Soreau R 1918 Par CR 166, 67-69 (F46, 50)   [word]
Spasskii I G 1952 Ist M Isl 5, 267-420 (MR16, 433)   [Russ]
Treutlein P 1877 Boncomp 10, 589-595   (F9, 17)
Venkatachacam I R 1951 M Stu 18, 79-82 (Z43, 1)   [Hindu, Patiganita]
Yeldham F A 1927 Arc Sto 8, 318-329 (F53, 13)   [XVII, England]
Yi-Yun Y 1951 MT 43, 402-404 [China]

## Accounting

See also Business, Economics, Finance.

Bentley H C + 1934 *Bibliography of Works on Accounting by American Authors,* Boston, 219 p [1796-1900]
Brown R 1905 *A History of Accounting and Accountants,* Edinburgh
DeRoover R 1944 Acc Rv 19, 381-407
Hatfield H R 1966 J Ind Eng 17(6)
Murray David 1930 *Chapters in the History of Book Keeping, Accountancy and Commercial Arithmetic,* Glasgow, 525 p

## Acoustics

Cherbuliez V 1871 Bern NG, 1-28
(F3, 12) [velocity of sound]
Loudon J 1901 Can Pt (2)7, 43-54
[Canada, XIX]
Mach E 1892 Prag DMG, 12-18
(F24, 53)
Miridonova O P 1964 Vop IET 16,
85-90 [Galileo, Mersenne]
Robel E 1891-1900 *Die Sirenen.
Ein Beitrag zur Entwickelungs-
geschichte der Akustik,* Berlin,
4 vol (F23, 41; 26, 66)
(see also ZM Ph 40, 61)
Tyndall John 1903 *Sound,* NY
(repr *Science of Sound* 1964
Citadel, 480 p)
Whewell W 1857 *History of the
Inductive Sciences,* London,
Book VIII

## Aerodynamics

See also Hydrodynamics.

Karman T v 1954 *Aerodynamics.
Selected Topics in the Light of
Their Historical Development,*
Ithaca NY, 203 p
Nemenyi P 1933 Naturw 21, 708-709
[Goettingen]
Wilson E B 1918 AMM 25, 292

## d'Alembert principle

Krbek F v 1953 Greifsw 2, 15-22
(MR14, 832; Z52, 2)

## Amateurs

Brousseau A 1961 M Mag, 311-315
Coolidge J L 1949 *The Mathematics of
Great Amateurs,* Oxford, 219 p

## Analysis and synthesis

Bopp 1907 Ab GM 20, 87-314 [XVII]
Crescini Angelo 1965 *Le origini del
metodo analitico,* Udine, 399 p
Duhamel J M C 1885 *Des méthodes dans
les sciences de raisonnement,*
3rd ed, Paris, Pt 1, 39-68
Hofmann J E 1963 Mathunt, 5-37
Mansion P 1902 Mathesis (2)2, 266-273
(F33, 53)
Matthes C J 1832-1833 Leyd ALB
Robinson R 1936 Mind 45, 464-473
(Z15, 52) [Greek geom]
Tannery P 1904 Int Con H 12, 219-229
(F35, 58)
Vetter Q 1923 Cesk Mysl 19, 147-158

## Anthropology

See also in Ch. 4 General: Theory.

Barnes H E 1965 *An Intellectual and
Cultural History of the Western
World,* 3rd ed (Dover)
Guenther S 1917 Mun Si, 111-125
(F46, 42)
White L A 1947 Phi Sc 14, 289-303
[antho approach to math]
Wilder R L 1950 Int Con 258-271
[cultural basis of math]
- 1953 AMSB 59, 423-448
- 1960 in *Essays in the Science of
Culture,* ed by G. E. Dole and
R. L. Carneiro, NY
- 1968 *Evolution of Mathematical
Concepts-An Elementary Study*
(Wiley), 237 [anthropological
approach]

## Applications

See also specific fields of
application.

Bell E T 1937 *The Handmaiden of the
Sciences,* Baltimore, 224. p
Bense Max 1946 *Konturen einer
Geistgeschichte der Mathematik,*

Applications   (continued)

- *Vol. 1, Die Mathematik und die
  Wissenschaft,*   Hamburg, 144 p
Bliss G A 1933 AMM 40, 472-
Bochner S 1963 Sci 141, 408-411
  [crises]
Bosmans H 1903 Rv Q Sc (3)3, 318-343
  (F34, 4) [sci]
Boyer C B 1959 Sci 130, 22-25
  ["inutility"]
Brown E 1917 AMSB (2)23, 213-230
  [nat sci]
Fort T 1940 AMM 47, 605 [sci]
*Frechet M 1955 *Les Mathématiques
  et la Concret,*   Paris, 438 p
*Freudenthal H 1968 Ed  St 1, 1-246
  [educ]
Fry T C 1964 Sci 143: 934-938
  [industry]
Giannelli B 1950 Archim 2, 16-20
  (Z36, 3)
Kaehler E 1941 DMV 51(2), 52-63
  (Z26, 98) [ast, phys]
Karpinski L C 1929 SSM 29, 126-132
  [sci]
Kline Morris 1959 *Mathematics and
  the Physical World,*   NY, 491 p
Piazzolla-Beloch M 1930 *La
  matematica in relazione alle sue
  applicazione e al suo valore
  educativo,* Ferrara 22 p  (F56, 2)
Rautenstrauch W 1937 Scr Lib 3
  [social devel]
Schaaf W L 1955 MT 48, 166-168
- 1956 MT 49, 475-476 [1700]
Steinhaus H 1956 Pol Ac Rv 1(4)
  (MR19, 108) [Poland]
Tannery P 1885 BSM (2)9, 311-324
  [anc geom]
Van Der Waerden B L 1955 Ens M 1, 44-
  55 (Z66, 3) [anc]
Walker J J 1890 LMSP 22, 4-18
Waschow H 1932 Arc Or 8, 127-131
  [Babyl]
Wiener N 1951 in *Structure, Method
  and Meaning, Essays in the honor
  of Henry M. Sheffer,* NY, 91-98
*Wigner E P 1960 CPAM 13, 1-14
 Woodward R S 1899 AMSB (2)6, 133-163
  [XIX]

517 / Epimathematical Topics

- 1900 Sci (NS)11, 41-51, 81-92
  [XIX, bib]


Architecture

See also Art, Engineering.

Dehio G 1895 *Ein Proportionsgesetz
  der antiken Baukunst und sein
  Nachleben im Mittelalter und in
  der Renaissance,*   Strassburg
Lockyer J N 1893 Nat 48, 55-58
  (F25, 91) [Egypt]
- 1893 Nat 48, 417-419  (F25, 91)
  [Egypt, Greek]
*Scholfield P H 1958 *The Theory of
  Proportion in Archicture* (Cambridge
  U P), 155 p  [bib]
Smith D E 1925 *Mathematica Gothica,*
  Paris, 30 p [cathedrals]
Williamson R S 1942 Nat 150, 460-461
  (F68, 3) [Saggare graph]
Wycherley R E 1949 *How the Greeks
  Built Cities,* Lon, 248 p


Arithmetization

Bourbaki Nicolas 1960 *Eléments
  d'histoire des mathématiques,*
  Paris, 36-
Klein F 1895 Gott N(1), 82-91
  (tr 1896 AMSB (2)2, 241-  )
- 1896 ZMN Unt 27, 143
*Luzin N N 1934 Bolsh S E 22, 622-
  642
Medvedev F A 1965 *Razvitie teorii
  mnozhesto v XIX veke,* Mos,
  34-77
Miller G A 1925 Am NASP 11, 546-
  548  (F51, 11)
- 1925 Sci 62, 328
Pierpont J 1899 AMSB (2)5, 394
- 1899 Prace MF 10, 249
Prag A 1931 QSGM (B)1, 381-412
  [Wallis]

## Astrology

See also Mysticism.

Ball W W R 1920 *Mathematical Recreations and Essays,* 9th ed, London Ch. 19

Camden C 1933 Isis 19, 26-73 [XVI]

Graubard M 1958 Osiris 13, 211-257

Haebler A 1879 *Astrologie im Alterthum,* Zwickau (F11, 46)

Mensinga J A M 1871 *Ueber alte und neuere Astrologie,* Berlin (F4, 3)

Pannekoek A 1932 Himmelsw 42, 13-19 57-66 [ast]

Pingree D 1963 Isis 54, 229-246 (Z112, 241) [ast, Hindu, Persia]

## Astronomy

See also Astrology, Celestial mechanics, Cosmology, Epicycle, Geodesy, Geophysics.

Abetti Giorgio 1952 *The History of Astronomy,* NY, 350 p (MR14, 343: Z41, 1)

Berry Arthur 1961 *Short History of Astronomy* (Dover)

Bruns H 1914 DMV 23, 12-28 (F45, 89) [Ptolemy to Newton]

Datta B 1938 Archeion 21, 28-34 (F64, 11; Z19, 100) [numb thy]

Dittrich A 1939 Ber Ab (2) (F65, 6) [Maya]

Doig Peter 1951 *A Concise History of Astronomy,* NY, 331 p (MR13, 809)

Dreyer John L E 1953 *History of Astronomy from Thales to Kepler* (Dover)

Eves H 1961 MT 54, 625-626 [Gulliver]

Forni G 1923 Pe MI (4)3, 1-10 [recent]

Gandz S 1950 Arc In HS (NS)3, 835-855 (MR12, 311; Z39, 3) [Maimonides]

Gilbert P 1882 Rv Q Sc 11, 353-393 [earth rotation]

Hanson N R 1960 Isis 51, 150-158 (Z108:2, 245) [epicycles]

Heath T L 1937 Scr 5, 215-222 [Greek]

Hogrebe J 1934 ZMN Unt 65, 23-28 (F60, 9) [anc Nordic]

*Houzeau J C + 1964 *Bibliographie générale de l'astronomie jusqu'en 1880,* London, 3 vols, 3082 p (16,000 entries, new index and additions by Dewhirst)

Karpinski L C 1941 Scr 9, 139-154 [Copernican]

*Kirchberger P 1921 *Mathematische Streifzuege durch die Geschichte der Astronomie,* Leipzig, 58 p (F48, 32)

Kline M 1954 M Mag 27, 127-139 [XVI-XVIII, Copernicus, Kepler]

Lilley S 1949 Nat 163, 468-469 [Laplace]

Lockyer J Norman 1964 *Dawn of Astronomy* (MIT P)

Mansion P 1899 ZM Ph 44(S), 277 [anc, geom]

Martin T H 1871 Boncomp 4, 464-466 (F3, 3) [re Sédillot]

Morehouse D W 1931 Sci (2)75, 27-32 [influence on math]

Moutton F R 1911 Sci (NS)33, 357 - 364 [influence on math]

Neugebauer O E 1932 Naturw 20, 169-170 [Babyl]

- 1941 in A E Speiser + ed *Studies in the History of Science* (U Pennsylvania P) (MR2, 306) [anc]

- 1942 Am J Phil 58, 455-459 (MR4, 65) [Plutarch]

- 1942 APST (NS)32, 209-250 (MR3, 25) [Egypt]

- 1943 ASP 55, 145-146 (MR5, 57) [ms]

- 1945 Isis 36, 10-15 (MR7, 105) [Babyl]

- 1945 JNES 4, 1-38 (MR6, 141) [anc]

- 1946 ASP 58, 17-43, 104-142 (MR7, 353) [anc]

- 1947 J Cun St 1, 143-148 (MR9, 73) [Sumerian]

- 1948 AMSB 54, 1013-1041 (Z33, 341) [anc]

Astronomy (continued)

- 1948 APSP 92, 136-138
  [conics]
- 1954 APSP 98, 60-89 (MR15, 383)
  [Babyl]
- 1956 Scr 22, 165-192 [anc, mid]
- 1957 Dan M Med 31(4) (MR21, 749)
  [Babyl]
- 1960 APST 50(2) (MR23A, 269)
  [Byzantine]
- 1960 Dan H 39(1) (Z102, 244)
  [Greek]
- 1960 *Egyptian Astronomical*
  *Texts. I: The Early Decans*,
  London, 144 p (Z113, 1)
Pannekoek A 1941 Eudemus 1, 1-14
  (MR2, 114) [Babyl]
- 1951 Lon As Mo N 111, 347-356
  (MR13, 611) [anc]
Rufus W C 1944 Lon As QJ 38, 143-
  153 (MR5, 253) [Greek]
Sedillot L A 1872 Boncomp 5, 294-296
  (F4, 5) [re Martin]
- 1872 Boncomp 5, 306-317 (F4, 4)
  [anc]
Shankar-Shukla K 1945 Benar MS
  (NS)7(2), 9-28 (MR9, 169)
  [Hindu]
Startsev P A 1961 *Ocherki istorii*
  *astronomii v kitae*, Mos, 156 p
  (MR23A, 3)
Subbotin M F 1959 Vop IET 7, 58-66
  (MR22, 618) [Euler]
Tannery P 1893 Bord Mm (4)1
  (F25, 92) [anc]
*Tanzi A 1912 *Confronto materatico*
  *fra il sistema tolemaico e*
  *copernicano*, Larino, 21 p
  (F43, 82)
Toomer G J 1963 Centau 9, 254-256
  (MR29, 219) [Tamil]
Turner Herbert H 1904 *Discoveries in*
  *Celestial Mechanics, Astronomical*
  *Discovery*, London, 236 p (repr
  1963 U Calif P)
van Der Waerden B L 1941 Eudemus 1,
  23-48 (MR2, 306) [Babyl]
- 1943 Leip Ber 95(1), 23-56
  (MR12, 577; F68, 8) [anc]
- 1944 Leip Ber 96, 47-56
  (MR8, 189) [Heraclides]

- 1949 JNES 8, 6-26 (MR10, 173)
  [Babyl]
- 1952 Mun Si, 219-232 (MR14, 831;
  Z51, 2) [Greek, Hindu]
- 1954 Mun Si, 159-168 (MR16, 985)
  [Byzantine]
- 1961 Arc HES 1, 107-121
  (MR27, 3) [Hindu]
- 1965 *Die Anfaenge der Astronomie.*
  *Erwachende Wissenschaft II*,
  Groningen, 327 p (MR33, 1199)
Vorontsov-Belyaminov B A 1960
  *Ocherki istorii astronomii v*
  *SSSR*, Mos, 227
Wolff R 1890 *Handbuch der Astronomie,*
  *ihrer Geschichte und literature,*
  Zurich, 2 vols (repr 1967)
Woolard E W 1942 Wash Ac 32, 189-216
  (MR3, 258) [bib]
Zelbr K 1896 ZM Ph 41, 121, 153
  [shortest twilight, bib]

Automata

See also Computer.

Anon 1962 *Simulation of Thought*
  *Processes by Computers*,
  Arlington, Va, 36 p [bib]
*Feiglebaum Edward + 1963 *Computers*
  *and Thought* (McGraw-Hill)
  [anthol]
Minsky M 1961 IRE HFET 2(Mar), 39-55
  [bib]
- 1966 Sci 156 (Sept)
*Moore E F 1964 *Sequential Machines:*
  *Selected Papers*, Reading, Mass
  [*bib]
Shannon C E 1958 AMSB 64, 123-129
  (MR19, 1084) [Von Neumann]
Simmons P L + 1962 IREECT 10, 462;
  11, 535 [bib]
*Von Neumann J 1958 *The Computer and*
  *the Brain* (Yale)

Axioms

See also Foundations, Philosophy.

Baumgart J K 1961 MT 54, 155-160 [alg]

Axioms (continued)

Chatelet A ed 1959 *La méthode
  axiomatique dans les mechaniques
  classiques et nouvelles,* Paris
  (Gauthier-Villars)
Dingler H 1931 Archeion 13, 1-10
  (F57, 12) [Greek]
Fritz K v 1955 Arc Begrf [Greek]
Gould S H 1962 M Gaz 46, 269-290
  [Euclid]
*Henkin L 1967 *Formal systems, and
  models of formal systems,* Enc Phi
  [bib]
Kalmar L 1953 Magy MF  (MR15, 383)
  [non-Euc geom]
Lacombe D 1949 Thales 6, 37-58
  (MR14, 1049) [Greek]
Miller G A 1925 Ens M, 24, 59-69
  (F54, 48)
- 1925 Int Con 7(2), 959-967
  (F54, 48)
Moore E H 1909 AMSB 16, 41
Scholz H 1930 Bla D Phi 4, 259-278
  [anc]
Szabo A 1956 Hun Antiq 4, 109-152
  (MR21, 230; Z74, 2) [Greek]
- 1957 M Lap 8, 8-36, 232-247
  (MR21, 230) [Greek]
- 1959 Ist M Isl 12, 321-392
  (MR23A, 692)
- 1959 M Lapok 10, 72-121
  (MR22, 618) [Greek]
- 1960 Arc HES 1, 37-106  (Z93, 2)
- 1962 Osiris 14, 308-369
  (MR30, 870) [Greek]
- 1964 Scr 27, ]7-48, 113-139
  (MR20, 218, 1068) [Greek]
Yanovskaya S A 1958 Ist M Isl 11, 63-
  96
Wang Hao 1957 JSL 22(2) [arith]


Ballistics

See also Mechanics, Military.

*Burington R S 1943 AMM 50, 404
  [bib]
 Charbonnier P 1928 *Essais sur
  l'histoire de la balistique,*
  Paris

Dugas R 1953 Rv Sc 91, 83-89
  (Z53, 246) [XVII]
Hall A R 1952 *Ballistics in the
  Seventeenth Century. A Study in
  the Relation of Science and War with
  Reference Principally to England,*
  Cambridge, 185 p
Itakura K 1963 Jp Stu HS (2), 136-145
  [Japan]
Mandryka A P 1954 Mos IIET 1, 146-192
  (Z59, 11; 60, 11) [Maievskii]
- 1958 *Ballistic investigations of
  Leonard Euler,* Moscow-Leningrad,
  185 p (MR21, 1169; Z95, 3)
  [Euler]
- 1960 Mos IIET 34, 241-263
  (MR33, 1208) [W L Kraft]
*Ocagne M d' 1928 Rv Q Sc (4)29, 239-
  253  (F55, 5)
Rothe R 1938 Tec Gsch 27, 41-47
  (F64, 920) [XX, Cranz]
Williams K P n.d. Field Art 19, 128-
  135  [XVI]


Billion

Miller G A 1931 Sch Soc 33, 591
  [word]


Biology

See also Genetics, Medicine.

Carruccio E 1936 Pe M (4)16, 37-54
  (F62, 4)
Mendenhall W 1958 Biomtka 45, 521-
  543 [life testing etc]
Ostwald W 1911 Rv Mois 12, 257-272
  [Arrhenius, Van't Hoff]
Pearson E S 1965 Biomtka 52, 3-18
  [biometry]
Williams H B 1927 AMSB 33, 273-293


Business

See also Accounting, Economics,
  Finance, Insurance, Operations
  research.

Business    (continued)

Bjornbo A A 1912 Bib M 12, 97-132
  193-224   [Florence, mid]
Davis N Z 1960 J H Ideas 21, 18-48
  [XVI, France]
Gardia de Zuniga Eduardo 1941
  *Origines del calculo mercantil,*
  Montevideo, 7 p
Kaefer K 1941 Zur U Hand 68
  (MR9, 484)
Rigobon P 1902 *Studi antichi e
  moderni intorno alla technica dei
  commerci,* Bari, 38 p  (F33, 53)
Shirk J A G 1940 MT 32, 203-208
Smith D E 1917 AMM 14, 257-265
  [econ]
- 1926 Isis 8, 41-49   [first
  commercial arith]


Calculating machines

See also Abacus, Computers,
  Harmonic Analysis, Instruments,
  Machines, Slide rule.

Anon 1931 Lar Men 8, 715-717
  [Pascal-Hollerith]
- 1942 Nat 150, 427   [Pascal]
- 1942 Sci L Bib S(582)
  [bib 1937-1942]
- 1961 Ist M Isl 14, 551-586
  [mss]
Brauner L 1926 Bei GTI 16, 248-260
Chapman S 1942 Nat 150, 427, 508-
  509 [Pascal]
Couffignal Louis 1933 *Les machines
  à calculer, leurs principes, leur
  évolution,* Paris, 88 p
Favier Jean + 1952 *La Méconographie,*
  Chapelle-Montligeon, France,
  267 p
Feldhaus F M 1918 Ges Bl TI 5, 149-
  151 [Pascal]
Fettweis E 1936 Schw Sch 22, 757-759
  (F62, 1009)
- 1936 *Urgeschichte und Rechenmaschine,* Rhein-Ruhr, 259-260
  (F62, 1010)
- 1939 Scientia 65, 273-281 (F65, 4)
  [cone]

Flad Jean-Paul 1958 Chiffres 1, 143-
  148  (Z82, 10)  [Schickard]
- 1963 *Les trois premières machines à
  calculer Schickard (1623)- Pascal
  (1642)  Leibniz (1673),* Paris, 27 p
  (Z115, 242)
Gradstein S 1963 Philip TR 24 (4/5),
  101-170  [Pascal]
Haga E J 1962 SSM 62, 197-205
Hartree Douglas R 1953 *Calculating
  Instruments and Machines* (U I11 P),
  138 p
Kiro S N 1957 Vop IET (4), 169-171
  [eqs]
Lilley S 1942 Nat 149, 462-465
Locke L L 1924 AMM 31, 422-429 [US]
- 1926 Type Top  [first mult]
- 1932 Scr 1, 147-152
- 1932 Scr 1, 315-321   [Leibniz]
Loria G 1920 Scientia 28, 77-93,
  "35-53"
Maistrov L E 1961 Ist M Isl 14,
  349-354  (Z118, 11)  [arithmometer,
  Chebyshev]
Mehmke R 1894 DMV 3, 59-62
  (F25, 96)
- 1895 Prace MF 6, 177-182 (F26, 68)
*- 1901 Enc MW 1(6)   [bib]
Michel F + 1898-1905 Intermed 5, 126,
  240; 6, 18-19, 252; 7, 133; 12, 14,
  152  (Q1299)
Michel H 1947 J Suis Hor (July-Aug),
  307-316
Ocagne, M d' 1893 Brx SS 17(A),
  84-86  (F25, 1913)
- 1930 Par CR 190, 1163-1164
  (F56, 19)  [Pascal]
- 1931 Brx SS (A)51, 23-27
  (F57, 1287)
Pearcey T 1965 Datamat 11(3), 37-38
Radovskii M I 1961 Ist M Isl 14, 551-
  586  (Z118, 7)
Reuleaux F 1892 *Die Sogenannte Thomas-
  sche Rechenmaschine. Fuer Mathe-
  matiker, Astronomen, Ingenieure,
  Finanzbeamte, Versicherungs-
  Gesellschaften und Zahlenrechner
  ueberhaupt,* 2nd ed, Leipzig, 67 p
  (F24, 58)
Runge C 1905 Int Con 3, 737-738
  (F36, 54)  [Leibniz]

## Calculating machines   (continued)

Sadovskii L E 1950 UMN (NS)(2),
57-71  (MR12, 69)  [Russ]
Taton René 1949 *Le calcul mécanique,*
Paris, 126 p
- 1963 Rv Hi Sc Ap 16, 139-160
Trinks F 1927 Braun Mo 14, 249-275
(F54, 46)  [Pascal]
Turch J A V 1921 *Origin of the
Modern Calculating Machines,*
Chicago, 196 p
Unger F A 1899 ZM Ph 44(S), 515-535
(F30, 61)
Wigand K 1963 Praxis 5, 44


## Calendar

See also Chronometry, Finger
reckoning.

Anon 1888 Vat NL Mm 1, 1-68
[300 anniv Gregorian calendar]
Barberena S H 1890 *Tratado elemental
del calendario musulman,* San
Salvador, 103 p  [Arab]
Bertrand J 1884 BSM (2)8, 8-19
[Egypt]
Bowman M E 1950 *Romance in Arithmetic.
A History of our Currency, Weights
and Measures and Calendar,*
(U London P)
Boyer J + 1899 Intermed 6, 201
(Q914)  [Baradelle, Macquart,
Meynier]
Bushell W F 1961 M Gaz 45, 117-124
[reform]
Cajori F 1929 Archeion 9, 31-42
[finger reckoning]
Dozy R ed 1873 *Le Calendrier de
Cordoue de Ponnée 961,* Leyden
(F5, 9)
Flemming Gustav 1869 *Ein Beitrag zur
Geschichte des Kalendars,* Altenburg
(F2, 5)
Geiger K 1935 Bay Bl GR 71, 133-137,
172-177 (F61, 952)  [Easter, Gauss]
Goldstein B R 1966 Isis 57(1), 116
[anc, metonic cycle]
Kennedy E S 1963 Fr SMB 27, 55-
[Hebrew, Al-Kwarismi]

Kinsella J + 1935 MT 27, 340-343
[Maya]
Kravitz S 1961 Recr M Mag (4), 22-25
(Z102, 246)  [mid]
Levy L + 1901 Intermed 8, 303; 11,
216 (Q2128)  [week]
Lindhagen A 1912 Ark MAF 7(23), 41 p
- 1914 Ark MAF 9 (36)   [mid]
- 1918 Ark Maf 13(1)  [Gmunden]
- 1918 Ark MAF 13(2), 21p
- 1922 Ark MAF 16(1)   [mid]
Lithberg Nils 1953 *Computus med
saerskild haensym till Runstaven
och den Bergenliga Kalendern,*
Stockholm, 326 p
Lockyer J N 1892 Nat 45, 487-490; 46,
104-107, 47, 32-35, 228-230
(F24, 56)  [year]
Loewy A 1917 DMV 26, 304-322
(F46, 12)  [Gauss, Hebrew]
Lynn W T 1897 Nat 56, 180-181
(F28, 52)
Mahler E 1892 Wien Si 101, 337-353,
1685-1693 (F24, 56)  [Babyl]
Maistrov L E + 1960 Ist As (6)279-
298  (Z115, 241)  [anc]
Matzka W 1881 Prag Ab (6)10(5)
(F13, 45)  [reform]
Melzi C M 1895 Vat NLA 48, 93-109
(F26, 67)
Orasanu C D 1925 Gaz M 30, 201-205,
324-327  (F51, 13)
Pogo A 1932 Isis 17, 6-24  [anc]
Poncet G 1887 Astmie 6, 378-382
[date of New Year]
Resnikoff L A 1943 Scr 9, 191-195
[Hebrew]
Sanford V 1952 MT 45, 198, 204
["computus"]
- 1952 MT 45, 336-369  [month]
- 1953 MT 46, 41
Schwarz A 1872 *Der juedische
Kalendar historisch und astro-
nomisch untersucht,*  Breslau
(F4, 3; 5, 9)
- 1874 Mo Ges W Ju 23, 375
(F5, 9)
Stael-Holstein A v 1935 Mon Ser 1,
277-314  [Tibet]
Tannery P 1902 Intermed 9, 67
- 1910 Intermed 17, 32

## Calendar (continued)

Vigarie E 1895 Intermed 2, 81
[Gauss]
Wilmart A 1934 Rv Bene 46, 41-69
[England]
Yeldham F 1928 Archeion 9, 324-326
[finger reckoning]

## Cartography

See also Geography.

Adams O S 1925 *Elliptic Functions Applied to Conformal World Maps,* Wash D C (US Dept Commerce 112)
[C S Peirce]
Carslaw H S 1924 M Gaz 12, 1-7
(F51, 14) [Mercator]
Chambertin William 1947 *The Round Earth in Flat Paper,* Wash DC (National Geographic Society)
Craig Thomas 1882 *A Treatise on Projections* (USCGS), 261 p
Delevsky J 1942 Isis 34, 110-117
(MR4, 65) [Mercator]
Diller A 1941 Isis 33. 4-7
(Z61, 2) [Ptolemy]
Eisele C 1963 APSP 107, 299-307
[C S Peirce]
Fiorini M 1889 It Cart An 3-4, 1-9
["curiosita"]
Frischauf J 1891 *Beitraege zur Geschichte und Construction der Karten-Projectionen,* Graz, 14 p
(F23, 40)
Hartnack W 1939 Mit R Land 15, 133-146, 207-222 (F65, 1089)
[Germany]
Hopfner F 1946 Wien Anz 83, 77-87
(Z61, 2) [Marinos, Ptolemy]
Lemaire G 1907 Intermed 14, 167
(Q3179) [journals]
Lemaire G + 1906 Intermed 13, 122ⱼ 219; 15, 50 (Q2974)
[first publication]
- 1907 Intermed 14, 36 [XIX, Italy]
Schaaf W 1953 MT 45, 440-443
[bib]
Wislicenus W F 1901 Bib M (3)2, 384
[Langrenus]

## Celestial mechanics

See also Astronomy.

Armellini G 1922 Pe MI (4)2, 464-474
[planets]
Eilander M 1949 Sim Stev 27, 16-51
(MR11, 571)
Herz Norbert 1887-1894 *Geschichte der Bahnbestimmung von Planeten und Kometen,* Leipzig, 2 vols (F25, 95)
Kriloff A N 1924 Lon As Mo N 85, 640-656 [comets, Newton]
Leite D 1915 Porto Ac 10, 65-73
[comets]
Ryabov Y 1959 *Elementary Survey of Celestial Mechanics,* Mos
(repr 1961 Dover)
*Van der Kamp P 1964 *Elements of Astromechanics* (Freeman)
Woolard E W 1940 M Mag 14, 179-189
(F66, 6) [planets]

## Chronometry

See also Calendar, Pendulum, Sun dials.

Baillie G H 1929 *Watches. Their History, Decoration and Mechanism,* London, 384 p
Ball W W R 1920 *Mathematical Recreations,* Ch. 22
Barneck A 1932 *Die Grundlagen unserer Zeitrechnung,* Leipzig und Berlin
Basserman-Jordan E V ed 1920-1925 *Die Geschichte der Zeitmessung und der Uhren,* Berlin-Leipzig
Berthoud F 1802 *Histoire de la mesure du temps par les horlages,* Paris, 2 vols
Dittrich A 1938 Scientia 63, 211-218
(F64, 12) [Maya]
Drecker J 1925 *Zeitmessungen und Sterndeutung in Geschichtlicken Darstellung,* Berlin, 188 p
(F51, 13)
Gandz S 1952 Osiris 10, 10-34
[Hebrew]

Chronometry  (continued)

Gerland E 1904 Bib M (3)5, 234-247
[pendulum]
Ginzel Friedrich K 1906-1914
*Handbuch der mathematischen und
technischen Chronologie. Das
Zeitrechnungswesen der Volker,*
Leipzig, 3 vols  (F43, 71)
Goldstein B R 1962 Scr 27, 61-6
[mid, table]
Guenther S 1876 *Vermischte Unter-
suchungen zur Geschichte der
Mathematischen Wissenschaften,*
Leipzig, Ch. 7, 308-343
[Huygens, pendulum clock]
Hoppe E 1919 Ham MG 5(7), 261-269
[anc]
Ideler C L 1925 *Handbuch der Matemati-
schen und technischen Chronologie,*
Breslau, 2 vols  (2nd ed 1883)
Kubitschek W 1928 *Grundriss der
antiken Zeitrechnung,* Munich, 241 p
(F54, 11)
Lamer H 1931 Bay Bl G 67, 340-341
(F57, 1288)
Marcolongo R 1937 Scientia 61, 16-23,
82-92  (F63, 817)
Meyer E 1904 *Aegyptische Chronologie,*
Ber Ab, 210 p
Petrie F 1933 Nat 132, 102
[pendulum]
Pierucci M 1932 Nuo Cim (2)10,
xvii-xviii [elec clock anniv]
Redl G 1929 Byz Z 29, 168-187
[Psellos]
- 1929-1930 Byz Par 4, 197-236; 5,
229-246
Sanford V 1959 Ari T 6, 322-323
[standard time]
Saporetti A 1893 Bln Mm (5)3, 145-
152  (F25, 95) [solar and
mean time]
Schaaf W L 1953 MT 46, 115-117  [bib]
Smith D E 1928 MT 21, 253-258
[time and math]
Staehlin O 1932 Bay Bl G 68, 13-15
(F58, 15)  [Roman]
Stewart G + 1954 Clas Week 47, 113-
118  [Roman]
Weinrich H 1934 ZMN Unt 65, 224-228
(F60, 8)  [Greek water clocks]

Wiedemann E + 1918 Leop NA 103,
163-202 [Archimedes]
Wittstein A 1888 Hist Abt 33, 96-97
[Greek]
- 1892 ZM Ph 39, 81-89  [water clock]
Zubov V P 1953 1953 Ist M Isl 6, 196-
212  (MR16, 660; Z53, 196)
[Kirik Novgorodets]
Zuckermann B 1882 *Materialien zur
Entwickelung der altjuedischen
Zeitrechnung,* Breslau  [Hebrew]


Coins

See also Postage stamps, Metrology.

Johnson R A 1932 Scr 1, 183-184
Smith D E 1925 AMM 32, 444-450
(F51, 1)  [portrait, medals]


Communication

See also Bibliography (Ch. 5),
Criticism, Education, Notation,
Plagiarism, Printing, Terminology.

Brun V 1950 Nor M Tid 32, 12-26
(Z40, 1)
Cajori F 1924 MT 17(2), 87-93
[notation]
Karpinski L C 1938 Sch Soc 48, 338-
[popularization]


Competetion and prizes

See also Plagiarism.

Ackermann A 1929 MZ 30, 320
[Ernst-Abbe prize]
Anon 1898 Kazn FMO (2)8, 96
[Lobachevskii prize]
- 1899 Par CR 129, 1064, 1166
[prizes]
- 1914 BSM (2)38, 31-32
[Weierstrass prize]
- 1915 BSM (2)39, 72  [notice]
- 1918 Crelle 149, 18
[Alfred-Ackerman-Teubner prize]

Competition and prizes (continued)

- 1925 Bln Rn 29, 191-192  (F51, 36)
  [Adolfo Merlani prize]
- 1925 M Ann 94, 176 [Ernst Abbe
  prize]
- 1926 AMSB 32, 40-41  [Chauvenet
  prize]
- 1934 Clct MS 25, 197-198
  [K Ganesh Prasad Prize in
  hist math]
- 1934 Nat 133, 110  [Fields prize]
Biermann K R ed 1960 Ber Ab (3) 75 p
  [election to Berlin Academy]
*Biermann K 1964 Ber Hum MN 13(2),
  185-199 [Berlin Academy]
Birkhoff G 1965 AMM 72, 469-473
  [W. L. Putnam Compet]
Bortolotti E 1926 Bin Rn 30, 51-62
  (F52, 14) [Ferrari, Tartaglia]
- 1927 It SPS 15, 163-180
  (F53, 10) [challenges]
- 1927 It UM 6, 23-27  (F53, 10)
  [XVI]
- 1934 Bln Sto SM 12  (F60, 826)
  [Cardan]
Bush L E 1961 AMM 68, 18-
  [W L Putnam compet]
- 1965 AMM 72, 474-483  [W L Putnam
  compet]
Fehr H 1933 Ens M 31, 279  (F59, 45)
  [Fields prize]
Graham E 1931 MT 24, 181-183
  [debates vs exams]
Klein F 1901 M Ann 55, 143
  [Beneke prize]
Koestlin J 1890 Die Baccalaurei und
  Magistri der Wittenberger phil-
  osophischen Facultaet 1538-1546
  und die oeffentlichen Disputationen
  derselben Jahre aus der Facultaets-
  matrikel veroeffentlicht (Halle U),
  24 p
Kurschak J 1923 MP Lap 30, 1-15
  [Julius Koenig prize]
Laisant C A + 1897-1900 Intermed 4,
  257; 5, 144, 183; 6, 117; 5, 66;
  7, 17, 55  (Q1079; Q1231)
  [prize questions]
Liagre 1879 Bel Bul (2)48  (F11, 29)
  [Belgium]

Lipka S 1940 MP Lap 47, 13-36
  (F66, 2) [Julius Koenig prize]
Loria G 1930 BSM (2)54, 245-254
  (F56, 19)  [Fermat, Frénicle]
Mayor J R 1949 MT 42, 283-289
  [usefulness]
Mittag-Leffler G 1916 Crelle 147,
  53 [Royal Swedish prize]
Noether M + 1908 Paler R 26, 145-151
  (F39, 49) [Guccia medal]
Pagliano C 1902 Bo Fir 1, 94
  [Ferrari, Tartaglia]
Pasquale L di 1958 Pe M (4)36,
  175-198  (Z83, 244) [Ferrari,
  Tartaglia]
Poincaré H 1925 Rv M Sp 36, 1-6
  (F51, 25) [1873 compet]
Putnam W L 1921 Harv G Mag (Dec)
  (repr in AMM 72, 472-273)
  [undergrad]
Rados G 1904 MP Lap 15, 73-93
  [Bolyai prize]
- 1906 BSM (2)30, 103-128
  (F37, 34) [Bolyai prize]
Reif F 1961 Sci 134, 1957-
Richardson R G D 1923 AMSB 29, 14
  (F49, 24) [Cole prize]
Rychlik K 1961 Cas M 86, 76-89
  (MR23, 697) [Prague Academy 1834]
Sanders S T 1934 M Mag 8, 49-50
Schaaf W L 1957 MT 50, 70-71
  [popularization]
Shoemaker R W 1959 MT 52, 453-456
  [Toledo, Ohio]
Suter H 1889 Bib M (2)3, 17-22
  [XVI]
Szasz G + ed 1968 Contests in
  Higher Mathematics, Budapest,
  260 p
Timerding H E 1914 Die Verbreitung
  mathematischen Wissens und mathe-
  matischer Auffassung, Leipzig
  (in Klein F + ed Die Kultur der
  Gegenwart (III)(1)(2))
Van Vleck E B 1922 AMSB 28, 42-44
  [Bocher prize]
Vasiliev A 1898 Rv M 6, 62
  [Lobachevskii]
Watson G N 1934 M Gaz 18, 5,
  [Cambridge tripos, Gauss]
White H S 1925 AMSB 31, 289
  [Cole prize]

## Computers

See also Automata, Calculating
machines, Instruments, Monte
Carlo methods, Numerical
methods.

Beach Ann F + 1954 *Bibliography on
Use of IBM Machines in Sciences,
Statistics, and Education,* NY,
60 p (2nd ed 1956)
* Bemer R W 1969 An Rv AP 5, 151-238
[Algol, bib]
Berkeley E C 1949 *Giant Brains or
Machines That Think,* NY, 286 p
Bernstein Jeremy 1966 *The Analytical
Engine* (Random House) [Babbage]
* Bowden B V 1953 *Faster Than Thought,*
London, 435 p
- 1960 Think 260, 28-32 [Babbage]
* Burks A W 1966 in Von Neumann J,
*Theory of Self-Reproducing
Automata,* (U Ill P), 1-28
[bib, Von Neumann]
Computer Abstracts 1959- London
(Bureau of Information)
Davis Ruth M 1965 Datamat 11(1),
24-28
* - 1966 Adv Comp 7 [bib processors]
Freudenthal H 1953 *Machine
pensantes,* Paris 16 p
Heller Jack + eds 1965 AMM 72
(2)(2), 156 p (Slaught Papers 10)
Kobori A 1952 JHS (22)
Larrivee J 1958 MT 51, 469-473,
541-544
Maurer W D 1968 *Programming* (Holden-
Day) [termin]
* Morrison P & E 1961 *Charles Babbage
and His Calculating Engines* (Dover)
Page S B 1963 *A Select Bibliography
on Digital Computer Education*
(Hertfordshire County Council)
28 p
Patterson G W 1960 Franklin 270, 130-
137 (MR23B, 534) [first elect]
Ransom W R 1952 M Mag 27, 205-207
Rees M 1958 MT 51, 162-168
[impact]
Sackman H 1967 *Computer System
Science and Evolving Society. The*

*Challenge of the Man-Machine
Digital Systems* (Wiley) [bib]
Schaaf W L 1952 MT 45, 110-112
*Serrell R + 1962 IRE (May) [bib]
Sippl Charles J 1967 *Computer
Dictionary and Handbook* (Sams),
766 p
Smith T S 1967 in Kranzberg M + ed
*Technology in Western Civilization*
Oxford, Vol 2, Ch.20
Spencer D D 1968 MT 61(1), 65-75
Stevens M E 1953 *Bibliography on
Electronic Information Processing
...,* (National Bureau Standards),
12 p
Takeda K 1952 JHS (22)(May), (23),
6-10
Tarjan R 1958 M Lap 9, 6-18
(Z80, 9) [Von Neumann]
Thomas Shirley 1965 *Computers:
Their History, Present Applica-
tions, and Future* (Holt), 135 p
Van Dam A 1966 Adv Comp 7
[displays]
Watson G W + 1960 SSM 60, 87-94
Youden W W 1965 *Computer Literature
Bibliography* (National Bureau
Standards 266)

## Conical refraction

See also Optics.

Sarton G 1932 Isis 17, 154-170
[Hamilton, Lloyd]

## Coriolis theorem

See also Mechanics.

Dugas R 1941 Rv Sc 79, 267-270
(Z60, 13)
Freyman L S 1956 Mos IIET 10, 213-
244 (Z74, 245) [proofs]

## Cosmology

See also Astronomy.

## Cosmology (continued)

Andrissi G L 1940 It UM Con 2, 908-911 (F68, 4) [geocentric]
Broecker W 1952 Stu Gen 5, 325-328 (Z47, 1) [Greek]
Duhem Pierre M M 1913-1959 *Le systeme du monde. Histoire des doctrines cosmologiques de Platon a Copernic,* Paris, 10 vols (repr 1954-1959 Hermann)
Forsyth T M 1932 Phi Lon 7(1), 54-61 [Newton, Plato, Whitehead]
Lampariello G 1961 Archim 13, 121-133 (Z94, 4)
Sikora J 1958 New Schol 32, 61-72
Whitrow G J 1940 M Gaz 24, 159-164

## Counterexample

See also Heuristic.

Gelbaum B R + 1964 *Counterexamples in Analysis* (Holden-Day), 218 p

## Cranks

See also Hoaxes.

Chrystal G 1965 *Perpetual Motion,* Enc Br [reprint]
De Morgan Augustus 1915 *A Budget of Paradoxes,* 2nd ed, Chicago, 2 vols
Greenblatt M H 1965 Am Sc (Dec) (see 1935 Inda Ac 45, 206-210) [legal value of pi etc]
*Gruenberger F J 1964 Sci 145, 1413-
Karpinski L C 1915 Sci 42 (19 Nov), 729- [re De Morgan 1915]

## Criticism

See also Communication.

Bieberbach L 1934 Ber Si (18/20), 351-360 (Z9, 388) [style]
Dickson L E 1923 AMM 30, 318 [book reviews]

Eneström G 1904 Bib M (3)5, 298-304 (F35, 51) [reviews]
Rosenbaum R A 1953 MT 53, 35-38 [Sylvester]

## Cryptanalysis

Andree R V 1952 MT 45, 503-509
- 1952 Scr 18, 5-16 [bib]
Anon 1943 AMM 50, 345-346 [bib]
Ball W W R 1960 *Mathematical Recreations,* Ch. 14
Chase P E 1859 M Mo Runkl 1, 194
Mendelsohn C J 1938 Scr 6, 157-168 [Cardan]
Mitchell U G 1919 AMM 26, 409-413
*Pratt Fletcher 1939 *Secret and Urgent: the Story of Codes and Ciphers,* Indianapolis
Simonov R A 1963 Bulg FM 6(39), 199-206 (MR29, 1) [XIV-XVI, Slavic]
Smith D E 1918 AMSB 24, 82-96 (F46, 8) [Wallis]

## Crystallography

See also Symmetry (Ch. 2)

Hilton Harold 1903 *Mathematical Crystallography and the Theory of Groups of Movements,* Oxford, 274 p (repr 1963 Dover) [bib]
Neumann C 1917 Leip Ab 33, 195-458 (F46, 5) [F Neumann]
Schoenflies A 1896 Chic Cong, 341-349 [group]
Sohncke L 1879 *Entwickelung einer Theorie der Krystallstruktur,* Leipzig, 255 p

## Cybernetics

See also Automata, Computer, Information theory.

Anon 1951 LHL Bib (3) [bib]
Feigenbaum E A + ed 1963 *Computers and Thought,* NY [bib]

## Cybernetics    (continued)

Fuchsman + 1965 Sci 150, 1666-1667
*Graham Loren R 1965 in *The State of Soviet Science* (MIT P), 3-18 [Russ]
- 1967 in Fischer G ed, *Science and Ideology in Soviet Society*, (Atherton), 83-106
Rapaport A 1959 in *American Handbook of Psychology*, NY
Unger G 1966 Ex Med IC (136) [Finsler]
Wiener N 1948 *Cybernetics--or Control and Communication in the Animal and the Machine*, NY

## Decidability

See also Foundations.

*Davis Martin 1965 *The Undecidable..*, NY, 440 p [bib, source book]
Kleene S 1952 *Introduction to Metamathematics*, NY, 136-141, 432-439
Tarski A 1953 *Undecidable Theories*, Amsterdam

## Definitions

See also Foundations.

Bortolotti E 1930 Pe M (4)10, 85-
Kaminski S 1958 Stu Log 7, 43-69 (MR20, 5119) [Hobbes]
Metzler W H 1913 MT 5, 143-144
Mogenet J 1947 Brx SS (1)61, 235-241 (MR9, 169) [anc]
Pastore A 1912 Tor FMN 47, 201-217, 478-494 [Aristotle]

## Demography

See also Geography.

Bartlett M S + 1951 Cam PSP 47, [bib]
Bodenheimer F S 1958 Arc In HS 11, 27-28 (Z95, 2)

Feller W 1941 An M St 12 [bib]

## Economics

See also Accounting, Business, Finance, Insurance, Operations research.

Brand Heinz W 1961 *The Fecundity of Mathematical Methods in Economic Theory*, Dordrecht
Castro G de 1955 Portu IA 11, 49-60 [decision]
Gale David 1960 *The Theory of Linear Economic Models* (McGraw-Hill) [bib]
Kantorovich L V 1959 Mangt Sci 6, 366-422 (MR23B, 2053) [planning]
Maillet E 1906 Intermed 13, 121, 122 (Q2973)
Masè-Dari E 1935 *Un precursore della econometria. Il saggio di Giovanni Ceva "De re numaria" edito in Mantova nel 1711*, Modena, 59 p (F62, 1035)
Nemchinov V S ed 1964 *The Use of Mathematics in Economics* (MIT P) [Russ]
Newman Peter 1968 *Readings in Mathematical Economics* (Johns Hopkins U), 2 vols [bib]
Read C B 1959 MT 52, 124 [living costs XIX]
Roy R 1933 Ecmet 1, 13-22 [Cournot]
Samuelson P A 1952 Am Ec Rv 42(2)
Scott A D 1951-1953 Lon St (A)114, 372-393; 116, 177-185 [appl of stat]
Shaw I B 1934 M News Let 8, 128-131 (F60, 840)
Smith D E 1917 AMM 24, 221-223
Theocharis Rheginos D 1961 *Early Developments in Mathematical Economics* (Macmillan), 142 p [bib to 1837]
Tinter Gerhard 1966 in Krupp S R ed, *The Structure of Economic Science*, Englewood Cliffs, 114-128
Weinberger O 1938 Nap Pont 59, 75 p (F64, 919)

Economics    (continued)

Zematis Z 1963 Lit M Sb 3(1), 289-
313  (Z117:, 249  [Revkovski]


## Education: Mathematics

See also Countries (Ch. 4),
Heuristic, Problems.

Argles M 1964 *South Kensington to
Robbins; An Account of English
Technical and Scientific Education
since 1851,* London, 178 p
Bateman H 1913 MT 5, 147-153
[England]
Cajori F 1915 Monist 25, 495-530
[Oughtred]
- 1920 MT 13, 57-62  [Greek]
*Cagan E J 1963 AMM 70, 554-560
[XIX, XX]
Courant R 1938 AMM 45, 601-607
[Germany to 1933]
Dainville R P F de 1956 XVII(30),
62-68  [XVII]
Dunnington G W 1937 M Mag 11, 318-
323  [Thibaut],
*Eisenhard L P 1945 *The Educational
Process* (Princeton U P)
Eneström G 1907 Bib M (3)7, 252-262
[mid]
Guldberg A 1905 Ens M 7, 433
[Norway]
Guenther Siegmund 1887 *Geschichte
des mathematischen Unterrichts
im deutschen Mittelalter bis zum
Jahre 1525,*  Berlin, 408 p
Hatzidakis N 1934 Inblk Con, 239-241
(F61, 957)
Ingalls E E 1954 MT 47, 409-410
[George Washington]
Jones P S 1967 AMM 74, 38-55
Lankov A V 1951 *K istorii razvitiya
peredovykh idei v russkoi metodike
matematiki,* Mos, 151 p [to 1915]
Lietzmann W 1930 ZMN Unt 61, 289-300
(F56, 4)  [Germany, 1903-1905]
- 1933 ZMN Unt 64, 101-105  (F59, 6)
[International Math Ed Comm]
Lorey W 1926 Ber MG 25, 54-68
[Klein]

Lynch J A 1939 MT 32, 61-64
[Herbart]
Marijon A 1934 Ens Pub (Oct), 115-126
[XVII, XVIII]
Natucci A 1932 Pe M (4)12, 173-179
(Z4, 194)  [XVI]
Ogura K 1932 *History of Mathematical
Education,* Tokyo, 360 p
(in Japanese)
Pahl F 1913 *Geschichte des natur-
wissenschaftlichen und mathe-
matischen Unterrichts,* Leipzig,
377 p  (F44, 42)
Piazzolla-Beloch M 1930 *La matematica
in relazione alle sue applicazione
e al suo valore educativo,* Ferrara,
22 p  (F56, 2)
Schaaf W L 1941 *Bibliography of
Mathematical Education,* Forest
Hills, NY  [since 1920]
- 1955 MT 48, 348-351
Schuelke A 1905 ZMN Unt 36, 22-23
(F36, 49)  [1811]
Simons L G 1924 MT 16, 340-347
[Dutch text 1730]
Smith D E + 1912 *Bibliography of the
Teaching of Mathematics 1900-1912*
Washington DC, 95 p
Wieleitner H 1927 Unt M 33, 312-315
(F53, 36)


## Education: Elementary

Bockstael P 1960 Isis 51, 315-321
[first arith texts]
Bradley A D 1934 Scr 2, 235-241
[1727]
Clason R G 1968 *Number Concepts in
Arithmetic Texts of the US from
1880 to 1966,* Ann Arbor
Galanin D D 1915 *Istoriya metodi-
cheskikh idei po arifmetike v
Rossii...XVIII veka,* Mos, 252 p
Grosse H 1901 *Historische Rechen-
buecher des 16 und 17 Johrhunderts
und die Entwicklung ihrer Grundge-
danken bis zur Neuzeit. Ein
Beitrag zur Geschichte der Methodik
des Rechenunterrichts...,* Leipzig
183 p

Education: Elementary (continued)

Ionides S A 1938 Engr B(May) [anc]
Karpinski L C 1933 Scr 2, 34-40
[texts]
Laisant C A 1908 Rv Sc (5)9, 449-454
Rappaport D 1965 SSM 65, 25-33
Ray L D 1890 Edu 10, 615-623 [XVII]
Read C B 1962 MT 55, 127-129
[1825]
- 1963 MT 56, 538-540
[improvement?]
Robson A 1949 M Gaz 33, 81-93
[XVII-XIX]
Sleight E R 1936 M Mag 10, 193-199
- 1937 M Mag 11, 310-317
[Pestalozzi, US]
Steinweller F 1899 Kurzer Abriss der
Geschichte des Rechenunterrichts,
Leipzig, 48 p (F30, 38)
Stoy H 1876 Zur Geschichte des
Rechenunterrichts, Jena (diss)
(F6, 27)
Williamson R S 1928 M Gaz 14, 128-
133 [XIX]
Yeldham Florence A 1936 The Teaching
of Arithmetic Through 400 Years,
1535-1935, London, 143 p

Education: Geometry

Buchanan H E 1929 M Mag 3, 9-18
Cajori F 1910 AMM 17, 181-201
[XVIII, XIX]
Coleman Robert 1942 The Development
of Informal Geometry (Teachers
College, Columbia U), 178 p
[XIX-XX, *bib]
Itard J 1936 Ens Sc 9, 193-198
[Legendre]
Kokomoor F W 1928 Isis 10, 21-32
376- [XVII]
Laumann J 1930 Unt M 36, 294-297
(F56, 4) [XVII]
- 1930 Unt M 36, 317-320
[XVII, XVIII]
Loria G 1893 Pe MI 8, 81 [Euclid]
Miller G A 1907 SSM 7, 752-755
[XIX]
- 1908 SSM 7, 752-755 [XIX]

Miller M 1935 ZMN Unt 66. 80-83
(F61, 7)
Metzler W H 1908 MT 1, 42-43
[Perry movement]
Mock G D 1963 MT 56, 150-153
[Perry movement]
Piaget Jean + 1948 La géométrie
spontanée de l'infant, Paris, 516
Segre C 1954 MT 47, 162-166
[Italy]
Stamper A W 1909 Loria 15, 10-11
* 1913 A History of the Teaching of
Elementary Geometry, with Reference
to Present-Day Problems, NY, 173p
[*bib]
Wolff G 1937 M Gaz 21, 82-98
[Germany]

Education: Teaching history of
mathematics

See also in Ch. 4 General: Textbooks;
in Ch. 5 Collections.

Amodeo F 1906 Bib M (3)6, 387-393
[Naples]
- 1908 Bib M (3)8, 403-410
[Naples 1905-1907]
- 1907 Bib M (3)8, 403-410
(F38, 56) [Naples]
Bashmakova I G + 1958 Ist M Isl 11,
185-192 (MR23A, 582) [Moscow U]
Bonolis A 1905 Pe MT (3)3, 103
[Russia]
Braunmuehl A v 1895 Bib M (2)3(4),
89 [Munich]
- 1902 Bib M (3)3, 403-404 (F33, 52)
[Munich]
Eneström G 1904 Bib M (3)5, 63-67
(F35, 50) [univ]
Favaro A 1887 Bib M (2)1, 49-54
[Padua U]
Lazzeri G 1905 Pe MI (3)2, 145-162
(F36, 48) [geometry, Pisa U]
Loria G 1909 Int Con (4)3, 541-548
(F40, 48)
- 1910 Arc GNT 1, 9-18 (F41, 48)
- 1937 M Gaz 21, 274-275
Mansion P 1900 Bib M (3)1, 232
[Gand U]

## Education: Teaching history of mathematics (continued)

*Natucci A 1960 Pe M (4)38, 228-233
(MR22, 1129)
Richeson A W 1934 Scr M 2, 161-165
[US courses]
Sarton G 1954 *The Study of the History of Mathematics* (Dover)
Smith D E 1898 Bib M 13 [Michigan]
Stott W 1915 Sci Prog 19, 204-217
Struik D J 1947 Isis 32, 123-124
[MIT]
Tannery Paul 1907 Rv Mois 16(April)
7 p
Tyler H W 1909 Bib M (3)10, 48-52
(F46, 48) [MIT]
White H S 1909 AMSB 15, 325
Wieleitner H 1927 Unt M 33, 312-315
[primary sources]
- 1927 Weltall 26, 186-191
[primary sources]
Yanovskaya S A 1958 Ist M Isl 1,
193-208 (MR23A, 582)

## Education: Using history in teaching mathematics

Anon 1885 Fiz M Nauk (A)1, 1-16,
97-121
Barwell M 1913 M Gaz 7, 72-79
Bockstaele P 1949 Nov Vet 27, 15 p
Bunt L N H 1954 *Geschichte der Mathematik als Thema fuer das Gymnasium,* Groningen-Djakarta,
2 vols (Z60, 2)
Carmichael R D 1914 SSM 13, 684-694
Chistyakov V D 1959 *Istoricheskie ekskursy na urokakh matematiki v srednei shkole,* Minsk, 95 p
Collins J V 1894 Sci 23, 44
Cook A J 1938 MT 30, 63-65
[Canada]
Eells W C 1925 MT 18, 296-297
Eneström G 1913 Bib M (3)14, 1-8
(F44, 41)
Frajese A 1950 It UM (3)5, 337-342
(Z38, 1)
Freitag H T + 1957 MT 50, 220-224
Funk J C 1921 MT 14, 405 [hist notes in texts?]

Gabriel R M 1937 M Gaz 21, 106-116
Gebhardt M 1912 *Die Geschichte der Mathematik in mathematischen Unterricht der hoeheren Schulen Deutschlands,* Leipzig-Berlin,
163 p (also ICTM Deu 3(6))
Gibbins N L 1954 MT 47, 488-489
Hallerberg Arthur E + 1969 *Historical Topics for the Mathematics Classroom,* Wash DC (NCTM), 524 p
Harmeling H 1964 MT 57, 258-259
Hassler J 1929 MT 22, 166-171
Heppel G 1893 Nat 48, 16-18
Hofmann J E 1961 Mathunt 7(3), 4-20
[entire number (3) also]
Jones P S 1957 MT 50, 59-64
Krueger R L 1954 MT 47, 408
Lilliacus L A 1920 AMM 27, 61-63
Malygin K A 1958 *Elementy istorizma v prepodavanii matematiki v srednei shkole,* Mos, 240 p
*Mendelssohn W 1912 ZMN Unt 43, 1-9
[using primary sources]
Miller B I 1922 MT 15, 416-422
Mitchell U G 1938 M Mag 13, 22-29
(F64, 2)
Molodshii V N 1953 *Elementy istorii matematiki v shkole,* Mos, 36 p
Servaty I 1906 SSM 5, 557
[Greek]
Shevchenko I N 1958 *Elementy istorizma v prepodavanie matematiki,*
Mos Ped (92)
Simons L G 1923 MT 16, 94-101
Whitrow G J 1932 M Gaz 16, 225-227
Wiltshire B 1930 MT 23, 504-508
Worden G F 1935 SSM 34, 361-371
[XV, XVI, XVII]

## Education: Pedagogy

Bell E T 1935 AMM 42, 472 [bib]
Bronshtein I N 1950 Ist M Isl 3,
171-194 [Lobachevskii]
Carver W B 1923 AMM 30, 132
[puzzles]
De Morgan Augustus 1831 *On the Study and Difficulties of Mathematics,*
London
Feierabend R L + 1959 *Psychological Problems and Research Methods in*

Education: Pedagogy   (continued)

*Mathematical Training,* St. Louis
(Washington U)
Itard J 1952 Rv Hi Sc Ap 5, 171-175
[La Chapelle]
Kropp Gerhard 1948 *Beitraege zur
Philosophie, Paedagogik und
Geschichte der Mathematik,* Berlin,
103 p (Z30, 1)
Lankov A V 1951 *K istorii razvitiya
peredovykh idei v russkoi
metodike matematiki,* Moscow
Meyer W F 1899 DMV 7, 147
Miller G A 1922 Sch Soc 15, 13-15
(F48, 41) [popular teachers]
*Moise E E 1965 AMM 72, 407-412
[motivation]
Nordgaard M A 1939 Scr 6, 88-94
[La Cour]
Palmer E G 1913 SSM 12, 692-693
[anal geom]
Parker E T 1965 AMM 72, 1127-1128
[using special cases]
Wilder R L 1953 AMSB 59, 423-448


Education: Secondary

Ball W R 1910 M Gaz 5, 202-205
[XVII]
Bass W S 1906 SSM 5, 712-716
Betz W 1950 MT 48, 377-
Breslich Ernst R 1950 in *A Half
Century of Science and Mathematics
Teaching* (Central Association of
Science and Mathematics Teachers)
[bib]
Bushell W F 1947 M Gaz 31, 69-89
[XIX-XX]
Dainville F de 1951 Fr Mod 19, 193-
196 [XVII, France]
- 1954 Rv Hi Sc Ap 7, 6-21,
109-123 [XVI, XVII]
- 1956 XVII(30), 62-68 [XVII]
Desforge 1936 Ens Sc 9(84), 97-117
[Hadamard]
Goldziher K 1912 Int Con (5)2, 608-
610 [bib 1900-1912]
Grundel F 1928-1929 *Die Mathematik
an den deutschen hoeheren Schulen,*

Leipzig, 2 vols  (F54, 8; 55, 3)
Heym C 1873 ZMN Unt 4, 427-429
(F5, 43)
Hooker J H 1895 M Gaz (4)  (F26, 47)
[old texts]
Lietzmann W 1921 ZMN Unt 52, 122-124
(F48, 40)
Mansion P 1895 Brx SS 19, 101
[XVII]
McDonough H B + 1934 MT 27, 117-127,
190-198, 215-224, 281-295  [USA]
Pahl F 1899 *Die Entwicklung des
mathematischen Unterrichts an
unserem hoeheren Schulen,*
Charlottesburg, 30 p  (F30, 35)
Paige D 1964 SSM 64, 195-201
[gen math]
Rollett A P 1963 M Gaz 47, 299-306
[England]
Siddons A W 1936 M Gaz 20, 7-26
[England, XIX-XX]
- 1956 M Gaz 40, 161-169
[England, XX]
Simons L G 1934 Scr 2, 165
[alg text]
Sister M Constantia 1961 MT 54, 261-
265
Smith D E 1926 NCTMY 1, 1-31  [XX]
Vogeli B R 1968 *Soviet Secondary
Schools for the Mathematically
Talented* (NCTM)
Willoughby Stephen S 1967 *Contem-
porary Teaching of Secondary
School Mathematics* (Wiley), Ch.1,
1-28
Wooton William 1965 *SMSG. The Making
of a Curriculum* (Yale U P),
182
Wren F L + 1935 MT 27, 117-127, 190-
198, 215-224, 281-295  [USA]
Zaremba S 1920 Wiad M 24, 97-107


Education: Technology

*Davidson P S 1968 MT 61, 509-524
[bib teaching aids]

## Education: Undergraduate research

Anon 1956 *Report on a Conference on Undergraduate Mathematics Curricular, held at Hunter College, October 12-13, 1956* (Yale U)
Calloway J M 1957 AMM (Feb), 141-142
Griffin F L 1930 AMM 37, 46-54
- 1942 AMM 49, 379-385
- 1951 AMM 58, 322-325
Lindquist Clarence B 1964 *Honors Programs for Superior Undergraduate Mathematics Students,* Wash DC (Office of Education-56015)
May K O 1948 AMM 65, 241-246
- 1968 AMM 75, 71-74
May K O + 1962 *Undergraduate Research in Mathematics,* Northfield, Minn (Carleton College)
*Simons L G 1941 Scr 8, 165-175
*Sleight E R 1941 AMM 48, 696-697
Wilansky A 1961 MT 54, 250-254
Wilder R L 1959 by L Henkin + ed, *The Axiomatic Method,* Amsterdam, 474-488

## Education: University

See also in Ch. 4, Cities and Universities.

Anon 1892 *Verzeichnis der seit 1850 an den deutschen Universitaeten erschienenen Doctor-Dissertationen und Habilitationsschriften aus der reinen und angewandten Mathematik,* Munich
Bohlmann G 1899 DMV 6(2), 91-110 (F30, 47) [calc, Euler]
Boyer C B 1946 MT 39, 159-167 [calc texts]
*Calkins R D 1966 Sci 152, 884-
Karpinski L C 1935 *An Exhibition of Early textbooks on College Mathematics,* Ann Arbor
- 1938 Scr 6, 133-140 [first year]
Klein F 1904 DMV 13, 267-275 [Germany]
- 1907 *Vortraege ueber den mathematischen Unterricht an den hoeheren Schulen,* Leipzig, 245 p
Likholetov I I + 1955 Ist M Isl 8, 127-480 (MR17, 1) [Moscow]
Lorey W 1916 *Das Studium der Mathematik an den Deutschen Universitaeten seit Anfang des 19. Jahrhunderts,* Berlin, 443 p
Mueller C H 1904 Ab GM 18, 51-143 (F35, 52) [XVIII, Goettingen]
Sanford V 1923 MT 16, 206-214 [first year]
Schoenflies A 1908 DMV 17, 22-35 [stat data]
Simons L G 1936 Scr 4, 207-219 [calc, US, XIX]
Webber W P 1927 M Mag 2, 10-11 [first year]

## Elasticity

Albenga G 1956 Tor FMN 90, 567-576 (Z73, 3)
*Freudenthal A M 1951 AMR 4, 394-396 [bib]
Frocht M M 1952 AMR 5 337-340 [bib]
Gaedecke W 1910 Intermed 17, 22 (Q3594) [bib, surface of elast]
Kaktseev E N 1959 Mos IIET 22, 214-239 (Z95, 4) [XIX, XX, Russia]
Love A E H 1927 *A Treatise on the Mathematical Theory of Elasticity* (Cambridge U P), 4th ed, 643 p (1st ed 1892-3, 2nd 1906, 3rd 1920) (repr 1944 Dover) [hist intro and notes]
Lure A I 1964 *Three-Dimensional Problems of the Theory of Elasticity* (Interscience) [bib]
Marcolongo R 1907 Nuo Cim (5)14, 371-410 (F38, 70) [XIX, Italy]
Nowinski J 1955 Roz Inz 3, 57-78 (MR17, 2) [survey]
Oldfather W A + 1933 Isis 20, 72-160 (F59, 27; Z7, 388) [Euler]
Todhunter I 1886-1893 *A History of the Theory of Elasticity and Strength of Materials...,* (Cambridge U P), 2 vols in 3 (F25, 88) (repr 1960 Dover; MR 22, 8784)

## Elasticity (continued)

Truesdell C 1952-1953 J Rat Me An 1, 125-300; 2, 593-616 [*bib]
- 1960 Acou SAJ 32, 1647-1656 (MR23B, 1715
- 1960 *The Rational Mechanics of Flexible or Elastic Bodies 1638 -1788,* Zurich, 435 p (Euler Opera (2)11(2)) (MR24, 221)
Vacca G 1916 Linc At (5)25(1), 30-37

## Electricity

Bertelli T 1888 Boncomp 20, 481-542 [anc, mod]
Gliozzi M 1933 Archeion 15, 202-215 [Franklin, Watson]
Graetz L 1907 *Elektrizitat und ihre Anwendung,* Stuttgart, 838 p
Schaefer C 1931 Nat 128, 339-341 (F57, 33) [Gauss]
Tricker R A R 1965 *Early Electro-dynamics: The First Law of Circulation* (Pergamon), 227 p (see Science 154, 497) [source book]
Windred G 1930 Phi Mag (7)10, 905-916 (F56, 8) [early a.c.]
- 1932 Isis 18, 184-190

## Engineering

See also Aerodynamics, Architecture, Linkages, Technology.

Allen D N de G 1954 *Relaxation Methods in Engineering and Science* (McGraw Hill), 266 p [bib]
Anon 1956 Mos IIET 7, 263 p
Brauer E 1908 DMV 17, 39-46 [Euler]
Cherepashinskii M 1889 Mos UZ(8), 1-33 [mech of construction]
Cressy E 1937 *A Hundred Years of Mechanical Engineering,* Lon, 340 p
Dirks H 1861 *Perpetuum mobile or Search for Self-motive Power During the Seventeenth, Eighteenth and Nineteenth Centuries,* London 600 p

Fitz-Patrick J + 1905 Intermed 12, 171, 286 (Q2942) [cisterns]
Henneberg L 1894 DMV 3, 567-599 (F25, 85) [frameworks]
Hering K 1907 Ab GM 23 [steam engine]
Ocagne M d' 1914 Rv GSPA 25, 469-474
Schaaf W 1951 MT 44, 54-57
Schmidt W 1902 Bib M (3)3, 337-341 [anc, Steam boilers]
Timoshenko S P 1953 *History of Strength of Materials...,* NY, 462 p (MR14, 1050; Z53, 246)
Todhunter I 1886-1893 *A History of the Theory of Elasticity and of the Strength of Materials...* (Cambridge U P), 2 vols in 3 (F25, 88) (repr 1960 Dover; MR 22, 8784)
Van den Broek J A 1947 AJP 15, 309-318 (MR9, 74) [Euler]
Von Mises R 1926 ZAMP 6, 263-264 (F52, 39) [foundries]

## Epicycle

See also Astronomy.

Aaboe A 1963 Centau 9, 1-10 (MR29, 218) [Greek]
Boyer C B 1948 Isis 38, 54-56 (Z30, 100) [Copernicus, ellipse, Lahire]

## Equality

See also Equivalence.

Cajori F 1923-1924 AMSB 29, 107; 30, 387 (F49, 23; F50, 19) [notation]
- 1923 Isis 5, 116-125; 6, 507-508 [nota]
Catalan E 1895 Intermed 2, 86, 115
MacDuffee C 1936 MT 29, 10-13 [equivalence]
Riquier C 1895 Rv Met Mor 3, 269 [axioms]
Tanner R C H 1962 Brt JHS 1-2(2) [inequality]

Equality   (continued)

Van Engen H 1958 MT 51, 42
   [meaning]


Equivalence

Couturat L 1901 *La logique de*
   *Leibniz d'après des documents*
   *inédits,* Paris, 313-315
Frattini G 1895 Pe MI 10, 153
   [polyg]


Ethics

Bredvold L I 1951 in *The Invention*
   *of the Ethical Calculus in the*
   *Seventeenth Century,* ed by R F
   Jones  (Stanford U P)
Lecat Maurice M 1923 *Probité*
   *scientifique,* Louvain,  24 p


Exhaustion (method of proof)

See also Integration  (Ch. 2).

Brusotti L 1952 Pe M (4)30, 241-248
   (MR14, 609; Z48, 241)
Christensen S A 1931 M Tid A, 101-
   130  (F57, 1291) [Archimedes,
   Euclid]
Hjelmslev J 1934 M Tid B, 33-39
   (Z9, 388)  [Euclid]
- 1934 M Tid B, 57-67  (F60, 8;
   Z10, 385)  [Archimedes, reals]
Yushkevich A P 1948 in *Trudy*
   *soveshchaniya po istorii*
   *estestvoznaniya (1946)* (Ak. Nauk
   SR SR), 173-182)  (MR1950, 572)
- 1964 Koyre 1, 635-653


Existence

See also Foundations, Philosophy.

Becker Oskar 1927 *Mathematische*
   *Existenz,* Halle, 377 p

Benacerraf P + 1964 *Philosophy of*
   *Mathematics: Selected Readings*
   (Prentice-Hall), Part 2
Beth E W 1951 Brt SHS 1, 165
Boussinesq J 1915 Par CR 161, 45-47
   (F45, 95)
Niebel Eckhard 1959 *Untersuchungen*
   *ueber die Bedeutung der Geometri-*
   *schen Konstruktion in der Antike*
   Cologne, 147 p  (Z113, 1) [bib]
Zeuthen H G 1896 M Ann 46, 222-
   [constr]


Fashion

See also Sociology of Science.

Curtiss D R 1937 AMM 44, 559-566


Fifty

Riedel E 1929 Clas J 25, 696-698
Roscher W H 1917 Leip Ph Ab 33


Finance

See also Accounting, Business,
   Economics, Insurance.

Archibald R C 1945 M Tab OAC 1(10),
   401-402  [compound interest tables]
De Morgan A 1861 Engl cyc (Arts and
   Sciences)  7, col 1008  [bib]
Eneström G 1896 Sv Ofv 53, 41
   [XVII, Halley, Witt]
Hagstroem K.-G 1931 Int Con (8)5,
   427-441  (F57, 1293)  [Rome]
Hodge W B 1857-1861 Assur Mag 6,
   301-; 7, 311-; 8, 68-; 9, 61-;
Laurent H 1899 Intermed 6, 138-139
   (Q1393)  [Leibniz]
Poole R L 1912 *The Exchequer in the*
   *Twelfth Century,* Oxford, 206 p
Stengers J 1941 Bel Rv PH 20, 573-588
   [value of ancient money]
Waller Zeper C M 1937 *De oudste*
   *intrest tafels in Italie, Frankrijk*
   *en Nederland met een herdrak van*

## Finance (continued)

*Stevin's "Tafeln van interest",*
Amsterdam, 191 p

## Finger reckoning

Cajori F 1926 Isis 8, 325-327
  [America]
- 1928 Arc Sto 9, 31-42 [calendar]
Carnoy A 1946 Museon 59, 557-570
Eastlake F W 1880 Chi Rv 9, 249,
  319 [China]
Fischer A 1934 Islamica 6, 48-57
  [Arab]
Jones P S 1955 MT 48, 153-157
Marre A 1864 Annali 6, 93
- 1868 Boncomp 1, 309-318
  (F1, 8) [anc]
Noury P 1932 Bude Bu 36, 40-45
  [Roman]
Richardson L J 1916 AMM 23, 7-13
Ruska J 1920 Islam 10, 87-119
  [Arab]
Seeger R J 1961 Ari T 8, 339-344
Stapleton H E 1956 Int Con HS 8,
  1103 [anc]
Yeldham A 1928 Archeion 9, 325-326
  (F54, 5) [calendar]

## Force

Chelini D 1873 *Interpretazione
  geometrica di formole essenziali
  alle scienze dell'estensione,
  del moto e delle forze...,*
  Bologna  (F5, 47)
Timerding H E 1908 DMV 17, 390-405
  (F39, 71)

## Formalism

Brouwer L E J 1913 AMSB 20, 81-96
Freudenthal H 1959 Nieu Arch (3)7,
  1-19 (Z85, 6)
*Kreisel G 1958 Dialect 12, 346-372
Von Neumann J 1964 in Benacerraf P +,
  *Philosophy of Mathematics,*

*Selected Readings* (Prentice-Hall),
  40-54

## Foundations

See also Axioms, Definitions,
  Existence, Paradoxes, Philosophy,
  Proof, Variable, Zeno paradox.

*Becker Oskar 1954 *Grundlagen der
  Mathematik in geschichtlicher
  Entwicklung,* Freiburg-Munich, 433 p
  (MR16, 433) (2nd ed 1964; MR30,
  373)
*Benacerraf P + 1964 *Philosophy of
  Mathematics, Selected Readings*
  (Prentice-Hall), Part 1
Beth Evert W 1952 Brt JPS 3
  [prehistory]
Bouligand G 1956 *Le repérage de la
  pensée mathématique, sa part dans
  la recherche,* Paris, 35 p
Cassina U 1961 *Critica dei principi
  ...,* Rome
Cavailles Jacques 1938 *Méthode
  axiomatique et formalisme,* Paris
  (Hermann), 3 vols (Act Sc Ind
  608-610; Le progrès de l'esprit
  9-11) [probl of found, ax and
  formal syst, non-contrad of arith]
Dehn M 1937 QSGM (B)4, 1-28
  (F63, 8; Z16, 196) [anc, phil]
Eves Howard + 1958 *An Introduction
  to the Foundations and Fundamental
  Concepts of Mathematics,* NY, 363 p
  [pop]
Freudenthal 1957 Nieu Arch (4)5,
  105-142
Hasse H + 1928 *Die Grundlagenkrisis
  der griechischen Mathematik,*
  Charlottenberg, 72 p (F54, 13;
  Z9, 97)
Hatcher W S 1968 *Foundations of
  Mathematics* (Saunders)
Jourdain P E B 1910 QJPAM, 324-352
  [logic]
Kleene S C 1965 Sci 146, 427
- 1952 *Introduction to Metamathe-
  matics,* Amsterdam
Kneebone G T 1963 *Mathematical Logic*

Foundations    (continued)

*and the Foundations of Mathematics*
(Van Nostrand), 435 p
Konigsberger L 1896 *Hermann von
Helmholtz's Untersuchungen ueber
die Grundlagen der Mathematik und
Mechanik,* Leipzig
Lietzmann Walther 1927 *Aufbau und
Grundlage der Mathematik,*
Leipzig, 89 p
*Menger K 1937 Phi Sc 4, 299-336
*Moore E H 1902 NCTMY 1   (repr 1967
MT 60, 360-)
Mordukhay-Boltovskoy D D 1928 Rostov
U(15), 35-129   (F54, 3)
Mostowski Andrej 1966 *Thirty Years
of Foundational Studies* (Barnes
and Noble), 180 p   (also Acta
Phi F17)
Schmidt A 1950 Enc M W 1(1)2
[bib]
Szabo A 1958 St It FC 30, 1-51
[anc]
- 1959 Ist M Isl 12, 321-392   [anc]
*Tarski A 1956   *Logic, Semantics,
Metamathematics,* Oxford
Van Dantzig D 1957 Nieu Arch (3)5,
1-18   (Z77, 9)   [Mannoury]


Fractions

See also Division, Egyptian fractions
(Ch. 4), Numeration.

Bobynin V V 1890 Bib M (2)4, 109-112
[Egypt]
- 1896 Bib M, 97
- 1899 Bib M (2)13, 81-85 (F30, 38)
- 1911 Kagan (535), 177-184
(F42, 55) [anc]
- 1913 Kagan (585-587) [anc]
Bruins E M 1952 Indagat 55
[Egypt, 2/n]
Chace A B 1931 Archeion 13, 40-41
[Egypt]
Chakravarti G 1934 Clct Let 24(9),
59-76 (F62, 1021) [Hindu]
Cimpan F T 1965 Iasi UM (NS)11(B),
9-13 (MR33, 7222) [Greek]

Eneström G 1914 Bib M 14, 41-54
[Jordanus Nemorarius]
Gandz Solomon 1945 in *Louis Ginsburg
Jubilee Volume,* (American Academy
for Jewish Research) 1, 143-157
[Hebrew]
Gillain O 1928 Mathesis 42, 405-413
(F54, 10) [Egypt]
Gillings R J 1959 Au JS 22, 247-250
(MR21, 1315) [Egypt, table]
Guérard 1838 JMPA 3, 483   [mid]
Hayashi T 1930 Toh MJ 33, 292-327
[Japan]
Jones E 1960 Ari T 7, 184-188
[decimals]
Kowalewski G 1938 Deu M 3, 698-700
[Egypt]
Lay L C 1958 MT 51, 466-468
Medovoi M I 1959 16 p IET (8),
101-106 [Abul Wafa]
Miller G A 1931 SSM 30, 881-883
[inverting denominator]
- 1932 SSM 31, 138-145
Neugebauer O 1926 *Die Grundlagen der
aegyptischer Bruchrechnung,* Berlin,
51 p
- 1930 QSGM (B)1, 425-483
[Babyl, sexag]
- 1932 QSGM (B)2(2), 199-210
[Babyl, sexag]
Niebecker E 1930 Arc GMNT 13, 82-92
[Egypt]
Padoa A 1933 Pe M 13, 87-98
Sachs A 1946 JNES 5, 203-214
[Babyl]
Sarfatti G 1958 Tarbiz 28, 1-17
[Hebrew]
Stoltenberg H L 1956 Gies Hoch 25,
130-137 (MR21, 750; Z71, 242)
[anc]
Tannery P 1886 Bib M, 235-236
[Greek]
Thompson H 1914 Anc Eg, 52-54
[Byzantine]
Thureau-Dangin F 1931 Rv Assyr 28,
9-12 [Babyl, Sumerian]
Van der Waerden B L 1938 QSGM (B)4,
359-382 [Egypt]
Vogel K 1931 Archeion 13, 42-44
(Z1, 321) [Egypt]
Yeldham F A 1927 Arc Sto 8, 313-329
[mid]

## Generalization

Blumberg H 1940 AMM 47, 451-462
Cajori 1929 AMM 36, 431-437
  (F55, 5) [in geom]
Candido G 1935 Pe M (4)15, 58-62
  (F61, 7) [Chasles, Fagnano,
  Pappus, Stewart]


## Genetics

See also Biology.

Ewens W J 1969 *Population Genetics*,
  London [bib, glossary, notes]
Kennedy H C 1965 Bio Sci 15, 418
  [Hardy-Weinberg law]
Stern C 1943 Sci 97, 137-
  [Hardy-Weinberg law]
*Sturtevant H H 1965 *A History of
  Genetics* (Harper)
Talacko J V 1966 Cz Am Con 3, 72 p
  [Mendel]


## Geodesy

See also Surveying, Tides.

Andrade Corvo J De 1881 Lisb JS
  (F13, 44) [arc length]
Baccardi J 1929 Rv Sc 67, 129-137
  (F55, 589)
Boersch O 1889 *Internationale
  Erdmessung. Geodaetische
  Litteratur auf Wunsch der
  Permanenten Commission im Central
  Bureau zusammengestellt*, Berlin,
  234 p
Bosmans H 1899 Brx SS 24(2), 111
  [Snell]
Brocard H + 1906-1908 Intermed 13,
  142; 14, 36, 51, 167; 15, 106,
  112, (Q3178) [XIX, J. de
  Géométres 1848- Jadanza]
Cajori F 1929 AMSB 35, 13-14
  (F55, 26) [mountain heights]
Cramer W 1877 *Beitraege zur
  Geschichte der Vorstellung von
  der Gestalt der Erde*, Barri
  (F9, 26)

Decourdemanche J-A 1913 J Asi (11)1,
  427-444 [Arab, Greek, Hindu,
  meridian arc]
- 1913 J Asi (11)1, 669-673
  [Babyl arc]
Diller A 1949 Isis 40, 6-9 [anc]
Eneström G 1890 Bib M 4, 64 (Q30)
Fischer I 1967 MT 60, 508-516
Fleet J F 1912 Lon Asia, 464-470
  [Hindu]
Fuhrmann A 1877 Civiling 23, 1-16
  [F9, 25)
Gelcich E 1899 ZM Ph 44(S), 107
  [distance at sea]
G F S 1930 Franklin 210,  684-686
  (F56, 8) [mountain height]
Gigas E 1958 Stu Gen 11, 47-62
  (Z79, 4)
Giletta L 1880 Batt 18, appendix
  (F12, 850) [bib]
Gore J H 1889 *A Bibliography of
  Geodesy*, USCS (1887), 311-512
  [bib]
- 1893 *Elements of Geodesy*, 3rd ed,
  NY
Guenther S 1883 *Die neueren
  Bemuehungen um schaefere Bestimmung
  der Erdgestalt*, Berlin
- 1890 Bib M (2)4, 73-80
  [Jacob's staff]
Humbert P 1939 Sphinx 9, 151-152
  (F65, 13) [meridian arc]
Koetter F 1893 DMV 2, 75-154
  (F25, 84)
Lambert W D 1936 Wash Ac 26, 491-
  506 (F62, 1011) [Clairaut, Newton,
  Stokes]
Mahnkopf 1932 Z Vermess 61, 23-33
  [Lower Saxony]
Marotte F 1931 Ens Sc 4, 269-275
  [Eratosthenes, Hipparchus]
Martin T H 1879 Par Mm 29
  (F11, 46) [Greek]
Mineo C 1911 Batt 49, 355-375
  [ellipsoid]
Mueller F J 1918 *Studien zur
  Geschichte der theoretischen
  Geodaesie*, Augsburg, 211 p
  (F46, 49)
Neugebauer O 1938 *Ueber eine Methode
  zur Distanzbestimmung Alexandria-
  Rom bei Heron*, Copenhagen, 26 p

## Geodesy   (continued)

Nobile V 1954 Linc Rn (8)16,
426-433 (MR16, 434; Z56, 243)
[Galileo]
Perrier G 1937 Rv Sc 75, 277-285
(F63, 3)  [survey]
- 1949 *Wie der Mensch die Erde
gemessen und gewogen hat. Kurze
Geschichte der Geodaesie,*
Bannberg, 190 p  (Z37, 1)
Read C B 1961 SSM 61, 18  [longitude]
Riccardi P 1879 Bln Mm (3)10,
431-528  (F11, 43)  [XIX, Italy]
- 1885 Bln Mm (4)4, 441-506, (4)5,
585-682  [Italy]
- 1887 Mod Mm 8, 63-68 [diameter]
Schmidt Fritz 1935 *Geschichte der
geodaetischen Instrumente und
Verfahren im Altertum und
Mittelalter,* Neustadt, 400 p
(Z12, 243)
Schub P 1932 Scr 1, 142-146
[ms in the Bastille]
Stackel P 1893 Leip Ber, 444
[geodesics]
Todhunter J 1873 *A History of the
Mathematical Theories of Attrac-
tion and the Figure of the Earth,
from the Time of Newton to that
of Laplace,* London, 2 vols
(F5, 52)  [to 1833]
Vitterbi A 1908 Pe M (3)23(5), 17-30
[geodesics]
Wittstein A 1892 ZM Ph 37 201-210
(F24, 58)  [old globes]
Wurschmidt J 1912 Arc MP (3)19,
315-320  [Arab, Gerbert,
instruments]
Zanotti-Bianco O 1893 Bib M (2)7,
75-79  (F25, 90)  [latitude]
- 1904-1919 Tor FMN 39, 539-565; 40,
18; 42, 25-46, 129-153; 43, 282-305,
331-353; 55, 118-131  (F39, 72)
[shape]

## Geography

See also Cartography, Demography,
Geodesy.

Beaujeu J 1963 in Taton R ed, *Ancient
and Medieval Science,* 323-330
Guenther S 1877-1879 *Studien zur
Geschichte der Mathematischen und
Physikalischen Geographie,* Halle,
6 vols  (F9, 26; 10, 32; 11, 44)
- 1899 ZM Ph 44(S), 125
[Cusanos]
Humbert P 1938 Ciel Ter 54, 233-238
(F65, 1088)  [coords, XVII]
- 1939 Ciel Ter 55, 81-86
(F65, 1088)  [latitude, XVII]
Neugebauer O 1942 APST (NS)32,
251-263  (MR3, 257)  [anc]
Schaefer H W 1873 *Die astronomischen
Geographie der Griechen bis auf
Eratosthenes,* Flensburg  (F5, 2)

## Geology

Brocard H 1895-1897 Intermed 2, 211,
328; 3, 39, 275; 4, 55 (Q108)
[bib]

## Gnomon

Allman G J 1887 Bib M (2)1, 22
[word]
Gandz S 1930 Am AJRP (2), 23-38
[Hebrew]
Dussavo R 1928 Syria 9, 80
[Syria]
Levy H L 1939 Am J Phil 60, 301-306
[Aulius Gillius]
Moreau F 1926 Brx SS 45(1), 149-153
(F52, 13)  [Arab]

## Golden section

See also Fibonacci sequences.

*Archibald R C 1918 AMM 25, 232-235
[bib]
- 1920 in Jay Hambridge, *Dynamic
Symmetry,* 146-157  (F47, 32)
*Baravalle H v 1948 MT 41, 22-31
[SMSGRS 9]  [bib]

## Heat

See also Fourier series (Ch. 2).

Burr A C 1933 Isis 20, 246-259 [thermal conduct]
Cherbuliez V 1871 Bern N G, 291-325 (F3, 11)
Gibson G A 1889 Edi MSP 7, 5-8 [since 1811]
Gradara E 1920 Rass MF 1, 8-17, 43-45 [thermodynamics]
Kowalski J 1892 Prace MF 3, 143-178 (F24, 54) [thermodynamics]
Langley S P 1889 Am JS (3)137, 1-23 [radiant heat]
Mach E 1900 *Die Principien der Waermlehre Historisch-Kritisch Entwickelt,* 2nd ed, Leipzig
Ostrogradskii M V 1953 UMN 8(1), 103-110 (Z51, 4)
Trevor J E 1920 AMM 27, 258 [thermodyn]
Vimercati G 1873 Boncomp 6, 61-65 (F5, 48) [Caldaie, heat in water]


## Heuristic

See also Analysis and synthesis, Counterexamples, Problems.

Bellman R + 1960 AMM 67, 119-133
Bieberbach L 1934 Ber Si 20, 351-360
Borel E 1909 Rv Mois 7, 360-362 [Poincaré]
*- 1936 Organon I, 33-42
*Bouligand G 1963 Par CR 256, 4138-4142 [Descartes, Leibniz, Euler]
Broglie L de 1951 *Savants et découvertes,* Paris
Carmichael R D 1922 AMSB 28, 179-210 [alg and transc]
Crombie A C 1963 *Scientific Change,* NY
*Fehr H + ed 1912 *Enquête sur la méthode de travail des mathématiciens,* Paris
*Ghiselin Brewster ed 1952 *The Creative Process. A Symposium* (U Calif P)

Gould S H 1955 AMM 62, 473-476 (Z65, 243) [Archimedes]
*Hadamard Jacques 1949 *The Psychology of Invention in the Mathematical Field,* Princeton, 158 p (repr 1954 Dover)
Halmos P R 1958 Sci Am (Sept) [pop] (repr in Kline 1968 Anthology--See Ch. 5)
Hermite C 1907 Mathesis (3)7, 145-146 (F38, 24) [observation]
Hoelder Otto 1924 *Die Mathematische Methode,* Berlin, 573 p
Loria G 1941 Archeion 23, 360-363 [language]
Martin L J 1965 *Psychological Investigations in Creativity. A bibliography (1954-1965),* Greensboro, NC
Poincaré H 1948 Sci Am (Aug) (repr in Kline 1968 anthology)
Polya G 1945 *How to Solve It* (Princeton U P) 268 p
1954 *Induction and Analogy in Mathematics* (Princeton U P), 2 vols [bib]
- 1962 *Mathematical Discovery* (Wiley), 2 vols
Popper Karl R 1959 *The Logic of Scientific Discovery,* London
Stein S K 1961 Sci Am 204, 148-158
Stuyvaert M 1923 *Introduction à la méthodologie mathématique,* Gand, 258 p
*Taton René 1957 *Causalité et accidents de la découverte scientifique,* Paris
*- 1957 *Reason and Chance in Scientific Discovery,* NY, 1yl p
*Van der Waerden B L 1954 Elem M 8, 1-28, 121-129; 9, 1-9, 49-56


## Hoaxes

See also Cranks.

Schaaf W L 1939 Scr 6, 49-55

## Hodograph

Anon 1910 Enc Br (11th ed)
Child J M 1915 Monist 25, 615-624
[Hamilton]

## Hydrodynamics and hydrostatics

See also Aerodynamics.

Appell P E 1914 Enc SM 4(5)(2),
102-208
Auerbach F 1881 *Die theoretische
Hydrodynamik nach dem Gange
ihrer Entwicklung in der neuesten
Zeit,* Braunschweig, 171 p
Dolgov A N 1933 *Nachala gidrostatiki.
Arkhimed, Stevin, Galilei,
Pascal,* Mos-Len, 403 p
Frankl F I 1950 UMN 5(4), 170-175
(Z36, 145) [Euler]
Maltézos C 1931 Athen P 6, 61-68
(Z2, 243) [clypsedra]
Nemenyi P F 1962 Arc HES 2, 52-86
(MR32, 4962; Z109, 5)
Oseen C W 1916 Nyt Tid 27(B), 2-20
(F46, 4)
Prandtl L 1957 *Fuehrer durch die
Stroemungslehre,* Braunschweig
Puppini U 1943 Bln Mm (9)10, 75-86
(MR9, 485; 485; Z61, 5)
[Bernoulli]
Riabouchinsky D 1946 Par CR 222,
426-428 (MR7, 354) [compl anal]
Rouse Hunter + 1957 *History of
Hydraulics* (Iowa Inst Hydraulic
Research) 269 p (repr 1963 Dover)
Schiller L ed 1933 *Drei Klassiker
der Stroemungslehre: Hagen,
Poiseuille, Hagenbach,* Leipzig,
97 p (F59, 848)
Timerding H E 1908 DMV 17, 84-93
[Euler]
Truesdell C 1953 AMM 60, 445-458
(MR15, 89)
Tyulina I A 1957 Vop IET (4), 34-45
(Z98, 7) [Euler]

## Induction (mathematical)

See Recursion.

## Information theory

See also Cybernetics.

Anon 1953 *Bibliography on Communica-
tion Theory,* (Union Internationale
des Telecommunications)
Gilbert, E N 1966 Sci 152, 320-
[bib]
Gnedenko B V 1957 in *Transactions of
the First Prague Conference on
Information Theory...,* 21-29
[bib Russ]
Goldstine H H 1961 Sci 133, 1393-1399
*Khinchin A I 1957 *Mathematical
Foundations of Information Theory*
(Dover)
Stumpers F L 1953 *A Bibliography of
Information Theory, Communication
Theory, Cibernetics* (Mass Inst
Tech Res Lab Electronics), 48 p
Von Neumann J 1932 *Mathematical
Foundations of Quantum Mechanics,*
Berlin, Ch. 5
Weiss P 1968 Ueberbl 1, 115-131
Zwicky F + 1967 *New Methods of
Thought and Procedure,* NY, 135-157,
163-199 [bib]

## Instruments: Arithmetical

See also Abacus, Calculating
machines, Computer, Finger
reckoning, Proportional compass,
Slide rule.

Anon 1959 Sci Am 200(4), 62 [anc]
Bespamyatnykh N D 1957 *K istorii
schetnykh instrumentov v Rossi v
XIX veka,* Minsk
- 1966 Ist M Isl 17, 281-288
[calcn]
Cheng D Chin Te 1925 AMM 32, 492-499
[China, rods]
Favaro A 1880 Ven I At (5)5, 495-521
(F12, 33) [Fuller]

## Instruments: Arithmetical (continued)

Hartree Douglas R 1953 *Calculating Instruments and Machines,* (U Ill P), 138 p

Horsburgh Ellice M ed 1914 *Napier Tercentenary Celebration Handbook of the Exhibition of Napier's Relics and of Books, Instruments, and Devices for Facilitating Calculation,* Edinburgh, 350 p ["modern instruments and methods of calculation"]

- 1916 *Modern Instruments and Methods of Calculation,* London, 343 p

Kargin D I 1959 Mos IIET 25

Kennedy E S 1950 Isis 4, 180-183 [al-Kashi]

- 1951 Am OSJ 71, 13-21 [Arab]

- 1951 Scr 17, 91-97 [XV, ast]

Ocagne M d' 1893 Brx SS 17(A), 84-86 (F25, 1913)

Sanford V 1950 MT 43, 292-294 [tallies]

Vanhee L 1926 AMM 33, 326-328 [China, Napier rods]

Wood F 1954 MT 47, 535-542 [sector compass]

Wright R R 1955 MT 48, 250 [re Wood 1954]

## Instruments: Geometrical

See also Constructions, Linkages.

Artobolevskii I I 1960 Int Con HS 9, 425-426 (Z114, 8) [curve drawing]

- 1960 Arc In HS 13, 87-93 [curve drawing]

- 1960 Vop IET (10), 129-131 (Z117, 9) [curve drawing]

Bosmans H 1920 Rv Q Sc, 13 p [geom, Hulsius]

Court T H + 1935 Phot J, 54-66 (F61, 8) [descri geom]

Curtze M 1895 Hist Abt 40, 61-165 (F26, 67)

## Instruments: Mathematical

See also Harmonic analyzer, Logic machines.

Anon 1936 Sci 83, 79 [Smith collection at Columbia U]

Dickstein S 1893 Wszech, 313-315 (F25, 97) [exhibit in Munich]

Dyck W 1893 *Einleitender Bericht ueber die Mathematische Ausstellung in Muenchen,* Allg Z (3 Nov)(S) (F25, 96) (repr DMV 3, 39-56; ZMN 25, 141-153)

Favaro A 1873 *Beitraege zur Geschichte der Planimeter,* (Allg Bauz) (F5, 52)

Frame J S 1945 M Tab OAC 1, 337-353

*Galle A 1912 *Mathematische Instrumente* (Teubner), 187 p [bib]

Gelcich E 1885 Hist Abt 30, 1-6 [Suardi]

Hellman C D 1932 Isis 17(1), 125-153 [Bird, XVIII]

Lopshits A M 1963 *Computation of Areas of Oriented Figures* (Heath), Ch. 2 [planimeter]

Maddison F R 1963 Hist Sci 2, 17-50 [bib, early]

Michel Henri 1935 Brx S Arc B (2), 65-79 [XVI, Belgium]

- 1939 *Introduction à l'étude d'une collection d'instruments anciens mathématiques,* Antwerp, 105 p (F65:1, 3)

Smith D E 1930 *Catalogue of Mathematical Instruments,* NY Ind MUS 1, 58

*Stibitz George R 1967 Mathematical Instruments, Enc Br

## Instruments: Scientific and technical

See also Astrolabe.

Anon 1891 Nat 44, 308-309 (F23, 42) [China, mariner compass]

Anthiaume A + 1910 BSM 34, 15-23 [astrolabe, Rouen museum]

Bedini Silvio A 1964 *Early American*

Instruments: Scientific and Technical (continued)

*Scientific Instruments and Their Makers,* Wash D C
Berthelot 1890 Par CR 111, 935-941 [hydrostatic balance etc]
Breithaupt G 1931 Z Instr 51, 256-259 (Z1, 323) [spirit level]
Breusing A 1890 *Die nautischen Instrumente bis zur Erfindung des Spiegelssextanten,* Bremen, 46 p
Curtze M 1896 Bib M, 65 [mid, surveying]
Destombes M 1960 Physis Fi 2, 197-210 [Arab, ast]
Eisenhart C 1949 Sci 110, 343- [design]
Eneström G 1888 Bib M (2)2, 32 (Q19) [measuring]
Fiorini M 1895 *Erd- und Himmels-globen, ihre Geschichte und Konstruktion,* Leipzig
Fretwell M B 1937 MT 30, 80-83 [thermometer]
Glaisher J 1851 *Philosophical Instruments and Processes Depending Upon Their Use,* Lewisham
Gore J H 1887 USCS (1887), 311-512 [ast, bib, geodesy, physics]
Govi G 1888 Boncomp 20, 607-622 [micrometer]
Hudson D R 1946 AJP 14, 332-336 [sector]
Kiely E R 1947 *Surveying Instruments - Their History and Classroom Use,* NY, 424 p (NCTMY 19) [bib]
Laussedat 1894-1904 Intermed 1, 23; 2, 324-325; 11, 73-74 (Q59) [recording]
Latham M 1917 AMM 24, 162-168 (F46, 52) [Astrolabe]
Luedemann K 1931 Z Instr 51, 136-144 (Z1, 114) [spirit level]
M 1932 Deu Opt Wo 18, 9-11 [thermometer]
Maistrov L E 1963 in *Voprosy istorii fiziko-matematicheskikh nauk,* Moscow
Maistrov L E ed 1968 *Nauchnye pribory,* Mos, 160 p + 278 plates [bib, selected items in Russ museums]

Maltezos C 1922 BSM (2)46, 313-316 [clepsydra]
Martin H 1871 Boncomp 4, 165-238 (F3, 2) [optical]
Millàs Vallicrosa J 1932 Isis 17, 218-258 (F58, 29) [quadrant]
Ocagne P Maurice d' 1905 *Le calcul siplifié par les procédés mécaniques et graphiques. Histoire et description sommaire des instruments et machines à calculer, tables, abaques et nomogrammes,* Paris, 234 p
Rome A 1927 Brx SS (A)47(3), 78-102 [astrolabe, Pappus]
- 1927 Brx SS (A)47 (4), 129-140 [Pappus]
Sanford V 1953 MT 46, 41 [level]
Schimpff W E 1941 MT 34, 320-331 [cross-staff]
Schmalzl Peter 1929 *Zur Geschichte des Quadranten bei den Arabern,* Munich, 142 p
Schmidt F 1935 *Geshichte der geodaetischen Instrumente und Verfahren im Altertum und Mittelalter,* Neustadt, 400 p [F61, 941)
Schmidt W 1903 Bib M (3)4, 7-12 [anc, tunneling]
Sloley R W 1926 Anc Eg, 65-67 [anc surveying]
Stone E N 1928 Wash UM 4, 215-242 [anc surveying]
Suter H 1895 Bib M (2)9, 13-18 (F26, 67) [Jacob staff]
Thorndike L 1955 Scr 21, 136-137 [ast, Bianchini]
Weissenborn H 1888 Bib M (2)2, 37 [mid, measuring]
Worrell W H + 1944 Scr 10, 170-180 [quadrant, Maridini]
Wuerschmidt J 1912 Arc MP (3)19, 315-320 [Arab, Gebert, geodesy]
Yates R C 1944 MT 37, 23-26
Zinner E 1956 *Deutsche und nieder-laendische Instrumente des 11 bis 18 Jahrhunderts,* Munich

Insurance

See also Finance.

Archibald R C 1945 M Tab OAC 1(10),
   402-403 [table]
Berkeley E C 1937 Am I Actu R 26,
   373; 27, 167 [Euler diagrams]
Biermann K-R 1955 Fors Fort 29, 205-
   208 (Z65, 244) [Leibniz]
Biermann K-R + 1959 Fors Fort 33,
   168-173 (Z85, 4) [Leibniz,
   de Witt]
Bohren 1911 Intermed 18, 55 (Q1359)
   [de Moivre, table]
Braun P 1925 Geschichte der Lebens-
   versicherung und des Lebens-
   versicherungstechnik, Nuernberg,
   433 p (F51, 37)
Brocard H 1911 Intermed 18, 272
   [de Witt]
De Morgan A 1854 Assur Mag 4, 185-
   [Baily, Barrett]
Du Pasquier L G 1909 Zur NGV 54, 217-
   243 (F40, 17) [Euler]
- 1910 Zur NGV 55, 14-22 (F41, 59)
   [Euler]
Eneström G 1896 Sv Ofv 53, 157
   [de Witt]
- 1898 Amst Verz 3, 263-272
   (F29, 30) [de Witt]
Farren E J 1851 Assur Mag 1, 40-
Hendriks F 1852-1854 Assur Mag 2,
   121-222-; 3, 93-; 4, 58-, 119-,
   300-
Henderson R 1923 Actuar SA 24, 1-
   [biogs]
Hennigsen V 1965 Soc Actu T 17(1), 227-
- 1969 Soc Actu T 21, 591-
   [Soc Actuaries]
Herman J R 1949 Actuar S A 50, 59-
Hoffman F L 1911 ASAQP 12, 667-760
   [USA]
Hohaus R A 1949 Soc Act T 1, 10-
   [Soc of Actuaries]
Houtzager Dirk 1950 Hollands lijf-en
   losrenteleningen voor 1672,
   Schiedam, 203 p
Kiepert L 1894 DMV 4, 116
Linder A 1936 Lon St (NS)99, 138-141
   (F62, 23) [D Bernoulli, Lambert]

Loewy A 1927 in J H Lamberts
   Bedeutung fuer die Grundlagen des
   Versicherungswesens, Berlin, 280-
   287 (F54, 31)
Lorey W 1937 Z Versich 37, 230-234
   (F63, 795) [XVIII, Goettingen]
- 1950 Deu Vers M 39-50 (MR15, 90)
   [Klein, Laplace]
Macaulay T B 1900 Actuar SA 6, 400-
   [Canada]
Macdonald W C 1915 Actuar SA 15, 2-
   [Actuarial Soc]
O'Donnell T 1936 The History of Life
   Insurance, Chicago
Peach P 1950 in Proc. Cleveland
   Conf. on Acceptance Sampling
   (Amer Stat Ass), 1-11
Pedoe A 1929 Actuar SA 30, 14-
Pierson I E 1901 Actuar SA 7, 1-
   [USA]
Quiquet A 1894 Apercu historique sur
   les formules d'interpolation des
   tables de survie et de mortalité,
   Paris
Raynes H E 1964 A History of British
   Insurance, London
Schaaf W L 1956 MT 49, 41-43
   [Halley, tables]
Sofonea T 1955 Versich 32, 57-69
   (Z65, 245) [Gauss]
- 1957 Verzek 34, 87-104 (Z77, 242)
   [Euler]
- 1958 Verzek 35(S), 62-79
   (Z83, 244) [Leibniz]
Todhunter R 1901 Institute of
   Actuaries' Textbook of the
   Principles of Interest, London,
   232 p (4th. ed 1937) [*bibs in
   various editions]
Toja G 1913 It SPS 7, 257-280
Van Haaften M + 1922 Verzekerings-
   Bibliografie, 1910-1920,
   Amsterdam, 98 p
Van Schevichaven S R J 1897 Bouwstof-
   fen voor de geschiedenis van de
   levensverzekeringen en lijfreuten
   in Nederland, Amsterdam, 370 p
   (F28, 39)

## Intuitionism

See also Philosophy.

Bockstaele Paul 1949 *Het intuitioni-
  sine bij de Franse Wiskundigen*,
  Brussels [Lebesgue]
Brouwer L E J 1952 So Af JS 49,
  139-146
Dresden A 1913 AMSB 20, 81-
  [and formalism]
- 1924 AMSB 30, 31-34  [Brouwer]
Fraenkel A A 1954 Isr RCB 30, 283-289
Heyting A 1954 *Les fondement des
  mathématiques. Intuitionism.
  Théorie de la démonstration*,
  Paris  [*bib]
*- 1956 *Intuitionism: An Introduction*,
  Amsterdam  [bib]
- 1960 in Nagel E + eds, *Logic,
  Methodology and Philosophy of
  Science* (Stanford U P), 194-197
- 1964 in Benacerraf P + *Philosophy
  of mathematics, Selected Readings*
  (Prentice-Hall), 42-49
- 1966 *Intuitionism*  (North Holland)
  [*bib]
Kleene S C + 1965 *The Foundations of
  Intuitionistic Mathematics*,
  Amsterdam
Larguier E H 1940 Scr 7, 69-78
  [Brouwer]

## Joseph game

  = Joseph problem = Josephus
  game  = Josephspiel

Ahrens W 1913 Arc Kul 11, 129-151
Curtze M 1895 Bib M (2)9, 33-42
  (F26, 53)
Smith David E 1925 *History of
  Mathematics* (Ginn), 541-544 [bib]

## Jurisprudence

See also Political science.

Halstead F 1949 Scr 15, 238-240

Loria G + 1933 Scr 1(3), 276

## "Kutsch"

Brocard H 1914-1915 Intermed 21, 199;
  22, 81  (Q1756)  [origin of the
  word]

## Least action

d'Abro A 1952 *The Rise of the New
  Physics*, 2nd ed  (Dover)
Brunet P 1938 *Etude historique sur
  le principe du la moindre action*,
  Paris, 114 p  (Act Sc Ind 693)
  (F64, 917)
Dugas R 1942 Rv Sc 80, 51-59
  [Maupertuis]
Geronimus J L 1954 *M W Ostrogradski.
  zum Prinzip der kleinsten Wirkung*,
  Berlin, 112 p  (260, 11)
Graf J H 1889 *Der Mathematiker
  Johann Samuel Koenig und das
  Princip der kleinsten Action*,
  Bern 46 p
Grigoryan A T 1956 Int Con HS 8,
  160-162  [XIX]
Helmholtz H v 1887 Ber Si, 225-236
Jourdain P E B 1912-1913 Monist
  22, 285-304, 414-459; 23, 277-293
Mayer A 1877 *Geschichte des Princips
  der Kleinsten Action*, Leipzig
  (F9, 23)
- 1878 Boncomp 11, 155-166  (F10, 28)
Polak L S 1935 Len IINT 5, 155-181
  (F61, 951)  [Lagrange]
- 1957 Mos IIET 17, 320-362
  (Z115, 242)  [Euler]
Vedel P 1888 Zeuthen (5)6, 13-22

## Least resistance

Bolza O 1913 Bib M (3)13, 146-149
  [Newton]
Lecat M 1913 Intermed 20, 166
  (Q4197)

## Lever

Lentzen V F 1932 Isis 17, 288-289
(Z4, 193) [Archimedes]
Pihl M 1950 M Tid B, 123-127
[MR12, 311)
- 1955 Nor M Tid 3, 148-156, 183
(MR17, 445)
Vailati G 1904 Loria 7, 33-39
[Archimede]

## Linguistics

Delavenay E + 1960 *Bibliography of
Mechanical Translation,* The Hague,
69 p  (Janu Lin M 11)
Geiger Lazarus 1868-1872 *Ursprung
und Entwicklung der menschlichen
Sprache und Vernunft,* Stuttgart,
2 vols
Ginsburg S 1966 *The Mathematical
Theory of Context Free Languages*
(McGraw-Hill), 232 p [bib, notes]
Hays D G + 1965-1966 *Computational
Linguistics: Biography 1964, 1965*
(Rand Corp), 2 vols
Kiefer F 1968 *Mathematical Linguis-
tics in Eastern Europe*  (American
Elsevier), 179 p
*Kotz S 1966 Sci Prog 54, 591
[mach tr]
MacDonald L A 1925 MT 18, 284-295
Mounin George 1964 *La Machine à
Traduire. Histoire des problèmes
linguistiques,* The Hague, 209 p
(Janu Lin M 32)
Papp Ferenc 1966 *Mathematical
Linguistics in the Soviet Union,*
The Hague, 165 p   (Janu Lin M 40)
[bib]
Schroder E 1897 Int Con 1, 147
[pasigraphy]
- 1898 Monist 9, 44-62 [pasigraphy]
Steinthal H 1890-1891 *Geschichte der
Sprachwissenschaft bei den Griechen
und Roemern, mit besonderer
Ruecksicht auf der Logik,* 2nd ed,
Berlin, 2 vols
Struve P G 1918 Pet Iz (6)(13),
1317-1318

Woodger J H 1952 *Biology and
Language* (Cambridge U P), 377 p

## Linkages

Archibald R C 1934 Scr 2, 293  [bib]
*Artobolevskii I 1964 *Mechanisms for
the Generation of Plane Curves,*
NY (Macmillan)
Aubry A 1900 Prog M (2)2, 337
[conics]
Delaunay N 1900 ZM Ph 44, 101-171
(F31, 17)  [Chebyshev]
*Ferguson E S 1966 *Straight Line
Mechanism,* MH En ST
Gebbia M 1891 Politcn 39, 778-782
(F23, 40)
Liguine V 1883 BSM (2)7, 145-160
[bib]

## Literature

See also Poetry.

Jones P S 1949-1950 MT 42, 401-402;
43, 164
Lord R D 1958 Biomtka 45, 282
[De Morgan, stat]
Slaught H E 1934 AMM 41, 167  [anc]

## Logarithmic paper

Beatty L B 1939 Isis 30, 95-96
Saussure R de 1894 Rv Sc 2, 743-750
[nomography]

## Logic diagrams

Duerr K 1949 Int Con Ph 10 (1948)2,
720-721  (Z31, 2)  [Euler, logic]
Gardner Martin 1958 *Logic Machines,
Diagrams and Boolean Algebra,* NY,
(repr 1968 Dover)
Keynesm J M 1906 *Studies and
Exercises in Formal Logic,* 4th ed
(Macmillan), 243 [Lambert]
More T 1959 JSL 24, 303-304
[Venn diagrams]

Logic diagrams (continued)

Sister M Stephanie 1963 MT 56, 98-
101 [Venn diagrams]
*Venn John 1894 *Symbolic Logic,*
2nd ed, London [Venn diagrams]


Logic machines

Gardner M 1958 *Logic Machines and
Diagrams, and Boolean Algebra,* NY
(repr 1968 Dover)
*- 1967 *Logic machines,* Enc Phi [bib]
Harley R 1879 Mind 4, 192-
[Stanhope demonstrator]
Mays W 1951 Elec Eng 23, 126-133,
278 [Jevons]
Mays W + 1951 Mind 60, 262-
[Jevons]
Tarjan R 1962 in W Hoffman ed,
*Digital Information Processors,*
NY [*bib]


Logicism

See also Philosophy.

Carnap R 1964 in Benacerraf P + ed,
*Philosophy of Mathematics,
Selected Readings* (Prentice-Hall),
31-41
Luchins E + 1965 Scr 27, 223-243


Machines

See also Computer, Calculating
machines, Instruments, Logic
machines.

Bogolyubov A N 1964 *Istoriya mekaniki
Mashin,* Kiev [Chebyshev]
Chapius A + 1949 *Les automate.
Figures artificielles d'hommes et
d'animaux. Histoire et technique,*
Neuchatel, 425 p
Khramoi A V 1956 *Ocherk istorii
razvitiya avtomatiki v SSSR.
Dooktyabrskii period,* Mos, 219 p

Laussedat 1895 Intermed 2, 324
[registers]
Lilley S 1942 Nat 149, 462-465
[math mach]
With E 1873 *Les machines. Leur
histoire, leur description, leur
usage,* Paris, 367 p


"Mathematics"

For Mathematics as a whole see
Philosophy, General (Ch. 4).

Bar A 1915 Intermed 22, 37 (Q4189
Bochner Solomon 1966 *The Role of
Mathematics in the Rise of Science,*
Princeton, 24-
Couturier C + 1900 Intermed 7, 252


Mechanics

See also d'Alembert principle,
Ballistics, Differential equations
(Ch. 2), Engineering, Force,
Gravity and free fall, Hodograph,
Hydrodynamics, Least action,
Least resistance, Lever, Linkages,
Rotation, Three body problem.


Mechanics: General

Anon 1959 *Variatsionnye printsipy
mekaniki. Sbornik statei,* Mos,
932 p
1963 *Ocherki istorii matematiki i
mekhaniki. Sbornik statei,* Mos,
272 p
Bobylev D K 1892 *Kratkii istoricheskii ocherk otkrytiya osnovnykh
printsipov i obshchikh zakonov
teoreticheskoi mekhaniki,* Pet, 36 p
Boll M 1943 *Les étapes de la
mécanique,* Paris, 127 p
Borel E 1943 *L'évolution de la
mécanique,* Paris, 227 p
Bybleinikov F D 1956 *O dvizhenii.
Iz istorii mekhaniki,* Mos, 212 p
Dijksterhuis E J 1923 C Huyg 3,
86-101 [ax]

Mechanics: General   (continued)

- 1924 *Val en Worp*, Paris, 474 p
Duehring E 1887 *Kritische Geschichte der allgemeinen Principien der Mechanik*, Leipzig, 638 p
Dugas Rene 1946 Rv Sc 84, 67-74 (MR8, 189) [kinetic energy]
- 1950 *Histoire de la mécanique*, Neuchatel, 649 p (MR14, 341; Z41; 1) (Engl tr 1955 Neuchatel; MR18, 982; 1957 NY; Z77, 241)
Duhem P M M 1905 *L'évolution de la mécaniques*, Paris, 348 p
- 1905-1906 *Origines de la statique*, Paris, 2 vols
Girvin H F 1948 *A Historical Appraisal of Mechanics*, Scranton, 284 p
Grigoryan A T 1962 *Ocherki razvitiya osnovnykh ponyatii mekhaniki*, Mos, 274 p [*bib]
Haas A E 1914 *Die Grundgleichungen der Mechanik, dargestellt auf Grund der geschichtlichen Entwicklung*, Leipzig, 221 p
Heun K 1903 DMV 12, 389-398 [tec]
Hoschek J 1968 Ueberbl 1, 193-213 [kinematics]
Ivanovskii M P 1957 *Zakony dvizheniya ...*, Mos, 127 p [discovery]
Johnson W W 1891 NYMS 1, 129-139 [ax]
Jouquet E 1908 *Lectures de mécanique, La mécanique enseignée par les auteurs originaux*, Paris, 2 vols
Klein H 1872 *Die Principien der Mechanik historisch und kritisch dargestellt*, Leipzig, 120 p
Koyre A 1955 APST (NS)45(4), 329-395 [free fall from Kepler to Newton]
Kramar F D 1964 Kazk Ak (12), 51-55 (MR30, 372) [geom]
Kragelskii I V + 1956 *Razvitie nauki o trenii. Sukhoe trenie*, Mos, 235 p [friction]
Krbek F 1954 *Grundzuege der Mechanik. Lehren von Newton, Einstein, Schroedinger*, Leipzig, 184 p
Lampe E 1917 Ber MG 16, 73-84 (F46, 42)

Lange L 1886 *Die geschichtliche Entwicklung des Bewegungsbegriffes und ihr voraussichtliches Endergebniss*, Leipzig
Liguine U 1876 Nou An (2)15, 499-501 (F8, 30) [kinematics]
MacGregor J G 1890 Nov Sco P(2)1, 460 [work]
Mach E 1889 *Die Mechanik in ihrer Entwickelung kritisch dargestellt*, 2nd ed, Leipzig [3rd ed 1897]
- 1902 *Mechanics and its Evolution*, London
- 1907 *The Science of Mechanics*, 3rd ed, Chicago [6th ed 1960 Open Court; MR22, 929]
Moiseev N D 1949 *Ocherki razvitiya teorii ustoichivosti*, Mos, 663 p
- 1961 *Ocherki razvitiya mekhaniki*, Mos, 478 p
Moody E A + ed 1952 *The Medieval Science of Weights*, Madison, 438 p
Polak L S 1960 *Variatsionnye printsipy mekhaniki, ikh razvitie i primechaniya v fizike*, Mos, 599 p
Redtenbacher F 1879 *Geist und Bedeutung der Mechanik und geschichtliche Skizze der Entdeckung ihrer Principien*, Munich (F11, 42)
Ruehlmann M 1885 *Vortraege ueber Geschichte der technischen Mechanik und der damit in Zusammenhang stehenden mathematischen Wissenschaften*, Leipzig, 565 p
Schell W 1895 Karlsruh 11, 136 [geom]
Schieldrop E B 1931 Nor M Tid 13, 1-13 [classics]
Sorel G 1903 Rv Met Mor 11, 716-718
Truesdell C A 1956 Phy Bl 12, 315-326 ["innere Druck"]
- 1958 ZAM Me 38, 148-157 (MR20, 129; Z80, 7)
- 1959 AMR 12, 75-80 (MR20, 7410) [materials]
- 1964 ZAM Me 44, 149-158 (Z117:2, 246) [momentum]
- 1968 *Essays in the History of Mechanics* (Springer), 384 p (reprints)

## Mechanics: General   (continued)

Tyulina I A + 1962 *Istoriya mekhaniki,*
 Mos, 228 p
Voss A 1901 Enz MW 4(1), 1-121
Zarankiewicz K 1958 *Kartki z*
 *dziejow mechaniki,* Warsaw, 186 p
Zeiliger D N 1898 *Ocherk razvitiya*
 *mekhaniki v tekushchem stoletii,*
 Kazan, 18 p

## Mechanics: Through XVII

Boussinesq J 1915 Par CR 161, 21-27,
 45-47, 65-70 [Aristotle]
Boutroux P 1921 Rv Met Mor 28, 657-
 688 [before Newton]
Clagett Marshall 1956 Osiris 12, 73-
 175 (Z72, 245) [Gerard]
- 1959 *The Science of Mechanics in*
 *the Middle Ages,* Madison, 711 p
De Vries H 1938 N Tijd 25, 232-252
 (F64, 918) [forces, torque]
Dijksterhuis E J 1928 C Huyg 7,
 161-180 [Huygens]
*Drake S + 1969 *Mechanics in Sixteenth*
 *Century Italy* (U Wis P), 428 p
 [bib]
Drake S 1964 AJP 32(8), 601-608
 [inertia]
Dugas R 1949 Rv Sc 87, 195-204
 (MR12, 577) [Descartes]
- 1954 *La mécanique au XVIIIe*
 *Siècle,* Neuchatel, 620 p
 (MR16, 659)
Duhem P 1905-1906 *Les orignes de la*
 *statique,* Paris, 2 vols
- 1906 Par CR 143, 809 [Torricelli]
Dupont P 1963 Tor FMN 98, 397-417
 (MR29, 807) [Gruebler 1884
 curvature of polar]
- 1963 Tor FMN 98, 489-513
 (MR29, 807) [Burmester, graph]
Enriques F 1921 Pe M (4)1, 77-94
 [relative motion, Greek]
- 1931 Milan Sem 4, 1-5 (F57, 14)
 [Greek]
Favaro A 1899 Ven Mm 26 [Galileo]
Fokker A D 1921 Ned Vs G 23(5)
 [inertial syst]

Galli M 1954 It UM (3)9, 289-300
 (MR16, 434) [Galilieo]
Grigoryan A T + 1957 Ist M Isl 10,
 671-765 (Z100, 7) [anc]
Gukovski M A 1935 Len IINT 7, 105-
 128 [Leonardo da Vinci]
- 1947 *Mekhanika Leonardo da Vinchi*
 Mos-Len, 815 p
Hall A R 1961 Arc HES 1, 172-178
 (Z95, 3) [Descartes]
Maier A 1940 *Die Impetus theorie der*
 *Scholastik,* Vienna, 178 p
 (F67, 968)
Moody E A 1951 Sci Mo 72, 18-23
 (MR12, 311) [mid]
Pennacchietti G 1889 *Gl'Italiani*
 *nella storia della meccanica,*
 Catania
Reidemeister K + 1933 *Christian*
 *Otter (1598-1660) Mechanismen,*
 Halle, 28 p (Z7, 387)
Vailati G 1897 Bo SBM, 21-22
 (F28, 49) [Euclid, lever]
- 1907 Bib M (3)8, 225-232
 (F38, 69) [moments]
Ver Eecke P 1931 Mathesis 45, 84-86
 (F57, 16) [Greek statics]
- 1933 Scientia 54, 114-122
 [Pappus]
Vicaire E 1894 Brx SS 18(1), 97-98
 [inertia]
Wiedemann E 1911 Bib M (3)12, 21-39
 (F42, 65) [Thabit ibn Qurra]

## Mechanics: From XVIII

Abraham Ralph + 1967 *Foundations of*
 *Mechanics* (Benjamin) [portraits]
Alekseev V P ⌐ 1936-1955 *Matematika*
 *i mehanika v izdaniyakh Akad. Nauk*
 *SSSR. Bibliografiya,* Mos, 2 vols
 (MR17, 813) [to 1947]
Anon 1945 Pri M Me 9, 185-192
 (MR7, 106) [Russ]
Aubry A 1927 Mathesis 41, 251-261
 (F53, 2) [vector]
- 1928 Ens M 27, 106-123 (F54, 8)
 [Steven, vector]
Bar A 1914-1915 Intermed 166-167;
 22, 182-183 (Q4190)
 ["rational mechanics"]

Mechanics: From XVIII (continued)

Basset A B 1894 Nat 49, 529-530
Bateman H 1942 Scr 10, 51-63
  [Hamilton]
Cartwright M L 1952 M Gaz 36, 81-88
  (MR13, 810) [no-lin vib]
Costabel Pierre 1951 Rv Hi Sc Ap 4,
  267-293 (MR14, 831)
  [Encyclopédie, XVIII]
- 1960 Leibniz et la dynamique.
  Les textes de 1692, Paris, 128 p
  (Z93, 5)
Cristescu N 1966 Ro Rv M 11, 1057-
  1074 [bib, Rumania]
Dugas R 1951 Rv Sc 89, 75-82
  (MR13, 198) [Poincaré]
Duhem P 1905 Par CR 140, 525
  [virtual displacements]
Dupont P 1963 Tor FMN 98, 537-572
  (MR29, 638) [D'Alembert, cen
  rot, Euler, Poisson, Poncelet]
Foppl A 1903 Die Mechanik im
  nuenzehnten Jahrundert, Munich
Freiman L S 1957 Vop IET (4), 164-
  167 (Z80, 4) [Euler]
Geppert H 1933 Ueber Gauss'
  Arbeiten zur Mechanik und
  Potentialtheorie, Berlin, 61 p
  (repr Werke 10(2)(7))
Geronimus Ya L 1952 Ocherki o
  rabotakh korifeev russkoi
  mekhaniki, Mos, 519 p
  (MR15, 275)
Gheorghita S I 1966 Ro Rv M 11, 1103-
  1122 [bib, Rumania]
*Goldstein H 1950 Classical Mechanics
  (Addison Wesley) [bib]
Golubev V V 1950 Pri M Me 14, 236-
  244 (Z37, 2) [Kovalevskaya]
- 1955 Ist M Isl 8, 77, 126
  (MR17, 1) [Mos U after 1917]
Grigoryan A T 1960 Int Con HS 9, 473-
  478 (Z114, 7) [non-Euc geom,
  Russ]
- 1960 Scientia 95, 347-350
  [non-Euc geom, Russ]
- 1961 Ocherki istorii mekhaniki v
  Rossii, Mos, 289 p [*bib 33 p]
- 1967 Evolyutsiya mekhaniki
  v Rossii, Mos, 167 p

Henneberg L 1894 DMV 3, 567-599
  [statics]
Honol S 1950 Glasn MPA (2)5, 21-32
  (MR12, 311) [Stay, Boscovitch,
  abs motion]
Kilmister C W 1967 Lagrangian
  Dynamics, NY, 136 p
Korner T 1904 Bib M (3)5, 15-62
  [XVIII]
Kosmodemyanskii A A 1964 Ocherki
  po istorii mekhaniki, Mos, 456 p
  (Z119, 10)
Laue M v 1948 Naturw (5)35, 193-196
  [Lange]
Lebesgue + 1922 Ens M, 209
  [Poincaré, Roberval, rel mot]
Marcolongo R 1909 Int Con (4)3,
  488-499 [Filippis, Lagrange]
Mikami Y 1914 Nieu Arch (2)11(1),
  76-78 [Mayeno, vectors]
Polak L S 1956 Int Con HS 8, 174-177
  [XIX, XX]
Putyata T V + 1952 Diyalnist
  vidatnikh mekhanikiv na Ukraini,
  Kiev
Truesdell C 1957 Ens M 3(2), 251-262
  [Euler]
- 1960 Arc HES 1, 3-36 (Z96, 3)
  [XVII]
- 1966 Six Lectures on Modern
  Natural Philosophy (Springer), 117 p
Tyulina I A 1955 Ist M Isl 8, 489-
  536 (MR17, 1) [XVIII, XIX,
  Moscow U]
Verchunov V M 1959 Mos UM Me 13(6),
  77-89 (Z86, 4) [Lobachevskii]
Wangerin A 1890 Ueber die Anziehung
  homogener ellipsoide. Abhandlung
  von Laplace, Ivory, Gauss, Chasles,
  Dirichlet, Leipzig, 118 p
Weyrauch J 1874 ZM Ph 19, 361-390
  (F6, 43) [graphical statics]
Wierzbicki W 1959 Kwar HNT 4, 595-
  604 [moments]

Medicine

See also Biology.

Brocard H 1902 Intermed 9, 186
  [doctor mathematicians]

Medicine    (continued)

Niewenglowski G 1867 *Les mathémati-*
*ques et la médecine,* Paris
Sanford V 1952 MT 368, 372  [XVI]


Meetings

See also Cities (Ch. 4),
  Organizations.

Bouligand G 1938 Rv Sc, 221-226
Brocard H 1910 Intermed 17, 33
  (Q2620)  [XVII]
Estève Madeleine 1956 *Répertoire des*
  *congrès...,* Paris (Inst H.
  Poincaré),  94 p
Gregory Winifred ed 1938 *Inter-*
  *national Congresses and Conferences*
  *1840-1937,*   NY
Laisant C A + 1898 Intermed 5, 81,
  175 (Q212)
Mittag-Leffler G 1926 Hels Comm 3(6)
  (F52, 35)
Wavre R 1948 in Le Lionnais F ed,
  *Les grands courants de pensée*
  *mathématique,* Marseille, 298-303
  [list]


Meteorology

Anon 1889 *Bibliography of Meteorology.*
  *A Classified Catalogue of the*
  *Literature of Meteorology from the*
  *Origin of Printing to 1887,* Wash
  DC, 390 p
- 1955 *50 Jahre Grenzschichtforschung,*
  Braunschweig, 499 p
- 1960 *Problema pogranichnogo sloya*
  *i voprosy teploperedachi,* Mos
- 1966 *Weather and Climate*
  *Modification, Problem and*
  *Prospects,* Wash DC, 2 vols
  (NAS-NRC 1350)
Frisinger Howard H 1965 Am Met B
  46(Dec)  [thunder, lightning]
- 1965 *The Role of Mathematics and*
  *Mathematicians in the Development*
  *of Meteorology to 1800...,*
  (U Michigan thesis)

Guenther S 1887 Bib M (2)1, 65-69
Harrington M W 1894 Smi R, 249-270


Metric system

See also Metrology.

Bigourdan G 1901 *Le Système*
  *métrique des Poids et Mesures,*
  Paris  [bib]
Brocard H 1922 Intermed (2)1, 9-10
  (F48, 46)  [Spencer]
Doublet E 1931 Rv Sc 69, 225-230
  (F57, 1289)
Gliozzi M 1932 Tor FMN 67, 29-50
  (F58, 32)  [precursors]
Guth J 1900 Cas MF 29, 121
  [anniv]
Hellman C D 1936 Osiris 1, 314-340
  [Legendre]
Johnson J T ed 1948 *The Metric*
  *System...,*  NCTMY 20
Karpinski L C 1921 Sci 53, 156-157
Kellaway F W 1944 M Gaz 28, 104-106
  [British and metric]
Kennelly Arthur E 1928 *Vestiges of*
  *Pre-metric Weights and Measures*
  *Persisting in Metric-System*
  *Europe 1926-1927,* NY, 189 p
  (F54, 45)  [bib]
Lehmann C F 1889 Ber Ph 7, 81-101
  [Babyl]
Schreiber E W 1929 MT 22, 373-381
  (F55, 589)  [facts for teachers]


Metrology

See also Coins, Metric system,
  Stamps.

Aaboe A + 1964 Koyre 1, 1-20
  [anc qualitative]
Airy W 1918 *On the Ancient Trade*
  *Weights of the East,* London, 36 p
Allotte de la Fuije 1930 Rv Assyr 27,
  65-71 [Babyl]
Anon 1901 Intermed 7, 165
  [XVII, Spain]
- 1936 Nat 137, 890-892 [Egypt]

Metrology     (continued)

Asimov Isaac 1960 *Realms of Measure*, NY (Fawcett) [pop]

Aurès A 1891 *Memoires, redigé pour completer la détermination des mesures agraires de longeur et de superfice autrfois en usage chez les assyriens*, Nimes, 102 p

Barbarin P + 1900-1902 Intermed 7, 373; 9, 104 (Q1751) [anc weighing]

Barny A 1863 *Traité historique des poids et mésures et de la vérification*, Paris

Baron R 1957 Isis 48, 30-32 [XII]

Berriman A E 1953 *Historical Metrology. A New Analysis of the Archaeological and Historical Evidence Relating to Weights and Measures*, Lon-NY, 350 p

Bowman M E 1950 *Romance in Arithmetic. A History of Our Currency, Weights and Measures and Calendar* (Univ London Press)

Boyer C B 1942 Sci 95, 553 [Babyl, thermometry]

Brandis J 1866 *Das Muenz-Maass- und Gewichtsystem in Vorderasien ...*, Berlin

Brunschvig R 1937 Alger Ori 3, 74-88 [capacity, XVII]

Chih K C 1939 M Gaz 23, 268-269 [China, length]

Clavero y Guervos 1902 Intermed 9, 11, 128 (Q1516) [XVII, Spain]

Cramer G F 1937 AMM 45, 344-347 [Maya length]

Decourdemanche J A 1909 *Traité pratique des poids et des mesures des peuples anciens et des Arabes*, Paris, 152 p (F40, 69)

Delesalle H 1965 Rv Hi Sc Ap 18, 305-308 [length]

Depman I Ya 1956 *Vozniknovenie mer i sposobov izmereniya velichin*, Mos, 136 p

Devignot 1909 Intermed 16, 130-131 (Q3373) [France XVII]

Eisenhart C 1963 USNBS(C)67, 161 [bib, calibration]

Ellis Brian 1966 *Basic Concepts of Measurement* (Cambridge U P), 220 p [bib]

Eneström G 1896 Bib M (2)10, 31-32, 64, 96, 120 [Arab coins]

Forbin V 1913 Nat Par (2), 287-288 [capacity, Hebrew]

Glanville S R K 1935 Lon RIGBP 29, 10 [Egypt]

Glazebrook R 1931 Nat 128, 17-28 (F57, 49) [standards]

Guerra F 1960 J H Med 15, 342-344 [anc America]

Guilhiermoz P 1919 Par EC Bib 80, 5-100 [mid]

Hallock W + 1906 *Outlines of the Evolution of Weights and Measures and the Metric System*, NY

Helmholtz Hermann v 1930 *Counting and Measuring*, NY, 73 p

Hemmy A S 1938 Iraq 5, 65-81 [Greek, Persia]

Hinz Walther 1955 *Islamische Masse und Gewichte, Umgerechnet ins metrische System*, Leiden, 66 p (Z66, 244)

Hultsch F 1862 *Griechische und roemische Metrologie*, Berlin

Humilis 1915 Intermed 22, 56 (Q1851) [length]

Kiel A 1895 *Geschichte der absoluten Masseinheiten*, Bonn, 18 p (F26, 66)

Kinkelin H 1873 Basl V 5, 147- [mid]

Kisch Bruno 1965 *Scales and Weights* (Yale U P), 297 p

La Ponce de 1858 *Mémoire et documents sur les déterminations de la mesure longimetrique du mille romain...*, Tours, 20 p

Lemaire G + 1906 Intermed 13, 120, 217 (Q2964)

Lenormant F 1873 *Essai sur un document mathématique chaldéen, et à cette occasion, sur le système des poids et mesures de Babylone*, Paris (F5, 1)

Lepsius C R 1877 *Die Babylonisch-Assyrischen Laengenmasse nach der Tafel von Senkereh*, Berlin, 144 p

Metrology (continued)

- 1884 *Die Langenmasse der Alten,*
  Berlin
Lysenko V I 1959 Ist M Isl 12, 161-
  178 [XVIII, Russia]
Machabey Armand 1954 *Aspects de la*
  *métrologie an XVIIe siècle,* Paris,
  18 p (Z60, 8)
Moody Ernest A + 1952 *The Medieval*
  *Science of Weights,* Madison,
  448 p (MR14, 1049; Z49, 145)
Moors B P 1904 *Le système des poids,*
  *mesures et monnaies des Isralites*
  *d'après la Bible,* Paris
Neugebauer O 1927 Gott Ab 13, 1
  [sexagesimal]
Nilsson N G 1936 Lychnos 1, 235-247
  [ships]
Oppert 1875 *Etalon des mesures*
  *assyriennes,* Paris
Paulme D 1942 Rv Sc 80, 219-226
  [Africa]
Petrie W M Flinders 1926 *Ancient*
  *Weights and Measures...,* London,
  58 p
- 1938 J Eg Arch 24, 180-181
  [Egypt]
Sachs A J 1944 Am SORB (96), 29-39
  (MR6, 141) [Babyl]
- 1947 J Can St 1, 67-71
  [Babyl, tables]
Schirmer O 1957 Bay Jb OK, 17-50
  (Z80, 6) [Arab]
Scott N E 1942 NY Met MA, 70-75
  [Egypt]
Segre A 1944 Am OSJ 64, 73-81
  [Assyrian, Babyl, Persian]
Sleight E R 1945 M Mag 19, 236-243
  [tables]
Thureau-Dangin F 1932 Rv Assyr 29,
  189-192 [Babyl "ga"]
- 1935 Rv Assyr 32, 1-28 (F61, 10)
  [Babyl]
- 1937 Rv Assyr 34, 80-86
  (F63, 6) ["ga"]
- 1938 Rv Assyr 35, 156-157
  [Babyl]
Ver Eecke P 1936 Brx SS(A)56, 6-16
  [Greek]
Vogel K 1931 Z Aeg SA 66, 33-35
  [Egypt]

Voronets A M + 1928 *O merakh i schete*
  *drevnosti,* Mos-Len, 3 vols
Warren Charles 1913 *The Early Weights*
  *and Measures of Mankind* (The
  Palestine exploration fund, London),
  135 p
Yu F B 1873 *Grecheskaya i rimskaya*
  *metrologiya,* Mos, 70 p
Zupko Ronald Edward 1968 *A Dictionary*
  *of English Weights and Measures*
  *from Anglo-Saxon Times to the*
  *Nineteenth Century* (U Wisconsin P),
  224 p [*bib (30p), chronology of
  laws]


Military

See also Ballistics, Cryptanalysis.

Berthelot D 1902 J Sav, 116
  [Leonardo da Vinci]
Hierholtz M R 1923 Ens M 23, 210-212
Morse M 1943 Sci Mo 56, 50-55
  [total war]
Neugebauer O 1932 QSGM (B)2, 305-310
  (Z6, 1) [Babyl]
Peters T 1933 QSGM (B)2, 352-363
  (Z7, 147) [fortification, Otter]
Schaaf W L 1945 Scr 11, 57-74
  [bib, recent]
Van Wijk H H 1935 Ind Mil Ti, 457-
  466, 682-697 (F61, 3)
  [artillery]
Waschow H 1933 Unt M 39, 368-373
  (F59, 14) [Babyl]


"Million"

Struik D J 1931 Ned Taal 50(2), 173-
Vosterman Van Oijen G A 1868 ZM Ph
  14, 22-25 (F1, 12)


Mistakes

(incorrect results, arguments,
  assumptions, and conjectures)

Archibald R C 1936 AMM 43, 487-
  [re Lecat 1935]

## Mistakes (continued)

Aubry A 1911 Fr AAS, 24    [list]
- 1912 Fr AAS 41, 57-64
Ball W W R 1960 *Mathematical
  Recreations,*  Chs 2-3
Betsch Christian 1926 *Fiktionen in
  der Mathematik,* Stuttgart, 396 p
*Dickson Leonard E 1919-1923 *History
  of the theory of numbers,*
  Washington, 3 vols (repr 1934 NY)
*Dubnov Y S 1963 *Mistakes in Geometric
  Proofs* (Heath) [elem]
 Emch A 1937 M Mag 11, 186-189
  [rejected papers]
 Epstein S S 1893 Bern N G 183-192
 Favre L 1912 Fr AAS 41, 64-66
 Genocchi 1880 Tor FMN 15, 803
  [Euler, Legendre]
 Goellnitz E 1936 DMV 46, 19-21
  (F62, 25) [Gauss]
 Heller S 1954 Mun Abh 63, 3-39
  [Archimedes]
 Hermelink H 1953 Arc In HS 25, 430-
  433 [Z52, 1] [Archimedes]
*Lecat Maurice 1935 *Erreurs de
  mathématiciens des origines à nos
  jours,* Brussels-Louvain, 179 p
  (F61, 1; Z10, 385) [bib]
 Lietzmann W 1923 *Trugschluesse*
  (Teubner) 54 p
 Lietzmann W + 1923 *Wo Steckt der
  fehler?* 3rd ed, Berlin, 48 p
 Lodge O J 1889 Nat 40, 273 [chem]
 Lorey W 1933 Crelle 170, 129-132
  (F59, 26) [Euler, primes]
*Mackie J L 1967 Fallacies in
  (Enc Phi) [bib]
 Maillet E + 1906-1914 Intermed 13,
  248-252; 21, 83-85; 22, 14-15
  (Q2855)
 Miller G A 1913 AMM 20, 14-20
  [group]
- 1923 Sci Mo 17, 216-228
  (F49, 21)
- 1936 Sci 84, 418-419
  [Webster's Dictionary]
 Pickering S 1889 Nat 40, 343-344
  [chem]
 Queneau R 1964 Koyre 1, 475-580
  [numb thy]

Ritter E 1868 Boncomp 1, 245
 (F1, 9)
Sang E 1890 Nat 42, 593

## Models (theoretical)

Freudenthal H ed 1961 *The Concept
  and Role of the Model in Mathematics
  and Natural and Social Sciences,*
  Dordrecht, 204 p
Suppes P 1960 Synthese 12, 287-301

## Music

See also Art.

Amir-Moez A R 1956 Scr 22, 268-270
 (SMSGRS 8) [scales]
Archibald R C 1924 AMM 31, 1-25
- 1924 Argosy 50, 135-142
Barbour J M 1940 Scr 7, 21-31
 (F66, 1187) [musical log]
Brown J D 1968 MT 61, 783-787
 [XVII-XX]
Coxeter H S M 1962 Can Mus J 6(2),
 13-24
- 1968 MT 61, 312-320
Dale H H 1945 Nat 156, 193-194
 [Newton]
Gonggrijp B 1916 N Tijd 4, 204-208
 [harmony]
*Helm E E 1967 Sci Am (Dec), 93-103
 [Pythagoras]
Hiller L + 1965 Sci 150, 161-169
 [bib, electronic]
*Mode E B 1962 M Mag 35, 13-20
 [SMSGRS 8]
Smith D E 1927 Tea Col R 28, 1
*Tannery P 1902 Bib M (3)3, 161
 [Greek]
Van der Waerden B L 1943 Hermes 78
 [Pythagoras]
Van Deventer C M 1929 Euc Gron 6,
 49-64   (F55, 599) [anc]

## Mysticism

See also Astrology, Numerology,
  Religion.

## Mysticism (continued)

Brunschwicg L 1937 *Le rôle du pythagorisme dans l'évolution des idées,* Paris (Act Sc Ind 446, 25 p
Cassirer E 1940 Lychnos, 248-265 (F66, 15)
Fettweis E 1932 Archeion 14, 207-220 (F58, 7; Z6, 1)
Junge G 1948 Clas Med 9, 183-194 (Z37, 290) [num thy]
Mahnke D 1937 *Unendliche Sphaere und Allmittelpunkt. Beitraege zur Genealogie des mathematische Mystik,* Halle, 260 p (repr 1966 Stuttgart) (F63, 12)
Mahnke D + 1949 Isis 32, 131-133
Naber H A 1914 *Meetkunde en Mystick,* Amsterdam, 99 p (F45, 78)
Sayle A H Bibcl Arc 4, 302-314 [Babyl]
Shaw J B 1941 Scr 8, 69-77 [Pascal]
Sister Alice Irene 1931 MΓ 24, 139-150 [spiral]

## Navigation

Bertelli T 1893 Vat NL Mm 9(1), 77-178, 9(2), 131-218 (F25, 89)
Formaleoni 1788-1789 *Storia filosofica e politica della navigazione,* Venice, 2 vols
Gelcich E 1899 ZM Ph 44(S), 105-111 (F30, 60)
Moreno N 1914 Arg SC 77, 5-48 [aerial]
Nilsson N G 1936 Lychnos, 235-250 (F62, 1016)
Photinos N G 1955 *"Nautika" Der Beitraeg der Griechen zur Entwicklung der theoretischen Nautik,* Athens, 172 p (Z68, 3)

## Nim

Archibald R C 1918 AMM 25, 132-142
*Ball W W R 1938 *Mathematical Recreations,* 36-40

Coxeter H S M 1953 Scr 19, 142-143 [Wythoff game similar to Nim]
Hardy G H + 1938 *An Introduction to the Theory of Numbers,* Oxford
Uspensky J V + 1939 *Elementary Number Theory,* NY

## Nomography

See also Graphical methods.

Boulanger G R 1949 *Contribution à la théorie générale des abaques à plan superposé,* Brussels-Paris
Glagolev A A 1951 in *Nomograficheskii sbornik* (Mos U), 7-24 (MR14, 523)
Lallemand C 1922 Par CR, 82-88
Luckey P 1923 Unt M 29, 54-59
- 1927 ZMN Unt 58, 455-465 (F53, 2)
Ocagne P Maurice d' 1899 *Traité de Nomographie,* Paris
* 1905 *Le calcul simplifié par les procédés mécaniques et graphiques. Histoire et description sommaire des instruments et machines à calculer, tables, abaques et nomogrammes,* Paris, 234 p
1907 Rv GSPA 18, 392-395 (F38, 61)
- 1922 Rv GSPA 32, 620-623
- 1926 Brx SS 46, 55-66 (F52, 3)
- 1928 Rv GSPA 39, 625
1929 Archeion 11, xiv-xv (F55, 4)
- 1929 Rv GSPA 40, 325-329
Pesci G 1908 Zarag Fac 2, 153-160, 233-240 [Luyando]
Soreau R 1922 Rv GSPA 32, 518-523
Weyrauch J 1874 ZM Ph 19, 361-390 (F6, 43) [mech]

## Notation (symbols)

See also Terminology (words), Numeration, Plus and minus signs, x.

Aldrich V C 1932 Monist 42, 564-576 (F58, 4)
André D 1909 *Des notations mathé-matiques. Enumération, choix et*

## Notation (symbols) (continued)

*usage,* Paris, 519 p (F40, 49)
Anon 1939-1940 AMM 46, 233; 47,
107
Baidaff B I 1928 B Ai Bol 1, 1-24,
33-35
Benedict S R 1910 SSM 9, 375-384
[alg]
Cajori Florian 1915 Nat 94, 643
[trig]
- 1921 AMSB 27, 453-458 [calc,
Leibniz, Newton]
- 1922 M Gaz 11, 136-143 [mult]
- 1923 AMM 30, 65-66 [angles]
- 1923 AMSB 29, 12 (F49, 22)
[mult]
- 1924 AMM 31, 65-71 [alg, Rahn]
- 1924 AMSB 30, 13 (F50, 19)
- 1924 An M (2)25, 1-46 (F50, 7)
[calc]
- 1924 Int Con (7)2, 929-936
(F54, 48)
- 1924 Int Con (7)2, 937-941
(F54, 48)
- 1924 Isis 6, 391-394
- 1925 AMM 32, 414-416 [USA]
- 1925 Isis 7, 412-429 [Leibniz]
- 1925 MT 17, 87-93 [unification]
- 1928-1929 *A History of Mathematical
Notations,* Chicago, 2 vols
Cazanacli P V 1932 Tim Rev M 12, 101-
102 (F58, 993) [divis, mult]
Clifton R B 1865 Cam PST 11, 213-
[DeMorgan]
Collins J V 1923 MT 16, 157-161
[defects]
Davies T S 1833 Bath Bris 8
Dordea T A + 1931 Tim Rev M 11,
14-17
Efendi S Z 1898 J Asi (Jan)
(F29, 28) [alg]
Feldhause Franz M 1953 *Geschichte
des technischen Zeichens,*
Wilhelmshaven, 109 p
Freudenthal H 1960 Bulg FM 3(36),
36-49 (MR23, 583)
Gandz S 1932 QSGM (B)2, 81-97
[Arab, Greek]
G B M 1908 Nat 77, 347-348 [arith]
Green M G 1959 Ari T 6, 215-216

Hofmann J E 1948 Experien 4, 364-366
[Euler]
Lattin H P 1933 Isis 19, 181-194
[Bubnov]
Le Paige C 1892 Brx SS 16(A), 56-57;
(B) 70-83 (F24, 43) [arith]
Lorey W 1930 ZMN Unt 61, 221-225
(F56, 4)
Luquet G H 1929 J Ps NP 15. 733-762
Mahnke D 1913 Bib M 13, 250-260
[Leibniz]
Malmsten C J 1883 Nor M Tid, 136-140
[re Lalande 1882]
Neville E H 1950 Adv Sc 7, 119-130
(Z37, 289)
Peano G 1906 Intermed 13, 18
- 1915 Scientia 18, 165-173
Pelseneer J 1932 Isis 17, 259
(F58, 34) [XV I]
Pogorzelski H A 1958 M Mag 33, 184
[Descartes]
Sarton G 1937 Isis 27, 328 [alg]
Smith D E 1932 Scr 1, 304
[nabla, del]
Tannery P 1904 Bib M (3)5, 5-8
(F35, 53) [Greek, subtraction]
Wieleitner H 1926 Arc Sto 7, 29-33
[Bombelli]

## Numeration

See also Arithmetic (Ch. 2), Billion,
Egyptian fractions (under Civilizations in Ch. 5), Fifty, Finger
reckoning, Million, Seven, Terminology.

## Numeration: General

Agostini A 1952 Archim (4-5), 213-
214 [place value]
Berdelle C + 1900 Intermed 8, 227
[base 64]
- 1905-1908 Intermed 12, 103, 234,
255; 14, 275; 15, 232 [bases]
Bertin G 1878 Bibcl Arc 7, 370-389
[Assyrian]
Boyer C B 1944 Isis 35, 153-168
(MR5, 253; Z60, 2)

## Numeration: General  (continued)

- 1959 MT 52, 127-129  [Egypt]
Bubnov N M 1908 *Arifmeticheskaya samostoyatelnost evropeiskoi Kultury,*  Kiev
Cajori Florian 1925 AMM 32, 414-416 [American Indian]
- 1928 *A History of Mathematical Notations,* 1, 1-70  [bib]
Chaykivskii M 1908 Lemb Sam 12, 4
Campbell W B 1937 AMM 44, 238 p (F63, 2)  [ordinals]
Cantor M 1858 Deu NA Ber 34, 135-
Datta B 1931 Scientia 50, 1-12 (F57, 6; Z18, 385)  [place value]
Eells W C 1912 Bib M (3)13, 218-222 (F43, 65)  [American Indian]
- 1913 AMM 20, 263-272, 293-299 [American Indian]
Frolov B A 1965 Sib TN 9(3) [prehistoric]
Gandz S 1930 Isis 14, 189-214 [abacus, Hebrew, Knot]
- 1933 Am AJRP  [Hebrew]
Ganguli S 1930 Clct MS 22, 99-102 [Babyl, Maya, place value]
Ginsburg J 1932 Scr 1, 60-62 [XIV]
- 1932 Scr 1, 87  ["cuento"]
Glathe A 1932 *Die chinesischen Zahlen,* Tokyo, 52 p  (F58, 994)
Gordon C H 1934 Rv Assyr 31, 53-60 (F60, 818)  [Nuzi tablets]
Grotefeld 1803-1804 Gott Anz (149); (178); (60), (11)
Guitel G 1956 Int Con HS 8, 52-56 [Aztec, Egypt]
Gunoermann G 1899 *Die Zahlzeichen,* Giessen, 50 p  (F30, 36)
Hartner W 1943 Paideuma 2, 268-326 [anc and mod]
Herskovits M J 1939 Man 39, 154-155 [Kra]
Himly Deu Morg Z 18, 292-381
Hopper V F 1938 *Medieval Number Symbolism,* NY, 253 p  (F64, 909) [mid]
Householder A S 1937 AMM 44, 463-464 (F63, 2)  [words]
Humboldt A v 1829 Crelle 4, 205-231 [place value]

Jacobs H v 1896 *Das Volk der Siebener-Zahler,*  Berlin
Karpinski L C 1911 AMM 18, 97-102
Keller H 1932 Scr 1, 66-67 [Hebrew]
Klingenheben A 1926 Afr USK, 17 [Africa]
Koca H J 1960 MT 53, 191
Koppelman D 1929 Int Arc Et 30, 77-94 [South America]
Laki K 1960 Wash Ac 50(4), 1-11 (MR22, 1; Z102, 243)  [base six]
Langley E M 1924 M Gaz 12, 468-269 [mixed Arabic and Roman]
Lehmer D H 1933 M Mag 7, 8-12
Locke L 1923 *The Ancient Quipu or Peruvian Knot Record* (American Museum of Natural History), 84 p (F49, 23)
Loefler Eugen 1912 Ulm VMN 15
- 1912 *Ziffern und Ziffernsysteme der Kultervoelker in alter und neuer Zeit,*  Leipzig, 97 p  (F43, 65)
- 1918 *Ziffern und Ziffernsysteme. 1 Teil. Die Zahlzeichen der alten Kulturvoelker,* 2nd ed, Leipzig, 58 p  (3rd ed 1928)  (F46, 39)
- 1919 Unt M 25, 85
- 1919 *Ziffern und Ziffernsysteme. 2. Teil. Die Zahlzeichen im Mittelalter und in der Neuzeit,* 2nd ed, Leipzig, 60 p  (F47, 22)
- 1920 Braun Mo, 540-548
Loewy A 1930 *Ueber die Zahlbezeichnung in der juedischen Literatur,* Berlin, 11 p
Lorey W 1940 Scientia 67, 68-75 (F66, 4)  [anc]
Marre A 1899 Tor FMN 34, 447-458 (F30, 38)  [Madagascar, Philippines]
Martin T H 1864 *Les signes numéreux et l'arithmétique chez les peuples de l'antiquité et der moyen-age,* Rome
*Menninger Karl 1934 *Zahlwert und Ziffer. Aus der Kulturgeschichte unserer Zahlsprache und der Rechenbrett,* Breslau, 365 p (F60, 1; Z8, 98)
- 1957-1959 *Zahlwort und Ziffer. Bd. I. Zahlreihe und Zahlspreche. Bd. II Zahlschrift und Rechnen,*

Numeration: General   (continued)

2nd ed, Goettingen, 2 vols
(Z82, 5)
*- 1969 *Number Words and Number
Symbols* (MIT Press), 480 p
(tr of 1957-1959)
Mercier G 1933 J Asi 222, 303-322
[Libya]
Mirick G R + 1925 MT 18, 465-471
[bases]
Mordell Phineas 1922 *The Origin of
Letters and Numerals,* Philadelphia,
71 p
Ninni A P 1889 Ven I At (6)7, 679-
686 [anc]
Oettel H 1961 Mathunt 7(3), 21-42
Picard E 1894 Intermed 1, 219
[anc]
Picton J A 1875 Liver LPS 29, 69-
116  (F7, 18)
Pihan 1860 *Expose des signes de
numération usites chez les
peuples orientaux anciens et
modernes,*  Paris
Pott 1847 *Die quinaere und
vigesimale Zaehlemethode bei
Voelkern aller Weltheile,* Halle,
312 p
- 1868 *Die Sprachverschiedenheit in
Europa an den Zahlwoertern
nachgewiesen, sowie die quinaere
und vigesimale Zaehlmethode,*
Halle
Richeson A W 1933 AMM 40, 542-546
(Z8, 99)  [Maya]
Ruska J 1922 Arc GNT 9, 112-126
Saalschutz L 1892 Konig Ph 5, 4-9
(F25, 57)  [anc]
Salyers G D 1954 M Mag 28, 44-48
(MR16, 207)  [Mayan]
Sánchez-Perez J A 1935 Al And 3, 97-
125  [XVI, Morocco]
Sanford V 1951 MT 43, 368-370
Savary A 1808 Par SASPN 2, 71-
Sethe Kurt 1916 *Von Zahlen und
Zahlworten bei den alten
Aegeptern...,* Strassburg, 155 p
[Egypt]
Shvetsov K I 1952 Matv Shk (2), 8-12
(MR13, 809)  [Slavonic]

Simeon 1867 Mex Arc CS 3, 523
[Mexico]
Simonov R A 1963 Chislovye gramoty
na bereste XIII-XIV vv. i nekotorye
voprosy istorii Kirillovskoi
numeratsii, in *Khilyada i Sto
Godini slavyauska pismenost*
Sophia
- 1964 O nekotorykh osobennostyakh
numeratsii, primenyavsheisya v
kirillitse in *Istochnik ovedenie
i istoriya russkogo yazika,*
Moscow
Slaught H E 1928 MT 20, 303-309
Smeltzer Donald 1953 *Man and Number*
London, 114 p  (repr 1958 NY)
[pop]
Smith D E 1919 *Number Stories of Long
Ago,* Boston, 143 p  [pop]
Smith D E + 1927 AMM 34, 258-260
[Arab]
- 1937 *Numbers and Numerals,* NY
(Teachers Coll)
Struik D 1964 MT 57, 166-168
[Kensington stone]
Sushkevich A K 1948 Mat v Shk 4, 1-16
[anc]
Tannery P 1895 Intermed 2, 214 (Q147)
[anc]
Taylor E B 1873 *Primitive Culture,*
London
Thureau-Dangin F 1928 Rv Assyr 25,
119-121 [Sumerian, ternary]
Unger E 1931 Fors Fort 7, 263
(F57, 1289)  [anc]
Vetter Q 1955 Leip MNR 5, 131-132
(MR18, 453)  [anc]
Villiers Melius de 1923 *The Numeral-
Words; Their Origin, Meaning,
History and Lesson,*  London, 124 p
(F49, 23)
Wertheimer M 1950 Numbers and
Numerical Concepts in Primitive
Peoples in W. D. Ellis ed, *A Source
Book in Gestalt Psychology*
(Humanities Press)
Wieleitner H 1917 ZMN Unt 48, 213-215
(F46, 46)  [Maya, zero]
Willers F A 1949 *Zahlzeichen und
Rechnen im Wandel der Zeit,*
Berlin-Leipzig, 84 p  (Z36, 145)

## Numeration: General (continued)

Woodruff C E 1909 AMM 16, 125-133
Zhukovskaya L P 1964 K istorii
  bukvennoi tsifiri i alfabitov u
  Slavyan, in *Istochnikovedenue i
  istoriya russkogo yazyka,* Moscow


## Numeration: Arabic

Abbott N 1948 Lon Asia, 277-280
Anon 1936 AMM 43, 99
Bayley E C 1883 Lon Anthr 14, 335:
  15, 1
Beckingham C F 1940 Lon Asia, 61-64
Benedict S R 1914 *A Comparative
  Study of the Early Treatises
  Introducing Into Europe the Hindu
  Art of Reckoning* (U Mich Thesis),
  132 p (also 1916 Concord, N.H.)
  (F45, 94; 46, 56)
Boev G P 1956 Mos M Sez, 165
Buddhue J D 1941 Sci Mo 52, 265-267
Bubnov N M 1908 *Proiskhozhdenie i
  istoriya nashikh tsifr,* Kiev, 197 p
  (= *Issledovaniya po istorii
  nauki v Evrope* 1 (2)) (F39, 53)
- 1924 Zagr BI, 102-113 (F50, 11)
Cajori F 1919 Sci Mo 9, 458-464
- 1925 AMSB 31, 113 (F51, 41)
- 1925 MT 18, 127-133 (F51, 5)
Campagne Maurice 1904 *De l'emploi
  des schiffres dits arabes au
  moyen-age,* Agen, 42 p (F35, 71)
Cantor M 1862 Heid NMV 3, 5-
Carra de Vaux 1917 Scientia 21,
  273-282 (F46, 54)
Ceretti H 1909 Rv FMSN 19, 503-517
  (F40, 51)
- 1910 Rv FMSN 21, 49-63 (F41, 55)
Clark W E 1929 in *Indian studies in
  honour of L. R. Lanman,* Cambridge,
  217-236
Coèdes G 1931 Lon SOAS 6, 323-328
Colin G S 1933 J Asi 222, 193-215
  [Greek]
Das S R 1927 Indn HQ (March), 44 p
Datta B 1926 AMM 33, 220-221
  (F52, 10)
- 1926 Benar MS, 7-8, 9-23 (F52, 10)
  [al-Biruni]

- 1928 Indn HQ 3, 536-540
  (F54, 5)
- 1933 Clct MS 24, 193-218
  (F59, 838; Z7, 386)
Davies T S 1833 Bath Bris 8
Decourdemanche J A 1912 Rv Eth Soc
  (Mar/Apr)
DeMilt C 1948 SSM 47, 701-708
Destombes M 1960 Physis Fi 2, 197-210
  (Z108:2, 245) [ast]
- 1962 Arc In HS 15, 3-45
  (Z109, 237)
Dubois-Maissoneuve 1808 Par SASPN 2,
  51-
Dziobek O 1911 Prometh (1104-1107),
  7 p (F42, 54)
Eneström G 1896 Bib M (2)10, 31-32,
  64, 96, 120 (F27, 2) [coins]
Friedlein Gottfried 1861 *Gerbert,
  die Geometrie des Boethius, und
  die indischen Ziffern,* Erlangen
Gandz S 1931 Isis 16, 393-424
  [abacus]
Ganguli S 1927 AMM 34, 409-415
  [Aryabhata]
- 1932 AMM 39, 251-256; 389-393
  (F 58, 988; F59, 19; Z5, 148; 6,
  3, 145) [Hindu, place value]
- 1933 AMM 40, 25-31, 154-157
  (F59, 19; Z5, 148; 6, 3, 145)
  [Hindu, place value]
Ginsburg J 1917 AMSB 23, 353-357,
  366-369 (F46, 45)
Goldschmidt V 1932 *Die Entstehung
  unserer Ziffern,* Heidelberg,
  51 p (F58, 4; Z7, 386)
Goodstein R L 1956 M Gaz 40, 114-129
  [arith]
Hagstroem K-G 1931 *Sagan om de tio
  Teeknen,* Stockholm, 147 p
  (Z2, 377; Z57, 1286) [digits]
Hill G F 1915 *The Development of
  Arabic Numerals in Europe...,*
  Oxford, 125 p
Horn-d'Arturo G 1925 Bln Oss 1, 187-
  204 (F51, 5)
Irani R A K 1952 Scr 18, 92-93
  [sexagesimal]
- 1955 Centau 4, 1-12 (MR17, 117;
  Z67, 245)
Iyengar A A K 1926 Indn MSJ 16(S), 24
  (F52, 37)

## Numeration: Arabic   (continued)

Karpinski L C 1910 Bib M (3)11, 121-124 (F41, 53) [Arab]
- 1912 Bib M (3)13, 97-98 (F43, 65) [Hindu]
- 1912 Sci 35, 969-970
Kaye G R 1907 Beng Asi J 3, 475 [Hindu]
- 1911 Beng Asi J 7, 801
- 1911 Indn Antq, 50-56
- 1918 Scientia 24, 53-55 (F46, 54) [Hindu]
Kleinwachter 1883 Chi Rv 12, 1
Langley E M 1924 M Gaz 12, 468-469 [Roman numerals]
Lattin H P 1933 Isis 19, 181-194 (Z7, 386; F59, 6) [Bubnov]
Lazzeri G 1904 Pe MIS 8, 3-7 (F35, 55)
Levey M + 1950 Isis 41, 196
Loeffler E 1911 Arc MP 19, 174-178 (F43, 64)
Marr N Ya 1930 K voprosy Len ZKV 5
Miller G A 1929 Indn MSJ 18, 121-125 (F55, 4)
- 1929 Sch Soc 29, 390-391 (F55, 588)
- 1929 Sch Soc 30, 431-432
- 1933 Sci (NS)78, 236-237
Mingana A 1937 Lon Asia, 315-316
Nagl A 1889 Hist Abt 34, 129-146, 161-170 [XII]
Narducci H 1883 BSM (2)7, 247-256 [Gobar]
Nau F 1894 J Asi 10(16), 219-228 [Severus] [First mention in West]
North G 1792 Archaeol 10, 360 [England]
Rey A 1935 Rv Et Grec 48, 525-539 [Greek]
Ross A C 1938 Indn Ar SM (57)
Saidan A S 1966 Isis 57, 475-490 [Alhazen, Arab, arith]
Sanford V 1951 MT 44, 135-157
- 1955 Ari T 2, 156-158
Sarton G 1934 Isis 22, 224 [early use]
Sherman C P 1923 MT 16, 398-401
Sleight E R 1939 MT 32, 243-248
- 1942 MT 35, 112-116

Smith D E + 1911 *The Hindu-Arabic Numerals*, Boston
Smith D E 1917 AMSB 23, 366
*- 1918 AMM 25, 99-108 [Abraham ben Ezra]
Thorndike L 1949 Isis 32, 301-303 (Z38, 146) [mid]
Turner E R 1912 Pop Sci Mo 81, 601-603
Vogel K 1960 Int Con By 11, 660-664 (Z94, 2) [Byzantine]
Vogel K ed 1963 *Mohammed ibn Musa Alkhwarizmi Algorismics: Das Fruheste Lehrbuch zum Rechnen mit indischen Ziffern,* Aalen, 41 p (MR20, 1; Z118:1, 246)
Vowles H P 1934 Nat 134, 1008 [intro to Europe]
Weissenborn H 1893 BSM (2)17, 47-50 [Gerbert]
Welborn M C 1932 Isis 17, 261-263 [Ghubar numer]
Whitrow G J 1959 Int Con HS 9, 614 (Z114, 5)
Woepcke F 1855 Tortolin 6, 321-
- 1859 *Sur l'introduction de l'arithmétique indienne en Occident...,* Rome, 72 p
- 1863 *Memoire sur la propagation des chiffres Indiens,* Paris (J Asi 1)
Wright G G N 1952 *The Writing of Arabic Numerals,* London, 446 p

## Numeration: Binary

Archibald R C 1918 AMM 25, 132-142 [Cardan, nim, Russian peasant meth]
Jones P S 1953 MT 46, 575-577
Li Shu-Tien 1959 ACMC 2(9), 28-29 (Z93, 2) [Octal]
Shirley J W 1951 AJP 19, 452-454 (MR13, 420; Z44, 243) [before Leibniz]
*Vacca G 1904 Int Con H 12, 63-67 (F35, 56)

## Numeration: Decimal

See also Metric system, Percent.

## Numeration: Decimal (continued)

Adams C W 1947 Isis 37, 68
  [instru]
Anon 1837 *Nouvelle arithmetique
  décimale...augmentée d'un precis
  historique...*, Paris
- 1949 Isis 40, 119 [anc numismatics]
Archibald R C 1945 M Tab OAC 1(10),
  400-401 [sexag]
Astier R 1899 Rv Sc (4)11, 501
Calvert H R 1936 Isis 25, 433-436
  (F62, 5) [metrology]
Cantor M 1856 ZM Ph 1, 65-
- 1858 ZM Ph 3, 325
Chasles M 1838-1839 Par CR 6, 678-;
  8, 72-; 9, 447-, 463-
Dickson L E 1919 Dickson 1, Ch 6
  [periodic deci]
Forno D M 1929 M Mag 3, 5-8
Friedlcin G Z 1864 ZM Ph 9, 73-
Gerhardt C I 1842 Arc MP 2, 427-
Gerson-Levy 1839 Metz 21, 336-
  [re Vincent 1839]
Gliozzi M 1932 Tor FMN 67, 29-50
Gore J H 1891 Am JS (3)41, 22-28
  (F23, 40) [XVII, metrology]
Hagstroem K G 1931 *Sagam om de 10
  tecknen*, Uppsala, 145 p
Hondard J C + 1896 Rv Sc (4)6, 214
Sarton G 1937 Isis 26, 304-305
  (F63, 4) [Hindu]
- 1950 Osiris 9, 581-601
  (Z39, 241)
Schade P 1908 DMV 17, 272-274
  (F39, 52)
Sedillot L A 1865 Vat NLA 18, 316-
Struik D J 1925 Linc Rn (6a)2,
  305-307 [Paul of Middleburg]
Terquem O 1853 Nou An 12, 195-
  [bib]
Thureau-Dangin F 1932 Rv Assyr
Vincent A J H 1839 JMPA 4, 261-
  (See Gerson-Levy) [abacus]

## Numeration: Decimal fractions

Boyer C B 1962 MT 55, 123-127
  [Viete]
Cajori F 1923 Arc Sto 4, 313-318
  [deci pt, Pitiscus]

- 1923 MT 16, 183-187
Christensen S A 1885 Zeuthen (5)4,
  149-152
Durand D B 1936 Isis 25, 134-135
  [re Sarton 1935]
Eneström G 1909 Bib M (3)10, 238-243
  (F40, 52)
Gandz S 1936 Isis 25, 16-45
  (Z15, 53) [Bonfils, XIV]
Ginsburg J 1928 AMM 35, 347-349
  [deci pt]
- 1932 Scr 1, 84-85 [deci pt]
Glaisher J W L 1873 Brt AAS 43, 13-
Gravelaar N L W A 1898 Nieu Arch
  (2)4(1), 54-73 (F30, 37)
Hunrath K 1892 ZM Ph 37, 25-28
  (F25, 60)
Karpinski L C 1917 Sci 45, 663-665
  [deci pt]
Ling W 1956 Int Con HS 8, 13-17
  [China]
Lorey W 1935 Unt M 41, 58-60
  (F61, 6)
Mautz O 1921 Basl V 32, 104-106
  [Buergi]
Muirhead R F 1935 M Gaz 19, 42-43
  (F61, 28) [deci pt]
Nyberg J A 1939 SSM 38, 59
  [deci pt]
Sanford V 1922 MT 14, 321-333
  [Stevin]
- 1953 MT 45, 71, 73 [deci pt]
Sarton G 1934 Isis 21, 241-303
  (See Durand 1936) [Stevin]
- 1935 Isis 23, 152-244 [Stevin]
Smith D E 1910 T Coll B (1)5, 11-21
  (F41, 55)
Struik D J 1959 MT 52, 474-478
  [Stevin]
- 1959 Wis Nat Ti 33, 65-71
  [al-Kashi, Stevin]
Vetter Q 1932 AMM 39, 511-514
  [Czech]
- 1932 Cas MF 61, R 113- R 118
  (F58, 41)
- 1935 Math Cluj 9, 304-309
  (F61, 943) [deci pt]

## Numeration: Duodecimal

Beard R H 1955 MT 48, 332-333
Essig Jean 1955 *Douze, notre dix futur. Essai sur le numération duodécimale et un système métrique condordant,* Paris, 169 p
Houzeau 1897 Rv Sc (4)7, 240 [clocks]
Terry G S 1938 *Duodecimal Arithmetic,* London, 407 p

## Numeration: Greek

Beaujouan G 1950 Rv Hi Sc Ap 3, 170-174 [XII-XVI]
Dow S 1952 Am J Arch 56, 21-23 [Greek]
Friedlein G 1869 *Die Zahlzeichen und das elementare Rechnen der Griechen und der Roemer und des christlichen Abendlandes von 7. bis 13. Jahrhunderts,* Erlangen
Heichelheim F 1927 Deu Morg Z 6, 78-81 [Greek]
Schultze E 1937 Archeion 19, 179-191 (Z17, 289) [Greek]
Shaw A A 1931 AMSB 37, 826 (F57, 49) [Armenia]
Tannery P 1899 Intermed 1, 139 [Greek]

## Numeration: Octal

Berdelle C + 1900-1904 Intermed 7, 5, 168, 370-371; 8, 168; 9, 299; 11, 241 (Q1716) [XIX, Colenne]
Collins Mark 1926 *On the Octaval System of Reckoning in India,* Madras, 28 p
Gelin E 1896 Mathesis (2)6, 161
Phalen H R 1945 AMM 56, 461-465 [Hugh Jones]
Tingley E M 1934 SSM 34, 395-399

## Numeration: Quinary

Bork F 1932 Orient Lz 35, 89-91
Latham R G 1860 *Opuscula. Essays,* *chiefly Philological and Ethnographical,* London
Murdoch J 1890 Am Anthro 3, 38- [Eskimo]
Pott 1847 *Die quinaere und vigesimale Zaehlsmethode bei Voelkern aller Weltteile,* Halle, 312 p
- 1868 *Die Sprachverschiedenheit in Europa an den Zahlwoertern nachgewiesen, sowie die quinaere und vigesimale Zaehlmethode,* Halle

## Numeration: Roman

Barrett J A S 1907 Edi SP 27, 161-182
Beajouan G 1961 Cah Civ Me, 159-169
Behafy E 1948 Louv Ped (1 Nov), 29-37
Bombelli R 1876 *Studi archeologico-critici circa l'antica numerazione italica ed i relativi numeri simbolici,* Rome (F9, 17)
Friedlein G 1868 Boncomp 1, 48-50 (F1, 2)
- 1869 *Die Zahlzeichen und das elementare Rechnen der Griechen, Roemer und des Christlichen Abendlandes vom 7. bis 13. Jahrhundert,* Erlangen (F2, 2)
Ionescu I 1925 Gaz M 30, 254-255 (F51, 38)
Jones P S 1954 MT 47, 194-195, 345 [large]
Langley E M 1925 M Gaz 12, 468-469 (F51, 38)
Mommsen Theodor 1868 *History of Rome,* NY 1, Ch 19
Piccoli G 1933 Pe M 13, 144-150 (F59, 18) [Etruscan]
Sanford V 1952 MT 44, 403-404
Shaw A A 1938 M Mag 13, 127-128 (F64, 24)
Smith D E 1926 Scientia 40, 1-8, 17-26, 69-78 (F52, 9)
Tregear E 1888 Nat 38, 565
Vincent A J H 1842 Par CR 14, 43
Zangenmeister K 1887 Ber Si, 1011-1028

## Numeration: Sexagesimal

Cajori F 1922 AMM 29, 8 [Babyl]
Gillings R J + 1964 Au JS 27, 139-141 (MR30, 1) [Babyl]
Hincks E 1855 Dub Ac T 22(6), 406-407 [first Babyl tablet]
Hoppe E 1909 Arc MP (3)15, 304-313; 16, 278-279 [cyclotomy]
- 1927 Arc Sto 8, 449-458 [cyclotomy]
Hultsch F 1881 Hist Abt 26, 38-39 (F13, 46) [cyclotomy]
- 1904 Bib M (3)5, 225-233 [Euclid]
Irani R A K 1952 Scr M 18, 92-93 (MR12, 121) [Arab, numer]
Kewitsch G 1909 Unt M 15, 122-128 (F40, 51)
- 1910 Arc MP (3)16, 277 (F41, 52)
- 1911 Arc MP (3)18, 165-168 (F42, 50)
- 1915 Arc MP (3)24, 286-287 (F45, 66)
Lewy H 1949 Am OSJ 69, 1-11 (Z37, 290)
Loeffler E 1910 Arc MP (3)17, 135-144 (F41, 52) [Babyl]
Miller G A 1931 Sch Soc 33, 58-59
Neugebauer O 1927 Gott Ab 13, 1-55 [metrology]
- 1928 Archeion 9, 208-216 (F54, 10)
- 1931 QSGM (B)1, 452-457, 458-463 [Babyl, frac]
Pelseneer J 1934 Mathesis 48, 71-72 (F61, 5) [Thureau-Dangin]
Sachs A 1946 JNES 5, 203-214 (MR8, 1) [Babyl, frac]
Seidenberg A 1965 Arc HES 2, 436-440 [Sumerian]
Sidersky D 1913 J Asi 2, 713-715 [Babyl]
Stael-Holstein v 1935 Mon Ser 1, 277-314 [Tibet]
Studnicka F J 1881 Cas MF 10, 87 (F13, 46)
Thureau-Dangin F 1930 Rv Assyr 27, 73-78
- 1932 *Esquisse d'une histoire du système sexagésimal,* Paris, 80 p (F58, 7; Z5, 241)

## 565 / Epimathematical Topics

- 1933 Rv Assyr 25, 115-118; 26, 43 (F59, 833)
- 1939 Osiris 7, 95-141 (MR1, 129; F65, 4; Z22, 294)

## Numerology

See also Fifty, Mysticism, Plato number.

Anon 1914 Braun Mo, 463-477 (F45, 70) [eight]
Barit J 1968 MT 61, 779-783
Bell E T 1933 *Numerology,* Baltimore (F59, 826)
- 1946 *The Magic of Numbers* (McGraw-Hill), 418 p
Borchardt L 1922 *Gegen die Zahlenmystik au der grossen Pyramide bei Gise,* Berlin
Heiberg J L 1925 M Tid B, 1-6 (F51, 5)
Hopper V F 1939 *Medieval Number Mysticism,* NY
Loria G 1913 Isis 1913, 313 Ens M 15, 193-201
Schultze E 1939 Archeion 21, 346-350 (F65:2, 1078) [Greek, nine, seven]
Staehle Karl 1931 *Zie Zahlenmystik bei Philon von Alexandreia,* Leipzig-Berlin, 98 p (Z3, 98)
Weinrich O 1916 Rel Ges VV 16, 1

## Operations research

*Anon 1958 *Comprehensive Bibliography on Operations Research,* NY (O.R. Group at Case Inst)
Batchelor James H 1959 *Operations Research: An Annotated Bibliography* 2nd ed (St Louis U P)
Boldyreff A W + 1960 Op Res 8, 798-860 (MR23 (B), 1591) [military]
*Gazis D C 1967 Sci 157, 273- [bib, traffic flow]
Haack L B 1950 in George Wash Univ Logist Papers, App 1, Quart Prog Rep 2 [inventory control]
Hammer P L + 1968 *Boolean Methods in Operations Research and Related*

Operations research   (continued)

Areas (Springer-Verlag), 329 p
[*bib]
Lewis R + 1956 Nav RLQ 3, 295- 303
[inventory control]
*McCloskey J F + 1954 Operations
Research for Management, Baltimore
Miles Herman W 1962 Operations
Research. An ASTIA Bibliography,
Arlington, Va, 132 p
Riley V 1956 in Operations Research
Management II (Johns Hopkins U P),
App A, 541-556 [bib]
Saaty Thomas L 1959 Mathematical
Methods of Operations Research
(McGraw-Hill) [bib]
Stroller David S 1964 Operations
Research: Process and Strategy,
Berkeley, 159 p
Whitin T M 1950 in George Washington
U Logistics Papers, App 1,
Quart Prog Rep 1 [inventory
control]


## Optics

See also Conical refraction, Rainbow.

Boyer C B 1959 The Rainbow from
Myth to Mathematics, NY, 376 p
Brocard H + 1912-1909 Intermed 12,
242; 13, 120, 217-219; 14, 68, 16,
25 (Q2964) [lunette anallatique]
Cohen I B 1940 AJP 8, 99-106
(MR3, 258) [interference]
Duhem P 1916 Rv Met Mor 23(1), 37-91
[Malebranche]
Emch A 1952 Scr M 18, 31-35
(Z47, 4) [catacaustic]
Guenther S 1874 Erlang Si (F6, 45)
Herzberger M 1932 Z Instr 52, 429-
435, 485-493, 534-542 (F58, 986)
Hoppe E 1907 DMV 16, 558-567
(F38, 13) [Euler]
Houstoun R A 1924 Edi MSN 23, 15-16
[Kepler, refraction]
Jadanza N 1896 Tor FMN (2)46, 253
Kahlbaum G W A 1889 Basl V 7, 884-888
[Newton]

Kramer P 1882 Ab GM 4, 235-278
[Descartes]
Lejeune Albert 1947 Brx SS (1)61,
28-47 (Z31, 1) [Archimedes]
- 1948 Arc In HS 28, 598-613
(MR11, 150) [Euclid]
- 1948 Euclide et Ptolémée, deux
stades de l'optique géométrique
grecque, Louvain, 196 p (Z39, 241)
- 1948 Isis 38, 51-53 (Z29, 385)
[Archimedes, reflexion]
- 1957 Recherche sur la catoptrique
grecque d'apres les sources
antiques et médiévales, Brussels,
199 p [bib]
Lemaire G + 1907 Intermed 14, 20, 91
[refraction]
Lohne J A 1966 Nor M Tid 14, 5-25
(MR34, 1140) [Fermat, Leibniz,
Newton]
Loria G 1942 Ens M 38, 250-275
(MR4, 65) [Euler]
Martinez G 1942 It UM Con 2, 617-627
(F68, 2)
Mugler C 1964 Dictionnaire
historique de la terminologie
optique des Grecs. Douze siècles
de dialogues avec la lumière,
Paris, 459 p (MR33, 1198)
Neurath O 1915 Arc GNT 5, 371-389
(F45, 87)
Rohr M v 1933 Naturw 21, 39-43
(F59, 10) [stereoscope]
Ruoss H 1894 ZM Ph 39, 1
Smith S 1924 Can T(3)(3)18, 227-230
[diffraction]
Southall J P C 1922 Franklin 193(5),
609-626 [aplanatic surf]
Steward G C 1932 M Gaz 16, 179-191
[Hamilton]
Vacca G 1940 It UM (2), 71-73
(MR3, 97) [Archimedes]
Wiedemann E 1890 Pogg An (2)39, 470-
473 [vision]
- 1890 Pogg An (2)39, 565-576
[Arab]
Winter H J J + 1951 Isis 42, 138-142
(Z42, 241) [Nasir al-Din]
Wolffing E 1902 Bib M (3)3, 361-382
[Fresnel surf]

## Organizations

See also Cities (Ch. 4), Meetings.

Ionescu I 1930 Gaz M 36, 211-212
(F56, 9) [list with info]
Kaminer L V + 1956 Vop IET 2, 326-334
[international org]
Oranzhereeva V F 1968 *Nauchno-
tekhnicheskie obshchestva SSSR.
Istoricheskii ocherk,* Mos, 455 p
Ornstein Martha 1913 *The Role of
Scientific Societies in the
Seventeenth Century,* Chicago
(3rd ed 1938)
Sparn E 1930 B Ai Bol 3, 117-120
[survey]

## Paper folding

Archibald R C 1918 AMM 25, 95-96
Hess A H 1956 M Mag 29, 217-221
(SMSGRS 10) [bib]
Row T Sundaro 1893 *Geometric
Exercises in Paper Folding,*
Madras (rev repr 1901 Chicago
repr 1966 Dover)

## Paradoxes

See also Foundations, Philosophy,
Zeno paradoxes.

*Beth E W 1959 *The Foundations of
Mathematics...,* Amsterdam
Brocard H + 1897-1899 Intermed 4,
108, 176; 6, 201 (Q923)
Delevsky J 1952 Rv Phi Fr E, 196-
222
Quine W V 1962 Sci Am (Ap)
Ruestow A 1910 *Der Luegner,* Leipzig
Thomas I 1958 JSL 23, 133-134
(MR21, 484) [infinity, mid]
Van Heijenoort J 1967 Logical
Paradoxes, Enc Phi [bib]
Weaver W 1938 AMM 45, 234-236
[Carroll]

## Pendulum

Bertelli P T 1873 Boncomp 6, 1-44
(F5, 48) [XVII]
Genocchi A 1883 Boncomp 15, 631-637
[Foucault]
Gerland E 1878 Pogg An (2)4, 585-613
(F10, 28)
Guenther S 1873 Erlang Si (F5, 48)
[Foucault]
- 1873 Erlang Si (F5, 51)
[Huygens]
Petrie F 1932 Anc Eg (Dec), 110-111
[Egypt]
- 1933 Nat 132, 102 (F59, 45)
Roethig O 1879 Hist Abt 24, 153-159
(F11, 42) [Foucault]
Safarik 1873 Prag Si, 51-57
(F5, 51)
Sestini G 1935 Pe M (4)15, 262-275
(F61, 942)
Zoellner F 1872 Leip Ber 24, 183-193
(F4, 29)
Zwerger M 1889 *Der Schwingungsmittel-
punkt zusammengesetzter Pendel.
Historisch-kritische Untersuchung
nach den Quellen bearbeitet,*
Munich, 135 p

## Percent

Datta B 1927 AMM 34, 530-531
(F53, 8)

## Perspective

See also Art, Descriptive geometry
(Ch. 2), Projective geometry
(Ch. 2).

Amodeo F 1933 It SPS 21 (2), 102-103
(F59, 858) [XVI]
- 1933 Nap Pont 62, 105-149, 517-519
(F59, 828; Z7, 50)
- 1933 Nap Pont 63, 52 p (F59, 829)
[XVII, France]
- 1934 It SPS 22(2)(A), 146-149
(F61, 941) [XVII, France]
- 1935 It SPS 23(2)(A), 1-6
(F61, 941) [XVIII]

Perspective   (continued)

Casotti M W 1953 Pe M (4)31, 73-103
  [Vignola]
Cassina U 1921 Pe M (4)1, 326-337
  [pt at inf]
Cremona L 1865 Rv It SLA 5, 226-231,
  241-245
Da Villa M 1937 Nap FM Ri (4)7,
  157-164 (F63, 803) [Pelerin]
Delbruck Richard 1899 Beitraege zur
  Kenntnis der Kunst, Bonn
Doehlemann K 1905 ZM Ph 52, 419
  [Van Eycks]
- 1912 Rep Kunst 34, 392-422,
  500-535 [Holland, art]
Doxiadis C A 1937 Ber MG 36, 46-49
  (F63, 8) [Greek]
Flocon Albert + 1970 La Perspective
  (Presses Universitaries), 128 p
Havelka F 1966 Olom UPF 21, 5-22
  [instruments]
Ivins William M 1938 On the
  Rationalization of Sight.  With
  an Examination of Three
  Renaissance Texts on Perspective,
  NY, 53 p [Viator]
Jones P S 1951 AMM 58, 597-606
  [B Taylor]
Laussedat A + 1899 Intermed 6, 102;
  10, 73 (Q149) [inverse problem]
Loria G 1933 Nap Pont 62, 506-509
  (F59, 828) [XV, XVI]
Luce J H 1953 Rv Hi Sc Ap 6, 308-321
  [Vitruvius]
Mercier P A 1950 Experien 6, 278-280
  [XV]
Mueller R 1914 DMV 23, 406-418
  [painting]
Pirenne M H 1952 Brt JPS 3, 169-185
  [Leonardo da Vinci]
Pittarelli G 1918 Loria (2)20(1),
  65-76 [word]
Poudra Noel G 1859 Histoire de la
  perspective ancienne et moderne...,
  Paris (also 1864)
Riccardi P 1889 Bib M (2)3, 39-42
  [re Poudra]
Rocquigny G de + 1897 Intermed 5, 96,
  163 (Q1222)
Taylor C 1881 Mess M (2)5, 112-113
  (F13, 43) [Greek]

Ten Doesschate G 1938 Bij Gesch 18,
  116-128
Vetter Q 1948 Int Con HS 5, 244-251
  [Czech]
Walcher-Casotti M 1953 Pe M (4)31,
  73-103 (Z51, 2) [Barozzi]
Wieleitner H 1915 Bib M 14, 320-335
  [Murdoch]
- 1920 Rep Kunst 42, 249-262
  [painting]
Wolff G 1915 ZMN Unt 46, 263-269
  (F45, 77)
- 1936 Unt M 42, 235-242  (F62, 5)
Zubov V P 1954 Mos IIET 1, 219-248
  (Z60, 6)  [Leonardo da Vinci,
  Vitello]


Philosophy

See also Ethics,  General: Theory
  (Ch. 4).

Beth E W 1942 Ned Ps 36, 80-83
  (MR8, 2) [Leibniz]
Enriques F 1912 Scientia 11 (21, 22)
  1-17, "3-20" [thy knowledge]
Jourdain P E B 1915 Monist 25,
  633-638
Kennedy E S 1942 M Mag 16, 290-298
  [XVIII-XX]
Keyser C J 1933 Scr 1(3), 185-203
Kliem F 1929 Humanismus und Mathe-
  matik, Breslau, 64 p (F55, 9)
Loeffler Eugen 1912 Der Anteil der
  mathematische Wissenschaften an
  der Entwicklung der menschlichen
  Geisteskultur, Mannheimer General
  Anzeiger (S)(2)(Feb)
Perkeo 1932 Deu Opt Wo 18, 120-121
Tabulski A 1868 Ueber den Einfluss
  der Mathematik auf die geschicht-
  liche Entwicklung der Philosophie
  bis auf Kant, Jena-Leipzig, 55 p
Whitehead A N 1925 in his Science
  and the Modern World, NY


Philosophy of Mathematics

See also Existence, Formalism,
  Foundations, Generalization,

Philosophy of Mathematics (continued)

Geometry (Ch. 2), Infinity (Ch. 2), Intuitionism, Logicism, Mysticism, Number, Space, Time.

*Abro A d' 1939 *The Rise of the New Physics*, 2nd ed (Dover), 982 p [section on intuitionism and formalism]

Barket S F 1967 Geometry, Enc Phi

*Benacerraf P + ed 1964 *Philosophy of Mathematics, Selected Readings*, (Prentice-Hall) [bib]

Bense M 1939 *Geist der Mathematik. Abschnitte aus der Philosophie der Arithmetik und Geometrieg* Munich, 173 p

Bernays P 1935 Ens M 34, 52-69 (tr in Benacerraf 1964) [Platonism]

*Beth E W 1944 *The Philosophy of Mathematics from Parmenides to Bolzano*, Antwerp (in Dutch)

*- 1955 *Les fondement logiques des mathématiques*, 2nd ed, Paris, 256 p (1st ed 1950)

* 1959 *The Foundations of Mathematics--a Study in the Philosophy of Science*, Amsterdam (tr of 1955) [bib]

Borel E 1912 Rv Mois (10 Aug), 218-227 [Brunschvicg, infinity]

Bouligand G 1949 Arc In HS 2, 291-302 [epistomology]

*- 1957 Arc In HS 36, 105-111

- 1957 Arc In HS 36, 105-111 (Z88, 4) [epistomology]

Boutroux Pierre 1913 Rv Met Mor 21, 107-131
1914-1919 *Les principes de l'analyse mathématique: Exposé historique et critique*, Paris, 2 vols

- 1920 *L'ideal scientifique des mathématicians*, Paris, 274 p (Ger tr 1927 Leipzig)

Bowne G D 1966 *The Philosophy of Logic 1880-1908*, The Hague, 157 p [bib]

*Brunschvicg Leon 1912 *Les étapes de la philosophie mathématique*, Paris, 602 p

Cajori F 1929 AMSB 35, 13 (F55, 26) [Barrow, Hobbes, Wallis]

- 1929 MT 22, 146-151 (F55, 588) [Barrow, Hobbes Wallis]

Cellucci Carlo 1967 *La filosofia della matematica*, Bari, 323 p

*Cooper J L B 1961 Smi R, 323-335 (also Smi P 4483)

Cornford F M 1932 Mind 41, 37, 173-190 [Plato]

Couturat Louis 1905 *Les principes des mathématiques, avec un appendice sur la philosophie des mathématiques de Kant*, Paris, 320 p (Ger tr 1908 Leipzig)

Dehn M 1937 QSGM (B)4, 1-28 (F63, 8; Z16, 196) [anc, foundations]

Dutordoir 1899 Brx SS 24(1), 52 [Aristotle]

Enriques F 1912 Int Con 1, 67-79 [influence]

- 1912 Scientia 12, 59-79, 172-191

Foraker F A 1919 MT 11, 196-198 [non-Euclidean geom]

Gonseth F 1926 *Les fondement des mathématiques. De la géométrie d'Euclide a la rélativité générale et à l'intuitionisme*, Paris, 260 p

Grey Gerhard 1968 *Einfuehrung in die philosophischen Grundlagen der Mathematik*, Hannover, 116 p [idealist]

Hawkins David 1964 *The Language of Nature...* (Freeman), 372 p

Jourdain P 1912 Int Con (5)2, 526-527 [Fourier]

Kennedy H C 1957 M Mag 32, 207-208 [Aristotle]

Kluge F 1935 *Aloys Mueller's Philosophie der Mathematik und der Naturwissenschaft*, Leipzig, 120 p (F61, 954)

Kolman E + 1931 UB Marx (5), 363-379 [Hegel]

Kreisel G 1967 in Schoenman R ed, *Bertrand Russell, Philosopher of the Century*, 201-272

## Philosophy of Mathematics (continued)

Krienelke K 1909 *J H Lamberts Philosophie der Mathematik,* Berlin

Kropp Gerhard 1948 *Beitraege zur Philosophie, Paedagogik und Geschichte der Mathematik,* Berlin, 103 p

Linsky L 1967 Referring, Enc Phi [Frege, Russell, Strawson]

Maupin G 1902 *Opinions et curiosités touchant la Mathématique,* Paris,

Maurer A 1959 Medvl St 21, 185-192 [Thomist]

Maziarz Edward A + 1968 *Greek Mathematical Philosophy* (Frederick Ungar), 271 p

*Mehlberg Henryk 1962 The present situation...in *Logic and Language, Studies Dedicated to Prof. Rudolf Carnap,* Dordrecht, 69-103 .[1930-1960]

Miller G A 1943 Sci 98, 38039

Mita K 1948 *Philosophical Questions of the History of Mathematics,* Tokyo, 225 p (in Japanese)

Murata T 1958 CMUSP 6, 93-114; 7, 57-63; 7, 100-116; 8, 81-96 (MR58, 1247) [French empiricism]

Myers G W 1928 MT 20, 93-100 [infl of phil]

Nekrassoff P A 1905 Mos MO Sb 25, 3 [Bugaev, Russia]

*Parsons C 1967 Math, Foundations of, Enc Phi [XIX, XX, bib]

Renyi A 1954 Hun Act M 5(S), 21-42 [non-Euc geom]

Robinet A 1961 Rv Hi Sc Ap 14, 205-254 [Malebranche]

Rybkin G F 1952 UMN 7(2), 123-144 (Z48, 243) [Osipovskii, Ostrogradskii]

Schaaf W L 1953-1955 MT 46, 515-516, 521-524; 48, 264-266 [what is math, bib, quotes]

Speiser A 1926 Crelle 157, 105-114 [Euler, Riemann]

Steklov V A 1921 *Teoriya i praktika v issledovaniyakh Chebysheva,* Pet (repr 1946 UMN 1(2)(12), 4-11)

Struik D J 1937 Sci Soc 1, 81-101, 545-549 [Marxism]

Synge J L 1957 *Kandelman's Krim: A Realistic Fantasy,* London, 175 p (MR21, 1)

Wang Hao 1959 Eighty years..., in *Logica, Studia Paul Bernays Dedicata,* Neuchatel, 262-293 [1879-1959]

Wedberg A E C 1955 *Plato's Philosophy of Mathematics,* Stockholm

White L A 1947 Phi Sc 14, 289-303 [anthropological]

Wieleitner H 1923 Arc G Med 15, 27-32 [nature of m]

Yanovskaya S A 1959 Matematicheskaya logika i osnovaniya matematika, in *Matematika v SSSR za sorok let 1917-1957,* Mos, vol 1, 13-120

Zaharia R O 1930 Tim Rev M 10, 125-127 (F56, 11) [Euclid]

## Philosophy of Science

See also Scientific method, General: Theory (Ch.4), Time.

Abro A d' 1939 *The Decline of Mechanism,* NY

Beth E W 1948 *Symbolische Logik und grundlegung der exakten Wissenschaften,* Bern, 27 p (Bibliographische Einfuehrung in das Studium der Philosophie 3) [bib]

Carnap R 1939 *Foundations of Logic and Mathematics* (U Chic P), 79 p (Int. Enc. Unified Sci 1(3))

Crombie A C ed 1961 *Scientific Change,* Oxford (repr 1963 NY)

Dijksterhuis E J 1956 *Die Mechanisierung des Weltbildes,* Berlin-Goettingen-Heidelberg, 601 p (MR21, 231)

Enriques Federico 1906 *Problemi della scienza,* Bologna, 597 p (tr 1914 *Problems of Science,* Chicago, 408 p)

Freistadt H 1956-1957 Phi Sc 23(Ap); 24 (Jan)

Lakatos I + 1968 *Problems in the*

## Physics (continued)

- 1912 Scientia 11 (21-22), 275-292
  [XIX]
Magie William F 1963 *A Source Book
in Physics,* Cambridge, Mass
Majorana Q 1934 Nuo Cim (2)11, 1-4,
48-66 [IX, mid, phys]
Picard E 1908 BSM (2)32, 141-159
  [math and phys]
- 1908 Rv GSPA 19, 602-609
- 1930 Math Cluj 4, vii-xv, 1-32
Poggendorf J C 1879 *Geschichte der
Physik,* Berlin, 937 p
  (repr 1964 Leipzig)
Poincaré H 1897 Act M 21, 331
  [math and phys]
- 1897 AMSB 4, 247
- 1897 Int Con 1, 81
Przibram K ed 1963 *Schroedinger-
Planck-Einstein-Lorentz: Briefe
zur Wellenmechanic* (Springer-
Verlag), 74 p (MR29, 5531)
Ravetz J 1961 Isis 52, 7-20
  (MR23, 271; Z106, 2) [XVIII]
Reichenbach Hans *From Copernicus to
Einstein* (Wisdom Library)
Rosenfeld L 1956 Nuo Cim S (10)5(S),
1630-1669 (MR19, 519)
  [velocity of light]
Schuster A 1882 Nat 25, 397
Study Eduard 1923 *Mathematik und
Physik...,* Braunschweig, 31 p
van Deventer C M 1928 Euc Gron 5,
177-192 [Greek]
Walden P 1933 Fors Fort 9, 347-348
  [atomic thy]
Whittaker Edmund T 1910 *A History of
the Theories of Aether and
Electricity...,* London, 489 p
  (repr 1951-1954 NY, 2 vols: MR14,
1; 1960; MR27, 899)
- 1942 Edi SP (A)61, 231-246
  (MR4, 65) [Aristotle, Newton,
Einstein]
- 1958 *From Euclid to Eddington.
A Study of Conceptions of the
External World,* NY, 221 p
  (Z84, 243)
Wilson C 1956 *William Heytesbury:
Medieval Logic and the Rise of*

*Mathematical Physics* (U Wisconsin
P), 231 p (MR18, 267)
Zubov V P 1960 Int Con HS 9, 626-629
  (Z114, 5)

## Plagiarism

See also Competition.

Agostini A 1925 Arc Sto 6, 115-120
  [Pacioli]
Graf J H 1899 ZM Ph 44(S), 113-122
  (F30, 51) [Le Clerc, Ozonam]
Loria G 1905 Loria 8, 65
  [J J Rousseau, Rossignol]
- 1938 Archeion 21, 62-68
  [Torricelli]
Miller G A 1926 Sci 64, 204-206
  [Guldin]
Pittarelli G 1909 Int Con (4)3,
436-440 [P della Francesca,
Pacioli]
Weaver J H 1916 AMSB 23, 131-133
  [Pappus]

## Plato number

(Nuptual, Geometric, Ideal number)

Adam J 1891 *The Nuptial Number of
Plato...,* 79 p (F23, 5)
Bertauld 1886 Boncomp 18, 441-450
Demme C 1887 Hist Abt 32, 81-99,
121-132
Diès l'Abbé *Le nombre de Platon,*
Paris, 144 p
Hultsch F 1882 Hist Abt 27, 42-60
Junge G 1949 Clas Med 10, 18-38
  (Z41, 337)
Laird A G 1918 *Plato's Geometrical
Number and the Comment of Proclus,*
Madison, 29 p
Martin T H 1856 Rv Arch 13
Petrie R 1911 Mind 20, 252-255
  (F42, 51)
Young G C 1923 Ens M 23, 212-213

## Plus and minus signs

See also in Ch. 2: Addition, Negative.

Cajori F 1913 Sci 28, 51-52
- 1923 AMSB 29, 195 (F49, 23) [minus]
Cantor M 1894 Intermed 1, 119
Cazanacli P V 1932 Tim Rev M 12, 88-89 (F58, 993)
De Morgan A 1842 Phi Mag 20, 135-230- [Leonardo da Vinci]
- 1864 Cam PST 11, 203-212
Eneström G 1894 Sv Ofv 51, 243-256 (F25, 59)
- 1849 Bib M (2)13, 105-106 (F30, 39) ["plus"]
Glaisher J W L 1921 Mess M (2)51, 1-148 (F48, 38) [Germany]
Wilhelm F 1893 ZMN Unt 24, 168-170 (F25, 60)

## Poetry

See also Literature.

Archibald R C 1933 Scr 1, 364 [about math]
- 1939 Sci 89, 19-26, 46-50
Birkhoff George D 1933 Aesthetic Measure (Harvard U P)
Buchanan S 1929 Poetry and Mathematics, NY, 197 p (repr 1962) (F55, 619)
Smith D E 1926 MT 19, 291-296
- 1934 The Poetry of Mathematics and Other Essays, NY, 94 p (Scr Lib
White A 1891 Acad Syr 6, 265

## Political science

See also Jurisprudence.

Grazia A de 1953 Isis 44, 42-51 (Z50, 243) [de Borda XVIII]
Rebiere A + 1896 Intermed 3, 124 (Q686) [statistics]

## Postage stamps

*Boyer C B 1949 Scr 15, 104-114
Johnson R A + 1932 Scr 1, 183-184
Larsen H D 1955-1956 MT 48, 277-280; 49, 395-396
Schaaf W L 1956 MT 49, 289-290
*- 1968 J Rec M 1(4), 195-216

## Precocity

See also Prodigies, Psychology of mathematics.

Coupin H 1932 Nat Par, 177-179
Sanders S T 1932 M Mag 7, 21-22

## Printing mathematics

See also Communication.

Cassie W + 1893 Nat 47, 8-9, 227
*Chaundy T W + 1954 The Printing of Mathematics, Oxford
*Danielson W A 1966 Adv Comp 7
*Peano 1916 Tor FMN 51
Phillips Arthur 1956 Setting Mathematics, London (Monotype Corp), 32 p (Also 1956 in the Monotype Recorder 40(4))
Schaaf W L 1956 MT 49, 554
Sintsov D 1898 Kazn FMO (2)8, 48

## Problems

See also Heuristic, Hilbert problems (Ch. 2), Kirkman problem (Ch, 2), Pothenot problem (Ch. 2), Pursuit problems, Recreations.

Ayyanagar A A K 1924 Indn MSJ 15, 214-223 (F50, 19) [Hindu]
Botts T 1961 MT 58, 496-500, 596-600 [gen]
Callandreau É 1949 Célèbres problèmes Mathématiques, Paris, 478 p (MR11, 571)
Chistyakov V D 1962 Sbornik starinnykh

## Problems   (continued)

*zadach po elementarnoi matematike
s istoricheskimi ekskursami i
podrobnymi resheniyami*, Minsk,
201 p

Doerrie H 1965 *100 Great Problems
of Elementary Mathematics, Their
History and Solution* (Dover)
(tr of *Triumph der Mathematik*,
Wuerzberg, 1958)

Eves H + ed 1957 *The Otton Dunkel
Memorial Problem Book*
(AMM 64(S), 7)

Hasse Helmut 1955 *Proben mathematischer Forschung in allgemein-
verstaendlicher Behandlung*,
Frankfurt
- 1967 *Proben mathematischer
Forschung* (Salle), 108 p

Hettner G 1908 *Alte Mathematische
Probleme und ihre Klaerung in
neunzehnten Jahrhundert*, Berlin,
18 p (F35, 53)

Jones P S 1957 MT 50, 442-444
[Baby1]

Kempner A J 1936 AMM 43, 467
["unsolvable"]

Kubota T 1932 Jp MASE 14, 151-160
[elem geom]

Maroger A 1951 *Les trois étapes du
problème Pythagore-Fermat. La
récurrence, l'art des reciproques*,
Paris, 107 p   (Z43, 5)

Miller G A 1915 Sci Mo (Oct), 93-97
[classic]

Milne J J 1930 M Gaz 15, 142-144

*Plenielj Josip 1964 *Problems in the
Sense of Riemann and Klein*, NY
(Interscience)

Popov G N 1929 *Pamyatniki matematicheskoi stariny v zadachakh*, Mos-
Len, 60 p
- 1938 *Sbornik istoricheskikh
zadach po elementarnoi matematike*,
Mos-Len, 215 p

Sanford V 1927 T Col Cont (251)
(F54, 44)  [alg]
- 1951 MT 44, 196-197
[lion in well]
- 1952 MT 45, 119-120, 368, 372
[Fibonacci]

Schreiber E W 1930 SSM 29, 818-824

Smith D E 1917 AMM 24, 64-71
(F46, 52).  [dioph anal, pursuit]

Tietze Heinrich 1949 *Geloeste und
ungeloeste Mathematische Probleme
aus alter und neuer Zeit*,
Munich, 2 vols   (MR11, 571)
(2nd ed 1959)
- 1965 *Famous Problems of Mathematics:
Solved and Unsolved Mathematical
Problems from Antiquity to Modern
Times* (Graylock), 383 p
(tr of Tietze 1959)

Trigg C W 1957 AMM 64, 3-8
[trig in AMM]

Weinreich H 1930 ZMN Unt 61, 204-208
[alphabetical puzzles]

## Prodigies

See also Precocity.

Archibald R C 1918 AMM 25, 91-94

*Ball W W R 1892 *Mathematical
Recreations and Essays*, Ch.13

Barlow Fred 1951 *Mental Prodigies:
An enquiry into the Faculties of
Arithmetical, Chess and Musical
Prodigies...*, London, Secs 1 and 4
[bib]

Bidder G P 1856 Lon ICE 15, 251-
[autobiog]

Binet 1892 Rv Deux M 111, 905-924
[Inaudi]

Bobynin V V 1895 Fiz M Nauk 12, 277-
300  (F26, 48)

Boyer J 1951 Lar Men (Sept), 709-710

Cauchy A 1840 Par CR 11, 952
[Herd boy]

Darboux G 1902 Mathesis (3)2, 167-170
(F33, 35)  [Bertrand]
- 1892 Par CR 114, 275-, 528-, 578-
[Inaudi]

Emch A 1909 Inc Con 4(3), 538-540
[J J Winkler]

Galle A 1917 Gott N 4, 1-24
[Gauss]

Keyser C J 1937 Scr 5, 83-94
[Pascal]

McCreery L 1933 M Mag 7, 4-12

## Prodigies (continued)

*Mitchell F D 1907 AJ Ps 18, 61-143
Newcomb S 1885 Sci 5, 106-108
[Georgia USA]
Scripture E W 1891 AJ Ps 4, 1-63

## Proof

See also Analysis and Synthesis,
Exhaustion, Foundations, Logic,
Rigor.

Balic K 1936 Wiss Weis 3, 191-217
[mid]
Bobynin V V 1910 Kagan (515)
Bosmans H 1925 Sphinx Oe 20, 113-119
[indirect]
Gericke H 1962 MP Semb 8, 129-141
(Z108:2, 244)
Lakatos I 1963 Brt JPS 14
Szabo A 1953 Hun Antiq 1, 377-410
[Greek]
- 1954 Hün Antiq 2, 17-62
[Greek]
*- 1958 Maia (NS)10, 106-131
(Z82, 9) [anc]
Van Der Waerden B L 1957 Bel SM 9,
8-20 [anc]
Ver Eecke P 1930 Mathesis 44, 382-
384 (F56, 13) [Greek, indirect]

## Proportional compass

See also Instruments.

Boffito Guiseppe 1931 *Il primo
compasso proporzionale costruito
da Fabrizio Mordente e la
Operatio Cilindri di Paolo dell.'
Abbaco,* Florence, 31 p
Favaro A 1908 Ven I At 67, 723-739
(F39, 75)
Marcolongo R 1931 Nap FM Ri (4)1, 7-
15 (F57, 22) [Boffito, Leonardo
da Vinci]
Pinette L 1955 MT 48, 91-95
Van Cittert P H 1947 Ned Ti N 13,
22 p

## Psychology

Luce R D + 1963 *Handbook of Mathematical Psychology,* NY
Miller George A 1964 *Mathematics and
Psychology* (Wiley), 295 p
[readings, bib]

## Psychology of mathematics

See also Precocity, Sociology of
mathematics, Heuristic.

Adams C W 1946 Isis 36, 166-169
[age]
Bieberbach L 1934 Ber Si, 10-20,
351-360 [style]
- 1934 Unt M 40(7), 236-243
[personality]
Bobynin V V 1895 Fiz M Nauk 12,
273-276 (F26, 48) (in Russian)
[geom intuit]
Chaslin Phillippe 1926 *Essai sur le
mécanisme psychologique de la
mathématique pure,* Paris, 280 p
Dennis W 1966 J Geron 21, 1-8
[age]
Eiduson B T 1966 Am Sc 54, 57-
[productivity]
Fettweis E 1951 Psy Fors 23, 391-398
[geom and arith]
- 1953 Scientia 88, 235-249
[arith]
Galton Francis 1869 *Hereditary
Genius...,* London, 396 p
(rev eds 1871, 1891)
- 1874 *English Men of Science:
Their Nature and Nurture,*
London, 283 p
Hardy G H 1934 Nat 134, 250 [types]
Hofmann J E 1955 Arc In HS 33, 339-
350 (Z68, 5) [genius]
Hugh-Hellmuth H V 1915 Imago 4, 52-68
[erotic and math]
Lakatos I 1963 Brt JPS (May)
Loria G 1923 Arc Sto 4, 144-155
[political activity]
- 1924 Scientia 35, 10-21
Lotka A J 1926 Wash Ac 16, 317-
[productivity]

## Psychology of mathematics (continued)

*Mannoury Gerritt 1947 *Les fondements psycholinguistiques des mathématiques,* Neuchatel
Moebius P J 1900 *Ueber die Analoge der Mathematik,* Leipzig (2nd ed 1907)
Ostwald W 1909 *Groesse Maenner,* Leipzig
Pickering E C 1908-1909 Pop Sci Mo 72; 74 [heredity, selection to Academies]
Roe A 1965 Sci 150, 313- [age]
Sanders S T 1932 M Mag 7, 3-4
Speiser Andreas 1932 *Die Mathematische Denkweise,* Zurich, 137 p (2nd ed 1946; 3rd ed 1952)
*Staeckel P 1912 *Hermann Grassmann. Ein Beitrag zur Psychologie des Mathematikers* (Int Monat)
Studnicka F J 1873 Cas MF 2, 57 (F5, 53) [genius]
- 1879 Cas MF 8, 85- (F11, 56) [genius]
Toulouse Edouard 1910 *Henri Poincaré* Paris, 204 p
Tsanoff R A 1949 *The Ways of Genius,* NY
Van der Waerden B L 1953-1954 Elem M 80, 121-129: 9, 1-9, 49-56
Vera Francesca 1947 *Psicogenesis del razonamiento matematico,* Buenos-Aires, 200 p
Wagner R 1861 Gott Ab 9, 59-
Weidman D R 1965 Sci 149, 1048 [emotional perils]

## Pursuit problems

Bernhart A 1954 Scr 20, 125-141 (Z55, 243)
- 1957 Scr 23, 49-65
- 1958 Scr 24, 23-50
- 1959 Scr 24, 189-206 (MR21, 5968)
Clarinval A 1957 Arc In HS 38, 25-37
Neville E H 1931 M Gaz 15, 436
Puckette C C 1953 M Gaz 37, 256-260

Sanford V 1951 MT 44, 516-517
Smith D E 1917 AMM 24, 64

## Quantum physics

Bligh N M 1929 Sci Prog 22, 619-632
Konovalov V M + 1960 Vop IET (10), 35-41 (Z117, 9)
Van Hove L 1958 AMSB 64, 95-99 (MR19, 1131) [Von Neumann]

## Rainbow

See also Optics.

Boyer Carl B 1952 Isis 43, 95-98 (Z47, 5) [Descartes]
*- 1959 *The Rainbow, From Myth to Mathematics,* NY, 376 p

## Recreations

See also Arithmomachia, Joseph game, Nim, Paper folding, Problems, String figures.

Ahrens Wilhelm 1910-1918 *Mathematische Unterhaltungen und Spiele* Leipzig (Teubner), 2 vols [*bib 56 p]
- 1918 *Altes und neues aus der Unterhaltungs mathematik,* Berlin
Anon 1935 *Congrès International de Récréation Mathématique. Comptes Rendus du premier (deuxieme) Congrès,* Bruxelles (Sphinx), 131 p 103 p
Bachet de Meziriac C G 1612 *Problèmes plaisans et délectables qui se font par les nombres,* Lyon (2nd ed 1624; 5th ed 1884 Paris; abr ed 1905; repr 5th ed 1959; MR21, 5546; Z89, 4)
Ball W W R 1892 *Mathematical Recreations and Essays,* London (9th ed 1920, 10th ed 1922, rev of 10th by H S M Coxeter 1938, repr 1960, 1962; Fr tr 1908) (Editions

## Recreations (continued)

differ by both additions and deletions. Some parts are indexed in this bib)

Boyer J 1899 Intermed 6, 112; 7, 228 (Q619) [bib]

Cajori F 1944 AMSB 30, 387 (F50, 19) [Despian]

Curtze M 1895 Bib M (2)9, 37-42, 77-88 (F26, 52) [XIV]

- 1895 Bib M 105-114 [Germany]

*Ghersi Italo 1921 *Matematica Dilettevole Curiosa*, 2nd ed, Milan, 756 p

Kowalewski Gerhard 1930 *Alte und neue mathematische Spiele*, (Teubner), 145 p

- 1938 *Der Keplersche Koerper und andere Bauspiele*, Leipzig, 65 p

Lacombe Jacques 1792 *Dictionnaire encyclopédiques des amusants des sciences mathématiques et physiques*, Paris

- 1792 *Dictionnaire des jeux mathématiques*, Paris

Lucas Eduard 1883-1894 *Récréations Mathématiques* (Gauthier-Villars), 4 vols

Perdrizet P 1931 Fr Arch Ov 30, 1-16 [icosahedron]

Richards J F C 1946 Scr 12, 177-217 (Z61, 5) [Pythagorean game]

Schaaf William L 1955 MT 48, 100-101 [H. F. McNeish, "lobster-quadrille"]

- 1955 MT 48, 166-168

- 1955 MT 48, 351-352 ["area of a mathematician"]

*- 1963 *Recreational Mathematics. A Guide to the Literature*, 3rd ed, Washington D C (NCTM), 143 p (also 1955, 1958)

Schubert Hermann 1898 *Mathematical Essays and Recreations* (Open court), 149

Simons L G 1923 MT 16, 94-101 [teaching]

Smith D E 1912 Tea Col R 13, 385-395 [number games, rhymes]

## Recursion

Biermann K-R 1956 Arc In HS 35, 233-238 [J Bernoulli]

Bussey W H 1917 AMM 24, 199-207 (F46, 52) [origin]

- 1918 AMM 25, 333-337 [Fermat]

Cajori F 1908 AMSB 15, 407

- 1918 AMM 25, 197-201 (F46, 53) ["induction"]

Cantor M 1902 ZMN Unt 33, 536 (F33, 53)

Eves H 1960 MT 53, 195-196 [Fermat]

Freudenthal H 1953 Arc In HS 22, 17-37 (MR14, 1049; Z50, 242) [Pascal]

Kaminski S 1958 Stu Log 7, 221-241 (MR20, 5120) [origin]

Lorey W 1921 ZMN Unt 52, 205-209 (F48, 40)

Stamatis Evangelos 1953 *Der Schluss von der Volstaendigen Induktion bei Euklid*, Athens, 6 p (Z51, 242)

Vacca G 1909 AMSB (2)16, 70-73 (F40, 49) [Maurolycus]

- 1910 Loria 12, 33-35 (F41, 49)

- 1911 Rv Met Mor 19, 30-33

## Relativity

Abro A d' 1927 *The Evolution of Scientific Thought from Newton to Einstein*, NY (2nd ed 1950 Dover)

Birkhoff G D 1925 *The Origins, Development and Influence of Relativity*, NY, 185 p

Dixon E T 1930 M Gaz 15, 1-4 (F56, 11) [Euclid]

*Einstein A + 1938 *The Evolution of Physics* (Simon and Schuster)

Enriques F 1930 It SPS 18(1), 411-413 (F56, 4) [non-Euc geom]

Frankfurt V I 1961 *Ocherki po istorii spetsialnoi teorii otnositelnosti*, Mos, 195 p [bib]

Hilbert D 1909 Gott N, 72-101 (F40, 36) [Minkowski]

Hoppe E 1923 ZMN Unt 54, 181-184 [Euler]

Relativity    (continued)

Infeld L 1955 Naturw 42, 431-436
    (MR17, 2)
- 1955 Ren M Ap 13, 270-281
    (MR17, 813)
Ivanenko D D 1957 Mos IIET 17, 389-
    424  (Z112, 243)
Kopff A 1924 Scientia 35, 397-406
    [deviation of light]
Kratzer A 1955 MP Semb 4, 171-182
    (MR17, 337)  [anniv]
Kuznetsov B G 1959 Printsip
    otnositelnosti v antichnoi,
    klassicheskoi i kvantovoi fizike,
    Mos, 232 p  (MR22, 3533)
Lampariello G 1963 Bari Sem
    (85-86-87), 17 p  (MR32, 3797)
Lecat M 1924 Bibliographie de la
    relativité, Brussels
Linder A 1931 Helv CM 3, 148-150
    (F57, 35)  [Schlaefli]
Marcolongo R 1933 Nap Nav 2, 5-47
    (Z7, 389)
Palatini A 1919 Scientia 26, 195-207,
    "59-72"  [special]
- 1919 Scientia 26, 277-289,
    "96-109"  [general]
Pierpont J 1923-1924 AMM 30, 425;
    31, 26  [Einstein, Riemann]
Polvani Giovanni 1955 Il moto della
    terra, filo storico della
    relatività, in Cinquant'anni di
    relativita 1905-1955, 3-28
    (Z67, 247)
Reynolds C N 1925 AMM 32, 74    [geom]
Staroselskaya-Nikitina O A 1957
    Vop IET (5), 39-49 (Z80, 5)
    [Poincaré]
Synge J L 1958 Edi MSP 11, 39-55
    (Z83, 245)  [Whittaker]
Weyl H 1951 Naturw 38, 73-83
    (MR12, 577)  [anniv]
*Whittaker E 1953 A History of the
    Theories of Aether and Electricity,
    NY, vol 2
*Williams L Pearce 1968 Relativity
    Theory: Its Origins and Impact on
    Modern Thought (Wiley)

Religion

Archibald R C 1936 AMM 43, 35-37
    [Wallis]
A S T 1817 Theologimena Arithmeticae,
    Leipzig
Brown B H 1942 AMM 49, 302
    [Diderot, Euler]
Gerardin A 1911 Intermed 18,
    133-136 (Q3580)
Goldberg O 1946 Scr 12, 231-232
    [numbers in Bible]
Hagenbach R R 1851 Leonhard Euler
    als Apologet des Christenthums,
    Basle
Hankin E H 1921 Sci Prog 16, 654-
    [Quakers]
Iwanicki Joseph 1936 Morin et les
    démonstrations mathématiques de
    l'existence de Dieu, Paris, 144 p
Lagrange C 1912 Bel BS, 422-437
    (F43, 63)  [numbers in Bible]
- 1925 Bel BS (5)10, 489-491
    (F51, 37) .
Miller G A 1912 SSM 11, 60-63
    [idolatry]
Schaaf W 1954 MT 47, 43-45
Scholz H 1934 Bla D Phi 8, 341-361
Smith D E 1921 AMM 28, 339
- 1921 MT 14, 413-426
Tannery P 1896 An Phi Chr 34
Tenca L 1960 Physis Fi 2, 84-89
    [Grandi]


Rigor

See also Proof.

*Hahn H 1954 Sci Am 190, 84-91
    [SMSGRS 5]
Jourdain P E B 1912 Int Con (5)2,
    526-527 [Fourier]
Kemeny J G 1961 MT 54, 66-74
Pierpont J 1928 AMSB 34, 23-53
    [Bolzano]
Schaaf W L 1951 MT 44, 259-261
Szabo A 1960 Arc HES.1, 37-106   [anc]
Vygodsky M Ya 1948 Int Con HS (1946),
    182-190  [XVIII]

## Rotation

Dupont P 1963 Tor FMN 98, 331-354
  (MR29, 639)
Galli M 1955 It UM (3)10, 77-96
  (MR16, 781) [centrifugal,
  Galileo]
Gilbert P 1878 Brx SS 2, 255-350
  (F10, 29)
Marcolongo R 1906 Loria 9, 1
  [Composition, XVIII]
Polubarinova-Kochina P Ya 1949
  SSSR Tek, 626-632 (MR11, 707)
Siacci F 1878 Boncomp 11, 217
  (F10, 30) [Gilbert]
Vasconcellos F de 1912 Porto Ac 7,
  4-45, 65-83 129-159

## Science

See also Applications, Philosophy,
names of sciences.

*Bernal J D 1954 Science in History
  London, 991 p (repr 1965 NY)
Bortolotti E 1932 Scientia 52, 273-
  286 (F58, 987; Z5, 241)
- 1940 It UM Con 2, 899 (F68, 4)
Bridgman P W 1950 Phi Sc 17, 63-73
Clagett Marshall ed 1959 Critical
  Problems in the History of Science,
  Madison, 569 p (MR21, 1023)
*Dampier- Whetham William C D 1929
  A History of Science and its
  Relations with Philosophy and
  Religion, Cambridge
Daumas Maurice 1957 Histoire de la
  science (Encyclopédie de la
  Pléiade), Paris, 1904 p
  [esp pp. 537-714]
Enriques F + 1937 Compendio di
  storia del pensiero Scientifico
  dall'antichità fino ai tempi
  moderni, Bologna, 487 p
  (F63, 794)
Feldhaus F M 1906 Geschichte der
  groessten technischen Erfind-
  gungen, Leipzig

- 1904 Lexikon der Erfindungen und
  Entdeckungen auf den Gebieten der
  Naturwissenschaften und Technik,
  Heidelberg
- 1906 Geschichte der groessten
  technischen Erfindungen,
  Leipzig
Forbes R J + 1963 A History of
  Science and Technology (Penguin),
  2 vols [through XIX]
Freudenthal H 1946 5000 Years of
  Science, Groningen
Goldsmith Maurice + eds 1964 The
  Science of Science, London
Loria Gino 1925 Pagine di storia
  della scienza, Turin, 166 p
  (F51, 2)
*Mason Stephen F 1962 A History of
  the Sciences, NY (Collier)
Pledge H T 1939 Science Since 1500,
  NY-Lon, 357 p (F65, 1077)
*Ross S 1962 An Sc 18, 65-85
  ["scientist"]
*Russell Bertrand 1924 Icarus, or the
  future of Science, London
*Taton René ed 1957-1964 Histoire
  générale des Sciences (Presse
  Universitaire), 4 vols
- 1963-1966 A General History of the
  Sciences (Basic Books), 4 vols
  (tr of 1957-1964) [bib]
Whewell William 1837-1838 History of
  the Inductive Sciences, London,
  3 vols (3rd ed 1847; Ger tr 1840-
  1841)

## Science and society

Anon 1945 The Social Impact of
  Science: A Select Bibliography
  (US Senate Comm on Military
  Affairs) Wash D C
Bernal J S 1939 The Social Function
  of Science (Routledge)
*- 1954 Science in History, London,
  991 p (repr 1965 NY)
Crowther J G 1941 The Social
  Relations of Science, London
Dupree A Hunter 1957 Science in the
  Federal Government: History

Slide rule (continued)

Jones P S 1953 MT 46, 501-503  [US]
Mehmke R 1902 ZM Ph 47, 489
   [Germany]
- 1902 ZM Ph 48, 134  [slider]
- 1902 ZM Ph 48, 317-318
   ["Soho-rules"]
Rohrberg A 1916 ZMN Unt 47, 338-344
   (F46, 46)
Sleight N 1946 M Mag 20, 11-20
   [educ]
Thompson J E 1930 A Manual of the
   Slide Rule: Its History, Principle
   and Operation, London, 227 p
Wormser 1919 Mit GMNT 19, 6-9
   [slider]


Social sciences

See also Economics, Political
   science, Sociology.

Fisher I 1930 AMSB 36, 225-243
Frechet M 1948 Ecmet 6(1), 51-53
Freudenthal H 1950 Int Con HS 6,
   162-171  [Quetelet]
- 1951 Arc In HS 30, 25-34
   (MR12, 577; Z43, 5)  [Quetelet]
Gibbs J W 1930 AMSB 36, 225-243
Granger G G 1956 La mathématique
   sociale du Marquis de Condorcet,
   Paris
Van Dantzig D 1950 Euc Gron 25, 203-
   232 (MR11, 707)  [Pascal]


Sociology of mathematics

See also Competition, Cranks,
   Fashion, Meetings, Organizations,
   Science and society, Women.

Curtiss D R 1937 AMM 44, 559
   [fashion]
Dresden A 1942 AMM 49, 415-429
   [migration of maths]
Emch A 1937 M Mag 11, 186-189
   (F63, 817)  [rejections]
Fisher C S 1966 Arc HES 3(2), 137-
   159 [death of invariant theory]

581 / Epimathematical Topics

Iznotskov I 1885 Kazn OEFM 3, 88-93
   [folk math]
Loria G 1924 Arc Sto 4(2), 144-155
   [maths in politics]
Miller G A 1934 Sci Mo (July), 40-45
   [folk math]
*Oblath R 1955 M Lap 6, 221-240
   [Gauss]
Sergescu P 1939 Sphinx Br 9, 145-151
   (F65, 13)  [revolutionaries]
Severi F 1935 Rv Sc 73, 581-589
   [latin spirit?]
Struik D J 1942 Sci Soc 6, 58-70
   (MR3, 257)
Tariste V 1915 Intermed 22, 117-118
   (Q4320)  [vegetarianism]
Taton R 1949 Thales 5, 43-49
   (MR13, 2)  [diffusion, Monge]


Sociology of Science

Barber B 1961 Sci 134, 596-
   [resistence to change]
Barzun Jacques + 1957 The Modern
   Researcher, NY (Harcourt  Brace)
Borel 1922 Rv Par 29, 850-860
   [sci property]
*Hagstrom W O 1964 Social Pr 12, 186-
   195    [anomy]
Malkin I 1963 Scr 26, 339-346
   [Alexander I, engineering, Russia]
Merton R K 1957 Am Soc Rev 22,
   635-659 [priorities]
- 1961 APSP 105, 470-486
   [duplications]
Weinberg Alvin M 1967 Reflections on
   Big Science (MIT Press)


Space (physical)

See also Cosmology, Spaces (Ch. 2),
   Time.

Abellanas P 1951 Zarag Ac (2)6, 9-26
   (MR12, 420)  [and geom]
Alexits G 1957 Ber FIM 1, 85-91
   (MR19, 573)
Borel E 1926 Space and Time, London,
   [bib]
Cantoni C 1904 Rv Met Mor 12, 305-
   319 [à priori]

## Space (physical) (continued)

Cornford F M 1936 in *Essays in honour of Gilbert Murray*, Lon, 215-225

Dingler Hugo 1923 *Das Problem der absoluten Raumes in historisch-Kritischer Behandlung*, Leipzig, 50 p

Doehlemann K 1924 An Phi 4, 369-384 [relativity]

Efros J J 1917 *The Problem of Space in Jewish Medieval Philosophy*, NY

Fleckenstein J-O 1958 Stu Gen 11, 29-34 (Z79, 242)

Freudenthal H 1955 Ned NGC 34, 82-95 (MR17, 117) [XIX-XX]

- 1961 in *Essays on the Foundations of Mathematics Dedicated to Professor A H Fraenkel*, Jerusalem, 322-332 (MR29, 564)

- 1965 MZ 63, 374-405 [Helmholtz, Riemann]

Gonseth F 1945 *La géometrie et la problème de l'espace*, Neuchatel, 160 p (2nd ed 1955)

Gruenbaum A 1963 *Philosophical Problems of Space and Time*, NY

Mach Ernst 1906 *Space and Geometry* (Open Court), 148 p (repr from Monist April 1901, July 1902, October 1903)

Mansion P 1895 Brx SS 19, 56

Newcomb S 1897 AMSB (2)4, 187

Tummers J H 1962 Dialect 16, 56-60 [and geom]

Vicaire E 1894 Brx SS 18(2), 283-301 [abs motion]

Wheeler John A 1968 *Einsteins Vision* (Springer), 108 p [bib]

## String figures

Amir-Moez A R 1965 *Mathematics and String Figures*, Ann Arbor (Edwards), 35 p

Ball W W R 1920 *Mathematical Recreations*, Ch. 18

- 1960 *String figures and Other Monographs*, NY (Chelsea), 411 p

## Sun dials

Arvanitakis G 1934 Intbalkn 159-162 (F61, 948) [Arab]

Barovlis P 1959 Deu Geod (26), 1-2 (Z95, 1)

Davidian M L 1960 Am OSJ 80, 330-335 [al-Biruni]

Drecker J 1925 *Die Theorie der Sonnenuhren*, Berlin, 112 p (F51, 13)

Lewis F M + 1936 MT 29, 295-303

Schoy K 1923 *Ueber den Gnomonschatten und die Schattentafeln der arabischen Astronomie*, Hannover, 29 p (F49, 23)

## Surveying

See also Geodesy, Pothenot problem (Ch. 2).

Cantor M B 1875 *Die roemischen Agrimensoren und ihre Stellung in der Geschichte der Feldmesskunst*, Leipzig

Dieperink J W 1928 *De techniek van het landmetan in een tweetal tijdperken der geschiedenis*, Wagennugin, 20 p (F54, 8)

Dunnington G W 1954 Scr 20, 108-109 [Gauss]

Fuije A de la 1915 Rv Assyr 12 [anc]

Gandz S 1929 QSGM (B)1, 255

Hofmann J E 1941 Deu M 6, 576-585 [Gauss, Lecoq]

Karpinski L C 1927 SSM 26, 853-855 [Roman]

*Kiely E R 1947 *Surveying Instruments. Their History and Classroom Use*, NY (NCTMY 19) [bib]

Longraire de + 1896-1909 Intermed 3, 209; 6, 179; 16, 175 (Q776) [Greek, Roman]

Mikami Yoshio 1943 Toh MJ 49, 223-242 (MR8, 497) [Kuroda]

- 1947 *Study of Land Surveying in Japan*, Tokyo, 218 p

Mortet V 1896 *Un nouveau texte des traités d'arpentage et de*

Surveying (continued)

geometrie d'Epaphroditus et de
    Vitruvius Rufus, Paris
Rossi G 1877 Groma e squadro ovvero
    storia dell'agrimsura italiana
    dei tempi antichi al secolo XVII,
    Turin (F9, 24)
Schippers H K 1935 Zutphen 4, 110-113
    (F61, 949) [XVII J P Dou, J.Sems]
Schmidt W 1903 Bib M (3)4, 234-237
    [groma]
Sloley R W 1926 Anc Eg, 65-67
    [instrument]
Stoeber E 1877 Die roemischen
    Grundsteuer-Vermessungen nach
    dem lateinischen Texte des.
    gromatischen Codex, insbesondere
    des Hyginus, Frontinus und
    Nipsus..., Munich (F9, 25)
Stone E N 1928 Wash ULL 4, 215-242
    [instruments, Roman]
Thulin C 1911 Zur Ueberlieferungs-
    geschichte des Corpus agrimensorum,
    Goeteborg, 69 p (F42, 71)
Vetter Q U 1919 Cas MF 48, 27-37,
    206-220 [levelling]
Worrell W H 1944 JNES 3, 91-100
    [Arab]


Tables

Aaboe A 1965 J Cun St 19, 79-86
    (MR33, 6) [Greek]
Abbud F 1962 Isis 53, 492-499
    (Z118, 246) [ast, al-Shatir]
Archibald R C 1943 M Tab OAC 1, 33-44
    [trig]
- 1943-1934 Scr 11, 213-245; 12,
    15-51
- 1945 M Tab OAC 1(10), 402-403
    [ins]
- 1946 M Tab OAC 2, 146 [Rudolphine]
- 1955 M Tab OAC 9, 62-63 (Z64, 1)
    [log base 10]
Bjoernbo A A 1909 in Festskrift til
    H G Zeuthen, Copenhagen, 1-17
    (F40, 64) [al-Khwarizmi, trig]
- 1965 Brendan T MT 58, 141-149
    [Ptolemy, trig]

Bryan N R 1922 AMM 29, 392-394
    [integrals]
Caley A + 1874 Brt AAS, 1-175
    [bib]
- 1875 Brt AAS, 305-336
- 1913-1914 Sphinx Oe 8, 50-60,
    72-79; 9, 8-14 [bib]
Cazalas G 1932 Rv Assyr 29, 183-188
    (Z6, 2) [Babyl]
Davis Harold T + 1949 A Bibliography
    and Index of Mathematical Tables,
    Evanston 311 p (2nd ed 1962
    A. Fletcher)
*- 1965 Mathematical Tables, Enc Br
    [bib]
Dittrich A 1933 Cas MF 63, 82-96
    [Babyl, trig]
Dupré A M 1938 Isis 29, 43-48
    [normal prob intgl]
Ellis A J 1881 Lon RSP 31, 398-413
    [radix method]
Fletcher A + 1946 An Index of
    Mathematical Tables, London, 459 p
Gandz S 1936 Isis 25, 426-431
    (Z15, 53) [Babyl, reciprocals]
- 1936 Isis 25, 426-431 (Z15, 53)
    [Babyl, reciprocals]
Glaisher J W L 1875 Cam PSP 5,
    386-392 (F7, 27) [Hohenberg]
- 1920 QJPAM 48, 151-192 [log]
Goldstein B R 1964 Scr 27, 61-66
    (MR29, 1) [mid, time]
Greenwood J A + 1962 Guide to
    Tables in Mathematical Statistics
    (Princeton U P), 1072 p
Henderson James 1926 Bibliotheca
    tabularum mathematicarum. Being a
    Descriptive Catalogue of Mathe-
    matical Tables (Cambridge U P),
    208 p
- 1930 M Gaz 15, 250-256 [log]
Hutton Charles 1785 Mathematical
    Tables, Lon, 550 p [hist preface]
Ibadov R I 1968 Uzb FMN 12(2),
    23-26 (MR37, 2564) [trig]
Irani R A K 1952 Scr 18, 92-93
    [sexagesimal mult]
Lehmer D H 1941 Guide to Tables in
    the Theory of Numbers, Wash D C,
    191 p
Loria G 1930 Geno ASL 9, 9-16
    (F56, 6, 798) [numerical]

Tables    (continued)

Lupton S 1913 M Gaz 7, 147-150, 170-
173 [radix method]
Mehmke R 1899 DMV 7(1), 123-126
(F30, 61) [Tables Commission]
Mercer S A B 1928 Toro ROM 7, 3-6
(F59, 833) [Babyl]
Neugebauer O 1934 Dan M Med 12(13),
1-14 (Z9, 97) [Babyl]
Owen D B 1962 Handbook of Statistical
Tables, Reading, Mass [bib]
Pendlebury R 1874 Mess M (2)4, 8-11
(F6, 15)
Simons L G 1932 Scr 1, 305-308
Smith D E 1908 Bib M (3)9, 193-195
(F39, 55) [Greek]
- 1922 AMM 29, 62-63 [XIV]
Stahlman W D 1960 Int Con HS 9,
593-605 (Z114, 5) [Ptolemy]
Sueltz B A 1965 MT 58, 446-447
[Babbage, legibility]
Suter H 1914 Die Astronomischen
tafeln des Muhammed Ibn Musa al
Chwarizmi..., Copenhagen, 290 p
Van den Berg F J 1892 Nieu Arch 19,
211-215 (F24, 43) [oldest
reckoning table]
Zinner E 1936 Osiris 1, 747-774
(Z14, 147) [Tabulae Toletanae]


Technology

See also Computer, Engineering,
Instruments, Navigation, Pendulum,
Telegraphy.

Barker A 1920 M Gaz 10, 86-91
[weaving]
Ferguson E S 1968 Bibliography of
the History of Technology (MIT P)
*Hole F 1966 Sci 153, 605-
[prehist]
Luckert H J 1937 DMV 47, 242-250
(F63, 817)
*Stodola A 1897 Int Con 1, 260
[applic to math]
Tannery P 1887 BSM (2)11, 17-28
[Euclid]
Usher A P 1954 A History of Mechan-
ical Inventions, Cambridge, 462 p

Telegraphy

Eichensieg A 1890 Mit Artil 21, 489-
524 (F22, 1261) [optical]
Feyerabend Ernst 1933 Der Telegraph
von Gauss und Weber im Werden der
elektrischen Telegraphie, Berlin,
228 p [bib]
Korn A 1941 Scr 8, 93-97
[picture transmission]


Terminology

See also Notation, Numeration,
Semantics.

Bar A 1914 Intermed 21, 166 (Q4189)
["general mathematics"]
Bentley Arthur F 1932 Linguistic
Analysis of Mathematics,
Bloomington, 311 p
Blondel A 1933 Par CR 197, 1555-1556
[and innovation]
Busch Wilhelm 1933 Die deutsche
Fachsprache der Mathematik.
Ihre Entwicklung und ihre
wichtigsten Erscheinungen mit
besonderes Rucksicht auf J. H.
Lambert, (Giessen U Thesis), 39 p
(Z7, 387)
Clark J R 1926 MT 19, 343-348
[classical antecedents]
Court N A 1937 AMM 44, 316
Eneström G 1912 Bib M 12, 180-181
["ratio subduplicata"]
Godefroy M 1914 Intermed 21, 106-107
(Q1752) [neologisms]
Greene E E 1962 MT 55, 484-489
[loan words]
Householder A S 1937 AMM 44, 463
Jones P S 1954 MT 47, 195-196
Karpinski L C 1924 SSM 24, 162-167
[elem geom]
Kasner E 1937 Scr Lib 3, 55-72
(F63, 2) [neologisms]
- 1938 Scr 5, 5-14 (also in Scr Lib 3,
55-72)
Kotz S 1965 Am Statcn (June), 16
[Russia, statistics]
Loria G 1919 Ens M 20, 237-244

Terminology   (continued)

- 1941 Archeion 23, 360-363
  [heuristic value]
Miller G A 1919 AMM 26, 290-291
- 1940 Sci 91, 571-572
  [alg laws]
Moulton J K 1946 MT 39, 131-133
Mueller F 1899 DMV 7, 159   (F30, 33)
  [French, German]
- 1899 ZM Ph 44(S), 303  [German]
Mugler Charles 1958 *Dictionnaire
  historique de la terminologie
  géométrique des Grecs,* Paris,
  2 vols
Mulcrone T F 1958 MT 51, 184-190
  [educ]
- 1968 MT 61, 475-478 ["inflection"]
Schirmer A 1912 *Der Wortschatz der
  Mathematik nach Alter und Herkunft
  untersucht,* Strassburg (Deu Wort
  Z14), 89 p   (F43, 81)
Smith D E 1935 Scr 3, 291-300
  (F61, 940: Z13, 193) [XVIII-XX]
Stewart L 1949 MT 42, 99-101,
  [Slang]
- 1951 MT 44, 19
Thureau-Dangin F 1931 Rv Assyr 28,
  195-198  (F57, 10)  [Babyl]
Tucker A N 1952 MT 45, 271-272, 275
Vailati G 1903 Rv M 8, 57-63
Van der Willen D D 1948 *Vremde
  Woorden in de Wiskunde,* Groningen


Thousand

Cajori F 1912 Bib M (3)12, 133-134
  [Spain]
- 1922 AMM 29, 201-202
  [Spain, Portugal]
Ginsburg J 1933 Scr 1, 264-265
  [Spain]
- 1932 Scr 1, 264-265


Three body problem

Anon 1923 Act M 39, 257-258
  (F49, 10)
Birkhoff G 1915 Paler R 39, 265-334
  [restricted]

Brown E W 1901 AMSB (2)8, 103-113
- 1913 Int Con (1912)1, 81-92
Cantor 1905 Paler R 19, 305-308
  [Weierstrass]
Cayley A 1862 Brt AAS (1862), 184-
  252
*Gautier Alfred 1817 *Essai historique
  sur le problème des trois corps...,*
  Paris, 295 p
Hadamard J 1913 Rv Mois 15, 91-96,
  385-418  [Poincaré]
Hall A 1901 As J 21, 113-114
Hill G W 1896 AMSB 2, 125-136
Houzeau J C + 1882 *Bibliographie
  générale de l'astronomie..,*
  Brussels 2, cols 539-569
Kulirich E 1891 *Zur Geschichte des
  mathemstischen Dreikoerperproblems,*
  (Hall U Thesis)
Kurth R 1957 Arc M 8, 381-392
  [Lagrange]
Laplace P S 1827 *Mécanique céleste* 5
Lovett E O 1909 Sci (NS)24, 81-91
- 1911 QJPAM 42, 252-315
Marcolongo R 1919 Scientia 26, 102-
  112, "17-27"
* - 1919 *Il problema dei tre corpi da
  Newton ai nostri giorni,* Milan,
  166 p  [*bib]
Moulton F R 1904 Univ Chicago
  Decennial Publics 8, 119-143
- 1902 *An Introduction to Celestial
  Mechanics* (Macmillan), 399 p
  (2nd ed 1914)
- 1914 Pop As 22(4)
Seydler A 1886 Cas MF 15, 7, 75, 102
Schwarzschild K 1904 DMV 13, 145-156
Whittaker E T 1889 Brt AAS (1889),
  121-160


Tides

See also Geodesy.

Aiton E J 1956 An Sc 11, 206-223
  (MR19, 624)  [Bernoulli, Euler,
  Newton]
- 1955 An Sc 11, 337-348
  (MR19, 624)  [Descartes]
Bateman H 1943 M Mag 18, 14-26 (MR5,
  57; Z60, 1)  [influence on math]

## Tides (continued)

Drake S 1961 Physis Fi 3, 185-194
[Galileo]


## Time

See also Chronometry, Philosophy.

Gunn J A 1929 *The Problem of Time,
An Historical and Critical
Study,* London
Heath L R 1936 *Concept of Time,*
Chicago
La Harpe J de 1945 in *Festschrift
zum 60 Geburtstag von Prof. Dr.
A. Speiser,* Zurich, 128-137
(MR7, 353) [Greek]
Lecat M 1911 Intermed 18, 157
(Q3799)
Vasiliev A V 1924 *Space, Time,
Motion,* London (Chatto and Windus),
255 p
Windrid G 1933 Isis 19, 121-153, 20,
192-319 (Z7, 389) [math time]
- 1935 M Gaz 19, 280-290
[imaginary time]


## Types, theory of

Barone Francesco 1953 *Il
neopositivismo logico, Filosofia
della scienza II,* Turin, 418 p
Church A 1939 Erkennt 9, 149-
[Schoeder]


## Variable

See also x.

Bortolotti E + 1927 Arc Sto 8, 49-63,
64 [Bombelli]
Dickson H 1967 *Variable, Function,
Derivative,* Gothenburg
Eneström G 1898 Intermed 5, 180
[unknown]
Gandz S 1932 QSGM (B)2(2), 81-97
[Arab, Greek]

Hevesi J 1953 Int Con HS 7, 366-371
[fun]
*Luzin N N 1934 *Differentsialnoe
izchislenie,* Bolsh SE
Miller G A 1925 Ens M 24, 59-69
- 1925 Int Con 7(2), 959-967
(F54, 48)
Wends C 1899 Intermed 6, 85 (Q1321)
[words]
Wieleitner H 1927 Arc Sto 8, 64
[notation]


## Women

See also Sociology of mathematics.

Archibald R C 1918 AMM 25, 136-139
Bernard Jesse 1964 *Academic Women,*
(Penn State U P), 356 p [bib]
Bower J W 1943 MT 36, 175-178
[effect of war]
Coolidge J L 1951 Scr 17, 20-31
(MR13, 1; Z43, 5) [lists six]
Dubreil-Jocotin Marie-Louise 1948
in Le Lionnais, *Les grands
courants de la pensée mathématique,*
Marseille, 258-269
Eells W C 1957 MT 50, 374-376
[XIX US Doctoral Thesis]
Eneström G 1893 Bib M 7, 96
(F25, 11-12) [living women maths]
- 1896 Bib M, 73 [bib]
Jones P S 1957 MT 50, 376-378 [XX, USA]
Joteyko J 1964 Rv Sc (5)1, 12-15
(F35, 48)
Kramer E E 1958 Scr 23, 83-95
(MR20, 945; Z83, 245) [re
Coolidge 1961, six more]
Loria G 1901 *Donne Matematiche,* 2nd
ed, Mantova (F33, 47)
- 1903 Rv Sc (4)20, 385-392
(F34, 42)
- 1905 Rv Sc (5)1, 338-340
(F35, 48)
Mattfield Jacquelyn A + 1965
*Women and the Scientific Profession,*
Cambridge, Mass
Ménage 1796 *Abrégé de l'Histoire
de la vie des Femmes Philosophes,*
Paris

## 4

HISTORICAL CLASSIFICATIONS

In this chapter publications are listed under five major historical classifications: General; Ancient civilizations; Time periods; Countries and regions; Cities, organizations, and universities.

GENERAL

See also Collections (Ch. 5),
Problems (Ch. 3).


General: Articles

Anon 1940 AMM 47, 107-108
  [selected countries]
Barrau J A 1909 *Ove de ontwikkelings-*
  *wijze der Wiskunde,* Delft, 29 p
  (F40, 54)
Belankin J 1909 Kiev U Iz 1(3)(1c),
  280-300  (F39, 2)
Bernoulli G II 1799 Ber Mm
  [anecdotes]
Bobynin V V 1892 Bib M (2)6, 110-115
  (F24, 4)  [Europe]
- 1896 Fiz M Nauk 3, 11-122
  [survey]
- 1910 Kagan (515), 272-281
  (F41, 63)
Bortolotti E 1932 Scientia 52, 273-
  286
- 1942 It UM Con 2, 899  (Z26, 289)
Bosmans H 1902 Rv Q Sc (3)1, 659-682
  (F33, 8)
- 1909 Rv Q Sc (3)16, 639-651
  (F40, 4)
- 1911 Rv Q Sc 70, 330-342
  (F42, 3)
Brown Ernest W 1923 Mathematics, in
  Woodruff L L ed, *The Development*
  *of the Sciences* (Yale U P), 1-42
Cantor M 1904 Int Con (3), 497
Carnahan W H 1950 MT 41, 159-164
  [and culture]
Court N A 1948 MT 41, 104-111
De Vries H 1938 N Tijd 26, 193-230
  (F64, 916)
Eneström G 1879 Zeuthen (4)3, 113-
  118, 161-165  (F11, 30, 39)
Fano G 1895 *Uno sguardo alla storia*
  *della matematica. Discorso,*
  Mantova, 34 p  (F26, 3; 27, 1)
  (=Virgil AM, 3-34)
Ferroni P 1789? It SS 7, 319
  [anecdotes]
Friedlein G 1870 Boncomp 3, 303-306
  (F2, 25)

Gericke H 1951 MP Semb 2, 71-97
  (MR12, 577; Z42, 1)
- 1954 Gies Hoch 23, 116-126
  (Z58, 2)
*Gonzalez M O 1950 Cub SCPMR 2, 150-
  158  (MR13, 420)  [modern]
Guenther S 1876 Hist Abt 21, 57-64
  (F8, 29)  [misc]
Henry C 1887 Boncomp 20, 389-404
  [misc]
Hoppe E 1911 Unt M 17, 106-112
  [founding]
Kinsella J J + 1947 MT 40, 355-358
  [pop]
Lampariello G 1940 It UM (2)2, 467-
  481  (F66, 7)  [surv]
Lietzmann W 1940 ZMN Unt 71, 25-30
  (F66, 6)  [legends]
Loewy A 1932 Fors Fort 8/9, 81-82
Loria G 1921 Scientia 29, 169-184,
  253-262  (F48, 38)  [nations]
- 1926 Arc Sto 7, 4-17  [questions]
- 1932 Geno ASL 11, 245-324
  (F58, 1)  [survey]
- 1933 Linc Rn (B)17, 768-775
  (Z7, 147)
Miller G A 1903 Sci (NS)17, 496-499
- 1915 AMM 22, 299-304
*- 1931 Toh MJ 34,
- 1932 Indn MSJ 19, 225-231
  [recent]
- 1934 Sch Soc 39, 211-212
  1938 M Mag 12, 388-392
  (F64, 24)  [legends]
- 1938 Sch Soc 48, 278-279
  (F64, 3)  [mistakes]
- 1939 M Mag 13, 272-277
- 1940 M Mag 14, 144-152
- 1940 MT 32, 209-211
- 1941 M Mag 15, 234-244
- 1943 M Mag 17, 13-20
  (MR4, 65)
- 1943 M Mag 17, 212-220
- 1943 M Mag 17, 341-350
  (MR5, 57)
- 1944 M Mag 18, 67-76
- 1944 M Mag 18, 261-270
  (MR5, 253)
- 1945 M Mag 19, 64-72
  (MR6, 141)
- 1945 M Mag 19, 286-293  (MR6, 253)

General: Articles   (continued)

- 1947 M Mag 21, 48-55
Miller N 1945 Queen Q 52(1), 22-30
Novak V 1932 Cas MF 61, R137-R141
*Ore O 1941 in Woodruff L L ed,
    *Development of the Sciences*,
    2nd series (Yale U P), 1-51
Piani D 1824 Bln OSN, 191-
Robinson A 1968 Composit 20, 188-193
    (MR37, 21)
Rome A 1929 Rv Q Sc, 250-275
- 1931 Rv Q Sc (4)24, 279-305
    (F59, 825; Z3, 97)
Sageret J 1906 Rv Sc (5)6, 577-585
    [origin]
Saltykow N 1935 Ens M 33, 214-220
    (F61, 1)
- 1935 Slav Cong 2, 255-256
    (F61, 1)
Schaaf W L 1931 MT 23, 496-503
    [and gen hist]
- 1956 MT 49, 291-292   [scope]
- 1961 Ari T 8, 5-9
Scott J F 1948 Phi Mag, 67-91
    (MR10, 174)   [to 1800]
Shaw I B 1934 M Mag 8, 31-37, 128-131
Slessenger W W O + 1950 M Gaz 34,
    82-83   (MR12, 311)
Smith D E 1937 in Keyser C J + ed,
    *Scripta Mathematica Forum
    Lectures*,  NY (Scr Lib 3)
Struik D J 1949 in *Philosophy of the
    Future*, NY
*Sullivan J W N 1925 in Singer C ed,
    *Chapters in the History of Science*,
    London, Ch 4   [Europe]
- 1932 M Gaz 16, 243-253   (F58, 993)
    [and culture]
Tropfke J 1928 ZMN Unt, 193-206
    (F54, 16)
Vetter Q 1935 Math Cluj 9, 304-
Wargny C 1913 Mathesis (4)3, 186
Wolf R 1869 Boncomp 2, 313-342
    (F2, 15)
Wolf W 1932 Verm  R 9, 164-170, 177-
    182   (F58, 993)
Wren F L 1957 MT 50, 361-371

General: Books to 1800

Baldi da Urbino 1707 *Cronica de
    matematici overo epitome dell'
    istoria delle vite loro*, Urbino
Biancani Giuseppe (Blancanus) 1615
    *De natura mathematicarum scientiarum
    tractatio, atque clarorum mathe-
    maticorum chronologia*, Bologna
Dechales Claude F M 1690 *Cursus seu
    mundus mathematicus. Pars I:
    tractatus proemialis, de progressu
    matheseos et illustribus mathe-
    maticis*, Leiden
Fenn J 1768 *A History of Mathematics*,
    Dublin
Frobesius J N 1750 *Historica et
    dogmatica ad mathesis...*,
    Helmstadt, 298 p
Heilbronner Johann C 1739 *Versuch
    einer mathematischen Historie*,
    Frankfurt, 204 p
- 1742 *Historia matheseos universae
    ...*, Leipzig
Kaestner Abraham G 1796-1800
    *Geschichte der Mathematik...*,
    Goettingen, 4 vols
Krafft Georg  W 1753 *Institutiones
    geometriae sublimioris*, Tuebingen
Minto Walter 1788 *An Inaugural
    Oration, on the Progress and
    Importance of the Mathematical
    Sciences...*, Princeton.., Trenton
Montucla Jean E 1758 *Histoire des
    mathématiques...*, Paris, 2 vols
Montucla Jean E + 1799-1802 *Histoire
    des mathématiques...*, Paris,
    4 vols  (rev of 1758)  (Ital tr
    1879)  (repr 1960 Paris)
Severien 1775 *Historia...ciencias
    esactas...*, Madrid
Voss Gerhard Johann (Bossius) 1650
    *De universa matheseos natura et
    constitutione liber, cui sub-
    jungitur chronologia mathematicorum*,
    Amsterdam

General: Books 1800-1849

Bossut Charles 1802 *Essai sur*

General: Books 1800-1849 (continued)

l'histoire générale des mathématiques, Paris (Ital tr 1802, Ger
tr 1804; Engl tr 1803, 1810)
- 1810 Histoire générale des mathématiques depuis leur origine
jusqu'à l'année 1808, Paris, 2 vols
Drobisch M W 1837 Quaestionum
mathematico-psychologicarum,
Leipzig
Duersch Max A 1834 Mathematische
Denkuebungen...mit der geschichte
der Mathematik, Nuernberg
Franchini P 1821 Saggio sulla storia
delle matematiche corredato di
scelte notizie biografische ad uso
della gioventu, Lucca
- 1823 Supplemento al Saggio sulla
storia delle matematiche, Lucca
Gerstenbergk H 1848 Geschichte der
Mathematik in uebersichte Umrissen
nebst Nomenclatur der beruehmstesten Mathematiker von der aeltesten
bis auf die neuesten Zeiten,
Eisenberg
Poppe J H M 1828 Geschichte der
Mathematik seit der aelteste
bis auf der neueste Zeit,
Teubingen
Sallustj C de 1846 Storia dell'
origine e de progressi delle matematische di piu autori riunita in
Commentari a forma di cronaca ad
uso dei giovani studenti, Rome
Reynaud M 1823 Problémes et
développements sur diverse
parties des mathématiques, Paris,
407 p


General: Books 1850-1899

Arneth A 1852 Die Geschichte der
reinen Mathematik in ihrer
Beziehung zur Geschichte der
Entwickelung des menschlichen
Geistes, Stuttgart, 291 p
Ball Walter W Rouse 1888 A Short
Account of the History of Mathe-
matics, London, 487 p (2nd rev ed

1893, 546 p; 4th ed 1908; repr
1960 Dover; Fr transl 1906-1907)
- 1895 A Primer of the History of
Mathematics, London, 158 p
(F26, 2)
Bobynin V V 1896 Ocherki istorii
razvitiya matematicheskikh nauk na
Zapade, Mos, 129 p
Cantor Moritz 1863 Mathematische
Beitraege zum Kulturleben der
Voelker, Halle
Guenther S 1881 Beitraege zur
Geschichte der Mathematik, Ansbach
(F13, 43)
*Hankel H 1869 Die Entwichelung der
Mathematik in den letzten Jahr-
hunderten, Tuebingen (F2, 15)
Hoefer F 1874 Histoire des Mathé-
matiques depuis leurs origines
jusq'ua Commencement du dix-
neuvième siècle, (F6, 37)
(4th ed 1895; F26, 45)
La Cour P 1888 Historisk mathematik,
Copenhagen, 382 p
Lavrov P L 1867 Ocherk istorii
fisiko-matematichesikh nauk,
Pet, 330 p
Marie M 1883-1887 Histoire des
sciences mathématiques et physiques,
Paris, 12 vols (F15, 1)
Ofterdinger L F 1867 Beitraege zur
Geschichte der Mathematik bis
zur Mitte des 17. Jahrhunderts,
Ulm
Suter H 1872-1875 Geschichte der
Mathematischen Wissenschaften,
Zurich, 2 vols [thru XVIII]
- 1873-1875 Geschichte der
mathematische Wissenschaften,
Zurich, 2 vols (F4, 24; 5, 37)
[to 1700]    (Russ tr 1876 (vol.1),
1905)
Wolf R 1869 Matériaux inédits pour
l'histoire des mathématiques,
Rome
Vashchenko Zakharchenko Mikhail Y
1882 Kharakter razvitiya matematicheskikh nauk u razlichnykh
narodov drevnevo i novogo mira
do XV veka, Kiev
- 1883 Istoriya Matematiki, Kiev, 695
695 p

General: Books 1900-1949

Árrighi G L 1905 *La storia della matematica in relazione con lo sviluppo del pensioro,* Turin, 146 p (F36, 45)
*Bell E T 1945 *The Development of Mathematics,* 2nd ed NY, 650 p (MR8, 1)
Bense Max 1946-1949 *Konturen einer Geistesgeschichte der Mathematik,* Hamburg, 2 vols
Belyankin I I 1908 *Kratkii ocherk istorii razvitiya matematiki ot drevneishikh do nashikh dnei,* Kiev, 20 p
Boev G P 1947 *Besedy po istorii matematiki,* Saratov, 103 p
Bouligand G + 1936 *L'évolution der sciences physiques et mathématiques,* Paris, 206 p (F62, 1042)
Boutroux P 1914 *Les principes de l'analyse mathématique. Exposé historique et critique,* Paris, 558 p
Boyer Jacques 1900 *Histoire des mathématiques,* Paris, 260 p (F31, 2)
Cajori F 1919 *History of Mathematics,* 2nd ed, NY (See Miller G A 1919-1923 SSM 19, 830-835; 20, 300-304; 23, 138-149)
Cipolla M 1949 *Storia delle matematica. Dai primordi a Leibniz,* Mazara, 170 p (Z41, 337)
Enriques F 1938 *La matematiche nella storia e nella cultura,* Bologna, 339 p (F64, 902; Z21, 2)
Frajese Attilio 1949 *Attraverso la storia della matematica,* Rome, 234 p (Z41, 337)
Freudenthal Hans 1946 *5000 Jaren Internationale Wetenschap,* Groningen
Gambioli D 1902 *Breve sommario della storia delle matematiche colle due appendici sui matematici italiani e sui tre celebri problemi geometrici dell'antichita,* Bologna, 241 p (F33, 3)
Giorgi Giovanni 1948 *Compandio di storia delle matematiche,* Turin, 148 p
Gnedenko B V 1946 *Kratkie besedy o zarozhdenii i razvitii matematiki,* Mos-Len, 39 p
*Guenther Siegmund 1908 *Geschichte der Mathematik, I. Tiel, Bis Cartesius,* Leipzig (F39, 1) (continued in Wieleitner 1911-1921)
Kropp Gerhard 1948 *Beitraege zur Philosophie, Paedogogik und Geschichte der Mathematik,* Berlin, 103 p (MR10, 174) [Lalouvère]
Loria Gino 1929 *Dalla tavola pitagorica alle equazioni integrali,* Milan, 51 p
Lebedev V I 1916-1919 *Ocherki po istorii tochnykh nauk,* Mos, 5 vols
- 1929-1933 *Storia delle matematische,* Turin, 3 vols
Maupin G 1902 *Opinions et curiosités touchant la mathématique,* Paris
Milhaud Gaston 1911 *Nouvelles études sur l'histoire de la pensée scientifique,* Paris
Ogura K 1935 *Studies in the History of Mathematics,* Tokyo, 350 p (F62, 1042)
*Pelseneer J 1935 *Esquisse du progrès de la pensée mathématique. Des primitifs au IX Congrès International des Mathématiciens,* Paris, 160 p (Z10, 385)
Popov G N 1920 *Istoriya matematiki,* Mos, 235 p
- 1925 *Ocherki po istorii matematiki,* Mos-Len, 163 p
Sergescu P ed 1949 *Histoire des mathématiques,* Paris (Act Sc Ind)
Sheremetevskii V P 1940 *Ocherki po istorii matematiki,* Moscow, 179 p
Sturm Ambros 1904 *Geschichte der Mathematik,* Leipzig, 152 p (F35, 1; 37, 1) [also 1906, 1911]
- 1911 *Geschichte der Mathematik bis zum Ausgange des 18. Jahrhunderts,* 2nd ed, Leipzig, 155 p (F42, 2) (3rd ed 1917)

General: Books <u>1900-1949</u> (continued)

Tannery J + 1903 *Notions de mathé-
matiques. Notions historiques,*
Paris, 362 p (Ger tr 1909 as
*Elemente der Mathematik;* Russ tr
1914 as *Osnovnye ponyatiya)*
Timerding H E 1914 *Die Verbreitung
mathematischen Wissens und
mathematischer Auffassung,* Leipzig
Ulehla J 1901 *Geschichte der
Mathematik,* Prague (F35, 47)
Vera Francesco 1946 *Breve historia
de la matematica,* Buenos Aires
Wargny C 1913 *Historie des las
matematicas,* Santiago, 375 p
(F44, 38)
Wieleitner H 1911-1921 *Geschichte
der Mathematik,* Berlin, 2 vols
(Vol 1= Guenther 1908) (2nd ed
1922-1923; repr 1939)
- 1925 *Der Gegenstand der Mathematik
im Lichte ihrer Entwicklung,*
Leipzig, 61 p (F51, 2)
Yakobson A B 1927 *Ocherki istorii
tochnykh nauk,* Mos, 204 p
*Zeuthen H G 1902 *Elements d'histoire
des mathématiques,* Paris

General: Books from 1950

*Aleksandrov A D + 1956 *Matematika,
ee soderzhanie, metody i znachenie,*
Mos, 3 vols
- 1963 *Mathematics Its Content,
Methods, and Meaning* (MIT Press),
(tr of 1956)
Boev G P 1956 *Lektsii po istorii
matematiki, Ch. I: Do nachale 18-
go veka,* Saratov, 281 p
*Bourbaki Nicolas 1960 *Eléments
d'histoire des mathématiques,*
Paris, 277 p (Ital tr 1963; MR33,
2514; Russ tr 1963; 2nd ed 1969)
Carruccio Ettore 1958 *Matematica
e logica nella storia e nel
pensiero contemporaneu,* Turin
[Engl tr 1964 Chicago]
Cassina Ugo 1961 *Della geometria
egiziana alla matematica moderna,*

Rome, 541 p (Z102, 243)
*Daumas Maurice ed 1957 *Histoire de
la Science* (Encyclopédie de la
Pléaide #5) (Gallimard), 1952 p
*Hofmann Joseph E 1953-1957
*Geschichte der Mathematik,* Berlin,
3 vols [to Fr Revo] (2nd ed 1963)
- 1957 *The History of Mathematics,*
NY, 132 p (MR19, 107) [to 1650]
- 1959 *Classical Mathematics,*
NY, 159 p [1650-1790]
Hosoi S 1955 *A History of Mathe-
matical Concepts in the East and
the West,* 2nd ed, Tokyo, 255 p
(Z67, 245)
Kofler E 1956 *Z dziejow matematiki,*
Warsaw, 279 p
Krbek Franz v 1952 *Eingefangenes
Unendlich. Bekenntnis zur
Geschichte der Mathematik,*
Leipzig, 338 p (MR14, 341;
Z47, 241)
Loria Gino 1950 *Storia delle matemati-
che dall'alba della civiltà al
secolo XIX,* 2nd ed, Milan, 3 vols,
1010 p (MR12, 69; Z39, 1)
Manwell A R 1959 *Mathematics Before
Newton,* London, 56 p
Mercier André 1964 *Antikes und
modernes Denken in Physik und
Mathematik,* Berlin, 154 p
(MR29, 218)
Meschkowski Herbert 1961
*grosser Mathematiker: Ein Weg zur
Geschichte der Mathematik,*
Braunschweig, 103 p (MR24, 571)
- 1964 *Ways of Thought of Great
Mathematicians,* San Francisco
(tr of 1961)
*Reidemeister K 1957 *Raum und Zahl,*
Berlin
Rey Pastor Julio + 1951 *Geschichte
der Mathematik,* Buenos Aires,
388 p (Z45, 289)
Ritchie A D 1958 *Studies in the
History and Methods of the Sciences,*
Edinburgh, 235 p [examples from
math]
Scott J F 1958 *A History of Mathe-
matics From Antiquity to the
Beginning of the Nineteenth Century,*
London, 276 p (Z79, 3)

General: Books from 1950 (continued)

Struik D J 1948 *A Concise History of Mathematics,* 2nd ed (Dover), 299 p (many editions and translations, each with some revisions)
*Wilder R L 1968 *Evolution of Mathematical Concepts. An Elementary Study* (Wiley)
Zhatykov O A 1951 *Mathematics and its Development,* Alma-Ata, 82 p (in Kazak)
- 1959 *From Simple Counting to Machine Mathematics,* Alma-Ata, 272 p (in Kazak)

General: Chronologies

Anon Am Q Reg 6, 16
Archibald R C 1921 AMM 28, 423 [8 key dates]
Baehr U 1955 *Tafeln zur Behandlung chronologischer Probleme,* Karlsruhe, 75 p
Buquoy George L v 1929 *Chronologische Auszug aus der Geschichte der Mathematik,* Leipzig
Darmstaedter Ludwig 1908 *Handbuch zur Geschichte der Naturwissenschaften und der Technik* 2nd ed (Springer), 1264 p
Daumas Maurice ed 1957 *Histoire de la Science* (Gallimard), 1952 p
Edwards E J 1936 *An Illustrated Historical Time Chart of Elementary Mathematics for Senior and Secondary Schools, Training Colleges and Universities,* London, 5 charts
Eves H 1961 MT 54, 452-454 ["time strip"]
Feldhaus Franz M 1904 *Lexikon der Erfindungen und Entdeckungen...,* Heidelberg, 144 p
*Houzeau J C 1876 Brx Ob, 52, 124 (F8, 31) [ast]
*Little Charles E 1900 *Encyclopedia of Classified Dates,* NY (Funk & Wagnall's) 1461 p
Mansion P 1899 *Mélanges Mathématiques III,* Paris

- 1905 Mathesis (3)5, 5-6, 33, 57-58, 89, 113-114, 145, 169, 201-202 (F36, 2)
*Mueller Felix 1892 *Zeittafeln zur Geschichte der Mathematik, Physik und Astronomie bis zum Jahre 1500, mit Hinweis auf die Quellen-Litteratur,* 109 p (Teubner), 109 p (F24, 3)
Rosenberger F 1882-1884 *Die Geschichte der Physik in Grundzuegen mit synchronistichen Tabellen der Mathematik, der Chemie und beschreibenden Naturwissenschaften,* Braunschweig, 2 vols

General: Elementary mathematics

Bortolotti E 1950 Enc M El (3)2(58), 539-570 (Z37, 289)
Cajori Florian 1896 *A History of Elementary Mathematics with Hints on Methods of Teaching,* NY (2nd rev ed 1917, 3rd ed 1930) (Russ tr 1917 Odessa; Japanese tr 1928; F54, 44)
Fink K 1890 *Kurzer Abriss einer Geschichte der Elementar-Mathematik mit Hinweisen auf die sich anschliessen den hoeheren Gebiete,* Tuebingen, 279 p [Engl tr 1900 Open Court]
Franchetti G 1893 *Cenni storici sulle matematiche elementari,* Sassari (F25, 82)
Lietzmann W 1928 *Ueberblick ueber die Geschichte der Elementarmathematik,* 2nd ed, Leipzig, 87 p (F54, 2)
Loria G 1896 Pe MI 11, 1
- 1898 Pe MI 14, 19
Miller G A 1909 AMM 16, 177-179
- 1925 Ens M 24, 59-60
- 1925 Sci (NS)61, 491-492
*- 1925 Sci Mo 21, 150-156 (F51, 11)
- 1933 Sci Mo 37, 398-404
- 1933 SSM 32, 838-844
- 1934 Sci Mo 39, 40-45 (F60, 816) [folk math]

General: Elementary mathematics
(continued)

- 1935 M Stu 3, 121-126   (F61, 940)
- 1936 Sci Mo 17, 230-235 (F62, 3)
Sanford Vera 1930 *A Short History of
Mathematics*, Boston, 412 p
*Smith D E 1923-1925 *History of
Mathematics. Vol I: General
Survey. Vol 2: Special Topics*,
NY, 618 + 737 p (Repr 1958 Dover)
(MR19, 1029) See Miller G A 1924
SSM 24, 939-947)
Tropfke Johannes 1902-1903 *Geschichte
der Elementar-Mathematik*, Leipzig,
2 vols
- 1921-1922 *Geschichte der Elementar-
Mathematik in Systematischen
Darstellung...*, 2nd ed, Berlin,
3 vols
- 1930-1940 *Geschichte der Elementar-
Mathematik in Systematischer
Darstellung*, 3rd ed, Berlin, 5 vols
(Z16, 145)
Wieleitner H 1925 Weltall 24, 128-134
(F51, 6)

General: Popularizations from 1900

Anon 1961 MT 54, 365-367
[rhymed hist]
Barmard Douglas S P 1968 *It's All
Done by Numbers* (Hawthorn), 128 p
*Bell E T 1934 *The Search for Truth*,
Baltimore, 289 p
*- 1937 *The Handmaiden of the
Sciences*, NY-Baltimore, 224 p
- 1938 *The Queen of the Sciences*,
NY, 138 p (F64, 1)
*- 1951 *Mathematics, Queen and
Servant of Science*, NY, 457 p
Beman W W + 1900 Nation 71, 314-315
Bioche Charles 1914 *Histoire des
Mathématiques*, Paris, 99 p
(F45, 1)
Boll Marcel 1947 *Les étapes des
mathématiques* (Press Univ.
de France), Paris, 127 p
- 1958 *Histoire des mathématiques*,
Paris, 127 p

Bortolotti E 1932 Scientia 273-286;
(S), 133-146
Bosteels G 1960 *La vie des nombres*,
Namur, 141 p
Boyer Lee E 1947 *Mathematics, a
Historical Development*, NY, 465 p
(2nd ed 1955)
Boyev T P  1947 *Besedi po istorii
matematiki*, Mos
Brown E W 1921 Sci Mo 12, 385-413
*Brun V 1964 *Alt er tall...*, Oslo,
239 p
Burgess E G 1925 SSM 24, 264-272
Colerus Edmont 1937 *Von Pythagoras
bis Hilbert*, Berlin, 364 p
(F63, 1)  (Ital tr 1939; Fr tr
1943; Hung tr)
- 1957 *Vom Einmaleins zum Integral*,
Hamburg, 403 p (Fr tr 1952,
Engl tr 1957)
Dantzig Tobias 1930 *Number: The
Language of Science...*, London,
271 p (4th ed 1954; editions
differ)
*Dedron  P + 1959 *Mathématiques et
mathématiciens*,  Paris, 442 p
Dehn M 1932 Scientia, 125-140; (S),
61-74
- 1943 AMM 50, 357-360, 411-414
(Z60, 5)
- 1944 AMM 51, 25-31, 149-157
(Z60, 5)
Denjoy A 1912 Rv Mois 13, 67-78
Depman I Ya 1950 *Iz istorii mate-
matiki*, Mos-Len, 116 p
[for schools]
- 1954 *Rasskazy o matematike*, Len,
144 p
De Vries H 1906 *Mathesis en
mathematica*, Delft, 28 p
(F37, 35)
Dziobek Otto 1962  *Mathematical
Theories*,  NY, 300 p
Fazzari G 1907 *Breve storia della
matematica. Dai tempi antichi al
Medio Evo*, Milan-Palermo-Naples,
267 p  (F38, 4)
Frajese Attilio 1962 *Attraverso la
storia della matematica*, Rome,
282 p  (MR30, 1141)
Frankland W B 1902 *The Story of
Euclid*, London, 173 p

General: Popularizations from 1900
(continued)

Freebury H A 1958 *A History of Mathematics for the Secondary Schools,* London, 208 p (also 1961 NY; MR23, 581)

Friederichs K O 1965 *From Pythagoras to Einstein* (Random House)

Gambioli D 1929 *Breve sommario della storia delle matematiche,* 2nd ed, Palermo, 236 p (F55, 587)

Hille E 1951 M Mag 26, 127-146

Hogben Lancelot 1955 *Man Must Measure,* London, 70 p

Jervey J P 1929 Milit Eng (May), 195-199

Keyser C J 1947 *Mathematics as a Culture Cue and Other Essays,* NY, 284 p (Z29, 2)

Kinsella J + 1948 MT 40, 355-358

Klimpert R 1902 *Storia della geometria ad uso dei dilettanti di matematica e degli alunni delle scuole secondarie,* Bari

*Kline Morris 1953 *Mathematics in Western Culture* (Oxford U P), 496 p (MR15, 769)

- 1959 *Mathematics and the Physical World,* NY, 159 p

Kramer Edna E 1951 *The Mainstream of Mathematics* (Oxford U P), 321 p

Kreitner J 1956 Myrin P 5, 3-26

Larrett Denham 1926 *The Story of Mathematics,* Lon-NY, 87 p

Malsch F 1928 *Geschichte der Mathematik,* Leipzig, 115 p (F54, 1)

Milhaud G 1911 Rv M Sp 21, 209-212

Miller G A 1921 Sci Mo 12, 75-82

- 1926 Sci Prog (Jan), 436-438
- 1928 Sci Mo 26, 295-298 (F54, 3)
- 1933 Sci Mo 37, 398-404 1941 Sci (NS)93, 235
- 1943 Sci (9 July), 38-39

Muir Jane 1961 *Of Men and Numbers...,* NY, 249 p (Z113, 1)

Ocagne Maurice d' 1955 *Histoire abrégée des sciences mathématiques,* Paris, 406 p (MR17, 117)

*Ore Oystein 1941 in *Development of the Sciences* Second Series, Ed by L. L. Woodruff (Z60, 1)

Pelletier Jean-Louis 1949 *L'age des mathématiques,* Paris (Z41, 337)

Perès J 1930 *Les sciences exactes,* Paris, 196 p (F56, 797)

Pledge H T 1949 *Science since 1500. A Short History of Mathematics, Physics, Chemistry, Biology,* NY, 359 p (MR10, 667)

Reid Constance 1963 *A Long Way from Euclid,* NY, 301 p

Rogers J 1966 *Story of Mathematics for Young People* (Random House)

Schaaf W L 1948 *Mathematics, Our Great Heritage,* NY

Sergescu P 1932 Cuget Cla 5, 386-395 (F59, 857)

- 1933 *Les Sciences mathématiques,* Paris, 182 p

Shaw H A + 1963 *The Story of Mathematics,* London (Edward Arnold)

Smith D E 1906 *History of Modern Mathematics,* 4th ed, NY, 81 p (F37, 1)

*- 1937 Scr Lib (3) [chronology]

*Speiser Andreas 1955 *Die Geistige Arbeit,* Basel-Stuttgart, 216 p

*Sullivan John W N 1925 *History of Mathematics in Europe From the Fall of Greek Science to the Rise. of the Conception of Mathematical Rigour* (Oxford U P)

Taton R 1946 *Histoire du Calcul,* Paris, 123 p

Turnbull Herbert W 1961 *The Great Mathematicians,* 4th ed, NY (MR24, 4)

Viviana R H 1917 Ed Rv 53, 30-43

*Whitehead A W 1925 in his *Science and the Modern World,* Ch 2

Willerding Margaret F 1967 *Mathematical Concepts, A Historical Approach,* Boston, 126 p

*Wolff Peter 1963 *Breakthroughs in Mathematics* (Signet Science Library)

## General: Quantitative Studies

See also General: Theory

Folta J + 1965 Act HRNT 1, 1-35
Holton G 1962 Am Ac Pr 91(2)
(Spring), 362-399 (repr in S R
Graubard +, *Excellence and
Leadership*... (Columbia U P)
[growth, quality]
*Lipetz Ben-Ami 1965 *The Measurement
of Effectiveness of Scientific
Research*, Carlisle, Mass [bib]
May K O 1966 Sci 154, 1672-1673
[growth]
- 1968 Isis 59(4), 363-371
[quantity and quality]
Novy L + 1965 Prag Dej 11, 25-55
Price Derek J de S 1961 *Science
Since Babylon* (Yale U P)
- 1963 *Big Science, Little Science*
(Columbia U P)
White H S 1915 Sci (NS)42, 105-113
[growth]

## General: Reference

See also Ch. 5.

Archibald Raymond C 1949 *Outline of
the History of Mathematics* (AMM
56(1)(2)), 114 p (repr 1966
Johnson) (corrections AMM 56,
492-497)
*Becker Oskar + 1951 *Geschichte der
Mathematik*, Bonn, 340 p (MR14, 341)
(Fr tr 1956) [*Index]
*Berzolari L + eds 1930- *Enciclopedia
delle matematiche elementari*,
Milan, 3 vols (= Enc M El)
*Cantor M 1880-1908 *Vorlesungen zur
Geschichte der Mathematik*, 4 vols,
Leipzig [To 1799] (see Bib M for
correction of over 2000 errors)
(repr 1965, NY) (MR33, 3863)
Darmstaedter L 1908 *Handbuch zur
Geschichte der Naturwissenschaften
und der Technik*, 2nd ed, Berlin,
1276 p (repr 1960)
*Meyer W F + eds 1896- *Encyklopedie*

*der mathematishe Wissenschaften*...,
(Rich historical and bibliographic
information)
Mieli A 1925 *Manuale di storia della
scienza*, Rome, 599 p (F51, 1)
*Molk J + eds 1904- *Encyclopédie des
sciences mathématiques*... (Even
richer tr and rev of Meyer 1896-)
*Sarton George 1927-1948 *Introduction
to the History of Science*,
Baltimore, 3 vols in 5
Tropfke J 1930-1940 *Geschichte der
Elementar-Mathematik*..., 3rd ed,
Berlin, 5 vols
*Uccelli A 1941 *Enciclopedia storica
delle science e delle loro
applicazioni. Le scienze fische
e matematiche* Milan, 730 p
(F67, 968)
*Wieleitner II 1939 *Geschichte der
Mathematik*..., Berlin, 2 vols

## General: Textbooks for courses

Of course many others are suitable, but we list here only those
intended as texts.

See also General: Books, Popularizations.

Boev G P 1956 *Lektsii po istorii
matematiki*, Saratov
*Boyer C B 1968 *A History of
Mathematics*, NY, 711 p
Carruccio Ettore 1951 *Curso di
storia delle matematiche presso
la Facoltà di Scienze dell'
Università di Torino*, Turin, 578 p
Eves Howard 1953 *An Introduction to
the History of Mathematics*, NY,
437 p (MR15, 89) (rev ed 1964:
MR29, 638) [for school teachers]
Otradnykh F P 1951 *Kurs istorii
matematiki. Lektsii 1-24*, Mos
Rybnikov K A 1960-1963 *Istoriya
matematiki*, Mos, 2 vols Moscow,
190 p (MR23, 692; Z98, 5)
Werner Carl 1840 *Kurzer Entwurf
einer Geschichte der Mathematik*,
Pasewalk

General: Theory

See also above Quantitative studies,
and in Ch. 3, Anthropology,
Heuristic, Philosophy, and in
Ch. 5, HISTORIOGRAPHY, REFERENCE
MATERIALS

Bobynin V V 1887 *Issledovaniya po
istorii matematiki,* Mos, 18 p
*Bochner Salomon 1963 Sci 141, 408-411
[crises]
- 1965 J H Ideas 26, 3-24 [why
growth]
1966 *The Role of Mathematics in
the Rise of Science* (Princeton
U P), 368 p
- 1969 *Eclosion and Synthesis.
Perspective on the History of
Knowledge* (Benjamin)
*Bouligand G 1952 Arc In HS 31, 230-
233
Bouligand Georges + 1957 *Hommage à
Gaston Bachelard,* Paris, 73-81
Boutroux P 1909 Scientia 6, 1-20
- 1920 *L'idéal scientifique des
mathématiciens,* Paris
- 1920 Rv Mois 20(14)(120), 604-621
Boyer C B 1959 Sci 130(3 July)
[utility]
Candido G 1935 Pe M 15, 58-62
[coincidence, priority]
*Eneström G 1904 Bib M (3)5, 1-4
(F35, 49) [regularities]
*- 1905 Bib M (3)6, 1-8
(F36, 47) [hist hyp]
*Gnedenko B V 1960 Int Con HS 9,
472 (Z114, 5) [concept
formation]
- 1966 Ist Met EN 5, 5-14
(MR34, 4085) [sci]
*Hadamard J 1958 Calcutta Math. Soc
Golden Jubilee (1), 11-14
(MR27, 899; Z102, 6) [current]
Halsted G B 1893 Tex Ac 1(2), 89
[new interpretation of old]
Karpinski L C 1921 SSM 20, 821-828
[rel betw arith and geom]
- 1937 Isis 27, 46-52 [progress?]
Klein F 1893 in his *Lectures on
Mathematics*...1893, NY (Macmillan)
41-50, [rigor]

Kline M 1954 Conflu 3, 196-206
*Kolmogorov A N 1926-1947 *Matematika,*
in Bolsh S E
*Kuhn Thomas S 1962 *The Structure of
Scientific Revolutions,* Chicago
(2nd rev ed 1970)
- 1962 Sci 136(June), 760-
Loewy A 1933 Fors Fort 9, 81-82
(F59, 4)
Loria G 1927 Scientia 41, 321-332;
41(S), 127-137 (F53, 1)
[hist laws]
- 1953 in Underwood E A ed, *Science,
Medicine and History* (Oxford U P),
2 vols
Meyer W F 1930 Unt M 36, 322
(F56, 35) [anc vs mod]
- 1931 Ham MG 7, 1-9 (Z2, 277)
[anc vs mod]
Miller G A 1917 AMM 24, 453
[the obsolete]
- 1926 Am NASP 12, 537-540
[hist assumptions]
- 1928 Am NASP 14, 214-217
[harmony]
- 1928 Scientia 44, 81-88
(F54, 4) [intangible advances]
- 1930 SSM 29, 954-960
- 1931 Am NASP 17, 463-466
(F57, 6) [hist laws]
- 1931 Toh MJ 34, 230-235
(F57, 4)
- 1933 Sci Mo 37, 398-404
(F59, 827)
- 1938 Sch Soc 17, 275-277
Mitchell U G 1933 MT 26, 296-301
[rel to human progress]
Nevanlinna R 1950 Zur NGV 95, 1-22
(MR11, 571)
Pavate D C 1933 Bombay 2(2), 188-195
(F61, 957) [rel to hist ideas]
Pelseneer J 1955 Isis 46, 95-98
(MR16, 985) [planning research]
Rosenthal F 1950 Osiris 9, 555-564
(MR16, 660) [Al-Asturlabi,
As-Samawol on progress]
*Rybnikov K A 1954 Ist M Isl 7, 643-
665 (MR16, 781) [creative and
critical periods]
*Scriba C J 1967 Bei GWT (9), 54-80
[problems]

General: Theory (continued)

Slaught H E 1934 AMM 41, 167-174
[uneven devel]
Smith D E 1921 Scientia 29, 417-429
[epochs]
*Toulmin S E 1967 Am Sc 55(4), 456-
[science]
*Voss A 1912 in *Die Kultur der
Gegenwart,* Leipzig, vol 2
White L A 1947 Phi Sc 14, 289-303
[anthropological approach]
- 1953 AMSB 59, 423-448
Yoneyama K 1926 Tok S Ph S 35, 399-
412 [discovery]


OLD CIVILIZATIONS

See also Topics, esp. Metrology,
Arithmetic, Numeration, etc, in
Chs. 2-3.


African

Kluge T 1940 *Die Zahlenbegriffe der
Voelker Amerikas, Nordurasien der
Munda und Palaienafrikaner; ein
dritter Beitrag zur Geistes-
geschichte der Menschen,* Berlin,
736 p (F66, 4)
Pauline D 1942 Rv Sc 80, 219-226
[metrology]
Raum O F + 1938 *Arithmetic in Africa*
(Evans Bros)
Zaslavsky C 1970 Mt 63(4), 345-356
[*bib]


American Indian

Bradley A D 1956 Scr 22, 275-280
[arith]
Cajori F 1925 AMM 32, 414-416
(F51, 5) [numeration]
Eells W C 1913 AMM 20, 263-272,
294-299 [numer]
- 1913 Bib M 13, 218-222 [numer]
Fettweis E 1929 Arc GMNT 11, 342-
344 (F55, 7)

Guerra F 1960 J H Med 15, 342-344
[metrol]
Kluge T 1940 *Die Zahlenbegriffe der
Voelker Amerikas, Norduraisien der
Munda und Palaienafrikaner; ein
dritter Beitrag zur Geistes-
geschichte der Menschen,* Berlin,
736 p (F66, 4)
Lietzmann W 1931 ZMN Unt 62, 155-160
[bib, Peru]
Spence L 1912 Rv Sc 50 [calendar]
Wren F L + 1934 SSM 33, 363-372
[North Amer]


Arab

See also TIME PERIODS: Middle ages,
Numeration (Ch. 3), Manuscripts
(Ch. 5).

Ahrens W 1917 Islam 7, 181-250
[magic sq]
Amodeo A F 1912 Nap Pont (2)17(4),
13 p (F43, 7) [importance]
Anon 1888 Bib M (2)2, 63
[translations]
*Arnaldez R + 1963 in Taton R ed,
*Ancient and Medieval Science,*
NY, 385-421
Baudoux C 1935 Brx Con CR 2, 73-75
(Z13, 193) [Syrian Euclid]
- 1937 Archeion 19, 70-71
(F63, 8) [Euclid, Ishaq]
Baqir T 1950 Sumer 6, 39-54
(MR12, 69) [Euclid]
Brandely G B 1873 Boncomp 6, 65-68
(F5, 7) [re Hankel 1872]
Brocard H + 1901 Intermed 7, 244
- 1903 Intermed 10, 144
[translations]
- 1903 Intermed 10, 171 [alg]
Carra de Vaux B 1899 Bib M (2)13,
33-36 (F30, 38) [arith]
- 1900 Bib M (3)1, 28
[Archimedes, Heron, machines,
Philon]
- 1921 *Les penseurs de l'Islam.
T. II: Les geographes, les sciences
mathématiques et naturelles,* Paris

Arab    (continued)

- 1948 Rv Hi Sc Ap 1, 206-212
  (MR11, 419) [magic sq]
Ceretti U 1900 Rv FMSN 2, 97; 3, 107
  [formulas]
Chasles M 1855 Par CR 40, 782-
  [re Woepke]
- 1865 Par CR 60, 601-
  [re Woepke]
Clagett M 1954 Osiris 11, 359-385
  [mid tr, hyperb]
Colin G S 1933 Hesperis 16, 151-155
  [Morocco, mult]
Datta B 1932 Benar MS 14, 7-21
  (Z9, 98) [Sanscrit, Persian]
- 1932 Clct MS 24, 193-218
  [numeration]
Decourdemanche J-A 1913 J Asi (11)1,
  427-444 [geodesy]
Destombes M 1960 Physis Fi 2, 197-
  210 [instruments]
Emine M 1899 Intermed 6, 39
Faddegon J M 1932 J Asi 220, 139-148
  [abacus]
- 1933 Arc Sto 14, 372-391
  [mid]
Fischer A 1934 Islamica 6, 48-57
  [finger reckoning]
Furiani G 1924 Z Sem Ver G 3(1), 27-
  52 [Syrian Euclid]
Gandz S 1927 AMM 34, 308 [abacus]
- 1932 QSGM (B)2, 81-97 (Z5, 3)
  [Greek]
- 1933 Isis 22, 220-222 (F60, 825)
  [rule of three]
- 1937 Osiris 3, 405-557 (F63, 797)
  [quad eq]
- 1937 Scientia, 249-257 [quad eq]
Garbers K 1943 Z Gesm Nat 9, 21-34
  (Z28, 1) [hist math]
Goldstein Bernard R 1967 *Ibn al-
  Muthanna's Commentary on the
  Astronomical Tables of al-
  Khwarizmi...* (Yale U P), 403 p
Hankel H 1872 Boncomp 5, 343-401
  (F4, 5)
Haskins C H 1925 Isis 7, 478-485
  [Europe]
Heiberg J L + 1910 Bib M (3)11, 193-
  208 (F41, 66) [parab, mirror]

Hermelink II 1959 Sudhof Ar 43, 351-
  354 [magic sq]
- 1961 Isis 52, 417 [geom, notat]
Irani R A K 1952 Scr 18, 92-93
  (Z46, 1) [sexagesimal, table]
Kapp A G 1934 Isis 22, 150-172
  [Euclid, al-Qifti]
- 1935 Isis 23, 54-59 (F61, 13)
  [Euclid]
Kennedy E S 1951 Am OSJ 71, 13-21
  (MR13, 1; Z45, 146) [instru]
- 1952 Isis 43, 42-50 (Z47, 3)
  [al-Kashi, instru]
Kennedy E S + 1961 Sudhof Ar 45, 85
  [notat]
- 1965 MT 58, 441-446 [ast]
Kokomoor F W 1936 MT 29, 224-231
  [mid]
Krause M 1936 QSGM (B)3, 437-532
  (F62, 1019; Z15, 289) [bib, mss]
Levey M + 1962 Ens M (2)8, 291-302
  (Z118, 246) [arith, ibn Labban]
Loria G + 1898 Intermed 5, 240
  [tr from Arabic]
- 1928 Archeion 9, 161-166 [Greek]
- 1929 Archeion 11, 15-22 (F55, 8)
  [miss]
- 1930 Bo Fir 11, 23-29
- 1930 Il Bol di Mat, 12, 13-14
  [Apollonius ms]
Luckey P 1948 Fors Fort 24, 199-204
  (MR11, 149; Z30, 337)
  [arith, alg]
- 1948 M Ann 120, 217-274 (Z29, 385)
  [binom thm, roots]
- 1948 Oriental 17, 490-510
  (Z37, 290)
Marre A 1866 Vat NLA 19, 362- [ms]
Massignon L 1933 Archeion 14, 370-371
  (F59, 18) [arith]
Merrifield C W 1866 LMSP 2, 175-177
Meyerhof M 1933 Archeion 15, 1-15
  (F59, 18) [Alexandria]
Mieli Aldo 1939 *La science Arabe et
  son role dans l'evolution
  scientifique,* Leyde (repr 1965)
- 1942 Archeion 24, 224-245
Moreau F 1926 Brx SS 45(1), 149-153
  (F52, 13) [gnomon, XIV]
Mydorge C 1884 Boncomp 16, 514-527
  [geom]

Arab (continued)

Nallino C A 1911 Bib M 12, 277-282
[ast, mid]
Plooij E B 1950 *Euclid's Conception
of Ratio and his Definition of
Proportional Magnitudes as
criticized by Arabian Commentators,*
Rotterdam, 71 p
Prell H 1961 Deu Morg Z 110, 26-42
(Z102, 245) [metrol]
Qifti Ibn al 1903 *Tarib Al-Huhama,*
Leipzig, 518 p (ed by J. Lippert,
A. Mueller) [biog coll]
Quadir M A 1950 Islam Rv, 33-34
Renaud H P J 1932 Isis 18, 166-183
[crit Suter 1900]
Rescher N 1964 *The Development of
Arabic Logic,* Pittsburgh [*bib]
Reynolds J H 1931 Nat, 913-914
(Z3, 99) [ebn Jounis, table]
Rodet L 1884 Boncomp 16, 528-544
[problems]
Rosenfeld B A 1958 Sov Vost (3),
101-108; (6), 66-76
Ruska Julius 1917 *Zur aeltesten
arabischen Algebra und Rechenkunst,*
Heidelberg, 125 p (Heid Si Ph)
- 1920 Islam 10, 87-119, 154-156
[finger reckoning]
Russell J C 1932 Isis 18, 14-25
[England, Hereford, XII]
Sanchez Perez José A 1927 *Biografías
de matemáticos arabes que
floreceiron en Espana,* Madrid,
164 p
- 1933 Rv M Hisp A (2)8, 86-96
(F59, 839)
- 1949 *Arithmetik in Rome, India
and Arab Lands,* Madrid-Granada,
263 p (in Spanish) (Z41, 338)
Sarton G 1933 Isis 20, 260-262
(F59, 19) [arith]
- 1943 Isis 34, 513 [Pythagorean
thm]
Schirmer O 1957 Bay Jb OK, 17-50
(Z80, 6) [metrol]
Schmalzl Peter 1929 *Zur Geschichte
des Quadranten bei den Arabern,*
Munich, 142 p
Schoy Karl 1917 ZMN Unt 49, 73-79
[chronom]

- 1923 Isis 5, 364-399 [trig]
- 1923 *Ueber den Gnomonschatten und
die Schattentafeln in der arabischen
Astronomie,* Hannover, 27 p
- 1926 Isis 8, 254-263
Sédillot Louis P E A 1837 *Recherches
nouvelles pour servir à l'histoire
des sciences mathématiques chez
les orientaux, ou notice de
plusiers opuscules mathématiques
qui composent le manuscript 1104
de la bibliographie du roi,*
Paris [cubic eq]
- 1845 *Matériaux pour servir à
l'histoire comparée des sciences
mathématiques chez les Grecs et
les Orientaux,* Paris
- 1853 *De l'algèbre chez le
Arabes,* Paris (J Asi (12))
- 1863 *Courtes observations sur
quelques points de l'histoire des
l'astronomies et des mathématiques
chez les orientaux,* Paris, 32 p
- 1868 Boncomp 1, 217-222
(F1, 5) [Bagdad]
- 1871 Boncomp 4, 401-418 (F3, 3)
- 1875 Boncomp 8, 63-78 (F7, 3)
[contributions]
Smith D E 1926 AMM 33, 28-31
[Schoy]
Smith D E + 1918 AMM 25, 99-108
[Ben Ezra]
- 1927 AMM 34, 258-260 [numer]
Steinschneider M 1863 Annali 5, 54-
- 1865 ZM Ph 10, 456-
- 1869 *Al-Farabi (Alpharabius). Des
Arabischen Philosophen Leben und
Schriften, mit besondere Ruecksicht
auf die Geschichte der grieschen
Wissenschaften under den Araben,*
Pet, 278 p (= Pet Mm (7)13(4))
- 1869 Deu Morg Z 24, 325-392
(F2, 4) [tr from India]
- 1872 Boncomp 5, 427-534 (F4, 4)
(also Rome 1874) [Baldi]
- 1891 Bib M (2)5, 41-52, 65-73
(F23, 4)
- 1892 Bib M (2)6, 53-62 (F24, 55)
[Ptolemy]
Studnika F J 1875 Cas MF 4 (F8, 12)
[cubic eq]

Arab (continued)

*Suter H 1892 Ab GM6, 1-89   [bib]
 - 1892 ZM Ph 37, 1-25
 - 1893 ZM Ph 38, 41-58, 126-127,
   101-185  (F25, 7, 14)
 - 1895 Die Araber als Vermittler der
   Wissenschaften in deren Uebergang
   vom Orient in den Occident, Aarau,
   31 p (F27, 5)
 - 1896 Bib M (2)10, 31-32, 64, 96,
   120  (F27, 2) [regula cecis]
 - 1897 Bib M (2)11, 83-86
   (F28, 2) [biog]
 - 1898 Bib M (2)12, 73-78
   (F29, 3) [Berlin mss]
 - 1899 Bib M (2)13, 86-88, 118-119
   (F30, 3)  [biog]
*- 1900 Die Mathematiker und
   Astronomen der Araber und ihre
   Werke, Leipzig, 286 p  (Ab GM 10)
   (F31, 4)
*- 1902 Ab GM 14, 155-185    (F33, 11)
   [suppl to 1900)
 - 1902 Bib M (3)3, 350-354
   [biog, "Liber augmenti et
   diminutionis"]
 - 1904 Int Con (3), 556
 - 1907 Bib M (3)8, 23-36  (F38, 65)
   [geom]
 - 1922 Beitraege zur Geschichte
   der Mathematik bei den Griechen
   und Arabern,  Erlangen, 116 p
   (F48, 28)
Thaer C 1942 Hermes 77, 197-205
   [Euclid]
Thureau-Dangin F 1931 Rv Assyr 28,
   23-25 ([angles]
Toqan Quadri 1948 Das grundlegende
   Buch der Arabes ueber Mathematik
   und Astronomie, Cairo, 267 p
   (Z41, 338)
Turetsky M 1924 MT 16, 29-34
   [permutations]
Vera Francisco 1947 La Matematica
   de los musulmanos espanoles,
   Buenos Aires
 - 1948 Pol SM An 21, 94-98
   (MR10, 174) [Toledo, tr]
Vernet J + 1950 Gac M (1)2, 78-82
   (Z36, 145)  [transf of coords]

Wiedemann E 1876 Pogg An 159, 656-658
   (F8, 12)  [sci]
 - 1879 Boncomp 12, 873-876
   (F11, 43)
 - 1882 Boncomp 14, 718-720
 - 1909 Arc GNT 1, 216-217  (F41, 3)
   [biog]
 - 1910 Erlang Si 41, 109   [biog]
 - 1918 Erlang Si (48/49), 1-15
   [al-Biruni, propor]
 - 1922 Z Instr 42, 114-121  [biog]
 - 1928 Erlang Si S8, 219-224  [mech]
Wieleitner H 1918 Unt M 24, 8-10
   [oldest]
 - 1925 Arc Sto 6, 46-48   [eqs]
Winter H J J 1953 Arc In HS 6, 171-
   192  (Z53, 338)
 - 1956 Centau 5, 73-88   [optics, ms]
Woepcke Franz 1850 Par CR 31, 715-
   [mss]
 - 1851 L'algèbre d'Omar Alkayami,
   Paris, 127 p
 - 1852 Notice sur une théorie ajoutée
   par Thabit ben Korrat a l'arithmétique speculative des Grecs,
   Paris, 12 p  (J Asi (12))
 - 1853 Extrait du Fakhri, traité
   d'algèbre par...Alkarkhi, précédé
   d'un mèmoire sur l'algèbre
   indeterminé chez les Arabes,
   Paris, 152 p
 - 1860 Recherches sur l'histoire des
   sciences mathématiques chez les
   Orientaux, d'apres des traités
   inédits des Arabes et des Persans,
   Paris  (also 1855 J Asi (S)5)
 - 1862 Notices sur quelques manuscrits Arabes relatifs aux mathématiques et recémment acquis par
   la bibliotheque impériale, Paris,
   31 p
 - 1864 Passage relatifs à des
   sommations de serie de cubes.
   Extrait de deux manuscrits Arabes
   inédits du British museum de
   Londres.  Cote N CCCCXVII et
   CCCCXIX des manuscrits orientaux
   N 7469 et 7470 des manuscrits
   additionnels, Rome, 25 p  (also
   1864 with same title, Rome, 39 p
   = Annali 5(3))

Arab (continued)

Worrell W H 1944 JNES (Ap), 91-100
[surveying]
Wuerschmidt J 1928 Erlang Si 60, 127-
154 (F54, 19) [trig]
Yushkevich A P 1951 Ist M Isl 4, 455-
488
- 1956 Int Con HS 8, 156-159
Yushkevich A P + 1960 in *Sowjetische
Beitraege zur Geschichte der
Naturwissenschaft,* Berlin, 62-160
[bib]
Zeki Salih 1911 *Athar-i-Baqiya, a
History of Arabic Mathematics,*
Istanbul, 512 p (in Turkish)


Asian (Near East and Central Asia)

See also Persian (below).

Faddegon J M 1932 Archeion 14, 372-
391 (F58, 15) [mid]
Grigoryan A T + eds 1966 *Fiziko-mate-
maticheskie nauki v stranach
Vostoka,* Mos, 358 p (MR34, 5608)
Mamedov K M 1960 Az FMT 6, 31-36
[mid]
Matvievskaya G P 1962 *K Istorii
matematiki Srednei Azii IX-XV
vekov,* Tashkent, 125 p (MR37, 2565)
Rozenfeld B A + 1958 Sov Vost (3),
101-108; (6), 66-76 [mid]
Taneri Kemal Zülfü 1958 *Türk
matematikcileri,* Istanbul, 104 p
Toth I 1955 Ro Sov (3)9(3), 74-83
(MR17, 337) [mid]
Yushkevich A P 1951 Ist M Isl 4,
455-488 (MR14, 523; Z43, 243)
[mid]
Yushkevich A P + 1960 in *Sowjetische
Beitraege zur Geschichte der
Naturwissenschaft,* Berlin, 62-168
(MR28, 221; Z99, 242) [bib, mid]
- 1960 Vost IINT 1, 349-421
(Z112, 241) [mid]
- 1963 *Die Mathematik der
Laender des Ostens im Mittelalter,*
Berlin, 160 p
Yusupov N 1933 *Ocherki po istorii
razvitiya arifmetiki na Blizhnem
Vostoke,* Kazan

Aztec

See also COUNTRIES: America.

Lemarie G 1907 Intermed 14, 224
(Q3125) [surveying]


Babylonian

See also Sumerian below.

Aaboe A 1955 Centau 4, 122-125
[Hipparchus]
- 1964 J Cun St 18, 31-34
[ast tab]
- 1965 Centau 10, 213-231 [ast]
- 1965 J Cun St 19, 79-86
[recip, sq, tab]
Agostini A 1959 Pe M (4)37, 201-212
(MR22, 1593; Z95, 1)
Allotte de la Fuije 1930 Rv Assyr 27,
65-71 [metrol]
- 1932 Rv Assyr 29, 11-19 (Z5, 1)
[tab AO 6456]
Anastasiades A S 1939 Gree SM 19,
189-203 (Z22, 1; 65:1, 4)
Anon 1928 Art Arch 26, 145-146
[mult tab]
- 1937 Sci Am 157, 311
Archibald R C 1929-1931 Sci 70, 66-67;
71, 342; 73, 68
- 1936 Isis 26, 63-81 (Z15, 147)
- 1936 MT 29, 209-219
(F62, 1012) [recent wk]
Aurès A 1891 *Traité de mètrologie
assyrienne ou étude de la
numeration at du systeme métriques
assyriens considéré dan leur
details, dans leur rapports et
dan leur ensembles,* Paris, 98 p
Becker O 1961 Praxis 3, 58-62 [pi]
Berriman A E 1956 M Gaz 40, 185-192
(MR18, 268) [quad eq]
Bertin G 1879 Bibcl Arc 7, 370-389
[numer]
- 1889 Nat 40, 237, 261, 285, 360
[ast]
Bilfinger G 1888 *Die babylonische
Doppelstunde: eine chronologische
Untersuchung,* Stuttgart
Bobynin V V 1893 Bib M (2)7, 18-21

Babylonian (continued)

  (F25, 57)   [numer]
Bortolotti E 1934 Bln Mm (9)1, 79-94
  (F60, 819)   [cubic eq]
- 1935 Bln Mm (9)2, 27-51
  (Z12, 242)   [alg]
- 1935 Pe M (4)15, 220-229
  (F61, 942)   [irrat, root]
- 1936 Pe M .(4)16, 65-81
  (F62, 7; Z13, 337)
- 1936 Pe M (4)16, 129-143, 225-241
  (F62, 7; Z16, 145)   [quad]
- 1937 Archeion 19, 192-195
  (F63, 797; Z17, 289)   [alg]
- 1937 Scientia 61, 9-15, 61(S),
  1-6   (F63, 5; Z15, 289)
  [gen meth, phil]
- 1938 Scientia 63,,305-312
  (F64, 903; Z18, 339)
- 1940 Bln Mm (9)7, 77-97
  (MR9, 483; Z26, 97; F66, 1185)
Boyer C B 1942 Sci 95, 553
  [thermometry]
Bruins E M 1948 Amst P 51, 332-341
  (Z30, 97)   [root]
- 1949 Amst P 52, 629-632
  (MR11, 149; Z33, 49)   [Pythagorean
  triples]
- 1950 Amst P 53, 1025-1033
  (MR12, 577; 1033,Z39, 1)
- 1950 Int Con HS 6, 121-122
  [recent wk]
- 1950 Rv Hi Sc Ap 3, 301-314
  (MR12, 577)
- 1952 Par PDC (D)H, 29 p
  (MR14, 1049)   [recent wk]
- 1953 Indagat 56, 412-422
  (MR15, 383)   [tria]
- 1955 Indagat 58, 16-23
  (MR16, 659; Z67, 243)   [geom]
- 1956 Int Con HS 8, 73-74   [geom]
- 1957 Janus 46, 4-11   [area]
- 1957 M Gaz 41, 25-28; 42, 212-213
  (MR21, 749)   [Pythagorean triples]
- 1959 Euc Gron, 131-159
- 1959 Janus 48, 5-23 [reg polyg]
- 1959 Praxis 1, 89-95
- 1959 Sim Stev 33, 38-60
  (MR21, 5156)   [geom of plummet]
- 1962 Physis Fi 4, 277-341
  (Z117:2, 241)

Bruins E M + 1961 *Textes mathé-
  matiques de Suse,* Paris, 146 p
  (Z101, 4)   (MR23A, 269; Z101, 4)
Cajori F 1922 AMM 29, 8-10
  [sexagesimal]
Cazalas 1932 Rv Assyr 29, 183-188
  (F58, 8)   [tab AO 6456]
David A 1928 Rv Assyr 25, 99-100
  (F59, 834)   [area, vol]
Decourdemanche J A 1913 J Asi (11)1,
  669-673   [geodesy]
Dittrich A 1933 Cas MF 63, 17-29
  (Z8, 337)   [ast]
- 1934 Cas MF 63, 82-96
  (F60, 819)   [tab, trig]
- 1938 Cas MF 67, 216-221
  (F64, 904)   [lunar tab]
Falkenstein A 1936 *Archaische Texte
  aus Uruk,* Berlin, 76 p (F62, 1011)
- 1953 Saeculum 4(2), 125-137
  (Z51, 2)
Fogelya K 1937 Kubicheskie uravneniya
  u vavilonyan, in Neugebauer O,
  *Lektsii po istorii antichnykh
  matematicheskikh nauk,* Mos-Len
Frank Carl 1929 *Strassburger
  Keilschrift texte in sumarischer
  und babylonischer sprache,*
  Berlin, 56 p  (F55, 593)
Gandz S 1929 Isis 12, 452-481
  [angle, Hebrew]
- 1936 Isis 25, 426-431
  (Z15, 53)   [recip, tab]
- 1936 Isis 26, 82-94
  (F63, 7; Z15, 147)
- 1937 It UM Con 1, 528-531
  (F64, 5)   [quad eq]
- 1937 Osiris 3, 405-557
  (Z18, 195) [quad eq]
- 1940 Isis 31, 405-425
  (MR7, 105; Z60, 4)
- 1947 Isis 32, 101-115   (MR8, 189)
  [isop, quad eq]
- 1947 in *Studies and Essays Offered
  to George Sarton,* NY, 449-462
- 1948 Osiris 8, 12-40
  (MR10, 667; Z32,241)   [numb thy]
Ganguli S 1930 Clct MS 22, 99-102
  (F56, 801)   [numer]
Gillings R J 1953 Au JS 16, 54-56
  (MR 15, 275) [errors in tablets]

Babylonian   (continued)

- 1954 Au MT
Gillings R J + 1964 Au JS 27, 139-141
   (MR30, 1) [recip, sexag, tab]
Goetze A 1946 JNES 5, 185-202
   [numb]
Gordon C H 1934 Rv Assyr 31, 53-60
   [numer, Nuzi]
Gundlach K-B + 1963 Ham Sem 26, 248-
   263 (MR28, 576) [quad eq]
Hilprecht H V 1906 Mathematical,
   Metrological and Chronological
   Tablets from the Temple Library
   of Nippur (U Pennsylvania), 88 p
   (F38, 4)
Hincks E 1787 Dub Ac T 22, 406-412
   [mythology]
Hofmann J E 1936 Muns Semb 8, 65-69
   (F62, 6) [sum of sq]
Huber P 1955 Isis 46, 104-106
   (MR16, 985; Z64, 1)
- 1957 Ens M (2)3, 19-27
   (MR21, 749; Z77, 4)
Jeremias Alfred 1908 Das Alter der
   babylonischen Astronomie,
   Leipzig, 64 p
Jones P S 1957 MT 50, 162-165, 442-
   444, 570-571 [recent wk: pi,
   polyg, probl, quad, trapezoid,
   zero]
Karpinski L C 1917 AMM 14, 257-265
   [alg]
- 1926 AMM 33, 325-326 (F52, 5)
   [recent wk]
- 1936 Am JSLL 52, 73-80
Kugler F X 1900 Die Babylonische
   Mondrechnung..., Freiburg, 214 p
Lenormant F 1873 Essai sur un
   document mathématique chaldéen, et
   a cette occasion, sur le système
   des poids et mesures de Babylone,
   Paris, 148 p (F5, 1)
Lepsius C R 1877 Die Babylonish-
   Assyrischen Laengenmasse nach der
   Tafel von Senkereh,  Berlin, 144 p
Levey M 1958 Ens M 4, 77-92
   [Kamil, alg]
Lewy Hildegard 1945 Am SORB(98),
   25-27 [metrol]
- 1947 Am OSJ 67, 305-320 (MR9, 169)

- 1949 Oriental 18   [metrol]
Lidonicci A 1933 Pe M (4)13, 4-5,
   310 [re Neugebauer]
Loffler E 1910 Arc MP (3)17,  135-
   144 [sexag]
Lorey W 1935 Unt M 41, 165-166
   (F61, 28) [recip, tab]
Loria G 1936 Archieon 18, 63-65
   [arith]
Lucas J D 1890 Rv Q Sc 28,  450-483
   [ast]
- 1891 Rv Q Sc 39, 513-541
   (F23, 43) [ast]
Mahler E 1886 Wien Si 93, 455-469
   [ast]
- 1892 Wien Si 101, 337-353, 1685-
   1693 [F24, 56] [calendar]
Martiny G 1932 Fors Fort 8/9, 122-123
   [ast]
Mercer S A B 1928 Toro ROM 7, 3-6
   (F59, 833)   [mult tab]
Miller G A 1930 Sci 72, 601-602;
   73, 68, 476-477
- 1935 Sch Soc 42, 438-439
   (F61, 10)
Neugebauer Otto 1929 QSGM (B)1(1),
   67-80
- 1929 QSGM 1, 81-92 [circle]
- 1930 QSGM (B)1, 120-130  [arith]
- 1930 QSGM (B)1, 183-193
   [sexag]
- 1931 Arc Or 7, 90-99  (F57, 1210)
   [approx, root]
- 1931 QSGM (B)1, 452-457, 458-463
   (Z2, 378) [fra, sexag]
- 1932 QSGM (B)2, 199-210  (Z5, 1)
   [sexag]
- 1932 QSGM (B)2, 305-310  (Z6, 1)
   [military]
- 1933 Gott N 1(43), 316-321
   (F59, 14; Z7, 386) [cubic]
- 1934 Dan M Med 12(13)
   (Z9, 97) [tab]
- 1934 QSGM (B)3, 106-114  (Z10, 97)
- 1934 Scr 2, 312-315  (Z10, 98)
- 1935 An Orient 12, 235-258
   (Z12, 242) [ratio]
- 1935 Mathematische Keilschrift-
   texte, I & II QSGM (A)3, 580 p
   (Z12, 97)
- 1936 M Tid B, 1-9 (F62, 5)

Babylonian   (continued)

- 1937 M Tid B, 17-21   (F63, 7;
  Z16, 145)   [Greek]
- 1937 *Mathematische Keilschrift-
  Texte. T. III* (QSGM (A)3), 92 p
  (Z15, 147)
- 1937 Scientia 61, 151-157; 61(S),
  70-75   (F63, 5; Z16, 145)
- 1938 M Tid, 1-10   (F64, 7)
- 1948 in *Studies and Essays
  presented to R. Courant on his 60th
  Birthday,* NY, 265-275   (MR9, 483)
  [dating]
- 1949 Oriental 18, 423-426
  (Z41, 337)   [re H. Lewy]
- 1951 APSP 95, 110-116 (MR13, 1;
  Z45, 289)   [ast]
- 1954 APSP 98, 60-89   (Z59;1, 2)
  [ast]
- 1952 *The Exact Sciences in
  Antiquity* (Princeton U P), 207 p
  (MR13, 809)   (2nd ed 1957; repr
  1962)
- 1963 APSP 107, 528-   [anc, mid]
Neugebauer O + 1932 QSGM (B)2, 291-
  297 (Z6, 1) [approx, roots]
- 1945 *Mathematical Cuneiform Texts*
  (American Oriental Series 29),
  New Haven, 187 p   (MR8, 1)
  [U S collections]
Oppenheim L 1938 Orient Lz 41, 485-
  486 [metrol]
Oppert 1875 *Etalon des mesures
  assyriennes,* Paris,
*Price D J de S 1964 Centau 10, 219-
  231 [Pythagorean triples]
Przyluski M J 1932 Brx U Rv 37,
  283-294   (F57, 1291)
  [influence on Greece, India]
Raik A E 1953 Perm M 8(1), 31-63
  (Z68, 1)   [quad]
- 1955 Perm M 8(3), 11-14 (Z68, 1)
  [quartic]
- 1958 Ist M Isl 11, 171-182
  (MR23, 692; Z100, 6)
- 1959 Ist M Isl 12, 271-318
Reisner G 1896 Ber Si, 417
  [metrol]
Rey A 1940 Thales 4, 227-234
  (MR9, 73; Z61, 1) [re Thureau-
  Dangin]

Sachs A J 1944 Am SORB (96), 29-39
  (MR6, 141; Z60, 4)   [metrol]
- 1946 JNES 5, 203-214   (MR8, 1;
  Z60, 4)   [frac]
- 1947 J Cun St 1, 219-240
  (MR9, 169)   [recip, sexag]
Sachs A J + eds 1955 *Late Babylonian
  Astronomical and Related Texts,*
  Providence
Sanchez Perez J A 1943 *La aritmetica
  en Babilonia y Egipto,* Madrid,
  72 p   (MR6, 253)
Sarton G 1940 Isis 31 398-404
  (MR7, 105; Z60, 3; F66, 1185)
Sayle A H n d Bibcl Arc 4, 302-314
  [mysticism]
Schamberger J 1953 Z Assyr (nf)16(50),
  224-229   (Z51, 2)
Scheil F V 1934 Rv Assyr 31, 171-172
  (F60, 818)   [metrol]
Scheil F V + 1896 *Ein altbabylonisher
  Felderplan,* Leipzig
Schott A 1933 QSGM (B)2, 364-368
  (Z7, 146)
Segre A 1944 Am OSJ 64, 73-81
  [metrol]
Speleers L 1925 *Recueil des inscriptions de l'Asie Intérieure des
  Musées Royaux du Cinquantenaire
  à Bruxelles,* Bruxelles
Thureau-Dangin F 1921 Rv Assyr 18,
  123-142 [metrol, numer]
- 1928 Rv Assyr 23, 187-188   [trig]
- 1931 Rv Assyr 28, 23-25 (F57, 1289)
  [angle]
- 1931 Rv Assyr 28, 85-88   [ast]
- 1931 Rv Assyr 28, 111-114
  (F57, 1289)   [angle, chronom]
- 1931 Rv Assyr 28, 195-198
- 1932 Rv Assyr 29, 1-10   (F58, 8)
- 1932 Rv Assyr 29, 59-66
  (Z5, 241)
- 1932 Rv Assyr 29, 21-28, 77-88,
  131-142, 188-192   (F58, 9; Z5,
  241; 6, 1)   [pi]
- 1932 Rv Assyr 29, 109-119   (Z5, 241)
- 1933 Rv Assyr 30, 144, 183-188
  (F59, 835)
- 1934 Rv Assyr 31, 49-52
  [fraction, root]
- 1934 Rv Assyr 31, 61-69   (F60, 818)
  [geom]

Babylonian (continued)

- 1935 Rv Assyr 32, 1-28    (Z11, 193)
  [vol]
- 1935 Rv Assyr 32, 188    [termin]
- 1936 Rv Assyr 33, 27-48
  (F62, 8)  [quad]
- 1936 Rv Assyr 33, 55-61
  (F62, 5)  [re Neugebauer]
- 1936 Rv Assyr 33, 65-84
- 1936 Rv Assyr 33, 161-168,
  180-184  (F62, 8)
- 1937 Archeion 19, 1-11
  (Z16, 195)  [alg]
- 1937 Rv Assyr 34, 9-28
  (F63, 6)
- 1938 Rv Assyr 35, 104-106
  (F64, 8)
*- 1938 Textes mathématiques
  Babyloniens, Leiden, 283 p
  (F64, 7)
- 1940 Rv Assyr 37, 1-10
- 1947 in Receuil offert à la
  Mémoire de Halil Edhem, (MR9, 73;
  Z32, 193)
Vaiman A A 1960 Ist M Isl 13, 379-382
  (MR24, 570)  [descr geom]
- 1961 Sumero-Vavilonicheskaya
  Mathematika, Moscow, 278 p
  (Z107, 242)
Van der Waerden B L 1940 Leip Ber 92,
  107-114   (MR2, 114; F66, 7)
  [ast]
- 1948 Leyd Ex OL 10, 414
  [bib, chron]
- 1951 JNES, 25
- 1957 Zur NGV 102, 39-60
  (MR21, 483)  [ast]
- 1960 Zur NGV, 97-114
  (MR27, 268)  [ast, Egypt]
Vaschenko-Zakharchenko M E 1881
  Istoricheskii ocherk matemati-
  cheskoi literatury khaldeev, Kiev,
  30 p
Veselovskiy I N 1955 Mos IIET 5,
  241-303  (Z68, 1)
Vetter Q 1922 Casopis 51, 271-278
  (F48, 37)  [arith]
- 1923 Arc Sto 4, 233-240
  (F49, 16)  [arith]
- 1927 Cas MF 51  [arith]

- 1936 Osiris 1, 692-702
  (Z14, 145)
Vogel K 1933 Arc Or 8, 22
  [pyramid]
- 1933 Unt M 39, 75-81
  (F59, 13)  [quad]
- 1934 Mun Si, 87-94  (F60, 7)
  [cubic]
- 1935 Bay Bl G 71, 16-29
  (F61, 942)
- 1936 Osiris 1, 703-717  [quad]
- 1937 Bay Bl G 71, 16-29
- 1937 Int Cor (10)2, 277-278
  (F63, 20)
- 1958 M Nach 18, 377-382  (MR21, 483;
  Z80, 2)  [Sumer or Akkadia?]
- 1959 Vorgriechische Mathematik, II:
  Die Mathematik der Babylonien,
  Hannover, 94 p  (Hannover Z)
  (Z86, 2)
- 1960 MP Semb 7, 89-95  (MR23, 583)
Vygodskii M Ya 1940 UMN 7, 102-153
  (MR1, 289; F66, 7; Z23, 193)
- 1941 UMN 8, 293-335
  (MR3, 97; Z60, 3)
- 1959 Ist M Isl 12, 393-420
  (Z100, 242)  [zero]
- 1967 Arifmetika i algebra v
  drevnem mire, 2nd ed, Mos, 367 p
Washchow H 1932 Arc Or 8, 127-131
  (F58, 9)
- 1932 QSGM (B)2(3), 211-214  [tria]
- 1932 QSGM (B)2, 298-304  [series]
- 1933 Arc Or 8, 215-220
  (F59, 836)  [appl]
- 1933 Unt M 39, 368-373
  [milit]
Wieleitner H 1930 Fors Fort 6, 319-
  320  (F56, 800)  [recent wk]
- 1930 in Historische Studien und
  Skizzen zur Natur- und Heilwissen-
  schaft Festgabe Georg Sticker,
  Berlin, 11-17  (F59, 832)
  [sexag]


Byzantine

Bierman K-R 1964 Deu LZ 85(2), 150-
  151  [re Hunger 1963]
Heiberg J L 1887 Dan Ov, 88-96

Byzantine    (continued)

- 1888 Hist Abt 33, 161-170
- 1899 ZM Ph 44(S), 163
Hunger H + 1963 Ost P Denk 78(2)
    (Z118, 248)  [XV]
Kotsakis D 1958 *Astronomy and*
    *Mathematical Sciences During the*
    *Byzantine Period,*  Athens, 16 p
    (Z84, 242)
Stephanides M 1923 Athena 35, 206-218
- 1933 Archeion 14(4), 492-496
    (F59, 22)  [Renaissance]
Tannery P 1884 BSM (2)8, 263-277
- 1906 BSM (2)29, 59
    [ephemerides]
- 1920 *Les science exactes chez les*
    *Byzantins,*  Paris
Theodorides J 1963 in Taton R ed,
    *Ancient and Medieval Science,*
    NY, 440-452
Van der Waerden B L 1954 Mun Si,
    159-168  (Z64, 1)  [sun tab]


Chinese

Anon 1955 *Iz istorii nauku i*
    *tekhniki Kitaya* (Sbornik statei)
    184 p (Mos IIET 8)
- 1895 Nat 52, 622-623  (F26, 67)
- 1966 *Sung  Yüan shu hsüeh shih*
    *lun wen chi,* Peking, 306 p
    (MR34, 2403) [Sung and Yüan
    periods]
Barbour J M 1933 AMM 40, 69-73
    [XVI, pi]
Banachiewicz T 1909 Wars SPTN 2, 7-10
    [primes, Fermat]
Barde R 1952 Arc In HS (NS)5, 234-
    281  (MR14, 523)  [I-King]
Berezkina E I 1960 Ist M Isl 13,
    219-230  (MR25, 4)  [Sun-Zi Suan-
    Jong]
- 1960 Vost IINT (1), 34-55
    (Z118, 245)  [Nine Books]
Bertrand J 1869 J Sav, 317-, 464
Biernatzki K L 1856 Crelle 52,
    59  [arith, based on Wylie]
- 1863 Nou An 2, 529
Biot J B 1841 J Asi (3)11

*Bodde D 1939 Am OSJ 59, 200-
Bosmans H 1903. Brx SS 27, 122-124
    [XVII, ast]
Cajori F 1903 AMM 10, 35  [zero]
Cammann S 1960 Am OSJ 80, 116-124
    [magic sq]
Carus P 1907 *Chinese Thought* (Open
    Court)
Chang Chia-Chu 1962 *Shen Kua,*
    Shanghai, 259 p
Cheng Chin-Te D 1925 AMM 32, 499-504
    [abacus]
Ch'ien Pao-Tsung ed 1963 *The Ten*
    *Mathematical Manuals,* Peking, 2 vols
Chih K C 1939 M Gaz 23, 268-269
    [metrol]
Chistyakov V D 1960 *Materialy po*
    *istorii matematiki v Kitae i Indii,*
    Mos, 167 p
Datta B 1930 Clct MS 22, 39-51
    (F56, 806)  [Brahmagupta, Suan-shu]
Eastlake F W 1880 Chi Rv 9, 249-319
    [finger reck]
Edgar J H 1936 Chi WBRS 8, 170
    [Tibet, numer]
Edkins J 1886 Pekin OSJ 1, 161
    [place value]
Endo T 1897 Tok M Ph 8, 19
    [I-King]
Faddegon M 1932 Archeion 14, 372-391
    (Z6, 145)  [current wk]
Ferguson J C 1941 Mon Ser 6, 357-382
    [metrol]
Franke O 1928 Deu Morg Z 7, 155-178
    [Leibniz]
Fujiwara M 1940 Toh MJ 46, 284-308;
    47, 35-57  (F66, 12, 13)
- 1940-1941 Toh MJ 47, 309-321; 48,
    78-88  (MR2, 306; Z25, 293; 26, 98)
    [Korea]
- 1941 Send CSJ, 64-84  (MR2, 306)
    [Wazan]
Gauchet L 1914 Toung Pao 15, 531-550
    [root]
- 1917 Toung Pao 18, 151-174
    [sph trig XIII]
Glathe A 1932 *Die chinesischen*
    *zahlen,* Tokyo, 47 p
Goodrich L C 1948 Isis 39, 64-65
    (Z30, 1)  [circle]
Halsted G B 1908 AMM 15, 84    [pi]

Chinese (continued)

Hayashi T 1909 Tok M Ph (2)5, 43-57 (F40, 65) [pi]

Ho Peng-Yoke 1965 Oriens Ex 12, 37-53 [V]

Jones P S 1956 MT 49, 607-610

Karpinski L C 1935 SSM 34, 467-472

Kaucky J 1963 MF Cas 13, 32-40 (MR27, 691) [XIX combin]

Kershaw F S 1930 Bost MFAB 28, 7-8 [metrol]

Lacouperie A T de 1882 Lon Asia (NS)14, 781- [I-King]

- 1892 The Oldest Book of the Chinese, London (repr of 1882 + bib)

Leavens D H 1920 AMM 27, 180-183 [abacus]

Lemaire G + 1906 Intermed 13, 72 [I-King]

Li Yan 1954-1955 Collection on the History of Chinese Mathematics, Peking, 5 vols (in Chinese)

- 1955 History of Chinese Mathematics, 2nd ed, Shanghai (in Chinese)

- 1956 Int Con HS 8, 70-72 [interpol]

- 1956 Mos M Sez 4, 246-248 (Z112, 1)

- 1957 Mathematics in China in the XIII-XIV Centuries, Peking, 77 p

- 1958 Essays on the History of Chinese Mathematics, Peking, 2 vols (in Chinese)

Li Shu-tien 1959 ACMJ 6, 102-110 (MR20, 1137) [abacus]

Ling W 1956 Int Con HS 8, 13-17 [dec frac]

Ling Wang + 1955 Horner's Method in Chinese Mathematics: Its Origins in the Root-Extraction Procedures of the Han Dynasty, Leiden (repr from Toung Pao 43(5), 345-401)

Loria G 1916 Rm Sem 4 (F46, 40)

- 1920 Mathes It 12, 63-70

- 1922 Arc Sto 3, 141-149

Loria G + 1921 Sci Mo 12, 517-521

Ma C C 1928 AMM 35, 29-30 (F54, 8) [root]

Mao Tsao-ben 1959 This Was Invented in China, Moscow, 160 p (MR23, 134)

Marakyeva A V 1930 Mery i vesy v Kitae, Vladivastok

Matthiessen H F L 1874 ZM Ph 19, 270-271 (F6, 9) [alg]

- 1880 Crelle 91, 254-

- 1881 Hist Abt 26, 33-37 (F13, 33) [Sun-tse]

Mei J-C 1963 Ko Hs Chi 6, 1-10 [Nine Chapters]

Mikami Yosio 1909 Arc MP (3)15(1), 68-70 [re M Cantor]

- 1910 Mathematical Papers From the Far East, (Ab GM 28), 235 p (F41, 39)

- 1910 Bib M (3)10, 193-200 [circle sq]

- 1911 Arc MN 32(3) [frac]

- 1911 Arc MP (3)18, 209-219 (F42, 54) [re M Cantor]

- 1911 DMV 20, 380-393 (F42, 54) [abacus]

- 1913 The Development of Mathematics in China and Japan (Ab GM 30), 347 p (repr 1961 Chelsea)

- 1926 Toyo Gak 15, 431-484; 16, 49-109

- 1934 Sugaku: Shima shiso, Tokyo, 64 p [gen hist]

- 1941 Archeion 23, 211-226 [XVI, chronol]

- 1961 The Development of Mathematics in China and Japan, NY, 374 p (Z114, 242)

Natucci A 1948 Pe M (4)26, 153-156 (Z32, 97) [circle sq, vol]

*Needham Joseph 1959 Science and Civilization in China. Vol 3: Mathematics and the Sciences of the Theorems and the Earth (Cambridge U P), 924 p (MR25, 569) [*bib]

- 1962 London As QJ 3, 87-98 (Z105, 1) [ast]

- 1964 Sci Soc 28, 385-408

Noda Churyo 1933 Chou pi suan ching no Kenkyu. Kyoto, 183 p [commentary on ancient text]

Nyien T G 1957 Shux Jinz 3, 335-339 (MR20, 1243) [re Li Yen]

Petrucci R 1912 Toung Pao (Oct) [alg]

Qian Bao-zong 1957 Zhongguo shuxueshi hua, Peking

Chinese (continued)

Raik A E 1961 Ist M Isl 14, 467-472
(Z118:1, 1) [Nine Books, Vol]
Rodet L 1880 Fr SMB 8, 158-169
(F12, 8) [abacus]
Rohrberg A 1936 Unt M 42, 34-37
[instr]
Shafranovskaya T K 1961 Vost IINT
(1), 56-63 (Z118, 245)
[compass]
Sarton G 1950 Isis 41, 51 [mid]
Saussure L de 1907 Rv GSPA 18, 135-
144 [ast]
- 1921 Rv GSPA 32, 729-736
[cosm]
Schlegel G 1877 Leip As 21 (F9, 28)
["uranographie"]
Sedillot L A 1868 Boncomp 1, 161-
166 (F1, 4) [ast]
Smith D E 1912 Pop Sci Mo 80,
547-601
- 1931 Sci Mo 33, 244-250
(F57, 19) [open hist probls]
**Startsev** P A 1961 Ocherki istorii
astronomii v Kitae, Mos, 156 p
(MR23, 4)
Struik D J 1963 MT 56, 424-432
[bib]
- 1964 Euc Gron 40, 65-79
(MR30, 201)
Takeda K 1953 JHS (26)
[Ming dynasty]
- 1954-1955 Kagak Ken (28/29)
(Z59, 6)
Tong C P + 1939 Report on Chou-King's
Tower for the Measurement of the
Shadow of the Sum, Shanghai, 129 p
Trigg C W 1960 M Mag 34, 107-108
(MR23, 4) [geom]
Tsien P T 1956 Act As Si 4, 193-211
(MR20, 620) [calendar]
- 1957 Vop IET (5), 164-165
(Z84, 2)
Ueno K 1922 Toh MJ 21, 138-147
[mult tab]
Vacca G 1905 Loria 8, 96-102
(F36, 3)
Van Gulik R H 1928 Euc Gron 5, 104-
109
Van Hee L 1912 Toung Pao 13, 291-300
[alg]

- 1913 Rv Q Sc 24, 574-587
[XIII, alg]
- 1914 Toung Pao 15, 111-164 [bib]
- 1914 Toung Pao 15, 182-184 [zero]
- 1914 Toung-Pao 15, 203-210
[Diophantine anal]
- 1914 Toung Pao 15, 454-457 [log]
- 1920 Toung Pao 20, 51-60
[Hai-Tao Souan-King]
- 1924 AMM 31, 235-237
[Hsia-Hou Yang]
- 1926 AMM 33, 326-328 [Napier rods]
- 1926 AMM 33, 502-506
- 1926 Arc Sto 7, 18-24 (F52, 11)
- 1926 Isis 8, 103-118 [Yuan Yuan]
- 1932 QSGM (B)2, 255-280 [III]
- 1937 Isis 27, 321 (F63, 12)
[alg]
- 1939 Isis 30, 84-88 (F65, 1081)
[Euclid]
Vogel Kurt ed 1968 Neun Buecher
Arithmetischer Technik, Ost Kl
(NF)4, Braunschweig (Vieweg), 160 p
Wang Ping 1962 in Proceedings,
Second Biennial Conference
(International Association of
Historians of Asia), Taipei
[Wylie]
Wittfogel K A 1938 Z Soz 7, 90-122
- 1964 Pensée 114, 3-73
Wong G H C 1963 Chun Ch HP 2(2),
169-180 [pre-Jesuit]
- 1963 Isis 54, 29-49
[Ming-Ch'ing dynasties]
Wylie A 1852 Nort Chi H
- 1897 Chinese Researches, Shanghai
(repr 1936 Peiping) [incl Wylie
1852]
Yabuuchi K 1960 Int Con HS 9, 617-
621 (Z114, 5) [ast]
Yabuuchi Kiyoshi ed 1963 Studies in
the History of Medieval Chinese
Science and Technology, Tokyo,
540 p
Yen Tun-Chieh 1958 Ko Hs Chi 1, 20-
28 [Euler]
Yushkevich A P 1955 in Iz istorii
nauki i tekhniki Kitaya, Moscow,
130-135
- 1955 Ist M Isl 8, 539-572
(MR17, 1; Z66, 4)

Chinese    (continued)

- 1956 Shux Jinz 2, 256-278
  (MR20, 738)


Egyptian: General

See below Egyptian: Fractions,
  Moscow papyrus, Rhind papyrus.

Adamo M 1962 Cagliari 32, 101-183
  (MR29, 879) [arith]
Ameline 1934 Rv GSPA 45, 243-247,
  (323-324) (F60, 840) [ell]
Anning N 1954 MT 47, 37-40
  [geom prog]
Antoniadi E M 1934 L'astronomie
  égyptienne depuis les temps
  les plus reculés jusqu'à la fin
  de l'époque alexandrine, Paris,
  167 p
*Archibald R C 1927 Bibliography of
  Egyptian Mathematics, Oberlin
  (MAA), 86 p (Supplement 1929, 23 p)
  (also in Chace 1927-1929
- 1930 Sci 72, 39
Baillet J 1892 Le papyrus mathématique d'Akhmin, Paris (Mem. de
  la Mission archéologique
  française au Caire 9)
Bertrand J 1884 BSM (2)8, 8-19
  [calendar]
Bilfinger G 1891 Die Sternstafeln in
  den aegyptischen Koenigsgraebern
  von Biban el Moluk, Stuttgart,
  80 p (F23, 43)
Bobynin V V 1895 Fiz M Nauk 12(4),
  301-340 (F26, 47) [Akhmin pap]
- 1897 Fiz M Nauk 13, 77-80
  (F28, 53) [arith]
- 1910 Pet MNP 11, 1-51, 11(2)(10),
  290-328 (F41, 1) [Hyksos]
- 1922 Matematika drevnikh egiptyan
  (po papirusu Rinda), Mos, 138 p
Boev G P 1950 Ves Drev I, 196-201
  [area, vol]
Bonner C 1930 J Eg Arch 16, 6-9
Borchardt Ludwig 1922 Gegen die
  Zahlenmystik an der grossen
  Pyramide bei Gise, Berlin, 40 p

- 1926 Laengen und Richtungen der
  vier Grundkanten der grossen
  Pyramide bei Gise, Berlin, 21 p
  (F52, 5)
Bortolotti E 1935 Bln Mm (9)2, 27-51
  (Z12, 242) [alg]
Boyer C B 1959 MT 52, 127-129
  [numer]
Britzelmayr W 1942 Allg St Ar 31,
  177-178 (F68, 3) [mean]
Brugsch H K 1849 Numerorum apud
  veteres Aegyptios Demoticorum
  doctrina..., Berlin, 37 p
- 1855 Deu Morg Z 4, 492-499
- 1865 Z Aeg SA 3, 65-70, 77
- 1870 Z Aeg SA 8, [Edfu temple]
Bruins E M 1945 Indagat 48, 206-210
  (MR7, 353) [pi]
- 1952 Indagat 14, 81-91
  (MR14, 1; Z46, 1) [arith]
Cantor M 1884 Wien Si 90, 475-477
- 1893 ZM Ph 38, 81-88 [Greek pap]
Cassina U 1940 Pe M (4)22, 1-29
  (MR8, 1; Z28, 98) [geom]
- 1942 It UM Con 2, 897-898
  (MR8, 497; F68, 3) [geom]
Cramer 1931 Neues L 38, 52-74
  (Z2, 243)
Davies T S 1850 Mech Mag 53, 150-154,
  169-174 [geom]
Dawson W R 1923 Sci Prog 19, 50-60
Delaney A A 1963 Ari T 10, 216
  [arith]
Delitzsch 1878 Z Aeg SA
Devéria T 1862 Rv Arch 6, 253-263
  [numer]
Dijksterhuis E J 1932 Euc Gron 8, 49-
  74 (F58, 11)
Drenckhahn F 1934 Unt M 40, 96-104
  [circle, pyr]
Dunton M + 1966 Fib Q 4, 339-354
Durach F 1929 Mittelaltreiche
  Bauhuetten und Geometrie,
  Stuttgart, 62 p (F57, 1298)
Ebert R 1896 Dres Isis 1, 44
  [arith tab]
Eisenlohr A 1877 Ein mathematische
  handbuch der alten Aegypter
  (Papyrus Rhind des British Museum)
  uebersetzt und erklaert...,
  Leipzig, 280 p (2nd ed 1897)

Egyptian: General   (continued)

Eyth M v 1901 Ulm VMNJ 10, 1
   [Cheops pyr]
Favaro A 1879 *Sulla interpretazione*
   *matematica del Papiro Rhind*
   *pubblicato ed illustrato dal*
   *Prof. Augusto Eisenlohr*, Modena
   (F11, 1)
Fortova-Samalova Pavla 1963
   *Egyptian Ornament*, London
   (University College)
Gandz S 1930 QSGM 1, 255-277
   [surveying]
   1947 in *Studies and Essays in the*
   *History of Science and Learning*
   *Offered in Homage to George*
   *Sarton...*, NY, 449-462 (MR8, 497)
Giles R P 1964 MT 57, 552-555
   [pyr vol]
Gillain O 1927 *La science égyptienne.*
   *L'arithmétique au Moyen Empire*,
   Bruxelles-Paris, 342 p
Gillings R J 1958 M Gaz 42, 212-213
   [Pytha triples]
Glanville S R K 1936 Lon RIGBP 29,
   10-40 [metrol]
Goodspeed E J 1903 AMM 10, 133-135
   [Ayer pap]
Gottheil R 1933 Archeion 15, 232-238
Gram H 1706 *De origine geometriae a*
   *apud Aegyptias...*, Copenhagen,
   44 p
Guggenbuhl L 1950 MT 58, 630-634
   [bib]
- 1965 MT 58, 630-634 [bib]
Heegard P 1931 Nor M Tid 13, 108-118
   (F57, 11; Z4, 2)
- 1937 Int Con (1936)2, 276-277
   (F63, 20) [Oslo, paps]
Heiberg J L 1900 Dan Ov 2, 147
   [paps]
Hultsch F 1895 *Die Elemente der*
   *Aegyptischen Teilungsrechnung*,
   Leipzig, 192 p (F26, 66)
- 1901 Bib (3)2, 177
Isely L 1894 Neuc B 20, 23
- 1895 Arc Sc Ph 33, 587
Jablonski E 1915 BSM (2)39, 166-168
   (F45, 95) [pi]
Jacobi C G J 1849 Ber Ber (Aug) 222-
   226 [Ptolemy geog]

- 1850 Ber Ber, 426 [arith]
Jaromilek A 1890 Ost IAW 15, 188,
   193, 203 (F22, 1259) [Cheops pyr]
Jéquier G 1922 in *Recueil dédie à*
   *la mémoire de Champollian*,
   Paris, 467-482 [numeration]
Jomard E F 1819 *Notice sur les signes*
   *numériques des anciens Egyptiens...*,
   Paris, 31 p
Junge G 1926 Unt M 32, 240-244
   (F52, 7) [pyr vol]
Karpinski L C 1917 AMM 24, 257-265
   [alg]
- 1923 Isis 5, 20-25 [Mich pap 621]
- 1926 AMM 32, 41 (F51, 38)
Kielland Else C 1955 *Geometry in*
   *Egyptian Art*, London
Kleppisch K 1927 *Willkuerliche oder*
   *mathematishe Unterlegung beim Bau*
   *der Cheopspyramide?*, Munich-Berlin,
   38 p
Lange H O + 1940 Dan H 1(2), 92 p
   (MR2, 114) [Carlsberg pap]
Laver J-P 1944 Chro Eg 19, 166-176
   [geom, pyrs]
Lepsius K R 1855 Ber Ab, 69-114
   [Edfu temple]
- 1965 Ber Ab H 1-56 [metrol]
Lockyer J N 1892 Nat 45, 296-299,
   373-375 (F24, 57) [ast]
- 1893 Nat 48, 417-419 (F25, 91)
   [archit]
Loria G 1892 Bib M (2)6, 97-109
   (F24, 43) [arith]
- 1893 Bib M 79-90 [arith]
- 1932 BSM (2)56, 236-247 (F58, 11;
   Z2, 241)
Luckey P 1930 ZMN Unt 90(4), 145-158
   [pyr vol]
- 1932 ZMN Unt 63, 389-391
   [pyr vol]
- 1933 Isis 20, 15-52 (F59, 15)
   [geom]
- 1933 ZMN Unt 63, 389-391
   [pyr vol]
Lundsgaard Erik 1945 *Aegyptisk*
   *matematik*, Copenhagen, 39 p
   (MR1948, 73; Z61, 1)
Luria S 1933 Len IINT 1, 45-70
   (F61, 945) [Greek geom]
Mahler E 1894 Wien Si 103, 832-844
   (F25, 90)

Egyptian: General  (continued)

- 1904 MP Lap 13, 30-53, 128-142
  (F35, 47)
Mau J + 1960 Arc Pap 17, 10 p
  (Z95, 1)  [Berlin pap]
Mehmke R L 1932 Bei GTI 21, 115-122
  [tec]
- 1935 Sci 81, 152 [pyth thm]
Montel P 1947 BSM (2)71, 76-81
  (Z33, 241)  [Giza pyr]
- 1947 Par CR 224, 1741-1743
  (Z29, 1)  [Cheops pyr]
Neugebauer Otto 1929 *Die Grundlagen
  der aegyptischen Arithmetik in
  ihres Zusammenhang mit der 2:n
  Tabelle des Papyrus Rhind,* Munich
- 1930 Arc GMNT 13, 92-99  (F56, 803)
  [re Vogel 1929]
- 1930 QSGM (B)1, 301-380
  [arith]
- 1931 QSGM (B)1(4), 413-451
  (Z2, 323)  [geom]
- 1932 Z Aeg SA 68, 122-123  [metrol]
- 1942 Act Ori 19, 138-139
  (F68, 4)  [Sothic period, chronom]
Newman James R ed 1956 *World of
  Mathematics,* NY, 170-178
Nichols I C 1929 M Mag 3, 10-13
Paravey de 1826 *Essai sur l'origine
  unique et hiéroglyphique des
  chiffres et des lettres de toms
  des peuples,* Paris
Peet T E 1922 Manc Mr 65(2), ix-xi
- 1923 *The Rhind Mathematical
  Papyrus, British Museum 10057.
  Introduction, transcription,
  Translation and Commentary,*
  London-Liverpool, 136 p
- 1931 J Eg Arch 17, 100-106
  (F57, 11; Z2, 5)  [geom]
- 1931 J Eg Arch, 17, 154
  [re Struve 1930]
- 1931 Manc JRL 15, 409-441
  (F57, 1290; Z2, 378)
Petrie F 1938 J Eg Arch 24, 180-181
  [metrol]
Quintino G di S 1825 *Saggio sopra il
  sistema de'Numeri presso gli
  antichi Egiziani,* Turin, 17 p
Raik A E 1958 Ist M Isl 11, 171-181

- 1959 Ist M Isl 12, 271-295, 295-318
  (MR24, 570)
*Rey A 1926 Rv Syn Sc 41, 19-62
  (F52, 37)
- 1927 Rv Syn Sc 43, 27-35
Robbins F E 1929 Clas Phil (Oct),
  321-329  [Mich pap 620]
- 1934 Isis 22, 95-103  (Z10, 243)
  [Mich pap 4966]
Roeber F 1854 *Beitraege zur Erforschung
  der geometrischen Grundformen in
  dem alten Tempeln Aegypter...,*
  Dresden, 60 p
Rodet L 1878 Fr SMB 6, 139-149
  (F10, 1)  [pap]
Sanchez Perez J A 1943 *La aritmetica
  en Babilonia y Egipta,* Madrid, 72 p
Sarton G 1936 Isis 25, 399-402
  (F62, 9)  [metrol]
Schack-Schackenburg 1900 Z Aeg SA 38,
  138-; 40, 65
Schoy C 1926 *Graeco-Arabische
  Studien nach mathematischen Hand-
  schriften der Vizekoeniglichen
  Bibliothek zu Kairo...,* Copenhagen
  (See Isis 8, 21-40)
Scott N E 1942 NY Met MA, 70-75
  [metrol]
Sethe K 1916 *Von Zahlen u Zahlwoerten
  bei den Aegypten und was fuer
  andere Voelker u Sprachen daraus
  zu lernen ist...,* Strasburg, 152 p
  (Strasb WG 25)
Seyfarth G 1829 Leip Lit Z (220)
  (12 Sept), 1753-1756  [numer]
Simon M 1905 Arc MP (3)9, 90,
  102-103, 181, 303  (F36, 2)
- 1905 Int Con (3), 526-535
  (F36, 2)
Stern C + 1966 Ari T 11, 254-257
  [mult]
Stoley R W 1922 Anc Eg, 111-117
Touraeff B 1917 Anc Eg, 100-102
  [pyr vol]
Tsinzerling D 1925 Len Ac (6)19,
  541-568  (F51, 38)  [geom]
- 1939 Mat v Shk 2, 5-20; 3, 3-15
Vacca G 1909 Loria 11, 65-67
  [circle sq]
Van der Waerden B L 1937 Leip Ab
  89, 171-172 (F63, 797; Z18, 339)
  [arith]

Egyptian: General   (continued)

- 1947 Amst P(B)50, 536-547
  (MR9, 73)  [tab]
- 1947 Amst P(B)50, 782-788
  (MR9, 169)  [tab]
Veselovskii I N 1948 Mos IIET 2, 426-
  498  (MR11, 150)  [Greek]
Vetter Q 1923 Bo Fir 2(2), xxxxiii-
  xxxix  [arith prog]
- 1925 Clas Phil 20, 309-312
  [Mich pap 621]
- 1927 Cas MF 54, 281-283
  (F51, 10)  [trig]
- 1935 Math Cluj 9, 304-309
  [cubic eq pyr vol]
Vogel Kurt 1929 Arc GMNT 11, 386-407
  (F55, 7)
- 1929 *Die grundlagen der aegyptis-*
  *chen Arithmetic in ihrem*
  *Zusammenhang mit der 2/n Tabelle*
  *des Papyrus Rhind,* Munich, 211 p
- 1930 Archeion 12, 126-162
  (F56, 10)  [alg]
- 1930
  [Mich pap 620]
- 1930 J Eg Arch 16, 242-249
  (F57, 1291)  [pyr vol]
- 1930 Z Aeg SA 66, 33-35
  [metrol]
*- 1958 *Vorgriechische Mathematik,*
  *Teil I. Vorgeschichte und*
  *Aegepten,* Hannover, 80 p
  (MR20, 944)
Vygodskii M Ya 1967 *Arifmetika i*
  *algebra v drevnem mire,* 2nd ed,
  Mos, 367 p
Weyr E 1884 *Ueber die Geometrie der*
  *alten Aegypter,* Vienna
Wieleitner H 1925 Arc Sto 6, 46-48
  (F51, 6)  [eq]
- 1925 Unt M 31, 91-92  [pyr vol]
- 1925 ZMN Unt 56, 129-137
  (F51, 6)
Williamson R S 1942 Nat 150, 460-461
  (Z28, 98)  [Saqqara graph]
Winter J W 1936 *Michigan papyri III:*
  *Papyri in the University of*
  *Michigan Collection,* Ann Arbor,
  408 p  (F62, 1017)

Egyptian: Fractions

Botts T 1965 MT 58, 596-
Bruins E M 1952 Indagat 55, 81
Carnahan W H 1960 SSM 60, 5-9
Chale A B 1931 Archeion 13, 40-41
  (F57, 12)
Bobynin V V 1899 ZM Ph 44(S)3
Bortolotti E 1932 Bln Mm (8)9, 85-93
  (F58, 11)
- 1933 Pe M (4)13, 2-3, 151-152
Gillain O 1928 Mathesis 42, 405-413
Gillings R J 1955 Au JS 18, 43-49
- 1959 Au JS 22, 247-250
Ianovskaia S A 1947 Mos IIET 1, 269-
  282
Kowalewski G 1938 Deu M 3, 698-700
  (F64, 9; Z19, 387)
Neugebauer Otto 1926 *Die Grundlagen*
  *der aegyptischen Bruchrechnung,*
  Berlin, 42 p  (F52, 4)
- 1929 Z Aeg SA 64, 44-48
Niebacker E 1930 Arc GMNT 13, 82-92
  (F56, 802)
Padoa A 1933 Pe M (4)13, 2-3, 87-98
Ransom W 1961 MT 54, 100-101
Ringler F 1929 MN Bl 23, 115-121
  (F57, 1290)
Simon M 1907 Arc MP (3)12, 377
  (F38, 56)
Van der Waerden B L 1938 QSGM (B)4,
  359-382  (F64, 9)
Vasconcellos F de 1921 Portu APS
- 1924 Lisb ISA 2
Vetter Q 1921 Prag V (14), 1-25
- 1923 Prag V 52, 169-177
Vogel K 1931 Archeion 13, 42-44
  (F57, 12)
Yanovskaya S A 1947 Mos IIET 1, 269-
  282  (MR12, 69)

Egyptian: Moscow papyrus

Bortolotti E 1935 Pe M (4)15, 87-92
  (F61, 10)  [pyr, sph]
Dijksterhuis E J 1930 Euc Gron 7,
  140-148  (F56, 802)
Gunn B + 1929 J Eg Arch 15, 167-185
  (F57, 1290)  [geom]
Karpinski L C 1923 Sci 57, 528-529

Egyptian: Moscow papyrus (continued)

Neugebauer O 1931 QSGM (B)1, 541
Struve W W 1930 QSGM (A)1,
Struve W W + 1930 *Mathematischer
Papyrus des Staatlicher Museums
der schone Kunste in Moskau,*
Berlin (Isis 16, 148-155)
Thomas W R 1931 J Eg Arch 17, 50-52
(F57, 1291)
Vetter Q 1933 J Eg Arch 19, 16-18
(F59, 836; Z7, 49)


Egyptian: Rhind papyrus

Brocard H + 1901 Intermed 7, 127
Budge E A Wallis 1899 Facsimile of
the Rhind Mathematical Papyrus in
the British Museum, Nat 59, 73-74
(F30, 36) [London]
Chace A B + eds 1927-1929 *The Rhind
Mathematical Papyrus,* Oberlin,
2 vols [*bib]
Faulkner R 1932-1933 *The Papyrus
Bremner-Rhind (British Museum
10188),* Brussels
Favaro A 1880 Mod Mm 19, 89-145
(F12, 1)
Gillings R J 1961 MT 54, 97-100
[probl 28, 29]
- 1962 MT 55, 61-69 [probl 1-6]
Glanville S R K 1927 J Eg Arch 13,
232-239 [London]
Guggenbuhl L 1964 MT 57, 406-410
[NY fragments]
Hein R 1951 Ost Si M 316-322
(Z45, 145) [frac]
Mansion P 1888 Brx SS (A)12, 44-46
[tab]
Miller G A 1906 SSM 5, 567-574
[pap]
- 1931 AMM 38, 194-197
(F57, 11; Z1, 321) [pap]
Newman James R 1952 Sci Am 187,
24-27 [pap]
Noguera Barrenche R 1956 Studia
1(10), 63-72 (MR19, 107) [pap]
Perepelkin J J 1929 QSGM 1, 108-112
[probl 62]
Révillout E 1882 Rv Egque 2, 292,
304 [Rhind=a student notebook]

Sarv J 1926 Dorpat U (A)11(5), 13 p
(F52, 5)
Shenton W F 1929 AMM 36, 409
Slaught H E 1929 AMM 36, 409-410
Smith D E 1927 AMM 34, 445
Vogel K 1961 Praxis 3, 267
(Z118, 245)


Etruscan

See also Roman (below).

Deroy L 1954 Latomus 13, 51-55
[numer]
Piccoli G 1933 Pe M (4)13, 144-150
(Z6, 337) [numer]


Greek

See also Byzantine (above), Roman
(below), Euclid (Ch. 1), Analysis
and Synthesis (Ch. 3), etc.

*Aaboe A 1964 *Episodes From the
Early History of Mathematics,*
(Blaisdell, Singer), Ch 2
(also New Math Library 13) [pop]
Allman G J 1889 *Greek geometry from
Thales to Euclid,* Dublin, 244 p
Anon 1885 Athm 1, 442 [re Gow 1884]
- 1885 *Recueil de travaux d'erudition
classique dediés à la mémoire de
Charles Graux,* Paris
Artom E 1924 Pe M (4)2, 326-342
[quad eq]
Baccou Roger 1951 *Histoire de la
science grecque de Thales à
Socrate,* Paris, 257 p (MR14, 523;
Z45, 290)
Bashmakova I G 1958 Ist M Isl 11,
225-438 (MR23, 582; Z96, 1)
- 1961 Ist M Isl 14, 473-490
(Z118:1, 246)
Becker Oskar 1965 *Zur Geschichte der
griechischen Mathematik,* Darmstadt,
482 p (MR34, 5610)
- 1966 *Das mathematische Denken der
Antike,* 2nd ed, Goettingen, 131 p
Bergh P 1886 Hist Abt 31, 135
[polyg]

Greek    (continued)

Blaschke W 1939 Gree SM 19, 167-174
    (F65, 2)
- 1953 *Griechische u anschauliche
    Geometrie,* Munich, 59 p
Bobynin V V 1892 Bib M (2)6, 1-3
    (F24, 4)
Boegel K 1930 *Aus dem mathematischen
    Schriften der Griechen,* Leipzig,
    32 p  (F56, 804)
Bosch F 1931 DMV 41, 59-72 (F57, 13)
    [irrat]
Boyer C B 1951 Scr 17, 32-54
    (Z44, 245)
- 1954 Scr 20, 30-36, 143-154
    (MR16, 433; Z55, 243)
    [anal geom]
Bruins E M 1948 Amst P 51, 332-341
    (MR9, 483)  [sq root]
- 1953 *Fontes mathesos: Hoofdpunten
    van het prae-Griekse en Griekse
    Wiskundigdenken,* Leiden,
    168 p  (MR14, 831)
- 1959 Janus 48, 5-23  [reg polyg]
Brunschvieg L 1937 *Le role du
    pythagorisme dans l'évolution
    des idées,* Paris, 25 p
Bunt L N H 1954 *Van Ahmes tot
    Euclides,* Groningen-Djakarta,
    171 p
Cajori F 1918 Sci 48, 577-578
    ["invisible lines"]
Cajori F + 1923 SSM 22, 463-464,
    715-717  [tan]
Cantor M 1869 Hist Abt 14, 29-30
    (F2, 2) [re Bretschneider]
- 1893 Hist Abt 38, 81-87
    (F25, 56)  [pap]
Cantzler R F B 1831-1832 *De
    Graecorum arithmetica dissertati-
    unculae,*  Greifswald
Carrara B 1903 Rv FMSM 4(2), 3-13
    [constr]
Cherniss Harold 1945 *The Riddle of
    the Early Academy* (U Calif P),
    109 p  (MR7, 353)
Chiari A 1923 Bo Fir 19, LXV-LXX
    [sophists]
Christensen A A 1895 Nyt Tid (B)6,
    52, 84  [circle sq]

Clagett Marshall 1955 *Greek Science
    in Antiquity,* NY
Cohen Morris R + 1948 *A Source Book
    in Greek Science* (McGraw-Hill),
    601 p  (MR10, 419)
Dantzig Tobias 1955 *The Bequest of
    the Greeks,* NY, 191 p    (MR17, 337;
    Z66, 243) [pop]
Davies T S 1850 Mech Mag 53, 150-154,
    169-174 [geom]
Decourdemanche J-A 1913 J Asi (11)1,
    427-444 [geod]
Dehn M 1943-1944 AMM 50, 357-360,
    411-414; 51, 25-31, 149-157
    [-VI to VI]
De Vries H 1926 N Tijd 14, 124-144
    [proj geom]
*Diels H 1903 *Die Fragmente der
    Vorsokratiker,* Berlin
Dijksterhuis E J 1928 Euc Gron 4,
    134-174   (F54, 13)
- 1930 Euc Gron 7, 97-112  [numb]
*- 1934 Gids 98(3), 189-209
    (F60, 821)
- 1936 Gids 100, 153-174
    (F62, 1015)
Dingler H 1931 Archeion 13, 1-10
    (Z1, 321)    [axiom]
Dow S 1952 Am J Arch 56, 21-23
    [num]
- 1958 Clas Phil 53, 32-34
    [arith, numer]
Drieberg F V 1819-1821 *Die Arithmetik
    der Griechen,* Leipzig, 2 vols
Edwards W M 1931 M Gaz 15, 449-460
    (F57, 1291)
Eells W C 1911 AMM 18, 3-14
    [quad eq]
Enriques Federigo 1923 Pe M (4)3,
    73-88 [geom]
- 1927 *L'évolution des idées
    géometriques dans la pensée
    grecque: point, ligne, surface,*
    Paris, 56 p
- 1929 Ham Sem 7, 70-81  (F55, 7)
- 1931 Milan Sem 4, 1-5  (F57, 14)
    [mech]
- 1936-1938 *Histoire de la pensée
    scientifique,*  Paris, vols 1-6
Enriques F + 1939 Act Sc Ind (845)
    (F65, 1080)  [ast]

Greek (continued)

Evans G W 1927 AMM 34, 354-357
[ratio]
Farrington Benjamin 1955 Greek
Science (Penguin Books), 320 p
Finger F A 1831 De primordiis
geometriae apud Graecos,
Heidelberg
Fotheringham J K 1932 QSGM (B)2(1),
28-44 [Chaldean ast]
Frajese A 1942 It UM (2)4, 49-60
(MR7, 353) [Thales, geom]
- 1949 Archim 1, 41-47
(MR12, 310)
- 1951 Archim 3, 98-104
(MR13, 1) [continuity, geom]
Frank E 1923 Plato und die
sogenannten Pythagoreer, Halle,
409 p
Frenkian A M 1953 Buc U Pol (3),
9-18 (Z53, 195)
- 1958 Ro SSM (NS)2(50), 5-18
(Z89, 3)
Fritz K v 1955 Arc Begrf 1, 13-103
(Z65, 241) [Archai]
Gandz S 1932 QSGM (B)2, 81-97
(Z5, 3) [lettering figures]
Gericke H 1964 MP Semb 11, 144-162
(MR30, 869) [phys]
Gerstinger H + 1932 Griechische
literatische Papyri I, Vienna,
170 p (Z4, 193)
Gillings R J 1955 M Gaz 39, 187-190
(MR17, 117; Z64, 241)
[oriental infl]
Gow J 1884 A Short History of Greek
Mathematics (Cambridge U P), 322 p
(repr 1968 Chelsea)
Guggenheimer H 1951 Methodos 150-164
[logic]
Guissani C 1894 Lom Gen (2)27, 435-
450 (F25, 54)
Gutenaecker J 1928 Kreis-Messung des
Archimeses von Syrakus nebst dam
dazu gehoerigen Kommentare des
Eutokius von Askalon, Wurtemberg,
166 p
Hallo R 1926 Deu Morg Z 5, 55-67
[numer]
Hasse H + 1928 Kantstu 33, 4-34
(F54, 11) [found crises]

- 1933 Gree SM 14(1), 61-97; 14(2),
65-86 (F59, 837) [found crisis]
Hauser G 1955 Geometrie der Griechen
von Thales bis Euklid..., Luzern,
176 p
Hawkes H E 1898 AMSB (2)4, 530-535
[arith]
Heath Thomas L 1920 M Gaz 10, 289-301
*- 1921 A History of Greek Mathematics,
Oxford, 2 vols
- 1923 M Gaz 11, 248-259 (F49, 24)
[geom, infinitesimal]
- 1924 Nat 111, 152-153
[infinitesimal]
- 1931 A Manual of Greek Mathematics,
Oxford, 568 p (F57, 12)
(repr 1963, Dover)
- 1938 Scr 5, 215-232 (F64, 904)
[ast]
Heiberg J L 1881 J Cl Phll 11(S),
357-399 (F13, 5) [philology]
- 1883 Hist Abt 28, 121-129
- 1884 Philolog 43(2) [and Roman]
- 1896 Dan Ov, 77-93 (F27, 2)
- 1927 Dan H 13(3), 107 p
(F54, 11)
- 1912 Naturwissenschaften und
Mathematik im klassischen Altertum
(Teubner), 106 p (F43, 3) (2nd
unchanged ed 1920; F47, 23; Engl tr
1922; Ital tr 1924; F50, 11; Russ
tr 1936)
- 1922 Mathematics and Physical
Science in Classical Antiquity,
London, 110 p
- 1925 Geschichte der Mathematik und
Naturwissenschaften in Altertum
Munich, 129 p
Heiberg J L + 1907 Bib M (3)8, 118-134
(F38, 59)
Heichelheim F 1927 Deu Morg Z 6, 78-
81 [numer]
Heidel W A 1940 Am J Phil 61, 1-33
(F66, 9) [Pyth]
Heller S 1956 Centau 5, 1-58
- 1958 Ber Ab (6)
Hemmy A S 1938 Iraq 5, 65-81
[metrol]
Hoffmann J J I 1817 Ueber die
Arithmetik der Griechen, Mainz
Holzwart K Israel 1892 Das System der
attischen Zeitrechnung auf neuer

Greek   (continued)

*Grundlage,* Frankfurt, 34 p
(F24, 56)
Hoppe E 1911 *Mathematik und
Astronomie in Klassichen Altertum,*
Heidelberg
- 1920 Ham MG 5, 289-304   (F47, 26)
Hultsch F 1862 *Griechische und
roemische Metrologie,* Berlin
- 1864 *Scriptores metrologici
Graeci,* Leipzig
- 1879 Hist Abt 24, 41-42   (F11, 29)
[termin]
Humbert P 1930 Rv Q Sc (4)18, 381-389
(F56, 10)
Huxley G L 1959 *Anthemius of Tralles.
A Study in Later Greek Geometry,*
Cambridge, 62 p   (Z113, 1)
*Itard J 1963 in Taton R, *Ancient and
Medieval Science...,* NY, 273-304
Ivor Thomas ed 1939-1941 *Selections
illustrating the History of
Greek Mathematics,*  Cambridge,
Mass, 2 vols
Jasinowski B 1936 Act Sc Ind (395),
9-19  (F62, 1015)
- 1942 Archeion 24, 455-456
Joja A 1960 Buc U Log 3(1), 7-47
(MR24A, 220)  [logic]
Junge G 1926 DMV 35; 66-80, 150-172,
251-268  (F52, 6)
*Karpinski L C 1934 Isis 22, 104-105
[alg]
*- 1937 Isis 27, 46-52  [anal geom]
Karpinski L C + 1929 Sci 70, 311-314
[Mich pap 620, alg eq]
Kaye G R 1919 Scientia (Jan), 1-4
Klein J 1934 QSGM 3(1), 18-105
[alg, arith]
- 1936 QSGM (B)3, 122-235
(Z10, 248)
Koveos M K 1954 Gree SM 29, 101-117
(Z59, 6)  [sorites]
Kutorga M S 1873 *O schetakh u
drevnikh grekov.  Istoriya slova
"kameshek,"* Mos, 26 p
Lacombe D 1949 Thales 6, 37-58
[axiom]
Langer R E 1941 AMM 48, 109-125
(MR1941, 305; Z60, 5)
[Alexandria]

Lasserre François 1964 *The Birth of
Mathematics in the Age of Plato,*
Larchmont, NY, 191 p
Lazzeri G 1905 Pe M 8(S), 33-37
(F36, 50)  [arith]
Lejeune A 1948 Louv HPRT (3)(31)
(MR11, 419)  [optics]
Levey M 1958 Ens M (2)4, 77-92
(MR20, 945)  [alg, Babyl]
Lockyer J N 1893 Nat 48, 417-419
(F25, 91)  [archit]
Longaire de 1896 Intermed 3, 209
(Q776)  [surveying]
Loria Gino 1890 *Il periodo aureo
della geometria greca. 'Saggio
storico,* Torino   79 p
- 1893 Bib M (2)7, 79-89
(F25, 56)  [arith, Egypt]
- 1893-1902 *Le scienze esatte nell'
antica Grecia,* Modena,  5 vols
(also Mod Mm (2)10; (2)11; (2)12)
(F26, 5; 33, 6)
- 1894 Batt 32, 28-57  (F25, 56)
[arith, Egypt]
- 1906 Intermed 13, 39  [Jabobi]
- 1912 Int Con (5)2, 518-525
[sq root]
- 1912 Mathes It 4  [trig]
- 1914 *Le scienze esatte nell'antica
Grecia,* 2nd ed, Milan, 974 p
(F45, 1)
- 1928 Archeion 9, 161-166  (F54, 14)
[Arab]
- 1928 Bo Fir (2)7, i-v  (F54, 13)
[Archimedes, Thales]
- 1929 *Histoire des sciences mathématiques dans l'antiquité
hellénique,*  Paris, 219 p
(F55, 2)
- 1939 Porto Fac 24, 65-69
(MR1, 33)  [arith]
Losev A F 1928 *Dialetika chisla u
Plotina,*  Mos, 196 p
Luders L 1811 *Geschichte der Mathe-
matik bei den alten Volkern.  Oder
Pythagoras und Ypatia,* Altenburg-
Leipzig
Lure S Ya 1932 QSGM 2, 106-185
(Z5, 1)  [infinitesimal]
- 1933 Len IINT 1, 45-70  (Z7, 49)
[Egypt infl]

Greek   (continued)

- 1935 Len IINT 4, 21-46
  (F61, 945) [approx]
- 1935 *Teoriya beskonechno malykh u
  drevnikh atomistov*, Mos-Len, 197 p
Markovich Z 1961 Scientia (6)55,
  37-41  (MR24, 127)
  [Plato, Aristotle]
Marotte F 1935 Ens Sc 8, 134-140
  [geom]
Messedaglia A 1891 Linc Rn (4)7(1),
  495-526  (F23, 44) [ast]
Metzler W H 1915 MT 7, 84-85
  [late]
Meyerhof M 1933 Archeion 15, 1-15
  [Alexandria]
Michel Paul H 1939 *Introduction à
  l'étude d'une collection
  d'instruments anciens de mathématiques*, Anvers, 105 p
  (MR1, 33)
- 1950 *De Pythagore à Euclide*,
  Paris, 699 p  (MR12, 69)
Mieli A 1912 Rv Fil 4(3), 10 p
  [Cyrene]
Milhaud G 1893 *Leçons sur les
  origines de la science grecque*,
  Paris
- 1897 Rv Met Mor 5, 419-442
  (F28, 53) [geom]
- 1900 *Les philosophes geométres de
  la Grèce*, Paris,  388 p
- 1906 *Etudes sur la pensée
  scientifique chez les Grecs et
  chez les modernes*,  Paris
- 1934 *Les philosophes géométries
  de la Grèce. Platon et ses
  prédécesseurs*,  2nd ed, Paris
Miller G A 1923 Indn MSJ 15, 153-155
  [quad eq]
- 1923 Sch Soc 23, 621-622
  (F49, 17) [hist error]
- 1925 SSM 24, 284-287
  [weaknesses]
- 1926 Scientia 39, 317-322 39(S)
  99-103  (F52, 6) [weaknesses]
- 1931 AMM 38, 496-500
  (Z3, 98) [pre-Greek]
- 1941 Sci 94, 89-90
  [pre-Euclid]

Mineur A 1945 Bel BS 31, 683-710
  (MR7, 74) [geom]
Mondolfo R 1934 *L'infinito nel
  pensiero dei Greci*, Florence,
  439 p  (F60, 822)
Mueller J H T 1860 *Beitrage zur
  Terminologie der griechischen
  Mathematiker*, Leipzig
Mugler Charles 1957 Antiq Cl 26,
  331-345  [geom, optics]
- 1959 *Dictionnaire historique de la
  terminologie géometrique des
  Grecs*, Paris, 2 vols  (MR25, 966;
  Z88, 241)
Narducci E 1880 Boncomp 13, 369-378
  (F12, 12) [Alexandria]
Nagl A 1914 *Die Rechentafel der
  Alten*, Vienna  [arith, tab]
Nesselmann G H F 1842 *Die algebra
  der Griechen*, Berlin
Neugebauer Otto 1937 Int Con (1936)1,
  157-170  (F63, 799)
- 1951 Centau 1, 266-270  (MR13, 1)
  [sun tab]
- 1953 Arc In HS 34, 166-173
  [Babyl, Hindu]
*- 1952 *The Exact Sciences in
  Antiquity* (Princeton U P), 207 p
Niebel Eckhard 1959 *Untersuchungen
  ueber die Bedeutung der
  geometrischen konstruktionen der
  Antike*, Koeln, 147 p  (Z113, 1)
Northrop F S C 1936 in *Philosophical
  Essays for A. N. Whitehead*,
  1-40  (F62, 1015) [phil]
Ofterdinger L F 1844 Arc MP 5, 102-
  [ms]
- 1860 *Beitraege zur Geschichte der
  griechischen Mathematik*, Ulm
Osborn J 1963 MT 56, 540-545
  [late]
*Pauly-Wissowa 1894-1938 *Pauly's
  Real Encyklopaedie der Classischen
  Altertumswissenschaft*,
  Stuttgart, 57 vols
Plakhowo N 1906 Intermed 13, 197
Reidemeister K 1939 *Die Arithmetik
  der Griechen*, Hamburg-Leipzig,
  31 p  (MR1, 129; F65;2, 1079)
- 1949 *Das exakte Denken der Griechen.
  Beitraege zur Deutung von Euklid*,

Greek (continued)

*Plato, Aristoteles,* Hamburg, 108 p
(MR13, 419; Z36, 2)
Rey Abel 1924 *Histoire des sciences
exactes et naturelles dans
l'antiquite greco-romaine,* Paris
1933 *La jeunesse de la science
Grecque,* Paris, 544 p (Z7, 49)
- 1935 Act Sc Ind (217)
(F61, 944; Z11, 193) [V]
- 1936 Rv Phi Fr E 121, 338-371
(F62, 1042) [logic, V]
- 1939 *La Maturité de la Pensée
Scientifique en Grèce,* Paris,
595 p (MR1, 289)
- 1946 *L'apogée de la science
techniques greque. Les sciences
de la nature et de l'homme.
Les mathématiques d'Hippocrate
à Platon,* Paris, 331 p (Z60, 5)
- 1948 *L'apogée de la science
technique greque. L'essor de la
mathématique,* Paris, 332 p
(Z41, 337)
Reymond Arnold 1923 Ens M 23, 257-268
- 1924 *Histoire des sciences exactes
et naturelles dans l'antiquité
gréco-romaine,* Paris, 246 p
(F50, 18) (2nd ed 1955; Z67, 244)
Richter A 1828 *Des Apollonius von
Perga zwei Buecher vom Raumschnitt.
Ein Versuch in der alten Geometrie,*
Halberstadt, 120 p
Rixecker H 1961 Mathunt 7(3), 43-45
[use of classic texts in schools]
Robbins F E 1934 Isis 22, 95-103
(Z10, 243) [arith]
Robinson R 1936 Mind 45, 464-473
(F62, 12) [geom]
Rome A 1932 Brx SS (A)52, 271-274
(Z6, 145) [trig]
- 1933 Antiq Cl 2, 177-192
[trig]
Rudio F 1908 Zur NGV 53, 481-484
(F39, 69) [termin]
Ruffini E 1923 Arc Sto 4, 78-92
(F49, 17) [Aristotle, geom]
Sambursky S 1956 Osiris 12, 35-48
(Z71, 243) [possible prob]
Sanchez Pérez J A 1947 *Die Arithmetik

*in Griechenland,* Madrid, 260 p
(MR10, 124; Z41, 337; 60, 15)
Sarton G 1936 Isis 24, 375-381
(Z13, 193) [Minoan]
- 1959 *A History of Science.
Hellenistic Science and Culture in
the Last Three Centuries B.C.*
(Wiley), 576 p (repr 1965)
Schmidt M 1906-1912 *Kulturhistorische
Beitrage zur Kenntnis des griechi-
schen und roemischen Altertums,*
Leipzig, 2 vols
Scholz H 1934 Gree SM 15, 40-65
(F60, 822) [irrat]
Schultze E 1937 Archeion 19, 179-191
(F63, 799) [numer]
Schweigger J S C 1847 Arc MP 9, 115-
[constr]
Sciacca M F 1936 Logos Nap 19, 85-104
(F62, 10) [inf]
Sedillot P E A 1845-1849 *Materiaux
pour servir a l'histoire comparée
des sciences mathématiques chez
les Grecs et les Orientaux,* Paris,
2 vols [Arab]
Seidenberg A 1959 Scr 24, 107-122
(MR22, 259) [peg and cord]
Sengupta P 1931 Clct Let 21
[sph ast]
Shaw A A 1937 M Mag 11, 3-7
[metrol]
Shimomura T 1956 Jp APS 1, 1-31
(Z72, 243)
Skeat T C 1936 Mizraim 3, 18-25
[tablet]
Smith David E 1909 Bib M (3)9,
193-195 [mult tab]
- 1923 *Mathematics (Our Debt to
Greece and Rome,* Boston, 185 p
[pop]
- 1932 AMM 39, 425 (Z5, 241)
[Vienna pap]
Steele A D 1936 QSGM (B)3, 287-369
(Z14, 146) [circle, line]
Stenzel J 1924 *Zahl und Gestalt bei
Platon und Aristoteles,* Leipzig-
Berlin, 152 p
- 1932 Int Con (1932)1, 324-335
(F58, 12; Z6, 337) [intuition]
- 1933 Antike 9, 142-158
(F59, 837)

Greek (continued)

- 1933 Fors Fort 9, 94-95
  (F59, 17)
- 1934 Scientia 55, 169-176
  (F61, 944) [phil]
Sturm A 1899 ZM Ph 44(S), 485-490
  (F30, 2)
Suter H 1922 *Beitraege zur Geschichte
  der Mathematik bei den Griechen
  und Arabern,* Erlangen, 116 p
  (F48, 28)
Szabo A 1954 Hun Antiq 2, 17-62,
  247-254, 377-410
  [Dialektik, Parmenides]
- 1956 Act Antiq 4, 109-152
  [deduction]
- 1957 M Lap 8, 8-36, 231-247
  (Z112, 241) [deductive]
- 1958 Maia (NS)10, 106-131
  (MR21, 484)
- 1959 M Lap 10, 72-121
  (Z114, 5) [axiom]
- 1960 Arc HES 1(1), 37-106
  (MR24, 2) [Euclid axioms]
- 1960 Maia (NS)12(2), 89-105
  (Z93, 2) [termin]
- 1962 Osiris 14, 308-369
  [axiom]
- 1963 M Ann 150, 203-217
  (MR27, 1077) [Pyth]
- 1964 Scr 27, 27-48, 113, 139
  [axiom]
- 1967 Greek Dialectic and Euclid's
  Axiomatics in Lakatos I ed,
  *Problems in the Philosophy of
  Mathematics,* Amsterdam, 1-8
- 1969 *Anfaenge der griechischen
  Mathematik,* (Munich-Vienna),
  494 p
Tannery Paul 1881 Bord Mm (2)4, 161-
  195 (F13, 30) [arith, Heron]
- 1886 BSM (2)10, 303-314
  [Academy]
- 1887 *La Géométrie grecque, comment
  son histoire nous est parvenue et
  ce que nous en savons,* Paris
- 1887 *Pour l'histoire de la
  science hellène: De Thales a
  Empedocle,* Paris (2nd ed 1930)
- 1902 Bib M (3)3, 161 [music]
- 1904 Bib M (3)5, 5-8 [subt]

- 1905 Bib M (3)6, 225-229
  [arith, pre-Euclid]
- 1912-1920 *Mémoires scientifiques,*
  Toulouse-Paris, 4 vols
Taylor C 1881 Mess M (2)5, 112-113
  (F13, 43) [persp]
Taylor T 1816 *Thecretic Arithmetic
  in Three Books...,* London
  [anthology of classic authors]
Thomas Ivor ed 1939-1941 *Selections
  Illustrating the History of
  Greek Mathematics* (Harvard U P),
  2 vols (MR1, 33; 3, 257; 13, 419)
  [from Thales to Pappus]
Tod M N 1913 J Helle St (June)
  [numer]
Toeplitz O 1935 Muns Semb 6, 4-18
  (F61, 11) [historiog]
Treutlein P 1883 Hist Abt 28, 209-
  227 [geom]
Vacca G 1936 It Attuar 7, 231-234
  (F62, 12) [prob]
Vailati G 1905 Int Con 3, 575-581
  (F36, 62) [axiom]
Van Der Waerden B L 1940 M Ann 117,
  141-161 (Z22, 294) [Zeno]
- 1948 M Ann 120, 127-153
  (MR9, 483) [numb thy, Pythagoreans]
- 1950 *Ontwakende Wetenschap,*
  Groningen
- 1961 *Science Awakening* (Wiley)
  [tr of 1950]
Ver Eecke P 1931 Mathesis 45, 84-86
  (Z1, 322) [mech]
- 1931 Mathesis 45, 352-355 [mech]
- 1933 Scientia 54, 114-121
  [mech, Pappus]
- 1940 *Les opuscules mathématiques
  de Didyme, Diophane et Anthemios,*
  Paris, 102 p
Veselovskii I N 1948 Mos IIET 2,
  426-498 (MR11, 150)
Vogel K 1936 Mun Si, 357-472
  (F62, 1015; Z16, 196)
- 1954 MP Semb 4, 122-130
  (MR16, 207) [alg, frac]
- 1931 ZMN Unt 62, 266-271
  (F57, 14) [alg]
Vygodskii M Ya 1967 *Arifmetika i
  algebra v drevnem mire,* 2nd ed,
  Mos, 367 p
Witting A 1934 Unt M 40, 53-54
  [quartic]

Greek   (continued)

Zeuthen H G 1912 Sk Kong 2, 3-13
   (F43, 3)
- 1913 Dan Ov (6), 431-473
   (F44, 3)  [pre-Plato]
   1916 Dan Sk (8)1, 199-382
   (F46, 47)
   1917 *Sur la réforme qu'a subie la*
   *mathématique de Platon à Euclide*
   ..., Copenhagen (F46, 47)


Hebrew

See also Middle ages (under
   TIME PERIODS).

Anon 1936 Scr 4, 61-65 [mid ms]
Carlebach J 1910 *Levi ben Gerson*
   *als Mathematiker,* Berlin, 240 p
   (F41, 3)
Dupont-Sommer P 1963 in Taton R ed,
   *Ancient and Medieval Science,*
   NY, Part 1, Ch 3
Efros J J 1917 *The Problem of Space*
   *in Jewish Medieval Philosophy,*
   NY
Feldman W M 1931 *Rabbinical Mathe-*
   *matics and Astronomy,* London,
   250 p  (F57, 1292; Z4, 5)
Forbin V 1913 Nat Par (2), 287-288
   [metrol]
Fraenkel A A 1947 *The Jewish*
   *Contribution to Mathematics and*
   *Astronomy,* Tel Aviv, 32 p
   (Z29, 194)
- 1960 Scr 25, 33-47 (MR22, 1341;
   Z100, 7)
Frank E 1963 Intercom 2, 5-7
   [chronom]
Gandz S 1927 AMM 34, 80-86
   [anal]
- 1929 Heb UC An 6, 247-276
   [mult]
- 1929 Heb UC An 6, 263-276
   [Mishnat-ha-middot]
- 1929 Isis 12, 452-481
   [angle, Babyl]
- 1930 Isis 14, 189-214
   [abacus, knot, numer]

- 1930 Am AJRP (2), 23-38
   [gnomon]
- 1932 *The Mishnat Hammidot, the firs*
   *Hebrew Geometry about 150 C. E. and*
   *the Geometry of ibn Musa-al*
   *Khowarizmi, the first Arabic*
   *Geometry (c. 820) representing the*
   *Arabic version of the Missatha-*
   *Middot,* Berlin, 104 p (QSGM(A)2)
   (Z5, 242)
- 1933 Am AJRP 4, 53-112
   (Z7, 386)  [numer]
- 1934 Isis 22, 220-222
   [rule of three]
- 1939 Am AJRP 9, 5-50
   (MR1, 129)
- 1941 Jew Q Rv 31, 383-404 [arith]
- 1945 in Louis Ginsburg Jubilee
   Volume (American Academy for
   Jewish Research, NY) 1, 143-157
   [frac]
- 1949 Jew Q Rv 39, 259-280
   [calen]
- 1949 Jew Q Rv 40, 157-172, 251-277
   [calen]
- 1951 Isis 42, 267-268 [numer]
- 1952 Jew Q Rv 43, 177-192, 249-
   270 [calen]
- 1952 Osiris 10, 10-34 (Z47, 2)
   [chronom]
- 1954 Jew Q Rv 44, 305-325  [ast]
Goldberg Anna 1901 *Die juedischen*
   *Mathematiker und die juedischen*
   *anonymen mathematischen Schriften*
   ..., Frankfurt, 12 p  (F32, 2)
Golovensky D I 1932 Scr 1, 53-55
   (Z5, 242)  [max, min]
Gottheil R J H 1930 Jew Q Rv 21, 75-
   84 [calen]
Groenman A W 1929 Act Ori 7, 241-274
   [numb]
Guggenheimer A 1953 Int Con Ph 11
   [dialectic logic]
- 1966 in *Confrontations with*
   *Judaism,* Philip Longworth ed,
   London, 171-196  [logic]
Guenther S 1876 *vermischte Unter-*
   *suchungen zur Geschichte der*
   *Mathematischen Wissenschaften,*
   Leipzig, Ch. 6
Hasofer A M 1966 Biomtka 54, 316-321
   [prob]

Hebrew (continued)

Keller H 1932 Scr 1, 66-67
  (Z5, 242) [numer]
Kennedy E S 1963 Scr 27, 55-59
  [calen, Al-Khwarizmi]
Loewy A 1930 *Ueber die zahlbezeich-
  nung in der juedischen Literatur*,
  Berlin, 11 p (F56, 803)
Mahler E 1884 Hist Abt 29, 41-43
  [irrat]
- 1884 Z Realsch 9, 465-471
- 1886 Hist Abt 31, 121-132
Moors B P 1904 *Le système des poids,
  mesures et monnaies des Israelites
  d'après la Bible*, Paris
Neverov S L 1904 Kiev U Izv (8)
  [geom]
Rabinovitch N L 1969 Biomtka 56,
  437-441 [prob]
Read C B 1964 SSM 64, 765-766
  [pi]
Renan E 1877 His Lit Fr 27
  [XIV, France]
Riccardi P 1893 Bib M (2)7, 54-56
  (F25, 5)
Robbins F Egleston 1898 *The Patric
  papyrus. Hieratic Papyri from
  Kahun and Curob*, London
Rubin S 1873 *Sod Hasfiroth
  (Secrets of Numbers)*, Vienna
Sacerdote G 1893 Rv Et Juiv
  [Simon Motot]
Sarfatti G 1958 Tarbiz 28, 1-17
  [frac]
- 1959 Leshonen 23-24, 156-171,
  [Mishnat Hammidot]
- 1960 Leshonen, 73-94
  [Mishnat Hammidot]
Sarton G 1931 Int His Sc 2(1), 208-
  209 [Mishnat Hammidot]
Schapira H 1880 Ab GM 3, 3-56
  (F12, 3) [Mishnat Hammidot]
Simon I 1963 in Taton R ed,
  *Ancient and Medieval Science*,
  NY, 453-467
Simon M 1891 *Grundzuege des
  juedischen Kalenders*, Berlin,
  39 p [(F23, 43)
*Spero S W 1967 Intercom 9(1), 15-19
  [bib]

Steinschneider M 1864 Heb Bib 7, 16 p
  [Mishnat Hammidot]
- 1867 ZM Ph 12, 1- [Savasorda]
  1880 ZM Ph 25(S), 57-128
  (F12, 5) [Savasorda]
  1889 Bib M (2)3, 35-38
  [mid, mss]
  1893 Bib M (2)7, 51-54
  (F25, 5)
  1893 Bib M (2)7, 65-72, 73-74,
  105-112 (F25, 4, 5)
  1893 *Die haebraischen Uebersetzungen
  des Mittelalters*, Berlin
  1894 Bib M (2)8, 37-45, 79-83,
  99-105 (F25, 4)
  1895 Bib M (2)9, 19-28, 43-50,
  97-104 (F26, 8)
  1896 Bib M (2)10, 33-42, 77-83,
  109-114 (F27, 2)
  1897 Bib M (2)11, 13-18, 35-42,
  73-82, 103-112 (F28, 2)
  1898 Bib M (2)12, 5-12, 33-40,
  78-89 (F29, 3)
  1899 Bib M (2)13, 1-9, 37-45,
  97-104 (F30, 3)
  1899 ZM Ph 44(S), 471-483
  (F30, 3)
  1901 Bib M (3)2, 58-76
  (F32, 2)
  1901 *Die Juedischen Mathematiker
  in die Juedischen anonymen mathematischen Schriften*, Frankfurt
- 1964 *Mathematik bei den Juden*
  (Olms), 219 p (repr of 1893-1899)
  (MR33, 3866)
Steinschneider M ed 1858-1882
  *Hebraeische Bibliographie*, Berlin
  [=Heb Bib]
Steveson P A 1965 SSM 65, 454 [pi]
Szechtman J 1960 Scr 25, 49-62
  (MR24, 127; Z100, 243)
Thiele E D 1951 *The Mysterious
  Numbers of the Hebrew Kings*,
  Chicago
Thomson W 1934 Isis 20, 274-280
  (rev of Gandz 1932)
Thorndike L 1938 Isis 29, 69-71
  [ast]
Turetsky M 1923 MT 16, 29-34
  [combin]
Vogel K 1937 Z Gesm Nat 2/3, 88-93
  (F63, 4) [anc, mid]

Hebrew (continued)

Zuckermann B 1878 *Das Mathematische im Talmud,* Breslau (F10, 4)
- 1882 *Materialen zur Entwickelung der alt juedischen Zeitrechnung,* Breslau

Hindu

(A vague category including the mathematics of ancient India)

Ayyangar A A 1939 M Stu 7, 1-16 (F65:1, 6) [Bakhshali ms]
Bag A K 1966 Indn JAS 1, 98-106 (MR37, 2562) [series, trig]
Bakhmutskaya 1960 Ist M Isl 13, 325-334
*Balagangadharan K 1947 M Stu 15, 55-70 (MR10, 667; Z41, 338) [bib]
Ballabh R 1962 Isis 53, 502-504 (Z112, 241) [hist res in India]
Boncompagni B 1869 Boncomp 2, 153-206 (F2, 4) [al-Biruni]
Brocard H 1908 Intermed 16, 66 [Laplace]
Brun V 1919 Nor M Tid 1, 33-40
Burgess Ebenezer 1935 *Translation of the Suryasiddlianto. A Text-book of Hindu Astronomy* (P. Gangooly ed) Calcutta, 460 p
Burrow R 1792 Asi R 3, 145- [arith]
Cajori F 1919 Sci Mo 9, 458
Cantor M 1877 Hist Abt 22, 1-23 (F9, 19) [Greek infl]
- 1904 Arc MP (3)8, 63-72 (F35, 4) [oldest]
Cattaneo P 1938 Bo Fir (2)17, 60-62 (F64, 906) [dioph eq]
Chakravarti G 1932 Clct MS 24, 79-88 (Z6, 3) [combin]
- 1933 Bhan ORIA 14, 87-102 (Z6, 145)
- 1934 Clct Let 24(6), 1-18 [series]
- 1934 Clct Let 24(7)(a), 19-22 (F62, 1024) [area, termin]
- 1934 Clct Let 24(7)(b), 23-28 (F62, 1023) [area]

- 1934 Clct Let 24(8), 29-58 (F62, 1021) [sq root]
- 1934 Clct Let 24(9), 59-76 (F62, 1021) [frac]
Chistyakov V D 1960 *Materialy po istorii matematiki v Kitae i Indii,* Mos, 167 p
Colebrooke H T 1817 *Algebra with Arithmetic and Mensuration from the Sanscrit of Brahmagupta and Bhaskorn,* London (2nd ed rev by H C Benerji, Calcutta 1927)
- 1927 *Translation of the Lilavati,* 2nd ed, Calcutta, 323 p (F57, 1293)
Collins Mark 1926 *On the Octaval System of Reckoning in India,* Madras, 28 p
Coomaraswamy A K 1933 Isis 19, 74-91 [tr probls]
Das S R 1928 AMM 35, 535-540 (F54, 21) [anal geom, ast]
- 1928 AMM 35, 540-543 [eq of time]
Datta B 1926 AMM 33, 449-454 [zero]
- 1927-1929 Alla UMAB 1-2, 62 p
- 1927 AMM 34, 420-423 ["root"]
- 1927 Beng Asi J 22, 26-42 [pi]
- 1928 AMM 35, 520-529 (F54, 20) [abacus]
- 1928 Beng Asi J 23, 261-267 [arith]
- 1928 Clct MS 19, 87-94 (F54, 20) [Pell eq]
- 1929 Indn HQ 5, 479-512 [Ganita]
- 1929 Clct MS 21, 1-60 (F56, 808) [Bakhshali ms]
- 1929 Clct MS 21, 115-145 (F56, 807)
- 1929 AMSB 35, 579-580
- 1930 QSGM (B)1, 113-119 [geom]
- 1930 QSGM (B)1, 245-254 [Jaina cosmol]
- 1931 AMM 38, 317-376 [root]
- 1931 AMM 38, 566-572 (F57, 17) [zero]
- 1931 Archeion 13, 401-467 (Z4, 4) [dioph anal]

Hindu (continued)

- 1932 *The Science of the Sulba. A Study in Early Hindu Geometry* (U Calcutta), 255 p (Z7, 387)
- 1933 Beng Asi J (2)26, 283-290 (F57, 1293) [polyg]
- 1941 Archeion 23, 78-83 [XVI, chronol]

Datta B + 1935-1938 *History of Hindu Mathematics. A Source Book. Pt. I. Numerical Notation and Arithmetic. Part II. Algebra.* Lahore, 281+330 p (F64, 906) (repr 1962 Madras-Lon; MR26, 1142)

De Boer F 1884 *De wiskunde der Indiers,* Leiden

Decourdemanche J-A 1913 J Asi (11)1, 427-444 [geod]

Delbos L 1892 BSM (2)16, 93-112 (F24, 44)
- 1893 BSM (2)17, 145-172 (F25, 94) [ast]

Devaraja 1944 *Kuttakarasiromani* (Anandasrama Sanskrit Series 125), Poona, 53 p (MR7, 105)

Dixit R R 1951 Scientia 45, 315-318 (MR13, 420) [ast]

Drenckhahn F 1936 DMV 46, 1-13 [circle sq]

Filliozat J 1963 in Taton R ed, *Ancient and Medieval Science,* NY, Part 1, Ch. 4

Ganguli S 1929 Isis 12, 132-145 [crit of Kaye]
- 1931 Indn MSJ 19, 110-120 (Z5, 149) [dioph anal]
- 1932 AMM 39, 251-256, 389-393 [numer]
- 1932 Indn MSJ 19, 153-168 [dioph anal]
- 1932 Scr 1, 135-141 [irrat]
- 1933 AMM 40, 25-31, 154-157

Goblet d'Alviella 1926 *Ce que l'Inde doit à la Grèce,* Paris, 153 p

Gupta P C S 1918 Clct MS 10, 73-80 [numb thy]

Gurjar L V 1942 Bombay (NS)10(5), 6-10 (MR6, 253) [root 2]
- 1942 Bombay (NS)10(5), 11-16 (MR6, 253) [pi]

- 1947 *Ancient Indian Mathematics and Vedha,* Poona, 208 p (MR9, 73)

Hoernle A F R 1883-1888 Indn Antq 12, 89-90; 17, 33-48, 275-279 [Bakhshali ms]

Hofmann J E 1953 MP Semb 3, 193-206 (MR15, 591) [pi]

Hromadko F 1876 Cas MF 5, (F8, 3) [Lilavati, arith]

Iyer R V 1954 Scr 20, 58-63 [abacus]

Kamalamma K N 1948 Clct MS 40, 140-144 (MR10, 667; Z41, 338) [dioph anal]

Karpinski L C 1913 AMSB (2)19, 294 (F44, 59) [numer]
- 1919 AMM 26, 298-300
- 1928 Scientia 43, 381-388 (F54, 20)

Kaye G R 1907 Beng Asi J 3, 475 [numer]
- 1907 Beng Asi J 3, 482 [numer]
- 1910 Bib M (3)11, 289-299 (F41, 2)
- 1912 Beng Asi J (NS)8(9), 349-361 [Bakhshali ms]
- 1915 *Indian Mathematics,* Calcutta, 75 p (See Mitra 1916)
- 1918 Scientia 24, 53-55 [numer]
- 1919 Beng Asi J 15, 153-189 [ast]
- 1919 Isis 2, 326-356
- 1919 Scientia 25, 3-16 [Greek infl]
- 1927 *The Bakhshali Manuscript. A Study in Medieval Mathematics,* Calcutta, 162 p (See Ganguli 1929)

Kern H 1863 Lon Asia 20, 371-387

Kokomoor F W 1937 MT 29, 224-231 [mid]

Krishnaswamy Ayyangar A A 1923 Indn MSJ 15, 214-223 [puzzle]
- 1933 M Stu 1, 1-18 (F59, 839)
- 1945 Mysore (A)5, 101-115 (MR7, 105; Z60, 1)

Labban K ibn 1965 *Principles of Hindu Reckoning,* Madison

Levi B 1908 Bib M (3)9, 97-105 (F39, 70) [geom]

Loria G 1934 Archeion 15, 395-407 (F60, 819; Z8, 337) [geom]

Hindu (continued)

- 1935 Bo Fir (2)14, i-iv
  (F61, 12)
Loria G + 1937 Mathesis 51, 327-330
  (F63, 801) [alg eqs]
Lucas E 1876 Boncomp 9, 157-164
  (F8, 2) [arith]
Marre A 1880 Cron Cien 3, 153-155,
  177-178 (F12, 1) [arith]
Mehta D M 1932 Theory of Simple
  Continued Fractions, Bhavagnar,
  168 p
Mitra 1916 Hindu Mathematics,
  Calcutta (Modern Review Office)
  (Crit of Kaye 1915)
Mohan B 1958 Banar JSR 9(1), 19-28
  (MR22, 769) [prog]
Mukhopadhyaya D 1930 Clct MS 22, 121-
  132 (F56, 808) [moon]
Mueller C 1929 Ham Sem 7, 173-204
  (F55, 8) [Sulvasutra]
- 1925 ZMN Unt 56, 234-236
  (F51, 38) [chronom]
Munjala 1944 Anand San (123), 32 p
  (MR7, 105)
Naraharayya S N 1923 Indn MSJ 15(2),
  105-113 [trig tab]
Neugebauer O 1952 Osiris 10, 252-276
  (MR14, 523) [Tamil ast]
Pingree D 1959 Am OSJ 79, 282-284
  (Z102, 244) [ast]
- 1968 APST 58 [ast tabs in USA]
Pizzagalli A M 1932 Scientia 51, 431-
  439 (F58, 17) [sci]
Playfair J 1790 Edi ST 2 [ast]
- 1798 Edi ST 4 [trig]
Przyluski M J 1932 Brx U Rv 37,
  283-294 (F57, 1291) [Babyl]
Rajagopal C T 1949 Scr 15, 201-209
  (MR11, 572; Z41, 338)
Rajagopal C T + 1949 Beng Asi J 15(1)
  (MR11, 572) [series, trig]
- 1951 Scr 17, 65-74 (MR13, 1)
  [Gregory series]
- 1952 Scr 18, 25-30 (MR14, 121)
  [pi]
Ranjan Das L 1932 Pe M (4)12, 133-140
  (F58, 19) [combin]
Renou Louis + 1953 L'inde classique.
  Manuel des études indiennes,
  Paris-Hanoi

Ross Allan S C 1938 The "Numeral-
  Signs" of the Mohenjo-Daro Script,
  Delhi, 25 p (=Indn Ar SM 57)
Sanchez Perez J A 1949 Arithmetic in
  Rome, India and Arab Lands,
  Madrid-Granada 263 p (in Spanish)
  (Z41, 338)
Sarkar B K 1918 Hindu Achievements
  in Exact Science, NY, 96 p
  (F46, 54)
Sarton G 1931 Int His Sc 2(1), 214-
  215 [Bakhshali ms]
- 1936 Isis 25, 323-326 [deci]
Schmidt O H 1944 Isis 35, 205-211;
  36, 46 (MR6, 253) [ast]
- 1952 Centau 2, 140-180 (Z49, 289)
  [ast]
Seal R 1915 The Positive Sciences of
  the Ancient Hindus, London, 302 p
  (F45, 3)
Sedillot L A 1875 Boncomp 8, 457-468
  (F7, 2) [re F Hoefer]
Sen S N 1966 A Bibliography of
  Sanskrit Works on Astronomy and
  Mathematics: 1: Manuscripts,
  Texts, Translations, and Studies
  (National Institute of Science of
  India), New Delhi, 281 p
Sengupta P C 1928 Clct Let 21(4)
  (F57, 1295) [sph trig]
- 1931 DMV 40, 223-227
  (F57, 18; Z2, 325) [calc]
- 1932 Clct MS 24, 1-18 [ast]
- 1944 Sci Cult 9, 522-526
  (MR6, 141) [ast]
Shankar Shukla Kripa 1955 Ganita 5,
  129-136 (MR18, 710; Z59:1, 3)
- 1955 Ganita 5, 149-151
  (MR18, 710; Z60, 3)
- 1966 Gantia 17, 109-117
  (MR37, 2563) [factoring]
- 1957 The Surya-Siddhanta with the
  Commentary of Paramesvara, Lucknow,
Singh A N 1936 Archeion 18, 43-62
  (F62, 15) [mid]
- 1936 Math Cluj 12, 102-115
  (Z15, 289) [roots]
- 1936 Osiris 1, 606-628 (F62, 17)
  [series]
- 1937 Int Con (1936)2, 275-276
  (F63, 20) [magic sq]

Hindu      (continued)

- 1951 Alla UMAB 16, 11-30
  (MR18, 710)
Smith D E 1913 Isis 1, 197-204
  [geom]
Smith D E  + 1911 The Hindu Arabic
  Numerals, Boston-London
Somayajulu D A 1934 M Stu 2, 12-21
  (F60, 820)  [trig]
- 1935 M Stu 3, 60-65
  (F61, 347)  [ast]
Spottiswoode W 1860 Lon Asia 17,
  221-
Staal J F 1967 in Prior A N,
  Formal Logic, 2nd ed, Oxford
  (1st ed 1953)
Strachey L 1813 Bija Ganita or the
  Algebra of the Indu, London
Subramani Iyer H 1938 M Stu 5,
  96-99  (F64, 907; Z19, 100)
  [ast]
Suter H 1905 Int Con (3), 556-561
  (F36, 3)
- 1908 Bib M (3)9, 196-199
  (F39, 70)  [surf area]
Thibaut G 1875 Beng Asi J 44
  [Sulvasustras]
- 1889 The Panchasiddhantantika of
  Varaha Mihira, Benares
- 1899 Astronomie, Astrologie, und
  Mathematik, Strassburg
Toomer G J 1963 Centau 9, 11-15
  (MR27, 1077)  [Tamil ast tab]
Uvanovic D 1936 Osiris 1, 652-657
  (F62, 15)  [and Europe]
Vallauri C M 1923 Arc Sto 4
Van der Waerden B L 1955 Zur NGV 100,
  153-170  [ast, dioph anal]
Vanecek J S 1881 Cas MF 10, 60, 127
  (F13, 1)  [geom]
Van Wijk W E 1938 Decimal Tables
  for the Reduction of Hindu Dates
  from the Data of the Surya-
  Siddhanta, The Hague, 40 p
Vashchenko-Zakharchenko M E 1882
  Istoricheskii Ocherk matemati-
  cheskoi literatury indusov,  Kiev
Venkatachalam Iyer R 1950 M Stu 18,
  79-82  (MR13, 420; Z43, 1)
  [abacus]

Venkatraman A 1948 M Stu 16, 1-7
  (MR11, 572; Z41, 338)
Vermeire M 1925 Rv Q Sc (4)8, 309-
  342  (F51, 15)  [ast]
Wiedemann E 1912 Mit GMNT 11, 252-
  255  [circle]
Woepcke Franz 1859 L'Introduction de
  l'Arithmetique indienne en
  Occident,   Rome
  1860 Sur l'histoire des sciences
  mathématiques chez les orientaux,
  Paris
Yajima S + 1955 JHS (33), 36-38


Japanese

For modern Japan, see under COUNTRIES.

Anon 1954-1957 A History of Japanese
  Mathematics in the Days Before
  Meijiera (1868-1911), (Japanese
  Academy), 3 vols  (Z77, 7)
Azuma R 1927 Jp MASE 9, 22-23
Beal E C 1950 Isis 41, 303
  [materials in US Lib Congress]
Berson 1891 Tou Mm (9)3, 268-271
  (F23, 36)  [geometric arith]
Bobori A 1964 Int Con HS (10)
  [infinitesimal]
Bourgeois L 1913 Rv Mois 16, 129-160
  (F44, 38)
Claudel 1875 La théorie des
  parallèles selon les géomètres
  japonais, Brussells  (F7, 18)
Endo T 1895 Tok M Ph 7, 103
  [cycloid, length]
- 1895 Tok M Ph 7, 123  [sph, vol]
- 1906 Tok M Ph 3, 72-74
  [ell, length]
- 1906 Tok M Ph  [Seki Kowa]
- 1918 History of Japanese Mathematics,  Tokyo, 702 p
Fujisawa R 1900 Int Con (2), 379-393
- 1902 Tok MPH 7, 88  [intgl]
- 1902 Int Con (2), 379-393
  (F32, 2)
Fujiwara M 1939 Jp Ac P 15, 101-104
  (F65, 1081)  [approx, dioph anal]
- 1939 Jp Ac P 15, 114-115
  (F65, 1081  [approx]

Japanese  (continued)

- 1939 Toh MJ 46, 123-134, 135-141,
  295-308  (F65, 8)  [Wazan]
- 1940 Toh MJ 46, 295-308, 47,
  49-57  (MR1, 289; F66, 13)
- 1940 Toh MJ 47, 322-338 (MR2, 306;
  F66, 13; Z26, 98) [approx, Newton]
- 1942 Toh MJ 49, 90-105
  (Z60, 1)  [Tanaka]
- 1954-1957 *History of Japanese
  Mathematics Before the Meiji Era,*
  Tokyo, 3 vols  (in Japanese)
Goryachkin V 1930 Gorsk Ped 7(2),
  43-58
Greenstreet W J 1906 M Gaz 3, 268-270
  (F36, 3, 37, 3)
Haga Y 1954 Kagak Ken (29)
  (Z60, 6)  [Shinpo-ken]
Hagiwara T 1885 Tok M Ph 3, 172
  [bib]
Harrison I + 1950 MT 43, 271-272
Harzer P 1905 Brt AAS 75, 325-329
  (F36, 3)
- 1905 *Die exakten Wissenschaften
  in alten Japan,*  Kiel, 39 p
  (also DMV 14, 312-339)  (F36, 3)
Hayashi G 1937 Toh MJ 43, 127-132
  (Z17, 290)  [Malfatti, probl]
Hayashi Tsuruichi 1895 Tok M Ph 7,
  60  [geom]
- 1901 Tok M Ph 9, 1  [trapezium]
- 1902 Bib M (3)3, 273-275
  [XVII, XVIII, pi]
- 1904 Nieu Arch (2)6, 296-361; 7,
  105-112, 113-  (F35, 4; 36, 3)
- 1905 Bib M (3)6, 323
  (F36, 53)  [Tait probl]
- 1905 Nieu Arch (2)7(1), 42
  [ast, Holland]
- 1906 AMM 13, 171  [conics]
- 1906 Nieu Arch (2)7(3), 232-237
  [bib, Holland]
- 1906 Tok M Ph  [Seki Kowa]
- 1909 Nieu Arch (2)9, 39-41, 42-48
  (F40, 4)  [Holland]
- 1910 Tok M Ph 25, 254-271
  (F41, 58)  [det]
- 1912 Batt 50, 193-211 [det]
- 1912 Tok M Ph 6, 144-152
  [Seki Kowa]

- 1912 Toh MJ, 204-206  [Casey thm]
- 1914 Toh MJ 6, 188-231  (F45, 95)
  [cont frac]
- 1916 Toh MJ 10, 15-27  [numb thy]
- 1918 Toh MJ 14, 127-151  [Tojutsu]
- 1919 Toh MJ 16, 26-40  (F47, 28)
  [Jinkuki, Kaisanki]
- 1919 Toh MJ 16, 299-333  (F47, 28)
  [Wasan]
- 1922 Toh MJ 21, 148-190  [log]
- 1920 Toh MJ 18, 302-308
  [Kemmochi, sph]
- 1924 Toh MJ 23, 64-67  [torus]
- 1924 Toh MJ 24, 185-189
  (F50, 6)  [bib]
- 1925 Toh MJ 26, 406-407
  [bib]
- 1925 Toh MJ 26, 408-419
  [trig tab]
- 1929 Toh MJ 30, 235-255  (F55, 592)
- 1929 Toh MJ 30, 506-527  (F55, 9)
  [Hasegawa]
- 1930 Toh MJ 33, 180  [bib]
- 1930 Toh MJ 33, 161-179
  (F56, 14)  [Hasegawa]
- 1931 Toh MJ 33, 292-327
  (Z1, 323)  [fract, integer]
- 1931 Toh MJ 33, 328-365
  (F57, 20)  [combin]
- 1931 Toh MJ 34, 145-185
  (F57, 20)  [eqs]
- 1931 Toh MJ 34, 349-396
  (F57, 20)  [max]
- 1932 Toh MJ 36, 182-188
  (Z6, 3)  [Ohara]
- 1932 Toh MJ 35, 171-226
  (Z4,,5)  [figurate numb, fin
  diffc, series]
- 1932 Toh MJ 35, 345-397
  (F58, 27)  [binom thm, eq, series]
- 1932 Toh MJ 35, 398-410
  (F58, 28; Z5, 4)
- 1932 Toh MJ 36, 135-181 (F58, 28)
  [reg polyg]
- 1933 Toh MJ 36, 346-394
  (F59, 839)  [intgl]
- 1933 Toh MJ 36, 395-397
  (F59, 840; Z6, 145)  [Ohara]
- 1934 Indn MSJ 20, 178-181
  (F60, 10)  [XIX, geom]
- 1934 Toh MJ 39, 125-179 (F60, 15)
  [curve, surf]

Japanese (continued)

- 1934 Toh MJ 39, 148, 406-424
  (Z9, 389)
- 1935 Toh MJ 40, 317-369
  (F61, 13) [alg]
- 1935 Toh MJ 40, 490-522
  (Z11, 194)
- 1935 Toh MJ 41, 249-264
  (Z12, 98)
- 1936 Toh MJ 41, 290-307
  (Z13, 338)
- 1936 Toh MJ 42, 1-31
  (Z14, 147)
Hirayama A 1939 Toh MJ 45, 377-404
  (F65, 8; Z21, 196)
  [Jinkoki, Kaisanki]
Hitomi C 1924 Tok S Ph S 33, 865-873
- 1925 Tok S Ph S 34, 170-176,
  370-375
- 1926 Tok S Ph S 35, 269-272
  [abacus]
Hiyama S 1923 MT 16, 359-365
Ichida A 1933 Toh MJ 37, 199-201
  (F59, 21) [Iwata, Ohara]
Itakura K 1963 Jp Stu HS (2),
  136-145 [ballistics]
Kamiya H 1930 Toh MJ 32, 365-372
  (F56, 14)
Kato H 1940 Toh MJ 47, 279-293
  (MR2, 306; Z26, 98) [catenary]
- 1941 Toh MJ 48, 1-24
  (Z25, 293) [Seki Kowa]
Kawakita C 1895 Tok MJ 7, 81
  [schools]
Kikuchi D 1895 Tok M Ph 7, 24-26
  (F26, 55) [area circle]
- 1895 Tok M Ph 7, 114
  [Ajima, length]
- 1895 Tok MJ 7, 47-53
  (F26, 56) [pi]
- 1895 Tok M Ph 7, 107
  [pi, series]
Knott C G 1892 Edi MSP 11, 167 [arith]
Kobori A 1957 Les étapes historiques
  des mathématiques du Japon (Par
  PDC (D)44), 21 p (MR23, 134)
- 1959 Sugaku 10, 145-147
  (MR25, 389)
Loria G 1906 Loria 9, 65-72
  (F37, 3)

Maruyama K 1953 JHS(27), 25-
  [tablets]
Michiwaki Y 1965 Nagao TC 2, 1-12
Mikami Yoshio 1906 DMV 15, 253-262
  (F37, 2) [re P. Harzer]
- 1906 Bib M (3)7, 364-366
  (F37, 5) [western infl, XVII]
- 1910 Mathematical Papers from the
  Far East, Ab GM 28, 1-226
- 1910 Nieu Arch 9, 231-234
  [ast, Holland]
- 1910 Nieu Arch 9, 301-304
  [Holland surveying]
- 1910 Tok M Ph (2)5, 372-392
  (F41, 65) [circle-principle]
- 1911 DMV 20, 381-393
  [abacus, China]
- 1911 Nieu Arch 9(4), 370-372
  [Holland, surveying]
- 1911 Nieu Arch (2)9, 373-386
  (F42, 3) [re Hayashi]
- 1911 Toh MJ 1, 98-105
  [fin diffc]
- 1912 The Development of Mathematics in China and Japan,
  Ab GM 30, 383 p (repr 1961
  Chelsea)
- 1912 Arc MP (3)20, 1-10, 183-186
  (F43, 6)
- 1912 Bib M (3)12, 225-237
  [ell length]
- 1912 Nieu Arch 10(1), 61-70
  [ast, Portuguese]
- 1912 Nieu Arch 10, 71-74
  [XVII, ast]
- 1912 Nieu Arch 10(2), 233-243
  [Buddhist, European ast]
  1913 Porto Ac 8, 210-216
  (F44, 47) [approx]
- 1913 Porto Ac 8, 5-14
  [Portuguese ast]
- 1913 Toh MJ 3, 29-37
  (F44, 52) [hyper, parab]
- 1913 Tok M Ph (2)7, 157-170
  (F44, 51) [pi]
- 1913 Tok M Ph (2)7, 46-56
  (F44, 51) [pi]
- 1914 Nieu Arch (2)11(1), 1-19
  [Kiell, ast]
- 1914 Arc MP (3)22, 183-199
  (F45, 3)

Japanese (continued)

- 1914 Isis 2, 9-36 [det]
- 1914 Toh MJ 5(3, 4)
- 1915  Toh MJ 7, 71-73  [circle]
- 1919 Toh MJ 15, 289-296
  (F47, 28)  [Casey thm]
- 1922 Shig Zas 29(3)
- 1929 Toh MJ 32, 173-192
  (F55, 592)  [inversive geom]
- 1930 Tok M Ph (3)12, 43-63
  (F56, 809)  [Yenri]
- 1934 in *Science in the Edo period*
  (Tokyo Science Museum), 25-49
  (F60, 823, 825)
- 1947 *Japanese Mathematics Viewed
  from the Point of View of Japanese
  Culture,* Tokyo, 200 p
- 1947 *History of Mathematics in
  Japan,* Tokyo, 244 p
  1947 *Study of Land Surveying in
  Japan,* Tokyo, 218 p
- 1951 JHS (19), 33-38
  [own work]
Minoda T 1940 Toh MJ 47, 99-109
  (Z23, 196) [Seki]
- 1941 Toh MJ 48, 174-184
  (MR7, 353; Z25, 293)  [Araki]
Mizuki Kozue 1928 *Nihon sugaku shi
  (History of Japanese Mathematics)*
  Tokyo, 532 p
Nagaoka H 1891 Tok M Ph 4, 90
  [surveying]
Ogura Kinnosuke 1935 *Sugaku-shi
  Kenkyu* (Studies in the History of
  Mathematics (1)10), Tokyo 340 p
- 1938 Chuo Kor 53, 31-47
  (F64, 920)
*- 1938 Kaizo 20, 107-121
  [mid, decline]
Okamoto Noribumi 1931 *A Catalogue
  of Japanese Mathematical
  Literature,* Tokyo, 822 p
- 1932 *Wasan tosho mokuraku
  (Catalogue of Books on Japanese
  Mathematics)* Tokyo, 835 p
Ota S-I 1953 JHS(27), 23-
  [bib, educ hist]
Shimizu Y 1924 Toh MJ 23, 53-63
  [torus]
Shinomiya A 1912 Toh MJ 2, 30-31
  [series]

- 1913 Toh MJ 4, 54-67   (F44, 46)
  [series]
- 1920 Jp MASE 2, 7-11  [geom]
Smith D E 1911 Toh MJ 1, 1-7
  (F42, 3)  [Western attitudes]
Smith David E + 1914 *A History of
  Japanese Mathematics,* Chicago
Sudo T 1954 Tok GCE 4, 165-177
  (Z67, 245)
- 1955 Tok GCE 5, 67-82, 179-189
  (Z67, 245)
Suzuki K 1934 in *Science in the Edo
  period,* Tokyo, 51-85  (F60, 825)
  [ast, calen]
Takeda U 1912 Toh MJ 2, 74-99,
  182-207  [Wada intgl tab]
Tanaka Tetsukichi 1936 *Ka etsu no
  ni okeru sugaku (Mathematics in the
  Kaga Feudal Clan),* Kawazawa, 134 p
W J G 1905 M Gaz 3, 268-270
Yanagihara K 1913 Toh MJ 3, 87-95
  (F44, 52)  [geom]
- 1914 Toh MJ 6, 120-123   (F45, 97)
  [Pyth triples]
- 1918 Toh MJ 14, 305-324
  [arith prog]
- 1924 Toh MJ 24, 128-135
Yendo M 1919 Jp MASE 1, 47-54
  [termin]


Korean

Edkins J 1898 Korean R 5  [numer]
Fujiwara M 1940 Toh MJ 47, 309-321
  (MR2, 306; Z61, 4)
- 1941 Toh MJ 48, 78-88  (MR7, 353)
Pogyom Kini 1960 Das Koreanische
  Zahlensystem, in Riekel August,
  *Koreanica,* Baden-Baden (Lutzeyer),
  85-94


Mayan

See America (under COUNTRIES).

Andersen R 1931 As Nach 242, 125-126
  (Z1, 323)  [calen]
Biermann K-R 1961 Ber Mo 3, 456-462
  (Z108, 245)  [calen]

Mayan    (continued)

Cramer G F 1938 AMM 45, 344-347
    (F64, 12) [metrol]
Dittrich A 1936 Ber Ab (3), 39 p
    (F63, 797) [chron]
- 1938 Scientia 63, 211-218
    (F64, 12) [chronom]
    1939 Ber Ab (2), 47 p
    (MR1, 130) [ast]
- 1943 Ber Ab (10), 51  p
    (MR8, 189) [ast]
Ganguli S 1930 Clct MS 22, 99-102
    (F56, 801) [numer]
Guthe C E 1931 Sci (2)75, 271-277
Kinsella J + 1935 MT 27, 340-343
    [calen]
Ludendorff H 1931 Ber Si (1/2), 8-19
    (Z1, 114)
- 1931 Ber Si 3, 40-62
    (Z1, 114)
- 1933 Ber Si, 4-49    (F59, 21)
    [ast]
- 1933 Ber Si, 772-798    (F60, 9;
    Z8, 338)    [ast]
- 1934 Ber Si, 40-45    (F60, 9)
    [ast]
- 1934 Fors Fort 10, 101-102
    (F60, 9) [ast]
- 1936 Ber Si, 65-88
    (Z13, 338) [ast]
- 1940 Ber Ab(6), 60 p    (MR5, 57)
    [ast]
- 1940 in Festschrift fuer Elis
    Stroemgen, Copenhagen, 143-162
    (MR3, 98; F66, 5) [ast]
- 1942 Ber Ab(16), 36 p
    (MR8, 189) [ast]
Makemson M W 1942 Pop As 50, 6-15
    (F68, 6) [calen]
- 1943 Carn CAAH (546), 185-221
    (MR6, 253; Z60, 2) [ast, table]
- 1946 Vass Ob Pu(5), 75 p
    (MR8, 305; Z60, 2)
Morley S G 1946 The Ancient Maya
    (Stanford U P)
*- 1915 An Introduction to the
    Study of Maya Hieroglyphs,  Wash
    DC (USGPO) [numer, zero]
    Richeson A W 1933 AMM 40, 542-546
    (F59, 22) [numer]

Salyers G D 1954 M Mag 28, 44-48
    [numer]
Sattherwaite Linton 1947 Concepts
    and Structures of Maya Calendrical
    Arithmetics,  Philadelphia, 168 p
    (Z36, 1)
Sanchez George I 1961 Arithmetic in
    Maya, Austin, Texas, 74 p
    (MR23, 433)  [bib]
Thomas C 1900 Smi Eth R 19, 853
Thompson J E S 1942 Carn CAAH (528),
    37-62    (MR6, 253)    (repr in
    Contributions to American
    Anthropology and History 36)
Wieleitner H 1917 ZMN Unt 48, 213-
    215  (F46, 46) [zero]


Persian

See also Asian.

Arne J T 1912 Orien Arc 2, 122-127
    [metrol, Sweden]
Datta B 1932 Benar MS 14, 7-21
    (F58, 17)
Hamadanizadeh J + 1965 MT 58, 441-
    446  [XI]
Kennedy E S + 1963 Am OSJ 83,
    316-  [chronom]
Rodet L 1884 BSM (2)8, 245-245
    [alg]
Schoy Carl 1927 Die trigonometrischen
    Lehren des persischen Astronomer
    ...Al-Biruni,  Hannover, 120 p
Segre A 1944 Am OSJ 64, 73-81
    [metrol]


Peruvian

See also America    (under COUNTRIES).

Locke L L 1912 Am Anthro 14,
    325-332    [quipu]
- 1923 The Ancient Quipu or Peruvian
    Knot Record (American Museum of
    Natural History), 84 P    (F49, 23)
    1932 Scr 1, 37-43  (Z5, 242)
    [abacus]

## Roman

See also Etruscan, Greek (above),
Middle Ages (under TIME PERIODS
below), Numeration (Ch. 3).

Behafy E 1948 Louv Ped (1 Nov), 29-
37 [numer]

Blume F + ed 1848-1952 *Die Schriften
der roemischen Feldmesser*, Berlin
2 vols [in Latin with comment]

Cantor Moritz 1875 *Die roemische
Agrimensoren und ihre Stellung in
der Geschichte der Feldmeskunst.
Eine Historische-mathematische
Untersuchung*, Leipzig (F8, 8)

Fettweis E 1929 Arc GMNT 11, 342-344
(F55, 7)    [Amer Indian]

Fields M 1934 MT 26, 77-84

Fischer 1846 *Roemische Zeittaffeln*,
Altona

Friedlein G 1869 *Die Zahlzeichen und
das elementare Rechnen der
Griechen und der Roemer und des
christlichen Abendlandes von
7 bis 13 Jahrhunderts*, Erlangen,
169 p

Gillet + 1903-1906 Intermed 10,
156, 272; 13, 197 (Q2612)
[arith]

Heiberg J L 1884 Philolog 43(2)

Hultsch F 1862 *Griechische und
roemische Metrologie*, Berlin

La Ponce de 1858 *Mémoire et
documents sur les déterminations
de la mesure longimetrique du
mille romain...*, Tours, 20 p

Longaire de + 1899 Intermed 6, 179
(Q776) [surveying]

Loria G 1928 *Archimed, la scienza
che domina Roma*, Milan

Noury P 1932 Bude Bu (36), 40-45
[numer]

Richardson L J 1915 AMSB (2)22, 7
(F45, 96)

Reymond A 1924 *Histoire des sciences
exactes, et naturelles dans
l'antiquité gréco-romaine*, Paris
246 p  (F50, 18)

Roesler M 1925 Engl Stu 59, 161-172
[vigesimal]

Sanchez Perez J A 1949 *Arithmetic in
Rome, India and Arab Lands*,
Madrid-Granada, 263 p  (in Spanish)
(Z41, 338)

Schmidt M C P 1914 *Kulturhistorische
Beitraege zur Kenntnis der
griechischen und roemischen
Altertums*, Leipzig, 285 p
(F45, 62)

Slotty F 1937 Arc Or Pra 9, 379-404
[Etruscan]

Smith David E 1923 *Mathematics (Our
Debt to Greece and Rome)*  Boston,
185 p  [pop]

Stoeber E 1877 *Die roemischen
Grundstener-Vermessungen nach dem
lateinischen Texte des gromstischen
Codex, insbesondere des Hyginus,
Frontinus und Nipsus...*, Munich
(F9, 25)

Stahlin O 1932 Bay Bl GR  68, 13-15
[numer]

Stone E N 1928 Wash UM 14, 215-242
[surveying]

Studnicka F J 1875 Cas MF 4
(F8, 12) [frac]

Terquem A 1885 *La science Romaine
a l'époque d'Auguste.  Etude
historique d'après Vitruve*, Paris

Thulin C 1913 *Corpus Agrimensorum
Romanorum*, Leipzig

## Sumerian

See also Babylonian  (above).

Belaiew N T 1927 Newcomen 8, 120-153
(F59, 834) [metrol]

- 1928 Beo Ark 2, 187-222  [metrol]

Bork F 1937 Arc Or 11, 369-372
(F63, 4)

Bortolotti E 1940 Bln Mm (9)7, 77-97
(MR9, 483) [Babyl]

Frenkian A M 1953 Buc U Pol (3),
9-18  (Z53, 195)

- 1957 Ro SSMP (NS)1(49), 17-32
(Z79, 4)

- 1957 Ro SSMP (NS)1(49), 281-294
(MR21, 484)

- 1958 Ro SSMP (NS)2(50), 5-18
(MR21, 893)

Sumerian (continued)

Goetze A 1951 Sumer 7, 126-155
Langdon S 1931 Lon Asia, 593-596
  [numer]
Thureau-Dangin F 1931 Rv Assyr 28,
  9-12 [(F57, 1289) [frac]


TIME PERIODS


Prehistoric mathematics

A C H 1897 Nat 55, 229-230
  (F28, 38) [neolithic]
Bortolotti E 1941 It UM (2)3, 395-
  401 (Z25, 145) [prog]
Brun Viggo 1963 *Alt er tall...*,
  Oslo, 239 p
Cordrey W A 1939 MT 32, 51-60
Fettweis E 1931 Scientia 49, 423-436
  (F57, 4; Z2, 5) [ethnology]
- 1937 Anthropo 32, 277-283
- 1956 Scientia (Jan), 1-15
  [megalithic]
Hawkins Gerald S 1965 *Stonehenge
  Decoded,* NY, 202 p [bib]
Kalmus H 1964 Nat 202, 1156-1160
  [animals]
Kluge T 1938 *Die Zahlbegriffe der
  Australier, Papua und Bantuneger
  nebst einer Einleitung ueber
  Zahl...,* Berlin-Steglitz, 305 p
  (F64, 902)
- 1940 *Die Zahlenbegriffe der
  Voekler Americas, Nordeurasiens,
  der palaioafricaner...,* Berlin-
  Steglitz, 736 p (F66, 4)
Lietzmann W 1934 Isis 20, 436-439
  (Z8, 98) [geom]
- 1934 ZMN Unt 65, 313-319
  [geom]
Lloyd D B 1965 MT 58, 720-723
Loefler E 1912 *Ziffern und
  Ziffernsysteme der Kulturvoelker
  in alter und neuer Zeit,*
  Leipzig, 97 p
Lorey W 1940 Scientia 67, 68-75
  (Z22, 294) [numer]
Ludendorff H 1940 Ber Ab, 1-60

(F66, 5; Z25, 145) [ast]
Maistrov L E 1957 Ist M Isl 10,
  595-616 [Russia]
Struik D J 1948 Sci Am 179, 44-49
  [stone age]
- 1964 MT 57, 166-168
  [Kensington Stone]
Thom A 1961 M Gaz 45, 83-92
  [megalithic geom]
Vetter Q 1926 *Jak se pocitalo c
  merilo na usvite Kultury,*
  Prague, 144 p
- 1955 Leip MNR 5, 131-132
  (MR18, 453) [bronze age, numer]
Vogel K 1939 Muns Semb 13, 105-128
  (F65, 3)
- 1939 Fors Fort 15, 95-97
  (F65, 1077)


Ancient mathematics

See also OLD CIVILIZATIONS.

Aaboe Asger 1964 *Episodes from the
  Early History of Mathematics,*
  NY, 133 p (MR28, 576)
Adamo M 1953 Cagliari 22 (Z52, 1)
  [geom]
Almeida e Vasconcellos F 1925
  *Historia des matematicas
  na antiquidade,* Paris-Lisbon,
  677 p (F52, 37)
Amodeo F 1907 Batt 45, 73-81
  (F38, 5)
Anon 1918 AMM 25, 453 [oldest
  math work]
- 1939 AMM 46, 234 [club topic]
Archibald R C 1930 Sci (NS)71,
  109-121, 342; 72, 39 (F56, 60)
  [pre-Greek]
Arton E 1925 Pe M (4)5, 255-264
  (F51, 11) [metrol]
Autonne L 1894 Rv GSPA 5, 561-563
Bashmakova I G 1956 Sov Cong 1, 228-
  229
- 1961 Ist M Isl 14, 473-490
  [I-IV]
- 1963 Ist M Isl 15, 37-50
Becker Oskar 1957 *Das Mathematische
  Denken der Antike,* Goettingen,

Ancient mathematics (continued)

128 p (MR19, 107; Z79, 241)
(2nd ed 1966)
- 1957 *Zwei Untersuchungen zur antiken Logik,* Wiesbaden, 55 p
Bell E T 1944 Kan Cit UR 11, 35-41
Betz W 1922 MT 15, 283-293
Bobynin V V 1889 Bib M (2)3, 104-108
- 1896 *Ocherki istorii donauchnogo perioda razvitiya arifmetiki,* Mos, 48 p
Boschenski J M 1951 *Ancient Formal Logic,* Amsterdam
Bortolotti E 1936 Osiris I, 184-230 [alg]
- 1938 Bln Mm (9)5, 147-159 (F64, 904) [infinitesimal, infinity]
Brambilla A 1896 *Saggio di storia della ragioneria press i populi antichi,* Milan, 50 p
Bretschneider C A 1870 *Die Geometrie und die Geometer vor Euklides,* Leipzig (F3, 1)
Bruce R E 1937 Scr 5, 117-121, 181-185, 245-250 [Sicily]
Bruins E M 1949 Amst P 52, 161-163 (MR10, 419) [calcn]
- 1953 *Fontes Mathesos. Grundzuege de vorgriechischen und griechischen mathematishes Denkens,* Leyden, 180 p (Z52, 244)
- 1959 Euc Gron 34, 131-159 [alg]
Brunet Pierre + 1935 *Historie des sciences. Antiquité,* Paris, 1224 p
Bubnov N M 1911 *Podlinnoe sochinenie Gerberta ob abake ili sistema elementarnoi arifmetiki klassicheskoi drevnosti,* Kiev, 508 p
- 1916 *Zabytaya arifmetika klassicheskoi drevnosti. Drevnii abakkolybel sovremennoi arifmetiki,* Kiev, 132 p [abacus]
Bunt L N H 1954 *Van Ahmes tot Euclikes,* Groningen-Djakarta, 178 p
Cantor M 1863 *Mathematische Beitraege Zum Culturleben der Voelker,* Halle
Carruccio E 1949 Archim 1, 177-180 (Z33, 337)

Dagobert E B 1955 MT 48, 557-559
De Hairs E 1929 Euc Gron 6, 88-98
Dehn M 1943-1944 AMM 50, 357-360, 411-414; 51, 25-31, 149 (MR5, 57, 253) [-VI to VI]
Dijksterhuis E J 1935 Gids 99(2), 209-230, 337-349; 99(3), 41-58 (F61, 942) [pre Hellenic]
Elachich V 1912 *Kak schitali lyudi v drevnie vremena,* Pet, 47 p
Enriques F + 1932 *Storia del pensiero scientifico. Vol. I. Il mondo antico.* Bologna-Milan, 682 p (F58, 2)
Fafara J 1884 *Historyczny zarys matematiki u starozytnych,* Tarnopol, 52 p
Favaro A 1875 *Saggio di cronografica dei matematici dell'antichità (A. 600 a.c.- A. 400 d.c.),* Padua (F7, 1)
Fazzari G 1899 Pitagora 3, 76, 98; 4(1), 4, 59, 78; 4(2), 28, 72
- 1905 Pitagora 11, 130 [alg, arith] 1905 *Breve storia della matematica dai tempi antichi al medio-evo,* Milan, 268 p
- 1923 *Kratkaya istoriya matematiki s drevneishikh vremen, konchaya srednimi vekami,* Mos, 214 p (tr from Italian)
Fettweis Ewald 1923 *Wie man einstens rechnete,* Leipzig, 56 p
- 1954 Z Ethn 79, 182-192 [geom]
Flauti V 1852 Nap Barbo 1, 17- [geom]
Frajese Attilio 1951 *La matematica nel mondo antico,* Rome, 160 p (Z45, 289)
Frenkian A M 1953 Buc U Pol 2(3), 9-18 (Z53, 195)
- 1957 Ro SSMP (NS)1, 17-32 (Z79, 4)
- 1957 Ro SSMP (NS)1, 281-294 (MR21, 484; Z109, 2) [Greek]
- 1958 Ro SSMP (NS)2, 5-18 (MR21, 893; Z89, 3)
Friedlein G 1869 *Die Zahzeichen und das elementare Rechnen der Griechen und Roemer und des*

Ancient mathematics (continued)

*christlichen Abenlandes vom 7 bis 13 Jahrhundert,* Erlangen, 169 p

Gandz S 1937 Scientia 62, 249-257 (F63, 798) [quad eq]

- 1947 in *Studies and Essays in the History of Science and Learning Offered in Homage to George Sarton on the Occasion of his Sixtieth Birthday,* NY, 449-462 (MR8, 497)

Gercke A + 1933 *Einleitung in die Altertumswissenschaft* (Teubner), 78 p (F59, 16)

Gerstinger H 1933 Fors Fort 9, 142-143 (F59, 16) [ms]

Ghirshman R 1934 Rv Assyr 31, 115-119 (F60, 819) [anc Iran]

Guenther S 1888 *Geschichte der antiken Naturwissenschaft* (Handbuch der Klassischen Altertums-wissenschaft in Systematischer Darstellung, I Mueller ed) (F21, 1255)

- 1893 *Abriss der Geschichte der Mathematik und der Naturwissen-schaften im Alterthum,* Munich (F25, 1913)

Hagstroem K-G 1932 *Les préludes antiques de la théorie des probabilités,* Stockholm, 54 p (F58, 986)

Hankel H 1874 *Zur Geschichte der Mathematik im Alterthum und Mittelalfer,* Leipzig, 414 p (F6, 1)

Heiberg J L 1929 Dan H 15(1) (F55, 8)

Hogben 1935 Antiq (June), 190-194

Hoppe E 1911 Unt M 17, 106-112 (F42, 51)

- 1911 *Mathematik und Astronomie im Klassischen Altertum,* Heidelberg, 454 p (F42, 48)

Karpinski L C 1929 Sci 70 622-627 (F59, 825) [recent research]

Kolman E Ya 1956 Sov Cong 1, 232 [open questions]

- 1958 Ist M Isl 11, 159-170 (MR23, 582; Z100, 242)

*- 1961 *Istoriya matematiki v

*drevnosti,* Mos, 235 p [Part 1 of Kolman E + Yushkevich A P, Mat. do epokhi vozrozhdeniya] (MR24, 570; Z114, 241)

Lefebvre B 1920 *Notes d'histoire des mathématiques,* Louvain, 162 p

Lietzmann W 1928 *Aus der Mathe-matik der Alten. Quellen zur Arithmetik, Planimetrie, Stereo-metrie und zu ihren Anwendungen,* Leipzig, 74 p (F54, 11)

Littrow C v 1870 *Ueber das Zueruck-bleiben der Alten in den Natur-wissenschaften,* Vienna (F2, 3)

Lorenzen P 1960 *Die Entstehung der exacten Wissenschaften,* Berlin

Loria Gino 1894 Batt 32, 28-57 [arith]

Lueders L 1811 *Geschichte der Mathematik bei den alten Voelkern,* Altenburg-Leipzig

Lure S Ya 1932 QSGM (B)2(2), 106-185 [infinitesimals]

- 1934 Len IINT 2, 297-303 [recent res]

- 1935 *Teoriya beskonechno malykh u drevnikh atomistov* (Len IINT 2(5)), Mos-Len, 199 p (Z12, 243)

Marcolongo R 1932 It SPS 20(1), 244-257 (F58, 7)

- 1932 Scientia 51, 21-34; (S), 19-32

Martin T H 1864 *Les signes numeraux et l'arithmétique ches le peuples d l'antiquite,* Rome

Martino A S 1842 *Sopra un'antica misura del centipondio: memoria storico fisico-geometrica,* Catania

McGee W J 1888 Am Anthro 1, 646

Menge R + 1881 *Antike Rechenaufgaben* Leipzig (F13, 45)

Meyer W F 1931 Ham MG 7, 1-19 (F57, 13) [anc vs mod]

Milhaud G 1911 Rv M Sp 21, 209-212 (F42, 51)

Miller G A 1931 Sci Mo 33, 419-423 (F57, 11)

- 1931 AMM 38, 496-500 (F57, 11; Z3, 98)

- 1932 Sch Soc 35, 833-834

- 1933 Sci (NS)77, 366

- 1938 Sci 87, 576-577 (F64, 8) [-2000 to -500]

Ancient mathematics (continued)

Neuburger A 1930 *The Technical Arts and Sciences of the Ancients,* London

Neugebauer Otto 1929 Ham Sem 7, 107-124 (F55, 7) [pre-Greek]
- 1931 Erkennt 2, 122-134 (F57, 10; Z2, 243)
- 1932 QSGM (B)2, 1-27 (F58, 9) [alg]
- 1933 Fors Fort 9, 503-504 (F59, 13)
- 1934 *Vorlesungen ueber Geschichte der antiken mathematischen Wissenschaften,* Vol. I, Berlin, 212 p (Z10, 97) (no vol 2; repr 1969; Russ tr 1937)
- 1938 Act Ori 17, 169
- 1948 AMSB 54, 1013-1041 (MR10, 419; Z33, 341) [ast]
  1952 *The Exact Sciences in Antiquity,* (Princeton U P), 207 p (MR13, 809) (2nd ed 1957; repr 1962)
- 1956 Scr 22, 165-192 (MR19, 722) [ast, mid]

Pauly-Wissowa 1894-1939 *Pauly's Real Encyclopadie der classischen Altertumswissenschaft,* Stuttgart

Pokorny E J 1959 *Niedyskrecje naukowe. Babilon, Egipt, Hellada,* Warsaw

Prell H 1958 Dres THWZ 7, 133-143 (Z87, 241)

Ramus Petrus 1569 *Scholae mathematicae,* Paris, vols 1-3 (3rd ed 1627 Frankfurt)

Rehm A + 1933 *Einleitung in die Altertumswissenschaft,* (Exakte Wiss. Bd. 2. H.5), Leipzig-Berlin, 78 p (Z6, 2)

Reymond Arnold 1924 *Histoire des sciences exact et naturelles dans l'antiquité gréco-romaine,* Paris, 238 p

Richardson L J 1916 AMM 23, 7-13 [arith]

Rome A 1930 Brx SS 40, 97-104 [combin]
- 1931 Rv Q Sc (4)20, 279-305 (F57, 1286)

Saalschuetz L 1892 Konig Ph 5, 4-9 (F24, 43) [numer]

*Sarton George 1927 Int His Sc 1
- 1936 Isis 24, 375-381 [Crete]
- 1952 *A History of Science. Ancient Science Through the Golden Age of Greece* (Wiley), 663 p (repr 1964)
- 1954 *Ancient Science and Modern Civilization* (U Nebraska P)

Seidenberg A 1962 Arc HES 2, 1-40 [arith]
- 1962 Arc HES 1, 488-527 [geom]

Simon M 1909 *Geschichte der Mathematik in Altertum in Verbindung mit antiker Kulturgeschichte,* Berlin, 418 p (F40, 1)

Slaught H E 1934 AMM 41, 167-174 (F60, 1; Z9, 97)

Smith D E 1936-1937 Scr 4, 111-125; 5, 1-17 [alg]
- 1938 AMM 45, 511-515

Steinschneider M 1867 *Intorno ad otto manoscritti arabi di mat. possedute dal Ch. sig. g. Libri,* Rome, 23 p

Sundwall J 1932 Hels Hum 4(4), 10 p (F58, 11)

Szabo A 1958 St It FC 30(1), 1-51 (MR21, 484; Z82, 7) [found]
- 1959 Ist M Isl 12, 321-392 (Z109, 3)

Tannery Paul 1885 BSM (2)9, 104-120
- 1893 *Recherches sur l'histoire de l'astronomie ancienne,* Paris
- 1912-1920 *Mémoires Scientifiques,* Paris, Vols 1, 2, 4

Taton R ed 1957 *La Science Antique et Mediévale,* Paris
- 1963 *Ancient and Medieval Science,* NY (tr of 1957)

Toeplitz O 1925 Antike 1, 175-203 (F51, 4)

Tseiten G G 1932 *Istoriya matematiki v drevnosti i v srednie veka,* Moscow-Leningrad

Van der Waerden B L 1950 *Ontwakende Wetenschap,* Groningen (Russ tr 1959)
*- 1954 *Science Awakening,* Groningen, 306 p (MR16, 1)
- 1955 Ens M 1, 44-55 [appl]

Ancient mathematics   (continued)

- 1957 Bel SM 9, 8-20
  (Z89, 2) [proof]
Vekerdi L 1963 Magy MF 13, 133-150
  (MR29, 638) [pre-Euclid]
Ver Eecke P 1937 Mathesis 11-14
  [locus]
Vogel Kurt 1930 Clas Phil 25, 373-
  375 [Mich pap 620, alg]
- 1931 ZMN Unt 62, 289-295
  (F57, 10) [pre-Greek]
- 1959 Vorgriechische Mathematik,
  Hannover, 2 vols  (Z80, 5; 86, 2)
Voronets A M + 1928 O merakh i
  schete drevnosti, Mos-Len, 3 vols
Vorwahl H 1930 Allq Verm 42, 113-
  117 (F56, 9) [metrol]
Vygodskii M Ya 1941 Arifmetiki i
  algebra v drevnem mire, Mos-Len,
  252 p (2nd ed 1967, 367, p)
  (MR37, 1215)
*Wussing Hans 1962 Mathematik in der
  Antike: Mathematik in der Periode
  der Sklavenhaltergesellschaft
  (Teubner), 305 p (MR25, 389;
  Z101, 5)
*- 1965 Mathematik in der Antike,
  2nd ed, Leipzig, 245 p
  Zeuthen H G 1893 Forelaesning over
  Mathematikens Historie. Oldtid og
  Middelalder, Copenhagen, 282 p
  (rev ed 1949; MR11, 149)
  1896 Geschichte der Mathematik im
  Altertum und Mittelalter.
  Vorlesungen, Copenhagen, 351 p
  (F26, 2) [tr of 1893] (Fr tr
  1902; Russ tr 1937)
- 1912 Die Mathematik im Altertum
  und im Mittelalter, Leipzig

Middle Ages  VI? - XIV?

See also particular centuries below,
  COUNTRIES and OLD CIVILIZATIONS.

Arrighi G 1956 Int Con HS 8, 103-
  105 [mss]
- 1956 Int Con HS 8, 106-108
  [Lucca]

Artz F B 1954 The Mind of the
  Middle Ages, AD 200-1500. An
  Historical Survey, NY [*bib]
Autonne L 1894 Rv GSPA 5, 561-563
Balić K 1936 Wiss Weis 3, 191-217
  (F62, 1042) [proof]
Baron R 1957 Isis 48, 30-32
  (Z79, 241) [geom, instrum]
Beaujouan Guy 1947 Recherches sur
  l'histoire de l'arithmétique au
  Moyen Age, Paris, 22 p
- 1948 Rv Hi Sc Ap 1, 301-313
  (Z31, 2) [XI-XII]
Björnbo A A 1909 Arc GNT 1, 385-394
  (F41, 2)
- 1912 Bib M (3)12, 97-132, 193-224
  [Florence mss]
Bobynin V V 1888-1889 Fiz M Nauk 7; 8
Boehner P 1952 Medieval Logic, An
  Outline of its Development from
  1250 to ca. 1400, Chicago
Bosmans H P R 1926 Brx SS 44(1), 458-
  462 (F52, 13) [Louvain, mss]
Bruins E M 1958 Euc Gron 34, 131-159
  (MR22, 769) [alg]
Cantor M 1865 ZM Ph 10, 1- [mss]
Clagett Marshall 1952 Isis 43, 36-38
  (Z47, 1) [Archimedes]
- 1952 Osiris 10, 587-618
  [Archimedes]
- 1953 Isis 44, 16-42
  [Adelard, Euclid]
- 1953 Isis 44, 371-381  (MR15, 383)
  [bib, mss]
- 1954 Osiris 11, 359-385
  (MR16, 659) [Arab, hyperb]
- 1955 in Essays in Medieval Life
  and Thought (Columbia U P),
  99-108
- 1959 Isis 50, 419-429
  (Z94, 2) [Archimedes]
  1959 The Science of Mechanics in
  the Middle Ages (U Wisconsin P,
  Oxford U P), 740 p  (MR22, 1852)
Clagett M + 1958-1959 Manuscr 2, 131-
  154; 3, 19-37 [bib, mss]
Cowley E B 1923 in Vassar Mediaeval
  Studies (Yale U P), 379-405
  [Italy, mss]
Crombie A C 1953 Medieval and Early
  Modern Science: Vol 1: Science in

Middle Ages   (continued)

   the *Middle Ages: V-XIII Centuries*
   (Harvard U P) (repr 1959 Anchor)
- 1961 *Augustine to Galileo,* London,
   2 vols
Curtze M 1895 Bib M (2)9, 37-42,
   77-88, 105-114
- 1895 Arc MP (2)13, 388-406
   (F26, 52) [Codex latinus
   Monacensis 14908]
- 1895 ZM Ph 40(S), 75-142
   (F26, 61) [Munich ms 14836]
- 1896 Bib M, 1 [Euclid]
- 1900 Bib M (3)1, 51 [phys]
- 1902 *Urkunden zur Geschichte der
   Mathematik in Mittelalter und der
   Renaissance,* Ab GM 12/13, 1-628
   (F33, 10)
Daly J 1964 Manuscr 8, 3-17 [mss]
- 1965 Manuscr 9, 12-29 [mss]
Drobisch M W 1852 Leip Ab 1, 431-
   [Florence]
Eneström G 1891 Bib M (2)5, 32 (Q33)
   (F23, 4) [astrolabe, mss]
- 1902 Bib M (3)3, 243
   (F33, 54) [series]
- 1906 Bib M (3)7, 252-262
   (F37, 2) [edu]
- 1907 Bib M (3)7, 252-262 [calc]
- 1908 Bib M (3)9, 265-266
   (F39, 58) [residues]
- 1910 Bib M (3)11, 90
   (F41, 65) [geom]
- 1914 Bib M 14, 83-84 [cube roots]
Feddegon J M 1932 Archeion 14, 372-
   391 (F58, 15) [Asia]
Funkhouser H G 1926 Osiris 1, 260-262
   [anal geom]
Friedlein G 1869 *Die Zahlzeichen und
   das elementare Rechnen der griechen,
   Roemer und des christlichen
   Abendlandes von 7 bis 13
   Jahrhundert,* Erlangen (F2, 2)
Goldstein B R 1964 Centau 10, 129
   [numb thy]
Guilhiermoz P 1919 Par EC Bib 80,
   5-100 [metrol]
Guenther S 1880 Leop 16 (F12, 4)
- 1887 *Geschichte des mathematischen
   Unterrichts im deutschen Mittel-*

*alter bis zum Jahre 1525,* Berlin
   423 p [edu, Germany]
Hankel Hermann 1874 *Zur Geschichte
   der Mathematik in Alterthum und
   Mittelalter,* Leipzig
Haskins C H 1924 *Studies in the
   History of Mediaeval Science,*
   Cambridge, Mass (2nd ed 1927)
Heiberg J L 1890 Hist Abt 35, 41-58,
   81-100
Hopper Vincent F 1938 *Medieval
   Number Symbolism,* NY, 253 p
d'Irsay S 1933 Arc Sto 15, 216-231
   [universities]
Junge G 1935 ZMN Unt 60, 117-119
   (F61, 15) [probls]
Kennedy E S + 1956 AMM 63, 80-83
   (Z72, 244) [approx]
- 1962 MT 55, 286-290
   (MR26, 449) [ast]
Lefebvre B 1920 *Notes d'histoire des
   mathématiques,* Louvain, 162 p
Loria G 1929 *Storia delle matematiche
   1, Antichità-medio rinascimento*
   Turin, 487 p (F55, 1)
Mamedov K M 1960 Az FMT (6), 31-36
   (MR24, 127; Z81, 243) [Asia]
Martin Henri 1864 *Les signes numeraux
   et l'arithmétique chez les peuples
   de l'antiquité et du Moyen Age,*
   Rome
Matthiessen L 1870 ZM Ph 15, 41-47
   (F2, 25) [false position]
Millás Vallicrosa José Mariá 1931
   *Assaig d'historia de les idees
   fisiques i matemàtiques a la
   Calalunya Medieval,* Barcelona,
   364 p
- 1953 Int Con HS 7, 451-454
   [Peter IV, ast tab]
Moody E A + 1952 *The Medieval
   Science of Weights,* Madison, 448 p
Muggli J 1953 Am Ben Rv 4, 34-46
   [Benedictines]
Murdoch J 1964 Koyre 1, 416-441
   [cong, cont, superposition]
Natucci A 1931 Pe M (4)11, 69-83
   (F57, 31) [circle, sph]
Ogura K 1938 Kaizo 20, 107-121
   (F64, 920) [Japan]
Rome A 1931 Rv Q Sc (4)20, 279-305
   (F57, 1286)

## Middle Ages (continued)

Rozenfeld B A + 1958 Sov Vost (3),
101-108; (6), 66-76 [Asia]
Sarton G 1938 Osiris 5, 41-245
[incunabula]
*Schrader D V 1967 MT 60, 264 [arith]
Schramm M 1965 Hist Sci 4, 70-103
[function]
Schreider S N 1959 Ist M Isl 12,
558-688 [alg, Jordanus Nemorarius]
Semyonov L 1949 Arm Aff 1, 80-81
[Armenia VII-XIII]
Sergescu P 1937 in Conferences de la
Reunion internationale des
Mathematiciens, Paris, July 1937,
Bris (F64, 909)
- 1937 Sphinx 7, 183-184 (F63, 13)
[French]
- 1939 Flambeen 22, 405-416
Severi F 1951 Archim 3, 45-55
(MR13, 1) [IX]
Singh A N 1936 Archion 18, 43-62
(Z14, 50) [Hindu to XII]
Sister Marie Stephen 1962 MT 55,
291-295 [Albert the Great]
Sokolow N P 1894 Kiev UFMO (F26, 49)
[arith, scholatics]
Steinschneider M 1866 ZM Ph 11, 235-
Studnicka F J 1903 Prag Si (6)
(F25, 59)
Sullivan J W N 1925 The History of
Mathematics in Europe from the
Fall of Greek Science to the
Rise of the Conception of Mathe-
matical Rigor, Oxford, 109 p
Suter H 1887 Die Mathematik auf die
Universitaeten des Mittelalters,
Zurich, 60 p (F21, 1256)
- 1905 Bib M (3)6, 112 (F36, 52)
["regula coeci"]
Taton R ed 1957 La Science Antique
et Médiévale, Paris
- 1963 Ancient and Medieval Science
From the Beginnings to 1450, NY
Thorndike L 1927 Speculum 2, 147-159
- 1956 Isis 47, 391-404 (Z72, 245)
[Vatican mss]
- 1958 Isis 49, 34-49 (Z80, 5)
[Vatican mss]
- 1959 Isis 50, 33-50
(MR20, 1041) [Italy, mss]

Toth I 1955 Ro Sov (3)9(3), 74-83
(MR17, 337) [Asia]
Tseiten G G 1932 Istorii Matemati-
cheskii v drevnosti i v srednie
veka, Moscow-Leningrad
Vashchenko-Zakharchenko 1883 Bord Mm
(2)5, 259-291 (F15, 2)
Veratti B 1859 De matematici
anteriori all'invenzione della
stampa, Modena
Vetter Quido 1935 Slav Cong 2, 263-
264 (F61, 2) [Czech, Jesuits]
Vogel K 1954 MP Semb 4, 122-130
[alg]
Wappler H E 1899 ZM Ph 45, 7
(Dresdensis C80, Lipsiensis 1470)
[Dresden, Leipzig, mss]
Weissenborn H 1888 Gerbert. Beitraeg
zur Kenntniss der Mathematik des
Mittelalters, Berlin, 257 p
Wieleitner H 1914 Bib M (3)14, 150-
168 [series]
Yeldom Florence A 1926 The Story of
Reckoning in the Middle Ages,
London
- 1927 Archion 8, 313-317
(F53, 12) [frac]
*Yushkevich A P 1961 Istoriya mate-
matiki v srednie veka, Mos, 448 p
(MR28, 220)
*- 1964 Geschichte der Mathematik im
Mittelalter, Leipzig, 454 p
(tr and rev of 1961)
Yushkevich A P + 1960 in Sowjetische
Beitraege zur Geschichte der
Naturwissenschaft, Berlin, 62-168
(MR28, 221) [Asia, bib]
- 1960 Vost IINT 1, 349-421
(Z112, 241) [Asia, bib]
Zeuthen H G 1893 Forelaesning over
Mathematikens Historie, Oldtid og
Middelalder, Copenhagen, 282 p
(rev ed 1949; MR11, 149)
- 1896 Geschichte der Mathematik im
Altertum und Mittelalter,
Copenhagen, 351 p (Fr tr 1902;
Russ tr 1937) [tr of 1893]
- 1912 Die Mathematik in Altertum und
im Mittelalter, Leipzig, 103 p
(Die Kultur der Gegenwart 3(1))
Zubov V P 1959 Isis 50, 130-134
(Z87, 241)

Middle Ages   (continued)

Zubov V P + 1963 Ist M Isl 15, 51-
72


Renaissance   XIII? - XVI?

Alter G 1958 Prag Rozp 68(11), 77 p
(MR21, 484)  [ast, Delmedigo,
Gans]
Amodeo F 1908 Batt 46, 91-108
(F39, 4)
Beaujouan G 1956 Int Con HS 8, 84-
88  [arith, France]
Curtze Max 1902 Ab GM 12/13, 1-628
[sources, mid and ren]
Perguson Wallace K 1948 *The
Renaissance in Historical Thought*,
Boston, 442 p  [alg]
Loria Gino 1929 *Storia delle mate-
matiche 1*,  Turin, 487 p
(F55, 1)
- 1930 Pe M (4)10, 152-154
(F56, 14) [alg]
Rudio F 1892 *Ueber den Antheil der
Mathematischen Wissenschaften an
der Kultur der Renaissance*,
Hamburg, 33 p (Sammlung gemeinver-
staendlicher wissenschaftlicher
Vortraege  (NF)(6)142)  (F24, 7)
Stephanides M 1933 Archieon 14,
492-496  [Byzantine]  (F59, 22)
Wightman W P D 1962 *Science in the
Renaissance*  (Hafner), 2 vols


Centuries:  XI

Bubnov N M 1913 *Abak i Boetsii.
Lotaringskii nauchnyi podlog XI
veka*, Pet, 353 p
Hofmann J E 1942 Ber Ab (8)
(F68, 8)
- 1942 Leip Ab, 1-19
(MR8, 189; Z27, 194) [XI, angle]
Lattin H P 1948 Isis 28, 205-225
[Munich ms 14436]

Centuries:  XII

Cipolla M 1935 Esercit (2)8, 1-13
(F61, 948)  [Italy]
Curtze M 1897 Mo M Phy 8, 193
[geom]
- 1898 Ab GM 8, 1-27  (F29, 28)
[arith]
Eneström G + 1904 Bib M (3)5,
312-316
Karpinski L C 1921 Isis 3, 396-413
[arith]
Millas-Vallicrosa J M 1960 Int Con HS
9, 33-65  (Z114, 5)  [tr]
Steinschneider M 1867 ZM Ph 12, 1-44
Thomas I 1958 JSL 23, 133-134
(MR21, 484)  [inf]
Van Wijk W E 1951 Int Con HS 6, 133-
139  (MR17, 1)  [calcn]
Vetter Quido 1956 *Co mohli stavitelé
v cechách na konci XII, XIII
stoleti znat o geometrii*,  Prague,
8 p  [architects and geom]


Centuries:  XIII

Beaujouan G 1954 in *Homenaje a
Millas Villacrosa 1*, 93-124
(MR16, 781) [U Paris arith XIII-
XIV]
Boffito G 1937 Fir Querc 36, 23 p
[folk math]
Fontes M 1897 Tou Mm (9)9, 382
[John of London, Maharn-Curia]
Loria G 1921 Arc Sto 3(1), 32-33
- 1922 Archeion 3, 32-33
(F48, 3)
Tannery P ed 1897 Anglès R, *Traité
du quadrant (Montpellier, 13e
siècle*,  Paris, 84 p  (F28, 54)
Taylor E G R 1960 Navig 13, 1-12
[navigation]
Ten Doesschate G 1948 *Rolduc als
middeleeuwsche voorpost der wis-,
natuur- en geneeskund in de
Nederlanden*,  Lochem, 154 p
(MR14, 1)  [bib XIII]
Thorndike L 1957 Scr 23, 61-76
(MR20, 945; Z80, 6)
[mss XIII-XIV]

## Centuries: XIII (continued)

Waters E G R 1928 Isis 11, 45-84
[arith]


## Centuries: XIV

Curtze M 1894 Bib M (2)8, 107-115
(F25, 78) [geom]
- 1895 Bib M, 77 [arith recr]
- 1896 Bib M, 43 [phys]
Eneström G 1898 Bib M (2)12,
19-22 (F29, 33) [geom, Norway]
Ginsburg J 1932 Scr 1, 60-62
(Z5, 242) [unknown mathn]
Karpinski L C 1929 Archeion 11, 170-
177 [arith, Jacob of Florence]
Karpinski L C + 1935 Isis 23, 121-
152 [arith]
Kinkelin H 1863 Basl V 3, 511-
Michalski K 1920 Krak BI, 59-88
[Oxford, Paris]
Narducci H 1883 BSM (2)7, 247-256
[Vatical ms, numer]
Procissi A 1954 It UM (3)9, 300-
326, 420-451 (Z57, 3)
[alg, Canacci]
Sanchez-Perez J A 1956 Gac M 8, 12-20
(Z71, 243)
Sarton G 1947 Int His Sc 3
(MR9, 484)
Sergescu Pierre 1947 Le développe-
ment de l'idée de l'infini mathé-
matique au XIV e siècle
(Univ de Paris), 15 p
Smith D E 1922 AMM 29, 62-63 [tab]
Vetter Q 1928 Archeion 9, 175-176
[eqs]
Wilson Curtis 1956 William
Heytesbury (U Wisc P), 219 p
Wittstein A 1895 ZM Ph 40, 121-125
[Jean de Linières, mss]


## Centuries: XV

Agostini A 1951 It UM (3)6, 231-240
(MR13, 420; Z43, 243) [arith]
Beaujouan G 1956 Int Con HS 8, 84-88
[arith, France, XIV-XV]

Berchet G 1880 Ven I At (5)6, 639-642
(F12, 7) [G. Leardo]
Bosmans H 1925 Brx SS, 458-462
[Louvain, mss]
Curtze M 1894 Arc MP (2)13, 388-406
[Monacensis 14908, China]
- 1894 ZM Ph 40(S), 31-74
[alg, Germany]
- 1895 Bib M (2)9, 1-8 (F26, 63)
- 1895 Bib M (2)9, 37-42, 77-88, 110-
114 [numb thy]
- 1898 Ab GM 8, 29-68 (F29, 34)
[vol]
Eneström G 1907 Bib M (3)8, 96-97
(F38, 8) [Italy]
- 1910 Bib M (3)11, 353-354
(F41, 56) ["radix relate"]
Hunger H + 1963 Ein byzantinischer
Rechenbuch des 15, Jahrhunderts,
Graz-Vienna-Cologne, 127 p
(MR27, 1078)
Jordan W 1898 Opus Palatinum.
Sinus- und Cosinus-Tafeln...,
Hannover-Leipzig
Kennedy E S 1949 Scr 17, 91-97
[ast, instru]
Lorey W 1933 Unt M 39, 120-122
[Erfurt U]
Rath E 1912 Bib M (3)13, 17-22
(F43, 67) [arith, Germany]
- 1914 Bib M 14, 244-248
[arith, Germany]
Sleight E R 1943 MT 35, 112-116
[arith, England]
Smith David E 1928 Le comput manuel
de Magister Anianus (Documents
scient. du XVe siècle, 4)
Paris, 107 p
- 1932 Am Bib SP, 143-171 [infl]
Studnicka F J 1892 Prag V, 100-104
[Kristans v Prachatic]
- 1893 Praf V, 14-
[Kristans v Prachatic]
Szily v K + 1895 MNB Ung 12, 134-143
[Georgius de Hungaria]
Thorndike L 1926 AMM 33, 24-28
[J. Adam, arith]
Torner J 1914 Esp SM 4, 33-36
[Andres de Li]
Vogel Kurt 1954 Die practica des
Algorismus ratisbonensis. Ein

Centuries: XV    (continued)

Rechenbuch der Benediktinerabtei
St. Enmeran aus der Mitte des 15.
Jahrhunderts..., Munich, 294 p
- 1960 Int Con HS 9, 610-613
  (Z114, 6)  [Benedictines]
Wappler E 1890 Ab GM 5, 147-169
  [Dresden mss]
           -1900 ZM Ph 45, 47-


Centuries: XVI

Bonelli M L 1959 Physis Fi 1, 127-148
  [Salernitano]
Bortolotti E 1925 Pe M (4)5, 147-192
  [alg, Bologna]
- 1933 Int Com HB, 268-283
  (Z7, 147)  [Italy]
Brandicourt V 1893 Pop Sci Mo 44,
  106-110  [curiosities]
Cantor M 1876 Boncomp 9, 183-187
  (F8, 28)  [arith, Kuckuck]
Curtze M 1895 ZM Ph 40, 161
Davis N Z 1958 Renais N 11, 3-10
  [France]
- 1960 JH Ideas 21, 18-48
  (MR21, 893)  [finance, France]
Drake S + ed 1969 Mechanics in
  Sixteenth Century Italy
  (U Wis P), 428 p  [bib]
Eneström G 1894 Bib M (2)8, 33-36
  (F25, 7)  [Spain]
- 1902 Bib M (3)3, 355-360
  [Andreas, Germany]
Fontes M 1899 Tou Bu 2, 202-
  [alg, arith]
- 1900 Tou Bu 3, 283
  [alg, arith, Scheubel]
- 1901 Tou Mm (10)1, 119-
  [alg, arith]
Giesing J 1879 Stifel's Arithmetica
  Integra. Ein Beitrag zur Geschichte
  der Arithmetik des 16 Jahrhunderts,
  Doebeln,  96 p
Grosse H 1901 Historische Rechen-
  buecher des 16 und 17 Jahrhunderts
  und die Entwicklung ihrer Grundge-
  danken bis zur Neuzeit...,
  Leipzig, 183 p

Hellman W 1895 Ueber die Anfaenge
  des mathematischen Unterrichts an
  den Erfurter evangelischen Schulen
  im 16. und 17. Jahrhundert,
  Erfurt, 16 p  (F26, 47)
Ionescu I 1934 Gaz M 39, 386-392
  (F60, 827)  [Germany, Loriti,
  Vuolph]
- 1935 Gaz M 40, 338-340  (F61, 17)
  [Germany, Lossius, Medler]
Johnson F R 1942 JH Ideas 3, 94-106
  [Hood]
Karpinski L C 1913 Bib M 13, 223-228
Loria G 1931 Storia delle matematiche.
  II: I Secoli XVI e XVII, Turin,
  595 p  (F57, 1)
Natucci A 1932 Pe M (4)12, 173-179
  [alg, edu]
Playoust C 1907 Rv GSPA 18, 550-552
  (F38, 8)  [France]
Rey A 1937 Act Sc Ind (531), 27-32
  (F63, 818)
Rey Pastor J 1926 Los Matemáticos
  Espanoles del siglo XVI,
  Madrid, 164 p  (F52, 37)
Russo F 1959 Rv Hi Sc Ap 12, 193-
  208  [alg]
Smeur A J E M 1960 The Sixteenth
  Century Arithmetics Printed in the
  Netherlands,  The Hague, 175 p
  (Z114, 6)
Smith D E 1917 An Med H 1, 125-140
  [medicine]
- 1930 A History of Modern Culture.
  The Great Renewal, 1543-1687,
  London, 683 p  (F56, 2)
Strong E W 1936 Procedures and
  Metaphysics. A Study in the
  Philosophy of Mathematical-Physical
  Science in the Sixteenth and
  Seventeenth Centuries (U Cal P)
Struik D J 1936 Isis 25, 46-56
  [Netherlands]
Suter H 1873 Geschichte der
  mathemtischen Wissenschaften.
  I Th.: Von den aeltesten Zeiten
  bis Ende des XVI Jahrhunderts,
  2nd ed, Zurich.
- 1889 Bib M (2)3, 17-22
  [competition, Leipzig]
*Taylor E G R 1954 The Mathematical

Centuries: XVI   (continued)

*Practitioners of Tudor and
Stuart England* (Cambridge,U P)
Treutlein P 1877 ZM Ph 22(S), 1-
Vorsterman van Oijen G A 1870
   Boncomp 3, 323-376  (F2, 11)
   [Holland, surveying]
Wappler E 1900 Hist Abt 45, 47-56
   (F31, 5)
Wolf A 1935 *A History of Science,
   Technology and Philosophy in the
   16th and 17th Centuries*, London,
   719 p  (F61, 2)
*Zeuthen H G 1903 *Forelaesninger
   over Mathematikens Historie. II:
   16 de og 17 de Aarhundrede*,
   (Ger tr 1903; Russ tr 1961)
*- 1903 *Geschichte der Mathematik
   im 16. und 17. Jahrhundert*,
   Leipzig, 434 p  (Ab GM 17)
   (repr 1966 Johnson)


Centuries: XVII

Allégret 1868 *Mélanges scientifiques
   et littéraires. Pascal, Viète,
   Newton, et Leibnitz. Liberté
   du calcul*, Clermont-Ferrand
   (F1, 15)
Amodeo F 1902 Nap Pont 31 (16), 32 p
   [Naples 1650-1732]
Anon 1960 *La science au seizième
   siècle*, Paris, 344 p
   (Colloque international Royaumont
   1957)  (Z101, 244)
Aubry A 1911 Porto Ac 6, 82-89
   (F42, 60)  [calc]
Ball W W R 1910 M Gaz 5, 202-205
   (F41, 51)  [edu]
Bell A E 1950 *Christian Huygens and
   the Development of Science in the
   Seventeenth Century*, London,
   220 p
Belyi Yu A + 1959 Ist M Isl 12,
   186-188 [Elizarev, geom, Russia]
Berenguer P A 1895 Prog M 5, 116
   [Spain, geom]
Bopp K 1929 *Drei Untersuchungen zur
   geschichte der mathematik*, Berlin

Bortolotti Ettore 1928 *Studi e
   ricerche sulla storia delle mate-
   matiche nei secoli XVI e XVII*,
   Bologna
- 1939 It UM 1, 351-371
   (MR1, 33) [calc, Italy]
Dainville R P F de 1956 XVII(Jan)(30),
   62-68    [edu]
De Morgan A 1862 *Contents of the
   Correspondence of Scientific Men
   of the 17th Century*  (Oxford U P)
Dickstein S 1894 Bib M (2)8, 24
   (F25, 7)
Dugas René 1953 Rv Sc 91, 83-89
   (MR16, 207) [ballistics]
- 1954 *Le mécanique au XVII*e *siècle*,
   Paris, 620 p  (Z57, 242)
   [Engl tr 1958 Neuchatel]
Eneström G 1892 Bib M (2)6, 32, 64,
   96, 120 [Parseval]  (F24, 5)
- 1896 Sv Ofv 53, 41
   [finance, Halley, Witt]
- 1911 Bib M (3)12, 135-148
   (F42, 59) [series]
Fueter E 1938 Fors Fort 14, 381-382
   (Z19, 388)
Fleckenstein J O 1948 Arc In HS 2,
   76-138 [Varignon]
*Gelcich E 1882 Ab GM 4, 191-231
   [anal geom, Ghetaldi]
Hallema A 1925 Euc Gron 1, 122-128,
   161-193 (F51, 7) [arith]
Hofmann J E 1950 MP Semb 1, 220-255
   (Z41, 340) [calc]
- 1957 *Geschichte der Mathematik.
   Zweiter Teil. Von Fermat und
   Descartes bis zur Erfindung des
   Calculus und bis zum Ausban der
   neuen Methoden*, Berlin, 109 p
   (MR19, 518)
- 1959 *Classical Mathematics:
   A Concise History of the Classical
   Era of Mathematics*, NY, 159 p
   (MR22, 1341)
Ionescu I 1933 Gaz M 39, 44-47
   (F59, 843) [Italy, edu]
- 1934 Gaz M 39, 163-167, 201-204
   (F60, 831) [Germany]
- 1935 Gaz M 40, 193-196
   [Germany, Trew]
- 1937 Gaz M 43, 173-176  (F63, 807)
   [England, John Karsey]

## Centuries: XVII  (continued)

Kokomoor F W 1928 Isis 10, 367-415
[geom]
- 1928 Isis 11, 85-110 [edu, geom]
Kropp G 1959 MN Unt 12, 23-26
(Z107, 246) [intgl]
Launay L de 1936 Rv Sc 74, 449-455
(F62, 20) [France]
Lietzmann W 1909 MN Bl 6, 57-61
(F40, 67) [geom]
Loria Gino 1926 Scientia 40, 205-216,
40(S), 61-70
- 1931 *Storia delle matematiche Vol.
II: I Secoli XVI e XVII*, Turin,
595 p (Z3, 97)
- 1941 Archeion 23, 1-35
[acta Eruditorum]
Mansion P 1895 Brx SS 19(A), 101-105
(F26, 50) [alg, edu]
- 1895 Mathesis (2)5(S), 3-6
(F26, 50) [alg]
Marian V 1957 Clug UBBN 1, 83-89
[arith]
Merton R K 1938 Osiris 4, 360-362
- 1939 Sci Soc 3, 3-27
Mikami Y 1907 Bib M 3, 364-366
[Japan, Engl infl]
Miller G A 1925 Indn MSJ 16, 123-129
(F51, 7)
Molodshii V N 1950 Ist M Isl 3, 431-
466 (Z41, 340) [numb]
Ornstein Martha 1928 *The Role of
Scientific Societies in the
Seventeenth Century*, 3rd·ed,
Chicago
Patterson B C 1939 Isis 31, 25-31
[Jaeger]
Podetti F 1913 Loria 15, 1-8
[prop]
Richeson A W 1937 M Mag 11, 165-171
[mss]
Riessen 1893 *Ein ungedrucktes
Rechenbuch aus dem Jahre 1676*,
Gluckstadt
Rigaud S P 1841 *Correspondence of
Scientific Men of the Seventeenth
Century*, Oxford, 2 vols
(repr 1963)
Schamhardt H C 1926 Euc Gron 3, 172-
186

Sergescu P 1942 Ro An (3)17, 419-
439 [J Sav]
- 1947 Arc In HS 1, 60-99
(MR9, 485) [J Sav]
Stamm E 1936 Wiad M 40, 1-216
(F62, 22) [Poland]
Strong E W 1936 *Procedures and
Metaphysics. A Study in the
Philosophy of Mathematical-
Physical Science in the Sixteenth
and Seventeenth Centuries*,
Berkeley, 308 p (F63, 794)
Suter H 1875 *Geschichte der math
Wissenschaften. Zweiter theil,
vom Anfange des 17ten bis Ende
des 18ten Jahrhunderts*, Zurich
(F7, 15)
Whiteside D T 1961 Arc HES 1, 179-388
(MR24, 2; Z99, 244)
*Wieleitner Heinrich K 1911 *Geschichte
der Mathematik von Cartesius bis
zur Wende des 18 Jahrhunderts*,
Berlin (= Vol 2 of following)
(Russ tr 1960)
*- 1924-1925 *Die Geburt der modernen
mathematik, historisches und
grundsaetzliches...*, Karlsruhe,
2 vols
Wolf R 1889 Bib M (2)3, 33-34
Zeuthen Hieronymus G 1903 Dan Ov,
553-572 (F34, 2) [XVI-XVII]
- 1903 *Geschichte der Mathematik im
XVI und XVII Jahrhundert*,
Leipzig (Russ tr 1963; Z8, 99)

## Centuries: XVIII

Bashmakova I G 1966 Ist M Isl 17,
317-323 [alg]
Biot J B 1803 *Essai sur l'histoire
générale des sciences pendant la
révolution française*, Paris
Bobynin W W 1908 Pet MNP 1, 1-50
(F39, 69) [geom]
Boyer C B 1950 Scr 16, 221-258
- 1960 Scr 25, 11-31 [Fr Revo]
Cajori F 1927 AMM 34, 122-130
(F53, 20) [Frederick the Great]
Christensen S A 1895 *Matematikens
Udvikling i Danmark og Norge i det
XVIII Aarhundrede*, Odense

Centuries:XVIII (continued)

Darboux G 1879 BSM (2)3, 206-228
(F11, 21) [France]
Delambre J B J 1810 *Rapport
historique sur les progrès des
sciences math depuis 1789...,*
Paris (repr 1966)
Dianni J 1963 Kwar HNT 8, 367-382
[Poland ms]
Fuss P H 1825-1843 *Correspondance
mathématique et physique de
quelques célèbres geométres du
XVIII^e siecle,* .. St Petersburg
(MR37, 1220) (repr 1968 Johnson)
[Bernoulli, Euler]
Goldsmith N A 1954 MT 47, 253-259
[England]
Guenther S 1911 *Geschichte der
Mathematik. Von Cartesius biz
zur Wende des 18 Jahrhundert,*
Leipzig
Hofmann J E 1957 *Geschichte des
Mathematik. III: Von den Ausein-
andersetzungen um der Calculus
bis zur Franzoesischen Revolution,*
Berlin, 107 p (MR19, 518)
Ionescu I 1934 Gaz M 40, 52-57, 148-
151 (F60, 831) [Germany, Martius,
Meissner]
- 1935 Gaz M 40, 193-196, 392-395
(F61, 20) [G. Sarganecks, ∟
L. Unterberger]
Koerner T 1904 Bib M (3)5, 15-62
(F35, 65) [mech]
Lichtenstein L 1939 Wiad M 46,
1-86; 47, 1-86
(F65, 1089) [Laplace]
Loria Gino 1899 Ab GM 9
[serial sources]
- 1899 Bib M 10 [curves]
- 1899 ZM Ph 44(S), 243
- 1929 Scientia 45, 1-12 (F55, 5)
- 1933 *Storia delle matematiche.
III: Dall'alga del secolo XVIII al
Tramonto del Secolo XIX,* Turin,
607 p (F59, 1)
- 1941 Archeion 23, 1-35 (MR3, 98)
[Acta Eruditorum]
Mascart J 1910 Intermed 17, 58
(Q3610)

Miller G A 1925 Indn MSJ 16(1), 123-
129 [elem math]
Miller M 1961 Dres Vert 8, 237
(Z93, 8) [Germany, Russia]
Molodshii V N 1953 *Osnovy ucheniya o
chisle v XVIII veke,* Mos, 180 p
Mueller C H 1904 Ab GM 18, 56, 141
[edu, Germany]
- 1909 Arc GNT 1, 61-77 (F40, 48)
[bib, Germany]
*Nielsen Niels 1930 *Géomètres français
sous la Révolution,* Copenhagen-
Paris, 251 p
- 1935 *Géomètres français du dix-
huitieme siècle,* Paris-Copenhagen,
437 p
Otradnykh F P 1954 *Matematika XVIII
veka i akademik Leonard Euler,*
Mos, 39 p (MR16, 552; Z59:1, 8)
Read C B 1963 SSM 63, 447-450
[notebooks]
Richeson A W 1937 M Mag 11, 221-230
[England ms]
Schaaf W L 1955 MT 48, 489-490
- 1956 MT 49, 290-291
Scott J F 1948 Phi Mag (S), 67-91
(Z30, 337)
Sergescu P 1940 Ro An (3)16(2)
(MR2, 306) [Fr Revo]
Smirnov V I + 1954 *Mikhail Sofronov
...,* Mos-Len, 54 p (Z59, 9)
Suter H 1875 *Geschichte der math.
Wissenschaften. II: Vom Anfange des
17ten bis Ende 18ten Jahrhunderts,*
Zurich (F7, 15)
Taton R 1951 Rv Hi Sc Ap, 255-266
[Encyclopedie]
*- 1952 Act Sc Ind (1166), 103-116
Taton R + ed 1964 *Enseignement et
diffusion des sciences en France
au XVIII siècle,* Paris
Toplis J 1805 Phi Mag 20, 25-
Truesdall C 1960 Arc HES 1(1), 3-36
(MR22, 1129) [mech]
Tyulina I A 1955 Ist M Isl 8, 489-
536 (MR17, 1) [XIX, mech, Moscow]
Vygodsky M Ya 1948 Int Con HS (1946),
182-190 [rigor]
Wolf A 1938 *A History of Science,
Technology and Philosophy in the
Eighteenth Century,* Lon, 814 p
(2nd ed 1952)

Centuries: XVIII   (continued)

Yushkevich A P 1949 Mos IIET 3, 45-
116  (MR11, 572)  [Euler, Russia]


Centuries:  XIX

*Adhémar R d' 1912 Rv Q Sc (2)20,
177-218
Anon 1934 Nat 133, 1-2
Bell E T 1931 *The Queen of the*
*Sciences,* NY, 138 p  [pop]
- 1940 *The Development of Mathe-*
*matics,* NY, 596 p  (MR2,  113)
- 1942 Kan Cit UR9 (Winter), 82-86
*Bernal J D 1953 *Science and Industry*
*in the Nineteenth Century,*
London
Bobynin V V 1898 *Biografii znamen-*
*itykh matematikov 19-go stoletiya,*
Mos, 5 vols
*- 1901 Fiz M Nauk 1, 205
[hist of math bib]
Bochner Salomon 1966 *The Role of*
*Mathematics in the Rise of Science*
Princeton, 396 p  (MR33, 2497)
Boutroux P 1909 Scientia 6, 1-20
Cajori F 1918 Sci 48, 279-284
[plans for hist]
Cavaignac E 1931 Scientia 49, 249-
254  (Z1, 322)  [since Comte]
Clariana y Ricart L 1892 Barc Mm 2,
539-
Czuber E 1895 ZMN Unt 26, 610-619
(F26, 4)
Deltheil R 1931 Tou Mm (12)8,
339-350  (F57, 1286)
Donati D + eds 1932 *L'Europa nel*
*secolo XIX. 3(1): Le scienze*
*teoriche,* Padova, 819 p  (F58, 3)
D'Ovidio Enrico 1889- *Uno sguardo*
*alle origini ed allo sviluppo*
*della makmatica pura,*  Turin
Eels W C 1927 AMM 34, 141-142
(F53, 3) [anniv]
- 1928 AMM 35, 437   (F54, 4)
[anniv]
Enriques Federigo 1912 Scientia
12, 172-191
- 1938 *Le Matematiche nella storia*
*e nella cultura,* Bologna

646 / Historical Classifications

Fink K 1890 *Geschichte der Mathematik*
Tuebingen   [bib]
Foote G A 1954 Osiris 11, 439-454
[England sci]
*Galdeano Z G de 1899 Prog M (2)1,
45, 77  [org]
- 1907 *Exposicion sumaria de las*
*teorias matematicas,* Zaragoza,
207 p
Guenther S 1876 *Vermischte Untersuch-*
*ungen zur Geschichte der Mathe-*
*matische Wissenschaften,* Leipzig
- 1901 ZMN Unt 32, 227
- 1901 *Geschichte der anorganischen*
*Naturwissenschaften im neunzehnten*
*Jahrhundert,*  Berlin, 1013 p
Guimaraes R 1900 *Les mathématiques*
*en Portugal au XIX siècle,*
Coimbre
Hadamard J 1912 *Notice sur les*
*travaux mathématiques de 1884 à*
*1912,* Paris, 150 p  (F43, 6)
Hankel H 1869 *Die Entwickelung der*
*Mathematik in dem letzten*
*Jahrhundert,* Tuebingen  (F2, 15)
*Hille E 1953 M Mag 26, 127-146
(MR14, 523)
Jourdain P E B 1917 Scientia 22,
245-255  [Fourier]
Klein F 1893 in his *Lectures on*
*Mathematics...1893,* NY(Macmillan),
41-50  [axiomat, rigor]
- 1926-1927 *Vorlesungen ueber die*
*Entwicklung der Mathematik im 19*
*Jahrhundert,* Berlin, 2 vols
(Russ tr 1937)
- 1927 Naturw 15, 5-11, 43-49
[France]
Kolmogorov A N 1943 Lobachevskii i
matematicheskoe myshlenie devyat-
nadtsatogo veka, in *Nikolai*
*Ivanovich Lobachevskii*  (Sbornik),
Mos-Len, 87-
Lampe E 1899 *Die reine Mathematik in*
*den Jahren 1884-1899*  Berlin
Le Lionnais F ed 1948 *Les grands*
*courants de la pensée mathématique,*
. Marseille
Lepage T 1932 Brx U Rv 37, 376-384
(F57, 1287)  [anal]
Levi-Civita T 1912 Scientia 11, 275-
292  [phys 1892-1912]

Centuries: XIX (continued)

Loria Gino 1900 Geno U An (1900-
  (1901)  (F31, 4)
- 1919 Arc Sto 1, 39-47  (F47, 32)
  [re Cajori]
- 1929 Scientia 45, 83-91, 107-115,
  225-234, 297-306; 46, 27-36, 63-
  71, 77-86, 153-162  (F50, 5)
- 1933 Storia delle matematiche.
  Vol. III. Dall'alba del secolo
  XVIII al tramonto del secolo XIX,
  Milan
- 1950 Storia delle matematiche dall'
  alba della civiltà al secolo XIX,
  2nd ed, Milan, Chs 37-44
Macfarlane Alexander 1916 Lectures on
  Ten British Mathematicians of the
  Nineteenth Century, NY, 148 p
- 1919 Lectures on ten British
  Physicists of the Nineteenth
  Century, NY, 144 p
Mandelbrojt S 1943 Can Rv Tri 29,
  253-258  [France]
Mangoldt H v 1900 Bilder aus der
  Entwicklung der reinen und
  angewandten Mathematik Waehrend
  der neunzehnten Jahrhunderts,
  ..., Aachen, 22 p  (F31, 13)
Marie M 1887-1888 Histoire des
  sciences mathématiques et
  physiques, Paris, vols 10-12
Marmery J Villin 1895 Progress of
  Science, London, 357 p  (F26, 45)
Meserve H G 1928 MT 21, 336-343
  (F54, 33)  [1828]
Miller G A 1924 Sci 59, 1-7 (F50, 12)
  [USA]
- 1938 Sci (21 Oct), 375-376
  [1900]
More W 1962 M Gaz 46, 27-29
  [early]
Mueller Felix 1911 Der Mathematische
  Sternenhimmel des Jahres 1811,
  Leipzig (Teubner), 30 p
*Nagel E 1939 Osiris, 7, 142-
  [geom, logic]
 Newcomb S 1893 Nat 49, 325
- 1893 NYMS 3, 95
 Ocagne M d' 1927 Rv Q Sc 12, 257-283
  [France]

Papp D + 1958 Las ciencias exactas
  en el siglo XIX, Buenos Aires,
  311 p
Pelseneer J 1936 Int Con (9), 161 p
  (F63, 817)
Picard E 1895 Paler R 9, 150
  [recent]
- 1900 Sur les développements, depuis
  un siècle, de quelques théories
  fondamentales dans l'analyse
  mathématique, Paris (Talk at
  Clark University)  (also Wiad M4,
  173-; 1904 BSM (2)28, 267-278,
  282-296)
- 1902 BSM (2)26, 37-53  (F33, 47)
- 1917 Les sciences mathématiques
  en France depuis un demi-siècle,
  Paris, 31 p
Pierpont J 1904 AMSB (2)11, 136-159
  (F35, 5)
- 1905 in Rogers Howard J ed,
  Congress of Arts and Sciences...,
  NY 1, 474-496  [by fields]
Prasad Ganesh 1933-1934 Some Great
  Mathematicians of the Nineteenth
  Century..., Benares, 2 vols
Read C B 1962 MT 55, 127-129
  [alg, arith]
Rey Pastor Julio 1951 Le matematica
  Superior. Métodos problemas del
  Siglo XIX, Buenos Aires, 349 p
  (Z45, 292)
Riccardi P 1890 Bln Mm (4)10, 635-
  651 [Italy, bib]
Ricci G 1961 Pe M (4)39, 201-235
  (F107, 247)
- 1962 Parma Rv M (2)3, 213-241
  (Z117, 7)
*Rogers Howard J ed 1905 Congress
  of Arts and Science..., NY 1,
  474-517, 535-558
Rosanes J 1904 DMV 13, 17-30
  (F35, 8)
Segre C 1905 AMSB (NS)10, 443
Sergescu P 1933 Les sciences mathé-
  matiques, Paris, 182 p
Simon M 1905 Ueber die Entwicklung
  der Elementar-Geometrie im XIX
  Jahrhundert, Leipzig
Simonart F 1950 Bel BS 36, 1010-1025
  (Z40, 289)

Centuries: <u>XIX</u> (continued)

*Smith David E 1906 *History of
Modern Mathematics,* 4th ed (Wiley),
80 p [bib]
*Staude O 1902 DMV 11, 280
*Taton René ed 1961 *Le XIXe Siècle*
(Presse Universitaire, Paris)
*- 1965 *Science in the Nineteenth
Century,* NY
Thomson J A 1906 *Progress of Science
in the Century,* London
Tyulina I A 1955 Ist M Isl 8, 489-536
(Z68, 6) [XVIII, mech, Moscow]
Van t'Hoff J H 1900 Naturw R 15,
557-
White H S 1916 Sci (NS)43, 583-592
Wieleitner Heinrich 1923 *Geschichte
der Mathematik. II: Von 1700 bis
zur Mitte des 19. Jahrhunderts,*
Berlin, 154 p
Woodward R S 1900 AMSB 6, 133-163
- 1906 Sci (NS)11, 41-51, 81-92
Zorawski K 1899 Prace MF 10, 48


Centuries: <u>XX</u>

Aleksandrov P S 1955 Ist M Isl 8, 9-
54 (MR17, 1) [Moscow]
Behnke H 1956 *Der Strukturwandel der
Mathematik in der ersten Haelfte
des 20 Jahrhunderts,* Koeln-
Opladen, 87 p
Berzolari L 1934 Esp Prog 1, 25-28
(Z12, 243) [alg]
Birkhoff G D 1938 in *AMS Semi-
centennial Publications* (AMS)2, 2
270-315 (F64, 3)
Crone C 1923 *Matematisk forening
gennem 50 aas,* Copenhagen, 91 p
(F49, 23)
Fantappie L 1929 Logos Pal 12, 192-
194 (F57, 1350)
*Hille E 1953 Am Sc (Jan), 106-
Jourdain P 1915 Sci Prog 10,
37-40, 114-120, 276-281, 431-436,
614-620
Kac Mark + 1968 *Mathematics and
Logic: Retrospect and Prospects,*
NY

Keyser C J 1939 Scr 6, 81-87
[current]
*Loève Michel 1965 *Integration and
Measure,* Enc Br
Loria G 1932 Int Con (1928) 6, 421-
426 (F58, 1)
Michal A D 1940 Sci (22 Mar), 289-
290
Miller G A 1928 Sch Soc 28, 363-364
(F54, 4) [algebraization]
Novy L + 1966 Prag Dej 11, 25-55
Pacotte J 1925 *La pensée mathématique
contemporaine,* Paris, 128 p
(F51, 2)
Pascal E 1910-1929 *Repertorium der
hoeheren Mathematik,* Leipzig,
5 vols
*Patri A 1950 Lib Chron 17(1)
Pelsener J 1935 *Esquisse du progrès
de la pensée mathématique. Des
primitifs au IX Congrès inter-
national des mathématiques,*
Paris, 160 p
Picard E 1905 AMSB 11, 404-426
(See also AMSB 14, 444-446)
(Russ tr 1912 Kharkov]
- 1906 *La science moderne et son
état actuel,* Paris, Ch 1
- 1913 *Das Wissen der Gegenwart in
Mathematik und Naturwissenschaften,*
Leipzig-Berlin, 196 p
- 1937 BSM (2)61, 41-47 (F63, 20)
Poincaré 1906 AMSB 12, 240-260
Prasad G 1920 Benar MS 2, 81-116
(F47, 34)
Read C B 1960 MT 53, 463-466
Saude O 1902 DMV 11, 280-292
Schwarz L 1963 Parana MB 6, 31-36
(Z112, 2) [France W. War II]
Sergescu P 1935 Slav Cong 2, 357-
[Roumania]
Severi F 1929 It SPS 17, 85-1115
(F55, 589)
- 1951 Archim 3, 45-55 [group]
Tannery P 1930 *Mémoires scientifiques.
X: Sciences modernes. Generalités
historiques. 1892-1930.*
Toulouse, 498 p (F56, 814)
Taton R ed 1966 *Science in the
Twentieth Century,* NY
Titchmarsh E C 1932 Sci Prog (90),
189-196 [recent]

Centuries: XX    (continued)

Tricomi F G 1959 Tor Sem 19, 5-17
   (MR23, 272) [since 1930]
Vasiliev A V 1928 M Obraz, 1-2
Vitali G 1931 It SPS 19(2), 315-327
   (F57, 4)
Volterra V 1906 Rv Mois 1, 1-20
   [bio-and soc sic]
Voss S 1914 *Die Beziehungen der
   Mathematik zur Kultur der
   Gegenwart,* Leipzig  (F. Klein ed,
   Die Kultur der Gegenwart 3(1)(2))
Warzewski T 1955 Shux Jinz 1(4),
   801-820
Weyl H 1951 AMM 58, 523-553
   (MR13, 420)
White H S 1915 Sci (NS)42, 105-113
Whittaker E 1950 Sci Am 183,
   40-42 [bib]
Young W H 1924 Int Con 1, 155-164
   (F54, 48)
- 1925 LMSP (2)24, 421-434
   (F51, 7)
- 1926 LMSP 26, 412-426


COUNTRIES AND REGIONS


America

See also below Latin America,
   United States of America, and
   under OLD CIVILIZATIONS above,
   American Indian, Aztec, Mayan,
   Peruvian.

Brasch F E 1939 Am Antiq(Oct)
   [Newton]
- 1942 APSP 86, 3-12   (MR4, 65)
   [J A Logan, Newton]
Cajori Florian V 1928 *The Early
   Mathematical Sciences in North
   and South America,* Boston, 155 p
   (F54, 21)
- 1927 Isis 9, 391-401   [arith]
Karpinski L C 1924 Colonial
   American Arithmetics in *Biblio-
   graphical Essays, a Tribute to
   Wilberforce Eames,*  242-248

- 1926 Sci 63, 193-195
   [earliest arith]
- 1936 Isis 26, 151-154
- 1940 *Bibliography of Mathematical
   Works Printed in America Through
   1850*  (U Mich P), 723 p (MR1, 290)
- 1941 Scr 8, 233-236 (MR4, 181)
   (S to 1940)
- 1954 Scr 20, 197-202 (S to 1940)
   (MR16, 433)
Sleight E R 1935 M Mag 10, 9-12
   (F61, 957) [early ariths]
Smith D E 1921 *The "Sumario
   Compendioso" of Brother Juan Diez,
   The Earliest Mathematical Work of
   the New World,*  Boston, 65 p
- 1921 AMM 28, 10-15 [first printing]
- 1933 Toh MJ 38, 227-232
Struik D J 1964 Post II (42), 1-5
   [Holland XVII-XVIII]


Argentina

Babini J 1964 Isis 55, 82-85
   (Z117, 250) [Balbin, serial]
Dassen Claro C 1924 *Las Matematicas
   en la Argentina,* Buenos Aires,
   140 p [bib]
Loria G 1938 Arg UM Pu 1   (F64, 901)
   [Spain]
Medici H J y Cabrera E 1953 *El
   desarrollo de la matematica en la
   Argentina...,* Buenos Aires, 63 p


Armenia

Anon 1957 Arm FM 10(5), 3-18
   (MR20, 1042; Z79, 5) [after 1917]
Leroy M 1936 in *Mélanges Franz
   Cumont,*  785-861   [Euclid]
Petrosyan G B 1959 *Matematika v
   Armenii v drevnikh i srednikh
   vekakh,* Erevan, 436 p
   (in Armenian)
Petrosyan G B ed 1967 *Voprosy
   istoria nauki,* Erevan, Armyanskoi
   SSR, 113-123
Semyonov L 1950 Arm Aff 1, 80-81
   (MR12, 577) [VII-XIII]

Armenia    (continued)

Shaw A A 1936 M Mag 10, 287- 289
    (F62, 1037)  [geom, trig]
- 1937 M Mag 11, 117-125
    (F62, 1037)  [alg, arith]
- 1939 M Mag 13, 368-372
    (F65, 3)  [numer]
Tumanyan T G 1953 Ist M Isl 6, 659-
    671  (MR16, 660)  [Euclid]


Austria

See also Germany.

Sadowski A 1892 Die oesterreichische
    Rechenmethode in paedagogischer
    und historischer Beleuchtung,
    Koenigsberg
Simon O 1902 Ens M 4, 157  [edu]
Weiss E A 1940 Deu M 5, 262-265
    (Z23, 194)  [Kepler]


Azerbaijan

Agaev B R 1967 in Petrosyan G B ed,
    Voprosy istorii nauki, Erevan,
    122-124
Agaev G N 1965 Az FTM (3), 3-9
    (MR33, 9)  [mech]
Guseinov A I + 1964 Az FTM (3), 3-17
    (MR29, 1069)


Belgium

See also Flanders, and under CITIES,
    Brussels, Liege, Louvain.

Alliaume M 1930 Rv Q Sc (4)18, 267-
    305  (F56, 9)  [after 1930]
Anon 1887 Notices biographiques et
    bibliographiques concernant les
    membres, les correspondants et les
    associés 1886, Brussels, 613 p
    [Belg Acad]
Belfroid J 1939 Et Class 8, 367-378
Bockstaele P B 1967 Janus 53(1), 1-16
    [XIX]

650 / Historical Classifications

- 1967 Janus 54, 228-235  [Jonghe]
Errera A + 1932 Mathesis 46 (S)
    (F58, 991)  [1830-1930]
Folie F 1881 Bel Bul (3)2, 661-678
    (F13, 45)  [ast]
Gloden A 1949 Hemecht (2), 12-36
    [Luxemburg]
- 1950 Brx Con CR 3, 31, 33
    [numb thy]
Godeaux Lucien 1915 Loria 17, 33-36
    (F45, 6)  [XVI]
- 1943 Esquisse d'une histoire des
    sciences mathématiques en Belgique,
    Brussels, 60 p  (Z28, 337)
- 1946 Leig Bul 15, 152-162
    [pre XX]
- 1950 Bel SM 3, 32-40
    (MR13, 197; Z40, 290)  [recent]
- 1950 Lieg Bul 514-524
    (MR13, 1)  [current]
- 1959 Rv GSPA 66, 6-16
    [geom, recent]
Goethals J V 1837-1838 Lettres
    relatives a l'histoire des sciences
    en Belgique, Brussels, 4 vols
Mailly E 1876 Bel Bul (2)42, 475-479,
    675-676  (F8, 26)  [XVIII]
Mansion P 1873 Boncomp 6, 277-312
    (F5, 38)  [1872]
- 1874 Les mathématiques en Belgique
    en 1872, Rome  (F6, 37)
- 1877 Boncomp 10, 471-542  (F9, 29)
    [1871-1875]
- 1907 Esquisse de l'histoire des
    mathématiques en Belgique,
    (Mathesis 7)
- 1907 Rv Q Sc 61, 270-285
Monchamps 1886 Bel Mm Cou 39, 1-643
    [Descartes]
Quetelet L A J 1835 Brt AAS  [XIX]
- 1837 Cor M Ph 9, 1-
- 1864 Histoire des Sciences Mathé-
    matiques et physiques chez les
    Belges, Brussels
- 1866 Sciences mathématiques et
    physiques chez les Belges au
    commencement du XIXe siècle,
    Brussels, 754 p
Tilly M de 1872 Rapport séculaire
    sur les travaux mathématiques de
    l'académie royale de Belgique
    (1772 bis 1872),  Brussels (F4, 26)

## Belorussia

See also Russia and Tomsk (CITIES).

Vetter Q 1958 Ist M Isl 11, 461-514

## Brazil

Anon 1956 Braz I Bib, 246 p  [bib]
Nachbin L 1956 Parana M 3, 28-41
  (MR19, 1150)  [recent]
- 1961 Parana MB 4, 22-28  (Z98, 9)

## Belgium

Sourek A V 1905 Ens M 7, 257  [edu]

## Canada

Keeping E S 1967 *Twenty-One Years of
  the Canadian Mathematical Congress*
  (U Alberta), 22 p
Tory H M 1939 *A History of Science
  in Canada* (Ryerson P)

## China

See also Mongolia and under
  OLD CIVILIZATIONS.

DeFrancis John 1962 MT 55, 251-255
  [current]
Hua Loo-Keng 1953 SSSR Vest 6, 14-20
  (MR15, 276  [current]
- 1959 Sci Sin 18, 565-567
  (also AMSN 6(7), 724-730)
  [1948-1958]
- 1960 UMN 15(3), 193-201
  (MR22, 928)  [1950-1960]
Hua Loo-Keng + 1959 Sci Sin 8(11),
  1218-1228  (MR22, 928)
  [1949-1959]
Liu S-C 1958 Sci Tec Ch (3), 163-174
  [Theses at NW U in China]
Look K H 1959 Sci Sin 8, 1229-1237
  (repr 1960 AMSN 7(2), 155-163)
  [complex anal 1948-1958]

Min S-H 1955 Shux Jinz 1(2), 397-402
  [numb thy]
Stone M H 1961 in S H Gould ed,
  *Sciences in Communist China*
  (AAAS Pub 68), 617-630
- 1961 AMSN 8, 209-215  (MR22, 1341)
  [1949-1960]
Su B + 1959 Sci Sin 8, 1238-1242
  (repr 1960 AMSN 7(2), 163-168)
  (MR22, 928)  [diff geom]
Tsao Chia Kuei 1961 *Bibliography of
  Mathematics in Communist China
  During the Period 1949-1960*
  (AMS), 83 p
Van S I 1958 Shux Jinz 4, 306-312
  (MR20, 738)  [bib]
Yushkevich A P 1956 Shux Jinz 2(2),
  256-278
Yuan T'ung Li 1963 *Bibliography of
  Chinese Mathematics, 1918-1960,*
  Washington, 164 p
Zen H C 1931 Pac Aff 4, 479-487

## Croatia

Nice V 1955 Zagr BS 2, 69-72
  (Z65, 244)

## Czechoslovakia

See also Slovakia below, and under
  CITIES, Prague.

Anon 1913 Cas MF 42, 273-324
  (F44, 41)  [50 anniv Czech Soc
  Math Phys]
- 1955 *Prazhskii universitet
  Moskovskomu universitetu, sbornik
  v chest yubileya, 1755-1955,* Prague
- 1960 Aplik M 5, 159-169
  (MR22, 424)  [appl]
- 1960 Cas M 85, 129-132
  (MR22, 770)  [current]
Balada F 1959 M Prosv 4, 95-110
  (Z88, 6)  [Czech Soc Math Phys]
Fabian F 1959 Ber Hum MN 5,
  699-703  [stat]
Gaiduk Yu M + 1968 Tbil M Me 129,
  412-428  [USSR-Czech, XVIII-XX]

Czechoslovakia   (continued)

Jarnik V 1955 Cz MJ 5(80), 291-307
   (MR17, 813; 18, 86)  [1945-1955]
Kaderavek F 1950 Prag V (15)
   (MR12, 311) [descr geom]
Kavan Jiri 1909 *Katalog Knihovny
   Jednoty ceskych mathematiku,*
   Prague, 227 p
Novy Lubos 1961 *Dejiny exaktnich ved
   v ceskych zemich do konce 19
   stoleti,*  Prague, 431 p
Peprny L 1902 Cas MF 31, 49-73
   (F33, 54)
- 1903 Cas MF 32, 57-66
Rychlik K 1961 Cas M 80, 76-89
   (Z119, 10)  [Acad prize 1834]
Studnika F J 1876 Cas MF 5
   (F8, 27)  [1856-1876]
- 1884 *Bericht ueber die mathe-
   matischen und naturwissenschaft-
   lichten Publicationen der Koenig-
   lich Boehmischen Gesellschaft der
   Wissenschaften Waehrend ihres
   hundertjaehrigen Bestandes   I.
   Heft, Abhandlungen der erten
   Periode betreffend,* Prague
- 1898 Prag V 6(9), 512-514
   (F29, 29)
- 1898 *Uebersicht der Studiums
   der exakten Wissenschaften bei Uns,*
   Prague, 41 p  (F29, 23)
Svec A 1960 Cas M 85, 389-409
   (MR23A, 231) [diff geom]
Teige K 1893 Cas MF 22  (F25, 15)
Vesely F 1962 *100 lit Iednoty
   ceskoslovenskych matematikii a
   fysiku,* Prague
Vetter Quido 1921 Arc Sto 2, 198-201
   (F48, 39)
- 1921 Pe M 1, 380-382
- 1923 AMM 30, 47-58
   1924 Bo Fir (2)4, i-x, xxxiii-xl
- 1925 Bo Fir (2)4, xiii-xxi
   (F51, 1)
- 1926 Cas MF 55, 165-171
   [duplation, meditation]
- 1932 AMM 39, 511-514  [dec frac]
- 1932 Int Con 6, 499-501
   (F58, 36)  [Hayck]
- 1934 Int Con HS (3), 171-179
   (F62, 1033)

- 1935 *Relations mathématiques entre
   les pays Tchèques et les pays de la
   Péninsule Ibérique, l'Amérique et
   l'Extrème Orient,* Lisbon, 8 p
   (repr of 1934)
- 1944 Scr 12, 141-146  [current]
- 1948 Arc In HS, 684-696
   [stat]
   1948 Int Con HS, 244-251
   [persp]
- 1958 Ist M Isl 11, 461-514
   (MR23, 583; Z96, 2)  [to Biela Gora]
- 1958 Prag Dej 4, 80-95
   (Z112, 242)  [to 1620]
- 1961 Ist M Isl 14, 491-516
   (Z115, 241)  [to XVIII]
Wydra Stanislaus 1778 *Historia mate-
   seos in Boemia et Moravia culta,*
   Prague


Denmark

See aslo Scandinavia below.

Christensen S A 1895 *Matematiken
   Udvikling i Danmark og Norge i det
   18.  Aarhundrede,* Odense, 270 p
   (F26, 45)
- 1909 in *Festskrift til H. G.
   Zeuthen,* Copenhagen, 18-26 [Euclid]
Larsen L M 1952 M Tid A, 1-21
   (Z48, 242)  [arith]
Nielsen Niels 1910 *Mathematikeni
   Danmark 1801-1908,* Copenhagen
- 1912-1916 *Mathematiken i Danmark
   1528-1800,* Copenhagen, 2 vols
Steen A 1873 Zeuthen (3)3, 161
   (F5, 38)  [since 1800]


England

See also Scotland and under CITIES,
   Cambridge, London, Oxford,
   Spitalfields Math. Soc.

Archibald R C 1929 M Gaz 14, 379-400
   [serials]
Clagett M 1954 Isis 45, 269-277
   (Z55, 3)  [Alfred, Euclid]

England (continued)

Cowley E B 1923 AMM 30, 189-193
(F49, 7) [1810]
DeMorgan A 1837 Notices of English
Mathematical and Astronomical
Writers Between the Norman Conquest
and the Year 1600, London
- 1852 Phi Mag (4)4, 321-330
[infinitesimals]
Dugas René 1953 De Descartes à
Newton par l'école anglaise,
Paris, 19 p (MR16, 433)
Elliott E B 1899 LMSP 30, 5-23
(F30, 29)
*Goldsmith N A 1953 MT 47, 253-259
[XVIII]
Halliwell J O 1841 Collection of
Letters Illustrative of the
Progress of England from the
Reign of the Queen Elizabeth
to that of Charles II, London
(repr 1965)
Henel Heinrich 1934 Studien zum
altenglischen Computus,
Leipzig, 104 p
Howarth O J R 1931 The British
Association for the Advancement of
Sciences. A Retrospect 1831-
1931, London
Le Lionnais F 1954 Osiris 11, 40-49
(MR16, 659) [XIX]
Loria G 1914 Isis 1, 637-654
- 1914 M Gaz 7, 423-427; 8, 12-19
(F45, 58)
Macfarlane Alexander 1916 Lectures on
Ten British Mathematicians of
the Nineteenth Century (Wiley)
MacKay J S 1893 Fr AAS 2, 303
[serials]
Mathews G 1915 Nat 95, 219-295
(F45, 58)
Maynard K 1932 Isis 17, 94-126
Pedersen O 1963 Centau 8, 238-262
Piaggio H T H 1931 M Gaz 15,
461-465 [Forsyth, Hardy, Hobson]
Richeson A W 1937 M Mag 11, 165-171
[XVII ms]
*- 1946 M Stu 14, 49-57
(MR9, 485) [1750-1850]
Scott J F 1948 Phi Mag, 67-90
[thru XVIII]

Shenton W F 1928 AMM 35, 505-512
(F54, 21) [Euclid]
Sleight E R 1942 M Mag 16, 243-251
(Z60, 7) [arith]
Taylor E G R 1953 Geog Mag (Dec),
409-416 [XVII]
- 1954 The Mathematical Practitioners
of Tudor and Stuart England
Cambridge, 454 p (MR16, 659;
Z57, 241)
- 1966 The Mathematical Practitioners
of Hanoverian England, Cambridge,
519 p
Wilkinson T T 1854 Manc Mr 11, 123-
[geom]
- 1856 Lancas HS 8, 75- [geom]
Wilson F P 1935 M Gaz 19, 343-354
(F61, 957) [XVII]


Estonia

See under CITIES, Tartu.

Lusisaj 1948 Matematika v SSSR 1917-
1947, 1028-1030
Ryago G 1955 Tartu M 37, 74-105
(MR17, 697) [U Tartu]


Ethiopia

Lakshmi Bai C 1952 M St 20, 84-85
(Z47, 2) [mult]


Finland

See also Scandinavia.

Dahlbo J 1897 Bib M, 116


Flanders

See also Belgium.

Bockstaele P 1948 Euc Gron 23, 204-
208 (Z30, 100)

France

See also under CITIES, Ecole Normale, Ecole Polytechnique, Paris, Strassbourg.

Adhémar R d' 1901 Rv Q Sc (2)20, 177-218 [XIX]

Amodeo F 1934 It SPS 2, 146-148 [XVII, persp]

Anon 1895 Nat 52, 637-638 (F26, 43) [anniv Acad]

- 1935 Nat 136, 1037-1038 (F61, 957) [XIX]

- 1960 *Iz istorii frantsuzkoi nauki,* Mos, 182 p (Z102, 247) [collection]

Appell Paul 1915 in *La Science Française (Exposition Universelle ...de San Francisco)* Paris 1, 77-91 (repr 1915 Paris 19 p

Archibald R C 1935 Scr 3, 193-196 [Gauss]

d'Argenson A R de V + 1782 *De la lecture des livres françois... science mathématiques...seizième siècle,* Paris, 389 p

Biggeri C 1947 Arg SC 143, 264-295 (Z20, 100)

Biot J B 1803 *Essai sur l'histoire générale des sciences pendant la révolution francaise,* Paris

Boatner C H 1936 Osiris 1, 173-183 (F62, 24) [Fr Revo]

Bocher M 1914 Rv I Ens 67, 20-31 (F45, 57)

Bouligand G 1941 Rv GSPA 51, 116-122 (F67, 970) [current]

Boyer C B 1960 Scr 25, 11-31 (MR22, 928) [Fr Revo]

Brunet P 1931 *L'introduction des theories de Newton en France au XVIII siècle,* Paris, 362 p

Carmichael R D 1919 AMSB 25, 180-184 (F47, 31) [1859-1919]

Chasles M 1870 *Rapport sur les progrès de la géométrie,* Paris

*Crosland M 1967 *The Society of Arcueil,*(Harvard U P)

de Dainville F 1954 Rv Hi Sc Ap 7, 6-21 (MR15, 770) [edu, XVI, XVII, XVIII]

- 1954 Rv Hi Sc Ap 7, 109-123 (MR15, 923)

Deniker J 1895 *Bibliographie des travaux scientifiques (sciences mathématiques, physiques et naturelles)...,* Paris (F26, 45)

Doublet E 1910 Rv Mois 9, 460-466 [Fr Revo]

Drach J 1903 *Histoire des Sciences Mathématiques en France, au XIX Siècle,* Leipzig

Dupin C 1835 Par CR 1, 564- [XIX]

Emch A 1935 AMM 42, 382-383 (F61, 21) [Gauss]

Estanave E 1902 BSM (2)26, 201-216 232-248, 272-280 (F33, 10) [XIX]

- 1903 *Nomenclature des thèses de sciences mathématiques...,* Paris 52 p (=Estanave 1902)

Fricker M + 1902 Intermed 9, 180 [Franche-Comte]

Geppert H 1935 Scr 3, 285-286 [Gauss]

Henry C 1882 Boncomp 15, 48-71 [arith, geom]

Houghtaling A E + 1927 MT 19, 179-183 [arith]

Karpinski L C 1932 Int Con 6, 455-458 (F58, 30) [arith]

Kusnetzov B G + 1960 *Frantsuzkaya nauka i sovremennaya fizika,* Mos

Lebeuf J 1841 *L'état des sciences en France depuis la mort du Roi Robert arrivée en 1031 jusqu'a celle de Philip-le-Bel arrivée en 1314,* Paris

Lenoir R 1929 Rv Syn Sc 47, 55-74 [Fr Revo]

*Mandelbrojt S 1943 Can Rv Tri 29, 253-258 (MR5, 57; Z61, 5) [XIX]

Marie A 1892 *Catalogue des thèses de science soutenues en France de 1810 a 1890 inclusivement,* Paris, 264 p (F24, 3)

Maupin G 1898 *Opinions et curiosités touchant la mathématique d'après*

France (continued)

*les ouvrages français des 16<sup>e</sup>,
17<sup>e</sup>, et 18<sup>e</sup>,* Paris
Mortet V 1908 Bib M (3)9, 55-64
(F39, 56) [arith]
Nielsen N 1927 Franske matematikere
under revolutionen, in *Festskrift
udgivet af København universitet,*
Copenhagen, 1-114
- 1929 *Géomètres français sous la
révolution,* Copenhagen-Paris,
259 p (F55, 609)
- 1935 *Géomètres français du
dix-huitième siècle,* Copenhagen,
437 p (F61, 3)
Ocagne M d' 1925 Rv Q Sc (4)8,
155-170 (F51, 4)
- 1927 Rv Q Sc (4)12, 257-283
(F54, 34) [XIX]
Picard E 1917 *Les sciences mathématiques en France depuis un demi
siècle,* Paris, 26 p (F46, 44)
[1857-1917]
- 1917 BSM (2)41, 237-260
(F46, 43) [XIX]
Pierpont J 1899 AMSB (2)6, 225
[edu]
Playoust C 1907 Rv GSPA 18, 550-552
[XVI]
Schwartz L 1951 Can Cong 2 (1949),
49-67 (MR13, 197; Z42, 243) [XX]
Sedillot L A 1870 Boncomp 3, 107-170
(F2, 17)
Sergescu Pierre 1933 *Les sciences
mathématiques,* Paris, 168 p
[XIX]
- 1937 Sphinx 7, 183-184 (F63, 13)
[mid]
- 1940 Ro An (3)16, 513-560
(F66, 1187) [Fr Revo]
- 1940 *Mathématiciens francais du
temps de la révolution française,*
Bucharest, 47 p
- 1941 *Quelques dates remarquables
dans l'évolution des mathématiques en France,* Timisoara,
32 p (Monog M4) (MR3, 98)
- 1942 Ro An (3)17(9) (MR4, 65)
[J Sav]
- 1952 Par BS Rou 1, 5-17 (MR13, 809)
[Rumania]

Simons L G 1931 Isis 15, 104-
[USA]
Taton René 1951 *La géomètrie
projective en France de Desargues
à Poncelet* (U Paris), 21 p
- 1965 *Les origines de l'Academie
Royale des Sciences,* Paris,
60 p
Taton René ed 1964 *Enseignement et
diffusion des sciences en France
au XVIII siècle,* Paris, 780 p
Thibault V 1958 M Mag 32, 79-82
[XIX, geom]
Varigny H de 1895 Nat 52, 644-650
(F26, 43) [anniv]
Walusinski G 1957 Ens M (2)3, 289-
297 (MR20, 738)
Watson E C 1939 Osiris 7, 556-587
(F65, 1089) [Academy]

Georgia (country)

Tskhakaya D G 1941 Tbil MI 9, 207-215
(MR14, 65; Z60, 11) [XVIII]
- 1944 Tbil MI 13, 207-219 (Z60, 6)
[trig]
- 1948 Tbil MI 16, 277-288
(MR13, 809) [XVII, XVIII]
- 1949 Tbil MI 17, 315-340
(MR13, 1) [alg, arith]
- 1959 *Istoriya matematicheskikh nauk
v Gruzin s drevnikh vremen do
nachala XX veka,* Tbilisi, 218 p

Germany

See also Austria and under CITIES,
Berlin, Bremen, Breslau, Erfurt,
Freiberg, Giessen, Goettingen, etc.

Anon 1892 *Verzeichniss der seit 1850
an Deutschen Universitaeten
erschienen Doctor-Disserationen und
Habilitationen-Schriften aus der
reinen und angwandten Mathematik,*
Munich, (DMV), 38 p
- 1936 Deu M 1, 389-420 (F62, 26)
[current]
Bartholomaei 1868 ZM Ph 13(S), 1-44
(F1, 12) [XVII, Weigel]

Germany (continued)

Busch W 1933 Gies Phil 30, 1-37
   (F59, 825)   [Lambert, termin]
Curtze M 1875 Hist Abt 20,.57-60,
   113-120   (F7, 16)   [XV]
- 1894 ZM Ph 40(S), 31-74
   [XV, alg]
Draeger M 1942 Deu M 6, 566-575
   (F68, 2)   ["Mathematik und
   Rasse"]
Dyck W 1893 Nat 48, 150-152
   (F25, 97)   [DMV]
Erman W + 1904 Bibliographie der
   deutschen Universitaten,
   Leipzig
Ersch Johann S 1813 Handbuch der
   deutschen literatur seit der
   mitte des achtzehnten jahrhunderts, 2(1), Amsterdam, 772 p
   (2nd ed 1828 Leipzig)
Fettweis E 1933 ZMN Unt 64(8), 353-
   360   [edu]
- 1935 Archeion 17, 64-75
   (F61, 5)   ["Arithmetik, Rasse
   und Kultur"]
- 1937 Scientia 62, 13-21 (F63, 3)
   ["nichteuropaeisher Rasse"]
Fraenkel A A 1967 Lebenskreise, aus
   den Erinnerungen eines juedischen
   Mathematikers, Stuttgart
   [facist period]
Gerhardt C J 1867 Ber Mo, 41-
   [alg]
- 1870 Ber Mo, 141 (F2, 6)   [alg]
- 1877 Geschichte der Mathematik in
   Deutschland, Munich, 319 p
   (F10, 24)
Gericke H 1966 DMV 68(2), 46-74
   (MR34, 1143)   [DMV]
Gnedenko B V + 1954 UMN (NS)9(4),
   133-154   (MR16, 433)   [current]
Graf J H 1889-1890 Geschichte der
   Mathematik und der Naturwissen-
   schaften in bernischen Landen vom
   Wideraufbluehen der Wissen-
   schaften bis in die Neuere Zeit,
   Bern-Basle, 2 vols
Grundel F 1929 Die Mathematik an
   den deutschen hoeheren Schulen.
   Teil II: Vom Anfang des 18.
   Jahrhunderts bis zum Anfang des 19.

Jahrhunderts, Leipzig, 154 p (F55, 3)
Guenther S 1875 Hist Abt 20(1)
   (F7, 16)   [XV]
- 1887 Geschichte des mathematischen
   Unterrichts in den deutschen
   Mittelalter bis zum Jahre 1525,
   Berlin
- 1900 Ens M 2, 237   [edu]
Gutzmer A 1904 BSM (2)31, 38-39
   [DMV]
- 1904 Geschichte der Deutschen
   Mathematiker Vereinigung in ihrer
   Begruendung bis zur Gegenwart,
   Leipzig, 67 p   (F35, 46)
- 1909 DMV 10(1), 1-49   [DMV]
Hofmann J E 1948 in Naturforschung
   und Medezin in Deutschland 1939-
   1946, Wiesbaden 1, 1-9
   (MR11, 149)
- 1954 Mun Abh 62 (MR15, 923; Z56, 2)
   [edu]
Keefer 1915 MN Bl 12, 137-140
   (F45, 68)   [arith]
Klein F 1893 in his Lectures on
   Mathematics..., 94-109
   [ed, research]
- 1904 DMV 13, 347-356   [edu]
- 1911 The development of mathe-
   matics in German universities in
   Evanston Colloquium, NY
Lampe E 1914 in Mathematik in
   Deutschland Unter kaiser Wilhelm
   II, ed by S Korte, Berlin 3,
   69-92 (F45, 56)
Lietzmann Walther 1940 Frueh-
   geschichte der Geometrie auf
   germanischen Boden, Ferdinand
   Hirt, 94 p   (Z23, 386)
- 1941 Fors Fort 17, 384-385
   (Z26, 97)   [re 1940]
Lorey Wilhelm 1916 Das Studium des
   Mathematik in dem deutschen
   Universitaeten seit Anfang des 19
   Jahrhunderts, Leipzig, 343 p
- 1938 Der deutsche Verein zur
   Foerderung des math. und naturw.
   Unterrichts. 1891-1938,
   Frankfurt 165 p   (F64, 4; Z19,
   389)
Mueller F 1899 ZM Ph 44(S), 303-333
   (F30, 33)   [termin]

Germany (continued)

Pietzker F 1901 Ens M (3)2, 77
[edu, XIX]
Pinl M 1969 DMV 71(4), 167-228
[1933-1945]
Rath E 1912 Bib M (3)13, 17-22
[XV]
Schaaf W L 1956 MT 49, 403-405
[Nazism]
Schimank H 1941 Ham MG 8(3), 22-45
(MR3, 258; Z24, 243) [arith]
Sudhoff K 1922 Hundert Jahre Deut-
scher Naturforscher-Versammlungen,
Leipzig
Unger F 1888 Hist Abt 33, 125-145
[arith]
Vahlen K T 1936 Deutsche Mathematik.
In Auftrage der deutschen For-
schungsgemeinschaft, Leipzig, 112 p
Villicus F 1897 Geschichte der Rech-
en-kunst..mit besonderer Ruecksicht
auf Deutschland und auf Oesterreich,
3rd ed, Vienna, 120 p
Vogel K 1949 in Festschrift Maximili-
ansgymnasium Muenchen, 231-277
[1842, arith]
Wappler E 1899 ZM Ph 44(S), 537-554
(F30, 38) [alg]
Weiss E A 1937 Deu M 2, 379-386
(F63, 2) [folk math]
Winter E J 1958 Die deutsch-russische
Begegnung und Leonhard Euler....,
Berlin, 204 p
Wolff G 1937 M Gaz 21, 82-98
(F63, 20) [edu, geom]


Greece

See also Greek under OLD
CIVILIZATIONS.

Blaschke W 1939 Gree SM 19, 167-174
(Z22, 1)
Sakellariou N 1935 Gree SM 16, 60-75
(F61, 3) [1830-1930]


Holland

See also Amsterdam under CITIES.

Anon 1885 Register, naar eene weten-
schappelijke verdeeling op de
werken von het wiskundig genootschap "een onvermoeide arbeid komt
alles te boven" gedurende het
tijdsverloop van 1818-1882,
Amsterdam, 445 p
- 1902 Ned Arch (2)7, 126
[100 anniv Dutch Asso Sci]
- 1911 Register op de Wiskundige
Opgaven uitgegeven door het Wiskundig Genootschap "Een onvermoiede
arbeid alles te boven" gedurende
het tijdsverloop van 1875-1910,
Amsterdam, 370 p (F42, 44)
- 1911 Catalogus van boeken in Noord-
Nederland verschenen van de vrveg-
sten tijd tot op heden. IX: Wis-en
natuurkunde..., The Hague, 58 p
Beth E W 1947 N Tijd 35, 100-104
Bierens de Haan D 1874 Amst Vs M (2)8,
57-99, 163-223 (F6, 40)
[source material]
- 1875 Amst Vs M 9, 1-40, 90-112,
323-370 (F7, 25)
- 1876 Amst Vs M 10, 161-206
(F8, 25)
- 1878 Boncomp 11, 383-452
(F10, 15) [1663]
- 1879 Amst Vs M 14, 180-187
(F11, 15)
- 1881 Amst Vs M 16, 1-44 (F13, 47)
- 1881 Boncomp 14, 519-630, 677-718
[XVI-XVIII, bib]
- 1881 Ned Arch 16, 443-462
(F13, 47)
- 1882 Boncomp 15, 225-312, 355-438
[XVI-XVIII, bib]
- 1883 Bibliographie néerlandaise
historique scientifique des ouvrages importants mathématiques et
physiques, Rome, 424 p
(repr 1965)
- 1883 Amst Vs M 18, 218-301
- 1884 Amst Vs M 19, 1-38, 78-84
249-295; 20, 102-195, 197-246
- 1885 Amst Vs M (3)1, 224-244
- 1886 Amst Vs M (3)3, 69-119
- 1887 Amst Vs M (3)3, 65-67
- 1887 Amst Vs M (3)3, 68-78
- 1887 Amst Vs M (3)3, 79-81
- 1887 Amst Vs M (3)3, 82-96

Holland (continued)

- 1891 Bib M (2) 5, 13-22
  [bib]
- 1892 Amst Vs M (3) 9, 4-47
  (F24, 5)
- 1893 Amst Vh (1) 2 (1), 3-60
  (F25, 20)
Bockstaele P 1960 Isis 51, 315-321
  (MR22, 769) [arith]
Burger C P 1908 *Amsterdamsche reken-*
  *meesters en zeevaartkundigen in de*
  *zestiendeeeuw*, Amsterdam, 238 p
  (F39, 11)
Cardinaal J 1900 Ens M 2, 318
  [edu]
De Groote H L V 1960 Sci Hist 2,
  161-172 [arith]
Dijksterhuis E J 1941 Arc Mu Tey
  (3) 9, 268-342 [XVI-XVII]
Gloden A 1960 Int Con HS 9, 467-471
  (Z118- 3) [calc]
Hayashi T 1909 Nieu Arch 9 (1), 39-
  41 [Japan]
Korteweg D J 1894 *Het bloeitijdperk*
  *der wiskundige wetenschappen in*
  *Nederland*, Amsterdam
Mikami Y 1910 Nieu Arch 9, 231-234,
  301-304 [ast, Japan, surveying]
Simons L G 1923 Mt 16, 340-347
  [1730]
Smeur A J E M 1960 *De Zestiende-*
  *eeuwse Nederlandse Rekenboeken*,
  The Hague, 175 p
Struik D J 1926 Ned HIR 6, 109-122
  [W. Gillesz]
- 1936 Isis 25, 46-56 (F62, 20)
  [XVI, bib]
- 1943 in Barnouw A J + ed, in *The*
  *Contribution of Holland to the*
  *Sciences*, NY, 281-295
- 1958 *Het land van Stevin en Huygens*,
  Amsterdam, 148 p (MR19, 1030)
- 1964 Post II (42), 1-5
  [America; XVII-XVIII]
Van der Blij F Enc Gron 24, 65-78
  (Z31, 145)
Van Schevichaven S R J 1897 *Bouwstof-*
  *fen voor de geschiedenis van de*
  *levensverzekeringen en lijfreuten*
  *in Nederland*, Amsterdam, 370 p
  (F28, 39)

Zinner Ernst 1956 *Deutsche und*
  *nederlaendische astronomische*
  *Instrumento des 11-18. Jahrhunderts*,
  Munich, 689 p (Z70, 4)


Hungary

See also under CITIES, Budapest.

Anon 1953 *Vazlatos a magyar mate-*
  *matika ujkori toerteneteboel...*,
  Budapest, 67 p (MR17, 2)
Gaspar Ilona 1930 *A magyar mathe-*
  *matikai irodolom bibliograviaja*
  *1901-1925*, Budapest, 97 p
Hars J 1938 Deb M Szem 14, 1-164
  (F64, 911; Z18, 340) [arith]
Jelitai J 1937 Archeion 18, 350-354
  (F63, 16)
- 1937 Int Con (1936) 2, 279
  (F63, 20)
- 1937 M Mag 12, 125-130 (F63, 808)
  [to 1830]
Keresztesi M 1935 Deb M Szem 9,
  1-197 (F62, 1042) [termin]
Ligeti B 1953 *A magyar matematika*
  *toertenete a XVIII szazad vegeig.*
  *A matematikai szakkoeroek szamara*,
  Budapest, 40 p (MR17, 2)
Rado T 1932 AMM 39, 85-90
  [current]
Szenassy B 1954 Deb U Act (1), 5-28
  (MR17, 813) [1830-1867]
- 1955 Deb U Act (1), 3 (MR17, 813)
Szily, K V 1888 MNB Ung 6, 211-223
  [XVIII-XIX]
Szily K V + 1894 MNB Ung 12, 134-143
  (F25, 58) [Magister Georgius]
Szinnyei Jozsef + 1878 *Magyarorszag*
  *termeszettu-domanyi es mathematikai*
  *koenyveszete 1472-1875*, Budapest,
  1008 columns [bib]
Szuecs A 1933 Int Con H 5 (19),
  284-291
Woyciechowsky J V 1932 Deb M Szem
  (Z5, 242) [P Sipos]

## India

See also Hindu (OLD CIVILIZATIONS),
and Calcutta (CITIES)

Anon 1929 Indn MSJ 17 (S) (F55, 26)
[current]
Bhabha H J 1951 Can Cong 2, 42-44
(Z42, 243) [recent]
Bose P + 1963 *Fifty Years of Science
in India: Progress of Statistics*,
Calcutta, 50 p
Een N R ed 1952 *Progress of Science
in India. Sec 1: Mathematics*,
New Delhi, 94 p
Hardy G H 1937 AMM 44, 137-155
(Z16, 145) [Ramanujan]
Masani P 1963 AMM 71, 671-676 [edu]
Seth Broj Raj 1963 *Fifty Years of
Science in India: Progress of
Mathematics*, Calcutta, 44 p
Vaidyanathaswamy R 1951 Alla UMAB
15, 33-37 (Z42, 243) [current]

## Ireland

See also Dublin (CITIES).

Colthurst J 1943 M Gaz (Oct), 166-170
[phys]
Purser J 1902 Brt AAS, 499-511
(F33, 9) [XIX]
*- 1902 Nat 66, 478

## Italy

See also Sicily and under CITIES
Academia dei Lincei, Bologna,
Florence, Livorno, Milan, Modena,
Naples, Padua, Palermo, Pisa,
Turin.

Agostini A 1950 Archim 2, 41-43
(Z35, 148) [alg, arith]
- 1951 Archim 3, 168-173
(MR13, 1; Z42, 242) [calc]
Amaldi V 1911 It SPS 5, 981-987
[geom, XIX]

Amici N 1906 *Matematici, fisici,
astronomi delle Marche, Macerata*
Amodeo F 1902 Nap Pont 32(2)7(9), 30
[edu, univ]
- 1902 Nap Pont 32 (2) 7 (11), 64
[Fratelli di Martino a Vito
Caracelli]
Anon 1816 Bib It 1, 225-, 358-;
2, 60-
- 1930 It Lm 9, 313-315 (F56, 9)
- 1960 *Cinquante ans de progrès
scientifique, 1907-1956*, Milan
210 p
Bariola P 1897 *Storia della ragioneria italiana*, Milan
Bibliografia matematica italiana 1950
(Unione matematica italiana)
Bjornbo A A 1903 Bib M (3) 4,
238-245 (F35, 5) [S. Marco mss]
- 1905 Bib M (3) 6, 230-238
(F36, 5)
Bompiani E 1931 AMM 38, 83-95
(F57, 4)
Bortolotti Ettore 1919 *Italiani
scopritori e promotori di teorie
algebriche*, Modena, 102 p
- 1928 *Studi e ricerche sulla storia
della matematica in Italia nei
secoli XVI e XVII*, Bologna, 254 p
(F54, 23)
- 1929 Pe M (4) 9, 161-166
(F55, 601) [geom alg]
- 1933 Int Com HB 5 (19), 268-283
[XVI]
- 1939 It SPS 2, 11-33 (Z22, 297)
[current]
- 1939 It SPS 27 (6), 511-520
(F65, 1079) [geom]
- 1939 It SPS 27 (3), 598-617
(F65, 1084; Z21 194)
Cavallaro V G 1938 Bo Fir 34,
LXI-LXII [triangle]
Cerboni G 1886 *Elenco cronologica
delle opere di computisteria e
ragioneria venuto alla luce in
Italia dal 1202 sino al presente*,
3rd ed, Rome
Cerruti V 1907 It SPS 1, 94-107
(F38, 46)
Cimmino G + 1937 It SPS 25 (2), 5-28
(F63, 817) [current]

Italy   (continued)

Cossali P 1797-1799 *Origine, tras-*
*porto in Italia, primi progressi*
*in essa dell' Algebra Storia*
*critica di nuove disquisizioni*
*analitiche e metafisiche arric-*
*chita.* Parma, 2 vols
Erasmo G d' 1935 Nap FM Ri (4)5,
3-6 (F61, 957) [1934]
Favaro A 1889 Bib M (2) 3, 113-115
[bib]
Fichera G 1949 Iasi IP 4, 63-107
(MR12, 1; Z49, 290) [1940-1945]
Gambioli D 1903 *Breve sommario della*
*storia delle matematiche colle due*
*appendici sui matematici italiani*
*e sui tre celebri problemi geo-*
*metrici dell'antichita,* Bologna,
241 p (F33, 3)
Gini Corrado + 1939 *I contributi*
*italiani al progresso della statis-*
*tica,* Rome, 153 p
Hortis A 1922 *La Riunioue degli Sci-*
*enzatti Italiani prima delle Guerre*
*d'Independenza,*   Citta de
Castello
Karpinski L C 1910 Bib M (3) 11,
209-219 (F41, 56) [XV, alg]
- 1929 Archeion 11, 331-335 [arith]
Lauricella G 1911 It SPS 5, 217-236
[recent, anal]
Lauriola L 1960 Civ Mac 8, 3-7
(MR22, 618)
Libri Guillaume 1838-1841 *Histoire*
*des sciences mathématiques en*
*Italie depuis la renaissance des*
*lettres jusqu'a la fin du dix-*
*septieme siècle,* Paris, 4 vols
(2nd ed 1 65 Halle)
- 1843 *Storia delle scienze mathe-*
*matiche in Italia,* Milan
(abr tr of 1838-1841)
Loria G 1908 Parola 1, 407-417
(F39, 5)
- 1909 Int Con (4) 3, 402-421
(F40, 3)
- 1916 Linc Rn (5) 25, 264-270
(F46, 41)
- 1932 It SPS 20 (1), 384-397
(F58, 986)
- 1932 Pe M (4)12, 1-16 (F58, 30)

- 1938 Math Cluj 14, 155-179
(F64, 3; Z20, 196)
- 1943 Bo Fir   [Liguria XV-XVIII
Mania B 1939 It SPS 2, 34-57
(Z22, 297) [current]
Mieli A 1923 *Gli scienziati Italian*
*dall'inizio del medio evo ai nostri*
*giorni,* Rome, 2 vols
- 1929 Arc GMNT 11, 292-308
(F55, 3)   [XVIII-XIX
Narducci E 1887 Boncomp 19, 335-640
Petrucci G B ed 1932 *L'Italia e la*
*scienza,* Florence, 408 p
Picone M 1953 *On the Mathematical*
*Work of the Italian Institute for*
*the Application of Calculus During*
*the First Quarter Century of its*
*Existence,* Rome, 37 p
Poletti L 1951 Parma Rv M 2, 417-434
(MR14, 121) [prime tab]
Prosper V R 1893 Prog M 3, 41-43
[logic]
Riccardi Pietro 1870-1880 *Biblioteca*
*matematica italiana dall a origine*
*della  stampa  ai primi anni del*
*secolo XIX,* Modena, 1475 p
(repr 1952 Milano) (F4, 26; 5, 37;
7, 16)  (see 1893)
- 1890 Bib M (2) 4, 56  [bib]
- 1890 Bln Mm (4) 10, 635-651
[bib XIX]
- 1893 *Correzioni ed agguinte,* Modena,
56 p  (Boncomp (7) 14)  (repr 1928)
- 1896 Bln Mm 6, 755-
- 1897 Bln Mm 7, 371-
Scorza-Dragoni G 1929 Bo Fir (2) 8,
41-53 [alg geom]
- 1935 It SPS 23(2)(A),   131-144
(F61, 957) [current]
Segre B Annali (4) 11, 1-16
(Z5, 242)  [geom]
Sergescu P 1953 Rv GSPA 60, 101-113
(MR 14, 1049) [post mid]
Severi F 1930 Bo Fir (2) 9, 121-124
(F56, 878)
- 1931 It SPS 19 (1), 189-203
(F57, 5)
Sobrero L+1936 It SPS 24, (2), 3-18
(F62, 1007)  [current]
Tosi A 1959 Pe M (4) 37, 78-85
(Z88, 244)  [Brera]

## Italy    (continued)

Volterra V 1909 Int Con (4) 1, 55-
65  (F40, 3)    [XIX]
Wilson E B 1904 AMSB 11, 77-
[geom]

## Japan

Anon 1922 Nat (2 Mar), 287-288
Kabori Akira 1956 Int Con HS (8),
10-12    [from XVII]
- 1957 Les étapes historiques des
mathématiques au Japon, Paris,
21 p
Mikami Y 1910 Mathematical Papers
from the Far East, Leipzig, 305 p
- 1911 AMM 18, 123-134  [edu]
Miller G A 1905 Sci 22, 215-216
Ogura Kinnosuke 1956 Kindai nikon-
nosu, Tokyo, 272 p
Pusey H C 1945 MT 38, 172-174
[Japanese-Americans]
Smith D E  + 1914 A History of
Japanese Mathematics, Chicago,
295 p

## Kazakstan

Anon 1960 Bibliografiya izdanii
akademii nauk Kazakhskoi SSR 1932-
1959, Alma Ata (Ak Nauk KSSR),
1-53
Zhatykov O A 1957 in Nauka v Kazakh-
stane za sorok let sovetskoi
vlasti, Alma Ata, 260-280
- 1958 Kazk M 1, 5-24  [bib]

## Latin America

See also America, Argentina, Brazil,
Peru.

Cajori F 1928 The Early Mathematical
Sciences in North and South
America, Boston, 156 p (F54, 21)
Karpinski L C 1947 Scr 13, 59-63
(MR 9, 74; Z29, 386)  [bib to 1850]

Pastor J R 1954 Modern Mathematics
in Latin America, Montevideo
(MR16, 985)

## Latvia

Lusis A Ya 1948 in Matematika V SSSR
1917-1947, Moscow, 1023-1030
(Z37, 291 41, 342)  [bib]
- 1958 Lat U Rak 20 (3), 5-20
(MR20, 1243)  [1948-1958]

## Lithuania

See Vilnius (CITIES).

## Luxemburg

Gloden A 1949 Hemecht (2), 12-36
Belguim

## Mongolia

Baranovskaya L S 1954 Mos IIET 1,
53-84  (MR16, 659; Z59: 1, 9;
60, 9)

## New Zealand

Duncan E R 1956 Ari T 3, 137-142
[arith]

## Norway

See also Scandinavia.

Anon 1911 in Sartryk av Festskrift
Matematikken av E. Holst, Kristi-
ania, 151 p (F42, 48)
[K. Fredriks U 1811-1911]
Brun V 1962 The Art of Calculation
in Old Norway Until the time of
Abel, Oslo Bergen, 125 p (Z114, 7)
(in Norwegian, English summary)

Norway (continued)

Christensen S A 1895 *Matematikens Udvikling i Danmark og Norge i det XVIII Aarhundrede*, Odense
Guldberg A 1905 Ens M 7, 433 [edu]
Piene K 1937 Nor M Tid 19, 52-68 (F63, 20) [edu after 1800]
Tambs Lyche R 1935 Nor VSF 7, 26*-33* (Z11, 194) [edu]
- 1952 Sk Kong (1949), xxi-xxxi (MR14, 523; Z47, 244)
Thalberg O M 1943 Nor M Tid 25, 65-75 (Z28, 195) [Norwegian Math Soc]

Peru

Popov G N 1922 *Kultura tochnogo znaniya v drevnem Peru*, Pet, 72 p

Poland

See also under CITIES, Krakow, Wroclaw.

Anon 1901 *Catalogue de la littérature scientifique polonaise*, Krakow
Banachiewicz F + 1903 Prace MF 24, 247-295 [bib]
Biernacki M 1958 Hels M (A)1(251) (MR20, 738) [fun thy]
Dianni Jadwiga + 1957 *z dziejow polskiej matematyczney*, Warsaw, 138 p (Z87, 5)
- 1963 *Tysiac Lat Polskiej Mysli Matematycznej*, Warsaw, 285 p [bib]
Dickstein S 1892 Wars NA Tr 3, 184-187 [bib]
- 1895 Kosmos 20, 352-358 (F26, 3) [bib 1873-1892]
- 1925 Bo Fir 5, i-vi (F52, 36) [bib, hist math]
Dickstein S + 1894 *Polnische mathematische Bibliographie des XIX. Jahrhunderts. Probeheft*, Krakow, 32 p (F25, 11)
- 1901-1902 Prace MF 12, 285; 13, 373-375 [bib 1899]

Folkierski W 1895 Prace MF 6, 151 [Polish maths in Paris 1870-1882]
Golab S 1947 It UM (3)2, 244-251 (MR9, 485; Z30, 338) [current]
- 1947 Zyc Nauk 3, 79-91
Grabowski J 1913 Krak BI (A), 63-64 (F44, 45)
- 1913 Krak Roz 53, 401-415 (F44, 45)
Jordan Z A 1945 *The Development of Mathematical Logic and Logical Positivism in Poland Between the Two Wars*, Oxford
Kotarbinski T 1959 *La Logique en Pologne*, Rome
Kuratowskii K 1955 UMN (NS)10(3), 217-221 (MR17, 2) [Acad]
- 1960 Wiad M (2)3, 199-216 (MR22, 1130) [Inst of Math]
Kuzawa M G 1967 MT 60, 383
- 1968 *Modern Mathematics. The Genesis of a School in Poland*, New Haven, 143 p [bib]
Loriowa Marija 1934 *Materjaly do bibljografji pismiennictwa Kobiet polskich (do roku 1929)*, Lwow, 282 p
Marczewski Edward 1948 *Ruzwoj Matematzki u Polsce*, Krakow, 47 p (MR11, 150; 17, 813)
Merczyng H 1907 Krak Roz 47, 199-218 (F38, 58) [edu]
Pawlikowska Z 1966 Wiad M (2)9, 23-43 (MR33, 1213) [termin]
Sierpinski W 1957 Glasn MPA (2)12, 125-132 (MR20, 620)
- 1959 Pol Rv 4, 1-13 (MR21, 893; Z85, 6) [Warsaw school]
Sikorski Roman + 1955-1956 *Polska bibliografia Analyczna. Matematyka*, Warsaw, 2 vols
Stamm E 1935 Wiad M 40, i-v, 1-216 (Z12, 388) [XVII]
- 1935 *z historji matematyki XVII W W Polsce*, Warsaw, 328 p
Steinhaus H 1956 Pol Ac Rv 1(4) (MR19, 108) [appl]
Wazewski T 1960 Wiad M (2)3, 217-221 (MR21, 1130) [Inst of Math]
Zebrawski T 1873 *Bibliografija pismiennictwa polskiego z dzialu Matematyki i Fizyki oraz ich zastosowan*, Krakow, 617 p (F5, 38)

## Poland   (continued)

- 1886 *Dodatki,* Krakow, 162 p
Zygmund A 1951 Can Cong (2), 3-9
   (MR13, 197; Z42, 243)


## Portugal

Almeida F de Y V 1925 *Historia des
   Matematicas na Antiguidade,* Lisbon
Bensaude Joaquim 1912 *L'astronomie
   nautique au Portugal à l' époque
   des grandes découvertes,* Berne
Cabreira A 1905 *Quelques mots sur
   les Mathématiques en Portugal,*
   Lisbon, 72 p   (F36, 41)
- 1910 *Les Mathématiques en Portugal.
   Deuxième défense des travaux de
   Antonio Cabreira,* Lisbon, 157 p
   (F41, 39)
Frick B M 1945 Scr 11, 327-339
   [arith]
Garcao-Stockler F 1819 *Ensajo his-
   torico sobre origem e progreso das
   matematicas em Portugal,* Paris,
   168 p
Guimaraes Rodolphe 1900 *Les mathématiques en Portugal au XIX^e
   siècle,* Coimbra, 167 p   (F31, 4)
   [*bib]
- 1909 *Les mathématiques en Portugal,*
   2nd ed, Coimbra, 660 p   (F40, 2)
   [bib]
- 1911 *Les mathématiques en Portugal,
   Appendice II,* Paris, 107 p
Loria G 1919 Scientia 26(87), 1-9
   (F47, 27)
Mikami Y 1912 Nieu Arch 10, 61-70
   [ast, Japan]
Miller G A 1910 AMM 17, 231-233
Ribeiro Dos Santos A 1912 Lisb Mr 8,
   148-229
Sarton G 1935 Isis 24, 106   [Coimbra]
Teixeira F Gomes 1906-1908 *Obras
   sobre mathematica publicadas por
   orden do governo portuges,* Coimbra,
   5 vols   (F39; 44; 40, 42)
- 1924 Ens M 23, 137-142
- 1925 Vat NL Mm (2)8, 331-351
   [XV-XVIII]

-1934 *Historia des matematicos en
   Portugal,* Lisbon, 306 p   (F62, 1028)


## Rumania

See also Bucharest, Iasi (CITIES).

*Andonie George St 1965-1967 *Istoria
   matematicii in Romania,* Bucharest,
   3 vols   [bib, ports]
Anon 1959 Cluj MF 10, 7-15
   (MR23, 272)   [since 1944]
- 1959 Ro Rv M 4, 337-340
   (MR22, 424)   [since 1944]
Cimpan F T 1955 Iasi M (1)6(1/2),
   143-153   (Z67, 247)   [alg]
Cimpan F T + 1955 Iasi UM (NS)1(1),
   400-414   (Z68, 6)   [edu]
Glivich A N 1958 Ist M Isl 11, 563-
   582   (Z95- 4)   [USSR]
Ionescu I 1930 Gaz M 36, 289-291
   (F56, 23)   [anal geom]
- 1935 Gaz M 40, 486-491   (F61, 953)
   [geom]
- 1935 Gaz M 41, 20-27   (F61, 951)
   [arith]
- 1935 Gaz M 41, 169-171 (F61, 958)
   [geom termin]
Jacob C 1959 Gaz M (B) 10(4), 195-206
   [XVIII, France]
Lazarin A 1903 Gaz M 8, 173-174
Marian V 1936 Gaz M 41, 515-517
   (F62, 3)   [arith tab XVIII]
- 1937 Gaz M 43, 239-243   (F63, 810)
   [arith]
- 1937 Gaz M 42, 449-451, 508-511
   (F63, 15)   [arith XVIII]
- 1938 Gaz M 43, 289-293   (F64, 20)
   [arith]
Nicolescu M ed 1966 Ro Rv M 11(9)(S),
   1041-1164
Pompeiu D 1934 Buc EP 5(1), 58-60
   [Borel]
Popa I I 1938 Archeion 31, 69-73
   (F64, 18)   [arith]
- 1958 Ist M Isl 11, 533-562
   (Z96, 3)
Sergescu Petre 1933 Int Com HB 5(19)
- 1934 *Matematica la Romani,* Valenii-
   de-Munte, 40 p

## Rumania (continued)

- 1934 Tim Rev M 13, 111-115, 123-127
  (F60, 825)
- 1935 Slav Cong 2, 257-262 (F61, 28)
  [XX]
- 1935 Thales 1, 45-47 [XVIII, XIX]
  1937 Le développment des sciences
  mathématiques en Roumaine,
  Bucarest, 61 p (F63, 15; Z17, 280)
- 1939 Sphinx Br 9, 145- [Fr Revo]
- 1939 Gaz M 44, 618-621 (F65, 17)
  [France 1870-1877]
- 1952 Par BS Rou 1, 5-17 [France]
Stoilov S 1956 UMN (NS)11(4),
  207-225
Villa M 1956 It UM (3)11, 591-593
  (MR18, 710) [diff geom]

## Russia

See also Soviet Union, Ukraine, and
  under CITIES, Kazan, Leningrad,
  Lwow, Moscow, Odessa, Tashkent.

Adamantov D L 1904 Kratkaya istoriya
  razvitiya matematicheskikh nauk s
  drevneishikh vremen i istoriya
  pervonachalnogo ikh razvitiya v
  Rossii, Kiev, 102 p
Aleksandrov P S 1947 Mos UZ 1(91), 4
  [XIX, XX]
Anon 1872-1891 Ukazatel russkoi
  literatury po matematike...,
  Kiev (Obshchestvo estestvoispytate-
  lei)
- 1904-1917 Russkaya bibliografiya po
  estestvoznaniya i matematike, St.
  Petersburg, (Byuro mezhdunarodnoi
  bibliografii), 5 vols
- 1945 Fisiko-matematicheskie nauki
  in Akademiya nauk za 220 let.
  Ocherki po istorii Akamii nauk,
  Mos, 78 p
- 1949 Voprosy istorii otechestvennoi
  nauki, Mos-Len, Pt 1, 63-236
Bapst G 1897 Par Cr 124, 1135
  [Poncelet]
Baranov P A 1912 M Obraz (1), 36-39
  (F43, 67) [painting]

Belozerov S E 1956 Mos M Sez,
  229-230 [compl anal]
Belyi A + 1959 Ist M Isl 12, 185-244
  (MR23, 693; Z100, 245) [XVII, geom]
Bobynin V V 1886-1893 Ocherki istorii
  razvitiya fisiko-matematicheskikh
  Znanii v Rossii, Mos, 324 p
  (Fiz M Nauk 1-11) (Fr tr 1888)
- 1886-1900 Russkaya fiziko-mate-
  maticheskaya bibliografiya...,
  Mos, 3 vols (F23, 2)
- 1895 Fiz M Nauk 13, 1-24, 50-67
  (F26, 43) [first org]
- 1896 Matematiko-astronomicheskaya
  i fizicheskaya sektsii pervykh
  devyati sezdov russkikh estestvoi-
  spytatelei i vrachei, ikh tseli i
  deyatelnost. Mos, 169 p
- 1899 Ens M 1, 77 [edu]
- 1899 Fiz M Nauk 1, 35, 71
  [first serial]
- 1900 Fiz M Nauk (2)(1), 100-111,
  138-147, 161-166 (F31, 5)
Chalilov Z I 1957 Az Iz (10), 25-38
  (Z77, 7)
Chernyaev M P 1955 Rostov Pe (3),
  5-18 [edu]
Chichiy O F 1957 Ist M Isl 10, 617-
  638 (Z99, 245) [first arith]
Dakhiya S A 1956 Ist M Isl 9, 537-612
  [edu, serials]
Deaux R 1940 Mathesis 54, 139-142
  [Magnitskii]
Delone B N 1948 Matematiki i ee
  razvitie v Rossii, Mos, 16 p
  (MR17, 117)
Delone B N + 1963 in Ocherki istorii
  matematiki i mekhaniki, Mos
Depman I Ya 1955 Ist M Isl 8, 620-629
  [XVIII, geom]
- 1955 Ist M Isl 8, 630-635
  [Paris, Bunyakovskii, Zateplinskii]
- 1949 Mos IIET 3, 378-380
  [XVIII, geom]
Dinze O V + 1936 Matematika v izdani-
  yakh Akademii Nauk SSSR 1728-1935,
  Mos-Len, 335 p [bib]
Dunnington G W + 1935 Scr 3, 356-358
  [Gauss]
Fel S E 1952 Mos IIET 4, 140-155
  [XVIII, geom, Peter I]

Russia   (continued)

Gayouk Yu M 1959 Ist M Isl 12, 245-
  258  (Z101-249)  [Jacobi]
Geronimus Ya L 1952 *Ocherki o
  rabotakh korifeev russkoi
  mekhaniki,* Moscow, 520 p
Gnedenko B V 1946 *Ocherki po istorii
  matematiki v Rossii,* Mos-Len, 247 p
  (MR9, 484)
- 1948 Mos IIET 2, 390-425  [prob]
- 1957 AMM 64, 389-408  [edu]
Grigorian A T 1960 Scientia 95, 347-
  350  [mech, non-Euc]
- 1961 *Ocherki istorii mekhaniki v
  Rossii,* Mos, 291 p  (Z107, 245)
Gussov V V 1953 Ist M Isl 6, 355-475
  (Z53, 339)  [cylinder fun]
Ioffe A F ed 1945 *Ocherki po Istorii
  Akademii Nauk Fiziko-Matematicheskie
  Nauki,* Mos-Len 78 p  (MR7, 106;
  Z60, 13)
Kiro, S N 1958 Ist M Isl 11, 133-158
  (MR23, 582; Z96, 1)  [meetings XIX]
Kolman E 1957 Vop IET (4), 15-25
  (Z99, 4)  [Euler]
Kuznetsov I B 1961 *Lyudi russkoi
  nauki,* Moscow, 600 p
Lankov A V 1951 *K istorii razvitiya
  peredovykh idei v russkoi metodike
  matematiki,* Mos 151 p
Leibman E B 1961 Ist M Isl 14, 393-
  441  (Z109, 239)  [org]
Lezhneva O A 1955 Mos IIET 5, 94-135
  (Z68, 6)  [XIX, physics]
Lure S Ya 1934 Len IINT 3, 273-311
  [bib hist math]
Lysenko V I 1959 Ist M Isl 12, 161-
  178  [XVIII, metrol]
Maistrov L E 1957 Ist M Isl 10, 595-
  616  (Z100, 244)  [anc, termin]
Malkin, I 1963 Scr 26, 339-346
  [Alexander I, eng, soc of sci]
Mordukhai-Boltovskoi D D 1926 Rostov
  U 8, 111-118  (F52, 38)  [current]
Prudnikov V E 1953 Ist M Isl 6, 223-
  237  (Z52, 3)  [encyclopedia]
- 1953 Matv Shk (2), 12-15
  (MR14, 833)  [Magnitskii]
- 1956 *Russkoe pedagogi-matematiki
  XVIII-XIX vekov,* Moscow, 640 p
Sadovskii L E 1950 UMN (NS)5(2),
  57-71  (Z36, 5)  [machine math]

Saltykow N 1930 Slav Cong (1), 81-91
  (F56, 9)
Severi F 1950 Archim 2, 177-182
  (MR12, 577; Z38, 147)
Shtokalo I Z + 1966 *Istoriya
  otechestvennoi matematiki,* Kiev,
  4 vols
Smirnov V I + 1954 *Mikhail Sofronov
  ..,* Mos-Len, 54 p  (Z59, 9)
  [XVIII]
Spasskii I G 1952 Ist M Isl 5, 269-
  420  (Z49, 289)  [abacus]
Sushkeich A K 1951 Ist M Isl 4, 237-
  451  (Z44, 245)  [XIX, XX, alg]
Vasiliev A V 1921 *Matematika v Rossii,*
  Len, 72 p  (F48, 40)
Vucinich A 1960 J H Ideas 21, 161-
  179
Volkovski D 1905 Kagan (409), 9-13
  (F37, 33)  [serials]
Winter E J 1958 *Die deutsch-russische
  Begegnung und Leonhard Euler...,*
  Berlin, 204 p
Yushkevich A P 1947 Mat v Shk (3/4)
  [edu, XVIII-XIX]
- 1948 Mos IIET 2, 567-572
  [Archimedes, Euclid]
- 1949 Mos IIET 3, 45-116
  [XVIII, Euler]
- 1957 Mos IIET 17, 300-311
  (Z117, 243)  [XVII, numb thy]
- 1957 in *Istoriya estestvoznaniya v
  Rossii,* Mos, vol 1, pt 1, 26-48,
  215-222; pt 2, 33-89; vol 2, 41-222
Zubov V P 1950 Ist M Isl 3, 407-430
  [XV, infinitesimals]
- 1952 UMN (NS)7(3), 83-96 (MR14, 121)
- 1956 *Istoriografiya estestvennykh
  nauk v Rossii,* Mos, 567 p
- 1956 Int Con HS 8, 89-102
  [infinitesimal]
- 1957 Rv Hi Sc Ap 10, 97-109
  [infinitesimal]


Scandinavia

See also Denmark, Finland, Norway,
  Sweden.

Mittag-Leffler G + 1910 DMV 19, 152
  [Stockholm Cong 1909]

## Scandinavia (continued)

- 1920 Sk Kong 4 (F47, 30)
  [Stockholm Cong 1916]
Tricomi F 1940 Saggiat 1, 155-160
  (Z25, 2)

## Scotland

See also England and under CITIES,
St. Andrews.

Gibson G A 1927 Edi MSP (2)1, 1-18,
  71-93 (F53, 13; 54, 30)
  [to 1800]
Sleight E R 1944 M Mag 18, 305-314
  [before 1700]
- 1945 M Mag 19, 173-185 [1669-1746]
Turnbull H W 1940 Lon RSNR 3, 22-38
  [Gregory, Royal Society]
Wilson Duncan K 1935 *The History of
  Mathematical Teaching in Scotland
  to the End of the Eighteenth Century,* (London U P), 107 p [bib]

## Sicily

See also Italy.

Bruce R E 1938 Scr 5, 117-121, 181-
  185, 245-250 (F64, 12, 909)
Calapso R 1953 It UM Con (4)1, 274-
  286 (Z50, 2)

## Slovakia

See also Czechoslovakia.

Koutsky K 1935 Slav Cong 2, 250-251
  (F61, 28)  [XVIII, XIX]

## South Africa

Gill D 1903 So Af AAS 1, 17-36
Metcalfe C 1904 So Af AAS 2, 33

## Soviet Union

See also Russia, Azerbaijan, Belorussia, Estonia, Georgia, Kazakh-
  stan, Latvia, Lithuania, Mongolia,
  Ukraine.

Anon 1932 *Mekanika v SSSR za 15 let,*
  Mos-Len, 174 p
- 1932 *Matematika v SSSR za 15 let
  1917-1932,* Mos-Len, 239 p
  1932 Allg Verm 44, 29-30 [metrol]
  1934 *Bibliografiya matematicheskoi
  literatury (1/1/1930-1/6/1934),*
  Moscow, 72 p
  1936-1957 Matematika v izdaniyakh
  Akademiya Nauk.  Bibliograficheskii
  Ukazatel, Moscow, 3 vols
  [1728-1952]
  1938 *Matematika i estestvoznanie v
  SSSR.  Ocherki razvitiya matematicheskikh nauk za 20 let,* Mos-Len,
  1013 p
- 1948 *Matematika v SSSR za tridtsat
  let 1917-1947,* Mos-Len, 1044 p
- 1950 *Mekhanika v SSSR za 30 let
  1917-1947, Sbornik statei,* Mos-Len,
  416 p
- 1952 *Istoriya estestvoznaniya v
  SSSR,* vol 3, 119- 199
  1958 *Matematika i fizika....,*
  Simferopol, 397 p (Krym Ped 29)
- 1959 Matematika v SSSR za sorok let
  1917-1957, Mos, 2 vols (MR22, 1130)
  [*bib]
Baikov A A 1944 *Twenty-five years of
  the Academy of Science of the USSR,*
  NY (American Russian Institute)
Bikadze A V 1958 Shux Jinz 4, 583-585
  (MR20, 738)  [1918-1958]
Chen K-K 1958 Shux Jinz 4, 586-590
  (MR20, 738)
Egorov D F 1928 *Uspekhi matematiki v
  SSSR,* Len, 12 p
*Forsythe George E 1956 *Bibliography
  of Russian Mathematical Books,* NY
  (Chelsea), 108 p [1930-1956]
Joravsky D 1960 Am Ac Pr 89, 562-580
  (MR22, 260)  [recent]
Lapko A F + 1957 UMN (NS)12(6), 47-130
  (MR19, 1029)  [meetings]

Soviet Union (continued)

- 1958 UMN 13(5), 121-166
  (MR20, 1042) [meetings]
- 1967 UMN 22(6), 13-140
  [1918-1923]
La Salle J P 1961 AMSN 8, 25-29
  (MR22, 2052)
Lavrentev M A 1948 SSSRM 12, 411-416
  (Z30, 2)
Lukomomskaya, Anna M 1957 *Bibliogra-
  ficheskie istochniki po matematike
  i mekhanike, izdannye v SSSR za
  1917-1952*, Mos, 355 p
- 1963 *Bibliograficheske istochiniki
  po matematike i mekhanike izdannye
  v SSSR za 1953-1960*, Mos
Turan P 1950 M Lap 1, 243-266
  (MR12, 311) [numb thy]
Varga O 1951 M Lap 2, 190-218
  (MR13, 611) [diff geom]
Vekua I N 1958 Prag Pok 3, 402-409
  (MR 21, 231; Z82, 242)


Spain

Anon 1895 Bib M (2)9, 120 (F26, 3)
  [to 1500]
- 1927 Rv M Hisp A (2)2, 154-156
  (F53, 36) [org]
Archilla y Espejo + 1888 *Discursos
  leidos ante la real Academia de
  Ciencias,* Madrid
Berenguer P A 1895 Prag M 5, 116-121
  (F26, 15) [XVII, geom]
Diéguez D F 1925 Rv M Hisp A 7, 145-
  148 (F51, 2) [crit of lit]
Echegaray J 1916 *Conferencias sobre
  fisica matematica,* Madrid, Curso
  de 1910 a 1911, 394 p; 1911 a 1912,
  582 p; 1912 a 1913, 531 p; 1913 a
  1914, 550 p (F46, 31)
Eneström G 1894 Bib M, 33-36 [XVI]
Feliv e Vegués F 1905 *Algunos trabajos
  matematicos,* Barcelona, 234 p
  (F36, 45)
Fontès J 1899 Rv Pyr 11, 16 p
  (F30, 4) [XVI]
Galdeano Z G de 1894 Prog M 4, 47
  [XVI]

- 1899 Ens M 1, 6 [XIX]
Karpinski L C 1936 Osiris 1, 411-420
  (Z14, 147) [arith]
Loria G 1919 Scientia 25, 79-85,
  99-108, 353-359, 441-449
- 1938 Arg UM Pu 1, 24 p [Argentina]
Mailly E 1868 Arc MP 48, 376
  (F1, 5)
Millas Vallicrosa Jose M 1931 *Assaig
  d'historia de les idees fisiques
  e matematiques a la Catalunya
  medieval,* Barcelona, 351 p
- 1949 *Studies on the History of
  Spanish Science,* Barcelona, 499 p
  (in Spanish) (Z40, 1)
Perott J 1882 Boncomp 15, 163-170
  [XVI, arith]
Rey-Pastor D J 1913 *Los matematicos
  espanoles del siglo XVI,* Oviedo,
  75 p (F44, 4)
- 1926 *Los matematicos espanoles del
  siglo XVI,* Madrid, 164 p
Sanchez Perez Jose A 1921 *Biografias
  de matematicos arabes que floreceiron en Espana,* Madrid, 164 p
Toledo L O de 1911 Esp SM 1, 62-66
  (F42, 50)
- 1915 Esp SM 4, 35-38, 175-177
  [XVII, log tab]
Vallin A F 1893 *Discursos leidos ante
  la real academia de ciencias
  exactas, fisicas y naturales en la
  recepcion publica,* Madrid, 337 p
Vera Francisco 1929-1933 *Historia de
  la matematica en Espana,* Madrid,
  4 vols (F55, 600; 59, 838)
- 1942 Archeion 24, 403-427
  (MR4, 181) [XVI, chronol]
- 1947 *Le matematica de los musul-
  manes espanoles,* Buenos Aires,
  236 p
- 1948 *Los judios espanoles y sus
  contribucion a las ciencias exactas,*
  Buenos Aires, 251 p
- 1948 Pos SM An 21, 94-98 (Z37, 1)
  [Toledo]
Vicuna G 1890 Bib M (2)4, 33-36
  [XVI, XVII]

## Sweden

See also Scandinavia.

Andersson T 1925 Nor St Tid 4, 1-24
(F51, 36) [stat]
Dahlin E M 1875 *Bidrag till de mate-*
*matiska vetenskaperns historia i*
*Sveriga foere 1679* Upp U (F7,17)
Eneström G 1886 Bib M, 45-47, 92-95,
140-141 [non-Swedish authors pub
in S]
Hueltmann F W 1869-1871 Dillner 2,
57, 105; 3, 7, 49, 241; 4, 5, 97
(F2, 15) [arith]
- 1875 Dillner 5 (F7, 17)
Nordenmark N V E 1933 Ark MAF 24(2)
[XVIII, ast]
Vanas E 1954 Lychnos, 141-164

## Switzerland

See also under CITIES, Basle, Geneva,
Zurich.

Anon 1923 Ens M 22, 291-300, 369-380
(F49, 22) [Math Soc]
Fueter Eduard 1941 *Geschichte der*
*exakten Wissenschaften in der*
*Schweizerischen Aufklaerung (1680-*
*1780),* Aarau-Leipzig, 336 p
(MR4, 65; Z25, 292)
- 1941 *Grosse Schweizer Forscher,*
2nd ed, Zuerich
Graf J H 1889 *Geschichte der Mathe-*
*matik und der Naturwissenschaften*
*in bernischen Landen vom Wiederauf-*
*bluehen der Wissenschaften bis in*
*die neuere Zeit,* Bern-Basel, 2 vols
[XVI, XVIII]
- 1896 Bern NG, 287-292 (F27, 2)
- 1902 Bern NG, 138-149, 245-254
- 1903 Bern NG, 96
- 1904 Bern NG, 196
- 1905-1906 *zur Geschichte der mathe-*
*matischen Wissenschaften an der*
*ehemaligen Akademie und der Hoch-*
*schule ,* Bern NG (S) (F37, 10)
- 1911 Bern NG, 277-281
[Micheli du Crests]

Guggenbuhl L 1961 MT 54, 363-365
[anniv Math Soc]
Isely L 1902 *Histoire des sciences*
*mathématiques dans la Swisse*
*Francaise,* Neuchatel, 215 p
(F33, 9)
Plancherel M 1961 Ens M (2)6, 194-
218 (Z94, 4) [1850-1950]
Polya G 1960 MT 53, 552-558 [edu]
Rudio F + 1908 Zur NGV 53(4),
605-611
- 1909 Zur NGV 54, 463-480, 505-506
- 1910 Zur NGV 55, 541-548 (F41, 10)
[Euler coll works]
- 1912 Zur NGV 56(4), 552-557
[Euler]
- 1913 Zur NGV 58, 431-437, 452-453
- 1914 Zur NGV 59, 564-571
- 1915 Zur NGV 60, 643-646
- 1916 Zur NGV 61, 726-733, 736-737
- 1917 Zur NGV 62, 690, 719-722
- 1918 Zur NGV 63, 566
- 1919 Zur NGV 64, 837, 837-841,
849-850, 855-861
- 1920 Zur NGV 65, 605, 616-619
Schinz H + 1922 Zur NGV 67, 396-399,
407-413, 422-423
- 1923 Zur NGV 68, 551-554, 593-596
- 1924 Zur NGV 69, 308-313, 326-341
- 1925 Zur NGV 70, 319-321
Wolf R 1845 Bern NG, 121-, 137-
- 1846 Bern NG, 161-, 209-
- 1847 Bern NG, 68-, 101-, 129-, 161-
- 1848 Bern NG, 46-, 217-, 269-
- 1858-1862 *Biographien zur Kultur-*
*geschichte der Schweiz,* Zurich,
4 vols
- 1879 *Geschichte der Vermessungen*
*in der Schweiz...,* Zurich
(F11, 43)

## Turkey

Adnan A H 1938 Archeion 21, 35-61
(F64, 11) [to XVII]
Husny H 1932 Int Con (8)6, 507-509
(F58, 30)
Taneri Kemal Zuelfue 1958 *Tuerk*
*matematikcileri,* Istanbul, 104 p

## Ukraine

See also Kharkov, Kiev, Rostov (CITIES).

Anon 1957 Ukr IM 9, 355-358
(Z77, 242) [1917-1957]
Gnedenko B V + 1956 Ist M Isl 9,
405-426 (Z70, 5)
- 1956 Ist M Isl 9, 477-536
(Z70, 242) [prob]
Gratsiyanskaya L N 1961 Ist M Zb
2, 148-155 [anc, Ukraine]
Putyata T V + 1952 *Diyalnist vidat-*
*nikh mekhanikiv na Ukraini,* Kiev
Shcherban O N 1952 Uk Do, 423-430
(MR20, 1042)
*Shtokalo I Z 1948 Uk Obcis (10),
5-40 (MR 12, 1) [1917-1947]
- 1958 *Naris rozvitku matematiki na*
*Ukraini za 40 rokiv Radnyanskoi*
*vladi,* Kiev, 81 p
Shtokalo I Z + 1963 *Ukrainska mate-*
*matichna bibliografiya 1917-1960,*
Kiev, 382 p
Sintsov D 1938 Geom Zb 1, 1-8
(F64, 920) [geom 1917-1937]

## United States of America

See also America, and under CITIES,
Amer Math Soc, Dartmouth, Harvard,
Johns Hopkins, Math Asso Amer,
Natl Coun Tea Math, New Engl Asso
Tea Math, Princeton, Waukasha Math
Soc, West Virginia U.

Anon 1926 AMM 33, 318 [Dict Nat Biog]
- 1939-1940 AMM 46, 233; 47, 107
[biog]
Anon ed 1938 *Semicentennial Addresses*
*of the American Mathematical*
*Society (AMS),* 315 p (Vol 2 of
*Semicentennial Publications*)
Birkhoff G D 1938 Sci 88, 461-467
[1888-1938]
Bradley A D 1937 Scr 5, 45-51
[Penn German arith]
Cairns W D 1944 AAASB 3, 1 [edu]
Cajori Florian 1890 *The Teaching and*
*History of Mathematics in the*
*United States,* (USGPO), 440 p

- 1891 Colo Stu 2, 39-47 (F23, 35)
[dioph anal]
- 1892 Edu 12, 170
- 1925 AMM 32, 414-416 [notat]
- 1928 AMM 35, 300 [notat]
Cochrane Rexmond C 1966 *Measures for*
*Progress: A History of the Nation-*
*al Bureau of Standards,* Wash D C,
703 p
Coolidge J L 1926 AMM 33, 61-76
[Adrain]
Dorwart H L 1948 Scr 16, 181-185
(Z39, 243) [early ms]
Dresden A 1922 AMSB 28, 303-307
[1897-1922]
Fields J C 1932 in *Fifty Years in*
*Retrospect: Anniversary Volume of*
*the Royal Society of Canada 1882-*
*1932,* Ottawa-Toronto, 107-112
Finkel B F 1940-1943 M Mag 14; 15;
16; 17 [math serials]
Fiske T S 1904 AMSB (2)11, 238-246
(F36, 44) [recent]
Ginsburg J 1933 Scr 1, 275
[early European notice]
Jones P S 1944 MT 37, 3-11
[early geom]
- 1953 MT 46, 341-342 [flag]
- 1955 MT 48, 333-338 [first math]
- 1956 MT 49, 30-33
- 1957 MT 50, 376-378 [XX, women]
Karpinski L C 1924 Sch Soc 19
[arith]
- 1940 *Bibliography of Mathematical*
*Works Printed in America Through*
*1850,* (Oxford U P)
- 1941 Scr 8, 233-236
[suppl to 1940]
- 1945 Scr 11, 173-177
[suppl to 1940]
Keyser C J 1902 Ed Rv 346-357
(F33, 9) [recent]
Lady C H 1948 MT 40, 38 [arith 1810]
Locke L L 1924 AMM 31, 422-429
[calculating machines]
Lowan A N 1949 Scr 15, 33-63
[Natl Bur Stan]
Martin Artemus 1900 *Notes on the*
*History of American Text-Books on*
*Arithmetic,* Wash D C
Meserve H G 1929 MT 21, 336-345 [1829]

United States (continued)

*Miller G A 1923 Sci 59, 1-7
[1850 to date]
- 1924 Sci (NS)59, 1-7
- 1934 Sci 80, 356-357 (F60, 833)
- 1935 M Stu 3, 8-11 (F61, 3)
[early]
- 1936 SSM 35, 292-296
- 1940 Sci (NS)92, 216-217
[first printed]
Phalen H R 1946 AMM 53, 579-582
[first prof]
Read C B 1941 SSM 40, 516-517
[arith 1813]
Reingold Nathan ed 1964 *Science in
Nineteenth Century America: A Documentary History,* NY, 339 p
Richeson A W 1936 M Mag 10, 73-79
[arith, Coleburn]
Rocquigny G de 1906 Intermed 13, 102
(Q970) [first math]
Schreiber E W 1941 MT 34, 76-77
[attitudes]
Simons Lao G 1924 *Introduction of
Algebra into American Schools in
the Eighteenth Century,* Wash D C,
80 p
- 1931 Isis 15, 104-123 (also 1939
Scr Lib 4) [XVIII, France]
- 1932 Scr 1, 29-36 [alg 1837]
- 1934 Scr 2, 294-295 [XVII]
- 1935 Scr 3, 355-356 [XVIII, alg]
- 1936 *Bibliography of Early American
Textbooks on Algebra,* NY, 68 p
(Scr Stu 1) [to 1850]
- 1936 M Mag 10, 188 [arith]
- 1936 Osiris 1, 584-605 (also Scr
Lib 4) (F62, 4) [geom XVIII]
- 1937 Scr 4, 207-219 [calc]
Slaught H E 1920 AMM 27, 443-451
Smith D E 1933 Scr 1, 277-285
[serials]
- 1933 Toh MJ 38, 227-232
(F59, 5; Z8, 99) [early]
- 1934 Scr 2, 221-223 [ms XVII]
Smith D E + 1934 *A History of Mathematics in America Before 1900,*
Chicago, 219 p (F60, 3; Z8, 338)
Struik D J 1956 Sci Mo 82, 236-240
[Ticonderoga]

- 1957 *The Origins of American Sci-
ence (New England),* NY, 442 p
(MR19, 825) [repr 1962 as *Yankee
Science in the Making]*
- 1959 Sci 129, 1100- [XVIII, XIX]
Sueltz B A 1944 MT 36, 183-185
[arith 1819]
Yates R C 1937 AMM 44, 194-201
[Sylvester, Virginia]


Yugoslavia

See also Croatia.

Nikolitch G M 1939 Wiad M 45, 115-
154; 46, 123-138 (F65, 9) [ast]
Stipanic E 1953 Beo MF 2, 159-168
(Z53, 340) [Getaldic]


CITIES, ORGANIZATIONS, UNIVERSITIES

Accademia dei Lincei

Picone M 1966 *L'Accademia nazionale
dei Lincei,* 2nd ed, Rome, 55 p
Drake S 1966 Sci 151, 1194-1200
Favaro A 1887 Boncomp 20, 95-158
Gabrieli G 1939 Linc Mor (6)7, 123-
530 (F65, 1084)


American Mathematical Society

Anon AMSB 45, 1-30 [50 anniv]
Archibald Raymond C 1938 *A Semicen-
tennial History of the American
Mathematical Society 1888-1938
with Biographies and Bibliographies
of the Past Presidents* (AMS), 268 p
- 1939 AMSB 45, 31-46 (F65: 1, 16)
*Birkhoff G D 1938 Sci (18 Dec),
461-467
Fiske T S 1905 AMSB 11, 238
Kline J R 1943 AAASB 2, 36-38
Mc Clintock E 1894 AMSB (2)1, 85-94
(F26, 44)
Meder A E 1938 Sci 88, 230-232
Richardson R G D 1924 Am CAS
11(1)(20), 131-133

## Amsterdam

Anon 1879 *Feestgave van het wiskundig genootschap te Amsterdam...,*
Haarlem (F11, 13)
Matthes J C 1879 *Feestrede ter ge-
legenheid van het honderdjarig
bestaan van het wiskundig genootschap...,* Amsterdam (F11, 13)


## Basle

Carathéodory C 1945 in *Festschrift
zum 60 Geburtstag von Prof Dr
Andreas Speiser,* Zurich, 1-18
(MR7, 354) [calc var]
Fleckenstein J O 1959 *L'école mathématique baloise des Bernoulli à
l'aube du XVIII siècle,* Paris,
21 p
Speiser A 1939 *Die Basler Mathematiker, 117 Neujahrsblatt,* Basel,
51 p (F65, 1083; Z21, 195)
Spiess D 1945 Basl V 56, 86-111
(MR7, 354; Z60, 12)


## Berlin

Biermann K-R 1960 *Vorschlaege zur
Wahl von Mathematiker in die
Berlin Akademie,* Berlin, 75 p
(also Ber Ab (3)) (Z94, 4)
- 1964 Ber Hum MN 13(2), 185-198
[competition]
- 1964 Schr GNTM, 11-20
- 1967 Ber Mo 9(3), 216-222
[Ber Acad]
Biermann K-R + eds 1960 *Deutsche
Akademie der Wissenschaften zu
Berlin, Biographischer Index der
Mitglieder,* Berlin, 248 p
Lamla Ernst 1911 *Geschichte des
Mathematischen Vereins 1861-1911,*
Berlin
Schroeder Kurt 1953 in *Bericht ueber
die Mathematiker-Tagung in Berlin
vom 14, bis 18, Januar 1953. An-
laesslich der Einweihung der neuen
Raume der drei Mathematischen
Institut der Humboldt-Universitaet*

*zu Berlin,* Berlin, 1-4 (Z52, 3)
Vasilev A V 1882 *Prepodavanie chistoi
matematiki v Berlinskom i Leiptsig-
skom universitetakh,* Kazan, 25 p
Yushkevich A P + 1959-1961 *Die Berliner und die Petersburger Akademie
der Wissenschaften in Briefwechsel
Leonhard Eulers,* Berlin, 2 vols


## Bologna

Agostini A 1923 Pe M 3, 1-4 [teaching]
Bortolotti Ettore 1921 *Lo studio di
Bologna ed il rinnovamento delle
scienze matematiche in Occidente,*
Bologna, 25 p
- 1925 Pe M (4)5, 147-192 (F51, 6)
[XVI, alg]
- 1926 Rm Sem (2)3, 62-64
(F52, 14) [XVI, alg]
- 1928 *L'école mathématique de Bologne,*
Bologna, 75 p (F54,23)
- 1933 *La scuola matematica di Bologna
nei secoli XVI e XVII,* Bologna, 30 p
(extract from *Bologna nella storia
d'Italia)*
- 1947 *La storia della matematica
nella Università di Bologna,*
Bologna, 226 p
Gherardi S 1846 Bln No Co 8, 519-
- 1871 Arc MP 52, 65-204 (F3, 6) [U]
Riccardi P 1879 Boncomp 12, 299-312
(F11, 13) [U]
Tricomi F G 1962 Bln Rn P 250(11)9(1),
116-119 (Z106, 3) [XIX]


## Bremen

Wietzke A 1928 Bremen NV 27, 125-142
(F57, 1303) [Gauss]


## Breslau

See also Wroclaw.

Hoffmann H 1934 Schles GZ 68, 107-117
(F60, 828) [Jesuit U]
Sturm R 1911 DMV 20, 314-321 [U]

Breslau  (continued)

1911 in *Festschrift Univ Breslau,*
434-440  (F42, 43)  [XIX U]


Brussels

Mailly E 1883 Bel Mm Cou 34; 35
  [Acad]
- 1886 Bel Bul (3)12, 786-794
  [orgs]
- 1887 Bel Mm Cou 40, 48 p
  [French control]


Bucharest

Nicolescu M 1956 M Lap 7, 18-25
  (MR20, 848)  [anal chair U]


Budapest Mathematical Society

Koenig D 1941 MP Lap 48, 7-32
  (Z24, 244)


Calcutta Mathematics Society

Anon 1909 Rv GSPA 20, 977
Sen N 1928 Clct MS 19, 151
  (F54, 48)  [20 anniv]


Cambridge, England

Anon 1932 Nat 130, 117-119
  (F58, 990)  [Lucasian profs]
Ball W W R 1889 *A History of the
  Study of Mathematics at Cambridge,*
  Cambridge
- 1912 M Gaz 6, 311-323
- 1921 *Cambridge Notes Chiefly Con-
  cerning Trinity College and the
  University,* Cambridge
Bushell W F 1960 M Gaz 44, 172-179
  [tripos]
Clark J W 1892 Cam PSP 7, i-
  [C Phil Soc]

Forsyth A R 1935 M Gaz 19, 164-172
  [tripos]
Gunther R T 1937 *Early Science in
  Cambridge,* Oxford, 513 p
  (repr 1969 Dawsons)
Pearson K 1936 M Gaz 20, 27-36
  [XIX, tripos]
Pedoe D 1964 AMM 71, 666-670
  [tripos]
Watson E C 1938 Scr 6, 101-106
  [XIX]


Dartmouth College

Brown Bancroft H 1964 *Belzaleel
  Woodward and his Successors, A
  Brief Account of Mathematics at
  Dartmouth College 1769-1960,*
  Hanover, N. H.  (multilith)


Dublin

Maxwell C 1946 *A History of Trinity
  College Dublin 1591-1892,* Dublin
McConnell A J 1944 Dub Ac P 50, 75-
  81  (MR6, 253; Z60, 13)  [1800-
  1850]


Ecole Normale (Paris)

Anon 1895 *Le Centenaire de l'Ecole
  Normale, 1795-1895,* Paris, 746 p
  [port]


Ecole Polytechnique (Paris)

Anon 1894-1897 *Ecole Polytechnique
  Livre du centenaire, 1794-1894*
  Paris 3 vols  [port]
- 1894 Nat 50, 82-83  (F25, 97)
Klein F 1927 Naturw 15, 5-11, 43-49
  (F53, 24)  [XIX]
Launay L de 1933 *Un grand Français,
  Monge, fondateur de l'Ecole
  polytechnique,* Paris, 280 p
Ocagne M d' 1931 Par EP (2)29, 55-86
  (F57, 5)  [Acad]

## Ecole Polytechnique (Paris) (cont'd)

- 1931 Rv Q Sc (4)20, 5-37
  (F57, 1305) [XIX]
Pinet G 1894 *Histoire de l'Ecole
  Polytechnique*, Paris-Liege, 400 p
  (F25, 97)
Taton R 1964 Koyre 1, 552-564
  [anal geom]

## Erfurt

Lorey W 1933 Unt M 39, 120-122
  (F59, 23) [XV]

## Florence

Procissi A 1938 It UM Con 1, 450-452
  (F64, 16)
Targioni-Tozzetti G 1780 *Notizie
  sugli aggrandimenti delle scienze
  fisiche, accaduti in Toscana nel
  corso degli anni XL nel secolo
  XVII*, Florence, 3 vols

## Freiburg

Gericke H 1955 *Zur Geschichte der
  Mathematik an der Universitaet
  Freiburg...*, Freiburg, 88 p
  (MR17, 445; Z65, 244)

## Geneva

Anon 1901 Rv GSPA 12, 1041

## Giessen

Lorey W 1935 Gies Hoch 10, 47-75
  (F61, 18)
- 1936 DMV 46, *10-11* (F62, 27)
  [Leibniz]
- 1937 Gies Hoch 11(3), 46-50
  (F63, 18)
- 1937 Gies Hoch 11(3), 54-97
  (F63, 18) [XIX-XX]

## Goettingen

Anon 1902 *Festschrift zur Feier des
  hundertfunfzigjahrigen Bestehens
  der koniglichen Gesellschaft der
  Wissenschaften zu Goettingen*,
  Berlin
Klein F 1908 Int Woch 2, 519-532
  (F39, 46) [sci org]
- 1909 DMV 18, 130-132 [air flight]
Mueller Conrad 1904 *Studium zur
  Geschichte der Mathematik, inbeson-
  dere des mathematischen Unterrichts
  an der Universitaet Goettingen im
  18 Jahrhundert* (Goettingen U Thesis),
  Leipzig (also in Ab GM 18)
Mueller C H ed 1904 Ab GM 18
  [educ, XVIII]
Wuestenfeld F 1887 *Die Mitarbeiter an
  den Goettingischen gelehrten Anzei-
  gen in den Jahren 1801 bis 1830*,
  Goettingen, 87 p

## Hamburg Mathematical Society

Anon 1881 Ham MG (1), 8-16 (F13, 18)
Lony G 1940 Ham MG 8(1), 7-41
  (Z23, 196; F66, 20)
Riebesell P 1928 Ham MG 6, 395-397
  (F54, 29)

## Harvard University

Badger Henry Clay 1888 *Mathematics
  Thesis of Junior and Senior Classes
  1782-1839* (Harvard U P), 14 p
Coolidge J L 1924 Harv Alu B 26,
  372-378
- 1943 AMM 50, 347-356
  (Z61, 5) [XVIII-XX]
Simons L G 1925 AMM 32, 63-70 [1730]
Young E J 1880 *Subjects for Masters
  Degrees in Harvard College 1655-
  1794*, Cambridge, Mass

## Heidelberg

Bopp K 1931 in *Taetigkeitsbericht der
  Math Fachschaft an der Univ*

## Heidelberg (continued)

*Heidelberg,* 18-19 (F57, 1311)
Christmann F 1925 *Studien zur Geschichte der Mathematik und der Mathematischen Unterrichts in Heidelberg* (Heidelberg U Thesis), 168 p (F51, 38)

## Iasi

Ionescu I 1929 Gaz M 35, 21 (F55, 622)
Myller A 1935 Rv St Adam 22, 3-9 [educ, U]
Popa I 1935 Rv St Adam 21, 104-111 [seminar]
- 1936 Nat 25, 35-39 [U]
- 1960 *Dezvoltarea matematicii in Contributii la istoria dezvoltarii Universitatii din Iasi 1860-1960, Bucarest* 2, 7-39

## International Mathematical Union

Fehr H 1933 Ens M 31, 276-278 (F59, 45)

## Jena

Haussner R 1911 DMV 20, 47-56 [institute at U]

## Jesuits

Stein P G 1941 Milan Sem 15, 129-146 (MR9, 485)

## Johns Hopkins University

Anon 1894 *Bibliographia Hopkinsiensis 1876-1893,* Baltimore, 64 p

## Karlsruhe

Lorey W 1947 ZAM Me (25/27), 142-144 (Z32, 98) [XX]

## Kazan

Gagaev B M 1960 Kazh Uch Z 120(7), 67-86 [U]
Laptev B Z 1959 Ist M Isl 12, 11-58 (MR23, 697; Z98, 9) [XX]
- 1960 Kazn Uch Z 120(7), 24-60 [XX]

## Kharkov

Anon 1933 Khar Me M (4)6, 79-80; (4)7, 87 (F59, 44) [1931-1932]
Bakhmutskaya E Ya 1955 Khar PI Tr 5, 185-191 [XIX, XX]
Marchevskii M N 1956 Ist M Isl 9, 613-666 (Z70, 5) [XIX, XX]
- 1956 Khar MUZ (4)24
Ossipova J + ed 1908 *The Physico-Mathematical Faculty of the Imperial University... 1805-1905,* Kharkov, 611 p (in Russian) (F40, 4)
Psheborskii A P 1911 *The Mathematical Society at the University of Kharkov (1879-1904),* Kharkov, 26 p. (F42, 42) (in Russian) (repr from a 100 Jubilee vol on U of K)
Sintsov D M 1908 *The Chair of Pure and Applied Mathematics at the University of Kharkov During its First Hundred Years (1805-1905),* Kharkov, 72 p (in Russian) (F39, 50)
- 1915 Khar U Za (1), 1-48; (2), 49-84 [bib, 1895-1905]

## Kiev

Dobrovolskii V O 1961 Ist M Zb 2, 57-67 (MR33, 1211) [alg]
- 1964 in *Z Istorii Vichiznyanogo Prirodoznabstva,* Kiev, 115-143
Gratsianskaya L N + 1959 in *Istoriya Kiivskogo universitet,* Kiev

## Krakow

Dianni Jadwiga 1963 *Studium mate-
matyki na uniwersytecie jagiel-
louskim do polowy XIX wiegu*,
Krakow, 236 p
Goloba Stanislawa ed 1964 *Studia z
Dziejow Katedr Wydziatu Matematyki,
Fizyki,... Univ. Jagiellouskiego*,
Krakow
Nozicka F 1958 Prag Pok 3, 373-377
(Z82, 242) [Univ]

## Leipzig

Vasilev A V *Prepodavanie chistoi
matematiki v Berlinskom i Leipt-
sigskom universitetakh*, Kazan,
25 p

## Leningrad (=St. Petersburg)

Anon 1931 *Na Leningradskom mate-
maticheskom fronte* (sbornik dokla-
dov), Mos-Len, 42 p
Bernstein S N 1940 Len MZ 10, 3-11
(MR2, 114) [probab]
Delone B N 1947 *Peterburgskaya shkola
teorii chisel*, Mos-Len, 421 p
(MR19, 1029)
Depman I Ya 1960 Ist M Isl 13, 11-106
(MR23, 584) [math soc]
Galchenkova P I 1961 Ist M Isl 14,
355-392 (Z109, 239) [XIX]
Yushkevich A P + 1959-1961 *Die Ber-
liner und die Petersburger Akademie
Wissenschaft in Briefwechsel L
Eulers*, Berlin, 2 vols
- 1964 Int Con HS [probab]

## Liége

Godeaux L 1953 Bel BS (5)19, 1412-
1423 (F59, 828) [geom]

## Livorno

Agostini A 1951 It UM Con 3, 50-51

(Z45, 147)

## London

Anon 1933 Nat 131, 896-897 [U]
Collingwood E F 1965 New Sci 25,
95-97 [100 anniv LMS]
- 1966 LMSJ 41, 577-594
(MR33, 5429) [LMS]
Glaisher J W L 1926 LMSJ 1, 51-64
(F52, 35) [LMS]
Zeuthen H G 1925 Nor M Tid 7, 47-49
(F51, 34) [LMS]

## Louvain

Bosmans H P R 1926 Brx SS 44(1),
458-462 [mid mss]
Gilbert P 1927 Rv Q Sc 12, 17-46
Lefebvre B 1929 Rv Q Sc 15, 29-57
[XV]
Mailly E 1877 Bel Oc 27 (F9, 12)
[XVIII, Zach]
Simonart F 1927 RvQ Sc 12, 73-100
(F54, 34) [U, 1427-1927]

## Lwow

Zajaczkowski W 1894 *History of the
Polytechnical School in Lwow
(1844-1894)*, Lwow, 169 p (in
Polish) (F25, 97)

## Marburg

Gerling C L 1842 Arc MP 2, 212-

## Mathematical Association
## (Great Britain)

Wilson C J M 1920 M Gaz 10, 239-247
[early hist]

## Mathematical Association of America

Allendoerfer Carl B + ed 1967

## Mathematical Association of America
(cont'd)

AMM 74(1)(2), 109 p
(Fiftieth Anniv Issue)
*Anon 1968 CUPMN 2 [CUPM survey]
Cairns W D 1916 AMM 23, 1-6
[organization]
Finkel B F 1931 AMM 38, 305-320
[AMM early history]
Ford L R 1946 AMM 53, 582 [AMM]
Karpinski L C 1920 Isis 3, 490-491
[AMM 1894-1919]
Slaught H E 1914 AMM 21, 1 [AMM]
*- 1915 AMM 22, 251-253 [formation]
*- 1915 AMM 22, 289-292 [motivation]
- 1927 AMM 34, 225 [sections]

## Milan

Masotti Arnaldo 1963 *Matematica e
matematici nella storia di Milano
da Severino Boezio a Francesco
Brioschi,* Pavia, 32 p + 35 ill
(=Milan Sem 33, 1-28) (MR28, 401)
[*facs]

## Modena

Bonacini C 1932 Int Con 6, 511-515
[observatory]
Bortolotti E 1918 Lonia (2)20, 1-11
(F46, 49)

## Moscow

Aleksandrov P S 1955 Ist M Isl 8,
9-54 (68, 6) [XX]
Aleksandrov P S + 1948 Ist M Isl 10,
9-42 [XX]
- 1957 UMN (ns)12(6), 9-46 (MR19,
1027; Z79, 4)
[Math So]
Golubev V V 1955 Ist M Isl 8, 77-126
(Z68, 6) [mech]
Likholetov I L + 1955 Ist M Isl 8,
127-480 (Z68, 6) [XIX, Univ]
Nekrasov P A 1904 *Moskovskaya filo-
sofsko-matematicheskaya shkola i ee*

*osnovateli,* Mos, 249 p
Shevolev F Ya 1966 Ist Met En 5,
62-74 (MR33, 7232) [MathSoc]
Volkovski D 1908 Kagan (459)
[teachers orgs]
Vygodskii M Ya 1948 Ist M Isl 1,
141-183 [XIX]
*Yushkevich A P 1948 Ist M Isl 1,
43-140 [XVIII, XIX]

## Muenster

Lorey W 1934 Muns Semb 5, 15-43
(F60, 832)
- 1935 Muns Semb 6, 110-120
(F61, 22) [E Heiss]
1937 Muns Semb 10, 124-133
(F63, 18) [Killing, Lilienthal]

## Munich

Mueller F J 1927 Z Vermess 26, 59-75
(F54, 47)

## Naples

Amodeo F 1898 Nap FM Ri (3)4, 102
- 1902 Nap Pont 32, 28- [XVII]
- 1905 Nap Pont (2)34, 1-60
(F36, 4) [institutes to 1800]
- 1905 *Vita matematica napoletana,*
Pt 1, Naples, 224 p (F36, 4)
(2nd ed 1924)
- 1922 Nap Pont (2) 27, 12-32
[institutes 1825-1860]
- 1924 Nap Pont 54, 31 p (F50, 8)
[rare books]
- 1924 *Vita matematica napoletana,
studio storico,* Pt 2, Naples,
391 p (F50, 7)
- 1934 *Argonenti trattati nella
scuola matematica Napoletana (1615-
1860),* Naples (F60, 11)
- 1935 It SPS 23(2) (A), 172
(F61, 957) [1651-1860]
Colangelo F 1833-1834 *Storia dei filosofi e dei matematici Napoletani e
delle loro dottrine, da Pitagorici
sino al secolo XVII dell'eravolgare,*

Naples (continued)

Naples, 3 vols
Pezzo P del 1899 Nap Pont 29, 6 p
[contest on hist of N math 1732-
1801]
Roubee P 1915 Nap Pont (2)45(20)(10),
9 [scuola di ponti e strade]

National Council of Teachers of
Mathematics

Austin C M 1929 MT 21, 204-213
Schreiber E W 1945 MT 38, 372-376

New England Association of Teachers
of Mathematics

Osgood W F + 1924 MT 17, 94-103
Ransom W R 1953 MT 46, 333-336

Nuremberg

Doppelmayr Johann G 1730 *Historische
Nachricht von den Nuernbergischen
Matematicis und Kuenstlern*,
Nuremberg, 360 p

Odessa

Kiro S N 1956 Odess UTM 146(6),
89-109 [U pub]
- 1961 Ist M Zb 2
[Odessa New Russian Univ]

Oxford

Gunther R T 1921 *Early Science in
Oxford*, Oxford, vol 1, 407 p
(repr 1967 Dawsons)
- 1922 *Early science in Oxford,
Vol 2: Mathematics*, London 101 p
(F48, 45)
James T E 1933 Nat 131, 716-717
[old Ashmolean]
Michalski K 1920 Krak BI, 59-88
[XIV phil]

Padua

Favaro .A 1878 Boncomp 11, 799-801
(F10, 23) [U]
- 1880 Pado Riv 30, 119-126
(F12, 28) [XIV, XV, XVI]
- 1919 Arc Sto 1, 151-152
[U 1222-1922]
Tenca L 1953 Ven I At 111, 83-102
(MR15, 923) [1700+]

Palermo

Franchis M de 1914 Paler RS 9, 1-68
(F45, 40) [Circolo M. di P]
- 1930 It SPS 18(1), 356-364
(F56, 9)

Paris

See also Ecole Normale, Ecole
Polytechnique.
Beaujouan G 1954 in *Homenaje a Millas-
Villacrosa* 1, 93-124 (MR16, 781)
[arith XIII-XIV]
Brocard H 1897 Intermed 4, 8 (Q733)
[monuments]
Cajori F 1929 AMM 36, 162
[Inst H Poincaré]
Folkierski W 1895 Prace MF 6, 151-175
(F26, 43) [Acad]
Fréchet M 1929 AMSB 35, 198-200
[Inst H Poincaré]
Lebesgue H 1922 Rv Sc, 249-262
[College de France]
Lemaire G 1909-1910 Intermed 16,
172; 17, 20-21 (Q3591) [XVI, U]
Michalski K 1920 Krak BI, 59-88
[XIV phil]
Sédillot L A 1869 Boncomp 2, 343-368,
387-448, 461-510 (F2, 17)
[Collège de France]
Sergescu P 1938 *Les mathématiques à
Paris au moyen age*, Paris, 16 p
(Réunion internationale des math.
à Paris, July 1937) (F64, 909)
Smith David E 1921 AMM 28, 62-63
[shrines]

Paris (continued)

- 1923 AMM 30, 107-113, 166-174
- 1924 *Historical Mathematical Paris,*
  Paris, 48 p

Pisa

*Ricci G 1951 Parma Rv M 2, 155-174
  (MR13, 420) [1848-1948]
Sansome G 1947 It UM (3)2, 135-139
  (Z31, 146) [XIX]

Prague

Seydl O 1951 Prag V (7) (MR15, 275)
  [Coll of St Clemens]

Princeton

Archibald R C 1932 Scr 1, 167-168
  (F58, 51) [Inst for Adv Study]
Bassi A 1938 Tor Sem, 65-66
  (F64, 23)
- 1939 Pe M (4)19, 57-79 (Z21, 3)

Rostov

Belozerov S E 1953 Ist M Isl 6, 247-
  352   (MR16, 659; Z53, 340)

Saint Andrews University

Anon 1911 *Saint Andrews University
  Memorial Volume,* Saint Andrews

Spitalfields Mathematical Society

Anon 1901 Nat 64, 478   [1717]

Strassbourg

Fréchet M 1920 Rv Mois 21(124),
  337-362

Sydney, Australia

Turner I S 1956 Au HSJP 41(6),
  245-266 [U Sydney]

Tartu

Meder A 1928 Arc GMNT (2)11, 62-67
  (F54, 33) [Gauss]

Tashkent

Zijaev K G 1964 Tash Tec M (19), 5-19
  (MR33, 5431) [1934-1964]

Tomsk

Krulikovskii N N 1967 *Istoriya razvitiya matematiki v Tomske,*
  Tomsk, 145 p (MR37, 18)

Turin

Gorresio G 1878 Tor An 1, 15-22
  (F10, 16) [academy]
Terracini A 1957 It UM (3)12, 290-298
  (Z77, 8) [Cauchy]

Ulm

Ofterdings L F 1866 *Beitraege zur
  Geschichte der Mathematik in Ulm
  bis zur Mitte des 17 Jahrhunderte,*
  Vienna, 12 p

Vilnius

Bespamjatnyh N D 1963 Karel Ped 14,
  49-69 (MR30, 871) [1803-1832]
Bielinski J 1890 MF 2, 265-432
  [Univ, bib]
Khmelevskii B 1962 Lit M Sb 2(2),
  319-342 (MR27, 899) [trig]
Zhemaitis Z 1962 Lit M 56 2(2),
  289-317 (MR27, 899; Z117, 249)
  [1579-1832]

Waukasha Mathematical Society

McAdam J 1968 Delta 1, 46-48


West Virginia University

Gould H W 1966 West Va AS 38, 123-
 [bib]


Wroclaw

See also Breslau.

Steinhaus H 1960 Bulg FM 3(36), 127-
 145  (MR23, 433)


Wurzburg

Buchner M ed 1932 *Aus der Vergangen-
 heit der Universitaet Wurzburg.
 Festschrift zum 350 jahrigen
 Bestehen der Universitaet, im
 Auftrage von Rektor und Senat,*
 Berlin (Springer), 807 p


Zurich

Anon 1897 *Festschrift der Naturforschenden Gesellschaft in
 zuerich 1746-1898,* Zurich
Fueter E 1946 in *Festschrift zur 200-
 Jahr-Feier der Naturforschenden
 Gesellschaft in Zuerich, 1746-1946,*
 135-137
Gagliardi E 1932 Fors Fort 8; 9,
 178-179

5

INFORMATION RETRIEVAL

In this chapter are listed reference
materials and publications on sources
and methods of information retrieval.
The main topics are: Bibliography;
Collections; Historiography; Information systems; Libraries; Manuscripts;
Museums, monuments, exhibits; Reference materials.

BIBLIOGRAPHY

Only bibliographies of general
interest are listed here. Others
are under topics throughout this
bibliography.

## Bibliography: History and criticism

Ahrens W + 1908 DMV 17, 339-340
  (F39, 19) [F Mueller, Hagan, Fuss]
Archibald R C 1928 Int Con 6, 465-472
Brocard H 1895 Par CR 120, 248
  [French Academy Catalog]
De Morgan A 1853 On the difficulties
  of correct description of books, in
  the *British Almanac & Companion*
* Eneström G 1890 Bib M (2)4, 37-42
  - 1897 Bib M (2)11, 65-72
    [F28, 1]
  - 1897 Wiad M 1, 186-198
  - 1898 Int Con 1; 281-288 (F29, 2)
  - 1900 Bib M (3)1, 480-484
    (F31, 3)  [Royal Soc Cat]
  - 1910 Bib M (3)11, 227-232
    (F41, 47) [Valentin]
Favaro A 1891 Rv M 1, 72-77
  (F23, 2)
Fehr H 1904 Ens M 6, 476-481
  (F35, 46) [Int Con 3]
Forster-Morley H 1906 Intermed 14(S),
  1 [international catalog]
Gerardin A 1912 Int Con 2, 444-446
Graf J F 1900 Bib M (3)1, 250
Jourdain Ph 1917 Sci Prog 11, 686-690
Laisant C A 1900 Bib M (3), 246-249
  (F31, 3)
Loria G 1912 Loria 14, 1-5, 61
Struyk Adrian 1955 MT 48, 58-59
Valentin G 1885 Bib M (2)2, 90-92
  [plans for a 100,000 title bib]
  - 1900 Bib M (3)1, 237-245 (F31, 3)
  - 1905 Ost Bib 9, 76-89
  - 1910 Bib M (3)11, 153-157 (F41, 47)
Wolffing E 1898 Wurt Sch 5, 417, 459
  - 1899 Wurt MN (2)1, 45
Young J W A 1904 AMM 10, 186

## Bibliographies of bibliographies

*Besterman Theodore 1954-1956 *A World

681 / Information Retrieval

*Bibliography of Bibliographies,* 3rd
  ed, Geneva, 4 vols. (Look under
  subtopics)
Vallée Leon 1883-1887 *Bibliographie
  des bibliographies,* Paris (Terquem)
West C J + 1923 *List of Manuscript
  Bibliographies in Astronomy, Mathematics and Physics,* Wash DC (NRC)

## Bibliographies of the history of mathematics

Anon 1903 *List of Works on the
  History of Mathematics (including
  works printed before 1800) in the
  New York Public Library,* NY
  (NY Pub Lib)
  - 1930- *International Bibliography of
    Historical Science,* Paris [lists
    all historical publications from
    1926, classified]
  - 1949 *Istoriya estestvoznaniya.
    Literatura, opublikovannaya v SSSR
    1917-1947,* Ak Nauka
  - 1955 *Istoriya estestvoznaniya.
    Bibliograficheskii ukazatel.
    Literatura opublikovannaya v SSSR
    1948-1950,* Mos, 396 p
  - 1957 *Bibliograficheskie istochniki
    po matematike i mekhanike, izdannye
    v SSSR za 1917-1952,* Mos-Len
  - 1953- *Index zur Geschichte der
    Medizin, Naturwissenschafften u.
    Technik,* Munich
*Bobynin V V 1901 Fiz M Nauk 1, 267-
  285 [XIX]
Boyer Carl B 1961 *Classics in the
  History of Mathematics,* (Lehigh U),
  32 p
Christensen S A + 1889 Bib M (2)3,
  75-83 [Denmark]
Curtze M 1902 *Urkunden zur Geschichte
  der Mathematik,* Leipzig
De Hairs E 1928 Euc Gron 5, 229-236
  [biog coll]
Dickstein S 1889 Bib M (2)3, 43-51
  [Poland]
  - 1890 Prace MF 2, 247-257 [Poland]
Dijksterhuis E J 1931 Euc Gron 7,
  185-195 [Review of 40 histories]

Bibliographies of the history of
mathematics (continued)

- 1950 Scr 16, 43-59 (MR12, 69;
  Z37, 291) [Holland 1930-1947]
Dobrovolskii V A + 1963 Ist M Zb 4,
  10-36 [Ukraine 1850-1960]
Enestroem G 1889 Bib M (2)3, 1-14
  [Sweden 1667-1888]
- 1889 Sv Ofv 46., 489-502 [Sweden]
Favaro A 1892 Bib M (2)6, 67-84
  (F24, 4) [Italy]
Gnedenko B V + 1963 Ist M Isl 15,
  11-36 [Russia 1956-1961]
Harig Gerard ed 1960 *Sowjetische
  Beitraege zur Geschichte der
  Naturwissenschaft,* Berlin, 250 p
  (Z99, 241)
Holst E 1889 Bib M (2)3, 97-103
  [Norway]
Jones P S 1953 MT 46, 500-501
Josephson Aksel G S 1911 *A List of
  Books on the History of Science*
  (John Crerar Lib) Chicago
  (Supplement 1916, repr 1967)
Koldomasova G V 1966 Ist Met EN 5,
  231-273 [Moscow libraries]
*Lemaire G 1907 Intermed 14, 225
  [history in elementary texts]
*Loria Gino 1946 *Guida allo studio
  della storia delle matematiche,*
  2nd ed, Milan (Hoepli), 385 p
  (1st ed 1916)
Luria S J 1934 Len IINT 3, 273-312
  (F61, 957) [Russ]
Mendelevich G A + 1956 Vop IET 1, 324
  [Russia, Theses 1949-1954]
Mueller F 1913 ZMN Unt 44, 461-463
  (F44, 43) [survey recent work]
Ocagne M d' 1935 Par EP (2)34,
  223-236 (F61, 957) [science]
*Read Cecil B 1966 SSM 66, 147-179
  (Covers hist articles in AT, M Gaz,
  MT, Scr, SSM, M Mag] (earlier
  versions: 1952 Wichita U Stu 25;
  1959 SSM 59, 689-717)
Riccardi P 1897-1898 Bln Mm (5)6,
  755-775; 7, 94, 371-425 [Italy]
  (F28, 37; 29,3)
Rocquigny G de 1895-1899 Intermed 2,
  143; 5, 250; 6, 108 (Q178)
  [treatises < 1800]

Rome A 1930 Rv Q Sc (4)18, 333-350
  (F56, 3)
Smith D E 1933 Scr 1(3), 204-
  [general histories]
Suter H 1890 Bib M (2)4, 97-106
  [Swiss]
Teixeira F G 1890 Bib M (2)4, 91-92
  [Portugal]
Uzakov J K + 1964 Tash Tec Mi 19,
  152-160 (MR33, 5422)
Vera Francisco 1935 *Los historidores
  de la matematica española,* Madrid
Vicuna G 1890 Bib M (2)4, 13-21
  [Spain]
Yushkevich A P 1948 in *Matematika v
  SSSR 1917-1947,* 993-1010, 1011-
  1020 (Z38,145)
- 1958 Ist M Isl 11, 11-46
  (MR62, 582; Z96, 1)
Zubov Vasilii P 1956 *Istoriografiya
  estestvennykh nauk v Rossii (XVIII-
  pervaya polovina XIX veka),* Mos

Bibliographies of Mathematics

Anon 1964- *New Publications (AMS)*
  =AMSNP)
Archibald R C 1932-1935 Scr 1,
  173-181, 265-274, 346-362; 2, 75-85,
  181-187, 282-292, 363-373; 3, 83-91
  (F58, 3; 60, 3, 815; 61, 4)
Baillet A 1725- *Jugements des savants
  sur les principaux ourvrages des
  savants,* Amsterdam
*Beughem Cornelius 1688 *Bibliographia
  Mathematica et artificiosa novis-
  sima perpetuo...,* Amsterdam, 526 p
Bobynin V V 1886-1900 *Russkaya fiziko-
  matematicheskaya bibliografiya,*
  Moscow, 3 vols [from printing to
  1816]
Brocard H 1896 Intermed 3, 279 (Q489)
  [Germany]
Dinze O V + 1936 *Matematike v izdani-
  yakh Akademii nauk 1728-1935,*
  Mos-Len
Eells W C 1923 AMM 30, 1-8; 318-321
  [10 "most important" books]
*Erlecke A 1873 *Bibliotheca Mathematica*
  Halle, 307 p (F5, 609; Hist Abt
  18, 1-4)

## Bibliographies of Mathematics (cont'd)

Ersch Johann S 1793-1807 *Allgemeines repertorium der literatur...1785-1790, 1791-1795, 1796-1800*, Jena-Weimar, 3 vols
- 1803 *Literatur der Mathematik, Natur- und Gewerbekunde, seit der Mitte des 18 Jahrhunderts bis auf die neueste Zeit*, Amsterdam-Leipzig, (1), 1-56

Estève Madeleine 1931 *Répertoire des congrès et ouvrages collectifs...*, Paris (Inst. H. Poincaré), 94 p

Heilbronner J C 1742 *Historia matheseos universae...*, Leipzig, 924 p

Kaestner A G 1796-1800 *Geschichte der Mathematik...* Goettingen, 4 vols

Koenigsberger Leo + 1877-1879 *Repertorium der literarischen Arbeiten aus dem Gebiete der reinen und angewandten Mathematik*, Leipzig, 2 vols [abstracts]

Loeffeholz-Colberg Friedrich von 1873 *Forstliche Chrestomathie...*, vol 3(2), Berlin, 859-1092

Mehmke Rudolf ed 1897-1899 *Verzeichnis von Abhandlungen aus der angewandten Mathematik...*, ZM Ph (S) (cont in ZM Ph)

Miller G A 1916 *Historical Introduction to Mathematical Literature*, NY, 316 p

Mueller Felix 1892 *Zeittafeln zur Geschichte der Mathematik, Physik und Astronomie bis zum Jahre 1500 mit Hinweis auf die Quellenliteratur*, Leipzig, 106 p
*- 1909 *Fuehrer durch die mathematische Literatur, mit besonderer Beruecksichtigung der historisch wichtigen Schriften*, Leipzig, 262 p (=Ab GM 27)

*Mueller Johann W 1820* Auserlesene mathematische Bibliothek, *Nuernberg, 288 p*
- 1822-1825 *Repertorium der mathematischen Literatur in alphabetischer Ordnung*, Augsburg, 3 vols

Murhard F W A 1797-1798 *Bibliotheca mathematica*, Leipzig, 5 vols, (another vol in 1803)

Okamoto 1931 *A Catalogue of Japanese Mathematical Literature*, Tokyo, 822 p (F57, 48)

*Parke N G III 1947 *Guide to the Literature of Mathematics and Physics*, NY-Lon, 205 p (Z29, 2) (2nd rev ed 1958)

*Pemberton John E 1964 *How to Find Out in Mathematics*, Oxford-NY (Pergamon-Macmillan), 158 p (rev 2nd ed 1969, 193 p)

Rogg Ignaz 1830 *Handbuch der mathematischen Literatur vom Anfange der Buchdruckerkunst...*, Tuebingen, 641 p

Schaaf W L 1954 Scr 20, 209-212 [curiosities]
- 1957 MT 50, 388-389 [paperbacks]

Scheibel Ephraim 1769-1798 *Einleitung zur mathematischen Buecherkenntnis*, Breslau, 19 parts

Sintsov D + 1915 Khar U Za 4, 1-24

Slaught H E 1907 Sch Rv 14, 679

Sohncke Ludwig A 1854 *Bibliotheca Mathematica...*, Leipzig, 388 p

Struik D J 1947 Scr 15, 115-131 [Selected English tr from 1200]

*Woelffing Ernst 1903 *Mathematischer Buecherschatz...*, Leipzig, 452 p (= Ab GM 16)

## Bibliographic indexes

See also the general serial indexes and the index volumes published by individual journals.

*Bibliografia matematica italiana*, 1950-, Rome (Unione Matematica italiana)

Carr G S 1886 *A Synopsis of Elementary Results in Pure Mathematics*, London, 935 p [the book index includes an index of journal papers from 1800!] (repr 1970 Chelsea)

*International Catalogue of Scientific Literature*, 1902-1917, (Royal Society of London) (Section A on Math, 14 vols)

## Bibliographic indexes (cont'd)

*Repertorium commentationum a socie-
tatibus litterariis editarum,
secondum disciplinarum indinem
digessit 1801-1822*, Goettingen,
7 vols (J D Reuss ed) (repr 1961)
[papers in academy proceedings to
1800]

*\*Royal Society Catalogue of Scientific
Papers 1800-1900*, 19 vols + 4 vol
index, 1867-1908, (Cambridge U P)
(repr 1965 Johnson)

Zuchold Ernst A 1851-1869 *Bibliotheca
historico-naturalis et physico-
chemica et mathematica...*,
Goettingen, 17 vols

## Abstracting and review journals

*Bulletin des sciences mathématiques
et astronomiques*, 1870-, Paris
(G Darboux ed) (=BSM)

*Bulletin signalétique*, 1941-, Paris
(Centre national de la Recherche
Scientifique) (*Bulletin analytique*
1941-1955) (=BS)

*Bulletin universel des sciences et de
l'industrie*, 1823-1831, Paris
(Ferussac ed)

Computer Abstracts, 1957-, Bureau of
Information London (=Comp Ab)

*Computer Mathematics*, 1969-, Washing-
ton DC (Cambridge Communications
Corp) (con of Cumulative Computer
Abstracts) (=Comp M)

*Jahrbuch ueber die Fortschritte der
Mathematik und ihrer Grenzgebiete*,
1868-1940, Berlin (=F)

*Mathematical Reviews*, 1940-,
(American Math Society) (=MR)

*Polytechnische Bibliothek*, 1866-1917,
Leipzig [math, phys, chem]
(=Poly Bib)

*Referativny Zhurnal*, 1953-, Moscow
(=RZ)

*Répertoire bibliographique des sci-
ences mathématiques*, 1889-1916,
Paris (F25, 8; 28, 1; AMSB (2)1,
186-189) [packets of cards, index
vols]

*Revue semestrielle des publications
mathématiques*, 1893-1934 (Societé
mathématique d'Amsterdam) (=RS)

*Statistical Theory and Method Ab-
stracts*, 1959-, London (International
Statistical Institute) (1959-1963
as International Journal of Abstracts.) (=Sta TMA)

Terquem O 1855-1861 *Bulletin de
bibliographie, d'histoire et de
biographie mathématique*, Paris
(=Terq Bul)

## Bibliography of rare books

See also MANUSCRIPTS.

Archibald R C 1918 AMM 25, 36-37
[oldest math work]

Anon 1940 AMM 47, 108 [incunabula]

Halliwell J O 1841 *Rara mathematica*,
London

Junk W 1900-1939 *Rara historico-
naturalia et mathematica*, Berlin,
3 parts, 295 p (F44, 60) (repr 1962)

Karpinski L C 1932 Scr 1, 63-65
[at U Mich]

Klebs A C 1939 Osiris 4, 1-359
(F64, 2) [incunabula]

Mueller F 1910 DMV 19, 163-178,
197-225 (F41, 48) [incunabula]

- 1911 Dres Isis 51-57 (F42, 50)
[incunabula]

Smith D E 1908 *Rara Arithmetica*,
Boston [Plimpton Lib]

- 1925 AMM 32, 287-294
[copies owned by famous]

- 1939 *Addenda to "Rara Arithmetica"*
..., Boston, 60 p (F65:2, 1077)

Thornton John L + 1962 *Books, Libraries and Collectors*, 2nd ed, London

## Bibliography of serials

Allen E S 1929 Sci 70, 592-594

Ahrendt M H 1955 MT 48, 588-589
[new journals]

Archibald R C 1929 M Gaz 14, 379-400
[England]

- 1931 AMM 38, 436-439 [new journals]

Bibliography of serials   (cont'd)

1932 AMM 39, 185-187 (F58, 993)
[new journals]
- 1933 Scr 1, 275  [Italy]
- 1928 M Gaz 14, 379-400  [England]
Babini J 1964 Isis 55, 82-85
   (Z117, 25)  [Argentina]
Barnes S B 1936 Osiris 1, 155-172
   (F62, 3)  [XVII]
Bellavitis G 1878-1880 Ven I At
   (5)4, 247-278, 357-388, 1069-1081,
   1099-1121; 5, 299-345
   (F10, 23; 12, 37)  [review of
   journals]
Bobynin W W 1899 Fiz M Nauk (2)1,
   35-54, 71-75  (F30, 33)
   [first Russian]
Bolton H C 1897 Catalogue of Scientific Technical Periodicals
   1665-1895, 2nd ed, Wash DC, 1247 p
   (repr 1965)
Cavallaro V G 1930 Bo Fir (2)9,
   xlix-lix (F56, 798)  (Italy)
Finkel B F 1940-1943 A history of
   American mathematical journals.
   M Mag 14, 197-210, 261-270, 317-328
   383-407, 461-468; 15, 27-34, 83-96,
   121-128, 177-190, 245-247, 294-302,
   357-368, 403-418; 16, 64-78, 188-
   197, 284-289, 341-344, 381-391; 17,
   21-30  [USA]
Fitz-Patrick J + 1905-1906 Intermed
   12, 172; 13, 28, 151  [Journal des
   Sci Math 1872, A. Labosne]
Hart D S 1875 Analyst H 2, 131-138
   (F7, 17)  [USA]
Hoffmann J C V 1899 ZMN Unt 30,
   142-150  (F30, 32)  [ZMN Unt]
Laronde 1908 Intermed 15, 131, 179
   [slavic]
Mackay J S 1893 Fr AAS 2, 303-308
   (F25, 11)  [England]
Miller G A 1916 Intermed 23, 201-202
   (F46, 34)  [num thy, geom curiosa]
- 1923 SSM 22, 276-280 [new journals]
Morgan Katherine L 1944 Journals
   Dealing With the Natural, Physical
   and Mathematical Sciences Publish-
   ed in Latin America, Wash DC
   (Pan American Union)

Mueller Felix 1902 Ber MG 1, 17-19
   (F33, 51)  [importance of serials]
- 1903 DMV 12, 427-444
   [title abbreviations]
- 1904 Int Con H 12, 105-113
   (F35, 51)
- 1909 Arc GNT 1, 61-77 (F40, 48)
   [Germany XVIII]
Rocquigny G de + 1901 Intermed 7,
   229; 9, 27 (Q1207)  [oldest]
Sarton G 1914 Isis 2, 125-161
Smith D E + 1918 Union List of Math-
   ematical Periodicals.  Wash DC
   (Department of Interior, Board of
   Education 9) (USGPO)  (F46, 37)
Smith D E 1933 Scr 1, 277-285
   [early USA]
Smith W 1932 MT 25, 85-86
Sparn E 1932 B Ai Bol 5, 36-38
   (F58, 993)  [valuable collections]
Thirion J 1895 Rv Q Sc, 58 p
   (F26, 68)  [Fr Long]


COLLECTIONS

See also Biographical Collections
   (Ch 1).

Anthologies on history of mathematics

*Aaboe A 1964 Episodes from the Early
   History of Mathematics (Blaisdell,
   Singer)  (also New Math Library
   13)  [anc, Babyl]
Anon 1956 Voprosy istorii estestvoznaniya i tekhniki, Moscow
- 1960 Istorii Estestvoznaniya i
   Tekhniki, Sbornik Nauchnykh Trudov
   I, Erevan (Izd A N Armyanskoi SSR),
   199 p [Armenian with Russian
   summary]
- 1962 Voprosy istorii i metodiki
   elementarnoi matematiki, Dushanbe
   (Uch Zap Dushanbinskogo gos ped
   in-ta 34)
   1963 Ocherki istorii matematiki i
   mekhaniki, Moscow (Inst Hist Sci
   Tech, Ac Sci), 271 p
- 1963 Voprosy Istorii Fiziko-
   Matematicheskikh Nauk, Moscow,
   523 p

Anthologies on history of mathematics (cont'd)

- 1968 *Rechenpfennige,* Munich (Forschungsinst d deu Museums f d Gesch d Wiss u d Technik), 239 p

Bopp K 1929 *Drei Untersuchungen zur Geschichte der Mathematik,* Berlin, 66 p (F55, 606)

Cassina Ugo 1961 *Dalla geometria egiziana alla matematica moderna,* (Edizioni Cremonese), 543 p (MR29, 2941)

Curtze M + 1899 *Festschrift zum siebzigsten Geburtstage Moritz Cantors,* Leipzig (Teubner), 675 p (=Ab GM 9)

Douglas J + ed 1941 Galois Lectures (Scr Lib 5)

Enriques F ed 1924 *Questioni reguardante le matematiche elementari,* Bologna, 4 vols (1st ed 1912-1914)

Franchetti G 1893 *Cenni Storici sulle matematiche elementari,* Sassari, 68 p (F25, 1912)

Giorgi Giovanni 1948 *Compendie di storia delle mathematiche,* Torino

Guenther S 1876 *Vermischte Untersuchungen zur Geschichte der Mathematischen Wissenschaften,* Leipzig, 352 p (F8, 30) (repr 1968 Wiesbaden)

Guenther Siegmund + 1909 *Festschrift Moritz Cantor...,* Leipzig, 204 p

Hutton C 1812 *Tracts on Mathematical Subjects,* London, 3 vols

Keyser C J 1935 *Mathematics and the Question of Cosmic Mind and Other Essays,* NY (Scr Lib 2) [pop]

Keyser C J + ed 1937 *Scripta Mathematica Forum Lectures,* NY (Scr Lib 3), 97 p [pop]

Klein Felix 1893 *Lectures on Mathematics... 1893,* NY (Macmillan), 109 p (repr 1911, Amer Math Soc) (Fr tr 1898 Hermann)

Klein F + ed 1912-1914 *Die Kultur der Gegenwart,* Part III, Section 1: Math. 3 vols: III 1(1), III 1(2), III 1(3), Leipzig

Kline M 1968 *Mathematics in the Modern World.  Readings from the Scientific American,* (Freeman) [*pop]

Laugwitz Detlauf ed 1968 *Ueberblicke Mathematik* 1 Mannheim (Bibliographisches Institut), 213

Lebesgue Henri 1958 *Notices d'histoir des mathématiques avec une intro de Mlle L Felix,* Genève (Ens M)

*Le Lionnais F ed 1948 *Les grands courants de la pensée mathématique,* Marseille (Cahiers du Sud)

Milhaud G 1911 *Nouvelles études sur l'histoire de la pensée scientifique,* Paris, 235 p

Neugebauer O + 1941 *Studies in the History of Science,* Philadelphia (U of Penn), 123 p

*Newman James R ed 1956 *The World of Mathematics,* NY, 4 vols (MR18-1, 453) [pop]

Petrosyan G B ed 1967 *Voprosy istorii nauki,* Erevan, 399 p

Sarton George 1948 *The Life of Science* (Henry Schuman), 118 p (repr 1960 Indiana UP)

Simons L G 1939 *Fabre and Mathematics and Other Essays,* NY (Scr Lib 4), 101 p [pop]

Smith D E 1934 *The Poetry of Mathematics and Other Essays,* NY (Scr Lib 1)

Tannery P 1912-1934 *Memoires Scientifiques,* Toulouse, 13 vols

Witting A + 1913 *Beispiele zur Geschichte der Mathematik.  Ein methodisch-historisch Lesebuch,* Leipzig, 69 p (F44, 43) (repr 1923)

Young J W A ed 1911 *Monographs on Topics of Modern Mathematics Relative to the Elementary Field* (Longmans) (repr 1955 Dover)


Source books

Ahrens Wilhelm 1904 *Scherz und Ernst in der Mathematik.  Geflugelte und ungeflugelte Worte.* Leipzig 522 p [pop]

Source books  (cont'd)

Archibald R C 1916 Mathematical
Quotation Books.  AMSB 22, 188-192
(F45, 59) [bib]

Bellman Richard ed 1961 *A Collection
of Modern Math Classics,* NY (Dover),
304 p (MR23, A789) [XX]

Bellman Richard + 1964 *Selected
Papers on Mathematical Trends in
Control Theory,* (Dover), 200 p

Benacerraf P + 1964 *Philosophy of
Mathematics, Selected Readings*
(Prentice-Hall)

Copi Irving M + ed 1967 *Contemporary
Readings in Logical Theory,* NY,
344 p

Dieck W 1920 *Mathematisches Lesebuch,*
Sterkrade, 5 vols

Fadiman C 1962 *The Mathematical
Magpie,* NY [pop]

Leitzmann W 1929 *Aus der neueren
Mathematik.  Quellen zum Zahl-
begriff und zur Gleichungslehre
...,* Leipzig, 78 p [XVI-XIX]

Marks R W 1964 *The Growth of Math
from Counting to Calculus,*
(Bantam Matrix), 217 p

- 1964 *Space, Time, & the New Math,*
(Bantam Matrix)

Maupin G 1898 *Opinions et curiosités
touchants la mathématique d'après
les ouvrages français des 16e,
17e et 18e siècles,* Paris, 207 p
(F29, 25)

Midonick Henrietta 1965 *The Treasury of Mathematics,* (Philosophical
Library), 820 p

Mikami Yoshio 1910 *Mathematical
Papers from the Far East,* Leipzig,
229 p (=Ab Gesch M 28)

Moritz R E 1914 *Memorabilia mathematica,* NY, 417 p (repr Dover
1958)

Newman James R 1956 *The World of
Mathematics,* 4 vols, NY, 2535 p
(MR18(1), 453) [pop]

Rapport S + ed 1963 *Mathematics,*
NY, 351 p [pop]

Rebière A 1898 *Mathématiques et
mathématiciens...,* Paris, 566 p
(F28, 37) [pop]

- 1923 *Pages choisies des savants
modernes,* Paris, 582 p  (F42, 48)

Reingold Nathan ed 1964 *Science in
Nineteenth Century America.  A
Documentary History,* NY, 339 p

Schaaf W 1957 MT 30, 229
[quotations]

Schmidt M C P 1901 *Realistische
Chrestomathie,* Leipzig

Schubert Hermann 1898 *Mathematical
Essays...,* Chicago (Open Court),
149 p

Smith David Eugene 1929 *A Source Book
in Mathematics,* NY, 701 p (repr
1959 Dover)  (MR21(2), 892)

Sierpinski W + 1967 *Congruence of
Sets and Other Essays,* (Chelsea),
(reprints of Sierpinski, Klein,
Runge, Dickson)

Speiser A 1925 *Klassische Stuecke
der Mathematik,* Zurich, 170 p
(F51, 2)

Struik D J ed 1969 *A Source Book in
Mathematics, 1200-1800,* (Harvard
U P), 427 p  (MR39, 11)

*Taylor Richard ed 1837-1852 *Scienti-
fic Memoirs, Selected from the
Transactions of Foreign Academies
of Science, and Learned Societies,
and from Foreign Journals,* London,
4 vols (repr 1966 Johnson)

*Wieleitner Heinrich 1927-1929 *Mathe-
matische Quellenbuecher,* 4 vols,
Berlin  (Russ tr 1935)

Witting Alexander + 1913 *Beispiele
zur Geschichte der Mathematik,*
(Teubner), 61 p [X-XVI]


HISTORIOGRAPHY

This term is here used with the
meaning "the body of techniques,
theories, and principles of historical research and presentation;
methods of historical scholarship"
(The Random House Dictionary 1966)

History of historiography

See also Bibliographies of history
of mathematics.

History of historiography (cont'd)

Anon 1885 Fiz M Nauk (A) 1, 195-216
[devel of hist of math]
- 1934 Clct MS 25, 197-198
[Ganesh Prasad prize]
*Becker O + 1951 in *Geschichte der
Mathematik*, Bonn, 254-261
Beer J L 1966 Res Mngt 9, 101-107
[func of hist in indust research]
Bobynin V V 1886 *Proiskhozdenie,
razvitie i sovremennoe sostoyanie
istorii matematiki*, Mos, 49 p
- 1888 Bib M (2)2, 103-110 [Russ]
- 1901 Fiz M Nauk 1, 205 [XIX]
Bosmans H 1907 Rv Q Sc (3)11, 636-
649 (F38, 4) [math and ast]
- 1908 Rv Q Sc (3)13, 319-335
(F39, 3) [ast]
- 1908 Rv Q Sc (3)14, 646-662
(F39, 3)
Bouligand G 1952 Arc In HS 5(31),
230-233 (Z82, 5)
Boyer C B 1964 MT 57, 242-253
[hist mistakes]
Brocard H + 1905-1906 Intermed 12,
129, 255; 13, 117-118, 202
[Lucas, Chasles, hoax]
Cajori F 1918 Sci 48, 279-284
[plans for gen hist of XIX]
*Cantor M 1900 Int Con (2), 27-42
[historians]
Eneström G 1887 Bib M (2)1, 3-7
[recent]
- 1900 Bib M (3)1, 1
[hist math journal]
- 1915 Bib M (3)14, 336-340
[J Molk]
Garbers K 1941 Mit GMNT 40, 306-307
[Arab]
Ginsburg B 1933 Archeion 15, 27-33
[value]
Gloden A 1948 Lun SNBM, 117-123
(Z37, 291) [most eminent historians]
Gnedenko B V + 1958 Ist M Isl 11,
441-460 (Z95, 1) [value]
- 1959 Wiad M (2)3, 50-64
(MR23, 3; Z91, 5) [value]
Guenther S 1876 *ziele und Resultate
der neueren math-historischen
Forschung*, Erlangen (F8, 32)

Loria G 1916 It SPS 8, 19 p
[current]
- 1950 It UM (3)5, 165-170
(MR12, 311; Z37, 1)
Miller G A 1927 Am NASP 13, 611-613
[Klein]
- 1938 M Stu 6, 71-73 (F64, 23)
[XX]
- 1938 Sci 88, 375-376 (F64, 3)
[1900]
Mueller F 1913 ZMN Unt 44, 461-463
(F44, 43) [recent]
Ostwald W 1910 Rv Mois 9, 513-525
[sci and hist sci]
- 1928 Arc GMNT 10, 1-11
Popa I + 1959 Gaz MF (A) 11(64),
641-646 (Z87, 241) [Ukraine]
Read C B 1960 MT 53, 463-466 [XX]
Rybkin G F + 1956 Mos M Sez,
104-105 [Russ recent]
Rybnikov K A O 1958 Ist M Isl 11,
209-224 (MR23, 582; Z95, 1)
*Rybnikov K A 1960 in *Istoriya Mate-
matiki I*, Mos, 5-16 [value]
Sarton George 1914 Isis 1, 577-589
[current]
Shimodaira K 1965 Jp Stu HS 4, 20-27
[Japan recent]
Stott W 1915 Sci Prog (Oct), 204-217
[value for math research]
*Whitrow G J 1932 M Gaz 16, 225-227
(F58, 3) [value]
Yushkevich A P 1958 in *Sbornik.
Matematika v SSSR za XL let*, Mos,
vol 1
Zubov V P 1956 Mos IIET 15, 277-322
[Bobynin]


Historiographic methods

See also under GENERAL, Quantitative
Studies, Theory (Ch. 4).

Amodeo F 1909 Int Con (4)3, 557-562
(F40, 48) [need for archive]
*Barzun Jacques + 1957 *The Modern
Researcher*, NY, 399 p
Bell E T 1945 Scr 11, 308-316
[projects]
Bierens De Haan 1878 Amst Vs M 12,
1-160 (F10, 12) [sources XII-XVII]

Historiographic methods   (cont'd)

Cantor M 1903 Bib M (3)4, 113-117
   (F34, 44)
De Morgan A + 1847 *Arithmetic Books
   From the Invention of Printing to
   the Present Time,* London
Dijksterhuis E J 1954 MP Semb 4,
   106-121   (MR16, 207)
Dingle Herbert 1947 *Inaugural Lectures,* London
Eneström G 1902 Bib M (3)3, 1-6
   (F33, 50)   [periods]
- 1902 Bib M (3)3, 226
   [calender of mathematicians]
- 1903 Bib M (3)4, 1-6   (F34, 43)
   [cultural specialized history]
- 1903 Bib M (3)4, 225-233 (F34, 44)
- 1904 Int Con H 12, 215-217
   (F35, 49)
- 1905 Bib M (3)6, 97-100   (F36, 47)
   [need for archive]
- 1907 Bib M (3)8, 1-12 (F38, 55)
- 1908 Bib M (3)9, 1-14 (F39, 51)
   [criticism]
- 1909 Bib M (3)10, 1-14 (F40, 47)
- 1910 Bib M (3)11, 1-10   (F41, 47)
- 1912 Bib M (3)12, 1-20 (F42, 49)
   [sources]
- 1912 Bib M (3)13, 1-13   (F43, 62)
   [how to discourage poor work]
Enriques Federigo 1934 *Signification
   de l'histoire de la pensée scientifique,* Paris, 68 p
Gnedenko B V 1958 Ist M Isl 11, 47-62
   (MR23, 582; Z96, 1) [problems]
*- 1963 Ist M Isl 15, 73-96 [problems]
Gnedenko B V + 1957 Ukr IM 9, 359-368
   (Z77, 241) [problems]
Gruebaum A 1964 Sci 143, 1406-
   [phil and hist]
Hofmann J E 1940 Deu M 5, 150-157
   (MR2, 113; F66, 3; Z23, 385)
Hofmann J H 1948 MN Unt 1, 63-67
*Karpinski L C 1938 Sch Soc 48, 338-
   340 [re L. Hogben, popularization]
Kuhn Thomas S 1962 *The Structure of
   Scientific Revolutions* (U Chicago
   P), 172 p   (2nd rev ed 1970)
Lecat M 1913 Intermed 20, 107
   [publications announced but unpub-
   lished]

Lefebvre B 1907 Rv Q Sc (3)12,
   594-626   (F38, 3)
- 1910 Rv Q Sc (3)17, 264-279,
   606-615   (F41, 2)
- 1911 Rv Q Sc (3)19, 600-616; (3)20,
   572-586   (F42, 3)
Lohne J A 1967 Hist Sci 6, 69-89
   [corruption of Newton diagrams]
*Loosjes Th P 1967 *On Documentation
   of Scientific Literature,* London,
   165 p  [*bib]
Loria Gino 1892 It Con Sto 5, 17 p
   (F24, 41)
- 1893 Bib M (2)7, 39-46   (F25, 54)
- 1908 Bib M (3)9, 227-236 (F39, 52)
   [planned research guide]
- 1908 Int Con H (4)3, 540-548
   (also Arc GNT 1)
- 1912 It SPS 6  [hist a sci?]
- 1916 *Guida allo studio della
   storia delle matematiche,* Milan,
   244 p  (2nd ed 1946)  [*bib]
- 1925 Arc Sto 6, 169-171 (F51, 1)
   [sources]
- 1926 Arc Sto 7, 4-17 (F52, 1)
- 1932 Archeion 14, 198-206
   (F58, 1; Z6, 1)
- 1933 Linc Rn (6)17, 768-775
   (F59, 2)  [anc, mid]
- 1947 Int Con HS (5), 51-52
   [universal history]
Luca F de 1857 Giambat 4, 451-
Mayer G A 1935 SSM 35, 977-983
   [mistakes]
Merton R K 1967 Sci 159, 56-63
   [Matthew or halo effect]
Metzger H 1933 Archeion 15, 34-44
   [thinking historically]
*Miller G A 1907 AMM 14, 193
   [library use]
- 1910 Ed Rv 39, 403-406
- 1921 Isis 4, 5-12
   [different modes of hist]
- 1926 Am NASP 12, 537-540 (F52, 16)
   [postulates in hist]
- 1936 SSM 35, 977-983 (F62, 1009)
   [mistakes]
- 1938 Sch Soc 48, 278-279 [mistakes]
Neugebauer P V 1937 As Nach 261, 152-
   199, 376-426, 460-523 (F63, 817)
   [tables for chronology]

Historiographic methods   (cont'd)

Radl E 1933 Scientia 54, 309-315
Rosenberger F 1899 ZM Ph 44(S), 361
   [value]
Sarton George 1931 *The History of
   Science and the New Humanism,* NY
- 1936 *The Study of the History of
   Mathematics,* Cambridge
- 1952 *A Guide to the History of Science,* Waltham , Mass
- 1957 *Study of the History of Mathematics and the Study of the History
   of Science* (Dover)  (MR19, 519)
   (enlarged from 1936)
*Scriba C J 1967 Bei GWT (9), 54-80
- 1968 Ueberbl 1, 9-33
Shelton J B 1959 MT 52, 563-567
   [chart]
Smith D E 1925 AMM 32, 444-450
   (F51, 1)  [coins, portraits, medals]
- 1930 Clct MS 20, 135-138  (F56, 3)
   [projects]
- 1930 Clct MS 20, 135-138
Stimson Dorothy ed 1962 *Sarton on the
   History of Science* (Harvard U P)
Vailati G 1897 *Sull'importanza delle
   ricerche relative alla storia
   delle scienze. Prolusione a un
   corso sulla storia della meccanica,*
   Turin, 22 p  (F28, 37)
Vetter Q 1918 Prag V 3, 1-52
   (F47, 22)
- 1922 Bo Fir (2)1, 33-42, 65-74
   (F48, 37)
- 1929 *O metodyce historji matematyki,*
   Warsaw, 19 p  (Wiad M (S))
Vogel K 1953 Arc G Med 37, 161-169
   (Z50, 1)  [mistakes]
Vygodsky M A 1930 Est Marks (2/3)
Zeuthen H G 1905 Int Con (3), 536-542
   (F36, 48)  [mistakes]


INFORMATION SYSTEMS

Knight G Norman 1969 *Training in Indexing* (MIT P), 219 p.  (A course
   of the Soc. of Indexers)
Lancaster F W 1968 *Information Retrieval Systems. Characteristics,
   Testing, and Evaluation* (Wiley)

Loosjes T P 1967 *On Documentation of
   Scientific Literature,* London,
   165 p  [survey, bib]
Vickery B C 1970 *Techniques of Information Retrieval,* London, 262 p
   [survey, bib]


LIBRARIES

Anderton B + 1901 *Catalogue of the
   Books and Tracts on Pure Mathematics in the Central Library,* New-
   castle-upon-Tyne, 49 p  (F32,946)
Anon 1890 *Katalog der auf Hamburger
   Bibliotheken vorhandenen Literatur
   aus der reinen und angewandten
   Mathematik und Physik. Herausge-
   geben von der mathematischen
   Gesellschaft in Hamburg anlaesslich
   ihres 200 jaehrigen Jubelfestes,*
   Hamburg, 88 p  (F25, 11)
   [supplements 1893, 1906]
- 1903 *List of Works on the History
   of Mathematics in the New York
   Public Library,* NY
- 1921 BSM 45, 89-90  (F48, 47)
   [Mittag-Leffler lib]
- 1958 *Union List of Periodicals on
   Mathematics and Allied Subjects in
   London Libraries,* London (U London
   Library), 61 p
Candido G 1942 It UM 2, 841-885
   (Z27, 4)  [Florence]
Diaraujo Bento V F 1885 *Catalogo de
   mathematica da bibliotheca publica
   municipal,* Porto, 115 p
Dresden A 1946 AMSB 52(5)(2), 173 p
   [AMS lib]
Favaro A 1894 Ven I At (7)6, 526
   [Eneström lib]
Frick B M 1936 Osiris 1, 79-84
   [Columbia U D. E. Smith lib]
Groenfeldt Stanislaus 1915 *Systematisk
   foesteckning oefven G. Mittag-
   Lefflers matematiska bibliotek,*
   Upsala, 352 p  (F45, 61)
Hiemenz K 1907 *Katalog des mathemat-
   ischen Lesezimmers der Universitaet
   Goettingen* Leipzig (Teubner), 235 p
Karpinski L C 1921 AMM 28, 484
   [Ziwet lib]

LIBRARIES (cont'd)

- 1933 Scr 1, 63-65  (U Michigan)
Kragemo H B 1928 Nor M Tid 10,
  156-158  (F54, 32)
Mueller F 1906 DMV 15, 430-434
  [Wertheim lib]
Munsterberg M 1942 Isis 34, 140-142
  [Bowditch coll, Boston Pub lib]
Plimpton G A 1928 Sci 68, 390-395
  (F54, 48)  [Plimpton lib]
- 1932 Int Con (8)6, 433-442
  (F58, 28)  [Plimpton lib]
Sarton G 1952 A Guide to the History
  of Science, Wattham, Mass, 260-289
Refior S R 1924 MT 17, 269-273
  [Columbia U, D. E. Smith lib]
Sanchez Peres José A 1929 Las mate-
  maticas en la biblioteca del
  Escorial, Madrid, 371 p  (also
  Mad Mm (2)7)
Sauvenier-Goffin Elisabeth 1958 Les
  sciences mathématiques et physiques
  à travers le fonds ancien de la
  bibliothèque de l'université de
  Liège, Louvain, 172 p
Smith D E 1925 AMM 32, 287-294, 393-
  397, 444-450  [Columbia U, D. E.
  Smith coll]
Theo + 1913-1914 Intermed 20, 267;
  21, 96  [Paris]
Vetter Q 1929 Archeion 11, LIII-LVII
  (F55, 9)  [Prague]


MANUSCRIPTS

Included here are all unprinted
  sources.  See also in Ch. 3, Coins,
  Postage stamps, and below MUSEUMS.

Manuscripts: General

*Anon 1962 Conference on Science
  Manuscripts, Isis 53(1), 1-157
  [bib]
Bruins E M 1960 Int Con HS (9), 436-
  438  (Z114, 5)  [Codex Constantinopolitanus palatii Veteris No 1]
Simons L G 1945 Scr 11, 247-262
  [autograph letters, D E Smith
  coll]

691 / Information Retrieval

Smith D E 1921-1922 AMM 28, 64-65,
  121-123, 166-168, 207-209, 254-255,
  303-305, 368-370, 430-435; 29, 14-
  16, 114-116, 157-158, 209-210, 253-
  255, 297-300, 340-343, 394-395
  [autographs]
- 1925 AMM 32, 393-397  [oriental]
Thorndike L 1948 Osiris 8, 41-72
  (MR10, 667; Z32, 242)  [ast, math]


Manuscript collections

Bjoerno A A 1903-1911 Bib M (3)4,
  238-245, 326-333; 6, 230; 12, 97-
  132, 193-224  (F42, 50)
  [San Marco, Florence]
Bortolotti E 1923 Esercit 3, 68-91,
  161-169  [algebra, Bologna]
Bosmans R P 1927 Brx SS 47 (A) 1,
  14-19  (F53, 16)  [London]
Bradley A D 1940 Scr 7, 49-58
  (F66, 1187)  [Schwenkfelder hist
  lib]
Brendel M 1902 DMV 12, 61-63
  [Gauss, Goettingen]
Favaro A 1885 Bib M 44-46
  [Libri-Ashburnham coll]
Gurland Jonas 1866 Kurze Beschreib-
  ung der Math, Ast und Astrologie
  hebraischen Handschriften der
  Firkowitsch'schen Sammlung in der
  Kayserlichen Oeffentlichen Bib-
  liothek zu St Petersburg, St
  Petersburg, 61 p (Neue Denkmaeler
  der juedischen Literatur (2))
Krause M 1936 QSGM(B) 3, 437-532
  [Arab, bib, Istanbul]
Matoevskaya G P 1965 Uzb FMN 9, 72-74
  (MR33, 1203)  [Uzbek Acadamey]
Millas Vallicrosa J 1935 Manuscripts
  catalans de caracter astronomie a
  la Bibliotheca nacional de Madrid,
  Barcelona, 11 p
Narducci E 1862 Catalogo di manoscrit-
  ti ora possedutti da D. B. Boncom-
  pagni, Rome (2nd ed 1892; F24, 41)
Steinschneider M 1890 Bib M (2)4,
  65-72  [Amplonian coll]
Thorndike L 1956 Isis 47, 391-404
  (MR18, 453)  [Vatican]
- 1957 Warb Cour 20, 112-172  [Paris]

692 / Information Retrieval

Manuscript collections   (cont'd)

Wilson J M 1911 M Gaz 5, 24-27;
  6, 19-27  [geom, Worcester
  Cathedral lib]
Wohlwill E 1895 *Galilei betreffende*
  *Handschriften der Hamburger*
  *Stadtbibliotheck,* Hamburg, 77 p
  (F26, 46)  [Galileo, Hamburg lib]


Manuscript guides

Anon 1959 *National Union Catalogue*
  *of Manuscript Collections* (Library
  of Congress)
Clagett M 1953 Isis 44, 371-381
  [mid]
Clagett M + 1958 Manuscr 2, 131-154
  [bib, mid]
- 1959 Manuscr 3, 19-37  [mid]
Hamer P M ed 1961 *A Guide to Archives*
  *and Manuscripts in the US-- Com-*
  *piled for the US National Historical*
  *Publications Commission* (Yale U P)
Richardson E C ed 1933 *A Union World*
  *Catalogue of Manuscript Books,* NY
  (Wilson), 5 vols
Richeson A W 1939 M Mag 13, 183-188
  (F65, 2)  [USA]
Rozenfeld B A 1966 *Physical-Mathe-*
  *matical Science in the East,* Moscow,
  256-289  (MR34, 5615)
  [Arab, Persian, USSR]


MUSEUMS, MONUMENTS, EXHIBITS

Baxandall D 1926 *Catalogue of the*
  *Collections in the Science Museum,*
  *South Kensington, With Descriptive*
  *and Historical Notes and Illustra-*
  *tions, Mathematics, I Calculating*
  *Machines and Instruments,* London,
  85 p
Fuchs F 1957 Deu Mus 25, 3
  [Germany, physics, 1905-1933]
Guenther S 1893 Deu NA Ver 2, 32
  (F25, 97)  [Germany instruments]
Gunther R T 1935 *Handbook of the*
  *Museum of the History of Science in*

the *Old Ashmolean Building,* Oxford,
  161 p  (F63, 817)
Moore L 1933 MT 26, 482-486
  [chic world's fair]
Rosskopf MF + 1949 MT 42, 187-191
  [math dept exhibit]
Sarton G 1952 *A Guide to the History*
  *of Science,* Waltham, Mass, 260-289
Smith David E 1924 *Historical Mathe-*
  *matical Paris* (Presses Universtaires)
- 1936 Sci (NS)83, 79-80
  [Columbia U instruments]
Wade H T 1909 Sci Am 101, 10


REFERENCE MATERIALS

See also in Ch. 4, General: Reference.

Reference: History and criticism

*Anning N 1921 AMM 28, 439
  [math in gen enc]
Anon 1926 Pe M (4)6, 46-48
  (F52, 39)  [Enc Italiana]
Biggiogero G 1933 Pe M (4)13, 2-3,
  69-73  [Enc Italiana]
Eells W C 1961 MT 54, 255-260
  [first dict]
Eneström G 1904 Bib M (3)5, 398-406
  (F35, 44)
- 1905 Int Con (3), 546-550
  (F36, 46)  [hist in math enc]
Erlecke A 1873 *Bibliotheca mathe-*
  *matica. I: Die enzyklopaedisch-*
  *mathematische Literatur umfassend,*
  Halle
Hadamard J 1940 Math Cluj 16, 1-5
  (F66, 7; Z25, 293)  [Enc française]
Halsted G B 1899 AMM 6, 7  [Enc MW]
Jourdain Ph 1919 Sci Prog 13(52),
  648-651  [Enc Br]
Karpinski L C 1912 Sci 35, 29-31
  [hist math in Enc Br]
Krazer A 1925 DMV 33, 125  (F51, 40)
  [enc]
Levey M 1952 Isis 43, 257-264
  (Z47, 242)  [enc of Savasorda]
Loria G 1934 Int Con HS 3, 146-160
  (F62, 1035)  [d'Alembert, Enc
  méthodique]

Reference: History and criticism (cont'd)

Marty F 1940 Thales 4, 257-261
[Enc française]
May K O 1966 AMM 73, 687-688
[Enc Br]
Meyer W F 1899 ZM Ph 44(S), 293-299
(F30, 30) [Enc MW]
Miller G A 1917 AMM 24, 106-109
[New International Enc]
- 1917 AMM 24, 453-456 (F46, 32)
[Davie and Peck dict]
- 1918 AMM 25, 383 [enc dict]
- 1924 Sci 60, 382-384 (F50, 12)
[New International Enc]
- 1936 Sci 84, 418-419 (F62, 1042)
[errors in Webster]
- 1935 Sci (13 Sept), 248-249
[defs of math in Engl dicts]
- 1938 Sci 88, 375 [hist errors in
Enc MW]
- 1939 Sci 90, 512-513
[errors in Enc Br]
*- 1939 Sci Mo 48, 268-271
[hist in Enc MW]
- 1940 Sci 91, 289-290
[revis of Enc MW]
*Peano G 1901 Rv M 7, 160-172
[dict]
Prudnikov V E 1953 Ist M Isl 6, 223-
237 [Chebyshev, Ostrogradskii,
Bunyakovskii in enc dict]
Sarton G 1921 Isis 4, 39-40
[Enc SM]
Simons L C 1935 Scr 3, 38-43
(F61, 28) [Oxford Engl Dict]
Taton R 1951 Rv Hi Sc Ap 4, 255-266
(MR14, 831) [Enc méthodique]
Vanhee L 1926 Arc Sto 7, 18-24
[Chinese]
Worms de Romilly 1920 Intermed 27,
109-110 (Q5083) (F47, 33)
[dicts]
Wrede E F 1812 Enzyklopaedisch-
szientifische Literatur. Heft 3:
enzyklopaedisch-mathematische Lit-
eratur enthaltend, Leipzig, 385 p
Youssoufian + 1895-1899 Intermed 2,
110, 347, 404; 5, 127, 250; 6, 108
(Q246) [exist of math dicts?]

Reference: General works of mathematical interest

Listed here are only a few most useful ones.

Bolshaya Sovetskaya Entsiklopediya,
1956 Mos, 51 vols (and other
editions)
The Century Dictionary and Cyclopedia,
1889-1906 NY, (and other editions)
[math and logic by C S Peirce]
Diderot Denis + 1751-1781 Encyclopédie, ou Dictionnaire raisonné
des Sciences, des Arts et des
Métiers, Paris et Neufchatel, vols
70-72 (also Encyclopédie méthodique
mathématiques, Paris, 1784-1789,
3500 p
Encyclopedia Americana, 1970 NY
(and other editions)
Encyclopedia Britannica 1910 11th ed
(All editions are useful)
Encyclopédie Française 1937 Paris,
vol 1
Harris John 1716-1744 Lexicon Techni-
cum, 3rd ed, London, 3 vols
McGraw Hill Encyclopedia of Science
and Technology, 1960 NY
Penny Encyclopedia, 1833-1846 London
[Articles by De Morgan]
Zedler Heinrich 1731-1750 Grosses
vollstaendiges universal Lexicon
aller Wissenschaften und Kuenste
welche bishere..., Halle-Leipzig,
64 vols + 4 vol suppl

Reference: Dictionaries

A chronological list of selected
mathematical dictionaries.

Dasypodius 1573 Lexicon, seu Dic-
tionarium mathematicarum...,
Strasboug
Vital H 1668 Lexicon mathematicum,
astronomicum, geometricum, Paris
Ozanam 1691 Dictionnaire mathématique
ou..., Paris, 850 p
Chauvin Etienne 1692 Lexicon Philo-
sophicum, Rotterdam [anc termin]

Reference: Dictionaries (cont'd)

Ralphson J 1702 *A Mathematical
Dictionary*, London
Harris John 1704 *Lexicon Techni-
cum...*, London, 2 vols (repr 1966
Johnson)
Wolff Christian 1716 *Mathematische
Lexikon...*, Leipzig (2nd ed 1732)
Stone Edmund 1726 *A New Mathematical
Dictionary*, London, (2nd ed 1743,
520 p)
Anon 1734 *Vollstaendiges matemat-
isches...Lexicon*, Leipzig, 2 vols
Savérien Alexandre 1752 *Dictionnaire
universel de mathématiques et de
physique*, Paris, 2 vols
Belidor 1754 *Dictionnaire portatif
de l'ingénieur...*, Paris
Walter Thomas 1762 *A New Mathematics
Dictionary...*, London
Hutton Charles 1795-1796 *A Mathemat-
ical and the Philosophical Diction-
ary:...*, London, 2 vols (2nd ed
1815)
*Kluegel Georg S + 1803-1847 *Mathe-
matisches Woerterbuch...*, Leipzig,
5 vols + 4 suppl vols
*Barlow Peter 1813 *A New Mathematical
and Philosophical Dictionary...*,
London, 6 vols
Mitchell J 1823 *A Dictionary of the
Mathematical and Physical Sciences
...*, London, 576 p
Montferrier A S de 1838 *Dictionnaire
des sciences mathématiques...*,
Paris-Brussels, 2 vols (2nd ed
1840)
Bunyakovskii V Ya 1839 *Leksikon
chistoi i prikladnoi matematiki*,
Pet, vol 1 (A to D), 462 p
(all published)
Montferrier A S de 1845 *Dictionnaire
des sciences mathématiques...*,
2nd ed, Paris, 3 vols
Davies Charles + Peck W G 1855 *Math-
ematics Dictionary and Cyclopedia
of Mathematical Science*, NY, 592 p
(and other editions)
Hoffmann L + 1857-1867 *Mathematisches
Woerterbuch*, Leipzig, 7 vols

Sonnet H 1868 *Dictionnaire des mathématiques appliquées*, Paris,
1482 p (2nd ed 1895)
Tannery Jules + ed 1896 *Dictionnaire
universel...*, Paris
*Naas Joseph + 1961 *Mathematisches
Woerterbuch*, Berlin, 2 vols,
2007 p
Ditkin V A ed 1965 *Tolkovyi slovar
matematicheskikh terminov*, Moscow,
534 p
James Glenn + 1968 *Mathematics Dictionary*, (Van Nostrand), 3rd ed,
446 p

Reference: Handbooks, Encyclopedias

A selected chronological list to 1912

Psellus Michel 1556 *Liber de quatuor
mathematicis scientiis: arithmetica
musica, geometria et astronomia*,
Basle-Wittemburg
Ramus Petrus 1569 *Scholarum mathemat-
icarum*, Frankfurt
Alsted J H 1611 *Elementale mathemat-
icum...*, Frankfurt, 342 p
Hérigone Pierre 1634-1642 *Cursus
mathematicus...*, Paris, 6 vols
Bussey C 1635 *Encyclopaedia mathe-
matica Collegii Claromontani
Parisiensis Societatis Jesu*, Paris
Psellus Michel 1674 *Compendium mathematicum*, Leiden
Schott Kaspar 1674 *Cursus mathematicus
...*, Frankfurt, 660 p
Dechales Claude F M 1690 *Cursus seu
Mundus Mathematicus*, Lyon, 4 vols
[bib]
Sturm Leonhard C 1707 *Kurtzer Begriff
der gesamten Mathesis*, Nuernberg
(2nd ed 1710)
*Wolf Christian von 1710 *Anfangs-
gruende aller mathematischen Wissenschaften*, Halle, 4 vols
Kaestner A G 1758-1766 *Anfangsgruende
der Mathematik*, Goettingen, 4 vols
*Gherli O 1770-1777 *Gli elementi
teorico-pratica delle matematiche
pure*, Modena, 7 vols

Reference: Handbooks, Encyclopedias
(cont'd)

Rosenthal G E 1794-1797 *Encyklopaedie der mathematischen Wissenschaften: ihrer Geschichte und Literatur,* Gotha

Buesch J G 1795 *Encyclopaedie der mathematischen Wissenschaften,* Hamburg, 560 p

Schloemilch O 1879-1881 *Encyklopaedie der Naturwissenschaften. Part 1: Mathematik,* Breslau, 2 vols

Wolf 1881 *Wolf's naturwissenschaft-lich-mathematisches Vademecum,* Leipzig, 252 p

*Carr George S 1886 *A Synopsis of Elementary Results in Pure Mathe-matics,* London-Cambridge, 968 p

Hagen Johann G 1891-1905 *Synopsis der hoeheren Mathematik,* Berlin, 3 vols

*Peano G ed 1894-1908 *Formulaire de mathématique,* Turin, 5 vols

Meyer W F + ed 1896-1935 *Encyklo-paedie der mathematischen Wissen-schaften mit Einschluss ihrer Anwendungen,* Leipzig, 7 vols in numerous parts and fascicules

Pascal E 1898-1900 *Repertorio di matematica superiori,* Milan, 2 vols

- 1900-1902 *Repertorium der hoeheren Mathematik,* Leipzig, 2 vols (tr of Pascal 1898)

Henke R + 1904 *Schloemilchs Handbuch der Mathematik. Zweite Auflage,* Leipzig, 3 vols (F34, 1023)

Molk J + 1904- *Encyclopédie des sciences mathématiques pures et appliqués,* [tr and revision of Enc MW, never completed]

Grave D A 1912 *Entsiklopediya matematiki. Ocherk ee sovremennogo sostoyaniya,* Kiev

APPENDICES

1

CONVENTIONS

## SPECIAL SYMBOLS

Centuries are indicated by Roman numerals.

A minus sign preceding a year or century indicates B.C.; otherwise A.D. is implicit.

In the case of multiple authorship, only the senior author's name is given, followed by "+".

An asterisk "*" in the Bibliography means a recommendation; in the Coded List of Serials it means that the journal is cited in the Bibliography.

In the Coded List of Serials, "//" means that the journal ceased publication in the last year indicated.

## ABBREVIATIONS

Below are listed the abbreviations used in this volume. In a few cases the same word has more than one abbreviation. Where the context assures a distinction, the same abbreviation may be used for quite different words. For example, "Coll" stands for "College," and "coll" stands for "collection" and related words. A single abbreviation may stand for different parts of speech closely related to the same word, inflected forms of a word, or synonyms in different languages with approximately the same spelling. Where confusion might result, the word is spelled out in full. Although these procedures will not please those who demand consistency, we think that they work satisfactorily. We hope that users will call attention to errors and counter-examples to this claim. Note that we do not use a final period to indicate an abbreviation.

Abh = Abhandlung
abr = abridgement
abs = absolute
abst = abstract
Abt = Abteilung
acad = academy, academia, etc.
acc = accounting
accad = accademia
actu = actuary
adj = adjunct, adjoint
adm = administration
adv = advisory, advancement
adva = advancement
ag = agriculture
akad = akademiya
alg = algebra
allg = allgemein
Amer = America
AMS = American Mathematical Society
Amst = Amsterdam
anal = analysis, analytic
anc = ancient
angew = angewandte
ann = annal
anniv = anniversary
anthol = anthology
anthro = anthropology
Anw = Anwendung
Ap = April
app = appendix
appl = applications, applied,
     applicata, etc.
approx = approximation
arch = archive, archivio, etc.
archeol = arch(a)eology
archit = architecture
arith = arithmetic
asso = associative, association
ast = astronomy
astroph = astrophysics
Au = Australia
Aug = August
autobiog = autobiography
autog = autograph
ax = axiom, axiomatics
Azer = Azerbaidzhanskoi


b-a = beaux arts
Babyl = Babylonian
Bei = Beitrage
Belg = Belgium

Ber = Bericht
betw = between
bib = bibliography
bibt = bibliotheque, bibliothek, etc.
bin = binary
binom = binomial
biobib = biobibliography
biog = biography
biol = biology
birat = birational
bis = bisect
bk = book
b-l = belles-lettres
bol = boletim, bolletino, etc.
Br = British
bull = bulletin
bur = bureau
bus = business


C = continued, continuation
Cal = California
calc = calculus
calcn = calculation
calc var = calculus of variations
calen = calendar
Can = Canada
cartog = cartography
categ = category
cel = celestial
cen = center
cent = century
Cesk = Ceskoslovensko
ch = chapter
char = character, characteristic
chem = chemistry
Chic = Chicago
chim = chimie
Chin = China
chron = chronology
chronom = chronometry
cien = ciencia
circ = circle, circular
civ = civilization
cl = class
co = company
coef = coefficient
col = column
coll = collection
Coll = College
Colo = Colorado
combin = combinatorics, combinations

comm = commutative
commi = committee
commis = commission
commu = communication
comp = computer
compet = competition
*con. See cont.*
cond = condition
conf = conference
confed = confederation
cong = congruent
Cong = Congress
conj = conjugate
Conn = Connecticut
consig = consiglio
constr = construction
cont = continued, continuous,
    continuity
contr = contribution
conv = convergent
coord = coordinate, coordinates
corp = corporation
corres = correspondence
cosm = cosmology
coun = council
CR = comptes rendues
crit = criticism, critical
crypt = cryptography
cryst = crystallography
CUPM = Committee on the Undergrad-
    uate Program in Mathematics

d = de, der, di, etc.
Dec = December
deci = decimal
def = definition, definite
dept = department
deriv = derivative
descr = descriptive
det = determinant
deu = deutscher, etc.
devel = development
diag = diagonal
dict = dictionary
dif = differential
diffc = difference
dim = dimension
Dioph = Diophantine
discon = discontinuous
discov = discovery

diss = dissertation, thesis
dist = distance
distr = distribution, distributive
div = division, divisibility
diverg = divergence
dokl = doklady
dual = duality
Dub = Dublin
dup = duplication
dyn = dynamics

ec = école
econ = economics
ed = editor, edition, edited
Edin = Edinburgh
edu = education
elast = elasticity
elec = electricity
elem = elementary
ell = ellipse, elliptic
enc = encyclop(a)edia
eng = engineering
Engl = English
enl = enlarged
eq = equation, equal, equality
esp = especially
est = estestvoznaniya
est = estadisticas
etc = and so on
ethn = ethnology
Euc = Euclidean
exist = existence
exp = experiment(al)
expo = exponent

f = fuer
fac = faculty, faculté, faculdade, etc.
facs = facsimile
Fak = Fakultet
Feb = February
fil = filosofia, filozoficzny,
    filosoficheskii, etc.
fin = finite
fis = fisica
fiz = fizyk, fizica, etc.
Fl = Florida
formu = formula
Fors = Forschung
found = foundations

Fr = France
frac = fraction
fun = function
fund = fundamental
fys = fysik


GB = Great Britain
Gemein = Gemeinschaft
gen = general
genzn = generalization
geod = geodesy, geodesic
geog = geography
geol = geology
geom = geometry
geomag = geomagnetism
Ger = Germany
Ges Abh = Gesammelte Abhandlungen
Gesch = Geschichte
Gesell = Gesellschaft
gior = giornale
gos = gosudarstvennyi
gov = government
grav = gravity
grp = group


hersgbn = herausgegeben
hist = history
historiog = historiography
homog = homogeneous
Hun = Hungary
hydrost = hydrostatics
hyp = hypothesis
hyperb = hyperbola


i = in, im
id = identity
Ill = Illinois
im = imeni
imp = imperial
incl = includes, including
ind = industry
indep = independence
indet = indeterminate
induc = induction
inf = infinity
infl = influence
info = information
ing = ingegnero, Ingenieur

ins = insurance
inscr = inscription
inst = institute
instru = instrument
internat = international
interpol = interpolation
inters = intersection
intgl = integral
intl = international
intro = introduction
invar = invariant
Ire = Ireland
irrat = irrational
irreg = irregular
isop = isoperimetric
issl = issledovanie
ist = istituto
Ital = Italy
iter = iteration
izv = izvestiya


j = journal, jornal, etc.
Jab = Jahrbuch
Jabr = Jahrbuecher
Jan = January
Jasb = Jahresbericht


k = kaiserlich, koeniglich, etc.
Kan = Kansas
Kl = Klasse
Kommis = Kommission
kon = koninklijke, etc.
Ky = Kentucky


l = la, le, les, etc.
La = Louisiana
lab = laboratory
Lancs = Lancashire
lang = language
lect = lecture
Len = Leningrad
lett = lettere, lettre
lib = library
lim = limit
lin = linear
lit = literature, litteraire, etc.
log = logarithm
Lon = London

MAA = Mathematical Association of
    America
mach = machine
mag = magazine
magn = magnetism
Mar = March
Mass = Massachusetts
mat = matematik, matematisk, etc.
math = mathematics, mathematical,
    mathematische, etc.
mathn = mathematician
max = maximum
Md = Maryland
meas = measure
mec = mechanics, etc.
med = medicine, etc.
mekh = mekhanika
mem = memoir
mens = mensuration
meteorol = meteorology
meth = method
metr = metric
metrol = metrology
Mich = Michigan
mid = middle ages
milit = military
min = minimum
minist = ministerio, ministerstvo,
    ministry, etc.
Minn = Minnesota
misc = miscellaneous
MIT = Massachusetts Institute of
    Technology
Mitt = Mitteilungen
M-L = Moscow and Leningrad
Mo = Missouri
mod = modern
mon = monthly, monatlich, etc.
Monatsh = Monatsheft
Monatsschr = Monatsschrift
Monb = Monatsbericht
monog = monograph
Mont = Montana
Mos = Moscow
mot = motion
MR = Mathematical Reviews
ms = manuscript
mt = mountain
mult = multiplication
multil = multilinear
mus = museum

Nac = Nachricht
nacl = nacional
NAS = National Academy of Sciences
    of the USA
nat = nature, natural
natl = national
Naturf = Naturforschung
naturhist = naturhistorischen
Naturw = Naturwissenschaft
naz = nazionale
NB = New Brunswick
NC = North Carolina
NCTM = National Council of Teachers
    of Mathematics, Washington DC
n.d. = no date
necr = necrology
ned = nederlandsch
neg = negative
newsl = newsletter
NF = neue Folge = NS
NH = New Hampshire
NJ = New Jersey
nomog = nomography
nor = normal
nota = notation
nou = nouveau
Nov = November
NRC = National Research Council of
    the USA
NRC Can = National Research Council
    of Canada
NS = new series
NSF = National Science Foundation
num = numerical
numb = number
numer = numeration
NY = New York
NZ = New Zealand


o = of
O = Ohio
obs = observatory
obshch = obshchestvo
Oct = October
Oest = Oesterreich
off = office
Okl = Oklahoma
Ont = Ontario
oper = operation
opt = optics

ordin = ordinary
org = organization
otd = otdelenie

P = Press
pap = paper, papyrus
parab = parabola
paral = parallel
part = partial
patho = pathological
ped = pedagogy, pedagogicheskii, etc.
Penn = Pennsylvania
perf = perfect
perio = periodic
perp = perpendicular
persp = perspective
Pet = St Petersburg
phil = philosophy
phys = physics
Pol = Poland
polyg = polygon
polyh = polyhedron
polyn = polynomial
polyt, polit = polytechnique,
            politechnische, etc.
pop = popularization, popular
port = portrait
postu = postulate
pot = potential
princ = principle
prir = prirodno
prob = probability
probab = probability
probl = problem
proc = proceeding
prof = professor, profession
prog = progression, program
program = programming
proj = projective, projection
prop = proposition
propor = proportion
prov = province
prz = przyrodniczy
pseud = pseudonym
PSR = Padomj Socialistiskas Repub
psych = psychology
pt = point
Pt = Part
pub = published, publisher,
    publishing, publication, etc.

pyr = pyramid
Pyth = Pythagorean

q = question
quad = quadratic
quadril = quadrilateral
quan = quantity
quar = quarterly

R = royal, real
rat = rational
re = relating to
rech = recherche
recip = reciprocal, reciprocity
reck = reckoning
reconstr = reconstruction
recr = recreation
recur = recursion
reg = regular
rel = relative, relation
renai = renaissance
rend = rendiconti
rep = report
repr = reprint
repres = represent
rept = report
repub = republic
res = research
rev = review, revue, etc.
revis = revision
revo = revolution
RI = Rhode Island
ric = ricerche
riv = rivista
rot = rotation
RPR = Republica Popular Romane
Russ = Russia

s = south
S = supplement, Beiheft
Samm = Sammlung
sb = sbornik
sch = school, Schule, etc.
sched = schedule
Schr = Schrift
Schw = Schweiz
sci = science
sec = secondary

sect = section
sek = sektora, sektsii, etc.
sem = seminar
sep = separately
Sept = September
seq = sequence
ser = series, seriya
sexag = sexagesimal
sez = sezione
Sib = Siberskii, etc.
sim = similar
sing = single, singular
Sitz = Sitzungsbericht
Skand = Skandinavisk
skr = skrift
soc = society, social
sociol = sociology
sol = solution
soob = soobshchenie
sov = soviet
sp = special (issue)
Sp = Spanish
spec = special
sph = sphere
sq = square
SSR = Soviet Socialist Republic
SSSR = USSR
stan = standard
stat = statistics
stud = study
subst = substitution
subt = subtraction
suff = sufficient
summ = summary
sup = superior
suppl = supplement
surf = surface
surv = survey, surveying
susp = suspended
sym = symmetry
symp = symposium
synth = synthetic, synthesis
syst = system, systematic

t = the
tab = table, tabulation
tan = tangent
tea = teacher, teaching
tec = technology, technik, etc.
tek = tekhnicheskii, tekhnika, etc.

teor = teoreticeskoi
termin = terminology
theor = theoretical
thm = theorem
thy = theory
tid = tidsskrift
tijd = tijdschrift
top = topology
tr = translation
transa = transactions
transc = transcendental
transf = transformation
tria = triangle
trig = trigonometry
trilin = trilinear

u = und
U = University
uch = uchenye
ue = ueber
Uk = Ukraine
um = umiejetnosci
undergrad = undergraduate
univ = university
Unt = Unterricht
U P = University Press
US, USA = United States of America
USGPO = United States Government
        Printing Office
USSR = Union of Soviet Socialist
        Republics

v = von
Va = Virginia
val = value
var = variation, variable
Ver = Verein, vereeniging
Verh = Verhandlungen
Vero = Veroeffentlichungen
verw = verwandte
vet = vetenskaps, etc.
vib = vibration
vid = videnskabers, etc.
vol = volume
vs = versus, against
Vt = Vermont

w = with
Wash DC = Washington, DC
wet = wetenschappen
Wis = Wisconsin
wisk = wiskundig
Wiss = Wissenschaft
wk = work, werke
Wochschr = Wochenschrift
wyd = wydzial


yr = year


z = zu, zur, etc.
zap = zapiski
zh = zhurnal
Zt = Zeitschrift


## TRANSLITERATION

Wherever possible we follow the original spelling. In particular we try
to follow the capitalization rules
of each language, though these are
often not well defined. However, we
omit diacritical marks and render
the German umlauted o, a, u by oe,
ae, ue. Regardless of the original
language, we have preferred transliteration to translation, since the
latter often creates problems of
identification, especially for journals. For Chinese we follow what appears to be the most widely accepted
usage. For Russian we use the Li-

brary of Congress system (which suggests an approximately correct sound
to English readers without the use
of special symbols), except that we
omit hard and soft signs. The following table shows our transliteration
with alternatives from other systems.

| | | |
|---|---|---|
| а | a | |
| б | b | |
| в | v | (w) |
| г | g | |
| д | d | |
| е, ё | e | (ye, je; ye, yo, e) |
| ж | zh | (z, sh) |
| з | z | |
| и, й | i | (y; j, i) |
| к | k | |
| л | l | |
| м | m | |
| н | n | |
| о | o | |
| п | p | |
| с | s | (ss) |
| т | t | |
| у | u | |
| ф | f | (ff) |
| х | kh | (h, ch) |
| ц | ts | (c, tz, z) |
| ч | ch | (c, tsch) |
| ш | sh | (s, sch) |
| щ | shch | (sc, stsch) |
| ы | y | |
| ь | | (', j) |
| э | e | (e, e) |
| ю | yu | (you, ju, u) |
| я | ya | (ja, a) |

## 2

CODED LIST OF SERIALS

The following is a list of serials cited in this bibliography (preceded by *) plus other mathematical journals that have come to our attention in examining many sources. A "mathematical journal" is defined as one whose primary content is mathematical in the broadest sense or one that is frequently cited in a mathematical abstracting service, a mathematical bibliography, or a history of mathematics. We do not pretend to be complete, and completeness is not meaningful because of the vague boundaries just described. We hope that readers will supply us with additional journals as well as with corrections.

Serials in the list are arranged alphabetically (letter-by-letter) according to the abbreviations used in this bibliography. Cross-references from countries, cities, and alternative names are included in the alphabetical order. These cross-references are not complete, but are intended to be sufficiently suggestive.

Abbreviations have been chosen short enough to be used by computers and by scholars in note-taking and long enough to be suggestive of the full title. Existing standard abbreviations (for example those used in *Mathematical Reviews*) are much too long for our purpose, and existing computer codes for serial titles are too short. We have found that a satisfactory compromise can be made with abbreviations using no more than eight characters.

The reader who wishes to find the abbreviation of a given journal can easily do so if he keeps in mind our criteria for establishing them:

Where a journal has a short distinctive name such as *Isis* (and blessed be those who chose such names!), the full name or a contraction is used. If the journal has a distinctive name not involving an organization, an abbreviation is chosen that involves all significant parts of the name while placing it in approximately the same position in an alphabetical ordering as it would have if the full name were used. For example, the *American Journal of Mathematics* is abbreviated "AJM." If a journal has a very well-established nickname, this is used. For example, the *Journal fuer die reine und angewandte Mathematik* is abbreviated "Crelle." Where, as is usually the case, the journal name contains some synonym of "journal" and some uninformative words referring to an organization, location, or nationality, the abbreviation begins with an indication of the most characteristic and stable feature of the journal name. This may be an abbreviation of the organization (for example, "AMS" for the American Mathematical Society) or a country or city (for example, "Lon RS" for the British Royal Society). Accordingly, if a journal does not have a distinctive name, the reader should look under the organization or the city or country of publication. Generally the city is more stable than the country (because cities do not move with national boundaries!), and for this reason in most cases the city has been chosen in preference to other geographical designations. For example, the academies of science located in Berlin appear under "Ber" and not under a national name, which would have been Prussia at one time, Germany at another, and now either East or West Germany, depending on the organization.

For each journal are given the full title (words within titles are abbreviated: see list of abbreviations, page 700), name variations in parenthesis (insofar as we have been able to find them), dates ("//" indicates cessation of publication), places of publication, publisher if not evident from the title, relations with other journals, and additional notes. An asterisk "*" appears before journals cited in this Bibliography.

When supplying additions and corrections, please include our code and all other information you have, written out in full without abbreviation.

*AAAS             Proc of the Amer Asso for the Adva of Sci.   (Am As P)
                        1848-.  Wash DC, etc.

*AAASB          Bull of the Amer Asso for the Adva of Sci.  1942-1946//
                        Lancaster, Penn.

*Aach THJ       Jb der Rheinisch-Westfalischen Tec Hochsch Aachen.
                        (Jahrb TH Aachen) 1921-1934//  Essen.

*Aarbok, arsbok, etc. See organization, city, region, country, etc.*

Ab Dok          Abh des (oesterreichischen) Dokumentationszentrums fuer
                        Tec und Wirtschaft. 1951-.  Vienna.

*Ab GM          Abh zur Gesch der math Wiss, mit Einschluss ihrer Anw.
                        (A z G d M) 1877-1912//  C of Hist Abt.  S to ZM Ph.
                        Leipzig.

*Ab G Med       Abh zur Gesch der Med. 1902-1906//  Breslau.

Ab G Med N     Abh zur Gesch der Med und Naturw. 1934-.  Berlin.

*Ab GN Med     Abh zur Gesch der Naturw und der Med. 1922-1925//  Erlangen.

*Abhandlungen. See Ab; also organization, city, region, country, etc.*

*Abhandlungen ueber den math Unt. See ICTM Deu.*

Ab Model       Abh zur Samm math Modelle. 1907-1911//  Leipzig.

*Abo. See Turku.*

Ab Pri Ges     Abh einer Privatgesell in Boehmen, zur Aufnahme der Math,
                        der Vaterlandgesch und der Naturgesch. 1775-1784//  Prague.

*Abruz BSS     Bol della soc storica Abruzzese. 1889-.  Abruzzi.

*Ab WGWL       Abh zur Wissenschaftsgesch und Wissenschaftslehre.
                        1951-1952//  Bremen.

*Academie, academy, accademia, akademie, etc. See also city, region,*
                        *country, etc.*

*Academy       The Academy (& Literature). 1869-1916//  London.

*Acad Syr      Academy. A J of Sec Edu. 1886-1892//  Syracuse, NY.
                        Asso Academic Principals of the State of NY.

Acc Mag        Accountant's Magazine. 1897-.  Edinburgh.  Inst of
                        Chartered Accountants.

*Acc Rv        The Accounting Review. 1926-.  Ann Arbor, Mich.

Acir At        Atti e rend dell(a) (R) accad di sci, lett e arti (degli
                        Zelanti) (e dello studio) di Acireale.  (Acireale Ac At;
                        Rend e mem 1900) 1890-.  Acireale.

*ACMC          Communications of the Asso for Computing Mach. 1958-.  NY.

*ACMJ          J of the Asso for Computing Mach. 1954-.  NY.

*Acou SAJ      J of the Acoustical Soc of Amer. 1929-.  NY.

*Acta mathematica acad sci hungarica(e). See Hun Act M.*

*Act Antiq     Acta antiqua. 1951-.  Budapest.  Acad sci hun (Magyar
                        tudomanyos akad).

*Acta Phi F    Acta philosophica fennica. 1935-.  Helsinki.

*Acta physica acad sci hungarica(e). See Hun Act Ph.*

Acta PM        Acta physico-medica. 1727-1754//  C of Eph MPG.
                        C as Leop NA.  Nuremberg.

*Act Ari        Acta arithmetica. 1935-1939, 1958-.  Warsaw.  Akad nauk
                        inst mat.

*Act As Si     Acta astronomica sinica.  (T'ien wen hsueh pao)
                        1953-1966//  Peking.

Act Bioth     Stichting voor theor Biol van Dier en Mensch, verbonden aan
                        de Rijksuniv te Leiden, Ser A, Acta biotheoretica.  1935-.

Act Br Si      Acta brevia sinensia. 1943-1945// C in Ling USB & in
                Sci Tec Ch. London. British Council.
Act Cry         Acta crystallographica. 1948-. Copenhagen.
Act Erud       Acta eruditorum. 1682-1732// C as Nov Acta. Leipzig.
Act Erud O    Opuscula omnia actis eruditorum Lipsiensibus inserta, quae ad
                universam mathesim, physicam, medicinam, anatomiam, chirurgiam
                et philologiam pertinent. 1740-1746// Venice.
Act Erud S    Acta eruditorum suppl. 1692-1734// C as Nov Acta S.
                Leipzig.
Act Ge Ca     Acta geodetica & cartographica sinica. 1957-. Peking.
*Act HRNT     Acta historiae rerum naturalium nec non technicorum. 1954-.
                Prague. Czech Acad Sci.
*Act HSNM     Acta historica scientiarum naturalium & medicinalium.
                (Acta hist sci nat med) 1942-. Copenhagen.
Act Inf         Acta informatica. 1971-. Springer Verlag.
*Act M          Acta mathematica. 1882-. S in Bib M. Stockholm, Uppsala.
Act Me          Acta mechanica. 1965-. C of Ost Ing. Vienna, NY.
Act Me Si     Acta mechanica sinica. 1957-. Tr as Act Me Tr. Peking.
                Inst of Mechanics.
Act Me Tr     Acta mechanica sinica. 1966-. Tr of Act Me Si. NY.
                Consultants Bur.
Act M Si      Acta mathematica sinica. (Shu hsueh hsueh pao) 1951-1966//
                Tr as Chi M. Peking, Shanghai.
*Act Ori       Acta orientalia. 1922-. Lund. Soc orientales batava,
                danica, norvegica.
*Act Ph Aus    Acta physica austriaca. 1947-. Vienna.
*Act Phch     Acta physicochimica URSS. 1934-1937// Moscow. Akad nauk
                SSSR.
*Act Phi Fe.*    *See Acta Phi F.*
Act Ph Po     Acta physica polonica. 1932-1940, 1946-. C of Pol To Fiz.
                Warsaw. Polska akad nauk inst fiz.
Act Ph Si     Acta physica sinica. 1933-1966//? Peking.
Act PSMC      Acta polytechnica (scandinavica), math and computing machine
                ser. 1956-. Stockholm.
*Act PSPM     Acta polytechnica, phys & appl math ser. (Acta polyt Stockh;
                Acta polytch scandinav) 1947-. Stockholm.
Act Sal         Acta salmanticensia, ser ciencias. 1952-. Salamanca.
*Act Sc Ind    Actualites scientifiques & industrielles. 1931-.
                C of Con Ac Sci. Paris. Hermann.
*Actuar SA     (Papers &) Transas of the Actuarial Soc of Amer. 1889-1949//
                C as Am Actu. NY, Chicago.
*Actuar St     Actuariele Studien. 1960-. The Hague. Ned ver ter bevor-
                dering van het levensverzekeringwesen.
Acu             Acustica, Intl J of Acoustics. 1951-. Incl Aku B. Zurich,
                Stuttgart.
*ADB            Allgemeine deutsche Biographie. 1875-1912// Leipzig.
*ADM. See ADB.*
*Adv Comp     Advances in Computers. 1960-. NY. Academic P.
*Adv M         Advances in Mathematics. 1961-. See also Shux Jinz. NY.
                Academic P.
Adv Ph         Advances in Physics. 1952-. S to Phil Mag. London.

*Adv Sc          Advancement of Science. 1939-. C of Brt AAAS. London.
AEG              AEG-Mitt, tec-lit Abt. 1917-. C of Ber Elek. Berlin.
                 Allg Elec-Gesell.
Aeq M            Aequationes mathematicae. 1968-. Waterloo, Ont; Basel.
Aer J            The Aeronautical Journal. 1897-1922// C as Lon Aer. London.
*Aerot           L'aerotecnica, notiziario tec del minist de aeronautica e
                 atti dell'asso ital de aerotecnica. 1920-. Rome.
*Aer Q           The Aeronautical Quarterly. 1949-. London. R Aeronautical
                 Soc.
AFAS.  See Fr AAS.
Africa.  See Afr, So Af, Alger; also organization, city, region, etc.
Afr NSSMP        Bull de la soc. des scis de l'Afrique du nord, scis maths &
                 phys. Algiers.
*Afr USK         Afrika und Uebersee: Sprachen, Kulturen. (Zt fuer
                 Kolonialsprachen; Zt fuer Eingeborenensprachen) 1910-.
                 Berlin, etc.
Agra.  See also Indn Ag Ou.
Agram.  See Zagreb.
Agra U           Agra Univ J of Res Sci. 1952-. Agra.
AIAAJ            AIAA Journal. 1963-. Easton, Penn. Amer Inst of Aero-
                 nautics & Astronautics.
AIEEJ            J of the Amer Inst of Elec Engs. 1920-1930// C of AIEE Pr.
                 C as El Engin. NY.
AIEE Pr          Proc of the Amer Inst of Elec Engs. 1905-1919// C of
                 AIEE Tr. C as AIEEJ. NY.
AIEE Tr          Transa of the Amer Inst of Elec Engs. 1887-1904// C as
                 AIEE Pr. NY.
AIPCHP           Newsletter of the Cen for the Hist & Phil of Phys. 1964-.
                 NY. Amer Inst of Phys.
*AJM             Amer J of Math. (Amer J) 1878-. Baltimore, Md.
*AJP             Amer J of Phys. 1940-. C of APT. NY.
*AJ Ps           Amer J of Psych. 1887-. Ithaca, NY.
Akademiya nauk.  See location (BSSR, SSSR, Uzb, etc); also journal title
                 (Avtomat, Vych Sist, etc).
Aktu Ved         Aktuarske vedy pojistna mat a mat stat. 1929-. Prague.
*Aku B           Akustische Beihefte. 1951-1965// Incl in Acu. Zurich.
*Aku Zh          Akusticheskii zhurnal. 1955-. C of SSSR Aku. Tr as
                 Sov Ph Ac. Moscow. Akad nauk SSSR.
*Al And          Al-Andalus, rev de las escuelas de estudios arabes de Madrid
                 y Granada. 1933-. Madrid.
Albania.  See Tirana.
*Alex UAB        Bull of the Fac of Arts, Alexandria Univ. (Majallat kulliyat
                 al-adab, jami'at al-iskandariya) 1943-. Alexandria.
Alger A          Publications scientifiques de l'univ d'Alger, ser A, sci
                 math. 1954-. Algiers.
Algeria.  See Alger.
Alger Ori        Anns de l'inst d'etudes orientales de l'univ d'Alger.
                 1934-. Algiers.
*Alg Log         Algebra i logika, seminar. 1962-. Tr as Alg Log Tr.
                 Novosibirsk. Akad nauk SSSR, sibirskoe otd inst mat.
Alg Log Tr       Algebra & Logic. 1968-. Tr of Alg Log. NY. Plenum P.

Algor          Algorytmy. 1962-. Warsaw. Pol akad nauk. Prace inst
               maszyn mat.
Algo Yaz       Algoritmy i algoritmichesie yazyki. 1967-. Moscow. Akad
               nauk SSSR, vychislitelnyi tsentr.
Alg U          Algebra universalis. 1971-. Winnipeg. Univ of Manitoba.
               Birkhaeuser.
*Alla UMAB     Bull of the Allahabad Univ Math Asso. 1928-. Allahabad.
*Allg St Ar    Allg statistisches Archiv. 1890-1944, 1949-. Munich.
               Deu Stat Gesell u Arbeitsgemein f gemeindliche Stat.
*Allg Verm     Allg Vermessungs-Nachrichten. 1889-1943, 1950-. Pt of
               Z Vermess 1943. C of Bildmess. Berlin, Liebenwerda.
*Allg Z        Allgemeine Zeitung. (Muenchener Allg Z) 1798-. Tuebingen,
               Augsburg, etc. Beilage merged with Int Woch 1909.
 Alma-Ata. See Kazk, Khalyk.
*Altpr Mo      Altpreussische Monatsschr, der neuen preussischen Provinzial-
               Blaetter. 1864-1923// Koenigsberg.
 Alzate. See Mex Alz.
*Am Ac Mm      Mem of the Amer Acad of Arts & Scis. 1833-. Boston.
*Am Ac Pr      (Proc) J of the Amer Acad of Arts & Scis. (Daedalus) 1846-.
               Boston.
 Am Actu       Transas of the Soc of Actuaries. 1949-. C of Actuar SA.
               Chicago, Ill.
*Am AJRP       Procs of the Amer Acad for Jewish Res. 1928-. NY.
*Am Anthro     American Anthropologist. 1888-. Lancaster, Penn.
*Am Antiq      Amer Antiquarian Soc Procs. 1812-. Worcester, Mass.
*Am Ben Rv     Amer Benedictine Rev. 1950-. Newark, NJ.
*Am Bib SP     Papers of the Bibliographical Soc of Amer. 1904-. Chicago.
*Am CAS        Casualty Actuarial (and Stat) Soc of Amer Procs. 1914-. NY.
 Am Ch         The Amer Chemist, a Mon J of Theor Chem. (Am C)
               1871-1877// NY.
*Am C Phil     Amer Catholic Phil Asso Procs. 1926-. Baltimore, Md.
*Am Dialog     American Dialog. 1964-. NY.
*Am Doc        Amer Documentation. 1950-1969// C as ASISJ. Amer Doc Inst.
*Am Ec Rv      Amer Economic Rev. 1911-. Ithaca, NY.
 American. See A, Am, US; also city, state, region, etc.
 American Acad of Arts & Scis. See Am Ac.
 American Actuarial Soc (Actuarial Soc of Amer). See Actuar SA.
 American Asso for the Advancement of Sci. See AAAS.
 American Documentation. See Am Doc, ASIS.
 American Inst of Aeronautics and Astronautics. See AIAA.
 American Inst of Electrical Engineers. See AIEE.
 American Inst of Physics. See AIP.
 American Mathematical Soc. See AMS.
 American Soc for Information Sci. See ASIS.
 American Society of Mechanical Engineers. See JA Me.
 American Statistical Asso. See ASA.
 Am ERJ         Amer Engineer & Railroad J. 1893-1913// NY.
*Am Geol SB     Bull of the Geol Soc of Amer. 1890-. Rochester, NY.
 Am Geoph U     Transas of the Amer Geophys Union. 1920-. Wash DC. NRC.
*Am I Actu R    The Record of the Amer Inst of Actuaries. 1909-1948//
                Chicago, Ill; Milwaukee, Wis.

*Am J Arch Amer Journal of Archaeology. 1885-1896, ser 2 1897-.
 Concord, Mass.
*Am J Phil Amer J of Philology. 1880-. Baltimore, Md. Johns Hopkins U.
*Am JS Amer J of Sci. (Amer J of Scis & Arts; Silliman J) 1818-.
 New Haven, Conn.
*Am JSLL Amer J of Semitic (Hebraic) Langs & Lit. 1884-1941//
 C as JNES. Chicago, Ill. Univ of Chicago.
*Am J Surg Amer J of Surgery. 1890-. NY.
*AMM The Amer Math Monthly. 1894-. Buffalo, NY; etc.
*Am Met B Bull of the Amer Meteorol Soc. 1920-. Easton, Penn.
*Am NASBM Natl Acad of Scis Biog Mems. 1876-. Wash DC.
*Am NAS Mm Mems of the Natl Acad of Scis. 1866-. Wash DC.
*Am NASP Procs of the Natl Acad of Scis (of the USA). 1915-. Wash DC.
*Am Nep Amer Neptune. 1941-. Salem, Mass.
*Am NRCB Bull (Pubs, Circular, Repr Ser, etc) of the Natl Res Council.
 1919-1951// Wash DC. Natl Acad of Scis.
*Am Opt J J of the Optical Soc of Amer. 1917-. NY, Philadelphia.
 See also Rv S Inst, Opt S Am.
*Am OSJ Amer Oriental Soc J. 1843-. Boston, Mass.
Amoy UASN Univ Amoiensis, acta scientiarum naturalium. (Siameny-
 dasiue-siebao tseyzhany kestiuebany; Hsia-men ta hsueh hsueh
 pao) 1957-. Amoy, China.
*Am Phlg TP Amer Philological Asso Transas & Procs. 1869-. Boston, Mass.
*Am Q Reg Amer Quarterly Register. (Quar J; Register & J) 1827-1846//
 Boston, Mass. Congregational Edu Soc.
*AMR Applied Mech Reviews. 1948-. NY.
*AMSB Bull of the Amer Math Soc. 1895-. C of NYMS. Providence,
 RI, etc.
*Am Sc Amer Scientist. 1942-. C of Sigma Xi Quar. New Haven, Conn;
 Easton, Penn. Soc of Sigma Xi.
AMS Col Amer Math Soc Colloquium Pubs (Lectures). 1903-. NY.
*AMS Contents of Contemporary Math Journals. See Contents.*
Am Seis Bull of the Seismological Soc of Amer. 1911-. Berkeley, Cal.
AMSM Mems of the Amer Math Soc. 1950-. Providence, RI.
*AMSN Notices of the Amer Math Soc. 1954-. Providence, RI.
AMSNP New Publications. 1964-. Providence, RI. AMS.
*Am SORB Bull of the Amer Schools of Oriental Studies. 1919-.
 S Hadley, Mass. Amer Sch of Oriental Res in Jerusalem.
*AMSP Procs of the Amer Math Soc. 1950-. Providence, RI.
*AMST Transas of the Amer Math Soc. 1900-. Providence, RI.
Amst Arch Archief uitgegeven door het wiskundig genootschap. 1856-
 1874// C of Verslagen Wiskundig Genootschap Amst. C as
 Nieu Arch. Amsterdam.
*Am Statcn Amer Statistician. 1947-. C of ASA Bu. Wash DC.
*Amsterdam. See Amst; also Indagat, Mengelw, Ned.*
*Amst Inst Het Instituut. (Amst I) 1841-1846// Amsterdam.
Amst IN Vh Nieuwe Verh der errste Kl van het K Ned Inst van Wet...te
 Amst. 1827-1852// Amsterdam.
Amst I Vh Verh der Eerste Kl van het K Ned Inst van Wet...te Amst.
 (Amst Vh; Amst Vh Ak) 1812-1825// Amsterdam.

| | |
|---|---|
| Amst Ja | Jaarbock van de K Akad van Wetenschappen. (Jaarb Akad Amst) 1857-1936// C as Ned Ja. Amsterdam. |
| *Amst L Med | (Verslagen en) Mededelingen der K nederlandse Akad van Wet, Afdeeling Letterkunde. 1855-. Amsterdam. |
| *Amst Luch | Natl Lucht- (en Ruimte)vaartlaboratorium Rapport, Verlagen en Verhandelingen. (Natl Aero- (& Astro)nautical Res Inst Rept) 1921-. Amsterdam. |
| Amst Maan | Maanblad voor Natuurwetenschappen. (Mbl Nt) 1871-. Amsterdam. Genootschap ter Bevordering van Natuur- Genees- en Heelkunde, Sectie voor Natuurwetenschappen. |
| Amst MCR | Math Centrum Amsterdam Rekenafdeling. 1963-. Amsterdam. |
| Amst MC Ra | Math Centrum Amsterdam Rapport. 1954-1963// C as Amst MCZW. Amsterdam. |
| Amst MCS | Amsterdam Math Centrum Scriptum. 1948-. Amsterdam. |
| Amst MCTW | Math Centrum Amsterdam Afdeling toegepaste Wiskunde. 1948-. Amsterdam. |
| Amst MCZW | Math Centrum Amsterdam Afdeling zuivere Wiskunde. 1963-. C of Amst MC Ra. Amsterdam. |
| *Amst NWNV | Nieuwe wis- & natuurkundige Verh van het Genootschap te Amst. (Amst N Ws Ntk Vh) 1844-1854// Amsterdam. |
| *Amst P | K Akad v Wet te Amst Afdeeling Natuurkunde. (Proc o t Sect o Sci; Proc Amsterdam; Amst Ak P) 1899-1950// C as Amst PA, & as Amst PB, & as Indagat. Tr of Amst Vs G. Amsterdam. |
| Amst PA | K ned Akad v Wet Procs (o t Sect o Scis), Sect A, Math Scis. 1951-. Pt C of Amst P. Tr from Amst Vs G. Vols 54-57 (1951-1954) = vols 13-17 Indagat. |
| *Amst PB | K ned Akad v Wet Procs, Ser B, Phys Scis. 1951-. Pt C of Amst P. Amsterdam. |
| *Amst PV | Processen-verbaal van de Gewone Vergaderingen der K Akad v Wet, Afdeeling Natuurkunde. (Amst Ak Wet P) 1865-1884// C in Amst Vs M. Amsterdam. |
| *AMS Trls | Amer Math Soc Translations. 1949-1954, NS 1962-. Providence. |
| *Amst Verz | Archief voor de Verzekeringswetenschap en aaverwante Vakken. (Arch Verzek) 1895-1918// C as Verzek. Gravenhage. Vereeniging v wisk Adviseurs bij Ned Maatschappijen v Levensverzekering. |
| *Amst Vh | Verh der K Akad van Wet. (Amst Ak Vh) 1854-1892// C in Amst Vh I. Amsterdam. |
| *Amst Vh I | Verh der K ned Akad v Wet Afdeling Natuurkunde, eerste Sect. 1892-. C of Amst Vh. Amsterdam. |
| *Amst Vs G | K Akad v Wet te Amst Verslag v d gewone Vergaderingen (d wis- en natuurkundige Afdeeling)(v d Afdeeling Natuurkunde). (Verslagen d zitlingen; Kon Ak Amst; Amst Ak Versl; Kon Ned Akad Weten Versl Afd Natuur) 1892-. C of Amst Vs M. Tr in Amst P, & in Amst PA, & in Indagat. Amsterdam. |
| *Amst Vs M | Verslagen & Mededeelingen d K Akad v Wet te Amsterdam. 1853-1892// C as Amst Vs G. Amsterdam. |
| Analog Bu | Info Bull, Intl Asso for Analog Computations. (Bull d'info d l'asso intl pour l calcul analogique). C as Analogiq. Brussels. |

| | |
|---|---|
| Analogiq | Anns d l'asso intl pour l calcul analogique. (Procs o t Intl Asso for Analog Comp) 1958-. C of Analog Bu. Brussels. |
| Analyst A | The Analyst. NS 1814// C of Analyst M. Philadelphia. Ed: Robert Adrain. |
| Analyst H | The Analyst, a Mon J of Pure & Appl Math. 1874-1883// Des Moines, Iowa. C as An M. |
| Analyst M | The Analyst, or Math Museum. 1808// C as Analyst A. Philadelphia. Ed: R Adrain. |
| *Anand San | Anandasrama Sanscrit Ser. 1888-. Poona. |
| *An Ast Met | Annuario astro-meteorologico. 1883-. Venice. |
| An Astph | Anns d'astrophysique. 1938-. Paris. |
| *Anc Eg | Ancient Egypt (& the East). 1914-1935// London. |
| *An Eug | Anns of (Eugenics) Human Genetics. 1925-. London. Galton Lab. |
| *Angelic | Angelicum. 1924-. Rome. |
| An Geof | Annali de geofisica, riv d ist naz d geofis. 1948-. Rome. |
| An Hy MH | Annalen der Hydrographie & maritimen Meteorologie. (Deu hydrographische Zt; Ger Hydrographic J; Zt f Seewarte & Meerkunde) 1875-. Berlin, Hamburg. Deu Seewarte. |
| *An In Hist | Anns intls d'histoire. 1901-1902// Paris. Congres intl d'hist comparee. |
| Ankara M | Communications d l fac d scis d l'univ d'Ankara, ser A, math-phys-ast. 1938-. Ankara. |
| *An M | Annals of Mathematics. (A Mth) 1884-. C of Analyst H. Princeton, NJ; etc. |
| *An Med H | Annals of Medical History. 1917-. NY. |
| *An Mines | Annales des mines. 1816-. C of J Min. Paris. |
| An M Log | Annals of Math Logic. 1969-. Amsterdam. North-Holland Pub. |
| *An M St | Annals of Math Statistics. 1930-. Baltimore, Md; etc. |
| *An M Stu | Annals of Mathematics Studies. 1940-. Princeton, NJ. |
| | *Annalen, annales, annali, annals, anales, etc. See An; also organization, city, region, country, etc.* |
| | *Annalen der Physik und Chemie. See Pogg An.* |
| | *Annales de l'inst d'etudes orientales. See Alger Ori.* |
| | *Annales de math pures et appl. See Gergonne.* |
| | *Annales nouvelles de math. See Nou An.* |
| | *Annales polonici math. See Pol An M.* |
| *Annali | Annali di mat pura & applicata. (Brioschi An; A Mt) 1858-. C of Tortolin. Rome, Milan, Bologna, etc. |
| | *Annali de scienze mat e fis. See Tortolin.* |
| | *Annali di fis, chim, e mat...Milan. See Majocc.* |
| | *Annals of Human Genetics. See An Eug.* |
| Annam B | J of the Annamalai Univ, Part B, Sci. 1932-. Annamalainagar. |
| *An Natphi | Ann der Naturphilosophie (& Kulturphil). 1902-1921// Leipzig. |
| *Annee Sci | L'annee scientifique & industrielle, ou expose mensuel des travaux scientifiques. 1856-1913// Paris. |
| | *Annuaire, etc. See also organization, city, region, country, etc.* |
| *Annuaire | Annuaire des mathematiques. 1901-1902// Paris. One number only. |
| | *Annuaire...bureau des longitudes. See Fr Long.* |
| *An Orient | Analecta orientalia. 1931-. Rome. Pontificio ist biblico. |

*An Ph            Annals of Physics. 1957-.  NY.
*An Phi           Annelen der Philosophie & philosophischen Kritik.  1919-1930/
                  C as Erkennt, & as J Un S.  Leipzig.
*An Phi Chr       Annales de philosophie chretienne.  1830-1913//  Paris.
*An Physiq        Annales de physique.  1914-.  Paris.
 An Ponts         Annales de ponts & chaussees.  1831-.  Paris.
 An Rad           Annales de radioelectricite.  1945-.  Paris.
*An Rv AP         Annual Rev of Automatic Programming.  1960-.  Pergamon P.
*An Sc            Annals of Sci, a Quar Rev for the Hist of Sci since the Renai
                  1936-.  London.
*An SENS          Annales scientifiques de l'ecole normale sup.  1864-.  Paris.
 An Telcom        Annales de telecommunications.  1946-.  Paris.
*Anthropo         Anthropos, ephemeris internationalis ethnologica & linguistic
                  1906-.  Vienna, etc.
*Antike           Die Antike, Zt fuer Kunst & Kultur des kl Altertums.  1925-.
                  Berlin, Leipzig.
*Antiq            Antiquity.  1927-.  Gloucester.
*Antiq Cl         L'antiquite classique.  1932-.  Brussels, etc.
 An Vers          Annalen des gesammten Versicherungswesens.  1870-1920//
                  Leipzig.
*Aplik M          Aplikace matematiky.  1956-.  Prague.  Cesk akad ved.
 APMEPB           Bull de l'asso de professeurs de math de l'enseignement
                  public.  1922-.  Paris.
 App Anal         Applicable Analysis, an Intl J.  1971-.  NY.  Gordon &
                  Breach Pub Co.
*Appleton         Appleton's Popular Sci Monthly.  1872-1900//  C as
                  Pop Sci Mo.  NY.
 *Applied Math & Mech.  See Pri M Me.*
 App Stat         Applied Statistics.  1952-.  London.  R Stat Soc.
*APS Lib B        Amer Phil Soc Library Bull.  1943-1959//  Philadelphia, Penn.
*APS Mm           Amer Phil Soc Memoirs.  1935-.  Philadelphia, Penn.
*APSP             Procs of the Amer Phil Soc (for Promoting Useful Knowledge).
                  1838-.  Philadelphia, Penn.
*APST             Transas of the Amer Phil Soc.  (TA) 1769-.  Philadelphia.
*APSY             Amer Phil Soc Yearbook.  1937-.  Philadelphia, Penn.
 APT              Amer Physics Teacher.  1933-1940//  C as AJP.  Lancaster, Pen
 *Arbeitsgemein f Fors d Landes Nordrhein-Westfalen.  See Nor Wes AF.*
 Arc Anth         Archiv fuer Anthropologie.  (Arch f Ap) 1866-.  Braunschweig
*Arc Artil        Archiv f die Artillerie- & Ingenieur-Officiere des deutschen
                  Reichsheeres.  1835-1897//  Berlin, Posen.
*Arc Begrf        Archiv f Begriffsgesch, Bausteine zu einem hist Woerterbuch
                  der Philosophie.  1955-.  Bonn.
 Arc El           Archiv f Elektrotechnik.  1912-1945, 1948-.  Berlin, etc.
 Arc El Com       Archive for Electronic Computing.  1966-.  Springer Verlag.
*Arc Fr Hi        Archivum franciscanum historicum.  1908-.  Florence.
*Arc G Med        Archiv f Gesch der Medizin.  1907-1928//  C as Sudhof Ar.
                  Leipzig.
*Arc GMNT         Archiv f Gesch der Math, der Naturw & der Technik.
                  1927-1931//  C of Arc GNT.  C as QSGM.  Leipzig.
*Arc GNT          Archiv f die Gesch der Naturw & der Tec.  1908-1922//  C as
                  Arc GMNT.  Leipzig.
 *Arc G Phi.  See Arc Phi.*

*Archaeol     Archaeologia, or Misc Tracts relating to Antiquity.  1770-.
              London.  Soc of Antiquaries.
*Archaeological.  See also Archeol.*
*Archeion     Archeion, archivo de historia de la ciencia.  1928-1943//
              C of Arc Sto.  C as Arc In  HS.  Santa Fe, Argentina.  Intl
              Acad of the Hist of Sci.
*Archeological Survey of India.  See Indn Ar.*
*Arc HES      Archive for Hist of Exact Scis.  1960-.  Berlin.  Springer
              Verlag.
*Archim       Archimede, riv per gli insegnanti e i cultori di mat pure ed
              appl.  1949-.  C of Bo Fir.  Florence.
*Archims      Archimedes, Anregungen & Aufgaben f Lehrer, Schueler, &
              Freunde der Math.  1949-.  Regensburg.
*Arc His SJ   Archivum historicum soc Jesu.  1932-.  Rome.
*Archiv, archive, archief, arkiv, etc.  See Arc;  also organisation, city,
              region, or country, etc.*
*Archives for the History of Sci & Tec.  See Len IINT.*
*Archiv fuer wiss Kunde von Russland.  See Erman Arc.*
*Arc In HS    Archives intls d'hist des scis.  1947-.  C of Archeion.  See
              also Int Ac HS.  Paris.  Acad intl d'hist des scis.
*Arc M        Archiv der Mathematik.  (Archives of Math; Archives mathe-
              matiques)  1948-.  Basel, Karlsruhe, Stuttgart, etc.
*Arc Me Sto   Archiwum mechaniki stosowanej.  1959-.  Warsaw.
              Pol akad nauk, inst podstawowych problemow techniki.
 Arc Met      Archiv f Meteorol, Geophys, & Bioklimatologie, Ser A,
              Meteorol & Geophys.  1948-.  Vienna.
*Arc MLG      Archiv f math Logik & Grundlagenforschung.  1950-.  Stuttgart.
*Arc MN       Archiv for Mathematik og Naturvidenskab.  1876-.  Oslo, etc.
*Arc MP       Archiv der Math & Phys mit besonderer Beruecksichtigung der
              Beduerfnisse der Lehrer....  (Grunert Arch; Hoppe Arch)
              1841-1920//  Leipzig.
*Arc MP Lit   Archiv der Math & Phys, literarische Bericht.  1841-1900//
              Leipzig, etc.
*Arc Mu Tey   Archives du musee Teyler.  1866-.  Haarlem.
 Arc MWSF     Archiv f math Wirtschafts- & Sozialforschung.  1935-.
              Stuttgart, Berlin, Leipzig.
*Arc Or       Archiv f Orientforschung (Keilschriftforschung).  1923-1944,
              1945/51-.  Berlin.
*Arc Or Pra   Archiv orientalni.  1929-.  Prague.  Cesk orientalni ustav.
*Arc Pap      Archiv f Papyrusforschung & verwandte Gebiete.  1900-.
              Leipzig.
*Arc Phi      Archiv f Gesch der Philosophie.  (Arch f syst Phil & Soziol;
              Phil Monatsh)  1868-1931//  Berlin.
*Arc Ph Par   Archives de philosophie.  1923-1951, 1955-.  Paris.
 Arc RAM      Archiv der reinen & angewandten Math.  1795-1800//  Leipzig.
 Arc RMA      Archive for Rational Mech & Anal.  1957-.  Berlin.
              Ed: Truesdell.
*Arc Sc Ph    Archives des scis phys & nats.  1846-1947//  C of Bib Un.
              Geneva, Paris, Lausanne.
*Arc Sto      Archivio di storia della sci.  1919-1927//  C as Archeion.
              Rome.  Ed: Aldo Mieli.
*Arc Teyler.  See Arc Mu Tey.*
 Arc T Mess   Archiv f tec Messen.  1931-1944, 1947-.  Munich, Berlin.

Arc Verz.   See Amst Verz.
*Arduo        L'arduo, riv mensile di sci, fil & storia.  1921-1922//
              Bologna.
Argentina.  See Arg;  also B Ai, La Pla, Physi Arg, Tucum U.
*Arg IM Ros   Pubs del inst de mat.  1939-.  Rosario.  Univ nacl del
              litoral, fac de ciencias mat (fis-quim y nat).
*Argosy       The Argosy.  1872-.  Sackville, NB.
*Arg SC       Anales de la soc cientifica Argentina.  (An Soc Argentina)
              1876-.  Buenos Aires.
*Arg UM Pu    Union matematica Argentina publicaciones.  (Publ Un Mat Arg)
              1936-1941//  C as Arg UMR.  Buenos Aires.
*Arg UMR      Revista de la union mat Argentina.  (Rev Un Mat Arg)  1941-.
              C of Arg UM Pu.  Buenos Aires.
*Aristot      Procs of the Aristotelian Soc.  1911-.  London.
*Ari T        The Arithmetic Teacher.  1954-.  Wash DC.  NCTM.
 Ark As       Arkiv f Astronomi.  1950-.  C from Ark MAF.  Stockholm.
 Ark Fys      Arkiv f Fysik.  1949-.  C from Ark MAF.  Stockholm.
*Arkhimed     Arkhimedes.  1949-.  Helsinki.
 Ark M        Arkiv f Matematik.  1949-.  C from Ark MAF.  Stockholm.
*Ark MAF      Arkiv f Mat, Ast och Fys.  1903-1949//  Pt C of Sv Bi, & of
              Sv Ofv.  C as Ark M, & and as Ark As, & as Ark Fys.  Uppsala.
*Arm Aff      Armenian Affairs.  1949-1950//  NY.  Armenian Natl Coun
              of Amer.
 Arm D        Doklady, akad nauk armyanskoi SSR.  (Aikakan SSR gitutiunneri
              akad zekuitsner)  1944-.  Erevan.
*Arm FM       Izv akad nauk armyanskoi SSR, ser fiz-mat nauk.  1957-.
              C of Arm FMET.  Erevan.
 Arm FMET     Izv akad nauk armyanskoi SSR, ser fiz-mat, est, i tek nauki.
              1948-1957//  C as Arm FM, & as Arm Tek.  Erevan.
 Arm IMM      Soobshcheniya inst mat i mekh, akad nauk armyanskoi SSR.
              1947-.  Erevan.
*Arm Ped FM   Sbornik nauchnykh trudov, ser fiz-mat.  1959-.  Erevan.
              Armyanskii gos ped inst.
 Arm Tek      Izv akad nauk armyanskoi SSR, ser tek nauk.  1957-.  C of
              Arm FMET.  Erevan.
*Art Arch     Art & Archeology.  1914-1934//  Concord, NH.  Archeol Inst
              of Amer.
 Art Intel    Artifician Intelligence, an Intl J.  1970-.  Amer Elsevier
              Pub Co.
*Aryan Pth    The Aryan Path.  1930-.  Bombay.
 ASA Bu       Amer Stat Asso Bull.  1935-1947//  C as Am Statcn.  Wash DC.
*ASAJ         J of the Amer Stat Asso.  1922-.  C of ASAQP.  NY, etc.
 ASA Pub      Pubs of the Amer Stat Asso.  1888-1918//  C as ASAQP.
              Boston, Mass.
*ASAQP        Quar Pubs of the Amer Stat Asso.  1920-1921//  C of ASA Pub.
              C as ASAJ.  Boston, Mass.
Asiatic Soc of Bengal.  See Beng.
*Asi R        Asiatick Researches, or Transas of the (Bengal) Soc....
              (As Researches)  1788-1836//  Calcutta.
*ASISJ        J of the Amer Soc for Info Sci.  1970-.  C of Am Doc.  Wash DC
*ASISN        Newsl of the Amer Soc for Info Sci (Amer Doc Inst).  1961-.

| | |
|---|---|
| ASISP | Procs of the Amer Soc for Info Sci (Amer Doc Inst). 1963-. |
| *As J | The Astronomical J. 1849-. Albany, NY; Boston, Mass; New Haven, Conn. |
| As Ja | Astronomischer Jahresbericht. 1899-. Berlin. |
| As J Ab | Astronomical Journal. Wash DC. Abst from As Zh. |
| *As Nach | Astronomische Nachrichten. 1821-. Altona, Berlin, Kiel, etc. |
| *ASP | Pubs of the Ast Soc of the Pacific. 1889-. San Francisco. |
| Assek Ja | Assekuranz-Jahrbuch. 1881-1943// Basel. |
| *Association* de professeurs de math.... *See APMEPB.* | |
| *Association intl pour le calcul analogique. See Analog.* | |
| *Association R des actuaires belges. See Bel Actu.* | |
| *As So GB | Transas of the R Asiatic Soc of GB & Ireland. (As S T) 1827-1835// London. See also Lon Asia, & Bom RAS. |
| *Assur Mag | The Assurance Mag & J of the Inst of Actuaries. 1830-1867// London. |
| *Astmie | L'astronomie, rev d'ast populaire.... (As) 1882-1894// Paris. Flammarian. |
| *Astmie Fr | Astronomie, bull de la soc ast de France. 1911-. C of Fr S Ast. Paris. |
| Astnaut | Astronautica acta. 1955-. Vienna. |
| Astph | Astrophysics. 1965-. Tr of Astrofiz. NY. |
| Astph J | The Astrophysical Journal. 1895-. Chicago, Ill. |
| Astph Nor | Astrophysica norvegica. 1934-. C of Chr Obs. Oslo. Norske vid-akad. Inst Theor Astrophys, Oslo Univ. |
| Astrofiz | Astrofizika. 1965-. Tr as Astph. Erevan. Akad nauk armyanskoi SSR. |
| *Astronomical, astronomique, astronomischer, etc. See As, Ast; also Bode As Jb.* | |
| *Astronomical Society of the Pacific. See ASP.* | |
| *As Zh | Astronomicheskii zhurnal. 1928-. C of Rus As Zh. Tr as Sov As. Moscow. Akad nauk SSSR. |
| Ataturk | Ataturk univ yayinlari, arastirmalar serisi matematik. 1959-. Erzurum. |
| *Athena | Athena. 1899-. Athens. |
| *Athen P | Praktika tes akademias athenon. 1926-. Athens. |
| *Atlantic | Atlantic. (Atlantic Monthly) 1857-. Boston, Mass. |
| *Atlas | Atlas. 1961-. NY. |
| *Atti. See organization, city, region, country, etc.* | |
| Au AAS | Rept of the...Australasian Asso for the Adva of Sci. (Aust As Rp) 1888-1928// C as Au NZAAS. Sydney. |
| *Au HSJP | R Australian Hist Soc J & Procs. 1901-. Sydney. |
| *Au JAS | Australian J of Applied Science. 1950-. Melbourne. |
| *Au JP | Australian J of Physics. (Au J Sci Res, Phys Sci). 1948-. Melbourne. |
| *Au JS | The Australian J of Science. 1938-. Sydney. |
| *Au J St | The Australian J of Statistics. 1959-. Sydney. |
| *Au MSJ | J of the Australian Math Soc. 1959-. Sydney, Brisbane. |
| *Au MT | Australian Math Teacher. 1945-. Sydney. |
| Au NZAAS | Rept of the Australian & NZ Asso for the Adva of Sci. 1929-. C of Au AAS. Sydney. |
| *Ausland | Das Ausland. (Ausl) 1828-1893// Munich, Stuttgart. |

*Australia.* See Au; also Melbrne, NSW, Tasm, Vict.
*Austria.* See Ost; also Act Ph Aus, Innsbr, Wien.

| | |
|---|---|
| Aus Unt | Aus Unterricht & Forschung. 1923-. C of Wurt Sch. Stuttgart. |
| Aut Con | Automatic Control. 1968-. Tr of Av Vych. Faraday P. |
| Aut Doc ML | Automatic Documentation & Math Linguistics. 1967-. Tr of Nau Tek In. Faraday P. |
| *Aut RC | Automation & Remote Control. 1956-. Tr of Avto Tel. Pittsburgh, Penn. Consultants Bur. |
| Avtomat | Avtomatika. 1948-. Kiev. Akad nauk UkSSR, inst elektrotek. |
| *Avto Tel | Avtomatika i telemekh. 1936-1942, 1946-. Tr as Aut RC. Moscow. Akad nauk SSSR. |
| Av Vych | Avtomatika i vychislitelnaya tek. (Avtomaticheskoe upravlenie i vychislitelnaya tek) 1958-. Tr as Aut Con. Moscow. |
| Az D | Doklady akad nauk Azerbaidzhanskoi SSR. (Az SSR elmler akad meruzeler) 1945-. Baku. |
| *Az FMT | Izv akad nauk Azer SSR, ser fiz-mat i tek nauk. 1936-1945// C as Az FTM. Baku. |
| *Az FRT | Kheberleri akad nauk Azer SSR, fiz-rijazijjat ve tek elmleri ser. (Izv ser fiz-mat i tek nauk) 1958-. Baku. |
| *Az FTM | Izv akad nauk Azer SSR, ser fiz-tek i mat nauk. 1958-. C of Az FMT. Baku. |
| Az IFM | Trudy inst fiz i mat, akad nauk Azer SSR. 1948-1960// C of Az M. C as Az IMM. Baku. |
| *Az IMM | Trudy inst mat i mekh, akad nauk Azer SSR. 1961-. C of Az IFM. Baku. |
| *Az Iz | Izv akad nauk Azer SSR. 1936-1957// Baku. See also Az F. |
| Az M | Trudy sekt mat, akad nauk Azer SSR. 1945-1948// C of Az T. C as Az IFM. Baku. |
| Az T | Trudy akad nauk Azer SSR. 1933-. Pt C in Az M. Baku. |
| Az U | Uch zap Azer gos univ Elmi Eserler. 1955-1958// C as Az UFMK. Baku. |
| Az UFM | Trudy gos Azer univ im S M Kirova, ser fiz-mat. 1940-1948// Baku. |
| Az UFMK | Uch zap Azer gos univ...Kirova, ser fiz-met i khim nauk. 1959-. C of Az U. Baku. |

*Babes-Bolyai univ.* See Cluj.

| | |
|---|---|
| *Bagd CS | Majallah, kulliyat alulum jami'at Baghdad. (Jami'at Baghdad, kulliyat al-adab wa-al-'ulum; Bull, Coll Arts & Sci) 1956-. Baghdad. |
| B Ai BM | Boletin matematico. 1928-. Buenos Aires. |
| *B Ai Bol | Boletin matematico elemental. (Bol mat B Aires) 1930-1934// Buenos Aires. |
| *B Ai Ci EFN | Anales de la acad nacl de cien exactas, fis y nat. 1935-1947// Buenos Aires. |
| B Ai El | Revista de matematicas y fisicas elementales. 1916-1918// C as Rv M Hisp A. Buenos Aires. |
| B Ai Sem | Bol del sem mat (argentino). 1928-. Buenos Aires. Ed: I Rey Pastor. |

B Ai UCT  Pub de la fac de cien exactas, fis y nat, univ de Buenos
Aires, ser B (cien-tec). 1929-. Buenos Aires.

B Ai UF  Contributiones cientificas de la fac de ciencias exactas,
fis y nat, univ de Buenos Aires, ser (B) fis. 1950-.
Buenos Aires.

B Ai UM  Contribuciones cientificas de la fac de ciens exactas y nats,
univ de Buenos Aires, ser (A) mat. 1950-. Buenos Aires.

B Ai US  Cursos y sems de mat, fac de ciens exactas y nats, univ
nacl de Buenos Aires. 1959?-1961// Buenos Aires.

*Baku. See Az.*

*Banaras. See also Benares.*

\*Banar JSR  The J of Sci Research of the Banaras Hindu Univ. 1950-.
Benares.

*Barbonica. See Nap.*

Barc Bo  Bol de la R acad de cien y artes de Barcelona. (Barcel Ac Bl)
1892-. Barcelona.

*Barcelona. See also Collect M, Chron Cien.*

\*Barc Mm  Mems de la R acad de cien y artes de Barcelona. (Barcel Ac
Mm) 1835-. Barcelona.

Barc Sem  Univ seminario mat. 1948-. Barcelona.

\*Bari Sem  Conferenze del sem de mat dell'univ di Bari. 1954-. Bari.

Baroda  J of the Maharaja Sayajirao Univ of Baroda. 1951-. Baroda.

\*Basl V  Verh der Naturfors Gesell in Basel. 1854-. Basel.

\*Bath Bris  Bath & Bristol Mag, or Western Miscellany. 1832-1834// Bath.

\*Batt  Giornale de mat (di Battaglini) ad uso (per il progresso)
degli studenti delle (nelle) univ ital. (G Mt; Giorn d mat
Batt; Napoli giorn d mat; Giornale d mat). 1863-. Naples.

Baumgart  Zt f Phys, Math, & verw Wiss. 1826-1842// Vienna.
Ed: Baumgartner.

\*Bay Bild  Bayerisches Bildungswesen. 1927-. Munich.

\*Bay Bl G  Blaetter f das Gymnasialschulwesen. 1864-1935// Munich.
Ver bayerischer Philologen (Gymnasiallehrerverein).

\*Bay Bl GR  Blaetter f das bairische Gymnasial- & Realschulwesen.
(Bayer Blat f d Gym) 1864-. Munich.

Bay Geom  Zt des bayerischen Geometervereins. C as Bay Verm. Wuerzburg.

\*Bay Jb OK  Jahresbericht Oberrealschule mit Knabenmittelschule. Bayreuth.

*Bayerisch. See also Mun, Deu, Oberbay.*

Bay Verm  Zt des Ver der hoeheren bayerischen Vermessungsbeamten.
C of Bay Geom. Wuerzburg.

Beh Sc  Behavioral Sci. 1956-. Baltimore, Md; Ann Arbor, Mich.

\*Bei GTI  Bei zur Gesch der Tec & Ind, Jab der Ver deu Ingenieure.
1909-1932// C as Tec Gsch. Berlin.

\*Bei GWT  Bei zur Gesch der Wiss & der Tec. 1961-. Wiesbaden. Deu
Gesell f Gesch d Wiss & d Tec.

*Beitragen, bijdragen, etc. See Bei, Bij; also organization, city, etc.*

Bel Actu  Bull asso R des actuaires belges. 1896-1915, 1919-1935,
1937-. Brussels.

\*Bel An  Annales de l'acad R des scis, des lettres.... 1835-.
Brussels.

\*Bel Anr  Annuaire de l'acad R des scis, des lettres.... (Jaarboek van
de K belgische Akad; Ann de l'acad de Brux). 1835-. Brussels.

*Bel Biog       Biographie nationale. 1866-. Brussels. Acad R des scis, des lettres....

*Bel BL       Bulletin, acad R des scis, des lettres..., cl des lettres.... 1899-. Brussels.

*Bel BS       Bulletin, acad R de Belgique, cl des scis. (K belg Acad Mededelingen van de Kl d Wet; Bull Acad Brux; Belg Bull Sc) 1832-. Brussels.

*Bel Bul       Bull, acad R des scis, des lettres, & des b-a de Belg. 1832-1898// Brussels.

*Belgium. See also Benelx, Brug, Brx, Lieg, Louv, Rv Q Sc.*

*Belgrade. See Beo.*

*Belleten.       Belleten. 1937-. Ankara.

*Bel Let Mm       Mems, acad R de Bel, cl des lett & des scis morales & politiques, ser 2. 1906-. Brussels.

Bel Log       Bull interieure d cen belge d rech d logique. 1954-1958// C as Log An.

Bell Sy J       Bell System Tec J. 1922-1941// NY.

*Bel Mm Cou       Mems couronnes & autres mems..., acad R de Belg. 1840-1904// Brussels.

*Belorussia, Byelorussia. See BSSR.*

Bel SM       Bull de la soc math de Belg. 1947-. Brussels.

*Bel Vla Jb       Jaarboek, k vlaamse Acad v Wet...v Belgie. 1928-. Ghent.

*Bel Vla Me       K vlaamse Acad v Wet (Verslagen en) Mededeelingen. 1939-. Antwerp.

*Bel Vla Ve       Verhandlung k vlaamse Acad v Wet.... 1942-. Brussels.

*Benares. See also Banaras.*

*Benar MS       Procs of the Benares Math Soc. (Bharata ganita parisad) 1909-1934, 1939-1947// C as Ganita. Benares.

*Benelx HS       Benelux Cong of Hist of Sci. 1960-. Luxembourg, etc. Also pub in Janus.

*Beng Asi J       J (and Procs) of the Asiatic Soc of Bengal. 1832-. C of Beng Asi P. Calcutta.

Beng Asi P       Procs of the Asiatic Soc of Bengal. (Beng As S P) 1865-1904// C as Beng Asi J. Calcutta.

Beo Ak       Glas srpske kralcevske akad. (Belgrad Ak) 1886-1933?// C as Beo M Ph. Belgrade.

Beo Ak GPM       Glas srpski akad nauka, otd prir-mat. 1946-. Belgrade.

*Beo Ak N       Glasnik srpske akad nauka. 1949-. Belgrade.

*Beo Ak N Zb       Zbornik radova, srpska akad nauka. 1949-. Belgrade.

Beo Ak PM       Predavanja, srpska akad nauki umetnosti, odeljenje prir-mat nauk. 1964-. Belgrade.

*Beo Ark       Sem (inst) im N P Kondakov, sb statei po arkheologii.... (Annales; Recueil d'etudes) 1927-1940// Belgrade.

*Beo DMF       Vesnik drustva mat i fiz narodne repub Srbije. (Bull soc math phys Serbie) 1949-1963// C as MV. Belgrade.

*Beo IM       Srpske akad nauka, mat inst. (Acad serb d sci, pubs d l'inst math; Belgrade Acad Serb Sci Pub Math) 1947-. C of Beo U. See also Beo IMS, Beo MI. Belgrade.

Beo IMS       Posebna izdanja, mat inst. (Special Pubs) 1963-. Belgrade.

*Beo MF       Savez drustava mat i fiz FNRJ, nastava mat i fiz u srednjoj skoli. 1952-. Belgrade.

| | |
|---|---|
| Beo MI | Srpska akad nauka, mat inst, zb radova. 1951-. Belgrade. See also Beo IM, Beo IMS. |
| Beo MN | Glas srpska akad nauka, cl scis maths & nats, NS. (Bull acad serbe des scis arts) Scis maths 1952-. Scis nats 1950-. C of Beo M Ph. _ Belgrade. |
| *Beo M Ph | Bull d l'acad d scis maths & nats, A: scis math & phys. 1933-1940// C of Beo Ak. C as Beo MN. Belgrade. |
| Beo RNI | Zap russkago naucnago inst v Belgrade. 1930-. Belgrade. Russkii nauchnyi inst. |
| *Beo U | Pubs maths de l'univ de Belgrade. 1932-1947// C as Beo IM. Belgrade. |
| Beo U El | Publikacije electrotehnickog fak, univ u Beogradu, ser mat i fiz. 1956-. Belgrade. |
| *Ber Ab | Abh d deu (preussischen) Akad d Wiss z Berlin, Kl f Math, Phys, & Tec. 1804-1944, 1955-. Berlin. |
| *Ber Ab H | Abh d deu Akad d Wiss z Berlin, philologische & hist Abt. 1908-. Berlin. |
| Ber As J | (Berliner) astronomisches Jahrbuch. 1773-. Berlin. Ed: Encke. |
| *Ber Ber | Ber ueber d z Bekanntmachung geeigneten Verh d k preussische Akad d Wiss z Berlin. (Berl B) 1836-1855// Berlin. |
| *Ber CPA | Berliner Studien f classische Philologie u Archaeologie. 1883-1894, 1894-1898// Berlin. |
| *Ber FIM | Schriftenreihe d Forschungsinst f Math bei d deu Akad d Wiss z Berlin. 1957-. Berlin. |
| Berg Mus | Bergens Museums Aarbog, Afhandl og Aarsberetning. 1886-1947// C as Berg U Nat. Bergen. |
| Berg U | Aarbok for Univ i Bergen, Mat-naturv Ser. 1960-. C of Berg U Nat. Bergen. |
| Berg U Nat | Bergen Univ Arbok, naturv Rekke. 1948-1959// C of Berg Mus. C as Berg U. Bergen. |
| Ber HARS | Hist d l'acad R d scis & b-l d Berlin avec les mems tires d registre d cette acad. 1745-1769// C of Ber Misc. C as Ber Mm. Berlin. |
| *Ber Hel Rm | Deu Akad d Wiss z Berlin, Inst f hellenistisch-roemische Phil, Veroeffentlichung. 1953-. Berlin. |
| *Ber Hum MN | Wiss Zt d (Humboldt-) Univ Berlin, math-naturw Reihe. 1952-. Berlin. |
| *Berichte, etc.* | *See organization, city, region, or country.* |
| *Berichte und Mitteilungen.* | *See ICTM.* |
| Ber Jb | (Preussische) Akad d Wiss Jahrbuch. 1939-. C of Ber Si. See also Ber Mo. Berlin. |
| *Berks HR | Historical Rev of Berks County. 1935-. Reading, Penn. Hist Soc Berks Co. |
| *Berk SMSP | Procs of the...Berkeley Symp on Math, Stat, & Prob. 1945-. Berkeley, Cal. |
| Ber Met | Vero d meteorol Inst d Univ Berlin. 1936-. Berlin. |
| *Berg MG | Sitz d Berliner math Gesell. 1901-. Vols 1-18 = S to Arc MP. Berlin. |
| Ber MI | Schr d math Sem & d Inst f angew Math an d Univ Berlin. 1932-. Leipzig. |

| | |
|---|---|
| Ber Misc | Miscellanea berolinensia ad incrementum scientiarum, ex scriptis societatis edita. 1710-1743// C in Ber HARS. Berlin. |
| *Ber Mm | Mems d l'acad R d scis (& b-1) d Berlin. (Berl Mm; Berl Mm Ac; Nouveaux Mem) 1770-1804// C of Ber HARS. Berlin. |
| *Ber MN Mit | Math-naturw Mitteilungen. 1883-1895// Pt of Ber Si. Berlin. |
| *Ber Mo | Monb d k preussischen (deu) Akad d Wiss z Berlin. (Berl Ak Mb; Berl Mb; Berl Monatsber). 1856-1881, C as Ber Si, then 1959-. Berlin. |
| *Bern NG | Mitt der naturf Gesell Bern. 1843-. Bern. |
| *Ber Ph | Verh d phys Gesell i Berlin. 1882-1898// C as Deu Ph G. |
| Ber Sam | Samm d deu Abh welche i d k Akad d Wiss z Berlin vorgelesen worden i d Jahren 1788-1803. 1789-1803// Berlin. |
| Ber SANBN | Samm ast Abh, Beobachtungen & Nac. 1793-1808// Berlin. Akad Wiss. |
| *Ber Si | Sitz d k (deu) Akad d Wiss z Berlin, Kl f Math, Phys & Tec. 1882-. C of Ber Mo. C as Ber Jb, and as Ber Mo. Berlin. |
| Ber Versi | Vero d Berliner Hochschulinst f Versicherungswiss. 1939-1942// Berlin. |
| *Bhan ORIA | Bhandarkar Oriental Res Inst Annals. 1918-. Poona. |
| Bib Brit | Bibliotheque britannique, ou Recueil extrait d ouvrages anglais periodiques & autres, pt d scis & arts. (Bb Brit) 1796-1815// C as Bib Un. Geneva. |
| *Bibcl Arc | Transas of the Soc of Biblical Archaeol. 1872-1893// London. |
| Bib Fis E | Bibliotheca fisica d'Europea. 1788-1791// C as Brugnat. Pavia. |
| *Bib Graec | Bibliotheca graeca. 1716-1729//? Hamburg. |
| *Bib Hum | Bibt d'humanisme & renai, travaux & documents. (Bibl Renaiss) 1941-. Paris. C of Humanisme & renaissance. |
| *Bib It | Bibt italiana, ossia gior d lett, sci.... (Bb It) 1816-1856// Milan. |
| *Bibliogr | Bibliograph. 1900-. |
| *Bibliographical Society of America. See Am Bib.* | |
| *Bib M | Bibt mathematica, Zt f Gesch d Math. 1884-1914// S to Act M. Stockholm, Leipzig. |
| Bib M Ross | Bibliographia math rossica. 1896-1900// S to Kazn FMO. Kazan. |
| *Bib Un | Bibt universelle d sci, b-1 & arts.... (Bb Un) 1816-1845// C of Bib Brit. C as Arc Sc Ph. See also Schw NGCR. Geneva. |
| *Bihar ORS | J of the Bihar & Orissa Res Soc. 1915-. Bankirpore. |
| *Bijdragen tot de natuurkundige Wetenschappen. See Hall Bij.* | |
| *Bij Gesch | Bijdragen tot de Geschiedenis der Geneeskunde. 1921-. Haarlem. |
| Bij Phi | Bijdragen Tijd voor Phil en Theol. (Bijd philos theol Fac Nederl Jezuieten) 1938-1943/45// Nijmegen. |
| *Bildmess | Bildmessung & Luftbildwesen, Zt d deu Gesell f Photogrammetrie. 1926-1938// C of Allg Verm. C in Allg Verm. Berlin-Gruenewald. |
| *Biog Bl | Biographische Blaetter. 1895-1896// C as Biog J. Berlin. |
| *Biog J | Biographisches Jab & deu Nekrolog. 1896-1913// C of Biog Bl. C as Deu Biog. Berlin. |

*Biog Slov     Biograficheskii slovar deyatelei est i tec, 2 vols, Moscow, 1958.

*Biomtcs      Biometrics, (Bull/J of the Biometric Soc). 1945-. Blacksburg, Va.

*Biomtka      Biometrika. 1901-. London. Biometrika Trust.

Bioph J       Biophysical J. 1960-. Richmond, Va.

*Bio Sci      Biological Sci. (Seibutsu Kagaku) 1949-. Tokyo.

BIT          (BIT) Nordisk Tid for Info Behandling. 1961-. Copenhagen.

*Bla D Phi    Blaetter f deu Phil. 1927-1944// Berlin. Deu phil Gesell.

*Blaetter. See Bla; also organization, city, region, country, etc.*

*Blaetter f d Gymnasialschulwesen, etc. See Bay Bl.*

*Bla Versi    Blaetter f Versicherungs-math & verw Gebiete. 1928-1944// S to Z Versich. Berlin. Deu Ver f Versicherungswiss.

Bln Comm     Commentarii d bononiensi scientiarum & artium inst & acad. 1731-1791// Bologna.

*Bln Mm      Atti d accad d sci d ist d Bologna, mem d cl d sci fis. (Memorie d accad) 1850-. Bologna.

*Bln No Co    Novi commentarii acad sci inst bononiensis. (Bologna N Cm) 1834-1849// Bologna.

Bln OS       Opuscoli scientifici. (Bologna Opusc Sc) 1817-1823// Bologna.

*Bln OSN     Nuova collezione d'opuscoli scientifici. (Bologna Opusc Sc N Col) 1824-1825// Bologna.

*Bln Oss     Pubblicazioni d osservatorio ast d R univ d Bologna. 1921-. Bologna.

*Bln Riv     Rivista bolognese d sci, lett, arti & scuole. 1867-1870// Bologna.

*Bln Rn      Rend d sessioni d accad d sci d ist d Bologna. 1829-1879, 1896-. Bologna.

*Bln Rn P    Atti d accad d sci d ist d Bologna, cl d sci fis, rendiconti. 1908-1914// Bologna.

*Bln SA Ing   Atti d R scuola d applicazione per gli ingegneri in Bologna.

*Bln Sto SM   Studi & mem per la storia d univ d Bologna. 1907-1942, 1956-. Bologna.

*Bo Bolog    Il bol.di mat e di sci fis & nat. 1900-1917// Bologna.

Bode As Jb   Astronomisches Jahrbuch. 1776-1829// Berlin. Ed: Bode.

*Boehmisch... See Prague.*

*Boethius     Boethius, texte & Abh z Gesch d exakten Wiss. 1962-. Wiesbaden.

*Bo Fir      Bolletino di mat, gior sci-didattico per l'incremento d studi mat nelle scuole medie. (Bo Mat Firenze) 1902-1943, 1947-1948// C of Loria. C as Archim. Florence, Genoa, etc.

*Bo Geno. See Bo Fir.*

*Bogota. See Colom.*

*Bohemia. See Prag.*

*Boklen Mitt. See Wurt MN.*

*Bollettino, boletin, bulletino, etc. See Bo, Bu; also Boncomp, Loria; also organization, city, region, country, etc.*

*Bologna. See Bln, It.*

*Bolsh SE     Bolshoi sovetskaya entsiklopediya. Various editions.

*Bombay      J of the Univ of Bombay. 1932-. Bombay.

*Bom RASJ      J (and Procs), R Asiastic Soc of GB & Ireland, Bombay Branch
1841-1923, 1925-.  Bombay.
*Boncamp       Bull (bol) di bib e di storia delle sci mat (e fis) pubblica
da B Boncampagni.  1868-1887//  Rome
Bonn           Bonner mathematische Schriften.  1957-.  Bonn.  Univ math In
*Bononiae, -ensis.  See Bln.*
*Borchardt J.  See Crelle.*
*Bord Mm        Mems de la soc des scis phys & nats de Bordeaux.  1854-.
Bordeaux.
*Bord PV        Proces verbaux des seances de la soc des scis phys & nats de
Bordeaux.  1894-.  Bordeaux.
*Bord U         Pub de l'univ de Bordeaux.  1937-.  Bordeaux.
Bo Roma        Bollettino di mat.  (Boll Mat Roma)  1902-1943, 1947-1948//
Rome, Bologna.
*Bo SBM        Bol di storia e bib mat.  1897//  C in Loria.  S to Batt.
Naples.
*Bost MFAB     Boston Mus of Fine Arts Bull.  1903-.  Boston, Mass.
*Bras Ac        Anais (anaes) da acad (soc) brasileira de ciencias.  1929-.
C of Bras Ac Rv.  Rio de Janeiro.
Bras Ac Rv     Rev da acad (soc) brasileira de ciencias.  1917-1922, 1926-
1928//  C of Bras Rv Sc.  C as Bras Ac.  Rio de Janeiro.
Bras Est       Rev brasileira de estatistica.  1940-1942, 1946-.  C of
Rv Ec Est.  Rio de Janeiro.  Inst brasileira de geog e
estatistica.
*Bras I Bib    Inst brasileiro de bib e documentacao, bib econ-social.
1950-.  See also Bras I Bol.  Rio de Janeiro.  Conselho
nacl de pesquisas.
Bras I Bol    Inst brasileiro de bib e documentacao, bol informativo.
1955-.  See also Bras I Bib.  Rio de Janeiro.  Conselho
nacional de pesquisas.
Bras Rv Sc     Revista de sciencias.  1920-1928//  C of and C as Bras Ac Rv.
Rio de Janeiro.  Acad brasileira de sciencias.
*Brat Act M    Bratislava univ prirodovedecka fac, acta mathematica.  (Acta
facultatis rerum nat univ Comenianae)  1956-.  Bratislava.
*Bratislava.  See also MF Cas.*
*Braun Mo      Braunschweiger GNC (Grimme Natalis & Co) Monatschr.
Braunschweig.
*Braun WG      Abh der braunschweigischen wiss Gesell.  1949-.  Braunschweig.
*Brazil.  See Bras; also Notas Rio, Parana, Rv Ec Est, Sao Pau, Sum Bras.*
*Bremen NV     Abh herausgegeben vom naturw Verein zu Bremen.  (Bremen Ab)
1868-.  Bremen.
*Brescia       Commentari d ateneo d Brescia.  1812-.  Brescia.  Ateneo d
sci, lett e arti.
*Breslau.  See also Wroclaw.*
*Bresl U Ch    Chronik, Breslau univ.  1886-1916//  Breslau.
Bretagne      Bull d l soc sci d Bretagne, scis math, phys, & nat.  1921-.
Rennes.
*Brioschi.  See Annali.*
Br JM St Ps    British J of Math & Stat Psych.  1947-.  London.  Br Psych Soc
*Brno AS       Prace brnenski zakladny ceskoslovenske akad ved.  (Acta acad
sci cechoslovenicau basis brunensis; AAS naturalium moravo-
silesiacae)  1948-.  C of Brno SS.  Brno.

| | |
|---|---|
| Brno Sb U | Sb vysokeho uch tec v Brne. 1926-. Brno. |
| Brno SS | Prace moravske prirodovedecke spolecnosti Brno. (Acta soc sci natur moraviae) 1924-1947// C as Brno AS. Brno. |
| *Brno U | Spisy prirodovecke fac univ v Brne. (Trudy estestvenno-istoriceskogo fac univ Brno; Pubs d 1 fac d scis d l'univ Masaryk a Brno) 1921-. Brno. Univ J E Purkyne. |
| *Brt AAS | Rept of the (Annual Meeting of the) Br Asso for the Adva of Sci. (Brit Ass Rept; B A Rp) 1831-1916, 1919-1938// C as Adv Sc. London. |
| Brt AAS Ta | Math Tables. 1931-. London. Br Asso for the Adva of Sci. |
| *Brt Alm C | Br Almanac & Companion. 1828-. London. |
| *Brt As J | J of the Br Ast Asso. 1890-. London. |
| *Brt JHS | Br J for the Hist of Sci. 1962-. C of Brt SHS. London. |
| *Brt JPS | Br J for the Phil of Sci. 1950-. Edinburgh. |
| *Brt J Psy | Br J of Psychology. 1904-. Cambridge. |
| *Brt Mus Q | Br Museum Quarterly. 1926-. London. |
| *Brt Q Rv | Br Quarterly Review. 1845-1886// London. |
| Brt SHS | Br Soc for the Hist of Sci Bull. 1949-1961// C as Brt JHS. London. |
| *Bruenn. See Brno.* | |
| *Brug Fm An | Annales, soc d'emulation de Bruges. 1839-. Bruges. Soc... pour l'etude d l'hist & d antiquites d 1 Flandre. |
| Brugnat | Gior de fis, chim & storia nat. (G fis-med) 1792-1796, 1808-1827// C of Bib Fis E. Pavia. Ed: Brugnatelli, etc. |
| *Brussels, Bruxelles. See Brx.* | |
| Brx CERO | Cahiers du centre d'etudes de recherche operationnelle. 1958-. Brussels. |
| *Brx Con CR | Congres natl des scis, CR. 1930-. Brussels. |
| *Brx Ob Anr | Annuaires d l'Obs d Bruxelles. 1834-1900, 1914-. (1901-1913 Ann ast & Ann meteorologique) Brussels. |
| *Brx S Arc B | Bull, soc R d'archeol de Bruxelles. 1928-. Brussels. |
| Brx SBL | Bull d l'acad R d sci & b-l d Bruxelles. 1835-1879// Brussels. |
| Brx Solvy | Cahiers d centre d math & d stat appl aux scis sociales d l'inst d sociologie Solvay. 1959-. Brussels. Univ libre. |
| *Brx SS | Anns d 1 soc sci d Bruxelles, ser I, sci math (ast) & phys. 1875-1914, 1919-. Brussels, Louvain, etc. |
| *Brx U Rv | Revue d l'univ d Bruxelles. 1895-1939, 1948-. Brussels. |
| BS | Bull signaletique, I, math pures & appls. 1956-. Paris. Min d l'edu natl, cen natl d 1 rech sci. |
| *BSM | Bull des scis maths. (Darboux bull) 1870-1876, 1877-. Paris. |
| BSSR Do | Dokl akad nauk BSSR. 1957-. Minsk. |
| *BSSRFMT | Trudy inst fiz i mat. 1956-. Minsk. Akad nauk BSSR. |
| BSSRFMV | Vestsi akad navuk BSSR, ser fiz-mat navuk. (Izv....) 1965-. Minsk. |
| BSSRFT | Vestsi akad navuk BSSR, ser fiz-tek navuk. 1956-. Minsk. |
| *BSSR Ped | Uch zap Grodnenskii gos ped inst, ser mat. Minsk. |
| BSSRU | Uch zap, Belorusskii gos univ im V I Lenin. 1939-. Minsk. |
| *Bu An Beh | Bull of Animal Behaviour. 1938-. London. |
| *Bu As | Bull astronomique. 1884-. (1919-1927 pt 2: Rv gen d travaux ast) Paris. |

Buc Disq      Disquisitiones mathematicae & physicae. 1940-1948// C as
Ro IM. Bucharest. Inst R de cercetari stiintifice al
Romaniei.

*Buc EP      Bull de math et de phys pures & appls de l'ecole polyt Roi
Carol II. 1929-1956?// C as Buc I Poly. Bucharest.

*Bucharest.*    *See also Cuget Cla, Gaz M, Numerus, Ro.*

Buc I Poly      Buletinul inst polit Bucuresti (Gheorghe Ghenghin-Dij).
1956-. C of Buc EP. Bucharest.

Buc SS      Buletinul soc de sciinte fiz (chim si mineralogia) din
Bucuresci-Romania. (Bull d l soc d sci B-R; Bucarest s sc bl)
1892-1910// C as Ro SRS. Bucharest.

*Buc U Log      Analele univ (C I Parhon) Bucuresti, ser acta logica.
1958-. Bucharest.

Buc UMF      Analele univ (C I Parhon) Bucuresti, ser stiintele nat, mat-
fiz. 1953-. Pt C of Buc U Pol. Bucharest.

*Buc U Pol      Revista univ (C I Parhon) sia polit Bucuresti, ser stiintelor
nat. 1952-1955// C as Buc UMF. Bucharest.

*Buc U Rev      Revista universiara mathematica. (Rev Univ Buc) 1929-.
Bucharest.

*Budapest.*    *See Bud, Hun, Magy.*

*Bude Bu      Asso Guillaume Bude bull. 1923-. Paris.

*Bud UM      Annales univ sci Budapestinensis de Rolando Eotvos nominatae,
sect math. (Tudomany-egyetem, annales, sect math) 1958-.
Budapest.

*Buenos Aires.*    *See Arg, B Ai, Rv.*

Bu Geod      Bull geodesique. 1922-1943, 1945-. Paris. Union geodesique
& geodesique intl, sect d geodesie.

Bu Geoph      Bull (Izv) Acad of Sci USSR, Geophys Ser. 1957-. Tr of
SSSR Gf. Wash DC, NY, London.

*Bu II St      Bull mensuel d l'off permanent, inst intl d stat. 1920-1932//
C as Rv II St. The Hague.

*Bukh PIUZ      Bukharskii gos ped inst, uch zap, ser fiz-mat nauk. 1958-.
Bukhara.

Bulg Abst      Absts of Bulgarian Sci Lit, Math, Phys, Ast, Geoph, Geod.
Sofia. Bulg Acad Sci.

*Bulgaria.*    *See also Sof.*

Bulg Do      Dokl Bolgarskoi akad nauk. (CR d l'acad bulgare d sci, scis
maths & nats) 1948-. Sofia.

Bulg F      Izv na bulgarskata akad na naukite, otd fiz-mat i tekh nauki,
ser fiz. 1950-1957// C as Bulg FI. Sofia.

Bulg FI      Izv na fiz insts aneb. (Bulgarski akad na naukite tsentralna
lab po geodeziia izv) 1958-. C of Bulg F. Sofia. Bulgarska
akad na naukite, otd za mat i fiz nauki.

*Bulg FM      Fiz-mat spisanie, fiz inst, mat inst, bulg akad na naukite.
1958-. C of Bulg FMD. Sofia.

Bulg FMD      Fiz-mat spisanie, bulgarska fiz-mat druzhestvo. 1950-1958//
C of Sof Sp FM. C as Bulg FM. Sofia.

Bulg M      Izv na mat inst, bulg akad na naukite, otd za mat i fiz nauki.
1953-. Sofia.

Bulg Sb      Sb na bulgarskata akad na naukite. (Recueil d l'acad bulgare
d scis) 1911. Sofia.

Bulg Spis    Spisanie na bulgarskata akad na naukite, knig 2, klon prir-
             mat. (Revue d l'acad bulgare d scis) 1911-1945// Sofia.
*Bulletin, buletinul, etc. See Bo, Bu; also organization, city, region,*
    *country, etc.*
*Bulletin de bibliographie. See Terq Bul.*
*Bulletin des sciences mathematiques. See Ferussac.*
*Bulletin des sciences historiques. See Int Com HB.*
*Bulletin mathematique. See Mos UBM.*
*Bulletino di bibliographia. See Boncamp, Loria.*
*Bulletin scientifique roumain. See Par BS Rou.*
Bu M Bioph    Bull of Math Biophysics. 1939-. S to Psychmet. Chicago,
              Ill; Colo Springs; Univ Mich, etc.
Bu M St       Bull of Math Stat. 1947-. Fukuoka.
*Bund (Universitaet). See Gott UBM.*
*Bureau of American Ethnology. See Smi.*
Burma RS      J of the Burma Res Soc. 1911-. Rangoon.
*Bu Sc M Ast  Bull des scis math (& ast). 1870-. Paris.
*Bu SMP El    Bull des scis math & phys elementaires. 1895-1906// Paris.
*Bu Thom      Bull thomiste. 1924-. Paris.
Byz Bost      Byzantion, Intl J of Byzantine Studs. 1924-. (1924-1939
              = Byz Par) Paris; Boston, Mass.
*Byz Par      Byzantion, rev intl des etudes byzantines. 1924-1939// Paris.
              See also Byz Bost.
*Byz Z        Byzantinische Zt. 1892-1913, 1919-1920, 1923-. Leipzig.
              Akad der Wiss, Muenchen.


Cadiz         Periodico mensual de ciens mat y fis. (Cadiz period m ci)
              1848-. Cadiz.
*Cagliari     Rend del sem della fac de sci della univ di Cagliari.
              1931-. Cagliari.
*Cah Civ Me   Cahiers d civ medievale X-XII siecles. 1958-. Univ d Poitiers.
*Cah Dr Hom   Cahiers des droits de l'homme. (Bull officiel, Ligue d d d
              l'h & d citoyen) 1901-. Paris.
*Cahiers. See also organization, city, region, or country.*
*Cah Mond     Cahiers d'hist mondiale. 1953-. Paris. Commis intl pour
              une hist d devel sci & culturel d l'humanite.
*Cah Phy      Cahiers d physique. 1941-. Paris.
*Cah Quin     Cahiers de la quinzaine. 1900-1914, 1923-1936// Paris.
*Cah Ratnl    Cahiers rationalistes. Paris.
*Cah Rhod     Cahiers rhodaniens. 1949-. Lyon. Univ, inst d math.
*Calc Au Ci   Calculo automatica y cibernetica. 1952-. Madrid.
Calcolo       Calcolo. 1964-. Rome. Asso ital per il calcolo automatico.
*Calcutta. See Clct.*
*Calif Ph     Univ of California Pubs in Phil. 1904-. Berkeley, Cal.
*Calif Pu     Univ of California Pubs in Math. 1912-. Berkeley, Los Angeles.
*Calif Sta    Univ of California Pubs in Stat. 1949-. Berkeley, Los Angeles.
Cam Analy     Mems of the Cambridge Analytical Soc. (Camb Mm Anal S)
              1813// Cambridge.

*Cam Dub          Cambridge & Dublin Math J.  (CD; Cam & Dub MJ)  1846-1854//
                  C of Cam MJ.  C as QJPAM.  Cambridge.
*Cam HJ           Cambridge Historical J.  1923-1957//  London.
 Cam Misc         Cambridge Miscellany of Math, Phys, & Ast.  1842//
                  Cambridge, Mass.
 Cam MJ           Cambridge Math J.  (CM)  1837-1845//  C as Cam Dub.  Cambridge.
*Cam PSP          Procs of the Cambridge Phil Soc.  (Camb Proc)  1843-.
                  Cambridge.
*Cam PST          Transas of the Cambridge Phil Soc.  1822-.  Cambridge.
*Cam Tract        Cambridge Tracts in Math & Math Phys.  1905-.  Cambridge.
 Canada.  See also Nov Sco, Ott, Queb, Queen, Toro.
*Can As           J of the R Ast Soc of Canada.  1906-.  Toronto.
*Can Cong         Can Math Congress Procs & Meetings.  1948-.  Toronto.
*Can JM           Can J of Math.  (CJM; J canadien de math)  1949-.  Toronto, etc.
*Can MB           Can Math Bull.  (Bull can de math)  1958-.  Toronto, etc.
                  Can Math Cong.
*Can Mus J        Can Music J.  1956-.  Toronto.  Can Music Council.
 Can Notes        Notes, News & Comments, Can Math Congress.  (Nouvelles &
                  Commentaires, Soc math du Can)  1969-.  Montreal.
*Can PT           Procs & Transas of the R Soc of Can.  1883-.  Ottawa.
*Can Rv Tri       Revue trimestrielle canadienne.  1915-1954//  Montreal.
*Can T            (Procs &) Transas of the R Soc of Can, sect III, chem, math,
                  & phys sci.  1882-.  Ottawa, Montreal.
 Canton          Pub of the Dept of Math, Sun Yat Sen Univ.  1937?//  Canton.
 Canton U        Acta scientiarum nat univ Sunyatseni.  1966-.  Canton.
 Caracas.  See Venzl.
 Carl PTMA        Repertorium f physikalische Tec, f math & ast Instrumenten-
                  kunde.  (Carl Rpm; Exner Rpm)  1865-1891//  Munich.
*Carn CAAH        Carnegie Inst of Wash, Contrs to Amer Anthro & Hist (Archaeol).
                  1931-.  Wash DC.
*Carn Pub         Carnegie Inst of Wash Pubs.  1902-.  Wash DC.
 Carus            Carus Math Monographs.  1925-1934, 1941-.  Chicago, etc.
                  Math Asso of Amer.
 Cas F            Casopis pro pestovani fysiky.  1951-.  Pt C of Cas MF.
                  Intl ed in Cz J Ph.  Prague.
*Cas M            Casopis pro pestovani matematiky.  1951-.  Pt C of Cas MF.
                  Contents from Cz MJ.  Prague.  Cesk akad ved.
*Cas MF           Casopis pro pestovani mat a fis.  1872-1950//  C as Cas F,
                  & as Cas M, & as Cz MJ, & as Cz J Ph.  Incl Vestnik J C M F,
                  Cas, Zt zur Pflege d Math & Phys...auf Studirende d Mittel-
                  & Hochschulen.  Prague.  Jednota Cesk math a fis.
 Casualty Actuarial Soc of Amer.  See Am CAS.
 Catania.  See also Matiche, Esercit.
*Cat Atti         Atti d accad gioenia d sci nat in Catania.  1825-.  Catania.
*Cat Bol          Bol d sedute d accad gioenia d sci nat in Catania.  (Boll
                  Acc Gioenia)  1907-.  Catania.
*Cat ISECA        Annuario d R Ist sup d sci econ e commerciali d Catania.
                  1932/1933-.  Catania.
*Centau           Centaurus, intl mag of hist of sci & med.  1950-.  Copenhagen.
 Center for the Hist and Phil of Physics.  See AIPCHIP.
 Centre d'etudes de recherche operationnelle.  See Brx CERO.

Cernau       Buletinul fac d stiinte din Cernauti. 1927-1937// Cernauti.

*Cesk Mysl      Cesca mysl. 1900-. Prague. Filosoficks jednota.

*Cescoslovenska academie ved. See Aplik M, Brno, Cas, Cz, Kyb Prag, Prag.*

*Ceylon. See Math Pera.*

*Charkiv, Charkow. See Khar.*

*Chelini       In memoriam Dominici Chelini, collectanea math nunc primum edita cura § studio di Cremona e E Beltrami. 1881// Mediolani.

*Chernovtsy. See Cernau.*

Chiba        J of the Coll of Arts § Sci, Chiba Univ. (Chiba daigaku, bunrigakubu) 1952-. Chiba.

*Chic Cong     Math Papers Read at the Intl Math Congr Held in Connection with the Worlds Columbian Exposition in Chicago, 1893, 2 vols, AMS, 1896 (repr 1904).

Chiffres      Rev fr d traitement d l'info, chiffres. 1958-. Paris. Asso fr d calcul.

Chile SS      Actes de la soc sci du Chile. (Chile S Sc Act) 1891-. Santiago. Soc cien d Chile.

Chil U       Anales de la univ de Chile. (Chile A Un; Santiago d Chile Un A) 1843-. Santiago.

Chi M        Chinese Mathematics. 1962-. Tr of Act M Si. Providence, RI. AMS.

Chi MS       J of the Chinese Math Soc. 1936-. Shanghai.

*China. See also Act As Si, Act Br Si, Act Ge Ca, Act Me Si, Act M Si, Act Ph Si, Amoy, Canton, Chun Ch HP, Ko Hs Chi, Ko Hs TP, Ling, Mon Ser, Nort Chi, Pekin, Ryojun, Sci Sin, Sci Tec Ch, Shang, Shu Hs, Shux, Tsing H.*

*Chi Rev      China Review. 1872-1901// Hongkong.

Chi SA       Sci Absts of China, Math § Phys Sci. 1958-. Peking.

*Chi WBRS     J of the West China Border Res Soc. 1922-1945// Chengtu.

*Chr Avh      Forhandlingar i videnskabs selskabet i Christiania. (F Kristiana) 1858-1924// C as Oslo Avh. Christiania.

*Christiania. See also Nor, Oslo.*

Chr Obs      Pub des Universitaetsobservatoriums in Christiania. 1931-1934// C as Astph Nor. Christiania.

*Chro Eg      Chronique d'Egypte, bull periodique d 1 fondation egyptologique reine Elisabeth. 1925-. Brussels. Mus R d'art § d'hist.

*Chronicle, chronik, etc. See also organization, city, region, country, etc.*

Chr Sk       Skrifter udgivne af Videnskabsselskabet i Christiania, math-naturv Kl. 1894-1924// C as Oslo Sk. Christiania.

CHUM         Computers § the Humanities. 1966-. NY.

*Chun Ch HP   Chung chi hsueh pao. (Chung Chi J) 1961-. Hongkong.

*Chuo Kor     Chuo koron (Central Review). 1887-. Tokyo.

*C Huyg       Christiaan Huygens, intl math Tijdschrift. 1921-1940// C from 1947 in Sim Stev. Groningen, Amsterdam.

Chymia       Chymia, Annual Studs in the Hist of Chem. 1948-1967// C as Hi Stu Ps. Philadelphia, Penn.

*Ciel Ter     Ciel § terre, bull mensuel d 1 soc belge d'ast, d meteorol § d phys d globe. 1881-1914, 1919-. Brussels.

Cien Lis      Ciencia, rev d cultura cientifica. Lisbon.

Cien Mad      Las ciencias. 1934-. Madrid.

Cimento        Il cimento, riv di sci, lett, & arti.  1844-1855//  C as
               Nuo Cim.  Turin.
*Circulaire*.   *See ICTMC*.
*Ci Tecnol     Ciencia y tecnologia.  1951-.  Wash DC.
*Civiling      Der Civilingenieur, Organ d saechsischen Ingenieur- &
               Architekten-Vereins.  1854-.  Leipzig.
*Civ Mac       Civilta dalle macchine.  1953-.  Rome.  Soc finanziaria
               meccanica.
*Clas J        The Classical J.  1905-.  Chicago.  Classical Asso of the
               Middle West & South, Classical Asso of New England.
*Clas Med      Classica & mediaevalia.  1938-.  Copenhagen.
*Clas Phil     Classical Philology.  1906-.  Univ Chicago.
*Clas Q        Classical Quarterly.  1907-.  London.
*Clas Week     The Classical Weekly.  1907-.  Brooklyn, NY.  Classical
               Asso of the Middle States & Maryland.
*Clct Let      J of the Dept of Letters, Univ of Calcutta.  1920-.  Calcutta.
*Clct MS       Bull of the Calcutta Math Soc.  1909-1960//  Calcutta.
Clct Sc        J of the Dept of Sci, Univ of Calcutta.  1919-1933; NS 1937-.
               Calcutta.
*Clct St       Bull, Calcutta Stat Asso.  1947-.  Calcutta.
*Clebsch An*.   *See M Ann*.
Cluj IP        Buletinul stiintific al inst polit din Cluj.  (Min invatamin-
               tului si culturii inst polit Cluj, lucrar stiintifice)
               1958-.  Cluj.
*Cluj MF       Studii si cercetari (stiintifice/de matematica)(si fiz),
               filiala Cluj, acad repub pop Romine.  1950-.  Cluj.
*Cluj Sem      Seminarul de mat, monograf mat, sem mat univ Cluj.  1929-.
               Cluj.
Cluj SS        Buletinul soc de stiinte din Cluj.  (Bull d l soc d sci d
               Cluj, Roumanie) 1921-.  Cluj.  Inst d arte grafice
               "Ardealul."
*Clum UBBM     Studia universitatis Babes-Bolyai, ser math-phys.  1962-.
               Cluj.
*Cluj UBBN     Buletinul universitatilor V Babes si Bolyai, ser stiintele
               nat.  1957-.  Cluj.
CM Ph          Communications in Math Physics.  1965-.  NY.
*CMUSP.  Commentarii mathematici univ Sancti Pauli.  See Paul UCM*.
Coim I         Instituto, j scientifico e litterario.  1852-.  C as
               Porto Ac.  Coimbra.
Coim U         Rev d fac d cien d univ d Coimbra.  1931-.  Coimbra.
*Collection, collezione, etc.  See also organization, city, region, or*
               *country*.
*Collect M     Collectanea mathematica.  1948-.  Barcelona.  Consejo sup
               d investigaciones cientificas, univ d Barcelona.
*Colliers      Colliers Encyclopedia.  Various editions.
*Colloq M      Colloquium mathematicum.  1948-.  Warsaw.
*Colo CPG      Colorado College Pubs, Gen Ser.  1904-.  Colorado Springs, Colo
*Colo CPS      Colorado Coll Pubs, Sci Ser.  1904-1926//  Colorado Springs.
*Cologne.  See Nor Wes*.
*Colombia.  See also Rv M El, Stvdia*.
Colom Ci       Rev d l acad colombiana d ciens exactas, fis, quims, y nats.
               1950-.  Bogota.

*Colom Cul    Rev de cultura moderna, univ nacl de Colombia. 1944-.
        Bogota.
Colom M    Rev colombiana d mat. 1968-. C of Rv M El. Bogota.
*Colo Stu    Colorado Coll Studs, Papers Read before the Colo Coll Sci
        Soc. 1890-1903// Colorado Springs, Colo.
*Comenianae. See Brat.*
*Commentarii, etc. See organization, city, region, or country.*
*Communications in Math Phys. See CM Ph.*
*Communications on Pure & Appl Math. See CPAM.*
Comp    Computing, Arch for Elec Comp. (Arch fuer Elek Rechnung)
        1966-. C of MTW. Vienna.
Comp Ab    Computer Abstracts. (Computers; Computer Bib) 1957-.
        London. Bureau of Tec Info.
Comp Appl    Computer Applications. 1969-. C of Cum Comp. Wash DC.
        Cambridge Communications Corp.
Comp Aut    Computers & Automation, Cybernetics, Robots, Automatic
        Control. 1953-. NY.
Comp B    The Computer Bull. (Lon Comp Grp Bull) 1957-. London.
        Br Comp Soc.
Comp BR    Computers & Biomedical Research. 1967-. Academic P.
Comp El    Computer Electronics (Hardware). 1959-. C of Cum Comp.
        Wash DC. Cambridge Communications Corp.
*Comp J    The Computer J. 1958-. London. Br Comp Soc.
Comp M    Computer Mathematics. 1969-. C of Cum Comp. Wash DC.
        Cambridge Communications Corp.
*Composit    Compositio mathematica. 1934-. Groningen.
Comp R    Computing Reviews. 1960-. NY.
Comp SHVB    Computer Studies in the Humanities & Verbal Behavior.
        1968-. The Hague.
Comp Soft    Computer Software. 1969-. C of Cum Comp. Wash DC.
        Cambridge Communications Corp.
*Comptes rendues. See organization, city, region, or country.*
*Computing Machinery. See ACMC, ACMJ.*
Con Ac Sci    Conferences d'actualites scis & inds. 1929-1931// C as
        Act Sci Ind. Paris.
*Conference, etc. See also organization, city, region, or country.*
*Conferenze e prolusione. See Parola.*
*Conflu    Confluence, An Intl Forum. 1952-. Cambridge, Mass. Harvard
        Univ.
*Congregational Education Soc. See Am Q Reg.*
*Congres, congress, etc. See also Int Con; also city, region, country, etc.*
Connaiss    Connaissance d temps ou d mouvements celestes. 1679-. Paris.
Conn Mm    Mems of the Conn Acad of Arts & Scis. 1810-1816//
        New Haven, Conn.
*Conn T    Transas of the Conn Acad of Arts & Scis. 1866-. New Haven.
Cont CV    Contrs to the Calculus of Variations. 1930-. Chicago, Ill.
        Univ of Chicago P.
Cont DE    Contrs to Differential Equations. 1963-. NY.
*Contemp R    Contemporary Review. 1866-. London.
Contents    Contents of Contemporary Math Journals. 1969-. AMS.
*Contribuciones cientificas. See B Ai UF, B Ai UM; also organization, city,*
        *region, or country.*

*Contributions to Education.  See T Col Cont.*

*Copenhagen.  See Dan;  also Act HSNM.*

*Copp Mit      Mitt Coppernicus-verein f Wiss & Kunst.  1878-.  Thorn.

*Cor M Ph      Corresp math & physique.  (Quetelet cor mth)  1825-1839//
              Ghent, Brussels.

*Correspo      Correspondant.  1843-1869, 1872-1933, 1935-1937//  Paris.

 CORS         CORS J, The Can Opers Res Soc J.  1963-.  Toronto.

*Cosmos.  See Kosmos.*

*Courtauld Institute.  See Warb Cour.*

*CPAM         Communications on Pure & Appl Math.  1948-.  NY, London.

*CR.  For comptes rendues, see Par CR, city, region, country, etc.*

*Cracow.  See Krak.*

*Crelle       J f d reine & angewandte Mathematik.  (J Math Berlin; J f M;
              Borchardt J; J f d r u a M)  1825-.  Berlin.

 Crim Ped     Izv Krymskogo ped inst im M V Frunze.  1927-.  Simferopol.

*Crim UMLP    Procs of the Math Lab of the Crimean Univ.  (Procs Math Lab
              Simf)  Simferopol.

*Croatia.  See Zagr.*

*Cron Cien    Cronica cientifica, rev intl d ciencias.  1878-1892//
              Barcelona.

*Cuadernos de estudio.  See Lim U.*

*Cuba.  See also Hab.*

*Cub SCPMR    Rev de la soc cubana de ciens fis y mats.  1942-.  Havana.

*Cuget Cla    Cuget clar.  Valerii-de-munte, Romania.  (A series of popular
              books by univ pop Nicolai Iorga, affiliated with univ
              Bucharest.)

 Cum Comp    Cumulative Computer Absts.  C in Comp Appl, Comp El, Comp M,
              & Comp Soft.  Wash DC.  Cambridge Communications Corp.

*Curr Biog    Current Biography.  1940-.  NY.

 Curr Con    Current Contents.  Philadelphia, Penn.  Inst for Sci Info.

 Cuyo U      Univ nacl d Cuyo, fac d cien, rev mat cuyana.  1955-.
              San Luis, Cuyo, Philippine Isls.

 Cybtica     Cybernetica.  1958-.  Namur.

 Cybtics     Cybernetics.  1965-.  Tr of Kibernet.  NY.

*Cz Am Con    Czechoslovak Soc of Arts & Sci of Amer, Absts of Papers.
              1962-.  Wash DC.

*Czechoslovakia.  See also Cesk;  also Act HRNT, Aplik M, Brat, Brno, Cas,*
              *Olom, Prag, Slov, Stroj ZI.*

 Cz Ec Pap    Czechoslovak Econ Papers.  1961-.  Prague.  Czech Acad Sci.

*Czernowitz.  See Cernau.*

 Cz J Ph     Czechoslovak J of Phys.  1952-.  Pt C of Cas MF.  Prague.
              Czech Acad Sci.

*Cz MJ        Czechoslovak Math J.  (Casopis pro pestovani mat)  1951-.
              Pt C of Cas MF.  ≠ Cas M.  Prague.  Czech Acad Sci.

DAB      Dictionary of Amer Biography, 21 vols, 1928-1937. 10-vol ed 1946.

*Daedalus. See Am Ac Pr.*

*Dan H      Historisk-filologiske meddelelser. (Medd Kob) 1917-. Copenhagen. K danske videnskabernes selskab.

*Dan H Sk      Skrifter, dansk videnskabernes selskab, historisk og filosofisk afdeling. (Mem d l'acad d scis & d letts d Danemark, sect letts) 1823-1936// Copenhagen.

*Dan M Med      Matematiskfysiske meddelelser udgivet af det K danske vid selskab. (Danske vid selsk mat-fys medd) 1917-. Copenhagen.

Dan M Sk      Matematiskfysiske skrifter udgivet af det K danske vid selskab. 1956-. Copenhagen.

*Dan      Oversigt over det K danske vid selskab forhandlingar. (Bull d l'acad R Danemark; Kjob ov) 1814-. Copenhagen.

*Dan      Skrifter det K danske vid selskab. (Kiob Dn vd selsk skr) 1745-1938// C as Biologiske skr. Copenhagen.

*Darboux bull. See BSM.*

*Datamat      Datamation. 1960-. NY; Chicago, Ill.

*Deb M Szem      Kozlemenyek, matematikau szeminariumabol. (Mitt math sem univ Debrecen) 1927-. Debrecen.

*Debrecen. See also Pub M.*

*Deb Szeml      Debreceni szemle. Budapest.

*Deb      Acta univ Debreceniensis de Ludovico Kossuth nominatae. 1954-. Budapest.

     Anns de l'ec polyt de Delft. 1885-1897// Leiden.

     Delta. (Math Delta) 1968-. Waukesha, Wis. Math Soc.

Delta      Delta-epsilon. 1960-1967// Northfield, Minn. Carleton Coll.

     See Gree SM.

*Denmark. See Dan, Nor.*

     Deu biographisches Jab. 1914-1929// C of Biog J. Berlin. Verbande der deu Akademien.

     Deu geodaetische Kommis bei d bayerischen Akad d Wiss, Reihe B, angew Geodaesie. 1952-. Munich.

*Deu      Zt d Vereins deu Ingenieure. (Zeit V D I) 1857-. Berlin. Deu Literaturzeitung f Kritik d intl Wiss. (Deu Lit Z V) 1880-. Berlin, etc.

     Deu Mathematik. 1936-1944// Berlin. Ed: Th Vahlen.

*Deu      Zt d deu morgenlandischen Gesell. 1847-. Leipzig. See also Z Assyr.

     Abh & Ber d deu Museums. 1929-. Berlin.

*Deu      Ber ueber d Versammlung d deu Naturf & Aerzte. (D Nf B; D Nf Vsm B) 1822-1883// Jena.

*Deu      Mitt d Gesell deu Naturf & Aerzte. 1924-. Duesseldorf.

*Deu      Tageblatt d...Versammlung, Gesell deu Naturf & Aerzte. (D Nf Tbl) 1836-. Leipzig.

*Deu      Verh d Gesell deu Naturf & Aerzte. 1890-. Leipzig. Neue deu Forschung, Abt mat. 1934-. Berlin.

*Deu      Deu optische Wochenschrift. 1914-1940// Berlin.

*Deu      Verh d deu physikalischen Gesell. 1899-. Braunschweig. Deu Forsgemein Kommis f Rechenanalogen Literaturteile, Titel v Vero & Analog- & Ziffernrechner & ihre Anw. 1954-. Wiesbaden.

Deu St Ab     Deu statisches Zentralblatt. 1909-1943// Leipzig.
         Deu Stat Gesell.
*Deutsche Akad der Wissenschaft zu Berlin. See Ber; also M Op Stat.*
Deu Versi     Vero d deu Ver f Versicherungswissenschaft. 1903-. Berlin.
*Deu Vers M     Blaetter d deu Gesell f Versicherungsmathematik. 1950-.
         Wuerzburg. Deu Aktuarverein.
*Deu Wort Z     Zt f deu Wortforschung. 1900-1914, 1959-. Strasbourg.
Devons     Rept & Transas of the Devonshire Asso for the Adva of Sci,
         Lit, & Art. (Devon As T) 1862-. Plymouth, London.
*Dialect     Dialectica, rev intl d phil d l connaissance. 1947-.
         Lausanne, Neuchatel.
Dickson     Dickson, L E, A Hist of the Thy of Numbers, 3 vols, 1919,
         Wash DC, Carnegie Inst Pub #256.
Dif Eq     Differential Equations. 1965-. Tr of Dif Ur. NY.
Dif Ur     Differentsial'nye uravneniya. 1965-. Tr as Dif Eq. Minsk.
         Akad navuk BSSR.
*Dillner     Tidskrift for mat och fys. 1868-1871, 1874// Uppsala.
         Ed: G Dillner.
Discr M     Discrete Math. 1971-. Amsterdam. North-Holland Pub Co.
Diskret     Diskretnyi analiz, sb trudov. 1963-. Novosibirsk. Akad
         nauk SSSR, Sibirskoe otd, inst mat.
*Disquisitiones mathematicae. See Buc Disq.*
Diss M     Dissertationes mathematicae. 1966-. C of Roz M. Warsaw.
         Polska acad nauk, inst mat.
*DMV     Jasb d deu Mathematiker-Vereinigung. 1890-1944, 1950-.
         Leipzig.
DMV Erg     Deu Math-Vereinigung, Ergaenzungsbaende. 1906-1914,
         1929-1930// Leipzig.
*DNB     Dictionary of Natl Biography, 63 vols, 1885-1900 & later eds.
*Doklady. See also organization, city, region, or country.*
*Doklady     Dokl akad nauk SSSR. (CR d l'acad d sci d l'URSS; Dokl
         Rossiiskoi akad nauk) 1922-. Incl Eur ed. See also Dokl F
         & Dokl M. Moscow, Leningrad.
Dokl F     Dokl akad nauk SSSR, otd fiz. 1937?-. Tr as Sov Ph Do.
         Moscow.
Dokl M     Dokl akad nauk SSSR, otd mat. 1937?-. Tr as Sov M.
         Moscow, Leningrad.
*Dokumentationszentrums. See Ab Dok.*
*Dominica     Dominicana, A Quar Treating Theol, Phil, & the Arts.
         1916-. Wash DC.
*Dorpat. See also Tartu.*
*Dorpat U     Eesti vabariigi Tartu ulikooli trimetused, A, math, phys, med
         (Acta & commentationes univ Dorpatiensis) 1921-. Tartu.
*Dres Isis     Sitz & Abh d naturw Gesell Isis in Dresden. 1861-. Dresden.
*Dres THWZ     Wissenschaftliche Zt d tec (Hochschule/Univ) Dresden.
         1952-. Dresden.
*Dres Verk     Wiss Zt d Hochschule f Verkehrswesen Friedrich List in
         Dresden, d anw math Methoden im Transport- & Nachrichtenwesen
         1953-. Dresden.
Dub Ac Mi     Minutes of Procs, R Irish Acad. (Absts of Min, R I A)
         1869-. Dublin.

| | |
|---|---|
| *Dub Ac P | Procs of the R Irish Acad, Sect A. 1841-. Dublin. |
| *Dub Ac T | Transas of the R Irish Acad. (Trans Dublin) 1781-1906// Dublin. |
| Dub IAS | Communications of the Dublin Inst for Adva Studies, Ser A. 1943-. Dublin. |
| *Dub Rv | Dublin Review. 1836-. London, etc. |
| *Dub S(A) | The Sci Procs of the R Dublin Soc, Ser A. 1877-1957, 1959-. Dublin. |
| *Dub S(B) | The Sci Procs of the R Dublin Soc, Ser B. 1877-. Dublin. |
| Duke MJ | Duke Math J. 1935-. Durham, NC. |
| *Dunkerq | Mems d l soc dunkerquoise pour l'encouragement d scis, d letts, & d arts. (Dunkerque Mm S Encour) 1853-. Dunkirk. |

| | |
|---|---|
| Earthqke | Bull of the Earthquake Res Inst. 1926-. Tokyo. Univ. |
| *East Indies.* | *See Sing.* |
| Ec Com Cyb | Economic Computation & Economic Cybernetics Studs & Res. 1966-. Bucharest. Cen of Econ Comp & Econ Cybernetics. |
| *Ec Hi | Economic History. 1926-1940// London. R Economic Soc. |
| *Ec J | The Economic J. 1891-. London. R Economic Soc. |
| *Ecmet | Econometrica. 1933-. Chicago, Ill; Colo Springs, Colo; etc. Econometric Soc. |
| Ecmica | Economica. 1921-. London Sch of Economics. |
| *Ecole polytechnique, Paris. See Par EP.* | |
| *Ecole polytechnique Roi Carol II. See Buc.* | |
| *Edi Actu | Transas of the Fac of Actuaries. 1901-. Edinburgh. |
| *Edi MSN | Edinburgh Math Notes. (Procs of the Edinburgh Math Soc) 1903-1917, 1923-1926, 1928-. Edinburgh. |
| *Edi MSP | Procs of the Edinburgh Math Soc. 1883-. Edinburgh. |
| *Edinburgh. See also Acc Mag.* | |
| *Edi PJ | The Edinburgh Phil J. 1819-1826// Edinburgh. |
| *Edi SPA | Procs of the R Soc of Edinburgh, Sect A, Math & Phys Sci. 1888-. Edinburgh. |
| *Edi ST | Transas of the R Soc of Edinburgh. (TE) 1783-. Edinburgh. |
| *Ed M | (L')Education mathematique. 1897-. Paris. Ed: H Vuibert. |
| *Ed Rv | Educational Rev. 1891-1928// C in Sch Soc. NY, etc. |
| *Ed Screen | Educational Screen (& Audio-visual Guide). 1922-. Chicago. |
| *Ed St | Educational Studies in Math. 1968-. Dordrecht. Ed: H Freudenthal. |
| *Edu | Education. 1880-. Boston, Mass. |
| Edu T | Educational Times. 1847-1923// See also MQSET. London. |
| Eg Ac | Procs of the Egyptian Acad of Sci. 1945-. Cairo. |
| *Eg IB | Bull d l'inst egyptien. (I Egypt Bll) 1859-. Cairo. |
| *Egypt. See also Alex, Anc Eg, UAR.* | |
| *Eidgenossischen tech hochsch. See Zur ETH.* | |
| Ek Mat | Ekonomika i mat metody. 1965-. Moscow. |
| Ek M Ob | Ekonomicko-mat obzor. (Economic-math rev) 1965-. Prague. Czech Acad Sci. |

*Elect Eng      Electronic Engineering.  (Television; Television & Short-wave;
                Electronics & Television & Short-wave World)  1928-.  London.
 Elek Dat       Elektronische Datenverarbeitung, Fachberichte ueber programmge
                steuerte Maschinen & ihre Anw.  1959-.  Braunschweig.
*Elek Nach      Elektrische Nachrichtentechnik.  1924-1944//  Berlin.
*Elek Rech      Elektronische Rechenanlagen.  1959-.  Munich.
 Elementa       Elementa, tids for elementar mat, fys och kemi.  1917-1937//
                Stockholm.
*Elem M         Elemente der Math.  (Rev d math elementaires;  riv d mat
                elementare)  1946-.  Basel, Zurich.  Short biogs in Beihefte,
                1947-.
*El Engin       Electrical Engineering.  1931-.  C of AIEEJ.  NY.  Amer Inst
                of Elec Eng.
*Elish Mit      J of the Elisha Mitchell Sci Soc.  1883-.  Chapel Hill, NC.
                Univ of NC P.
 *Emperor Franz Joseph Academy.  See Prag BI.*
 *Emperor Nicolas II Inst.  See Wars PI.*
*Enc Br         Encyclopedia Britannica.  Various editions.
*Enc M El       Enciclopedia d matematiche elementari.  1930-.  Milan.
                Eds: L Berzolari, G Vivanti, D Gigli.
*Enc Met        Encyclopaedia metropolitana.  1817-1845.  2nd ed 1848-1858.
                London.
*Enc MW         Enzyklopaedie der math Wiss....  (Teubner)  1898-.  Leipzig.
*Enc Phi        The Encyclopedia of Phil.  1967.  NY.  Macmillan Pub Co.
*Enc SM         Encyclopedie d scis mathematiques.  1904-1915.  Tr & revision
                of Enc MW, never completed.  Paris.
*Endeav         Endeavour.  1942-.  London.
 Engl Abs       English Absts of Selected Articles from Sov Bloc & Mainland
                China Tec Js, Ser 1, Phys & Math.  1961-.  Wash DC.
 *England.  See also Bath, Br, Brt, Cam, L, Lancas, Leeds, Leic, Lincolns,*
                *Lon, Manc, Newcomen, Nort Br, Not Qu, Ox, Peterbor, Warb Cour.*
*Engl Cyc       English Cyclopaedia.  1854-1870.  London.  Ed: Charles Knight.
*Engl HR        The English Historical Rev.  1886-.  London.
*Engl Stu       Englische Studien.  1877-1944//  Leipzig.
*Engr B         Engineers' Bull.  1918-.  Denver, Colo.  Soc of Engineers.
 Engr Cyb       Engineering Cybernetics.  1963-.  Tr from Tek Kib.  NY.
                Inst of Elec & Electronics Engineers.
 Enig           Enigmatical Entertainer & Math Repository (Associate).
                1827-1830//  London.
 Enqu Bos       The Enquirer, or Math & Phil Repository.  1811-1814//
                Boston, Mass.
 Enqu Lon       Enquirer, or Lit, Math, & Phil Repository.  1811-1813//
                London.
*Ens M          L'enseignement mathematique.  1899-.  Geneva, Paris.
*Ens M Mon      Monographie de l'enseignement mathematique.
 *Ens MP.  See Beo MF.*
*Ens Pub        Enseignement public, rev pedagogique.  1878-1942//  Paris.
*Ens Sc         Enseignement scientifique.  1927-.  Paris.
 *Enzyklopaedie.  See Enc MW.*
 Epinal         Anns d l soc d'emulation d dept d Vosges.  (Epinal Vosg A)
                1831-.  Epinal.

| | |
|---|---|
| *Epist Tec | Episteme kai techne. Athens. |
| Erg AM | Ergebnisse d angew Math. 1952-. Berlin. |
| *Ergebnisse.* | *See also organization, city, region, country, etc.* |
| Erg Ex N | Ergebnisse d exakten Naturw. 1922-. Berlin. |
| Erg MG | Ergebnisse d Math & ihrer Grenzgebiete. 1932-. Berlin, Gottingen, Heidelberg. |
| *Erg MK | Ergebnisse eines math Kolloquiums. 1928-1936// Leipzig. See also Not DMC. |
| Ericsson | Ericsson Technics. 1933-1940, 1942-. Stockholm. |
| *Erkennt | Erkenntnis. 1930-1939// C of An Phi. C as J Un S. Leipzig. Gesell f wiss Phil Berlin & d Ver Ernst Mach in Wien. |
| *Erlang Si | Sitz d phys-med Sozietat zu Erlangen. (S B phys med Soz; Erl Ber) 1864-. Erlangen. |
| Erman Arc | Archiv f wiss Kunde von Russland. (Erman Arch Rs) 1841-1867// Berlin. Ed: Erman. |
| *Ernst-Moritz-Arndt-Univ.* | *See Greifsw.* |
| Ertekez | Ertekezesek a math osztaly korebol. (Mem on Math Subjects, Hung Acad Sci; Mag tud ak etk, math) 1867-1894// Budapest. Kiadja a Magyar tudomanyos akad. |
| *Erzurum.* | *See Ataturk.* |
| *Esercit | Esercitazioni matematiche. (Note ed esercitazioni) 1921-. Catania. Circolo mat d Catania. |
| *Esp Prog | Anales d 1 asoc espanola para el progreso d 1 ciencias. Madrid. |
| *Esp SM | Rev soc mat espanola. 1911-1917// Madrid. |
| *Essling J | Jab f Gesch d oberdau Reichstaedte. 1955?-. Esslingen. |
| Estad | Estadistica. 1943-. Wash DC. |
| Estad Esp | Estadistica espanola. 1958-. Madrid. |
| Est FMT | Eesti NSV teaduste akad, toimetised, fuusika-mat tehnikateaduste seeria. (Izv akad nauk Est SSR, ser fiz-mat i tekh nauk) 1956-1966// Tallinn. |
| *Est Marks | Estestvoznanie i marksizm. 1929-. Moscow. Kommunisticheskaya akad, sekt est i tochnykh nauk. |
| *Estonia.* | *See also Dorpat, Tartu, SSSR.* |
| *Et Class | Les etudes classiques. 1932-. Namur. |
| *Euc Gron | Euclides, tijd voor d didaktiek d exact vakken. (Nieuw tijd voor wiskunde bijvoegsel) 1924-. Groningen. |
| Euc Mad | Euclides, rev mensual d ciencias exactas, fis, quim, nat, y appl tec. 1941-. Madrid. |
| *Eudemus | Eudemus, An Intl J Devoted to the Hist of Math & Ast. 1941// Copenhagen; Providence, RI. Brown Univ. Eds: R C Archibald, O Neugebauer. |
| Eureka | Eureka, The Archimedeans' J. 1939-. Cambridge. |
| *Exc Med IC | Excerpta medica, intl congress ser. Amsterdam. |
| *Exner Rpm.* | *See Carl PTMA.* |
| *Ex oriente lux.* | *See Leyd Ex OL.* |
| *Experien | Experientia. 1945-. Basle. |

F     Jab ueber d Fortschritte d Math. (Ftschr Math; JFM)
      1868-1942// 68 vols. Merged with RS, 1935-. Berlin.
*Facultes des sciences et grandes ecoles. See Mars M.*
*Fam Volk   Familie & Volk, Zt f Genealogie & Bevoelkerungskunde. 1952-.
      C of Genealogie & Heraldrie.
Faraday D   Discussions of the Faraday Soc. 1947-. London.
Faraday T   Transas of the Faraday Soc. 1947-. London.
*Fennica. See Hels.*
*Ferra U An   Annali univ Ferrara. 1936-. Ferrara.
Ferra UM   Annali d univ d Ferrara, NS, sez 7, sci mat. 1950/1951-.
      Ferrara.
Ferussac   Bull d scis math, ast, phys, & chim. (Ferussac Bll Sc Mth)
      1824-1831// Paris.
*Fib Q    The Fibonacci Quarterly, A J Devoted to the Study of Integers
      with Special Properties. 1963-. St Mary's Coll, Cal.
      Fibonacci Asso.
*Field Art   The Field Artillery J. 1946-. Wash DC; Baltimore, Md.
*Filos    Filosofia. 1950-. Turin.
*Filo Tor. See Filos.*
*Finland, Finska. See Acta Phi F, Arkhimed, Hels, Turku.*
*Fir Atti   Atti d R accad (econ-agraria) d georgofili (d Firenze).
      (Firenze Ac Georg At, etc) 1818-. Florence.
*Fir Querc   Pubb d collegio alla Querce d Firenze. 1901?-. Florence.
Fiz M Ezh   Fiziko-mat ezhegodnik. 1900-1902//? Moscow.
*Fiz M Nauk   Fiz-mat nauka v ikh nastoyashchem i proshedchem. (Fiz-mat
      n v khod ikh razvitaya; D phys-math Wiss; J d reinen & angew
      Math Ast & Ph; Les sci phys-math dans 1 marche d leur
      developpement) 1885-1904// Moscow. Ed: V V Bobynin.
Fiz Tverd   Fiz Tverdogo tela. 1959-. Tr as Sov Ph Sol. Moscow,
      Leningrad.
Fiz Zh    Fisicheskii Zhurnal. (J of Phys Moscow) 1939-1947//
      C of Tek Ph. Moscow. Akad nauk SSSR.
*Flambeau   Le Flambeau, rev belg d questions politiques & litts.
      1918-. Brussels.
*Flammarian rv d'ast. See Astmie.*
*Formosa. See Taiwan.*
*Fors Fort   Forschungen & Fortschritte, Korrespondenzblatt (Nachrichten-
      blatt) d deu Wiss & Tec. (Forsch Fortsch dtsch Wiss; FuF)
      1925-. Tr in Res Prog. Berlin.
*Fors G Op   Forschungen z Gesch d Optik. 1928-1930, 1935-. S to Z Instr.
      Berlin.
Fors Ing   Forschungen im Ingenieurwesen. (Fors Gebiete Ingenieurwesens)
      1930-1944, 1949-. Duesseldorf.
Fors Log G   Forschungen z Logik & z Grundlegung d exakten Wiss. 1937-.
      Leipzig.
Fort Min   Fortschritte d Mineralogie, Kristallographie & Petrographie.
      1911-. Stuttgart.
*Fortschritte. See also F.*
Found Phy   Foundations of Phys, an Intl J Devoted to the Conceptual
      Bases of Mod Nat Sci. 1970-. NY. Plenum P.

*Fr AAS        Assoc francaise pour l'avancement d scis nats, actes d congres.
               1872-. See also Sciences. Paris.
Fr AASB        Bull d l'asso fr pour l'avancement d scis. 1901-1935// C of
               Fr AASI. C as Sciences. Paris.
Fr AASI        Intermediare d l'afas. 1896-1900// C in Fr AASB. Paris.
Fr Actu        Bull trimestriel d l'inst d actus frs. 1890-. Paris.
*Fraenkis      Gesell f fraenkische gesch Vero. 1907-. Wuerzburg.
*Fragen.* See organization, city, country, region, etc.
Fr Air         Pubs scis & tecs d minist d l'air. 1930-. Paris.
Fr Air BST     Pubs scis & tecs d minist d l'air, bull d services tecs.
               1937-. Paris.
*Fr Air NT     Notes techniques, dir d inds aeronautiques, minist d l'air.
               1942-. Paris.
*France.* See also Bord, Bretagne, Dunkerq, Epinal, Gren, Lyon, Mars, Metz,
               Nancy, Par, Rouen, Rv, Strasb, Tou.
*France, inst de.* See Par.
*Frankel's Zeitschrift.* See Mo Ges W Ju.
*Franklin      J of the Franklin Inst. 1826-. Philadelphia, Penn.
*Frank PVJ     Jasb d physikalischen Ver z Frankfurt a M. 1845-.
               Frankfurt am Main.
*Fr Arch Or    Inst francais d'archeol orientale (de Cairo), mems. 1902-.
               Cairo.
Fr CNRS        J d rech d centre natl d l rech sci. 1946-. Paris.
Fr CTHSM       Bull sci d centre d travaux hist & sci, pt II, math. 1956-.
               Paris. Minist d l'edu natl.
Freib ITP      Mitt aus d Inst f theor Phys & Gloph d Bergakad Freiberg.
               Freiberg.
*Freib NG      Ber d naturf Gesell z Freiburg (in Breisgau). 1855-.
               Freiburg.
*Friedrich-Schiller-Univ.* See Jena UWZ.
*Fr Long       Annuaire...publie par l bur d longitudes, avec d notices scis.
               1799-. Paris.
*Fr Mod        Francais moderne, rev d synthese & d vulgarisation
               linguistiques. 1933-. Paris.
Fr ONERA       Office natl d'etudes & d rechs aeronautiques, pub. 1948-.
               Paris.
Fr ON Met      Memorial d l'office natl meteorol d France. 1949-. Paris.
*Fr SMB        Bull d l soc math d France. 1873-. Paris.
Fr SMCC        Communications & conferences, soc math d Frence. 1872-.
               Paris.
Fr SMCR        CR d seances & conferences d l soc math d France. 1912-.
               Paris.
*Fr Soc Ag     Mems d'ag, d'econ, rurale & domestique publies par l soc
               d'agriculture. (Fr S Ag Mm) 1801-. Paris.
*Fujihara Memorial Faculty of Engineering.* See Tok Keio.
*Fukuoka.* See also Bu M St.
Fukuo U        Bull of Fukuoka Gakugei Univ, III, Nat Scis. (Fukuoka
               gakugei daigaku kiyo) 1951-. Fukuoka.
Fun An Ap      Functional Analysis & Its Appls. 1967-. Tr of Fun An Pri.
               NY. Plenum P.
Fun An Pri     Funktsionalnyi analiz i ego prilozheniia. 1967-. Tr as
               Fun An Ap. Moscow.

| | |
|---|---|
| *Fund M | Fundamenta mathematicae. 1920-. Warsaw. Polska akad nauk. |
| *Fun Ekv | Funkcialaj ekvacioj, ser internacia. (Fako d l funkcialaj ekvacioj Japana mat soc) 1958-. Kobe, Tokyo. |
| Fys Tid | Fisisk tidsskrift. 1902-. Copenhagen. Selskabet for naturlaerens udbredelse. |

| | |
|---|---|
| *Gac M | Gaceta matematica. 1949-. Madrid. |
| *Gaea | Gaea, Natur und Leben. 1865-1909// C in Naturw R. Leipzig. Ed: H J Klein. |
| *Ganita | Ganita. 1950-. C of Benar MS. Lucknow. |
| *Gauss Mt | Mitteilungen der Gauss Gesellschaft. 1964-. Goettingen. |
| *Gaz M | Gazeta matematica. 1895-1934// C as Gas MFA. Bucharest. |
| *Gaz MA | Gazeta mat, ser A. 1964-. C of Gaz MFA. Bucharest. Soc d stiinte mat din RPR. |
| *Gaz MFA | Gazeta mat si fiz (ser A). 1934-1963// C of Gaz M. Pt C as Gaz MA. Bucharest. Soc d stiinte mat si fiz d RPR. |
| *Gaz MFB | Gazeta mat si fiz, ser B. 1954-. C of Rv M Fiz. Bucharest. Soc d stiinte mat si fiz din RPR. |
| *Gaz M Lisb | Gazeta d matematica. 1940-. Lisbon. |
| Gen D | The Gentleman's Diary or Mathematical Repository. 1721-1840/ London. Eds: O Gregory, T Leybourn. |
| Genetica | Genetica, Ned tid voor Erfelijkheids-an Afstammingsleer. 1919-. 's Gravenhage. |
| Genetics | Genetics. 1916-. Princeton, NJ; Baltimore, Md. |
| Genet Res | Genetical Research. 1960-. NY, London. |
| *Geno AL | Atti d accad Ligure d sci e lett. (Att Accad Ligure) 1941-. C of Geno AS. Genoa, Rome. |
| Geno AS | Atti d soc d sci e lett d Genova. 1936-1940// C of Geno ASL. C as Geno AL. |
| *Geno ASL | Atti d soc ligustica d sci e lett. 1890-1935// C as Geno AS. Pavia, Genoa. |
| *Geno U | Atti d R univ d Genova. (Genova Un At) 1869-. Genoa. |
| *Geno U Ann | Annuario d R Univ d Genova. |
| Gen Top Ap | Gen Topology & Its Appl. 1971-. Amsterdam. North-Holland Pub Co. |
| *Genv Arc | Archives des sciences. 1948-. 1949- incl Genv CR. Soc d phys & d'hist nat d Geneve. Geneva. |
| *Genv CR | CR d seances d l soc d phys & d'hist nat d Geneve. 1884-. S to Genv Arc. Geneva. |
| *Genv Mus | Les musees de Geneve. 1944-. Geneva. |
| | *Geofizicheskii sbornik. See Len Sb Gf.* |
| *Geog Mag | Geographical Mag. 1935-. London. |
| Geometra | Geometra. 1946-. Turin. Coll d geom d prov d Torino. |
| *Geom Zb | Geometrichnii zbirnik. 1938-. Kharkov. Naukov-doslednii inst mat i mekh pri Kharkovskomu derzavnomu univ. |
| | *Georg-Augustus-Univ. See Gott.* |
| | *Georgia, Georgian SSR. See Gruz, Tbil.* |
| GE Rev | General Electric Rev. 1903-1958// Schenectady, NY. |

Gergonne          Anns d math pure & appls.  (Gergonne A Mth)  1810-1933//
                  C as JMPA.  Nimes, Paris.
*German R         Germanic Review.  1926-.  NY.
 Germany.    See Deu; also Aach, Bay, Ber, Bonn, Braun, Dres, Erlang, Essling,
                  Fraenkis, Frank, Freib, Gies, Gorlitz, Gott, Greifsw, Hall,
                  Halle, Ham, Hannov, Heid, Jena, Karlsruh, Leip, Mainz, Marb,
                  Mitdeu, Mun, Muns, Nor Wes, Nurnb, Oberbay, Pots, Regensb,
                  Rostock, Sarav, Schwa, Ulm, Wurt, Wurz.
*Ges Bl TI        Geschichtsblaetter f Tec & Industrie.  1914-1927//  Berlin.
 Gesellschaft.    See organization, city, region, or country.
*Gesnerus         Gesnerus, Vierteljahrsschr f Gesch d Med & d Naturw.  1943-.
                  Zurich.
*Giambat          Il Giambattist-Vico, gior scientifico.  (Giamb Vico)  1857-.
                  Naples.
*Gids             De Gids.  1837-.  Amsterdam.
 Gi Ec St         Gior d economisti e riv d stat.  1910-1938//  Milan.
*Gies Hoch        Nac d Giessener Hochschulgesell.  1918-.  Giessen.
*Gies Phil        Giessener Bei zur deu Philologie.  1921-.  Giessen.
*Gies Sem         Mitt (aus) d math Sem (d Univ) Giessen.  1881-.  Giessen.
 Gies Sem B       Justus Liebig Hochschuler math Sem, Mitt Beiheft.  1952-.
                  Giessen.
 Gifu             Sci Report of the Fac of Lib Arts & Edu, Gifu Univ (Nat Sci).
                  (Gifu daigaku gakugeigakubu kenkyu hokoku).  1953-.  Gifu.
*Gi Let Ven       Gior d letterati d'Italia.  1710-1730//  C as Oss Lett.
                  Venice.
*Gi M Fin         Gior d mat finanziara, riv tec d credito e d previdenza.
                  (G mat fin)  1919-.  Turin.
 Gioenia.    See Cat.
 Giornale.   See also organization, city, region, or country.
 Giornale de matematiche...studenti...univ italiane.  See Batt.
 Giornale (fisico-medico) di fisica, chimica, e storia naturale, Pavia.
                  See Brugnat.
 Glas, glasnik.   See Glasn; also organization, city, region, or country.
*Glasg M          Procs of the Glasgow Math Asso.  1952-.  Glasgow.
 Glasg PS         Procs of the Phil Soc of Glasgow.  (Glasg P Ph S; Glasg Ph
                  S P)  1841-.  Glasgow.
 Glasn MPA (Glasnik mat-fiz i ast, ser II).  See Zagr Gl.
 Gleangs          Gleanings in Science.  1829-1831//  Calcutta.
*Goress Ja        Philosophisches Jahrbuch.  1888-1942, 1946-1951, 1953-.
                  Fulda.  Founded Goress-Gesell.
*Gorlitz          Abh d naturf Gesell z Gorlitz.  (Gorl Ab)  1827-1842//
                  Gorlitz.
 Gotb B           Goeteborgs K vetenskaps- och vitterhetssamhaelles
                  handlingar, femte folyden, ser B, mat och naturvet skr.
                  1850-.  Goteborg.
 Gott Ab          Abh d (K) Gesell (Akad) d Wiss z (in) Goettingen, math-naturw
                  Kl.  1838-1940//  C in Gott Ab MP, & in Gott Ab H.
                  Goettingen.
*Gott Ab H        Abh d Gesell d Wiss z Goettingen, phil-hist Kl.  1895-.
                  Pt C of Gott Ab.  Goettingen.

Gott Ab MP    Abh d Akad d Wiss i Goettingen, math-phys Kl.  1940-.
               C from Gott Ab.  Goettingen.

*Gott Anz      Goettingische gelehrte Anzeigen.  (Goett Zt v gelehrte
               Sachen; G Anz v gel Sachen)  1739-.  Berlin; Goettingen.
               K Gesell d Wiss Goettingen.

*Gott Bei      Bei zur Gelehrtengeschichte Goettingens.

Gott Cm       Commentationes (recentiores) societatis regiae scientiarum
               Gottingensis.  (Gott commentarii; Gott C; Novi commentarii)
               1751-1837//  C in Ber Ab.  Goettingen.

*Gott Jb       Jab Gesell (Akad) d Wiss Goettingen.  (Nachrichten;
               Geschaeftlichen Mitt; Nac Jasb)  1894-1944//  Pt of Gott N.
               Goettingen.

*Gott N        Nac v d (K) Gesell (Akad) d Wiss z Goettingen.  (Gott gelehrte
               Anzeigen; ...& d Georg-Augusts-Univ; ...soc regia sci
               Gottingensis; Gott Nachr)  1845-1938, 1945-.  II, math-phys
               Kl 1893-1938, 1945-.  See also Gott Jb.  Goettingen.

*Gott Ph HN    Gesell d Wiss i Goettingen, philologisch-hist Kl Nac.
               1894-.  Goettingen.

Gott Stu       Goettinger Studien.  1845-1847//  Goettingen.

*Gott UBM     Mitt Univ Bund.  1918-.  Goettingen.

Gott U Jb     Georg-Augusts-Univ Goettingen, math-naturw Fak, Jab.
               1920-1924//  Goettingen.

Gott Vorl     Math Vorlesungen Georg-Augusts-Univ Goettingen.  1907-1914//
               Goettingen.

*Grande En    La grande encyclopedie, 1886-1902, 31 vols.  Paris.

*Grande Rv    Grande revue.  1897-.  Paris.

*Greece.  See also Athen, Inblk, Intbalkn, Rv.*

*Gree SM     Bull d l soc math d Grece.  (Deltion, Hellenike math
               hetaireia)  1919-.  Athens.

*Greifsw      Wiss Zt d (Ernst-Moritz-Arndt-) Univ Greifswald, math-naturw
               Reihe.  1951/1952-.  Greifswald.

Greifsw N    Mitt aus d naturw Ver von Neu-Vorpommern & Rugen in Greifswald.
               (N-Vorp Mt)  1869-.  Berlin.

Gren Annl    Anns d l'univ d Grenoble, NS, sect sci-med.  1924-.  Grenoble.

*Gren IF      Anns d l'inst Fourier.  1949-.  Grenoble.  Univ d Grenoble.

*Griefsw(ald).  See Greifswald.*

*Grodnen gos ped inst.  See BSSR Ped.*

Grund MW     Grundlehren d math Wiss in Einzeldarstellung.  1921-.  Berlin.

*Grunert Archive.  See Arc MP.*

*Gruz.  See also Tbil.*

Gruz PIT     Gruzinskii politek inst im V I Lenina, trudy.  (Izv...)  1934-.
               Tbilisi.

*Gruz Soob    Soobshcheniya akad nauk Gruzinskoi SSR.  (Bull Acad Sci
               Georgian SSR)  1940-.  Tbilisi.

Gruz Vich    Trudy vychislitelnogo tsentra, akad nauk Gruzinskoi SSR.
               1960-.  Tbilisi.

*Guld Pass    Gulden passer, le compas d'or.  (Bull trimestriel d l soc
               d bibliophiles anversois)  1878-1883, 1923-.  Antwerp.
               Vereeniging d antwerpsche bibliophilen.

*Haarlem. See also Ned.*
Haar Nat          Natuurkundig verhandelingen van de (bataafsche) hollandsche
                  maatschappij der wetenschappen te Haarlem.  (Haarl Nt Vh
                  Mtsch)  1799-.  Haarlem.
*Habana. See also Cub.*
Hab               Mems d 1 fac d ciencias, univ d 1 Habana, ser mat.  1963-.
                  Havana.
*Hall Bij         Bijdragen tot de natuurkundige wetenschappen.  1826-1832//
                  Amsterdam.  Ed: Hall.
*Halle NG         Ber d naturf Gesell zu Halle.  (Halle Nf Gs B; Naturf Ges
                  Halle)  1856-.  Halle.
*Halle UM         Wiss Zt d Martin-Luther-Univ, Halle-Wittenberg, math-naturw
                  Reihe.  1953-.  Wittenberg.
*Hall Mgr         Hallische Monographien.  1848-1951//  Halle.  Univ.
*Ham M Einz       Hamburger math Einzelschriften.  1923-.  Leipzig, Hamburg.
*Ham MG           Mitt d math Gesell zu (in) Hamburg.  1889-.  Hamburg.
*Ham Sem          Abh aus d math Sem d Univ Hamburg.  (Hamb Abh; Abh Math Sem)
                  1921-1943, 1947-.  Hamburg.
*Hannov Z         Zt d Architekten- & Ingenieurvereins zu Hannover.  1855-.
                  Hannover.
*Hansischen Univ. See Leip Sem.*
*Harpers          Harper's Magazine.  1850-.  NY.
*Harv Alu B       Harvard (Alumni) Bull.  1898-.  Cambridge, Mass.
*Harv G Mag       Harvard Graduates' Magazine.  1892-.  Boston.
*Harv J Asi       Harvard J of Asiatic Studies.  1936-.  Cambridge, Mass.
                  Harvard-Yenching Inst.
*Harv Lib B       Harvard Univ Library Bull.  1875-1894//  Cambridge, Mass.
*Harv Reg         Harvard Register.  1827-1828, 1880-1881//  Cambridge, Mass.
*Havana. See Cub, Hab.*
*Heb Bib          Hebraeische Bibliographie, Blaetter f neuere & aeltere Lit
                  d Judenthums.  1858-1882//  Berlin.
*Heb UC An        Hebrew Union Coll Annual, J of Jewish Lore & Phil.  1924-.
                  Cincinnati, O.
*Heid Abh         Abh d Heidelberger Akad d Wiss, math-naturw Kl.  1910-.
                  Heidelberg.
*Heid NMV         Verh d naturhist-med Ver zu Heidelberg.  (Heidl Nt Md Vh;
                  Heidl Vh Nt Md)  1857-.  Heidelberg.
*Heid Si          Sitz d Heidelberger Akad d Wiss, math-naturw Kl.  1909-1944,
                  1948-.  Heidelberg.
*Heid Si Ph       Sitz d Heidelberger Akad d Wiss, phil-hist Kl.  1910-.
                  Heidelberg.
*Heid Tat         Taetigkeitsbericht d math Fachschaft an d Univ Heidelberg.
                  1930-1931//?  Heidelberg.
*Hels Act         Acta societatis scientiarum fennica.  1839-1926//  Helsinki.
                  Finska vetenskaps societeten.
 Hels Act A       Societatis scientiarum fennica, acta A, opera phys-math.
                  1927-.  Helsinki.  Finska vetenskaps societeten.
*Hels Arsb        Arsbok-vuosikirja soc sci fennica.  1922-.  Helsinki.
*Hels Comm        Commentationes phys-math.  1922-.  Helsinki.  Soc sci
                  fennica; Finska vet soc.
*Hels Hum         Commentationes humanorum litterarum.  1922-.  Helsinki.

*Helsingfors*.   *See Helsinki*.
*Hels M        Suomalaisen tiedeakatemian toimituksia, sarja A, math (phys).
               (Annales acad sci fennicae, ser A, math) 1909-.  See also
               Hels Phy.  Helsinki.
*Hels Ofv      Ofversigt af finska vetenskaps-societetens forhandlingar.
               1853-.  Helsinki.
Hels Phy       Suomalaisen tiedeakatemian toimituksia, sarja A VI, phys.
               (Annales acad sci fennicae) 1957-.  C from Hels M.  Helsinki.
Hels Si        Sitz der finnischen Akad der Wiss.  (Proc of the Finnish
               Acad of Sci & Letts.  1908-.  Helsinki.
*Hels Tids     Finsk tidskrift for vitterhet, vetenskap, konst och politik.
               1876-.  Helsinki.
Helv Acta      Acta Helvetica, phys-math, bot-med....  1751-.  Basle.
*Helv CM       Commentarii mathematici helvetici.  1928-.  Zurich.
*Helvetia*.   *See also Switzerland*.
Helv Phy       Helvetica physica acta.  1928-.  Basle.  Soc phys helveticae
               commentaria publica.
*Helv SN (Actes d l soc helvetique d scis nats;  Atti d soc elvetica d sci*
               *nat; Verh d schw naturf Gesell).  See Schw NG.*
*Hemecht       Hemecht.  1949-.  C of Ons Hemecht, 1895-1939//  Luxembourg.
*Herd          Abh d Herder-Gesell & d Herder-Inst zu Riga.  1925-1935//
               Riga.
Heredtas       Hereditas, genetiskt arkiv.  1920-.  Lund.  Mendelska
               sallskapet i Lund.
*Hermath       Hermathena, a Ser of Papers on Lit, Sci, & Phil, by Members
               of Trinity Coll.  1873-.  Dublin.
*Hermes        Hermes, Zt f kl Philologie.  1866-1944, 1952-.  Berlin.
*Hesperis      Hesperis.  1921-.  Rabat.  Inst des hautes etudes marocaines.
*Himmelsw      Die Himmelswelt, Mitt (d Ver) von Freunden d ast & kosmischen
               Phys.  (Zt f Pflege d Himmelskunde & verw Gebiete) 1920-.
               Berlin.
*Himm Erde     Himmel & Erde.  1888-1915//  Leipzig, Berlin.
*Hindi Science Academy*.   *See Vijm*.
*Hinrichsen's Musical Yearbook*.   *See Mus Bk*.
Hiro A         J of Sci of the Hiroshima Univ, Ser A, (Math, Phys, Chem).
               1931-.  Hiroshima.
Hiro En        Mems of the Fac of Engineering, Hiroshima Univ.  (Hiroshima
               daigaku, kogakubu, kenkyu hokoku) 1952-1958//  Hiroshima.
Hiro RITP      Sci Reports of the Res Inst for Theor Phys, Hiroshima Univ.
               (Hiroshima daigaku, biron butsuri kenkyuju) 1961-.
               Hiroshima, Takehara.
Hirosak        Sci Reports of the Fac of Lit & Sci, Hirosaki Univ.  1954-.
               Hirosaki.
*His Lit Fr    Histoire litteraire d l France.  1733-.  Paris.
*His Relig     History of Religions.  1961-.  Chicago, Ill.  Univ of
               Chicago P.
*Hist Abt      Historische-literarische Abt.  (Literaturzeitung) 1859-1876//
               C as Ab GM.  S to Z M Ph.  Leipzig.
*Hist Math Studies*.   *See Ist M Isl*.
*Hist Sci      History of Science.  1962-.  Cambridge.

Hi Stu PS     Historical Studies in the Phys Scis. 1969-. C of Chymia.
           Philadelphia, Penn. Univ of Penn Pr.
Hitot JAS     Hitotsubashi J of Arts & Scis. 1960-. Pt C of Annals of the
           Hitotsubashi Acad. Tokyo.
HM     Historia mathematica. (1974- anticipated.) Toronto.
Hochfreq     Hochfrequenztechnik & Elektroakustik. 1932-. Leipzig.
*Hochschule fuer Verkehrswesen Friedrich List in Dresden. See Dres.*
*Hoffmann Zt. See ZMN Unt.*
Hokk M     J of the Fac of Sci, Hokkaido Univ, Ser I, Math. 1930-.
           Sapporo.
*Holland. See Netherlands.*
*Hoppe Archiv. See Arc MP.*
Hou Bla     La houille blanche, rev d l'ingenieur hydraulicien. 1902-.
           Grenoble.
*Hum Bio     Human Biology, a Rec of Research. 1929-. Baltimore, Md.
           Johns Hopkins P.
*Humboldt Univ. See Ber.*
*Hun Act M     Acta mathematica (acad sci) hungarica(e). (Hungarica acta
           mathematica) 1946-. Budapest.
Hun Act Ph     Acta physica (acad sci) hungarica(e). (Hungarica acta phys)
           1947-1949, 1951-. Budapest.
Hun Act T     Acta technica acad sci hungaricae. 1950-. Budapest.
*Hun Antiq. See Act Antiq.*
*Hungary. See also Act Antiq, Bud, Deb, Magy, MNB Ung, MP Lap, M Ter Er,*
           *Szgd.*
Hun St     J d l soc hongroise d statistique, rev trimestrielle. 1923-
           1943// Budapest. Magyar statistikai tarsasag.
*Hunt LB     Huntington Library Bull. 1931-1937// San Marino, Cal.
           Henry E Huntington Lib & Art Gallery.
Hutton     Hutton's miscellana mathematica. 1771-1775// London.


Iasi FT     Studii si cercetari stiintifice, fiz si sti tehn. 1956-.
           C from Iasi St. Iasi. Acad repub pop Romine, filiale Iasi.
*Iasi IP     Buletinul inst polit din Iasi. (Bull of the polyt inst
           Jassy; Izv Yasskogo pol inst) NS 1946-.
Iasi Ist     Studii si cercetari stiintifice istorie. 1956-. Iasi.
           Filialla Iasi acad repub pop Romine.
*Iasi M     Studii si cercetari stiintifice, mat. 1956-. Pt C of
           Iasi St. Iasi. Acad rep pop Romine, filiale Iasi.
Iasi St     Studii si cercetari stiintifice (ser 1, mat, fiz, chem).
           1950-1957// C as Iasi M, & as Iasi FT. Iasi. Acad repub
           pop Romine, filiale Iasi.
Iasi U     Annales sci d l'univ d Jassy. 1900-1936// Pt C as Iasi UM.
           Iasi.
*Iasi UM     Analele stiintifice ale univ Al I Cuza din Iasi, sect I, mat
           (phys, chem). 1937-. C of Iasi U. Iasi.
IBMJRD     IBM Journal of Research & Development. 1957-. NY.
Icarus     Icarus, Intl J of the Solar System. 1962-. NY.

| | |
|---|---|
| ICCB | ICC Bulletin. (Bull of the Provisional Intl Comp Cen) 1958- Rome. Intl Computation Cen. |
| ICSUR | ICSU Rev of World Sci. 1959-1964// Amsterdam. Intl Council of Sci Unions. |
| ICTM Am | Bull of the Amer Commis. 1909-1910// NY. Intl Commis on the Teaching of Mathematics. |
| ICTMB | Ber & Mitt, Intl Commis on the Teaching of Math. 1909-1917// Leipzig. |
| ICTM Ci | Circulaire ICTM. 1909-1911// Geneva. |
| *ICTM Deu | Abh ueber d math Unterricht im Deutschland.... 1909-1916// Leipzig, Berlin. |
| ICTM Hun | Schr d ungarischen Commis. 1911-1912// Leipzig. |
| ICTM It | Atti d sottocommis italiano. 1911-1912// Rome. |
| ICTM Rus | Rapport d l sous-commis russe. 1910-1912// Geneva. Intl Commis on the Teaching of Math. |
| ICTM Wien | Ber ueber d Unterricht in Oesterreich. 1910-1914// Vienna. |
| *IEE | Procs of the Inst of Electrical Engs. 1949-. C of IEEJ. London. 1949-1950 3 pts; 1951-1954 4 pts (I Gen, C as IEEJ(NS) Electronics & Power; II Power, C as Procs Pt A Power; III Radio & Communication, C as Procs Pt B Radio & Electron Eng incl Communication, C as Procs Pt B Electron & Communication; IV Inst Monographs, C as Procs Pt C Inst Monographs); 1963- (vol 110-) 1 pt. |
| IEEE | Procs of the Inst of Elec & Electronic Engs. 1963-. C of IRE. NY. |
| IEEE El Co | IEEE Transas on Electronic Computers. 1952-1967// NY. Inst of Elec & Electronic Engs. |
| IEEE Inf | IEEE (IRE) Transas on Info Theory. 1953-. NY. |
| IEEJ | J of the Inst of Electrical Engs, (A General, B Power Eng, C Communication Eng). 1889-1948// C of Lon Tel En. C as IEE. London. |
| *II St B | Bull d l'inst intl d statistique. 1886-1915, 1924-. Rome. |
| II St Rv | Rev d l'inst intl d statistique. 1933-. The Hague. |
| IJCIS | Intl J of Computer & Info Scis. 1972-. NY. Plenum P. |
| IJCM | Intl J of Computer Mathematics. 1964-. NY. |
| IJ Con | Intl J of Control. 1965-. London. |
| IJG Thy | Intl J of Game Thy. 1971-. Physica Verlag. |
| IJ Me S | Intl J of Mechanical Scis. 1960-. London, NY, Oxford. |
| IJMEST | Intl J of Math Edu in Sci & Tec. 1970-. London, etc. Wiley. |
| IJQC | Intl J of Quantum Chem. 1967-. Interscience Pub Co. |
| IJ Sys Sci | Intl J of Systems Sci. 1970-. Taylor & Francis Pub Co. |
| IJTP | Intl J of Theoretical Phys. 1968-. NY. Plenum P. |
| *Ill JM | Illinois J of Math. 1957-. Urbana, Ill. |
| Ill MN | Illinois Math Newsletter. 1961-. |
| Ilmen TH | Wiss Zt d tec Hochschule (f Elektrotechnik) Ilmenau. 1954/1955-. Ilmenau. |
| *Imago | Imago, Zt f Anwendung d Psychoanalyse. 1912-1937// Leipzig, Vienna. |
| *Imago Mun | Imago mundi, rev d hist d la cultura. 1953-. Buenos Aires. |
| IM Ap J | J of the Inst of Math & Its Applications. 1965-. London, NY. Academic P. |

IMI. *See Ist M Isl.*

IMN  Intl Math News. (Bull of the Intl Math Union; Intl math Nachrichten; Nac ost math Gesell; Nouvelles maths intls) 1947-. Vienna. Intl Math Union.

*Imperial University, Japan. See Tok.*

*Inblk Con  Actes d congres interbalkanique de math, 1934. 1935// See also Intbalkn. Athens.

*Inda Ac  Procs of the Indiana Acad of Sci. 1891-. Brookville, Indianapolis, Ind; etc.

*Indagat  Indagationes math ex actis quibus titulus, Procs of the Sect of Scis, Sect 1A, Math Scis. 1939-. Pt C of Amst P. Tr from Amst Vş G. Amsterdam. K ned akad v wet. Vols 13-17 = vols 54-57 of Amst PA.

*Inda U Stu  Indiana Univ Studies (Pubs). 1910-. Bloomington, Ind.

Inde GMNT  Index z Gesch d Med, Naturw, & Technik. 1953-. Munich.

*India. See Ind, Indn; also Agra, Alla, Anand, Aryan, Asi, Banar, Baroda, Benar, Bengal, Bhan, Bihar, Bom, Clct, Islam, Karnat, Madra, Mysore, Rajasthan, Rv, Sankh.*

*Indian J of Stat. See Sankh.*

Ind M  Industrial Math. 1950-. Detroit, Mich.

*Ind Mil Ti  Indisch militair tijd. 1869-1877, NS 1877-. Batavia, Bandoeng. Indische krijgskundige vereeniging.

Indn Ac A  Procs of the Indian Acad of Sci, Sect A. 1934-. Bangalore.

Indn Ag Ou  Procs of the Acad of Scis of the United Provs of Agra & Ouhd. 1931-1935// C of Indn NAS. C as Indn NASA. Allahabad.

*Indn Ag St  J of the Indian Soc of Ag Statistics. 1948-. New Delhi.

*Indn Antq  The Indian Antiquary, a J of Oriental Res.... 1817-1933// Bombay.

*Indn Ar Sm  Archaeological Survey of India, Mems, Sers Western, Southern Northern. 1874-. Bombay.

*Indn HQ  Indian Hist Quar. 1925-. Calcutta.

Indn ISJ  J of the Indian Inst of Sci. 1914-1938// C as Indn ISQJ. Bangalore.

Indn ISQJ  Quarterly J of the Indian Inst of Sci. 1938-1951// C of Indn ISJ. Bangalore.

*Indn JHS  Indian J of the Hist of Sci. 1966-. Calcutta. Natl Inst of Sci.

Indn JM  Indian J of Math. 1958-. Allahabad.

Indn J Me M  Indian J of Mech & Math. 1963-. Calcutta.

Indn J Ph  Indian J of Phys & Procs of the Indian Asso for the Cultivation of Sci. 1926-. Calcutta.

Indn JT Ph  Indian J of Theoretical Physics. 1953-. Calcutta.

*Indn MSJ  The J of the Indian Math Society (Club). 1909-1932, 1934-. C as M Stu. Madras.

Indn NAS  Procs of the Natl Acad of Scis of India. 1931/1932// C as Indn Ag Ou. Allahabad.

*Indn NASA  Procs of the Natl Acad of Sci, India, Sect A. 1936-. C of Indn Ag Ou. Allahabad.

Indn NISA  Procs of the Natl Inst of Sci of India, Pt A, Phys Sci. 1955-. New Delhi.

Indn NIST  Transas of the Natl Inst of Scis of India. 1935-. New Delhi.

Indn Ph MJ    The Indian Physico-Mathematical J. 1930-1937// Calcutta.

Indn RJFS    India Rev & J of Foreign Sci & the Arts. (I Rv) 1837-1838// Calcutta.

Indn St A    J of the Indian Statistical Asso. 1963-. Bombay.

Inf Cont    Information & Control. 1957-. NY.

Inf PJ    Info Processing J. 1962-. Wash DC. Cambridge Communication Corp.

Inf Pro Jp    Info Processing in Japan. 1961-. Tokyo.

Inf Pro L    Info Processing Letters. 1970/1971-. Amsterdam. North-Holland Pub Co.

Inf Sci    Information Scis. 1968-. American Elsevier Pub Co.

Inf SR    Info Storage & Retrieval. 1963-. Oxford, NY.

*Innsbr B    Ber des naturw-medizinischen Ver in Innsbruck. (Innsb Nt Md B) 1870-. Innsbruck.

*Institut, institute, instituto, istituto, etc. See I...; also organization, city, region, country.*

*Institut    L'institut, j universel d scis & d socs savantes en France & a l'etranger, sect I, scis math, phys, & nat. 1933-. Paris.

*Institut de France. See Par Mm Div.*

*Institut de sociologie Solvay. See Brx Solvy.*

*Institute of Physical & Chemical Research. See Tok IPCR.*

*Institute of Statistical Math. See Tok I.*

*Institut Fourier. See Gren IF.*

*Institut fuer angewandte Math (& Mech), deu Akad d Wiss. See Ber, M Op Stat.*

*Institut Henri Poincare. See Par HP.*

*Institution of Civil Engineers. See Lon ICE.*

*Instituto Gulbenkian. See Lisb.*

*Instituto Jorge Juan. See Mad IJJ.*

Instr Ab    Instrument Manufacture. 1957-. Engl absts of Pribor, Instrum. Wash DC.

Instr CS    Instru & Control Systs. 1959-. Pittsburgh, Penn.

Instrum    Instrument Construction. 1959-. Tr of Pribor. London.

Int Ac HS    Collection d travaux d l'acad intl d'hist d scis. 1948-. See also Arc In HS. Paris.

*Int Arc Et    Intl Archives of Ethnography. (Intl Archiv f Eghnographie; Archs intls d'ethnographie; Intl Gesell f Ethn) 1888-. Leiden.

Int Arc Ph    Intl Archiv f Photogrammetrie. 1938-1942// C in Photgram. Amsterdam.

*Intbalkn    Rev math d l'union interbalkanique. 1936-. See also Inblk Con. Athens. L'union interbalkanique d mathematiciens.

*Int Com HB    Bull Intl Commi of the Hist of Sci. (Bull Sci Hist) 1926-1943// Paris, Wash DC.

*Int Con    Procs (atti, actes, etc) of Intl Congresses of Mathematicians. (Int Congr Math; ICM) 1897-. Zurich, etc.

*Int Con By    Congres intl d etudes byzantines. 1924-. Rome, Istanbul, etc.

*Int Con H    Intl Congresses of Historical Scis. 1903-. Rome, etc.

*Int Con HS    Intl Congress of the Hist of Sci & Tec, Procs (actes, etc). (P I C H S) 1929-. Paris, London, Coimbra, Prague, etc.

*Int Con L    Intl Congress for Logic, Methodology & Phil of Sci. 1960-.

*Int Con Phy    Intl Congresses of Phil. 1900-. Paris, Geneva, etc.

*Int Con PS      Actes d cong intl d phil d scis (phil scientifique). 1936-.
*Int Con US      Intl Cong for the Unity of Sci. 1935-. (Cong intl d phil
                 sci) 1935-. Paris, etc.
*Intercom        Intercom. 1959-. NY. Asso of Orthodox Jewish Scientists.
*Intermed        Intermediaire d mathematiciens. (L'intermed math) 1894-
                 1925// C as Interm RM. Paris.
 Interm RM       Intermediaire d rechs maths. 1945-1949// C of Intermed.
                 Paris.
*International* Absts in Oper Res. *See Op Res Ab.*
*International* Asso for Analog Computation. *See Analog.*
*International* Congress. *See Int Con; also organization, city, region, etc.*
*International* Journal. *See IJ.*
*International* Journal of Abstracts, Statistical Theory and Method. *See
                 Sta TMA.*
*Int His Sc      Intro to the Hist of Sci. By George Sarton. 3 vols (1,1927;
                 2,1931; 3,1947-48). Baltimore, Md. Carnegie Inst of Wash.
*Int M Nach (Intl mathematische Nachrichten; Intl Math News; Nouvelles
                 mathematiques intls). See IMN.*
*Int Monat       Intl Monatsschrift f Wiss, Kunst & Tec. (Int Mschr Wiss)
                 1911-. C of Int Woch. Munich.
*Int MU Bul      See IMN.*
*Int Woch        Intl Wochenschrift f Wiss, Kunst & Tec. 1907-1911// C as
                 Int Monat. Munich.
 Inv M           Inventiones mathematicae. 1966-. Berlin, NY.
 Inzh Zh         Inzhenernyi zh. (Inzh zh mek tverdogo tela) 1961-. Moscow.
                 Organ otd tek nauk i inst mekh akad nauk SSSR.
*Iorga, Nicolai. See Cuget Cla.*
 Iow Ac          Procs of the Iowa Acad of Sci. 1875-. Des Moines, Iowa.
 Iow SCJS        Iowa State (Coll) J of Sci, (a Quar Rev). 1926-. Ames, Iowa.
*Iraq. See also Bagd, Sumer.*
*Iraq            Iraq. 1934-1940, 1946-. London. Br Sch of Archae in Iraq.
 Iraq SS         Procs of the Iraqi Sci Socs. 1957-. Baghdad.
*IRE (Inst of Radio Engineers). 1962- = IEEE.*
*IRE             Procs of the Inst of Radio Engineers (& Waves & Electrons)
                 1913-1962// C as IEEE. NY.
*IRECTT          IRE (IEEE) Transas on Circuit Thy. 1954-. NY. Inst of
                 Radio Engineers.
*IREECT          IRE Transas on Electronic Computers. 1952-1962// C as
                 IEEE El Co. NY.
*IREHFET         IRE Transas, Human Factors in Electronics. (IEEE...)
                 1960-. NY.
*Ireland. See Dub, Phi St Ire.*
*Iron. See Mech Mag.*
*Isis            Isis. 1913-. Bruges; Cambridge, Mass; etc. Hist of Sci Soc.
 Isis Oken       Isis, oder encyclopaedische Zt. 1817-1849// Jena.
*Islam           Der Islam. 1910-1943, 1947-. Strassburg.
*Islamica        Islamica, Zt f d Erforschung d Sprachen, d Gesch & d Kulturen
                 d islamischen Volker. 1924-1938// Leipzig.
*Islam Rv        Islamic Rev. (Muslim India & Islamic World; Islamic Rev &
                 Muslim India) 1913-. Woking.
 ISMA            Annals of the Inst of Stat Math. 1949-. Tokyo.
*Israel. See also Tarbiz.*

Isr JM        Israel J of Math.  1963-.  C from Isr RCB.  Jerusalem.
*Isr RCB      Res Council of Israel Bull.  1951/1952-1963//  Pt C as Isr JM.
              Jerusalem.
Istanb        Istanbul univ fen facultesi mecmuasi.  (Rev d 1 fac d scis
              d l'univ d'Istanbul)  1923-.  Istanbul.
*Istanb UB    Istanbul tek univ bulteni.  (Bull of the Tec Univ of Istanbul)
              1948-.  Istanbul.
*Ist As       Istoriko-astronomicheskie issledovaniya.  1954-.  Moscow.
  *Istituto superiore navale.  See Nap Nav.*
*Ist Met EN   Istoriya i metodologiya estestvennykh nauk.  (Hist & Meth of
              Nat Sci)  1960-.  Moscow.  Univ.
*Ist M Isl    Istoriko-matematicheskie issledovaniya.  (Trudi sem po ist mat
              Historical-mathematical Studies)  1948-.  Moscow, Leningrad.
*Ist M Zb     Istoriko-matematichnii zbirnik.  1959-1963//  Kiev.  Akad nauk
              URSR inst mat.
It Ac At      Atti d R accad d'Italia, mem d cl d sci fis, mat, e nat.
              1940-1944//  C of Linc Mr.  C as Linc Mr M, and as Linc Mr FC.
              Rome.
  *It Ac Mm (R accad d'Italia, mem d cl d sci fis, mat e nat).  See It Mm FM.*
It Ac Ren     Atti d R accad d'Italia, rend d cl d sci fis, mat & nat.
              1939-1943//  C of and C as Linc Rn.  Rome.
It Ac Sc      Atti d (R) accad italiana d sci....  (At Ac It)  1810-.
              Leghorn.
  *Italian Asso of Theor & Appl Mech.  See Meccan.*
  *Italy.  See also Abruz, Acir, Bln, Brescia, Cagliari, Cat, Ferra, Fir,
              Geno, Linc, Livo, Lom, Lucca, Mathes F, Mod, Nap, Pado,
              Paler, Parma, Pavia, Pelori, Physis Fi, Piac, Pisa, Rm,
              Romag, Rv, Siena, Tor, Triest, Vat, Ven, Virgil.*
*It Annu      Annuario d R accad d'Italia.  (Annu Accad Ital)  1930-.  Rome.
*It As        Mem d soc astronomica italiana.  1920-.  C of It Spet.
              Milan, Rome.
It Assic      Atti d ist naz d assicurazioni.  1929-1942//  Rome.
*It Attuar    Gior d ist italiano d attuari.  1930-.  Rome.
*It Cart An   Annuario d ist cartografico italiano.  1884-1889//  Rome.
*1t Con Sto   Atti d congresso storico italiano.  1885-.  Turin, etc.
*It Geog Bo   Soc geografica italiana, bollettino.  1896-.  Rome.
*It INFM      Mem d ist naz italiano, cl d fis e d mat.  (Bologna Mm I It)
              1806-1813//  Bologna.
It Mm FM      Mem d cl d sci fis, mat e nat, R accad d'Italia.  1930-1944//
              Incl in It Ac At 1940-1944.  Rome.
Itogi M       Itogi nauki, seriya matematicheskaya.  1962-.  Moscow.
It Riun S     Riunione d scienziati italiani, atti.  (At Sc It)  1839-.
              Pisa, etc.
It Spet       Mem d soc degli spettroscopisti italiani.  (Spet It Mm)
              1872-1919//  C as It As.  Palermo.
*It SPS       Atti d soc italiana per il progresso d scienzia.  (Atti Soc
              It Progr Sc)  1907-.  Rome.
*It SS        Mem d mat e d fis d soc italiana d sci (detta dei XL).  (Mem
              d Soc It; Mod S It Mm; Mod Mm S It; Rm S It Mm; Verona Mm
              S It; Verona S It Mm; Mem Mat Sci Fis Nat Soc It Sci; Soc It
              Sc detta d XL)  1782-1855, 1862-.  Rome, etc.

It SS An        Annuario d soc italiana d scienze. 1922-. Rome.
It Stat         Atti d riunione scientifica d soc ital d statistica. 1939-.
                Pisa, etc.
*It Sto At      Atti d soc ital d storia critica d sci med & nat. 1907-1922//
                C as Rv Sto Cr. Venice, etc.
*It UM          Boll d unione mat italiana. (Boll Un Mat Ital; Bol UMI)
                1922-. Bologna.
*It UM Con      Atti d congresso unione mat italiana. 1937-. Rome.
*It XL An       Annuario generale accad naz dei XL. Rome.
It XL Rn        Rend accad naz dei XL. 1950-. Rome.
*Ivan GPI       Uch zap Ivanovskii gos ped inst im D A Furmanova, Ivanovskoe
                mat obshchestvo. 1941-. Ivanov.
*Izvestiya, izvjesca, etc. See also organization, city, region, country.*
*Izvjesca, acad d scis & d arts d Slaves d sud d Zagreb. See Zagr Iz.*
*Iz VUZM        Izv vishchykh uchebnykh zavedenii matematika. 1958-. Kazan.
                Minist vishchego obrazovaniya SSSR.


*J Aer Sci      J of the Aerospace (Aeronautical) Scis. 1934-1962// C in
                AIAAJ. NY. Inst of Aer Sci.
*Jahrbuch, jaarbock, etc. See Jb; also organization, city, region, etc.*
*Jahrbuch ueber die Fortschritte der Mathematik. See F.*
*Jahresbericht. See Jasb; also organization, city, region, etc.*
*Jahresbericht Oberrealschule mit Knabenmittelschule. See Bay Jb OK.*
J Alg           J of Algebra. 1964-. NY.
JA Me           (Transas of the ASME, Ser E:) J of Appl Mechanics. 1935-.
                NY. Amer Soc of Mech Engs.
JA Me TF        J of Appl Mech & Tec Phys. 1965-. Tr of PMTF. NY. Faraday P.
*JAM Me         J of Appl Math & Mech. 1958-. Tr of Pri M Me. NY.
J Anal M        J d'analyse mathematique. 1951-. Jerusalem.
*Janu Lin M     Janua linguarum, studia memorial Nicolai Van Wijk dedicata,
                ser minor. 1956-. The Hague.
*Janus          Janus, rev intl d l'hist d scis, d l med, d l pharmacie & d l
                technique. 1896-1942; 1957-. Leiden.
JAP             J of Appl Physics. 1937-. NY.
*Japan. See Jp; also Chiba, Chuo Kor, Fukuo, Gifu, Hiro, Hirosak, Hitot,
        Hokk, Kagak Ken, Kagosh, Kanaz, Kitami, Kodai, Kuma, Kyo,
        Kyush, Miyaz, M Jp, Muror, Nagao, Nagoya, Nara, Niiga, Ochan,
        Okay, Osak, Paul U, Saita, Sasayama, Send, Shig, Shiz, Sugaku,
        Toh, Tok, Tokush, Toyo Gak, Waseda, Yam, Yok.*
J Approx        J of Approximation Thy. 1968-. NY. Academic P.
JA Prob         J of Appl Probability. 1964-. East Lansing, Mich.
*Jaros Ped. See Yaros.*
*J Asi          J asiatique. (J Asiat; Jou Asiat) 1822-. Paris.
*Jassy. See Iasi.*
J Atmos         J of the Atmospheric Scis. (J of Meteorology) 1944-.
                Boston, Mass.
*J Bot          J of Botany. (J Bt) 1863-. London.
*Jb Phi Pha     Jab f Phil & phaenomenologische Forschung. 1913-1930//
                Halle.

| | |
|---|---|
| J Ch Ph | J of Chemical Physics. 1933-. NY. |
| *J Cl Phll | J der classischen Philologie. |
| J Comb Thy | J of Combinatorial Thy. 1963-. NY, London. Academic P. |
| J Comp Ph | J of Computational Physics. 1966-. NY. Academic P. |
| J Comp S Sc | J of Computer and Systems Scis. 1967-. NY. Academic P. |
| *J Comp Sys | J of Computing Systems. 1952-. St Paul, Minn. Inst of Appl Logic. |
| *J Cun St | J of Cuneiform Studies. 1947-. New Haven, Conn. Amer Sch of Oriental Research. |
| J Cyb | J of Cybernetics, Transas of the Amer Soc of Cybernetics. 1971-. Wash DC. |
| JDE | J of Differential Equations. 1965-. NY. |
| JD Geom | J of Differential Geometry. 1967-. Bethlehem, Penn. Lehigh Univ. |
| J Doc | J of Documentation. 1945-. London. |
| *J Eg Arch | J of Egyptian Archaeology. (JEA) 1914-. London. Egypt Exploration Soc. |
| Jena Sem | Mathematisches Seminar zu Jena, Bericht. |
| Jena UWZ | Wiss Zt d Friedrich-Schiller-Univ, math-naturw Reihe. 1951-. Jena-Thueringen. |
| JEP | J d l'ecole polytechnique. 1796-. Paris. |
| *Jesuit SB | Jesuit Sci Bull. (Bul Am Ass Jes Sci) 1923-. Chestnut Hill, Mass. Amer Asso of Jesuit Scientists. |
| JETPL | JETP Letters. 1965-. Tr of Pisma v redaktsiyu, Acad of Sci, USSR. NY. |
| *Jeune Sci | Jeune scientifique. 1962-. Joliet, Quebec. Asso canadienne-francaise pour l'avancement d scis. |
| *Jew Q Rv | Jewish Quar Rev. 1888-1908, 1910-. London; Philadelphia, Penn. |
| *J Fl Me | J of Fluid Mechanics. 1956-. London. |
| J Fun Anal | J of Functional Analysis. 1967-. NY. Academic P. |
| J Gen | J of Genetics. 1910-. Cambridge; Chicago, Ill; etc. |
| *J Gen Psy | J of Genetic Psych. (Pedagogical Seminary & J of G P) 1891-. Provincetown, Mass; etc. |
| J Geoms | J des geometres. 1848-. Benais. Soc geom de France. |
| J Geoph R | J of Geophysical Research. 1896-. Wash DC. |
| *J Geron | J of Gerontology. 1946-. Springfield, Ill. |
| JH Ast | J for the Hist of Astronomy. 1970-. Cambridge. Univ. |
| *JHBS | J of the Hist of the Behavioral Scis. 1965-. Brandon, Vt. Intl Soc for the Hist of the Behavioral Scis. |
| *J Helle St | J of Hellenic Studies. 1880-. London. |
| *JH Ideas | J of the History of Ideas. 1940-. NY; Lancaster, Penn. |
| *JH Med | J of the Hist of Med & Allied Scis. 1946-. NY. |
| *JHS.* | *See Kagak Ken.* |
| *J Ind Eng | J of Industrial Engineering. 1949-. NY, etc. Amer Inst of Ind Engineers. |
| JM Ala | J of Math. 1962-. University, Ala. Alabama Math Foundation. |
| *JM An Ap | J of Math Analysis & Applications. 1960-. NY. |
| J Me | J de mecanique. 1962-. Paris. |
| *JM El | J d math elem a l'usage d tous les candidats aux ecoles d gouvernement & d aspirants au baccalaureat en scis. (Math Elementaire) 1877-1902// C from JM El Sp. See also JM Sp. Paris. Ed: J Gourget, G d Longchamp. |

| | |
|---|---|
| JM Elem | J d mathematiques elementaires. 1870-1872// Paris. |
| JM El Sp | J d math elem & speciales. 1880-1881// C in JM El. Paris. |
| JM El V | J d math elem. 1876-. Paris. Ed: H Vuibert. |
| J Me Ph At | J d mecanique & d phys d l'atmosphere. (J sci d l meteorol) 1949-. Paris. |
| J Me Ph Sol | J of the Mechanics & Phys of Solids. 1952-. London. |
| J Ml Ph | J of Mathematical Physics. 1960-. NY. |
| JM Me | J of Math & Mechanics. 1957-. C of J Rat Me An. Bloomington, Ind. |
| JM Met | J of Math. 1951-. Tokyo. Metropolitan Inst for Math. |
| *JMPA | J d math pures & appls. (Liouv J Mth) 1836-. C of Gergonne. Paris. Ed: J Liouville, etc. |
| *JM Ph | J of Math & Physics. 1921-1969// C as Stu Apl M. Cambridge, Mass. MIT. |
| JM Ps | J of Mathematical Psych. 1964-. NY. Academic P. |
| JM Pu Ap | Jornal d mat pura e aplicada. 1936// Sao Paolo. Univ. |
| JM Soc | J of Mathematical Sociology. 1971-. NY. Gordon & Breach Pub. |
| *JM Sp | J d math speciales a l'usage d candidats aux ec polyt. 1882-1902// Pt of JM El. Paris. Ed: Longchamp. |
| JM Spec | J d math speciale. 1869-1870// Montpellier. |
| J Multi An | J of Multivariate Anal. 1971-. Academic P. |
| *JNES | J of Near Eastern Studs. 1942-. C of Am JSLL. Chicago, Ill. |
| JNSM | J of Nat Scis & Mathematics. 1961-. Lahore. |
| J Num Thy | J of Number Theory. 1969-. NY. Academic P. |
| *John Rylands Library.* | *See Manc JRL.* |
| *Johns HAM | Johns Hopkins Alumni Mag. Baltimore, Md. |
| *Johns HC | Johns Hopkins Univ Circulars. 1879-. Baltimore, Md. |
| J Optim | J of Optimization Thy & Appls. 1967-. NY. Plenum P. |
| *Jornal de sciencias mathematicas e astronomicas.* | *See Teixeira.* |
| *Journal, etc.* | *See also organization, city, region, country, etc.* |
| *Journal fuer die reine und angewandte Mathematik.* | *See Crelle.* |
| *Journal of Applied Math and Mechanics.* | *See Pri M Me.* |
| *Journal of Meteorology.* | *See J Atmos.* |
| *Journal of Natural Philosophy, Chemistry, and the Arts.* | *See Nicholso.* |
| *Journal of Physics.* | *See Fiz Zh.* |
| *Journal of Spatial Mathematics.* | *See Sasayama.* |
| *Journal of Statistical Abstracts.* | *See Sta TMA.* |
| *Journal of the History of Science Japan.* | *See Kagak Ken.* |
| JPA Alg | J of Pure & Appl Alg. 1971-. Amsterdam. North-Holland Pub. |
| *Jp Ac P | Procs of the Imp Acad of (Tokyo) Japan. (Procs of the Japan Acad) 1912-. Tokyo. |
| *Jp APS | Annals of the Japan Asso for Phil of Sci. (Kagaku kisoro gakkai (kenkyu). 1955-. Tokyo. |
| Jp Asi So | Transas of the Asiatic Soc of Japan. (Jap As S T) 1872-. Yokohama. |
| *J Phil | J of Philosophy (Psych & Sci Methods). 1904-. NY. |
| *J Phi Ps SM.* | *See J Phil.* |
| *J Phlol | J of Philology. 1868-. London. |
| Jp Inf Pr J | J of the Inf Processing Soc of Japan. 1961-. Tokyo. |
| Jp J As Gf | Japanese J of Ast & Geophys, Transas & Absts. (Nippon temmongaku oyobi chikyu butsuri-gaku shuho) 1922-1945/47// Tokyo. Natl Res Council of Japan. |

Jp J Biost      Japanese J of Biostatistics.  (Seibutsu tokeigaku zasshi)
                1959-.  Tokyo.
*Jp JM          Japanese J of Mathematics.  1924-.  Tokyo.  Sci Coun of Japan.
Jp J Ph         Japanese J of Physics, Transas & Absts.  1922-1942//  Tokyo.
                Natl Res Council of Japan.
*Jp MASE        J of the Mathematical Asso of Japan for Secondary Education.
                1919-.  Tokyo.
*Jp MS          J of the Math Soc of Japan.  1948-.  C from Tok M Ph.  See
                also Jp MSP.  Tokyo.
Jp MSP          Math Soc of Japan Pubs.  (Nikon suggakkai) 1955-.  See
                also Jp MS.  Tokyo.
Jp ORS          J of the Operations Res Soc of Japan.  1957-.  Tokyo.
Jp PS           J of the Physical Soc of Japan.  1946-.  C from Tok M Ph.
                Tokyo.
Jp RSAR         Repts of Statistical Appl Res.  1951-.  Tokyo.  Union of
                Japanese Scientists & Engs.
Jp SMEB         Bull of the Japan Soc of Mechanical Engs.  (Bull JSME)
                1958-.  Tokyo.
Jp SMET         Transas of the Japan Soc of Mechanical Engs.  1935-.  Tokyo.
*J Ps NP        J d psychologie normale & pathologique.  1904-1915, 1920-.
                Paris.
*Jp Stu HS      Japanese Studies in the Hist of Sci.  (Nippon kagakusi gakkei)
                1962-.  Tokyo.
*J Rat Me An    J of Rational Mechanics & Anal.  1952-1956//  C as JM Me.
                Bloomington, Ind.
J Rec M         J of Recreational Mathematics.  1968-.  NY.
JRME            J for Res in Mathematics Education.  1970-.  NCTM.
*J Sav          J des savants.  1684-1792, 1816-.  Paris.
J Sc En R       J of Sci & Engineering Res.  1957-.  Kharagpur.
J Sci Mat       J d scis mathematiques.  1872//  Paris.  Ed: A Labosne.
J Sc Instr      J of Scientific Instruments.  1922-.  London.  Inst of Phys
                & Natl Phys Lab.
J Ship R        J of Ship Research.  1957-.  NY.
*JSL            J of Symbolic Logic.  1936-.  Asso for Symbolic Logic.
*J Sp Phil      J of Speculative Philosophy.  1867-1893//  St Louis, Mo.
*J Suis Hor     J suisse d'horlogerie.  1876-.  Geneva.
J Theo Bio      J of Theoretical Biology.  1961-.  London, NY.  Academic P.
JTP             J of Tec Physics.  1956-.  Eng absts of Zh TF.  See also PMTF,
                & Sov Ph TP.  Wash DC.
*Jugoslav.  See Yugoslavia.*
J Und M         J of Undergraduate Math.  1969-.  Greensboro, NC.  Guilford
                Coll.
J Un S          J of Unified Sci.  1930-1940//  C of Erkennt.  The Hague,
                Leipzig.
*Jurjew.  See Tartu, Dorpat.*
J Zos Kiok      J of Zosen Kiokai.  1959-.  Tokyo.

Kab FM          Uch zap, ser fiz-mat, Kabardino-Balkarskii gos univ.
                1957-. Nalchik.
*Kagak Ken      Kagakushi kenkyu. (Japanese J of the Hist of Sci; J Hist
                Sci Japan) 1941-. Tokyo.
*Kagan          Vestnik opytnoi fiz i elem mat. (Kagans bote; Spaczins Bote)
                1886-1917// C of Zh EM. Kiev, Odessa.
Kagosh U        Sci Repts of the Kagoshima Univ. (Daigaku bunrigakubu
                rika hooku) 1952-. Kagoshima.
*Kaiserliche Akademie. See city, region, country, etc.*
*Kaizo          Kaizo, la rekonstruo. 1919-. Tokyo.
*Kaliningrad. See Koenigsberg (Konig).*
Kalin Ped       Uch zap Kalininskii gos ped inst im M I Kalinina. 1926-.
                Kalinin.
Kanaz U         Sci Repts of the Kanazawa Univ. 1951-. Kanazawa.
*Kan Cit UR     Univ of Kansas City Rev. 1934-. Kansas City, Mo.
Kan MT          Bull Kansas Asso of Math Teachers. 1926-. Manhattan, Kan.
*Kantstu        Kantstudien, philosophische Zt. 1897-. Berlin, Hamburg.
                Kant-Gesellschaft.
*Karel Ped      Uch zap Karelskii ped inst, ser fiz-mat nauk. 1948-.
                Petrozavodsk.
*Karl-Marx Univ. See Leip.*
*Karlovy Univ. See Prag Karl.*
*Karlsruh       Verh d naturw Ver in Karlsruhe. (Karlsruhe Nt Vr Vh)
                1862-1935// Karlsruhe.
Karnat          J of the Karnatak Univ (Sci Sect). 1956-. Dharwar.
*Kasan, Kazan. See Kazn, Iz VUZM.*
*Kaunas. See also Lit.*
Kaunas U        Mems d l fac d scis d l'univ d Vytautas le Grand. (Vytauto
                disziojo univ, mat-gamtos fak darbai) 1922-. Kaunas.
*Kazk Ak        Vestnik akad nauk Kazakhskoi SSR. 1956-. Alma-Ata.
Kazk Asf        Trudy astrofiz inst, akad nauk Kazakhskoi SSR. 1961-.
                Alma-Ata.
Kazk Asf I      Izv astrofiz inst, akad nauk Kazakhskoi SSR. 1955-.
                Alma-Ata.
Kazk FM         Izv akad nauk Kazakhskoi SSR, ser fiz-mat nauk. 1963-.
                C of Kazk M Me. Alma-Ata.
*Kazk M         Trudy sektora mat i mekhaniki, akad nauk Kazakhskoi SSR.
                1958-. Alma-Ata.
Kazk M Me       Izv akad nauk Kazakhskoi SSR, ser mat i mekh. (Kazak SSR
                eylym akad khabarlary) 1947-1963?// C as Kazk FM. Alma-Ata.
Kazk U          Uch zap Kazakhskogo gos univ im S M Kirva. 1938-. Alma-Ata.
*Kazn FMO       Izv fiz-mat obshch pri Kazanskom univ. (Kazan Ges; Nachr
                d ph-mat Ges; Bull d l soc phys-math d l'univ d Kazan)
                1891-. C of Kazn OEFM. S in Bib M Ross. Kazan.
Kazn FMT        Izv Kazanskogo filiala, akad nauk SSSR, ser fiz-mat i tek
                nauk. 1948-1962// Kazan.
*Kazn OEFM      Procs of the Phys-Math Sect of the Naturalists' Soc of the
                Imp Univ of Kazan. (Kazan S Nt (Ps-Mth) P; Kasan Ber; Kaz
                univ obshch estestv) 1883-1890// C as Kazn FMO. Kazan.
*Kazn Uch Z     Uch zap Kazanskogo (gos) univ (im V I Lenina). (Kazan Mm Vn;
                Kazan Un Mm; Mem d Kasan; Izv i uch zap) 1834-1916, 1924-.
                Kazan.

Kazn U Iz     Bull of the Imp Univ of Kazan. 1865-1884// Kazan.

*Keio Univ. See Tok Keio.*

*Kexue Tongbau. See Ko Hs TP.*

Khalyk     Khalyk mugalimi. (Kazak Journal) 1947-. Alma-Ata.

Khar Geom     Naukovo-daslidnii inst mat i mech pri Charkivskomu derzavnomu univ, geom zbirnik. (Praci, sek geom, n-d inst mat i mech i kafedri geom...) 1938-. Kharkov.

*Kharkov. See also Teor Fun.*

*Khar Me M     Communications d l'inst d sci math & mec d l'univ d Kharkov & d l soc math d Kharkov. 1928-. Kharkov.

Khar M Iz     Kharkovskoe mat obshch izdanii. 1892-1908// Kharkov.

*Khar M So     Soobcheniya (i protokol) Kharkov univ, mat obshch. (Charkow Ges; Charikov soobsc mat obsc; Rapports d l soc math d Kh) 1879-1918// C as Khar MUZ. Kharkov.

*Khar MUZ     Uch zap, nauchno-issl kafedr (Ukrany), otd mat. (Ann soc des Inst Sav d Kark; Charikow Ann Sci) 1924-1926// C of Khar M So. C as Khar M Za. Kharkov.

*Khar M Za     (Uch) zap nauchno-issl inst mat i mekh i Kharkovskogo mat obshch. 1927-. C of Khar MUZ. Kharkov. Narodniia komissariat prosveshcheniia USSR, gos univ im A M Gorkogo.

*Khar PI Tr     Trudy (Kharkovskogo) politek inst (im V I Lenina). 1952-. Kharkov.

*Khar U Za     Mems d l'univ imp d Kharkov. (Uch zap...; Charikov ann univ) 1874-. Kharkov.

Kibernet     Kibernetika. 1965-. Tr as Cybtics. Kiev. Akad nauk UkRSR, otd mat, mekh i kibernetiki.

*Kiev. See also Uk, Vych MK.*

Kiev OEZ     Zap Kievskogo obshch estestvoispitatelei. (Sb nauchnykh obshch est Kiev; Recueil d travaux sci d l soc nat Kiev; Mems d l soc d naturalistes d Kiev) 1870-1929// Kiev.

Kiev Ped     Naukovi (nauchnye) zap, Kiivskii derzhavnii ped inst im O M Gorkogo. 1939-. Kiev. Minist osviti UkRSR.

Kiev Pol     Izv Kievskogo ordena Lenina politek inst. (Visti; nauchno-info biull; soobshchenniya; nauchno-issl rabote) 1939-. Kiev.

Kiev Stu     Kiivskii derzhavniia univ im T G Shevchenka, studenski naukovi pratsi (raboty). (Travaux sci d etudiants; Studentecheskie nauchnye raboty) 1934-. Kiev.

*Kiev UFMO     Otchet i protokol fis-mat obshch pri imp Kievskom univ. (Trav d l soc phys-mat d l'univ imp d Kiev; Nac d kais Univ z Kiew; Kiew phys math Ges) 1891-1916// S to Kiev U Iz. Kiev.

*Kiev U Iz     Kiev univ izv. (Bull univ imp Kiev; Anzeiger Univ Kiew) 1861-1919, 1935-1936// Kiev.

Kiev Ukaz     Kiev univ ukazatel Russoi lit po mat. 1899-1905// Kiev.

Kiev UNZ     Kiev univ naukovi zap. 1935-1940, 1946-. Kiev.

*Kiev Vis     Visnik Kiivskogo (gos) univ im T G Shevchenka, ser ast, mat ta mek. 1958-. Kiev.

*Kiob, Kjob. See Dan.*

*Kishinev. See also Mold IM, Tiras Ped.*

Kish U     Uch zap Kishinevskii gos univ. 1949-. Kishinev. Komitet vishego i srednego spetsialnogo obrazovaniya Soveta minist Moldavskoi SSR.

Kitami CT      Mems of the Kitami Coll of Technology.  Kitami.
*Klein Ann.*   *See M Ann.*
Kl EW          Klassiker d exakten Naturw.  1853-.  See also Ostw Kl.
               Leipzig.  Ed: W Ostwald.
Kodai Sem      Kodai Math Seminar Repts.  1949-.  Tokyo.
*Koeln.*   *See Nor Wes AF.*
*Koeniglich privilegierte berlinische Zeitung.... See Voss Z.*
*Ko Hs Chi     Ko-hsueh chi-lu.  (Science Record)  1958-.  Peking.
Ko Hs TP       K'o hsueh t'ung pao.  (Kaciue tunbao; Kexue tongbao; Nauchnyi
               vestnik; Scientia; Sci Bull)  1950-.  Peking.  See also Sci Sin.
*Kolom Ped     Uch zap Kolomenskii (gos) ped inst, ser fiz-mat.  1956-.
               Kolomna.
*Kolonialsprachen.  See Afr USK.*
*Kolozsvar.  See Cluj.*
*Kondakova.  See Beo Ark.*
*Kongelige norske videnskabers selskabs skrifter.  See Tron Skr.*
Konig Gel      Schr d koenigsberger gelehrten Gesell, naturw Abt.  1925-.
               Berlin, Halle, Koenigsberg.
*Konig Ph      Schr d phys-oekonomischen Gesell z Koenigsberg.  (Schriften
               Koenigsberg)  1860-.  Koenigsberg.
*Konig U As    Ast Beobachtungen auf d K Universitaetssternwarte in Koenigs-
               berg.  1815-1899//  Koenigsberg.
*Koninklijke, etc.  See organization, city, region, country, etc.*
*Korea.  See also Kyung.*
*Korean R      Korean Repository.  1892-1898//  Seoul.
Korea ST       Choson minchuchuy inmin konkhvaguk kvakhakvon khakpo.
               (Vestnik akad nauk Koreiskoi narodno-demokraticheskoi respub-
               liki, ser estestvoznaniia tekh)  1966-.  Pyongyang.
*Kosmos        Kosmos.  (Cosmos, the J of the Polish Soc of Naturalists
               Founded in Honor of Copernicus)  1875-1939, 1947-.  Lvov.
*Koyre         Melanges Alexandre Koyre.  1964//  Paris.
*Koz M Lap     Kozepiskolai mat (es fiz) lapok.  1893-1915, 1925-1948, 1950-.
               Budapest.
Krak Ak Pr     Krakow akad gornicza zaklad, mat prace.  1926-1931//  Krakow.
Krak BI        Bull Intl of the Polish Acad of Sci in Krakow (Sect A, Math
               Sci).  (Bull intl d l'acad...; Anzeiger...; Wydziai mat
               przyrodniczy; Krakower Anz; Crc Ac Sc Bull)  1889-.  Krakow.
Krak CR        Wyd mat prz Pol akad um, CR.  1928-.  Krakow.
Krak Mm        Wyd mat prz Pol akad um, mems, sect A, math.  1929-.  Krakow.
*Krakow.  See also Kwar.*
Krak Pam       Wyd mat prz Pol ak um, pametnik.  (Krak Denkschr)  1874-1899//
               Krakow.
*Krak Roz      Rozprawy wyd mat prz Pol ak um.  (Krk ak mt-prz rz (& sp);
               Krak Abh; Krak Ber)  1874-1939, 1946-1952//  C of Krak RTN.
               Krakow.
Krak RTN       Rocznik towarzystwa naukowego z univ Krakowskim Polaczonego.
               (Krk roczn tow nauk; Krk roczn uniwers; Anns of the Sci Soc
               Univ of Krakow)  1817-1872//  C as Krak Roz.  Krakow.
Krak UM        Zeszyty naukowe uniwersytetu jagiellonskiego, prace, mat (fiz
               chem).  1955-.  Krakow.
*Kralcevska Akad.  See Beo.*
*Kralovsje Ceske spolecnosti nauk.  See Prag.*

*Kristal       Kristallografiya. 1956-. Tr as Sov Ph Cr. Moscow.
               Akad nauk SSSR.
*Kristiania (Oslo). See Chr, Nor, Oslo.*
*Kronecker J. See Crelle.*
*Krym. See Crim.*
Kuma Ed        Mems of the Fac of Edu, Kumamoto Univ. (Kumamoto daigaku
               kyoikugakubu kiyo) 1953-1961// C in Kuma Ed M. Kumamoto.
Kuma Ed M      Mems of the Fac of Gen Edu, Kumamoto Univ, Sect 1, Nat Sci,
               Math. (Kumamoto daigaku, kyoyobu kiyo, shizen kagaku hen,
               sugaku) 1966-. C from Kuma Ed. Kumamoto.
Kuma J         Kumamoto J of Sci, Ser A, Math, Phys, Chem. 1952-. Kumamoto.
*Kwar Fil      Kwartalnik filozoficzny. 1922-. Krakow. Akad umiejetnosci.
*Kwar HNT      Kwartalnik hist nauki i techniki. 1956-. Warsaw.
Kyb            Kybernetik. 1961-. Berlin.
Kyb Prag       Kybernetiki. 1965-. Prague. Ceskoslovenska akad ved.
*Kyo CS        Mems of the Coll of Sci, Kyoto Imp Univ, Ser A, Math.
               1914-1960// C of Kyo CSE. C as Kyo JM. Kyoto.
Kyo CSE        Mems of the Coll of Sci & Eng, Kyoto Imp Univ. 1903-1914//
               C as Kyo CS. Kyoto.
Kyo Eng        Mems of the Fac of Eng, Kyoto Univ. 1914-. Kyoto.
Kyo En RI      Tec Repts of the Engineering Res Inst, Kyoto Univ. (Kogaku
               kenkyush) 1951-. Kyoto.
Kyo JM         J of Math of Kyoto Univ. 1960-. C of Kyo CS. Kyoto.
Kyo UB         Bull of the Kyoto Gakugei Univ (Ser B, Math & Nat Sci).
               1951-. Kyoto.
Kyung MJ       Kyungpook Math J. 1958-. Taegu (Taikyu), Korea.
Kyush AM       Repts of the Res Inst for Appl Mechanics, Kyushu Univ.
               1952-. Fukuoka.
Kyush IT       Bull of the Kyushu Inst of Tec, Math, Nat Sci. (Kyushu
               kogyo daigaku) 1955-. Tobata.
Kyush UA       Mems of the Fac of Sci, Kyushu Univ, Ser A, Math.
               1940-. Fukuoka.


*Lancas HS     Procs & Papers of the Lancashire & Cheshire Hist Soc. (Lanc
               Hist S T; Lanc T Hist S) 1849-1854// Liverpool.
La Pla Con     Contribucion al estudio d l ciencias fis y mats, ser mat-fis.
               1914-1931// La Plata. Univ nac d La Plata.
*La Pla Pub    Publicaciones, fac d ciencias fis-mats. 1901-. La Plata.
               Univ nac d La Plata.
*Lardomshistoriska samfundets arsbok. See Lychnos.*
*Lar Men       Larousse mensuel (illustre). 1907-1939, 1948-1957// Paris.
Lat FT         Latvijas PSR zinatnu akad, vestis fiz & tehn zinatnu ser.
               (Izv akad nauk Latviiskoi SSR, ser fiz i tekh nau) 1947?-.
               Riga.
Lat IF         Latvijas PSR zinatnu akad, fiz inst, raksti. (Trudy inst
               fiz, akad nauk Latv SSR) 1947-. Riga.
*Latomus       Latomus. 1937-. Brussels.
*Lat U Rak     Latvijas univ raksti (ser mat). (Acta univ latviensis; Lat-
               vijas augstskolas raksti) 1921-. See also Riga UZR. Riga.

Lat UUZ            Zinatniskie raksti Latviiskii gos univ. (Uch zap) 1949-.
                   Riga.
*Lat Ves           Latvijas PSR zinatnu akad, vestis. (Izv akad nauk Latv SSR)
                   1947-. Riga.
*Latvia*.  *See also Herd, Riga, SSSR.*
*Lebensbi          Lebensbilder aus Kurhessen & Waldeck. 1830-1930// Marburg.
                   Hist Kommis fuer Hessen & Waldeck.
Lebensv            J d Collegiums f Lebensversicherungswissenschaft. 1868-1871//
                   Berlin.
Leeds PS           Procs of the Leeds Phil & Lit Soc, Sci Sect. 1925-. Leeds.
*Leghorn*.  *See Livo.*
*Leic LPS          Transas of the Leicester Lit & Phil Soc. (Leic S T)
                   1835-. Leicester.
*Leiden*.  *See Leyden; also Act Bioth.*
*Leip Ab           Abh d (K) saechsischen Akad (Gesell) d Wiss z Leipzig, math-
                   naturw (phys) Kl. (Leip Ab Mth Ps; Leip Mth Ps Ab; Abh Math
                   Phys Kl Sachs Ak W L) 1848-. See also Leip Ph Ab. Leipzig.
*Leip As           Vierteljahrsschr d ast Gesell. (Ast Viert; Vjschr ast Ges)
                   1866-. Leipzig.
*Leip Ber          Ber ueber d Verh d (K) saechsischen Gesell (Akad) d Wiss,
                   math phys Kl. 1846-1944, 1949-. See also Leip PH, Leip Si.
                   Leipzig.
*Leip Lit Z        Leipziger Literatur-Zeitung. 1812-1834// Leipzig.
Leip Mag R         Leipziger Magazin f reine & angew Math. 1786-1788// Leipzig.
                   Eds: J Bernoulli, K F Hindenberg.
*Leip MNR          Wiss Zt d Karl-Marx-Univ Leipzig, math-naturw Reihe. 1951-.
                   Leipzig.
*Leip PH           Saechsische Akad d Wiss z Leipzig, Ber ueber d Verh, philo-
                   logisch-hist Kl. 1849-1944, 1949-. See also Leip Ber.
                   Leipzig.
*Leip Ph Ab        Abh, saechsische Akad d Wiss z Leipzig. 1850-. See also
                   Leip Ab. Leipzig.
Leip Sem           Abh aus d math Seminar d hansischen Univ. Leipzig.
Leip Si            Sitz d saechsischen Akad d Wiss z Leipzig, math-(naturw/phys)
                   Kl. 1849-1945, 1948-. See also Leip Ber. Leipzig.
*Leipzig*.  *See also Acta Erud, Deu.*
*Lemberg*.  *See also Lvov.*
*Lemb Sam          Zb mat-prir-likarckoi sek naukovogo tovarictva im Shevchenka.
                   (Sammelschr d math-naturw-aerztlichen Sect d Sevcenko-Gesell
                   d Wiss i Lemberg) 1897-. Lvov.
*Lemb Si           Sitz d math-naturw-aerztlichen Sekt, Ukraini Sevcenko-Gesell
                   d Wiss. (Procs, mat-pryrodopysno-likarska sekt, naukove
                   tovarystvo im Shevchenka) 1917-. Lemberg, NY.
Len F Kh           Vestnik Leningradskogo univ, ser fiz, khim. 1956-. Pt C of
                   Len Vest. Leningrad.
Len FMO            Leningrad fis-mat obshch, zh. 1926-1930// C of Pet MO.
                   Leningrad.
Len FN             Uch zap Leningradskii gos univ, ser fiz nauk. 1935-.
                   Leningrad.
*Len IINT          Trudy inst istorii nauki i tek. (Archiv, ist nauk tekhn, akad
                   nauk SSSR) 1933-1936// C of Len KIZ. Moscow, Leningrad.

*Leningrad Bulletin. See SSSR Iz.*

| | |
|---|---|
| Len KIZ | Trudy Kommissiia po ist znaniya. 1927-1930// C as Len IINT. Leningrad. |
| Len MV | Vestnik Len univ, ser mat, mek i ast. Pt C of Len Vest. 1956-. Leningrad. |
| *Len MZ | Uch zap Len gos univ, ser mat nauk. (Len State Univ Annals of Math) 1936-. Leningrad. |
| Len Ped | Leningradskii gos ped inst im A I Gerchena, uch zap. 1932-. Leningrad. |
| *Len Sb Gf | Geofiz sb. (Recueil d geophys, obs phys centrale d Russie) 1914-1930// Leningrad. |
| Len TI | Trudy Len tek inst im Lensoveta, trudy kafedr mekh fak. 1934-. Leningrad. |
| Len Vest | Vestnik Leningradskogo univ. 1946-1956// C in Len F Kh, & in Len MV. Leningrad. |
| *Len ZKV | Zap kollegii Vostokovedov. 1925?-. Leningrad. Akad nauk SSSR. |

*Leodiensis. See Lieg.*

| | |
|---|---|
| *Leop | Leopoldina, Mitt (amtliches Organ) d K Leopoldino-Carolinischen deu Akad d Naturforscher. 1859-1930// C in Leop NA. Dresden, Halle. |
| Leop MC | Miscellanea curiosa acad Caesarea-Leopoldina-Carolina. 1671-. Jena, etc. |
| *Leop NA | Nova acta acad Caesareae Leopoldina-Carolinae germanicae naturae curiosorum. (Verh (Abh) K Leop-Carol deu Akad d Naturf, etc) 1757-. Halle. |
| *Leshonen | Leshonenu. 1928-. Tel Aviv, etc. Va'ad haloshon haivrit be Eretz Yisrael. |
| *Let Fr Par | Lettres francaises. 1942-. Paris. |
| *Leyd ALB | Anns academiae Lugduno-Batavia. (Leijd A Ac) 1815-1875// Leiden. |
| *Leyd Ex OL | Jaarbericht (joarboek), ex oriente lux. 1933-. Leiden. Vooraziatisch-egyptisch gezelschap. |
| *LHL Bib | Linda Hall Lib Bibliographic Bull. 1950-. Kansas City, Mo. |
| *Lib Chron | Library Chronicle. 1933-. Philadelphia. Friends of the Univ of Penn Lib. |
| *Lib Civ | Libruni civitas, rassegna d attivita municipale. 1928-. Livorno. |
| *Lieg Bul | Bull d l soc R d scis d Liege. 1932-. Liege. |
| *Lieg Mm | Mems d l soc R d scis d Liege. 1843-. Brussels. Fondation univ d Belgique. |
| *Life Let | Life & Letters, & the London Mercury. 1928-1950// London. |

*Ligure. See Geno.*
*Ligustica. See Geno.*

| | |
|---|---|
| *Lim Ac | Actas d l acad nacl d ciencias (exactas, fis y nat). 1938-. Lima. |
| *Lim Rev | Rev d ciencias. (Rev cienc Lima) 1897-. Lima. |
| Lim U | Cuadernos d estudio, univ nacl d ing, inst d mat puras y aplic. 1963-. Lima. |
| Lin Alg Ap | Linear Algebra & Its Appl. 1968-. American Elsevier Pub Co. |
| *Linc At | Atti d R accad d Lincei. (Acc R d L; Rom Acc L) 1870-1883// C as Vat NLA, Linc Mr, Linc Mor, Linc Tr. Rome. |

*Lincei*. *See also It XL, Vat NL.*

| | |
|---|---|
| *Linc Mor | Atti d R accad (naz) d Lincei, cl d sci morali, memorie. 1876-1939// C of Linc At. Rome. |
| *Linc Mr | Atti d accad naz d Lincei, cl d sci fis, mat e nat, memorie. 1876-1939// C of Linc At. C as It Ac At, Linc Mr M, Linc Mr FC. Rome. |
| Linc Mr FC | Atti d accad naz d Lincei, mem, cl d sci fis mat e nat, sezione IIa, fis, chim, geol, paleontologia, e mineralogia. 1947-. C of Linc Mr. Incl in It Ac At 1940-1944. Rome. |
| *Linc Mr M | Atti d accad naz d Lincei, mem, cl d sci fis mat e nat, sezione I a, mat, mec, ast, geodesia e geofis. 1946-. C of Linc Mr, It Ac At. Rome. |
| *Lincolns | Lincolnshire Mag. 1932-1939// Lincoln. Lindsey Local Hist Soc. |
| *Linc Rn | Atti d R accad naz d Lincei (d Italia), rendiconti, cl d sci fis mat e nat. (Rom Acc L Rend) 1884-1891, 1925-1939, 1946-. C of Linc Tr Incl in It Ac Ren 1939-1943. Rome. |
| *Linc Rn Mo | Accad naz d Lincei, cl d sci morali, storiche e filologiche, rendiconti. 1892-1929// Rome. |
| Linc Sol | Atti d accad naz d Lincei, rend d adunanze (sedute) solenni. 1884-. Rome. |
| Linc Tr | Atti d R accad d Lincei, transunti. (Rm R Ac Linc T) 1877-1884// C of Linc At. C as Linc Rn. Rome. |
| Ling USB | Sci Bull, Lingnan Univ. 1930-. C of Act Br Si. Canton. |
| *Liouville J.* | *See JMPA.* |
| *Lisb A | Univ d Lisboa, rev d fac d ciencias, ser A, ciencias mats. 1950-. Lisbon. |
| Lisb Ann | Annaies d scis e lets, sci math, phys, hist-nat e med. (A Sc) 1857-1859// Lisbon. Acad R das sciencias. |
| Lisb Arq | Arquivo d inst Gulbenkian d ciencia, seccao A, estudos mat e fis-mat. 1963-. Lisbon. |
| Lisb Bol | Boletin d acad d scis d Lisboa. 1929-. C of Lisb JS. Coimbra. |
| Lisb H Mm | (Hist e) mems d acad R d scis d Lisboa. (Lisb Ac Sc Mm; Lisb Mm; etc) 1780-. Lisbon. |
| *Lisb ISA | Anais d ist sup d agronomia. 1920-. Lisbon. Univ tecnica. |
| *Lisb JS | J d scis math, phys e nat. 1866-1927// C as Lisb Bol. Lisbon. Acad R das sciencias. |
| *Lisb Mr | Mems d acad d ciencias d Lisboa, cl d ciencias. 1780-. Lisbon. |
| *Listener | Listener. 1929-. London. |
| Lit B | Lietuvos TSR mokslu akad, darbai, ser B. (Trudy akad nauk litovskoi SSR) 1955-. C of Lit V. Vilna. |
| *Literaturzeitung der ZM Ph.* | *See ZM Ph.* |
| *Lithuania.* | *See also Kaunas, SSSR, Viln.* |
| *Lit M Sb | Litovskii mat sbornik. (Vischie uchebnie zavedenkya Litovskoi SSR) 1961-. Vilna. |
| *Litoral.* | *See Arg IM Ros.* |
| *Lit UM | Lietuvos univ mat gamtos fak darbai. (Mems, fac sci, univ Lithuanie) 1922-. Kaunas. |
| Lit V | Lietuvos TSR mokslu akad, Vilna vestnik, zinynas. 1947-1952// C as Lit B. Vilna. |

| | |
|---|---|
| *Livo NAPS | Pubblicazioni sci a cura d accad navale. 1947-. Livorno. |
| *Livo Rv | Riv d Livorno, rassegna d attivita municipale a cura d comune 1952-. Livorno. |
| Ljub IM | Univ of Ljubljana, Inst of Math, Phys, & Mech, Pubs of the Dept Math. 1964-. Ljubljana. |
| *Ljubljana.* | *See also Matica Sl, Slov MFT.* |
| LMSB | Bull of the London Math Soc. 1969-. London. |
| *LMSJ | J of the London Math Soc. 1865-. London. |
| *LMSP | Procs of the London Math Soc. 1865-. London. |
| Lodz NMP | Zeszyty naukowe univ Lodzkiego, nauki mat prz, ser II. 1955-. Lodz. |
| Lodz SSL | Bull d l soc d scis & d lets d Lodz, cl III, scis math & nat. (Lodzkie towarzystwo naukowe, wydzial III, nauk mat-prz) 1946-. Lodz. |
| *Log An | Logique & analyse. 1958-. C of Bel Log. Louvain. |
| Logos Nap | Logos, riv internaz d filosofia. 1914-. Naples. |
| *Logos Pal | Logos, organo d bibt fil d Palermo. 1918?-. Naples. |
| Logos Rom | Logos, riv trimestrale d fil e d storia d filosofia. 1914-. Florence, Naples, Rome. |
| *Logos Tub | Logos, intl Zt f Phil d Kultur. 1910-1933// Tuebingen. |
| *Lom Gen | Ist Lombardo (accad) d sci e let, rend (parte gen e atti ufficiali). (Lomb Ist Rend) 1864-. Milan. |
| *Lom Mr | Mem d (R) ist Lombardo, accad d sci e let, cl d sci mat e nat, 1867-. Milan. |
| *Lom Rn M | Ist Lombardo, accad (R ist) d sci e let, rend (cl d sci mat e nat), sci mat, fis, chim, & geol, A. 1868-. Milan. |
| *Lon Actu | J of the Inst of Actuaries. (I Act J) 1850-. London. |
| Lon Aer | J of the R Aeronautical Soc. 1923-. C of Aer J. London. |
| *Lon Anthr | J of the R Anthropological Inst of GB & Ireland. 1871-. London. |
| *Lon Asia | J of the R Asiatic Soc. (As S J) 1834-. London. |
| *Lon Asia T.* | *See As So GB.* |
| Lon As Mm | Mems of the R Ast Soc. (As S Mm) 1822-. London. |
| *Lon As Mo N | Monthly Notices of the R Ast Soc. (As S M Not; Monthly Not) 1827-. London. |
| Lon ASON | Occasional Notes of the R Ast Soc. 1938-1941, 1947-1959// C as Lon As QJ. Incl in Obs 1941-1947. London. |
| *Lon As QJ | Quar J of the R Ast Soc. 1960-. C of Lon ASON. London. |
| *London.* | *See also LMS, Phil T.* |
| *London Computer Group Bull.* | *See Comp B.* |
| *London, Edinburgh, & Dublin Phil Mag & J of Sci.* | *See Phi Mag.* |
| *Lon ICE | Minutes of Procs of the Inst of Civil Engs.... (CE I P) 1837-1935// C in Lon ICEJ. London. |
| Lon ICEJ | J of the Inst of Civil Engs. 1935-. C of Lon ICE. London. |
| Lon IME | Procs (J) of the Inst of Mech Engs. 1847-. London. |
| *Lon Math Soc.* | *See LMS.* |
| *Lon RIGBP | Procs of the R Inst of GB. 1929-. C of Lon RIP. London. |
| *Lon RIP | Notice of the Procs at the Meetings of the Members of the R Inst.... (R I P) 1851-1928// C as Lon RIGBP. London. |
| *Lon Royal Society.* | *See also Phil T.* |
| Lon RSA | Procs of the R Soc, Ser A, Math & Phys Sci. 1905-. London. |

| | |
|---|---|
| *Lon RSBM | Biog Mems of the Fellows of the R Soc. 1955-. C of Lon RS Ob. London. |
| *Lon RSNR | Notes & Records of the R Soc of London. 1938-1941, 1945-. London. |
| *Lon RS Ob | Obituary Notices of Fellows of the R Soc. 1901-1904, 1932-1954// C as Lon RSBM. London. |
| *Lon RSP | Procs of the London R Soc. (R S P) 1854-. London. |
| Lon RSY | Yearbook of the R Soc of London. (R S Yearbook) 1896/97-. London. |
| *Lon SOAS | Sch of Oriental (& African) Studies, Bull. 1917-. London. Univ. |
| *Lon St A | J of the R Stat Soc, Ser A (Gen). 1838-1942, 1946-. London. |
| *Lon St B | J of the R Stat Soc, Ser B (Meth). 1838-. Incl C of Lon St Sup. London. |
| Lon St Sup | Suppl to the J of the R Stat Soc (Ind & Ag Res Sect). 1934-1941, 1945-1947// C in Lon St B. London. |
| *Lon Tim | London Times. 1785-. London. |
| *Lon Tim LS | London Times Literary Supplement. 1902-. London. |
| *Loria | Boll d bibliografia e storia d sci mat. 1898-1919// C of Bo SBM. C in Bo Fir. Parts in Batt. Turin, Geneva. |
| *Louvain. | See also Rv Neo. |
| *Louv HPRT | Recueil d travaux d'hist & d philologie. (Confs d'hist & d p) 1890-. Louvain. Univ catholique d Louvain. |
| *Louv ISPA | Anns inst sup d philosophie. 1912-1924// Louvain. Univ catholique, ecole St T d'Aquin. |
| Louv Ped | Bull d cercle ped d l'univ cath d Louvain. 1946-1956?// Louvain. |
| *Lub A | Anns univ Mariae Curie-Sklodowska, sect A, math. (Rczniki uniwersytetu Marii C-S, dzial A, mat) 1946-. Lublin. |
| Lub AA | Anns univ M Curie-Sklodowska, sect AA, phys & chem. 1953-. Lublin. |
| *Lucca | Atti d R accad Lucchese d sci, let, & arti. (Lucca At Ac) 1821-. Lucca. |
| *Luftf | Luftfahrt-Forschung. 1928-1932, 1933-. Munich, Berlin. Zentrale f tec-wiss Berichtswesen ueber Luftfahrtforschung. |
| *Lugduno-Batavia Acad. | See Leyd. |
| Lund Af | Lunds akademiens (akademis) afhandlingar. 1832-. Lund. |
| *Lund F | K fisiografiska sallskapets i Lund forhandlingar. (Procs of the R Physiographic Soc at Lund; Lund Univ Arsskr) 1931-. C of Lund UA. See also Lund U, Lund UMMS. Lund. |
| Lund Sem | Meddelanden fran Lunds univ mat sem. (Communications d sem math d l'univ d Lund) 1933-. Lund. |
| *Lund U | Acta univ Lundensis. (Lunds univ arsskrift, andra avdelningen) 1864-1904// C as Lund UA. Incl K fis sallsk Lund handlingar 1889-1907. Lund. |
| Lund UA | Lund univ aftdeining, aftdeining 2, med samt mat och naturvet. 1905-1924// C of Lund U. Lund. |
| Lund UMMS | Acta univ Lundensis, sect 2, med, math, sci rerum nat. 1964-. Lund. |
| *Lux Arch | Archives d l'inst grand-ducal d Luxembourg, sect d scis nat, phys & math. 1906-. Luxembourg. |

Luxembourg.  *See also Benelux, Hemecht.*

*Lux SNBM    Bull mensuel, soc d naturalistes luxembourgeois.  1907-.
             C of Lux SNCR.  Luxembourg.

Lux SNCR     CR d seances d l soc d naturalistes luxembourgeois.  1891-
             1906//  C as Lux SNBM.  Luxembourg.

*Lux SNM     Pub d l'inst R grand-ducal d Luxembourg, sect d scis nats
             et maths.  (Lux I R Pb)  1870-.  Luxembourg.  Soc des sci
             nat.

Lux SSN      Soc d scis nats d grand-duche d Luxembourg.  (Lux S Sc Mm;
             Lux S Sc Nt)  1853-1869//  Luxembourg.

Lvov.  *See also Lemb.*

Lvov Ped     Naukovi zap, Lvivskii derzhavnii ped inst.  (Nauchnye zap)
             1946-.  Lvov.

Lvov PID     Doklady, Lvovskii politek inst, ser fiz-mat.  (Dokl mat;
             dokl fiz)  1955-.  Lvov.

Lvov PINZ    Nauchnye zap, Lvovskii politek inst, ser fiz-mat.  1947-.
             Lvov.

Lvov UDP     Dopovidi ta povidomlennia, Lvivskii derzhavnii univ im
             I Franka.  (Inf Repts, Franka State Univ of Lvov)  1949-.
             Kharkov.  Minist kulturi UkRSR.

Lvov UNZ     Naukovi zap, Lvivskii derzhavnii univ im I Franka.  (Lvov
             univ uch zap)  1946-.  Lvov.

*Lvov U Vis  Visnik Lvivskogo ordena Lenina derzavnogo univ im Ivana
             Franka, ser mekh-mat.  1965-.  Lvov.

*Lychnos     Lychnos, lardomshistoriska samfundets arsbok.  1936-.
             Uppsala, Stockholm.  Swedish Hist Sci Soc.

*Lyon A      Annales d l'univ d Lyon, sect A, sci math & ast.  1891-.
             Lyon.

Lyon An      Annales d sci, phys & nat, d'ag, & d'industrie.  (Lyon A S Ag)
             1838-.  Lyon.

Lyon B       Annales d l'univ d Lyon, sect B, sci phys & chim.  1891-.
             Lyon.

Lyon Mm      Mems d l'acad d scis, b-l & arts d Lyon, cl d scis.  (Lyon
             Ac Mm (Sc))  1845-.  Lyon, Paris.


M (mathematical, mathematics).  *See also Mat, Math.*

Maa EM       Maanedsskrift for d elem math.  1886-1889//  C as Nyt Tid.
             Copenhagen.

Maa ML       Maandelykse math liefhebbery.  1754-1769//  Holland.

Maanblad, etc.  *See organization, city, region, country, etc.*

*MAA Stu     Math Asso of Amer Studies in Math.  1962-.  Englewood Cliffs,
             NJ.  Prentice Hall.

M Ab Mod     Math Abh aus d Verlage math Modelle von Martin Schilling.
             1899-1913//  Leipzig.

Macedonia.  *See Mak, Skopj.*

Mad IJJ      Mems d mat d inst Jorge Juan.  1946-.  Madrid.  Consejo sup
             d investigaciones cientificas.

*Mad Mm     Mems d l R acad d ciencias exactas, fis y nats d Madrid, ser ciencias exactas. 1850-. Madrid.

Madra B     J of the Madras Univ, B, Contrs in Math, Phys & Biol Scis. 1943-1957// Madras.

*Mad Rv     Revista d l R acad d ciencias exactas, fis y nats d Madrid. 1904-1935; 1940-. Madrid.

Mad UM     Univ d Madrid, pubs d l sect d mat d l fac d ciencias. 1956-. Madrid.

*Mag Nat     Magazin for naturvidenskaberne. (Mg Ntvd) 1823-1836// Christiania, Lund.

Magnet     Magnetohydrodynamics. 1966-. Faraday P.

Magy Alk     Magyar tudomanyos akad alkamazott, mat intezetenek kozlemenyi. 1952-1954// C as Magy MK. Budapest.

*Magyar. See also Hungary.*

Magy Evk     A Magyar (tudos tarasag) tudomanyos akademia, evkonyvei. 1833-. Budapest.

*Magy MF     A Magyar tudomanyos akad mat (termeszet)(es fiz) tudomanyok osztalyanak kozlemenyei. 1951-. Budapest.

Magy MK     A Magyar tudomanyos akad mat kutato intezetenek kozlemenyei. (Publication; Trudy) 1956-. C of Magy Alk. Budapest.

*Magy MT     Magyar tudomanyos akad ertesito, a math es termeszettudomanyi osztalyok kozlonye. (Rept of the Hungarian Acad, Communications of the Math & Nat Sci Sects; Mag ak ets math term; Ber ung Akad d Wiss) 1840-. Budapest.

*Maharaja Sayajirao Univ. See Baroda.*

*Maia     Maia, rivista d letterature classiche. 1948-. Rome, Messina.

*Mainz MN     Akad d Wiss & d Lit in Mainz, Abh d math-naturw Kl. 1950-. Wiesbaden.

*Majallat kulliyat.... See Alex.*

Majocc     Annali d fis, chim, & mat.... (Majocchi A Fis C) 1841-1850// Milan.

*Mak DMF     Bull d l soc d mathns & d physns d l repub pop d Macedoine. (Bilten na drustvota na math i fiz od narodna repub Makedonija) 1950-. Skopje.

*Man     Man. 1901-. London.

*Manchuria. See Ryojun.*

*Manc JRL     John Rylands Lib, Manchester, Bull. 1903-. Manchester.

*Manc Mr     Mems (& Procs), Manchester Lit & Phil Soc. (Manch Ph S Mm) 1785-. Manchester.

Manc NHMJ     Monthly J of Hist, Folklore..., Math, Metaphysics,.... 1889-. Manchester, NH.

*Manc Pr     Procs of the Lit & Phil Soc of Manchester. (Manch Lt Ph S P) 1857-1887// Manchester.

*Mangt Sci     Management Sci. 1954-. Baltimore, Md; Providence, RI. Inst of Management Sci.

*M Ann     Mathematische Annalen. (Klein Ann; Clebsch An) 1869-. Berlin, Goettingen, etc.

Manu M     Manuscripta mathematica. 1969-. Berlin.

*Manuscr     Manuscripta. 1957-. St Louis, Mo. Univ Lib.

*Marb Si     Sitz d Gesell z Beforderung d gesamten Naturw z Marburg. 1866-. Marburg.

*Marb U Mit      Mitt Universitaetsbund. 1930-. Marburg.

*Marie Curie-Sklodowska Univ. See Lub.*

*Mars Ann      Annales d l fac d scis d Marseille. (Pubs d l'univ d'Aix-Marseille) 1891-. Marseille.

Mars M      Bull math d facs d scis & grandes ecoles. (Pubs d l fac d scis d Marseille) 1934-. Bar-le-Duc.

*Martin-Luther-Univ. See Halle.*

*Masaryk Univ. See Brno.*

*Masc Buch      Die Maschinen-Buchhaltung, Anregung f leitende Personen zu zweckmaess Verwendung & Buchhaltungsmaschinen. 1926-. Thueringen.

*Matematikai es fizikai lapok. See MP Lap.*

*Matematikai es termeszettudomanyi ertesitoe. See M Ter Er.*

Mathcian      The Mathematician. 1843-1850// London. Ed: Davies, etc.

Mathc Lon      The Mathematician. 1745-1751// London.

*Math Cluj      Mathematica. 1929-. Incl Math Tim. Cluj. Sem d mat al univ Cluj. (Soc d st mat si fiz din RPR, filiala Cluj)

*Mathematica. See also Zutphen.*

*Mathematical, mathematics, etc. See M, Mat, Math.*

*Mathematical Reviews. See Math Rev, MR.*

*Mathematikunterricht. See Mathunt, M Obraz.*

*Mathematisch Centrum Amsterdam. See Amst MC.*

*Mathematisch-naturw Mittheilungen.... See Wurt MN.*

*Mathes F      Atti d soc ital d sci fis e mat Mathesis. 1909-. See also Mathes It. Bologna, Padua.

*Mathesis      Mathesis, recueil math a l'usage d ecs sps & d etablissements d'instruction moyenne. 1881-1916, 1920-. Paris, Ghent.

*Mathes It      Bol d (associazione) Mathesis. 1896-1920// Merged with Pe MI 1899-. See also Mathes F. Rome. Soc ital d mat.

*Mathes Po      Mathesis polska. 1926-1938// Warsaw.

Mathes Ti      Mathesis, tydschr voor wiskunde. 1928-1931// The Hague.

*Mathes Un      Mathesis universalis, Quellenschr z Geistgesch d exacten Wiss & d Kuenste. 1948// Halle.

Mathika      Mathematika, a J of Pure & Appl Math. 1954-. London.

*Math Mngl      Mathematica monongaliae, Mathematical Papers. 1961-. Morgantown, W Va. Univ of W Va.

Math N      Math Notes (of the Acad of Sci, USSR). 1967-. Tr of Mat Zam. Consultants Bur.

*Math Nat Anz. See M Ter Er.*

Math Pera      Mathematica. 1965-. Peradeniya. Ceylon Univ Math Soc.

*Math (Phil) Repository. See Enig, Enqu, Gen D.*

Math Rev      The Mathematical Rev. 1896-1899// Worcester, Mass.

*Math Scan      Mathematica scandinavica. 1953-. C of M Tid B. Copenhagen.

*Math Tim      Mathematica, Timisoara. Part of Math Cluj.

*Mathunt      Der Mathematikunterricht, Beitraege z seiner wiss & methodischen Gestaltung. 1955-. Stuttgart.

*Matica Sl      Matematica slovenska v Ljubljani, zb. (Zbornik znastvenih iu poucnih spisov; Slovenska sb....) 1899-. Ljubljana. Na svetlo daje slovenska.

*Matiche      Le matematiche. 1945-. Catania.

*Mat PA      Le matematiche (pure ed appl). 1901-1903// Citta d Castello.

| | |
|---|---|
| Mat Per | Matematika periodicheskii sb perevedor inostrannykh statei. 1957-. Moscow. |
| Matr Ten Q | The Matrix & Tensor Quar. 1950-. London, Kent. Tensor Club of GB. |
| Mat Schul | Math Schuelerzeitschrift. (Volk & Wissen) 1967-. Berlin. |
| *Mat v Shk | Matematika v shkole. 1937-1941, 1946-. Pt C of MF v Shk. Leningrad, Moscow. |
| Mat Zam | Mat zametki. 1967-. Tr as Math N. Moscow. Akad nauk SSSR, otd mat. |
| Max PI | Mitt aus d Max-Planck-Inst f Stroemungsfors. 1950-. Goettingen. |
| M Biosc | Math Biosciences (An Intl J). 1967-. NY. Amer Elsevier Pub. |
| M Cas | Matematicky casopis. 1967-. C from MF Cas. Bratislava. Slovenska akad ved. |
| *McGraw-Hill.* | *See MH.* |
| M Compan | The Mathematical Companion. 1828-1832// NY. |
| M Comp | Math of Computation. 1960-. C of M Tab OAC. Wash DC. |
| M Corresp | The Mathematical Correspondent Quar. 1804-1807// NY. |
| M Diary | The Math Diary, Containing New Res & Improvements in the Math. 1825-1832// NY. |
| Meccan | Meccanica, J of the Ital Asso of Theor & Appl Mech. 1966-. Pergamon P. |
| *Mech Mag | Mechanics' Mag, Mus, Register, J, & Gazette. (Iron) 1823-1893// London. |
| Mec Tr | MT, Mechanical Translation, Devoted to the Tr of Langs with the Aid of Machines. 1954-. Cambridge. |
| *Med. See Ab G Med, etc; also organization, city, region, country, etc.* | |
| *Mededelingen. See organization, city, region country, etc.* | |
| *Medvl St | Mediaeval Studies, a J of the Thought, Life, Letters & Culture of the Middle Ages. 1939-. Toronto. Univ of Toronto. |
| *Melanges. See also Koyre.* | |
| *Melanges | Melanges mathematiques d l'acad d scis d St Petersburg. 1849-. St Petersburg. |
| Melbrne | Procs of the R Soc of Victoria, NS. 1888-. Melbourne. |
| M El Mad | Matematica elemental, rev dedicada a l estudiantes d mat. 1932-. Madrid. Soc mat espanola, sem mat d consejo sup d investigaciones cientificas. |
| *M El Rm | Matematica elementare. 1922-1924// Rome. |
| *Memoires d'agriculture.... See Fr Soc Ag; also region, city, etc.* | |
| *Memoirs, memoires, memorias, etc. See also organization, region, etc.* | |
| *Memorial d l'office national meteorologique d France. See Fr ON Met.* | |
| *Memorie della R accad delle scienze, etc, Naples. See Nap FM At.* | |
| Mem SM | Memorial d scis mathematiques. 1925-. Paris. |
| Mem SP | Memorial d scis physiques. 1928-. Paris. |
| Mengelw | Mengelwerk van uitgeleezene en andere wis- en natuurkundige verhandelingen. (Amst Mengelwerk; Mengelwerk Wisk Vh) 1796-1816// Amsterdam. Genootschap d math wetenschappen. |
| Mens Est | Mensario d estatistica da producao. C as Rv Ec Est. Rio de Janeiro. |
| *Merchist | The Merchistonian. 1873?-. Edinburgh. Merchiston Castle School. |

*Messager d phys exp & d math elem.* See *Kagan*.
| | |
|---|---|
| *Mess M | Messenger of Mathematics. (Me; Mess Math) 1872-1929// C of Ox CD. C in QJM Ox. London, Cambridge. |
| Meteo | La meteorologie. 1854-. Paris. |
| Meteo Z | Meteorologische Zt. 1884-1944// Vienna. Gesell f M u d deu M Gesell. |
| *Methodos | Methodos, rev trimestrale d metodologia e d logica simbolica, 1949-. Milan. Centro ital d metodologia e anal d linguaggic |
| *Metrika | Metrika. 1958-. C of Mit M Stat, & of Sta Vier. Wuerzburg, |
| Metroec | Metroeconomica, rev intl d economica. 1949-. Bologna. |
| *Metron | Metron, Intl Rev of Statistics. 1920-. Rome. |
| *Metz | Mems d l'acad (R/imp) d Metz. (Metz Ac Mm) 1819-. Metz. |
| M Ex | Math Exercises. 1750-1753// London. |
| *Mex Alz Mm | Mems (y rev) d l soc cientifica Antonio Alzate. (Mex Soc Alzate) 1887-1931// C as Mex Alz MR. Mexico City. |
| Mex Alz MR | Mems y rev d l acad nacional d ciencias Antonio Alzate. 1931-. C of Mex Alz Mm. Mexico City. |
| Mex Arc CS | Archives d l commis sci d Mexique. (Mex Arch Com Sc) 1865-1869// Paris. Min d l'instr publique. |

*Mexico.* See also *Rv Mex*.
| | |
|---|---|
| Mex IM | Anales d inst d matematicas. 1961-. Mexico City. Univ nac (autonoma) d Mexico. |
| M Exp | Mathematical Expositions. 1940-. Toronto. |
| *Mex SM | Bol d l soc matematica Mexicana. 1943-. Mexico City. |
| *MF Cas | Matematicko-fyz casopis. 1953-. C of MF Sb. Pt C as M Cas. Bratislava. Slovenska akad ved. |
| M For | Mathematics Forum. 1958-. Chicago, Ill. |
| MF Sb | Matematicko-fyz sbornik. 1951-1953// C as MF Cas. Bratislava. |
| MF Sof | Mat i fiz. 1958-. Sofia. Minist na prosvetata i kulturata. |
| MF v Shk | Mat i fiz v (srednei) shkole. 1934-1936// Pt C as Mat v Shk. Moscow, Leningrad. |
| *M Gaz | The Mathematical Gazette. 1894-. London. |
| M Geom Phi | Math, Geom, & Phil Delights. 1792-1798// London. |
| Mgt Sc | Management Sci, J of the Inst of Management Science. 1954-. Philadelphia, Penn. |
| *MH En ST | McGraw Hill Enc of Sci & Technology. 1966// NY. |
| MHMMS | McGraw-Hill Modern Men of Science. 1966// NY. |
| Mich Hu | Univ of Michigan Studies, Humanistic Ser. 1904-. Ann Arbor. |
| *Mich MJ | The Michigan Math J. 1952-. Ann Arbor, Mich. |
| Mich Sc | Univ of Mich Studies, Sci Ser. 1914-. Ann Arbor, Mich. |

*Milan.* See also *Lom*.
| | |
|---|---|
| Milan At | Atti d (ist R/ist lombardo) d sci, let, ed arti. (Mil At I Lomb) 1858-1864// Milan. |
| Milan Gio | Gior d I R ist lombardo d sci, let, ed arti. (Mil G I Lomb) 1841-1856// Milan. |
| *Milan Sem | Rend d sem mat e fis d Milano. 1927-. Milan. |
| *Milan USC | Univ cattolica d sacro cuore, pubblicazioni, (1) scienze filosofiche. 1923-. Milan. |
| *Milit Eng | Military Engineer. 1920-. Wash DC. |
| *Mind | Mind, a Quar Rev of Psych & Phil, NS. 1892-. London, NY. Mind Association. |

*Minn CR        Minnemath Cen Repts. 1962-1969// Minneapolis. Univ of Minn.
M in Schul      Mathematik in der Schule. 1963-. Berlin. Volk & Wissen
                Volkseigner.
*Minsk. See BSSR.*
*Miscellanea Berolinensia. See Ber Misc.*
*Mit Artil      Mitt ueber Gegenstaende d Artillerie- & Genie-Wiss.
                1870-1919// Vienna.
*Mitdeu Lb      Mitteldeu Lebensbilder. 1926-1930// Magdeburg. Hist Kommi
                f d Provinz Saechsen....
*MITM Cont      Mass Inst of Tec Dept of Math Contributions. 1920-1932//
                Cambridge, Mass.
*Mit M Stat     Mitteilungsblatt f math Stat. 1949-1957// C in Metrika.
                Munich.
*Mit R Land     Mitt d Reichamts f Landesaufnahme. Berlin.
*Mitteilungen. See also organization, city, region, etc.*
Miyaz En        Mems of the Fac of Eng, Miyazaki Univ. 1956-. Miyazaki.
*Mizraim        Mizraim, J of Papyrology, Egyptology, Hist of Anc Law &
                Their Relation to Civs of Bible Lands. 1933-. NY.
M Jp            Mathematica japonicae. 1948-. Kobe.
*MLAAP          Mod Lang Asso of Amer Pubs. (PMLA) 1884-. NY.
*M Lap          Matematikai lapok, Bolyai Janos mat tarsulat. 1949-. C of
                MP Lap. Budapest.
M Lieb          Der math Liebhaber. 1767-1769// Hamburg.
M List          Matematicheskii list. 1879-1882// Moscow.
*M Mag          Mathematics Mag. (Natl M M; NMM) 1934-1945, 1947-. C of
                M News Let. Pocoima, Cal; Buffalo, NY.
*M Mag Mart     Mathematical Mag. 1882-1910// Erie, Penn; Wash DC.
                Ed: Artemus Martin.
M Mag Mitc      Math Mag, Mitchell & Moss. 1761// London.
M Mess          The Mathematical Messenger. (Sch Messenger) 1884-1901//
                Tyler, Tex. Ed: G H Harvill.
M Misc          The Mathematical Miscellany. (Mth Misc) 1836-1839// NY.
M Mo Lehr       Mathematisches Monatsblatt f Lehrer & andere Freunde d Math.
                1832-1833// Elberfeld. Ed: Schurmann.
*M Mo Runkl     The Mathematical Monthly. (Camb M Mth M; Camb U S Mth M)
                1858-1861// Cambridge, Mass; London. Ed: J M Runkle.
*M Nach         Mathematische Nachrichten. 1948-. Berlin.
*MN Bl          Mathematisch-naturw Blaetter. 1904-1933// Berlin, etc.
*MNB Ung        Math & naturw Ber aus Ungarn. (Math Naturw Ung; Ungar Ber)
                1882-1931// Ger ed of M Ter Er. Berlin, Budapest, Leipzig.
*M News Let     Mathematics News Letter. 1926-1934// C as M Mag.
                Baton Rouge, La. La State Univ P.
*MN Mit. See Wurt MN.*
*M Notae        Mathematicae notae, bol d inst d matematica. 1941-.
                Rosario. Univ nac d litoral, Rosario d Santa Fe.
*M Notes (Math Notes, a Rev of Elem Math & Sci, Edinburgh). See Edi MSN.*
*MN Unt         Der math & naturw Unterricht. 1948-. C of Unt M, ZMN Unt,
                Z Ph C Unt. Bonn.
*M Obraz        Matematicheskoe obrazovanie. (Math Unterr) 1912-. Moscow.
*Mod At         Atti e mem, (R) accad (naz) d sci lett & arti. (Atti Ac
                Modena) 1926-. C of Mod Mm. Modena.

*Modern Sc    Modern Schoolman, a Quar J of Phil. 1925-. St Louis, Mo.
    St Louis Univ.
*Mod Mm    Mem d R accad d sci, lett & arti d Modena. (Mod Ac Sc Mm)
    1833-1922// C as Mod At. Modena.
Mod Sem    Atti d sem mat e fis d univ d Modena. 1947-. Modena.
Mod SN    Atti d soc d naturalisti (e mat) d Modena. (Mod S Nt At)
    1866-. Modena.
*Mod U An    Annuario univ Modena. 1863-. Modena.
*Mo Ges W Ju    Monatsschr f Gesch & Wiss d Judenthums. 1851-. Breslau.
    Gesell z Foerderung d Wiss d Judenthums.
*Moldavia. See also Kish.*
Mold IM    Mat issledovania. 1966-. Kishinev. Akad nauk Moldavskoi
    SSR, inst mat.
*Molotovsk. See Perm.*
Mol Phy    Molecular Physics. 1958-. London.
Mo M    Monatshefte f Math. 1948-. C of Mo M Phy. Vienna.
*Mo M Phy    Monatshefte f Math & Phys. (Mh Math Phys) 1890-1944//
    C as Mo M. Vienna.
*Monatliche Correspondenz zur Befoerderung d Erd- & Himmelskunde.
    See Zach MC.*
*Monatsblatt.... See organization, city, region, country, etc.*
*Mondes    Les Mondes. 1863-. Paris. Ed: F N M Moigno.
*Monist    The Monist, a Quar J Devoted to the Phil of Sci. 1890-1936//
    Chicago, Ill; London. Ed: Paul Carus.
*Monog M    Monografii matematice. Cluj. Univ din Cluj, sect mat.
Mono M War    Monografje mat. 1932-. Warsaw.
*Mon Ser    Monumenta serica, J of Oriental Studies of the Catholic Univ
    of Peking. 1935-. Peking.
*Monthly Notices. See organization, city, region, country, etc.*
*Montv Bol    Bol d l fac d ing y agrimensura d Montevideo. 1935-.
    Montevideo.
*Montv Did    Fac d ing y agrimensur, pub didacticas d inst d mat y
    estadistica. 1943-. Montevideo. Univ.
*Montv H Ci    Univ d l republica, fac d humanidades y ciencias, rev.
    1947-. Montevideo.
Montv IM    Fac d ing y agrimensur, univ d l republica, pub d inst d mat
    y estadistica. 1943-. Montevideo.
M Op Stat    Mathematische Operationfors & Stat. 1970-. Berlin. Inst
    f angew Math & Mec d deu Akad d Wiss z Berlin.
Mos AIS    Moskovskii gos univ im M V Lomonosova, soob gos ast inst im
    P K Sternberga. 1948-. C of Mos AIT. Moscow.
Mos AIT    Moskovskii gos univ im M V Lomonosova, trudy gos ast inst im
    P K Sternberga. 1922-. C as Mos AIS. Moscow.
*Moscow. See also Len, SSSR, Vych.*
Mos DU    Trudy sem po teorii dif uravnenii s otklonyayushchinskya
    argumentom. 1962-. Moscow. Univ, druzhby narodov im
    Patrisa Lumumby.
Mos Fin    Nauchnye zap, Moskovskii finansovii inst. 1951-. Moscow.
*Mos IIE    Trudy inst istorii est. 1947-1952// C as Mos IIET. Moscow.
    Akad nauk SSSR.
*Mos IIET    Trudy inst istorii est i tekh. 1954-. C of Mos IIE. Moscow.

| | | |
|---|---|---|
| *Mos IINT | Transas, Inst Hist Sci Tech. Moscow. | |
| Mos ITAB | Byulleten inst teor ast. (...astronomischeskii inst) 1924-. Moscow. Akad nauk SSSR. | |
| Mos ITAT | Trudy inst teor ast. 1952-. Moscow, Leningrad. Akad nauk SSSR | |
| *Mos MO Sb* | *(Matematicheskie sb; Math Sammlung math Gesell; Rev math Mosc; Rec math Mosc). See M Sbor.* | |
| Mos MOT | Trudy Moskovskogo mat obshch. 1952-. Tr as Tra MMS. Moscow. | |
| *Mos M Sez | Trudy...vsesoyuznogo mat sezda. Moscow. | |
| *Mos Ped | Izv akad ped nauk RSFSR. 1945-. Moscow. | |
| Mos Ped Do | Dokl akad ped nauk RSFSR. 1957-. Moscow. | |
| Mos Ped Kr | Uch zap, Moskovskii oblastnoi ped inst im N K Krupskoi. 1937-. Moscow. | |
| Mos Ped Le | Uch zap, Moskovskii gos ped inst im V I Lenina. 1934-. Moscow. | |
| Mos Philom | J d l soc philomathique de Moscou. | |
| Mos SIN | Bull d l soc imp d naturalistes d Moscou. 1829-. Moscow. | |
| Mos Tek | Zap, Mos otdel imp Russ tekh obshch. 1878-1917// Moscow. | |
| Mos UBM | Bull math (d l'univ d Moscou), ser intl. 1927-. Sect A (math & mec) 1937-. Moscow. | |
| Mos UF As | Vestnik Mos univ, ser 3, fiz, ast. 1960-. Pt C of Mos UV. Tr as Mos UPB. Moscow. | |
| Mos UFM | Uch zap, (imp) Mos (ordena Lenina) (gos) univ, otd fiz mat. (Mosc Un Mm Ps Mth) 1880-1896// Moscow. | |
| Mos UM | Mathematika, uch zap, Mos gos univ, otd (fiz) mat. 1933-. Pt C of Mos UZ. Moscow. | |
| Mos U Me | Vop mekh. Moscow. Moskovskii gos univ im M V Lomonosova. | |
| *Mos UM Me | Vestnik Mos univ, ser 1, mat mekh. 1960-. Pt C of Mos UV. Moscow. | |
| Mos UPB | Moscow Univ Physics Bull. 1966-. Tr of Mos UF As. NY. | |
| Mos UV | Vestnik Moskovskogo univ. 1946-. Ser fiz-mat i est nauk 1950-1956// Ser mat (mekh) ast fiz kim 1956-1959// Pt C in Mos UF As, Mos UM Me. Moscow. | |
| *Mos UZ | Uch zap, Mos univ. (Wiss Ber (Nachrichten) Mosk Univ; Sci Mem of the (Imp) Univ of Mos) 1833-1932// C as Mos UM. Moscow. | |
| *Mos Ve Ten | Trudy sem po vektornomu i tenzornomu anal s ikh prilozheniyami k geom, mekh i fiz. (Abh Sem Vektor, Tensoranalysis) 1933-. Moscow. Univ, fiz-mat fak. | |
| *MP Ap | Le matematiche pure & appl, per mensile...ad uso d istruzione media e sup. 1901-. Citta d Castello. | |
| *MP Bib | Mathematisch-physikalische Bibliothek. 1911-. Leipzig. | |
| M Ph S | Math & Physical Scis. 1967-. Madras. Indian Inst of Tec. | |
| *MP Lap | Mathematikai es phys lapok (tarsulat). 1892-1918, 1921-1943// C as M Lap. Budapest. | |
| M Prog | Math Programming. 1971-. Amsterdam. North-Holland Pub Co. | |
| M Pros | Mat prosveshchenie. 1957-. C of M Prosv. Moscow. | |
| *M Prosv | Matematicheskoe prosveshchenie. (Mat, ee prepodavanie, prilozheniya i istoriya, sb statei po elem mat) 1933-1957// C as M Pros. Moscow, Leningrad. | |
| *MP Semb | Math-phys Semesterberichte. 1949-. Goettingen. Math Forschungsinst Oberwolfach. | |

MQSET      Math Qs & Solutions from the Educational Times. (Educ Times) 1864-1918// London.

*MR      Mathematical Reviews. 1940-. Ann Arbor, Mich. AMS.

M Rep      Math Repository. 1748-1753, 1795-1804, 1806-1835// London. Ed: T Leybourn. ≠ Gen D.

*M Sbor      Matematicheskii sbornik. (Math Sammlung Math Gesell; Rev Math Mosc; Rec Math Mosc) 1866-. Tr as MUSSR Sb. Moscow. Moscow Math Soc, and Akad Sci USSR.

M Spect      Mathematical Spectrum. 1968-. Oxford. Univ P.

*M Stu      The Math Student. 1933-. C of Indn MSJ. Madras, Poona.

M Stu J      The Math Student J. 1954-. Wash DC. NCTM.

M Surv      Math Surveys. 1943-. NY.

M Syst      Mathematical Systems Thy. 1966-. Berlin, Heidelberg, NY.

*MT      The Mathematics Teacher. 1908-. Syracuse, NY. NCTM.

*M Tab OAC      Mathematical Tables & Other Aids to Computation. 1943-1959// C as M Comp. Wash DC. NRC.

M Tchg      Mathematics Teaching. 1955-. Nelson, Lancashire. Asso of Teachers of Math.

*M Ter Er      Mathematikai es termeszettudomanyi ertesito. (Math Nat Anz, Ungar Akad Wiss) 1882-. Hungarian ed of MNB Ung. Budapest.

M Ter Koz      Math es termeszettudomanyi kozlemenyek, vonatkozolag a hazai visconyokva. 1861-. Budapest.

*M Tid A      Matematisk tid A udgivet af (Dansk) mat forening i Kobenhavn. 1919-1952// C of Nyt Tid. C as Nor M Tid. Copenhagen.

*M Tid B      Matematisk tid B udgivet af mat forening i Kobenhavn. 1919-1952// ‚C of Nyt Tid. C as Math Scan. Copenhagen.

*MTW      Math-tec-Wirtschaft, Zt f mod Rechentec & Automation. (Mitt...tec Hochschule, math Labor; ...Zt z Pflege d Beziehungen zwischen Math etc) 1954-1966// C as Comp. Vienna, Graz.

*Muenster.* *See Muns.*

*Mun Abh      Abh d (K) bayerischen Akad d Wiss, math-naturw Abt, II Kl. (Abh Muenchen; Munch Ak Ab) 1829-1945, 1948-. Munich.

*Munich, Muenchen.* *See also Bay.*

*Mun Jb      Jab d (K) bayerischen Akad d Wiss. 1912-. Munich.

Mun Mus KM      Vero d Fors-Inst d deu Mus f d Gesch d Naturw & d Tec, Reihe A, kleine Mitt. 1964-. Munich.

*Mun Si      Sitz d (K) bayerische Akad d Wiss z Muenchen. (Munch Ber) 1860-1944, 1947-. Math-naturw Abt 1870-. Math-phys Kl 1903-. Munich.

*Muns Semb      Semester-Ber z Pflege d Zusammenhangs v Univ & Schule in math Unterricht. 1933-. Bonn, Muenster.

Muns UMI      Schriftenreihe d math Inst d Univ Muenster.

*M Unt.* *See Mathunt.*

Muror IT      Mems of the Muroran Inst (Coll) of Tec. 1950-. Muroran.

*Mus Bk      Music Book. (Hinrichsen's Musical Yearbook) 1944-. London.

*Musee Teyler.* *See Arc Mu Tey.*

*Museon      Museon. 1882-. Louvain, etc.

*Muslim India.* *See Islam Rv.*

MUSSR Iz      Math of the USSR, Izv. 1968-. Tr of SSSRM. See also Math N. Providence, RI. AMS.

MUSSR Sb        Math of the USSR, Sbornik. 1968-.  Tr of M Sbor.
                Providence, RI.  AMS.
*MV             Matematichki vesnik, NS.  1964-.  C of Beo DMF.  Belgrade.
                Mat inst.
M Vis           Mathematical Visitor.  1877-1894//  Wash DC; Erie, Penn.
M Wirt          Mathematik & Wirtschaft.  1963-.  Berlin.
*Myrin P        Myrin Inst, Adelphi Coll, Procs.  1952-.  Garden City, NY.
Mysore          The Half-yearly J of Mysore Univ.  1927-.  Mysore.
*MZ             Mathematische Zt.  1918-1944, 1946-.  Berlin, Goettingen,
                Heidelberg.


*Nagao TC       Res Repts of the Nagaoka Tec Coll.  (Kenkyu kiyo, Nagaoka
                kogyo tanki daigaku, koto semman gakko)  1962-.  Nagaoka.
Nagoya CP       Coll Papers from the Math Inst.  1949-.  Nagoya.  Univ,
                Fac of Sci.
*Nagoya MJ      Nagoya Math J.  1950-.  Nagoya.
Nagoya Sp       J of Spatial Math of the Sasayama Res Room.  1958-.  Nagoya.
Nancy SBM       Bull mensuel d l soc d sci d Nancy.  1936-.  C of Nancy SLA.
                Paris.
Nancy SLA       Mems d l soc (R) d sci (let, & arts) d Nancy.  (Nancy Mm S Sc)
                1830-1870, 1929-1935//  C of Nancy SS.  C as Nancy SBM.
                Paris, Strasbourg.
Nancy SS        Bull d l soc d sci d Nancy.  (Nancy S Sc Bll)  1868-1928//
                C as Nancy SLA.  Nancy.
*Naniwa Univ.*  *See Osaka.  (Name changed 1955.)*
Nanta M         Nanta mathematica.  1966-.  C of Sing MS.  Singapore.
*Nap Barbo      Atti d (R) accad d sci e B-L, sez d soc R Barbonica.
                1819-1851//  Naples.
*Nap FM At      Atti d (R) accad d sci fis e mat d Napoli.  (Napoli Atti;
                Memorie...)  1780-1917, 1926-.  Naples.  Soc Barbonica.
*Nap FM Ri      Rend d accad d sci fis e mat.  (R C Accad Napoli)  1842-.
                Naples.
*Nap Nav        Ist univ navale, annali.  1932-.  Naples.
*Nap Pont       Atti d soc (accad) Pontaniana d Napoli.  (Soc R d Napoli;
                R accad Pontaniana d sci morali e politiche; accad naz d
                sc mor e pol)  1810-.  Naples.
Nara NS         J of Nara Gakugei Univ, Nat Sci.  (Nara gakugei daigaku kiyo)
                1951-.  Nara.
*N Arc          Nieuw archief voor wiskunde.  1875-.  Amsterdam, Groningen.
*Nat            Nature 1869-.  London.
*Nation         Nation.  1865-.  NY.
*Nationaal lucht- en ruimtevaartlaboratorium....  See Amst.*
*National Acad of Sci.  See Am NAS;  also country, etc.*
*National Bureau of Standards.  See USNBS.*
*National Council of Teachers of Mathematics.  See Ari T, MT, NCTM.*
*National Math Mag.  See M Mag.*
*National Research Council.  See Am NRC.*
*Nat Kult       Natur & Kultur.  1903-.  Munich.
*Nat Offen      Natur & Offenbarung.  1855-1910//  Muenster.
*Nat Par        La Nature.  1873-.  Paris.

Nat Prir        Nature. Engl absts of Priroda. Leningrad.
*Naturf GB      Naturfors Gesell Bamberg, Berichte. 1852-. Bamberg.
*Naturw         Die Naturwissenschaften. 1913-. C of Naturw R. Berlin,
                Goettingen, Heidelburg.
*Naturw R       Naturw Rundschau. 1886-1912// C as Naturw. Braunschweig.
*Naturw Wo      Naturw Wochenschrift. 1888-1920// Jena.
*Nauk Pri       Nauka i priroda. 1948-. Belgrade.
*Nauk Zh        Nauk i zhizn. 1934-. Moscow.
Nau Tek In      Nauchno-tekhnicheskaya informatsiya. Tr as Aut Doc ML.
*Navig          (Navigation) J of the Inst of Navigation. 1948-.
                Los Angeles, Cal.
Nav RLQ         Naval Res Logistics Quar. 1954-. Wash DC.
*NCTM (Natl Council of Teachers of Math). See also Ari T, JRME, MT.*
NCTMRN          NCTM Research Newsletter. 1968-. Columbus, O.
*NCTMY          NCTM Yearbooks. 1926-. NY.
*Ned Arch       Archs neerlandaises d scis exactes & nats. 1866-1933//
                Ser 3a C as Physica. Haarlem. Soc hollandaise d scis.
Ned As IB       Bull of the Ast Inst of the Netherlands. 1921-. Haarlem.
*Ned Ja         Jaarboek d K nederlandse akad v wetenschapen. 1927-. C of
                Amst Ja. Amsterdam.
*Ned NGC        Handelingen van het nederlansch natuur en geneeskundig
                congres. 1887-. Haarlem.
*Ned Ps         Algemeen nederlands tijd voor wijsbegeerte en psychologie.
                1907-. Amsterdam, Assen.
*Ned Taal       Tijdschrift voor nederlandsche taal- en letterkunde. 1900?-.
                Leiden.
*Ned Ti N       Nederlands tijd voor natuurkunde. 1934-. C of Physica.
                Utrecht.
*Netherlands.   See also Act Bioth, Amst, Benelux, Bij, Delft, Haar, Leyd,*
                *Rott, St Ned, Utr, Zutphen.*
Networks        Networks, an Intl J. 1971-. Amer Elsevier Pub Co.
*Neuc B         Soc neuchateloise d scis nats, bull. (Soc d scis nats de
                Neuchatel) 1844-. Neuchatel.
Neuc Mm         Mems d l soc neuchateloise d scis nats. (Mems soc d scis
                nats d Neuchatel) 1835-1914, 1938-. Neuchatel. Univ.
*Neuc U Mm      Mems d l'univ d Neuchatel. 1914-. Neuchatel.
Neuc U Sem      Pubs d sem d geom d l'univ d Neuchatel. 1958-. Neuchatel.
*Neue. See also N.*
*Neue Denkschrift...Schweizerischen Gesellschaft.... See Schw NG.*
*Neuvorpommern und Rugen. See Greifsw.*
*Newcomb Am J. See Am JM.*
*Newcomen       Newcomen Soc Transas. 1920/21-. London. Newcomen Soc for
                the Study of the Hist of Eng & Tec.
*New Eng        New Englander & Yale Rev. 1843-1892// C as Yale Rv.
                New Haven.
*New M Lib      New Mathematical Library. 1961-. NY; New Haven, Conn.
*New Schol      New Scholasticism. 1927-. Wash DC.
*New Sci        New Scientist. 1956-. London.
*Newsletter. See organization, city, region, country, etc.*
*New Yorkr      The New Yorker. 1925-. NY.
*New Zealand. See NZ.*
*Nicholso       J of Nat Phil, Chem, & the Arts. (Nicholson J) 1797-1813//
                London. Ed: Nicholson.

| | |
|---|---|
| *Nieu Arch | Nieuw arch voor wiskunde. 1875-. C of Amst Arch. Amsterdam. |
| Niiga M | Sci Repts of Niigata Univ, Ser A (Math). 1964-. Pt C of Niiga MPC. Niigata. |
| Niiga MPC | J of the Fac of Sci, Niigata Univ, Ser 1, Math, Phys, Chem. 1953-. Pt C as Niiga M. Niigata. |
| *Nippon. See Jp.* | |
| *N Jb MGP | Neues Jab f Mineralogie, Geol & Palaontologie. (N Jb Mn) 1833-1942// Stuttgart. |
| *N Jb W Jug | Neue Jab f Wiss & Jugendbildung. 1925-1937// Leipzig, Berlin. |
| *NMM. See M Mag.* | |
| *Nor As Tid | Nordisk astronomisk tidsskrift. 1920-. Copenhagen. |
| *Nordisk tid for info behandling. See BIT.* | |
| *Nor(di)sk vid akad. See also Oslo.* | |
| *Nordrhein-Westfalen. See Nor Wes.* | |
| *Nor | Norsk mat forenings skr. 1929-. Oslo. |
| *Nor M | Nordisk (Norsk) mat tid. 1919-. Incl C of M Tid A. Oslo. Norsk mat forening. |
| Nor St J | Nordic Stat J. 1929-1932// Engl ed of Nor St Tid. Stockholm. |
| *Nor St Tid | Nordisk stat tid. 1922-1932// Stockholm. |
| *Nort | North British Review. 1844-. Edinburgh. |
| *Nort | North-China Herald. 18??-1916// Shanghai. |
| *North Africa. See Afr N, Alger, Hesperis.* | |
| | Forhandlinger, det K norske vid selskab. (Norske vid selsk forh; Tron for) 1926-. Trondheim. |
| *Nor | Oversigt over videnskaps-selskabet moder i Kristiania. 1868-1924// Christiania. |
| *Norway. See also Berg, Chr, Oslo, Phy Norv, Sk, Tron.* | |
| *Nor Wes AF | Arbeitsgemeinschaft f Fors d Landes Nordrhein-Westfalen, wiss Abh. 1958-. Koeln. |
| Notas Lim | Notas d matematicas. Lima. Univ nac d ingenieria, inst d mat puras y aplicadas. |
| Notas Rio | Notas d matematica. 1942-. Rio de Janeiro. |
| *Not DJFL | Notre Dame J of Formal Logic. 1960-. Notre Dame, Ind. |
| Not DMC | Pub of the Univ of Notre Dame, Repts of a Math Colloquium, Sec Ser. (Rep math Colloqu, Indiana) 1939-1948// See also Erg MK. Notre Dame, Ind. |
| Not M Disc | Notes on Mathematical Discussions. 1940?-. London. Mathematical Discussion Group. |
| *Not Qu Lon | Notes & Queries. 1850-. London. |
| *Not Qu SD | Notes & Queries, Somerset & Dorset. 1888/89-. Sherborne. |
| *Nou An | Nou annales d math, j d candidats aux ecs polyt & normale. (N A Mth; Ann Nou Math; Terquem Annales) 1842-1927// Paris. First ed: Terquem. |
| *Nou Cor | Nou corres d math. (N C M) 1874-1880// Paris. |
| Nov Acta | Nova acta eruditorum. 1732-1776// C of Act Erud. Leipzig. |
| Nov Acta S | Nova acta eruditorum supplementa. 1735-1757// C of Act Erud S. Leipzig. |
| Novor | Zap Novorossiiskogo (Odesskoe) obshch estestvoispytatelei. (Mem Soc Nat Nouv-russ Est) 1872-. Odessa. |
| *Novor M | Zap mat otd Novorossiiskogo obshch est. (N Rs S Nt Mm (Mth); Odessa Ges) 1878-1900// Odessa. |

*Novosibirsk.*   *See Alg Log, Sib, Siber.*

*Nov Sco P     Procs (& Transas) of the Nova Scotian Inst of Sci. 1863-.
               Halifax.

*Nov Vet      Nova et vetera. 1926-. Fribourg, Switzerland.

*NSWRSJ      J & Procs of the R Soc of New South Wales. 1867-. Sydney.

*N Tijd       Nieuw tijd voor wiskunde. 1913-. Pt C of Zutphen.
               Groningen, Djakarta.

*NTM         NTM, Schriftenreihe (Zt) f Gesch d Naturw, Tec & Med. (ZNTM)
               1960-. Berlin.

Numerus      Numerus, rev lunara d mat elem pentru invatamantul secundar,
               teor, nor, commercial si ind. 1935/36-. Bucharest.

Num M       Numerische Mathematik. 1959-. Berlin, Goettingen, Heidelberg.

*Nuo Cim     Il nuovo cimento. 1855-. C of Cimento. Bologna, Pisa.

Nuo Cim S    Suppl al (del) nuovo cimento. 1948-. Bologna.

*Nuovi Lincei.*   *See Vat NL.*

*Nurnb NG    Abh d naturhist Gesell z Nuernberg. (Nuernb Nt Gs Ab)
               1852-. Nuernberg.

NYASA       Anns of the NY Acad of Scis. 1823-. NY.

NYAST       Transas of the NY Acad of Scis. 1881-. NY.

*Ny Ill Tid   Ny illustrerad tidning for konst, bildning och noje.
               1865-1900// Stockholm.

*NY Ind Mus   Industrial Museum of NY. 1930// NY.

*NY Met MA   Metropolitan Museum of Art Bull. 1905-1942, NS 1942-. NY.

*NYMS       Bull of the NY Math Soc, a Hist & Crit Rev of Math Sci.
               (Bu NY Soc) 1891-1894// C as AMSB. NY. Eds: Th S Fiske,
               H Jacoby.

*NYT         The New York Times. 1851-. NY.

Nyt Mag     Nyt magazin for naturvidenskaberne. 1836-1934// C of
               Mag Nat. C as Nyt Mag O. Christiania.

Nyt Mag O    Nytt magazin for naturwidenskapene. 1936-. C of Nyt Mag.
               Oslo.

*Nyt Tid     Nyt tidsskrift for matematik. 1890-1918// C of Tid M, & of
               Maa EM. C as M Tid A, & as M Tid B. Copenhagen.

NYU Cl       J of the Math Clubs. 1937-. NY. New York Univ.

NZJS        New Zealand J of Sci. 1958-. C of NZJST. Wellington.

NZJST       New Zealand J of Sci & Tec, Sect B, Gen Res. 1918-1958//
               C as NZJS. Wellington.

NZRSP       Procs of the R Soc of New Zealand. 1958-. Wellington.

NZRST      Transas of the R Soc of NZ, Gen. 1952-1961// Wellington.


*Oberbay     Oberbayerisches arch f vaterlaendische Gesch. 1839-.
               Munich. Hist Verein von Oberbayern.

*Obs         The Observatory, a (Monthly) Rev of Ast. 1877-. London,
               Hailsham.

Ochan U     Nat Sci Rept of the Ochanomizu Univ. 1951-. Tokyo.

Och Ist Zn   Ocherki po istoria znania. 1927-. Leningrad.

*Odessa.*   *See also Novor.*

Odess Mec   Trudy Odesskogo gos univ im I I Mechnikova. (Pratsi Odesk
               derzh u Mechnikova) 1935-. Odessa.

Odess U       Nachrichten v d (imp) Univ Odessa. 1865-1920, 1933-. Odessa.
*Odess UTM      Odessa, univ, trudy mat. 1935-. Odessa.
Odess UZ      Zapiski Odessa univ. 1867-1914// Ser fiz mat fac 1910-1919//
                Odessa.
*Oesterreich. See Ost.*
*Office national d'etudes et de recherches aeronautiques. See Fr ONERA.*
Okay MJ       Math J of Okayama Univ. 1952-. Okayama.
*Olmuetz. See Olom (Olomouc).*
*Olom Pri V     Vyosoka skola ped, sbornik, prir vedy. 1954-1959// Olomouc.
*Olom UPF     Sb praci prir fak univ Palackehu v Olomouci. 1954-. Prague.
                Statni ped nakl.
*Open Ct      Open Court, a Quar Mag. 1887-1936// Chicago, Ill.
*Op Phil Co     Opuscula philologica. 1887// Copenhagen. Filologisk-hist
                samfund.
Op Phil Li     Opuscula philologica. 1926-. Linz. Katholisch-akad
                Philologenverein in Wien.
*Op Res       Operations Research, the J of the Opers Res Soc of Amer. (OR)
                1952-1964// Baltimore, Md.
Op Res Ab     Intl Absts in Operations Research. 1961-. Baltimore, Md.
Op Res Q      OR, Operations Research Quar. 1950-. London.
Opt Ac        Optica acta (European J of Optics). 1954-. London, Paris.
*Optical Society of America. See Am Opt J, Opt S Am.*
Optik         (Optik) Zt f d gesamte Gebiet d Licht- & Elektronenoptik.
                1946-. C of Opt Run. Stuttgart.
Optik N       Optik, Nachrichtenblatt f d Optikerhandwerk. 1943-1944//
                C of Opt Run. See also Optik. Weimar.
Opt Run       Optische Rundschau, Zentralorgan f d Gesamtinteressen d deu
                Opt & Feinmech. 1910-1943// C in Optik, & in Optik N.
                Schweidnitz, Berlin.
*Opt S Am     J of the Optical Soc of Amer. 1917-. NY; Rochester, NY.
Opt Soc       Transas of the Opt Soc. 1899-1932// C in Lon Phy S. London.
Opusc M       Opuscula mathematica. 1937-. Krakow. Akad gornicza,
                zaklad mat.
*Opuscoli scientifici. See Bln OS.*
*Opuscula omnia actis eruditorum.... See Act Erud O.*
*OR (Operations Research). See Op Res.*
*Organon      Organon. 1936-1938// Warsaw. Pol Acad of Scis, Dept of the
                Hist of Sci & Tec; Union int d'hist & d phil d scis, div
                d'hist d scis.
*Orien Arc     Orientalisches arch, illustrierte Zt f Kunst, Kulturgesch &
                Voelkerkunde d Laender d Ostens. 1910-1913// Leipzig.
*Oriens Ex     Oriens extremus, Zt f Sprache, Kunst & Kultur d Laender d
                fernen Ostens. 1954-. Hamburg.
*Oriental      Orientalia, commentarii periodici d rebus orientis antiquii.
                1932-. Rome. Pontifico ist biblico.
*Orient Lz     Orientalistische Literaturzeitung. 1898-1944, 1952-.
                Berlin, Leipzig.
Osak IP       J of Math (J of the Inst of Polytec, Ser A, Math), Osaka
                City Univ. 1950-1963// C in Osak JM. Osaka.
Osak IST      J of the Osaka Inst of Sci & Tec (Kinki Univ), Pt I, Math &
                Phys. 1949-. Osaka.

Osak IU          Coll Papers from the Fac of Sci, Osaka Imp Univ, Ser A, Math.
                 1933-1948//  C in Osak JM, Osak Rep.  Osaka.
*Osak JM         Osaka J of Math (Osaka Math J).  1949-.  Pt C of Osak IU, &
                 of Osak IP.  Osaka.
Osak NS          Mems of the Osaka Univ of the Liberal Arts & Education,
                 Ser B, Nat Sci.  1952-.  Osaka.
Osak PUM         Bull of the Univ of Osaka Prefecture, Ser A, Eng & Nat Scis.
                 (Osaka furitsu daigaku) 1952-.  Osaka, Sakai.
Osak Rep         Annual Rept of Sci Works from the Fac of Sci, Osaka Univ.
                 1952-.  Pt C of Osak IU.  Osaka.
*Osiris          Osiris, Studies on the Hist & Phil of Sci & on the Hist of
                 Learning & Culture.  (Commentationes d sci & eruditionis
                 historia) 1936-.  Bruges.
  *Oslo (Christiania).  See also Chr, Nor.*
Oslo Avh         Avhandlinger utgitt av det Norske videnskaps-akademi i Oslo,
                 I, mat-naturv kl.  1925-.  C of Chr Avh.  Oslo.
Oslo Gf          Geofysiske publikasjoner utgitt av det Norske vid-akad i
                 Oslo.  1920-.  Oslo.
*Oslo Sk         Skr utgitt av det Norske vid-akad i Oslo, I, mat-naturv kl.
                 (Skrifter Oslo) 1925-.  C of Chr Sk.  Oslo.
Oss Lett         Osservazioni letterarie.  1737-1740//  C of Gi Lit Ven.
                 Venice.
*Ost Anz         Anzeiger, oesterreichische Akad d Wiss, math-naturw Kl.
                 1947-.  C of Wien Anz.  Vienna.
*Ost Bau         Oesterreichische Wochenschr f d oeffentlichen Baudienst.
                 1901-1920//  C of, & C as Ost Bau M.  Vienna.
*Ost Bau M       Oesterreiche Monatsschr f d oeffentlichen Baudienst.
                 1895-1900, 1921-1924//  See also Ost Bau.  Vienna.
*Ost Bib         Oesterreichischer Ver f Bibliothekswesen, Mitt.  1897-1910//
                 Vienna.
Ost Denk         Denkschriften d oesterreichischen Akad d Wiss, math-naturw
                 Kl.  1947-.  C of Wien Denk.  See also Ost P Denk.  Vienna.
Ost Gesch        Vero d Kommis z Gesch d Math & d Naturw, Sitz d oester-
                 reichischen Akad d Wiss, math Kl.  1947-.  Vienna.
Ost IAW          Wochenschrift d oesterreichischen Ing- & Architektenvereins.
                 1876-1891//  C in Ost IAZ.  Vienna.  Ed: P Kortz.
*Ost IAZ         Zt d oesterreichischen Ing- & Architektenvereins.  1849-1938//
                 Incl C of Ost IAW.  C as Bau, & as Ost Ing Z.  Vienna.
                 Ed:  P Kortz.
*Ost Ing         Oesterreichisches Ingenieur-Archiv.  1947-.  C as Act Me.
                 Vienna.
Ost Ing Z        Oesterreichisches Ingenieur-Zeitung.  1958-.  C of Ost IAZ.
                 Vienna.  Oesterreichisches Ing- & Architekten-Verein.
  *Ost MG Nach.*  *See Int M Nach.*
*Ost P Denk      Denkschriften d oesterreichischen Akad d Wiss, phil-hist Kl.
                 1947-.  C of Wien Denk.  Vienna.
*Ost PZ          Oesterreichische polytec Zt.  1907-1918//  Vienna.
*Ost Si M        Oesterreichische Akad d Wiss, math-naturw Kl, Sitz, Abt II,
                 Math, Ast, Phys, Meteorologie & Tec.  1947-.  C of Wien Si.
                 Vienna.

| | |
|---|---|
| Ost Verm | Oesterreichische Zt f Vermessungswesen. 1903-1937, 1948-. Baden. Ver d oesterreichischen kk Vermessungsbeamten. Suppls: Mittblatt...; ...Gesell f Photogrammetrie. |
| *Ostw Kl | Ostwalds Klassiker d exakten Wiss. 1889-. See also Kl EW. Leipzig. |
| *Ott UR | Revue de l'univ d'Ottawa. 1931-. Ottawa. |
| Ox CD | The (Oxford, Cambridge & Dublin) Messenger (of Math), a J Supported by Junior Math Students of Three Univs. 1861-1871// C as Mess M. London, Cambridge. Eds: Whitworth, Casey, Challis, McDowell, Taylor, Turnbull. |

| | |
|---|---|
| *Pac Aff | Pacific Affairs. 1926-. Honolulu, NY. Inst of Pacific Relations. |
| Pac JM | Pacific J of Math. 1951-. Berkeley, California. |
| *Pado A Mm | Atti (e mem) d (R) accad (patavina) d sci, lett & arti (in Padova), cl d sci mat e nat. (Atti Acad Padova; Padova Ac At e Mm) 1884/85-1943, 1947-. C of Pado Riv. 1935- incl Atti accad sci veneto-trentino-istriana. Padua. |
| *Pado Pat.* | *See Pado A Mm.* |
| *Pado Riv | Riv periodica d lavori d I R accad d sci, let, & arti d Padova. (Padova Rv Period; Riv Per) 1851/53-1883/84// C in Pado A Mm. Padua. |
| *Pado Sem | Rend d sem mat d (R) univ d Padova. 1930-. Padua. |
| *Paideuma | Paideuma, Mitt z Kulturkund. 1938-. Frankfurt, Bamberg. |
| *Pak J Sc | Pakistan J of Sci (Scientific Research). 1949-. Lahore. Pakistan Asso for the Adva of Sci. |
| Pak St | Procs of the Pakistan Stat Asso. 1954/55-. Lahore. |
| *Palais de la decouverte.* | *See Par PDC.* |
| *Paler Bi | Annuario biografico d circolo mat d Palermo. 1884-1928// Palermo. |
| *Paler R | Rend d circolo mat d Palermo. (Palermo Rend) 1884-. Palermo. |
| *Paler RS | Rend d circ mat d Palermo, suppl. 1906-1914, 1919-. Palermo. |
| Paler (S) | Atti d accad d sci lett e arti d Palermo, pt I prima, sci. 1845-. Palermo. |
| Paler SM | Gior d mat e fis d scuola media. 1919-. Palermo. |
| Panjab U | Res Bull of the (East) Panjab Univ. 1950-. Hoshiarpur. |
| *Papers.* | *See organization, city, region, country, etc.* |
| *Papyrusforschung.* | *See Arc Pap.* |
| Par Actu | Bull d l'asso d actus diplomes d l'inst d sci financiere & d'assurances. 1890-. Paris. |
| Paramet | Parametr, czasopismo poswiecone nauczaniu mat. 1930-1932// Warsaw. |
| Parana FT | Bol d univ d Parana, fis teorica. 1961-. Curitiba. Conselho d pesquisas, inst d fis. |
| *Parana M | Anuario d sociedade paranaense d matematica. 1954-. Curitiba. |

```
*Parana MB      Bol d soc paranaense d mat.  1958-.  Curitiba.  Univ d Parana
                Escola d engenharia.
 Par Annu       Annuaire d l'inst d France.  1796-.  Paris.
*Par BS Rou     Bull sci roumain.  1952-.  Paris.  Inst univ roumaine
                Charles 1er; fundatia regala univ Carol I.
 Par CAM        Annales d conservatoire d arts & metiers.  (A Cons Arts et
                Met)  1861-.  Paris.
*Par CR         CR hebdomadaires d seances d l'acad d scis.  (Com ren d'acad
                des sci)  1835-.  1964- 8 sects.  Paris.
*Par EC         Paris ecole d chartes, positions d theses.  1846-.  Paris.
*Par EC Bib     Bibliotheque d l'ecole d chartes.  1839-.  Paris.
 Par EHE        Bibliotheque d l'ecole d hautes etudes, sect d scis nats.
                1869-1880//  Paris.
 Par ENS An.    See An SENS.
*Par ENSB       Bull d l'asso amicale d secours d anciens eleves d l'ec
                normale superieure.  Paris.
*Par EP         Journal (d l'ecole) polyt.  (Bull d travail fait a l'ec
                cen d travaux publics)  1795-.  Paris.
 Par EP Cor     Corres d l'ec polyt.  1804-1816//  Paris.
 Par Hist       Hists d l'acad d scis.  1666-1699//  C as Par Mm.  Paris.
 Par HP         Annales d l'inst Henri Poincare.  1930-.  Paris.
*Par IBL        Acad R d inscriptions & B-L, CR.  (...acad d inscrs &
                medailles; ...inst d Fr, cl d'hist & litt anc)  1663-.  Paris
 Par IBLHM      Hists & mems d l'acad d inscrs & B-L d Paris.  1717-1736//
                Paris.
 Par IHESM      Inst d hautes etudes scis, pubs maths.  1959-.  Paris.
 Par I Org      Org & reglemens d l'inst d scis, letts & arts.  1798-.  Paris
 Paris.  See also Act Sc Ind, Arc, Mem, Philom, Rv, etc.
 Paris ecoles normales, annales scientifiques.  See An SENS.
*Par I St       Pubs d l'inst d stat d l'univ d Paris.  1952-.  Paris.
*Parma Rv M     Rivista d mat d univ d Parma.  1950-.  Parma.
*Par Med        Paris medical, la semaine d clinicien.  1910-1950//  C in
                Rev d practicien, 1951-.  Paris.
*Par Mm         Mems d l'acad d scis d l'inst d France.  (Mem d 1 cl d scis
                math & phys d l'inst; Par mm d l'I; Mem l'acad sci l'inst
                France; Par ac sc mm; Mem ac sci ins imp France; etc)
                1699-1818//  Incl C of Par Hist.  Paris.
*Par Mm Div     Mems presentes par divers savants a l'acad d scis d l'inst
                d France.  (Par Mm Sav Etr)  1746-1786, 1806-1811, 1815-1914//
                C from Par RPRP.  Paris.
*Par Mor Tr     Acad d scis morales & publiques (politiques), rev d travaux
                & CR.  (Seances & travaux...)  1842-.  Paris.
*Par Not D      Notices & discours, acad d scis, Paris.  1924-.  Paris.
 Par off natl d'etudes & d rech aeronautiques.  See Fr ONERA.
*Parola         Parola.  (Conferenze e prolusioni)  1907-.  Rome.
*Par PDC        Les confs d palais d 1 decouverte.  Paris.
*Par Radio      Les cahiers d Radio-Paris.  1930-.  Paris.
 Par RPRP       Recueil d pieces qui ont remporte 1 prix....  1721-1777//
                C in Par Mm Div.  Paris.  Acad des scis.
 Par Rv.  See Rv Par.
*Par SASPN      Notices d travaux d 1 soc d amateurs d scis phys & nat d
                Paris.  (Par S Amat Tr; Par Tr S Amat)  1807-1808//  Paris.
```

```
*Par Sav CR      Comite d travaux hist & sci, CR d congres d socs savantes
                 d Paris & d depts, sect scis. 1883-. Paris. Gauthier-
                 Villars.
*Par St          J d l soc d stat d Paris. 1860-. Paris.
 Par TNSP        Pamietnik towarzystwa nauk scislych w Paryzu. (Par T Nauk
                 Sc Pam; Par Denkschr) 1871-1882// Paris.
*Par U An        Annales d l'univ d Paris. 1926-. Paris.
 Patavina. See Pado.
*Paul UCM        Commentarii math univ sancti Paulii. (Rikkyo daigaku
                 sugaku zassi) 1952-. Tokyo.
*Pavia U An      Annuario d R univ d Pavia. 1879-. Pavia.
 Pedagogical. See also organization, city, region, country, etc.
 Pedagogical Seminary. See J Gen Psy.
*Ped Sb          Pedagogicheskii sbornik. (Pedagog Sammlung) 1864-1917//
                 St Petersburg.
*Peirce Tr       Transas of the Charles S Peirce Society. 1965-. Amherst, Mass.
*Pekin OSJ       J of the Pekin Oriental Soc. 1885-1898// Peking.
 Pekin U         Acta sci nat univ Pekinensis. (Tszyzhany kesiue snebao)
                 1955-. Peking.
 Pelor FMN       Atti d soc peloritana d sci fis mat e nat. 1955-. Messina.
 Pelori          Atti d R accad peloritana. (Acc Peloritana) 1878-.
                 Messina.
*Pe M            Periodico d matematiche. 1924-. C of Pe MI. Bologna.
 Pe M Hun        Periodica mathematica hungarica. 1970-. Bolyai Math Soc.
*Pe MI           Periodico d mat per l'insegnamento sec. 1886-1918// C as
                 Pe M. Merged with Mathes It 1899-. Rome, Livorno.
*Pe MIS          Suppl al periodico d mat per l'insegnamento sec. 1898-1917//
                 Livorno.
*Pensee          La pensee, rev d rationalisme mod, arts, scis, phil.
                 1939, 1944-. Paris.
*Pentagon        Pentagon, the J of Kappa Mu Epsilon. 1941-. Albuquerque, NM.
 Periodical, periodico, etc. See Pe; also organization, city, region, etc.
 Periodicum. See Zagr Gl.
*Perm M          Uch zap, Permskii (Molotovski) gos univ im A M Gorkogo, ser mat.
                 (Scientific Memoires...) 1925-. Perm. Minist vyssego i
                 srednego spec obrazovaniya RSFSR.
*Perm Sem        Permski gos univ, mat semmariya, trudy. (Permi ouvrages sem
                 math de l'univ) 1927-1928// Perm.
*Perm Zh         Perm gos univ fiziko-mat obshch zh. 1918-1930// Perm.
 Peru. See also Lim, Notas Lim.
*Peru U Rv       Pontificia univ catolica d Peru, revista. (Revista univ
                 catolica Peru) 1931-1945, 1954-. Lima.
 Pet Acta        Acta acad sci Petropoli. 1777-1782// C of Pet Nov Co.
                 C as Pet Nov Ac. St Petersburg.
*Pet B           Bull d l'acad imp d scis d St Petersbourg. (Petersb Bull;
                 St Pet Ac Sc Bll; St Pet Bull Sc; Petersb Abh; Bull St
                 Petersb) 1860-1916// C of Pet BPM. C as Pet Iz.
                 St Petersburg.
 Pet BPM         Bull d l cl phys-math d l'acad imp sci d St Petersbourg.
                 (St Pet Ac Sc Bul; St Pet) 1843-1859// C of Pet BS.
                 C as Pet B. St Petersburg.
```

Pet BS   Bull sci pub par l'acad imp d scis d St Petersbourg. (St Pet Ac Sc Bll; St Pet Bll Ac Sc; St Pet Bull Sc) 1836-1842//   C as Pet BPM.   St Petersburg.

Pet Comm   Commentarii acad sci imp Petropolitanae.   1726-1746// C as Pet Nov Co.   St Petersburg.

*Peterbor   Precis of the Annual Rept of the Peterborough Nat Hist, Sci, & Archeol Soc.   1872/73-.   Peterborough.

*Peterman   Petermanns geog Mitt aus Justs Perthes' geog Anstalt. (Peterm Mt) 1855-1944, 1949-.   Gotha.   Nebst Ergaenzungsheften.

Pet Etr   Mems presentes a l'acad imp d scis d St Petersbourg par divers savants.   (St Pet Mm Sav Etr)   1831-1859// St Petersburg.

Pet F Kh   Zh Russkago fiz-khim obshch pri imp SPB univ, (a) fiz, (b) khim.   (J d phys-chem Gesell z St Petersburg; J soc phys chem russe; Rs Ps CSJ)   1869-1930//   Pt C as Zh ETF. St Petersburg.

*Pet Gor   Anns d l'inst d mines d l'imp Cath II.   1907-1911// St Petersburg.

Pet Iz   Izv Rossiskoi akad nauk Petrograd.   (Bull acad sci Russ) 1917-1928//   C of Pet B.   C as SSSR Iz.   Petrograd.

Pet Mel   Melanges math & ast d l'acad d scis d St Petersbourg. 1849-1919//   St Petersburg.

Pet Mm   Mems d l'acad imp d scis d St Petersbourg, cl d scis phys & maths.   (St Pet Ac Mm;   Mem;   Pet Abh;   Mem Sci Math Phy Nat) 1803-.   Classes 1897-.   C of Pet Nov Ac.   Russ ed: Pet Zap.   St Petersburg.

*Pet MNP   Zh minist narodnykh prosveshcheniya Sankt Petersburg. 1834-1917//   St Petersburg.

*Pet MO   Sanktpeterburgskoi mat obshch.   C as Len FMO.   St Petersburg.

*Pet Nov Ac   Nova acta acad sci imp Petropolitanae.   (St Pet Ac Sc N Acta) 1783-1802//   C of Pet Acta.   C as Pet Mm.   St Petersburg.

Pet Nov Co   Novi commentarii acad sci.   1747-1775//   C of Pet Comm. C as Pet Acta.   St Petersburg.

*Petrograd (St Petersburg, Leningrad).   See also Len.*

*Petrozavodsk.   See Karel.*

*Petrus No   Petrus Nonius.   (Anuario d hist d cien) 1937-.   Lisbon. Grupo Portugues da hist das ciencias.

Pet Zap   Zapiski imp akad nauk po fis-mat otd St Pet.   1862-1930// Russ ed of Pet Mm.   St Petersburg.

Phila ANS   J & Procs of the Philadelphia Acad of Nat Sci.   (J Ph Ac Nat Sc) 1817-1918//   Philadelphia.

*Philb Wi   Philobiblon, eine Zt f Buecherliebhaber.   1928-1940//? Vienna.

*Philippine Islands.   See Cuyo.*

*Philip TR   Philips Tech Review.   1936-.   Eindhoven.

Phil Mag   Philosophical Mag, a J of Theor, Exp & Appl Sci.   1945-. C of Phi Mag.   See also Adv Ph.   London.

*Philolog   Philologus, Zt f d klassichen Altertum.   1846-1943, 1947-1948, 1953-.   Stolberg.

Philol Q   Philological Quarterly.   1921-.   Iowa.

*Philom Bu (Nouveau) Bull d (scis d) l soc philomathique. (Par Bll  
     S Phim) 1791-1936// C of Philom Rp. C in Science. Paris.  
Philom PV Extraits d proces-verbaux d seances d l soc philomath.  
     (Par S Phim PV) 1836-1863// Paris.  
Philom Rp Rapports generaux (d travaux) d l soc philomath d Paris.  
     1788-1799// C as Philom Bu. Paris.  
*Phi Lon Philosophy. 1926-. London.  
*Philosophical Repository. See Enqu, Phi Rep.*  
*Philosophical Transactions. See Lon RS, Phil T.*  
*Philosophische Monatshefte. See Arc Phi.*  
*Philosophisches Jahrbuch. See Goress Ja.*  
Phil T Phil Transas of the R Soc of London. 1665-. London.  
Phil TA Phil Transas of the R Soc London, Ser A, Math & Phys Scis.  
     1887-. London.  
Phil T Ab Absts of the Papers Printed in the Phil Transas from  
     1700-1854. London. Royal Society.  
*Phi M Philosophia mathematica. 1964-. Dekalb, Ill, etc. Asso  
     for the Phil of Math.  
*Phi Mag London, Edinburgh, & Dublin Phil Mag & J of Sci. (Tilloch  
     Ph Mag) 1798-1944// C as Phil Mag. London.  
*Phi Nat Philosophia naturalis. 1950-. Meisenheim.  
*Phi Phen R Phil & Phenomenological Res. 1940-. Buffalo, NY. Intl  
     Phenomenological Soc, Buffalo Univ.  
*Phi Q Philosophical Quar. 1950-. St Andrews.  
Phi Rep Philosophical Repository. 1801-1804// Pt of Math Rep.  
     London.  
*Phi Rev The Philosophical Review. 1892-. Boston, Mass; Ithaca, NY;  
     etc.  
*Phi Sc Philosophy of Sci. 1934-1948, 1951-1953, 1955-1961//  
     Baltimore, Md.  
*Phi St Ire Philosophical Studies. 1951-. Maynooth. St Patrick's  
     Coll, Phil Soc.  
Photgram Photogrammetria, organe officiel d l soc intl d photogram-  
     metrie (offizielles Organ d intl Gesell f Photogrammetrie).  
     1938-1943// C of Int Arc Ph. C in Z Vermess. Amsterdam,  
     Berlin.  
*Phot J Photographic Journal. 1876-. London.  
*Phronsis Phronesis, a J for Anc Phil. 1955-. Assen.  
*Phy Bl Physikalische Blaetter. 1945-. Mosbach, Baden.  
Phy Fl The Physics of Fluids. 1958-. NY. Amer Inst of Physics.  
Phy L Physics Letters. 1962-. Amsterdam.  
Phy Norv Physica norvegica. 1961-. Oslo. Norske videnskaps-akademi.  
*Phy R The Physical Review. 1893-. Ithaca, NY. Cornell Univ.  
Phy RL Physical Review Letters. 1958-. NY.  
Phy R Sup The Phys Rev Suppl, a J of Experimental & Theor Physics.  
     1929// C as Rv Mod Ph. Minneapolis, Minn. Amer Phys Soc.  
Physi Arg Physis, revista d l soc argentina d ciencias nats. 1915-.  
     Buenos Aires.  
*Physica Physica. 1921-. C of Ned Arch. C as Ned Ti N.  
     Amsterdam, etc.  
*Physics. See also Fiz, P, Ph.*

*Physis Fi       Physis, revista d storia d sci. 1959-. Florence.
*Phys Math Wiss. See Fiz M Nauk.*
*Phy Today       Physics Today. 1948-. NY. Amer Inst of Physics.
*Phy Z           Physikalische Zt. 1899-. Leipzig.
*Piac IT An      Annuario d R ist tec d Piacenza. 1926-. Piacenza.
*Pi Mu EJ        The Pi Mu Epsilon J. 1949-. Norman, Okla. Okla Univ Dept
                 Math.
*Pisa SNS        Annali d R scuola nor sup d Pisa, sci fis e mat. (Ann di
                 Pisa) 1871-. Pisa.
*Pisa U          Annali d univ toscane. 1846-. Pisa.
*Pitagora        Il Pitagora, gior mat d scuole sec. 1895-. Palermo.
*Platon          Platon, hetaireia hellenon filologon. 1949-. Athens.
*PMLA. See MLAAP.*
*PMTF            PMTF, Zh prikladnoi mekh i tekh fiz. 1960-. T as JA Me TF.
                 See also Zh TF. Moscow.
*Pogg An         Annalen d Phys & Chem. (Annalen d Physik; Wiedemann Ann;
                 Poggendorff's Annalen) 1799-. Leipzig. Phys Gesell zu
                 Berlin.
 Poggen          Poggendorff, biog Woerterbuch. (Biog-lit Handwoerterbuch
                 z Gesch d exakten Wiss) 1863-. Leipzig.
*Pokroky. See Prag Pok; also organization, city, region, country, etc.*
*Pol Ac Rv       The Rev of the Polish Acad of Scis. 1956-. Warsaw.
*Poland. See also Act Ari, Act Ph Po, Algor, Arc Me Sto, Bresl, Fund M,
         Krak, Lodz, Lub, Mathes Po, Par TNSP, Poz, Prace M, Pr Vib,
         Prz Fil, QRSP, Roz, Schles, Slav, Stu Log, Stu M, Wars,
         Wroc, Zyc.*
*Pol An M        Polska akad nauk, annales polonici math. 1954-. C of
                 Pol SM An. Warsaw.
 Pol Bib         Pol akad nauk, osrodek bib i dokumentacji nauk, Pol bib anal.
                 1955-. Warsaw.
*Pol HPL         Studs & Materials on the Hist of Polish Learning, Ser C,
                 Math & .... 1953-. Warsaw. Pol Acad of Sci Res Cen, Hist
                 of Sci & Tec.
*Pol HST Mo      Monographs on the Hist of Sci & Tec. (Polsk Ak Um Hist Nauk
                 Pol Monografiach) 1948-. Krakow. Pol Acad of Sci, Comm on
                 the Hist of Sci.
 Pol HST Ss      Sources of the Hist of Sci & Tec. 1957-. Warsaw. Pol Acad
                 of Sci, Comm on the Hist of Sci.
*Politec         Il politecnico, gior d ingegnere, architetto civile ed
                 industriale. 1853-1937// Incl C of Politecn. Milan.
*Politecn        Il politecnico, repertorio mensile d stud appl alla
                 prosperita e coltura sociale. (Il polit) 1839-1844, 1860-.
                 C in Politec. Milan.
 Pol MAFK        Biulleten Polskii akad nauk, otd trete, mat, ast, fiz, khim,
                 geol, geog. 1952-. Tr of Pol MFCG. Warsaw.
 Pol MAPB        Bull d l'acad pol d scis, ser d scis maths, asts, & phys.
                 1958-. C of Pol MFCG. Warsaw.
 Pol MAPM        Mem d l'acad pol d scis, ser d scis maths asts & phys.
                 1929-. Warsaw.
 Pol Me          Pol bib analityczna, mech. (Pol Sci Absts, Mech) 1955-.
                 Warsaw. Polska akad nauk, inst podstawowych problemow tec.

| | |
|---|---|
| Pol MFCG | Bull, Pol akad nauk, wydzial III, nauk mat-fiz, chem, geol-geog. 1952-1957// C of Pol MP. C as Pol MA Ph. Tr as Pol MAFK. Warsaw. |
| Pol MP | Bull intl, Pol akad umiejetnosci, wydzial mat-prz, ser A, scis math. 1910-1951// C as Pol MFCG. Krakow. |
| *Pologne | La Pologne politique, economique, litt, & artistique. (Pologne) 1920-1934// Warsaw. Asso France-Pologne. |
| *Pol Rv | Polish Review. 1956-. NY. Pol Inst of Arts & Scis in Amer. |

*Polska akad nauk, Pol akad umiejetnosci, Polish Acad of Scis. See also Poland.*

| | |
|---|---|
| *Pol SM An | Annales d l soc pol d math. (Polskie towarzystwo mat, rozprawy; Rocznik Polski) 1922-1939, 1945-1952// C as Pol An M. Krakow. |
| Pol To Fiz | Polskie towarzystwo fiz, sprawozdania i prace. 1920-1931// C as Act Ph Po. Warsaw. |
| Poly Bib | Polytechnische Bibliothek, monatliches Verzeichnis d neu erschienenen Werke aus d Faechern d Math, Phys & Chem. 1866-1917//? Leipzig. |

*Pontaniana. See Nap.*
*Pontificio, etc. See Vat.*

| | |
|---|---|
| *Pop As | Popular Astronomy. 1894-. Northfield, Minn. |
| *Pop Sci Mo | Popular Sci Monthly. (Appleton's...) 1872-. NY. |

*Port Arthur, Manchuria. See Ryojun.*

| | |
|---|---|
| *Porto Ac | Annaes scientificos d acad polyt do Porto. 1905-1921// C of Coim I, & of Teixera. C as Porto Fac. Coimbra. |
| *Porto Fac | Annaes d fac d sci d Porto. (Anais Porto) 1922-. C of Porto Ac. Porto. |
| *Portu APS | Associasso portugeso para o progresso d scis, congresso. 1921// Porto. |

*Portugal. See also Coim, Lisb, Petrus No, Teixeira.*

| | |
|---|---|
| *Portu IA | Inst d actuarios portugueses, boletim. 1946-. Lisbon. |
| Portu M | Portugaliae mathematica. 1937-. Lisbon. |
| Portu Ph | Portugaliae physica. 1943-. Lisbon. |

*Posebna izdanja. See Beo IMS.*
*Posen. See Poz.*
*Poske Z. See Z Ph C Unt.*

| | |
|---|---|
| *Pos Rv | The Positivist Review. 1893-1923// C as Humanity. London. |
| *Post II | Post iucundam iuventutem. 1922-. Utrecht. Univ. |
| *Pots Ped | Wiss Zt d paed Hochschule Potsdam, math-naturw Reihe. 1954/55-. Potsdam. |
| Poz B | Bull d l soc d amis d scis & d letts d Poznan, ser B, scis maths & nats. 1925-. Poznan. |
| Poz Prac | Prace komisji mat-prz, Poznanskie towarzystwo przyjaciol nauk. 1921-1948// C as Poz Wyd. Poznan. |
| *Poz Rocz | Poznanskie towarzystwo przjaciol nauk, roczniki. (Soc sci d Poznan; Posener Gesell d Freunde d Wiss, Denkschr) 1860-. Poznan. |
| Poz UM | Uniwersytet im Adama Mickiewicza w Poznaniu, prace wydzialu mat, fiz i chem, ser mat. 1963-. Poznan. |
| Poz UZ | Zeszyty naukowe uniw im Adama Mickiewicza. 1956-. Poznan. |
| Poz Wyd | Wydzial mat-prz, komisja biol. 1948-. C of Poz Prac. Poznan. |

*Prace.  See also organization, city, region, country, etc.*
*Prace M   Roczniki Polskiego towarzystwa mat, ser I, prace mat,
       ser II, wiadomosci mat.  1955-.  Pt C of Prace MF.  Warsaw.
*Prace MF   Prace mat-fiz.  (Mat phys Abh Wars)  1888-1955?//  Pt C as
       Prace M.  Warsaw.
*Prag Ab   Rozpravy tridy math prir Kralovsje Ceske spolecnosti nauk.
       (Abh d K boehmischen Gesell d Wiss, math nat)  1775-1892,
       1927-.  Prague.
Prag BI ⁄   Acad d scis (d l'empereur Francois Joseph I), bull intl,
       resume d travaux presentes, scis maths & nats.  (Cesk akad
       ved a umeni Cisare Frantiska Josefa I; Prag Fr Jos Ac Sc
       Bll, Mth Nt)  1895-.  See also Prag Roz P.  Prague.
Prag Com M   Commentationes math univ Carolinae.  1960-.  Prague.  Univ,
       math-phys fac.  Founded E Cech.
*Prag Dej   Sb pro dejiny prir ved a techniky.  1954-.  Prague.
       Nakladatelstvi Cesk akad ved.
*Prag DMG   Mitt d deu math Gesell d Prag.  (Deu Mat Ges Prag)  1892-.
       Prague.
Prag Ja   Jasb d K boehmischen Gesell d Wiss.  (Bohm Gs Ws Jbr)
       1875-1917//  C as Prag Zpr.  German ed of Prag Zpr.  Prague.
Prag Karl   Spisy vydavane prir Karlovy univ.  (Acta fac rer nat Carol;
       Pub d 1 fac d sci d l'univ Char)  1923-.  Prague.
*Prag M Ap.  See Aplik M.*
*Prag Pok   Pokroky mat fys a ast.  1955-.  C of Prag Sov.  Prague.
Prag Rozh   Rozhledy mat-prirodovedecke.  1921-.  Prague.  Jednota Cesk
       mat a fys.
*Prag Roz P   Rozpravy Ceske akad (Cisare Frantiska Josefa) (pro) vedy,
       (Slovesnost) a Umeni (II cl).  (Roz Cesk Akad Ved)  1891-.
       Prague.  See also Prag BI.
*Prag Si   Sitz d K boehmischen Gesell d Wiss.  (Prag Ber)  1859-1917//
       C in Prag V.  Prague.
Prag Sov   Sovetska veda, mat-fyz ast.  1950-1955//  C as Prag Pok.
       Prague.  Cesk-Sovetsky inst.
Prag St ML   Prague Studies in Math Linguistics.  1966-.  Czech Acad
       of Scis.
*Prague.  See also Cas MF, Casopis, Cesk, Cz.*
*Prag V   Vestnik Kralovske Ceske spolecnosti nauk, trida mat prir.
       (Mem d 1 soc R (lettres &) sci Boheme)  1859-.  Incl C of
       Prag Si.  Prague.
Prag Zpr   Vyrocni zprava, Ceska (Kralovske) spolecnosti nauk.
       (Resume soc sci Boheme)  1918-.  C of Prag Ja.  Prague.
       Boehm ges wiss; Soc d letts & d sci Boh.
Prag ZZCM   Zpravy zednoty Cesk math.  1870-1872//  Prague.
*Praktika.  See organization, city, region, etc.*
*Praxis   Praxis d Math, Monatsschr d reinen & d angewandten Math i
       Unterricht.  1959-.  Cologne.
*Predavanja.  See organization, city, region, country, etc.*
*Preussische Akad d Wiss zu Berlin.  See Ber.*
Pribor   Priborostroenie.  1956-.  Tr as Instrum.  Moscow.  Minist
       mashinostroenie i priborostroenia SSSR.
Primary M   Primary Mathematics.  1963-.  C of Tea Arith.  Oxford.

*Pri Me Prikladna mekhanika. 1955-. Tr as Sov Ap Me. Kiev. Akad
nauk UkRSR, inst mekh (otd mat, mekh i kibernetiki).

*Pri M Me Prikladnaya mat i mekh. 1933-. C of Ves Me Pr M. Tr as
JAM Me. Moscow. Akad nauk SSSR, otd tekh nauk, inst mekh.

*Prin MS Princeton Math Ser. 1939-. Princeton, NJ.

*Priroda Priroda (pop est-istoricheskii zh). 1912-1919, 1921-.
Leningrad, Moscow. Akad nauk SSSR.

Prir Sof Priroda. 1952-. Sofia.

*Proceedings. See organization, city, region, country, etc.*

*Proces-verbaux, Processen-verbaal, etc. See organization, city, region, etc.*

Pro Cyb Problems of Cybernetics. 1960-. Tr of Pro Kib. London.

*Prog M El progreso mat, periodico d mat puras y aplicadas.
1891-1895, 1899, 1900// Madrid, Zaragoza. Dirs: E Zoel,
G d Galdeano.

*Prog OR Progress in Operations Res. 1961-. NY. Opers Res Soc Amer.

Prog T Ph Progress of Theor Phys. 1946-. Kyoto, Osaka.

Pro In Tra Probs of Info Transmission. 1965-. Tr of Pro Per In. NY.

Pro Kib Problemy kibernetiki. 1958-. Tr as Prob Cyb. Moscow.

*Pro Per In Probl peredaci info. 1965-. Tr as Pro In Tra. Moscow.
Akad nauk SSSR, lab sistem peredaci info.

*Proteus Proteus. 1931-1937// Berlin, Bonn. Thein Gesell f Gesch
d Naturw, Med, & Tec.

Proteus L Proteus, a J of Sci, Phil, & Therapy of Nature. 1931-1932,
NS 1947-. London.

Pr Vib Procs of Vibration Problems. 1959-. Warsaw. Pol Acad of
Scis, Inst of Basic Tec Probls (Inst podstawowych prob tec).

*Prz Fil Przeglad filozoficzny. 1921-. Warsaw.

*Prz St Przeglad stat, kwartalnik poswiecony nauce i potrzebom zycia.
1954-. Warsaw. Pol towarzystwo ekonomiczne.

*Psychmet Psychometrika, a J Devoted to the Devel of Psych as a Quant
Rational Sci. 1936-. S by Bu M Bioph. Chapel Hill, NC; etc.
Psychometr Soc.

*Psy Fors Psych Forschung, Zt f Psych. 1922-. Berlin.

*Publications, etc. See also organization, city, region, country, etc.*

*Pub M Pubs mathematicae. (Debrecen) 1949-. Debrecen.
Tudomany-egyetem, mat intezet.

*Pubs scis d ministere d l'air. See Fr Air.*

*Punjab. See Panjab.*

*Pyramide Die Pyramide, naturw Monatsschrift. 1951-. Innsbruck,
Munich.

Pyth Pythagoras. 1968?-. Groningen.

.Pyth E Pythagoras Engl Ed. 1968-. London. Ed: Bruckheimer,
Mansfield.


*QAM Quar of Appl Math. 1943-. Providence, RI. Brown Univ.

QJ Econ The Quar J of Economics. 1886-. Cambridge, Mass. Harvard
Univ P.

QJ Me AM The Quar J of Mech & Appl Math. 1948-. Oxford, London.

QJM Ox        The Quar J of Math, Osford.  1930-.  C of QJPAM, & of Mess M
              Oxford, London.
*QJPAM        Quar J of Pure & Appl Math.  (QJ Math Lon; QJM; Qua JM: Q)
              1857-1927//  C of Cam Dub.  C in QJM Ox.  London.
QRSP          Quar Rev of Sci Pubs.  1955-.  Ser C, Pure & Tec Scis, 1958-
              Warsaw.  Pol Acad of Scis, Distr Cen for Sci Pubs.
*QSGM         Quellen & Stud z Gesch d Math.  A, Quellen, 1929-1938//
              B, Stud, 1930-1938//  C of Arc GMNT.  Berlin.
QSG Ost       Quellen & Stud z Gesch Osteuropas.  1958-.  Berlin.  Akad
              Wiss, hist Abt.
Quad SSM      Quaderni d storia d sci e d medicina.  1963-.  Ferrara.
              Univ d studi.
*Quarterly, etc.  See also organization, city, region, country, etc.*
*Quatrns      Bull of the Intl Asso for Promoting the Study of Quaternions
              & Allied Systs of Math.  1900-1913//  Lancaster, Penn.
              Ed: A Macfarlane.
Queb AMB      Bull d l'asso math d Quebec.  1958-.  Montreal.
*Queen Q      Queen's Quarterly.  1893-.  Kingston, Ont.  Queen's Univ.
*Querce.  See Fir Querc.*
*Quest M Ap   Questioni d mat appl trattate nel II convegno d mat appl
              (Roma 1939).  (Quest appl math)  Bologna.  Consiglio naz
              d ricerche, comitato per l fis e l mat.
*Quetelet Cor Mth.  See Cor M Ph.*
*Quinzain     La quinzain, rev litt, artistique & sci.  1894-1907//  Paris.


RAAGNL        RAAG Newsletter.  Tokyo.  Res Asso of Appl Geometry.
RAAGRN        RAAG Res Notes.  Tokyo.  Res Asso of Appl Geometry.
*Raboty.  See organization, city, region, country, etc.*
*Rad.  See organization, city, region, country, etc.*
Rad El        Radiotekhnika i elektronika.  1956-.  Tr as Rad En El.
              Moscow.  Akad nauk SSSR.
Rad En El     Radio Engineering & Electronic(s) (Physics).  1957-.  Tr of
              Rad El.  NY.  IEEE.
*Rainer papyrus sammlung.  See Wien Papy.*
Rajasth       Procs of the Rajasthan Acad of Scis.  1960-.  Pilani, Jaipur.
Rajasth B     Rajasthan-Bharati.  1946-.  Bikaner.  Sadul R Res Inst.
*Raksti.  See organization, city, region, country, etc.*
*Rass MF      Rassegna d mat e fis, periodico mensile d inst Esu Ferraris.
              1920-1927//  Rome.
*Ratio        Ratio.  1957-.  Oxford.
RCAR          RCA Review.  1936-.  NY.  Radio Corp of America.
Realist       The Realist, a J of Sci Humanism.  1929-1930//  London.
Rec Aeron     Recherche aeronautique.  1948-1963//  C as Rec Aersp.  Paris.
Rec Aersp     La rech aerospatiale.  1963-.  C of Rec Aeron.  Paris.
              Off natl d'etudes & d rech aerospatiales.
*Record.  See organization, city, region, country, etc.*
*Recr M Mag   Recreational Math Mag.  1961-1964//  Idaho Falls, Idaho.
*Recueil d travaux d'hist & d philologie.  See Louv HPRT.*
Recueil M     Recueil math a l'usage d ecs specs & d etablissements d'in-
              struction moyenne.  Paris.

*Regeltec Regelungstechnik, Zt f Steuern, Regeln & Automatisieren.
    1953-. Munich.
*Regensb Ber d naturw Vereins z Regensburg. 1886-. Regensburg.
*Rel Ges VV Religionsgesch Versuche & Vorarbeiten. 1903-. Giessen.
*Renais N Renaissance News, a Quar Newsl Pub by Dartmouth Coll Library.
    1948-. Hanover, NH. Renaissance Soc of Amer.
*Rendiconti, etc. See also organization, city, region, country, etc.*
*Ren M Ap Rendiconti d mat e d sue appl. (RC Mat Univ Roma) 1940-.
    Rome. Univ, ist naz d alta mat.
*Repertorium f physikalische Technik.... See Carl PTMA.*
*Rep Kunst Repertorium f Kunstwissenschaft. 1876-1931// Berlin,
    Stuttgart, etc.
Rep Lit M Repertorium d lit Arbeiten aus d Gebiete d reinen & angew
    Math. (Rpm Mth) 1876-1879// Leipzig.
*Report, rapport, etc. See organization, city, region, country, etc.*
Res BMV Resultate aus d Beobachten d magn Ver. 1836-1841//
    Goettingen. Eds: Gauss, Weber.
*Res Mngt Research Management. 1958-. Easton, Penn. Ind Res Inst.
*Review, revue, revista, rivista, etc. See R, Rv; also organization,*
    *city, region, country, etc.*
*Revista d cien. See Lim Rev.*
*Revista d Sciencias. See Bras Rv Sc.*
*Revista matematica cuyana. See Cuyo U.*
*Revista universiara mathematica. See Buc U Rev.*
*Revue des maths elems. See Elem M.*
*Revue des sciences humaines. See Rv Hi Phi.*
*Revue rose. See Rv Sc.*
*Rheinisch-westfalischen.... See Aach THJ.*
*Riazan. See Ryaz.*
*Rice Pam The Rice Inst Pamphlet(s) 1915-. Houston, Tex.
Ricerca La ricerca, riv d mat pure & appl. (Ricerca Naples)
    1950-. Naples.
Rice Stu Rice Univ Studies. 1915-1961// Houston, Tex.
Ric M Ricerche d mat. 1952-. Naples. Univ, inst d mat.
*Ric Sc La ric sci & il progresso tec nell'econ naz. (Ricerca sci,
    Roma) 1930-. Rome. Minist d edu naz, consig naz d ric.
*Riga. See also Herd, Lat.*
Riga NV Correspondenzblatt d naturfors Ver z Riga. (Riga Cor-Bl)
    1845-. Riga.
*Riga univ. See also Lat U.*
Riga UZR Petera Stuckas Latvijas valsts univ, mat un dabas zinatnu
    fak, raksti. (Latvijas valsts univ zinantn raksti;
    Latviiskii gos univ im Petra Stucki, uch zap) 1929/30-.
    Riga. PSRS augstakas izglitibas minist. (Minist vyssego
    obrazovaniya)
*Riv Lemat Riveon lematematika, A Quar J Intended to Promote Math
    Res among Students of Math. 1946-1959// Jerusalem.
*Rjazan. See Ryaz.*
*Rm At. See Vat NLA.*
Rm ISP Fac d sci stat, demografiche e attuariali, ist d stat e ist
    d calcolo d probabilita. 1956-. Rome. Univ.

```
*Rm Sem        Sem mat d fac d sci d R univ d Roma, rendiconti.  (RC Sem
               Mat Univ Roma)  1913-1939//  C as Rm U Rn.  Rome.
*Rm U Rn       Rend d mat e d sue appl, R univ d Roma e R ist naz d alta
               mat.  (Rend mat sue appl, univ Roma ist naz alta mat)
               1940-.  C of Rm Sem.  Rome.
 Ro Acad An    Analele acad Romane.  1867-.  Incl Ro An 1880-1922.
               Bucharest.
*Ro An         Analele acad Romane, mem sect stiintifice.  1880-.  Incl in
               Ro Acad An 1880-1922.  Bucharest.
*Ro An MFC     Analele acad repub pop Romine, ser mat, fiz, chim.  1948-.
               Bucharest.
 Ro Bul        Bulitinul stiintific, acad repub pop Romine.  1948-.  C of
               Ro Bul St.  Pt C as Ro MF.  Bucharest.
 Ro bull scientifique.  See also Par BS Rou.
 Ro Bul St     Bull, sect stiintifica, acad Romana.  (Bull d 1 sect sci,
               acad Roumaine)  1912-1947//  C as Ro Bul.  Bucharest.
 Rock IPOP     Occasional Papers of the Fac & Friends of the Rockefeller
               Inst.  1958-.  NY.
 Rocky         Rocky Mountain J of Math.  1970-.  Missoula, Montana.  Univ
               of Montana, Rocky Mt Math Consortium.
 Ro Com        Comunicarile, acad repub pop Romine.  1951-1963//  Bucharest.
 Roczniki.  See organization, city, region, country, etc.
*Ro IM         Stud si cercetari mat.  (Mat trudy i issledovaniya; Etudes
               d rechs maths)  1950-1964//  C of Buc Disq, & of Ro SSM.
               C as Stu Cer M.  Bucharest.  Acad RPR, inst d mat.
 Ro I Me       Stud si cerc d mec (si metalurgie) apl.  1950-.  Bucharest.
               Acad repub pop Romine (inst d mec aplicata "Traian Vuia").
*Romag Sto     Atti e mem d R deputazione d storia patria per (1'Emilia e)
               1 Romagna.  1862-1947, NS 1948-.  Bologna, Ferrara, etc.
 Romania.  See also Buc, Cluj, Cujet Cla, Gaz M, Iasi, Math Cluj, Math Tim,
           Monog M, Par BS Rou, Rv Me A, Rv M Fiz, Tim.
 Rome.  See Rm, It, Vat;  also organization, etc.
 Ro MF         Bul stiintific, sect d stiinte mat si fiz, acad repub pop
               Romine.  1950-1957//  C from Ro Bul St.  Bucharest.
*Ro Mm Mo      Acad d stiinte din Romania, ser IIIa, mem si monog.
               Bucharest.
*Ro Rv M       Rev (Roumaine) d math pures & appl.  (Rev math pures appl;
               Zt f reine & angew math; J of pure & appl math)  1956-.
               C of Ro Rv MP.  Bucharest.  Acad d 1 repub pop Romine.
 Ro Rv MP      Revue d math & d phys.  1954-1955//  C of Sci dans 1 repub
               pop Roumaine.  C as Ro Rv M.  Bucharest.  Acad RPR.
 Rosario.  See Arg IM Ros.
 Ro SF         Bull d 1 soc Roumaine d phys.  (Bul soc Romane d fizica)
               1923-.  Pt C of Ro SRS.  C in Ro SSMP.  Bucharest.  Lab
               d'acoustique et opt, univ d Bucharest.
 Ro Sov        Analele Romano-sovietice.  1946-1949//  C as Ro Sov MF.
               Bucharest.  Acad repub pop Romane, inst d studii romano
               sovietic.
 Ro Sov MF     Analele Romino-sov, ser mat-fiz.  (Rumyno-Sov zap, fiz-mat
               ser)  1949-.  C of Ro Sov.  Bucharest.  Acad repub pop
               Romine, inst d stud Romino-sov.
```

| | |
|---|---|
| *Ro SP | Bul soc politecnice din Romania. 1887-. Bucharest. |
| Ro SRS | Buletinul soc Romane d stiinte. (Bull soc roum sci) 1911-1915// C of Buc SS. C as Ro SF, & as Ro SSM. See also Buc. Bucharest. |
| *Ro SSM | Pubs d 1 fac d scis d Bucarest, bull mat d 1 soc roumaine d scis. (BMSSM & phys d 1 repub pop Roumanie; Bul math soc Roum) 1916-1947// Pt C of Ro SRS. C as Ro IM, & as Ro SSMP. Soc Romana d stiinte, sect mat. Bucharest. |
| *Ro SSMP | Bull math d 1 soc d scis math & phys d 1 repub pop roumaine. (Soc stiinte mat din RP Rom) 1957-. C of Ro SSM, & of Ro SF. Bucharest. |
| Rostock U | Wiss Zt d Univ Rostock, math-naturw Reihe. (Wiss Zt, Heft 2) 1951-. Rostock. |
| *Rostov Pe | Rostovskii-na-Donu gos ped inst, fiz-mat fac, uch zap. 1949-. Rostov-on-Don. |
| *Rostov U | Izv severo-Kavkazskogo gos univ, Izv Donskogo gos univ. (Rostov-Don izv sev Kavk univ) 1918-1931?// Rostov-on-Don. |
| Rostov UZ | Uch zap, Rostovskii-na-Donu gos univ. 1934-1942, NS 1945-. Rostov-on-Don. |
| *Rott NV | Nieuwe verh v het bataafsch genootschap d proefondervinde-lijke wijsbegeerte te Rotterdam. (Rot N Vh) 1800-. Rotterdam. |
| Rouen Ac | Precis anal d travaux d l'acad d scis, B-L, & arts d Rouen. (Rouen Ac Tr; Rouen Tr Ac) 1744-1794, 1803-. Rouen. |
| *Royal.... | *See city or country; e g Dub, Lon, etc.* |
| Royal Asiatic Soc. | *See As So GB, Bom RAS, Lon Asia.* |
| Rozhlady. | *See city, organization, region, country, etc.* |
| *Roz Inz | Rozprawy inzynierskie, inst podstawowych probl tec, Pol akad nauk. 1953-. Warsaw. |
| Roz M | Rozprawy mat, inst mat, Pol akad nauk. 1952-1966// C as Diss M. Warsaw. |
| Rozprawy. | *See also organization, city, region, etc.* |
| RS | Rev semestrielle (des pubs maths). 1893-1934// Merged with F, 1935-. Amsterdam. Wiskundig genootschap. |
| *Rueb | Reubezahl. 1868-1875// Breslau, etc. |
| Rumania. | *See Ro.* |
| *Rus As Zh | Ruskii ast zh. 1924-1932// C as As Zh. Moscow. |
| *Rus MS | Russian Math Surveys. 1960-. Tr of UMN. London. Math Soc. |
| Rv Acous | Revue d'acoustique. 1932-. Paris. |
| *Rv Arch | Revue archeologique. 1844-. Paris. |
| *Rv Assyr | Rev d'assyriologie & d'archeol orientale. 1884-. Paris. |
| *Rv Bene | Rev benedictine. 1884-. Abbaye d Maredsous, Belgium. |
| *Rv Bourg | Revue bourguignonne. 1891-1914// Dijon. Univ d Dijon. |
| Rv Cat Ser | Riv d catasto e d servizi tec erariali. 1934-. Rome. Ist poligrafico d stato PV, minist d finanze. |
| Rv Cien. | *See Lim Rev.* |
| *Rv Cou Con | Revue d cours & confs. 1892-1940// Paris. |
| *Rv Deux M | Revue d deux mondes. 1829-. Paris. |
| Rv Ec Est | Rev d economica e estatistica. 1936-1940// C of Mens Est C as Bras Est. Rio d Janeiro. |
| Rv Ec St | Review of economic(s &) statistics. 1919-. Cambridge, Mass. |

Rv Ec Stu      Rev of Economic Studies. 1933-. London.
*Rv Enc        Rev encyclopedique, ou anal raisonee d productions 1 plus
                 remarquables dans 1 litt, 1 scis, & 1 arts. 1819-1835//
                 Paris.
*Rv Et Grec     Rev d etudes grecques. 1888-1939//? Paris. Asso pour
                 l'encouragement d etudes grecques en France. C of their
                 Annuaire.
*Rv Eth Soc     Rev d'ethnographie & d sociol. 1910-1914// Paris. Inst
                 ethnographique intl d Paris.
*Rv Et Juiv     Rev d etudes juives. 1880-. Paris. Soc d etudes juives.
*Rv Europ      Rivista europea, riv intl. 1868-1883// Florence.
*Rv Fil        Riv d filosofia. 1909-. C of Rv Fil Pav. Modena, Perugia.
                 Soc filosofica ital.
Rv Fil Neo      Riv d filosofia neo-scolastica, pub per cura d fac filosofia
                 d univ cattolica d Sacro Cuore. 1909-. Milan.
Rv Fil Pav      Rivista filosofica. 1899-1908// C as Rv Fil. Pavia, Rome.
*Rv FMSN       Rivista d fis, mat, e sci nat. 1900-1912// Pavia. 1927-,
                 Naples.
*Rv Fr         Revue d France. 1921-1939// Paris.
  *Rv francaise d traitement d l'info.* See Chiffres.
*Rv Fr Asq     Rev francaise d'astronautique. 1958-. Paris. Soc francais
                 d'astronautique.
Rv Fr OR       Revue francaise d rech operationelle. 1956-. Paris.
*Rv Gen IS     Revue gen intl sci, litteraire, & artistique. 1896-1898//
                 Paris.
*Rv GSPA       Rev gen d scis pures & appls & bull d l'asso francaise pour
                 l'avancement d scis. (Rev Gen Sci) 1890-1939, 1944-1947//
                 C in Philom Bu. Paris.
*Rv Hebdom     Revue hebdomadaire. 1892-1939// Paris.
*Rv Hi Phi     Rev d'hist d 1 phil (& d'hist gen d 1 civ). (Rev d scis
                 humaines, fac d letts d l'univ d Lille) 1927-. Paris, Lill
*Rv Hi Rel     Rev d l'hist d religions. 1880-. Paris. Mus Guimet.
*Rv Hi Sc Ap   Rev d'hist d scis & d leurs appls. (Rev Hist Sci) 1947-.
                 Paris. Sect d'hist d scis d cen intl d synthese.
*Rv HL Fr      Rev d'hist litt d 1 France. 1894-. Paris.
*Rv I Ens      Rev intl d l'enseignement. 1881-. Paris. Soc d l'en-
                 seignement sup.
*Rv II St      Rev d l'inst intl d stat. (Rev of the Intl Stat Inst)
                 1933-. C of Bu II St. The Hague.
*Rv In Gr It   Riv indo-greco-ital d filologia, lingua, antichita. 1916-.
                 Naples.
Rv It Sc Ec    Riv ital d sci econ (statistica). 1929-. Bologna.
*Rv It SLA    Riv ital d scienze, lett & arti. 1860?-1875?// Milan.
*Rv M         Revue d math. (Rivista d mat; Riv mat Peano) 1891-1906//
                 Turin.
Rv MCF        Rev d math commerciales & financieres. Vienna.
Rv Me A       Rev d mechanique appl. (Rev d scis techs) 1954-. Buchares
                 Acad d 1 repub pop roumaine.
*Rv M El      Rev d mat elem, fac d ciencias, univ nacl, univ d 1 Andes.
                 1952-1967// C as Colom M. Bogota.
*Rv Met Mor    Rev d metaphys & d morale. 1893-1942, 1945-. Paris.

| Rv Mex F | Rev mexicana d fis. 1952-. Mexico City. |
|---|---|
| Rv MF El | Rev d mat y fis elem. 1919-1924// Buenos Aires. |
| Rv M Fiz | Rev mat si fiz. -1954// C as Gaz MFB. Bucharest. |
| *Rv M Hisp A | Rev mat hispano-amer. 1919-. C of B Ai El. Madrid. |
| Rv M Ibalk | Rev math d l'union interbalkanique. 1936-. Athens. |
| *Rv Min Eng | Rev mineira (d engenharia). (Revta min Engenh) 1929-. Lisbon. |
| *Rv Mod Ph | Revs of Modern Phys. 1930-. C of Phy R Sup. NY. |
| *Rv Mois | Revue d mois. 1906-1914, 1919-1920// Paris. Ed: Borel. |
| Rv M Pit | Rev d mat pitagora pnetru uzul scoalelor secundare. Iltov. |
| Rv M Sant | Revista d mat. 1903-1906?// Santiago de Chile. |
| *Rv M Sp | Rev d maths speciales. (Rev math spec Paris) 1890-1914, 1920-. Paris. |
| Rv Neo | Rev neoscholastique (d phil). 1894-1914, 1919-1941// C as Rv Phi Lou. Louvain. Soc phil d Louvain. |
| Rv Opt | Rev d'optique theor & instrumentale. 1922-. Paris, Nancy. |
| *Rv Par | Rev d Paris. 1894-1939, 1945-. Paris. |
| *Rv pedagogique.* | *See Ens Pub.* |
| *Rv Phi | Rev d philosophie. 1900-1914, 1919-1939// Paris. |
| *Rv Phi Fr E | Rev phil d l France & d l'etranger. 1876-. Paris. |
| *Rv Phi LHA | Rev d philologie, d litt & d'hist anc. 1845-. Paris. |
| Rv Phi Lou | Rev phil d Louvain. 1945-. C of Rv Neo. Louvain. |
| *Rv Pyr | Rev d Pyrenees (& d l France meridionale). 1889-1914// Toulouse. Univ. Asso pyreneenne. |
| *Rv Q Sc | Rev d questions scis, pub par l soc sci d Bruxelles. 1877-1914, 1920-1939, 1946-. Bruxelles, Louvain. |
| *Rv Sc | Revue scientifique. (Rev d cours sc d l Fr & d l'etranger; La rev sci d l Fr & d l'etr; Rev rose; Rv Sc; Rv Scient) 1863-1954// C as Nucleus. Paris. |
| *Rv sciences humaines..., univ Lille.* | *See Rv Hi Phi.* |
| *Rv Sco Ind | Riv sci-ind d principali scoperte & invenzioni fatte nelle sci e nelle ind. (Rv Sc-Ind) 1869-1909// Florence. |
| Rv S Inst | Rev of Sci Instruments. 1930-. C from Am Opt J. Lancaster, Penn, etc. Amer Inst of Physics. |
| *Rv St Adam | Revista stiintifica 'V Adamachi'. 1923-. Iasi. |
| *Rv Sto Cr | Rev d storia critica d sci med e nat. (Rivista storia sci) 1910-1956// C of It Sto At. Faenza, Siena. Ital Soc for the Hist of Sci. |
| *Rv Syn | Revue d synthese. 1931-. C of Rv Syn Sc. Paris. |
| *Rv Syn Sc | Rev d synthese hist. 1900-1914, 1919-1930// C in Rv Syn. Paris. |
| *Rv Trim | Rev trimestral d mat. 1901-1906// Zaragoza. |
| Ryaz Ped | Ryazanskii gos ped inst, uch zap. 1939-. Ryazan. |
| *Rylands Library.* | *See Manc JRL.* |
| Ryojun CE | Mems of the Rvojun Coll of Eng. 1927-. Ryojun. |
| RZ | Referativnyi zh, mat referaty. 1953-. Moscow. Akad nauk SSSR, inst naucnoi info. |
| RZ Me | Referativnyi zh, mekh referaty. 1953-. Moscow. Akad nauk SSSR, inst naucnoi info. |

SAAB          Saab Tecnical Notes. 1951-. C of Svenska aeroplan A B
              Tec Notes. Linkoping. Svenska aeroplan aktiebolaget.
*Saar. See Sarav.*
*Sachsen und Thuringen. See Halle.*
*Saechsischen Akademie, Gesellschaft. See Leip.*
*Saeculum     Saeculum, Jab f Universalgesch. 1950-. Freiburg.
*Saggiat      Il saggiatore, riv mensile d attualita sci. 1951-. Turin.
*Saint Petersburg. See Len, Pet.*
Saita A       Sci Repts of the Saitama Univ, Ser A, Math, Phys, & Chem.
              1952-. Urawa.
*Sakharth SSR mecn akad math.... See Tbil MI.*
*Salamanca. See Act Sal.*
Samark U      Alishe Navoii nom Uzbek davlat univ asarlari. (Trudy
              Samarkandskogo gos univ im Alesha Navoi) 1935-. Samarkand.
              Minist vyshego i srednego spets obrazovaniya UzSSR.
*Sancti Paulii univ, Tokyo. See Paul UCM.*
*Sankh A      Sankhya, the Indian J of Stat, Ser A. 1933-. Calcutta.
*Sankh B      Sankhya, the Indian J of Stat, Ser B. 1933-. Calcutta.
Sao Pau CM    Cadernos d mat, Sao Paulo univ escola polyt. 1945-.
              Sao Paulo.
*Sao Paulo univ. See also JM Pu Apl.*
*Sao Pau SM   Bol d soc d mat d Sao Paulo. 1946-. Sao Paulo.
SA Ph         Sci Absts, Sect A, Phys Absts. 1898-. London.
*Saragossa. See Zarag.*
Sarav MN      Anns univ Saraviensis, math-naturw Fak. 1963-. C of
              Sarav N. Berlin.
Sarav N       Anns univ Saraviensis, naturw-scis. 1952-1963// C as
              Sarav MN. Saarbruecken, univ d Saarlaendes.
*Sarav PL     Anns univ Saraviensis, phil-letts. 1952-. Saarbruecken,
              univ d Saarlaendes.
*Sardinia. See Cagliari.*
*Sasayama     J of Spatial Math of the Sasayama Res Room. 1958-. Nagoya.
*Sat Rv       The Saturday Rev, Politics, Lit, Sci, & Art. 1855-1938//
              London.
*Sbornik, etc. See organization, city, region, etc.*
*Schles GZ    Zt d Ver f Gesch Schlesiens, Breslau. 1855-1943// Breslau.
*Schles Jb    Jasb d schlesischen Gesell f vaterlaendische Kultur.
              1850-. Breslau.
*Schloemilch. See Z M Ph.*
*Scholas      Scholastik, Vierteljahrsschr f Theologie & Phil. 1926-1944,
              1949-. Eupen, Esch, Freiburg.
*School of Oriental Studies. See Lon SOAS.*
*School Sci & Math. See SSM.*
*School Sci Rev. See SSR.*
*Schr GNTM. See NTM.*
*Schriften, skrifter, etc. See organization, city, region, country, etc.*
*Schriftenreihe des Forschungsinstituts. See Ber FIM.*
*Sch Rv       School Review, a J of Sec Edu. 1893-. Ithaca, NY; Chicago.

| | |
|---|---|
| *Sch Soc | School & Society. 1915-. Garrison, NY. Soc for the Adva of Education. |
| Sch Vis | School Visiter, Devoted to Practical Math, Examination Work, Notes, Queries, & Answers. 1880-1894// Versailles, Ohio. Ed: J S Royer. |
| *Sch Vit | Schola & vita. 1926-. Milan. Acad pro interlingua. |
| *Schwa Mer | Schwaebischer Merkur, eine politische Zeitung. 1785-. Stuttgart. |
| *Schw NG | Verh d schw naturfors Gesell. (Actes d 1 soc helvetique d scis nats; Atti d soc elvetia d sci nat; Eroeffnungsrede...; Kurze uebersicht...) 1817-. See also Schw NGCR. Aarau, Basel, Geneva, etc. |
| Schw NGCR | CR schw naturforschende Gesellschaft. 1880-1909// C from Schw NG, & from Bib Un. C in Schw NG. Aarau. |
| *Schw Sch | Schweizer Schule, Halbmonatsschr f Erziehung & Unterricht. Olten. Katholische Schul- & Erziehungsver d Schweiz. |
| *Schw Verm | Schw Zt f Vermessungswesen & Kulturtec (& Photogrammetrie). (Revue tec suisse d mensurations et ameliorations foncieres, d genre rural et d photogrammetrie; Schweiz geometerzeitung) 1903-. Winterthur. Schw Geometerverein (Soc suisse d geometres). |
| *Schw Vers | Mitt d Vereinigung schw Versicherungsmath. (Bull d l'asso d actuaires suisses) 1906-. Bern. |
| *Sci | Science. 1883-. Incl Sci Mo 1958-. NY; Wash DC. |
| *Sci Am | Scientific American. 1846-. NY. |
| *Sci Cult | Science & Culture. 1935-. Calcutta. |
| *Sciences | Sciences, rev (bull) d l'asso fr pour l'avancement d scis. 1936-. C of Fr AASB. Paris. Hermann. |
| *Scientia | Scientia, riv d scienza. 1907-. Bologna. |
| *Sci Hist | Scientiarum historia. 1959-. Antwerp. |
| Sci Inf N | Scientific Info Notes (News). 1959-1968// Wash DC. NSF. |
| *Sci L Bib S | Sci Lib Bib Ser. 1930-. London. Science Mus. |
| *Sci memoirs, selected...foreign academies...London. See Taylor Ms.* | |
| *Sci Mo | The Scientific Monthly. 1915-1957// C in Sci. NY. |
| *Sci Papers of the Inst...Tokyo. See Tok IPCR.* | |
| *Sci Prog | Sci Progress in the Twentieth Cent. 1906-. London. |
| *Sci Record. See Ko Hs Chi.* | |
| *Sci Sin | (Acta) Scientia sinica. (Kexue tongbao) 1952-1966?// See also Ko Hs TP. Peking. Academia sinica. |
| *Sci Soc | Science & Society, a Marxian Quar. 1936-. NY. |
| *Sci Tec Ch | Sci & Tec in China. 1948-. C in Act Br Si. Nanking. |
| *Sci Tec Pi | Sci & Tec, a Record of Lit Recently Added to the Carnegie Lib of Pittsburgh. 1946-. Pittsburgh, Penn. |
| *Scotland. See Edi, Glasg, Nort Br.* | |
| *Scr | Scripta math, a Quar J Devoted to the Phil, Hist, & Expository Treatment of Math. 1932-. NY. |
| *Scriptor | Scriptorium, revue intl d etudes rels aux manuscrits. 1946-. Brussels. |
| *Scr Lib | Scripta mathematica Library. 1934-. NY. Academic P. |
| *Scr Stu | Scripta mathematica Studies. 1936-. NY. Academic Pr. |
| *Seibutsu. See Bio Sci, Jp J Biost.* | |

*Seismological Society of Amer. See Am Seis.*
*Semester-Berichte. See organization, city, region, country, etc.*
Semigrp          Semigroup Forum. 1970-. NY. Springer Verlag.
*Seminar, etc. See organization, city, region, etc.*
*Send CSJ        J of the Sendai Intl Cultural Soc. Sendai.
*Serbia. See Beo.*
Sev Kav Pe       Uch zap, Severo-Osetinskii gos ped inst im K L Khetagurova.
                 (Izv Gorsk ped inst; Izv Sev Kav ped inst; Izv Gorsk inst
                 narod obraz) 1923-1937, NS 1938-. Dzandzhikan.
Shang SI         J of the Shanghai Sci Inst, Sect I, Math, Ast, Phys, Geophys,
                 Chem & Allied Scis. 1932-. Shanghai.
*Shevchenko Gesellschaft. See Lemb Si.*
*Shig Zas        Shigaku zasshi. (Zt f Geschichtswiss) 1889-. Tokyo.
Shiz UNS         Repts of Liberal Arts & Sci Fac, Shizuoka Univ, Sect Nat Sci,
                 Ser B. 1950-1965// Shizuoka.
Shu Hs           Shu hsueh chin chan t'sung shu. 1959-. C of Shux Jinz.
                 Peking.
Shux Jiao        Shuxue jiaoxue. (Shusiue tsziaosiue; Teaching Math) 1958?-.
*Shux Jinz       Shuxue jinzhan. (Shu hsueh chin chan; Shusieu tszinzhang;
                 Advancement in Math; Progress in Math) 1955-1959// C as
                 Shu Hs. Shanghai.
Shux Ton         Shuxue tonbau. (Shu hsueh t'ungpao; Bull of Math) 195?-.
                 Peking.
*SIAMJ           SIAM J of Appl Math. 1953-. Philadelphia, Penn. Soc for
                 Ind & Appl Math.
SIAMJ Con        SIAM J on Control. (J of the Soc for Ind & Appl Math, Ser A,
                 Control) 1963-. Philadelphia, Penn.
SIAMJNA          SIAM J on Numerical Anal. (J of the Soc for Ind & Appl
                 Math, Ser B, Num Anal) 1964-. Philadelphia, Penn.
*SIAMR           SIAM Review. 1959-. Philadelphia, Penn. Soc for Ind &
                 Appl Math.
Siber MJ         Siberian Math J. 1966-. Tr of Sib MJ. NY. Consultants
                 Bureau.
Sib FM Tr        Sibirskii fiz-mat inst, trudy. 1932-. Tomsk.
*Sib MJ          Sibirskii mat zhurnal. 1960-. Tr as Siber MJ. Moscow.
*Sib TN          Izv Sib otd akad nauk SSSR (ser tekh nauk). 1957-1962,
                 series 1963-. Novosibirsk.
*Sicily. See Acir, Cat, Paler.*
*Siena           Atti d accad (d sci d Siena detta) d fisio-critici. (Siena
                 At Ac) 1760-1930// C in sects. Siena.
*Silliman J. See Am JS.*
*Simferopol. See Crim UMLP, Tauric U.*
*Sim Stev        Simon Stevin, wis- en natuurkundig tijd. 1947-. C of
                 C Huyg, & of Zutphen, & of Wis Nat Ti. Groningen, Djakarta.
Sing MS          Majallah tahunan 'ilmu pasti. (Bull of Math Soc, Nanyang
                 Univ; Shu hsueh nien k'an) 1959-1965// C as Nanta M.
                 Singapore.
*Sinolog         Sinologica, Zt f chinesische Kultur & Wiss. 1947-. Basel.
*Sitzungsberichte. See organization, city, region, country, etc.*
*Sk Akt          Skandinavisk aktuarietidskrift. 1918-. Uppsala.
*Sk Kong         Skand matkongres. (D ... skand matkongres; CR d ... cong
                 d mathns scandinaves) 1909-. Oslo.

Skopj           Annuaire, fac d scis d l'univ d Skopje.  Skopje.
*Skopje.*  *See also Mak.*
Skopj PM        Godisenzbornik, prirodno-mat oddel, fil fak, Skopje univ
                (Macedonian Univ).  1948-.  Skopje.
*Sky Tel        Sky & Telescope.  1941-.  NY.
*Slav Cong      CR d premier cong d mathns d pays slaves, Warszawa, 1929.
                (Sprawozdanie z I kongresu math krajow slowianskych)
                1930//  Warsaw.
Slov DVT        Z dejin vied a tec na Slovensku.  (From the Hist of Sci &
                Tec in Slovakia) 1962-.  Bratislava.
*Slovenska.*  *See also MF Cas.*
Slov MFT        Razprave, razred za mat fiz in tec vede, ser A, mat, fiz in
                kemicne vede, Slov akad znanosti in umetnosti.  (Acad
                scientiarum et art slovenica, classis 3, math, phys, chem,
                dissertations) 1950-.  Ljubljana.
*SMF.*  *See Fr SM.*
Smi C As        Smithsonian Contributions to Astrophysics.  1956-.  Wash DC.
*Smi Eth R      Annual Report, Smithsonian Inst Bureau of Amer Ethnology.
                1879/80-.  Wash DC.
*Smi R          Annual Rept, Board of Regents of the Smithsonian Inst,
                Showing the Opers, Expenditures, & Condition of the Inst.
                (Smiths Rp) 1846-.  Washington DC.  US Govt Printing Office.
*So Af AAS      Rept of the S African Asso for the Adva of Sci.  1903-1908//
                C as So Af JS.  Cape Town.
*So Af JS       S African J of Sci.  1909-.  C of So Af AAS.  Cape Town.
So Af PS        Transas of the S African Phil Soc.  (S Afr Ph S T)
                1877-1909//  C as So Af RS.  Cape Town.
*So Af RS       Transas of the R Soc of S Africa.  1909-.  C of So Af PS.
                Cape Town.
*Social Pr      Social Problems.  1953-.  NY.
*Sociedad cientifica "Antonio Alzate."*  *See Mex Alz.*
*Societe philomathique....*  *See Philom; also city, region, etc.*
*Society for Ind and Appl Math.*  *See SIAM.*
Sociomet        Sociometry, a J of Research in Psych.  1937-.  NY.
Sof F           Godisnik na Sofiiskiya univ fiz fak.  (Annuaire d l'univ d
                Sofia, fac d phys) 1904-.  Sofia.
*Sof FMD        Jubileen sb na fiz-mat druzestvo v Sofija po Slucaj
                40-godisnija mu jubilej.  1939//  Sofia.
*Sofia.*  *See also Bulg, MF Sof.*
Sof M           Godisnik na Sofiiskija univ mat fak.  (Annuaire d l'univ
                d Sofia, fac d math) 1904-.  Sofia.
Sof Sp FM       Spisanie fiz-mat druzhestvo.  1916-1950//  Sofia.  Univ.
*South Africa.*  *See So Af.*
Sov Ap Me       Soviet Appl Mechanics.  1966-.  Tr of Pri Me.  NY.  Faraday P.
*Sov As         Soviet Astronomy.  1957-.  Tr of As Zh.  NY.  Amer Inst of
                Physics.
*Sov Cong       Transas of All Union Math Meetings of the Acad of Sci, USSR.
*Sov Cult.*  *See Survey.*
Sov M           Soviet Mathematics, Doklady.  1960-.  Tr of Dokl M.
                Providence, RI.  AMS.
*Sov Ph Ac      Sov Phys, Acoustics.  1955-.  Tr of Aku Zh.  NY.  Amer Inst
                of Physics.

*Sov Ph Cr     Soviet Phys, Crystallography. 1956-. Tr of Kristal. NY.
               Amer Inst of Physics.
Sov Ph Do      Soviet Phys, Doklady. 1956-. Tr of Dokl F. NY. Amer
               Inst of Physics.
Sov Ph J       Soviet Physics, JETP. 1955-. Tr of Zh ETF. NY. Amer Inst
               of Physics.
Sov Ph Sol     Soviet Phys, Solid State. 1959-. Tr of Fiz Tverd. NY.
               Amer Inst of Physics.
Sov Ph TP      Soviet Phys, Tec Phys. 1956-. Tr of Zh TF. See also JTP.
               NY. Amer Inst of Physics.
Sov Ph Usp     Soviet Phys, Uspekhi. 1958-. Tr of UFN. NY. Amer Inst
               of Physics.
*Sov Rv        Soviet Review, a J of Translations. 1960-. NY.
*Sov Survey*.  *See Survey*.
*Sov Vost      Sovetskoe vostokovedenie. 1955-1958// Moscow. Akad nauk
               SSSR, inst vostokovedeniya.
Spacetim       Spacetime. 1967-. Arlington, Tex. Univ of Texas.
*Spain. See Esp; also Al And, Barc, Cadiz, Estad Esp, Mad, Majocc,*
               *Rv M Hisp A, Sal, Zarag.*
*Spatsinski's Bote, Spaczin's Bote. See Kagan.*
*Speculum      Speculum, a J of Mediaeval Studies. 1926-. Cambridge, Mass.
               Mediaeval Acad of Amer.
*Sphinx        Sphinx. 1902?-. Chicago, Ill.
*Sphinx Br     Sphinx, rev mensuelle d questions recreatives. 1931-1939//
               Brussels.
*Sphinx Oe     Sphinx-Oedipe. 1906-1932// Nancy.
*Spisanie, Spisy. See organization, city, region, country, etc.*
*Srbije. See Beo.*
S Res In J     Stanford Research Inst J. 1964-. C of SRIJ. Menlo Park, Cal
SRIJ           SRI Journal. 1957-1963// C as S Res In J. Menlo Park, Cal.
               Stanford Res Inst.
*Srpska. See Beo.*
*SSM           School Science & Math. 1901-. Chicago, Ill.
*SSR           School Science Review. 1919-. London.
*SSSR. See also USSR, Sov, Russ.*
SSSR Aku       Trudy komissia po akustike. 1939-1955// C as Aku Zh.
               Moscow. Akad nauk SSSR.
SSSR Ener      Izv akad nauk SSSR, otd tec nauk, energetika i avtomatika.
               1959-. Pt C of SSSR Tek. Moscow.
SSSRF          Izv akad nauk SSSR, ser fiz. 1936-. Pt C of SSSR Iz.
               Moscow.
*SSSR Gf       Izv akad nauk SSSR, ser geofiz. 1952-. C of SSSRGG. Tr as
               Bu Geoph. Moscow.
SSSR Gf Tr     Trudy geofiz inst, akad nauk SSSR. 1948-. C of SSSR Seis.
               Moscow, Leningrad.
SSSRGG         Izv akad nauk SSSR, ser geog i geofiz. 1937-1951// Pt C of
               SSSR Iz. C as SSSR Gf. Moscow.
*SSSR Iz       Izv akad nauk SSSR. (Len Bull Ac Sc; Bull Acad Sc URSS;
               Bull Ac Sci Len) 1928-1935// C of Pet Iz. C as SSSRF,
               SSSRGG, SSSRM, etc. Leningrad.
*SSSRM         Izv akad nauk SSSR, ser mat. 1937-. Pt C of SSSR Iz.
               Tr as MUSSR Iz. Moscow.

| | |
|---|---|
| *SSSR Mash | Trudy sem po teor mash i mekh. 1947-. Moscow. Akad nauk SSSR, inst mashinovedeniya. |
| *SSSR Mek | Izv akad nauk SSSR, otd tec nauk, mekh i mashinostroenie. 1959-. Pt C of SSSR Tek. Moscow. |
| SSSR Met | Izv akad nauk SSSR, otd tec nauk, metallurgiya i toplivo. 1959-. Pt C of SSSR Tek. Moscow. |
| SSSR Opt | Optika i spektroskopiya. 1956-. Moscow, Leningrad. Akad nauk SSSR. |
| SSSR Seis | Trudy seismologicheskogo inst. (Pubs d l'inst seismologique) 1930-1946// C as SSSR Gf Tr. Moscow, Leningrad. Akad nauk SSSR. (Acad d scis d l'URSS) |
| *SSSR Stek | Trudy (fiz-) mat inst im V A Steklova. 1931-. Moscow, Leningrad. Akad nauk SSSR. See also Stekl MI. |
| *SSSR Tek | Izv akad nauk SSSR, otd tek nauk. (Bull d l'acad d scis d l'URSS, cl d scis tecs) 1937-1958// C as SSSR Ener, SSSR Mek, SSSR Met, Tek Kib. Moscow. |
| *SSSR Vest | Vestnik akad nauk SSSR. (J acad sci USSR) 1931-. Moscow, Leningrad. |
| SSSRVSB | Biulleten, vsesoyuznyi komitet po delam vysshei shkoly. 1936-1939// C as Ves Vy Shk. Moscow. |
| SSSRVSFM | Nauchnye dokl vysshei shkoly, fiz-mat nauki. 1958-. Moscow. Minist vysshego obrazovaniya SSSR. |
| SSSR Vych | Sb trudov vychislitelnogo tsentra, akad nauk SSSR. |
| *SSSR vychislitelnyi tsentr.* | *See also Vych, Algo Yaz.* |
| *Stal Ped | Uch zap Stalingradskogo gos ped inst im A S Serafinovicha. 1948-. Stalingrad. |
| Sta MR | Stat Methodology Revs. (Res & Stat Methodology) 1933-. New Brunswick, NJ. |
| *Sta Neer.* | *See St Ned.* |
| *Stanford Research Institute.* | *See SRI.* |
| Sta Res Mm | Statistical Res Mems. 1936-1938// London. Univ of London, Univ Coll, Dept of Stat. |
| Sta Rund | Stat Rundschau, Zt f stat Theorie & Praxis. (Stat obzor, rev pro stat theorii a praxi) 1920-. Prague. Stat Zentralamt. (Vydavatel ustredni stat urad) |
| *Stat Bln | Statistica. 1941-. C of Sup St. Bologna, Ferrara, Milan. Univ of Bologna. |
| Sta Tid | Statsokonomisk tid. 1887-. Oslo. Styret for statsokonomisk forening. |
| Statistn | The Statistician. (The Incorporated Statistician) 1950-. London. Inst of Statisticians. |
| Sta TMA | Stat Thy & Meth Absts. (Intl J of Absts; Stat Thy & Meths) 1959-. London, Edinburgh. Intl (Lon) Stat Inst. |
| *Sta Vier | Stat Vierteljahresschrift. 1948-1957// C in Metrika. Vienna. |
| Sta Wars | Statistica. 1929/30-. Warsaw. Szkola glowna gospodarstwa wiejskiego. |
| *Stekl MI | Trudy mat inst imeni V A Steklova. 1933-. Moscow, Leningrad. Akad nauk SSSR. See also SSSR Stek. |
| *Sterne | Sterne. 1921-. Stuttgart, Potsdam, Leipzig. |
| *Stichting voor theoretische Biologie....* | *See Act Bioth.* |

*St It FC     Studi ital d filologia classica. 1893-1915, NS 1920-.
         Florence.
*St Ned     Statistica (neerlandica). 1946/47-. The Hague, Leyden,
         Rijswijk. Vereniging voor Stat.
Stoc Ing     Handlingar, ingeniors vetenskaps akad. (Procs, R Swedish
         Inst for Eng Res) 1921-. Stockholm. Svenska bokhandels-
         centralen.
Stockholm. See also Sv.
Stoc TH     Handlingar, K tek hogskolans. (Transas of the R Inst of
         Tec) 1946-. Stockholm.
St Petersburg. See Len, Pet.
*Strand     Strand Magazine. 1891-. London.
*Strasb H     Schriften d strassburger wiss Gesell (in Heidelberg) (an der
         Univ Frankfurt). 1920-1923, 1928-1931// C of Strasb WG.
         Strassburg, etc.
*Strasb WG     Schriften, strassburger wiss Gesell. 1907-1919// C as
         Strasb H. Strassburg.
*Stroj ZI     Stroje na zpracovani info. (Cesk akad ved, ustav mat stroju;
         lab mat troju) 1953-. Prague.
Stu Apl M     Studies in Appl Math. 1969-. C of JM Ph. Cambridge, Mass.
         MIT Press.
Stu Cer M     Studii si cercetari mat. 1964-. C of Ro IM. Bucharest.
Studies, studi, etc. See organization, city, region, country, etc.
Studii si cercetari stiintifice.... See Iasi, Ro, Tim.
*Stu Gen     Studium generale, Zt f d Einheit d Wiss. 1947-. Berlin.
*Stu HI     Studies in the Hist of Ideas. 1918-. NY. Columbia Univ,
         Dept of Philosophy.
*Stu Log     Studia logica. 1953-. Warsaw. Polska akad nauk, komitet
         filozoficzny.
*Stultif     Stultifera navis. 1944-1957// Basel. Schw bibliophilen
         Gesell.
Stu M     Studia mathematica. (Lviv-Leopol, vidannja Lvivskogo
         derzavnogo univ im Ivana Franka) 1929-1940, 1948-. Lvov,
         Warsaw. Polish Acad of Scis.
Stu M Ober     Studia math, math Lehrbuecher Oberwolfach Math-Forschungs-
         inst. 1948-. Goettingen.
Stu SMH     Studia scientiarum mathematicarum Hungarica. 1966-.
         Budapest. Akad kiado.
Stuttgart. See Wurt.
*Stvdia     Stvdia, rev d l univ d Atlantico. 1942-. Barranquilla,
         Colombia.
*Sudhof Ar     Sudhoffs Arch f Gesch d Med (& d Naturw, d Pharmazie & d
         Math). 1929-. Wiesbaden.
*Sugaku     Sugaku. 1947-. Tokyo. Math Soc of Japan.
Sum Bras     Summa brasiliensis math. 1945-. S Paulo, Rio d Janeiro.
Sumer     Sumer, a J of Archaeology in Iraq. 1945-. Baghdad. Iraq
         Directorate-General of Antiquities.
Sun Yat Sen University. See Canton.
Suomi, Suomalaisen, Suomen Tasavalta. See Hels.
Sup St     Suppl stat ai nuovi problemi d politica, storia & economica.
         1935-1940// C as Stat Bln. Ferrara.

Sup VW            Suppl op "De vriend d wiskunde." 1887-1913// Culemborg,
                  Haarlem, Schoonhoven.
*Survey           Survey, a J of Sov & East European Studies. (Soviet Culture;
                  Soviet Survey; Survey, an Anal of Culture Trends in the Sov
                  Union) 1956-. London, Paris. Congress for Cultural Freedom.
*Sv Arsb          Kongl svenska vetenskapsakad arsbok. 1903-. Uppsala,
                  Stockholm.
Sv Avh            Kongl svenska vet avhandlingar i naturskyddsaerenden.
                  1938-. Stockholm.
*Sv Bi            Bihang till K svenska vetenskaps-akad handlingar. 1872-
                  1903// C as Ark MAF, etc. See also Sv Ofv. Stockholm.
*Sv Han           K svenska vet-akad handlingar. (Stock Ak Hndl; Stockholm
                  Handl) 1739-. Stockholm.
Sv Han Deu        Der K schwedischen Akad Abh. (Deu A Schw) 1739-1790//
                  German ed of Sv Han.
*Sv Ofv           Ofversigt af K vetenskaps-akad forhandlingar. 1844-1902//
                  C as Ark MAF, etc. See also Sv Bi. Stockholm.
*Sweden.* *See also Act PSMC, Act PSPM, Gotb, Lund, Nor, SAAB, Sk, Stoc, Upp.*
*Switzerland.* *See Helv, Schw; also Basl, Bern, Genv, J Suis Hor, Neuc, Zur.*
*Symp PM          Symp in Pure Math. 1959-. Providence, RI. Amer Math Soc.
*Syn Dor          Synthese, Intl J for Epistemology, Methodology, & Phil of
                  Sci. 1936-. Dordrecht. ≠ Synthese.
*Synthese         Synthese, an Intl J Devoted to Present-day Cultural & Sci
                  Life. 1949-. Bussom. ≠ Syn Dor, Rv Syn.
*Syracuse.* *See Acad Syr.*
*Syria            Syria, rev d'art oriental & d'archeol. 1920-. Paris.
                  Inst fr d'archeol d Beyrouth.
*Szgd Acta        Acta univ Szegediensis acta sci math. (A Szegedi egyetem
                  kozlemenyei; Acta sci math) 1942-. C of Szgd ALS. Szeged.
Szgd ALS          Acta litt ac sci, regiae univ ungaricae Francisco-Josephinae,
                  sect sci math. (Tudomanyos kozlemenyei, mathematikai
                  tudomanyok) 1922-1941// C as Szgd Acta. Szeged.


Tadz Do           Dokl akad nauk Tadzhikskoi SSR. (Dokladkhoi akad fankhoi
                  SS Tochikistoi) 1951-. Dushanbe, Stalinabad.
Tadz U            Uch zap Tadzhikskii gos univ. 1952-. Stalinabad.
*Taegu, Taikyu.* *See Kyung.*
*Tageblatt.* *See organization, city, region, country, etc.*
Tai Imp           Mems of the Fac of Sci & Ag, Taihoku Imp Univ. 1929-1942//
                  Pt C as Tai Imp Sc. Formosa.
Tai Imp Sc        Mems of the Fac of Sci, Taihoku Imp Univ. 1943// Pt C of
                  Tai Imp. C as Tai Nat Sc. Formosa.
Tai Nat Sc        Mems of the Fac of Sci, Natl Taiwan Univ. 1946-. C of
                  Tai Imp Sc. Taipei.
*Taiwan.* *See also Tamk.*
Tamk JM           Tamkang J of Math. 1970-. Tamsui, Taiwan. Tamkang Coll
                  of Arts & Sci.
*Tarbiz           Tarbiz, a Quar Rev of the Humanities. 1929-. Jerusalem.
                  Jerusalem Univ.

*Tartu. See also Dorpat.*
Tartu Ann     Anns soc rebus naturae investigandis in univ Tartuensi
              constitutae. 1934-. C of Tartu Est. Tartu.
*Tartu Est    Sitz d natfors Gesell bei d Univ Jurjew. (Protokoly
              obshch estestvoispytatelei pri imp univ Yurev) 1853-1933//
              C as Tartu Ann. Jurjew (Tartu).
*Tartu M      Tartu riikliku ulikooli toimetised mat-loodusteaduskonna
              toid. (Uch zap Tartuskogo gos univ, trudy est-mat fak)
              1945-. Tallin.
*Tashkent. See also Uzb.*
*Tash NO Tr   Trudy Turkestanskogo nauchnogo obshch pri sredne-Aziatskom
              gos univ. (Transas of the Sci Soc of Turkestan at Middle
              Asiatic Univ) 1923-1925//? Tashkent.
Tash Pe       Uch zap, fiz-mat, Tashkentskii gos ped inst im Nizami.
              1947-. Tashkent. Minist prosvescenija UzSSR.
*Tash Tec M   Tashkentskii tekstilnyi inst, kafedra vyssei mat, sb.
              Nauchno-issledovatelskih rabot, ser mat. 1953-. Tashkent.
*Tash UA      Acta univ Asiae mediae, ser V-a, math. (Trudy sredne-
              aziatskogo gos univ) 1929-. Tashkent.
Tash UB       Bjulletin sredneaziatskogo gos univ. (Bull d l'univ d
              l'Asie Centrale) 1923-. Tashkent. Izdatelstvo sredne-
              aziatskogo gos univ.
Tash UT       Trudy Tashkentskogo gos univ im V I Lenina, mat. 1929-.
              Tashkent.
Tasm PP       Papers & Procs of the R Soc of Tasmania. 1848-. Hobart.
*Tauric ML    Protokoly, mat lab, Krymskii (Tavricheskii) univ. (Procs
              of the Math Lab of the Crimean (Tauric) Univ) 1912-1921//
              Simferopol.
Tauric U      Procs of the Tauric Univ. Simferopol.
Taylor Ms     Sci Mems, Selected from the Transas of Foreign Acads &
              Learned Socs & from Foreign Journals. (Taylor Sc Mm)
              1837-1852// London.
*Tbilisi. See also Gruz.*
*Tbil MI      Sromebi, Sakharthvelos SSR mecnierebatha akad A Razmadzis
              sahelobis Thbilisis math inst. (Trudy Tbilisskogo mat inst
              im A M Razmadze; Trav inst math Tbilissi) 1937-. Tiflis.
              Akad nauk Gruzinskoi SSR.
*Tbil M Me    Trudy, ser mekh-mat nauk, Tbilisskii gos univ. (Tbilsis
              univ shromebi) 1936-. Tiflis.
*T Col Cont   Cóntributions to Edu. NY. Columbia Univ, Teachers Coll.
              ≠ Cont Ed.
*T Coll B     Teachers Coll Bull (on Higher Edu). 1930-1932// NY.
              Columbia Univ, Teachers Coll.
*Tea Col R    Teachers Coll Rec. 1900-. NY. Columbia Univ. Teas Coll.
Tec Cyb       Tec Cybernetics USSR. 1963-. Tr from Tek Kib. Wash DC.
              US Off of Tec Services.
*Tec Gsch     Technikgesch, Bei z Gesch d Tec & Ind. 1933-1941// C of
              Bei GTI. Berlin.
Technmet      Technometrics, a J of Stat for the Phys, Chem & Eng Scis.
              1959-. Princeton, NJ; Richmond, Va.
*Tec Rv       Technology Review. 1899-. Boston, Mass.

| | |
|---|---|
| Tec Trls | Technical Translations. 1959-1967// C in USRDR Ind, & in Tr Reg Ind. Wash DC; Springfield, Va. US Clearinghouse for Fed Sci & Tech Inf. |
| *Teixeira | Jornal d scis maths e asts pub pel Dr F Gomes Teixeira. 1868-1902// C as Porto Ac. Coimbra. |
| Tek Kib | Izv akad nauk SSSR, tek kibernetika. 1963-. Pt C of SSSR Tek. Tr in Engr Cyb, & in Tec Cyb. Moscow. |
| Tek Ph | Tech Phys of the USSR. 1934-1938// C as Fiz Zh. Leningrad. |
| Teletek | Teleteknik. 1950-. Copenhagen. |
| *Teletek E | Teleteknik (English). 1957-. Copenhagen. |
| *Television.* | *See Elect Eng.* |
| *Temeszvar.* | *See Timisoara.* |
| *Tensor | Tensor. 1938-. Tokyo, Sapporo. Hokkaido Univ, Dept Math, Tensor Soc. |
| Teor Fiz | Teoreticheskaya fizika. 1965-. Tr as Theor MP. Moscow. |
| Teor Fun | Teoria funktsii, funktsionalnyi analiz i ego prilozheniia. 1966-. Kharkov. |
| *Teoria | Teoria. 1953-. Madrid. |
| *Teor Ver | Teorija verojatnostei i ee primenenija. 1956-. Tr as Th Pr Ap. Moscow. Akad nauk SSSR. |
| *Ter Magn | Terrestrial Magnetism & Atmospheric Elec, an Intl Quar J. 1896-. Baltimore, Md; Cincinnati, O. Johns Hopkins Univ P; Ruter P. |
| Terq Bul | Bull d bib, d'hist & d biog math. 1855-1862// S to Nou An. Paris. Ed: O Terquem. |
| *Terquem Annales.* | *See Nou An.* |
| *Teubner.* | *See Enc MW.* |
| *Tex Ac | Transas of the Texas Acad of Scis. (Texas Ac Sc T) 1892-1944// Austin, Tex. |
| *Tex Q | Texas Quar. 1958-. Austin, Tex. Univ of Texas P. |
| *Thales | Thales, recueil annuel d travaux d l'inst d'hist d scis & d techs d l'univ d Paris. 1934-1937, 1948-. Paris. Ed: G Sarton. |
| *Theoria | Theoria, tid for fil och psych. (Theoria, a Swedish J of Phil & Psych) 1935-. Goteborg, Lund. |
| Theor MP | Theoretical & Mathematical Phys. 1970-. Tr of Teor Fiz. Consultants Bur. |
| The Prak | Theor-praktische Volksblatt f d Groessen-Lehre. 1831-. Ulm. |
| *Think | Think, a "Spotlight" on New Things & Thoughts in the World of Affairs. 1935-. NY. Intl Business Machines Corp. |
| *Thomist | Thomist, a Speculative Quar Rev. 1939-. Baltimore, Md. |
| *Th Pr Ap | Thy of Probability & Its Appls. 1956-. Tr of Teor Ver. Philadelphia, Penn. SIAM. |
| Th Pr Ap Ab | Thy of Probability & Its Appls, Abstracts. Wash DC. Dept of Commerce, Off of Tec Services. |
| *Tid El M | Tidskrift for elem mat, fys och kemi. 1917-. Stockholm. |
| *Tid M | Tid for math. (Ts Mth; Tidsskr) 1859-1889// C as Nyt Tid. Copenhagen. |
| Tid MF El | Tid for mat och fys, tillegnad den Svenska elem underwisningen. (Ts Mt Fys) 1868-1874// Uppsala. |
| *Tidskrift, Tijdschrift, etc.* | *See organization, city, region, country, etc.* |

*Tidskrift for matematik och fysik. See Dillner.*
*Tidsskrift for mathematik. See Zeuthen.*
*Tiflis. See Gruz, Tbil.*
Tijd BMW   Tijd d bevordering d math wetenschappen. 1823-1828//
     Purmerende. Wisk genootschap zu Amsterdam.
*Tijd Phi   Tijd voor phil. 1939-. Leuven.
*Tijdspgl   Tijdspiegel. 1844-1919// The Hague.
*Tim EP   Bull sci d l'ec polytec d Timisoara, CR d seances d l soc
     sci d Timisoara. (Institutul politehnic, bul sti si teh;
     Bulletin Timisoara) 1925-. Timisoara.
*Timisoara. See also Math Tim.*
Tim MF   Analele univ din Timisoara, ser stiinte mat-fiz. (Uch;
     Inst ped Temeszvar, lucrarile stiintifice, mat-fiz) 1958-.
     Timisoara.
*Tim Rev M   Revista mat din Timisoara. 1921-. Timisoara.
Tim ST   Studii si cercetari stiinte tehn. 1954-. Timisoara. Acad
     repub pop Romine, baza d cercetari stiintifice Timisoara.
Tirana U   Bul i univ shteteror te Tiranes, ser shkencat natyrore.
     1956-. Tirana.
*Tiras Ped   Uch zap Tiraspolskogo gos ped inst im T G Shevchenko (ser
     fiz-mat). 1956-. Kishinev.
*Toh MJ   The Tohoku Math J. (Tohoku sugaku dezassi) 1911-1942,
     1949-. Sendai.
Toh RIA   Sci Repts of the Res Insts, Ser A, (Phys, Chem & Metallurgy),
     Tohoku Univ. 1949-. Sendai.
Toh Sc   Sci Repts of the Tohoku Imp Univ, first ser (a) (Math, Phys,
     Chem (ast)). 1911-. Sendai.
Toh Tec   Tec Repts of the Tohoku Imp Univ. 1920-. Sendai, Tokyo.
*Tok But Gak Zas. See Tok S Ph S.*
Tok C En   J of the Coll of Eng, Imp Univ of Tokyo. 1908?-1921//
     C as Tok F En. Tokyo.
*Tok CGE   Sci Papers of the Coll of Gen Edu, Univ of Tokyo. (Tokyo
     daigaku, kyoyogakubu) 1951-. Tokyo.
Tok CS   J of the Coll of Sci, Imp Univ, Japan. (Tokyo J of Sci;
     Teikoku daigaku; Japan J) 1887-1925// C as Tok FSA. Tokyo.
Tok F En   J of the Fac of Eng, Univ of Tokyo, Ser B. 1922-1943,
     1952-. C of Tok C En. Tokyo.
Tok FSA   J of the Fac of Sci, Univ of Tokyo, Sect I, Math, Ast,
     Phys, Chem. 1925-. C of Tok CS. Tokyo.
Tok Gak   Bull of Tokyo Gakugei Univ. 1949-. Tokyo.
Tok IPCR   Sci Papers of the Inst of Phys & Chem Res. 1922-. C as
     Tok SRI. Tokyo.
Tok IST   Reports of the Inst of Sci & Tec, Univ of Tokyo. 1947-.
     Tokyo.
*Tok I Sta M   Procs of the Inst of Stat Math. (Tokesuri kenkinse ikho)
     1953-. Tokyo.
Tok I St MA   Annals of the Inst of Stat Math. (Tokei suri kenkyojo)
     1949-. Tokyo.
Tok IT   Bull of the Tokyo Inst of Tec (Ser B). 1950-. Tokyo.
Tok Keio   Procs of the Fujihara Memorial Fac of Eng, Keio Univ.
     (Keio gijuku daigaku kogakubu kenkyu hokoku) 1948-. Tokyo.

Tok Kyo A      Sci Repts of the Tokyo Kyoiku Daigaku, Sect A.  1930-.  Tokyo.
*Tok M Ph      Tokyo sugaku butsurigaku dwai kiji.  (Procs of the (Tokyo)
               Math-Phys Soc (o Japan); Tokyo Math Ges; Nip Sug-But Kij)
               1902-1945//  C as Jp MS, Jp PS.  See also Tokush J.  Tokyo.
*Tok S Ph S    J Tokyo Sch of Phys Sci.  (Tokyo But Gak Zas) 1892?-.  Tokyo.
Tok SRI        J of the Sci Res Inst of Tokyo.  (Rikagaku kenkyusho hokoku)
               1948-.  C of Tok IPCR.  Tokyo.
Tok U Sc D     Mems of the Sci Dept, Univ of Tokio, Japan.  (Tok Un Mm)
               1879-1885//  Tokyo.
*Tokush J      J of Gakugei, Tokushima Univ.  (J of Sci of the Gakugei
               Coll/Fac) 1950-.  Tokushima.
*Tokyo.  See Tok; also Jp, Kaizo, Kodai, Paul UCM, Toyo Gak.*
*Tomsk.  See also Sibirsk.*
Tomsk IU       Izv imperatorskago Tomskago univ.  1889-1929//  C as
               Tomsk M Me.  Tomsk.
Tomsk M Me     Trudy Tomskogo gos univ im V V Kuibyseva, ser mekh-mat.
               (Geom Sb) 1930-.  C of Tomsk IU.  Tomsk.
Tomsk UZ       Tomskii gos univ im V V Kuibyseva, uch zap.  1946-.  Tomsk.
Top            Topology, an Intl J of Math.  1962-.  Suppl 1964-.  Oxford,
               NY.
*Tor An        Annuario d'accad R d sci e d lett d Torino.  1878-.  Turin.
*Tor FMN       Atti d accad d sci d Torino, cl d sci fis, mat e nat.
               (Torino Atti) 1865-.  Turin.
*Tor Mm        Mems d R accad d sci, Torino.  (Miscellanea phil-math,
               melanges d phil et d math; Miscellanea taurinensia;  Mems
               d l'acad R des scis d Turin; Torino Mem Acc Sc; etc)
               1759-.  Turin.
Toro MPS       Papers Read before the Math and Phys Soc at the Univ of
               Toronto.  1890-.  Toronto. ·
*Toronto.  See also M Exp.*
*Toro ROM      Bull of the R Ontario Mus of Archae.  1923-.  Toronto.
Toro UAM       Univ of Toronto Studies, Appl Math Ser.  1935-.  Toronto.
Toro UM        Univ of Toronto Studies, Math Ser.  1915-.  Toronto.
*Tor Sem       Univ e politec d Torino rend d sem mat.  (Conferenze d fis
               e d mat) 1929-1938, 1950-.  Turin.
*Tortolin      Annali d sci mat e fis, compilati d B Tortolini.  (An;
               Tortolini A)  1850-1857//  C as Annali.  Rome.
*Tor U Ann     Annuario d R univ d Torino.  1954/55-.  Turin.
*Toscane.  See Pisa.*
*Tou Bu        Bull d l'acad d scis, inscriptions & B-L d Toulouse.  (Tou
               Ac Sc Bll) 1898-1899//  Toulouse.
*Tou FS        Anns d 1 fac d scis d l'univ d Toulouse pour 1 scis maths
               & 1 scis phys.  (Tou An) 1887-.  Toulouse.
*Tou Mm        Mems d l'acad d scis, inscriptions & B-L d Toulouse.
               (Toul Ac Sc Mm) 1782-.  Toulouse.
*Toung Pao     T'oung pao, ou archs concernant l'hist, 1 langues, 1 geog
               & l'ethnographie d l'Asie orientale.  1890-.  Leiden.
Tou Ob         Anns d l'obs ast, magn & meteorol d Toulouse.  (Toul Obs A)
               1873-.  Toulouse.
*Toyo Gak      Toyo gakuho.  1911-.  Tokyo.
*Tpilisi.  See Gruz, Tbil.*

| | |
|---|---|
| *Urk GMA | Urkunden z Gesch d Math in Altertums. 1907-. Leipzig. |
| *Urk GNG | Urkunden z Gesch d nichteuklidischen Geom. 1899-1913// Leipzig. |
| *Uruguay.* | *See Montv.* |
| *USA.* | *See A, Am, US; also Actuar SA, AMS, Berk, Berks, Bost, Calif, Carn, Chic, Colo, Conn, Duke, Franklin, Harv, Heb UC, Hunt L, Ill, Inda, Johns H, Kan, Kans, LHL, MAA, Manc, Mich, MIT, Myrin, New Eng, Not D, NY, Pac, Phila, Prin, Rice, Rock, Rocky, Sci Am, Sci Tec Pi, Smi, SRI, Syr, T Col, Tex, Vass, Virg, Wash, Wesl, Wich, Yale.* |
| *USCS | Repts of the Superintendent of the Coast (& Geodetic) Survey, Showing Progress of the Surv from Year to Year. (US Coast Sv Rp) 1851-. Wash DC. |
| USNBSA | J of Res of the Natl Bur of Standards, A, Phys & Chem. 1934-. C from USNBSJ. 1934-. Wash DC. |
| USNBSAM | Natl Bur of Standards, Appl Math Ser. 1948-. Wash DC. |
| *USNBSB | J of Res of the Natl Bur of Standards, B, Math & Math Phys. 1934-. C from USNBSJ. Wash DC. |
| *USNBSC | J of Res of the Natl Bur of Standards, C, Eng & Instrumentation. 1934-. C from USNBSJ. Wash DC. |
| USNBSD | J of Res of the Natl Bur of Standards, D, Radio Sci. 1934-. C from USNBSJ. Wash DC. |
| USNBSJ | Bur of Standards J of Research. 1928-1934// C as USNBSA, USNBSB, etc. Wash DC. |
| *USNI | Procs of the US Naval Inst. 1874-. Annapolis, Md. |
| *USO Ed CI | United States Off of Edu, Circ of Info. 1870-1903// Wash DC. |
| *Uspekhi.* | *See UMN.* |
| USRDR Ind | US Govt Res & Devel Repts Index. (Govt-wide Ind to Fed Res & Dev Grants) 1965-. Pt C of Tec Trls. Wash DC. US Clearinghouse for Fed Sci & Tec Inf. |
| *USSR.* | *See also Russ, Sov, SSSR; also Act Phch, Aku Zh, Alg Log, Arm, As Zh, Av, Az, BSSR, Bu Geoph, Bukh, Cernau, Crim, Diskret, Dokl, Dorpat, Est, Gruz, Ilmen, Inzh Zh, Ivan, Kab, Kalin, Karel, Kazk, Kazn, Khalyk, Khar, Kibernet, Kiev, Kish, Kolom, Konig, Lat, Lemb, Len, Lit, Lvov, Mold, Mos, Novor, Odess, Perm, Pet, Pribor, Priroda, Pro Kib, Pro Per, Rostov, Ryaz, RZ, Samark, Sib, Stal, Stekl, Tadz, Tash, Tauric, Tbil, Tik Kib, Teor, Tomsk, Tula, Turkm, UFN, Ukr, UMN, Ural, Uzb, Uzh, Viln, Vop, Vych, Yaros, Zh.* |
| USSR Comp | USSR Computational Math & Math Phys. 1962-. Tr of Zh Vych M. NY. Pergamon P. |
| *Utr Aant | Aanteekeningen van het verhandelde in de sectievergaderings van het prov Utrechtsch genootschap van kunsten e wetenschappen. (Utr Aant Prv Gn) 1845-1921// Utrecht. |
| *Uzbekistan.* | *See also Samark, Tash.* |
| Uzb Fan Do | Doklady akad nauk UzSSR. (Uzb SSR fanlar akad dokladlari) 1948-. Tashkent. |
| Uzb Fan Iz | Uzbek SSR fanlar akad akhboroti, Izv. 1940-1942, 1947-1956// C as Uzb FMN. Tashkent. |

| | |
|---|---|
| *Uzb FMN | Izv akad nauk UzSSR, ser fiz-mat nauk. (UzSSR fanlar akad ahboroti, fiz-mat fanlari ser) 1957-. Pt C of Uzb Fan Iz. Tashkent. |
| Uzb IM | Akad nauk Uzbekskoi SSR, trudy inst mat (i mekh) im V I Romanovskogo. 1946-. Tashkent. |
| Uzh UDS | Dokl i soobshcheniia Uzhgorodskogo gos univ, ser fiz-mat i khim. 1957-. Uzhgorod. |
| Uzh UNZ | Nauchnye zap, Uzhgorodskii gos univ. 1947-. Lvov, Uzhgorod. |


| | |
|---|---|
| *Varanasi*. | *See Banaras, Benares.* |
| *Varsovie*. | *See Warsaw.* |
| *Vass Ob Pu | Vassar Coll Observatory, Pubs. 1900-. Poughkeepsie, NY. |
| *Vat Act | Acta, pontificia acad sci. 1937-. C of Vat NLA. Rome, Vatican City. |
| *Vat An | Annuario d pontifica accad d sci. 1929-, NS 1936/37-. Rome, Vatican City. |
| *Vat Com | Pontificia acad sci, commentationes, ex aedibus acad in Civitate Vaticana. 1937-. C of Vat NL Mm. Rome. |
| *Vat NLA | Atti d accad pontificia d Nuovi Lincei, Roma. (Rm At; Rm At N Linc; Atti Pontificia Accad) 1847-1935// C as Vat Act. Pt C of Linc At. Rome. |
| *Vat NL Mm | Mem d accad pontificia d Nuovi Lincei, Roma. (Roma Mem Acc Nuovi Lincei; Mem d pontificia Ac d Sc NL) 1887-1935// C as Vat Com. Rome. |
| *Vector*. | *See Wektor.* |
| *Ven Atene | L'ateneo veneto, riv mensile d sci, lett & arti. 1812-. Venice. |
| Ven Atti | Atti d ateneo veneto. 1864-1881, 1921-1924// Pt of Ven Atene. Venice. |
| *Venezuela*. | *See Venzl.* |
| *Ven I At | Atti d adunanze d I R Ist veneto d sci, lett & arti. (Ven At; Ven Ist Atti; Att Acc Ven; Ven Atti Ist) 1841-. Venice. |
| *Ven ML | Ist veneto d sci, let & arti, atti, pt sec, cl d sci morali e lett. 1840-. Venice. |
| *Ven Mm | Mem d R ist veneto d sci, lett & arti. (Mem di Venez) 1843-. Venice. |
| *Ven MN | Ist veneto d sci, lett & arti, atti, pt sec, cl d sci mat e nat. 1841-. Venice. |
| *Venzl Ac | Repub d Venezuela, bol d l acad d ciencias fis, mat y nat. 1934-. Caracas. |
| Venzl FM | Coleccion cien fis y mat. 1965-. Caracas. Univ Central. |
| Venzl Ing | Univ central d Venezuela, bol d l fac d ing. 1954?-. Caracas. |
| *Verein, vereniging, etc*. | *See organization, city, region, country, etc.* |
| *Verein der hoeheren bayerischen Vermessungsbeamten*. | *See Bay Verm.* |
| *Verhandlungen, etc*. | *See organization, city, region, country, etc.* |
| *Vermessungsnachrichten*. | *See Allg Verm.* |

*Verm R          Vermessungstec Rundschau, Zt f d Vermessungswesen.  1924?-.
                 Hamburg.
*Veroeffentlichungen.  See organization, city, region, country, etc.*
*Versich         Das Versicherungsarchiv, Monatsblaetter f private & oeffent-
                 liche Versicherung.  1930-.  Vienna.
*Verslagen, etc.  See organization, city, region, country, etc.*
Vers Wisk        Versameling d wiskundige voorstellen.  (Recueil d qs math)
                 1811-1815, 1820-1836, 1841-1846, 1850-1854//  Amsterdam.
                 Wisk Genootschap.
*Verzek          Het verzekeringsarchief.  1920-.  C of Amst Verz.
                 's Gravenhage.  Bedrijfsgroep levensverzekeringen d ver v
                 verzekeringswet.
*Ves Drev I      Vestnik drevnei istorii.  (Rev d'hist anc)  1937-1941, 1945-.
                 Moscow.  Akad SSSR, inst istorii.
Ves Me Pr M      Vestnik mekh i prikladnoi mat.  1929-1931//  C as Pri M Me.
                 Leningrad.
Ves MN           Vestnik mat nauk.  1861-1863//  Vilnus.  Ed: M M Gusev.
*Ves St          Vestnik statistiki.  1919-.  Moscow.
*Vestnik, vesci, vestis, etc.  See also organization, city, region, country.*
*Vestnik jednoty Ceskoslov, mat a fys v praze.  Pt of Cas MF.*
*Vestnik opytnoi fiz i elementarnoi matematiki.  See Kagan.*
Ves Vy Shk       Vestnik vysshei shkoly.  1940-1941, 1945-.  C of SSSRVSB.
                 See also SSSRVSFM.  Moscow.
*Vict RS         (Transas &) Procs of the R Soc of Victoria.  (Vict RST;
                 Vict TRS)  1861-.  Melbourne.
*Vienna.  See Wien.*
*Vierteljahrsschr d ast Gesell.  See Leip As.*
*Vier W Phi      Vierteljahrsschr f wiss Phil.  1876-1916//  Leipzig.
Vijn PAP         Vijnana parishad anusandhan patrika.  (Res J of the Hindi
                 Sci Acad)  1958-.  Allahabad.
*Vilna.  See also Lit.*
Viln M Bul       Vilna Univ Sem Math Bull.  1938-1939?//  Vilna.
Viln Sem         Vilna univ sem ze skarbowosci i prawa skarbowego oraz ze
                 statystvki, prace.  1931-1934?//  Vilna.
Viln UT          Vilniaus valstybinis univ, mokslo darbai, mat, fiz ir chem
                 mokslu ser.  (Vilnjusskii gos univ, uch trudy, ser mat,
                 fiz i khim nauk)  1949-.  Vilna.  TSRS aukstojo mokslo
                 minist.  (Minist vyssego obrazovanija SSSR)
Virg JS          Virginia J of Sci.  1940-.  Blacksburg, Richmond, Va.  Va
                 Acad of Sci.
*Vjschr.  See Vierteljahrsschrift.*
*Vlaamse.  See Bel Vla.*
*Volgograd.  See Stalingrad.*
*Vop Fil         Voprosy filosofii.  1947-.  Moscow.  Akad nauk SSSR, inst fil.
*Vop F Ps        Voprosy filosofii i psikhologii.  1889-1917//  Moscow.
*Vop IET         Voprosy istorii est i tekh.  1956-.  Moscow.  Akad nauk SSSR,
                 inst ist est i tekh.
Vop Kos          Voprosy kosmogonii.  1952-.  Moscow.  Akad nauk SSSR.
*Voprosy.  See also organization, city, region, country, etc.*
*Voss Z          K privilegirte berlinische Zeitung v Staats- & gelehrten
                 Sachen.  (Vossische Zeitung)  1816-.  Berlin.

*Vost IINT      Izv istorii nauki i tekh v stranah vostoka. 1960-. Moscow.
                  Verlag d orientalischen Lit.

*Vychislitelnyi tsentr, akad nauk SSSR. See also SSSR Vych.*

Vych M          Vychislitelnaya mat. 1957-. C from Vych MVT, Vych MT.
                  Moscow. Akad nauk SSSR, vychislitelnyi tsentr.

Vych MK        Vychislitelnaya mat. 1966-. Kiev. Akad UkRSR.

Vych MT        Vychislitelnaya mat i tekh. 1953-1957// C in Vych M.
                  Moscow. Akad nauk SSSR, vychislitelnyi tsentr.

Vych MVT       Vychislitelnaya mat i vych tekh. 1953-. C in Vych M,
                  Vych T. Moscow. Akad nauk SSSR, inst tochoi mekh i
                  vych tekh.

Vych Sist      Vychislitelnye sistemy, sb trudov. 1962-. Novosibirsk.
                  Akad nauk SSSR, Sib otd, inst mat.

Vych T          Vychislitelnaya tekh. (Trudy...kafedra mat; Sb statei)
                  1955-. C from Vych MVT. Moscow. Vysshei tekh.

*Vysshei shkoly. See SSSRV, Ves Vy Shk.*

*Vytautas univ. See Kaunas U.*


*Warb Cour      J of the Warburg & Courtauld Insts. 1937-. London.
                  Univ of London, Warburg Inst.

*Wars Na Tr      Travaux d 1 soc d nat d Varsovie, CR d assemblees gens.
                  (Vars S Nt Tr (CR Ps C); Warschau phys sect; Varsava Prot
                  Obsc jest; Proces-verbaux d 1 soc d nat d Varsovie; Obshch
                  estestvoispitatelei Protokoly; CR & mem d 1 soc d nat a
                  l'uni TMP; Trav mat & Phys d Vars) (1) gen 1899-1913//
                  (2) bio 1889-1908// (3) fi kh 1889-1908// Warsaw.

Wars PI         Bull d l'inst polyt d l'emp Nicolas II a Varsovie. (Varsava,
                  Izv politechn inst) 1878-1914// Warsaw.

*Wars PI An      Ann d l'inst polyt d Varsov. (Nachrichten d Warschau Polyt
                  Inst)

Wars Pra M      Travaux d 1 soc d scis & d letts d Varsovie, (b) mat.
                  (Prace towarzystwa nauk warszawskiego; Tow nau War wydz III
                  prace) 1908-1919, 1930-. Warsaw. (Pol ak nauk)

Wars PZAM      Prace zakladu aparatow mat. 1959-. Warsaw. Pol akad nauk.

*Wars SPTN      Sprawozdania z posiedzen towarzystwa nauk Warszawskiego.
                  (CR soc sci Varsovie; Warsch Sitzungsber) 1908-1939//
                  Warsaw.

Wars U Iz       Bull d l'univ imp d Varsovie. (Varsava, Izv Univ; Warsch
                  Nachr) 1870-1915// Warsaw.

Waseda Mm      Mems of the Sch (Coll) of Sci & Eng, Waseda Univ, Tokyo.
                  1922-. Tokyo.

*Wash Ac        J of the Wash Acad of Scis. 1911-. Wash DC.

*Wash PS        Bull of the Phil Soc of Washington. (Wash Bull) 1871-1910//
                  Wash DC.

*Wash UB        Washington Univ Asso Bull. 1903-1917// St Louis, Mo.

*Wash ULL       Univ of Washington Pubs in Language & Lit. 1920-. Seattle,
                  Wash.

*Wash UM        Univ of Wash Pubs in Math (and Phys Sci). 1915-1952//
                  Seattle, Wash.

| | |
|---|---|
| *Wash U St | Washington Univ Studies, Sci Ser. 1913-1926// See also Wash U St S. St Louis, Mo. |
| Wash U St S | Washington Univ Studies, Sci & Tec. 1928-. See also Wash U St. St Louis, Mo. |
| *Wiad M | Roczniki polskiego towarzystwa mat, ser II, wiadomosci mat. (Wiad mat; Wiadom mat) 1877-. Warsaw. |
| *Wich U Stu | Univ Studies Bull. 1936-. Wichita, Kan. Municipal Univ. |
| Wiedemann Ann. | See Pogg An. |
| *Wien Anz | Anzeiger, math naturw Kl, Akad d Wiss z Wien. (Anzeiger Wien) 1864-1946// C as Ost Anz. Vienna. |
| Wien Denk | Denkschr d K Akad d Wiss, math-naturw Kl. 1850-1946// C as Ost Denk. Vienna. |
| *Wien Papy | Mitt, Wien Natlbibt, Mss. (Samm d papyrus erzherzog Rainer) 1886-1897, NS 1932-. Vienna. |
| Wien PI | Jab d kk polyt Inst in Wien. (Wien Jb Pol I) 1819-1839// Vienna. |
| *Wien Si | Sitz d K Akad d Wiss, math-naturw Kl, Abt IIa, Math, Ast, Phys, Meteorol, Tec. (Wien Ak Sb; Wien SB; Wien Sb; Wien Ber) 1848-1947// C as Ost Si M. Vienna. |
| *Wien Si PH | Sitz, K Akad d Wiss, phil-hist Kl. 1848-. Vienna. |
| Wisk Opg | Wiskundige opgaven met d oplossingen van het wisk genootschap ter apreuke voerende.... 1855-. Amsterdam, Groningen. |
| *Wisk Tijd | Wiskundig tijd. 1904-1921// Haarlem. |
| *Wis Nat Ti | Wis- en natuurkundig tijd. 1921-1945// C in Sim Stev. Ghent. Flaamisch nat geneeskundig cong. |
| Wiss | Wissenschaft, Sammlung naturw & math Monographien (E Wiedemann). 1904-. Braunschweig. |
| *Wiss An | Wissenschaftliche Annalen. 1952-1957// Berlin. Deu Akad d Wiss. |
| *Wiss Weis | Wissenschaft & Weisheit, Vierteljahrsschr f sys franziskanische Phil & Theologie in d Gegenwart. 1934-1943, 1949-. Freiburg i Br, Dusseldorf, etc. |
| Wolf Z. | See Zur NGV. |
| *Wort Sach | Woerter & Sachen. 1909-1933, 1938-1944// Heidelberg. |
| Wroc B | Prace, Wroclawskiego towarzystwo naukowego, ser B. (Travaux, soc d scis (& d letts) d Wroclaw) 1947-. Wroclaw. |
| Wroclaw. | See also Bresl. |
| *Wszech | Wszechswiat (organ o Polskie towarzystwo przyrodnikow im Kopernika). 1882-1914, 1928-. Warsaw. |
| *Wurt MN | Math-naturw Mitteilungen. (Boklen Mitt) 1884-1892, 1900-. Stuttgart. Math-naturw Ver i Wuerttemberg. |
| *Wurt Sch | (Neues) Korrespondenz-Blatt f d hoeheren Sch (Gelehrten- & Realsch) Wuerttembergs. 1854-1922// C as Aus Unt. Stuttgart. |
| Wurz Ber | Ber d phys-med Gesell z Wuerzburg, neue Folge. 1936-. C of Wurz Si, Wurz Ver. Wuerzburg. |
| *Wurz Jb | Wuerzburger Jbr f d Altertumswiss. 1946-. Wuerzburg. |
| Wurz Si | Sitz d phys-med Gesell z Wuerzburg. (Wurzb Ps Md Sb) 1859-1862, 1881-1923// C in Wurz Ver, Wurz Ber. Wuerzburg. |
| *Wurz Ver | Verh d phys-med Gesell z Wuerzburg. 1850-1860, 1868-1935// C in Wurz Ber. Incl C of Wurz Si. Wuerzburg. |

*XL. See ItSS, It XL, Linc*
*XVII         XVIIe siecle, bull d l soc d'etude d XVIIe siecle. 1949-. Paris.


*Yale Review. 1892-, NS 1911-. C of New Eng. New Haven, Conn.
Yam U Bu       Yamagata daigaku kie sidzen kagaku. (Bull Yamagata Univ, Nat Sci) 1950-. Yamagata.
*Yaros Ped      Dokl na nauchnykh konferentsiya, ser 3, mat, ast, fiz, tekh, Yaroslavskii gos ped inst. 1962-. Yaroslavl.
Yaros Pe M     Uch zap, ser mat, Yaroslavl gos ped inst. 1960-. Yaroslavl.
Yaros PUZ      Uch zap Yaroslavskii gos ped inst im K D Uchinskogo. 1944-. Yaroslavl.
*Yasskogo. See Iasi.*
*Yayinlari. See organization, city, region, country, etc.*
Yok DGNVO     Mitt d deu Gesell f Nat- & Voelkerkunde Ostasiens (in Tokio). (D Gs Ostas Mt) 1873-. Yokohama, Tokyo.
Yok MJ         Yokohama Math J. (Yok Mun Univ J, Ser D, Math) 1953-. Yokohama. Municipal Univ Fac of Arts & Sci.
Yok U M Ph    Sci Repts of the Yokohama Nat Univ, Sect 1, Math, Phys (Chem). (Yokohama kokuritsu daigaku) 1952-. Yokohama, Kamakura.
*Yugoslavia. See Beo, Ljub, Mak, Matica Sl, Skopj, Zagr.*
*Yurev. See Dorpat, Tartu.*


Z               Zentralblatt f Math & ihre Grenzgebiete. (ZBL) 1931-. Berlin, Goettingen, Heidelberg.
*Zach MC       Monatliche Correspondenz z Befoerderung d Erd- & Himmels-kunde. (Zach M Cor) 1800-1813// Gotha. Ed: von Zach.
*Z Aeg SA      Zt f aegyptische Sprache & Alterthumkunde. 1863-1943, 1954-. Leipzig. Deu morgenlaendische Gesell.
*Zagr Ac. See Zagr Iz.*
*Zagr BI        Bull intl d l'acad Yougoslave d scis & d B-A, cl d scis mat (& nat) (phys & tec). 1930-. C of Zagr Iz. Zagreb.
*Zagr BS        Bull sci, conseil d acad d l RSF d Yougoslavie. 1954?-. Zagreb.
*Zagr Gl        Glasnik mat-fiz i ast, ser 2. (Periodicum math-phys ast; Drustvo mat i fiz Hrvatske) 1943-1965// Zagreb. Soc hist-nat Croatica.
Zagr Gl M      Glasnik mat, ser 3. (Drustvo mat i fiz Hrvatske) 1966-. C of Zagr Gl. Zagreb.
Zagr Hrva     Hrvatska akad znanosti i umjetnosti, rad. 1941-1945// Zagreb.
*Zagr Iz        Izv o raspravama, mat prir razred, Jugosl akad znatnosti i um. (Bull Int Acad Yougosl) 1914-1929// C as Zagr BI. Zagreb.
Zagr Ljet     Ljetopis Jugoslavenske akad znanosti i umjetnosti. 1867/77-. Zagreb.
*Zagr Rad      Rad Jugoslavenske akad znanosti i umjetnosti, odjel za mat, fiz, i tehn nauk. (Rad Jug Akad Znan Umj; Rad jug aka zna i um) 1867-1940, 1946-. Zagreb.

Zagr Rasp        Jugoslavenska akad znanosti i umjetnosti, rasprave odjela za
                 mat, fiz, i tekh nauke. 1952-. Zagreb.
*ZAM Me          Zt f angew Math & Mech (ing wiss Forschungsarbeiten). (Z f
                 Angew Math; Z Ang M Mech) 1921-1944, 1947-. Berlin.
*ZAMP            Zt f angew Math & Phys. (J of Appl Math & Phys; J d math &
                 d phys appl) 1950-. Basel.
ZAP              Zt f angew Phys. 1948-. Berlin.
 Zapiski.   See organization, city, region, country, etc.
*Zarag Ac        Rev d l'acad d ciencias exactas, fis-quim y nat d Zaragoza.
                 1916-1937, 1946-. Zaragoza.
*Zarag Fac       Anales d l fac d ciencias, univ Zaragoza. 1907-1909//
                 Zaragoza.
 Zarag U An      Zaragoza univ anales. 1917-. Zaragoza.
*Z Assyr         Zt f Assyriologie & (verw Gebiete) (vorderasiatische Archeol).
                 1886-. C of Z Keil. See also Deu Morg Z. Berlin, Leipzig,
                 etc. Deu morgenlandische Gesell.
*Zastos M        Zastosowania mat. 1953-. Warsaw. Polska akad nauk, inst
                 mat.
 Z Astph         Zt f Astrophysik. 1930-1944, 1947-. Berlin, Goettingen,
                 Heidelberg. Deu phys Gesell.
*Z Bild K        Zt f bildende Kunst. 1866-1932// C as Z Kunst. Leipzig,
                 etc.
 Zbornik.   See also organization, city, region, country, etc.
 Zbornik znastvenih iu poucnih spisov.   See Matica Sl.
 Z eingeborenen-sprach.   See Afr USK.
 Zeitschrift.   See Z; also organization, city, region, country, etc.
 Zeitschrift f Physik, Math...Vienna.   See Baumgart.
 Zelanti.   See Acir At.
 Zen Did         Zentralblatt f Didaktik d Math. 1969-. Karlsruhe. Univ,
                 Zentrum f Didaktik d Math, intl math Unterrichtskommission.
*Z Ethn          Zt f Ethnologie. 1869-1943, 1949-. Berlin.
*Zeuthen         Tid f math. 1859-1889// Copenhagen. Ed: Zeuthen.
*Z Flug          Zt f Flugwissenschaften. 1953?-. Braunschweig.
*Z Geoph         Zt f Geophys. 1924-. Braunschweig, Vieweg, Wuerzburg.
*Z Gesm Nat      Zt f d gesamte Naturw einschliesslich Nat-phil & Gesch d
                 Naturw & Med. 1935-. Braunschweig, Vieweg.
*ZGNTM           Zt f Gesch d Naturw, Tek & Med. 1960-. Berlin.
 Zh EM           Zh elem mat. 1884-1886// C in Kagan. Kiev.
 Zh ETF          Zh eksperimentalnoi i teor fiz. (Fiz zh, ser A) 1931-. Pt
                 C of Pet F Kh. Tr as Sov Ph J. Moscow. Akad nauk SSSR.
 Zh PF           Zh prikladnoi fiziki. (J Appl Phys Moscow) 1924-1930//
                 C as Zh TF. Moscow.
 Zh TF           Zh tekh fiz. (Fiz zh, ser B; J Tech Phys Moscow) 1931-.
                 C of Zh PF. See also PMTF, JTP. Tr as Sov Ph TP.
                 Moscow, Leningrad. Akad nauk SSSR.
 Zhurnal.   See also city, organization, region, country, etc.
 Zh Vych M       Zh vychislitelnoi mat i mat fiz. 1961-. Tr as USSR Comp.
                 Moscow. Akad nauk SSSR.
*Z Instr         Zt f Instrumentenkunde (Vereinsblatt d deu Gesell f Mech &
                 Optik; Deu Mech Zt) 1881-1943, 1957-. Berlin.

| | |
|---|---|
| Z Keil | Zt f Keilschriftforschung & verw Gebiete.  1884-1885// C as Z Assyr.  Leipzig. |
| Z Krys | Zt f Krystallographie & Mineralogie (& Petrographie). 1877-1944, 1955-.  Leipzig. |
| Z Kunst | Zt Kunstgesch.  1932-1943, 1950-.  C of Z Bild K.  Berlin, Leipzig, Munich. |
| *Z Latein | Zt f lateinlose hoehere Schulen.  1889-1919//  Leipzig. |
| Z Mech | Zentralblatt f Mech.  1930-.  Berlin. |
| Z Met | Zt f Meteorologie.  1946-.  Berlin, Potsdam. |
| Z M Log Gru | Zt f math Logik & Grundlagen d Math.  1955-.  Berlin. Humboldt Univ,  Inst f math Logik. |
| *ZMN Unt | Zt f math & naturw Unterricht.  (Hoffmann; Z f math Unterricht)  1870-1943//  C in MN Unt.  Leipzig.  Founder: Hoffmann. |
| *ZM Ph | Zt f Math & Physik.  (As Math Leipzig; Schloemilch; Z Math Ps; Z)  1856-1917//  Leipzig.  Founder: Schloemilch. |
| *Z Natfor | Zt f Naturforschung.  1946-.  Tuebingen, Wiesbaden. |
| *Z Natok | Zt f Nationaloekonomie.  1929-.  Vienna.  Springer. |
| *ZNTM.  See NTM.* | |
| *Z Ph | Zt f Physik.  1920-1944, 1947-.  S to Deu Ph G.  Braunschweig, Berlin, Goettingen, Heidelberg.  Deu phys Gesell. |
| *Z Ph C Unt | Zt f d phys & chem Unterricht.  1888-1943//  C in MN Unt. Berlin.  Ed: Poske. |
| *Z Phi Fors | Zt f phil Forschung.  1946-.  Wurzach, etc. |
| Z Ph M | Zt f Phys & Math.  (Wiener Zt f Phys, Chem, & Min)  1826- 1832//  C in Baumgart.  Vienna. |
| Z Phon Spr | Zt f Phonetik (u allg) Sprachwiss (& Kommunikationsfors). 1947-.  Berlin. |
| *Z Realsch | Zt f d Realschulwesen.  1876-1920//  Vienna. |
| *Z Schw St | Zt f d schw Stat & Volkswirtschaft.  (J d statistique & rev econ suisse; Schw Z f V u S; Rev suisse d'ec pol et d stat) 1865-.  Basel, Bern.  Schw stat Gesell. |
| *Z Sem Ver G | Zt f Semitistik & verw Gebiete.  1922-1935//  Leipzig. Deu morgenlaendische Gesell. |
| *Z Soz | Zt f Sozialforschung.  1932-1939//  Frankfurt a M. |
| *Z T Ph | Zt f tec Phys.  1920-.  Leipzig, Barth.  Deu Gesell f tec Phys. |
| Zur ETH | Mitt aus d Inst f angew Math an d eidgenossischen tec Hochsch in Zurich.  1957-.  Zurich. |
| Zur NG Mt | Mitt d naturfors Gesell in Zur.  (Zur Mt)  1847-1856// C as Zur NGV.  Zurich. |
| *Zur NGV | Vierteljahrsschr d naturfors Gesell in Zurich.  (Wolf Z) 1856-.  C of Zur NG Mt.  Zurich. |
| *Zur U Hand | Handelswiss Seminar Mitt.  (Mitt Handelswiss Sem Univ Zur) 1922-.  Zurich.  Univ. |
| *Zutphen | Mathematica, tijd v 1932-1946, stud v d acten wisk L O en M O K en v Stud aan univ (pts A & B).  (Mathematica Zutphen) 1932-1946//  C in N Tijd & in Sim Stev  Zutphen, Leyden. |
| *Z Vermess | Zt f Vermessungswesen.  1872-1943, 1949-.  Incl C of Photgram.  See also Allg Verm.  Munich, Stuttgart.  Deu Geomvereins. |

| | |
|---|---|
| *Z Ver Wien | Zt f Vermessungswesen. Vienna. |
| *Z Versich | Zt f d gesamte Versicherungswiss. 1899-. S by Bla Versi. Berlin. Deu Ver f Versicherungswiss. |
| *Z Wahr | Zt f Wahrscheinlichkeitstheorie & verw Gebiete. 1962-. Berlin. |
| *Zyc Nauk | Zycie nauki, miesiecznik naukoznawczy. 1946-1952// C as Zyc Szkol. Krakow, Warsaw. |
| Zyc Szkol | Zycie szkoly wyzszej. 1953-. C of Zyc Nauk. Warsaw. Minist szkolnictwa wyzszego, Zarzad glowny, Zwiazek zawodowy nauczycielstwa polskiego. |

Lightning Source UK Ltd.
Milton Keynes UK
UKHW030615210722
406167UK00006B/612